Green Energy and Technology

Climate change, environmental impact and the limited natural resources urge scientific research and novel technical solutions. The monograph series Green Energy and Technology serves as a publishing platform for scientific and technological approaches to "green"—i.e. environmentally friendly and sustainable—technologies. While a focus lies on energy and power supply, it also covers "green" solutions in industrial engineering and engineering design. Green Energy and Technology addresses researchers, advanced students, technical consultants as well as decision makers in industries and politics. Hence, the level of presentation spans from instructional to highly technical.

Indexed in Scopus.

Indexed in Ei Compendex.

Nils Bullerdiek · Ulf Neuling · Martin Kaltschmitt
Editors

Powerfuels

Status and Prospects

 Springer

Editors
Nils Bullerdiek
Hamburg University of Technology
(TUHH), Institute of Environmental
Technology and Energy Economics (IUE)
Hamburg, Germany

Ulf Neuling
Hamburg University of Technology
(TUHH), Institute of Environmental
Technology and Energy Economics (IUE)
Hamburg, Germany

Martin Kaltschmitt
Hamburg University of Technology
(TUHH), Institute of Environmental
Technology and Energy Economics (IUE)
Hamburg, Germany

ISSN 1865-3529 ISSN 1865-3537 (electronic)
Green Energy and Technology
ISBN 978-3-031-62410-0 ISBN 978-3-031-62411-7 (eBook)
https://doi.org/10.1007/978-3-031-62411-7

This Springer imprint is published by the registered company Springer Nature Switzerland AG
The registered company address is: Gewerbestrasse 11, 6330 Cham, Switzerland

If disposing of this product, please recycle the paper.

Preface

Confronting climate change is one of the most important and urgent challenges facing humanity. If the global society does not find and implement suitable development pathways to substantially reduce greenhouse gas emissions in the coming years, the world as we know it today is unlikely to be preserved in the same way for future generations. The use of renewable sources of energy promises to be a cornerstone in the overall portfolio of possible reduction measures accessible to humans. Energy sources like solar radiation, wind power, hydropower, as well as biomass and geothermal energy, can basically be used in a climate-neutral way. Out of these options, especially photovoltaic systems and wind turbines, characterized by unexpected growth on a global scale within the last two decades, are able to provide cheap electrical energy. This energy can be used highly efficiently within global energy systems, and these options still show huge unexploited potential. Yet, our industrialized global society requires not only energy-rich electrons (i.e., electrical energy) but also "green" molecules characterized by a high heating value. This is particularly true for sectors like aviation and maritime transportation, which, due to technological constraints, rely heavily on fuels with high volumetric and gravimetric density. Moreover, certain production processes, for example in the chemical industry, still need "green" hydrocarbons as an educt (raw materials).

Addressing these demands and considering the goals for greenhouse gas reduction, hydrocarbon-based renewable energy carriers and feedstocks ("green" molecules) are becoming increasingly important in the years to come. The respective process pathways and chemical substances can be collectively referred to under the umbrella terms "Power-to-X" (PtX) and "powerfuels". These general terms encompass a broad spectrum of technologies capable of converting electricity, typically from renewable sources of energy, into various hydrocarbons as well as other synthetic "green" molecules. The viability, necessity, feasibility, and practicality of such Power-to-X and powerfuel technologies are the subject of intense debates on a global and especially European scale within industry, politics, research, and society, particularly on aspects like technical functionality, economic viability, environmental benefits, climate compatibility, and social acceptability. Additionally, in parallel to these discussions, research and development activities in the Power-to-X field are ongoing. Some processes are in the early stages of development, while others are more advanced, facing not only technological challenges but also issues related to scaling up and global implementation.

In this context, this book provides a comprehensive and in-depth overview of Power-to-X technologies and powerfuels. It aims to cover essential aspects such as the fundamentals of supplying the required electrical energy, hydrogen provision, and feedstock/carbon supply, as well as the main conversion pathways and technologies. This publication also addresses the use of Power-to-X products and powerfuels in key sectors dependent on these technologies. Additionally, it tackles the conditions necessary for the future market development of Power-to-X technologies. This is achieved through a comprehensive collection of various papers contributed by international experts in the diverse fields related to Power-to-X applications.

The successful publication of this book, which encompasses a broad spectrum of topics across the entire provision chain and necessary framework conditions of Power-to-X, was made possible thanks to the significant support of various highly engaged authors. Their expertise and detailed analysis have thoroughly addressed the diverse aspects of this field. Therefore, the editors would like to extend their sincere thanks to all authors and co-authors for their invaluable contributions. Their high motivation and strong commitment were key to providing such a comprehensive overview. This achievement would not have been possible without their relentless support and the incorporation of their extensive expertise. We also express our gratitude to the publisher for their constructive and straightforward collaboration.

Hamburg, Germany   Nils Bullerdiek
November 2024   Ulf Neuling
                                                                    Martin Kaltschmitt

Contents

Introduction

Nils Bullerdiek, Ulf Neuling, and Martin Kaltschmitt

Abstract As in recent decades, global population and average living standards are expected to continue to grow substantially in the years and decades to come. The consequence is that, even despite increasing exploitation of existing energy efficiency potentials and a continued shift toward more efficiently usable energy carrier (e.g., electrical energy), a significant increase of the global energy demand is expected in the coming decades. In this context, the transformation of the global energy system towards a more sustainable and climate-friendly energy supply is becoming imperative.

Keywords Power-to-X · Powerfuels · e-fuels · Synthetic fuels · Eelectricity-based fuels · Electrofuels · Hard-to-abate · Hard-to-decarbonize

1 Introduction

As in recent decades, global population and average living standards are expected to continue to grow substantially in the years and decades to come. The consequence is that, even despite increasing exploitation of existing energy efficiency potentials and a continued shift toward more efficiently usable energy carrier (e.g., electrical energy), a significant increase of the global energy demand is expected in the coming decades. So far, meeting this already quite substantial energy demand heavily relies on the use of a priori limited fossil fuel-based resources of natural gas, crude oil, and coal. The use of these resources poses severe challenges and drawbacks on global society and the fragile ecosystems of the planet. Besides many other undesirable effects, such an ongoing and even increased use of fossil fuel energy would further harm the local as well as the global environment. The latter is particularly true through the emissions of even further amounts of greenhouse gases (GHG), thereby undermining the chances to achieve the global climate goals defined and agreed within the Paris

N. Bullerdiek · U. Neuling · M. Kaltschmitt (✉)
Hamburg University of Technology (TUHH), Institute of Environmental Technology and Energy Economics (IUE), Hamburg, Germany
e-mail: kaltschmitt@tuhh.de

1

Agreement. In addition, locally effective emissions like nitrogen oxides (NO_x), sulfur dioxides (SO_2), and particulate matter released during the use of fossil fuel energy also contribute to negative social and environmental impacts.

Additionally, fossil fuel-based resources are limited and unevenly distributed across the planet. This unfortunate situation results in numerous frictions and severe political challenges. For instance, many cost-effective, exploitable fossil fuel resources are located in partly politically unstable regions, and only a limited number of global companies control substantial portions of the global natural gas, crude oil, and coal markets. This concentration of resources, as well as the existing market oligopoly, influences the development of global energy prices and impacts the security of a global cost-effective energy supply. Moreover, there is also uncertainty about whether the currently known fossil fuel reserves can be fully exploited in the years to come, considering the significant environmental and climatic constraints, which impact the future development of global human society. In particular, future generations may be affected as the consequences of the current energy system on the global climate may take years to decades to fully manifest.

Overall, this challenging situation does not guarantee a stable, reliable, and socially acceptable global energy supply, which is vital for the peaceful and prosperous development of global societies in the future. Given these pressing issues and challenges, the transformation of the global energy system toward a more sustainable and environmentally / climate-friendly energy supply becomes imperative. This shift is not only necessary to achieve global climate goals but also to align global economies with the unalterable planetary boundaries defined by nature.

In theory, transitioning from the current unsustainable energy system to a globally aligned "green" energy system in line with the United Nations Sustainable Development Goals (UN SDG) might appear straightforward and universally accepted within globalized economies. The necessary measures to achieve these goals have been well-known for decades and can easily be translated into clear and achievable transformation steps. However, in reality, such a shift is far more complex and highly controversial within society. In fact, transitioning to a fully "green", stable, and cost-effective global energy system is a comprehensive and time-consuming endeavor that requires collaborative efforts from stakeholders across technological, economic, environmental, political, scientific, and societal levels over decades. Additionally, such a process also demands substantial financial resources for the costly construction of a reliable infrastructure, which is likely to be provided only when long-term investment security can be assured.

On a technological level, the transition toward an energy supply system based on renewable sources of energy offers various approaches. These approaches can be used within a wide range of partly very diverse applications in the overall globalized economies. Up to now, the focus has been clearly on generating "green" electricity from renewable sources of energy such as solar radiation (via photovoltaic systems), wind power (via wind turbines), hydropower (via hydropower plants), and biomass (via biogas plants and/or thermal power plants based on solid biofuels). In addition to the provision of "green" electricity, bioenergy — and to some extend also geothermal and solar energy — are also utilized within the heating sector. Furthermore, for land transportation, biomass in the form of biofuels (e.g., bioethanol, biodiesel, biomethane) is partially used to power ground-based vehicles.

However, relying solely on the renewable energy options outlined above for an extensive integration of renewable energies into the global energy system in the next decades would lead to significant challenges and limitations. For instance, in terms of energetic use, certain sectors of the global economy, especially those that require propulsion systems with high power and high energy storage densities (e.g., aviation, maritime), cannot meet their energy demands based on currently market-ready technologies operating solely on electricity; so far the volumetric and gravimetric energy density of battery systems is way too low. For a significant share of such sectors this limitation is likely to apply even to fully electrified technologies developed in the upcoming decades. Thus, while there is a clear shift and an evident transition towards a more electric world, achieving fully electric applications across all sectors within the few decades remaining to meet global climate goals is highly improbable and truly unlikely. Furthermore, in several industrial production processes with significant markets, hydrocarbons are essential as an educt (as raw materials). This demand for a raw material that undergoes very diverse processing cannot be fulfilled through electrification. Additionally, while biomass can indeed serve as a renewable resource, the availability of sustainable biomass resources is limited, and organic matter is already in high demand across various sectors, applications, and markets.

Therefore, as an efficient way to utilize "green" energy to fulfill various duties within both industry and society, it remains crucial to transform significant portions of the global energy system toward a higher share of a direct use of "renewable" / "green" electricity. Most likely, this approach is also the most efficient and thus probably the most cost-effective option. This transformation will be particularly relevant for sectors where electrification is feasible, due to forthcoming developments in appropriate / adapted technologies and the further improvement of the respective infrastructure. Additionally, given anticipated global advancements, such approaches are expected to become economically viable in a timely manner. These sectors may encompass portions of the heating sector, large parts of the land-based transportation, as well as substantial segments of industrial production.

However, as previously mentioned, there are other sectors, particularly the so-called "hard-to-abate" or "hard-to-decarbonize" sectors like aviation and maritime transportation, which currently rely on other energy options (primarily carbon-based fuels such as kerosene or diesel) and propulsion systems distinct from fully electrified systems. In certain segments of these sectors, given the current state of technology and foreseeable advancements, it appears highly unlikely that commercially viable electrified propulsion systems and the necessary infrastructure developments will be able to electrify substantial portions of these sectors in the coming decades. For instance, regarding aviation, even if technological solutions emerge in the future to enable intercontinental flights using electrical energy, these options are likely to be available too late to significantly contribute to the greenhouse gas reduction goals set for 2045 and 2050 on a system level. Moreover, several processes are operated to produce carbon-based products primarily for use as materials rather than as energy carriers, (e.g., in the chemical industry). These hydrocarbon-based products have vast global markets that have shown substantial growth in recent years.

In these hard-to-abate and hard-to-decarbonize sectors and applications, renewable fuel options such as kerosene, diesel fuels, or primary chemicals like methanol can for example be derived from sustainable biomass sources. However, despite its renewable origin, biomass is an a priori limited resource and already in demand by various other sectors and applications, including food and fodder production, as well as industries like construction, furniture, pulp, and paper. Thus, while it is likely that a significant expansion of the use of sustainable biomass for energy purposes will be essential in achieving global climate objectives, nevertheless, with numerous applications across various sectors competing for the limited resource of biomass, it is highly improbable that the future demand for renewable energy and renewable molecules can be entirely met by sustainably provided organic matter (biomass) in cases where large-scale, economically viable direct electrification is not feasible in the foreseeable decades.

In this context, energy carriers and hydrocarbon feedstocks derived from renewable sources other than biomass are gaining significance for these hard-to-abate and hard-to-decarbonize sectors. The broad variety of respective provision chains, conversion pathways and technologies used are primarily categorized and summarized under the umbrella terms "Power-to-X" (PtX) and "Powerfuels". However, depending on the specific context a variety of other terms also exist (e.g., e-fuels, synthetic fuels, electricity-based fuels, electrofuels). In general, these terms encompass a wide range of technologies able to convert electricity — typically generated from renewable sources of energy due to their very low climate impact — into various synthetic / synthesized fuels and / or hydrocarbons. The respective process chains typically start with the production of hydrogen via water electrolysis. Depending on the desired fuel and / or energy carrier, sustainably sourced carrier molecules, such as carbon dioxide or nitrogen, are utilized to synthesize gaseous or liquid "green" molecules, which can be used in sectors like transportation or the chemical industry.

The topic of Power-to-X technologies and powerfuels is currently the subject of intense and often contentious discussions within industry, research, politics, as well as the overall society. These discussions primarily revolve around the practical and technical feasibility of Power-to-X processes and applications, their economic viability, the respective environmental benefits, the contribution to climate protection as well as the social acceptability. To a large degree, these options are seen as an essential element for the transition toward a sustainable energy supply in the coming decades and for achieving the global greenhouse gas reduction goals — especially in the hard-to-abate and hard-to-decarbonize sectors mentioned above. In many areas related to Power-to-X, research activities, as well as technological and conceptual developments, are still ongoing. As a result, some technologies, processes, and concepts are only in early stages of technological and / or commercial development, while others are more advanced but face the challenges of scaling up to commercially available large-scale units. Such an up-scaling is crucial to exploit economy of scale effects and to facilitate a large-scale technological ramp-up on a global scale — all within a highly competitive economic environment with a limited availability of only finite resources.

Within this context, the primary aim of this book is to provide a comprehensive overview of various aspects, diverse considerations, and different perspectives regarding the future role and utilization of Power-to-X pathways on a global scale. This encompasses the challenge of sourcing necessary educts / feedstock options, their conversion into different products and product groups, exploring the possibilities of using these electricity-based fuels / hydrocarbons in various markets, and establishing suitable framework conditions for viable and sustainable markets in the years to come. These objectives are achieved through a collection of articles contributed by experts actively engaged in the various fields related to Power-to-X. Figure 1 provides an overview of the structure and contents in this book.

With the overarching goal defined above, this book is specifically structured to offer a comprehensive overview of various Power-to-X pathways and powerfuels, along with their potential roles across different sectors of global society. Below, the main sections are briefly outlined.

- **Development of the Mobility Sector**. The first section focuses on the mobility sector, including ground-based mobility, aviation and maritime, discussing aspects of their overall development in the past and in future. At least the aviation and maritime sectors will probably face substantial technological challenges in implementing direct electrification, making Power-to-X options pivotal. For example, the aviation sector currently heavily relies on kerosene as a fuel option, which can be derived from various Power-to-X pathways. It is expected that this sector will continue to depend on liquid fuels in the coming decades. This necessitates the likely provision of such fuels through Power-to-X pathways, as there may not be enough biomass to meet the demand of a still-growing global civil aviation industry. In contrast, for maritime applications, a broader range of Power-to-X options may become feasible in the coming years and decades, such as hydrogen, methanol, and ammonia. The objective of this section is to provide an initial comprehensive overview and understanding of these sectors and their potential development. This includes topics related to the use and corresponding markets for renewable fuels, especially in the context of greenhouse gas (GHG) reduction goals, as implemented, for instance, by the European Union (EU) in the Green Deal.
- **Feedstock**. When discussing powerfuels, a pivotal aspect is the provision of the necessary energy and material feedstocks for their production. Therefore, the second section of this book addresses key aspects regarding the feedstocks / educts required for large-scale powerfuel production. This covers various fundamental aspects on the generation of electrical energy ("power") from renewable sources of energy. Following that, the provision of hydrogen based on water electrolysis is addressed in more detail. In this context, aspects related to freshwater resources are also discussed; the question of whether the availability of water limits "green" hydrogen production is tackled. Additionally, water treatment for the subsequent hydrogen production through electrolysis is explored. In the context of hydrogen

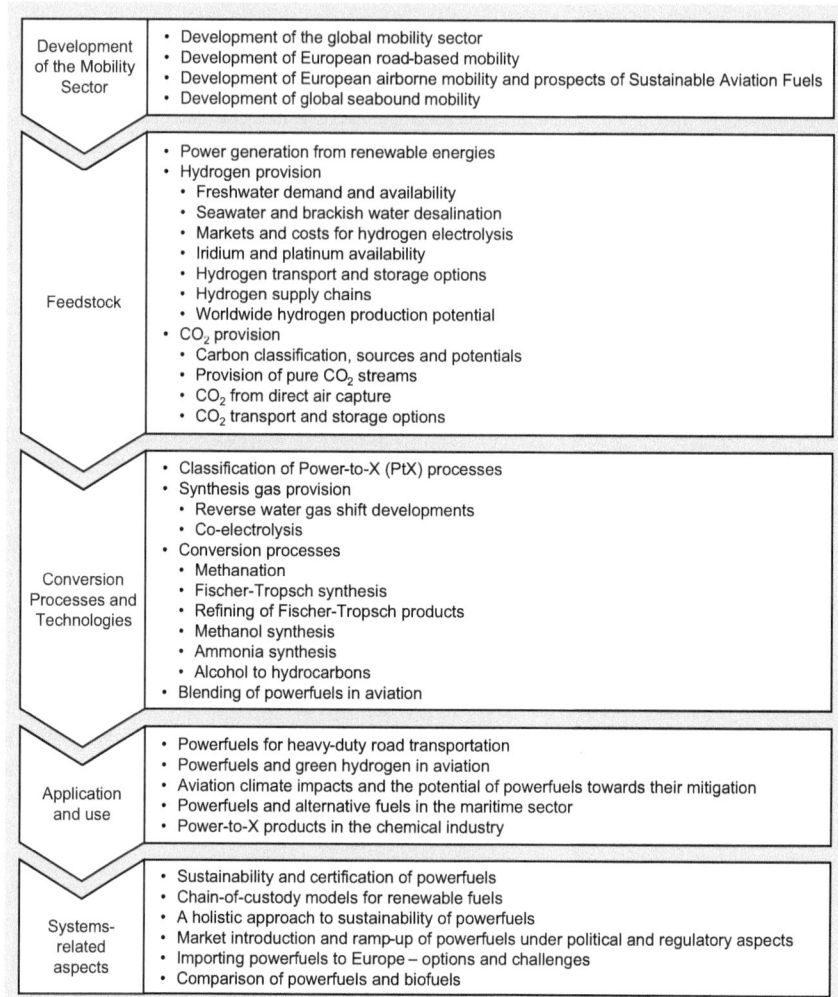

Development of the Mobility Sector	• Development of the global mobility sector • Development of European road-based mobility • Development of European airborne mobility and prospects of Sustainable Aviation Fuels • Development of global seabound mobility

Feedstock	• Power generation from renewable energies
	• Hydrogen provision
	  • Freshwater demand and availability
	  • Seawater and brackish water desalination
	  • Markets and costs for hydrogen electrolysis
	  • Iridium and platinum availability
	  • Hydrogen transport and storage options
	  • Hydrogen supply chains
	  • Worldwide hydrogen production potential
	• CO_2 provision
	  • Carbon classification, sources and potentials
	  • Provision of pure CO_2 streams
	  • CO_2 from direct air capture
	  • CO_2 transport and storage options

Conversion Processes and Technologies	• Classification of Power-to-X (PtX) processes
	• Synthesis gas provision
	  • Reverse water gas shift developments
	  • Co-electrolysis
	• Conversion processes
	  • Methanation
	  • Fischer-Tropsch synthesis
	  • Refining of Fischer-Tropsch products
	  • Methanol synthesis
	  • Ammonia synthesis
	  • Alcohol to hydrocarbons
	• Blending of powerfuels in aviation

Application and use	• Powerfuels for heavy-duty road transportation • Powerfuels and green hydrogen in aviation • Aviation climate impacts and the potential of powerfuels towards their mitigation • Powerfuels and alternative fuels in the maritime sector • Power-to-X products in the chemical industry
Systems-related aspects	• Sustainability and certification of powerfuels • Chain-of-custody models for renewable fuels • A holistic approach to sustainability of powerfuels • Market introduction and ramp-up of powerfuels under political and regulatory aspects • Importing powerfuels to Europe – options and challenges • Comparison of powerfuels and biofuels

Fig. 1 Overall structure of the book

production, material-related aspects regarding the construction and large-scale production of electrolyzers are discussed, such as the availability of platinum or iridium, being so far important trace elements for the electrodes used within an electrolyzer. Within a broader context, various aspects of hydrogen logistics and supply, including transportation and storage, are presented. The latter is particularly relevant in hydrogen production and provision since there are many different options for hydrogen storage and transport. The best choice depends heavily on the specific details of hydrogen production and usage, as well as the transportation distances involved and the local conditions. The selection of a specific hydrogen transport medium and storage method can also significantly impact the overall costs of the hydrogen supply chain. Furthermore, the framework conditions and opportunities for cost-effective hydrogen production are strongly dependent on the ability to provide large amounts of electricity from renewable sources of energy at scale, making it a highly location-dependent issue. In this regard, the analysis and presentation of hydrogen production potentials for different regions of the world are also considered. In addition to hydrogen provision, another focus is placed on aspects of carbon provision in the form of carbon dioxide (CO_2), which is particularly relevant when aiming for carbon containing powerfuels (hydrocarbons) as end products. A closed carbon loop is crucial for climate mitigation, so the CO_2 should originate from the atmosphere. Therefore, this section discusses additionally various potential CO_2 sources for Power-to-X production and the required technologies for a respective CO_2 supply.

- **Conversion Processes and Technologies**. In the third section, specific conversion technologies and processes for powerfuel production are described in detail. To produce carbon-containing Power-to-X products and product groups, a corresponding synthesis gas containing carbon, typically in the chemical form of carbon monoxide (CO), is required in addition to hydrogen. Therefore, two basic technologies for providing a carbon-containing synthesis gas are introduced: the reverse water gas shift approach and the co-electrolysis concept. Subsequently, various conversion processes currently discussed within the Power-to-X context are described in detail. Methanation plays a pivotal role in providing methane for different applications and sectors. Via the Fischer–Tropsch route, synthetic crude oils can be produced, which can then be further processed into a wide array of products based on the specific refining approach applied. In this context, the Fischer–Tropsch conversion route and the corresponding possibilities for further processing of the provided Fischer–Tropsch crude are discussed in detail. Furthermore, the synthesis of methanol, ammonia, and various other molecules are tackled. The same holds true for processing pathways to convert alcohols into higher (long chain) hydrocarbons. Methanol synthesis, for instance, offers the advantage of being an already established large-scale and commercially utilized process, making it a rather technologically feasible option within the Power-to-X context. A similar case applies to ammonia synthesis, where, depending on the final application, an advantage lies in not requiring carbon as a carrier molecule.

- **Application and Use**. Powerfuels present unique challenges and opportunities across various industries and sectors. In the fourth section of this book, a comprehensive perspective on the use of powerfuels and their diverse roles and potentials in different sectors is presented. This includes their use in heavy-duty road transportation, aviation, maritime as well as in the chemical industry. Powerfuels are considered a key option for integrating renewable energies into these sectors at scale, especially where a complete electrification is challenging. Depending on the specific powerfuel option (e.g., kerosene or diesel), there can be compatibility options or synergies with the already existing infrastructure on a global scale, which might be advantageous for a market ramp-up while minimizing the need for new investments. In the case of aviation, besides the use of kerosene-based aviation fuels, options for integrating "green" hydrogen into this sector are also presented. Additionally, aviation-related climate effects beyond CO_2-related greenhouse gas (GHG) emissions (referred to as non-CO_2 effects) and the potential influence of powerfuels on these effects are discussed. Finally, the integration of Power-to-X technologies and products within the chemical industry is analysed, highlighting their role to replace carbon-based materials currently derived from fossil fuel-based resources. Especially in this industry, appropriate technologies and processes are required to produce carbon-containing products (e.g., plastics) based on a sustainably sourced carbon, thus being in direct competition with the transport sector.

- **System-Related Aspects**. This section addresses overarching systemic aspects of the integration and the utilization of powerfuels. This includes considerations such as sustainability aspects, certification, reporting and accounting, which are pivotal in gaining acceptance and integrating powerfuels into the current energy landscape. In this section aspects related to market introduction, scale-up, and the import of powerfuels are also presented and political as well as regulatory frameworks that may hinder or facilitate their widespread adoption are examined. Finally, a comparative analysis of powerfuels and biofuels is conducted to evaluate their roles in transitioning towards a more sustainable energy system.

In summary, this book aims to provide a comprehensive understanding of Power-to-X technologies and powerfuels by addressing key aspects. These include their application in crucial sectors that will rely on these options, the essential energy and feedstock inputs, the most important conversion technologies, selected cross-sector applications, and the implications of widespread commercial use of powerfuels at scale.

Development of the Mobility Sector

Development of the Global Mobility Sector

Barbara Lenz

Abstract In light of climate change and the significant role traffic plays in anthropogenic greenhouse gas (GHG) emissions, mobility and transport are increasingly facing critical evaluation. Nonetheless, mobility remains a vital and indispensable aspect of people's daily lives. The same holds true for the transportation of goods, which occurs between companies and from companies to retailers and end consumers through global logistics, operating around the clock. Worldwide, many countries have set ambitious plans to ensure high-quality passenger mobility and freight transport, both for the present and the long term, while also addressing climate and environmental protection goals. This chapter aims to provide an overview of the current global transport situation, and the developments anticipated in the coming years and decades. It is clear that both segments will continue to grow, due to increasing demand, but also due to technical developments that will boost international trade in particular.

Keywords Mobility · Freight · Transport demand · Vehicle stock · Fuel consumption

1 The Need and Demand for Mobility and Transport from a Global Perspective

In view of climate change and the significant contribution that traffic makes to anthropogenic greenhouse gas (GHG) emissions, mobility, and transport are increasingly subject to critical scrutiny. Regardless, mobility is an important element in people's everyday lives and is therefore indispensable. The same applies to the transport of goods that takes place between individual companies and from companies to retail and end consumers by worldwide logistics around the clock.

B. Lenz (✉)
Humboldt University of Berlin, Berlin, Germany
e-mail: barbara.lenz@geo.hu-berlin.de

11

Worldwide, many countries have set ambitious plans to ensure passenger mobility and freight transport in good quality and to the necessary extent both for the present and in the long term, while also addressing climate and environmental protection requirements. Against this background, this chapter is intended to provide an insight into the current situation of global transport and into the developments that are expected in the next few years and decades. Both passenger and freight transport are considered. In the part on passenger transport, the focus is on everyday mobility, while the freight transport part is mainly concerned with long-distance transport.

2   Mobility of People

Mobility is a basic requirement for people to participate in life in society. This not only means access to gainful employment, but also to education, culture, and opportunities for political participation. The establishment and maintenance of interpersonal relationships are another important field, and in today's societies, it is difficult to achieve them without mobility [2]. Mobility is both a need and a necessity. If and how the demand for mobility is accomplished depends, on the one hand, from a person's individual resources and skills and, on the other hand, from the technical and infrastructural options that are available to satisfy the mobility need. It is important to note that satisfactory levels of mobility also make an important contribution to the achievement of sustainability goals, by means of providing access and connectivity as well as supply, economic development, and employment.

As would be expected, the volume of global traffic is growing significantly and continuously.[1] For all regions of the world together, i.e., Asia, Europe (European Economic Area and Turkey), Latin America, Middle East and North Africa, OECD-Pacific, Sub-Saharan Africa, emerging markets and North America (USA and Canada), this results for the year 2015 in a value of approximately 52,752 billion passenger kilometers (Table 1).

The data presented in Table 1 is additionally shown in the subsequent Fig. 1 for better comparison, in descending order.

As shown in Fig. 1, the highest traffic performance is achieved in Asia; the penultimate place on the list is Sub-Saharan Africa (SSA)—although in 2015 almost a billion people, or around 12% of the world's population, lived in this region. The

---

[1] One of the main sources of data on transport in an international comparison is the International Transport Forum (ITF) which is part of the OECD. Additional relevant in-formation comes from the International Energy Agency (IEA) publishes statistics for transport-related energy topics; Ren21, a UN organization, publishes with a focus on "renewable energies"; The US government's Energy Information Administration (EIA) repeatedly presents also an international perspective. The ITF compiles data in order to derive forecasts for future developments using models. However, these data do not allow for an in-depth comparison between the major global regions. There are large data deficits with regard to developing and emerging countries, and also with regard to traffic in large cities. Further difficulties can arise by different formats of published data, or varying definitions of the "same" object of data collection.

Table 1 Demand for passenger transport by world region 2015

World region	Year 2015 (billion pkm)
Asia	18,088
United States and Canada	12,870
European Economic Area (EEA)/Turkey	8,093
Latin America	4,431
Middle East and North Africa (MENA)	3,391
OECD-Pacific	2,490
Sub-Saharan Africa (SSA)	1,893
Transition countries	1,496
Total	52,752

pkm—passenger kilometers
Source ITF Transport Outlook [16]; ITF Transport Outlook 2021

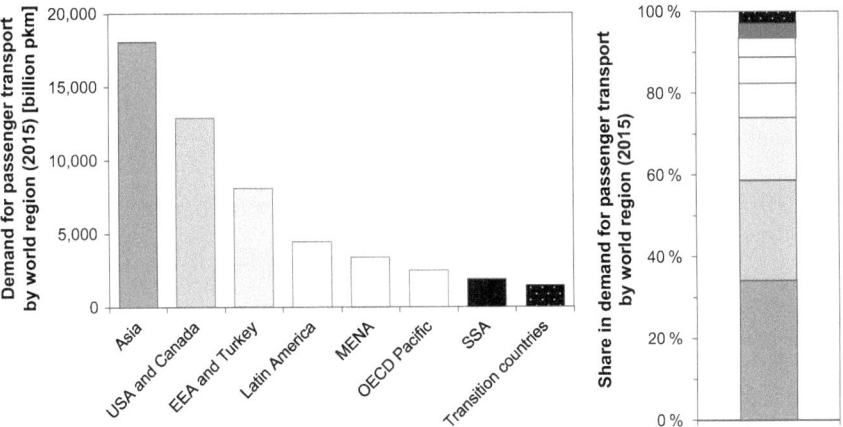

Fig. 1 Demand for passenger transport by world region 2015 in descending order (EEA—European Economic Area; MENA—Middle East and North Africa; OECD—Organization for Economic Cooperation and Development; pkm—passenger kilometers; SSA—Sub-Saharan Africa; USA—United States of America; data according to Table 1)

right part of the figure shows that the three regions—Asia, North America (USA and Canada), and the European Economic Area (EEA) with Turkey—account for about three-quarters of the passenger transport demand, while the remaining shares are distributed among the other world regions. When comparing the major regions, the high proportion of rail traffic in China and India is also noticeable (not explicitly shown in the figure). The railway infrastructure available for passenger trains in these countries has a significant impact on the use of rail-based travel in both local and long-distance transport.

Table 2 Global urban travel by mode group in 2015

World region	Year 2015 (billion pkm)
Private vehicle	10,480
Paratransit[a]	2,439
Public transport[b]	4,070
Active and micro mobility	3,103
Shared vehicle	316
Shared mobility	283
Total	20,691

pkm—passenger kilometers

Source ITF Transport Outlook [16]; ITF Transport Outlook 2021

[a] Paratransit: Public transport-like services operating under unclear regulatory frameworks. Paratransit is more common in developing countries where they serve a significant role in the transport system, operating in parallel to formal services. The term is also used in the United States and Canada to mean on-demand transport services, typically used by the elderly or those with mobility restrictions who find it difficult to use fixed-route systems. However, there services are not included in the Transport Outlook's definition of paratransit (ITF Transport Outlook [16], Glossary)

[b] Public transport: Public transport services served by bus, metro, tram, and rail (ITF Transport Outlook [16], Glossary)

Urban Travel/Transport. Around 40% of passenger transport worldwide occurs in cities (about 20,700 of about 52,750 pkm with regard to the year 2015). In urban traffic, the focus of the modal split is on the car; the term "modal split" here refers to the division of the kilometers traveled between private cars, public transport, sharing options (car and bike), and active individual mobility (bicycle and on foot) (Table 2).[2]

The data presented in Table 2 is additionally shown in the subsequent Fig. 2 for better comparison.

The high proportion of public transport is remarkable, at around 30%. This reflects the great importance of publicly available mobility options, especially in economically less developed countries, where this share is important compared to the respective modal split shares in industrialized countries. In 2017, the modal split share of local public transport in Germany, for instance, was in the metropolises around 20%, but went down to around 12% in the bigger cities and to 6–8% in the smaller cities (long-distance public transport not included ([22], p. 47).

Inter-urban Travel/Transport. Regional and inter-urban transport, i.e., the transport that occurs worldwide within countries at the regional level outside of cities and in rural areas as well as between cities, comprises the remaining 60% of passenger transport performance. This translates to about 32 billion passenger kilometers for the year 2015 (Table 3) (ITF Transport Outlook [15], p. 145).

---

[2] Another calculation of "Modal Split" is by number of trips by transport mode. Then the share of the car is usually lower, as many trips are made by foot or bike; They are made, however, rather on short journeys, while the car is used to a large extent on longer journeys.

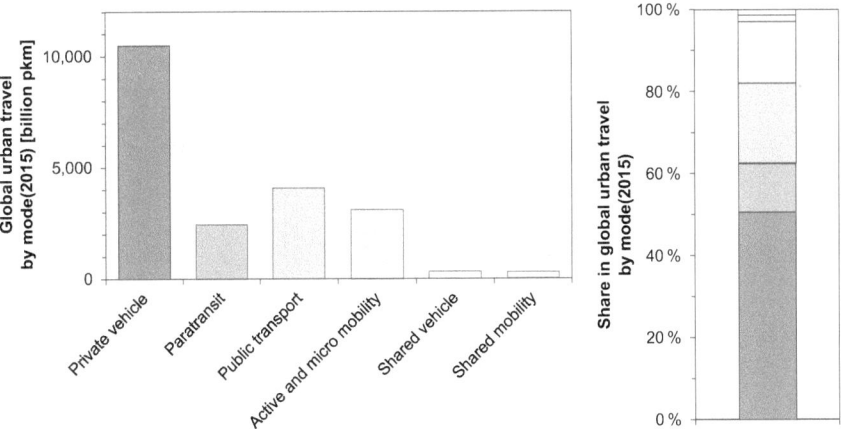

Fig. 2 Comparison of global urban travel by mode group in 2015 (pkm—passenger kilometers; *Source* Data according to Table 2)

Table 3 Global non-urban passenger transport by sub-sector in 2015	World region	Year 2015 (billion pkm)
	Regional	14,867
	Intercity surface	10,211
	Domestic aviation	2,510
	International aviation	4,489
	Total	32,077

pkm—passenger kilometers
Source ITF Transport Outlook [16]; ITF Transport Outlook 2021

The data presented in Table 3 is additionally shown in the subsequent Fig. 3 for better comparison.

Within the non-urban passenger transport category, regional transport accounts for nearly half; the share of ground-based traffic between cities and city regions is almost a third. The remainder of non-urban passenger transport consists of national and international passenger air transport, which makes up about 20%. Within these sub-sectors, about two-thirds come from international aviation and about one-third from domestic aviation.

2.1 Global Differences in Mobility

The amount of traffic that people generate in a region is not determined solely by their sheer number. How many trips people take in a day and how many kilometers they cover depends largely on the resources available at the individual level. Accordingly,

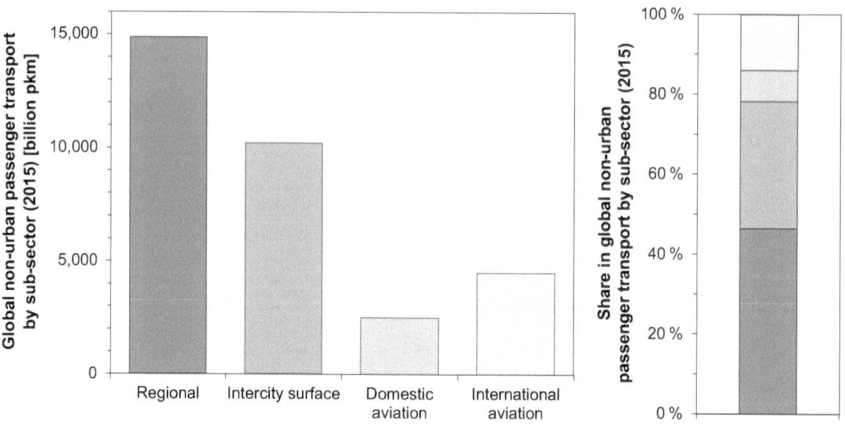

Fig. 3 Comparison of non-urban passenger transport by sub-sector in 2015 (pkm—passenger kilometers; *Source* Data according to Table 3)

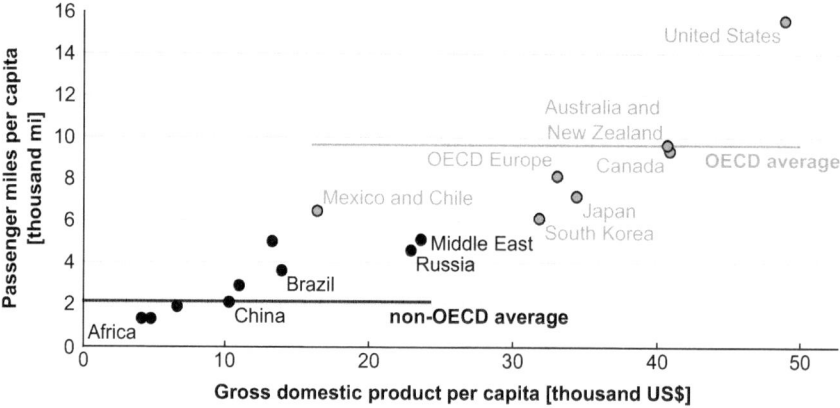

Fig. 4 Passenger miles and gross domestic product per capita in 2012. *Source* EIA [31]

there are significant statistical interrelationships between the number of passenger kilometers and the per capita income of a country or region: The number of kilometers traveled per person increases with the level of gross domestic product (GDP) per capita. While the industrialized countries have values between 7,000 and almost 16,000 km per person per year, in the poorer countries it is only 2,000 km per person or less (Fig. 4).[3]

---

[3] Transport planning and research differentiate between passenger kilometers and vehicle kilometers. The total number of passenger kilometers results from the distances traveled by a specific number of people. The total number of vehicle kilometers results from the distances the vehicles have traveled in traffic. So, if four people sit in a car and drive 10 km, this results in 40 passenger kilometers (pkm) and 10 vehicle kilometers (vkm).

Another significant factor influencing the extent and type of individual mobility is the mobility options available in a city or region. In order to be mobile and have access to facilities for everyday life, built infrastructure such as roads and pathways or railways and stops and stations are essential. This applies to all inland means of transport—walking, cycling, car, public road and rail transport. At the same time, transport infrastructures are a basic requirement for access to markets, both locally and regionally as well as nationally and internationally.

There are, of course, significant differences between countries and regions in the extent and quality of infrastructure networks. An industrialized country like Germany, for instance, has a network of roads (including about 400,000 km of municipal and urban roads) and highways totaling around 625,000 km, while it covers an area of almost 358,000 km^2 and has a population density of 238 inhabitants per km^2 (2021). This makes 7.5 m of road for every resident.

As expected, the situation in developing countries is very different. Considering the countries in Sub-Saharan Africa, for instance, the value of road length per capita can drop to less than half of the value for Germany, with—at the same time—a high proportion of 70 to 80% of unpaved roads in the overall road network (*Source* CIA (no date)). However, in such a comparison, the differences in the land surface, in the population density and the spatial distribution of the population must also be considered (*Source*; 16.05.2023). The average population density in sub-Saharan Africa is only 51 inhabitants per km^2; the total area of the 49 Sub-Saharan states covers 23.7 million km^2 and is therefore five times larger than Europe (*Source* Staatslexikon 2017).

In addition, there are significant differences between regions in the provision of mobility options—this applies especially to local and long-distance public transport and the quality of the service in terms of the frequency and reliability of connections. The share of investments in road and rail infrastructure in GDP can be used as a valuable indicator for assessing the quality of the infrastructure. In 2020, investment in national transport infrastructure—measured as a share of GDP—varied between 5.8% in China and 0.2% in Mexico. The investments relate to both road and rail infrastructure, while the respective shares differ greatly. In the European Economic Area (EEA) and Great Britain, for instance, public investment in rail was about 42% of the total national transport infrastructure, while it was 11% in North America in the year 2018. It has to be noted that investments as mentioned here usually include investment in new infrastructure as well as maintenance expenditure for existing infrastructure [14].

2.2 Vehicle Stock

Another essential component of individual mobility options is the private car. The appreciation of one's own car is reflected not least in the continuously increasing number of cars worldwide, which grew from 276 million vehicles to almost 1.3 billion vehicles between 1978 and 2022 [27]. The strongest driver of this increase is

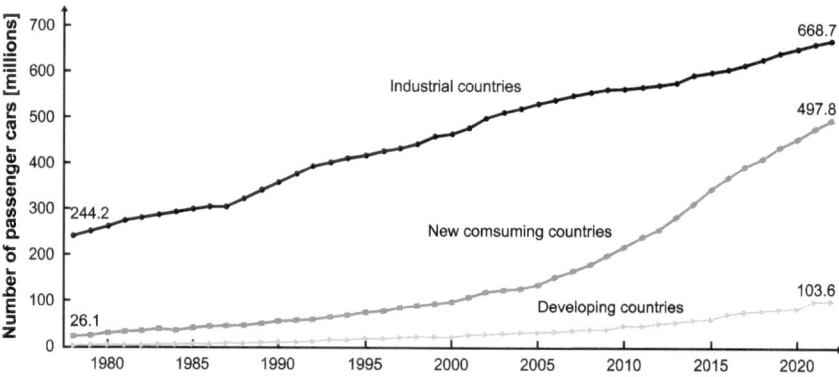

Fig. 5 Development of the global car stock in industrialized countries, new consumer countries and developing countries (1978–2022). *Source* Umweltbundesamt 2024 (Marktdaten Mobilität)

China, where car ownership has augmented dramatically since the mid-2000s, so that in 2022 the motorization rate is 119 cars per 1000 inhabitants [27]. This value is still comparatively low compared to the motorization rates in industrialized countries. In Germany, which ranked nineth in 2022 in Europe in terms of motorization rate (*Source* destatis 2023), there are 585 cars per 1000 inhabitants in 2022, and the trend has been rising for years.

From a global perspective, it continues to be the industrialized countries and the so-called "new consumer countries",[4] which are the triggers for the increase in the number of cars. A fourfold increase in the number of cars between 1978 and 2022 can also be seen in developing countries; however, this remains here at 104 million cars, which is 8.2% of the global stock, at a comparatively low level ([27]; Fig. 5).

In industrialized countries, the number of households with their own car is usually above average where the availability of public transport is rather limited and therefore easy access to services and jobs is not guaranteed. This applies particularly to the edges of urban areas and rural areas. In Germany, for example, 78% of all households had at least one car in 2017. In major cities, only 58% of households owned a car, while in small towns and villages the figure went up to 90% [22].

Such differences between cities and rural areas can be found worldwide, including the highly motorized USA with around 830 cars per 1000 inhabitants. In 2021, 92% of all US households owned at least one vehicle (*Source* Tilford and Megna 2023). However, many US American big cities and big cities worldwide have below-average motorization rates. The city of New York, for instance, has the lowest number of cars per household in the USA at around 46% (value for 2021). In contrast, however, there are a significant number of medium-sized cities with motorization

---

[4] Argentina, Brazil, China, India, Indonesia, Iran, Colombia, Malaysia, Mexico, Pakistan, Philippines, Poland, Russia, Saudi Arabia, South Africa, South Korea, Thailand, Turkey, Ukraine, Venezuela.

rates of almost 100%—the front runner in 2021 was the metropolitan area of Daphne-Fairhope-Foley in the state of Alabama with a motorization rate of almost 98% (about 250,000 inhabitants).

For the New York case, it is noteworthy that there is a similar picture compared to major European cities: While car ownership is very low in the central areas, it increases toward the outskirts of the city (Fig. 6). Still, regardless of the high car availability in the outer parts of the city, public transport is the most important means of transport for many commuters' way to work (*Source* NYCEDC 2018; see also Cortright [9]).

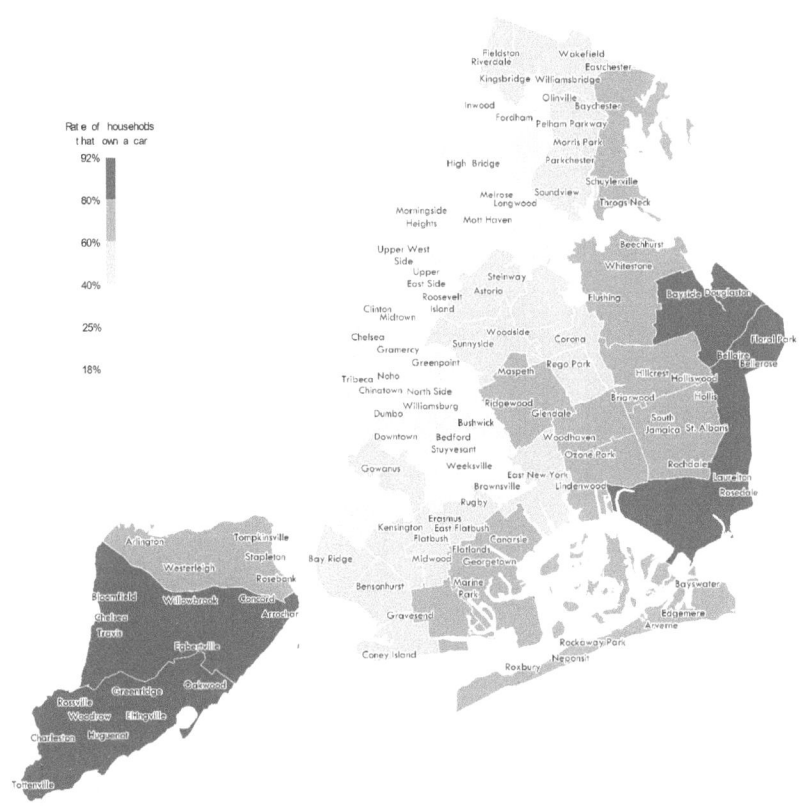

Fig. 6 New York City car ownership rates in 2018. *Source* NYCEDC 2018

2.3 Fuel Consumption and CO$_2$ Emissions

The transport sector's share of global final energy consumption was between 25 and 30% of total final energy consumption in 2018 (data according to World Energy Outlook 2022 and Renewables 2023 Global Status Report). Around 90% are provided on the basis of (fossil) petroleum, the remaining 10% is (fossil) natural gas or biofuels, or electrical energy and others. The major share, around 74%, of the energy is consumed by road transport, 12% by air transport, around 9% by shipping and 2% by rail (Fig. 7). These figures make clear the strong dependence that road transport currently has on the supply of petroleum-based fuels (i.e., petrol and diesel fuels).

The demand for fuels for transport has constantly grown. Between 2009 and 2019, around 2% more fuel was used for road transport every year. Meanwhile, the share of renewable energies in the mobility sector hardly increased; rather, it has remained largely constant over the years in the lower single-digit percentage range (Fig. 8).

By the end of 2022, the number of electrically powered vehicles in the global car fleet had reached around 27 million, with a share of around 70% being purely battery-electric vehicles (BEVs); this corresponds to 2.1% of the total stock (IEA 2023); This development was accompanied, of course, by a corresponding use of energy in the form of electrical energy. Given extensive government support, the increase in electric

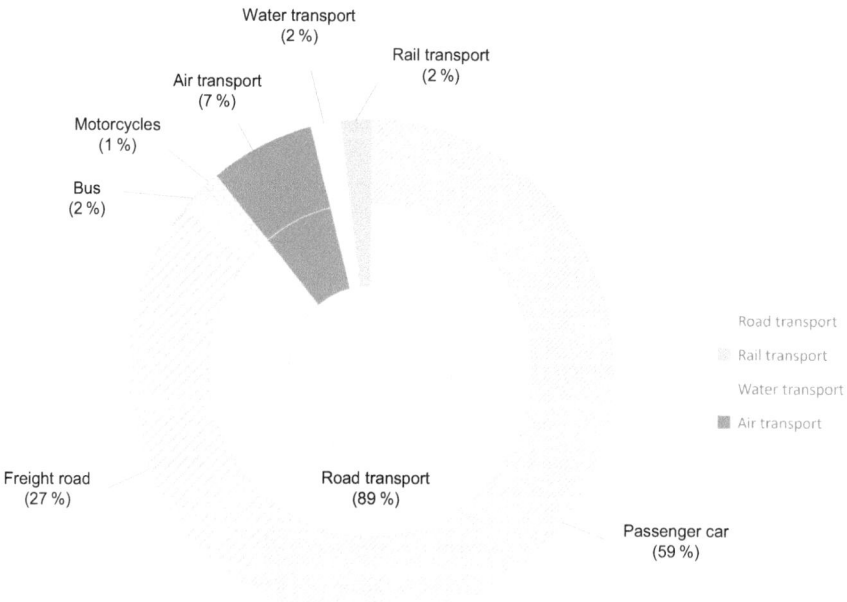

Fig. 7 Shares of energy consumption by modes and types of transport in total energy consumption for transport in 2018 (Passenger cars includes cars, sport utility vehicles and personal trucks; *Source* IEA 2018)

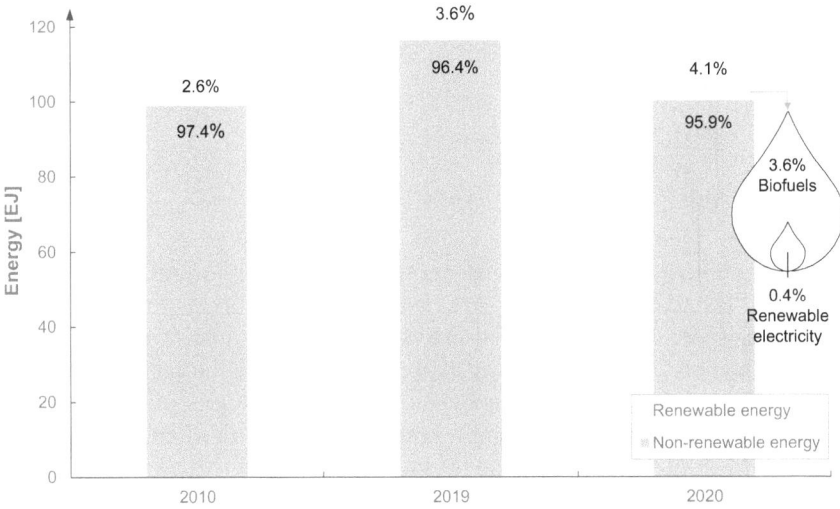

Fig. 8 Renewable share of total final energy consumption in transport, 2010, 2019 and 2020. *Source* Ren21, p. 40

vehicles was particularly pronounced in China, the USA, Germany, Great Britain, and France.

The number of vehicles worldwide is estimated to continue growing. However, a substitution of combustion engines is expected mainly in industrialized countries and not in all parts of the world. Against this background projections await that the demand for fossil fuels for use in the transport sector will increase in the non-OECD countries (Fig. 9).

The same applies to the use of electrical energy. Still, it should be considered in the current projections for 2050 that changes in policies or technological innovations

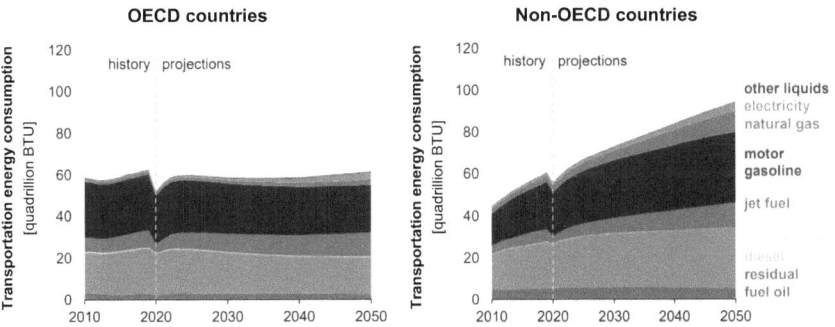

Fig. 9 Projected transportation energy consumption in OECD and non-OECD countries (light duty includes cars, vans, SUVs, pickup trucks, gross vehicle weight rating less than or equal to 10,000 pounds according to https://www.lawinsider.com/dictionary/light-duty-vehicles; 16.05.2023; *Source* EIA [32])

could only be incorporated to a very limited extent given the long projection period. Uncertainties in the projection also arise in view of the price development of liquid fuels.

2.4 Increase in Global Mobility

Traffic has been increasing worldwide for years. This applies to all areas of motorized transport, including air transport, and applies not only to passenger travel, but also to the transport of goods between individual companies and from companies to retailers and end consumers. A key reason for this increase in both passenger and freight transport is, from a global perspective, growing incomes and thus increasing consumption. Both everyday mobility and long-distance travel are increasing as a component of private consumption. More consumption inevitably also means that more and more goods, raw materials and means of production must be transported. This is reinforced by the constantly growing regional and global division of labor, which in turn entails extensive long-distance travel for work-related reasons.

Forecasts suggest further strong growth in all areas of transport. In the passenger transport segment, passenger kilometers are expected to triple between 2015 and 2050 (ITF [12], p. 26; Fig. 10).

There will be a shift in the shares of the world regions; China and India will gain significant importance in future. In addition, high rates of increase are expected in urban areas, not least in view of the global growth of urban centers. A slight decline in the use of private cars is forecast for OECD countries—resulting from an expansion of public transport and sharing mobility options, but also as a result of measures to reduce car traffic in cities—while in the non-OECD countries urban car traffic continues to grow (ITF [12], p. 29; Table 4).

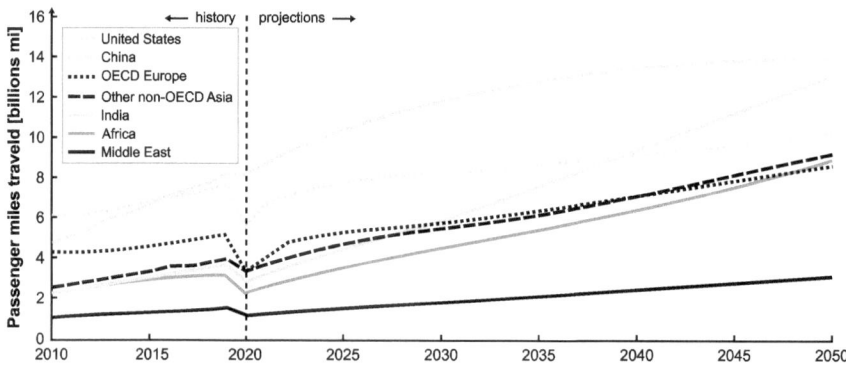

Fig. 10 Forecasts for growth in car traffic up to 2050 in selected regions of the world. *Source* EIA [32]

Table 4 Urban transport growth by mode

	2015–2030 (%)	2015–2050 (%)
OECD urban transport demand		
Private cars	− 0.9	− 0.2
Two and three wheelers	3.0	2.1
Bus and BRT	1.7	1.9
Rail and metro	2.7	2.2
Shared mobility (including all modes)	14.9	8.4
Non-OECD urban transport demand		
Private cars	− 0.2	0.4
Two and three wheelers	3.1	2.2
Bus and BRT	3.3	2.6
Rail and metro	4.7	3.6
Shared mobility (including all modes)	23.8	12.4

Current demand pathway, compound annual growth rates of passenger kilometers in percentages; BRT—bus rapid transit
Source ITF Transport Outlook [12], p. 29

The data presented in Table 4 is additionally shown in the subsequent Fig. 11 for better comparison.

The shift in shares on the global level is largely due to the increase in car ownership in new consumer countries and in developing countries. At the same time, it must be noted that a further strong increase in mobility is hardly possible for the industrialized countries. Their level of mobility already achieved is extremely high and the extent of individual mobility cannot be increased arbitrarily. Concrete limits arise for the individual because of her or his ties to physiological requirements and rhythms, as well as because of the time required to overcome space. This inevitable capping of demand at the individual level is referred to as the law of the "constant travel time budget". This describes the observation that the average (!) time spent on mobility is approximately the same worldwide at 60 to 80 min per day, and it has been rather stable over the years.

An analysis for Great Britain, for instance, shows how the mobility budget in terms of time spent remains the same over the years 1970 to 2008 [20]. There is also hardly any increase in the daily number of trips. However, the number of kilometers traveled per person is increasing significantly, primarily due to the better technical equipment of households with private vehicles, whose performance in terms of speed, comfort and capacity is continuously improving and for which more and better infrastructure is increasingly available (Fig. 12). This observation must be associated with strong variation within the major regions, in line with the diversity of technical and infrastructural mobility options as well as the population's supply of resources that make the use of these options possible [17].

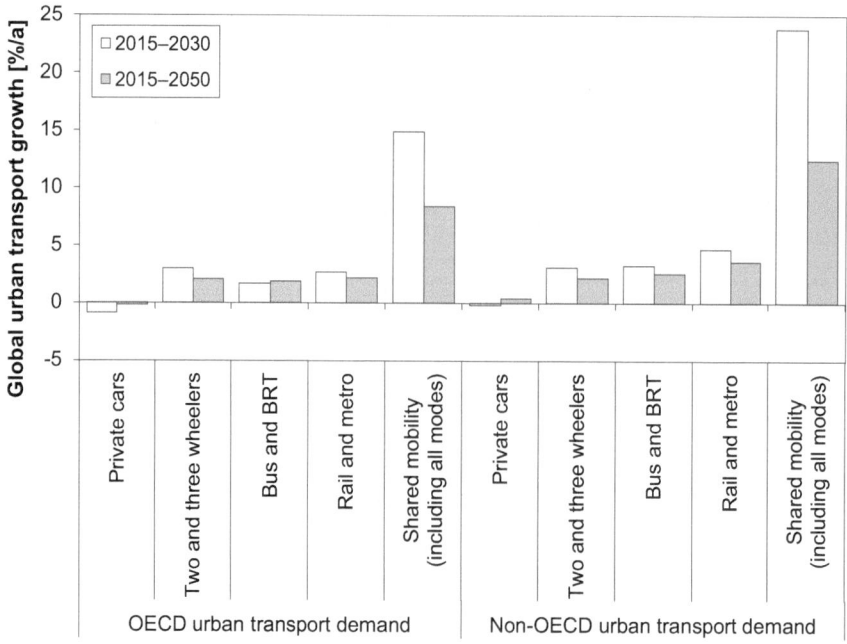

Fig. 11 Forecasts for growth in car traffic up to 2050 in selected regions of the world. BRT—bus rapid transit; *Source* EIA [32]

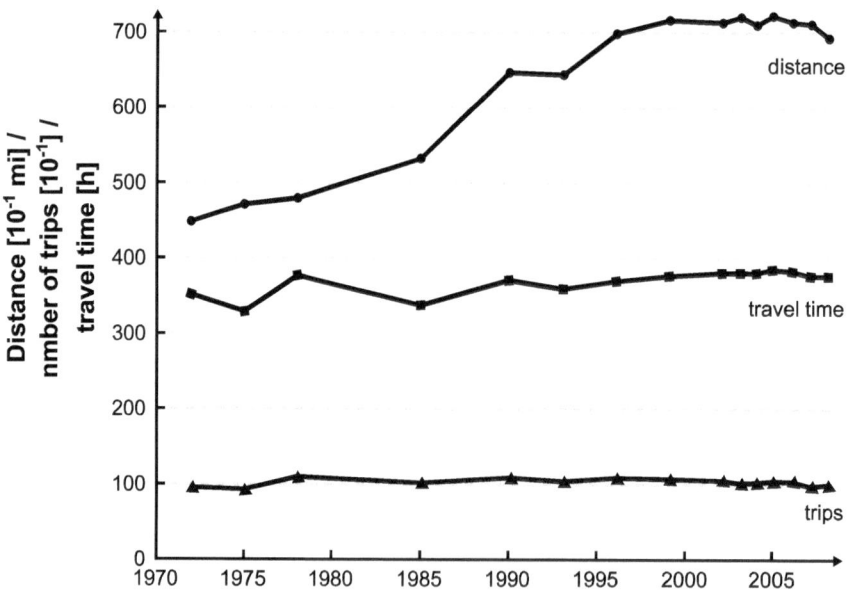

Fig. 12 Travel time (hours per person per year), distance (miles per person per year) and journeys (per person per year) in Great Britain from the 1970s to the 2000s [20] (*Source* Metz, p. 661))

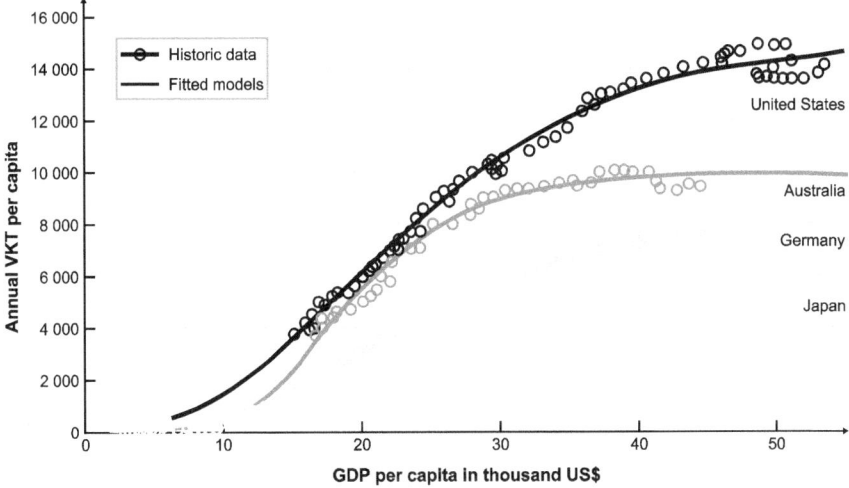

Fig. 13 Forecasted saturation levels for vehicle kilometers (VKT) (estimate based on vehicle kilometers plotted against GDP per capita PPP: *Source* Seum et al. [26], p. 12)

The fact that mobility cannot be increased indefinitely also means that above a certain level, the growth in per capita income has less and less of an impact on the amount of transport demand. Using the example of the BRIC countries (Brazil, Russia, India, and China), it was shown at what magnitude of GDP saturation occurs, measured by average passenger kilometers per inhabitant. Yet, there is no "global level" of saturation. Rather, clear differences between the individual countries come up: with the same GDP per capita, saturation is achieved in the USA at an annual average of 15,100 pkm compared to 6000 pkm in Japan ([26]; Fig. 13).[5]

In addition to economic aspects, other factors included in the model-based calculations were lack of alternatives, pro-car policies, and car culture as causes that affect the level of saturation to varying degrees.[6]

---

[5] The parallel development of GDP or income and mobility is not only a sign of people's ability to be mobile, but it also indicates that with increasing prosperity, new "living conditions" develop that require mobility—so, for instance, the opportunities for better training, a higher-quality job or specific medical care. In addition, mobility is becoming an important purpose in leisure time. In Germany, for example, a quarter of car mileage is spent traveling to leisure activities (excluding vacation trips). Similar values can be found in other European countries such as Italy, Denmark or Belgium (EU; URL: https://ec.europa.eu/eurostat/statistics-explained/index.php?title=Passenger_mobility_statistics; 16.05.2023).

[6] The importance of private cars must be regarded hand in hand with the growing cultural and social importance of mobility. The car has developed into the epitome of individual mobility, combined with a maximum of flexibility, both in terms of time of use (the private car is available 24/7) and accessibility to a wide variety of locations. At the same time, cars can be used to express social status. Numerous studies have examined to what extent and, above all, why young people in industrialized countries in particular use cars less. It turned out that the declining status value of one's own car is only part of the explanation, but that the lower car use is also largely dependent on the context: In Western Europe, for example, the proportion of young people who continue education

3 Freight Transport

3.1 Freight Transport Worldwide

Freight transport includes the transport of all kinds of goods between different locations by road, rail, water, air, or pipeline. Goods are transported from one company to other companies, or from a company to private consumers or commercial end users. In addition, freight transport occurs in the form of internal company transport, which is usually not recorded by statistics. The share of modes for freight transport depends on the distance and accessibility between the place of dispatch and the place of receipt. The available infrastructures, as well as the costs associated with their use and the operation of the respective means of transport, have a significant influence on the modal split of freight transport.

It was in the 1960es when a massive growth in freight transport—global and national—began, which has continued almost uninterrupted since then. The trigger for this increase at the international level was the onset of globalization in the production of industrial and agricultural products. Between 1950 and 2022, the amount of goods traded worldwide increased 45-fold. During the same period, the nominal value of traded goods increased four hundred times—from 62 billion US$ in 1950 to 24,715 billion USD in 2022 (*Source* WTO 2023).

Rising incomes in industrialized and emerging countries and population growth were and are the main drivers on the consumption side. The continuously growing demand for consumer goods brought about a fundamental change in the product structure of international trade. Manufacturing industries across all branches massively promoted the international division of labor. Long and often complex production chains emerged, that made it possible to use regional differences in production costs—especially for labor—to produce goods cost-effectively. This meant that the proportion of intermediate goods in the total amount of goods transported was constantly growing. The term "intermediate goods" refers to a category of manufactured goods that are precursors to a consumer good or an investment good (mainly machinery used for the production of other goods).

In the 1960s, world trade comprised around 24% manufactured goods, 21% agricultural products, and 52% fuels and mining products [5, 25, 29]. Today, the share of manufactured goods in global exports is almost two-thirds, around 60% are intermediate goods so that around half of all goods traded globally are intermediate goods. The strongest category in terms of value within intermediate goods in 2022 was "other industrial supplies" with a value of 4771 billion US$, while "food and

---

and training after leaving school is increasing and is often combined with residence in a bigger city. There, public transport mostly provides a high level of service, in addition the urban cycling network is well developed. This means that the need to own a car is relatively low [18]. The use of alternative means of transport such as buses, trams or bicycles also fits with a sustainability-oriented mindset, which is particularly true among younger people. In a different social environment, the car's promise of status can still be present among younger people if it conveys values such as autonomy and independence.

beverages" had the lowest value at about 472 billion US$ [34, 35] (Table 5). Current trade in intermediate goods predominantly happens between developed countries. This development was and is primarily driven by multinational enterprises that have above-average trade flows of intermediate inputs and, as a consequence, a higher ratio of foreign to domestic inputs [21].

The long and complex supply chains would have been unthinkable without a significant drop in transport costs (Fig. 14). This drop in costs was a result of the rapidly growing demand for international freight transport from the 1960s onwards, which led to an expansion of transport capacities associated with economies of scale and supplemented by fundamental technical innovations. For international trade in intermediate and final goods, containerized shipping was "one of the most important transportation revolutions" in the twentieth century ([10], p. 141). Its introduction started in the 1960s by routes between the USA and Europe as well as the USA and Japan. Transporting the goods in standardized containers provided considerable cost savings, as the previously necessary reloading operations during the transport from the starting point of the load to its actual destination, sometimes from ship to ship, but also from ship to train or to truck, were no longer necessary. This saved demurrage costs in ports, labor costs for reloading processes and storage costs. Due to these changes in the transport and logistics processes, but also due to higher ship speeds, the transport time was significantly reduced, for example, in traffic between Europe and North America by 5–10 days; today, a cargo ship from Europe to the US east coast takes 8–10 days. Estimates specify that, compared to the era before containerization, the lay time of break-bulk (non-container) cargo ships went from one-half to two-thirds of the ship's life (UNCTAD, Unitization of Cargo, 1970; cited from Hummels [10], p. 141).

Despite the numerous advantages, the dissemination of containerized shipping was relatively slow. This was due to the large, fixed costs of adoption for ships and containers. In addition, the expansion of container traffic could only take place at those ports that were equipped with specialized cranes, storage areas, and railheads. The transport of goods by container required new options for transport to and from the port by train and truck. The rapid development in the transport of goods since the 1960s and 1970s, made transport logistics becoming a key area for the international division of labor and trade [33].

Table 5 World merchandise export of intermediate goods by category in 2022

Category of intermediate goods	Merchandise export (billion US$)
Other industrial supplies	4,771
Parts and accessories (excl. transport equipment)	1,904
Parts and accessories (transport equipment)	1,057
Ores, precious stones	918
Food and beverages	472
Total	9,122

Source WTO [34, 35]

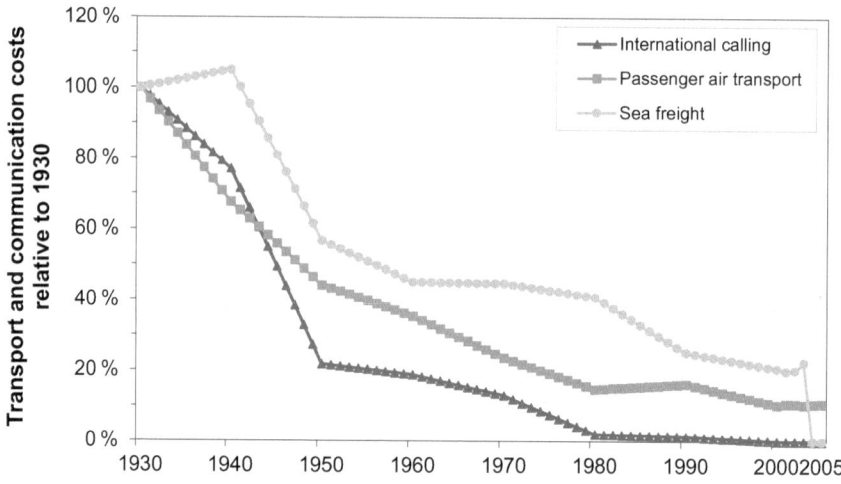

Fig. 14 Decline of transport and communication cost relative to 1930 (Sea freight corresponds to average international freight charges per ton; passenger air transport corresponds to average airline revenue per passenger mile until 2000 spliced to US import passenger fares afterward; international calls correspond to cost of a three-minute call from New York to London. *Source* Ortiz-Ospina et al. [23]; OurWorldInData 2023)

3.2   Inland Freight Transport

Parallel to global growth, there was a strong and sustained increase in inland freight transport, especially in the industrialized countries. In Germany, for example, domestic freight transport increased by 16% between 2000 and 2019, in absolute values from 440,000 million tkm to 476,000 million tkm. In the EU 27, this figure was even higher at 23%. The biggest winner was road freight transport with 31% (Table 6).

The modal split as presented in Table 6 points to the dominant role of the road as primary mode for inland goods transport. An important prerequisite for this was the expansion of the road infrastructure, which was driven forward in industrialized countries with the onset of motorization of road transport. In Germany, for example, the length of trunk roads increased from 26,900 km to 36,300 km between 1960 and

Table 6 Freight transport performance in EU 27 by mode of transport 2000/2019 (in million tkm)

	Road	Rail	Inland waterways	Pipelines	Sea	Total
2000	1,343.9	387.9	133.7	115.7	775.7	2,756.0
2019	1,764.8	407.9	139.7	101.0	979.1	3,392.5
2000–2019 (%)	+ 31.3	+ 5.2	+ 4.5	− 12.7	+ 26.2	+ 23.1

Source Eurostat, Transport in Figures 2022, Part 2 Transport. Chapter 2.2 Performance of Freight Transport in tkm

Table 7 Length of motorways and trunk roads in West Germany 1960–1989

Year	Motorways (km)	Trunk roads (km)
1960	2,500	26,900
1970	4,100	36,300
1980	7,290	39,590
1989	8,720	39,820
2022	13,172	37,810

Sources BPB [6], p. 231; BMDV [4]

1970; in 1960, the highways were 2,500 km, and in 1970, it was already 4,100 km (Table 7).

In most European countries, the construction of trunk roads and motorways only began after the Second World War, motivated by the beginning of motorization in households, but also by the rapidly growing freight traffic. As early as 1950, a declaration of intent was launched on the European level with the aim of a coordinated networking of concentrations of population and economy in Europe via roads that were reserved exclusively for motor vehicle traffic. However, the network of "European roads" was built only in 1975 with the concretization of the agreement [19]. Today, the European motorway network (EU27) covers around 82,000 km (Wikipedia 2023).

In the United States, the focus on building a nationwide road network connecting all states and urban regions was also after World War II. The 1956 plan called for expanding the existing trunk road network by around 66,000 km within 10 years and creating an interstate highway system that was further supplemented by the expansion of state and regional highways. In order to secure the financing, the "Federal Highway Act" and the "Highway Revenue Act" set aside the fuel tax to finance road construction so that 90% of the construction of the Interstate Highway System could be financed by the federal government. Downstream roads were also supported by the federal government [7]. The Interstate Highway System today stretches for almost 80,000 km.

In the emerging countries construction of roads and especially highways only began from the 1980s. Given the enormous speed in which highways are being built, China plays a special role, having established a network of over 160,000 km in length since 1988 (Fig. 15).[7]

---

[7] International comparisons of road infrastructure are difficult. They are useful, though feasible only to a limited extent, on the one hand due to the lack of comparability of roads with regard to their state of development and maintenance—e.g. paved or unpaved, number of lanes—but also because of the limited significance of the absolute total length of roads in countries of different size. The network density (km of roads per km^2 of area) might be used as a comparative value with some more significance. However, this value does not take the quality of roads into account and must in any case be considered in relation to the size and distribution of population and of significant economic activities. The same applies to the density value "inhabitants per km of road" which is also rather often used.

Fig. 15 Expressway network in China 2022. *Source* Alan Fan Pei (21.07.2022): China National Express-way Network; Wikimedia Commons License https://creativecommons.org/licenses/by-sa/4.0/deed.de; 06.09.2023

While road freight transport may appear to be the "norm" for many countries, there are significant differences. In Europe, for instance, countries with high road traffic shares include the Czech Republic, Poland and Hungary. At the same time, other countries show high percentages in freight transport by rail: in 2019 the Baltic countries Estonia, Latvia and Lithuania had shares of 67, 84, and 70%, respectively (Fig. 16).

Considerable differences in the modal split of domestic freight transport are also evident at the international level. Countries with low shares of rail freight transport (measured in tkm in 2019) include, for example, Japan with just under 9% or the UK with 8% (data for 2019; *Sources*: OECD Transport Statistics. OECD Transport Statistics 2020; Transport Statistics Great Britain 2020). High proportions can be found in Australia (67%) and Kazakhstan (48%), among others, but also in the USA, where 29% of goods are transported by rail (*Source* rosap Transportation Statistics Annual Report 2022). This quite high share in the USA is due to the traffic running on the approximately 150,000 km main routes of the rail freight network, on which seven private companies (so-called Class I railroads) offer their transport service. The highest proportions of modal split of 30% and more are achieved in the

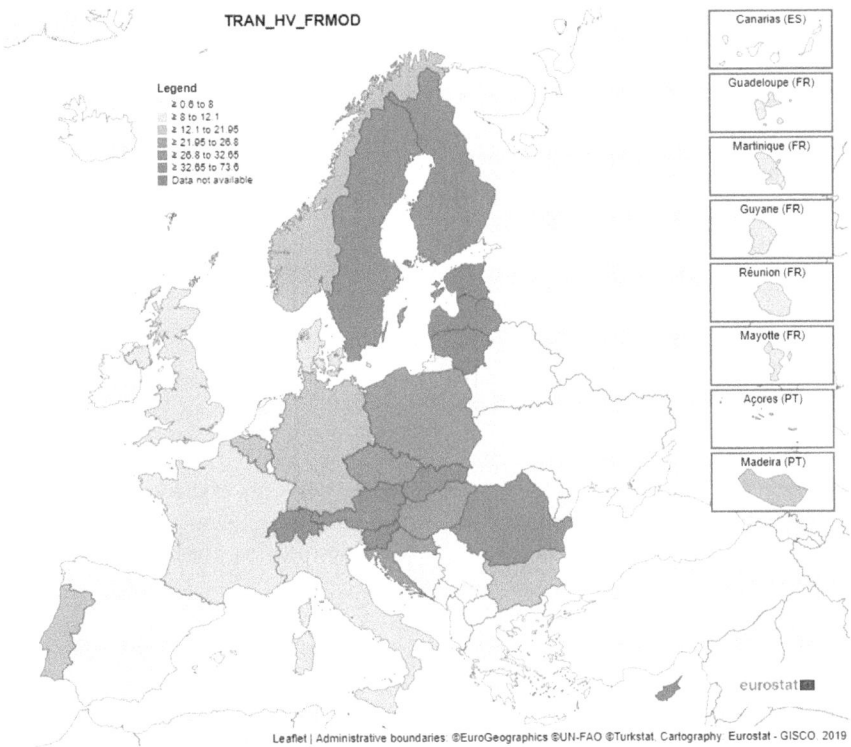

Fig. 16 Share (in percent) of rail in the modal split of freight transport in the countries of Europe (EU27) and Switzerland in 2019. *Source* Eurostat 2023

distance range between 1000–2000 miles (1600–3200 km) (*Source* US Department for Transportation [30]).

Through extensive investments in infrastructure and rolling stock, e.g., larger cars, double stack container railcars, and on-dock rail, the modal split of the railway in the USA could be continuously increased over the years, starting at 22% in 1980. One of the important growth drivers in rail freight transport in the USA today is intermodal transport, especially the combination of sea ships and trains. For rail transport in the USA, it is important to note the fundamental difference between freight and passenger transport: In contrast to freight transport, passenger rail transport is largely a state responsibility, and for many years the state did not provide significant investment in the passenger rail network [7].

For the global transport of goods between countries, ocean shipping is by far the most important mode. In 2022, 11 billion tons of goods were transported by sea. That is 80–90% of the total world trade and corresponds to almost doubling compared to 2005 with 6.76 billion tons (*Sources* UNCTAD [28]; OECD n.d.). This development is also reflected in the increase in freight transport performance in all freight segments of maritime transport in the period mentioned. The corresponding

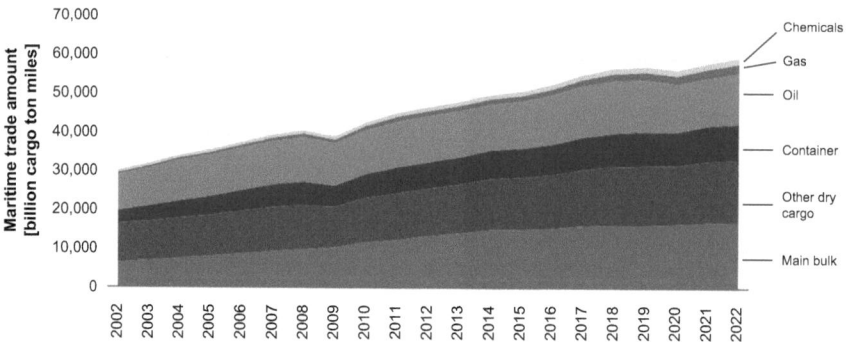

Fig. 17 International maritime trade 2002–2022 (billions of cargo ton-miles; a: includes iron ore, grain, coal, bauxite/alumina, and phosphate; b: estimated; c: forecast; *Source* UNCTAD [28], p. 5)

numbers make clear that the volume of maritime transport of crude oil and refined products has hardly changed over time. At the same time, however, the transport of raw materials and bulk goods as well as container transport have been increasing (Fig. 17).

In parallel with growing demand, transport capacities in all areas of maritime shipping increased between 1981 and 2022. The fleet available worldwide changed significantly between 1999 and 2022 in terms of both the number of ships and their cargo capacity. In 1999, the world's cargo carrying fleet comprised 46,002 ships of 778 million dwt[8] (*Source* Lloyd's Register Foundation World Fleet Register 2023). In 2022 there were almost 60,000 ships with a capacity of 2180 million dwt (*Sources* atlas-mag.net 2023; UNCTAD [28]).

The enormous change in ship sizes can be seen in the example of container ships, whose capacity has grown from 10,000 TEU in 2005 to about 25,000 TEU in 2023 (Fig. 18). However, the development in ship sizes seems to be reaching its limits. This applies, on the one hand, to the economies of scale, which cannot be increased indefinitely by the size of the ship alone, and, on the other hand, to the high costs, which in many countries are created by the public sector for the expansion of ports to allow loading and unloading of ships with increased volumes [13, 16].

To reach their final destination durable and consumable goods need delivery. End products are delivered to wholesalers and retailers almost exclusively by truck. In addition, direct delivery to the end customer has become ever more important in recent years; the usual mode then is the light commercial vehicle. Delivery is carried out by so-called CEP services (courier, express, parcel) that form a separate and very specific area of freight transport. CEP service providers carry piece goods with a relatively low weight and volume. An important feature of the CEP service is its speed of shipping goods.

---

[8] Deadweight tonnage (dwt) represents a ship's capacity to carry weight, excluding the vessel's own empty weight.

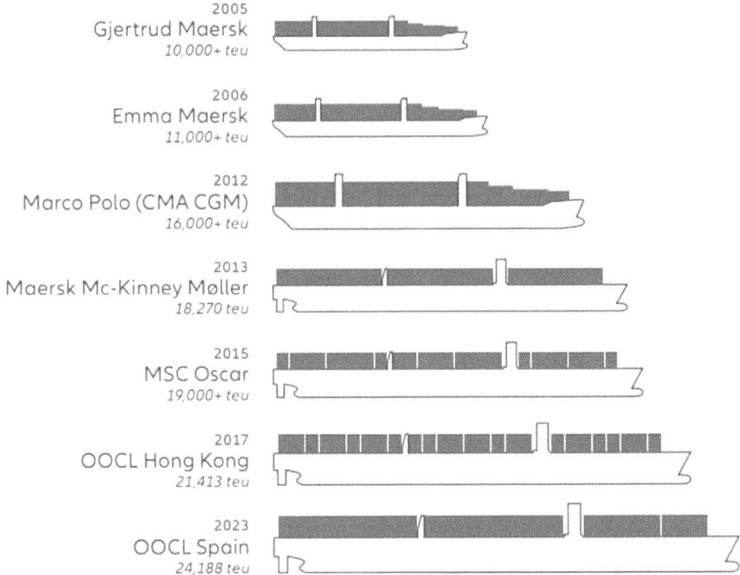

Fig. 18 Capacity of the largest container vessels 2005 to 2023. *Source* Allianz [1], p. 21

The CEP sector has grown almost explosively, especially with the expansion of e-commerce, boosted by in-shop shopping restrictions during the Covid19 crisis. While in 2014, 43 billion packages were delivered worldwide, this value had already doubled by 2018; in 2022 there were 161 billion packages with worldwide sales of 485 billion USD (*Source* Pitney Bowes Parcel Shipping Index 2018 und 2022 [24]). The leader in the number of packages was China with 111 billion parcels, followed by the USA with 21 billion parcels, Japan with 9 billion parcels, UK with 5 billion parcels, and Germany with 4 billion parcels. Delivery to the end customer is usually carried out using light commercial vehicles (maximum permitted total weight 3.5 tons). In larger cities, cargo bikes and delivery stations are increasingly used for CEP deliveries and deposition.

The origin of the CEP services were air courier companies that specialized in international delivery services. They emerged in North America in the late 1960s, Western Europe, and the Pacific Rim. Their service was to provide fast and reliable delivery services for urgent documents, such as finance, shipping, and engineering papers. The first companies founded included DHL and FedEx: Dalsey. Hillblom and Lynn started DHL in 1969 with rapid delivery of shipping and banking documents between the west coast of the United States and Hawaii, and Fred Smith started Federal Express in 1971. One essential prerequisite for the possibility of expanding the services was the liberalization of the postal services [8]. In Germany, for instance, the Deutsche Bundespost's monopoly on forwarding parcels up to 20 kg ended in 1976.

The rapid increase in demand for courier and express services motivated CEP companies very early to spend greatest attention on the development and implementation of measures to increase efficiency and effectiveness. The parcel services soon developed into "promoters and pacemakers of the whole logistics industry" for high-performance logistics ([33], p. 173). This was initially achieved by standardizing the transported goods, which included limiting the dimensions and weight of the packages. An important further step was the introduction of barcodes, to put into operation tracking and documentation of the packages from the consignor to the recipient along the entire transport chain. Tracking and documentation gave a strong push to internationalization of the forwarding companies as it was applicable not only for parcel shipping within the system of a CEP company nationally, but also in the global market.

In the early years of express services, companies used their own aircraft on international routes. This was later replaced by a hub and spoke air transport operation, which, however, was only possible where appropriate infrastructure was available. To meet the ever-growing demand, companies in the US built a network of cargo airplanes for long-distance transport that was independent of the belly freight capacity of passenger airplanes. To transport urgent documents and parcels in domestic commerce, US express companies set up their fleets of trucks and small aircraft [8]. As a consequence, the railway was replaced as a means of transport by trucks for long distances and light commercial vehicles for the last mile. At the same time, technological developments in aviation were of particular importance: Introduction of jet aircraft into commercial airline fleets in the 1960s reduced the costs and delays inherent in long distance commerce and set off a boom in international trade [8]. However, as mentioned earlier, a whole series of fundamental regulatory changes like the liberalization of postal services or trade agreements between states were still required, to start establishing a global cargo network in the 1980s ([8], Chap. 5).

3.3   The Future Demand in Freight Transport

Global trade in goods and thus the need for freight transport will continue to grow in the future and will have growth rates that are even higher than those of passenger transport. Global freight transport (measured in ton-kilometers) is expected to increase almost three-fold between 2015 and 2050—that would be around 330,000 billion tkm in 2050 [13, 16] (Table 8).

Maritime shipping then remains by far the most important area and is assumed to increase by 3.4 times. It will be increasingly supplemented by freight traffic on the inland waterways (expected growth by 3.7 times by 2050). The increase is also particularly high for air cargo (4.6-fold increase), which has already become an important transport route for higher-value and time-critical commercial goods in recent years. A growth rate of 3.0 times is expected for road freight transport and 2.3 times for rail freight transport (Table 9).

Table 8 Global freight by mode 2015, 2030 and 2050 (ITF Transport Outlook [13]; ITF Transport Outlook [15])

	Rail (billion tkm)	Road (billion tkm)	Sea (billion tkm)	Air (billion tkm)	Total (billion tkm)
2015	12,676	19,729	79,715	141	112,261
2030	19,796	31,539	130,694	319	182,348
2050	31,383	51,149	245,303	883	328,718

Table 9 Projected growth rates of freight transport demand

	2015–2030 (%)	2015–2050 (%)
Freight transport demand	3.1	3.4
Rail	2.7	2.5
Road	3.5	3.2
Inland waterways	3.4	3.8
Aviation	5.5	4.5
Sea	3.0	3.6

Source ITF Transport Outlook [12], p. 37

The main causes of the projected development are population growth and economic growth. The ITF assumes that despite recurring economic downturns and difficulties such as the Covid19 crisis or Russia's war against Ukraine, long-term growth in the global economy will occur. Accordingly, there will be an increase in the transport demand for goods [11]. While total volumes increase, a fundamental change in global freight transport patterns is expected: Model-based projections conclude that the importance of developing and emerging countries within the global economy will grow significantly; over the course of the 2030s, the share in global exports from developing and emerging countries will come to exceed that of developed countries. While at the same time the share of global trade within developing countries will rise by 18 percentage points, the share of trade within developed countries will fall from 40 to 21%. In addition, the importance of service sectors in world trade will become considerably more important than today [3].

Digitalization will influence the projected growth of freight transport in different ways. On the one hand, it will affect the demand side, where consumption is expected to continue to rise, not least through online retail worldwide. On the other hand, digitalization will help transporting goods more efficiently and make cost savings possible. This in turn can lead to a further deepening of the international division of labor.

The growth of freight demand will result in additional energy requirement is primarily due to the increase in road freight transport and to the growing use of heavy trucks (Fig. 19). However, as already mentioned above, growth will not be equally distributed globally. Rather, the ITF forecasts assume that the increase in

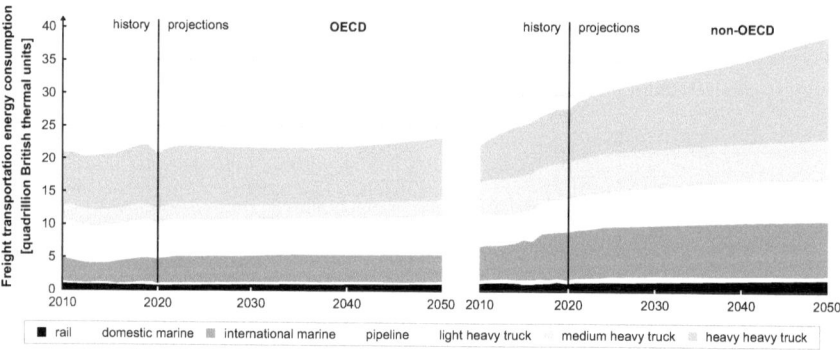

Fig. 19 Current and future freight energy consumption by mode for OECD and non-OECD countries. projection for 2050 [32] *Source*: EIA [32]

freight transport on the road will occur primarily in developing economies, this means especially in Africa (+435%) and also in Asia (+269%) (ITF [12]. p. 40).

To counteract the expected strong growth in road freight transport, several technical and operational measures are being discussed, especially with a view to the resulting CO_2 emissions. In the ITF scenarios for long-distance transport, the use of fuels based on renewable energies in heavy trucks and the use of high-capacity city vehicles are predicted to be particularly effective (ITF [12], p. 202).

4 Conclusion—More Transportation More CO_2

The ongoing growth in global traffic of both people and goods is not an assumption, but a fact. It is an expression of the satisfaction of needs that all people are equally entitled to. The main challenge is to design transport in such a way that the associated release of climate-impacting CO_2 emissions is minimized. Today, the share of transport in global emissions of CO_2 emissions that have an impact on the climate is around 20%, still, transport is emitting more and more CO_2 emissions in absolute terms (Fig. 20).

Almost all scenarios that are intended to show ways out of the high greenhouse gas contribution of transport emphasize that it will need considerable political ambition to implement effective measures. There is agreement that only with a mix of technological innovation, the timely implementation and scaling of the technological innovations which already exist today and with changing travel behavior it will be possible to tackle the gigantic task of efficient climate protection in the mobility sector. Continuing in "business as usual" mode would mean that the amount of GHG from transport carry on rising. The climate protection measures which were currently announced or already implemented by national governments are not sufficient to reduce CO_2 according to national and international objectives.

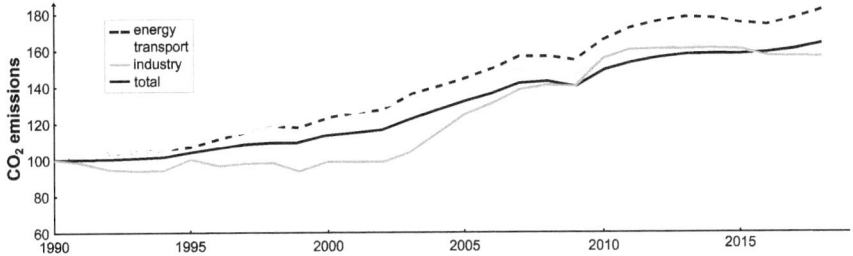

Fig. 20 Increase of global CO_2-emissions from fuel combustion by end-use sector, index 1990 = 100. [15] *Source* ITF 2021

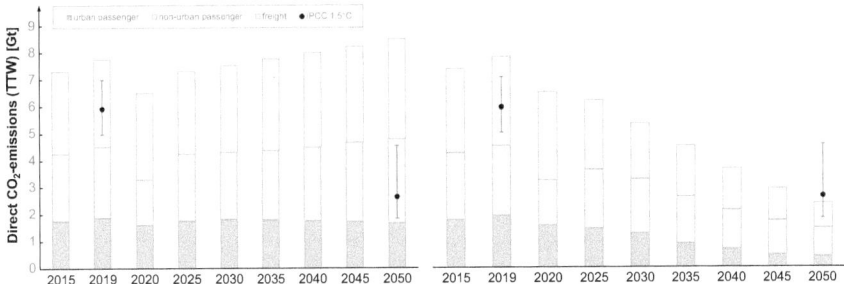

Fig. 21 CO_2-emissions for urban passenger. non-urban passenger and freight transport 2015 to 2050. ITF Scenarios "Recover"·(left) and "Reshape+" (right). [22] *Source* ITF

There are ways to achieve the target value stated by the IPCC for maintaining the 1.5 °C target (Fig. 21). However, this requires enormous additional efforts and changes, in particular the shift to more environmentally and climate-friendly means of transport, the improvement of energy efficiency and the rapid shift to alternative drives and renewable fuels. In fact, the commitments from large parts of politics, business, and society to more climate protection have become part of the everyday public discourse. But the slow or often complete lack of implementation raises doubts as to whether the global community will succeed in achieving the goal set out in the 2015 Paris Agreement.

References

1. Allianz (2023) Safety and shipping review 2023. https://commercial.allianz.com/content/dam/onemarketing/commercial/commercial/reports/AGCS-Safety-Shipping-Review-2023.pdf; 06.09.2023
2. Bartelheimer P (2007) Politik der Teilhabe. Ein soziologischer Beipackzettel. Friedrich-Ebert-Stiftung (Hrsg.). Fachforum: Analysen und Kommentare 1. Berlin

3. Bekkers E, Corong EL, Métivier J, Orlov D (2023) How will global trade patterns evolve in the long run? WTO Staff Working Paper, No. ERSD-2023-03, World Trade Organization (WTO), Geneva. https://www.wto.org/english/res_e/reser_e/ersd202303_e.pdf; 20.09.2023
4. BMDV (Bundesministerium für Digitales und Verkehr) (2023) Verkehr in Zahlen 2022/2023. https://bmdv.bund.de/SharedDocs/DE/Publikationen/G/verkehr-in-zahlen-2022-2023-pdf.pdf?__blob=publicationFile; 06.10.2024
5. BPB (Bundeszentrale für politische Bildung) (2016) Entwicklung des Warenexports nach Warengruppen. https://www.bpb.de/kurz-knapp/zahlen-und-fakten/globalisierung/52557/entwicklung-des-warenexports-nach-warengruppen/; 20.09.2023
6. BPB (Bundeszentrale für politische Bildung) (2022) Deutschland in Daten. Zeitreihen zur Historischen Statistik. Bonn. https://www.bpb.de/system/files/dokument_pdf/deutschland_in_daten_online_komplett.pdf; 06.10.2024
7. Bühler R, Kunert U (2008) Trends und Determinanten des Verkehrsverhaltens in den USA und in Deutschland. Endbericht. Forschungsprojekt im Auftrag des Bundesministeriums für Verkehr. Bau und Stadtentwicklung. Projektnummer 70(2008). https://www.diw.de/sixcms/detail.php?id=diw_01.c.94284.de; 09.09.2023
8. Campbell JI (2001) The rise of global delivery services. James Campbell Press, Washington D.C. https://www.jcampbell.com/docs-campbell/books/RoGDS_complete.pdf; 06.09.2023
9. Cortright J (2010) New York City's Green Dividend. Download via: https://www.zukunft-mobilitaet.net/866/umwelt/auto-sparen-autoverzicht-new-york-19-milliarden/; 16.05.2023
10. Hummels D (2007) Transportation costs and international trade in the second era of globalization. J Econ Perspect 21(3):131–154. https://pubs.aeaweb.org/doi/pdf/, https://doi.org/10.1257/jep.21.3.131; 20.09.2023
11. ITF (2023) ITF Transport Outlook 2023. OECD Publishing. Paris. https://www.oecd-ilibrary.org/transport/itf-transport-outlook-2023_b6cc9ad5-en; 16.05.2023
12. ITF (2019) ITF Transport Outlook 2019. OECD Publishing. Paris. https://doi.org/10.1787/transp_outlook-en-2019-en; 16.05.2023
13. ITF (2017) ITF Transport Outlook 2017. OECD Publishing. Paris. https://www.oecd-ilibrary.org/transport/itf-transport-outlook-2017_9789282108000-en; 16.05.2023
14. ITF (2022) Spending on transport infrastructure. ITF Statistics Brief. June 2022. https://www.itf-oecd.org/sites/default/files/docs/inland-transport-infrastructure-investment-brief-2022.pdf; 16.05.2023
15. ITF (2021) Demand for passenger transport by world region to 2050: under three different scenarios. Billion passenger-kilometres. In: ITF Transport Outlook 2021. OECD Publishing, Paris. https://stats.oecd.org/Index.aspx?DataSetCode=ITF_OUTLOOK_URB_PAX; 16.05.2023
16. ITF (2017) The impact of mega-ships. Case specific policy analysis. Paris. https://www.itf-oecd.org/sites/default/files/docs/15cspa_mega-ships.pdf; 16.05.2023
17. Joly I (2004) Travel time budget—decomposition of the worldwide mean. In: International Association of Time-Use Research, annual conference 2004, 27–29 Oct, Rome Italy. https://www.researchgate.net/publication/5087269_Travel_Time_Budget_-_Decomposition_of_the_Worldwide_Mean; 16.05.2023
18. Kuhnimhof T, Buehler R, Wirtz M, Kalinowska D (2012) Travel trends among young adults in Germany: increasing multimodality and declining car use for men. J Transp Geogr 24:443–450
19. Lenz, Barbara (2013): "Auto-Europa" – das europäische Fernstraßennetz und die Vertiefung regionaler Disparitäten. In: Gebhardt H, Glaser R, Lentz S (eds) Europa – eine Geographie. Berlin, Heidelberg, pp 432–435
20. Metz D (2010) Saturation of demand for daily travel. Transp Rev 30(5):659–674
21. Miroudot S, Lanz R, Ragoussis A (2009) Trade in intermediate goods and. OECD Trade Policy Working Paper No. 93 https://www.oecd-ilibrary.org/trade/trade-in-intermediate-goods-and-services_5kmlcxtdlk8r-en; 20.09.2023
22. Nobis C, Kuhnimhof T, Follmer R, Bäumer M (2019) Mobilität in Deutschland – Zeitreihenbericht 2002–2008–2017. Studie von infas, DLR, IVT und infas 360 im Auftrag des Bundesministeriums für Verkehr und digitale Infrastruktur (FE-Nr. 70.904/15). Bonn, Berlin. https://www.mobilitaet-in-deutschland.de; 16.05.2023

23. Ortiz-Ospina E, Beltekian D, Roser M (2018) Trade and globalization. Published online at OurWorldInData.org. https://ourworldindata.org/trade-and-globalization; 06.09.2023
24. Pitney Bowes Parcel Shipping Index 2018 und 2022. https://www.pitneybowes.com/con tent/dam/pitneybowes/us/en/shipping-index/pb_globalshippingindexinfographic_2022stats__ final.pdf; 07.09.2023
25. Rodrigue JP (2000) The geography of transport systems. Routledge.
26. Seum S, Schulz A, Phleps P (2019) The future of driving in the Brics countries (Study Update 2019). ifmo/DLR. https://www.ifmo.de/files/publications_content/2020/2019_ifmo_BRICS_ reloaded_en1.pdf; 16.05.2023
27. UBA (Umweltbundesamt) (2023) Marktdaten: Mobilität. 12.01.2023 https://www.umweltbun desamt.de/daten/private-haushalte-konsum/konsum-produkte/gruene-produkte-marktzahlen/ marktdaten-bereich-mobilitaet#globaler-autobestand-china-stockt-weiter-auf; 16.09.2023
28. UNCTAD (United Nations Conference on Trade and Development) (2022) Review of maritime transport 2022—navigating stormy waters. New York. https://unctad.org/system/files/official-document/rmt2022_en.pdf; 06.09.2023
29. UNCTAD (United Nations Conference on Trade and Development) (2020) Key statistics and trends in international trade 2019. Geneva. https://unctad.org/system/files/official-document/ ditctab2019d7_en.pdf; 20.09.2023
30. U.S. Department of Transportation. Bureau of Transportation Statistics (ongoing) Freight Facts and Figures. Washington, DC. https://www.bts.gov/browse-statistical-products-and-data/fre ight-facts-and-figures/value-tonnage-and-ton-miles-freight; 06.10.2024
31. U.S. Energy Information Administration (EIA) (2016) Annual passenger travel tends to increase with income. https://www.eia.gov/todayinenergy/detail.php?id=26192; 06.10.2024
32. U.S. Energy Information Administration (EIA) (2021) Chart library. Transportation. https:// www.eia.gov/outlooks/ieo/pdf/IEO2021_ChartLibrary_Transportation.pdf; 16.05.2023
33. Vahrenkamp R (2012): The rise of transportation and logistics in Europe 1950–2000. In: Cuadra-Montiel H (ed) Globalization—education and management agendas. InTech. https:// www.intechopen.com/chapters/38265; 06.09.2023
34. World Trade Organization (WTO) (2022) Evolution of trade under the WTO: handy statistics https://www.wto.org/english/res_e/statis_e/trade_evolution_e/evolution_trade_wto_e.htm#: ~:text=World%20trade%20values%20today%20have.the%20WTO%20was%20first%20esta blished.; 07.09.2023
35. World Trade Organization (WTO) (2022) Information note on trade in intermediate goods: fourth quarter 2022. https://www.wto.org/english/res_e/statis_e/miwi_e/info_note_2022q4_e. pdf; 21.09.2023

Development of European Road-Based Mobility

Karsten Wilbrand and Michael Schulthoff

Abstract The document provides a comprehensive analysis of the latest advancements in defossilization of the transport and mobility sector. It highlights progress in biofuels, e-mobility, and hydrogen and discusses the economic and technical challenges faced in the adoption of these technologies. The paper offers insights into the potential future developments and the impact of policy changes on decarbonization of the transport sector.

Keywords Biofuels · e-fuels · Hydrogen · e-mobility · Ethanol · E20 · Mobility · Transport · Decarbonisation · Defossilisation

1 Introduction

The 2015 Paris Agreement defined a bold ambition to limit global warming to below 2 °C above pre-industrial levels and pursue efforts to limit it to 1.5 °C—in part by pursuing net carbon neutrality by 2050. In response, many countries, industries, and individual organizations set targets to limit their carbon emissions and began developing plans on how to reach them. Nevertheless, further substantial efforts must be made to address climate change and achieve the goals defined within the Paris Agreement. The United Nations Environment Program notes, "On current unconditional pledges, the world is heading for a 3.2 °C temperature rise" [1].

The challenge is particularly pronounced in six so-called "harder-to-abate" sectors that, according to the International Energy Agency (IEA), currently account for around 32% of global CO_2 emissions (for a better overview, a cross-sectoral distribution of global CO_2 emissions in 2020 is shown in Fig. 1). These harder-to-abate industry sectors, namely cement, iron and steel, chemicals, shipping, and aviation

K. Wilbrand (✉)
Shell Global Solutions (Deutschland) GmbH, Hamburg, Germany
e-mail: karsten.wilbrand@shell.com

M. Schulthoff
Hamburg University of Technology (TUHH), Institute of Environmental Technology and Energy Economics (IUE), Hamburg, Germany

© The Author(s), under exclusive license to Springer Nature Switzerland AG 2025
N. Bullerdiek et al. (eds.), *Powerfuels*, Green Energy and Technology,
https://doi.org/10.1007/978-3-031-62411-7_3

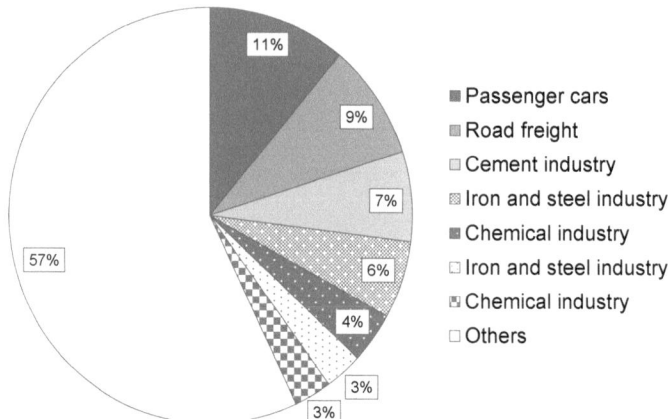

Fig. 1 Share of CO_2 emissions by different sectors [1]

share common characteristics, such as long asset lifespans, a strong dependency on energy carrier characterized by high energy densities, and a significant complexity and truly high challenges related to full electrification. As a result, a defossilization of these sectors will be slower, more investment-intensive, and technically more demanding than in other sectors. As defossilization happens most likely more rapidly elsewhere than in these harder-to-abate sectors, pressure and focus on these sectors are expected to increase in future years.

On a global level, the passenger car sector accounts for approximately 11% of worldwide CO_2 emissions, an even larger share than the heavy-duty road-freight sector (approximately 9%). The share in global CO_2 emissions of the cement, iron and steel industries is about 7% and 6%, respectively. For the global chemical industry, CO_2 emissions account for about 4% of the global CO_2 emissions, while these shares are in the order of magnitude of 3% for global aviation and maritime transport.

The two road-based sectors "passenger cars" and "road-freight" together emit ca. 20% of global CO_2 emissions, see Fig. 1. Total global CO_2 emissions in 2020 amounted to about 35.26 Gt CO_2 according to [2] (without land-use-change). These two road-based sectors hence account for ca. 7 Gt CO_2 per year.

Against this background, this chapter takes a closer look at how to defossilize the road-based mobility sector. It describes, assesses, and compares options to decarbonize and defossilize the road-based transport sector, with a focus on Europe, which has the most stringent targets globally to date. The chapter covers various energy carriers such as biofuels, gaseous fuels, hydrogen, and electrons for e-mobility.

2 Evolution of Road-Based Mobility

Road-based mobility has shown a tremendous increase in the past decades. Figure 2 shows exemplarily the continuous growth of registrations for passenger cars (i.e., other vehicles are not shown, such as busses and heavy-duty vehicles). The same is true for the annual car production, only slightly disturbed by the world financial crises in 2009 and the COVID-19 pandemic in 2020/2021. For 2021, the total number of global car registrations will be roughly 1.1 billion. Light and heavy-duty vehicles, as well as buses, add up to this.

With the (rapid) development of the transportation sector, also the energy demand of this sector has continuously increased over the past decades (despite successive, primarily technical, energy efficiency improvements). The corresponding development of the energy demand for the global transport sector between 1975 and 2022 is shown in Fig. 3. As can be seen, the increase in energy demand from 1975 (ca. 40 EJ/a) to 2022 (ca. 120 EJ/a) amounts to roughly a factor of three.

Furthermore, from the data shown in Fig. 3 it becomes evident that in the overall global transportation sector roughly 95% of the energy demand is covered by energy carriers derived from crude oil. Alternative fuels like natural gas and biofuels provide only a small share of this energy. Additionally, electricity as another alternative energy source is used to a minor extent. Even since the data shown in Fig. 3 also includes maritime and aviation transportation, the highest share of the energy demand shown is caused by ground-based transport.

The relatively small share of vehicles driven by alternative fuels within the operated fleets' shows regionally significant differences. For example, within the EU 27, now (2022) 13.7 million of a total of 270 million passenger cars are on the roads using a fuel outside of the "classical" crude oil-based fuels gasoline and diesel; this sums up

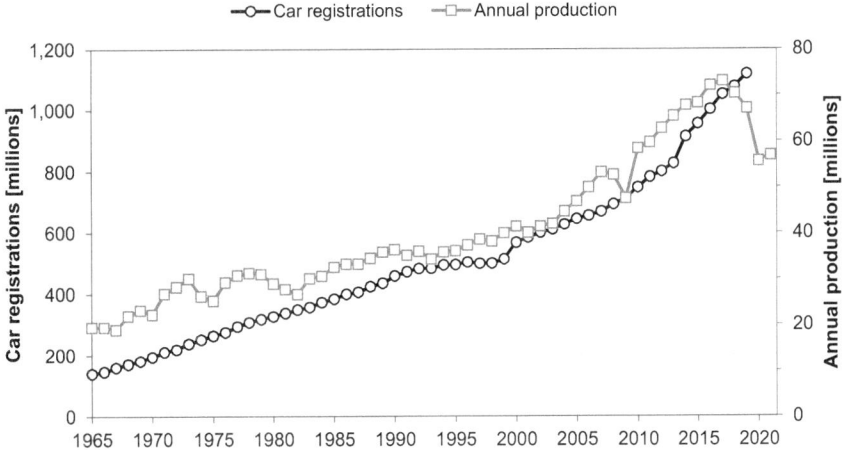

Fig. 2 Worldwide automobile production [3] and fleet 1965 till 2021 [4]

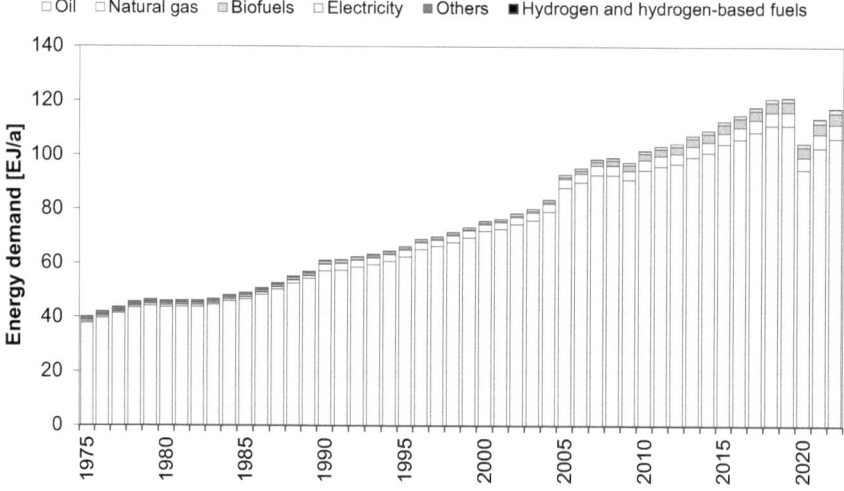

Fig. 3 Energy consumption in transport by fuel from 1975 to 2022 [5]

to about 5% (Fig. 4). Nevertheless, not all of these alternative fuels are "green," as for example gas driven vehicles are typically operated by methane from (fossil) natural gas. In 2022, the exponential growth of battery electric vehicles (BEVs) arose due to policies and subsidies within countries. Anyway, regarding emission-intensities of "alternative fuel production" within those countries, none of the alternative fuels can be considered "green" by default.

3 Taxonomy and Regulatory

3.1 Taxonomy

The regulatory perspective on different fuels classification in road transport is crucial to evaluate their alignment with the Paris Agreement and the climate aspirations of the European Union (EU). Here, low-carbon fuels, carbon–neutral fuels (including biofuels and carbon-based Renewable Fuels of Non-Biological Origin (RFNBO)), zero-carbon (e.g., hydrogen, ammonia), and zero-emissions fuels (e.g., green electricity) in the context of road transport are differentiated. Low-carbon und carbon–neutral fuels both emit CO_2 after the combustion, but in balance those fuels reduce the CO_2 emissions by utilizing carbon from the atmosphere or other sources.

Ammonia and hydrogen can be classified as zero-carbon fuels if used in an internal combustion engine, as they still emit nitrogen oxides (NO_x). If hydrogen is used in a fuel cell road vehicle, it can be considered zero-emissions in road transport, although it still emits water (H_2O), which can cause climate effects if released in the

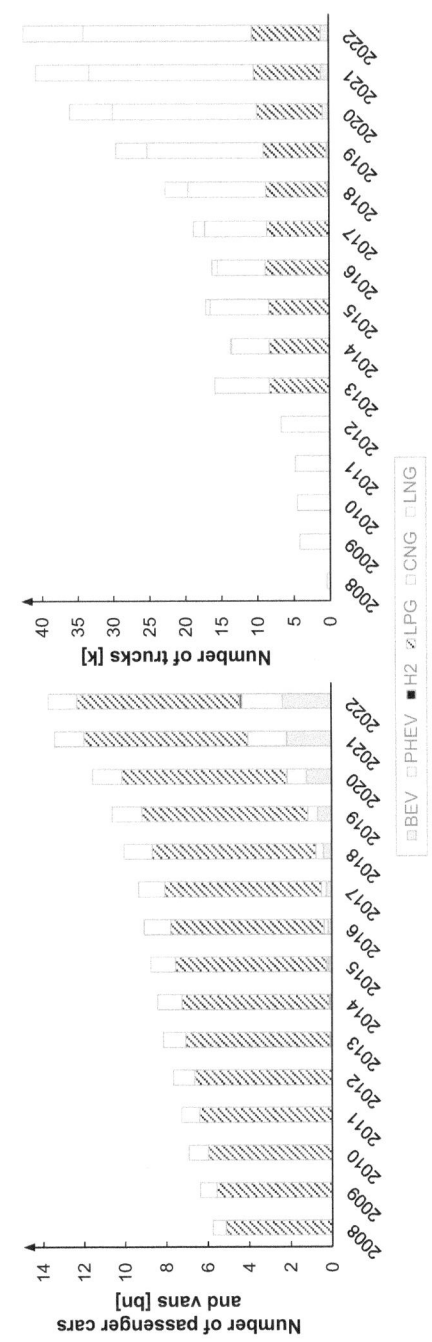

Fig. 4 Number or alternative fuels passenger cars & vans (left) and trucks (right) in EU 27 [6]. (BEV—battery electric vehicle; CNG—compressed natural gas; H₂—hydrogen; LNG—liquefied natural gas; LPG—liquefied petroleum gas; PHEV—plug-in hybrid electric vehicle)

atmosphere, e.g., by planes. If green electricity is used in a vehicle, it is also regarded as "zero-emissions".

Diverse low-carbon fuels, such as blends of conventional diesel and HVO diesel (hydrogenated vegetable oil (HVO) diesel) or FAME (fatty acids methyl esters), advanced ethanol blends, or compressed natural gas (CNG) or liquefied natural gas (LNG), based on fossil natural gas or on biomethane (e.g., from biogas), offer varying degrees of net carbon emissions reduction compared to conventional fuels, playing a role in CO_2 emission mitigation.

Carbon–neutral fuels, exemplified by carbon-based RFNBO like e-fuels, endeavor to counterbalance emitted carbon through removal from the atmosphere, theoretically achieving carbon neutrality. Biofuels, a subset of carbon–neutral fuels, are produced based on organic material and can thus contribute to a circular carbon cycle, as the carbon or CO_2 emissions released into the atmosphere during fuel combustion were previously absorbed from the atmosphere by feedstock growth through photosynthesis.

Zero-carbon fuels or zero-emission fuels, such as hydrogen and electricity, ensure the complete elimination of CO_2 emissions during usage, provided they are completely supplied based on renewable energies. While low-carbon and carbon–neutral fuels can partially align with climate objectives, they might not comprehensively meet the Paris Agreement's 1.5 °C global warming threshold. Zero-carbon fuels demonstrate a stronger alignment with the Paris Agreement and EU climate targets, offering a direct pathway to substantial decarbonization. However, challenges concerning technology readiness, infrastructure development, and cost may impede widespread zero-emission fuel adoption in road transport. As policies evolve and innovations emerge, a strategic combination of these fuel categories may be imperative to accomplish a comprehensive transition that resonates with international climate goals.

3.2 Regulatory Frameworks

In many regions of the world, a complex legislation framework is driving the development of alternative powertrains, significantly reducing transport-related CO_2 emissions. Within the EU, one of the most complex and demanding regulatory frameworks is being installed to reduce CO_2 as well as local emissions/pollutant emissions.

Figure 5 shows a selection of the most important legal measures. According to this, governmental regulation addresses the fuel provider, the vehicle producer, and the consumer. The latter is mainly burdened by various types of taxes intending to make transportation options characterized by high CO_2 emissions more expensive for the final consumer than more climate-sound possibilities. It currently implies separate regulations on fuel providers (i.e., on a well-to-tank level) and OEMs (Original Equipment Manufacturer/vehicle producer).

Figure 6 shows selected measures that are primarily implemented to be fulfilled by the fuel supplier on the one hand and by the OEMs on the other. Here, it becomes

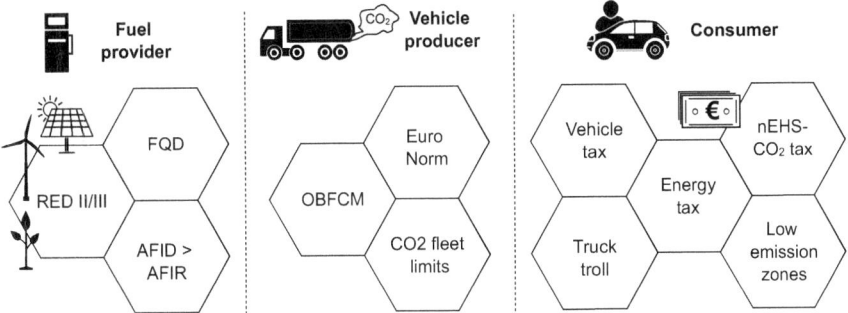

Fig. 5 Climate- and health-related policy landscape for cars and trucks within the EU in 2022 (AFID—Alternative Fuels Infrastructure Directive; AFIR—Alternative Fuels Infrastructure Regulation; OBFCM—On-board fuel consumption measurement; nEHS—nationales Emissions-Handels-System; FQD—Fuels Quality Directive; RED—Renewable Energy Directive) [7]

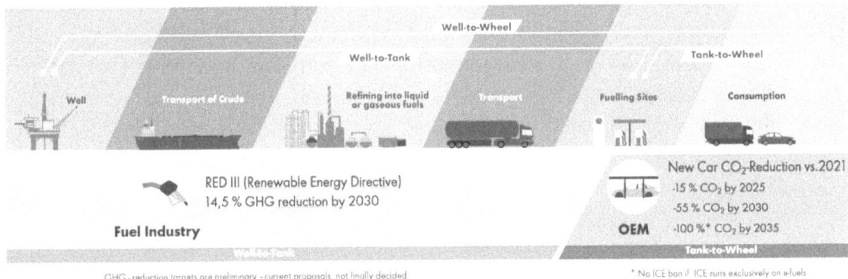

Fig. 6 Separate legislation for energy companies [8] and vehicle OEMs [9]

obvious that the strategy of the legal framework directs clearly to address the GHG intensity of the energy needed within the transport sector and the vehicle fleet. These measures can be understood as a combination to follow a GHG reduction pathway, taking the existing fleet still operated in the years to come into consideration.

Selected measures implemented by the EU are discussed in the following sections in more detail.

3.2.1 CO_2 Emission Performance Standards for Cars, Vans, and Trucks

On April 19, 2023, the European Parliament and the Council adopted Regulation (EU) 2023/851 with the following EU fleet-wide CO_2 emission reduction targets for new passenger cars and vans as compared to the 2021 target [9]. For heavy-duty trucks, the EU Commission proposed the following reduction targets (2023/0042(COD)) as compared to the reference period (1 July 2019–30 June 2020), as shown in Table 1.

Table 1 CO_2 emissions reduction target of EU for cars, vans (LDV) [9], and trucks (HDV) [10] (HDV—heavy-duty vehicle; LDV—light-duty vehicle)

Time period	Cars		Vans		Trucks
	CO_2 emissions reduction versus 2021	CO_2 emissions (gCO_2/km)	CO_2 emissions reduction versus 2021	CO_2 emissions (gCO_2/km)	CO_2 emissions reduction versus 2019
2020–2024	-/-	95	-/-	147	-/-
2025–2029	− 15%	80.75	− 15%	124.95	− 15%
2030–2034	− 55%	42.75	− 50%	73.5	− 45%
2035–2039	− 100%	0	− 100%	0	− 65%
2040	-/-	-/-	-/-	-/-	− 90%

Specific CO_2 emission targets are set annually for each vehicle manufacturer. These are based on the EU fleet-wide targets and consider the average mass of the manufacturer's new vehicles registered in a given year. The measurement of CO_2 emissions is focused on the vehicle tailpipe per the test protocol (NEDC[1] until 2021, WLTP[2] afterward). Hence, no CO_2 emissions are considered for battery electric and fuel-cell powertrains, respectively, types of energies driving those powertrains (i.e., electricity and hydrogen). Combustion engines operated on carbon-containing fuels (gaseous or liquid, regardless of their origin, fossil or renewable) emit tailpipe CO_2 emissions.

Consequently, an emission reduction target of 100% by 2035 versus the fleet value in 2021 will enforce the elimination of combustion engine powertrains running on carbon-containing fuels from the newly registered fleet by that time, per the actual legislation (to be reviewed in 2026). The only exemption will be new vehicles with combustion engines that run exclusively on RFNBO (Renewable fuel from non-biological origin), which means either on power fuels (also referred to as e-fuels) or green hydrogen. The exact regulations for how it will be controlled that the combustion engines cannot run on fossil or biofuels are not yet defined. These regulations are being discussed and developed while this book is written.

The first-ever EU-wide CO_2 emission standard for heavy-duty vehicles was adopted in 2019, and targets have been set for reducing the average CO_2 emissions from new trucks for 2025 and 2030 [1]. A review of this regulation aiming at an alignment with the ambition of the "Fit For 55" package proposal was initiated in late December 2021 via a public consultation. The fundamental policy decision related to the CO_2 tailpipe approach will also be crucial for the prospects of biomethane, liquid biofuels, and other low-carbon fuels for use in heavy-duty combustion engine powertrains. The consultation issued by the European Commission has requested viewpoints about the potential of recognizing the role of low-carbon and renewable fuels via different methods, for instance, a carbon correction factor, or a fuel crediting system. Furthermore, given non-carbon-containing fuels (e.g., hydrogen,

---

[1] NEDC: New European Drive Cycle.

[2] WLTP: Worldwide-harmonized light-duty test protocol.

H_2) in heavy-duty combustion engine powertrains, a clarification in the regulation of how non-avoidable CO_2 tailpipe emissions originating from combusted lubricant and the reducing agent for emission control (i.e., AdBlue) shall be treated would be necessary. At this moment, the concept of "zero-emission heavy-duty vehicles," provided by the incumbent CO_2 emission standard for heavy-duty vehicles (criterion: < 1 g CO_2/kWh), may play a helpful role in leaving a legal scope for non-carbon-containing fuels in heavy-duty combustion engine powertrains in the time beyond 2030.

Initial member states' discussions have revealed preferences regarding the ambition for the target levels of the CO_2 emission performance standards for cars and vans. There is a general understanding that the automotive sector must contribute to the increased overall EU climate ambition for 2030 and 2050. For a few delegations, the proposal is key for achieving the overall ambition and to fulfill their increased national targets under the "Effort Sharing Regulation." For a few delegations, the proposed ambition level is a source of concern, notably the 2035 target of 100% (concerning fleet-wide CO_2 emission reductions for new passenger cars), which some explicitly deem unacceptable. These delegations have highlighted a few considerations in this context, notably, the need to consider the different situations of Member States regarding the uptake of zero- and low-emission vehicles due to disparities in the roll-out of charging infrastructure and in purchasing power.

Besides the CO_2 emission performance standards, the EURO Norms define limits for non-CO_2 emissions, which contain harmful emissions like carbon monoxide, nitrogen oxides, methane, and others. Even though CO_2 emissions are not represented in the EURO Norm, its limits will also significantly impact the future of combustion engines and, therefore, the use of low-carbon and carbon–neutral fuels within the road transport sector. The upcoming next norm, "EU 7," is under trilogue negotiation, and final emission targets are expected to be defined by the end of 2023.

3.2.2 Renewable Energy Directive

The Renewable Energy Directive (RED) is the legal framework for developing renewable energy across all sectors of the EU economy. The last thoroughly revised RED (RED II) entered force in 2018. The European Commission (EC) proposed a revision of selected portions of the RED in July 2021 (RED III) as part of the package to deliver on the European Green Deal [11]. In line with the EU Climate Law, the targets and measures set in the revised directive should be ambitious enough to reduce greenhouse gas (GHG) emissions by at least 55% in 2030 compared to 1990 levels.

On an EU level, this includes an EU-wide binding increase to the preceding targets for the share of renewable energy in 2030 (32%) to an adapted target of 40% by 2030. It also includes the introduction of new targets and raising existing targets for various sectors of the EU economy (e.g., an indicative industry target of 1.1% annual GHG emissions reduction) [12]. The EC also aims at a more energy-efficient and circular energy system that facilitates electrification (i.e., the use of electricity based on renewable energies) and promotes the use of renewable and low-carbon fuels, including hydrogen, in sectors where electrification is not feasible yet.

The RED III allows member states to either implement a 14.5% GHG reduction target in the transport sector or a min 29% renewable energy target in 2030. While the RED II focusses on the road and rail transport, the transport sector in RED II includes not only road and rail, but also marine and aviation sectors. Multipliers of 2 for 2G advanced and RFNBO are proposed but not finally decided. For energy based quotas, the reference is the total final energy demand of the transport sector, while the GHG reduction target is based on the reference emissions calculated with a fossil reference value [12]. In addition, it introduces a 5.5% target for advanced biofuels and renewable fuels of non-biological origin (RFNBO). The directive retains the caps on food- and feed-based biofuels. The RED needs transposition into national legislation of EU Member States.

Table 2 shows the compliance options of the RED III in terms of the "Fit For 55" package proposal on the left-hand side. The mandatory targets and cap limits can be seen on the right-hand side (energy-based). Cap limits for sustainable food and feed crop-based biofuels are proposed, as well as for biofuels from used cooking oils (UCO) and animal fats (both caps are maintained from RED II). These compliance options can be used only to a limited volume, hence the cap. The use of advanced biofuels and renewable fuels from non-biological origin (RFNBO) are enforced by mandatory volumes; there are no cap limits. Hence, these renewable energy options would provide an "open door" to achieve compliance with the proposed RED III targets. Given the prospects of advanced biofuels and RFNBO in road transport in the mid to long term, the fundamental policy decisions related to the CO_2 emission performance standards for cars, vans, and heavy-duty vehicles are crucial. Investment potentials into these renewable energy options will not be fully unlocked if combustion engines for road transport are about to be phased out, regardless of the fuel's renewable footprint.

The following multipliers are proposed toward the overall 29% renewable energy target and all applicable sub-targets for either an energy target or GHG target:

- Multiplier of 2 for advanced biofuels, RFNBO, and waste oils;
- Multiplier of 4 for renewable electricity in vehicles;
- Multiplier of 1.5 for renewable electricity in rail;
- Multiplier of 1.2 for advanced biofuels and 1.5 for RFNBO in aviation and maritime sectors.

The European Commission has indicated that they target 30 million electric vehicles in Europe by 2030 [14], primarily light-duty vehicles (LDVs). They estimate a renewable electricity contribution of around 10–12% in the transport sector by 2030 (due to higher uptake of new electric vehicles driven by assumptions on vehicle standards). Combined with the proposed CO_2 emission performance standards for cars and vans, using "renewable" electricity for RED compliance holds the strongest prospects for road transport. Consequently, it can be observed that energy suppliers for road transport significantly invest in the buildup of electric vehicle (EV) charging

Table 2 Proposed targets of Renewable Energy Directive III (RED III) as of July 2023; energy based quotas based on total final energy consumption of transport sector of the respective year (1G—first generation; 2G—second generation; 3G—third generation, RFNBO—renewable liquid and gaseous fuels of non-biological origin; UCO—used cooking oil) [13]

Renewable type fuel	Examples	Minimum (energy based)	Maximum (energy based)
Food/feed crop-based biofuels (1G)	Biodiesel produced from rapeseed Ethanol produced from sugar beet or corn cane	-/-	7.0%
Biofuels from UCO and waste fats (2G) (RED Annex IX, part B)	Biodiesel from used cooking oils (UCO) and waste fats	-/-	1.7%
Advanced biofuels (2G) (RED Annex IX, part A)	Biofuels or biogas from forestry and agricultural waste, algae, or biomass from municipal and industrial sectors	5.5% (sub-target for RFNBO ≥ 1%)	-/-
RFNBO	Renewable fuel of non-biological origin, e.g., green hydrogen, e-fuels		-/-
Renewable electricity	Wind and solar energy	-/-	-/-

infrastructure and intend to grow future business through electricity supply for road transport.

4   Defossilization via Lower Carbon Energy Carriers

Various fuel options exist to achieve GHG emission reductions as required according to the RED III target. First of all, crop-based or first generation (1G) biofuels are the established fuel options feeding into the diesel and gasoline markets in the form of fatty acid methyl esters (FAME) in case of diesel and bio-ethanol in case of gasoline. Different regulations cap the use of feedstocks for both of these fuel options, thus overall RED compliance is not feasible based on these two options (also known as "blend wall"). Further options to contribute to RED compliance beyond the blend wall are renewable diesel from non-capped feedstock options, green hydrogen, green electricity, and some North-Western European countries also Bio-CNG and Bio-LNG.

- Renewable diesel is produced by the HEFA conversion process (hydro-processed esters and fatty acids) or the HVO (hydrotreated vegetable oils) process using oils and fats. These processes deliver a bundle of product streams, mostly straight-chain paraffinic hydrocarbons free of aromatics, oxygen, and sulfur, and have high cetane numbers. HEFA diesel and HVO diesel offer a number of benefits over fatty acid methyl esters (FAME), such as better storage stability and better cold flow properties. While the production can be very focused on diesel fuels according to EN15940, the process can also be optimized toward the production of kerosene (i.e., the production of renewable fuels for the aviation sector; so-called "sustainable aviation fuels" (SAF)). In the case of kerosene or SAF production, bio-naphtha is a sizeable side stream, while the diesel stream is usually rather small. While SAF will likely find increasing demand in aviation in the coming years, bio-naphtha could be used in the chemical industry or as a blend component in the gasoline pool for corresponding passenger cars. However, blending is limited due to the low octane number of the naphtha.
- Liquefied Natural Gas (LNG)—allowing for lower specific CO_2 emissions within combustion processes compared to fuels based on crude oil (like, e.g., gasoline and diesel), but is still based on natural gas and thus of fossil origin—has made the first step of reducing CO_2 emissions in the Commercial Road Transport (CRT) sector by introducing gas powered engines. This allows for a rather seamless introduction of bio-LNG (i.e., bio-methane) as a possible next step. Initially, the feedstock used for the production of bio-methane (biogas production) by anaerobic digestion was predominantly maize silage but recently shifted towards manure and agricultural residues.

This shift—from fuels based on food and fodder crops to fuels based on manure and agricultural residues—is aligned with the challenge seen by the regulator and, to a certain extent, by society for crop-based biofuels in their competition with food and fodder demand. Therefore, The EU strategy strongly focuses on biofuels based on organic residues, by-products, and wastes, which do not compete with food and fodder applications. Biofuels using such a feedstock are also called "advanced biofuels" or "second-generation (2G) biofuels".

The type of feedstock that is used for renewable products is becoming more and more relevant as regulations evolve. So far, three main categories of feedstocks have been identified: crop-based feedstock, waste-based feedstock (RED Annex IX Part B), and so-called "advanced" feedstock (RED Annex IX Part A) (Fig. 7).

The different types of biofuel products as shown in the figure above can be described as follows.

- First-Generation Biofuels. Crop-based biofuels are based on crops that can also be used to produce already existing products in the food and fodder industry and could be seen as in competition with food (also called "first generation biofuels" or "1G biofuels").
- Second-generation biofuels. Waste-based biofuels are based on organic by-products, like used cooking oil and animal fats (also called "second-generation biofuels" or "2G biofuels").

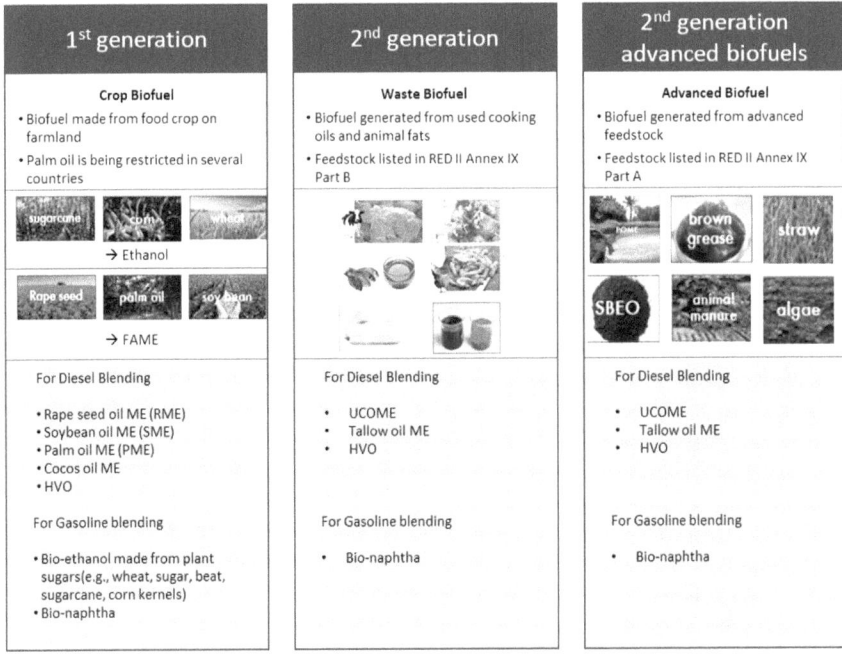

Fig. 7 Types of bio feedstocks and biofuel products for road applications (FAME—fatty acid methyl esters, HVO—hydrotreated vegetable oil; ME—methyl ester; UCOME—used cooking oil methylester)

- Second-generation advanced biofuels. Advanced biofuels are new biofuel products that are not considered to be in competition with food farmland nor harmful to animals in any way. For the production of advanced biofuels, by-products of current processes are used, such as forestry or agricultural residues that would not be used otherwise (also called "advanced second-generation biofuels" or "2G advanced biofuels").

Electric Vehicles (EV). While the biofuel pathways, as mentioned above, enable compliance with CO_2 emission reductions not only by application in new vehicles entering the market but also by leveraging the legacy fleet of vehicles in the market, the contribution of electricity charging is linked with the share of new battery electric vehicles (BEV) entering the market. As governments want to stimulate the development of the electric vehicle (EV) fleet, in many EU member states, the regulation allows for multiple counting of the electric vehicles (EV) contribution (the so-called multiplier varies per member state). This means that any kWh_{el} of electricity of renewable origin sold to a battery electric vehicle (BEV) can be counted multiple times toward the renewable energy mandate under the current RED II regulation. Of course, this only applies to so-called "green" electricity, i.e., electricity provided by using renewable energy sources, such as solar radiation or wind energy, in combination with photovoltaic plants or wind turbines. This "green" electricity can be applied in two ways: a direct line to a green electricity source like a photovoltaic power plant or a percentage of the commonly used grid that the government calculates to be "green" (ratio of green vs. gray electricity being supplied into the grid—in this case "gray electricity" means that it is produced from fossil fuels (e.g., coal or natural gas)).

Hydrogen. A similar situation applies to hydrogen: only a very small number of hydrogen vehicles are on the roads today. While the passenger car sector is focusing on battery electric vehicles (BEV) technology, vehicles that run on hydrogen are expected to gain market share in the heavy-duty sector due to their fast-refueling capability, delivering similar operational flexibility as today's diesel or LNG trucks. However, the timeline for market introduction is such that sizeable fleets of hydrogen trucks are expected only toward the end of the decade, so their contribution to CO_2 emission reductions and RED compliance by 2030 is most likely limited. In terms of RED compliance, basically only "green" hydrogen—i.e., from electrolysis using electricity of renewable origin or from nuclear energy (due to its ultra-low CO_2 emissions)—is applicable.

Ammonia. Also, the utilization of hydrogen derivatives like ammonia is gaining momentum in the market, predominantly as a fuel within the maritime sector but also as a hydrogen carrier for long-distance transport (i.e., to transport hydrogen bound in the ammonia molecule from a production site to distant locations for the use of hydrogen). Again, only "green" ammonia, utilizing "green" hydrogen produced by electrolysis with electricity from renewable sources of energy, is applicable for RED compliance, while others building on fossil feedstock are not—even when carbon capture and storage (CCS) technologies are used.

Methanol. Another fuel option considered in the maritime sector is methanol, as it is a rather "easy-to-handle" liquid fuel, and the large-scale supply infrastructure already exists in many ports globally. While methanol is produced today primarily from natural gas or coal (i.e., from fossil fuel energy), alternative biomass-based production pathways exist. Furthermore, synthetic production using "green" hydrogen and sustainably provided CO_2 to produce renewable methanol is also possible.

While many of these biomass-based or synthetic fuels are directly fit for purpose in some applications, alcohols like methanol and ethanol offer the opportunity to be further processed into products feasible for blending into high-demand target sectors. Ethanol, for example, can be converted based on a so-called "Ethanol-to-Jet" (ETJ) upgrading process into SAF [15], as well as by a Methanol- or Ethanol-to-Gasoline (MTG/ETG) process into an aromatic-rich naphtha stream for blending in high ratios into the gasoline pool [16–18].

In terms of CO_2 abatement costs (only fuel production side), a high-level overview would look like this from cheapest to most expensive fuel/energy carrier option (Fig. 8). Of these options, ethanol and FAME are commercially most mature so far. HVO is becoming more mature; however, HVO is characterized by some challenges related to the feedstock requirement. Mostly, crop and RED Annex IXb caps will be filled with (cheap) ethanol and FAME, and therefore, HVO will need to be more RED Annex IXa (i.e., uncapped) based.

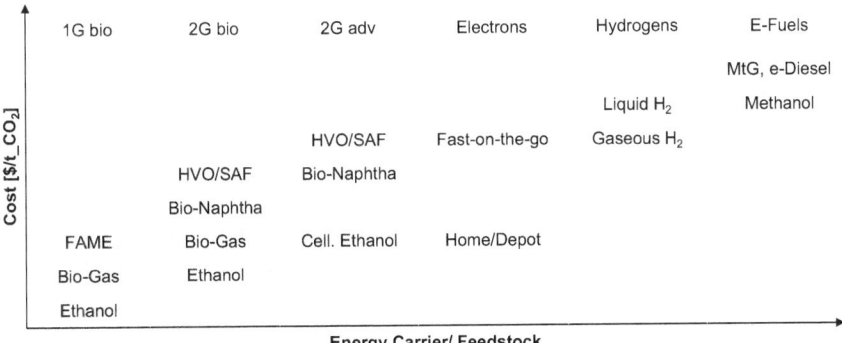

Fig. 8 Cost ladder of fuel options and energy carriers for defossilization purposes (adv—advanced; FAME—fatty acid methyl ester; HVO—hydrotreated vegetable oil; MtG—methanol-to-gasoline; SAF—sustainable aviation fuel) [19]

Bio-LNG, battery electric vehicles (BEV), and hydrogen are mostly dependent on the acceptance of the respective end-user. Hydrogen will, however, remain a relatively costly product in the years to come but offers carbon emission-free mobility (on a tank-to-wake level) combined with a number of customer benefits in comparison to battery electric vehicles (BEV) for certain use cases (e.g., a higher range, faster refueling, higher payload). Hydrogen as a fuel shows potentially significant cost reductions once the infrastructure has been set up and once large-scale (in the gigawatt (GW) scale) electrolyzer projects have come online.

With the focus of the EU regulation to promote a shift toward zero tailpipe CO_2 emissions vehicles, the demand for liquid fuels in road transportation will diminish over time. Assuming a phase-out of vehicles driven by internal combustion engine (ICE) vehicles powered by carbon-containing fuels, e.g., within the passenger car sector by 2035, given the announced 100% CO_2 reduction demand by new passenger car vehicle sales and an average lifespan of such vehicles of 11.5 years on EU roads [20], demand for liquid fuels will be there up to roughly the year 2050. Similar steps are being discussed for Commercial Road Transport (CRT), though it is realized that a technology change from internal combustion engine (ICE) powered trucks toward only battery electric vehicles (BEV) and hydrogen-powered trucks is much harder to realize than lighter vehicles. The demand for diesel-type fuels can therefore be foreseen well beyond 2050, especially since the age of heavy-duty trucks is even longer, with 13 years on average [20].

With new biofuel production facilities coming online this decade, increased availability of renewable fuel molecules can be expected. Those facilities might be produced for the decades to come to support a defossilization of the transport sector as a whole, particularly the harder-to-abate sectors of commercial road transport, aviation, and maritime sectors. Additionally, the demand for fuels within road transport will be declining due to a substitution of internal combustion engine (ICE) vehicles for battery electric vehicles (BEV) and hydrogen-powered trucks. Depending on how the competition for biofuels across the various transport sectors will turn out in the end, the share of renewable molecules within the liquid fuel market may grow relative and absolute over time, finally enabling a fully renewable liquid fuel for the remaining internal combustion engine (ICE) legacy fleet.

5 Lower Carbon Fuel Options for Existing Engines

Low-carbon fuels already contribute to CO_2 emission reductions within the EU today. For example, in Germany in the year 2021, biofuels lead to a reduction in CO_2 emissions of 11.1 Mt/a [21]. The difference of such options will be further described in the following, differentiated into diesel fuel options and gasoline fuel options.

5.1 Diesel Fuel Options

In the following, three low-carbon diesel fuel options that can be used in existing vehicle fleet engine technologies are described in more detail, namely "Blue Diesel," "HVO/GTL diesel blends" and "HVO 100".

Blue Diesel. A significant number of vehicles with diesel combustion engines will be operated in the foreseeable future. Only in Germany alone, about 20 million diesel vehicles are operated, consisting of passenger cars, trucks, buses, vans, and off-road vehicles. Similarly, within the EU, there are about 700,000 buses, several million trucks, and more than 25 million vans on the road today, primarily powered by diesel engines and with expected lifespans often exceeding ten years. Also, combustion technology is still present in the overwhelming share of new vehicles, typically around 95% for heavy-duty diesel vehicles [19]. Hence, given these circumstances, a need can be seen to provide fuel solutions for such vehicles that allow for mobility based on carbon-containing fuels and ICE-engines with a reduced climate impact (i.e., low-carbon fuels that meet the existing diesel specification EN 590). One example of such a fuel is so-called "Blue Diesel." This diesel type is a response to the question of how much CO_2 emissions can be avoided within the EN 590 specification while also improving fuel quality and setting new standards for sustainability. Figure 9 shows the fuel composition and product definition of Blue Diesel.

According to Fig. 9, this type of diesel fuel is characterized by a share of up to one-third of renewable components provided by sustainable sources. The remaining two-thirds are still fossil fuel-based diesel-compatible components. With this blend, about 20% well-to-wheel (WtW) CO_2 reduction can be achieved. Blue Diesel is on the market in a growing number of refueling stations across Europe.

HVO / GTL blends. Another option to provide "low-carbon" fuel options for existing diesel combustion engines is a blend of gas-to-liquid (GTL) diesel and HVO diesel. GTL diesel is a fossil synthetic paraffinic diesel made from natural gas.

R33 Blue Diesel - Composition

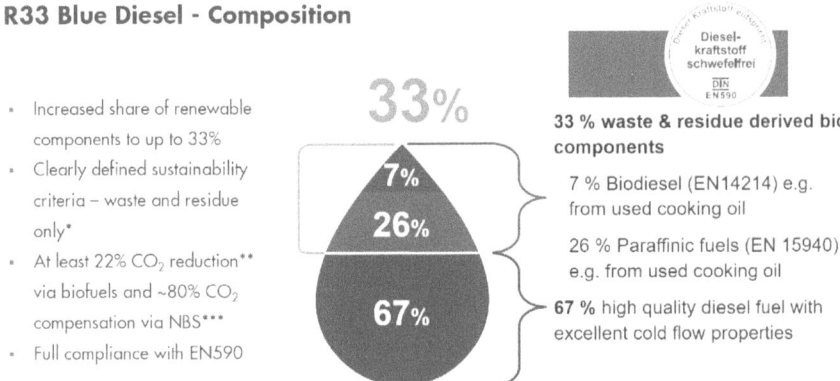

- Increased share of renewable components to up to 33%
- Clearly defined sustainability criteria – waste and residue only*
- At least 22% CO_2 reduction** via biofuels and ~80% CO_2 compensation via NBS***
- Full compliance with EN590

33%

7%
26%

67%

33 % waste & residue derived bio components

7 % Biodiesel (EN14214) e.g. from used cooking oil

26 % Paraffinic fuels (EN 15940) e.g. from used cooking oil

67 % high quality diesel fuel with excellent cold flow properties

Fig. 9 Fuel composition and product definition of Blue Diesel (* according to mass-balanced PoS/certified feedstock; ** well-to-wheel CO_2 reduction based on GHG value of PoS of used components with 95.1 g CO_2/MJ as reference for fossil diesel; *** compensation via VCS-certified nature-based solutions)

As shown in Fig. 10, by using a diesel blend of 45% v/v HVO with 55% v/v GTL a CO_2 emission reduction (WtW) of about 40% can be achieved.

HVO 100. A 100% HVO option (HVO100) is also available in some markets in the EU (e.g., in the Netherlands and in Scandinavia). Such a paraffinic diesel fuel (EN 15940 compliant) provides a further step along a pathway to provide non-fossil diesel fuel and can allow for CO_2 emission reductions of almost 90%, depending on the feedstock supply (Fig. 10). Data for the well-to-wheel (WtW) CO_2 emission impact are taken from [22] (Fig. 11).

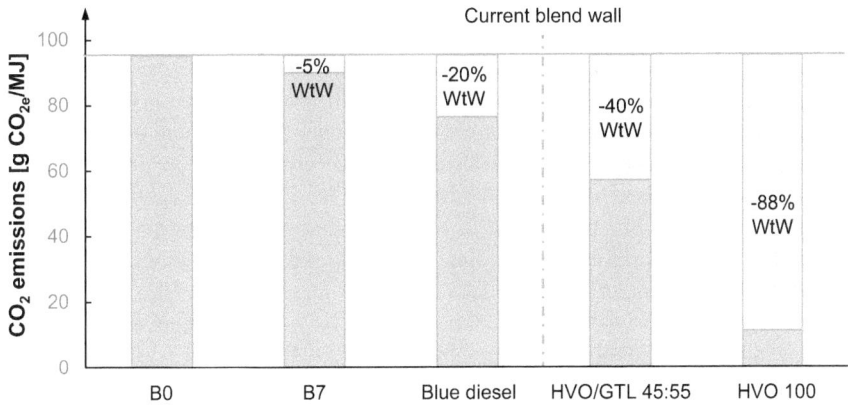

Fig. 10 CO_2 emissions reduction potential of Blue Diesel, HVO100, and a (volume-related) 45:55 blend of HVO and GTL (HVO data from [22] WtW report, WOHY1a pathway [5], GTL data from [22] WtW report and WtT reports [22]

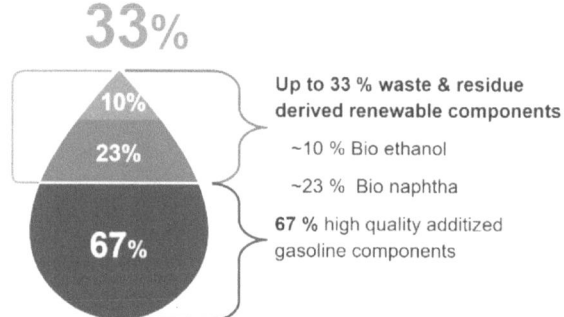

- Increased share of renewable components to up to 33%
- Clearly defined sustainability criteria – waste and residue only*
- At least 20% CO_2** reduction via biofuels and ~80% CO_2 compensation via NBS***
- Full compliance with EN228 RON95 E10 specification

Up to 33 % waste & residue derived renewable components

~10 % Bio ethanol

~23 % Bio naphtha

67 % high quality additized gasoline components

Fig. 11 Fuel composition and product definition of blue gasoline (* according to mass-balanced PoS/certified feedstock; ** well-to-wheel CO_2 reduction based on GHG value of PoS of used components with 93.3 g CO_2/MJ as reference for fossil gasoline; *** compensation via VCS-certified nature-based solutions)

5.2   Gasoline Fuel Options

In the following, two low-carbon gasoline fuel options that can be used in existing vehicle fleet engine technologies are described in more detail, namely "E20 gasoline" and "Blue Gasoline".

E20 Gasoline. When blended into a fossil-based base stock, an addition of 20% v/v of bio-ethanol to make an E20 grade results in Well-to-Wheel (WtW) GHG reductions of approximately 10%, compared to fossil gasoline (E0). This is illustrated in Fig. 12, assuming GHG emissions of 93.3 g CO_2/MJ as a comparator for fossil gasoline and ethanol typically available in the European market. Moving from E10 to E20 would result in an incremental 5% GHG savings, which would be a significant step integrated across the whole fleet.

Blue Gasoline. In the year 2021, so-called "Blue Gasoline" was launched by a German industry consortium (Shell, Volkswagen, and Bosch), where the gasoline equivalent of Blue Diesel was pursued, based on the same motivation to demonstrate that, also for gasoline fuels, the overall CO_2 emissions balance can be substantially improved for gasoline fuels while also improving overall fuel quality. Also, long-term low-carbon fuel solutions will be needed for gasoline cars with a stock of well above 70 million vehicles in Europe, and that still make up around 50% of new sales (particularly when including plug-in hybrid electric vehicle (PHEVs)) to achieve the ambitious CO_2 emissions reduction targets in the EU. Blue gasoline can deliver a CO_2 emission reduction of about 20% (WtW). This fuel was tested against numerous technical requirements. To achieve high product quality regarding particulate emissions (PN/PM), the E150 specification of "blue" gasoline was set to a minimum of 85% v/v, and the distillation end point was set to a maximum of 195 °C. Figure 11 shows the fuel composition and product definition of blue gasoline.

Again, the renewable molecules comprise roughly 33%, mainly composed of ca. 10%-points of bio-ethanol and ca. 23%-points of bio-naphtha. The remaining rest

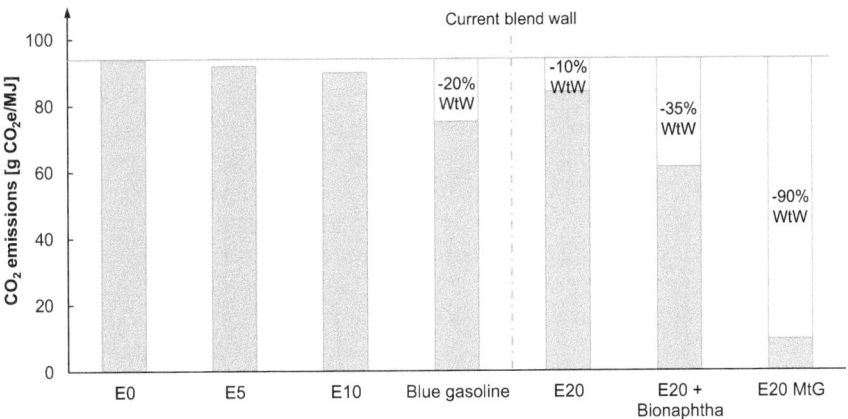

Fig. 12 Well-to-wheels (WtW) CO_2 emissions: illustration of E20 blue gasoline as part of the defossilation strategy

are gasoline components derived from fossil-based crude oil, contributing to meeting the given fuel standards. Additionally, clearly defined sustainability criteria need to be taken into consideration.

Blue gasoline comprises ca. 10% ethanol by volume (making it an E10 fuel) and about 23% v/v of a bio-naphtha, giving overall WtW GHG savings of 20%, as also illustrated in Fig. 12. The C3 Mobility Project shows that adding 10% v/v of bio-ethanol to synthetic gasoline produced from a "Methanol-to-Gasoline" (MtG) process, that it is possible to produce a fuel that meets the prevailing EN 228 gasoline specification [23]. Indeed, one of the great advantages of ethanol [24] is that it can be blended with synthetic streams with a poorer octane quality to make a 95 RON fuel.

Blue gasoline is an E10 fuel, as was the experimental gasoline produced as part of C3 Mobility. However, one could envisage a subsequent step in which E20 becomes an enabler for a further step in WtW reduction in which E20 is combined with a bio-naphtha produced by a methanol-to-gasoline (MtG) or an ethanol-to-gasoline (EtG) pathway. This step is illustrated in the far-right-hand side, which envisages a 35% reduction of WtW-GHG emissions from having E20 in conjunction with bio-naphtha. However, if a synthetic stream with a higher octane content compared to bio-naphtha is used (such as MtG or EtG naphtha/gasoline), then it might be possible to go further than this, and a 100% renewable content would be potentially possible with corresponding higher reductions in WtW-GHG-emissions.

6 Outlook

As in the previous sections shown, a variety of options and conversion pathways is available to defossilize road transportation applications. Some options (i.e., biofuels and e-fuels) can be produced in a quality that is very similar to fossil fuels and allows for their "drop-in" utilization, meaning that blending such components with currently used (conventional) fuel options allows to meet the required fuel norms and allows to feed them into the existing fuel infrastructure and vehicle fleet. This can help to defossilize applications where internal combustion engines (ICE) are used, which will most likely be in the vehicle fleets for the years and maybe decades to come. Other fuel options require additional infrastructure as well as new powertrains for their application within the road transportation segment, such as (bio-)LNG, the direct use of electricity, or hydrogen.

Regulations and mandates are needed to stimulate the transition toward "green" fuels and GHG-neutral transportation. Biofuels and bio-LNG are lower-cost options and are expected to play an important role in helping defossilize road transport in the near term and as bridging fuels until battery electric and hydrogen-powered vehicles become more mainstream. Options like Blue Diesel and HVO100 are fuels that can be used as "drop-in" fuels for existing diesel combustion engines and in the existing fuel infrastructure. For gasoline combustion engines, bio-ethanol is a blend component allowing for a reduction in WtW-GHG-emissions and delivering high octane, which enables the blending of lower-octane bio-naphtha or gasoline produced via a "Methanol-to-Gasoline" (MtG) pathway. The E20 fuel option is seen as a logical next step for gasoline combustion engines, as many newer vehicles on the roads are technically able to run on E20.

In the medium term, on-road transport is proposed to move to zero tailpipe CO_2 emissions in the EU (i.e., passenger cars and trucks will be more and more replaced by battery electric vehicles (BEV) and fuel-cell electric vehicles (FCEV)). Especially for heavier trucks with long-distance driving use cases, hydrogen is expected to play an important role, as the range of FCEV trucks is larger and refueling is faster than BEV trucks. Once hydrogen infrastructure has been built up and once fuel cell technologies and hydrogen storage systems are mass-produced—and thus are available at lower cost, it is expected that also in some passenger car segments (e.g., in the segment of sport utility vehicles (SUV) or for frequent long-distance driving) hydrogen can play a role.

With the transition of on-road transport toward a direct use of electricity and hydrogen, available biofuels can be used for the maritime and aviation sectors, as they will still rely on liquid and energy-dense fuels, most likely also in the coming decades. When powerfuels/e-fuels (synthetic fuels from renewable power) can be produced at a larger scale and, therefore, most likely at a lower cost, these fuels are expected to play a longer-term role, especially in aviation and the maritime sector. A possible mobility defossilization roadmap is shown in Fig. 13.

In [25] it has been shown that a mix of solutions enables faster defossilization of the transport sector rather than focusing on single or few solutions. However, another

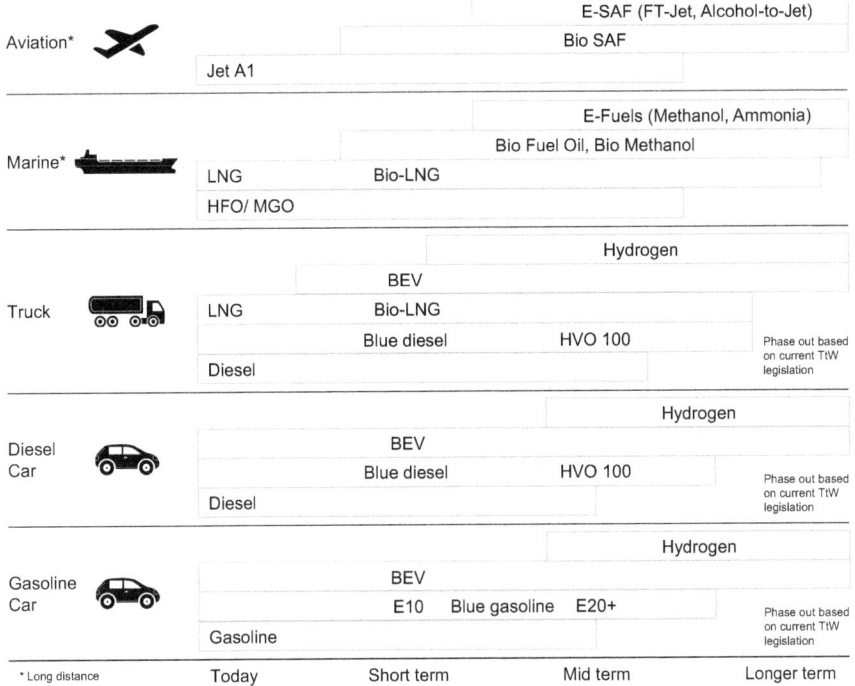

Fig. 13 Possible fuel defossilization roadmap for EU vehicle fleet [19]

most important factor is the response of the end-consumer that may affect policy-shaping of how certain sub-sectors of transport shall transition into a defossilized future with reduced climate impacts by GHG emissions.

Apart from defossilization through non-fossil energy carriers, a range of technologies may well impact road-based mobility in future, but their impact on the defossilization of mobility is hard to quantify. Some of these technology options are listed here for completeness:

- Car sharing and/or mobility-as-a-service may well lead to fewer vehicles on the roads. It is unclear, though, how strong the impact will be on driven kilometers and, hence, on GHG reduction.
- More automated or even fully autonomous vehicles are technically feasible, and developments by OEMs are making progress. However, it is unclear what the potential will be to reduce GHG emissions from road mobility.
- Battery swapping for cars and trucks is seen as an option to make BEV more acceptable for long-distance driving. However, potentially, a higher number of batteries will be needed, as batteries will not only be on cars, but additional batteries will be waiting for customers to swap at battery swapping stations. This means potentially a higher demand for battery materials and metals.

- Megawatt charging for trucks is being developed for fast charging. However, providing this amount of electrical power at truck ports is seen as a challenge.
- Catenary systems (overhead power lines) for battery electric trucks have been developed, but roll-out seems to be slow/limited, as the cost for infrastructure is very high, and another hen-egg-problem will be created.
- Parcel delivery with vans and trucks to homes is expected to grow further, maybe reducing the need for individual transport and hence potentially reducing GHG emissions from on-road transport.
- Digital AI-supported multi-modal mobility offerings will make it easier for customers to travel conveniently from A to B, potentially making lower GHG modes of mobility more acceptable and reducing the dependency of people to own a private car.

Road-based mobility and transport by passenger cars, vans and trucks is a major contributor to global greenhouse gas emissions. A range of options are available and have been presented in this chapter to defossilize the sector. In view of cost and time constrains, it is suggested to implement those options first that come at lower cost and can be implemented as quickly as possible, while at the same time not losing sight of the carbon–neutral long-term ambitions.

References

1. Shell, Deloitte (2022) Decarbonising road freight: getting into gear [Online]. https://www.shell.com/energy-and-innovation/the-energy-future/decarbonising-road-freight.html
2. Statista (2023) Global CO_2 emissions by year 1940–2022. Statista
3. IEA (2022). https://www.iea.org/data-and-statistics/charts/passenger-car-sales-2010-2022
4. Transportgeography (2022). https://transportgeography.org/contents/chapter5/road-transportation/automobile-production-fleet-world/
5. IEA (2023). https://www.iea.org/data-and-statistics/charts/energy-consumption-in-transport-by-fuel-in-the-net-zero-scenario-1975-2030
6. EAFO (2022) European alternative fuels observatory. https://alternative-fuels-observatory.ec.europa.eu/transport-mode/road/european-union-eu27
7. Schulthoff M, Kaltschmitt M, Balzer C, Wilbrand K, Pomrehn M (2022) European road transport policy assessment: a case study for Germany. Environ Sci Eur 34, Article 92. [Online]. Available: https://doi.org/10.1186/s12302-022-00663-7
8. ICCT (2022) Provisions for transport fuels in the European Union's finalized "Fit for 55" package—International Council on Clean Transportation (theicct.org)
9. Europa.eu (2022) CO_2 emission performance standards for cars and vans (europa.eu)
10. HD_Europa.eu (2022) https://climate.ec.europa.eu/eu-action/transport-emissions/road-transport-reducing-co2-emissions-vehicles/reducing-co2-emissions-heavy-duty-vehicles_en
11. European Commission (2021) Proposal for a directive of the European Parliament and of the Council amending Directive 2003/87/EC establishing a system for greenhouse gas emission allowance trading within the Union, Decision (EU) 2015/1814 concerning the establishment and. Brussels
12. European Parliament (2021) Renewable Energy Directive Proposal 2021/0218(COD)—14/07/2021. [Online]. Available: https://oeil.secure.europarl.europa.eu/oeil/popups/summary.do?id=1670384&t=e&l=en

13. ICCT (2023) Provisions for transport fuels in the European Union's finalized "Fit for 55" package (theicct.org)
14. Euractiv (2020) EU to target 30 million electric cars by 2030—EURACTIV.com
15. ASTM International (2020)Standard specification for aviation turbine fuel containing hydrocarbons—D7566—19b
16. Arthur MB (2015)"Chapter 4—Synthesis gas-based fuels. In: Renewable motor fuels—the past, the present and the uncertain future, 2015, pp 33–46
17. Siemens Energy (2021) Haru Oni: a new age of discovery. [Online]. Available: https://www.siemens-energy.com/global/en/news/magazine/2021/haru-oni.html. Accessed Feb 2022
18. Volkswagen Aktiengesellschaft (2021) Construction begins on world's first integrated commercial plant for producing nearly CO_2-neutral fuel in Chile. [Online]. Available: https://www.volkswagenag.com/en/news/2021/09/construction-begins-on-world-s-first-integrated-commercial-plant.html
19. Wilbrand K et al (2022) Pathways to meet renewable energy targets in transport. In: Vienna Engine Symposium 2022
20. ACEA (2021) Average age of the EU vehicle fleet, by country, 1 Feb 2021. [Online]. Available: https://www.acea.auto/figure/average-age-of-eu-vehicle-fleet-by-country/
21. Bundesamt für Landwirtschaft und Ernährung (2021) Evaluations und Erfahrungsbericht für das Jahr 2020 - Biomassestrom-Nachhaltigkeitsverordnung Biokraftstoff-Nachhaltigkeitsverordnung
22. Prussi M, Yugo M, De Prada L, Padella M, Edwards R (2020) JEC well-to-wheels report v5, EUR 30284 EN. Publications Office of the European Union, Luxembourg
23. Cracknell RF, Hemberger Y, Alayon L (2021)Low carbon mobility options for internal combustion. In: 11th Emission control conference, Dresden, Germany, 2021
24. Wang C, Chahal J, Jannsen A, Cracknell R, Xu H (2017) Investigation of gasoline containing GTL naphtha in a spark ignition engine at full load conditions. Fuel 194:436–447
25. Kramer U, Bothe D, Gatzen C, Pfannenschmidt A, Baum C, Schrogl F, Mahmood O—FVV Fuel Study IVb—Follow-up study: transformation of mobility to the GHGneutral post-fossil age—FVV Final report 1313/2022. https://www.fvvnet.de/fileadmin/Storys/Wie_schnell_g eht_nachhaltig/TB_R603_HT22_AB_1452_Future_Fuels_FVV_Fuel_Study_IVb_2022-10-06_final-web.pdf

Development of European Airborne Mobility and Prospects of Sustainable Aviation Fuels

Marc C. Gelhausen, Fabian Baier, Peter Berster, and Nico Flüthmann

Abstract There has been a strong growth in European and global demand for air transport services in the past decades, and there is likely to be a continuation of such growth in future, as indicated by major forecasts . European and global traffic development is closely interrelated because of the global nature of air travel, so that one has to look at both European and global traffic development. The air transport sector needs to reduce its environmental impact, especially if expected future air traffic volumes are considered, which is forecast to increase in Europe by more than 70% and almost double globally by 2050 compared to 2022. One way of achieving this is to employ so called sustainable aviation fuels (SAF), which also includes certain powerfuel option, to an increasing extent in future. This will be analysed in the following chapters. Other approaches to reduce the aviation related climate impact—which are not the focus of this study—include for example more passengers per flight, i.e. increasing load factors and / or employing larger aircraft, and the use of more efficient aircraft in terms of fuel consumption and emissions. A brief overview of the regulatory framework of air transport relating to the mitigation of climate change and the use of SAF is given, as well as the "Fit for 55" strategy of the European Union (EU) as a measure for mitigating such climate change. Subsequently, two scenarios up to the year 2050 are presented to show a possible impact of SAF on air transport volumes. Finally, a summary and the major conclusions of this analysis will be presented.

Keywords Air transport forecast · Air transport modelling · "Fit for 55" strategy · Mitigation of climate change · Sustainable aviation fuels (SAF)

M. C. Gelhausen (✉) · F. Baier · P. Berster · N. Flüthmann
German Aerospace Center (DLR), Institute of Air Transport, Hamburg, Germany
e-mail: marc.gelhausen@dlr.de

© The Author(s), under exclusive license to Springer Nature Switzerland AG 2025
N. Bullerdiek et al. (eds.), *Powerfuels*, Green Energy and Technology,
https://doi.org/10.1007/978-3-031-62411-7_4

65

1 Introduction

Global as well as European air traffic has been and remains highly concentrated on a relatively small number of important airports, the majority of which are facing capacity problems or will face such problems in the coming years [12]. An outstanding example is London Heathrow airport which is already operating at its maximum capacity and has been for about 20 years, with there being still no solution to the capacity crunch in sight [13]. Nevertheless, global air traffic is expected to continue to grow in the long term, albeit at a pace that likely differs greatly between Asia and the Middle East on the one hand, and Europe and North America on the other. For instance, as prosperity levels increase in these regions, demand has only just begun to grow in the last few decades and is now rapid in Asia. In North America, however, demand development is more mature as there is already a high level of propensity to fly, and the demand development shows signs of saturation, with relatively low growth compared to other markets. Demand in Europe, however, is also expected to grow continuously, albeit at a reduced rate, and the demand growth in the Middle East and Asia contributes to this growth significantly due to the global nature of air travel.

The overall growth in air traffic is coupled to increasing greenhouse gas (GHG) emissions, if the use of (fossil) petroleum-based aviation fuels is continued. With this respect, so called sustainable aviation fuels (SAF) are one measure to mitigate the greenhouse gases produced by aviation. The term SAF represents sustainable and non-conventional aviation fuels that are not fossil-derived. SAF can be produced from biological resources (i.e. biofuels) or non-biological resources (i.e. synthetic fuels). So far, SAF is be blended with conventional aviation fuel prior to its use. After proper blending, it can be used as a so called drop-in-fuel [18].

The outline of this chapter is as follows: In the next section the past development of air transport is presented. In Sect. 3, the model employed for the simulation of the effects of SAF on the future development of air transport and its environmental impact is briefly described. Section 4 covers the topic of the legal framework of the air transport system relating to the mitigation of climate change. In Sect. 5, SAF and the "Fit for 55 Strategy" of the European Union (EU) are described in more detail. This sets the scene for the air transport scenarios up to 2050 in Sect. 6:

- In the "Traditional Scenario", the use of 100% conventional jet fuel is assumed.
- The "SAF Scenario" represents blending conventional jet fuel with SAF according to the "Fit for 55" strategy on a global scale.

However, while a global adoption of this strategy may be unrealistic, the two scenarios cover a spectrum which most likely comprises the actual future development (as far as this is possible in a long-term forecast).

The focus of this chapter is on scheduled commercial passenger traffic, in that it makes up the bulk of global air traffic that is responsible for aviation related CO_2-emissions. Thus, air cargo traffic and so called business aviation are excluded.[1] This chapter is finally closed with a summary of the major findings and conclusions that can be derived from the analyses.

2 Air Transport Development in the Past (Up to 2021)

Figure 1 shows the passenger volume development between 1950 and 2021 on a global scale. The pattern of development is very similar for the European market, as the major global crises in Fig. 1 affected all world regions. Despite the periodic crises such as the oil crisis in the Seventies, 9/11 and the Global Financial Crisis in 2008/2009, long-term passenger volume development has been on a steady growth path. The COVID-19 crisis, however, is unprecedented in its extent, and especially in 2020 there was a 60% drop in passenger volume, an impact on air transportation that has never occurred before. Nevertheless, since 2021/2022, air traffic has been on a steep recovery path, and COVID-19 might only be considered as another major crisis in the long term in retrospect.

Figure 2 displays the development of passenger volume between 2006 and 2021 in terms of the seven world regions. There was a strong passenger volume growth, especially in Asia, which almost tripled between 2006 and 2019. There was also a substantial growth in the other world regions, but due to COVID-19 there was a sharp decline of passenger volume in all regions. This decline was extremely large in Asia, and the Asian region is, apart from the Southwest Pacific region, the only one which still declined in 2021. On the other hand, the recovery is decidedly strong in North America, Europe and South America. This leads to a situation where North America is the largest passenger market in 2021, marginally larger than Asia, which was the biggest one in 2019 by a large margin. The decline was almost the same in Europe as in Asia in absolute terms (a drop of about 0.8 billion passengers), but in Europe it was much larger in relative terms: passenger volume dropped to about 35% of pre-COVID-19 levels in 2020 (Asia: 50%) but, unlike Asia, European air transport was already on a recovery path in 2021.

In addition to the development of passenger volumes, Fig. 3 illustrates the corresponding flight volume development. Flight volume grew only very slowly in most regions, or even stagnated during the period 2006–2019. The only exception is Asia, where a strong growth in flight volume can be observed. However, it is still not as strong as the passenger volume growth for this region. In all regions, flight volume

---

[1] AIR cargo traffic volumes, where so-called integrators such as FedEx or UPS handle a large share of the global freight volume, are inherently difficult to model, because they are typically not included in air traffic databases such as OAG or Sabre AirVision MI, therefore requiring a different approach [3]. We have further excluded so-called business aviation, i.e. the use of small aircraft that are typically not operated on a particular schedule. Such traffic heavily depends on local circumstances [4].

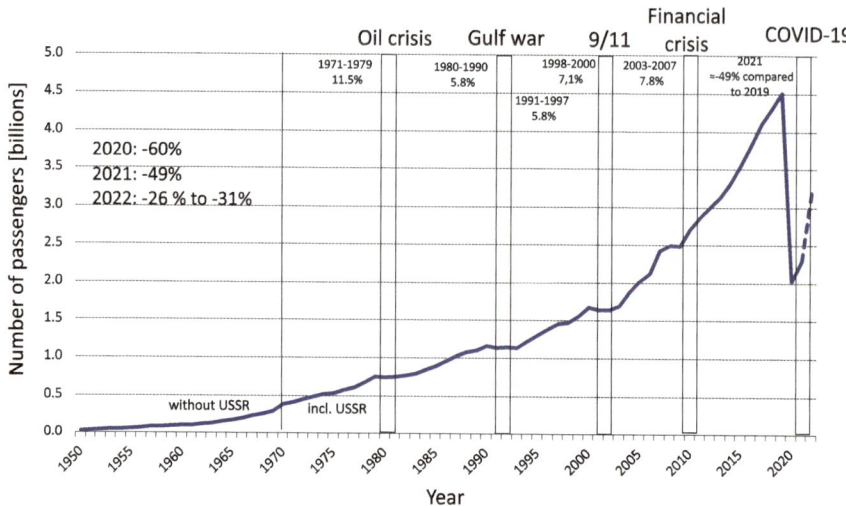

Fig. 1 Global air transport passenger volume 1950–2021 ([19]; USSR: Union of Soviet Socialist Republics)

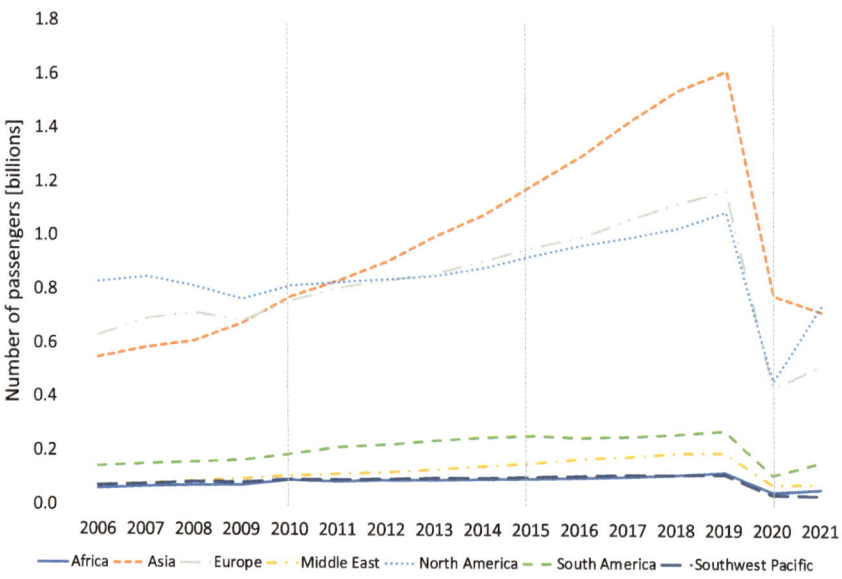

Fig. 2 Global air transport passenger volume by world region 2006–2021 [27]

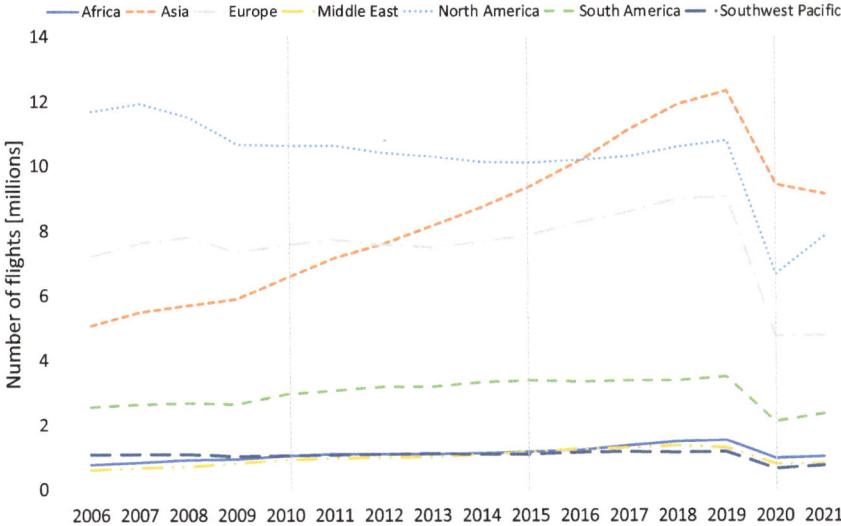

Fig. 3 Global flight volume by world region 2006–2021 [26]

development was weaker than passenger volume growth, leading to an increase in the number of passengers transported per flight. This was either achieved by raising load factors or increasing the seat capacity per flight, for example by employing larger aircraft. As load factors are well above 80% on a global level, and in some cases even approaches 90%, raising seat capacity is likely to be the major measure to serve an even larger passenger demand in the long term. However, there is only limited potential in raising load factors further. They technically cannot exceed 100% and, on a global level, with a value close to 90%, seems currently to be at a realistic upper bound.

A main driver of increasing passengers per flight is the economic considerations of airlines because it is more economical to transport more passengers per flight up to a certain limit as long as the level of flight frequencies remains attractive from the point of view of the air traveller—i.e. the possibility to take a flight for a certain connection in a flexible manner. If this is not the case, passengers may look for other options, depending on factors like trip purpose, destination and season of the year. This might be a problem for smaller airports and put them under pressure, because they may not be able to maintain an adequate level of flight frequencies if more passengers are transported per flight.

Another main driver of increasing passengers per flight lies in capacity constraints, especially with regard to airport (and airspace) capacities and at major hub airports like London Heathrow. Long-term airport capacity is mainly determined by the capacity of the runway system, as this part of the airport is typically difficult to enlarge, as thus adaptions represent major infrastructure expansions, where there is—amongst others—an involvement of public-stakeholders, which typically often oppose such plans due to an expected increase in noise and pollution emissions [32].

This is particularly the case in western countries, but to some degree in other countries as well. Even in Asian countries such as China, where airport enlargements are relatively easy to realise than, for example, in European countries, the forecast increase in passenger volume cannot be served without a substantial increase in the number of passengers per flight. Nevertheless, the barriers to enhance airport capacity, i.e. the runway system, are one of the highest in Europe, as can be seen by the example of London Heathrow: the airport has been operating at its capacity limit for about 20 years, and there is still no runway enhancement in sight—as a consequence, the air traffic management (ATM) had to be designed there in a rather stringent and efficient manner. Other major European hubs such as Paris Charles de Gaulle or Amsterdam Schiphol still have some capacity reserves but are fast approaching their limits.

Innovative usage concepts that have been developed can help to improve the capacity of existing runway systems, in particular those with a layout that is not optimal for maximum capacity such as closely spaced parallel runways [22]. Other airport infrastructure enlargement investments such as terminals and aprons can normally be realised during normal business activities of an airport and do not present a long-term capacity restriction. Other parts of the global air traffic network, such as en-route airspace capacity, are also expected to be less of a capacity problem in the long term [23]. The trend to more passengers per flight not only affects airports with a capacity shortage, but also those with ample capacity as well, because of connecting traffic, puts pressure on the smaller airports.

These developments lead to a different fleet structure in the longer term with the employment of larger aircraft and thus fewer flights per passenger (from an aircraft operator perspective), which will have effects on fuel consumption and emissions. This is particularly true for the European market because of the high utilisation of hub airports and the barriers to significant expansion of their capacity. The future fleet therefore cannot be extrapolated from the current fleet, adjusted by the growth in passenger volume. Rather, a different methodological approach is needed, one which is briefly described in the next section.

3   Overview of the Forecast Model

In this section, a brief overview of the model that is used to generate the scenarios presented in Sect. 6 is provided. A full description of the model can be found in [12, 13] and [14], but the major model elements including air transport demand, airport capacity and aircraft size forecast, are briefly discussed in a more technical manner in the next section.

Figure 4 illustrates the model approach. In the first step, the unconstrained passenger and flight forecast are established. This includes new nonstop flights, which become viable due to the increase in demand. There is, for example, some potential for new nonstop flights in the long-haul market, on routes that are currently served only by stopover flights [15, 33]. On the other hand, shorter routes of say up to 600 km may be also served by high-speed train instead. In Fig. 4, boxes with

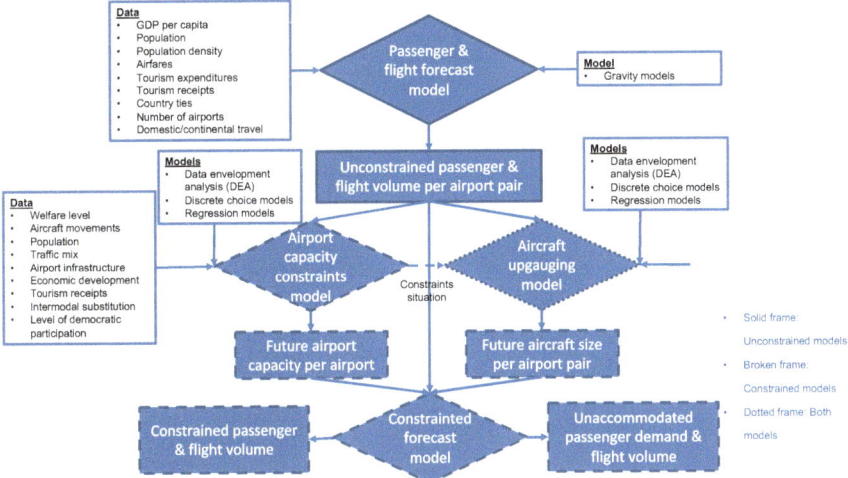

Fig. 4 Overview of the air transport forecast model [13]

solid frames refer to unconstrained models, while boxes with broken frames refer to constrained models that include the effects of limited airport capacity and related aircraft upgauging, i.e. more seats per aircraft. The box with dotted frames (aircraft upgauging) refers to both categories. Passenger and flight volume per airport pair is modelled by a gravity model. The three major drivers of passenger demand are

- **Real Gross Domestic Product (GDP) Per Capita**: the use of the term "real" means that it is inflation-adjusted to reflect actual purchasing power and is a measure of the wealth or income of the population per capita that can for example be spent on air travel. It is split into GDP per capita for the origin and the destination of a journey, to allow for more complex relationships.
- **Population**: the larger the population, the larger is the potential for air travel demand at a particular wealth or income level (i.e. real GDP per capita). Real GDP per capita and population taken together for origin and destination is the typical real GDP variable that appears in many air transport demand models.
- **Real Airfare Development**: the development of real, i.e. inflation-adjusted, airfares has a direct effect on passenger demand volume. Increasing real airfares reduces demand volume, while decreasing real airfares has a positive effect on passenger demand volume. In the past, real airfares declined by about 1.5% per year on a global level [12] due to organisational and technological innovations, i.e. better organisation, and the employment of more efficient aircraft. It is difficult to assess if this assumption will hold for the future, especially in the light of increases in kerosene and SAF prices. Fuel costs make up about 15–30% of the ticket price [17, 30, 31], so that this assumption is still applied in the "Traditional Scenario", but this assumption is adjusted accordingly in the "SAF Scenario", as SAF will most likely be (significantly) more expensive than kerosene.

The elasticities of these major drivers determine the overall unconstrained passenger volume development. Real GDP, i.e. real GDP per capita times population, has an elasticity of 1.31, so that an increase of 1% of real GDP increases passenger volume by 1.31%. Real airfares have an elasticity of -1.11, so that a decrease of 1% increases passenger volume by 1.11%. Both real GDP and airfares are thus elastic so that passenger volume reacts over-proportionally to changes in real GDP. However, there is one caveat: these drivers are measured in real terms, so effects of inflation need to be considered. One could argue that inflation reduces real airfares (given nominal airfares remain the same or rise more slowly than inflation), which is positive for demand development, but the GDP also needs to be considered. Inflation burns purchasing power, i.e. real GDP (given that nominal GDP remains the same, or rises more slowly than inflation), which is negative in terms of demand development. Real GDP has a greater impact on passenger volume than real airfares (1.31 compared to 1.11 in absolute values), so the total effect on demand is negative. An exact forecast of inflation over the next 20–30 years is not possible, even estimations are difficult to make and coupled with high uncertainties. Thus, it is assumed that inflation will return to its steady state in the long term, and that there is no long-term inflation effect in the forecast. This assumption is of course debatable, and therefore needs to be kept in mind when discussing the forecast results. However, the discussion shows that a long-term high-inflation scenario would most likely lead to lower passenger volumes.

After obtaining the unconstrained passenger and flight volume forecast for each airport pair, airport capacity constraints and aircraft upgauging are applied. The airport capacity constraints model contains an element that calculates current airport capacity for each airport using data envelopment analysis (DEA) and regression models and, on the basis of discrete choice theory, offers a model that estimates the probability of airport capacity expansion, if capacity is not sufficient to handle the forecast demand. Based on this probability, an expected delay regarding the realisation of a new runway is derived, if indeed and if it is possible at all.

Aircraft upgauging depends not only on the level of airport capacity constraints but also on various other factors such as passenger demand volume, flight distance and more. It affects constrained as well as unconstrained airports because of interdependencies in the global air traffic network [5]. The upgauging model belongs to both the unconstrained and constrained models (dotted frames). The model is implemented using data envelopment analysis (DEA) and regression models and incorporates factors such as passenger volume, flight distance and the constraints' situation at airports. The forecast result is the average number of passengers per flight ("aircraft size") for each airport pair. Combining future airport capacity and aircraft size per airport pair with the unconstrained passenger forecast yields the constrained forecast model. The forecast results are the constrained passenger and flight volume, as well as the lost passenger demand and restricted flight volume due to limited airport capacity.

Figure 5 displays the relationship between passenger demand volume, airport capacity and aircraft size on a very general level: given a (forecast) passenger demand, a minimum airport capacity and average aircraft size are needed so that this demand

Fig. 5 Relationship between
passenger demand, airport
capacity and aircraft size
[13]

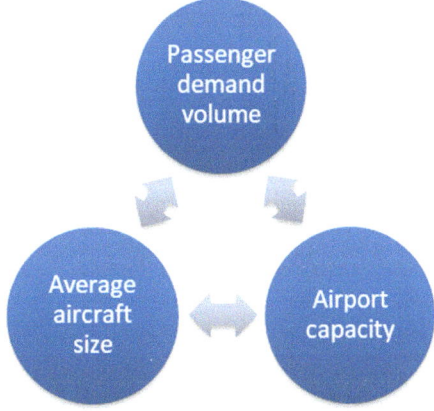

can be met. Both aircraft size and airport capacity limit the maximum number of passengers that can be handled. In this respect, aircraft size and airport capacity substitute for each other to some degree: if a particular aircraft size is not sufficient to serve a given passenger demand volume, increasing the size can compensate for the lack of airport capacity to meet that demand, at least within limits. For simplification, the influence of aircraft mix on airport capacity is neglected. The same applies for airport capacity: insufficient aircraft size can be substituted for by increased airport capacity, so that more flights, albeit with fewer average passengers per aircraft, can be handled. However, airport capacity is typically the bottleneck.

This relationship is particularly important in a world in which (future) airport capacity tends to be scarce. The bottom line is that if these interrelations between passenger demand, airport capacity and aircraft size are accounted for, adjustments to all three elements in a constrained forecast will be seen: there will be some degree of unaccommodated passenger demand, stronger growth in average aircraft size and less airport capacity expansion in future. An unconstrained forecast always assumes a best case scenario regarding the development of airport capacities, which means that potential bottlenecks are neglected.

4   Political and Legislative Framework

The legal framework for European environmental policy law is based upon the European Green Deal which was presented by the European Commission in 2019. The goal is to reduce net greenhouse gas emissions to zero by 2050 with specific interim emission targets being set during the transition period. Milestones are constantly being adjusted according to political goals and ambitions to mitigate the effects of climate change: there is a trade-off between environmental protection and economic power. The cost of climate change has become clearly visible, increasingly since

2020: there have been extreme weather events almost everywhere in Europe—heat and drought together with high wildfire activity on the one hand, and extreme rainfall with heavy floods on the other. The economic consequences of the COVID-19 crisis which started in 2020, and the escalation of the war between Ukraine and Russia in 2022 have pushed the European economy near to a recession, raising voices calling for a delay to additional environmental protection in line with restrictions for industries, whilst others view recent crises as a wake-up call to change the economic systems in Europe to more innovative, energy self-sustaining ones. Europe has the opportunity to take up a leading role in terms of establishing sustainable economies, stimulating innovation and attracting investment in such a way as to tackle the problems that are on the horizon.

In this current situation of political uncertainty, and with potentially substantial changes ahead, it has become rather difficult for companies and industries to anticipate and assess future market conditions, and thus knowing how to invest and transform themselves in such a way as to remain competitive. Such circumstances also apply to the market ramp-up situation for SAF. Political goals to achieve environmental targets (e.g. reducing greenhouse gases by 55% by 2030 according to the European Climate Law) therefore go hand-in-hand with setting incentives and supporting the aviation industry to make a change towards a more climate friendly operation. Appropriate proposals and procedures were presented in 2021 in the "Fit for 55" strategy by the European Commission. With regard to sustainable aviation fuels (SAF), related policies mainly target to support a production, distribution and use of SAF—a proposal which is also known as the ReFuelEU Aviation initiative [7]. In detail, the share of SAF (including advanced biofuels and synthetic fuels) is required to increase over time, thus reducing the usage shares of conventional kerosene and consequently limit aviation related greenhouse gas emissions over time, in accordance with the sustainability and greenhouse gas emission criteria which are part of the EU directives [8].

Starting in 2025 increasing minimum shares of sustainable fuel have to be admixed (2% in 2025, 6% in 2030, 20% in 2035, 34% in 2040, 42% in 2045 and 70% in 2050), whereby the shares have to contain a binding quote of synthetic fuels themselves. In order to avoid "tankering", e.g. operators order aircrafts to carry more conventional fuel than needed for flights to EU airports to skip uplifting the respective SAF shares, this practice will be limited. The law also includes airport and infrastructure directives that the supply of SAF has to be proactively mandatory where possible (targeting transport, storage and refuelling). The law of proof applies to airports and operators, respectively; efficient distribution of SAF on airports, aircrafts and routes thus underly competitive market mechanisms avoiding additional regulatory measurements.

The main goal of ReFuelEU is to hit the "Fit for 55" target for the aviation sector, as neither the EU emission trading system (EU ETS) nor the offsetting mechanism "CORSIA" or associations with the RED II programme have provided sufficient retrenchment regarding greenhouse gas emissions [7], as stated in a parliament resolution 2020 which serves as the legislative baseline for the call being in line with the EU's climate ambitions.

Most stakeholders in the aviation sector see the need for additional action, and thus support the ReFuelEU Aviation initiative. However, there are different views as how to achieve the goals, especially on the part of airlines. Supply-side SAF obligations are generally supported if fuel distribution can be handled in a flexible way, and the production of (non-biological) sustainable fuels is incentivised by the EU. Suggestions from airline representatives mainly consider SAF fuel price compensations, the additional investment needed for research and development as well as for the built up of SAF production plants, to remain competitive within a global aviation market, and an averaging clause for tank loads across (total annual) routes to avoid tank overloads if certain airports do not provide SAF. The IATA notes that production incentives need to be set by the EU to minimise higher prices for airlines and customers, while airport representatives criticise the compulsion for airports to provide SAF infrastructure and storage, rather favouring a competition approach (via a book-and-claim system). Practicability of the "proof-system" is yet to be seen and exceptions for individual airports being discussed in order to ensure competitive equality across Europe's airports.

Opposing cost (largely due to higher fuel prices) with benefit (pollution avoidance, "cleaning-cost") of the measurement, an efficiency of four to six times higher than the original investment is estimated economically [7].

5 Sustainable Aviation Fuels (SAF)

The umbrella term "sustainable aviation fuel" (SAF) encompasses a variety of synthetic kerosene options that can be produced from a wide range of biomass-based and non-biomass-based renewable feedstock options via a number of different conversion processes. While the former SAF options (biomass-based) are sometimes also referred to as biokerosene (among other terms), the latter (non-biomass-based) are also referred to as Power-to-Liquid (PtL) SAF (and more).

A look at the current production costs of SAF shows that the various types of SAF are associated with significantly higher production costs compared with the current and historic prices for conventional kerosene. Based on an extensive literature analysis, data on production costs for four main SAF types is considered. In this context, three biomass-based SAF options are presented, namely Hydroprocessed Esters and Fatty Acids (HEFA), Biomass-to-Liquid (BtL) and Alcohol-to-Jet (AtJ), as well as the Power-to-Liquid (PtL) SAF option based on renewable electricity. Figure 6 provides an overview of average values and value ranges for these four SAF. Accordingly, as of today the average production costs for HEFA SAF are the lowest, with a factor of 2.5 compared to conventional kerosene prices (between 2010 and 2021, the global range was primarily between 500 to 800 €/t). The production of PtL is by far estimated as the most cost-intensive, with a factor of about 5 compared to conventional kerosene prices. Furthermore, the analysis shows that the value range of information on production costs—and thus the uncertainty—is widest for PtL compared to the other SAF.

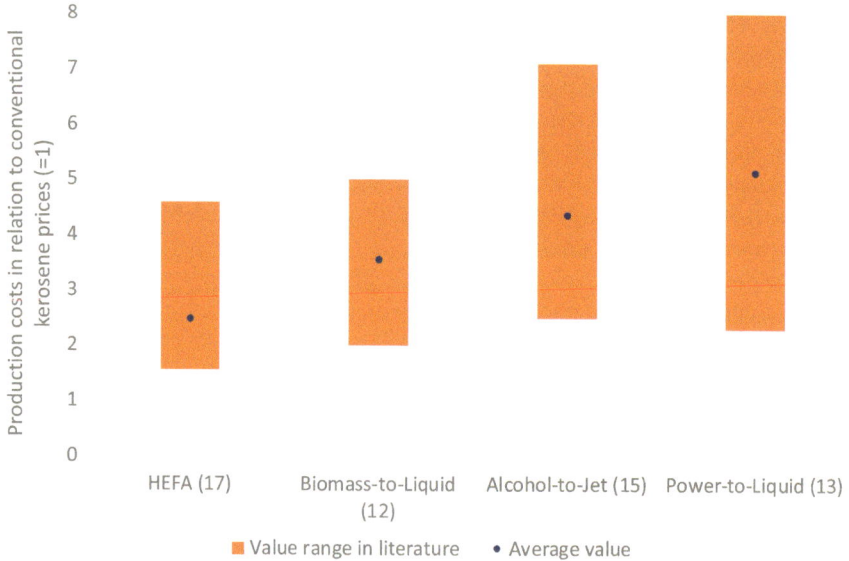

Fig. 6 Production costs estimations of SAF in relation to conventional (fossil) jet fuel prices (values in brackets indicate number of studies considered)

An analysis, where future SAF production costs are estimated, reveals that in general there are cost reductions that can be achieved for various SAF types. Especially the built up of plants to produce PtL SAF requires large investments, and the transition will take years up to decades [28]. While production costs for PtL SAF are currently the highest (or currently estimated to be the highest), they are often expected to be significantly up until 2050, if an increased production can be realised (Fig. 7). This is also the case for the other SAF types, but to a lesser extent.

Because of longer-term policy measures and expected reductions in SAF production cost, its share in aviation is expected to increase substantially. Based on an extensive literature analysis, Fig. 8 displays an overview of the possible SAF share in future air traffic. Here, studies from various actors (for example, IEA [26], ICAO [22], IATA [18], Eurocontrol [11], ATAG [2], NLR [33]) were analysed regarding estimates of future SAF shares. In addition to various SAF scenarios, mandates such as the ReFuelEU proposal from the EU (SAF share in 2030: 6%, 2040: 34%, 2050: 70%) or the Net Zero Carbon 2050 Resolution from IATA [16] (SAF share in 2030: 5%, 2040: 40%, 2050: 65%) are also included in this analysis. Looking at the dotted regression line in Fig. 8, it immediately becomes clear that estimates of future SAF shares in air traffic have a high degree of uncertainty. For example, in 2050—depending on the scenario and the underlying assumptions—SAF shares ranging between 5% and 100% are considered as possible, depending on the study.

Therefore, it is important to remember that many of these very optimistic SAF share projections were made in the case of normative aviation scenarios which have the goal of achieving net zero CO_2 emissions in 2050. As a result, many of the

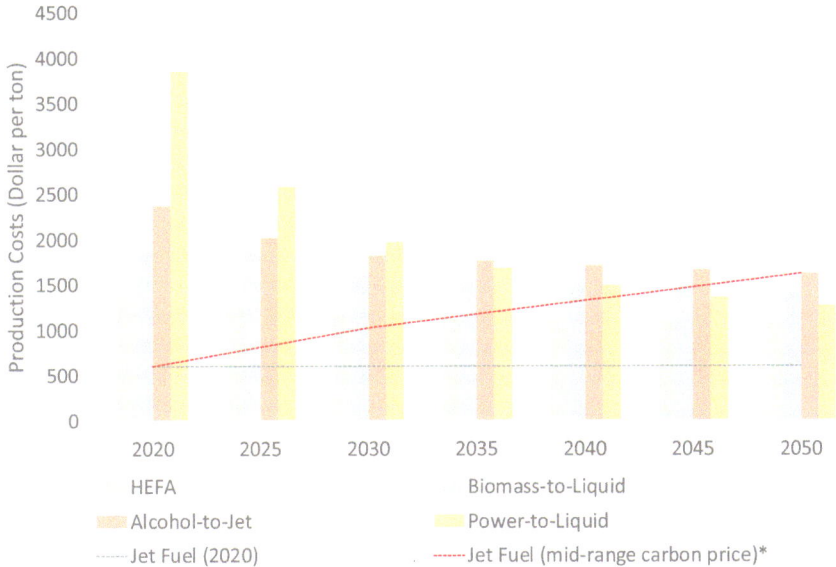

Fig. 7 Future SAF cost development ([24], mid-range carbon price based on [29])

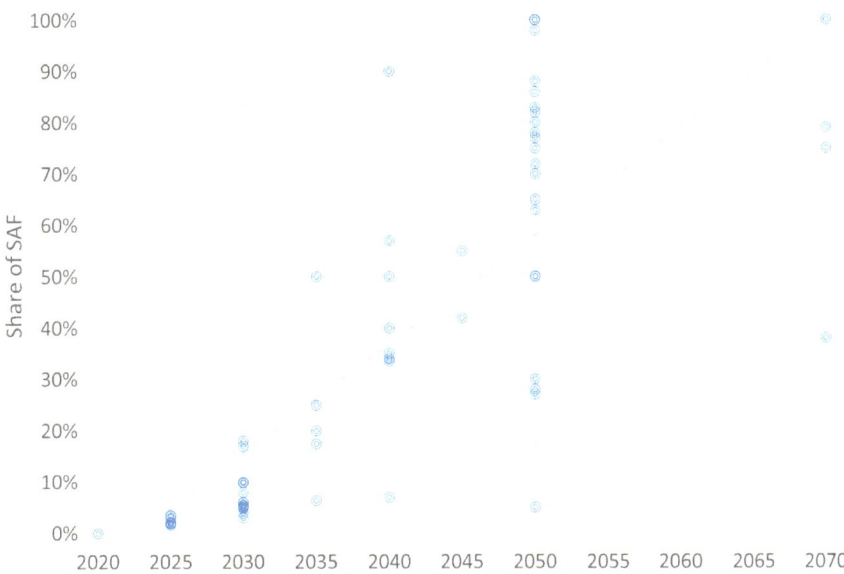

Fig. 8 Future possible SAF shares in aviation

assumptions are very ambitious, but also show the challenges of achieving net zero aviation.

For the scenarios to be presented in the next section a SAF share development as outlined by the ReFuelEU proposal, such that SAF shares will be 6% in 2030, 34% in 2040 and 70% in 2050, is assumed. On average, we assume a 20% share of fuel costs with regard to the airfare. In real terms, airfares will increase by 0.4% in 2025, rising to 13.2% in 2050 compared to kerosene jet fuel without any SAF (the development is proportional to the SAF shares according to the ReFuelEU proposal). The cost calculation is based upon the ReFuelEU proposal shares with regard to biomass-based SAF and synthetic SAF (PtL). For example, the share of biomass-based SAF types starts with a share of 4.8% in 2030 and reaches a share of 35% in 2050. The share of synthetic or PtL SAF increases from 1.2% in 2030 to 35% in 2050. The impact on airfares is almost negligible initially, but becomes more and more substantial over time. As an example, if the airfare for a given flight is 150 € in the case of no SAF in 2050, it is 169 € in the SAF case. Nevertheless, as shown in Fig. 7, cost differences between different SAF options are expected to decrease substantially in future, in particular for PtL, which becomes more and more important until 2050. Furthermore, a 100% shift of SAF-related costs to airfares is assumed, which is considered as a realistic but not a mandatory assumption. In the event that airlines do not fully shift the cost increase to airfares, this would dampen the demand reduction as SAF shares increase over time.

6   Impact of SAF on Future Traffic and Passenger Volume Development and CO_2-Emissions

In this section, the forecast results for the "Traditional Scenario", i.e. without the introduction of SAF and the "SAF Scenario", where SAF is being introduced globally according to the "Fit for 55" strategy of the EU are presented. Major inputs for the forecast are real GDP per capita, population and the development of real airfares. GDP per capita and population are retrieved from forecasts of the IHS Markit of S&P. For real airfares, an annual decline of 1.5% in real airfares is assumed, excluding the cost effects of SAF, which are treated separately. This value of 1.5% is based upon our own analysis of historical airfares and inflation data [12]. The results are presented for passenger volume, revenue passenger kilometres (RPK) and flights for the global and European levels. While the focus of this section is on the European level, the global results help to put them into perspective.

Figure 9 displays the results for global passenger volume development up to 2050. In 2022, the passenger volume is about 3.3 billion and is expected to grow up to over 13 billion by 2050 in the "Traditional Scenario" and 12 billion in the "SAF Scenario". Thus, demand will be about 10% lower in the SAF case as a result of the higher airfares. The corresponding compound annual growth rates (CAGR) are 5.1% for the "Traditional Scenario" and 4.7% for the "SAF Scenario". Despite the

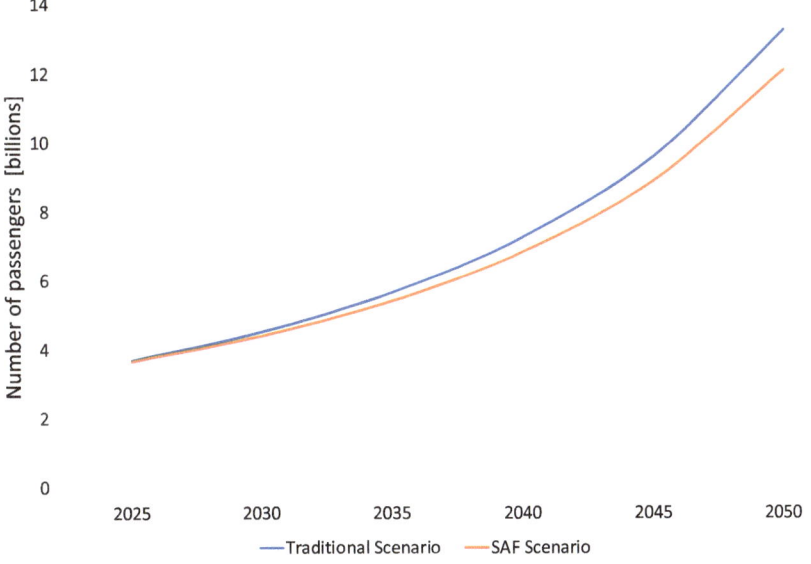

Fig. 9 Global forecast results of passenger volume up to 2050 for the "Traditional Scenario" and "SAF Scenario"

additional cost for SAF, there is a huge increase of passenger volume in both cases: it increases by a factor of 3.6 in the "SAF Scenario" and by a factor of 4 in the "Traditional Scenario".

Figure 10 shows the forecast results for the European passenger volume development up to 2050. This includes not only intra-European air travel, but also passengers between European airports and airports on other continents. In 2022, the European passenger volume is about 919 million, which is expected to grow up to 2.6 billion in the "Traditional Scenario" and 2.3 billion in the "SAF Scenario". As in the global case, demand will be about 10% lower in the SAF case in 2050 due to the higher airfares. The corresponding CAGRs are 3.5% for the "Traditional Scenario" and 3.1% for the "SAF Scenario". This is significantly less compared to the global development, but still almost triples (factor of 2.7) in the "Traditional Scenario", and more than doubles (factor of 2.4) in the "SAF Scenario". The major reason for the European passenger volume growing more slowly compared to the global development is a less favourable economic development measure in terms of GDP per capita and population. Scarce airport capacity at major European airports also plays a role, but this impacts more the number of flights and fleet structure.

Figure 11 illustrates the results for the global RPK volume development up to 2050. In 2022, the RPK volume is about 5,600 billion, which is expected to grow up to 22,400 billion (factor of 4.0) in the "Traditional Scenario" and 20,400 billion (factor of 3.6) in the "SAF Scenario". Thus, the RPK volume is forecast to be about 10% lower in the SAF case in 2050 due to higher airfares. The corresponding CAGRs are 5.0% for the "Traditional Scenario" and 4.6% for the "SAF Scenario" and the RPK

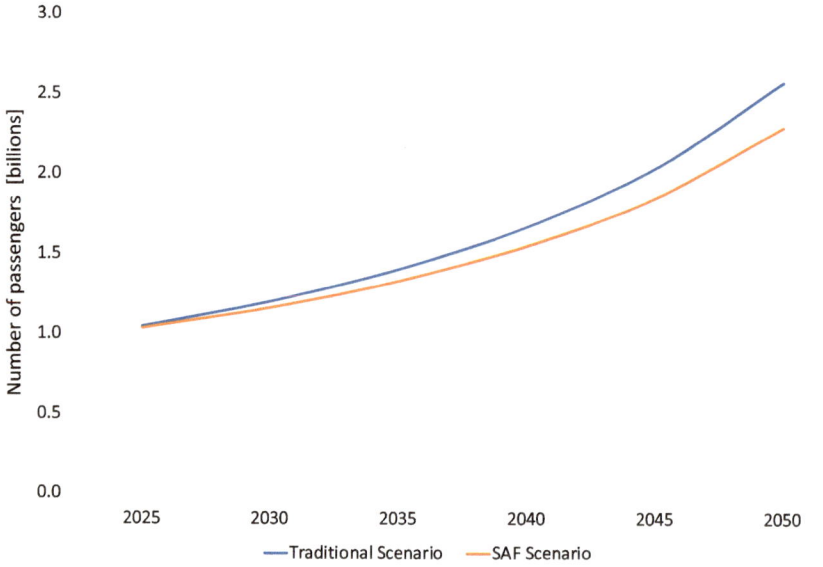

Fig. 10 European forecast results in terms of passenger volume up to 2050 for the "Traditional Scenario" and "SAF Scenario"

volume more than triples in the "SAF Scenario" and quadruples in the "Traditional Scenario".

Figure 12 shows the results for the European RPK volume development up to 2050. In 2022, the RPK volume is about 2040 billion, which is expected to grow up to over 6300 billion in the "Traditional Scenario" and 5600 billion in the "SAF Scenario". Thus, the RPK volume is forecast to be about 10% lower in the SAF case in 2050 due to higher airfares. The corresponding CAGRs are 3.9% for the "Traditional Scenario" and 3.5% for the "SAF Scenario", and the RPK volume more than doubles (factor of 2.6) in the "SAF Scenario" (factor of 2.6) and in the "Traditional Scenario" (factor of 2.9). As in the case of passenger volume, the European development lags behind the global development because of the weaker economic development and the capacity situation of major European hubs that is expected for the future.

Figure 13 displays the results for global flight volume development up to 2050. In 2022, flight volume was about 32 million, which is expected to grow up to 58 million in the "Traditional Scenario" and 55 million in the "SAF Scenario". The corresponding CAGRs are 1.8% for the "Traditional Scenario" and 1.6% for the "SAF Scenario" and are much lower than passenger and RPK volume growth. The rate at which the number of passengers per flight rises is approximately the difference between passenger and flight CAGR, i.e. 3.2% in the case of "Traditional Scenario" and 3.1% for the "SAF Scenario". As a result, a large share of the additional passenger volume up to 2050 will be handled by better employment and increased aircraft capacity instead of more flights. This leads to an increasing number of passengers carried per flight, and the employment of higher seat load factors and larger aircraft to

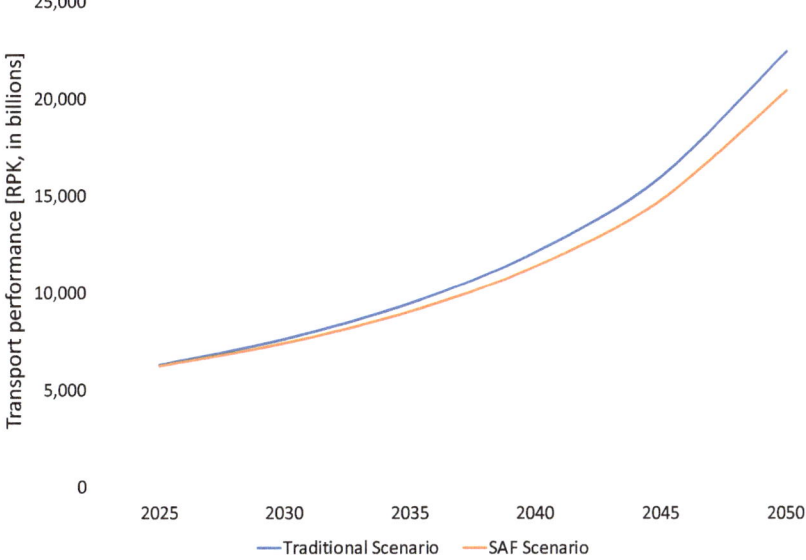

Fig. 11 Global forecast results of revenue passenger kilometres (RPK) volume up to 2050 for the "Traditional Scenario" and "SAF Scenario"

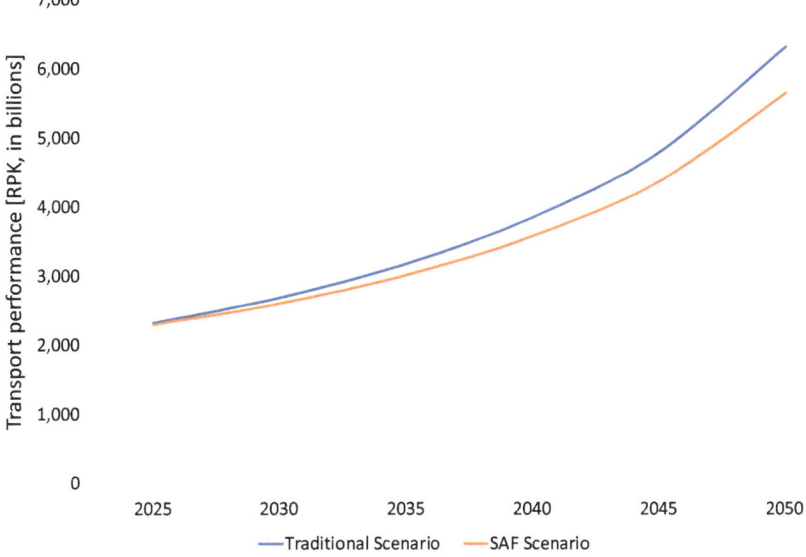

Fig. 12 European forecast results of revenue passenger kilometres (RPK) volume up to 2050 for the "Traditional Scenario" and "SAF Scenario"

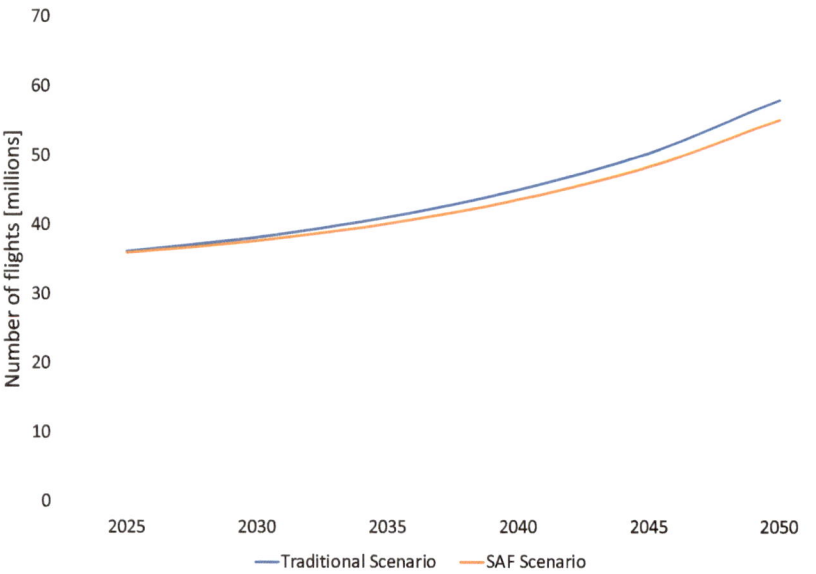

Fig. 13 Global forecast results of flight volume up to 2050 for the "Traditional Scenario" and "SAF Scenario"

accommodate passengers. As a result, the larger part of additional passenger volume up to 2050 will be handled by better employment and increased aircraft capacity instead of more flights. This is caused not only by the economic considerations of airlines, but because of the capacity constraints of major airports worldwide. This will even affect smaller airports with capacity reserves because of connecting traffic to capacity-constrained airports.

Figure 14 illustrates the results for the European flight volume development up to 2050. In 2022, flight volume was about 7.7 million, and this is expected to grow up to 12.3 million in the "Traditional Scenario" and 11.6 million in the "SAF Scenario". The corresponding CAGRs are 1.2% for the "Traditional Scenario" and 1.0% for the "SAF Scenario", and as in the global case are much lower than passenger and RPK volume growth. The rate at which the number of passengers per flight rises is approximately the difference between passenger and flight CAGR, i.e. 2.4% in the case of "Traditional Scenario" and 2.2% for the "SAF Scenario". As a result, a large share of the additional passenger volume up to 2050 will be handled by better employment and increased aircraft capacity instead of more flights. The relative difference between the flight volumes in the two scenarios is larger on a global level, because the capacity situation will be worse on the global level than on the European level in 2050. This is mainly due to the expected demand and flight volume development in Asia: there is a far stronger demand development to be expected compared to Europe and North America, which puts airport capacities under pressure. This leads to more passengers transported per flight, either by increasing load factors where possible, or increasing aircraft size. As a result, the number of passengers increases more on the global than

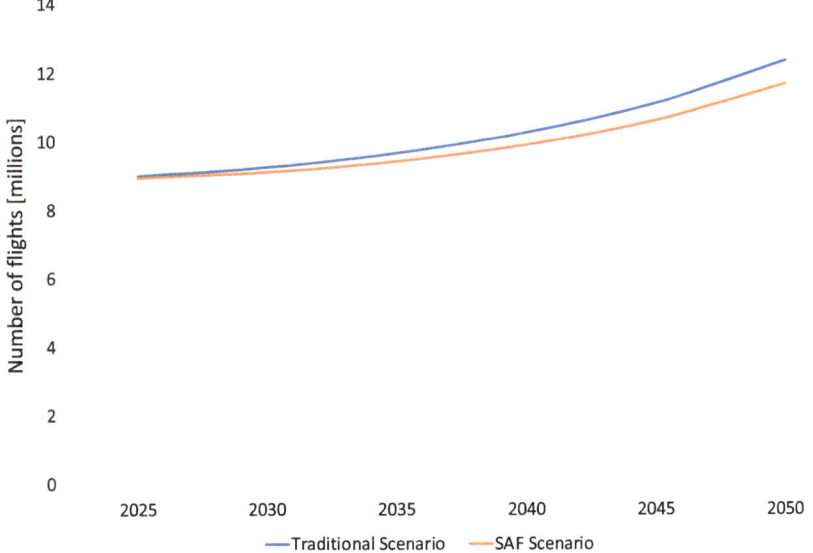

Fig. 14 European forecast results of flight volume up to 2050 for the "Traditional Scenario" and "SAF Scenario"

on the European level, while flight volume growth rates are not that much different. It is increasingly difficult to enlarge airport capacity substantially the larger an airport is, even in Asia. Furthermore, adding runways typically becomes increasingly inefficient, i.e. with each runway, less and less absolute capacity is added, because of interferences between (dependent) runways [12]. Another problem is that air traffic is heavily concentrated on a rather small number of (major) airports, and there are currently no signs that this will change substantially in future [10]. However, in future it might be the case that more flights are operated from neighbouring airports with capacity reserves to help mitigate the capacity crunch.

Increasing SAF share in aviation will have an impact on future air transport demand and, at the same time, significantly reduce CO_2-emissions resulting from aviation activities. In this study SAF share according to the "Fit for 55" strategy of the EU is assumed to increase over the course of time. Furthermore, a constant fuel consumption per RPK is assumed. Based on these assumptions, CO_2-reductions relative to the year 2025 are calculated regarding the "Traditional Scenario" for Europe and worldwide. The aim of this approach is to identify the direct impact of SAF on air transport related overall CO_2-emissions resulting from the use of "greener" fuel and the air transport demand (and also fuel) reduction resulting from higher airfares. As already shown in this section, fleet structure is also forecast to change in future, i.e. more passengers carried per flight leading to higher load factors and larger aircraft being employed. This tends to lower the specific fuel consumption and thus specific CO_2-emissions per RPK even further. Another area in which to reduce the specific fuel burn is the introduction of more fuel-efficient aircraft over time. This happens as older aircraft are retired, or additional aircraft are needed to meet the forecast passenger demand. However, fleet-related measures to reduce fuel consumption and CO_2-emissions are independent of SAF and are therefore excluded from this analysis. This analysis only focuses on the SAF-related effects, i.e. greener fuel and the (slight) demand reduction. Yet it must be kept in mind that the results presented are just an upper bound of CO_2-emissions development because effects that lower specific fuel consumption are not included in this analysis (e.g. different fleet composition or other measures that increase fuel efficiency).

Figure 15 presents the results for both scenarios on a global level. In the "Traditional Scenario", CO_2-emissions by 2050 increase up to a factor of 3.6 of the 2025 "Traditional Scenario" level. This is driven by the large increase of RPK between 2025 and 2050. In the "SAF Scenario", CO_2-emissions by 2050 remain virtually on the level of the 2025 "SAF Scenario", which is far lower. RPK volume increases until 2050 by a factor of 3.5 in the "Traditional Scenario" and 3.3 in the "SAF Scenario" of the 2025 "Traditional Scenario" level, which is only slightly lower. The high share of SAF in 2050 is mainly responsible for this result.

Figure 16 shows the results of the analysis on a European level. In the "Traditional Scenario", CO_2-emissions increase up to 272% by 2050 compared to the 2025 "Traditional Scenario" level which is again driven by the large increase in RPK. In the "SAF Scenario", CO_2-emissions increase by a factor of 0.74, i.e. they are reduced by 26%, by 2050 compared to the 2025 "Traditional Scenario" level, and thus decline slightly in the long term. In this case, the RPK volume is 30% percentage points

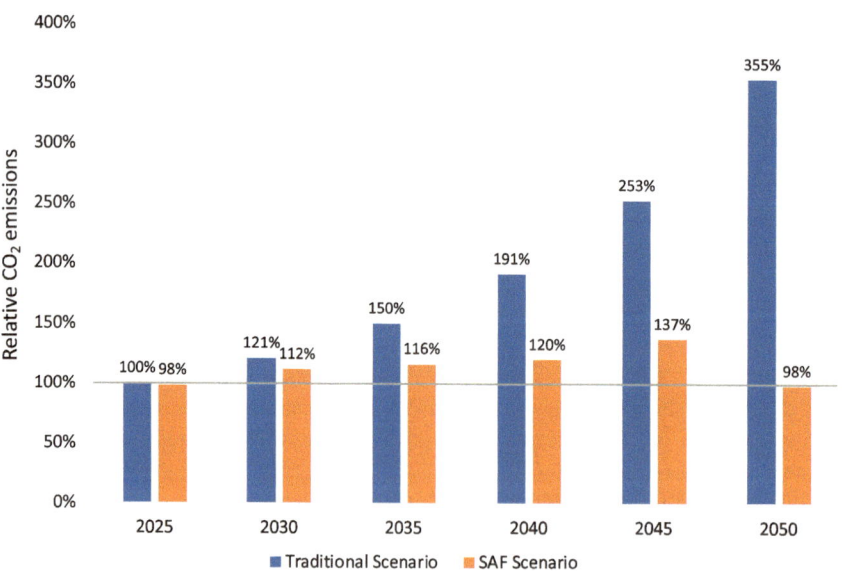

Fig. 15 Development of relative CO_2-emissions on a global level compared to the 2025 level of the "Traditional Scenario"

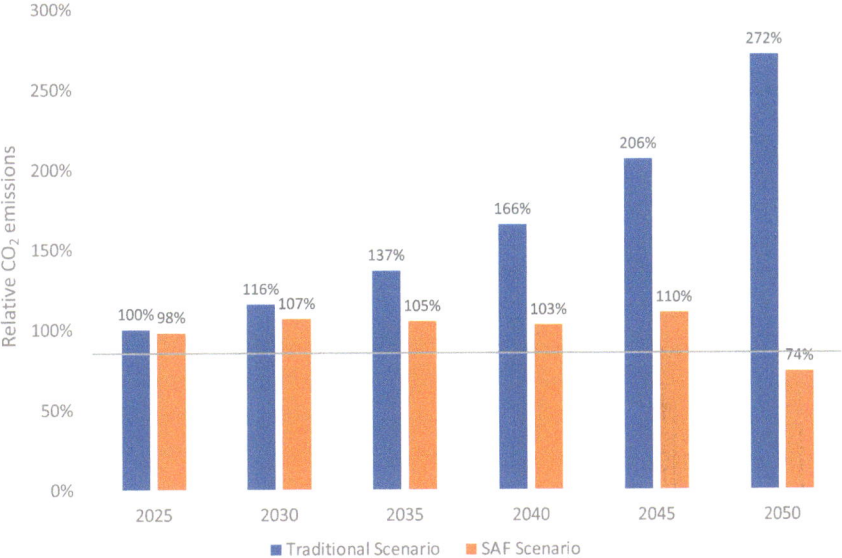

Fig. 16 Development of relative CO_2 emissions reduction potential by SAF on a European level

lower in 2050 (321% vs. 291% in 2050 compared to the 2025 level of the "Traditional Scenario"). As far as global results are concerned, it is the high share of SAF in 2050 which is responsible for this outcome.

As the analysis shows, the use of SAF has a big potential to decrease aviation related CO_2-emissions on a global and European level even though overall air traffic will most likely increase. There may be an increase in airfares due to SAF, which will reduce demand and thus CO_2-emissions additionally, but the bulk of CO_2-reduction comes from the actual employment of SAF, as they allow for reduced life cycle CO_2-emissions compared to conventional kerosene.

7 Summary and Conclusions

To mitigate the climate impact of the aviation sector, renewable aviation fuels, mostly referred to as "sustainable aviation fuels" (SAF), are pivotal. In addition to biomass-based SAF options, electricity-based or Power-to-Liquid (PtL) SAF options ("powerfuels") are also expected to play a central role in this. In this context, this study examines the effects of a continuous global SAF ramp-up on air travel demand and the resulting CO_2-emissions of the aviation sector. Two scenarios of traffic and passenger volume development between 2025 and 2050 are analysed. In the first scenario ("Traditional Scenario"), it is assumed that no SAF will be employed until 2050 (baseline), while in the second scenario ("SAF Scenario") a SAF share according to the "Fit for

55" strategy of the EU is assumed over the course of time. The "Fit for 55" strategy sets the target of SAF shares as follows: 6% in 2030, 34% in 2040 and 70% in 2050.

A substantial growth in air traffic and passenger volume is forecasted up to 2050: compared to 2022, the global RPK volume in aviation increases a factor of 3.9 in the "Traditional Scenario" and by 3.6 in the "SAF Scenario". Global flight volume rises by a factor of 1.64 in the first and 1.56 in the second scenario. For Europe, these values are slightly lower: the RPK volume is forecasted to increase by a factor of 2.9 in the "Traditional Scenario" and 2.6 in the "SAF Scenario". The corresponding flight volume growth is 73% and 63%, respectively.

The enormous difference between the RPK and flight volume growth results in large part from increasing capacity constraints at major airports worldwide. This will lead to a substantial increase in the number of passengers transported per flight. Nevertheless, even the ambitious goals of the "Fit for 55" strategy will have only a modest impact on forecasted air traffic and passenger volume growth. SAF share targets are low for the next few years and rise progressively until 2050. Such a strategy, i.e. a progressive ramp-up of SAF, will help to mitigate or "soften" the negative effects on passenger demand volume in the course of time.

Progressively increasing the SAF share used in commercial passenger aviation will influence CO_2-reductions. Here, the "Traditional Scenario", which involves only conventional jet fuel (i.e. kerosene of fossil origin) as aviation fuel, serves as a baseline (especially the values for the year 2025) and thus for a comparison to evaluate the effects of SAF in the second scenario. CO_2-emissions still increase until 2050 in the "SAF Scenario", but at a much slower rate than RPK. In 2050, global CO_2-emissions from aviation resulting from the "SAF Scenario" are 98% of the "Traditional Scenario" in 2025, i.e. remain almost constant in the long term. For Europe, this value is just 74%, so that CO_2-emissions are lower than the 2025 baseline levels. Consequently, while passenger volume more than triples on the global level, and more than doubles on the European level by 2050, CO_2-emissions remain relatively low if SAF is employed according to the "Fit for 55" strategy of the EU. The main reasons for this outcome are the high share of SAF in 2050. Demand reduction due to higher airfares due to the use of SAF plays only a small role. Therefore, SAF can be considered as a powerful measure to a greener aviation with relatively small side effects in terms of passenger demand. Overall, the model results indicate that even if the ambitious SAF ramp-up of ReFuelEU were realised not just in the EU but globally, it alone would not ensure the achievement of net zero CO_2-emissions by 2050 for global aviation. While SAF is a crucial component for achieving net zero CO_2-emissions in the aviation sector by 2050, the use of SAF alone is most likely not feasible and requires additional measures (e.g. fuel efficiency improvements by technological and operational enhancements).

The aim of this study is to identify exclusively the impact of SAF on CO_2-emissions and whether it is an efficient measure to achieve a more sustainable aviation. Thus, this study focuses just on SAF-related effects on CO_2-emissions. There are also fleet-related effects with potentially with have large effects on CO_2-emissions, in particular more passengers transported per flight by higher load factors and / or employing larger aircraft and introducing more fuel-efficient aircraft by technological improvements, fleet renewal and enlargement to meet this increased demand. As a result, the CO_2-emissions shown in this study can represent an upper boundary only. Future studies should address these fleet-related effects to address the total effect on CO_2-emissions. On the other hand, there are non-CO_2 effects of fuel burn, which are not considered in this study, and they have a potentially large impact on global warming.

References

1. Airbus (2022) Global market forecast 2022–2041. Airbus, Blagnac, France
2. Air Transport Action Group (ATAG) (2020) Waypoint 2050—balancing growth in connectivity with a comprehensive global air transport response to the climate emergency. Air Transport Action Group, Geneva, Switzerland
3. Baier F, Berster P, Gelhausen MC (2022) Global cargo gravitation model: airports matter for forecasts. Int Econ Econ Policy 19(1):219–238
4. Berster P, Gelhausen MC, Wilken D (2011) Business aviation in Germany: an empirical and model-based analysis. J Air Transp Manag 17(6):354–359
5. Berster P, Gelhausen MC, Wilken D (2015) Is increasing aircraft size common practice of airlines at congested airports? J Air Transp Manag 46:40–48
6. Boeing (2022) Current market outlook 2022–2041. Boeing, Seattle
7. European Union Parliament (2023) ReFuelEU aviation initiative—sustainable aviation fuels and the fit for 55 package. https://www.europarl.europa.eu/RegData/etudes/BRIE/2022/698900/EPRS_BRI(2022)698900_EN.pdf. Accessed 29 Aug 2023
8. European Union (2018) Directive 2018/2001 of the European Parliament and of the Council. https://eur-lex.europa.eu/legal-content/EN/TXT/PDF/?uri=CELEX:32018L2001&from=EN#page=29. Accessed 29 Aug 2023
9. Eurocontrol (2022) Aviation outlook 2050. Eurocontrol, Brussels, Belgium
10. Gelhausen MC, Berster P (2017) Domination of hub-and-spoke systems. In: Finger M, Button K (eds) Air transport liberalization: a critical assessment. Edward Elgar Publishing, Cheltenham, UK, pp 266–283
11. Gelhausen MC, Berster P, Wilken D (2013) Do airport capacity constraints have a serious impact on the future development of air traffic? J Air Transp Manag 28:3–13
12. Gelhausen MC, Berster P, Wilken D (2019) Airport capacity constraints and strategies for mitigation: a global perspective. Elsevier, New York, USA
13. Gelhausen MC, Berster P, Wilken D (2021) Post-COVID-19 scenarios of global airline traffic until 2040. Aerospace 8(19):300
14. Gelhausen MC, Grimme W, Junior A, Lois C, Berster P (2022) Clean Sky 2 technology evaluator—results of the first air transport system level assessments. Aerospace 9(4):204
15. Grimme W, Maertens S, Bingemer S, Gelhausen MC (2021) Estimating the market potential for long-haul narrowbody aircraft using origin-destination demand and flight schedules data. Transp Res Procedia 52:412–419
16. International Air Transport Association (IATA) (2021a) Net zero carbon 2050 resolution—fact sheet. International Air Transport Association, Montreal, Canada
17. International Air Transport Association (IATA) (2021b) IATA economics' chart of the week—shares of key cost items changed during the crisis. https://www.iata.org/en/iata-repository/publications/economic-reports/shares-of-key-cost-items-changed-during-the-crisis_/. Accessed 21 Nov 2022
18. International Air Transport Association (IATA) (2022) What is SAF? https://www.iata.org/contentassets/d13875e9ed784f75bac90f000760e998/saf-what-is-saf.pdf. Accessed 24 November 2022

19. International Civil Aviation Organization (ICAO) (2021) ICAO traffic statistics. ICAO, Montreal, Canada
20. International Civil Aviation Organization (ICAO) (2022) Report on the feasibility of a long term aspirational goal for international civil aviation CO_2 emission reductions—fuels sub group report. International Civil Aviation Organization, Montreal, Canada
21. International Energy Agency (IEA) (2021) Net zero by 2050: a roadmap for the global energy sector. International Energy Agency, Paris, France
22. Knabe F, Dreyzehner T, Korn B (2022) Super close runway operations (SupeRO): a new concept to increase runway capacity. In: Integrated communications, navigation and surveillance (ICNS). Conference—proceedings. Volume April 2022. Integrated communications, navigation and surveillance conference, ICNS 2022, Dulles, 5th April–7th April 2022
23. Kuenz A, Gelhausen MC, Grimme W, Knabe F, Mollowitz V, Störmer M (2022) Filling enroute airspace—where are the limits? In: AIAA/IEEE digital avionics systems conference—proceedings, Volume September 2022. 41st IEEE/AIAA digital avionics systems conference, DASC 2022, Portsmouth 18th September–22nd September 2022
24. McKinsey & World Economic Forum (WEF) (2020) Clean skies for tomorrow: sustainable aviation fuels as a pathway to net-zero aviation. Switzerland, Geneva
25. Netherlands Aerospace Center (NLR) (2021) Destination 2050: a route to net Zero European aviation. Netherlands Aerospace Center, Amsterdam, The Netherlands
26. Official Airline Guide (OAG) (2021) Market analysis. Reed Travel Group, Dunstable, UK
27. Sabre AirVision Market Intelligence (MI) (2021) Data based on market information data tapes (MIDT). Sabre, Southlake, UK
28. Scheelhaase J, Maertens S, Grimme W (2019) Synthetic fuels in aviation—current barriers and potential political measures. Transp Res Procedia 43:21–30
29. Shell (2021) Decarbonising aviation: cleared for take-off. Shell, London, UK
30. Traveller (2018) To what extent do fuel costs affect plane ticket prices. https://www.traveller.com.au/to-what-extent-do-fuel-costs-affect-plane-ticket-prices-h12by. Accessed 21 Nov 2022
31. US Department of Transportation (DOT) (2019) What the cost of airline fuel means to you. https://www.transportation.gov/administrations/assistant-secretary-research-and-technology/what-cost-airline-fuel-means-you. Accessed 21 Nov 2022
32. Wilken W, Berster P, Gelhausen MC (2011) New empirical evidence on airport capacity utilisation: relationships between hourly and annual air traffic volumes. Res Transp Bus Manag 1:118–127
33. Wilken W, Berster P, Gelhausen MC (2016) Analysis of demand structures on intercontinental routes to and from Europe with a view to identifying potential for new low-cost services. J Air Transport Manag 56(Part B):79–90

Development of Global Seabound Mobility

Stephan Krüger, Ann-Kathrin Lange, and Carlos Jahn

Abstract The importance of maritime shipping for world trade is paramount. The globalization of supply chains is only made possible to its current extent by shipping and can therefore be considered one of the most important transport systems in the world. In the future, the transport of goods is forecast to triple by 2050 compared to 2015. At the same time, due to its size, global maritime shipping today accounts for approx. 3% of global greenhouse gas (GHG) emissions. With the adoption of the "Strategy to Reduce Greenhouse Gas Emissions from Ships," the International Maritime Organization (IMO) has set a milestone in that its stated goal is to reduce GHG emissions from shipping to net zero emissions from 2008 levels by 2050. In addition to the efforts of the IMO, other political entities such as the European Union (EU) and interest groups from the shipping industry have taken measures that significantly underpin the declared goals. To achieve this, the transition of using conventional fuels consisting of heavy fuel oils (HFO) to climate-neutral and sustainable fuels must be undertaken. In this article, alternative fuels are considered and assessed in terms of their potential and availability to contribute to GHG reduction in the maritime sector. The article concludes with an analysis of the current state of renewable and low carbon fossil fuels in shipping and gives a forecast of how their distribution and development may look in the future.

Keywords Maritime freight transportation · Climate neutrality · Renewable fuels · Decarbonization of shipping

1 Introduction

Maritime shipping represents one of the most important elements of the global transportation system and thus forms the prerequisite for prosperity in many parts of the world. With the ongoing globalization of markets, the transport of goods by sea has increased enormously in recent decades. From the 1950s until the outbreak of the

S. Krüger · A.-K. Lange · C. Jahn (✉)
Hamburg University of Technology (TUHH), Institute of Maritime Logistics, Hamburg, Germany
e-mail: carlos.jahn@tuhh.de

© The Author(s), under exclusive license to Springer Nature Switzerland AG 2025
N. Bullerdiek et al. (eds.), *Powerfuels*, Green Energy and Technology,
https://doi.org/10.1007/978-3-031-62411-7_5

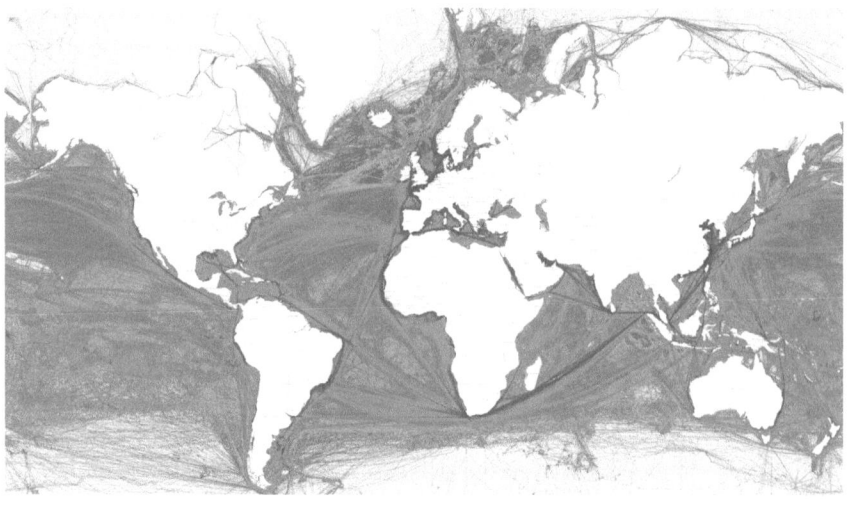

Fig. 1 Density map of global shipping 2020 (map taken from [40])

world financial crisis in 2008/2009, international trade grew at almost twice the rate of global economic activity as a whole. Today, based on weight, about 90% of cross-border trade in goods is transported by sea. More than 130 countries are involved in the seaborne transport of goods. The majority of transcontinental freight traffic takes place on main shipping routes (Fig. 1). The busiest routes worldwide are the entrances to the ports of Europe, East Asia (especially Shanghai, Singapore, and Hong Kong), and the USA. Straits lead to an additional concentration of shipping traffic. Neuralgic points are the Straits of Dover, Gibraltar, Malacca, Lombok, and Hormuz, but also the Cape of Good Hope on the southern coast of Africa. Traffic clusters in these straits make it vulnerable to disruption.

The routes shown in Fig. 1 were obtained by analyzing data from the Automatic Identification System (AIS)[1] of vessels in 2020. Maritime shipping can be roughly divided into three subsectors: (1) tanker shipping with goods such as oil, oil products, gas, or chemicals, (2) the transport of dry bulk and bulk general cargo, of which the five most important are iron ore, coal, grain, phosphate, and bauxite, (3) and other dry cargo, consisting of other bulk goods such as non-ferrous metal ores, feedstuffs, and fertilizers and, above all, goods of all kinds bundled into smaller transport units (Fig. 2). The latter are referred to as general cargo and are today mostly transported in containers and regularly by liner shipping, which operates fixed routes according to announced schedules.

---

[1] This is an automatic tracking system that improves the safety and guidance of vessel traffic by exchanging navigational and other vessel data. It was adopted as a mandatory standard by the IMO in 2000. The processing and use of data transmitted by AIS can significantly improve, among other things, collision prevention between vessels, monitoring of illegal fishing, and shore-based monitoring and guidance of vessel traffic. Furthermore, the transmitted data allow the comprehensive analysis of vessel traffic over a longer period.

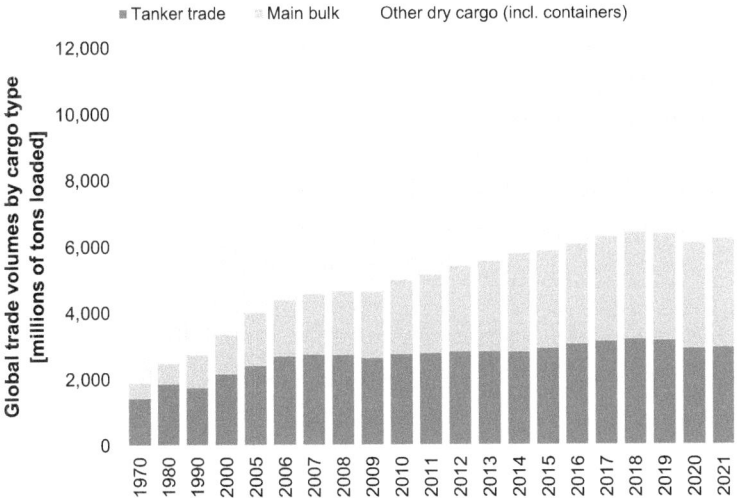

Fig. 2 International maritime trade by cargo type, selected years; tanker trade includes crude oil, refined petroleum products, gas, and chemicals [64]

Since 1980, the volumes of main bulk transport and container transport, in particular, have increased significantly. The latter is partly due to the now very high degree of containerization of general cargo. However, volumes for the other types of goods have also increased significantly since 1980. This has been made possible by factors such as the increase in world trade, the liberalization successes of the World Trade Organization (WTO), and China's accession to the WTO in 2001. More recently, only 2009 and 2020 show a decline in growth. In 2009, this was caused by the consequences of the global economic and financial crisis. The after-effects of the crisis in 2008 extended into 2010. In 2020, international maritime trade and global supply chains were affected by the impact of the COVID-19 pandemic. Global economic output fell by 3.5% and merchandise trade by 5.4%, while international maritime shipments declined by 3.8% to 10.65 billion metric tons. Overall, however, maritime transport managed to overcome the crisis. The short-term outlook is positive and the medium and longer-term prospects also offer the potential for further growth in principle. But despite this, the recovery will be determined by the further development of the pandemic and the associated closures and restrictions. A sustained recovery also depends on creating supportive macroeconomic and fiscal conditions while minimizing trade protectionism [63].

In the context of climate protection, which will continue to gain importance in the coming years, maritime freight transport offers many potentials. Compared to other modes of freight transport, it is characterized above all by its high mass efficiency and the associated relatively low emissions per ton kilometer transported. Figure 3 compares the four modes of freight transport by air, by road, by rail, and by water in terms of carbon dioxide (CO_2) emissions in grams per ton kilometer (tkm).

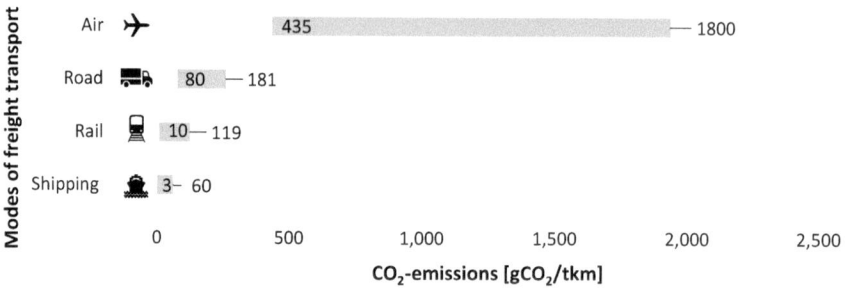

Fig. 3 Comparison of transport modes [27]

Figure 3 illustrates that CO_2-efficiency can be increased through increased multi-modal transport. Based on the specific circumstances, for example, the type of goods and overall available transport options, ocean shipping may present certain benefits over other transport methods in terms of GHG-emission reduction, as fewer CO_2-emissions are usually emitted per ton of freight transported.

Shipping is currently responsible for about 3% of global GHG emissions. As international efforts to combat climate change gain momentum, the shipping industry is coming under increasing pressure to become more climate-friendly. GHG emissions increased by about 10% between 2012 and 2018. Defossilizing the industry will require major investments in green technologies and alternative fuels. A growing number of ships are already converting to liquefied natural gas (LNG), while other alternative fuels are under development, including ammonia, hydrogen, and methanol, as well as completely electrically powered ships. However, some of the alternative fuels still need to be counted as fossil fuels. The use of alternative fuels could be associated with an increased risk of engine failures during the transition period [1].

Different organizations and alliances around the world have committed to climate neutrality in recent years—albeit with different time horizons. In 2018, the IMO passed a resolution agreeing for the first time on a sector target for mitigating GHG emissions. In this strategy, the international community agreed to reduce GHG emissions from shipping to net zero emissions from 2008 levels by 2050 [32]. This is supported and often exceeded by many industry players. For example, at the COP26 UN Climate Change Conference in Glasgow, more than 20 countries signed up for so-called Clydebank Declaration, which aims to drive forward the zero-emission target for shipping in the coming years. The declaration calls for the establishment of green shipping corridors with zero-emission shipping routes between two (or more) ports. At least six of these zero-emission corridors are to be established by the middle of the decade, with more to follow by 2030. In preparation for COP26, the "Call to Action for Shipping Decarbonization" was issued by representatives of the entire maritime ecosystem, including shipping, chartering, finance, ports, and fuel production. These stakeholders have recognized that the ambitious goal of climate neutrality can only be achieved through the intensive cooperation of all stakeholders, right down to shipbuilders and component manufacturers. In the meantime, more than 230 representatives of different industries are among the signatories [24].

2   Progressive Growth in Maritime Freight Traffic

In recent decades, the international division of labor and specialization has led to a global interdependence of supply chains. Politicians have provided the appropriate framework for this with the multilateral, rules-based trading system of the WTO. This has made it possible to open up markets worldwide, from which both the industrialized countries and the developing and emerging countries are still benefiting today. However, these positive developments were threatened by the COVID-19 pandemic and the trade disruptions it has caused. The pandemic has revealed vulnerabilities of supply chains in crises as well as risks of one-sided dependencies in supply relationships. Companies have already started to revise and diversify their supply chains based on the lessons learned from the crisis [48].

The development of maritime freight transport can be evaluated using various key performance indicators. In addition to the development of transport volumes and capacities of ships and ports, these also include the efficiency improvements achieved and their influence on the safety of ships and processes.

2.1   Development of GDP, World Merchandise Trade, and World Maritime Trade

There is a strong correlation between population and gross domestic product (GDP) growth on the one hand and maritime trade growth on the other. The population is projected to continue to grow considerably, but after 2050, population growth is expected to slow down. This is due in part to the composition of the population: many industrialized countries, but also economies such as China, will face an aging population, which could lead to a shift in consumption patterns—for example, from goods to services. Overall GDP is predicted to increase, but at a lower level than before. In the long term, there is a convergence of growth rates between industrialized and developing countries. The globalization process, based on labor cost differentials and massive outsourcing of production, which has driven maritime trade, has probably reached its limits [33]. World maritime trade remains highly dependent on the development of the world economy (Fig. 4).

Due to slower growth in the global economy and world trade, growth in international maritime trade stagnated in 2019, reaching its lowest level since the 2008/2009 financial crisis. After a modest 2.8% increase in 2018, volumes grew by only 0.5% in 2019. The subsequent COVID-19 pandemic further disrupted maritime traffic. However, the consequences turned out to be less severe than originally feared.

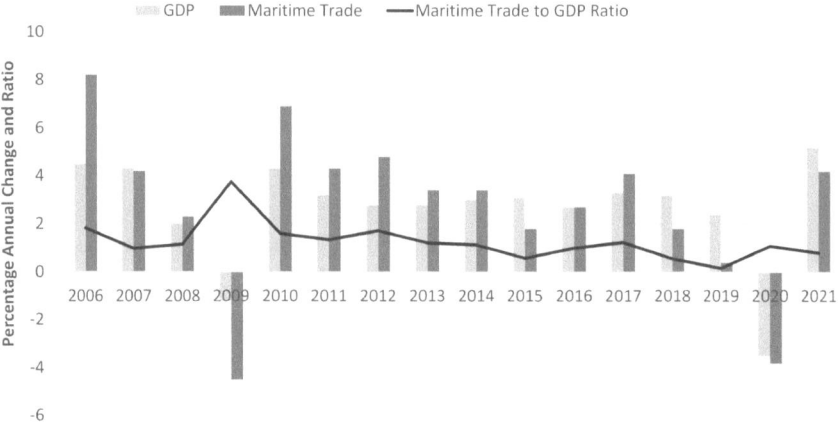

Fig. 4 International maritime trade, world gross domestic product (GDP), and maritime trade-to-GDP ratio from 2006 to 2021 [63]

Already in the second half of the year, volumes recovered in both container and dry bulk traffic. In tanker shipping, however, there has not yet been a full recovery. Two shipping segments that had previously enjoyed success were hit particularly hard by the pandemic in 2020: the cruise ship and car carrier markets. As a result, the cruise sector came to an almost complete standstill in 2020. Going forward, all shipping segments are expected to recover slowly [62, 63].

2.2 Growing Fleet

At the beginning of 2021, the world merchant fleet (ships of 300 gross tonnage (GT) and above) consisted of 58,000 vessels with a total capacity of 2.03 billion deadweight tons (dwt), an increase of 3.2% compared to the previous year (Fig. 5). Although this figure represents a smaller increase than in previous years, it continues the trend of rising transport capacity. The situation is different for the number of ships. This value fell for the first time after a long upward trend as of January 2021 [48].

In 2020, only 1280 merchant ships with a total capacity of 88 million dwt were delivered. In terms of transport capacity, this represents a year-on-year decline of 11.3%. During the same period, 492 merchant vessels with a total capacity of 22.4 million dwt were sold to shipbreaking yards, representing an increase of 42% year-on-year in terms of transport capacity. Against the backdrop of the pandemic, technical concerns, and the issue of which fuel to choose, there was a 27% year-on-year decrease in ordered tonnage based on 2020 as a whole [48].

In 2021, global orders picked up again. New orders recorded a 138% increase over 2020 levels in terms of compensated gross tons (CGT). In the case of container

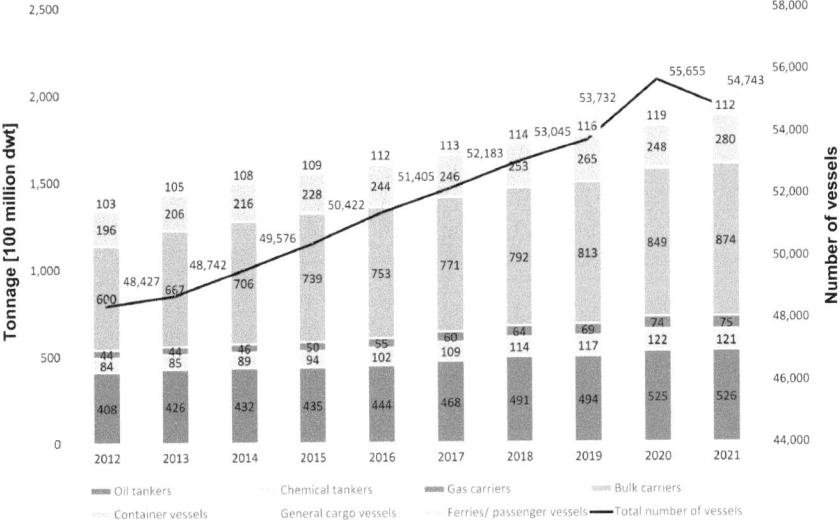

Fig. 5 Development of the world merchant fleet in numbers and tonnage from 2012 to 2021 [41–48]

vessels, this was the most stable order level since 2014. Order volumes also increased in the bulk carrier, LNG and LPG, and pure car carrier segments. However, measured by the size of the fleet, the level of orders placed is not yet at a structural peak. Despite the increase in order volume, orders continued to be concentrated in the major (Asian) shipyards. In this context, Chinese and Korean shipyards had a dominant share of orders (46% and 40% in CGT, respectively), while reported order volumes were lower for Japanese (8.5% market share) and even more so for European shipyards [56].

In addition to the development of the ship fleet in terms of volume, the price level at which new ships are purchased is also a useful indicator of the health of the shipbuilding industry. Clarkson's newbuilding price index can be used for this purpose. This is calculated by averaging the US$ per dwt of different ship types. While it should be noted that newbuilding prices vary by classification, the Clarkson Newbuilding Price Index indicates the general price trend in the industry (Fig. 6).

In 2022, newbuilding prices show the strongest increase in almost two decades, mainly due to the extremely high number of container vessels and LNG carriers ordered in the last 15 months. Nevertheless, prices did not exceed mid-2008 levels, despite 12 years of steady inflation increases. In addition, the waiting time for shipyards to order ships has increased significantly. It stood at 2.9 years at the beginning of 2022 [56].

Fig. 6 Clarkson newbuilding price index [9]

2.3 Vessel Size Growth

One of the most striking developments is the average increase in ship size. Larger ships reduce the costs related to cargo units for the crew, fuel, demurrage, insurance, maintenance, and upkeep of the ships by exploiting economies of scale. This has been particularly evident in container vessels over the past 20 years. For example, the largest container vessel in 2002 could only hold 8000 TEU (Twenty-foot-Equivalent Units), while in 2022 the largest vessel can hold almost 24,000 TEU (Fig. 7) [52].

The development of ship size is a stepwise process in this respect. The changes correspond to the introduction of a new container ship class by a shipping company (Maersk Line was until 2014 often the pioneer), quickly followed by others. Since the early 2000s, a larger share of the world's cargo volume has been carried by large container ships. These ships have a container capacity of more than 10,000 TEU.

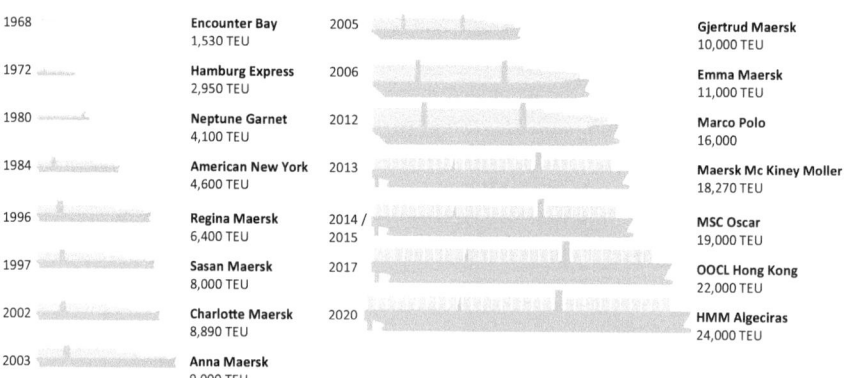

Fig. 7 Development of vessel sizes in container shipping (TEU: Twenty-foot-Equivalent Units [1])

Between 2011 and 2021, their share of transport capacity rose from 6 to almost 40% [1].

These larger ships have been made possible by technological advances and are part of broader business strategies to exploit economies of scale. However, this has led to oversupply on the major liner shipping routes, increasing pressure on port infrastructure and logistics. Furthermore, in a holistic view, it must be considered that larger container vessels also require more investment in port facilities including port access roads. The costs in the ports are therefore also based on the size of the ship. Furthermore, it should be borne in mind that very large ships entail some special risks. Delays have a significant impact on supply chains and their size means that they represent a massive concentration of risk. This results in disproportionately high costs due to the more complex response to accidents when a risk occurs. Port facilities and salvage equipment for large vessels are specialized and limited, while salvage and wreck removal are more expensive and often still uncharted territory [1].

The trend of ship size growth is expected to continue in the face of climate change and the introduction of GHG emission reduction targets in the industry. For example, despite the COVID-19 pandemic, orders for larger and larger ships have been increasing in recent years. These include, in particular, ships that run on alternative fuels [1]. Their characteristics and developments are described in more detail in Chap. 4.

2.4 Increase in Requirements for Capacities in Ports

The importance of well-functioning seaports for industrial activity, trade in goods, globalized production processes and economic growth has become increasingly clear in recent years. The world's ports handle more than 90% of the volume and more than two-thirds of the value of the global merchandise trade. As key nodes in global transportation chains, ports connect consumers and producers, provide access to markets, and support supply chains. Both port competition and the requirements to adapt to changes in the economic, institutional, legal, and operational landscape are creating intense pressure for ports. This is not limited to striving for higher levels of performance in terms of criteria such as operational optimization, cost reduction, time efficiency, and trade facilitation, but also specifically targets factors such as safety, resource conservation, environmental protection, and social inclusion. At the same time, several megatrends are impacting the port industry and the container port segment in particular. These trends include increasing concentration and consolidation in the liner shipping market, ship size growth, and the emergence of mega alliances. Therefore, achieving higher levels of port performance and private sector participation in container terminal operations has become key issues [57, 63].

Increasing sea transport volumes, combined with global economic developments over the last 20 years, has led to Asian ports making great leaps in throughput figures as a result of extensive investments and rebuilding, and new planning (Fig. 8). Only

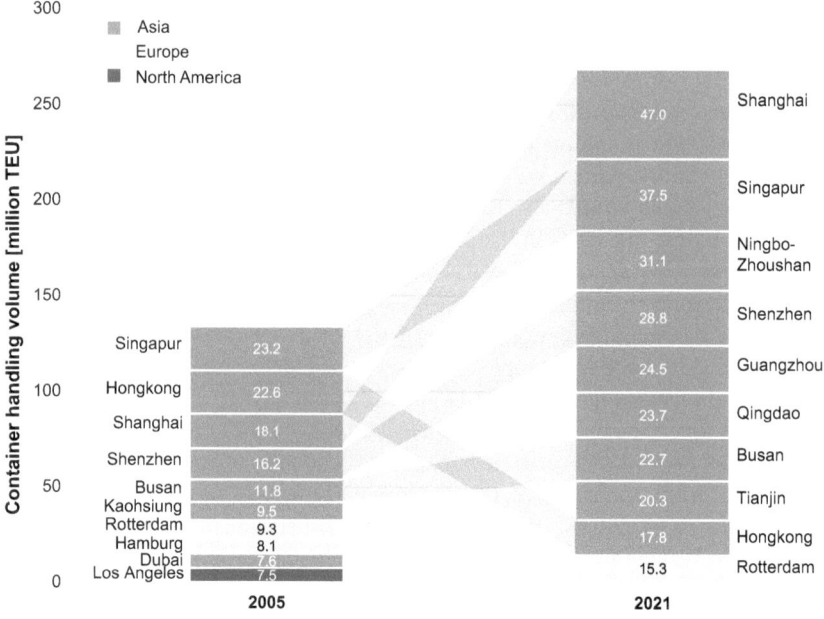

Fig. 8 Handling volume of the world's 10 largest container ports (in million TEU) [34]

Rotterdam, as a non-Asian port, is still represented in the top ten of the largest ports (according to their handling volume in TEU).

The goals set for climate neutrality in shipping will pose further challenges for ports. For example, they are responsible for enabling bunkering of alternative fuels. This includes creating technical solutions and corresponding capacities to receive, store, and deliver these fuels to ships as needed. Different characteristics of the various fuel options will require different technical solutions. In addition, since conventional fossil fuels will also have to be provided during the decades-long transition period until defossilization, storage and bunkering facilities for the diverse fuel alternatives will have to be kept ready in parallel. In addition to the technical questions, there are also questions about the availability of suitable areas in the ports and legal and safety requirements (lack of international standardization) that must be met.

2.5   Increase in Canal Passages

Maritime shipping canals, waterways artificially created and developed for economic reasons, are elementary routes of many shipping lines. They often shorten the transport distance considerably and thus reduce the transport time and, depending on the charges, also the transport costs. The most important shipping channels are the Suez Canal and the Panama Canal. The Suez Canal is located in Egypt between the port

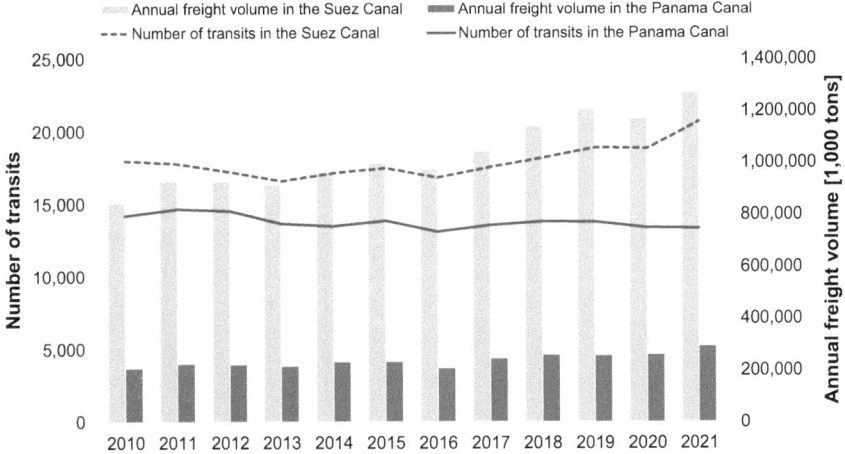

Fig. 9 Tonnage and number of transits in the Suez Canal and Panama Canal from 2010 to 2021 [23, 60]

cities of Port Said and Port Taufiq near Suez and connects the Mediterranean Sea with the Red Sea via the Isthmus of Suez. The Suez Canal has a length of 163 km and saves ocean shipping between the North Atlantic and the Indian Ocean the detour around Africa. The Panama Canal is an approximately 82 km long, artificial waterway that cuts through the Isthmus of Panama in Central America, thus connecting the Atlantic and Pacific Oceans for shipping. This eliminates the need to sail around Cape Horn or through the Strait of Magellan at the southern tip of South America. Approximately 15% of the world's traded goods pass through the Suez Canal and about 5% through the Panama Canal (Fig. 9). These important nodes of the global maritime transport network are very sensitive to any disruption, such as the accident of the 20,150 TEU container vessel Ever Given in the Suez Canal in 2021, and can cause shock waves in global supply chains when they occur [48].

Through ongoing deepening and widening projects, the Suez Canal reached a depth of 22.5 m in 2001 and 23.5 m in 2008. In 2014, a new widening project was carried out, increasing capacity to about 100 passages per day. The new Suez Canal was inaugurated in August 2015 at a cost of over $8 billion, and the depth was increased to 24 m. This expansion included a new 35 km section that allows simultaneous passage of ships in both directions. The canal has a handling capacity of up to 25,000 vessels per year (about 78 per day). In 2019, about 18,800 ships were handled, resulting in an average daily handling of about 55 ships [52].

Traffic volumes in the Panama Canal increased particularly sharply in the 2000s, prompting the decision to expand the canal with a new set of locks to accommodate significantly larger vessels. After the expansion was completed in 2016, tonnage surged while the number of transits did not change significantly [52]. The COVID-19 pandemic also impacted the Panama Canal with a short-term decline in passage numbers due to the slowdown in the global economy. Subsequently, the Canal, as well as all ports in the Americas, Asia, and Europe, experienced an increase in traffic that coincided with the peak season for container and LNG segments. To cope with the temporary increase in traffic, the canal made operational adjustments, such as increasing staffing levels and providing information on expected waiting times. Furthermore, a draft of approximately 15.2 m became possible at the beginning of the year 2020 [48].

Due to the expected developments in maritime shipping, such as the growth in ship size and the further increase in fleet size, it is to be expected that the utilization of the canals will continue to increase and that further adjustments may be necessary. However, this is also dependent on developments such as the increase in charges for canal transits or adjustments regarding climate protection.

2.6 Increase Safety on Vessels and Reduce Total Losses

Due to the high importance of the international shipping industry for world trade, the safety of ships is of crucial importance. The number of total losses of larger ships has generally decreased until 2020 (Fig. 10). However, in addition to the shortage of skilled labor, congested ports, old fleets, and extreme weather conditions are causing concern for the insurance and shipping industries. Furthermore, the switch to alternative fuels is creating new challenges.

Over the past four years, 38–72 major ships have been lost worldwide each year. While there were 38 total losses in 2022, 109 total losses were reported in 2013. At the same time, the number of ships sailing the world's oceans has risen continuously. However, the number of reported accidents alone rose by 16% to 3032 with the channel region seeing the highest number at 679. Engine damage and failure were the most common cause of shipping accidents worldwide. More than one in three incidents worldwide (1478) was due to machinery damage, followed by collisions (280) and fires (209). The frequency of fires increased to the highest total for a decade.

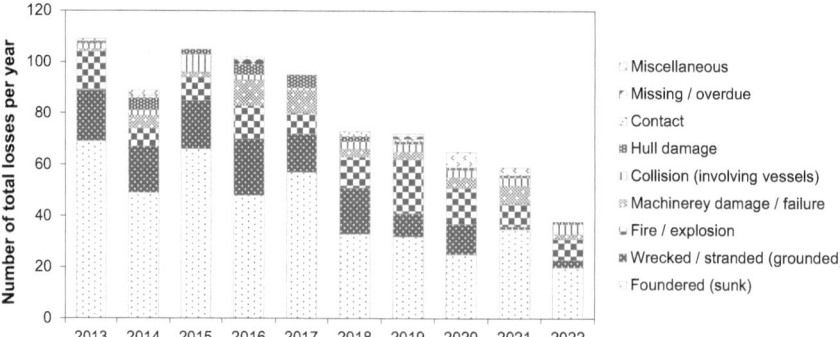

Fig. 10 Total losses by causes 2013 to 2022 [2]

Containers often burn when hazardous cargoes such as chemicals or batteries have not been declared correctly. Furthermore, the growing number of electric vehicles being transported increases the risk of fire [2].

2.7 Digitalization to Reduce GHG Emissions

Continuous technological development in recent years, especially in the field of artificial intelligence (AI), enables the use and improvement of a wide range of systems, e.g., autonomous control, route optimization through the use of AIS and weather data, and information exchange in the logistics chain, thus reducing waiting times. Volatile weather patterns due to climate change are increasingly impacting shipping. In the future, the shipping industry will need to be more proactive in responding to and mitigating the effects of extreme weather. More accurate weather forecasts and technologies will help shipping companies plan and take action to avoid losses, such as postponing sailings, seeking shelter, or calling another port [1].

In particular, the introduction of unmanned and autonomously operating cargo and passenger ships will have a significant impact on shipping and the role of nautical-technical onboard personnel. Although only a few ship automation projects have been completed to date, there is a discernible trend among shipping companies to increasingly install automated systems on ships and, soon, to also deploy unmanned ships controlled from shore stations. So far, the focus has been mainly on technological specifications and developments. However, little emphasis has been placed on the training of seafarers who will operate these future vessels. Therefore, it is necessary to first identify the new skills and competencies that the seafaring personnel of the future will need to operate highly automated ships either as crewmembers or in operations centers from shore [48].

This also applies to ports. To optimize logistics operations, port operators must be able to handle data efficiently. To this end, port logistics is increasingly relying on digitization—for exchanging information between customers, partners, suppliers, and other stakeholders, as well as for offering new services. Furthermore, AI methods and autonomous systems are also being used there at various points, e.g., in horizontal transport and the allocation of storage locations.

3 Frameworks and Initiatives Advancing Climate Neutrality

Environmental protection in maritime shipping is becoming increasingly important, as shipping has a high impact on the marine environment with its 54,753 vessels. Through maritime shipping, environmental damaging substances such as oil, chemicals, ship sewage, and ship waste find their way into the world's oceans. However, in addition to environmental and marine protection through shipping, the impact on the climate due to the use of fossil fuels in this sector is also becoming increasingly important in the context of environmental effects. Estimates assume that the seagoing ships operating worldwide consumed about 207 million tons of heavy fuel oil (HFO) in 2020 (IMO; [48]). Thus, ocean shipping was responsible for emissions of 1.05 billion tons of CO_2, which represented a share of approx. 3% of global GHG emissions in 2018 [29]. Currently, both the IMO, as the maritime umbrella organization of international shipping, and other political entities such as the European Union as well as industry stakeholders are taking further steps to achieve the goal of climate-neutral shipping close to 2050.

3.1 Policies of the IMO

The International Maritime Organization (IMO), a specialized agency of the United Nations (UN), with its 175 member states, has the task of providing maritime shipping with an international framework for regulating the specific areas of maritime safety, maritime security, and the protection of the marine environment. The rules and guidelines adopted by the IMO are approved by delegates of the member states and then come into force through their incorporation into national legislation.

As an organization, the IMO reserves voting rights for delegates representing nation states, but also allows other organizations such as classification societies, maritime industry trade associations, and non-governmental organizations to participate in its work. Even before the establishment of the IMO in 1958, international shipping was regulated in terms of safety measures as well as environmental issues. After its foundation, the IMO took responsibility for the further development of the SOLAS (Safety of Life at Sea) Convention of 1914, as well as the OILPOL (International Convention for the Prevention of Pollution from Oil) Convention of 1954. OILPOL was replaced in 1974 by MARPOL (International Convention for the Prevention of Pollution from Ships), which today constitutes the main part of the IMO environmental regulations. The discussion of environmental issues takes place in the responsible environmental committee MEPC (Marine Environmental Protection Committee) [30].

In 20 articles, the MARPOL Convention contains general obligations of the contracting states as well as procedural instructions and basic regulations [3]. Annexes I to VI regulate the practically relevant requirements for the prevention of various types of pollution arising in connection with the operation of ships and apply to 99% of the world's merchant tonnage. Since its introduction, it has contributed to a substantial reduction in pollution from international shipping. The following is an overview of Annexes I through VI:

- MARPOL Annex I—Regulations for the Prevention of Pollution by Oil.
- MARPOL Annex II—Regulations for the Prevention of Pollution by Noxious Liquid Substances.

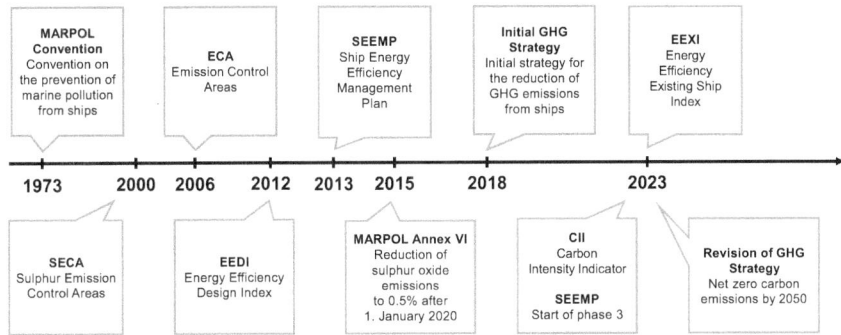

Fig. 11 Environmental regulations of the IMO International Maritime Organization [28, 29, 30, 32]

- MARPOL Annex III—Regulations for the Prevention of Pollution by Pollutants Carried by Sea in Packaged Form.
- MARPOL Annex IV—Regulation for the Prevention of Pollution by Sewage from Ships.
- MARPOL Annex V—Regulation for the Prevention of Pollution by Waste from Ships.
- MARPOL Annex VI—Convention on the Prevention of Air Pollution from Ships.

One of the IMO's challenges in addressing climate change is the ratification process that must be undertaken when making adaptations to conventions. When the IMO makes a convention or its adaptations, it does not necessarily enter into force immediately. In most cases, the convention must be ratified by a certain number of nation states, and those states together must represent a certain percentage of the world fleet by gross tonnage or some other threshold specified in the convention text. It then takes a year or significantly longer for the convention to enter into force [31]. This can take a considerable amount of time and demonstrates that even when IMO delegates can agree on the text of a convention, it does not immediately become a regulation. For example, the 2004 Ballast Water Convention was not ratified until September 2016 and entered into force in September 2017. This example shows how great the challenges are for the IMO in terms of efforts to achieve climate neutrality in shipping through legal hurdles [4].

Figure 11 provides an overview of selected measures introduced by the IMO to reduce GHG emissions from ships. The measures are contained in Annex VI of the MARPOL Convention and define how they are to be applied.

3.1.1 IMO Greenhouse Gas Strategy

Under the 1997 Kyoto Protocol to the United Nations Framework Convention on Climate Change (UNFCCC), the IMO was given responsibility for limiting GHG from international shipping that occurs outside national boundaries. In 2011, the IMO

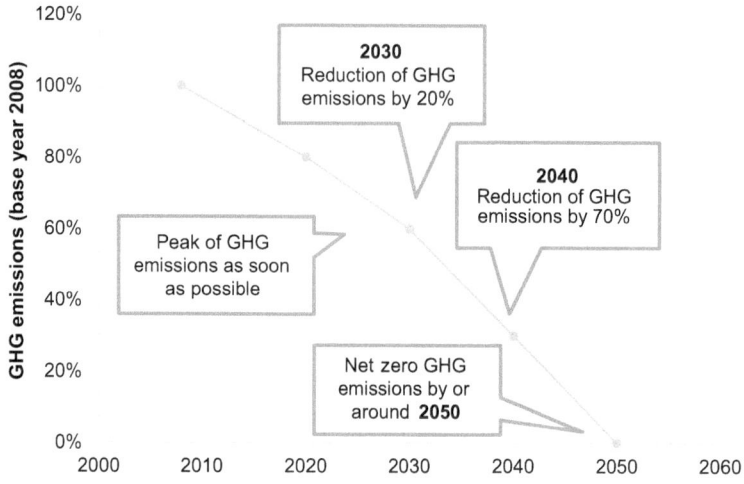

Fig. 12 IMO roadmap toward defossilization of shipping [32]

adopted its first mandatory GHG emissions requirements for ocean-going ships, the Energy Efficiency Design Index (EEDI), which will undergo its third adaptation by 2023. The EEDI is a measure used to determine the climate efficiency of new ships and is largely dependent on deadweight and ship type. Until the adoption of the fourth IMO GHG Strategy at the 72nd Marine Environment Protection Committee (MEPC) in April 2018, international shipping remained the only major sector not covered by a global GHG agreement. In its 72nd MEPC resolution, the IMO announced the agenda for decarbonization of the shipping industry, which can be seen as a historic milestone in the IMO's efforts to reduce GHG in shipping. In the fourth Greenhouse Gas Strategy, the emission reduction targets for shipping up to the year 2050 (Fig. 12) were specifically named [29] and revised in 2023 at the 80th MEPC [32].

The agenda defines the GHG emission reduction targets for international maritime transport as follows:

- The peak level of GHG emissions is to be achieved as soon as possible.
- A reduction of total annual GHG emissions from international shipping by at least 20%, striving for 30%, by 2030, compared to 2008.
- A reduction of total annual GHG emissions from international shipping by at least 70%, striving for 80%, by 2040, compared to 2008.

The IMO is aiming for net zero GHG emissions, i.e., climate neutrality, by the end of this century at the latest. However, due to as yet unforeseeable technical progress, climate neutrality may also be achieved significantly earlier. To achieve the GHG strategy targets, a subdivision is made into short-, medium- and long-term measures to reduce GHG emissions and additional measures to reduce the burden on developing countries. Table 1 provides an overview of the different measures in

Table 1 Measures of the GHG strategy [26]

Type	Year	Measure
Short term	2018–2023	New Energy Efficiency Design Index (EEDI) phases
		SEEMP phases
		Existing fleet improvement program
		Speed reduction
		Measures to address methane and volatile organic compounds (VOC) emissions
Medium term	2023–2030	Implementation of (fossil) low-carbon fuels and renewable fuels
		Further operational efficiency measures (e.g., EEXI, CII, operational efficiency standards)
		Market-based measures (e.g., integration into the EU emissions trading system)
Long term	2030+	Development and provision of renewable fuels

terms of their impact horizon. In 2023, the IMO greenhouse gas strategy was revised and is to be reviewed by 2028.

The short-term measures included in Table 1 are intended to ensure that the 2030 GHG strategy targets are met. Although these are technically challenging proposals, they are only a first step. It is expected that others will follow. Of particular importance are the medium-term efforts to promote the development of renewable fuels, as these are critical if the shipping industry is to meet its 2050 targets as well as the ultimate goal of becoming climate-neutral.

3.1.2 IMO Emission Reduction Targets

With the emission control areas, the maritime sector does not directly aim at reducing greenhouse gases. Rather, the emission types of sulfur oxides (SO_x) and nitrogen oxides (NO_x), which are very significant in the maritime sector, with their comparatively large environmental impacts, are to be limited. Therefore, this type of regulation is listed here in the context of essential IMO policy.

As can be seen in Fig. 13, a new limit value for the sulfur content of the fuel used on board was prescribed with the entry into force of the so-called IMO [28] rule and thus experienced a further lightening.

The maximum global allowable sulfur content was reduced from 4.5% over 3.5% in 2012 to 0.5% in 2020 for fuels used or carried on ships. This new limit was made mandatory by an amendment to Annex VI of the International Convention for the Prevention of Pollution from Ships (MARPOL). Because SO_x and nitrogen oxides NO_x are known to have negative impacts on human health, the environment, and the climate, consistent implementation of the 0.5% sulfur limit for all ships can lead to a variety of environmental and health improvements by reducing emissions. However,

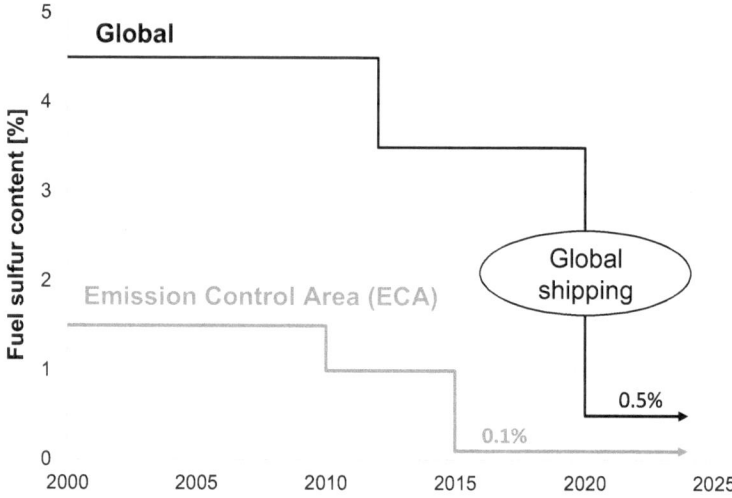

Fig. 13 Adjustment of marine fuel sulfur limits (*Source* by IMO [28])

exceptions apply to ships equipped with an approved "equivalent device" to meet the sulfur limit—such as an exhaust gas cleaning system (EGCS). These engineered devices are commonly referred to as scrubbers. They are capable of cleaning exhaust gases in such a way that even when HFO with a high sulfur content is burned, the prescribed emission limits can be met. For certain areas around the globe, known as Emission Controlled Areas (ECAs), this limit is lower. As of 2015, the mass-based limit is 0.1%. Emission Controlled Areas are located along the west and east coasts of the USA, the North Sea, and the Baltic Sea. While the ECAs around North America contain both SO_x and NO_x limits, the ECAs in Europe only limit SO_x emissions [28].

The Mediterranean Sea is expected to become an ECA in 2025, lowering the sulfur content of the fuel to a limit of 0.1%. In addition, countries such as Australia, Turkey, China, and several Central American countries have expressed their intention to work on NO_x emission regulations to designate their territorial waters as future ECAs. An overview of current and future ECAs is shown in Fig. 14 [36]. The figure does not distinguish between SECAs and ECAs.

In the Mediterranean, home to about 250 million people, emissions from ships have a significant impact on air quality, and about 70% of coastal cities in the Mediterranean are above recommended particulate matters (PM) 2.5 (particles smaller than 2.5 μm) pollution levels. Several studies, such as the 2018 IIASA, have shown that the creation of an ECA in the Mediterranean Sea would reduce sulfurous emissions by 95% and PM 2.5 emissions by 11%. If the ECA included NO_x limits, the study suggests that 10,000 premature deaths could be avoided annually by 2050 [10].

ECA's
Discussed ECA's

Fig. 14 Current and potential emission controlled areas [36]

3.2 European Union—Fit for 55 Activities

With the launch of the European Green Deal, the EU has committed itself to the goal of achieving climate neutrality by 2050, thereby committing to the Paris Climate Agreement. As a milestone, the EU has set a plan to reduce its GHG emissions by at least 55% by 2030 compared to 1990 levels. As part of the "Fit for 55" package of measures, current EU legislation is being revised and updated to bring it into line with the EU's climate targets.

The elementary components of the legal norm adjustments of the "Fit for 55" package consist of

- Ensuring a fair and socially equitable transition,
- maintain and strengthen innovation and competitiveness of the EU industry, while ensuring a level playing field vis-à-vis economic operators from third countries and
- underpin the EU's position as a leader in the global fight against climate change.

The "Fit for 55" package also explicitly addresses shipping with several measures [20]. These are presented below.

Expansion of the EU Emissions Trading Scheme. In April 2023, the European Council (EC) and the European Parliament (EP) agreed on a political provision on a reform of the EU Emissions Trading Scheme (EU ETS). Industries under the EU ETS will be required to reduce GHG emissions by 62% by 2030 compared to 2005 levels. The adaptation of existing provisions and the expansion of the scope of the EU ETS are to be achieved in particular by including shipping GHG emissions [20]. With the extension of the EU ETS to the maritime sector, GHG emissions from all large vessels (larger than 5000 gross tonnage (GT)) calling at EU ports, regardless of their flag, will be considered from 2024 onward. At a later stage, the EU ETS will also be extended to smaller vessels (400 to 5000 GT). The EU ETS will cover 50% of the GHG emissions generated when entering or leaving EU territory or 100% of GHG emissions from voyages within the EU. The EU ETS will first start to cover CO_2 emissions and will be extended to emissions of methane (CH_4) and nitrous oxide (N_2O) in 2026. These two greenhouse gases, alongside CO_2, have significant relevance in the maritime sector as well. GHG emissions from maritime transport fit into the overall cap of the EU ETS, which sets the maximum amount of greenhouse gases across all different sectors at EU level. The cap is adjusted (reduced) over time to contribute to the achievement of the EU climate target of net zero emissions by 2050 [17].

More Environmentally Friendly Fuels in Maritime Transport. Through the directive promoting the use of low carbon fuels (of fossil origin) and renewable fuels in maritime transport, known as the Fuel EU Maritime initiative, the EU aims to decrease the GHG intensity of vessels gradually, targeting a 2% reduction by 2025 and up to 80% by 2050. Therefore, the primary goal of the EU is to promote renewable fuels of non-biological origin (RFNBO) with a high potential for GHG reductions [22]. In doing so, the EU follows the principle of technology neutrality. Currently, the fuel mix in maritime transport is based almost entirely on fossil fuels (e.g., heavy fuel oil (HFO) and marine gasoil (MGO)). There are many reasons for this, including the lack of technological mature and globally viable alternatives. Further complicating factors are the long-term investment required to build vessels and uncertainty about future regulatory requirements.

Proportionally, fuel costs in the operating costs of vessels can range from 35% for small tankers to up to 53% for container and bulk carriers. Therefore, the volatility of fuel costs has a very high impact on the economic performance of ship operators. To date, the price differential between fossil marine and renewable fuels such as hydrogen, methanol, or ammonia remains high. To maintain competitiveness while preparing the sector for the intended transition to new fuels, clear and consistent commitments for the use of low-carbon fuels and renewable fuels on ships are needed [15, 18].

Alternative Fuels Infrastructure Development. In 2023, the Alternative Fuels Infrastructure Development regulation (AFIR) was adopted by the EU. The purpose of the AFIR measure is to ensure that there is an adequate publicly accessible infrastructure network for recharging of battery systems and refueling alternative fuel vessels. That includes enough hydrogen, methanol and ammonia refueling points as its intention to provide alternatives to stationary engine operation of vessels at berth. This measure is about ensuring full interoperability within the EU and that this infrastructure is user-friendly. Specifically, it requires that at least 90% of containers and passenger ships (larger than 5000 GT) at the largest seaports have access to shore-side power by 2030 [19, 21].

Moreover, seaports that are called at least 50 times by large passenger ships or 100 times by container ships must provide shore power for these ships by 2030. The measure plays an important role in accelerating the development of these infrastructures so that the use of zero- or low-emission ships is not impaired, a virtuous circle is set in motion, and shipping can thus significantly reduce its carbon footprint [16].

3.3 Clydebank Declaration

Beyond the initiatives spearheaded by the IMO and the EU at the political level, multiple countries voluntarily collaborated at Climate Summit 26 in Glasgow to introduce an additional initiative for significant climate protection advancements in the maritime sector. This initiative, known as the Clydebank Declaration, seeks to provide added momentum toward achieving the zero emissions goal in shipping in the upcoming years. The declaration envisages the establishment of "green" shipping corridors with zero GHG emission maritime transport routes between two (or more) ports. At least six of such corridors are to be established by the middle of the decade (2025), with more to follow by 2030. Signatory countries include Germany, the United Kingdom, the USA, France, Italy, the Netherlands, Sweden, Denmark, Norway, as well as Finland. For these "green" shipping corridors, partnerships are to be formed between the respective ports as well as shipping companies and other stakeholders to enable the defossilization of the entire transport chain. The necessary "CO_2-free fuels" and propulsion technologies must ensure that no additional greenhouse gases are released during production, transport, or consumption [61].

3.4 Call to Action for Shipping Decarbonization

In addition to the steps taken by policymakers to achieve the targets set, the private sector operating across the maritime value chain is also positioning itself. In its 2021 "Call to Action for Shipping Decarbonization," a coalition of more than 230 industry leaders and organizations from the shipping, freight, energy, finance, ports, and infrastructure sectors calls for greater collaboration with governments. The self-proclaimed "Getting to Zero Coalition" intends to make its fair contribution to defossilization by offering its expertise to guide the investments needed to reach the critical tipping points in defossilizing global supply chains and the world economy. The Getting to Zero Coalition not only recognizes the IMO's Greenhouse Gas Strategy as an important and necessary step in underpinning the Paris Climate Agreement. It sets a goal of using current technologies to power shipping entirely with Net Zero Energy Sources and Zero-Emission Vessels (ZEV) by 2050. The Getting to Zero Coalition sees the need and requirement for the development of these same ZEVs and Net Zero Energy Sources as trade volumes continue to grow. The first milestone in defossilization is seen as the provision of the first commercially viable zero-emission ships by 2030, which must be massively increased by 2040. Furthermore, this goal is flanked by the measure that a minimum of 5% of zero-emission fuels are used in international shipping by 2030. This will also support the development of necessary infrastructure, including production, distribution, storage, and bunkering. The Getting to Zero Coalition sees the Call to Action as an opportunity for collective action to ensure sufficient scalability to make the production of sustainable fuels competitive and also to create benefits for other energy-intensive industries. Signatory companies include major shipping companies. These companies, for example, acknowledged that achieving the long-term goal of decarbonization will require new fuel types and systemic change within the industry, including new legislation to finance net zero solutions. Some companies have additionally signed a bilateral partnership agreement in which they commit to developing zero-carbon technologies and solutions for the maritime industry, which will then be secured through in-house measures [24].

Hapag-Lloyd—Targeting Climate Neutrality by 2045. The shipping company Hapag-Lloyd has set itself the goal of being climate-neutral by 2045. By 2030, Hapag-Lloyd aims to have reduced the CO_2 intensity of its ships by 60% compared to 2008 and 30% compared to 2019. The path that Hapag-Lloyd is taking to this end involves extensive investments in the fleet of ships to make them dual fuel capable. The first of these ships was the "Brussels Express" with a capacity of 15,000 TEU in 2020. The container vessel is the largest conventional vessel of its kind to date that has undergone a retrofit to operate in LNG mode. At the same time, the shipping company has ordered 12 more ships with a capacity of more than 23,500 TEU each, which are scheduled to enter service from 2023 to 2024. Another type of fuel that Hapag-Lloyd intends to increase its use of as part of its target achievement strategy is biofuels. Biofuels have been in the test phase since 2020. These are made up of biological raw materials such as used cooking oil. Compared to conventional fuels, a reduction in GHG emissions of up to 80% can be achieved. Hapag-Lloyd has been certified to operate 16 of its A 19 (19,000 TEU) and A 15 (15,000 TEU) class ships with biofuels in 2021. Meanwhile, it is planned to convert further ship classes and the charter fleet to operate on biofuels [6].

A.P. Moller-Maersk—Targeting Climate Neutrality by 2040. A.P. Moller-Maersk, as one of the world's largest shipping companies, is going even further in its climate ambitions, taking the lead in the Getting to Zero Coalition. Maersk aims to be the world's most carbon–neutral shipping company by 2040, reaching the 2050 goal postulated by the Getting to Zero Coalition 10 years earlier. The roadmap calls for a 50% reduction in GHG in the shipping fleet and a 70% reduction in terminals by 2030, measured against 2020. Combined emission reductions are to be 35% and 50%. Maersk also aims to have at least 25% of its shipping fleet operating on green and sustainable fuels by 2030 [37]. Compared to other shipping companies, Maersk is not resorting to LNG as a bridge fuel but is skipping this technology by going straight to running its ships on carbon–neutral methanol [50]. In 2023, Maersk has put the first "green" methanol-powered ship into service [5]. The feeder vessel will have a capacity of approximately 2100 TEU and will operate in the North Sea and the Baltic Sea services. Although equipped with dual fuel engine technology and thus capable of operating on very low sulfur oil it is intended to run on carbon–neutral e-methanol or sustainable bio-methanol from the outset. The feeder ship represents the beginning of a transition to further green methanol-fueled ships [38]. In addition to the feeder vessel, Maersk has ordered 18 container ships with a capacity of 16,000 TEU each and dual fuel engine technology to be launched between 2024 and 2025 [5]. The newbuilds will mainly replace smaller, older vessels that are at their 30-year service end. At the same time, the introduction of methanol as a fuel offers alternative fuel suppliers the opportunity to gain insights into fuel needs and thus scalability of future fuel production. For security of supply of methanol-fueled ships, Maersk is creating facts and entering into various global partnerships with bio-methanol producers from China to Europe and North America. By the end of 2025, Maersk intends to have 730,000 tons of green methanol available annually. Maersk currently sees methanol as the most mature fuel technology [54].

LNG as a "Bridge Technology". Although the French shipping company CMA CGM is not part of the Getting to Zero Coalition, it has committed itself to the same goal of operating its ships in a climate-neutral manner by 2050. While Maersk has completely abandoned LNG as a bridging technology, CMA CGM is launching a major program to procure LNG-fueled container vessels [53]. By the end of 2024, the shipping company intends to have 44 LNG-fueled ships in its fleet [51]. CMA CGM made a start by announcing the construction of 9 LNG-powered 23,000 TEU container vessels in 2017 [58]. In addition, CMA CGM continues to invest heavily in alternative solutions such as wind-assisted propulsion, hydrogen engines, and green fuels. In 2019, the group became the first shipping company in the world to successfully test a biofuel made from 20% recycled vegetable oil and forestry waste. The company has further committed to meet at least 10% of its consumption with alternative fuels by 2023 [8].

The analysis so far has shown very clearly that the shipping industry as a whole is under increasing pressure to implement the Paris Agreement and reduce GHG emissions. Meanwhile, the world's largest shipping companies, such as Maersk, Hapag-Lloyd, and CMA CGM, are adapting their fleets to the circumstances to meet the internationally binding but also self-imposed targets for achieving emission reductions in the coming decades. This requires not only efforts to control their own GHG emissions, but also the communication with associated companies to complete their reporting and expectations of banks and carriers. As a consequence, carbon-intensive and non-transparent shipping companies could increasingly suffer from a loss of attractiveness in the charter and capital markets in the future.

4 Conversion to Low Carbon and Renewable Fuels

The main engines of the shipping fleet are still mainly powered by heavy fuel oil (HFO). However, the number of alternatively powered ships is constantly increasing, especially liquefied natural gas (LNG) is increasingly being used, among other things. In addition to LNG, methanol is also increasingly finding its way into the shipping industry as an alternative fuel. Furthermore, other alternative fuels such as hydrogen and ammonia, as well as propulsion systems such as fuel cells, are expected to be used to power the global shipping fleet. However, the introduction of these novel fuels must be supported by accompanying technical measures such as increased energy efficiency of ships to curb climate change. At the same time, measures aimed at defossilizing shipping include advancing digitalization, increased fleet utilization, and speed reductions.

According to DNV [14] Maritime Forecast 2050 study, the introduction of all alternative fuels for shipping is fraught with challenges and obstacles, with hurdles varying in severity depending on the fuel type. Typical major barriers common to all future fuels include the cost of the required machinery, fuel storage onboard ships, the additional need for storage space onshore, the low technical maturity of many propulsion technologies, limited fuel availability, and a lack of global bunkering infrastructure. Another issue voiced by the study authors regarding deployment is safety. The lack of rules and regulations makes the use of alternative fuels machinery and storage systems difficult. The reality of long transit voyages means that large amounts of energy must be stored onboard ocean-going ships. Therefore, the choice of alternative fuels for ocean shipping is currently still reduced to liquefied natural gas (LNG), liquefied petroleum gas (LPG), and few biofuel options. DNV [11] In the following, different alternative fuel options are discussed based on general aspects on their current usage and aspects on their integration within the maritime industry. Therefore, first so-called low carbon fuels are addressed, which are still of fossil origin, but can allow for a reduction in GHG emissions. Subsequently, renewable fuel options are presented, which are based on renewable energies and thus facilitate further GHG emissions reductions.

4.1 Low Carbon Fuels

As "low carbon fuel options," in the following liquefied natural gas (LNG) and liquefied petroleum gas (LPG) are considered. As the world transitions to cleaner and more sustainable energy sources, the importance of LNG and LPG increases. Both fuels offer specific advantages over heavy fuel oil (HFO) and marine gasoil (MGO), such as lower GHG emissions and improved fuel efficiency. Both these fuel options, namely LNG and LPG, are of fossil origin but allow for reduced CO_2 emissions when combusted. This is due to their higher hydrogen-to-carbon ratio. In this context, LPG is partially considered as attractive an energy source as LNG and could therefore also serve as a bridge technology in the further development toward renewable fuels [12].

Liquefied Natural Gas (LNG). Liquefied natural gas (LNG) is a clear, odorless, and non-toxic liquid. It is formed by cooling (fossil) natural gas down to about -162 °C. Within the maritime industry, it is increasingly being adopted as a "cleaner" alternative fuel option to conventional crude oil-based fuels, reducing GHG emissions and pollutants. Nevertheless, methane release (slip) must be considered when assessing the GHG reduction potential of LNG as a marine fuel. While LNG bunkering infrastructure is still limited, the supply situation for ships is improving significantly. Much of the LNG supply takes place in port by truck, and the same is true for supply to bunkering sites in ports. Delivery by rail would also be possible but is not currently practiced. The development of LNG bunkering vessels for seagoing vessels will therefore become increasingly important shortly. In recent years, LNG bunkering vessels entered service in the eastern Baltic Sea, Rotterdam, North America, and

Singapore. Furthermore, bunkering vessels for other locations such as the Western Mediterranean, Gulf of Mexico, Middle East, Singapore, China, South Korea, and Japan have recently been ordered or are under development [25, 65], [39, 59]. According to DNV's maritime advisory board, the expansion of the bunker fleet needs to be increased; otherwise, bottlenecks in the supply of LNG-fueled vessels may occur.

Liquefied Petroleum Gas (LPG). Liquefied petroleum gas (LPG) is a flammable hydrocarbon gas that is liquefied through pressurization, consisting primarily of propane and butane. In the maritime sector, just like LNG, LPG is gaining traction as an alternative fuel to conventional marine fuels. As it is the case for LNG (methane slip), this must be considered when considering whether unburned LPG could potentially escape into the atmosphere. At the same time, the use of LPG will significantly reduce sulfur emissions, as well as particulate matter emissions. The reduction in NO_X emissions depends on the technology used. For a two-stroke diesel engine, NO_X emissions can probably be reduced by 10 to 20% compared to heavy fuel oil (HFO). Therefore, to meet the standards, it will be necessary to equip a two-stroke diesel engine with an exhaust gas cleaning system (EGCS) or selective catalytic reduction systems. For four-stroke gasoline engines, the expected reduction in sulfur emissions is more significant and may be below IMO Tier III NO_X limits. Currently, there are three orders for newbuilds ahead of so-called Very Large Gas Carriers (VLGC) to operate on LPG, and four existing LPG tanker vessels have been converted to operate on LPG fuel in 2019. Bunkering infrastructure for LPG is relatively easy to build, as it can be constructed at existing LPG storage facilities or terminals by simply adding distribution facilities. Distribution to vessels can be done either from dedicated facilities or from dedicated bunkering vessels [11].

4.2 Renewable Fuels

In the following, various renewable fuel options currently under discussion for use in the maritime sector will be described. These primarily include methanol, different biofuel options, hydrogen, ammonia, and battery storage combined with corresponding propulsion systems. Unlike the aforementioned low carbon fuels (of fossil origin), the production of these fuel options for maritime use are discussed based on renewable energies.

Methanol. Methanol is a much-produced industrial alcohol obtained from synthesis gas. However, the mixture of hydrogen and carbon monoxide has so far mostly been produced from fossil raw materials. Two-stroke engines that run on methanol are already on the market and have proven themselves. The first methanol-powered ship, Stena Germanica, has been operated by Stena Lines since 2015. By the end of 2022, 21 ships will use methanol as their primary fuel. In addition to large shipping companies such as the aforementioned shipping company A.P. Moller-Maersk, smaller shipping companies such as North Sea Container Line (NCL) are also making the switch to methanol-powered ships. Likewise, fuel cell technology using methanol has already been demonstrated for smaller ships [11, 49].

Biofuels. Biofuel is a collective term for a range of energy carriers that are converted into liquid or gaseous fuels by converting primary biomass or biomass residues. There is a global lack of infrastructure and bunkering facilities for biofuels. The only place where biofuels can be bunkered is in certain ports in the Netherlands, Australia, or Norway. The handling of biofuels varies widely and depends on the chemical properties. Hydrotreated vegetable oil (HVO), basically a marine diesel option, for example, can in most cases be distributed through existing marine gasoil (MGO) and HFO (heavy fuel oil) distribution routes, although modifications are sometimes required. Using existing distribution systems for fatty acid methyl esters (FAME) presents a greater challenge. Due to the potential for oxidation of FAME and possible sedimentation, storage of FAME for more than six months should be avoided. Liquefied biogas (LBG), meanwhile, can take advantage of the expanding LNG infrastructure. Since methane is the main component of LNG, LBG should blend easily with LNG. The actual GHG emissions of a given biofuel are highly dependent on the type of feedstock used and the fuel production process. Therefore, reductions in GHG emissions can vary from low values such as 20% up to higher values of around 90%. In general, however, it can be stated that the use of biofuels in shipping is currently very limited. In a first trial, the shipping company CGM CMA is bringing biofuels into use by having 32 container vessels run on different blends of conventional HFO and biofuel. Some of the ships in the trial will run on B24 biofuel, which contains 24% cooking oil methyl ester and is expected to lead to a 21% reduction in GHG emissions [7, 11].

Hydrogen. Hydrogen is a widely used basic chemical material as an energy carrier and can be produced from various energy sources. Today, 95% of hydrogen is produced from fossil fuels, mainly natural gas (68%), but also petroleum (16%) and coal (11%). To date, only 5% of the hydrogen production is by electrolysis [11]. Since there is currently little demand for hydrogen fuel in the maritime sector, there is no distribution or bunkering infrastructure, but there could be options to access LNG distribution infrastructure in the future. The production of hydrogen by electrolysis is a well-known and commercially available technology that could be suitable for local production in ports under certain suitable circumstances and provided that sufficient (renewable) electricity can also be made available. Current barriers to the use of hydrogen as a marine fuel include lack of safety requirements, low maturity of the technology, required onboard storage space, and high investment costs. Demonstration projects have been conducted for both internal combustion engines and fuel cell systems. Also, in 2024, the first hydrogen-fueled bulk carrier is supposed to be commissioned by the Norwegian shipping company Egil Ulvan Rederi [55].

Ammonia. Due to safety-related challenges, as well as space, weight, and cost considerations associated with storing large quantities of fuel, such as LNG, LPG, and hydrogen on ships, there has been increasing interest in exploring alternative hydrogen-based energy sources such as ammonia. Major challenges include ammonia toxicity, its combustion characteristics, NO_x emissions, and potential ammonia slip. The development of engines capable of burning ammonia is expected to be ready for the market before 2030. An ongoing EU project, ShipFC, is scheduled to demonstrate a 2 MW ammonia-fueled solid oxide fuel cell in 2023 to retrofit an existing supply vessel (Viking Energy). Such demonstration and pilot projects are expected to significantly accelerate technology maturity. Some commercial applications are also expected, as several shipowners have announced plans for use on vessel types such as roll-on roll-off of cargo and passenger (RoPax) ships, tankers, and bulk carriers. Before 2020, four ammonia projects were initiated for vessels over 5000 dwt, but 10 new projects have started since then [11, 35].

Battery Propulsion Systems. Battery propulsion systems onboard ships represent a departure from previous approaches to energy use and distribution, while at the same time offering numerous new possibilities in ship propulsion technology. The reasons for this are the improved power supply systems for ships, which are increasingly electrified, and the improved battery technology, which is continuously improving and becoming more affordable. All-electric ships are a major advance in the development of propulsion systems, but are currently only feasible in limited applications such as ferries and short sea shipping. The feasibility of all-electric operation for ocean-going vessels is typically limited either by the size of the battery system required or by its cost. Further research and development is needed to significantly improve this technology. Currently, solid electrolyte technologies in marine applications are among the most promising technologies that still need further development. Nevertheless, they offer great potential for improving ship safety, should the battery technology succeed in meeting the stringent performance requirements of maritime transport. Shipboard battery systems require adequate charging infrastructure depending on the application, battery size, and required charging times can increase power demand. In principle, existing shore power infrastructure can be used to supply power to ships [14].

4.3   Status and Possible Developments

For the presented alternative fuel options, the subsequent sections will delve into their general development over recent years. Between 2018 and 2021, there's been a trend toward climate-friendly shipping, looking at the number of new ship orders equipped with alternative propulsion systems and the existing ships that have adopted alternative fuels. As depicted in Fig. 15, approximately 0.3% of the global ship fleet operating in 2018 were powered by alternative fuels.

It must be added, however, that in the database used it is not differentiated between ocean shipping and short sea shipping. Nevertheless, it allows for a general overview of the current situation of possible trends in shipping regarding the development toward climate neutrality. If Figs. 15 and 16 are compared with each other, the overall shares of ships to use alternative fuels are still low, both in terms of their shares in the existing fleet as well as in ship orders, but an emerging dynamic in the

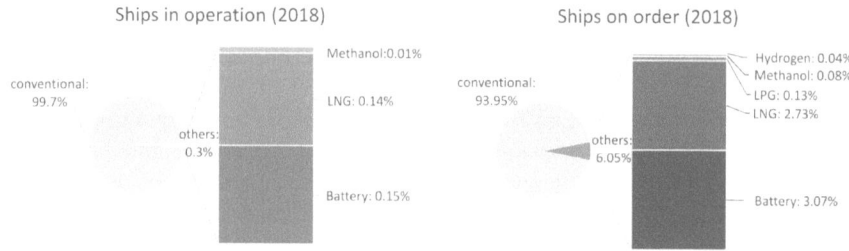

Fig. 15 Global fleet distribution by fuel type 2018 (*Source* DNV 2019)

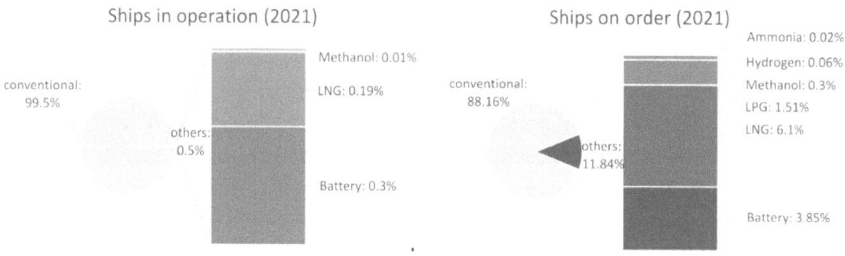

Fig. 16 Global fleet distribution by fuel type in 2021 (*Source* DNV [14])

current new ship orders can already be seen toward increasing shares of ships that run on the described alternative fuels.

In 2021, about 12% of the orders were already provided for new ships with alternative fuel systems. This means that the share of newbuilds with alternative fuels has doubled since 2018 [13, 14]. About the propulsion situation of new ships for the next few years, a global increase in LNG-fueled ocean-going vessels and batteries for fully electric or partially electric operation in short sea shipping seems likely. Apart from electrification in short sea shipping, fuels are currently still mainly based on low carbon fossil fuels such as LNG. For low fuel applications in maritime shipping, storage capacity in particular is a key challenge.

According to a study by [66], the two alternative fuels ammonia and hydrogen are considered as most promising when it comes to leading shipping toward climate neutrality. LNG is not expected to play a major role in the World Bank's view but rather plays a limited role. The study shows that the fuels ammonia followed by hydrogen can hold their own against other alternative GHG neutral fuels, such as biofuels in various forms and methanol, because of their many positive properties. Ammonia as well as hydrogen stand out, especially compared to the other fuels, due to the great strategic advantage that they can be produced in different ways. This is particularly important in the strategic positioning concerning the fuels of the future in terms of capacity limits and scalability. Furthermore, other parameters such as GHG emissions emitted throughout the life cycle of the fuels, profitability, and technical and safety requirements that arise when using these fuels have a very positive impact. For example, ammonia and hydrogen have similar technical requirements for the main engines of ships as heavy fuel oil (HFO). As a result, the technical changes to the main engine are minimal and do not require the replacement of the main engine or adjustments to the drive trains. In addition, ammonia and hydrogen are also compatible with new fuel cell solutions. The study indicates that potential is seen for both alternative biofuels and synthetically produced carbon-based fuels. However, compared to ammonia and hydrogen, supply constraints are seen due to the limited availability of and competing demand for biomass.

DNV points to challenges in existing fuel options across all scenarios in its Maritime Forecast to 2050 [14]. Like the World Bank, DNV sees the most promising carbon–neutral fuel in ammonia. However, unlike the World Bank, DNV sees methanol as the fuel of choice in the long term after ammonia. Furthermore, DNV points out that not all ships are likely to switch to the same fuel. Rather a diverse energy mix that includes both fossil and non-carbon fuels is expected, with fossil fuels being phased out by 2050. Zero-carbon fuels are being introduced both as drop-in alternatives and for use with specific technologies. Fossil LNG is projected to gain a significant share until further regulatory tightening occurs.

5   Summary of Challenges for the Decarbonization of Shipping

In the course of the approaches described so far, the goal emerges clearly. The defossilization of shipping is a major global project with numerous challenges. It involves much more than simply introducing new fuel options. However, alternative fuels and energy sources are needed to meet the 2050 targets, although all currently known alternative fuels and technologies have limitations and no truly zero-carbon fuels are widely available. In this respect, frameworks are set for innovation and targets to ensure the infrastructure and production for renewable fuels such as green hydrogen and ammonia in sufficient quantities. In summary, the challenges include the following four issues:

Marine and Bunkering Technologies. Technologies for the use of alternative fuels in marine operations are basically available. Various pilot projects for methanol-, ammonia- or hydrogen-powered ships demonstrate this, and the increasing orders for large ships with alternative propulsion systems point to the change here. Nevertheless, considerable research and development work is needed to convert the global merchant shipping fleet to synthetic fuels in the coming years. Ship concepts and bunkering solutions, including the many technical system components optimized for the use of these new fuels, must be developed and made ready for series production. Safety, reliability, and the achievement of economic cost levels for the procurement and operation of these systems are key targets.

Fuel Availability and Markets. The defossilization of shipping can only succeed if the fuels required for this, produced based on renewable energies, are available globally, in sufficient quantities, and at reasonable cost. The technologies for producing the alternative fuels do exist, but not yet on the industrial scale that will be needed for shipping in the future. There is a significant need for research and development in this area as well as for large-scale commercial technology roll-outs. The capacities of fuel production and distribution are to be built up with the commissioning of the corresponding ships. Ideally, supply and demand will gradually develop into a healthy market that ensures a global green fuel supply at an economical cost. But this is precisely where another risk occurs. The enormous investments required for fuel supply will hardly be covered by the initially subdued development of the markets, so support from policymakers will be necessary.

Port Bunkering Capacity for Alternative Fuels. The task of the ports here is to enable bunkering with synthetic fuels. This involves creating technical solutions and appropriate capacity to receive, store, and deliver these fuels to ships as needed. Different characteristics of the various fuel options will require different technical solutions. In addition, since conventional fossil fuels will also have to be provided during the decades-long transition period until decarbonization, storage, and bunkering facilities for the diverse fuel alternatives will have to be kept ready in parallel. In addition to the technical questions, there are also questions about the availability of suitable areas in the ports. Considerable funds will be needed for the immense task of the ports which will require enormous public support.

Investment in Alternative Fuel Ships. Investments by shipping companies in green fuel ships are still subject to numerous uncertainties regarding timing, scale, fuel options, and associated costs. Likewise, this is true for ports regarding the creation of storage and bunkering facilities and the potential alternative fuel producers to build up production capacity. A key challenge for national and international policymakers is therefore to reduce the market and technological uncertainties of decarbonization and create reliable prospects. To this end, research and development must be promoted and technologies developed for the distribution and use of alternative fuels for marine propulsion. In addition, incentives must be provided for the emergence of markets for low-carbon fossil and renewable fuels, for example, through investment subsidies. Furthermore, ports must be supported in the creation of storage and bunkering facilities, and standardized and binding regulations for defossilization must be created in fair competition worldwide.

For this to succeed, the stakeholders from politics, shipping, ports, and the energy and fuel industries must join forces. And, of course, the shipping industry and end customers worldwide will be called upon. Their willingness to bear the additional costs of climate neutrality in shipping resulting from the necessary technological change will be a key success factor for implementation.

References

1. Allianz Global Corporate & Speciality (2022) Safety and shipping review 2022. An annual review of trends and developments in shipping losses and safety. Edited by Allianz Global Corporate & Specialty SE. Munich, Germany

2. Allianz Global Corporate & Speciality (2023) Safety and shipping review 2023. An annual review of trends and developments in shipping losses and safety. Edited by Allianz Global Corporate & Specialty SE. Munich, Germany

3. BSH (2022a) MARPOL-Übereinkommen. Edited by Bundesamt für Seeschifffahrt und Hydrographie. Available online at https://www.bsh.de/DE/THEMEN/Schifffahrt/Umwelt_und_Sch ifffahrt/MARPOL/marpol_node.html

4. BSH (2022b) Übereinkommen und Umsetzung. Edited by Bundesamt für Seeschifffahrt und Hydrographie. Available online at https://www.bsh.de/DE/THEMEN/Schifffahrt/Umwelt_ und_Schifffahrt/Ballastwasser/Uebereinkommen_und_Umsetzung/Uebereinkommen_und_ Umsetzung_node.html

5. Bahtić F (2023) Maersk: world's 1st containership operating on green methanol delivered. In: Offshore Energy, 2023. Available online at https://www.offshore-energy.biz/maersk-worlds-1st-containership-operating-on-green-methanol-delivered/, checked on 7/13/2023

6. Behncke N, Beenken M (2022) HAPAG-LLOYD AG | Sustainability report 2021. Available online at https://www.hapag-lloyd.com/sustainability-report-2021/en/index.html, updated on 4/21/2022, checked on 9/5/2022

7. Biofuels Central (2022) CMA CGM with the support of the maritime and Port Authority of Singapore started biofuel bunkering for 32 ships. In: Biofuels Central, 2/24/2022. Available online at https://biofuelscentral.com/cma-cgm-maritime-port-authority-singapore-biofuel-bun kering-32-ships/, checked on 9/6/2022

8. CMA CGM (2022) CMA CGM launches global biofuel bunkering trial in Singapore. Available online at https://www.cma-cgm.com/news/4082/cma-cgm-launches-global-biofuel-bun kering-trial-in-singapore, updated on 9/5/2022, checked on 9/5/2022

9. Clarkson's Research LTD. (2022) Clarksons newbuilding price index. Available online at https://www.clarksons.com/research/, checked on 9/13/2022

10. Cofala J, Aman M, Borken-Kleefeld J, Gomez-Sanabria A, Heyes C (2018) The potential for cost-effective air emission reductions from international shipping through designation of further emission control areas in EU waters with focus on the Mediterranean Sea. Edited by International Institute for Applied Systems Analysis. Available online at https://previous.iiasa. ac.at/web/home/research/researchPrograms/air/Shipping_emissions_reductions_main.pdf

11. DNV (2019a) Assessment of selected alternative fuels and technologies (rev. June 2019). Edited by Det Norske Veritas. Available online at https://www.dnv.com/Publications/assess ment-of-selected-alternative-fuels-and-technologies-rev-june-2019--116334, updated on 9/6/ 2022, checked on 9/6/2022

12. DNV (2019b) Making LPG fuel an option for the shipping industry—DNV. Edited by Det Norske Veritas. Available online at https://www.dnv.com/expert-story/maritime-impact/Mak ing-LPG-fuel-an-option-for-the-shipping-industry.html

13. DNV (2019c) Martime Forecast to 2050. Energy Transition Outlook 2019. Edited by Det Norske Veritas. Available online at https://sustainableworldports.org/wp-content/uploads/ DNV-GL_2019_Maritime-forecast-to-2050-Energy-transition-Outlook-2019-report.pdf

14. DNV (2021) Maritime forecast to 2050. Energy transition outlook 2021. Available online at https://eto.dnv.com/2021/

15. European Commission (2021) Regulation of the European Parliament and of the Council on the use of renewable and low-carbon fuels in maritime transport and amending. Directive 2009/16/ EC. Available online at https://ec.europa.eu/info/sites/default/files/fueleu_maritime_-_green_ european_maritime_space.pdf

16. European Commission (2023a) New law agreed to deploy alternative fuels infrastructure. Available online at https://ec.europa.eu/commission/presscorner/detail/en/IP_23_1867

17. European Commission (2023b) Reducing emissions from the shipping sector. Available online at https://climate.ec.europa.eu/eu-action/transport-emissions/reducing-emissions-shipping-sector_en

18. European Council (2022) Fit for 55 package: Council adopts its position on three texts relating to the transport sector. Available online at https://www.consilium.europa.eu/en/press/press-releases/2022/06/02/fit-for-55-package-council-adopts-its-position-on-three-texts-relating-to-the-transport-sector/

19. European Council (2023a) Alternative fuels infrastructure: Council adopts new law for more recharging and refuelling stations across Europe. Available online at https://www.consilium. europa.eu/en/press/press-releases/2023/07/25/alternative-fuels-infrastructure-council-adopts-new-law-for-more-recharging-and-refuelling-stations-across-europe/
20. European Council (2023b) European Green Deal. Fit for 55. Available online at https://www. consilium.europa.eu/en/policies/green-deal/fit-for-55-the-eu-plan-for-a-green-transition/
21. European Council (2023c) Fit for 55: towards more sustainable transport. Available online at https://www.consilium.europa.eu/en/infographics/fit-for-55-afir-alternative-fuels-inf rastructure-regulation/
22. European Council (2023d) FuelEU maritime initiative: council adopts new law to decarbonise the maritime sector. Available online at https://www.consilium.europa.eu/en/press/press-rel eases/2023/07/25/fueleu-maritime-initiative-council-adopts-new-law-to-decarbonise-the-mar itime-sector/
23. Georgia Tech Panama Logistics Innovation and Research Center (2022) Statistics. Available online at https://logistics.gatech.pa/en/assets/panama-canal/statistics, checked on 9/13/2022
24. Global Maritime Forum (2021) Getting to zero coalition. Available online at https://www.glo balmaritimeforum.org/getting-to-zero-coalition, updated on 9/5/2022, checked on 9/5/2022
25. Habibic, Ajsa (2022) MOL's 12,000 cbm LNG bunker vessel launched in Singapore. In: Offshore Energy, 4/11/2022. Available online at https://www.offshore-energy.biz/mols-12000-cbm-lng-bunker-vessel-launched-in-singapore/, checked on 9/5/2022
26. ICCT (2018) The International Maritime Organization's initial greenhouse gas strategy. Edited by International Council on Clean Transportation. Available online at https://theicct.org/sites/ default/files/publications/IMO_GHG_StrategyFInalPolicyUpdate042318.pdf
27. IMO (2009) Second IMO GHG Study 2009. With assistance of Buhaug Ø, Corbett JJ, Endresen Ø, Eyring V, Faber J, Hanayama S, Lee DS, Lee D, Lindstad H. Edited by International Maritime Organization. London, UK
28. IMO (2020) Cutting sulphur oxide emissions. Edited by International Maritime Organization. Available online at https://www.imo.org/en/MediaCentre/HotTopics/Pages/Sulphur-2020.aspx, updated on 9/4/2022, checked on 9/4/2022
29. IMO (2021) Fourth IMO GHG Study 2020—full report and annexes. Edited by International Maritime Organization. London
30. IMO (2022a) Brief history of IMO. Edited by International Maritime Organization. Available online at https://www.imo.org/en/About/HistoryOfIMO/Pages/Default.aspx
31. IMO (2022b) Conventions. Adopting a convention, entry into force, accession, amendment, enforcement, tacit acceptance procedure. Available online at https://www.imo.org/en/About/ Conventions/Pages/Default.aspx, updated on 9/5/2022, checked on 9/5/2022
32. IMO (2023) Revised GHG reduction strategy for global shipping adopted. Edited by International Maritime Organization. Available online at https://www.imo.org/en/MediaCentre/ PressBriefings/pages/Revised-GHG-reduction-strategy-for-global-shipping-adopted-.aspx, updated on 7/21/2023, checked on 7/21/2023
33. ITF (2020) Future maritime trade flows: summary and conclusions. Edited by OECD Publishing. Paris (ITF Roundtable Reports, No. 178)
34. Janson M (2022) Die größten Containerhäfen - 2005 und 2021. Digitales Bild. Edited by Statista GmbH. Available online at https://de.statista.com/infografik/23337/umschlagvolumen-von-containerhaefen/, updated on 28. April, 2022, checked on 9/8/2022
35. Jendrischik M (2021) Viking Energy: Schiff der Reederei Eidesvik erhält weltweit erste Ammoniak-Brennstoffzelle. In Jendrischik PR, Leipzig, 3/3/2021. Available online at https://www.cleanthinking.de/viking-energy-schiff-der-reederei-eidesvik-erhaelt-weltweit-erste-ammoniak-brennstoffzelle/, checked on 9/6/2022
36. Kuehne und Nagel (2022) Sea freight emission control areas. Edited by Kuehne und Nagel. Available online at https://home.kuehne-nagel.com/-/knowledge/emission-control-areas, updated on 9/4/2022, checked on 9/4/2022

37. Maersk (2022a) A.P. Moller—Maersk accelerates net zero emission targets to 2040 and sets milestone 2030 targets. Available online at https://www.maersk.com/news/articles/2022/01/12/apmm-accelerates-net-zero-emission-targets-to-2040-and-sets-milestone-2030-targets, checked on 9/5/2022

38. Maersk (2022b) Maersk signs shipbuilding contract for world's first container vessel fueled by carbon neutral methanol. Available online at https://www.maersk.com/news/articles/2021/07/01/container-fueled-by-carbon-neutral-methanol, checked on 9/5/2022

39. Mandra, Jasmina Ovcina (2023) World's largest LNG bunkering vessel refuels first VLCC in milestone operation. In: Offshore Energy, 2023. Available online at https://www.offshore-energy.biz/worlds-largest-lng-bunkering-vessel-refuels-first-vlcc-in-milestone-operation/, checked on 9/4/2023

40. MarineTraffic (2021) Singapore top shipping centre for eighth year running. Density map of global shipping 2020. Available online at https://www.marinetraffic.com/blog/singapore-top-shipping-centre-for-eighth-year-running/, updated on 7/16/2021, checked on 9/12/2022

41. Marinekommando (2014) Marinekommando Jahresbericht 2014. Fakten und Zahlen zur maritimen Abhängigkeit der Bundesrepublik Deutschland. Edited by Marinekommando. Bundeswehr

42. Marinekommando (2015) Marinekommando Jahresbericht 2015. Fakten und Zahlen zur maritimen Abhängigkeit der Bundesrepublik Deutschland. Edited by Marinekommando. Bundeswehr

43. Marinekommando (2016) Marinekommando Jahresbericht 2016. Fakten und Zahlen zur maritimen Abhängigkeit der Bundesrepublik Deutschland. Edited by Marinekommando. Bundeswehr

44. Marinekommando (2017) Marinekommando Jahresbericht 2017. Fakten und Zahlen zur maritimen Abhängigkeit der Bundesrepublik Deutschland. Edited by Marinekommando. Bundeswehr

45. Marinekommando (2018) Marinekommando Jahresbericht 2018. Fakten und Zahlen zur maritimen Abhängigkeit der Bundesrepublik Deutschland. Edited by Marinekommando. Bundeswehr

46. Marinekommando (2019) Marinekommando Jahresbericht 2019. Fakten und Zahlen zur maritimen Abhängigkeit der Bundesrepublik Deutschland. Edited by Marinekommando. Bundeswehr

47. Marinekommando (2020) Marinekommando Jahresbericht 2020. Fakten und Zahlen zur maritimen Abhängigkeit der Bundesrepublik Deutschland. Edited by Marinekommando. Bundeswehr

48. Marinekommando (2021) Marinekommando Jahresbericht 2021. Fakten und Zahlen zur maritimen Abhängigkeit der Bundesrepublik Deutschland. Edited by Marinekommando. Bundeswehr

49. Maritime Executive (2022) Two methanol-fueled feeder ships to launch north sea green corridor. In: The Maritime Executive, 7/4/2022. Available online at https://maritime-executive.com/article/two-methanol-fueled-feeder-ships-to-launch-north-sea-green-corridor, checked on 9/6/2022

50. NGI (2021) Maersk orders eight methanol-powered vessels, Eschewing LNG as marine fuel—natural gas intelligence. Natural Gas Intelligence. Available online at https://www.naturalgasintel.com/maersk-orders-eight-methanol-powered-vessels-eschewing-lng-as-marine-fuel/, updated on 8/27/2021, checked on 9/5/2022

51. Nair S (2022) CMA CGM and Total energies introduce LNG bunkering operation in Marseille. In: Ship Technology, 1/24/2022. Available online at https://www.ship-technology.com/news/cma-cgm-totalenergies-marseille/, checked on 9/5/2022

52. Notteboom T, Pallis AA, Rodrigue J-P (2022) Port economics, management and policy. Routledge, Taylor & Francis Group, Abingdon, Oxon, New York, NY

53. Ovcina J (2020) CMA CGM sets sights on becoming carbon neutral by 2050. In: Offshore Energy, 6/3/2020. Available online at https://www.offshore-energy.biz/cma-cgm-sets-sights-on-becoming-carbon-neutral-by-2050/, checked on 9/5/2022

54. Prevljak NH (2022a) Maersk secures green fuel supply for 12 methanol-powered boxships. In: Offshore Energy, 3/10/2022. Available online at https://www.offshore-energy.biz/maersk-sec ures-green-fuel-supply-for-12-methanol-powered-boxships/, checked on 9/5/2022
55. Prevljak NH (2022b) World's first hydrogen-powered bulk carrier receives Enova funding. In: Offshore Energy, 4/8/2022. Available online at https://www.offshore-energy.biz/worlds-first-hydrogen-powered-cargo-ship-receives-enova-funding/, checked on 9/8/2022
56. SEA Europe (2022) SEA Europe shipbuilding market monitoring. Report No 53—March 2022. Edited by Shipyard's & Maritime Equipment Association (53)
57. Saxe S, Jahn C, Brümmerstedt K, Fiedler R (eds) (2017) Digitalization of seaports—visions of the future. Fraunhofer-Center für Maritime Logistik und Dienstleistungen; Fraunhofer IRB-Verlag. Fraunhofer Verlag, Stuttgart. Available online at http://www.bookshop.fraunhofer.de/buch/digitalization-of-seaports-visions-of-the-future/247608
58. Schuler M (2017) CMA CGM Opts for LNG fuel to power record-breaking 'megaships'. In: Unofficial Networks LLC, 11/7/2017. Available online at https://gcaptain.com/cma-cgm-opts-for-lng-fuel-to-power-record-breaking-megaships/, checked on 9/5/2022
59. Ship Technology (2022) Optimus LNG bunkering vessel, Estonia. In: Ship Technology, 1/27/2022. Available online at https://www.ship-technology.com/projects/optimus-lng-bunkering-vessel-estonia/, checked on 9/5/2022
60. Suez Canal Authority (2020) Suez Canal traffic statistics. Annual report 2019. Available online at https://www.suezcanal.gov.eg/English/Downloads/DownloadsDocLibrary/Naviga tion%20Reports/Annual%20Reports%E2%80%8B%E2%80%8B%E2%80%8B/2019.pdf, checked on 9/13/2022
61. UK Department for Transport (2021) COP 26: Clydebank Declaration for green shipping corridors. Edited by GOV.UK. Available online at https://www.gov.uk/government/public ations/cop-26-clydebank-declaration-for-green-shipping-corridors/cop-26-clydebank-declar ation-for-green-shipping-corridors, updated on 9/5/2022, checked on 9/5/2022
62. UNCTAD (2021) Review of maritime transport 2020. [S.l.]: Division on Technology and Logistics. United Nations Conference on trade and development
63. UNCTAD (2022) Review of maritime transport 2021. [S.l.]: Division on Technology and Logistics. United Nations conference on trade and development
64. UNCTAD (2023) Review of maritime transport 2022. United Nations, Geneva (Review of maritime transport/United Nations conference on trade and development, Geneva, 2022)
65. Vard (2022) Vard marine secures design contract for 7600 m3 LNG Bunker Vessels for Seaspan ULC—Vard Marine. Available online at https://vardmarine.com/vard-marine-secures-design-contract-for-7600m3-lng-bunker-vessels-for-seaspan-ulc/, updated on 5/24/2022, checked on 9/5/2022
66. World Bank (2022) Carbon revenues from international shipping: enabling an effective and equitable energy transition—technical paper. Available online at https://www.worldbank.org/en/topic/transport/publication/carbon-revenues-from-international-shipping

Hydrogen as a Feedstock

Power Generation from Renewable Energies

Jelto Lange and Martin Kaltschmitt

Abstract This chapter explores the critical role of renewable energy in the global transition toward a sustainable and climate-neutral future, particularly focusing on the production of synthetic electricity-based fuels, or "powerfuels". It begins by discussing the imperative of defossilizing the energy sector to meet global climate targets, emphasizing the increasing reliance on renewable energy sources like photovoltaics, wind power, and hydropower. The chapter highlights the challenges and opportunities associated with these renewable energy technologies, including their fluctuating nature and site-specific limitations. It provides an in-depth analysis of their technical and economic developments, current utilization, and global potential. The chapter concludes by emphasizing the need for an integrated approach in designing optimal renewable energy supply systems, combining different generation technologies, storage solutions, and demand management to achieve cost-effective and sustainable production of powerfuels. This integrated evaluation is essential for meeting climate protection goals while minimizing production costs in a rapidly evolving energy landscape.

Keywords Renewable energy · Energy transition · Power generation · Integrated energy systems

1 Introduction

Average global temperatures in the year 2022 were more than 1.15 °C above pre-industrial levels [1]. In order to limit this global warming to below 2 °C, various nations are increasingly taking action to considerably reduce greenhouse gas (GHG) emissions in the years to come [2]. The defossilization of the energy sector (i.e., its transformation from the use of fossil fuels toward renewable energies) is an important aspect of these reduction measures, as energy consumption is responsible for the

J. Lange · M. Kaltschmitt (✉)
Hamburg University of Technology (TUHH), Institute of Environmental Technology and Energy Economics (IUE), Hamburg, Germany
e-mail: kaltschmitt@tuhh.de

© The Author(s), under exclusive license to Springer Nature Switzerland AG 2025
N. Bullerdiek et al. (eds.), *Powerfuels*, Green Energy and Technology,
https://doi.org/10.1007/978-3-031-62411-7_6

majority of anthropogenic GHG emissions. Electricity generation based on renewable energies, which will replace other conventional (i.e., fossil) energy sources, will be the foundation of a large share of the necessary energy system transformation. Therefore, the power sector and its adaptation toward an electricity generation from renewable sources of energy will be a major requirement for the transition of global energy systems to a sustainable and climate-neutral future [3]. However, while some energy demands will most likely be electrified in the course of this development, in certain—so-called harder-to-abate—sectors the energy demand will likely not be met directly with electricity (e.g., long haul aviation, intercontinental shipping). In addition to the use of biomass-based fuels, electricity-based synthetic fuels represent another option for supplying the corresponding sectors with energy in a climate-neutral and non-fossil way. These electricity-based fuels are also known as "powerfuels." In the past, electricity used to be the most valuable form of secondary or final energy because it is very efficiently convertible into any other form of usable energy. As the energy transition progresses, however, electricity is also increasingly becoming a feedstock for further processing (i.e., power-to-X processes) in areas of application in which aspects beyond efficient convertibility play a greater role (e.g., volumetric and gravimetric energy density in aviation).

There are various options for power generation based on renewable energies. Recent history has shown an unprecedentedly dramatic development in photovoltaics and wind energy, with electricity generation costs falling by around 60–80% per decade [4, 5]. They are now among the cheapest options for electricity generation globally. However, due to their dependency on the local time- and site-dependent characteristics of solar radiation and wind, they supply electricity in a fluctuating manner. This can be challenging for power consumers in need of a stable and continuous power supply. Therefore, photovoltaics and wind power may incur a need for additional system components (e.g., energy storage) in order to achieve similar overall supply characteristics as conventional power plants. Compared to electricity from solar and wind, other renewable power generation options show more controllable supply characteristics (e.g., dammed hydropower, geothermal power plants). However, these options are generally characterized by a strictly limited global energy potential, as promising, economically viable sites are becoming increasingly scarce or are located far away from existing infrastructures and demand centers.

Thus, the design of optimal electricity supply systems based on renewable energies will increasingly become a multifactorial challenge, considering global, regional, and local influences. Especially for the production of synthetic "powerfuels," the choice of location and type of power generation will determine the availability and cost of the electricity supply. Both of these aspects have a significant impact on the achievable production costs of synthetic "powerfuels," which in turn play a key role in the market ramp-up of such electricity-based liquid energy carriers.

In order to elucidate the context of renewable power generation further, the following elaborations provide an overview over different electricity supply options based on renewable energies. For this purpose, Sects. 2, 3 and 4 give insights into power generation from photovoltaics, wind energy, and hydropower. Here, the first two are the technologies that have recently shown by far the most dynamic growth, and the latter is currently still the most developed technology for electricity generation from renewable energies. The different subsections address the energy resource and the physical working principle including availabilities, the technical and economic development focusing on levelized cost of electricity as well as the current utilization and global technical potential of each power generation option. For each of these aspects, in addition to a discussion of the status, an outlook on possible future developments is provided. Based on this, Sect. 5 compares the discussed options for power generation and qualitatively derives challenges and solutions for the identification and design of optimal supply configurations with regard to demand-side requirements (e.g., regarding the production of synthetic "powerfuels"). Finally, Sect. 6 closes this elaboration with a conclusion.

2 Photovoltaics

Since the mid-2010s, photovoltaics has advanced to become the strongest-growing technology for power generation globally. This was based on and accompanied by a drastic decrease of specific system installation cost [5]. Overall, photovoltaics has thereby emerged as an important technology for the system transformation toward a more sustainable energy supply.

2.1 Energy Resource and Physical Working Principle

The following section gives an overview of the physical working principle and the energy resource of photovoltaic power generation. After an initial discussion of the current situation, potential changes of the energy resource (i.e., due to climate change) and general trends regarding availability will be elucidated.

2.1.1 Status

Photovoltaics is defined as the direct use of solar radiation for the generation of electricity (direct current). It is based on the so-called photoelectric effect—the creation of free electrons within a solid body by the absorption of light. To make technical use of this effect for generating electricity, the solid body must be characterized by an internal electric field that prevents the immediate recombination of free electrons with remaining positive charges. For this purpose, photovoltaic cells (solar cells) are made up of a positive (p) doped base semiconductor material and a thin negative (n) doped layer. Within the interface between these two layers, an electric field develops due to diffusion of charges in the p-n-transition [6]. When a photon (e.g., from solar irradiation) is absorbed by the doped semiconductor material of a solar cell, electrons are elevated from the valence to the conduction band, making them able to flow freely within the material. Due to the internal electric field, the charges (i.e., the electron and the remaining positive charge) separate. By electrically connecting the top and bottom of the solar cell, an electric current and, thus, electric power can be generated [6]. The overall process of using the photoelectric effect for power generation is called the photovoltaic effect.

The technical design of such solar or photovoltaic cells and solar modules generally aims at maximizing the electricity yield. A metallic contact is applied over the entire surface of the rear side of the solar cell and a grid-like contact system, optimized to minimize shading losses, is applied to the irradiated side. Alternatively, full-surface transparent conductive layers can be used. To minimize reflection losses, anti-reflective layers for the photon-richest region of the solar radiation spectrum (green–yellow–red) can be applied to the solar cell surface [6].

The amount of electricity that a solar cell or a module can generate directly depends on the intensity of solar radiation reaching its surface. The cell's current and voltage (and thus its power) show nonlinear dependencies on cell temperature as well as direct and indirect solar radiation, among other factors. Nevertheless, a simplified estimation of the maximum power output of a photovoltaic system can be given based on Eq. (1) [7].

$$P_{pv} = \frac{G}{G_{STC}} P_{pv,STC}\left[1 + \gamma\left(T - T_{STC}\right)\right] \tag{1}$$

Accordingly, the main external influence on the power output of a photovoltaic system P_{pv} with a rated power output at standard test conditions $P_{pv,STC}$ (STC, $G_{STC} = 1000\,W\,m^{-2}$, $T_{STC} = 25\,°C$ [7]) is the global irradiance G in relation to the irradiance at standard test conditions G_{STC}. While also the difference between the temperature T and the temperature at standard test conditions T_{STC} plays a non-negligible role (depending on the temperature correction factor γ), the electricity output of a photovoltaic system is primarily determined by the intensity of global irradiance.

Besides the diurnal changes due to earth's rotation (i.e., day and night), this intensity strongly depends on the latitudinal location together with the altitude of a specific photovoltaic system, as it defines the distance, light has to travel through the atmosphere before reaching the surface of the respective solar module. While passing the atmosphere, light is scattered, which generally reduces the amount of usable energy. Therefore, the degree of latitude of a location as well as the relative tilt of earth's axis toward the sun (and thus the season) play an important role in harvesting solar irradiation. The slight variations in the energy reaching the earth's outer atmosphere (i.e., the solar constant) throughout the year have only a minor influence [8, 9].

Consequentially, it follows that photovoltaic power generation shows a diurnal pattern as well as an annual variation throughout the seasons. During the day, electricity generation is generally strongest when the sun is highest and as perpendicular as possible above the installed solar modules. Over the course of a year, the solar power yield is highest in the summer months. In addition, stochastic fluctuations of irradiation due to different weather conditions (e.g., clouds) influence power generation. A measure of this necessarily limited availability is the capacity factor expressing the ratio between actual electricity generation in a given period and the theoretical maximum electricity generation at rated power over the same period. Figure 1 shows two examples of the diurnal (left) and annual (right) variation of the distribution of this capacity factor for photovoltaic systems installed at different locations for the years 2011 to 2020 calculated on an hourly basis. The upper part shows the capacity factors for a location in northern Central Europe (50.0° N, 10.5° E) and the lower for a location in North Africa (33.2° N, 7.3° E).

The diurnal patterns indicate strong similarities for both locations. However, the capacity factors for the location in North Africa (bottom) have a narrower distribution and reliably reach higher values. This is in part due to lower annual variations for the location in North Africa. Closer to the equator, seasonal differences are less pronounced than at more northerly or southerly latitudes, such as the location in northern Central Europe. Thus, photovoltaic power generation is also reasonably strong in the winter months, the closer to the equator a system is installed. Further away from the equator, the intensity of solar irradiation changes more and more considerably throughout the year, resulting in increasingly stronger differences in power output for summer and winter months. In addition, solar radiation is generally weaker further away from the northern and southern tropics (i.e., Tropic of Cancer and Tropic of Capricorn). This results in lower average capacity factors closer to the poles. For the location in northern Central Europe, the overall mean capacity

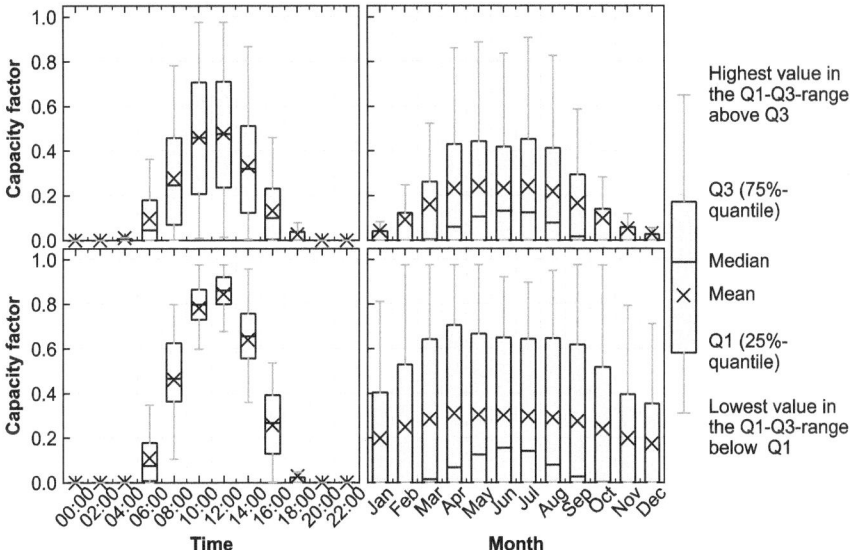

Fig. 1 Diurnal (left) and annual (right) distribution of the hourly capacity factor of solar photo-voltaics for a location in northern Central Europe (50.0° N, 10.5° E) (top) and in North Africa (33.2° N, 7.3° E) (bottom) for the years 2011 to 2020. Calculations are based on ERA5 data reanalysis [10]

factor amounts to roughly 15.1% while it is roughly 26.0% in the North African location. Some locations close to the equator with preferable weather conditions throughout the year (i.e., little cloudiness) even reach capacity factors close to 30% [5]. Today's global average capacity factor of new photovoltaic systems amounts to roughly 16.4% (three-year average) [5].

Thus, photovoltaic power generation shows strong variations over the day as well as throughout the year. Even though the values of these fluctuations may differ, the overall availability is generally limited regardless of the specific location.

2.1.2   Outlook

While new solar cell materials might play an increasingly important role in the future, the overall working principle of photovoltaics will expectedly not change in the years to come. Thus, fundamental changes in the availability of photovoltaic power generation are not to be expected, even though (partial load) efficiencies will likely still improve. Judging from recent development (i.e., almost stagnating capacity factors of new installations [5]), global average capacity factors will likely only increase slightly from their current value of around 16.4% [5]. However, locations with little solar irradiation might still reach capacity factors around 10% while exceptional locations will continue to achieve almost 30% [5].

Beyond these already visible trends, climate change could also have an impact on the availability of photovoltaics. Changing climate conditions could, e.g., affect the water content in the atmosphere, the formation of clouds, etc. and thereby influence the availability of the basic energy resource—i.e., the solar irradiation on the module's surface for subsequent electricity generation. Nevertheless, overall changes in photovoltaic resource quality are expected to be rather limited [11, 12]. With solar irradiation being the most influential factor for photovoltaic power generation, its potential changes are of special concern. On a global level, however, no corresponding change trends are currently discernible, with an increase in solar radiation in some regions and a decrease in others to be expected in the course of climate development [13]. Therefore, no generally increasing influence due to irradiation is currently discernible. Further, rising temperatures have a negative impact on photovoltaic efficiencies and, thus, power generation. A slight negative influence of climate development can be identified in this regard [12, 13]. Overall, the energy resource of photovoltaics is expected to be stable and provide similar, though possibly slightly lower, annual energy output [12, 13]. However, while annual electricity generation could be only slightly affected, seasonality could experience stronger changes. As with electricity generation, both increased and decreased regional seasonality is anticipated [14].

2.2 Technical and Economic Development

Over the last decade, photovoltaic power generation experienced a profound technical and economic development. This section aims at providing an overview of the recent progression, the current state as well as the potential for further development.

2.2.1 Status

Monocrystalline silicon modules make up more than 80% of photovoltaics market shares and are, thus, the most prominent cell-design; polycrystalline modules make up 15% and thin film modules 5% of the global production (values for 2021, [15]). Under standard test conditions (STC), these cell types are able to convert up to 26.7% (monocrystalline), 24.4% (polycrystalline), and roughly 22.4% (thin film) of the incident light into electricity. Over the last decade, these efficiencies improved by several percentage points [15]. For photovoltaic modules, efficiencies are generally around 2–4%-pts. lower [16].

Parallel to improving system efficiencies, the specific amount of material required for the production of photovoltaic cells and modules decreased continuously. While crystalline solar cells roughly needed 15 g of silicon per installed W in 2005, modern modules merely need around 2.5 g/W (in 2021, [15]). Due to these reduced material demands associated with energy-intensive production processes, the "invested" energy can be regained more than 20-fold over the systems technical lifetime depending on the location [15]. The future will most likely show additional reductions in material requirements while continuing to improve efficiencies. In addition, new material combinations for modules might arise, using cheaper and more abundant resources, while offering comparable energy yields [17].

Based on the recent global surge of photovoltaic installations, costs were drastically reduced in the past decade. The discussed improvements in material usage, mass production concepts, and generally optimized manufacturing have transformed photovoltaics from a rather expensive power supply technology to one of the cheapest options for power generation—and this is true not only among renewable energy technologies but also in comparison to fossil fuel-based power supply [5]. In 2022, global average levelized cost of electricity for newly installed photovoltaics amounted to 0.049 $_{2022}$/kWh and was, thus, roughly 80% lower than the cost in 2012, i.e., a decade earlier [5].

2.2.2 Outlook

Under the assumption of constant learning rates, capital expenditures will drop by additional 30 to 60% between 2020 and 2030 and by 50 to 75% between 2020 and 2050 [18]. Since no substantial changes in capacity factors are expected, this cost reduction would yield global average levelized cost of electricity of between roughly

0.026 and 0.045 $_{2022}$/kWh in 2030 and between around 0.016 and 0.033 $_{2022}$/kWh in 2050.

2.3   Utilization and Potential

Photovoltaic power generation recently saw a drastic increase in globally installed power capacities. The following section provides an overview of this (recent) development and elaborates on the remaining technical potentials for further expansion.

2.3.1   Status

Figure 2 shows the development of global photovoltaic capacities and electricity generation over the last decade. Total globally installed capacities recently (i.e., in 2022) surpassed the 1 TW-mark and global photovoltaic systems are now capable of supplying more than 1200 TWh of electricity annually [19].

Overall, photovoltaics grew tremendously throughout the last decade, increasing global capacities by a factor of more than seven. Hence, photovoltaic capacities grew on average by more than 22.6% annually [19]. At the same time, electricity

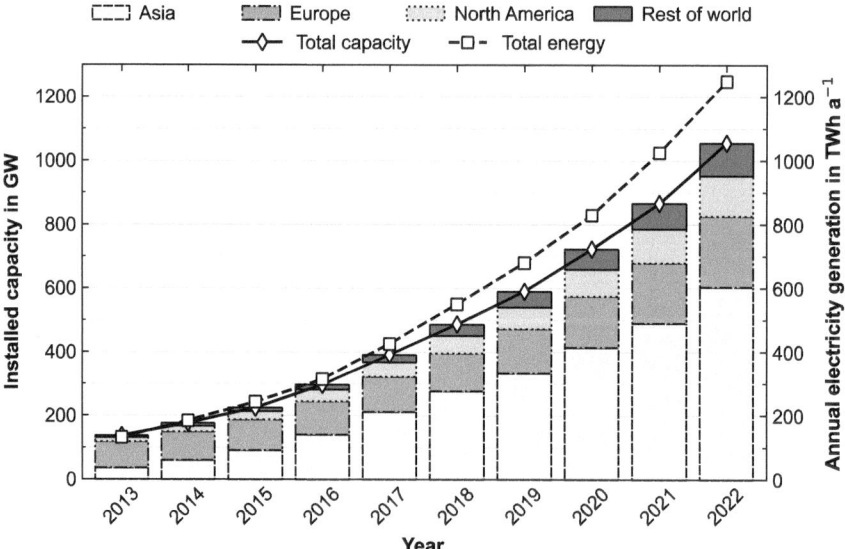

Fig. 2 Development of the installed capacity at year-end of photovoltaics in different regions (bars) and on a global level (solid line) as well as global annual electricity generation from photovoltaics (dashed line) [19]

generation grew by a factor of more than nine. The over-proportionate growth of electricity generation can be attributed to increasing installations in regions with higher solar irradiation. While the global north (e.g., Europe) financed the initial market uptake of photovoltaics, other regions adopted the technology once system costs fell to levels that were more or less competitive with the use of fossil fuels under the given local circumstances. Nowadays, photovoltaics is developed more commonly solely due to economic reasons as photovoltaic power generation often outperforms fossil fuel-based options, even without the inclusion of externalized environmental cost of the conventional power generation technologies.

2.3.2   Outlook

Judging from the recent development of photovoltaics, it can be expected that both global capacities and electricity generation from photovoltaic systems will strongly increase further in the coming years and decades. From a technical potential perspective, this growth trend will likely not face any limitations in the short to medium term.

With roughly 173,000 TW of continuous power, the amount of solar energy that reaches earth's surface per year exceeds the current annual primary energy demand as well as expected future primary energy demands by several orders of magnitude [20]. However, this energy is not fully technically usable. There have been numerous attempts to quantify the technical potential of electricity generation from solar energy resulting in different conclusions. According to these assessments, the quantified amount of technically realizable solar power generation ranges from between 1000 and 2000 EJ/a (280 to 560 PWh/a) [21–23] up to between 10,000 and 15,000 EJ/a (2800 to 4200 PWh/a) [24, 25]. Thus, even the most pessimistic assessment of the technical potential of electricity generation by solar radiation via photovoltaic systems alone leads to a value that clearly exceeds an estimated global primary energy demand of 750 EJ/a in 2050 [26]. In a more optimistic scenario, the global primary energy demand could be exceeded by a factor of 20. Moreover, the substitution potential for primary energy is likely even higher, as the photovoltaic potential is quantified as electrical energy, while the primary energy demand includes large shares of fossil fuels characterized by much lower conversion efficiencies for supplying the same amount of useful energy. Therefore, when comparing primary energy equivalents, the large surplus of technical solar potential becomes even more apparent.

Overall, compared to the magnitudes of expected global energy demand, solar energy is largely unlimited. Thus, human civilization could potentially operate solely based on the sun's energy available at the earth's surface. Nevertheless, since power generation based on solar irradiation shows strong fluctuations and is only available during daytime, a predominantly solar energy-based energy system would still be in need of various additional measures for dealing with the volatile availability (i.e., energy storage).

3 Wind Energy (On- and Offshore)

Like the use of solar irradiation, also the utilization of wind energy has recently shown a drastic increase in global energy market shares. These huge gains in market relevance are linked directly to technical and economic improvements (cost reductions) that have made the use of wind energy an increasingly promising option for a sustainable and affordable power generation at a multitude of locations and under strongly varying framework conditions.

3.1 Energy Resource and Physical Working Principle

In the following. the physical working principle and the energy resource of wind energy are discussed. The current situation as well as possible changes—both due to technology development and climate change—are elucidated.

3.1.1 Status

Wind forms due to pressure differences within the earth's atmosphere caused by differences in surface temperatures [6]. On a global scale, the latter result from varying solar inclinations and, thus, irradiation intensities over the latitude as well as the large-scale structures of land and water masses. Locally, they can be attributed to varying surface heat capacities (e.g., water, desert, forests, agricultural land, urban areas) that lead to temperature and air pressure differences also over shorter distances (e.g., coastal areas) as well as surface structure (e.g., mountain ranges) and roughness (e.g., cities, forests) [27]. Overall, these aspects result in an uneven distribution of the wind energy resource globally. Additionally, since solar radiation is the cause of wind, the wind energy resource exhibits similar temporal variations, i.e., characteristic diurnal and seasonal patterns—albeit with partially more stochastic properties.

Wind turbines allow for a technical use of the kinetic energy contained in moving air masses. Thus, the kinetic energy of the air within the atmosphere layer close to the earth's surface is the energy resource for on- and offshore wind power plants. The most common type of modern wind turbines uses the lift principle, in which the kinetic energy of the wind is converted into lift perpendicular to the direction of flow and consequently into rotary motion. This rotation is ultimately turned into electricity via a generator [6].

As the rotor extracts energy from the approaching moving air mass, the wind is slowed down and partially diverted past the rotor. Consequently, the energy extraction at the rotor is in a conflict between gaining energy from the wind and leaving enough energy for letting the slowed air pass. The optimal harvest factor (i.e., the optimized energy extraction considering these conflicting factors) is obtained for a relative wind speed ratio of 1/3 between the wind speeds behind and in front of the rotor and amounts to 16/27 [28]. However, this optimal ratio is usually not achievable in reality, so that the power coefficient c_P reaches values below the stated maximum. Furthermore, there are additional losses that define the overall system efficiency η_{sys}. Based on this, the power output of a wind turbine P_{wt} is given in Eq. (2) with the rotor area A_{rot} as well as the density ρ_{air} and velocity v_{air} of the wind.

$$P_{wt} = \frac{1}{2} c_P \, \eta_{sys} \, A_{rot} \, \rho_{air} \, v_{air}^3 \tag{2}$$

At higher wind speeds, the efficiency is of minor concern, since the output of the wind turbine is limited by the installed generator power. Thus, the efficiency is mostly relevant for partial load operation. A more important design parameter for increasing power output is the length of turbine blades, which defines the swept rotor area. The length of the rotor blades factors into the area to the second power and, thus, increases the power output of the wind turbine quadratically.

However, the most important factor for electricity generation is the wind velocity, as it influences the power output of a wind turbine to the third power. Therefore, the site selection (i.e., the wind resource) plays a crucial role, when it comes to maximizing the electricity generation of a wind farm. For this reason, the average wind power and mean capacity factors of wind turbines can differ strongly depending on the location.

Figure 2 shows two examples of the diurnal (left) and annual (right) distribution of capacity factors of wind turbines in different locations for the years 2011 to 2020. The upper part shows the capacity factors for a location in northern Central Europe (50.0° N, 10.5° E) and the lower part for a location in North Africa (33.2° N, 7.3° E) (Fig. 3).

It is apparent that the yearly average capacity factors differ clearly between the locations. It amounts to roughly 20.1% for the location in northern Central Europe and roughly 34.8% for the location in North Africa. Besides these overall differences, the locations depicted in Fig. 2 show strongly different diurnal and annual patterns. While wind is distributed almost equally over the day for the location in northern Central Europe (top), in the North African location it is much stronger during the night (bottom). Regarding annual variations, winter months in northern Central Europe show the strongest winds, while in the North African location, the winds are the strongest from April to June.

Therefore, the power output from wind turbines shows strong diurnal and annual variations, whose specific characteristics depend on the individual location. Furthermore, the annual amount of electricity generation depends strongly on the wind resource, which is highly site-dependent. On a global scale, these influences lead to average capacity factors for new onshore wind installations of roughly 37% (three-year average) being more than twice as high as for solar photovoltaics [5]. The average capacity factor for new offshore wind turbines in 2022 reached 42% and is, thus, even higher [5].

Even though wind energy generally has higher capacity factors than photovoltaics, it is also inherently limited, as the wind generally does not blow steadily throughout

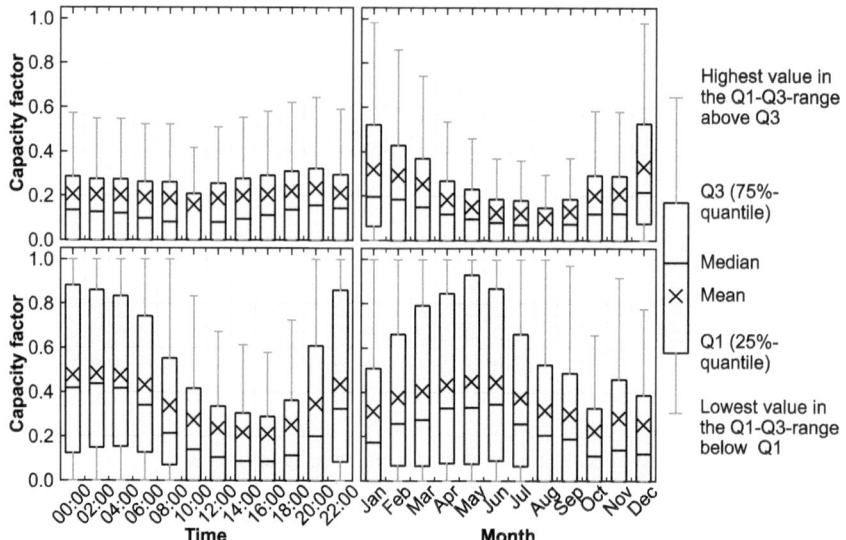

Fig. 3 Diurnal (left) and annual (right) distribution of the hourly capacity factor of wind turbines for a location in northern Central Europe (50.0° N, 10.5° E) (top) and in North Africa (33.2° N, 7.3° E) (bottom) for the years 2011 to 2020. Calculations are based on ERA5 data reanalysis [10]

the year. In addition, wind power can be challenging to integrate into power systems, as it can fluctuate strongly and rather stochastically. Thus, the design of a wind power-based energy supply configuration depends on a site-specific assessment of its potential and most likely on further system components (i.e., energy storage).

3.1.2 Outlook

Even though wind turbine designs that exploit the resistance principle (in contrast to the lift principle) are discussed every now and then, it is not to be expected that the mainstream working principle of exploiting wind energy will change in the near future. While new turbine designs using the lift principle (e.g., turbines with fewer blades) might be conceivable, the extraction of energy from the wind via aerodynamic lift will expectedly remain the dominant physical principle due to the higher overall conversion efficiencies.

Nevertheless, from today's perspective, a global development trend toward ever larger rotor areas and not necessarily proportionally larger generators (i.e., decrease in area-specific capacity) is expected, increasing global average capacity factors. For onshore wind, capacity factors of new turbines are expected to rise from a global average of 37% to 30 to 50% in 2030 and 32 to 58% in 2050. Regarding offshore wind, global average capacity factors are expected to increase to 36 to 58% in 2030 and 43 to 60% in 2050 [29].

Regarding the overall energy resource, no substantial changes are expected on the global scale due to climate change—locally or regionally, however, annual electricity generation from wind energy might increase or decrease by up to 10% [30]. While most regions are expected to see much smaller changes in annual power generation, overall there is a high degree of uncertainty regarding the increase or decrease in annual electricity output [13, 30]. Further, it is expected that seasonal patterns of wind energy availability are affected by changing wind patterns in the course of climate change [30]. Where seasonal differences are increased, this might complicate the integration of wind energy into energy systems. Beyond seasonal variation, supra-regional wind patterns might change; e.g., for Europe, it is expected that correlation lengths (i.e., the distance for which weather patterns stay similar) will increase by roughly 15% from current values of between 300 and 700 km [12]. This would reduce the potential for regionally balancing differing generation patterns and, thereby, increase wind energy integration cost [12].

However, the global wind energy resource is expected to be only slightly affected by climate change. Thus, wind power development will not face substantial changes of the energy resource. Nevertheless, it can be beneficial to include the potential for changing wind patterns on a local level into the planning of new wind power installations.

3.2 Technical and Economic Development

In the course of wind energy development, both technical and economic parameters have been improved continuously. This section provides an overview of the recent progress and current state of the techno-economic characteristics of wind energy and its further development potentials.

3.2.1 Status

With the most important wind turbine design factor for a high energy yield being the rotor diameter, recent development was strongly focused on increasing blade length and, thus, the swept rotor area [31]. Although this development was partly accompanied by a reduction of the rotor area-specific capacity, it also led to a strong increase of the electricity output per turbine. While wind turbines reached electrical output capacities in the two-digit kW-range in the beginning of commercial wind power utilization (in the early 1980s), modern offshore wind turbine models have electrical output capacities exceeding 10 MW, with rotor diameters above 200 m [29]. In addition to this development, hub heights increased to harness stronger and more consistent winds at greater heights above the earth's surface—this is especially important when the terrain has greater roughness and, thus, a stronger influence on the wind [29].

With the wind resource being the most important influence on annual electricity generation, a focus of wind power development lied on exploiting the most promising locations. In the wake of this development, offshore wind power recently showed substantial growth, as the unrestrained offshore winds generally promise higher wind speeds and an improved wind energy yield. Offshore wind development was the main driver for the increase of rotor and turbine sizes. Regarding the electrical power output, the largest dimensions of offshore wind turbines currently in development are in the range of 15 MW [32]. After initially being installed in shallow waters, offshore wind power is increasingly developed in higher depths. In the future, even floating offshore wind turbines for depths of up to 1000 m are expected to gain in importance and market shares [32]. If the current development trends regarding turbine dimensions continue, it is possible that wind turbines will exceed 20 MW of installed power in the coming years.

Market shares of wind energy increased strongly in the recent decades. Between 2000 and 2010, the average annual growth rate amounted to 26%/a; this growth reduced to roughly 11%/a between 2015 and 2020. Even though the relative growth declined, annual capacity additions currently exceed 50 GW/a. Thus, the globally installed wind power is still strongly growing [33]. Within this development, the share of offshore wind on total wind power installations increased but still plays a subordinate role in comparison to the global expansion of onshore wind [33]. In the context of increasing efforts to mitigate global climate change, wind power growth is expected to intensify further.

The development of wind energy has led and continues to lead to a sharp decline in system installation costs. Thereby, in 2022, global average levelized cost of electricity for new onshore wind installations at promising sites declined to 0.033 $_{2022}$/kWh, which is roughly 65% lower than a decade before [5]. For newly installed offshore wind systems, the weighted-average levelized cost of electricity amounted to 0.081 $_{2022}$/kWh in the year 2022. This represents a reduction by approx. 54% in comparison to 2012.

3.2.2   Outlook

Assuming constant learning rates, specific investment cost for onshore wind turbines will drop by another 10 to 30% between 2020 and 2030 and 15 to 40% between 2020 and 2050 [18]. For offshore wind utilization, cost might be reduced by 10 to 40% between 2020 to 2030 and around 20 to 50% between 2020 and 2050 [18]. When additionally considering a relative average increase in capacity factors of 15% and 25% by 2030 and 2050, respectively, global average levelized cost of electricity for new onshore wind turbines could fall to between 0.027 and 0.035 $_{2022}$/kWh until 2030 and to between 0.021 and 0.031 $_{2022}$/kWh by 2050. For offshore wind energy, levelized cost of electricity might fall to around 0.053 to 0.074 $_{2022}$/kWh until 2030 and to between 0.038 and 0.056 $_{2022}$/kWh by 2050.

3.3 Utilization and Potential

Throughout recent history, capacities for wind energy generation have been increased drastically. The following sections show both the recent development of capacities and the remaining technical potentials for further expansion.

3.3.1 Status

Figure 4 shows the installed capacities and annual electricity generation of wind energy in different regions as well as global cumulative values throughout the last decade. Globally installed capacities more than tripled over the depicted time span, resulting in a compound annual growth rate of more than 12%/a. While this is lower than for photovoltaic power generation, the global wind energy industry still saw an exceptional growth. An increasingly important driver of this development is the installation of offshore wind turbines gaining more and more importance in specific parts of the world. While total wind capacities grew by roughly 12%/a throughout the last decade, offshore wind capacities grew with a compound annual growth rate of more than 24%/a; global capacities of offshore wind turbines increased by a factor of almost nine throughout the last ten years [19].

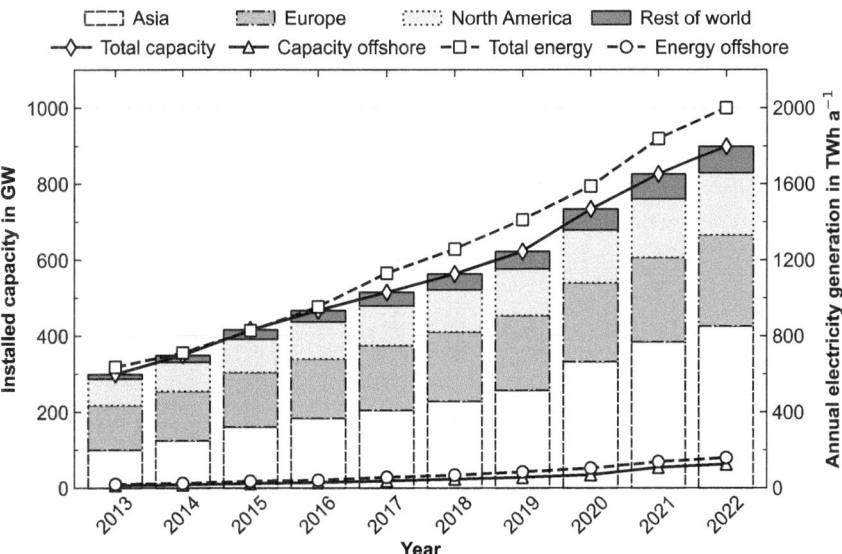

Fig. 4 Development of the installed capacity at year-end of wind energy in different regions (bars) and on a global level (solid line) as well as global annual electricity generation from wind energy (dashed line). Beyond total global capacities and energies (on- and offshore), capacities and energy are also plotted explicitly for offshore wind [19]

Since wind energy achieves much higher capacity factors than photovoltaics, the annual electricity generation from wind is much higher than that from photovoltaics, even though the installed electrical capacities of photovoltaic systems have recently surpassed those of wind energy on a global scale. Annual electricity generation from wind power recently surpassed 2 PWh/a [19].

Overall, wind power is, hence, already an important option for electricity generation in the energy industry. While its contribution to total global electricity generation still only amounts to roughly 8%, its growth by far outpaced the growth of global electricity consumption. Thus, wind energy both increases its absolute and relative relevance for electricity supply systems globally [34].

3.3.2   Outlook

With wind being caused by solar irradiation, the global amount of wind energy is smaller than the amount of the sun's energy reaching earth's surface. Nevertheless, wind power shows a large technical potential for the generation of electricity. This potential is estimated to lie between roughly 200 to 500 EJ/a (56 to 140 PWh/a) [21, 22, 24] and up to 3000 EJ/a (830 PWh/a) of electrical power [23]. Thus, in a rather pessimistic scenario, wind power could supply roughly 25% of the expected global primary energy demand in 2050 (only comparing electricity and primary energy—in reality, electric power would most likely substitute larger shares of primary energy due to higher conversion efficiencies). In a more optimistic case, wind power could provide four times the primary energy demand. In comparison to the amount of electricity currently generated by wind turbines on a global level (roughly 7.2 EJ/a [19]), this results in a possible increase of at least 2800%. Therefore, there are rather minor foreseeable limits to the overall expansion of wind power systems from a technical perspective. It is therefore to be expected that wind energy will make an important contribution to a more sustainable energy future and continue on its path of growth.

4   Hydropower

With regard to power generation based on renewable energies, hydropower is the most established technology in global energy systems. In the past, the development of hydropower plants was determined mainly by their economic performance, as they were by far the cheapest way to generate electricity at some locations. Nowadays, their potential to supply energy without operational GHG emissions is considered another advantage over conventional power plants.

4.1 Energy Resource and Physical Working Principle

The following sections explain the physical working principle and energy resource of hydropower and show how the availability of this energy resource might develop in the coming years on a global scale.

4.1.1 Status

Hydropower essentially uses the potential and kinetic energy of water to drive a turbine and, conclusively, a generator. Therefore, the amount of potential and kinetic energy of a certain volume of water directly determines the amount of electricity that can be generated. The power output of a hydropower plant P_{hy} depends on the mass flow rate $\dot{m}$ of water (e.g., in a river), the usable geographical difference in height Δh, the overall system efficiency η_{sys} as well as the acceleration due to gravity g, as shown in Eq. (3) [6].

$$P_{hy} = \eta_{sys}\dot{m}g\,\Delta h \qquad (3)$$

There are different types of hydropower plants. Dammed hydrofacilities store the water of a river in a reservoir with the help of a dam. Such dams allow for a more demand-oriented power generation; i.e., the stored amounts of water can be utilized whenever electricity is needed. However, the construction of these reservoirs usually leads to the flooding of land areas and the possible destruction of wildlife habitats. In addition, the use of the stored water leads to strongly fluctuating water levels, potentially harming the local (aquatic) flora and fauna.

So-called run-of-river plants, on the other hand, primarily use the energy of flowing water in a river without implementing a considerable water reservoir. Therefore, they are less able to provide electricity on demand since their power output depends on the currently available flow of water (i.e., runoff). This runoff might vary strongly throughout the year, e.g., due to seasonal patterns of rainfall or meltwater. Therefore, electricity generation from run-of-river power plants generally exhibits greater variability and less controllability compared to that of dammed hydropower facilities. Overall, while a dam and the resulting water reservoir enable and improve the control over the power supply, they usually cause more negative impacts on the environment (e.g., destruction of up- and downstream habitats, flooding of fertile land) than run-of-river power plants [6].

Depending on the local conditions and the technical plant type (i.e., whether a dam and a reservoir are parts of the plant), hydropower can reach high capacity factors of roughly 80%. Between 2010 and 2022, average capacity factors of new hydropower plants amounted to roughly 45%, with some plants reaching 23% while others achieved 80% and more [5]. Low capacity factors can, however, also be the consequence of power plant design decisions that aim at providing peak power as opposed to base load [5]. For pumped hydropower plants, capacity factors play a minor role but generally lie at considerably lower values [35].

4.1.2   Outlook

Hydropower is a long-established technology for electricity generation. Thus, no substantial changes in physical working principles are to be expected, even though the plant design might develop in the direction of smaller systems.

Like for the other options for electricity generation based on renewable energies, the development of the energy resource due to climate change (e.g., changing precipitation patterns) is still a factor to consider in the assessment of the future development of hydropower on a global scale. However, it is not currently expected that global hydropower potentials will be strongly influenced by climate change [34], with some estimations suggesting a slight net increase of global hydropower electricity output [11]. Regionally, however, generation potentials can be influenced quite significantly. Some regions could face a decline of hydropower generation by more than 20%, while others might experience an increase in the same order of magnitude. In many regions close to the equator, it is expected that the overall precipitation and, consequentially, electricity generation from hydropower will decrease, while it will most likely increase closer to the poles [12, 30]. In addition, seasonality could be affected, as regional snowfall and the associated meltwater could decrease, even if annual precipitation levels do not change significantly [12, 30].

Overall, like for photovoltaics and wind power, the energy resource for hydropower might not change strongly on a global scale but will still be affected locally. Thus, when installing new hydropower plants, it could be advisable to also consider the potential for changing local precipitation patterns. Nevertheless, on a global scale, it can be expected that average capacity factors stay in the order of magnitude of today's hydropower plants with possible minor reductions due to an increasing exploitation of less advantageous sites.

4.2   Technical and Economic Development

With hydropower being the longest established technology for electricity generation based on renewable energies, it is highly mature both technically and economically. Against this background, the following sections briefly summarize the current techno-economic characteristics of hydropower and elaborate on how they might change in the future.

### 4.2.1	Status

In 2022, hydropower was by far the most important option for supplying electricity based on renewable energies, despite the recent growth of photovoltaics and wind power. With hydropower being a mature technology that has been substantially optimized during the past decades, the system design will most likely not change considerably in the future. Nevertheless, for the coming years, a steady expansion of additional hydropower capacities is expected globally (expected growth of approx. 17% or 230 GW between 2021 and 2030) [36].

Having been built more than 40 years ago, many of the existing hydropower plants in operation are already quite old. Therefore, the future expansion of hydropower could be accompanied by a modernization of existing plants to a greater extent. This could be combined with an increase of installed capacities (and deliberately lower capacity factors) in order to enable an improved complementation of volatile power generation from photovoltaics and wind power [36]. Further hydropower development might also be possible in small hydropower plants (i.e., plants with an installed power of up to 10 MW), which recently showed noticeable annual growth rates (exceeding 3%/a) [37].

As favorable sites are increasingly being exploited, global average costs for electricity generation from hydropower plants have risen recently. Thereby, the weighted-average levelized cost of electricity increased from 0.041 $_{2022}$/kWh in 2012 to roughly 0.061 $_{2022}$/kWh in 2022 [5].

### 4.2.2	Outlook

Due to the advanced technological maturity, hydropower plants already reach extremely low levelized cost of electricity at favorable locations—even in comparison with conventional options for power generation. At the same time, there are little to no opportunities for a considerable further reduction in electricity generation costs. It is even likely, that these costs might increase for the ongoing development as new installation sites become more challenging and, thus, more cost intensive, since power plants are already in operation at locations that are rather "easy to access." Therefore, it has to be expected, that levelized cost from the beginning of the 2020s (0.052 $_{2022}$/kWh; three-year average) will likely not be undercut in the near future [5]. For the development of small hydropower plants, one generally has to expect even higher cost [37]. Overall, global average levelized cost of electricity for hydropower might, therefore, slightly increase in the future but still stay in a similar order of magnitude.

4.3 Utilization and Potential

Hydropower is a long developed energy technology. Recent progress and potentials for further development are briefly discussed in the following.

4.3.1 Status

Figure 5 shows the development of globally installed capacities for electricity generation from hydropower and annual electricity output of the existing hydropower plants in operation. In contrast to the "younger" renewable energies (i.e., photovoltaics and wind power), hydropower has not experienced an above-average growth recently. With compound annual growth rates of installed capacities at around 2%/a, the global hydropower capacities increased in the same order of magnitude as the demand for electrical energy [19, 34]. As the global average capacity factors decreased marginally, electricity generation from hydropower grew even slightly slower than the global electricity consumption [19]. Therefore, despite absolute growth in electricity generation, hydropower has a rather constant (to slightly decreasing) relevance for covering the global demand for electrical energy.

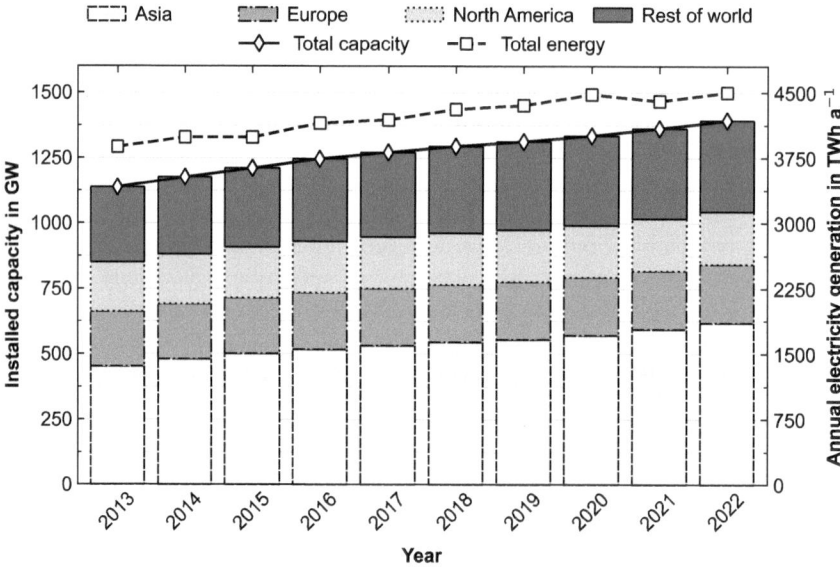

Fig. 5 Development of the installed capacity at year-end of hydropower in different regions (bars) and on a global level (solid line) as well as global annual electricity generation from hydropower (dashed line) [19]

4.3.2   Outlook

The technical potential for electricity generation based on hydropower has been narrowed down to around 50 EJ/a (13.9 PWh/a) by various studies [21, 22, 24, 25]. Upper and lower boundaries of the technical potentials are quantified to range from roughly 20 EJ/a to 110 EJ/a (5.6 PWh/a to 30.6 PWh/a) [38]. A share of 13 EJ/a of this potential is already utilized [23]. The largest untapped potential currently lies in Asia and amounts to roughly 22 EJ/a. In Europe, already more than 50% of the hydropower potential is exploited, leaving approximately 3.6 EJ/a of electricity generation for further expansion [23]. However, this potential is likely to be further reduced if environmental and sustainability criteria (i.e., nature conservation, landscape protection, biodiversity conservation) are taken into account. Overall, the global technical potential for electricity generation from hydropower is, thus, much more foreseeably limited than the technical potential of photovoltaics and wind power. Further, it is much lower than the expected primary energy demand in 2050, while suitable and economic sites for the installation of hydropower plants will increasingly become a scarce resource. Nevertheless, due to its ability to more controllably supply electricity, (dammed) hydropower will most likely still play an important role within sustainable power supply systems in the years to come.

5   Energy Supply from Renewable Energies

This section qualitatively discusses the techno-economic fundamentals for designing energy supply solutions based on renewable energies. To this end, Sect. 5.1 summarizes the key techno-economic criteria of the discussed power generation options highlighting their advantages and disadvantages for the design of supply configurations for energy-intensive (base load) processes (e.g., synthesis of "powerfuels"). Based on this compilation, various approaches to optimizing the supply side for specific applications are discussed in Sect. 5.2.

5.1   Comparison of Supply Options

The different options for electricity supply based on renewable energies exhibit distinctly different characteristics regarding availability, techno-economic performance and potentials, which are summarized below.

Power generation based on photovoltaics experienced an unprecedented development over the last two decades. In the course of this progress, photovoltaics has advanced to become one of the cheapest ways to generate electricity—even in comparison to conventional and, thus, far developed power generation technologies. Beyond that, the use of photovoltaics shows promising potentials for further technological and economic development. The techno-economic characteristics and the technical potential can be summarized as follows.

- Low generation costs with expected strong further reduction. Global average levelized cost of electricity generation currently (i.e., values for 2022) amount to roughly 0.049 $_{2022}$/kWh. Further reductions to around 0.036 $_{2022}$/kWh in 2030 and ca. 0.024 $_{2022}$/kWh by 2050 seem possible and likely.
- Lowest availability without expected improvements. Photovoltaic power generation shows strong variability and an inherently limited availability of the supplied electrical energy over time, reaching the lowest capacity factors of all discussed options for power generation based on renewable energies. Capacity factors of new installations amount to roughly 16.4%, reaching around 10% in less favorable (but potentially still economic) sites and almost 30% at very favorable locations. Values beyond 30% will hardly ever be achievable due to the lack of sunlight at night times. Thus, a stable and steady energy supply from photovoltaics is always in need of additional measures (e.g., energy storage).
- Largest technical potentials. Photovoltaic power generation can still draw on an enormous technical potential globally. Estimations for power generation range from around 280 PWh/a to 4 200 PWh/a, which is up to 20-fold the expected primary energy demand for the year 2050. Thus, due to its low cost and vast potentials, it can strongly contribute to an increasingly sustainable energy supply despite its limited power availability.

Wind power exhibits some parallels to photovoltaics. Despite its technological development being less drastic, wind power still experienced a strong progress during recent decades, becoming a highly competitive option for power generation. Additionally, a continued development of wind turbine technology and economics can be expected. The techno-economic characteristics and the technical potential can be summarized as follows.

- Low generation costs with expected further reduction. Average levelized cost of electricity generation amount to 0.033 $_{2022}$/kWh (onshore) and 0.081 $_{2022}$/kWh (offshore) (values for 2022). They will likely decline further in the years to come, potentially reaching 0.031 $_{2022}$/kWh (onshore) and 0.064 $_{2022}$/kWh (offshore) in 2030 and 0.026 $_{2022}$/kWh (onshore) and 0.049 $_{2022}$/kWh (offshore) in 2050.
- Moderate availability with expected improvement. Like photovoltaics, wind power generation shows strong (partially stochastic) fluctuations, currently reaching average capacity factors of 37% (onshore) and 42% (offshore) for new installations. However, turbine development trends will likely further increase these factors (e.g., through lower rotor area-specific capacities) to 30 to 50% (onshore) and 36 to 58% (offshore) by 2030 and 32 to 58% (onshore) and 43 to 60% (offshore) in 2050. Thus, wind power will show ever-higher availabilities, reducing the need for backup power generation for a stable and steady sustainable energy supply; some form of complementary (dispatchable) generation (or energy storage) will, however, still be needed to ensure a secure (base load) power supply.

- Large technical potentials. Global technical potentials for wind power generation are lower than for photovoltaics but still substantial. The exploitable electricity generation potentials are estimated to amount to between 56 and 443 PWh/a. Compared to the primary energy demands expected for the year 2050, wind power could realize a huge substitution of fossil fuels (even with pessimistic expectations).

Hydropower is a well-established power generation technology. Sites that promise economic exploitation are likely to be developed further, which will be the main driver for additional hydropower expansion in the future. The techno-economic characteristics and the technical potential can be summarized as follows.

- Low power generation costs with expected slight increase. Hydropower achieves relatively low average levelized cost of electricity of roughly 0.061 $_{2022}$/kWh (value for 2022). Since all system components are already highly developed and fully market mature, costs will likely not decrease further. Even slight increases of levelized costs are expectable, as further hydropower expansion has to focus on less favorable locations.
- Moderate to high availability without expected improvement. Hydropower allows moderate to high availability, reaching global average capacity factors of 45%. Due to strong influences of location and plant design (i.e., turbine rating), capacity factors of newly installed hydropower plants range from roughly 23% to 80%. No further improvements are expected regarding availability. However, hydropower generation is generally (at least partially) controllable, making it a potentially valuable complement to fluctuating renewables.
- Strongly limited technical potentials. With roughly 14 PWh/a of power generation as technical potential, the global development of hydropower is strongly limited. For the continued expansion, promising locations will become increasingly scarce so that less favorable sites will need to be exploited. Therefore, the overall expansion of hydropower is foreseeably limited.

The following figures compare the different characteristics of photovoltaics, wind energy and hydropower regarding capacity factors (Fig. 6), levelized cost of electricity (Fig. 7) as well as utilization and technical potential (Fig. 8).

As Fig. 6 shows, wind energy is the only technology with expected substantial development of capacity factors due to general design paradigms (reduced area-specific capacities) that lead to slightly higher utilization rates of the generator. While this development would lead to average capacity rates beyond current hydropower availabilities, favorable locations of (dammed) hydropower still show the highest capacity factors of the considered options for electricity generation based on renewable energies. Photovoltaics will stay the technology with the lowest capacity factors, as its availability is inherently limited.

Figure 7 shows current global average values of levelized cost of electricity for newly installed photovoltaic systems, wind turbines (on- and offshore), and hydropower plants. Photovoltaics is the technology with the largest potential for cost reduction; it will likely reach the lowest average levelized cost globally in the near

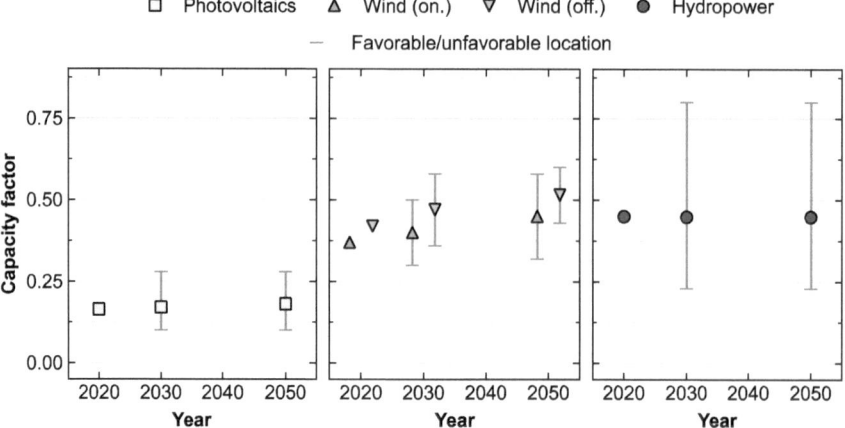

Fig. 6 Current and expected average capacity factors as well as ranges between minimum and maximum values of newly installed photovoltaics, wind energy onshore and offshore as well as hydropower plants ([5, 19, 29] and own assumptions)

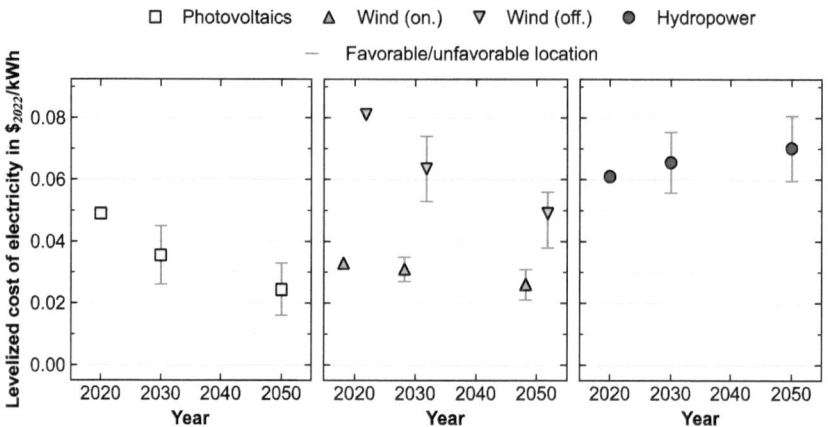

Fig. 7 Current and expected average levelized cost of electricity as well as ranges between minimum and maximum values of newly installed photovoltaics, wind energy onshore and offshore as well as hydropower plants ([5, 18] and own assumptions)

future. Currently, average levelized cost of electricity for photovoltaics are strongly influenced by less cost efficient small-scale systems (e.g., solar energy on residential rooftops). Onshore wind—exclusively built at utility scale—on average currently reaches lower levelized cost. However, the potential for further reductions seems to be slightly more limited. Nevertheless, strong cost reductions are expected also for offshore wind.

Against the background of recently increasing levelized cost of electricity for hydropower, no cost reductions are expected for this technology. Thus, the former

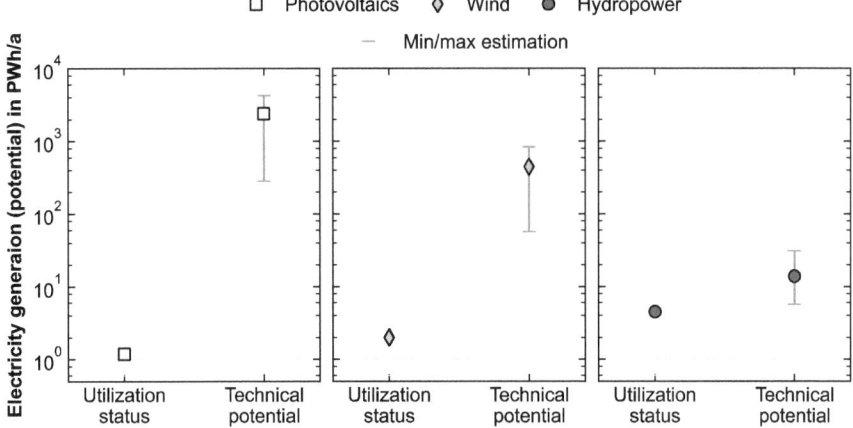

Fig. 8 Realized utilization and total technical potentials (including ranges between minimum and maximum values) for photovoltaics, wind energy (sum of onshore and offshore) as well as hydropower plants [19, 21–25, 38]

cheapest option for electricity generation based on renewables will foreseeably become a more expensive one. However, due to its controllability, it will likely still be an important option for sustainable, resilient, and reliable energy systems based on renewable energies. Further, all discussed technologies for electricity generation based on renewable energies generally achieve levelized generation costs below the cost levels of fossil fuels—especially when considering cost or penalties for CO_2 emissions [5].

Figure 8 finally summarizes current utilization levels and technical potentials of photovoltaics, wind energy, and hydropower. While hydropower is visibly still the most developed technology, technical potentials of photovoltaics and wind power are several orders of magnitude higher than those of hydropower. Future energy systems will, therefore, rely increasingly on photovoltaics and wind energy and inevitably less on hydropower.

This compilation of availabilities, techno-economic data, and utilization potentials emphasizes the strong differences that technologies for electricity supply from renewable energies exhibit. While photovoltaics and wind power show vast potentials for further global development, their operation is strongly dependent on the specific location and prevailing (local) weather conditions, leading to limited and uncontrollable availabilities. Even though (dammed) hydropower can be more tightly controlled regarding its power output, its potentials are strictly limited, preventing hydropower-based energy systems on a global scale. Further, while wind and solar energy will likely see ongoing strong reductions in levelized costs of electricity, generation costs are more likely to increase for hydropower.

Overall, these most prominent examples of electricity generation from renewable energies symbolically illustrate the trade-off between different system characteristics that is associated with the use of renewable energies. Other options for supplying electricity from renewable energies—such as geothermal energy and concentrated solar power—also exhibit individual availability characteristics, levelized cost of energy, and technical potentials. Regionally, these options can be promising technologies for electricity generation from renewable energies (i.e., locations with corresponding technology-specific potential), so that they are also being expanded worldwide, albeit at growth rates that are far below those of the renewable energies under discussion [5]. Nevertheless, it can be summarized that the challenge of creating a sustainable, reliable (i.e., available) and cost-effective energy supply is becoming increasingly complex in the course of the transformation toward a defossilized energy system. There are always various supply options that offer different advantages and disadvantages. Overall, the design of energy supply configurations, thereby, increasingly demands more holistic approaches considering not only different options for generating electricity but also incorporating the demand side and energy storage.

5.2   Optimal Supply Configurations

The optimization of energy supply configurations increasingly becomes a multifactorial problem. For the production of synthetic "powerfuels" based on renewable electricity, this plays an especially important role. As the costs of these fuels are largely determined by the costs of the input energy, the power generation costs of the overall supply configuration are of paramount importance. However, processes like the production of synthetic "powerfuels" are only partially flexible. While, e.g., the generation of hydrogen can be flexibilized (using hydrogen storage), synthesis processes are generally only controllable in narrow ranges and typically need a stable energy supply throughout the year. Thus, the optimal supply solution for such systems is not simply based on the cheapest power supply option, but rather the supply configuration with the overall lowest supply cost (i.e., the lowest total cost, that still allow a sufficiently demand-oriented supply of all process components). Against this background, different aspects that can be important for optimized supply configurations are briefly discussed below.

- If an option for controllable power generation from renewable energies with high capacity factors is available locally (e.g., hydropower at favorable locations), it might be the "easiest" option for decarbonizing power generation and, e.g., controllably supplying a processing plant. Due to the potential omission of energy storage and an optimized dimensioning of demand-side system components, the total cost of the processing plant including the supply system might be the lowest even at higher cost for electricity generation.
- When volatile generation is the only locally available option for a power supply based on renewable energies, a partial flexibilization of the demand side can be beneficial. A more flexible demand can more easily adapt to fluctuations on the supply side and, thus, ease the integration of volatile electricity generation. This enables the process management to operate with higher shares of

fluctuating renewable energies. However, for the same annual output, this leads to larger necessary demand-side system dimensions (e.g., higher installed electrolyzer capacities) to react more optimally to an increased or reduced availability of volatile power generation.

- Similarly, the supply side can be over-dimensioned in order to increase overall process utilization (e.g., installing a photovoltaic system with rated capacities strongly exceeding maximum demand-side load). As far as potential losses of surplus electricity result in less economic drawbacks than insufficient utilization rates of different process components, over-dimensioning is a cost-effective way of increasing the process capacity factor, without changing the capacity factor of the supply side. However, increasing process utilization is limited to times where power production from the particular technology for electricity generation is possible at all and, further, is accompanied by increasing total supply cost.
- If a certain process has only limited potential for flexibilization or if flexibilization is highly expensive, energy storage can offer a solution for the power supply of inflexible system components. Energy storage can help to integrate surplus energy when supply exceeds demand and provide electricity during times of limited availability from volatile renewables. The installation of storage facilities, however, necessarily increases total system cost (both due to the required energy storage installations and due to a potentially increased demand for input energy, as storage losses increase overall energy consumption).
- To reduce the necessity to flexibilize load or to implement storage options, different power supply technologies can be combined for an increase of overall supply-side availability. Provided the location allows the installation of a controllable renewable power source (e.g., hydropower), the combination with a potentially cheaper (volatile) supply option might be economically viable. The controllable power source then provides electricity complementary to the volatile supply. Such a design of the supply system is strongly cost-driven, as a combination only makes sense, if it provides benefits over a power supply solely based on the controllable option.
- If the specific location does not allow the installation of an option for controllably supplying electricity from renewable energies, it might be beneficial to combine systems using different fluctuating sources (i.e., photovoltaics and wind power). As solar irradiation and wind speed are strongly linked, the fluctuations of both technologies can complement each other. For example, there are locations with pronounced power generation from photovoltaics during the day and in the summer and stronger power supply from wind during the night and in the winter. At such locations, the combination of both options can strongly increase the overall availability of volatile power generation, which at least reduces the need for additional energy storage and/or demand flexibilization.

In sum, the design of power supply systems is strictly a question of minimizing total cost of the desired output of a specific application (e.g., "powerfuels"). As renewable energies are strongly site-dependent, similar supply tasks might be accomplished completely differently for various locations, depending on the prevailing framework conditions and overall performance potentials for different technologies as well as the cost of process flexibilization and energy storage.

6 Conclusion

The need to switch from conventional energy supply systems based on fossil fuels to the widespread utilization of (fluctuating) renewable energies is unavoidable from a climate protection perspective. The recent global development of energy supply systems is beginning to reflect this reality, as electricity generation from renewable energies is increasing at disproportionately high rates. Moreover, these trends are expected to intensify in the coming years.

However, electricity from renewable energies will likely not be able to defossilize all areas of the energy system. For the sectors for which direct electrification is not an option (i.e., long haul aviation), synthetic electricity-based fuels—so-called powerfuels—represent an alternative solution for the transformation of the energy system. In order to achieve this cost-effectively, supply configurations based on renewable energies must be identified that enable the production of "powerfuels" at the lowest possible cost.

There are various power generation options that can be part of such supply configurations, of which three (i.e., photovoltaics, wind power, and hydropower) are discussed here in more detail. These options have fundamentally different supply characteristics than fossil fuel-based conventional power plants, and their operating characteristics depend heavily on the site conditions. Newly installed photovoltaic systems, for example, have a very limited availability with average capacity factors of around 16% (i.e., roughly 1400 full load hours) and supply electricity in a volatile manner. While modern wind power plants achieve higher average global capacity factors of around 37% (onshore) and 42% (offshore), which are likely to increase further in the future, their electricity generation is also characterized by strong, partly stochastic fluctuations. The electricity generation from (dammed) hydropower plants, on the other hand, can be controlled much more directly. Although these plants do not necessarily achieve higher average annual availabilities, they enable complementary operation to fluctuating renewable energies. However, hydropower can draw on significantly lower remaining technical potentials for further expansion globally. At the same time, suitable economically viable locations are an increasingly scarce resource, which has recently led to an increase in hydropower generation costs. Photovoltaics and wind power, on the other hand, show falling levelized costs of electricity and (virtually) unlimited expansion potentials.

Consequently, there is no universally optimal generation technology for supplying "powerfuel" processes. Rather, the intelligent combination of different generation technologies in combination with energy storage and potential demand flexibilization must be used to find the best possible individual configuration that enables minimal generation costs on the product side. Thus, since electricity generation will become more dependent on the local circumstances during the transition from a fossil fuel-based power supply to more sustainable options based on renewable energies, the identification of suitable supply configurations will rely increasingly on a more integrated evaluation and assessment of available options and potentials.

While the definition of power supply configurations was focused almost exclusively on the identification of the cheapest generation technology in the past, the most successful system designs of the future will likely be those that strongly incorporate site-specific influences as well as multifactorial aspects of the overall supply task (i.e., supply, storage, demand). Thus, the most competitive production facilities (e.g., for the synthesis of "powerfuels") will likely need a pronounced expertise regarding the integrated evaluation of the overall energy supply system, reducing production cost while adequately addressing climate protection requirements.

References

1. World Meteorological Organization (WMO) State of the global climate 2022, Genf, 2023
2. Vereinte Nationen (2015) Übereinkommen von Paris - Sammlung völkerrechtlicher Verträge, Paris
3. Lange J, Hilgedieck J, Kaltschmitt M (2019) Renewable resources of energy for electricity generation—development trends and necessities within the overall energy system. Jordan J Mech Indust Eng 13(3):207–220
4. NREL (2021) Documenting a decade of cost declines for PV systems—NREL marks ongoing cost reductions for installed photovoltaic systems, while also establishing benchmark of PV-plus-storage systems. https://www.nrel.gov/news/program/2021/documenting-a-decade-of-cost-declines-for-pv-systems.html, 2021. Zuletzt geprüft: 04.04.2022
5. IRENA (2023) Renewable power generation costs in 2022, Abu Dhabi
6. Kaltschmitt M, Streicher W, Wiese A (2020) Erneuerbare Energien. Springer, Berlin Heidelberg
7. Rosell JI, Ibáñez M (2006) Modelling power output in photovoltaic modules for outdoor operating conditions. Energy Convers Manage 47(15–16):2424–2430
8. Häckel H (2016) Meteorologie, 8th edn, UTB Geowissenschaften, Ökologie, Agrar- und Forstwissenschaften, vol 1338, Verlag Eugen Ulmer, Stuttgart
9. Malberg H (2007) Meteorologie und Klimatologie - Eine Einführung ; mit 56 Tabellen, 5th edn. Springer-Lehrbuch, Springer, Berlin, Heidelberg
10. Copernicus Climate Change Service (C3S): ERA5: Fifth generation of ECMWF atmospheric reanalyses of the global climate—Copernicus Climate Change Service Climate Data Store (CDS), https://cds.climate.copernicus.eu/cdsapp#!/home. Zuletzt geprüft: 20.04.2022
11. Gernaat DEHJ, de Boer HS, Daioglou V, Yalew SG, Müller C, van Vuuren DP (2021) Climate change impacts on renewable energy supply. Nat Clim Chang 11(2):119–125
12. Schlott M, Kies A, Brown T, Schramm S, Greiner M (2018) The impact of climate change on a cost-optimal highly renewable European electricity network. Appl Energy 230:1645–1659
13. Narvaez G, Giraldo LF, Bressan M, Pantoja A (2022) The impact of climate change on photovoltaic power potential in Southwestern Colombia. Heliyon 8(10):e11122
14. Hou X, Wild M, Folini D, Kazadzis S, Wohland J (2021) Climate change impacts on solar power generation and its spatial variability in Europe based on CMIP6. Earth Syst Dynmics 12(4):1099–1113
15. Fraunhofer ISE (2023) Photovoltaics report. Freiburg
16. Green M, Dunlop E, Hohl-Ebinger J, Yoshita M, Kopidakis N, Hao X (2021) Solar cell efficiency tables (version 57). Prog Photovoltaics Res Appl 29(1):3–15
17. IRENA (2019) Future of solar photovoltaic—deployment, investment, technology, grid integration and socio-economic aspects (A Global Energy Transformation: paper), Abu Dhabi, 2019

18. Sens L, Neuling U, Kaltschmitt M (2022) Capital expenditure and levelized cost of electricity of photovoltaic plants and wind turbines—development by 2050. Renew Energy 185:525–537
19. IRENA (2023) Renewable energy statistics 2023. International Renewable Energy Agency, Abu Dhabi
20. Department of Energy: Top 6 things you didn't know about solar energy
21. Hoogwijk M, Graus W, Global potential of renewable energy sources—a literature assessment: background report
22. Krewitt W, Nienhaus K, Kleßmann C, Capone C, Stricker E, Grauss W, Hoogwijk M, Supersberger N, Winterfeld UV, Samadi S (2009) Role and potential of renewable energy and energy efficiency for global energy supply, climate change
23. IPCC (2011) Renewable energy sources and climate change mitigation—special report of the Intergovernmental Panel on Climate Change, Cambridge University Press, Cambridge
24. de Vries BJ, van Vuuren DP, Hoogwijk MM (2007) Renewable energy sources: their global potential for the first-half of the 21st century at a global level: an integrated approach. Energy Policy 35(4):2590–2610
25. Jacobson MZ, Delucchi MA (2011) Providing all global energy with wind, water, and solar power, part i: technologies, energy resources, quantities and areas of infrastructure, and materials. Energy Policy 39(3):1154–1169
26. bp (2022) Energy outlook 2022
27. Liljequist GH, Cehak K (2006) Allgemeine Meteorologie, 5th edn. Springer, Berlin, Heidelberg
28. Betz A (1920) Das Maximum der theoretisch möglichen Ausnützung des Windes durch Windmotoren, Zeitschrift für das gesamte Turbinenwesen
29. IRENA (2019) Future of wind—deployment, investment, technology, grid integration and socio-economic aspects
30. Solaun K, Cerdá E (2019) Climate change impacts on renewable energy generation. A review of quantitative projections. Renew Sustain Energy Rev 116:109415
31. IEA-ETSAP and IRENA (2016) Wind power—technology brief
32. U. S. Department of Energy (2021) Offshore Wind Market Report: 2021 Edition
33. Global Wind Energy Council (GWEC) (2021) Global wind report 2021, Brussels
34. Enerdata (2023) World energy & climate statistics—yearbook 2023. https://yearbook.enerdata.net/electricity/electricity-domestic-consumption-data.html, 2023. Zuletzt geprüft: 09.10.2023
35. International Energy Agency (2020) World energy statistics and balances 2020—(database)
36. International Energy Agency (2021) Hydropower special market report—analysis and forecast to 2030
37. Liu D, Liu H, Wang X, Kremere E (2019) World small hydropower development report 2019—global overview
38. Gernaat DEHJ, Bogaart PW, van Vuuren DP, Biemans H, Niessink R (2017) High-resolution assessment of global technical and economic hydropower potential. Nat Energy 2(10):821–828

Freshwater Demand and Availability for Green Hydrogen Production

Anne Rödl

Abstract Green hydrogen, produced from renewable energy sources (e.g., by electrolysis), is projected to play a pivotal role in the defossilization of a wide range of applications. Water is a fundamental component of this process, as hydrogen (H_2) production via electrolysis requires water (H_2O) as a raw material to be split into hydrogen and oxygen. However, freshwater is a scarce resource in many regions of the world, and densely populated areas are increasingly experiencing water scarcity. This trend is likely to be exacerbated by ongoing climate change and global population growth. From an economic perspective, sunny and dry regions of the planet seem particularly suitable for hydrogen production due to their abundant solar energy. However, water availability for hydrogen production depends on more than just physical supply; factors such as water quality, the regeneration of water reserves, and local social and environmental impacts of water withdrawal must also be considered. Therefore, this article examines the availability of water for green hydrogen production across various regions of the world.

Keywords Green hydrogen · Water demand · Freshwater demand · Water scarcity · Hydrogen

1 Introduction

So-called green hydrogen, produced from renewable sources of energy, e.g., via electrolysis, is projected to play a pivotal role for defossilation of a wide range of applications. This includes its direct use as an energy carrier, as well as its use for the production of subsequent fuel derivates like ammonia, methanol, and synthetic hydrocarbons such as kerosene and diesel. And water is a fundamental component of this process. The hydrogen (H_2) production via electrolysis requires water (H_2O)

A. Rödl (✉)
Hamburg University of Technology (TUHH), Institute of Environmental Technology and Energy Economics (IUE), Hamburg, Germany
e-mail: anne.roedl@sgs.com

© The Author(s), under exclusive license to Springer Nature Switzerland AG 2025   155
N. Bullerdiek et al. (eds.), *Powerfuels*, Green Energy and Technology,
https://doi.org/10.1007/978-3-031-62411-7_7

as a raw material to be split into its elemental components, hydrogen and oxygen, e.g., through electrolysis.

However, freshwater is a scarce resource in some areas of this earth. Many densely populated regions globally are currently experiencing water scarcity. In more and more parts of the world today, the water needs of people and nature can no longer be adequately met. Most likely, this trend could be exacerbated by ongoing climate change and a continued growth in global population, even though it is not yet known to what extent anthropogenic climate change contributes to the current extent of water scarcity [1].

Thus, also in the context of a large-scale production of green hydrogen, the questions of water availability and potential are crucial and these aspects move more into the focus of public discussion. From an economic point of view, the sunny and dry areas of our planet seem particularly suitable for green hydrogen production from solar energy due to the favorable solar energy supply [2]. Additionally, water potentials suitable for a potential hydrogen production do not solely relate to the physical availability of water in a specific region or area alone. Other factors, such as, e.g., water quality, the ability to regenerate or rebuild water reserves, and the local impacts of water withdrawal on society and nature need to be considered as well. Thus, also for the production and utilization of green hydrogen, it is necessary not only to consider the water demand and availability for the provision of hydrogen, but also to understand broader conditions and implications of its use to develop truly sustainable use strategies.

Against this background, the overall goal of this paper is it to have a closer look at the availability of water for green hydrogen provision in different parts of the world. Therefore, an overview of the methods and approaches used to describe water demand for specific products will be given first. The following section gives a more detailed overview of water availabilities in different regions and countries. Different indices and date used to assess general water availability, water scarcity, drought as well as the so-called "water stress" levels are discussed. Then, water requirements for the production of green hydrogen are analyzed for selected countries as well as for Germany. Finally, overall conclusions are drawn.

2 Basics

The issue of water availability is determined by various factors and thus rather location or region specific. General data, often available for countries or large watersheds on an annual basis, are not necessarily suitable for drawing conclusions on water availability on a local or regional level.

Another important issue is the quality of the available water. If the available water in a certain region is of poor or inadequate quality, it might not meet all of the water demands sufficiently. In addition, when determining water availability for specific applications (e.g., water electrolysis), existing water demands for other human or ecosystem purposes in the same area have also to be taken into consideration.

Additionally, while assessing the water demand, a differentiation between *withdrawal* or use of water and water *consumption* is needed.

- *Water withdrawal*: Withdrawal of water can be described as the amount of water that is taken from a specific freshwater source. In some cases, not all of the water is consumed, but is returned to the source or another waterbody in the same watershed [3–5].
- *Water consumption*: In contrast to water withdrawal, the amount of "consumed" water can be described as the amount of water not returned to the source; i.e., the water is definitively lost with regard to the respective basin or source. This can occur through evaporation, changes in water quality or incorporation of the water in a product that is then transferred to another region. In most cases, water consumption is the variable studied because it better reflects the impact of water use [3–5].

Water is generally considered a more or less renewable resource. Even if water is not newly created or additionally arrives on earth from space, water bodies regenerate qualitatively and quantitatively in different time periods from very fast to very slow. Very slowly (i.e., within geological time frames) regenerating water bodies are considered typically as fossil water resources. Fossil water is water from deep reservoirs formed millions of years ago and having often only a very weak or even no connection to surface water. If water is withdrawn from such water bodies, they regenerate only very slowly or—what is often more likely—not at all. Tapping such sources, realized especially often in arid regions, leads to the depletion of the resource and its irretrievable loss. Such fossil water resources are mainly used in desert countries like in North Africa and the Middle East. Compared to that, countries located within humid areas usually do not rely on the use of fossil sources. In addition, these fossil water resources are only partially fresh water; brackish or saline water is often present in the subsurface, requiring additional water treatment.

Based on these definitions selected basics of water assessment and water analysis are outlined below.

2.1 Water Availability

In most cases, when water availability is mentioned, it refers to the average potentially available renewable surface and groundwater supply. The so-called *total renewable water resource* (TRWR) is a statistical value describing this water availability in a certain region. The characteristic value is calculated for a 30-year period. The internal resource of a defined region is based on the difference between precipitation and evapotranspiration and the inflows from neighboring areas determined from the outflows

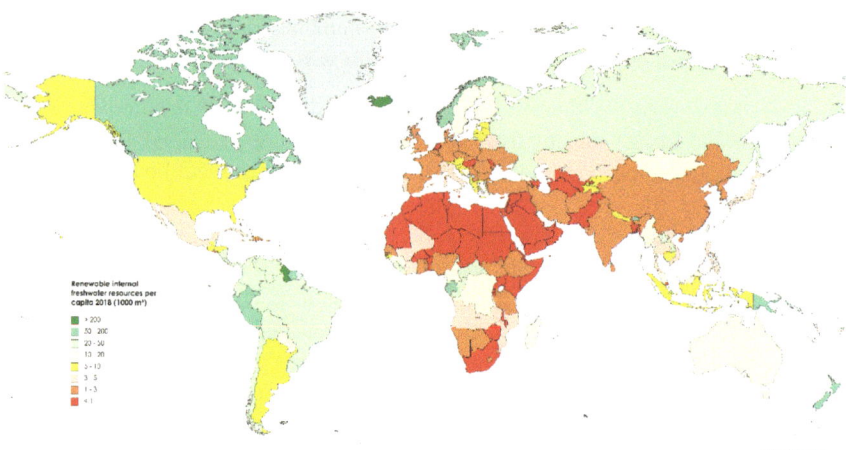

Fig. 1 Annual renewable internal freshwater resources per capita in 184 countries (in 1,000 m³) (data for 2019 retrieved from [7]; map created with mapchart.net [8] licensed under CC-BY-SA 4.0)

of gauges close to the border. The renewable water resources are subject to considerable annual changes fluctuating around the potential supply [6].[1] In Germany, for example, the average annually available renewable water in the period from 1991 to 2020 was about 176 billion m³/a. In the years 1961–1990, about 188 billion m³ of water were available each year [6]. In contrast, another database [17] indicates only approximately 107 billion m³/a for Germany in the year 2018. Per capita, that is about 1,300 m³.

Figure 1 shows the visualization of renewable internal freshwater resources per capita worldwide (data from [7]). Countries with few internal freshwater resources are shown in reddish colors, and countries with higher water resources available for each inhabitant are indicated in greenish colors.

Information on total water resources is generally not very useful for assessing water availability. Countries with large land areas tend to have huge water resources; smaller countries show typically fewer. Important in assessing water availability is the existing demand and the exact location within the country where this water demand occurs. In regions with very high population densities (e.g., metropolitan areas) or with specific geographic conditions, water availability may be severely limited, even if it is adequate on average for the country. Figure 2 illustrates this ratio of water availability per area and per capita for selected countries that are also of interest for hydrogen projects.

- Chile is a good example for such circumstances. In absolute terms, water seems to be a plentiful source (Table 1). Nevertheless, there might be regions with extreme

---

[1] The food and agriculture organisation AQUASTAT database [7] provides annual water resource information. But also national authorities like the German environmental agency [6] provide figures on water availability.

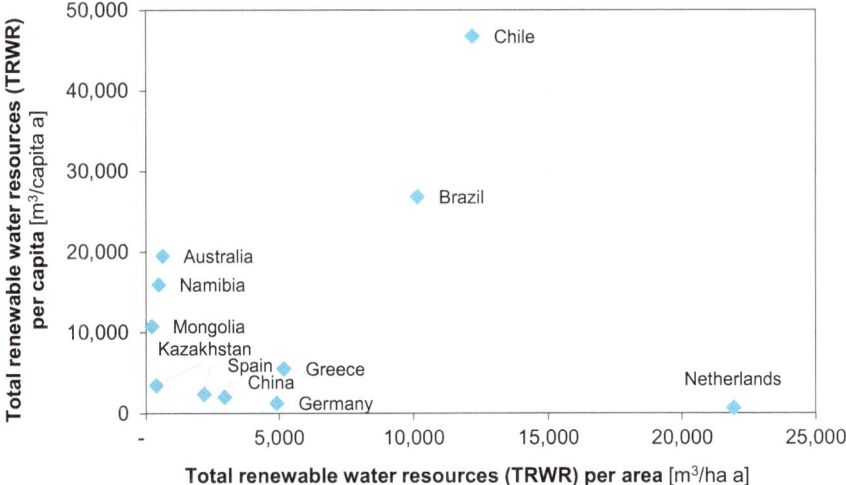

Fig. 2 Total renewable water resources (TRWR) per area and capita for selected countries (data based on Table 1)

water scarcity (see Fig. 3). Especially in the North, there are extremely arid areas, and, in the center, a lot of water is used for irrigation of agricultural land. The topography of the country leads on the one hand to large amounts of surface water running of the Andes Mountains, but large parts of this fresh water is quickly reaching the Pacific Ocean [9]. Therefore, in parts of the country, a certain share of fresh water is provided with the help of seawater desalination [10].

- Another example is the Netherlands. Usually, the country is not known to be a water scarce country, but it is densely populated and has a small territory (Table 1). But during the summer 2022, the Dutch government declared a water shortage because of the low water flow within the river Rhine. This caused serious problems for the shipping industry and resulted in seawater intrusion throughout the highly interconnected water system and weakening of the dike stability [11].

It is therefore advisable to assess water availability at least within a river basin or a watershed. But also, on river basin level, regional peculiarities can be found. However, in official statistics or databases, data is usually available only on a country level, and, at best, for major river basins, detailed information on water resources for almost every country in the world is provided, for example, by AQUASTAT [7].

Table 1 presents the total renewable water resource of selected countries representing areas interested in green hydrogen production facilities on a larger scale (Sect. 3.1).

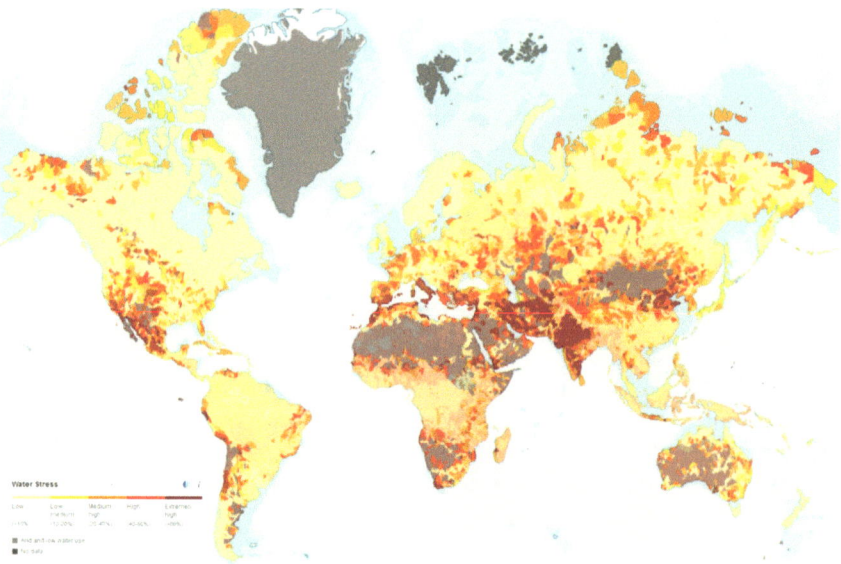

Fig. 3 World map showing water stress on regional level (WRI Aqueduct, accessed on 09.12.2022)

2.2 *Water Drought and Scarcity*

The local water availability can vary considerably due to numerous reasons. For example, a shortage of water can be caused both by a low natural availability of water or by excessive consumption. Additionally, a low availability can be due to climatic or infrastructural causes. Therefore, the differences between the two terms *drought* and *scarcity* are important and will be further explained.

- *Drought* is caused by climatic influences. There is a difference between permanent climate-induced drought in a region and acute drought periods. Drought usually results from a mismatch between precipitation and evapotranspiration [12]. In this case, the high demand of evapotranspiration cannot be adequately met by precipitation. Acute droughts are usually temporary and result from irregular fluctuations in precipitation intensity. The occurrence, duration, and severity of droughts are difficult to predict [13]. Average annual water availability can vary significantly between seasons. Droughts can occur in some months even in areas with averagely adequate water supplies. There are also differences in the regularity and reliability of precipitation between different areas with similar annual rainfall. Potential evapotranspiration is another factor that can lead to differences in the expression of dry seasons, in areas otherwise similarly supplied with water.
- *Water scarcity* often refers to a mismatch between demand and supply of fresh water, measured in physical terms [14]. Water scarcity means that there is not

enough water available to satisfy the needs of people or the environment. Nevertheless, there is no uniform definition of the term or a fixed method for determining the extent of water scarcity. One definition of the term water scarcity [15] describes water scarcity as the point at which the demands of all users, under the prevailing institutional arrangements, negatively affect water supply or water quality and the needs of all sectors, including the environment, can no longer be adequately met. This water scarcity has a physical as well as an economic or institutional dimension and mostly refers to human demands. Water scarcity can be caused by, e.g., a climatically induced water shortage, by poor infrastructure or by increased consumer demands. In most cases, water scarcity occurs in arid or semi-arid regions due to drought or climate fluctuations combined with strong population or economic growth. The mismatch due to overexploitation of ground or surface waters is then usually accompanied by deterioration of water quality [13].

A whole series of water scarcity or drought indices have been developed in recent years. They are designed quite differently and can be based on:

1. Water cycle related ratios (e.g., precipitation to evapotranspiration),
2. The relation of use and availability of water/the resource,
3. The competitive situation for water/a resource.

The criteria of the first group are used to form the so-called *drought indices*. The criteria of group two and three are used to determine the so-called *scarcity indices*.

2.2.1 Drought Indices

Climatic drought indices are a measure of long-term and large-scale climatically determined water conditions, e.g., within climate zones. They are not suitable for mapping of smaller areas within certain climatic zones. Most of these indices classify the degree of permanent drought using the parameters like precipitation, temperature, or potential evapotranspiration. Examples are the UNEP Aridity Index [16] or the Budyko Index [17] used to classify an area on a scale from hyperarid to humid.

Further indices for the classification of hydrological conditions in a watershed also include the yearly balance between runoff, precipitation, and evapotranspiration. In long-term equilibrium, runoff corresponds to the difference between precipitation and evapotranspiration. In other indices, irradiation and water balance of a catchment area are considered also. Evapotranspiration increases with increasing irradiation energy and precipitation. When irradiance is high and precipitation is low, soil dryness increases, while evapotranspiration decreases and runoff increases when irradiance is low.

2.2.2 Scarcity Indices

The other group of indices refer to the term *scarcity*, which generally means the limitation of a good in comparison with its needs. Scarcity can only occur when there is a need. Existing indices usually denote scarcity related to human use. Arid conditions often trigger water scarcity, but are not always the sole cause.

So far, there is no common agreement on a minimum amount of water that should be available for a person on a daily basis to meet all needs (drinking water, hygiene, consumption) [14]; i.e., no globally accepted benchmark exists. For example, the World Health Organization (WHO) assumes a need of up to 15 L/d to meet the minimum water requirements for drinking, hygiene, and for cooking in crisis situations [18]. However, under normal circumstances, this amount (15 L/d) does not meet the water demand of a modern human in homes with piped water. Other sources recommend 50 L/d [19], and a supply of 100 L/d is considered to be optimal [20]. The minimum requirement for the amount of daily consumption of a single person also depends on access to water and lifestyle [20]. Water use in Western societies is higher than in developing countries [21] and also related to household income [22].

Falkenmark Index Is one of these indicators and describes the size of a country's renewable water resource per inhabitant [23]. For example, the lower limit of water consumption in a modern society is 500 m^3 per capita and year. Due the fact that this figure includes not only water consumption of private households but also the needs of agriculture and industry, countries with high irrigation needs sometimes also consume above 1,000 m^3 per capita and year [24]. In a country located within the arid zone, at least about 400 m^3 per capita and year is needed for irrigation to ensure independent food supply.

Although the Falkenmark index has been one of the most widely used indicators to date, it is becoming less important with global food markets that can be used to meet demand. Another difficulty with this indicator is that it is calculated at country level and does not take into account small-scale differences in water consumption and availability [25]. Artificial water resources, such as desalinated seawater, are not taken into account either. In summary, using basic water needs as a metric for determining scarcity has only limited validity.

Social Water Stress Index (SWSI) The Falkenmark indicator has been further developed by combining it with the widely accepted Human Development Index (HDI) to create a Social Water Stress Index (SWSI) [26]. This evolved index takes into account a society's adaptive capacity, as expressed by the Human Development Index (HDI). As a result, water supply in highly developed countries with comparatively high population densities (e.g., United Kingdom, South Korea, and Belgium) is no longer classified as water stressed. This is because these countries are institutionally and economically capable of counteracting water scarcity.

Water Stress Index (WSI) Another scarcity indicator is the Water Stress Index (WSI). It expresses the ratio of total water use, including environmental water demand, to water availability in a specific area. The Water Stress Indicator (WSI) is interpreted as indicated by Table 2.

Withdrawal to Availability Index (WTA) The Water Resources Vulnerability Index [27], often referred to as the Withdrawal to Availability (WTA) Index [28], denotes the relationship between the overall annual water withdrawals within a country and its available water resources. Water withdrawals include domestic, industrial, irrigation, and livestock consumption and non-consumptive uses. Water scarcity is then determined based on critical withdrawal rates. For example, a withdrawal of more than 40% of the total renewable water resources means that there is a serious water shortage in the country. The Withdrawal to Availability Index (WTA) is commonly applied to characterize a country's management of its natural water resources [29]. A classification of the Withdrawal to Availability Index (WTA) can be made according to Table 3.

Table 1 Total renewable water resources (TRWR) in selected countries, land area, and water resources per capita. Data retrieved from AQUASTAT latest available data [7]

Country	TRWR (billion m³/a)	Total country area (1000 ha)	TRWR per area (m³/ha a)	TRWR per capita (m³/capita a)
Australia	492	774,122	636	19,521
Brazil	8,647	851,577	10,154	26,823
Chile	923	75,670	12,199	46,697
China	2,840	960,001	2,958	2,005
Denmark	6	4,292	1,398	1,040
Germany	176	35,758	4,922	1,281
Greece	68	13,196	5,183	5,538
Kazakhstan	108	272,490	398	3,469
Mauritania	11	103,070	111	88
Mongolia	35	156,412	222	10,790
Namibia	40	82,429	484	15,999
Netherlands	91	4,154	21,907	643
Oman	1.4	30,950	45	281
Saudi Arabia	24	214,969	112	70
Spain	111	50,594	2,194	2,379

Table 2 Water Stress Index (WSI) classification and implications

WSI value	WSI classification/implication
WSI > 1	Water shortage, resource depletion
0.6 > WSI < 1	Water stress, very high water withdrawal
0.3 > WSI < 0.6	Medium water stress, moderate water withdrawal
WSI < 0.3	No water stress, low water withdrawal

Table 3 Withdrawal to Availability Index (WTA) classification

WTA value	WTA classification
WTA < 10%	No water shortage
10% < WTA < 20%	Low water shortage
20% < WTA < 40%	Medium water shortage
WTA > 40%	Severe water shortage

Although the Withdrawal to Availability Index (WTA) is often used with regard to water resources, there are various critical aspects to its application:

- Technically, not all available water resources can actually be used by humans.
- The Withdrawal to Availability Index (WTA) does not take into account how much water is consumed and how much water is redirected unchanged after its use.
- Problems in accurately recording all water resources of a country.
- Seasonal differences in water availability are not taken into account. Water withdrawals during dry seasons can easily deplete water supplies in some regions, even if the annual average balance is uncritical [29, 30].
- Artificial resources, such as desalinated seawater or recycled water, which account for a large part of the water supply in some regions, are not recorded.

Additional Indices The concept of the Withdrawal to Availability Index (WTA) has been extended in various directions. For example, the Water Poverty Index [31] was created by additionally considering the supply security and access to drinking water of households. Also water treatment have been considered when assessing water availability [32].

As a further development, the Blue Water Scarcity Indicator have been introduced [33]. This indicator compares total surface and groundwater withdrawals within a watershed to the overall amount of water availability. The latter (available water) is calculated from the total runoff minus environmental water demand.

A similar approach was chosen to measure achievements with regard to target 6.4 of the United Nations (UN) sustainable development goals (SDGs), which aims to increase efficiency in water use and reduce water scarcity [34].

2.3   Water Stress Data and Maps

Different organizations provide data or maps of water stress using mostly the water-to-availability approach. They provide this data on different special levels.

The world resource institute (WRI) is providing a comprehensive tool to analyze and visualize water risks around the world. The open source data in the Aqueduct Water Risk Atlas [35] provides information about water stress for almost all river basins worldwide and additional information about water quality issues or regulatory risks. It is based on the Withdrawal to Availability Index (WTA) as an indicator of water stress (Fig. 3).

Figure 3 shows regional water stress at basin level as mapped in the Aqueduct Water Risk Atlas [35]. Dark red areas suffer from extremely high water stress. The lighter the color, the lower the water stress. Well visible are desert areas in Africa, Asia, and Australia and on the Pacific coast in the Americas. But even in Central Europe and around the Mediterranean, there is at least medium water stress in many regions.

The Water Risk Atlas [35] also provides more precise risk indicator values considering consumptive water use for many watersheds worldwide. The ratio of water consumption to available renewable water supply is calculated for this. Here, consumed water is not returned to the original water source. The risk indicator takes values higher than 25% in regions with medium risk for water stress, lower than 50% in areas with high risk, and higher than 75% in regions with extremely high risk for water stress.

The Food and Agricultural Organization (FAO) also provides maps displaying water stress [36]. Among others, they provide data and a map for Sustainable Development Goals (SDG) indicator 6.4.2, displaying the ratio between withdrawals for all economic activities and the total renewable freshwater by considering environmental flow requirements (37). High values of water stress mean more water users are competing for limited water supplies. Conversely, low water stress indicates a minimal potential impact on resource sustainability and little potential competition among users.

The Food and Agricultural Organization (FAO) also provides maps displaying water stress [36]. Among others, they provide data and a map displaying the ratio between withdrawals for all economic activities and the total renewable freshwater by considering environmental flow requirements [37]. The map is shown in Fig. 4, and the data is used among others for reporting Sustainable Development Goals (SDG) indicator 6.4.2.

Figure 4 shows regional water stress at basin level as mapped in the FAO Aquamaps [37]. The dark red color indicates a critical water situation in the catchment area. High values of water stress mean more water users are competing for limited water supplies. Conversely, low water stress indicates a minimal potential impact on resource sustainability and little potential competition among users.

All maps presented can just serve as a first and overall indication, while a more detailed assessment and evaluation of local water situation require further in-depth analyses. For example, it might be the case that local groundwater sources are fossil water reservoirs, which are not regenerating by precipitation. Additionally, groundwater in arid regions can be saline and need treatment before it can be used for further use [38].

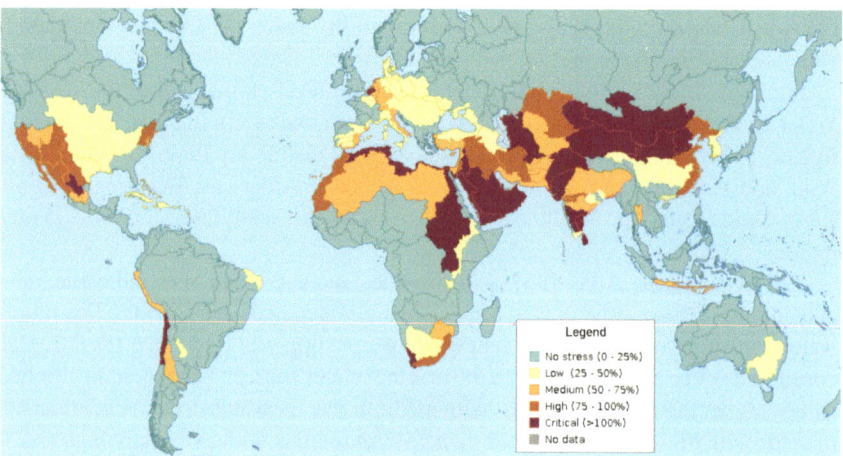

Fig. 4 Level of water stress (SDG indicator 6.4.2) by major river basin (https://data.apps.fao.org/aquamaps/; CC BY-NC-SA 3.0)

3 Water Demands for Hydrogen Production

Water is an essential resource for the production of green hydrogen via, e.g., water electrolysis assumed to play a pivotal role in future energy systems progressively shifting toward defossilization. The stoichiometric water demand to produce hydrogen by electrolysis amounts to 9 L/kg of hydrogen (L_{H2O}/kg_{H2}). It is higher than for steam methane reforming (4.5 L_{H2O}/kg_{H2}), as in this case water is also built by the reforming of methane consisting of four atoms of hydrogen per molecule.

However, the total water demand for hydrogen production not only results from the stoichiometric water demand, but water is also needed for additional process steps within the overall production pathway, such as for purification, cooling, and water disposal.

- **Water treatment**. Within most of the currently available electrolyzer technology, ultrapure water is required as a feedstock. Thus, due to losses during the treatment of the required water as a feedstock (e.g., desalination, demineralization), the overall demand for water as a primary feedstock increases as its quality decreases; i.e., the lower the water quality, the higher the overall specific water input demand. As a rough order of magnitude, around 20 to 40% of the input water might be wasted due to the respective treatment processes. Untreated salt or brackish water is not yet a viable source for hydrogen production by electrolysis, although research activities are ongoing to allow the direct use of, e.g., seawater within an electrolyzer. Also, if the water used is of very low quality, in most cases the wastewater streams (brine, etc.) must be treated before they are discharged to the environment.

- **Cooling water**. Water is also needed for cooling of the electrolyzer. This may require 30 to 40 kg_{H2O}/kg_{H2}, especially if the efficiency of the electrolyzer stack decreases over time and generates more heat and consequently demands increased cooling. An increase of the cooling demand by 40 to 70% during the life time of an electrolyzer stack can occur [39]. Also, compressors with intercooling might increase the overall water demand, as well as steam or cooling water demand for the conversion of hydrogen to a carrier such as liquid organic hydrogen carriers (LOHCs). However, cooling water is typically not polluted and might be returned into the respective water body from which it was taken. Nevertheless, the mere change in water temperature of the water used for cooling can lead to a thermal pollution of the water body if the often realized airborne water cooling is insufficient.

Taking these upstream water requirements into account, the production of hydrogen by electrolysis results in a significant higher water consumption compared to the sole stoichiometric demand. To further contextualize the above, Table 4 presents the key orders of magnitude of water consumption for various hydrogen categories. The figures in Table 4 are based on the assumption that good quality water is used and evaporative cooling is applied. When using sea water, brackish water, or waste water (in case of green hydrogen), the water demand increases further. Using air cooling or chiller systems instead of evaporative cooling reduces water demand considerably.

Looking further upstream in the green hydrogen supply chain, water use for power generation or power-generating infrastructure could also be considered. In the case of a power generation by photovoltaics (PV), minor amounts of water are needed for cleaning the solar panels or mirrors in case of concentrated solar power (CSP) plants.

Table 4 Total water demand, considering upstream processes, cooling, and water treatment [39]

Hydrogen category	Production process/technology	Water demand (L_{H2O}/kg_{H2})
Gray	Natural gas steam methane reforming	15–40
Blue	Natural gas reforming with CO_2 capture	18–44
Black/brown	Coal gasification	~ 70
Orange	Biomass gasification	~ 60
	Biomethane/biogas reforming	20–45
Green	Water electrolysis based on electricity generated from renewable energy sources (evaporative cooling)	60–95
	Water electrolysis based on electricity generated from renewable energy sources (air cooled)	< 18
Stoichiometric water demand (for comparison)		9

A cradle-to-grave water consumption to produce green hydrogen via proton exchange membrane (PEM) electrolysis and energy from wind or PV has been calculated to be in a range between 17 and 43 L_{H2O}/kg_{H2} [40].

3.1 Analysis of Selected Countries

In this section, the associated water demands for different countries willing to produce green hydrogen are analyzed. Therefore, the non-exhaustive list in Table 5 gives an overview over countries with large planned green hydrogen projects on a global scale (Status: End 2022). Only projects with more than 1 GW electric capacity installed within the electrolyzer have been considered. The number of large-scale projects per country and their planned total hydrogen output have been calculated based on [41]. Also, in the table, estimations are shown about the total water demand for these amounts of green hydrogen calculated using the specific water demand figures based on Table 4 (18, 60, and 95 L_{H2O}/kg_{H2}). For informative purposes additionally, the total renewable water resources are listed for every country.

Many of the hydrogen electrolysis projects recently announced are planned at locations close to the coast and in combination with seawater desalination plants. Only few projects are planned inland depending solely on local freshwater sources [41–43]. In most cases, information on the size of the power generation facilities, the capacity of the electrolyzers, and the amount of hydrogen or its derivatives produced annually are given for the projects enumerated here. Only in some cases the planned water supply concept is also reported (e.g., [44]).

For a better overview and comparison, the required amounts of water for the production of green hydrogen are graphically depicted in Fig. 5 for selected countries; these characteristics will be further explained below.

Based on the projected hydrogen production volumes across different countries, corresponding different water demands arise (Fig. 5, left). Considering a specific water demand of 60 L_{H2O}/kg_{H2}, the annual water requirement for the production of green hydrogen is around 800 million m³/a for Australia. For other countries shown, this water demand is considerably less and mostly below 200 million m³/a of water.

Compared to the total available renewable water resources (TWRW), only in a few countries total potential water demand for the planned electrolyzer projects seem to be critical (Fig. 5, right). These are Oman and the United Arab Emirates (UAE). For these countries the water demand for green hydrogen production sums up to approximately 5 to 20% for Oman and 5 to 30% for the United Arab Emirates (UAE) related to the total available renewable water resources (TWRW), depending on the specific water demand applied. In contrast, for most of the other countries presented, these shares are typically less than 1% of the total renewable water resources (TWRW), with Denmark being an exception where it approaches up to 2% related to the total available renewable water resources (TWRW). By interpreting these figures, one has to keep in mind, that for hydrogen projects planned in countries like Oman and Algeria, typically a respective sea water desalination facility is planned. Thus,

Table 5 Overview over planned large-scale green hydrogen projects exceeding 1 GW electric capacity, their targeted annual hydrogen output in 2035, and estimations about the total water demand (considering specific water demand estimates based on Table 4 and information about the total renewable water resources (TRWR) retrieved from FAO AQUASTAT country profile pages [45] (UAE: United Arab Emirates)

Country	Number of projects $\geq$ 1 GW	Projected total H_2 production in 2035 (t H_2/a)	Potential water demand for various estimates of specific water demand per kg H_2 (million m³/a)			Average TRWR (billion m³/a)
			18 L_{H2O}/ kg_{H2}	60 L_{H2O}/ kg_{H2}	95 L_{H2O}/ kg_{H2}	
Australia	17	12.8 M	231	770	1219	492
USA	4	4.2 M	75	250	395	3069
Mauritania	2	3.5 M	64	212	336	11.4
Kazakhstan	1	3.5 M	62	208	329	108.4
Netherlands	6	3.2 M	58	194	307	91
Chile	6	2.8 M	51	168	267	923.1
Oman	4	2.8 M	50	167	264	1.4
Argentina	2	2.5 M	45	150	237	876.2
Germany	5	2.3 M	42	139	221	176
China	4	2.0 M	37	124	196	2,840
Spain	4	1.7 M	31	103	163	111
South Africa	2	1.7 M	30	101	160	51.35
Denmark	4	1.5 M	27	90	143	6
India	3	788,000	14	47	75	1,911
Brazil	1	600,000	11	36	57	8,647
UAE	1	480,000	9	29	46	0.15
Namibia	2	470,000	8	28	45	39
Ireland	1	459,000	8	28	44	52
Ukraine	2	433,000	8	26	41	175.3
Norway	1	407,000	7	24	39	393
Egypt	2	360,000	6	22	34	57.5
Great Britain	1	260,000	5	16	25	147
Greece	1	250,000	5	15	24	68.4
Korea	1	208,000	4	12	20	69.7
Sweden	1	191,000	2	11	18	174

the values given above might be misleading because it is expected that the water desalination plant of such a project is typically designed to provide more freshwater than needed for the project to be on the safe side; thus, such a project can also help to improve the water supply security within a certain arid or desert area.

However, the comparison of the potential water demand with the average annual total renewable water resources (TRWR) in every country is only helpful to a limited

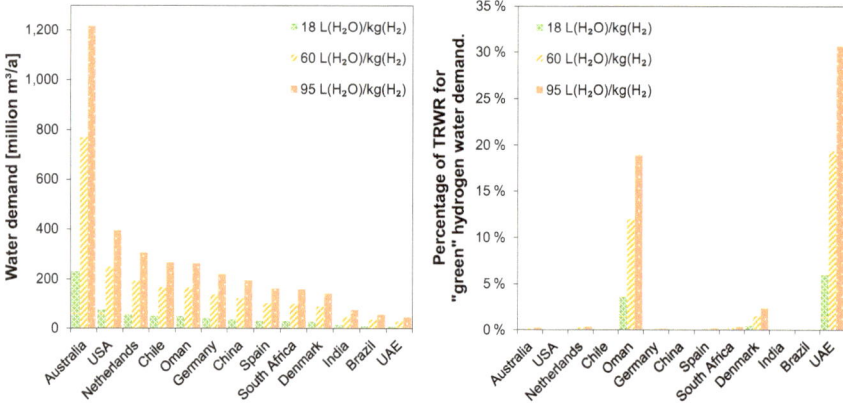

Fig. 5 Water demand estimates for green hydrogen supply scenarios for different countries (left) and percentage share of TRWR (right) for the year 2035 (TRWR: total renewable water resources; UAE: United Arab Emirates; data based on Table 5)

extend. In some countries (i.e., Oman, Saudi Arabia, and United Arab Emirates) desalinated water is used already to a certain extend and has to be used necessarily for hydrogen production. Additionally, the total available amount of water within a country is not meaningful on a local level if water resources are distributed unevenly within the country. The above comparison also completely ignores other uses of available water resources in agriculture, industry, or households. However, these already existing uses should definitely be considered. Therefore, looking at water demand together with the water stress indicators, as shown Fig. 3 or Fig. 4, could give a deeper insight.

Figure 6 serves this purpose. The graphic [37] visualize the water situation at locations of currently planned large-scale green hydrogen production projects [41]. The figure provides an overview of selected large-scale green hydrogen projects with electrolyzer capacities of over 1 GW_{el} currently under discussion. Thus, at the moment about 76 large-scale projects are in the planning stage with a total foreseen production volume of about 49 Mt/a of hydrogen [41].[2]

According to this, critical water situations prevail especially in Central Asia and the Arabic Peninsula, but also in North Africa and North America. Overall, the map indicates that in locations and countries where large-scale hydrogen projects (i.e., electrolyzer capacities) are intended to be implemented, there is often also at least a low level of water stress estimated. Specifically, in regions of Central Asia and the Arabic Peninsula, the water stress level ranges rather from medium to high. The following aspects provide a detailed overview of the conditions surrounding hydrogen projects in specific countries and regions.

---

[2] According to the IEA hydrogen database [41] currently around 740 hydrogen projects based on water electrolysis are in planning or conception stage worldwide in around 65 countries. Countries with most activities or lot of large-scale ambitions are Australia, Chile, Netherlands, Spain, and Germany for example.

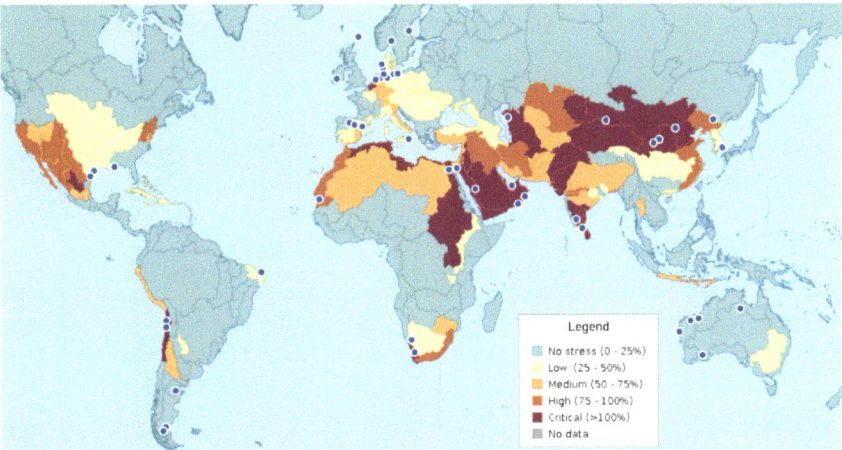

Fig. 6 Selection of planned large-scale hydrogen projects with electrolyzer capacities of more than 1 GW_{el} integrated in the water stress map of Fig. 4 (basic map retrieved from 37)

- **Australia**. Projects to ramp-up hydrogen production in Australia are in most cases planned in the Western and Northern part. A lot of these projects will be located close to the sea, planning to use desalinated water as a resource. But, according to Fig. 3, large parts of Australia suffer from water stress [35]. Thus, it has been reported that an electrolyzer project in South Australia with 6 GW of renewable electricity supply was cancelled in 2022 due to water supply concerns [46].
- Chile (South America). Chile wants to become one of the most important hydrogen producers in the coming years by initiating a lot of projects across the country [44]. Chile has a great renewable energy potential with potentially low levelized costs of hydrogen production. The planned project in the North and the Atacama Desert is designed with desalinated seawater supply. Especially in the Northern and central parts of Chile, medium or high water stress are displayed according to Figs. 3 and 4 [37].
- Oman and UAE (Arabic Peninsula). Projects in Oman or the United Arab Emirates (UAE) are planned at the coast using sea water sources. Other freshwater resources would definitely not be available, as the whole country suffers from severe water shortages. This is clear in Figs. 3 and 4, although they are based on different indices.
- **Kazakhstan and China (Central Asia)**. Larger local effects on freshwater sources might be caused by large-scale projects in Kazakhstan or inland China where no sea water sources are available. About 80% of the total Kazakhian area are deserts and steppes [47] with very low precipitation and high evapotranspiration. Surface water is unevenly distributed. In particular, the Mangystau region, where the large-scale green hydrogen project is planned [43], is mentioned [58] as a region with significant water deficit and hardly any freshwater.

- **Europe**. In Central Europe, the water situation is more relaxed in most cases. Nevertheless, the Rhine catchment is also characterized as a region with medium water stress. In some regions, the water situation has worsened in recent years during hot and dry summers. Furthermore, it has to be considered that the drinking water supply systems may not be sufficiently prepared to serve an additional demand for water for the production of green hydrogen. Already, communities around major cities such as Berlin, Frankfurt, or Hamburg are fighting for their right to retain control over the water sources in their area [48].

Overall, the above explanations show that the landscape of hydrogen projects varies significantly in terms of water supply situations and the associated challenges. Regions such as Central Asia, the Arabic Peninsula, North Africa, and North America are particularly affected by conditions of water stress. Both Chile and Australia are promoting large-scale hydrogen projects, despite the respective water stress challenges; coastal projects in these countries therefore usually propose utilizing desalinated seawater. Conversely, initiatives in regions like Kazakhstan or inland China, without seawater access, might strain local freshwater resources. In Central Europe, water supply might not be a critical issue, nevertheless, locally the situation might be different.

3.2 Focus: Germany

Based on estimates of specific water consumption for green hydrogen production (18 to 95 L_{H2O}/kg_{H2}), the resulting water demand for different scenarios of green hydrogen provision are estimated exemplarily for Germany. Figure 7 presents a range of water demand estimations for respective scenarios for Germany for the years 2030 and 2050. These values are provided for three specific water consumption values for green hydrogen production (18, 60, and 95 L_{H2O}/kg_{H2} based on Table 4). Within those German hydrogen scenarios, three distinct hydrogen demand projections are considered: 90 TWh_{H2} (2030 minimum value), 110 TWh_{H2} (2030 maximum and 2050 minimum value), and 400 TWh_{H2} (2050 maximum value) [49–51].

Based on these frame assumptions, the required water amounts for the year 2030 (minimum hydrogen projection of 90 TWH_{H2}) varies between 50 to 260 million m^3. Similar water demands result for the hydrogen amounts of 110 TWH_{H2} (maximum for 2030, minimum for 2050). For the maximum hydrogen projection of 400 TWH_{H2} in the year 2050, the absolute water demand varies significantly, ranging between about 200 and 1,140 million m^3.

However, while the required absolute amounts of water for the production of green hydrogen differ significantly between the analyzed years and specific water demands, even the 1,141 million m^3 of water required for green hydrogen supply for the year 2050 (maximum scenario) represents less than 1% of the annual renewable natural water supply in Germany (2018). Thus, based on these results, the required water

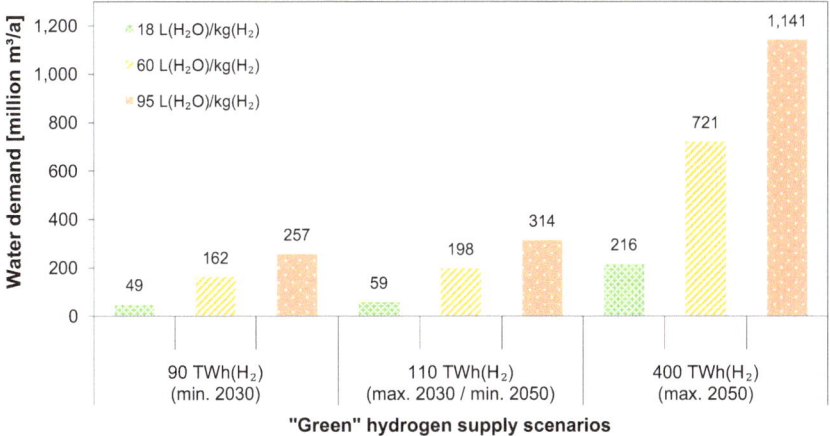

Fig. 7 Water demand estimates for green hydrogen supply scenarios in Germany (2030 and 2050). Values for hydrogen demand according to [49]

amounts for the supply of green hydrogen represent rather a non-critical proportion compared to the mentioned available amounts of water.

Still, it should be kept in mind that this total amount may be very unevenly distributed. Additionally, these water resources are usually not available without any treatment in drinking water quality. For comparison, in 2019, a total of 5,355 million m^3 of water was extracted by public water suppliers for drinking water supply in most cases from groundwater. The potential water demand for hydrogen production shown in Fig. 7 sums up to 1 to 21% of the public water production to be provided additionally to the currently produced amount.

Nevertheless, one has to compare the local production capacities with the water demand for hydrogen production. Then, a different picture might emerge. The development of the regional water production infrastructure may be costly and lengthy. Also, strong seasonal differences with droughts in summer might require a completely different assessment of the situation [52].

In the year 2019, a total of around 20 billion m^3 of water was withdrawn from the environment and used in Germany [6]. This corresponds to approximately 11.4% of the natural water supply in Germany amounting to about 176 billion m^3/a [6]. Additionally, large quantities of water are already being used for energy generation. For example, in the year 2019, the energy sector used 8.8 billion m^3 of water and 84.7% of all freshwater consumed by German companies was used for cooling of power generation and production facilities [53].

Considering not only water demand for cooling during power generation, but also for extraction of lignite, water consumption is much higher than statistics of water use for electricity generation display (e.g., [54]). Studies calculate that transforming energy system from a prevailing use of fossil fuels to high shares of green hydrogen would save huge amounts of water [4]. Comparing the maximum estimation in water demand, approx. 1.1 billion m^3 of water for the required amounts of green hydrogen in the year 2050, to the 8.8 billion m^3 consumed by the energy sector in Germany in the year 2019, this water demand can be considered less critical and seems like a fairly modest demand.

4 Conclusion

The use of so-called green hydrogen, derived from renewable energy source through water electrolysis, holds promise for diverse applications in order to reduce greenhouse gas emissions. Besides renewable electricity, also central to its production is the supply of water as a raw material. However, this resource is facing scarcity in many regions, which could be amplified by factors like climate change and population growth. Thus, with regard to the need of water for green hydrogen production, aspects on its availability, quality, and the broader implications need to be considered as well. Against this background, this article gives an insight into the challenge of fresh water supply for green hydrogen projects. It was shown that water should be considered as a relevant raw material, but the assessment of its availability depends very much on local environmental and economic factors.

Water scarcity is often not only due to a lack of natural water resources, but also has man-made components on both the supply and consumption side. Important are the climatic conditions and the scale of existing use in the agricultural, industrial, and domestic sectors of a country.

Additional water demand from green hydrogen production could be problematic in regions with existing water stress or weak water supply infrastructure. The use of desalinated water might be the only way to obtain enough raw material for hydrogen production in these regions without harming local communities. Instead, these projects have potential to improve the local situation if they are generously planned to allow for local water supply as well.

Also, if it seems that water availability might not be the major point to consider in planning green hydrogen projects the topic might get more important in the future or in particular regions of the world. The change of water demand is resulting from socio-economic changes and climatic change [1]. There are huge uncertainties in projecting future water availability or scarcity on local level [1] which should be considered also in the planning of green hydrogen projects.

References

1. Caretta MA, Mukherji A, Arfanuzzaman M, Betts RA, Gelfan A, Hirabayashi Y, Lissner TK, Lopez Gunn E, Liu J, Morgan R, Mwanga S, Supratid S (2022) Water. In: Pörtner H-O, Roberts DC, Tignor MM, Poloczanska E, Mintenbeck K, Alegría A, Craig M, Langsdorf S, Löschke S, Möller V, Okem A (eds) Climate Change 2022: impacts, adaptation and vulnerability. Working group II contribution to the sixth assessment report of the Intergovernmental Panel on Climate Change. Cambridge University Press, Cambridge, pp 551–712

2. Sens L, Piguel Y, Neuling U, Timmerberg S, Wilbrand K, Kaltschmitt M (2022) Cost minimized hydrogen from solar and wind—Production and supply in the European catchment area. Energy Convers Manage. https://doi.org/10.1016/j.enconman.2022.115742

3. Lohrmann A, Child M, Breyer C (2021) Assessment of the water footprint for the European power sector during the transition towards a 100% renewable energy system. Energy. https://doi.org/10.1016/j.energy.2021.121098

4. Beswick RR, Oliveira AM, Yan Y (2021) Does the Green Hydrogen Economy Have a Water Problem? ACS Energy Lett. https://doi.org/10.1021/acsenergylett.1c01375

5. Vanham D, Medarac H, Schyns JF, Hogeboom RJ, Magagna D (2019) The consumptive water footprint of the European Union energy sector. Environ Res Lett. https://doi.org/10.1088/1748-9326/ab374a

6. UBA (2022) Wasserressourcen und ihre Nutzung. https://www.umweltbundesamt.de/daten/wasser/wasserressourcen-ihre-nutzung#wassernachfrage

7. FAO (2022) AQUASTAT core database. https://www.fao.org/aquastat/en/. Accessed 12 Jan 2022

8. MapChart (2022) World map: simple. https://www.mapchart.net/world.html. Accessed 13 Dec 2022

9. FAO (2015) Country profile—Chile. https://www.fao.org/aquastat/en/countries-and-basins/country-profiles/country/CHL

10. FAO (2015) AQUASTAT Perfil de País - Chile. Organización de las Naciones Unidas para la Alimentación y la Agricultura, Roma, Italia. https://www.fao.org/3/ca0440es/CA0440ES.pdf. Accessed 26 January 2023

11. Toreti A, Bavera D, Acosta Navarro J, Cammalleri C, De Jager A, Di Ciollo C, Hrast Essenfelder A, Maetens W, Magni D, Masante D, Mazzeschi M, Niemeyer S, Spinoni J (2022) Water scarcity in the Netherlands August 2022, EUR 31176 EN. JRC, European Commission, Publications Office of the European Union, Luxembourg. https://publications.jrc.ec.europa.eu/repository/handle/JRC130436

12. Thornthwaite CW (1948) An approach toward a rational classification of climate. Geogr Rev 38:55–94

13. Pereira LS, Cordery I, Iacovides I (2002) Coping with water scarcity. technical documents in hydrology. Programm, International Hydrological, Paris

14. Chenoweth J (2008) Minimum water requirement for social and economic developement. Desalination 229:245–256

15. UN (2007) Coping with water scarcity. Challenge of the twenty-first century. World water day 2007. United Nations Water Scarcity Initiative (UN-water); Food and Agriculture Organisation (FAO) (2007)

16. UNEP (1992) World atlas of desertification. United Nations Environment Programme, London

17. Budyko MI (1958) The heat balance of the earth's surface. U.S. Department of Commerce, Weather Bureau, Washington D.C

18. Reed B, Reed B (2011) How much water is needed in emergencies. Technical notes on drinking-water, sanitation and hygiene in emergencies. Water, Engineering and Development Centre, Geneva

19. Gleick PH (1996) Basic water requirements for human activities. Meeting basic needs. Water Int 21:83–92

20. Howard G, Bartram J (2003) Domestic water quantity, Service, Level and Health. Organization. World Health, Geneva

21. FAO (2011) AQUASTAT

22. Crouch ML, Jacobs HE, Speight VL (2021) Defining domestic water consumption based on personal water use activities. J Water Supply Res Technol AQUA. https://doi.org/10.2166/aqua.2021.056

23. Falkenmark M (1989) The massive water scarcity now threatening Africa. Why isn't it being addressed? Ambio. https://doi.org/10.2307/4313541

24. Falkenmark M (1997) Meeting water requirements of an expanding world population. Philos Trans Royal Soc B: Biol Sci. https://doi.org/10.1098/rstb.1997.0072

25. White C (2012) Understanding water scarcity. Definitions and measurements (2012). Accessed 2013
26. Ohlsson L (2000) Water conflicts and social resource scarcity. Phys Chem Earth Part B: Hydrol Oceans Atmosphere (2000). https://doi.org/10.1016/S1464-1909(00)00006-X
27. Raskin P, Gleick P, Kirshen P, Pontius G, Strzepek K (1997) Water futures. Assessment of long-range patterns and problems. Stockholm Environment Institute
28. Brown A, Matlock MD (2011) A review of water scarcity Indices and methodologies. White paper. University of Arkansas
29. Anisfeld SC (2010) Water resources. Foundations of contemporary environmental studies. Island Press, Washington, DC
30. Rijsberman FR (2006) Water scarcity. Fact or fiction? Agricultural water management 80:5–22
31. Sullivan C (2002) Calculating a Water Poverty Index. World Dev. https://doi.org/10.1016/S0305-750X(02)00035-9
32. Vörösmarty CJ, Douglas EM, Green PA, Revenga C (2005) Geospatial indicators of emerging water stress. An application to Africa. AMBIO: J Hum Environ. https://doi.org/10.1579/0044-7447-34.3.230
33. Hoekstra AY, Chapagain AK, Aldaya MM, Mekonnen MM (2011) The water footprint assessment manual. Setting the global Standard. Network, Water Footprint, London (2011)
34. UN (2023) SDG indicator metadata. Goal 6: ensure availability and sustainable management of water and sanitation for all. https://unstats.un.org/sdgs/metadata/files/Metadata-06-04-02.pdf. Accessed 29 July 2023
35. WRI (2019) Aqueduct water risk atlas. https://www.wri.org/applications/aqueduct/water-risk-atlas/#/?advanced=false&basemap=hydro&indicator=w_awr_def_tot_cat&lat=30&lng=-80&mapMode=view&month=1&opacity=0.5&ponderation=DEF&predefined=false&projection=absolute&scenario=optimistic&scope=baseline&threshold&timeScale=annual&year=baseline&zoom=3. Accessed 14 Dec 2022
36. FAO (2022) Aquamaps. global spatial database on water and agriculture. https://data.apps.fao.org/catalog/dataset/aquamaps-ll. Accessed 30 Jan 2023
37. FAO (2018) Aquamaps. level of water stress (SDG 6.4.2) by major river basin. https://data.apps.fao.org/aquamaps/?share=f-2334e062-ed85-4ac2-a01c-286b97df20e0. Accessed 30 Jan 2023
38. Nilsson A, de Vivero G, Lopez Legarreta P, Day T, Pudlik M, Seyfang B, Dugarjav B (2021) Green hydrogen applications in Mongolia. Technology potential and policy options. Fraunhofer ISI, New Climate Institute, GIZ. https://www.isi.fraunhofer.de/content/dam/isi/dokumente/ccx/2021/NewClimate_Green_Hydrogen_Applications_in_Mongolia.pdf. Accessed 30 Jan 2023
39. Coertzen R, Potts K, Brannock M, Dagg B (2020) Water for hydrogen. https://www.ghd.com/en/perspectives/water-for-hydrogen.aspx. Accessed 17 Nov 2022
40. Altgelt F, Micheli M, Sailer K, Crone K (2021) Water consumption of powerfuels. Global Alliance Powerfuels, Berlin. https://www.powerfuels.org/news/news/new-publication-water-consumption-of-powerfuels/. Accessed 19 Jan 2023
41. IEA (2022) Hydrogen projects database. https://www.iea.org/reports/hydrogen-projects-database. Accessed 12 Jan 2023
42. Iberdrola (2020) Iberdrola builds the largest green hydrogen plant for industrial use in Europe. https://www.iberdrola.com/about-us/what-we-do/green-hydrogen/puertollano-green-hydrogen-plant. Accessed 30 Jan 2023
43. SVEVIND (2022) HYRASIA ONE. Green energy for the energy transition and decarbonization of industry. https://hyrasia.energy/
44. Gobierno de Chile (2021) Chile's Green Hydrogen Strategy and investment opportunities. Ministries of Energy and Mining. https://energia.gob.cl/sites/default/files/documentos/green_h2_strategy_chile.pdf
45. FAO (2021) AQUASTAT—FAO's global information system on water and agriculture. Country Profiles. https://www.fao.org/aquastat/en/countries-and-basins/country-profiles

46. IEA (2023) Global hydrogen review 2022. License: CC BY 4.0. International Energy Agency, Paris. https://www.iea.org/reports/global-hydrogen-review-2022. Accessed 30 Jan 2023

47. FAO (2012) AQUASTAT Country Profile – Kazakhstan. Food and Agriculture Organisation (FAO), Rome, Italy. https://www.fao.org/3/ca0366en/CA0366EN.pdf. Accessed 30 Jan 2023

48. Frey A, Schlömer O (2023) Hauptstadt auf dem Trockenen. Frankfurter Allgemeine Sonntagszeitung, 1 October 2023. https://www.faz.net/aktuell/wissen/trinkwassermangel-in-berlin-hauptstadt-auf-dem-trockenen-19203983.html#void. Accessed 28 Oct 2023

49. Deutscher Bundestag: Ausarbeitung Wasserstoffbedarf, WD 5-3000-024/22. Wissenschaftlicher Dienste des Deutschen Bundestags, Berlin. https://www.bundestag. de/resource/blob/894040/0adb222a2cbc86a20d989627a15f4bd8/WD-5-024-22-pdf-data.pdf. Accessed 19 Jan 2023

50. BMWi (2020) Die Nationale Wasserstoffstrategie. Bundesministerium für Wirtschaft und Energie, Berlin. https://www.bmwk.de/Redaktion/DE/Publikationen/Energie/die-nationale-wasserstoffstrategie.pdf?__blob=publicationFile. Accessed 19 Jan 2023

51. Wietschel M, Zheng L, Arens M, Hebling C, Ranzmeyer O, Schaadt A, Hank C, Sternberg A, Herkel S, Kost C, Ragwitz M, Herrmann U, Pfluger B (2021) Metastudie Wasserstoff - Auswertung von Energiesystemstudien. Studie im Auftrag des Nationalen Wasserstoffrats, Karlsruhe, Freiburg, Cottbus

52. Saravia F, Graf F, Schwarz S, Gröschl F (2023) Genügend Wasser für die Elektrolyse. Wieviel Wasser wird für die Erzeugung von grünem Wasserstoff benötigt und gibt es ausreichende Ressourcen? Deutscher Verein des Gas- und Wasserfachs e.V., Bonn. https://www.dvgw.de/medien/dvgw/leistungen/publikationen/h2o-fuer-elektrolyse-dvgw-factsheet.pdf. Accessed 7 Aug 2023

53. DESTATIS (2022) 85% der Wassernutzung in der Wirtschaft dienten 2019 der Kühlung von Anlagen. https://www.destatis.de/DE/Presse/Pressemitteilungen/Zahl-der-Woche/2022/PD22_34_p002.html;jsessionid=585F2E722C8C1569577D5ACD5C937F4C.live712

54. EUROSTAT (2022) Water abstracted by sector of use. https://ec.europa.eu/eurostat/databrowser/product/view/ENV_WAT_ABS. Accessed 19 Jan 2023

Seawater and Brackish Water Desalination

Muhammad Usman, Dolly Dipakkumar Sharma, and Mathias Ernst

Abstract The increasing scarcity of water due to the world's population growth, urbanization, industrial development, and climate change is an important topic of concern for many countries around the world. Various technical measures for improving the quantity and quality of water supply and water resources often rely on saline resources like seawater and brackish water, as these are available in abundance. The technical solutions as well as the capacities of technical desalination systems grow every year, making this water production technology increasingly reliable. In this context, various conventional thermal desalination technologies like multi-stage fraction distillation (MSF), multi-effect distillation (MED), mechanical vapor compression (MVC) as well as membrane-based methods like reverse osmosis (RO) are presented. For successful process implementation, pre- and post-treatment of the raw and desalinated water is essential. The membrane-based processes, in particular, are prone to fouling and/or scaling, while re-mineralization of the desalinated water is required. As research and development continue, new desalination techniques like membrane distillation (MD), capacitive deionization (CDI), forward osmosis (FO), the coupling of thermal and membrane desalination methods in a hybrid desalination system as well as brine management strategies are under development. Different aspects of the reducing the specific energy consumption in desalination are also discussed. Finally, insights into the current state of the desalination market and freshwater demand are given, and an outlook on future expectations related to freshwater consumption for green hydrogen production and the economic effect of providing freshwater using RO desalination for green hydrogen is presented.

Keywords Water scarcity · Desalination · Reduction of energy consumption · Brine management · Hybrid desalination systems · Desalination market

M. Usman · D. D. Sharma · M. Ernst (✉)
Hamburg University of Technology (TUHH), Institute for Water Resources and Water Supply (WWV), Hamburg, Germany
e-mail: mathias.ernst@tuhh.de

© The Author(s), under exclusive license to Springer Nature Switzerland AG 2025   179
N. Bullerdiek et al. (eds.), *Powerfuels*, Green Energy and Technology,
https://doi.org/10.1007/978-3-031-62411-7_8

1 Introduction

Water scarcity is currently one of the most concerned topics of global development taking into consideration the rapid population growth, urbanization, climate change, and industrialization along with a change in the usage pattern, agriculture, and living standard also contributing to the steadily growing water demand [1, 2]. About 71% of the global population lives under conditions of moderate to severe water scarcity, and about 66% live under severe water scarcity for at least one month of the year [1]. As a result of increasing urban population and more often climate change events, over-abstraction of freshwater is continue to grow. It is predicted that by 2050, 2 billion people in 44 countries will most likely suffer from scarcity of freshwater and out of which 95% may live in developing countries based on projected population data (AQUASTAT data [3]).

The total dissolved substances in freshwater resources is routinely quantified by measurements of salinity or total dissolved solids (TDS) [4]. The salinity is usually low for freshwater sources, at less than 0.5 g/L. In comparison, brackish water (BW) and seawater (SW) are saline waters (salt concentration is in the range of 0.5 to 30 g/L) and, therefore, not suitable for human consumption as drinking water due to associated health issues such as cardiovascular diseases, diarrhea, and abdominal pain [5]. Considering the freshwater scarcity and abundance of saline water (97% of water on the earth's surface), the desalination techniques to treat BW and SW are widely accepted to meet the growing demand of freshwater. Desalination of BW and SW is a reliable and drought-proof solution as it does not depend on the flow of river or filling state of reservoirs [6].

Figure 1 shows development for the desalination technology market since 1960. Thermal desalination is constantly growing; in 2022 the installed capacity of thermal desalination amounted about 28 Mm³/d. However, since the 2000s, there has been a significant rise in the development of membrane technology for desalination. By 2022, the installed capacity reached approximately 97.2 million m³/day (Mm³/d). Today, about 17,000 desalination plants are operated around the world (Table 1), out of which almost 45% are located in the Middle East and North Africa, due to rising water demands (annual population growth rates of about 7 to 9%), water scarcity and the availability SW [7]. About 80% of all desalination, plants are replying on RO membranes and the remaining 20% are thermal plants [3].

The global desalination capacity per region is presented in Fig. 2 distinguishing between SW, BW, and wastewater as three water sources. In all the regions, SW is the main water source for desalination with exception of North America where BW desalination accounts for about 73% (7.3 Mm³/d) of the regional capacity followed by 19% for wastewater (1.9 Mm³/d).

In 2016, the amount of desalinated water in the Middle East and North Africa (MENA) regions was approximately 7 billion m³ [9]. Apart from the MENA region, the US has been considering desalination and investing to meet the water supply requirement [9]. India and China have freshwater production from thermal and membrane-based desalination higher than the water scarcity threshold out of which

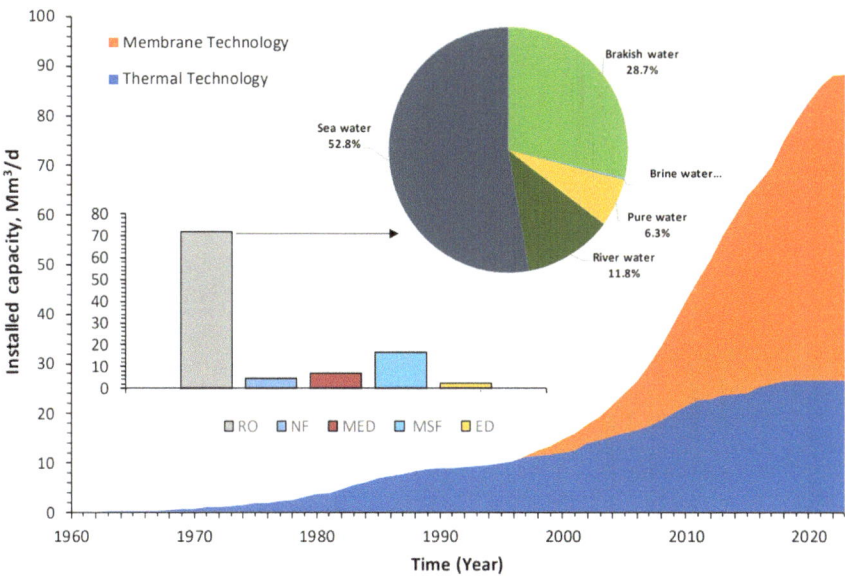

Fig. 1 Global desalination trend based on Desal Data, 2020. Taken from Dhakal et al. [3]; ED: electrodialysis; MED: multi-effect distillation; MSF: multi-stage flash; NF: nanofiltration; RO: reverse osmosis)

Table 1 World desalination capacity in 2020 (according to Global Water Intelligence [8])

No. of plants	Status of desalination plants	Capacity (m^3/d)
20,957	Total plants	115,625,178
3,823	Offline	7,193,546
16,860	In operation	97,305,664
274	Under construction	11,125,968
17,134	In operation plus under construction	108,431,632

75% water is used by industries. Currently installed desalination capacity in India and China is 2.87 Mm^3/d and 9.72 Mm^3/d, respectively [3]. As per data from 2020, the total desalination capacity in Algeria, Morocco, and South Africa is 2.76 Mm^3/d, 0.27 Mm^3/d, and 0.45 Mm^3/d, respectively. Out of which Algeria uses about 90% for municipal purposes, while the other two countries use 30% for industrial purposes [3]. The global water production from desalination has increased to more than 90 Mm^3/d in 2020 from 20 Mm^3/d in 2000 [10]. At the moment, more than half a billion people receive their daily drinking water by means of desalination [9].

As presented above, current desalination technologies include mainly thermal (evaporation/condensation) as well as membrane-based desalination for freshwater production. For both options the principal remains the same including the removal of salt and solids from the BW and SW. Thermal distillation was the only applied method

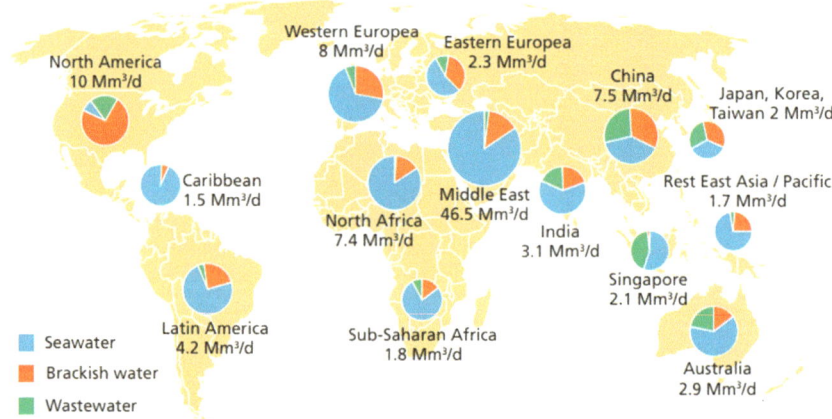

Fig. 2 Desalination capacities in different regions of the world per percentage capacity production of various water sources (seawater, brackish water, wastewater effluent) (Information from [8]. For example: North America desalinates water with a total capacity of 10 Mm³/d of which 73% is produced from brackish water, 19% from wastewater effluent and 8% from seawater

for desalination before and in the 1950s. However, pressure-driven membrane-based various desalination techniques were developed in the 1960s and treatment capacities increased tremendously in the 1970s. The development of membrane-based RO concept increased rapidly in the 1980s and gave strong competition to other traditional methods replying on thermal desalination. In 1981, RO systems are already applied for desalination in more than 63 countries [11, 12].

Today RO membrane-based desalination is the most prevalent technique due to its relatively low specific energy demand resulting in low cost of water production as compared to other technical approaches [6]. When fossil fuel-based energy is used to meet the energy requirement of a desalination plant, greenhouse gases (GHGs) are emitted along with emission of other substances (e.g., NO_x) harmful to humankind and the environment. In addition, limited reserves and market-dependent high prices of fossil fuel energy are another challenges determining the cost of produced potable water [13]. A possible solution is that desalination systems can be powered by electricity from renewable sources of energy (e.g., photovoltaic (PV) systems, wind turbines). Thus, this approach has been within the focus of many research projects. The production costs of electricity from PV are today clearly below 0.05 US$/kWh due to technological advancements resulting in a sharp decrease in capital costs [14]. Renewable energy driven desalination systems allow to produce potable water economically and with much lower GHGs emission than that of current fossil fuel driven desalination plants [15].

In addition, brine (the highly concentrated saline water) is generated as by-product of a desalination system and must be handled carefully for safe disposal. Beside this, desalination systems might have even more severe impacts on the environment other than GHGs emissions and brine generation, which are, e.g., consequences of applied chemicals (e.g., acid, caustic solutions, antiscaling agents).

Suitable sustainable solutions to reduce such environmental impacts together with the use of renewable energy may lead to more environment-friendly desalination technologies [16]. Recently, photovoltaic driven decentralized RO plants for BW desalination have been investigated for constant production capacity of potable water with much higher efficiency (from PV electricity to used hydraulic energy) and evaporation pond for reducing brine volumes [17].

Following the aforementioned aspects this article presents the technical basics of desalination and introduces various desalination processes as well as pre- and post-treatment processes. Moreover, various technological strategies related to the combination of different desalination processes, and further developments and improvements (especially evaluation and minimization of energy consumption) in SW desalination are presented. Finally, potential development of desalination, freshwater demand for production of green hydrogen energy using water electrolysis and impact of SW desalination green hydrogen economy are discussed.

2 Conventional Desalination

In this section, first an overview of desalination technologies is presented. This is followed by the description on desalination pre-treatment processes, membrane-based desalination technologies, and thermal desalination. Afterward, the processing of desalination products is described and energy aspects on membrane-based and thermal desalination are addressed.

Desalination separates saline waters such as SW and BW into two streams, the fresh water and the brine solution [18]. Such desalination systems can be divided based on utilized energy sources (thermal, mechanical, chemical, or electrical) or based on the principle of desalination such as evaporation/condensation, filtration/diffusion, and crystallization (Fig. 3) [19]. Thermal energy is the source of desalination techniques like multi-stage flash (MSF), multi-effect distillation (MED), membrane distillation (MD) and humidification–dehumidification (HDH); nevertheless, these options need also electrical process energy. Compared to that, RO, nanofiltration (NF), forward osmosis (FO) and mechanical vapor compression (MVC) rely mainly on electrical energy that is converted into mechanical energy (pressure). In addition, other desalination processes like, e.g., electrodialysis (ED) and capacitive deionization (CDI) need direct electrical energy supply to separate the salt ions within an electric field [20]. Ion-exchange resin (IXR) uses a chemical energy source and filtration as a desalination process different diffusion coefficients to separate ions from water [21]. However, thermal desalination techniques work on the evaporation and condensation principle of desalination operation and most membrane-based techniques work on filtration and diffusions as shown in Fig. 3.

Forward osmosis (FO), membrane distillation MD and capacitive deionization (CDI) are considered as innovative desalination methods.

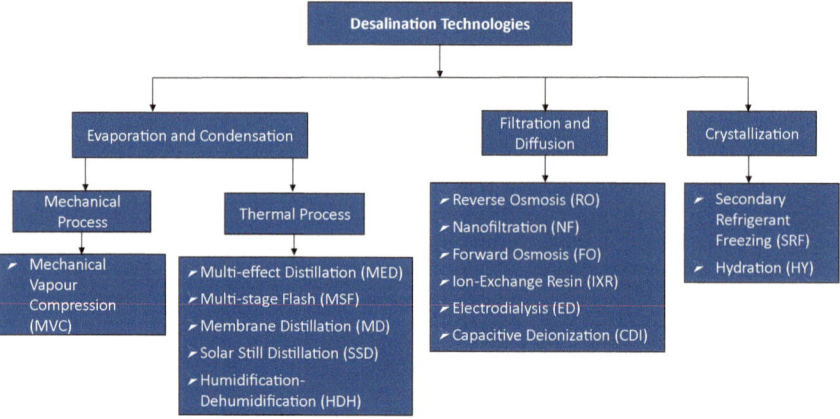

Fig. 3 Desalination technologies based on working operations. Adapted from Curto et al. [21]

- FO is a membrane separation process utilizing the osmotic pressure difference resulting from the solute concentration difference between the feed water and receiving draw solution. Accordingly, in FO water is transported into a high salt containing solution and must be extracted from this again.
- MD is a thermally driven membrane separation process in which salt separation is driven by phase changes. Water vapor is transported through a hydrophobic membrane at moderated temperatures by vacuum and condenses later as a liquid permeate stream. Due to advances in membrane materials, MD is an emerging method for the production of fresh water [22].
- CDI is a process system that removes ions (e.g., Na^+, Cl^-) from water using an electrical potential difference (electrical driving force on the ions) between a pair of electrodes (anode and cathode) made often of porous carbon. The positively charged anode adsorbs negatively charged ions (e.g., Cl^-) and the negatively charged cathode adsorbs positively charged ions (e.g., Na^+).

These innovative desalination methods may be a promising future opportunity to supply water in water-scarce regions and overcome the shortage due to climate change and population growth [23, 24].

Figure 4 shows mature conventional thermal or membrane-based desalination technologies. Among these methods, RO systems currently show the highest share of 65% produced water, followed by multi-effect distillation (MED) which contributes 21%, and multi-stage flash (MSF) as 7% of the total installed water production facilities [25].

For BW treatment thermal desalination techniques are rarely applied due to their relatively constant production costs at different source water salinities [26]. (RO concept was introduced to treat BW in the 1970s and with improved membrane materials and better peripheral equipment was later used also for SW desalination. Hence, it is currently the appropriate technique to treat both SW and brackish groundwater [3, 27]. A reason for this is that the energy requirement for the thermal processes

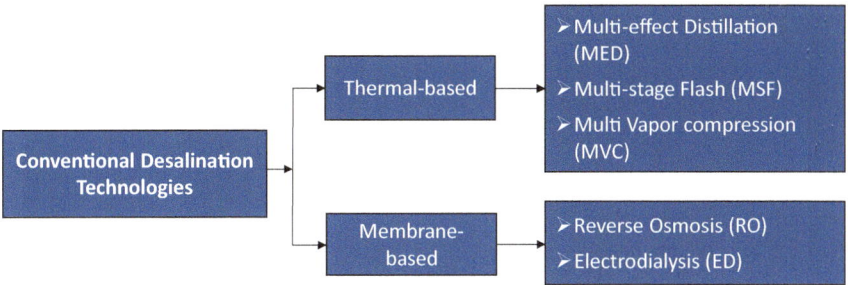

Fig. 4 Conventional desalination technologies. Adapted from Saavedra et al. [25]

is much higher and independent of the salt concentration, while the energy required for the RO concept depends on the salt concentration (osmotic pressure) [28, 29]. Thermal desalination is still widely popular in the Gulf states. RO systems leads the USA, Japan, and Spain markets, while Italy and the Caribbean Islands have balanced technologies [30]. About 70–80% of the world´s desalination uses RO and MSF technologies [31].

2.1 Desalination Pre-Treatment Process

An efficient pre-treatment system forms the backbone of a BW and SW desalination unit. Pre-treatment steps are usually common regardless of the type of desalination process. This is because high-quality feed water substantially enhances the productivity and operability of the plant, reducing downtime duration and frequency [16, 32].

In membrane-based desalination systems, for example, the aim of the pre-treatment steps is to prevent scaling of sparingly soluble salts such as calcium carbonate, gypsum, and barium sulfate over the membrane surface, fouling by organic water constituents and control microbial growth [33]. Hence, pre-treatment of salty water is important to mitigate particulate/colloidal fouling, inorganic fouling/scaling, biofouling, and organic fouling. Otherwise, membrane fouling will result in increased energy consumption and reduced life time of the membranes [32, 34]. Other implications due to membrane fouling include a decreased removal efficiency for salts, resulting in increased salt passage and thus deteriorate the permeate quality, an increase in membrane degradation leading to membrane replacement costs, and a decrease rate of water production due to extended downtime during chemical cleaning and membrane replacement [32]. All these implications lead to the increasing operational costs of the systems and thus higher water production costs.

In general, the conventional pre-treatment process can be categorized into two sections as physical and chemical pre-treatment.

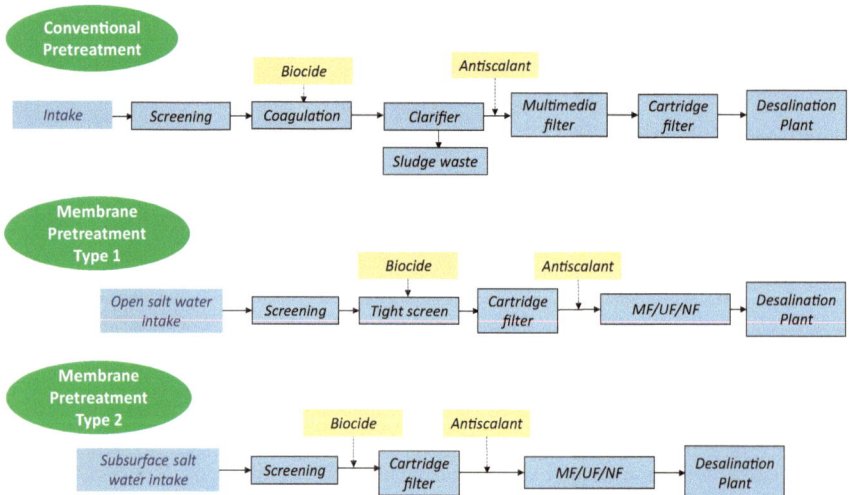

Fig. 5 Examples of conventional and membrane pre-treatment systems. Adapted on basis of Elsaid et al. [34]

- The physical process includes the basic mechanical screening of particulate matter through the screens and deep-bed/dual media filters.
- The chemical process involves the addition of scale inhibitors, possible coagulants and disinfectants [35].

Coagulation (using ferric chloride and polymer coagulants) is a prevalent conventional pre-treatment process to eliminate potential foulants such as aqueous particulate and colloidal matter by neutralizing the charge of particles/colloids. It is usually applied prior to media filtration, micro- or ultrafiltration membranes. As a result of filtration large aggregates of loosely bound suspended particles can be attached to the media filter or retained by microfiltration (MF) or ultrafiltration (UF) membranes (Fig. 5).

Microfiltration (MF), ultrafiltration (UF) and NF membrane processes are the most widely used pre-treatment systems for desalination because of their ability to remove colloidal particles, suspended matter and to some extent bacterial matter [36, 37]. Figure 5 presents conventional (made from screening, disinfection, coagulation, flocculation, sedimentation, filtration, addition of chemical agents called antiscalants) and membrane pre-treatments systems for desalination [38]. Even if particulate and colloidal fouling are well controlled by pre-treatment, the occurrence of sparingly soluble salts and biofouling may contribute to flux decline and remains a challenge that requires frequent chemical cleaning procedures [39–41]. Addition of antiscalants in thermal desalination processes is useful to curb the scaling of heat transfer surfaces, which is a major challenge in thermal desalination of BW and SW [42]. To mitigate membrane biofouling due to microbial growth on a membrane, biocides such as chlorine-based compounds such as chlorine gas, sodium hypochlorite, chlorine dioxide and/or chloramines are added in the feed water [43]. However, chlorine-based compounds my damage the membrane material, contribute to the formation of disinfection by-product. Thus, ultraviolet (UV) treatment is becoming increasingly popular as it does not produce many disinfection by-products [44]. Taking into account the quality of raw salt water, the source salt plays a major role in the selection of different pre-treatment steps. Figure 8 shows that tight screen is commonly exercised when open SW is to be desalinated for drinking water production.

2.2 Membrane-Based Desalination

RO, NF, and ED processes separate the sources of raw saline water into the permeate and the brine discharge [25]. Such pressure-driven membrane-based desalination techniques rely solely on electrical energy, eliminating the need for a thermal energy source [45]. The individual processes are explained in more detail below.

2.2.1   Reverse Osmosis (RO) Desalination

RO is a water purification process that separates water molecules from other substances such as inorganic salts by pushing them under hydraulic pressure through a semi-permeable RO membrane. The separation mechanism of RO involves solution-diffusion where water molecules dissolve into the membrane polymer material from the feed side, diffuse across the membrane, and then are desorbed at the permeate side. In practice, RO membranes retain 98–99% of the dissolved (monovalent and multivalent) inorganic ions from the feed stream into the concentrate, while the permeate can be considered as high-quality water [16].

The RO based processes can be grouped into brackish water RO (BWRO) plants where the salinity is in the range of 0.50–10 g/L and SWRO plants where the salinity is around 30 g/L. BWRO is further sub-grouped into low salinity BRWO and high salinity BRWO plants. While in the first case, feed water with a salinity between 0.50 and 2.5 g/L is processed, in high salinity BRWO plants water with a salinity between 2.5 and 10 g/L is processed [46]. SWRO has a water conversion factor of 35 to 45%, which means 35–45% of the feedwater is converted to usable permeate water and the remainder is drained as concentrate. In comparison, the WCF of BW from RO could be more than 90% [47].

Among membrane-based processes, RO desalination is considered the preferred option as it is easy to adapt, modular (plant size can be adjusted/expanded), and offers cost advantages in treating BW as well as SW. In general, the capital costs are approximately 25% less than for thermal techniques [19].

2.2.2   Nanofiltration (NF) Desalination

NF is another pressure-driven membrane separation process to retain salt ions using NF membranes and has gained widespread usage in the desalination industry owing to its ability to reject molecular or ionic species [48]. Key distinguishing characteristics of NF membranes are low rejection of monovalent ions, high rejection (> 99%) of divalent ions (e.g., Ca^{2+}, Mg^{2+}), and higher flux compared to RO membranes. NF works as a combination of convective, diffusive, and electrostatic transport mechanisms with more open membranes compared to RO membranes. NF has been used as a standalone membrane process for BW desalination with salinity up to 6 g/L. The salt rejection increases with decreasing molecular weight cut-off of the NF membrane and decreases with increasing salinity [49]. For SW desalination, NF desalination is typically not applied as a standalone membrane process due to lower salt rejection [49] but can be applied as a pre-treatment step to RO to remove the divalent ions (e.g., Ca^{2+}, Mg^{2+}) and reduce scaling potential by carbonates or sulfates on the RO feed site [50]. The advantages associated with NF-RO hybrid system in desalination include minimizing the scaling propensity of feed water, decreasing operating pressure, energy requirement and increasing membrane flux [51, 52].

2.2.3   Electrodialysis (ED) Desalination

Electrodialysis (ED) is a process where salts ions are separated from saltwater across semi-permeable ion-exchange (IEX) selective membranes under the effect of an applied electric field [53, 54]. Thus, ED is mainly based on selective movement of ions in solutions as a result of an electrical driving force (Fig. 6). The mandatory condition for an ED process to be executed is an alternating order of cation exchange (CEX) and anion exchange (AEX) membranes and electric field applied across the entire assembly. When electrical field is imposed, the positive ions migrate to the negative electrode or cathode. The negative ions migrate to the positive electrode, or anode. A CEX preamble membrane allows positive ions to pass, but blocks negative ions. An AEX permeable membrane does the opposite. It allows negative ions to pass, but blocks positive ions. As a result, saltwater that is flowing into the ED assembly is separated into desalinated (clean water) and concentrated streams (brine) [53, 54].

2.3   Thermal Desalination

The major conventional thermal desalination technologies are multi-stage flash (MSF), multi-effect distillation (MED), and mechanical vapor compression (MVC). They are technologically and operationally matured, globally commercialized and generally suitable for large water treatment capacities. Their operation is based on the utilization of thermal energy to evaporate the water and later condense the same for potable water provision. High salinity water may also be used as a source as the

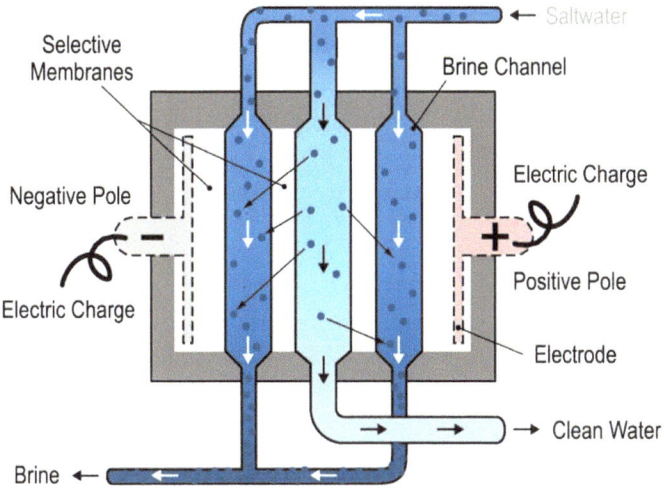

Fig. 6 Schematic diagram of electrodialysis (ED) system. Taken from Saavedra et al. [16]

process is rarely limited by maximum salinities and was realized at large capacities by wealth Middle East countries already decades ago [54].

Thermal desalination has five primary streams: (1) source water, (2) brine solution, (3) thermal energy/steam required for evaporation, (4) cooling water for condensation, and (5) distilled water with no salinity [55]. Figure 7 displays these streams of thermal desalination processes. The combustion of fossil fuel is conventionally used to satisfy the energy requirements for steam generation [56].

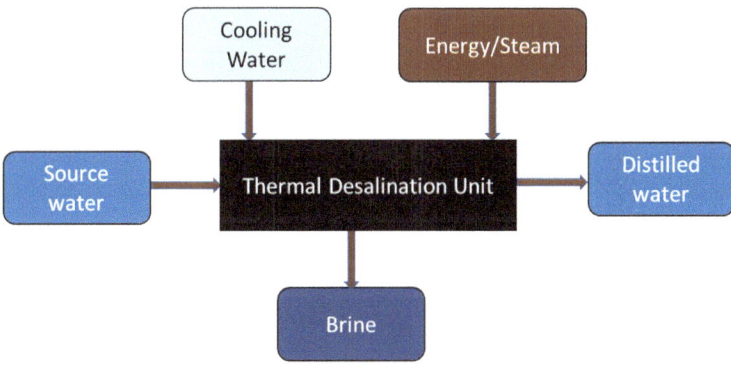

Fig. 7 Primary input and output streams in a thermal desalination process

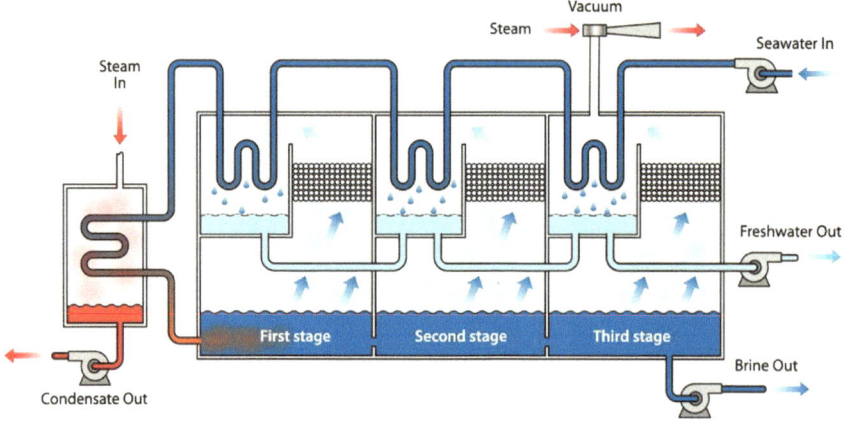

Fig. 8 The working principle of multi-stage flash distillation (MSF) plant. Taken from do Thi et al. [20]

2.3.1 Multi-stage Flash Distillation (MSF)

Multi-stage flash distillation (MSF) is known for its simple layout because the phase change is caused by a decrease in total pressure. In MSF, SW is converted to distillate water by flashing a portion of the source water into steam in multiple stages (Fig. 4). MSF typically function at high temperatures between 90 and 120 °C. The typical heat requirement is between 250 and 390 kJ/L product water [57]. Due to such high requirements in thermal energy, MSF desalination plants are reasonable in regions where significant amounts of affordable fuel energy or even waste energy is available [58].

Each section of an evaporator within a multi-stage flash distillation (MSF) is named "stage" and a conventional MSF plant may have 4–40 stages. The evaporation-cooling process is repeated at each stage and each next stage functions at a lower pressure and temperature level than the one before [25]. The multi-stage configuration allows to decrease the boiling point of the raw water compared to ambient conditions and its concentration in salt increases when leaving each stage [59]. Several times boiling is consequently achievable without providing extra heat next to the brine heater (Fig. 8). The concentration factor (ratio of the brine salinity to the feed salinity) is normally maintained at 1.5 to avoid scale formation on the heat exchanger surfaces [59].

The performance ratio (PR) and gained output ratio (GOR) are the two different performance indicators to assess the performance of such a multi-stage flash distillation (MSF) plant.

- The performance ratio is realized as the ratio between product water (in kg) and the steam input heat (in kJ) supplied to the system. The PR vary between 2.3 and 4.57 kg/kJ [39].

- The gained output ratio is a dimensionless parameter and realized as the mass ratio between product water and heating steam supplied to the system. For MSF plants, GOR values of are often between 8 and 10 with a practical maximum of 12, which corresponds to about 55 kWh/m^3 of thermal energy [39, 40].

2.3.2 Multi-effect Distillation (MED)

The MED works on the principle of heat transfer from the steam to the salty water, where the heat transfer leads to the condensation of steam into product water [25]. A MED system comprises of a series of airtight elements, named "effects". At each element, a sequence of successive condensation/evaporation processes take place in descending order of temperatures and pressures. The first effect with the highest pressure initiates the first boiling of the feed water with an external heat supply provided inside. The steam generated in one effect is applied as the energy source for the subsequent effect. While, on one side, the entering steam condenses, on the other side, the feed water boils, and thus generates additional steam. This process continues in each effect [41].

The MED was established in the late 1950s and early 1960s [27]. The capacity of MED plants in practice varies from 5,000 to 15,000 m^3/d [55]. MED has been exercised for SW desalting, but had a major practical challenge with the accumulation of scale on the boiler pipes [60]. In addition, evaporators with large surface areas are needed to reduce the temperature difference of the next effects.

The main difference between the MED and MSF processes is that in an MSF system steam is created through flashing, whereas in a MED system, evaporation of salty water is achieved through heat transfer from the steam in condenser tubes into the source water sprayed onto these tubes. This heat transfer also causes steam condensation to freshwater. In general, desalination plants based on MED are more cost-effective than these based on MSF [25].

2.3.3 Mechanical Vapor Compression (MVC)

A MVC system is basically a heat exchanger that is an evaporator/condenser. A vapor compressor is used to extract vapor produced inside the chamber. Due to the compression, the vapor increases its temperature and pressure. By raising the temperature and using a heat exchanger, the pressurized vapor can transfer heat to the saline water inside the chamber and produce vapor. After the preheating, the saline water is mixed with a brine recirculation flow. This solution is sprayed externally on the main heat exchanger inside the desalination unit.

The MVC essentially requires electricity to run the process; therefore, a small standalone desalination unit can be realized to satisfy a freshwater demand ranging from 100 to 3,000 m^3/d. [61]. MVC has been used to increase energy efficiency by extraction heating from recompressed vapors [25, 61].

The MED and MVC work on a similar heat transfer principle. The heat transfer occurs in the tubes within the multi-effect distillation (MED), while the vapor in is generated by spraying sources of water on the heat exchanger surface [21]. The MVC is used for small projects and MED is used for mid- and large-scale projects.

2.4 Desalination Products

Desalination is typically a multi-stage process that splits saltwater into two products: (a) the product stream (freshwater) and the brine (highly concentrated brine) as a by-product stream. The freshwater and brine streams are further processed for production of drinking water and mitigation of negative impacts of brine, respectively.

2.4.1 Post-Treatment of the Permeate

Desalination permeates are slightly acidic and contain very low buffering capacity due to the absence of salts. As a consequence, these permeates are soft waters. As such, the water may be aggressive and corrosive to infrastructure, resulting in adverse health and economic effects. Consequently, a post-treatment step in which the chemical stability, buffering capacity and minerals are adjusted, is invariably practiced [62]. In addition, post-treatment includes disinfection of the final product and to meet the local drinking water standards [33].

Three classical groups of post-treatment processes are available for stabilizing membrane permeates.

1. Processes that are based on direct dosage of chemicals (e.g., dosage of calcium hydroxide followed by carbon dioxide);
2. Processes that are based on mixing the desalinated water with other water sources, with or without further adjustment of water quality parameters;
3. Processes that center around dissolving calcium carbonate/calcite for alkalinity and calcium addition followed by pH adjustment using sodium hydroxide [63].

The product water must also be disinfected before entering the receiving water distribution system. Hence, the objectives of the desalination post-treatment include safeguarding human health and corrosion control in water distribution systems [64].

2.4.2 Post-treatment of the Brine Solution

A water recovery of 35–53% for RO systems using SW and BW, respectively, and up to 25% for thermal desalination systems has been quantified [65]. The rest of the stream is realized as brine. The generated brine is at least two times more concentrated in terms of TDS when compared to the feed water. The handling of the concentrate streams is a problem especially in RO desalination where undesired chemicals and antiscalants are used [33, 66]. The brine from thermal desalination shows comparatively high temperatures [2]. Some antiscalants like polyphosphonate-based antiscalants (used for inhibition of carbonate scaling on some types of membranes [67] and high brine salinity have adverse effects on various microbial communities and reduction in biodiversity as it leads to eutrophication [68, 69]. Various biotic and abiotic impacts include effects on chemical oxygen demand, salinity, turbidity, dissolved oxygen levels, physiological state, and metabolic rates of the organisms [70]. In addition, the impacts of a desalination plant on the marine environment depend on the chemical properties of the reject streams and the hydrographical and biological features of the receiving environment [71].

The mitigation measures of high salinity can be a dilution of the brine before discharge by the addition of cooling water of adjacent plant is a practice that can reduce the impacts [71]. Moreover, the discharge of the brine through the diffuser system can help in the reduction of the effects on marine life [66]. To avoid the high temperature of the discharge maximum heat dissipation by cooling towers of the reject stream can be helpful [71, 72]. Some major mitigation measure for pre-treatment of brine includes selection of suitable adsorption material, optimizing coagulant and flocculant dosage and use of biodegradable chemicals. All these steps eventually cause less harm to marine life and recovers valuable materials from the brine [34, 73].

To mitigate the effect of antiscalants, environmentally friendly antiscalants that are effective against the formation of inorganic scaling can be adopted. Application of natural or green antiscalants that are phosphorus and nitrogen free can be effective in controlling inorganic scales while preventing colloidal fouling and eutrophication. The chemical groups like acrylic or maleic based polymers, polyepoxysuccinic acid, polysaccharide-based polycarboxylates, and polyaspartic polymers are green biodegradable scaling inhibitors that can prevent the deposition of many inorganic salts including calcium carbonate, and calcium sulfate [74].

2.5   Energy Aspects of Membrane and Thermal Desalination

2.5.1   Energy Aspects of Membrane-Based Desalination

The transport properties of a membrane are determined by the membrane water permeability and operating pressure as driving force. Both of which influence energy requirements of a membrane system. The permeate flux is directly proportional to the operating pressure and is expressed by the Eq. 1 at constant temperature:

$$Q_p = A_m(\Delta P - \Delta \pi) \tag{1}$$

According to Zhu et al. [75], the operating pressure for RO membrane is generally calculated as

$$Q_p = S_m A_m (\Delta P - \sigma \Delta \pi), \tag{2}$$

where Q_p is membrane permeate flux, S_m is the RO membrane surface area, A_m is membrane water permeability, ΔP is the pressure difference between the pump inlet (P_f) and outlet (P_0), σ is the reflection coefficient (assumed 1 for high rejection membranes) and $\Delta \pi$ is osmotic pressure difference across the membrane. Fouling (the collimation of the membranes surface by particles, salts, organics, microorganisms) causes a loss in water flux and significant increase in operation and maintenance costs. Although most of the flux drop due to fouling can be avoided by suitable selection of pre-treatment processes for raw salty water. Another key equation of desalination processes is the water recovery (Eq. 3)

$$\text{Recovery} = \frac{Q_p}{Q_f} \times 100 = \left(\frac{Q_f - Q_c}{Q_f} \right) \times 100, \tag{3}$$

where Q_f, Q_c and Q_f are the feed, concentrate and permeate flow rates. The water recovery varies between 0 and 1 and is mainly restricted by the solubility product of present salts such as carbonates or sulfates.

The viability of a membrane system for production of potable water via desalination depends on the pump work and more specifically on the operating pressure required to achieve a designed permeate flux [76, 77]. The energy demand is expressed quantitatively as specific energy consumption (SEC, kWh/m^3) which is the electrical energy required to produce one cubic meter of product water and is given by the expression:

$$\text{SEC} = \frac{W_{pump}}{Q_p}, \tag{4}$$

where W_{pump} is the pump work and the mathematical expression for W_{pump} is given as

$$W_{pump} = \frac{\Delta P Q_f}{\eta_{pump}}, \tag{5}$$

where η_{pump} is the efficiency of the high pressure pump (HPP). Hence, the SEC can also be expressed in terms of water recovery as follows:

$$\text{SEC} = \frac{\Delta P}{\eta_{pump} \times \text{Recovery}} \tag{6}$$

Equation 6 shows that increasing the recovery, the SEC of the RO membrane system decreases. Nevertheless, it has negative impact on the concentrate factor (CF). CF is a function of the water recovery as shown in Eq. 7:

$$CF = C_c/C_f = \left(\frac{1}{1 - \text{Recovery}} \right) \qquad (7)$$

Assuming salt rejection efficiency η_{SR} of 1 or 100%. In SW desalination recovery rates are limited 50% and CF to < 2 to ensure the operational stability, but can increase to CF > 10 in high recovery BWRO [78, 79]. Table 2 shows the energy requirements along with main characteristics of membrane–based desalination.

Table 2 Main characteristics along with the energy consumption of conventional membrane-based desalination technologies

Method	Membrane-based desalination	
Technique	Reverse osmosis (RO)	Electrodialysis (ED)
Applicability	Commercial	Commercial
Electrical energy consumption (kWh/m^3)	Seawater (SW): 2–4 Brackish water (BW): 1.5–2.5	Normal: 2.6–5.5 Low TDS: 0.7–2.5
Advantages	• Low energy requirement, high product water quality • Lower operating cost, and investment cost compared to the thermal desalination process • Lower membrane cost compared to ED • High salt rejection (> 99% for NaCl) • Mature technique • Large-scale application	• Pre-treatment: Medium requirements • Electrical potential difference is required • Better efficiency at low salt concentrations • High recovery rate possible • Lower fouling or scaling possibility
Disadvantages	• Pre-treatment: High requirements • Prone to membrane fouling and scaling • High costs for membranes and chemicals • High external pressure required • In general need of chemical dosing (antiscalants, bactericidal additives), results in challenges for brine disposal • Possible membrane degradation by biofouling	• Prone to corrosion • Best suited for BW desalination • Ion removal only • No bacteria/pathogen removal • Regular electrode cleaning required • Thicker membranes required
References	[9, 12, 22, 26, 30, 36, 37]	

2.5.2 Energy Aspects of Thermal Desalination

A thermal desalination system is based on the evaporation/condensation process of water. With thermal energy input water vapor from the saline water source is formed/evaporated. This water vapor is condensed on a cooling surface to collect freshwater. A small portion of electrical energy is required for fluid transport and to create a pressure difference between feed and permeate side suitable for evaporation process. A general representation of a phase-change desalination process including an evaporator and a condenser is shown in below (Fig. 9).

The yield of any phase-change desalination process based on first law of thermodynamics can be written as:

$$\frac{m_f}{m_s} = \frac{\frac{Q_i - Q_l}{m_s} + (h_s - h_b)}{h_{L(T_e)} + (h_f - h_b)}, \tag{8}$$

where h_s, h_b and h_f is the specific enthalpy of saline water, brine and freshwater respectively, T_s, T_b and T_f is the temperature of saline water, brine and freshwater respectively, and Q is the heat transfer rate. Here, $\frac{m_f}{m_s}$ is process yield in kg/kg, Q_{in} is the rate of heat input and $Q_{out} = m_f h_{L(T_e)}$ is the heat rejection rate, where m_f is the freshwater production rate, latent heat of vaporization, and $h_{L(T_e)}$ is the latent heat of condensation at the evaporation temperature (T_e). The heat loss rate is $Q_{loss} = UAT$, where U is the heat transfer coefficient, A is the heat transfer area, and T is the temperature difference between the evaporation chamber and the ambient. Using Eq. 8, the freshwater production rate as a function of saline water feed rate and evaporation temperature can be predicted for a given energy input.

SEC in kJ/kg of product water/freshwater at a fixed evaporation temperature can be calculated as:

$$SEC\left(\frac{kJ}{kg}\right) = \frac{Q_{in}}{m_f} = \left[c_p(T_e - T_s) + h_{L(T_e)}\right] + \frac{Q_{loss}}{m_f}, \tag{9}$$

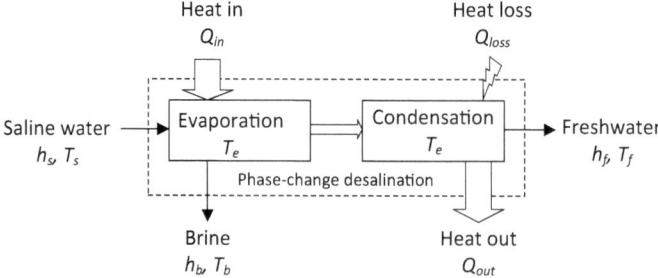

Fig. 9 Generic phase-change thermal desalination process with heat energy flows. Adapted from Gue et al. [80]

where c_p is specific heat of water at fixed pressure. According to above heat balance Eq. (9), SEC increases with rising evaporation temperature due to large sensible heat demand and increase in thermal losses [80]. Table 3 shows the energy requirements along with main characteristics of thermal desalination.

2.5.3 Theoretical Minimum Specific Energy Consumption for BW and SW Desalination

Two approaches are introduced to determine SEC_{min} of any SW desalination technology (Eqs. 10 and 11); this is regarded as one of the most important characteristic figures based on the states of the influent and effluent streams [84].

$$\text{Approach 1}\quad SEC_{min} = \int -d(\Delta G_{mix}) = \int -RT \ln a_w dn_w = \int \pi_{sw} V_w dn_w \quad (10)$$

$$\text{Approach 2}\quad SEC_{min} = \sum X_{f,out} - \sum X_{f,in} \quad (11)$$

X_f is defined according to Eq. 12.

$$X_f = (h - h^*) - T_0(s - s^*) + \sum_{i=1}^{n} w_i(\mu_i^* - \mu_i^0) \quad (12)$$

G_{mix} represents the Gibbs function for mixing, R and T is the gas constant and the temperature, respectively. a_w describes the activity of water, n_w reflects the number of moles of water, π_{sw} denotes the osmotic pressure of saline water, and V_w is the molar volume of water. $\sum X_{f,out} - \sum X_{f,in}$ denotes the sum of the flow exergy of outlet and inlet streams, respectively. w_i is the mass fraction, h, s and μ denotes the specific enthalpy, specific entropy, and chemical potential, respectively. Superscripts * and 0 denote the restricted dead state and the global dead state respectively.

- Approach 1 reveals that this theoretical minimum energy, which is independent of the desalination method, is realized when the separation occurs as a reversible thermodynamic process. Thus, the energy for the separation is equal in magnitude but opposite in sign to the free energy of mixing. This approach also exhibits a close relationship between the free energy of mixing and the osmotic pressure [85]. The rise in SW salinity increases minimum energy required for desalination.
- Approach 2 informs that SEC_{min} increases with a rising SW concentration and with increasing temperature. According to this, the effect of temperature on SEC_{min} is smaller than that of the SW concentration. The SEC_{min} of saline water treating 35 g/L salt content at 25 °C temperature with 50% recovery is 1.1 kWh/m³. This minimum energy requirements for production of potable water from saline water is the same as determined in other studies [85, 86].

Table 3 Main characteristics along with the energy requirements of conventional thermal desalination technologies

Method	Thermal desalination		
Technique	Multi-effect distillation (MED)	Multi-stage flash (MSF)	Mechanical vapor compression (MVC)
Applicability	Commercial	Commercial	Commercial
Thermal energy consumption (kJ/kg)	230–390	250–390	–
(Additional) electrical energy consumption (kWh/m^3)	1.5–5	3–5	7–12
Salient features	• Pre-treatment requirements: low • Higher TDS removal efficiency compared to RO • Effective waste heat utilization • Works at a lower temperature than MSF • Pilot plants in China and the Middle East	• Pre-treatment requirements: low • Simple operation • Higher TDS removal efficiency compared to RO • High water quality • High installed capacity: 75% of thermal desalination plants all over the world • Convenient option in regions with cheap heat energy availability	• Pre-treatment requirements: low • Higher TDS removal efficiency (< 10 mg/L) compared to RO • Effective for both BW and SW desalination • MVC needs electrical or mechanical energy only • Comparatively low energy requirements • Low maintenance requirements • Used at small municipals and resorts
Challenges	• Requires both electrical and thermal energy • Scaling of heat exchange surfaces at elevated temperatures ($T >$ 90 °C) • Hot brine discharge • Lower water recovery than membrane-based desalination	• Requires both electrical and thermal energy • High energy consumption • Prone to corrosion • High investment cost • Hot brine discharge • Lower water recovery than membrane-based desalination	• When MVC operated at T > 80 °C, pre-treatment required to prevent fouling of heat exchangers • Possible water contamination due to refrigerant • Low production capacity • When MVC operated at T < 80 °C, large compression work is required
Typical plant capacity (m^3/d)	5,000–15,000	50,000–75,000	100–3,000
References	[21, 45, 54, 55, 81–83]		

Today, for example, the SEC of RO systems treating 35 g/L salt content at 25 °C temperature with 50% recovery is 2.26 kWh/m^3 (SEC$_{min}$ is 1.1 kWh/m^3). This implies that the exergy efficiency of RO systems treating SW is 47.4% [84] and that 1.16 kWh/m^3 of additional energy is needed due to various non-ideal conditions.

3   Desalination Options at the Horizon

In this section, various emerging technologies and advancements in desalination are presented. First, novel water desalination technologies are described, followed by an introduction to hybrid desalination technologies. Afterwards further developments and improvements are addressed.

3.1   *Novel Water Desalination Technologies*

3.1.1   Forward Osmosis (FO)

FO represents a new membrane technology mainly for further concentrating salt loaded solutions driven by the osmotic pressure. However, it can also be applied for desalination. It works based on a dense hydrophilic semi-permeable membrane that separates two solutions with different osmotic pressures. The water molecules pass through the membrane from the low osmotic pressure side (feed solution) to high osmotic pressure (realized as the draw solution) due to natural osmotic pressure driving force and no external hydraulic pressure must be applied. Thus, the initial FO process involves no energy cost for transport of water molecules through the membrane except the energy required to circulate draw solution (DS) in the system to minimize external concentration polarization and membrane fouling. The osmotic driving forces in FO systems can be considerably higher than hydraulic driving forces in reverse osmotic systems [87, 88].

In FO, water permeation from the salty feed solution through the FO membrane dilutes the draw solution. The resulted dilute draw solution is then separated to a product water and solutes/salts in next step called draw solution regeneration step. Hence, SW desalination using FO process involves two steps: (1) FO unit and (2) regeneration step to concentrate the draw solution for reuse and to separate the desalinated water prior to any post-treatment.

There are two major challenges in the wider implementation of FO process. Firstly, selection of a suitable draw solution facilitating the mass transfer and secondly, determination of an energy-efficient method for recovery of draw solution in salt regeneration system [89]. The regeneration step can be thermal-based or membrane-based. Most studies employ RO systems in the draw solution regeneration step. Thus, this process is generally considered as a RO process with FO as pre-treatment [87, 88, 90].

FO has not been integrated with RO as a pre-treatment approach for regeneration of draw solution but also to protect severe scaling on the RO membrane as the membrane fouling and scaling are still the major limiting factors of the SWRO plants. Many studies have found FO as an effective fouling and scaling mitigation method. The FO membranes are typically designed to be selective toward water molecules, which enable them to reject salt, scaling ions as well as dissolved organics. In addition, the lack of hydraulic pressure-driven force forms the loosely loaded, visible salts precipitates/scales on the surface of FO membrane and thereby, the permeate flux of FO process is virtually not impacted by scaling [76, 87]. In comparison to FO membrane, high hydraulic pressure (greater than the osmotic pressure of the saline water) forms the scaling irreversible on RO membrane and compels to apply chemical cleaning to destroy salt precipitates. So, when FO process are applied as pre-treatment of RO process, FO membranes have shown low scaling, thus offering more stable operating conditions with longer and more sustainable water production without cleaning [76, 87, 88].

3.1.2 Capacitive Deionization (CDI)

CDI has emerged as an innovative desalination technology and has found its application in BW and SW desalination [91–93]. CDI is considered to be an alternative desalting technology to the thermal and membrane desalination technologies [92, 93]. This concept is an electro-sorption process that acts as a flow-through capacitor consisting of pairs of electrodes. In CDI systems, low external voltage (in the range of 1–1.4 V) is imposed to porous, high-surface area electrodes that are placed parallel at some distance and distanced is maintained by means of spacer and/or space compartment. Saline water flows between the electrodes. An external potential is applied to the electrodes and applied potential difference creates an electrical field that causes the cations (e.g., Na^+, Ca^{2+}) and anions (e.g., Cl^-, SO_4^{2-}) of the dissolved salts (e.g., NaCl, $CaSO_4$) to travel via electrostatic attraction (without charge transfer) toward the oppositely charged electrodes and subsequently, the ions are adsorbed temporarily into electrical double layer (EDL) of the charged electrode. Thus, the resulting effluent possesses significantly lower content of salt. The solution is continuously desalinated as it flows through successive pairs of porous electrodes [93]. When the porous electrodes become saturated, the electrodes are regenerated by reversing electrical polarity (electro-desorption), allowing the attracted salt ions to release into a brine stream. During current discharging/electro-desorption, the charges leaving the CDI cell can be leveraged to recover energy (analogously to the energy from a discharging electric capacitor) and considered as energy-efficient strategy to reduce the costs. A major challenge during electro-sorption/electrode regeneration is that the electrically sorbed salts ions are desorbed as brine stream. But at the same time, the oppositely charged ions from the brine stream are adsorbed on the electrodes. Therefore, the amount of salt that is electrically sorbed in the first desalination cycle is not desorbed completely during the electro-desorption step. The un-desorbed ions in the electrodes contributes in the decrease of the electro-sorption

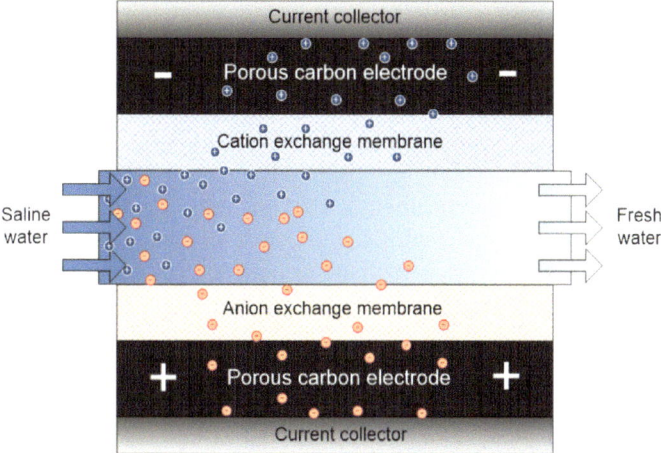

Fig. 10 Schematic elucidation of membrane capacitive deionization (MCDI) system. Taken from Folaranmi et al. [96]

performance of salts species onto electrode materials in the next desalination cycle [94, 95]. Recently, researchers have found a promising solution to solve this issue of lower electro-sorption efficiency in following electro-sorption cycles by integration of an ion-exchange (IEX) membrane on top of the porous electrodes (MCDI) (Fig. 10).

In most cases, porous carbon electrode is deposited or immobilized on a current collector whose function is to provide structural support to the chamber separators [93, 94, 96]. The membrane is positioned next to electrodes either as an unintegrated layer (with a thickness of 50–200 μm) or can be fabricated directly onto the electrode surface). In comparison to CDI systems without membranes having ion-exchange (IEX) properties, when the positive cell potential is applied to a saturated electrode during electro-desorption the cations from the anode are electrically desorbed in the membrane and pass through the cation exchange (CEX) membrane into water stream, while the incorporated cation exchange (CEX) membrane does not allow the anions to permeate through membrane and adsorb electrically on the electrode surface. Likewise, in the case of an anion exchange membrane (AEX), the cations will not penetrate through in the electro-desorption step. Thus, no electro-sorption occurs when ion exchange (IEX) membranes are integrated, leading to the availability of fully regenerated electrodes in the next electro-sorption step, and hence, no loss of electro-sorption efficiency takes place [94]. In addition to improve the electro-sorption efficiency, electrical energy used to charge electrodes during electro-sorption could be recovered during electro-sorption/current discharges. It has been reported through lab-scale investigations that up to 83% of the energy can be recovered when the charging and discharging of the electrodes are well-controlled using constant current conditions. In this case, the SEC of membrane capacitive deionization (MCDI) system can be 0.26 kWh/m^3 for BW desalination with water recovery of 80% and salt rejection of 80% [95, 97].

A suitable material for electrodes of CDI electrode should own a large surface area, large adsorption capacity, high porosity (both meso and micropores), relatively high electrical conductivity, electrochemical stability, adsorption–desorption kinetics, good wetting behavior (hydrophilicity), low cost, and scalability. Carbon electrodes possess most of these characteristics. Carbon electrode materials that have been tested for CDI are activated carbon, carbon aerogel, carbon cloth, templated nanoporous carbons, carbon nanofibers, carbon nanotubes, and graphene and graphene-based composites. Bio-based carbon materials that are primarily derived from biomaterials (many agricultural residues and wastes) have also been tested as electrode materials to minimize electrode costs. Such biomaterials have shown promising results [91].

Research and commercialization efforts on CDI have exponentially grown in the last decade but it is limited mostly to low salt content waters up to 10 g/L like BW due to lower salt removal capacity (measured as amount of salt removed per unit mass of electrode) [98]. CDI and MCDI have not been explored for the SW desalination due to the high salt content. In order to consider the MCDI for SW desalination, there is a need to first increase the salt removal capacity [99].

CDI is considered to be a cost-effective alternative to BWRO systems [98]. However, the energetic performance of CDI concepts has recently been compared with that of a BWRO system under constant working condition (2 g/L feed salinity, 50% water recovery, 75% salt rejection, and 10 L/(m^2 h) average water flux) and found that the SEC of the CDI concept (without energy recovery during electrodesorption) is eight times higher than of a RO system (without energy recovery from brine). To minimize the cost of CDI process, different efforts including use of bio-based carbon electrodes have been attempted; but they contribute insignificantly to the reduction of energy consumption of CDI concepts [100].

3.1.3 Membrane Distillation (MD)

A concept of combining thermal and membrane-based desalination has emerged as a new method called MD. It allows only water vapor to transfer through a hydrophobic porous membrane. This concept is thus based on a temperature-dependent process and requires a source of heat for the vaporization of feed water. One side of the hydrophobic porous membrane faces hot salty water and the other side faces cold product water. In this way, a temperature gradient is established between the membrane sides, creates a vapor pressure difference, and drives the water vapor to move through the hydrophobic membrane and condensed on permeate side of membrane to liquid water as product water. Thus, MD is a desalination method based on water evaporation and vapor filtration [22, 28, 101]. MD is considered as an attractive technology for BW and SW desalination as well as for the treatment of brine of SWRO and thermal desalination methods. The treatment of brine is possible because hydrophobic porous membranes in MD exhibit low sensitivity to concentration polarization [102]. Several benefits of MD include low operating temperatures compared to other concepts. Heat waste can be utilized and material requirements of membrane are lower compared to RO with respect to mechanical strength [22, 31, 101, 103].

Four MD configurations have been investigated to maintain the driving force on feed and permeates sides of the hydrophobic membrane. These are direct contact membrane distillation (DCMD), air gap membrane distillation (AGMD, vacuum membrane distillation (VMD and sweeping gas membrane distillation (SGMD) (Fig. 11).

- In DCMD, direct contact is established between condensing fluid (usually pure water) and permeate side of the membrane surface.

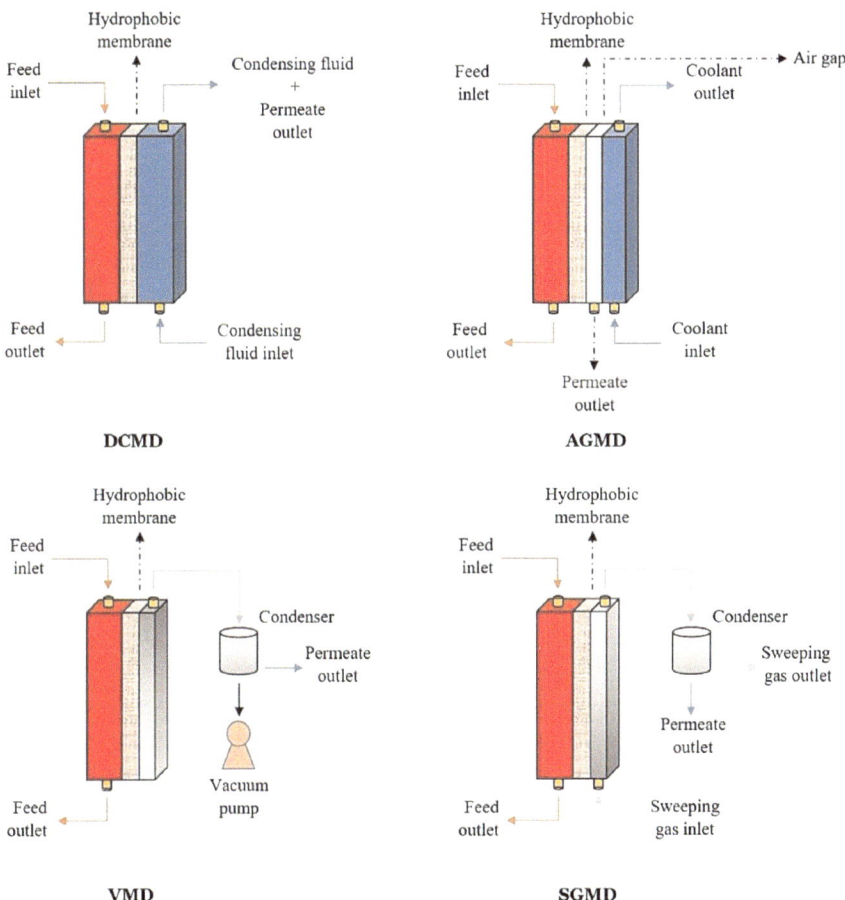

Fig. 11 Schematic elucidation of various membrane distillation configurations. Taken from Abdelrasoul [104]

- In comparison, a stagnant air layer is placed during AGMD between membrane surface on permeate side and condensing wall to reduce heat loss by conduction.
- In VMD configuration, a vacuum pump is employed to suck the permeated vapor molecules from the permeate side of the membrane and condensation occurs outside of the membrane module by means of an external condenser.
- In SGMD configuration, condensation of water vapor takes place outside of membrane module like VMD, but a cold inert gas is blown instead of vacuum permeate side of membrane module to sweep permeated vapor molecules.

All types of MD have proven to be an effective approach for SW desalination except SGMD which is found to be promising solution for BW desalination [104]. Among these configurations, DCMD is the most widely and comprehensively investigated due to its simple membrane module design and large permeate flux than other configurations [103–105].

Similar with other membrane separation processes including RO, fouling of MD is the principle challenge [103]. However, the effect of membrane fouling on MD performance is lower than other pressure-driven membrane processes. In MD, the fouling layer caused by organic matter, mostly suspended constituents on porous membranes, is mostly hydrated and does not thus hinder physically the transport of evaporated water vapor through the hydrophobic membrane pores. Instead, fouling layer helps to reduce the temperature at the interface between the membrane and the vapor and lowers the driving force for the transport of water vapor. Under laboratory investigations, MD proved to be suitable for high content salt water (120–160 g/L) when UF and coagulation was applied as pre-treatment and provided the best results in mitigating the membrane fouling during VMD [103, 105].

In recent time, new concept of application of super hydrophobic membranes (e.g., using surface deposition of fluorinated nanoparticles on membrane surface) inspired by lotus leaf has been investigated. The porous membranes are rated as super hydrophobic membranes when water contact angle is more than 150 °C (water contact angle is quantitative measure of wetting of a membrane material by water). Such membranes have shown promising results to overcome fouling issues owing to their much higher wetting resistance against various low surface tension organic compounds and self-cleaning properties and thereby, improved water vapor permeability [103, 106]. However, long-term tests should be carried out to confirm the results [101].

In 2021, a pilot MD system based on AGMD for small-scale SW desalination was investigated in Vietnam with aim to augment fresh water supply. The pilot MD system was operated with an evaporator inlet temperature of 70 °C and a coolant inlet temperature of 25 °C. The treatment capacity amounted to 46 L/h of product water. The SEC of this pilot MD system was measured to be 87 kWh/m^3, which results in production costs of approx. 20.5 US\$/m^3 of product water. Due to high water production costs, MD has not found large-scale industrial application although a number of pilot systems have been carried out. The major identified challenges include membrane scaling and fouling that contributes to underperformance of MD system due to low long-term stability [105]. In addition, lack of appropriate membrane module and high SEC are another challenges for commercialization and widespread application of MD [104].

3.2 Hybrid Desalination Technologies

Hybridization in desalination is considered to overcome the drawbacks of widely used conventional desalination methods by integrating several techniques to improve sustainability aspects. The main objective of coupling different desalination strategies is to maximize gross recovery, reduce the amount of brine discharge, minimize energy requirements and thus the environmental characteristics as well as specific costs [107]. Technically, there are two strategies to hybridize the desalination technologies. In homogenous hybrid desalination systems, the same types of desalination technologies either membrane or thermal-based are combined. On the other hand, the heterogeneous configuration of hybrid desalination technologies involves two different desalination technologies (e.g., a membrane with a thermal process [34]) such as MD.

3.2.1 Homogenous Hybrid Desalination Plants

Different homogenous hybrid desalination systems have been employed to minimize energy consumption for desalination and to increase water recovery rates [34, 108]. One of the widespread membrane hybrid couplings is a combination between NF and RO. NF has been integrated successfully with the RO process, mainly as a pre-treatment for SW. Such a pre-treatment of SW by NF within SWRO desalination plants resulted in such benefits as follows:

1. prevention of RO membrane fouling by the removal of turbidity and bacteria,
2. elimination of scaling in SWRO membrane by removal of scale forming Ca^{2+} and Mg^{2+} ions,
3. decreasing hydraulic pressure required to operate SWRO plants by reducing salinity of feed SW (TDS decreased by 30–60%) depending on the type of NF membrane and operating conditions [109].

Using NF within SWRO configurations, the recovery of hybrid SWRO plant can be increased to double (27.8–55.1%) [110], permeate flow can be increased by 48% [111], 3.5% of energy can be saved and water production cost can be reduced by about 0.18 US\$/m^3 compared to the conventional SWRO [112].

Recently, a new concept of hybridization of ED and RO has been studied [108] at the lab-scale for BW desalination. Electrodialysis was employed in the first step to desalinate BW and RO was considered to treat the highly concentrated brine of the electrodialysis system. This ED-RO hybrid configuration led to gross recovery of 87.5%. The SEC and the production costs were reduced to 0.73 kWh/m^3 and 0.37 US\$/m^3, respectively. The performance of this hybrid system in terms of energy consumption, water production, water recovery, and water cost is far better than those of standalone RO and ED systems for BW desalination. In case of SW desalination, standalone RO systems operate at limited recovery rates due to issues related to scaling and retentate salinity. In comparison to RO, ED systems can be operated at higher recovery rates, although this is also eventually limited by scale formation. The high recovery rates possible in electrodialysis combined with high salt removal rates in RO provide an opportunity for hybridization to produce high-quality permeate at high recovery and low SEC [109].

A hybrid FO-NF configuration has been proposed in a study for SW desalination. Solute rejection of the FO membrane was over 99.4% and NF was applied to regenerate the draw solution and produce desalinate. However, solute rejection of NF is depended on the concentration of dilute the draw solution. $MgSO_4$ and Na_2SO_4 are proposed as the most likely draw solutions for the hybrid process because rejection of these salts by NF was about 98%. The total dissolved salt of the desalinate was 113.6 mg/L, which is below the recommended guideline value for total dissolved salt set by the World Health Organization (WHO) for drinking water [113]. Compared to standalone SWRO plants, FO-NF system requires less hydraulic pressure and is less prone to membrane scaling [109]. This hybrid configuration seems to be an attractive option to reduce the SEC and to increase the water recovery ratio, but the SEC and the recovery of this configuration has not yet been determined. Another homogenous hybrid desalination system is FO-RO and has proven to be effective in mitigating RO membrane fouling and scaling.

3.2.2   Heterogeneous Hybrid Desalination Plants

Within heterogeneous hybrid desalination systems two or more desalination processes can be coupled. The major objective of heterogeneous hybridization is to take advantage of certain aspects associated with each desalination process. Similar to NF-RO homogenous hybridization, NF has been exploited mainly as a pre-treatment process for multi-stage flash to remove divalent ions with the objective of decreasing scaling issues on heat exchange surfaces in multi-stage flash at elevated temperature. A study communicated there was limited scaling problems on heat exchange surfaces due to 97% removal of calcium sulfate, 97% removal of barium sulfate, and 94.9% removal of calcium fluoride. Hybridization of NF with multi-stage flash enables multi-stage flash plant to operate at higher temperatures (> 120 °C). In consequence, 35–75% increase in distillate flow rate was achieved and energy consumption was further reduced [114].

Similarly, FO was investigated to minimize the concentration of divalent ions (e.g., Ca^{2+}, Mg^{2+}, SO_4^{2-}) in the SW to the thermal processes like multi-stage flash and multi-effect distillation. Pre-treatment of SW with FO has successfully reduced the scaling problems in heat exchange tubes of multi-stage flash and multi-effect distillation with clear advantage of increased distillate flow rate at same top brine temperature of 111 °C [115].

Moreover, new concepts on hybridization of MD with thermal and membranes processes such as MD-RO and FO-MD, ED-MD, MED-MD, and MSF-MD have been investigated to achieve combined benefits [31, 109]. In hybrid FO-MD configuration, MD was used to reconcentrate the diluted draw solution at 45 °C [109]. In another study on FO-MD, Na_3PO_4 was selected as the draw solute and 100% recovery of the draw solute was achieved using MD at a flux of 10.3 L/(m^2 h) [116]. The lab-scale studies on hybrid MED-MD and MSF-MD configuration demonstrate that the hot brine from multi-effect distillation or multi-stage flash may help to run the MD and thus, energy consumption of such hybridized desalination systems can be lower compared to conventional treatment [83, 101, 117, 118]. The optimum operating parameters for operation of MD with multi-effect distillation hot brine were 0.36 bar of vapor pressure difference, feed flow rate of 73.6 L/h, and permeate flow rate of 17.1 L/h [117].

3.3 Further Developments and Improvements

3.3.1 Reduction of Specific Energy Consumption

Many studies have shown that overall energy consumption of the SWRO plants is significantly lower than that of thermal desalination technologies. Nevertheless, there are continuous developments for lowering the energy requirements of the RO process. For instance, by developing better energy-efficient pumps, high-performance membranes, membrane module design to achieve a lower pressure drop in the membrane channel the SEC of the SWRO process has been decreased from 20 kWh/m^3 in 1970 to as low as 2 kWh/m^3 [85, 119]. However, SWRO is still an energy-intensive process as the calculated theoretical SEC_{min} for SW desalination using RO is around 1.1 kWh/m^3 at a salinity of 35 g/L with complete salt rejection and 50% recovery.

Upon detailed exergy analysis, major factor that contributes to larger SEC of SWRO than this theoretical SEC_{min} of 1.1 kWh/m^3 is the irreversible work of the high-pressure pumps (HPP) (0.70 kWh/m^3). In addition, 0.31 kWh/m^3 and 0.19 kWh/m^3 are additional SEC owing to inefficiency of the high-pressure pumps and booster pump, respectively. These energy losses can be improved by improving the efficiencies of the pumps. To reduce the energy losses due to irreversibility of the high-pressure pumps, three options have been exploited:

1. minimize irreversible work in the high-pressure pumps,
2. harvest osmotic energy from RO brine, and
3. decreasing the SW osmotic pressure.

To minimize irreversible work of the high-pressure pumps, two-stage RO configuration has shown the best results. It is possible to lower the SEC of the SWRO process by ca. 0.34 kWh/m^3 using two-stage RO configuration with moderate circulation pressure loss [120–123].

Harvesting Osmotic Energy from Reverse Osmosis Brine

As RO brine carries exceptionally large osmotic pressure. A new concept to decrease the SEC of SWRO systems has been reported by utilizing the harvested energy from RO brine. Conventional technologies such as pressure retarded osmosis (PRO) and reverse electrodialysis (RED) have been explored to harvest the osmotic energy from RO brine. Coupling of these energy-harvesting technologies with RO process can be an attempt not only to reduce the SEC of SWRO systems but also to minimize the negative impacts of brine on marine life and ecosystem [124, 125].

Pressure retarded RO was identified as a promising technology to produce clean and sustainable osmotic energy from salinity gradient. This approach allows water molecules to pass through a semi-permeable membrane from a low-salinity feed solution (e.g., river water, wastewater RO (WWRO) retentate, secondary treated wastewater) to a high salinity solution (e.g., SW, RO brine) against an applied hydraulic pressure [125, 126]. Osmotic pressure provided by high salinity solution draws water molecules through a semi-permeable membrane and gets diluted. The diluted high salinity solution then moves through a hydro-turbine to extract the converted hydraulic energy and thus electricity is generated [125]. Like FO, pressure retarded RO is also an osmotically driven membrane process but high salinity solution in the pressure retarded RO process is hydraulically pressurized and somewhat retards water permeation [126]. In pressure retarded RO, brine is used as high salinity solution to harvest the sustainable osmotic energy from RO brine. A power density of 4.6 W/m^2 of membrane area can be achieved when WWRO brine and SWRO brine at 20 bar are used as feed and high salinity solutions, respectively. The severe fouling of the feed solution (e.g., WWRO brine) on the membrane is a challenge and thus, pre-treatment of the feed solution should be considered. When WWRO brine was pre-treated with ultra-filtration and nano-filtration, power density of 6.6 and 8.9 W/m^2 of membrane area was estimated, respectively. The corresponding net power density for ultra-filtration and nano-filtration is 6.2 and 6.4 W/m^2 of membrane area, respectively. Thus pre-treatment of the SWRO brine can contribute to significant increase in power density [125]. The cost of electricity generation can lie between 0.005 to 0.024 US\$/kWh depending on the feed solution, salinity gradient, and size of membrane module [127]. It is concluded in a study that the SEC of RO can be reduced by 0.8 kWh/m^3 by harvesting osmotic energy using pressure retarded RO with 50% recovery. This translated into energy saving of 40% from the SEC of the SWRO plant [128]. According to theoretical minimum energy demand concept [124], around 0.2 to 0.3 kWh/m^3 (out of 0.7 kWh/m^3—a part of energy lost due to irreversible work in the high-pressure pumps) of energy can be saved by utilizing pressure retarded RO in SWRO plants.

Reverse electrodialysis (RED) is a technology to generate electricity from two solutions with different salinities using ion-exchange (IEX) membranes. A reverse electrodialysis process is analogous to a well-established electrodialysis process. Here, low and high salinity solutions flow alternatively through channels between stacked anion exchange (AEX) and cation exchange (CEX) membranes. Anions and cations travel in opposite directions from high to low-salinity through the stacked

AEX and CEX membranes. When an external load resistance is connected to the two electrodes, ionic diffusion from high salinity compartment to the low-salinity compartment generates an electrochemical potential recorded as electric current. The maximum power density of 3.0 W/m^2 membrane can be harvested when operated with 0.1 M NaCl as low-salinity solution and 5 M NaCl as high salinity solution. However, the power densities of reverse electrodialysis depend strongly on the kind of solution pairs and thus, should be considered prior to project planning [129]. Operation conditions (e.g., feed flowrate, feed temperature, feed concentration) and stack scaling up in reverse electrodialysis are regarded as important parameters affecting the power density. The power density can be increased by approximately 5 times by optimizing the RED system and operation conditions (e.g., improving membrane material, electrode, and electrolyte system, and decreasing the feed solution pressure drop, ion diffusion flux) [120, 130]. A study based on simulation results indicates that the RO-RED has the potential to reduce the SEC of SWRO by 0.7 kWh/m^3 [121]. One of the greatest challenges in wider application of RED is the low conductivity of currently available IEX membranes. In comparison to reverse electrodialysis, pressure retarded RO can achieve both greater energy efficiencies and higher power densities. The higher efficiency is credited to the capability of pressure retarded RO membranes to exploit more efficiently the salinity energy gradient to drive the transport of water molecules and better suppress the detrimental leakage of salts [122]. Thus, pilot-scale operation of RO-pressure retarded RO process has already been accomplished [123].

Decreasing the Osmotic Pressure of Seawater

If the osmotic pressure of the feed water decreases, e.g., by dilution of the feed stream of RO system, the specific energy requirements for desalination decrease as well. Three techniques for dilution of the feed stream are available, i.e., (1) dilution of feed by a low-saline concentrate stream, (2) dilution of feed by a partial permeate stream in the SWRO process, and (3) dilution of feed by exploiting an additional low salinity feed stream [124]. In dilution of feed by a low-saline concentrate stream, BWRO concentrate can be supplied to SWRO feed as the salt content of the BWRO concentrate is much lower than the SWRO feed. When feed with lower osmotic pressure is used, the minimum work required for desalination will be reduced as the RO system requires lower operating pressure. This technique is widely adopted in most two-pass SWRO systems [75]. In dilution of feed by a partial permeate stream in a single pass SWRO process, the first RO module produces pure permeate, while the second permeate is more saline. However, salt content of the second permeate is still lower than feed. The second permeate can be mixed with the feed to reduce the osmotic pressure of the feed water. This technique of feed dilution lowers the feed osmotic pressure as well as produces permeate with extremely low salt content because of diluted feed.

In contrast to aforementioned feed dilution techniques, the diluting feed water might be realized by the application of FO [131]. Principally, this approach shows two steps. For dilution of SW using FO, in the first step, the low saline water is extracted from wastewater using SW as draw solution (DS). In the second step, the diluted SW is used as feed stream of RO. Using FO-RO hybrid process, desalination can be done at lower SEC compared to standard SW. Some studies have concluded that SEC of the FO–RO hybrid configuration for SW desalination is approximately 1.3–1.5 kWh/m^3, indicating energy saving of 35% [131]. Recently, the SW dilution using FO and the corresponding theoretical energy consumption has been analyzed [124]. At a dilution of 33.3% and FO recovery of 50%, the theoretical minimum energy can be reduced from 1.1 to 0.66 kWh/m^3, indicating an energy saving of 0.4 kWh/m^3. Similarly, at dilution of 50 and 75%, a decrease in SEC of 0.6 and 0.9 kWh/m^3 was realized, respectively. However, FO recovery of 50% is not easy to implement practically in large-scale FO plants. Moreover, wastewater feed flow rate for FO process at 50% recovery in the first step should be double of the SW feed stream to achieve 50% dilution. In large-scale SW desalination plants, it can be extremely challenging to acquire such higher amount of wastewater to be used in the first step of FO-RO hybrid configuration to dilute the SW feed stream [124]. Further, serious bacterial biofilms and deposition of large molecular organic matter can be observed on the side of the FO membrane facing secondary treated wastewater [132]. In addition, additional SEC of FO (in the range of 0.1–0.2 kWh/m^3) for pretreatment of wastewater should be taken in account to estimate the overall SEC of FO-RO hybrid configuration. Also, the additional capital and operating cost of the FO process shall also be regarded. These limitations hinder wide spread application of the FO-RO hybrid process for SW desalination [124, 133]. It should be noted also that recycling treated wastewater for irrigation is a valuable reclamation technique, which cannot be done if wastewater becomes salty.

Development of New Reverse Osmosis Membrane Materials

In a classical RO membrane unit, high-pressure pumps (e.g., 55–70 bar) are needed for SW desalination. Accordingly, the energy demand as well as capital cost of these high-pressure pumps is significantly high. In the last decade, efforts have been made to develop mechanically stable and high-performance RO membranes aiming to decrease the energy demand (and capital cost) of SWRO [134]. It is estimated that the SEC of the SWRO could be decreased by 15% by tripling the water permeability of current RO membranes (e.g., 1 L/(m^2 h bar)) at same water recovery [135].

Vastly used polymeric SWRO membranes made of cellulose acetate (CA) and polyamide (PA) polymers are asymmetric (anisotropic) and are synthesized using the phase inversion (polymer precipitation) method where a polymer is transformed from a liquid to a solid state in a controlled manner. In such asymmetric membranes, a thin dense layer composes is supported on a more porous layer. As the active layer, the dense film defines the membrane performance (e.g., water flux and solute rejection characteristics), while the porous sublayer acts as a mechanical support [136]. Based on the structure, asymmetric RO membranes are typically classified into two distinct groups: (1) integrally skinned asymmetric membranes, and (2) thin film composite (TFC) membranes [137]. Within integrally skinned asymmetric membranes, both, the top skin layer and the porous sublayer are composed of the same material. The membrane consisted of a dense 200 nm skin layer on a thick microporous support [138]. The thin film composite membranes are essentially composites of two polymers that are casted on a woven or non-woven fabric (e.g., polyester web/ polysulfone) for mechanical support [138]. Transport through the polyamide (PA) layer is well described by the solution-diffusion model, in which permeants (e.g., water and solutes) partition into the dense polyamide (PA) layer and diffuse through it. The resultant permeability and selectivity (commonly known as perm-selectivity), of the membrane is attributed to differences in abilities and rates of species to dissolve into and diffuse through the polyamide membrane material [139]. Today, thin film composite (TFC)—polyamide (PA RO membranes are widely adopted by commercial SWRO membrane manufacturers because of their ability to achieve relatively higher water permeability at high salt rejection (e.g., 99.8%). Moreover, TFC PA RO membranes offer low-cost fabrication costs based on their technological maturity [138]. Despite of their stability over a broad range of temperature and pH values, thin film composite SWRO membranes still face some challenges [134]. The major challenge relevant for reduction of SEC is perm-selectivity.

RO materials with a high perm-selectivity are desirable because higher water permeability minimizes the applied hydraulic pressure and same selectivity [112]. However, above a certain water permeability, membranes cannot further reduce the energy consumption of SWRO because the hydraulic pressure must be higher than that of the osmotic pressure of the saline water [134]. The studies on modeling results concluded that SEC of the SWRO decreases by 12.5% when the water permeability is increased from 0.7 to 2.0 L/(m^2 h bar). As the water permeability increases from 2.0 to 3.0 L/(m^2 h bar), the decrease in SEC is only 2.5%. At same recovery, the water permeability above 3 L/(m^2 h bar) does not lead to further decrease in SEC because the thermodynamic limit is attained (minimum required hydraulic pressure to overcome the osmotic pressure of the saline water stream). Hence, the water permeability of greater than 3.0 L/(m^2 h bar) does not reduce the SEC of SWRO. However, SEC of the BWRO decreases by 80% as the water permeability increases from 0.7 to 4.0 L/(m^2 h bar) due to lower osmotic pressure in the feed water of the BW. These studies showed that there is minimum margin to reduce the SEC of SWRO by improving the water permeability when compared with BWRO (with salt content of 2 g/L) [140, 141]. Accordingly, there is more room to reduce the energy consumption of BWRO employing RO membranes with water permeability greater than 4.0 L/(m^2 h bar). In case of SWRO, the ideal SWRO membrane could be one with a water permeability of 3.0 L/(m^2 h bar) and almost complete salt rejection (> 99.8%) to achieve SEC$_{min}$ keeping in view the perspective of the thermodynamic limit [134].

Many membrane researchers over the years have developed the novel approaches to improve the perm-selectivity and antifouling properties of PA RO membranes. However, the focus of all of studies have been either application of nano-filtration (below 20 bar) or desalination of BW [112, 142]. In comparison, the operating conditions for employment of RO are much more stringent. Thus, breakthrough advancements in fabrication of thin film composite—polyamide membranes are not possible to apply in SWRO applications because the prerequisites (e.g., high salt rejection, water permeability and mechanical strength) essential for a SWRO process are not satisfied [134, 141].

Chemists have exploited more advanced membrane materials to further improve the performance of BW and SW desalination relying on membranes to the maximum extent [143]. Some membrane materials including graphene [144] and so-called "MXene" [145] have shown promising results with regards to higher salt rejection (nearly 100%) and high water permeability [143, 146]. For instance, graphene-based membranes have two to three order of magnitudes higher water permeability compared to conventional PA RO membranes due to its unique two-dimensional structure and fast mass transfer channels [135]. Although application of graphene-based membranes has been limited to the laboratory scale due to the expensive methods to synthesis them. Currently, researchers are looking at affordable methods to fabricate graphene-based membranes so that these membranes can be applied in pilot-scale applications [143].

For further energy reduction desalination membranes that remove solutes via molecular sieving have been developed to offer attractive water permeability and high rejection of monovalent [138, 147–149]. Carbon nanotubes (CNT) and carbon nanochannels, two-dimensional laminates, and nano-porous sheets haven been studied to develop molecular sieves for desalination membranes. In this mechanism of ion rejection, highly uniform, rigid pores that are smaller than the diameter of hydrated salt ions transport water and nearly completely reject ions by size exclusion [139]. For instance, modification of networked cellulose with (7 wt%) carbon nanotubes (CNTs) not only results in 93% higher water permeability and enhancement of mechanical properties but also show almost complete rejection (99.8%) of salts [148]. The increasing of perm-selectivity is attributed to the similarity between carbon nanotubes fluid transport properties and those of water transport channels in biological membranes [150].

Among other attempts to overcome the limitations of the solution-diffusion-based PA RO membranes, aquaporins have been incorporated into polymeric membrane for desalination purpose. Incorporation of aquaporins into polymeric membrane matrices can improve the perm-selectivity trade-off [151]. Aquaporins are pore-forming proteins and ubiquitous in living cells. Under the right conditions they form so-called "water channels" that can exclude ionic species. The biomimetic aquaporin membranes may achieve high water permeability and good selectivity due to its hydrophobic inner channel that rejects the ions and protons [152–154]. The most optimized aquaporin-bearing bilayer can exhibit a water permeability of 2.1 L/(m^2 h bar) [139]. However, the production costs of biomimetic desalination membranes could be unaffordable due to complex protein purification methods. The natural proteins can easily denature and loose functionality, hence it could affect the stability and create additional problems during membrane synthesis and use [155].

To overcome these limitations, a new concept of artificial water channels (AWCs), synthetic analogues to aquaporins, for the development of novel biomimetic

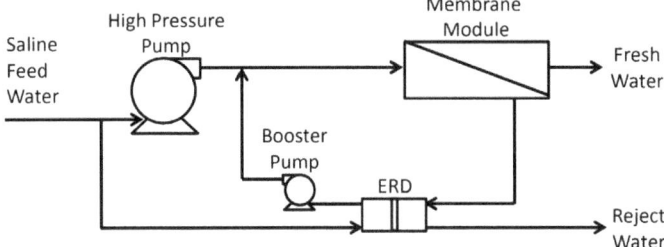

Fig. 12 Membrane desalination system coupled with dual work exchanger energy recovery system. Taken from Qiu and Davies [158]

membranes was proposed by incorporating the small carbon nanotubes into the polymeric RO desalination membranes that offer exceptionally high water permeability. For instance, a new kind of artificial water channels based on peptide-appended hybrid[4]arene (PAH[4]) has shown very strong monovalent ion rejection with a water-salt selectivity (ratio of water permeability to salt permeability) of 10^9, which is almost four orders of magnitude higher than that of modern thin film composite membranes [153]. Artificial water channels consist of unimolecular or self-assembled channels presenting an inner water conducting pore and an outer hydrophobic exterior in interaction with the lipid bilayer or polymeric membranes [155, 156]. Some artificial water channels have come close to the performance and selectivity of natural aquaporins by mimicking their structure. Protein-based membranes and artificial water channels incorporated membranes having high water-salt selectivity can play an important role in future of the desalination industry for producing high-quality water with low energy consumption [2, 157].

3.3.2 Brine Management Strategies

Enhancement of Energy Recovery

Roughly, 40–50% of the energy consumed in any desalination process is contained in the brine rejected by the RO plant [28]. This energy can be recovered by an equipment called "energy-recovery device" (ERD) and reused for pumping the saline water into the membrane system. Figure 12 shows a typical SW system equipped with an isobaric energy recovery system. Pre-treated SW is pumped to both the main high-pressure pump and the low-pressure inlet of the ERD unit. The pressurized, rejected brine (about 60% of feed SW) from the SWRO unit enters the energy-recovery device and pressurized SW leaves the energy-recovery device unit to the booster pump. The booster pump adds the small amount of pressure (to equalize the hydraulic pressure of the high-pressure pump) lost to friction in the energy-recovery device unit, the RO membranes and the associated piping. Fully pressurized SW from the booster pump merges with the SW from the high-pressure pumps.

In the absence of such an energy-recovery device, this energy contained in the brine is lost due to friction in the brine valve. Coupling of energy-recovery devices with membrane desalination systems can reduce the SEC of SW desalination by 25–40% [159]. The reduced SEC of current SWRO plants compared to former installations is mainly due to use of energy-recovery devices [119].

Energy-recovery devices can be divided into two different types namely, centrifugal and isobaric energy recovery systems based on their working principles [54, 82].

- In the centrifugal pump, the pressure in the brine is applied to the impeller, which converts it into rotational energy. This energy is then used to pump the saline feed into the plant. The first successful energy recovery systems were the centrifugal type, and most common is a Francis turbine, a Pelton wheel, and a hydraulic turbocharger [82, 159]. The energy recovery efficiency of the energy-recovery device is realized as change in feed pressure to the change in brine pressure. The energy recovery efficiencies of a Francis turbine, a Pelton turbine, and a hydraulic turbocharger are 77%, 88%, and 80%, respectively [119, 159, 160].
- The isobaric energy-recovery devices (also known as pressure-equalizing recovery system) works on the principle of pressure exchange, wherein the energy is directly transferred for pumping through a piston with no excessive energy loss due to the direct transfer of the brine pressure to the feed SW [82]. One major advantage is that its size does not affect the plant's scale and size [82, 160]. In comparison to centrifugal devices, the use of isobaric devices has increased over time due to their higher energy recovery efficiency [82]. The energy recovery efficiency of isobaric energy-recovery devices is between 95 and 97% [160] and up to 40% energy can be saved when SWRO plants are coupled with isobaric energy-recovery devices compared to standalone SWRO plants [119, 159]. Rotary and piston-driven isobaric energy-recovery devices are mostly common and available as commercial energy recovery systems. Rotary-driven isobaric energy-recovery devices such as the pressure exchanger (PEx) use a rotor with a series of ducts to convey the raw SW into direct contact with the pressurized brine. Piston-driven isobaric energy-recovery devices such as the dual work exchanger energy recovery use large cylinders with pistons and valves to pressurize and inject the SW into the membrane unit using the brine energy [161]. Such isobaric energy-recovery devices are usually applied to large-scale plants and the centrifugal energy recovery systems work better for small-scale plants that deals high-pressure concentrated SW. The flow rate of the single isobaric system is limited and hence, multiple energy-recovery devices are applied in parallel unit [159].

The SEC of a SWRO plant (43% recovery) with and without energy-recovery device are critically determined [159]. The SEC of SWRO without energy-recovery device is 5.57 kWh/m^3. The SEC can be decreased to 3.65 and 3.05 kWh/m^3 when SWRO is coupled with a Francis turbine and a pressure exchanger energy-recovery device, respectively. The energy contained in the brine was calculated to be 2.39

Table 4 Energy losses in the different constituents of a SWRO plant with and without energy-recovery device adapted to [159]

Energy loss (kWh/m^3) by	SWRO without energy-recovery device	SWRO with Francis turbine	SWRO with Isobaric chamber
High pressure pump including engine	1.25	1.15	0.68
Membrane	0.04	0.04	0.04
Energy-recovery device	–	0.57	0.08
Booster pump	–	–	0.51
Product water	0.79	0.79	0.79
Hydraulic fittings	1.10	1.10	> 1.10

kWh/m^3. Out of that brine energy, Francis turbine and pressure exchanger energy-recovery device enabled energy recovery of 1.82 and 1.98 kWh/m^3, respectively. In addition, the energy consumption of such a system is slightly lower (lower by 0.34 kWh/m^3) than of Francis turbine energy-recovery device [162]. However, the noise of pressure exchanger energy-recovery device is slightly higher than a Francis turbine energy-recovery device due to the high rotation speed of pressure exchanger device [162].

There are energy losses in commonly used energy-recovery devices and these energy losses could affect the SEC of SWRO. The energy losses are categories into energy losses in the brine, product water, due to RO membrane, and hydraulic losses. They are summarized in Table 4. As the energy recovery efficiency of the pressure exchanger device is higher than Francis turbine and hence, the size of the high pressure pump (HPP) in case of the pressure exchanger energy-recovery device is reduced because it has to manage less feed flow than in the case of the Francis turbine.

At lower feed flow rates, the energy losses in the HPP is lower (0.68 kWh/m^3). Compared to a Francis turbine (0.57 kWh/m^3), the energy losses in the pressure exchanger energy-recovery device is 0.08 kWh/m^3. But, there is an additional energy loss in the booster pump. Therefore, the energy losses due to membrane and product water are similar for the both energy-recovery device concepts. The losses in hydraulic fittings in the pressure exchanger energy-recovery device are more due to more hydraulic elements [159]. The energy losses can be reduced by improving the performance of different SWRO components.

Improvement of Environmental Aspects

Conventional brine management methods discharge brine directly into the environment (e.g., oceans, surface water, sewer discharge, evaporation ponds, and deep well injection) without any treatment. Hence, each of the above-stated brine management/

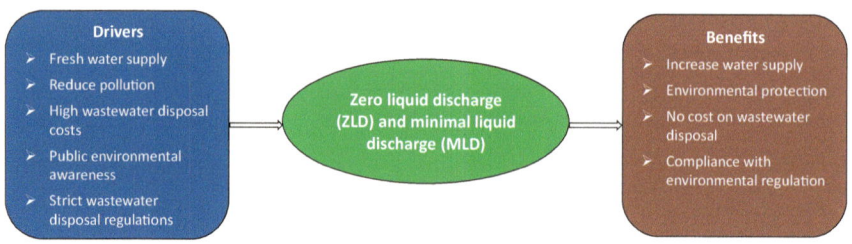

Fig. 13 Importance of zero liquid discharge (ZLD) and minimal liquid discharge (MLD) with drivers and benefits. Adapted from Tong and Elimelech [168]

disposal strategies has diverse range of negative environmental impacts (i.e., marine pollution, groundwater contamination, and soil salinization) [163, 164].

The development of an appropriate and efficient brine management that can be implemented in current desalination plants system has been the focus of many studies in the last decade. The thermal and membrane-based technologies/processes might be transferred into minimal liquid discharge (MLD) and zero liquid discharge (ZLD) systems. The target of these systems is to minimize (and completely eliminate) the volume of freshwater and liquid waste from brine, produce solid salts, recover useful resources like metals, and some chemicals like sodium hydroxide, chlorine, hydrogen, potassium chloride, hydrochloric acid, and pure salts. Figure 13 summarizes the drivers to adopt such system for brine management [165–167].

A regular zero liquid discharge system replying on thermal-based methods consists of brine concentrators and crystallizers. Brine from desalination is pretreated by adjusting pH values and filtration, and sent to a brine concentrator to achieve higher salinity (e.g., from 50 g/L to higher than 220 g/L) and the crystallizers separates salts from liquids through crystallization. Zero liquid discharge replying on crystallizers have promising results since their energy demand is not strongly related to the concentration of brine [169]. Nevertheless, energy demand of these crystallizers might be exceptionally large (52–70 kWh/m^3 includes both heat and electricity requirements) because these methods reply on phase changes [166]. Hence today, there is no full-scale desalination plant in the world that involves the zero liquid discharge crystallizer for brine management system [170].

In comparison to zero liquid discharge, up to 95% of the liquid water can be recovered in minimal liquid discharge system. The minimal liquid discharge system offers major advantage due to lower capital expenditures and energy demand compared to zero liquid discharge [165].

Membrane-based brine management systems have shown encouraging results for energy-efficient management of brine. However, the membranes used in brine management are different from conventional desalination membranes (e.g., RO membranes). Conventional RO membranes are not suitable for handling hypersaline solutions (e.g., brine) at high pressures (higher than 83 bar is not possible in case of commercial spiral wound RO membrane). Conventional RO membranes are optimally designed by selective layers for SW and SW desalination [166].

In 2021, bench-scale high-pressure membrane system employing flat-sheet RO membrane was designed, tested and found to be a durable system to withstand hydraulic pressures up to 150 bar (suitable for hypersaline brine at concentration of 117 g/L) with a rejection efficiency of around 98% [167]. Such high-pressure RO systems are expected to handle hypersaline brines with two- to three-fold lower energy consumption compared to thermal technologies. To maximize water recovery in such membrane systems, inorganic scaling needs to be studied comprehensively in hypersaline conditions. Moreover, the fundamental mechanisms and nature of membrane deformation associated with high-pressure RO are still lacking. Detailed investigations are needed to explore fully the potential of high-pressure RO system as well as thermal-based desalination for future brine management [167].

4 Outlook and Conclusion

4.1 Future Freshwater Demand and Seawater Desalination

The world is rapidly urbanizing. From 1950 to 2020, the global population living in cities increased from roughly 0.8 billion (29.6%) to 4.4 billion (56.2%) and is projected to reach 6.7 billion (68.4%) by 2050. Due to this rapid urbanization together with climate change and overall population growth, domestic water demand in urban regions is expected to increase by 50–80% over the next three decades [171, 172]. The greatest increment of 300% is expected to occur in Africa and Asia; an increase of 200% is expected for Central and South America. This growth is attributed to the increase in water supply services to urban settlements [173]. By 2050, the demand for water is thus expected to increase dramatically, but unequally across continents.

Based on facts released by the International Desalination Association (IDA) and the International Water Association (IWA) in 2022, about 1% of the world's freshwater comes from SW desalination in water scarcity-suffering regions, where more than 300 million people (about 4% of the global population) rely fully or partially on this source to fulfill their daily water needs. Countries like Uganda, Burundi, Nigeria, Somalia, Malawi, Eritrea, Ethiopia, Haiti, Tanzania, Niger, Zimbabwe, Afghanistan, Sudan, and Pakistan do not yet have established the desalination facilities to meet the demand and thus, can suffer severe water scarcity [3]. By 2050, more than half of the global population (ca. 57%) is expected to live in areas that suffer water scarcity for at least one month each year [174]. According to recent research on future global urban water scarcity, the global urban population facing water scarcity is projected to increase from 933 million (one-third of the global urban population) in 2016 to about 1,690 to 2,370 million people (one third to nearly half of global urban population) in 2050 [175].

The global desalination capacity in 2020 is approximately 97.2 Mm^3/d [23, 176, 177]. Of the global desalination capacity, reverse osmosis (RO) was producing about 69% (65.5 Mm^3/d). By 2050, the projected global desalination capacity, estimated based on logistic curves using historical data, would be 170 Mm^3/d, with RO accounting for about 80% of the total projected capacity [178]. Thus, the desalination capacity will increase almost two-fold by 2050, based on past growth trends [178].

These estimates indicate a clear dominance of seawater (SW) RO in future desalination market. Gao et al. [23] estimated the economic feasibility of desalination based on the decreasing SWRO costs and increasing water prices in different countries. By 2050, the global population dependent on SW desalination will increase by 3.2 times when compared with the population dependency on desalination in 2015. Hanasaki et al. [179] linked gross domestic product (GDP), aridity and proximity to the shore with demand for SW desalination. Based on these observed relationships, the production of desalinated water is projected to increase by up to 2.1 times before 2040. The demand is estimated to increase further by 6.7–17.3 times during the years 2041–2070.

4.2 Potential Development of Global Green Hydrogen Fuel and Water Demand

Over the past two years, requirements to meet climate targets (Paris Climate Agreement in 2016, which sets the goal to limit global warming to below 2 °C of preindustrial levels) have been intensified [180]. In addition to the tightening of the climate targets and the demand for net zero achievement by as early as 2045, there have been discussions on implementing and realizing these goals. Hydrogen has emerged as a promising climate-neutral energy carrier. Conventionally hydrogen is mostly produced by steam methane reformation (SMR) of natural gas and – in this case— referred to as "gray hydrogen" [181]. Hydrogen can also be produced via water electrolysis. The hydrogen produced by electrolysis is identified as "green hydrogen" if renewable energy is used for the water electrolysis [182].

The development of a green hydrogen economy is seen as a significant contributor to achieving the climate goals set by Paris Climate Agreement. In this sense, policymakers are working on roadmaps and deployment programs to support the development of a green hydrogen economy. In parallel, researchers are foresting the supply of green hydrogen, and evaluating the impact of green hydrogen on the fuel business, whereas companies are identifying potential future business opportunities [181, 183].

Today, the chemical industry and refineries are the main hydrogen consumers on the demand side. In 2020, around 45 Mt of hydrogen was dedicated to ammonia and methanol production in the chemical industry and nearly 40 Mt of hydrogen was consumed by refineries. Additionally, 5 Mt of hydrogen demand came from the steel industry for direct reduction of iron ore. Apparently, hydrogen demand in new applications stayed very low. For example, in transportation, less than 20 kt (0.02 Mt) of hydrogen was consumed in 2020 [184, 185]. According to the International Energy

Agency (IEA), hydrogen production in 2020 via water electrolysis accounted for only 0.03% of the total global hydrogen production and the installed electrolysis capacity in 2020 was estimated at approximately 0.3 GW_{el} [184].

Green Hydrogen Demands The International Renewable Energy Agency (IRENA) forecasted a green hydrogen demand of about 3 EJ in 2030, which corresponds to about 15 Mt_{H2} [186]. According to another study published by Hydrogen Council [187], a green hydrogen supply of 20 to 30 Mt_{H2}, assuming 4,000 operating hours of electrolyzers, is forecasted, leading to a required installed capacity of 200–250 GW_{el}.

To compare the effects of the different demand forecasts from the above-mentioned studies, the needed installed electrolyzer capacity for different operating hour scenarios using 15 Mt_{H2} and 25 Mt_{H2} (the average forecast of [187]) are shown in Table 5 [181]. Assuming this green hydrogen demand is produced via water electrolysis, 174 to 695 GW_{el} electrolyzer capacity must be installed by 2030.

Considering the current installed electrolyzer capacity of about 0.3 GW_{el}, it is clear that there is a huge gap to fill the demand expectations. To increase the installed electrolyzer capacity, the manufacturing capacity of the electrolyzers shall also be increased significantly. In 2022, global electrolyzer manufacturing capacity reached almost 11 GW_{el} per year, which is an increase of a more than 25% compared to 2021. Europe and China account for about two-thirds of global manufacturing capacity [188]. In a recent study [181], three market scenarios based on hydrogen electrolyzer manufacturing capacity were developed to evaluate different markets in the coming years.

- Market scenario 1: average manufacturing capacity growth rate of 10% p.a. from 2026 to 2030 (conservative estimation)
- Market scenario 2: average manufacturing growth rate of 40% p.a. from 2026 to 2030 (extrapolation of the average growth rate from 2021 to 2025)
- Market scenario 3: average growth rate of 200% p.a. from 2026 to 2030 (necessary to reach net zero climate target).

With the existing and announced capacities and an average annual growth rate of 10% p.a. between 2026 and 2030 (market scenario 1) an installed capacity of around 150 GW_{el} will be operational in 2030, which would not be sufficient to fill the projected demand of 15 Mt_{H2} from the IRENA study. With an average growth

Table 5 Installed electrolyzer capacities to produce 15 Mt_{H2} and 25 Mt_{H2} of green hydrogen depending on the operating hours

Operating hours (h/year)	Installed electrolyzer capacity (GW_{el}) to achieve 15 Mt_{H2}	Installed electrolyzer capacity (GW_{el}) to achieve 25 Mt_{H2}
2,000	417	695
4,000	208	348
8,000	104	174

rate of 40% p.a. between 2026 and 2030 (market scenario 2), the installed capacity in forecast scenario 1 will reach approximately 200 GW_{el} in 2030, which will be sufficient to fill the prospected demand from IRENA assuming average operating hours of 4,000 h/a (or more). Market scenario 3 assumes a growth rate of 200% for the manufacturing capacities starting in 2026. If these capacities are fully used and deployed two years later, the installed capacity will reach around 900 GW_{el} in 2030, which would actually exceed the IEA's scenario projection for 2030 for achieving the net zero climate target by 2045 [184]. This means it is not only necessary to foster demand for green hydrogen to build a green hydrogen economy. At the same time, hydrogen production technologies like water electrolyzers have to be installed to cover this demand.

Water Demand for Electrolysis The freshwater consumption for water electrolysis can be evaluated based on Eq. (13) [189] (Si).

$$\text{Freshwater consumption}(FWC) = (FWC_{mol} + FWC_{cool}) \\ + (FWC_{ele} + FCW_{mat} + FCW_{dis}), \quad (13)$$

where FWC_{mol} refers to direct molecular FWC to split water into hydrogen and oxygen; FWC_{cool} refers to direct cooling water consumption; FWC_{ele} refers to indirect water consumption embodied in electric energy consumption; FCW_{mat} refers to indirect water consumption embodied in material and equipment in electrolyzers; and FCW_{dis} refers to indirect water consumption embodied in waste disposals. Based on Eq. (13), the production of 1 kg of hydrogen consumes about 0.04 m^3 of water when solar photovoltaics (PV) is used, and 0.02 m^3 of water when wind electricity is used [189]. Assuming over 3 EJ of green hydrogen are produced globally in 2030 by water electrolysis, the freshwater demand would vary from about 0.3 to 1.0 billion m^3 per year (0.8 to 2.7 Mm^3/d), depending on the source of renewable energy (Table 6).

Assuming all of the hydrogen demand is met by water electrolysis using renewable energy sources (either solar PV or wind energy), the freshwater consumption is still small compared with the global figure of 2,770 billion m^3 per year for agriculture (the largest consumer), and 251 billion m^3 per year for industrial uses including fossil fuel energy production and power generation [190]. The freshwater requirements are quite low when all hydrogen in future will be produced using renewable energy.

Table 6 Freshwater demand to produce 15 Mt and 25 Mt of (green) hydrogen depending on the source of electricity

Hydrogen demand	Freshwater demand (billion m^3 per year) for water electrolyzer	
	Solar PV	Wind energy
15 Mt_{H2} (IRENA)	0.6	0.30
25 Mt_{H2} (average of [187])	1.0	0.50

4.3 Desalination of Saltwater for Green Hydrogen Production

Accessible freshwater makes up about 3% of the planet's water [79], and it is best to avoid creating any additional burden on freshwater usage, especially in areas where freshwater is scare and/or difficult to attain. However, almost all the remaining 99% (about 1.4 billion km^3) is SW, which can be purified through desalination processes before being used as an electrolysis feedstock. The leading desalination technology today is RO. Current state-of-the-art RO desalination plants, such as the Ashkelon plant in Israel, can achieve water recoveries of up to 50% [79], meaning that twice the amount of water desired for the water electrolyzers must be desalinated using the RO plant. To meet the green hydrogen demand of 15 Mt$_{H2}$ in 2030, maximum of 1.0 billion m^3 of water per year is required for water electrolyzers. This is around 2.5 ppb (2.5×10^{-7}%) of the world's available supply of SW each year, a negligible amount compared to the available SW resources.

Adding a RO desalination process to supply the water for the water electrolyzer increases the energy requirement of the life cycle of electrolytic hydrogen production, but this too is negligible in comparison to powering the electrolyzer itself. Overall, RO requires 2 to 4 kWh for each m^3 of clean water produced. For a global green hydrogen demand of 15 Mt, this yields an additional 0.0006–0.0011 EJ of annual energy required to perform RO for water electrolysis. This applies that, i.e., 0.019 to 0.038% of the minimum energy required to produce the hydrogen by electrochemical water splitting [190]. From an economic viewpoint, desalination by RO would add an energy cost of 0.53 to 1.50 US$/m^3 of clean water produced [79], which would add no more than 0.01 US$ to the cost of hydrogen production per kg [190]. Khan et al. [191] found that RO desalination would comprise a maximum of 0.1% of the energy requirement of water electrolyzers and add 0.02 US$ to the cost of hydrogen per kg. Therefore, even if desalination processes were integrated into green hydrogen production, the US Department of Energy (DoE) target to produce green hydrogen for less than 2.0 US$ per kg would still be within reach [192].

In conclusion, desalination is a potential solution to water scarcity and the supply of potable fresh water. The conventional desalination technologies, thermal and membrane-based desalination gives high water quality and are reliable methods. Pre-treatment of raw saline water is an integral part of both, thermal and membrane-based desalination methods for the removal of suspended particles, pathogens, inorganic salts and organic compounds. These impurities in raw saline water might cause not only membrane fouling/scaling but also possible scaling on heat exchange surfaces in thermal desalination. Pre-treatment systems relying on conventional (e.g., media filtration) and/or porous membrane filtration are employed to remove all particles before the main desalination system. Membrane processes (e.g., MF/UF/NF) are increasingly used as pre-treatment for thermal and RO desalination as an alternative to conventional filtration because membrane filtration offers a sustainable and cost-effective pre-treatment of saline water.

Over the past few decades, efforts have been made to advance thermal and membrane desalination processes to minimize some barriers (exceptionally high SEC of thermal desalination methods, RO membrane fouling and scaling). MD, CDI, and FO are novel desalination technologies that have been investigated in the last decade and have promising results. However, the up-scaling of these novel desalination methods needs to be investigated. The hybridization of well-established technologies (RO, ED, MSF, MED) with membrane filtration (e.g., NF) and novel desalination methods can be beneficial as it helps overcome the drawbacks (e.g., effective utilization of rejected brine, high energy consumption, low water recovery rates and salt rejection, membrane fouling) of standalone techniques. The concept of hybridization of desalination methods with proper selection of several heterogeneous and homogenous desalination systems could be suitable to utilize the benefits of each desalination method and achieve the objectives of pre-treatment systems.

The current SWRO are still energy-intensive (SEC is twice than thermodynamically demanded) and thus different strategies have been considered to lower the SEC. Minimizing the irreversibility in HPP, decreasing the osmotic pressure of SW and recover osmotic energy from RO brine are the main strategies that can decreases the SEC of SWRO. Two-stage RO configuration is recommended to minimize the irreversibility in high-pressure pumps. Two-stage RO configuration has the potential to decrease SEC of the SWRO process by more than 0.3 kWh/m^3. The RO feed stream is usually diluted to reduce the osmotic pressure of the SW. The dilution of the feed stream can be engineered by utilizing a recycle stream (by mixing the second permeate with FO feed stream) in the SWRO process and exploiting an additional feed stream. Utilizing a recycle stream in SWRO does not enable decrease of SEC. However, it could produce higher-quality product water than the conventional single-stage RO process. In comparison, dilution of feed stream by exploiting an additional feed stream using treatment of secondary treated municipal wastewater with FO could decrease the SEC of about 0.2 to 0.3 kWh/m^3 at FO recovery of 50%. This strategy of reducing the osmotic pressure of SW is an attractive solution to decrease SEC only when large amounts of secondary treated municipal wastewater are available. The recovery of osmotic energy from RO brine is suggested by integrating the RO and RED and the actual obtainable energy is approximately 0.14 kWh/m^3. If all the suggested strategies are successfully applied, the SEC of current SWRO could be decreased by 35%.

It is also estimated that SEC value of the SWRO could be reduced by 15% if the water permeability of the current RO membrane is increased by three times at same water recovery. Novel membrane materials (e.g., aquaporin proteins, carbon nanotubes, nanoporous graphene, and graphene oxide) have emerged as promising candidates for synthesizing high-performance RO membranes with high water permeability. These materials can potentially achieve relatively higher water permeability higher compared to the state-of-the-art TFC PA RO membranes. The SEC of the SWRO can be decreased further by utilization these high-performance membrane because the energy losses due to pressure drop across the RO membrane could be lowered, which might lead to the reduction of energy losses in the high-pressure pumps and hydraulic fittings. Today, the focus of research is on the development of sustainable cost-effective brine management systems based on high-pressure RO process because the energy demand of current brine management systems replying on thermal technologies is exceptionally high (52 to 70 kWh/m^3 includes both heat and electricity requirements). Such high-pressure RO systems are expected to handle hypersaline brines (>70 g/L) with two- to threefold lower energy consumption compared to thermal technologies. However, some challenges such as development of high-pressure RO apparatus as well as membrane deformation and compaction under high pressures (e.g., 150 bar) and its impact on membrane permeability need to be addressed.

Nevertheless, desalination is currently one of the most researched and applied techniques to meet the water demand. Compared with thermal desalination, RO desalination is the well-developed and well-optimized desalination method and has already takeover the thermal desalination market. This trend is expected to increase continuously all over the world due to low SEC of SWRO plants compared with thermal desalination plants.

References

1. Mekonnen MM, Hoekstra AY (2016) Four billion people facing severe water scarcity. Sci Adv 2:e1500323. https://doi.org/10.1126/sciadv.1500323
2. Ihsanullah I, Atieh MA, Sajid M, Nazal MK (2021) Desalination and environment: a critical analysis of impacts, mitigation strategies, and greener desalination technologies. Sci Total Environ 780:146585. https://doi.org/10.1016/j.scitotenv.2021.146585
3. Dhakal N, Salinas-Rodriguez SG, Hamdani J, Abushaban A, Sawalha H, Schippers JC, Kennedy MD (2022) Is desalination a solution to freshwater scarcity in developing countries? Membranes (Basel) 12. https://doi.org/10.3390/membranes12040381
4. McCleskey RB, Cravotta CA, Miller MP, Tillman F, Stackelberg P, Knierim KJ, Wise DR (2023) Salinity and total dissolved solids measurements for natural waters: an overview and a new salinity method based on specific conductance and water type. Appl Geochem 154:105684. https://doi.org/10.1016/j.apgeochem.2023.105684
5. Chakraborty R, Khan KM, Dibaba DT, Khan MA, Ahmed A Islam MZ (2019) Health implications of drinking water salinity in coastal areas of Bangladesh. Int J Environ Res Public Health 16. https://doi.org/10.3390/ijerph16193746
6. Voutchkov N, Kaiser G, Stover R, Lienhart J, Awerbuch L (2019) Sustainable management of desalination plant concentrate-desalination industry position paper-energy and environment committee of the International Desalination Association (IDA). In: The international desalination association world congress on desalination and water reuse, pp 1–32
7. Ashraf HM, Al-Sobhi SA, El-Naas MH (2022) Mapping the desalination journal: a systematic bibliometric study over 54 years. Desalination 526:115535. https://doi.org/10.1016/j.desal.2021.115535
8. Global water intelligence (2020). In: Microbial desalination cells for low energy drinking water. 32nd Worldwide Desalting Plant Inventory Media Analytics Ltd, p 214
9. Riera JA, Lima RM, Hoteit I, Knio O (2022) Simulated co-optimization of renewable energy and desalination systems in Neom, Saudi Arabia. Nat Commun 13:3514. https://doi.org/10.1038/s41467-022-31233-3
10. Williams J (2022) Desalination in the 21st century: a critical review of trends and debates. Water Altern 15(2):193–217

11. Khordagui HK (1999) Desalination. In: Environmental geology. Kluwer Academic Publishers, Dordrecht, pp 124–125, ISBN 0-412-74050-8
12. Naseer MN, Zaidi AA, Khan H, Kumar S, Owais MTB, Wahab YA, Dutta K, Jaafar J, Uzair M, Johan MR et al (2022) Desalination technology for energy-efficient and low-cost water production: a bibliometric analysis. Green Processing and Synthesis 11:306–315. https://doi.org/10.1515/gps-2022-0027
13. Sanna A, Kaltschmitt M, Ernst M (2019) PV-betriebene Umkehrosmoseanlage zur Meerwasserentsalzung – Modellierung und Analyse verschiedener Energieversorgungsvarianten. Chem Ing Tec 91:1853–1873. https://doi.org/10.1002/cite.201900095
14. Bakhshi-Jafarabadi R, Sadeh J, Dehghan M (2020) Economic evaluation of commercial grid-connected photovoltaic systems in the Middle East based on experimental data: a case study in Iran. Sustainable Energy Technol Assess 37:100581. https://doi.org/10.1016/j.seta.2019.100581
15. Nassrullah H, Anis SF, Hashaikeh R, Hilal N (2020) Energy for desalination: a state-of-the-art review. Desalination 491:114569. https://doi.org/10.1016/j.desal.2020.114569
16. Amy G, Ghaffour N, Li Z, Francis L, Linares RV, Missimer T, Lattemann S (2017) Membrane-based seawater desalination: present and future prospects. Desalination 401:16–21. https://doi.org/10.1016/j.desal.2016.10.002
17. Sanna A, Buchspies B, Ernst M, Kaltschmitt M (2021) Decentralized brackish water reverse osmosis desalination plant based on PV and pumped storage—Technical analysis. Desalination 516:115232. https://doi.org/10.1016/j.desal.2021.115232
18. Khawaji AD, Kutubkhanah IK, Wie J-M (2008) Advances in seawater desalination technologies. Desalination 221:47–69. https://doi.org/10.1016/j.desal.2007.01.067
19. Alkaisi A, Mossad R, Sharifian-Barforoush A (2017) A review of the water desalination systems integrated with renewable energy. Energy Procedia 110:268–274. https://doi.org/10.1016/j.egypro.2017.03.138
20. Do Thi HT, Pasztor T, Fozer D, Manenti F, Toth AJ (2021) Comparison of desalination technologies using renewable energy sources with life cycle, PESTLE, and multi-criteria decision analyses. Water 13:3023. https://doi.org/10.3390/w13213023
21. Curto D, Franzitta V, Guercio A (2021) A review of the water desalination technologies. Appl Sci 11:670. https://doi.org/10.3390/app11020670
22. Tijing LD, Woo YC, Choi J-S, Lee S, Kim S-H, Shon HK (2015) Fouling and its control in membrane distillation—A review. J Membr Sci 475:215–244. https://doi.org/10.1016/j.memsci.2014.09.042
23. Gao L, Yoshikawa S, Iseri Y, Fujimori S, Kanae S (2017) An economic assessment of the global potential for seawater desalination to 2050. Water 9:763. https://doi.org/10.3390/w9100763
24. Kundzewicz, ZW, Gerten, D (2015) Grand challenges related to the assessment of climate change impacts on freshwater resources. *J. Hydrol. Eng.* **2015**, *20*. https://doi.org/10.1061/(asce)he.1943-5584.0001012.
25. Saavedra A, Valdés H, Mahn A, Acosta O (2021) Comparative analysis of conventional and emerging technologies for seawater desalination: Northern Chile as a case study. Membranes (Basel) 11. https://doi.org/10.3390/membranes11030180
26. Sabry AS, Youssef YM, El-Kilany K. Productivity prediction of seawater reverse osmosis desalination plant using robust regression. In
27. Krishna HJ (2004) Introduction to desalination technologies. Texas Water Dev 2:1–7
28. Alkhudhiri, A, Hilal, N (2018) Membrane distillation—Principles, applications, configurations, design, and implementation. In: Emerging technologies for sustainable desalination handbook. Elsevier, pp 55–106. ISBN 9780128158180
29. Yadav A, Labhasetwar PK, Shahi VK (2021) Membrane distillation using low-grade energy for desalination: a review. J Environ Chem Eng 9:105818. https://doi.org/10.1016/j.jece.2021.105818
30. Al-Sahali M, Ettouney H (2007) Developments in thermal desalination processes: design, energy, and costing aspects. Desalination 214:227–240. https://doi.org/10.1016/j.desal.2006.08.020

31. Ghaffour N, Bundschuh J, Mahmoudi H, Goosen MF (2015) Renewable energy-driven desalination technologies: a comprehensive review on challenges and potential applications of integrated systems. Desalination 356:94–114. https://doi.org/10.1016/j.desal.2014.10.024
32. Henthorne L, Boysen B (2015) State-of-the-art of reverse osmosis desalination pre-treatment. Desalination 356:129–139. https://doi.org/10.1016/j.desal.2014.10.039
33. Kharraz JE (2020) Desalination as an alternative to alleviate water scarcity and a climate change adaptation option in the MENA region. Konrad-Adenauer-Stiftung, Morocco. ISBN 978-3-95721-811-7
34. Elsaid K, Kamil M, Sayed ET, Abdelkareem MA, Wilberforce T, Olabi A (2020) Environmental impact of desalination technologies: a review. Sci Total Environ 748:141528. https://doi.org/10.1016/j.scitotenv.2020.141528
35. Prihasto N, Liu Q-F, Kim S-H (2009) Pre-treatment strategies for seawater desalination by reverse osmosis system. Desalination 249:308–316. https://doi.org/10.1016/j.desal.2008.09.010
36. Brover S, Lester Y, Brenner A, Sahar-Hadar E (2022) Optimization of ultrafiltration as pretreatment for seawater RO desalination. Desalination 524:115478. https://doi.org/10.1016/j.desal.2021.115478
37. Wolf PH, Siverns S, Monti S (2005) UF membranes for RO desalination pre-treatment. Desalination 182:293–300. https://doi.org/10.1016/j.desal.2005.05.006
38. Kavitha J, Rajalakshmi M, Phani AR, Padaki M (2019) Pre-treatment processes for seawater reverse osmosis desalination systems—A review. J Water Process Eng 32:100926. https://doi.org/10.1016/j.jwpe.2019.100926
39. Al-Mutaz IS (2020) MSF challenges and survivals. DWT 177:14–22. https://doi.org/10.5004/dwt.2020.24908
40. Garg MC (2019) Renewable energy-powered membrane technology: cost analysis and energy consumption. In: Current Trends and future developments on (bio-) membranes. Elsevier, pp 85–110
41. Li C, Goswami Y, Stefanakos E (2013) Solar assisted sea water desalination: a review. Renew Sustain Energy Rev 19:136–163. https://doi.org/10.1016/j.rser.2012.04.059
42. Budhiraja P, Fares AA (2008) Studies of scale formation and optimization of antiscalant dosing in multi-effect thermal desalination units. Desalination 220:313–325. https://doi.org/10.1016/j.desal.2007.01.036
43. Kim YM, Kim SJ, Kim YS, Lee S, Kim IS, Kim JH (2009) Overview of systems engineering approaches for a large-scale seawater desalination plant with a reverse osmosis network. Desalination 238:312–332. https://doi.org/10.1016/j.desal.2008.10.004
44. Kim D, Jung S, Sohn J, Kim H, Lee S (2009) Biocide application for controlling biofouling of SWRO membranes—an overview. Desalination 238:43–52. https://doi.org/10.1016/j.desal.2008.01.034
45. Okampo EJ, Nwulu N (2021) Optimisation of renewable energy powered reverse osmosis desalination systems: a state-of-the-art review. Renew Sustain Energy Rev 140:110712. https://doi.org/10.1016/j.rser.2021.110712
46. Goh PS, Lau WJ, Othman M, Ismail AF (2018) Membrane fouling in desalination and its mitigation strategies. Desalination 425:130–155. https://doi.org/10.1016/j.desal.2017.10.018
47. Karagiannis IC, Soldatos PG (2008) Water desalination cost literature: review and assessment. Desalination 223:448–456. https://doi.org/10.1016/j.desal.2007.02.071
48. Sharma M, Meena N, Chauhan P, Nehra S, Pachwarya RB, Sharma R, Kumar D (2023) Desalination through nanofiltration technique. In: Nanofiltration membrane for water purification. Ahmad A, Alshammari MB (eds) Springer, Singapore, pp 141–156, ISBN 978-981-19-5314–9
49. Dach H (2008) Comparison of nanofiltration and reverse osmosis processes for a selective desalination of brackish water feeds. Université d'Angers
50. Ismail AF, Matsuura T (2022) Nanofiltration. In: Membrane separation processes. Elsevier, pp 61–68, ISBN 9780128196267
51. Srivastava AKA, Nair A, Ram S, Agarwal S, Ali J, Singh R, Garg MC (2021) Response surface methodology and artificial neural network modelling for the performance evaluation

of pilot-scale hybrid nanofiltration (NF) & reverse osmosis (RO) membrane system for the treatment of brackish ground water. J Environ Manage 278:111497. https://doi.org/10.1016/j.jenvman.2020.111497

52. Parlar I, Hacıfazlıoğlu M, Kabay N, Pek T, Yüksel M (2019) Performance comparison of reverse osmosis (RO) with integrated nanofiltration (NF) and reverse osmosis process for desalination of MBR effluent. J Water Process Eng 29:100640. https://doi.org/10.1016/j.jwpe.2018.06.002

53. Prajapati M, Shah M, Soni B, Parikh S, Sircar A, Balchandani S, Thakore S, Tala M (2021) Geothermal-solar integrated groundwater desalination system: current status and future perspective. Groundw Sustain Dev 12:100506. https://doi.org/10.1016/j.gsd.2020.100506

54. Voutchkov N (2013) Seawater desalination-costs and technology trends. In: Encyclopedia of membrane science and technology. In: Hoek EM, Tarabara VV (eds) Wiley, Hoboken. ISBN 9781118522318

55. Ghalavand Y, Hatamipour MS, Rahimi A (2014) A review on energy consumption of desalination processes. Desalination Water Treatment 1–16. https://doi.org/10.1080/19443994.2014.892837

56. Chen C, Jiang Y, Ye Z, Yang Y, Hou L (2019) Sustainably integrating desalination with solar power to overcome future freshwater scarcity in China. Glob Energy Interconnection 2:98–113. https://doi.org/10.1016/j.gloei.2019.07.009

57. Faibish RS, Konishi T (2003) Nuclear desalination: a viable option for producing freshwater. Desalination 157:241–252. https://doi.org/10.1016/S0011-9164(03)00403-X

58. Najafi FT, Alsaffar M, Schwerer SC, Brown N, Ouedraogo J (2016) Environmental impact cost analysis of multi-stage flash, multi-effect distillation, mechanical vapor compression, and reverse osmosis medium-size desalination facilities. In: 2016 ASEE annual conference & exposition

59. El-Ghonemy A (2018) Performance test of a sea water multi-stage flash distillation plant: case study. Alex Eng J 57:2401–2413. https://doi.org/10.1016/j.aej.2017.08.019

60. Zhang X, Hu D, Li Z (2014) Performance analysis on a new multi-effect distillation combined with an open absorption heat transformer driven by waste heat. Appl Therm Eng 62:239–244. https://doi.org/10.1016/j.applthermaleng.2013.09.015

61. Aybar HS (2002) Analysis of a mechanical vapor compression desalination system. Desalination 142:181–186. https://doi.org/10.1016/S0011-9164(01)00437-4

62. Birnhack, L, Lahav O (2018) Post-treatment of desalinated water—Chemistry, design, engineering, and implementation. In: Sustainable desalination handbook. Elsevier, pp 305–350. ISBN 9780128092408

63. Birnhack L, Lahav O (2007) A new post-treatment process for attaining Ca^{2+}, Mg^{2+}, SO_4^{2-} and alkalinity criteria in desalinated water. Water Res 41:3989–3997. https://doi.org/10.1016/j.watres.2007.06.007

64. Rehman AU, Zaini DB, Lal AB (2022) Post-treatment of desalinated water. In: Lal B, Nallakukkala S (eds) Gas hydrate in water treatment. Wiley, pp 289–301. ISBN 9781119866114

65. Macias-Bu L, Guerra-Valle M, Petzold G, Orellana-Palma P (2022) Technical and environmental opportunities for freeze desalination. Separation Purif Rev 1–10. https://doi.org/10.1080/15422119.2022.2098504

66. Frank H, Fussmann KE, Rahav E, Bar Zeev E (2019) Chronic effects of brine discharge form large-scale seawater reverse osmosis desalination facilities on benthic bacteria. Water Res 151:478–487. https://doi.org/10.1016/j.watres.2018.12.046

67. Gryta M (2012) Polyphosphates used for membrane scaling inhibition during water desalination by membrane distillation. Desalination 285:170–176. https://doi.org/10.1016/j.desal.2011.09.051

68. Belkin N, Rahav E, Elifantz H, Kress N, Berman-Frank I (2017) The effect of coagulants and antiscalants discharged with seawater desalination brines on coastal microbial communities: A laboratory and in situ study from the southeastern Mediterranean. Water Res 110:321–331. https://doi.org/10.1016/j.watres.2016.12.013

69. Petersen KL, Paytan A, Rahav E, Levy O, Silverman J, Barzel O, Potts D, Bar-Zeev E (2018) Impact of brine and antiscalants on reef-building corals in the Gulf of Aqaba—Potential effects from desalination plants. Water Res 144:183–191. https://doi.org/10.1016/j.watres.2018.07.009

70. Kress N (2019) Seawater quality for desalination plants. In: Marine impacts of seawater desalination. Elsevier, pp 35–52. ISBN 9780128119532

71. Lattemann S, Höpner T (2008) Environmental impact and impact assessment of seawater desalination. Desalination 220:1–15. https://doi.org/10.1016/j.desal.2007.03.009

72. Alameddine I, El-Fadel M (2007) Brine discharge from desalination plants: a modeling approach to an optimized outfall design. Desalination 214:241–260. https://doi.org/10.1016/j.desal.2006.02.103

73. Elsaid K, Sayed ET, Abdelkareem MA, Baroutaji A, Olabi AG (2020) Environmental impact of desalination processes: mitigation and control strategies. Sci Total Environ 740:140125. https://doi.org/10.1016/j.scitotenv.2020.140125

74. Pervov AG, Andrianov AP, Danilycheva MN (2018) Preliminary evaluation of new green antiscalants for reverse osmosis water desalination. Water Supply 18:167–174. https://doi.org/10.2166/ws.2017.106

75. Zhu A, Christofides PD, Cohen Y (2009) Minimization of energy consumption for a two-pass membrane desalination: effect of energy recovery, membrane rejection and retentate recycling. J Membr Sci 339:126–137. https://doi.org/10.1016/j.memsci.2009.04.039

76. Zaviska F, Chun Y, Heran M, Zou L (2015) Using FO as pre-treatment of RO for high scaling potential brackish water: energy and performance optimisation. J Membr Sci 492:430–438. https://doi.org/10.1016/j.memsci.2015.06.004

77. Singh R (2011) Analysis of energy usage at membrane water treatment plants. Desalin Water Treat 29:63–72. https://doi.org/10.5004/dwt.2011.2988

78. Pan S-Y, Haddad AZ, Kumar A, Wang S-W (2020) Brackish water desalination using reverse osmosis and capacitive deionization at the water-energy nexus. Water Res 183:116064. https://doi.org/10.1016/j.watres.2020.116064

79. Greenlee LF, Lawler DF, Freeman BD, Marrot B, Moulin P (2009) Reverse osmosis desalination: water sources, technology, and today's challenges. Water Res 43:2317–2348. https://doi.org/10.1016/j.watres.2009.03.010

80. Gude VG, Nirmalakhandan N (2009) Desalination at low temperatures and low pressures. Desalination 244:239–247. https://doi.org/10.1016/j.desal.2008.06.005

81. Chandwankar RR, Nowak J (2019) Thermal processes for seawater desalination: Multi-effect distillation, thermal vapor compression, mechanical vapor compression, and multistage flash. In: Lahnsteiner J (ed) Handbook of water and used water purification. Springer, Cham, pp 1–38. ISBN 978-3-319-66382

82. Voutchkov N (2013) Desalination engineering: planning and design. McGraw-Hill Professional

83. Zaragoza G, Andrés-Mañas JA, Ruiz-Aguirre A (2018) Commercial scale membrane distillation for solar desalination. npj Clean Water 1. https://doi.org/10.1038/s41545-018-0020-z

84. Park J, Lee S (2022) Desalination technology in South Korea: A comprehensive review of technology trends and future outlook. Membranes (Basel) 12. https://doi.org/10.3390/membranes12020204

85. Elimelech M, Phillip WA (2011) The future of seawater desalination: energy, technology, and the environment. Science 333:712–717. https://doi.org/10.1126/science.1200488

86. Werber JR, Deshmukh A, Elimelech M (2017) Can batch or semi-batch processes save energy in reverse-osmosis desalination? Desalination 402:109–122. https://doi.org/10.1016/j.desal.2016.09.028

87. Chekli L, Phuntsho S, Kim JE, Kim J, Choi JY, Choi J-S, Kim S, Kim JH, Hong S, Sohn J et al (2016) A comprehensive review of hybrid forward osmosis systems: performance, applications and future prospects. J Membr Sci 497:430–449. https://doi.org/10.1016/j.memsci.2015.09.041

88. Bamaga OA, Yokochi A, Zabara B, Babaqi AS (2011) Hybrid FO/RO desalination system: Preliminary assessment of osmotic energy recovery and designs of new FO membrane module configurations. Desalination 268:163–169. https://doi.org/10.1016/j.desal.2010.10.013

89. Sreedhar I, Khaitan S, Gupta R, Reddy BM, Venugopal A (2018) An odyssey of process and engineering trends in forward osmosis. Environ Sci: Water Res Technol 4 129–168. https://doi.org/10.1039/C7EW00507E

90. Darwish MA, Abdulrahim HK, Hassan AS, Mabrouk AA, Sharif AO (2014) The forward osmosis and desalination. Desalination Water Treatment 1–27. https://doi.org/10.1080/19443994.2014.995140

91. Elsaid K, Elkamel A, Sayed ET, Wilberforce T, Abdelkareem MA, Olabi A-G (2022) Bio-based carbon materials for capacitive deionization CDI desalination processes. In: Encyclopedia of smart materials. Elsevier, pp 402–410. ISBN 9780128157336

92. Gabelich CJ, Tran TD, Suffet IHM (2002) Electrosorption of inorganic salts from aqueous solution using carbon aerogels. Environ Sci Technol 36:3010–3019. https://doi.org/10.1021/es0112745

93. Suss ME, Porada S, Sun X, Biesheuvel PM, Yoon J, Presser V (2015) Water desalination via capacitive deionization: what is it and what can we expect from it? Energy Environ Sci 8:2296–2319. https://doi.org/10.1039/C5EE00519A

94. Wu Q, Liang D, Lu S, Wang H, Xiang Y, Aurbach D, Avraham E, Cohen I (2022) Advances and perspectives in integrated membrane capacitive deionization for water desalination. Desalination 542:116043. https://doi.org/10.1016/j.desal.2022.116043

95. Długołęcki P, van der Wal A (2013) Energy recovery in membrane capacitive deionization. Environ Sci Technol 47:4904–4910. https://doi.org/10.1021/es3053202

96. Folaranmi G, Bechelany M, Sistat P, Cretin M, Zaviska F (2020) Towards electrochemical water desalination techniques: a review on capacitive deionization, membrane capacitive deionization and flow capacitive deionization. Membranes (Basel) 10. https://doi.org/10.3390/membranes10050096

97. Porada S, Zhang L, Dykstra JE (2020) Energy consumption in membrane capacitive deionization and comparison with reverse osmosis. Desalination 488:114383. https://doi.org/10.1016/j.desal.2020.114383

98. Porada S, Zhao R, van der Wal A, Presser V, Biesheuvel PM (2013) Review on the science and technology of water desalination by capacitive deionization. Prog Mater Sci 58:1388–1442. https://doi.org/10.1016/j.pmatsci.2013.03.005

99. Tang K, Kim Y, Chang J, Mayes RT, Gabitto J, Yiacoumi S, Tsouris C (2019) Seawater desalination by over-potential membrane capacitive deionization: opportunities and hurdles. Chem Eng J 357:103–111. https://doi.org/10.1016/j.cej.2018.09.121

100. Qin M, Deshmukh A, Epsztein R, Patel SK, Owoseni OM, Walker WS, Elimelech M (2019) Comparison of energy consumption in desalination by capacitive deionization and reverse osmosis. Desalination 455:100–114. https://doi.org/10.1016/j.desal.2019.01.003

101. Ghaffour N, Soukane S, Lee J-G, Kim Y, Alpatova A (2019) Membrane distillation hybrids for water production and energy efficiency enhancement: a critical review. Appl Energy 254:113698. https://doi.org/10.1016/j.apenergy.2019.113698

102. Bindels M, Carvalho J, Gonzalez CB, Brand N, Nelemans B (2020) Techno-economic assessment of seawater reverse osmosis (SWRO) brine treatment with air gap membrane distillation (AGMD). Desalination 489:114532. https://doi.org/10.1016/j.desal.2020.114532

103. Abdel-Karim A, Leaper S, Skuse C, Zaragoza G, Gryta M, Gorgojo P (2021) Membrane cleaning and pre-treatments in membrane distillation—a review. Chem Eng J 422:129696. https://doi.org/10.1016/j.cej.2021.129696

104. Abdelrasoul A (2020) Advances in membrane technologies. IntechOpen. ISBN 978-1-78984-806-9

105. Hsieh I-M, Thakur AK, Malmali M (2021) Comparative analysis of various pre-treatments to mitigate fouling and scaling in membrane distillation. Desalination 509:115046. https://doi.org/10.1016/j.desal.2021.115046

106. Su C, Horseman T, Cao H, Christie K, Li Y, Lin S (2019) Robust superhydrophobic membrane for membrane distillation with excellent scaling resistance. Environ Sci Technol 53:11801–11809. https://doi.org/10.1021/acs.est.9b04362

107. Helal AM (2009) Hybridization—A new trend in desalination. Desalin Water Treat 3:120–135. https://doi.org/10.5004/dwt.2009.263

108. Generous MM, Qasem NA, Zubair SM (2021) An innovative hybridization of electrodialysis with reverse osmosis for brackish water desalination. Energy Convers Manage 245:114589. https://doi.org/10.1016/j.enconman.2021.114589

109. Ahmed FE, Hashaikeh R, Hilal N (2020) Hybrid technologies: The future of energy efficient desalination—A review. Desalination 495:114659. https://doi.org/10.1016/j.desal.2020.114659

110. Kaya C, Sert G, Kabay N, Arda M, Yüksel M, Egemen Ö (2015) Pre-treatment with nanofiltration (NF) in seawater desalination—Preliminary integrated membrane tests in Urla, Turkey. Desalination 369:10–17. https://doi.org/10.1016/j.desal.2015.04.029

111. Al-Sofi MA, Hassan AM, Hamed OA, Dalvi AGI, Kither MN, Mustafa GM, Bamardouf K (2000) Optimization of hybridized seawater desalination process. Desalination 131:147–156. https://doi.org/10.1016/S0011-9164(00)90015-8

112. Wang C, Wang Z, Yang F, Wang J (2021) Improving the permselectivity and antifouling performance of reverse osmosis membrane based on a semi-interpenetrating polymer network. Desalination 502:114910. https://doi.org/10.1016/j.desal.2020.114910

113. Tan CH, Ng HY (2010) A novel hybrid forward osmosis—nanofiltration (FO-NF) process for seawater desalination: draw solution selection and system configuration. Desalin Water Treat 13:356–361. https://doi.org/10.5004/dwt.2010.1733

114. Al-Shammiri M, Ahmed M, Al-Rageeb M (2004) Nanofiltration and calcium sulfate limitation for top brine temperature in Gulf desalination plants. Desalination 167:335–346. https://doi.org/10.1016/j.desal.2004.06.143

115. Altaee A, Mabrouk A, Bourouni K, Palenzuela P (2014) Forward osmosis pre-treatment of seawater to thermal desalination: high temperature FO-MSF/MED hybrid system. Desalination 339:18–25. https://doi.org/10.1016/j.desal.2014.02.006

116. Nguyen NC, Nguyen HT, Ho S-T, Chen S-S, Ngo HH, Guo W, Ray SS, Hsu H-T (2016) Exploring high charge of phosphate as new draw solute in a forward osmosis-membrane distillation hybrid system for concentrating high-nutrient sludge. Sci Total Environ 557–558:44–50. https://doi.org/10.1016/j.scitotenv.2016.03.025

117. Boubakri A, Hafiane A, Bouguecha SAT (2014) Application of response surface methodology for modeling and optimization of membrane distillation desalination process. J Ind Eng Chem 20:3163–3169. https://doi.org/10.1016/j.jiec.2013.11.060

118. Camacho L, Dumée L, Zhang J, Li J, Duke M, Gomez J, Gray S (2013) Advances in membrane distillation for water desalination and purification applications. Water 5:94–196. https://doi.org/10.3390/w5010094

119. Kim J, Park K, Yang DR, Hong S (2019) A comprehensive review of energy consumption of seawater reverse osmosis desalination plants. Appl Energy 254:113652. https://doi.org/10.1016/j.apenergy.2019.113652

120. Giacalone F, Papapetrou M, Kosmadakis G, Tamburini A, Micale G, Cipollina A (2019) Application of reverse electrodialysis to site-specific types of saline solutions: a techno-economic assessment. Energy 181:532–547. https://doi.org/10.1016/j.energy.2019.05.161

121. Li W, Krantz WB, Cornelissen ER, Post JW, Verliefde AR, Tang CY (2013) A novel hybrid process of reverse electrodialysis and reverse osmosis for low energy seawater desalination and brine management. Appl Energy 104:592–602. https://doi.org/10.1016/j.apenergy.2012.11.064

122. Yip NY, Elimelech M (2014) Comparison of energy efficiency and power density in pressure retarded osmosis and reverse electrodialysis. Environ Sci Technol 48:11002–11012. https://doi.org/10.1021/es5029316

123. Straub AP, Lin S, Elimelech M (2014) Module-scale analysis of pressure retarded osmosis: performance limitations and implications for full-scale operation. Environ Sci Technol 48:12435–12444. https://doi.org/10.1021/es503790k

124. Park K, Kim J, Yang DR, Hong S (2020) Towards a low-energy seawater reverse osmosis desalination plant: a review and theoretical analysis for future directions. J Membr Sci 595:117607. https://doi.org/10.1016/j.memsci.2019.117607

125. Wan CF, Chung T-S (2015) Osmotic power generation by pressure retarded osmosis using seawater brine as the draw solution and wastewater retentate as the feed. J Membr Sci 479:148–158. https://doi.org/10.1016/j.memsci.2014.12.036

126. Low JH, Zhang J, Li WP, Yang T, Wan CF, Esa F, Qua MS, Mottaiyan K, Murugan S, Aiman M et al (2023) Industrial scale thin-film composite membrane modules for salinity-gradient energy harvesting through pressure retarded osmosis. Desalination 548:116217. https://doi.org/10.1016/j.desal.2022.116217

127. Matsuyama K, Makabe R, Ueyama T, Sakai H, Saito K, Okumura T, Hayashi H, Tanioka A (2021) Power generation system based on pressure retarded osmosis with a commercially-available hollow fiber PRO membrane module using seawater and freshwater. Desalination 499:114805. https://doi.org/10.1016/j.desal.2020.114805

128. Prante JL, Ruskowitz JA, Childress AE, Achilli A (2014) RO-PRO desalination: an integrated low-energy approach to seawater desalination. Appl Energy 120:104–114. https://doi.org/10.1016/j.apenergy.2014.01.013

129. Tufa RA, Curcio E, van Baak W, Veerman J, Grasman S, Fontananova E, Di Profio G (2014) Potential of brackish water and brine for energy generation by salinity gradient power-reverse electrodialysis (SGP-RE). RSC Adv 4:42617–42623. https://doi.org/10.1039/C4RA05968A

130. Mehdizadeh S, Kakihana Y, Abo, T, Yuan, Q, Higa, M. Power Generation Performance of a Pilot-Scale Reverse Electrodialysis Using Monovalent Selective Ion-Exchange Membranes. *Membranes (Basel)* **2021**, *11*. https://doi.org/10.3390/membranes11010027.

131. Valladares Linares R, Li Z, Yangali-Quintanilla V, Ghaffour N, Amy G, Leiknes T, Vrouwen-velder JS (2016) Life cycle cost of a hybrid forward osmosis - low pressure reverse osmosis system for seawater desalination and wastewater recovery. Water Res 88:225–234. https://doi.org/10.1016/j.watres.2015.10.017

132. Kim D-H, Lee C, Nguyen T-T, Adha RS, Kim C, Ahn S-J, Son H, Kim IS (2021) Insight into fouling potential analysis of a pilot-scale pressure-assisted forward osmosis plant for diluted seawater reverse osmosis desalination. J Ind Eng Chem 98:237–246. https://doi.org/10.1016/j.jiec.2021.03.048

133. Park K, Heo H, Kim DY, Yang DR (2018) Feasibility study of a forward osmosis/crystallization/reverse osmosis hybrid process with high-temperature operation: modeling, experiments, and energy consumption. J Membr Sci 555:206–219. https://doi.org/10.1016/j.memsci.2018.03.031

134. Lim YJ, Goh K, Kurihara M, Wang R (2021) Seawater desalination by reverse osmosis: current development and future challenges in membrane fabrication—A review. J Membr Sci 629:119292. https://doi.org/10.1016/j.memsci.2021.119292

135. Cohen-Tanugi D, Grossman JC (2014) Water permeability of nanoporous graphene at realistic pressures for reverse osmosis desalination. J Chem Phys 141:74704. https://doi.org/10.1063/1.4892638

136. Duarte AP, Bordado JC (2016) Smart composite reverse-osmosis membranes for energy generation and water desalination processes. Smart composite coatings and membranes. Elsevier, pp 329–350. ISBN 9781782422839

137. Wang Y-N, Wang R (2019) Reverse osmosis membrane separation technology. In: Membrane separation principles and applications. Elsevier, pp 1–45. ISBN 9780128128152

138. Lee KP, Arnot TC, Mattia D (2011) A review of reverse osmosis membrane materials for desalination—Development to date and future potential. J Membr Sci 370:1–22. https://doi.org/10.1016/j.memsci.2010.12.036

139. Porter CJ, Werber JR, Zhong M, Wilson CJ, Elimelech M (2020) Pathways and challenges for biomimetic desalination membranes with sub-nanometer channels. ACS Nano 14:10894–10916. https://doi.org/10.1021/acsnano.0c05753

140. Cohen-Tanugi D, McGovern RK, Dave SH, Lienhard JH, Grossman JC (2014) Quantifying the potential of ultra-permeable membranes for water desalination. Energy Environ Sci 7:1134–1141. https://doi.org/10.1039/C3EE43221A

141. Werber JR, Deshmukh A, Elimelech M (2016) The critical need for increased selectivity, not increased water permeability, for desalination membranes. Environ Sci Technol Lett 3:112–120. https://doi.org/10.1021/acs.estlett.6b00050

142. Chowdhury MR, Steffes J, Huey BD, McCutcheon JR (2018) 3D printed polyamide membranes for desalination. Science 361:682–686. https://doi.org/10.1126/science.aar2122

143. Dai Y, Liu M, Li J, Kang N, Ahmed A, Zong Y, Tu J, Chen Y, Zhang P, Liu X (2022) Graphene-based membranes for water desalination: a literature review and content analysis. Polymers 14:4246. https://doi.org/10.3390/polym14194246

144. Shao C, Zhao Y, Qu L (2020) Tunable graphene systems for water desalination. ChemNanoMat 6:1028–1048. https://doi.org/10.1002/cnma.202000041

145. Zhang B, Gu Q, Wang C, Gao Q, Guo J, Wong PW, Liu CT, An AK (2021) Self-assembled hydrophobic/hydrophilic porphyrin-Ti3C2Tx MXene Janus membrane for dual-functional enabled photothermal desalination. ACS Appl Mater Interfaces 13:3762–3770. https://doi.org/10.1021/acsami.0c16054

146. Chen X, Feng Z, Gohil J, Stafford CM, Dai N, Huang L, Lin H (2020) Reduced holey graphene oxide membranes for desalination with improved water permeance. ACS Appl Mater Interfaces 12:1387–1394. https://doi.org/10.1021/acsami.9b19255

147. Shimura H (2022) Development of an advanced reverse osmosis membrane based on detailed nanostructure analysis. Polym J 54:767–773. https://doi.org/10.1038/s41428-022-00627-x

148. Ahmed FE, Hashaikeh R, Hilal N (2019) Fouling control in reverse osmosis membranes through modification with conductive carbon nanostructures. Desalination 470:114118. https://doi.org/10.1016/j.desal.2019.114118

149. Sabir, A, Falath, W, Shafiq, M, Gull, N, Wasim, M, .I. Jacob, K. Effective desalination and anti-biofouling performance via surface immobilized MWCNTs on RO membrane. *Chinese Journal of Chemical Engineering* **2022**. https://doi.org/10.1016/j.cjche.2022.06.027.

150. Noy A, Park HG, Fornasiero F, Holt JK, Grigoropoulos CP, Bakajin O (2007) Nanofluidics in carbon nanotubes. Nano Today 2:22–29. https://doi.org/10.1016/S1748-0132(07)70170-6

151. Tang CY, Zhao Y, Wang R, Hélix-Nielsen C, Fane AG (2013) Desalination by biomimetic aquaporin membranes: review of status and prospects. Desalination 308:34–40. https://doi.org/10.1016/j.desal.2012.07.007

152. Gong B (2018) Artificial water channels: inspiration, progress, and challenges. Faraday Discuss 209:415–427. https://doi.org/10.1039/c8fd00132d

153. Noy A, Wanunu M (2020) A new type of artificial water channels. Nat Nanotechnol 15:9–10. https://doi.org/10.1038/s41565-019-0617-5

154. Werber JR, Osuji CO, Elimelech M (2016) Materials for next-generation desalination and water purification membranes. Nat Rev Mater 1. https://doi.org/10.1038/natrevmats.2016.18

155. Strilets D (2021) Development of highly efficient artificial water channels for water desalination. Ecole nationale supérieure de chimie, Montpellier

156. Song W, Joshi H, Chowdhury R, Najem JS, Shen Y-X, Lang C, Henderson CB, Tu Y-M, Farell M, Pitz ME et al (2020) Artificial water channels enable fast and selective water permeation through water-wire networks. Nat Nanotechnol 15:73–79. https://doi.org/10.1038/s41565-019-0586-8

157. Park HB, Kamcev J, Robeson LM, Elimelech M, Freeman BD (2017) Maximizing the right stuff: the trade-off between membrane permeability and selectivity. Science 356. https://doi.org/10.1126/science.aab0530.

158. Qiu T, Davies PA (2012) Comparison of configurations for high-recovery inland desalination systems. Water 4:690–706. https://doi.org/10.3390/w4030690

159. Arenas Urrea S, Díaz Reyes F, Peñate Suárez B, La Fuente Bencomo JA, de. (2019) Technical review, evaluation and efficiency of energy recovery devices installed in the Canary Islands desalination plants. Desalination 450:54–63. https://doi.org/10.1016/j.desal.2018.07.013

160. Schunke AJ, Hernandez Herrera GA, Padhye L, Berry T-A (2020) Energy recovery in SWRO desalination: current status and new possibilities. Front Sustain Cities 2. https://doi.org/10.3389/frsc.2020.00009

161. Kadaj E, Bosleman R (2018) Energy recovery devices in membrane desalination processes. In: Renewable energy powered desalination handbook. Elsevier, pp 415–444. ISBN 9780128152447

162. Wang C, Meng P, Wang S, Song D, Xiao Y, Zhang Y, Ma Q, Liu S, Wang K, Zhang Y (2022) Comparison of two types of energy recovery devices: pressure exchanger and turbine in an island desalination project case. Desalination 533:115752. https://doi.org/10.1016/j.desal.2022.115752

163. Subramani A, Jacangelo JG (2014) Treatment technologies for reverse osmosis concentrate volume minimization: a review. Sep Purif Technol 122:472–489. https://doi.org/10.1016/j.seppur.2013.12.004

164. Giwa A, Dufour V, Al Marzooqi F, Al Kaabi M, Hasan SW (2017) Brine management methods: recent innovations and current status. Desalination 407:1–23. https://doi.org/10.1016/j.desal.2016.12.008

165. Semblante GU, Lee JZ, Lee LY, Ong SL, Ng HY (2018) Brine pre-treatment technologies for zero liquid discharge systems. Desalination 441:96–111. https://doi.org/10.1016/j.desal.2018.04.006

166. Park K, Kim J, Hong S (2022) Brine management systems using membrane concentrators: future directions for membrane development in desalination. Desalination 535:115839. https://doi.org/10.1016/j.desal.2022.115839

167. Davenport DM, Wang L, Shalusky E, Elimelech M (2021) Design principles and challenges of bench-scale high-pressure reverse osmosis up to 150 bar. Desalination 517:115237. https://doi.org/10.1016/j.desal.2021.115237

168. Tong T, Elimelech M (2016) The global rise of zero liquid discharge for wastewater management: drivers, technologies, and future directions. Environ Sci Technol 50:6846–6855. https://doi.org/10.1021/acs.est.6b01000

169. Panagopoulos A (2022) Brine management (saline water & wastewater effluents): sustainable utilization and resource recovery strategy through Minimal and Zero Liquid Discharge (MLD & ZLD) desalination systems. Chem Eng Process Process Intensif 176:108944. https://doi.org/10.1016/j.cep.2022.108944

170. Tsai J-H, Macedonio F, Drioli E, Giorno L, Chou C-Y, Hu F-C, Li C-L, Chuang C-J, Tung K-L (2017) Membrane-based zero liquid discharge: myth or reality? J Taiwan Inst Chem Eng 80:192–202. https://doi.org/10.1016/j.jtice.2017.06.050

171. Flörke M, Schneider C, McDonald RI (2018) Water competition between cities and agriculture driven by climate change and urban growth. Nat Sustain 1:51–58. https://doi.org/10.1038/s41893-017-0006-8

172. Garrick D, L de Stefano, Yu W, Jorgensen I, O'Donnell E, Turley L, Aguilar-Barajas I, Dai X, de Souza Leão R, Punjabi B et al (2019) Rural water for thirsty cities: a systematic review of water reallocation from rural to urban regions. Environ Res Lett 14:43003. https://doi.org/10.1088/1748-9326/ab0db7

173. Wada Y, Flörke M, Hanasaki N, Eisner S, Fischer G, Tramberend S, Satoh Y, van Vliet MT, Yillia P, Ringler C et al (2016) Modeling global water use for the 21st century: The Water Futures and Solutions (WFaS) initiative and its approaches. Geosci Model Dev 9:175–222

174. World Water Assessment Programme (2018) The United Nations World Water Development report 2018. United Nations Educational, Scientific and Cultural Organization, New York, United States

175. He C, Liu Z, Wu J, Pan X, Fang Z, Li J, Bryan BA (2021) Future global urban water scarcity and potential solutions. Nat Commun 12:4667. https://doi.org/10.1038/s41467-021-25026-3

176. Jones E, Qadir M, van Vliet MTH, Smakhtin V, Kang S-M (2019) The state of desalination and brine production: a global outlook. Sci Total Environ 657:1343–1356. https://doi.org/10.1016/j.scitotenv.2018.12.076

177. Eke J, Yusuf A, Giwa A, Sodiq A (2020) The global status of desalination: an assessment of current desalination technologies, plants and capacity. Desalination 495:114633. https://doi.org/10.1016/j.desal.2020.114633

178. Mayor B (2019) Growth patterns in mature desalination technologies and analogies with the energy field. Desalination 457:75–84. https://doi.org/10.1016/j.desal.2019.01.029

179. Hanasaki N, Yoshikawa S, Kakinuma K, Kanae S (2016) A seawater desalination scheme for global hydrological models. Hydrol Earth Syst Sci 20:4143–4157. https://doi.org/10.5194/hess-20-4143-2016

180. Glanemann N, Willner SN, Levermann A (2020) Paris climate agreement passes the cost-benefit test. Nat Commun 11:110. https://doi.org/10.1038/s41467-019-13961-1

181. Wappler M, Unguder D, Lu X, Ohlmeyer H, Teschke H, Lueke W (2022) Building the green hydrogen market—Current state and outlook on green hydrogen demand and electrolyzer manufacturing. Int J Hydrogen Energy 47:33551–33570. https://doi.org/10.1016/j.ijhydene.2022.07.253

182. Ursua A, Gandia LM, Sanchis P (2012) Hydrogen production from water electrolysis: current status and future trends. Proc IEEE 100:410–426. https://doi.org/10.1109/JPROC.2011.2156750

183. Gondal IA, Masood SA, Khan R (2018) Green hydrogen production potential for developing a hydrogen economy in Pakistan. Int J Hydrogen Energy 43:6011–6039. https://doi.org/10.1016/j.ijhydene.2018.01.113

184. IEA. Global hydrogen review 2021. Available online https://www.iea.org/reports/global-hydrogen-review-2021. Accessed on 30 Oct 2023

185. Albrecht U, Bünger U, Michalski J, Raksha T, Wurster R. International hydrogen strategies. Available online https://www.weltenergierat.de/wp-content/uploads/2020/10/WEC_H2_Strategies_finalreport.pdf. Accessed on 30 Oct 2023

186. IRENA. Global energy transformation: a roadmap to 2050. Available online https://www.irena.org/publications/2019/Apr/Globalenergy-transformation-A-roadmap-to-2050-2019Edition

187. Hydrogen Council. Hydrogen for net zero-A critical cost-competitive energy vector. Available online: https://hydrogencouncil.com/wp-content/uploads/2021/11/Hydrogen-for-Net-Zero.pdf. Accessed on 01 Nov 2023

188. IEA. Global hydrogen review 2022. Available online: https://www.iea.org/reports/global-hydrogen-review-2022. Accessed on 01 Nov 2023

189. Shi X, Liao X, Li Y (2020) Quantification of fresh water consumption and scarcity footprints of hydrogen from water electrolysis: a methodology framework. Renewable Energy 154:786–796. https://doi.org/10.1016/j.renene.2020.03.026

190. Beswick RR, Oliveira AM, Yan Y (2021) Does the green hydrogen economy have a water problem? ACS Energy Lett 6:3167–3169. https://doi.org/10.1021/acsenergylett.1c01375

191. Khan MA, Al-Attas T, Roy S, Rahman MM, Ghaffour N, Thangadurai V, Larter S, Hu J, Ajayan PM, Kibria MG (2021) Seawater electrolysis for hydrogen production: a solution looking for a problem? Energy Environ Sci 14:4831–4839. https://doi.org/10.1039/D1EE00870F

192. Peterson D, Vickers J, DeSantis D (2020) Hydrogen production cost from PEM electrolysis-2019. DOE hydrogen fuel cells program record 19009

Markets and Costs for Hydrogen Electrolysis

Paul Balcombe⬤, **Marian Chatenet**⬤, **Jonathan Deseure**⬤,
Helmut Schäfer⬤, **and Iain Staffell**⬤

Abstract Hydrogen is currently undergoing a renaissance, emerging as a potential key option in the global transition to sustainable energy systems, particularly for decarbonising hard-to-electrify sectors. Hydrogen electrolysis has rapidly reduced costs over the last decade, making large-scale hydrogen production more feasible. However, achieving cost competitiveness with conventional energy sources remains a challenge. This chapter investigates the current capital costs of electrolysers, focusing on the impact of technology, manufacturing scale, and regional production differences. Electrolyser costs vary widely, from $200 to $2,500 per kW, depending on the technology and production scale. Further cost reductions are achievable through manufacturing scale-up, technological improvements, and shifting production to regions with lower costs, such as China. The scale and pace of these potential future cost reductions for hydrogen production will determine whether hydrogen becomes a significant component of the coming low-carbon energy economy.

Keywords Hydrogen electrolysis · Capital costs · Green hydrogen · Manufacturing scale · Technology cost reduction · Global electrolyser market

P. Balcombe
Queen Mary University of London, School of Engineering and Material Science, London, UK

M. Chatenet · J. Deseure
University Grenoble Alpes (UGA), Institute of Engineering (Grenoble-INP), Grenoble, France

H. Schäfer
University of Osnabrück, Institute of Chemistry of New Materials (Electrochemical Energy and Catalysis Group), Osnabrück, Germany

I. Staffell (✉)
Imperial College London, Centre for Environmental Policy, London, UK
e-mail: i.staffell@imperial.ac.uk

© The Author(s), under exclusive license to Springer Nature Switzerland AG 2025
N. Bullerdiek et al. (eds.), *Powerfuels*, Green Energy and Technology,
https://doi.org/10.1007/978-3-031-62411-7_9

1 Introduction

Hydrogen is currently undergoing a renaissance, emerging as a potential key option for sustainable energy solutions. Recognizing this potential, major financial institutes are positioning themselves to advice on hydrogen [1–3] in anticipation of a growing commercial market. The European Union's 2020 hydrogen strategy signaled a step-change in commitment to the technology, establishing a target for 40 GW of electrolyzers installed over the coming decade [4]. The industry has responded with manufacturing scale-up and the advent of "gigafactories" [5, 6]—mirroring the GW-scale production plants for lithium-ion batteries.

For these plans to materialize and embed hydrogen as a mainstream part of the global energy system, it is critical that hydrogen achieves cost competitiveness against incumbent technologies. The two most important drivers of hydrogen cost are the capital cost (CAPEX) of the electrolyzer and operational costs in the form of input cost of electricity (OPEX). Both costs vary widely across regions, between technologies and over time.

Against this background, this chapter reviews the markets for hydrogen and anticipated scale-up of the industry. The focus is on current capital costs of electrolysis devices and the influence of components and manufacturing stages. Projected developments in capital costs over time and surveys on the drivers for potential cost reduction are reviewed. Finally, levelized costs of hydrogen production are presented, which includes both capital and operating cost.

2 Commercial Status of Hydrogen Electrolysis

As of today, electrolysis only provides around 1 to 2% of global hydrogen production, or around 7 Mt of hydrogen per year [7]. This share is set to increase though; Fig. 1 shows the global installed hydrogen electrolyzer capacity over time, and near-term projections from various sources. Global capacity has grown rapidly over the last decade, by an average of 32%/a since 2010. Alkaline electrolysis (AEL) was the most mature technology, forming over 90% of global capacity as recently as 2010. However, growth since then has only been 19%/a, whereas proton exchange membrane electrolysis (PEMEL) capacity has grown at 80%/a, overtaking the installed capacity of AEL in 2019. Aurora identifies over 200 GW of new electrolysis projects planned for delivery by 2040, [8] of which 85% is located within Europe. This suggests that the market will accelerate over the coming decade with 75% annual growth.

Markets for Hydrogen Widespread optimism about the prospects for hydrogen is not a new phenomenon [12–14]. Hydrogen technologies have been a faithful adherent to the Gartner hype cycle model [15], experiencing cycles of excessive expectations followed by disillusion and bankruptcies [12, 16]. However, the potential markets for hydrogen are changing compared to past hype cycles, as competition from other

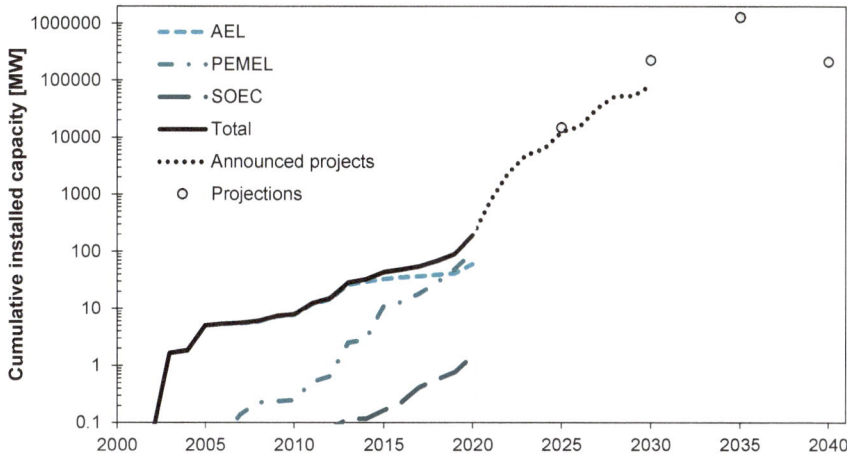

Fig. 1 Cumulative installed capacity of modern hydrogen electrolyzers, split by technology; with analysts' projections for future market size (AEL: alkaline electrolysis; PEMEL: proton exchange membrane electrolysis; SOEC: solid oxide electrolyzer cell; historical data from Buttler and IEA [9, 10], and future trajectories from Aurora and the ETC [8, 11]). Projections are from different organizations, so lower capacity in 2040 is due to different assumptions rather than indicating capacity falls after 2035

low-carbon technologies intensifies. In previous decades, passenger vehicles [17] and home-heating systems [18] were thought of as the leading sectors to be served by hydrogen. Their prospects are now seen as waning, as battery electric vehicles [19] and electric heat pumps [20] have gained early ground in the transition away from fossil fuels. Figure 2 shows three examples of analysts' expectations for where hydrogen will be competitive. Hydrogen is projected to be most competitive in sectors such as steel production, fertilizer and refining, heavy duty road transport (trucks and buses), shipping, or aviation and especially for long-duration electricity storage.

Major Manufacturers of Electrolyzers The global electrolyzer market is relatively consolidated. Buttler and Spliethoff surveyed the market in 2018, finding only 33 medium to large suppliers in total: 20 alkaline electrolysis (AEL) suppliers, 12 proton exchange membrane electrolysis (PEMEL) suppliers, and 1 solid oxide electrolyzer cell (SOEC) supplier [9]. This situation may change as the market is dynamic with acquisitions being common [22]. IRENA [23] and the ETC [11] and various market research firms discuss the main technology manufacturers. Some prominent examples are listed by technology in Table 1.

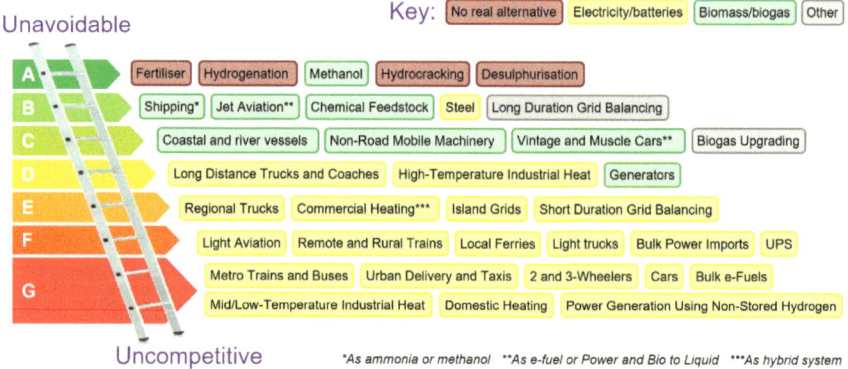

Fig. 2 Perceived competitiveness of hydrogen across different market sectors, visualized as the "hydrogen ladder". Adapted from Liebreich Associates [21], CC-BY 4.0)

Table 1 Non-exhaustive selection of major manufacturers of electrolyzers (AEMEL: anion exchange membrane electrolysis; AEL: alkaline electrolysis; PEMEL: proton exchange membrane electrolysis; SOEC: solid oxide electrolyzer cell)

AEL	PEMEL	SOEC	AEMEL
Asahi Kesei (Japan)	Cummins (US)[a]	Ceres (UK)	Enapter (Italy)
Cockerill Jingli (China)	Elogen (Germany)	Haldor Tøpsoe (Denmark)	
McPhy (France)	ITM power (UK)	Sunfire (Germany)	
Teledyne (US)	NEL (Norway)[a]	Toshiba (Japan)	
Thyssenkrupp (Germany)	Siemens (Germany)		
Tianjin Mainland (China)			
Yangzhou Chungdean (China)			

[a] Also manufacture alkaline electrolyzers

3 Capital Cost of Electrolyzers

A major diver of hydrogen production costs is the capital expenditures (CAPEX) associated with the electrolyzer. They play a defining role in the overall economic viability of hydrogen electrolysis. Thus, the cost of electrolyzers will be critically important to the success of this technology and its competitiveness against other routes to produce hydrogen and other low-carbon fuels. However, predicting the future cost range of electrolyzers is challenging for four main reasons:

1. Electrolysis technology is still at an early stage of large-scale commercial development (thus, data are not readily available).

2. Electrolyzer costs differ substantially by technology due to design and material requirements, as well as the maturity and scale of production.
3. Electrolyzer prices vary strongly based on country of manufacture, with a prominent disparity between China and the rest of the world.
4. Electrolyzer prices are changing rapidly as manufacturers increase their scale of production.

Considering these aspects and complexities, it is imperative to maintain a dynamic perspective when evaluating and estimating future hydrogen electrolysis costs and prices. As the industry matures and scales up, more definitive data on costs and efficiencies will emerge. Taking this into consideration, the subsequent section presents current cost ranges for electrolyzers in greater detail. Following this, possible cost developments for electrolyzers up to 2050 are discussed as well as primary drivers that could influence these cost dynamics in the future.

3.1 Current Capital Cost of Electrolyzers

Current estimates of electrolyzer costs vary by an order of magnitude from 200 to 2,500 $/kW of capacity. Values are differentiated by technology type, with estimates for AEL at 200 to 1,100 $/kW, PEMEL at 800–2,200 $/kW, and SOEC at around 2,200 $/kW [23–26]. The minimum cost for alkaline electrolyzers of 200 $/kW is noteworthy, a value cited in several organizations relating to claims of cost from recent Chinese manufacturing plants.

Influence of Materials and Components Alkaline and proton exchange membrane electrolyzers are relatively mature technologies, with several products commercially available at known prices. Solid oxide electrolyzers only surpassed 1 MW of capacity installed in 2019, so greater variation and uncertainty surrounds their costs. For more novel technologies (such as anion exchange membrane (AEM) and proton-conducting ceramics (PCC)), costs can only be speculated upon as large systems (more than 100 kW) have not yet been built.

Electrolysis systems consist of more than just the electrolyzer stack. Ancillary equipment, known as the balance-of-plant (BoP), include the power conditioning (transformer and rectifier to condition the DC supply), water treatment (purification and heating), and hydrogen conditioning (separation, drying and pressurization). All these components are mature technologies and used in a wide array of other industries and settings.

The cost contribution of the electrolyzer stack itself varies widely across literature, from 27 to 64%. Figure 3 shows a range of study estimates of the contribution to CAPEX from different electrolyzer components. For example, IRENA calculates that the stack contributes 45% of total system cost [23]. The remainder comes from the balance-of-plant (BoP) components: power supply (28%), water circulation (12%), hydrogen processing (11%), and cooling (4%) [23]. Mayyas and Mann similarly model the stack as contributing 40% of the total system cost [27], with the BoP share mostly coming from the power supply. The share from balance-of-plant (BoP) grows with scale of production, from 60% at 10 MW per year to 70% at 1 GW per year due to declining stack production costs [27]. IRENA [23] and ETC [11] also present breakdowns of AWE cost, giving 45% and 55% share, respectively, to the electrolyzer stack. Most of this cost is from manufacturing the diaphragm/electrode package, and the breakdown of BoP costs is similar to that for PEM electrolyzers.

Broadly as the capacity or production levels increase, the contribution from the stack increases as their largely modular design lends less favorably to economies of scale. Lower cost estimates are associated with larger capacity installations: Fig. 4 shows a breakdown of specific system costs for different electrolyzer system capacities. However, substantial cost reductions are achieved with the balance-of-plant

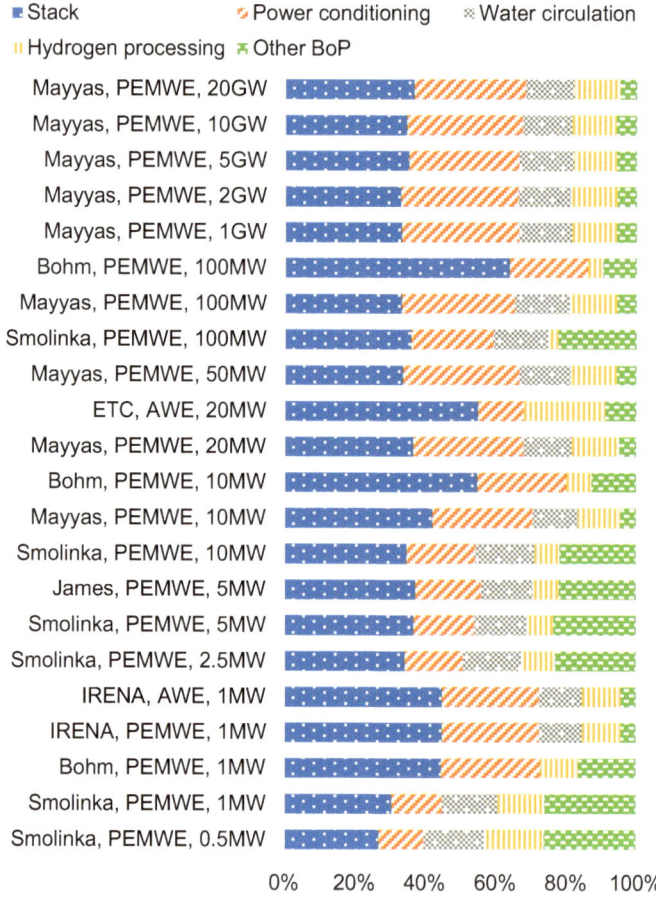

Fig. 3 Comparison of the cost contribution of different electrolyzer components (BoP: balance-of-plant; PEM: proton exchange membrane; data from: [11, 23, 24, 27–29])

(BoP), including hydrogen and water conditioning. There are differences in the literature on the cost contribution from different stack elements for PEMEL (Fig. 5).

There are very few publicly available inventories for electrolysis stacks to understand the contributing components of the costs, and there is a large variation across manufacturers and scales of production. NREL suggest that AEL stacks cost 100 $/kW (1 MW capacity, producing 10 to 20 units per year). The catalyst-coated membrane is typically the largest cost (23 to 47% of the total costs) due to use of iridium and platinum, whereas bipolar plates represent a high cost (9 to 51%)

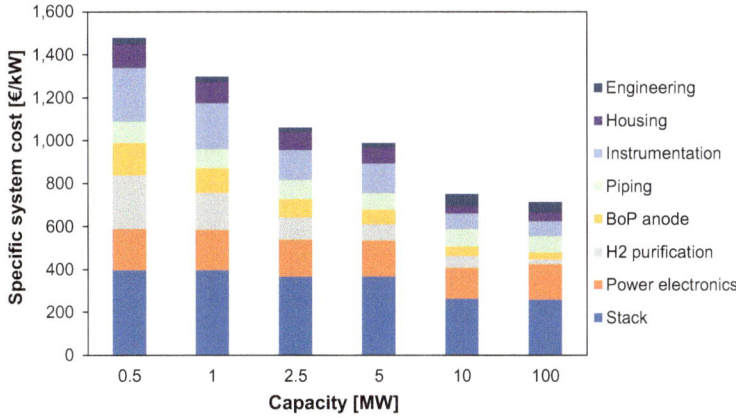

Fig. 4 Component contribution to PEMEL system cost at different capacities (BoP: balance-of-plant; PEMEL: proton exchange membrane electrolysis; data from [24])

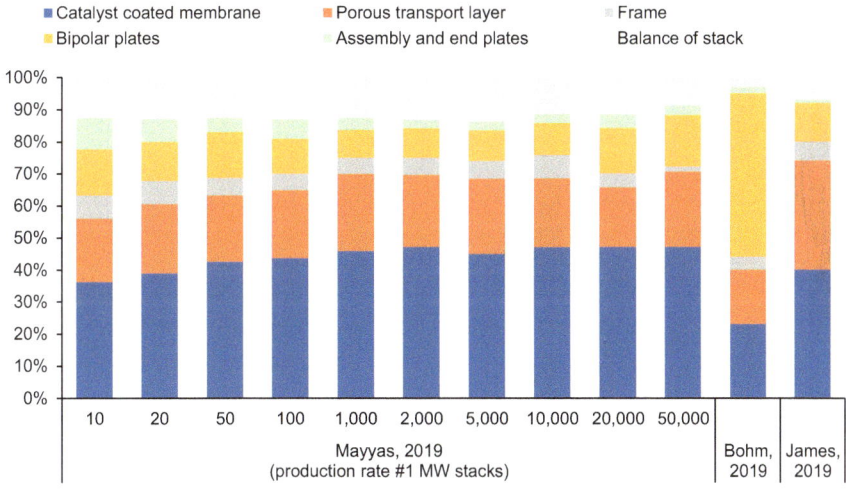

Fig. 5 Estimates of cost contribution of different PEM electrolyzer stack elements from three studies (PEM: proton exchange membrane; data from: [27–29])

depending on the material used: higher costs are associated with titanium plates, whereas lower costs may be from gold-coated steel manufacture [27].

3.2 Future Capital Cost of Electrolyzers

As new hydrogen production technologies are developed, established technologies undergo refinement and rapid scale-up of manufacturing, and there is widespread expectation that current prices will continue to fall. This has been observed widely across the energy sector, with prominent examples being solar photovoltaic (PV) panels [30], offshore wind farms [31], electricity storage systems [32], and hydrogen fuel cells [33]. Figure 6 illustrates both historical electrolyzer costs as well as near-term projections up to 2030 for alkaline, proton exchange membrane (PEM), and solid oxide electrolyzer cell (SOEC) electrolyzer technologies.

It is evident that costs have been rapidly falling in recent years. BNEF estimates that the CAPEX of large-scale electrolyzers fell by 40–50% in the five years to 2019 [34]. Specifically, AEL costs fell from 2,000 to 1,200 $/kW over the period, while PEMELs fell from 2,800 to 1,400 $/kW. According to the data presented in Fig. 6, further cost reductions for these technologies are also expected in the future, up to the year 2030. However, based on this data, these reductions might be smaller compared to the cost reductions of the past one to two decades.

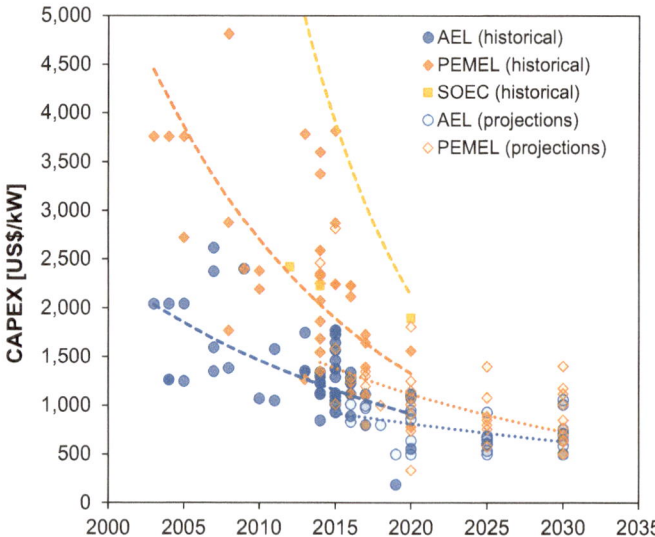

Fig. 6 CAPEX costs of electrolyzers, both historical and projections for alkaline, PEM and SOEC technologies (AEL: alkaline electrolysis; PEMEL: proton exchange membrane electrolysis; SOEC: solid oxide electrolyzer cell; data compiled from: [23–26])

To assess the competitiveness of hydrogen going forward, the following section assesses the evidence of projections of cost reductions, particularly related to experience curve analysis and expert elicitation.

3.2.1 Experience Curve Analysis

Experience curves are an empirical approach used to track the development of a product's price as a function of its cumulative installed capacity. For each doubling of installed capacity, historical prices are often observed to fall by a fixed percentage—known as the experience rate (ER) or learning rate. Product prices have been observed to relate to the experience by the following equations: [35]

$$P_n = P_{\text{base}} \left(\frac{X_n}{X_{\text{base}}} \right)^b \quad \text{and} \quad ER = 1 - 2^b \tag{1}$$

where P_n is the price of a specific unit, P_{base} is the price of a reference unit, X_n is the experience, X_{base} is the cumulative experience gained before the construction of the product, and b is an exponent. Experience can be represented by number of units, or more commonly by the installed production capacity (e.g., MW of electrolyzer).

Experience curves are well established within the energy sector for modeling future product prices [36, 37], and can be traced back to Wright's Law [38] from the 1930s. Solar PV panels are a prime example, with module prices falling by 23% for each doubling of cumulative installed capacity between 1976 and 2019 (i.e., $b = -0.38$) [39]. Experience rates for energy technologies typically lie in the region of 5 to 30% [32, 33, 40].

Neij argues that modular technologies such as electrolyzers should experience higher learning rates than monolithic products such as turbines [41]. Malhotra and Schmidt [42] show empirically that simple and standardized products such as solar panels or LED lights have higher learning rates (18–22%) than complex or customized/bespoke technologies such as conventional power plants or building insulation (3–5%). With electrolyzer stacks being modular assemblies of standard repeated units, electrolysis would appear to fit the "simple and standardized" group of technologies, which ought to experience the highest of these learning rates.

IRENA [23] and Saba et al. [43] surveyed previous studies of learning rates for electrolyzers (Table 2). As there are relatively few studies to date, these learning rates are compared to estimates for hydrogen fuel cell systems, which "can be adapted also to electrolyzers" [43]. Various studies have suggested that fuel cells have a comparable learning rate in the region of 15–21% [33, 44–47].

Böhm et al. [28] anticipate that the experience rate for electrolyzers will decline over time as cumulative production increases (Fig. 7). This would occur because the core components of the electrolyzer (catalyst layers, bipolar plates) are expected to have the higher learning rates than the generic components (flanges and pumps) and,

Table 2 Estimates for the learning rate for hydrogen electrolyzers (AEL: alkaline electrolysis; PEMEL: proton exchange membrane electrolysis)

Technology	Notes	Learning rate (%)	Reference
AEL	Hypothetical, 1977–1994	10	[48]
AEL	Observed, 1972–2004	18 ± 13	[49]
AEL	Observed, 1956–2014	18 ± 6	[32]
AEL	Projection for 2020–2030	9	[50]
PEMEL	Projection for 2020–2030	13	[50]

as these core components become cheaper, their impact on the overall system's rate of cost decline will weaken.

These estimated learning rates can be combined with a forecast for the future market size (in terms of GW of capacity installed) to create future cost projections. Schmidt et al. [32] provided an example of this, projecting the price of alkaline electrolyzers up to a cumulative capacity of 100 GW. When combined with a market projection, which is conservative in today's terms, this gives prices of 1,300 $/kW in 2030 and 970 $/kW in 2040 for alkaline electrolyzers.

ETC [11] provides another example yielding much lower costs: attaining 160 $/kW in 2030 and 80 $/kW in 2040 in their "optimistic scenario" (Fig. 8). This prediction uses an 18% learning rate, the same as in Schmidt et al., but yields much lower prices due to a lower reference price for electrolysis (825 $/kW in 2020 compared to 1,340 $/kW in Schmidt et al. [32]) and more optimistic scenario for market growth (3,300 GW installed by 2040 versus 270 GW in Schmidt et al. [32]). This comparison highlights the sensitivity of experience curve analyses to their specific assumptions.

Fig. 7 Development of experience rates for electrolysis stack modules as a function of cumulative production (AEL: alkaline electrolysis; PEMEL: proton exchange membrane electrolyzer; SOEC: solid oxide electrolyzer cell; based on data from Böhm et al. [28])

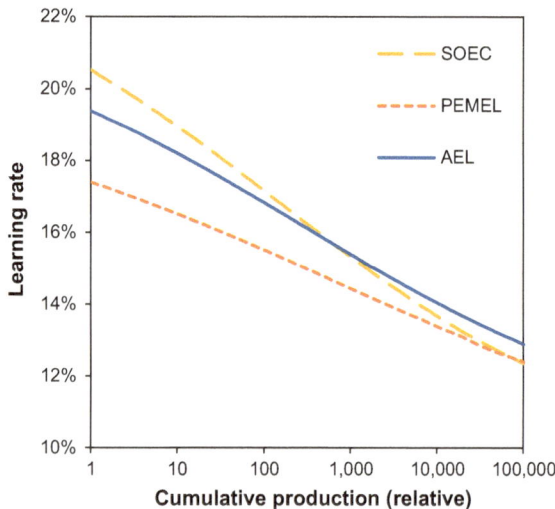

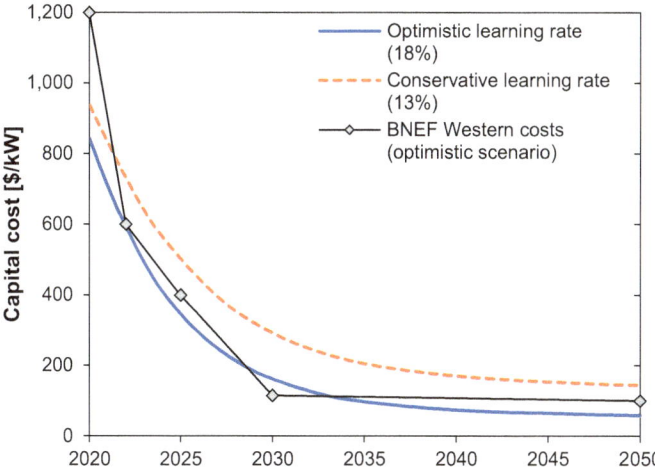

Fig. 8 Cost projections from ETC based on optimistic and conservative learning rates for electrolyzers (technology-neutral); compared to the BloombergNEF scenario for costs outside of China (data from [11])

In summary, high learning rates for electrolyzer stacks and, to a lesser extent, the full systems are expected, broadly in the range of 10–20%. This is likely to yield substantial cost reductions in the coming decade, which will then slowdown in the longer term.

3.2.2 Expert Elicitation Analysis

Due to the scarcity of empirical data, studies have compiled expert estimates of future electrolyzer costs up to 2030. Saba et al. [43] compiled a list of estimates for alkaline and PEM electrolyzer costs spanning back to the 1990s (Fig. 9). For both technologies, they see cost estimates falling and converging to below 1,000 €/kW after 2020.

Bertuccioli et al. [26] provided trajectories for AEL and PEMEL costs out to 2030, using expert elicitation with 22 people from industry and academia. The expert estimates for AEL systems cost fell from 1,100 $/kW (900–1,300 $/kW range) in 2015 to 700 $/kW (450–950 $/kW range) in 2030. For PEMEL, the estimates were 1,900 $/kW (1,450–2,350 $/kW range) in 2015 falling to 900 $/kW (300–1,500 $/kW range) in 2030.

Similarly, Schmidt et al. [51] conducted an expert elicitation with ten people from industry and academia to gage opinion on future cost reductions with both increased R&D funding and production scale-up (Fig. 10). These elicitations yielded similar ranges to those from Bertuccioli et al. albeit with narrower ranges in 2030. The experts estimated that increased R&D funding for water electrolysis could lower capital costs by 7–24% by 2030, with the weakest effect seen for AEL due to its

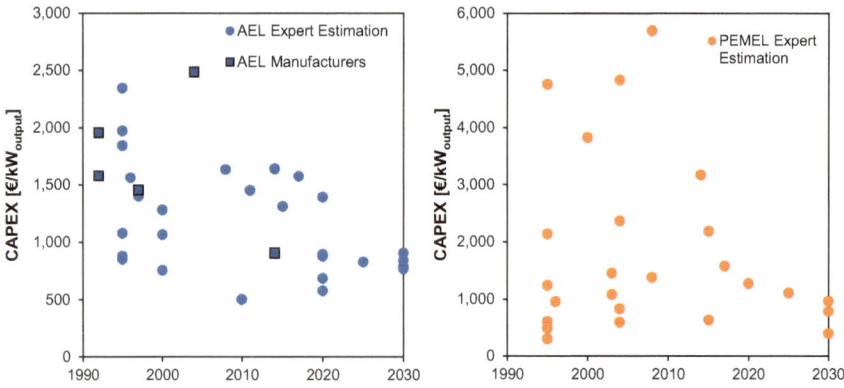

Fig. 9 Cost projections for alkaline and PEM electrolyzers surveyed from the literature (Currency in 2017 Euros; AEL: alkaline electrolysis; PEM: proton exchange membrane; based on data from Saba et al. [43])

maturity. Production scale-up was consistently thought to reduce costs by a further 22 to 29% across all technologies (Fig. 10).

3.3 Drivers of Cost Reduction

Cost reductions for electrolyzers are likely to be driven by a quickly maturing and growing market, namely: manufacturing scale-up and plant size increases, design improvements, and shifting production to world regions with lower production costs.

3.3.1 Electrolyzer Plant Size and Production Scale-Up

While electrolyzers spent several decades at the kW scale, the size of individual electrolyzer projects has increased markedly over the last decade as manufacturing supply-chains mature. Between 2010 and 2017, AEL systems increased in size from 120 kW to 2 MW on average (a factor of about 17), and PEMEL increased from 10 kW to 2.9 MW [10] (a factor of about 290). Projects are expected to increase by three orders of magnitude over the coming decade, with rapid scale-up from 1 to 5 MW in 2020 to 30 to 300 MW by 2025 [8, 10] (Fig. 11).

The impact of increasing plant size reduces system cost via economies of scale. As the capacity of the system increases, the material and energy requirement typically reduces per unit of production (Fig. 12).

Manufacturing scale-up also gives substantial potential for cost reduction. As with increasing plant size, increasing economies of scale in manufacturing can significantly reduce specific costs such as energy and material requirements and labor via increased automation and increased learning rates. For electrolyzers, a move away

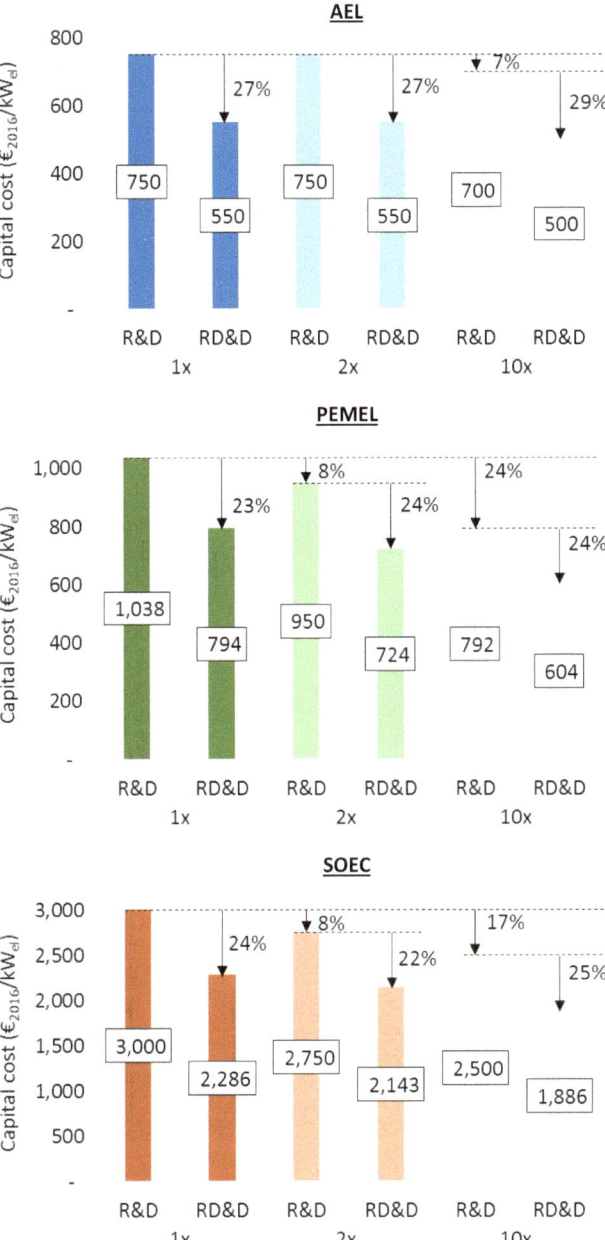

Fig. 10 Estimated capital costs for water electrolysis in 2030 from expert elicitations conducted by Schmidt et al. [51]. The median cost from all experts is given by technology (top to bottom). Each panel shows the relative impact of increased R&D funding (1×, 2×, 10×) by bars labeled R&D. This impact combined with production scale-up due to increased deployment is shown by bars labeled RD&D (AEL: alkaline electrolysis; PEMEL: proton exchange membrane electrolysis; R&D: research and development; RD&D: research, development and demonstration; SOEC: solid oxide electrolyzer cell)

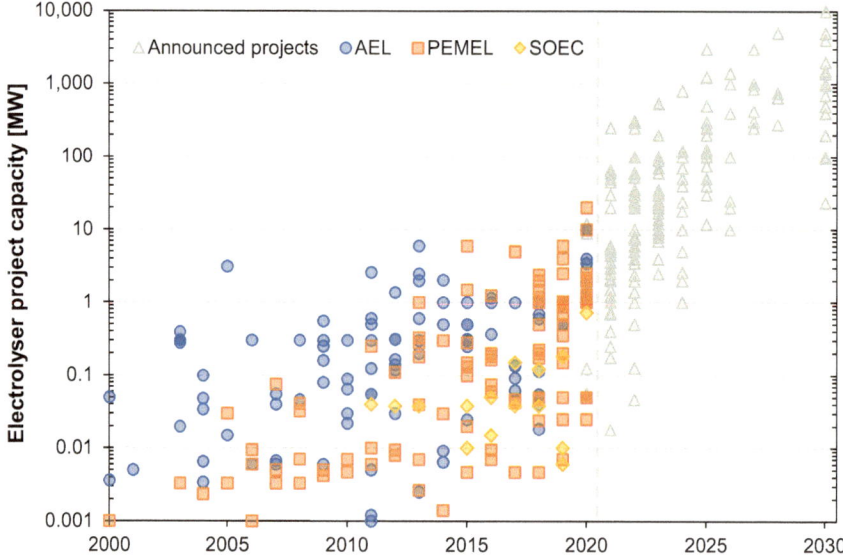

Fig. 11 Size of individual electrolysis plants commissioned over the last two decades, and announced by companies for construction during the next decade (AEL: alkaline electrolysis; PEMEL: proton exchange membrane electrolyzer; SOEC: solid oxide electrolyzer cell; compiled using data from IEA [10] and Aurora [8])

Fig. 12 Estimate of cost reduction associated with plant size increases for alkaline and PEM electrolyzers (PEM: proton exchange membrane; compiled using data from the IEA [7])

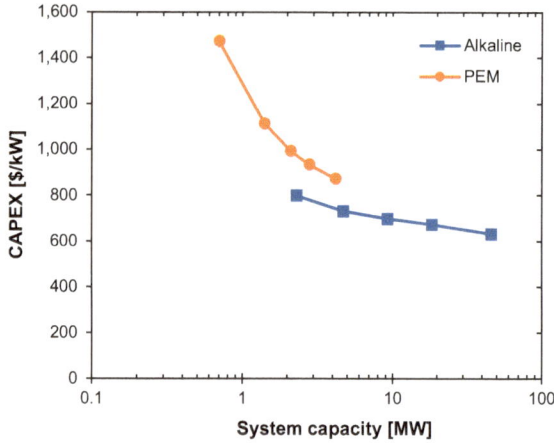

from manual stacking and connecting, toward high volume manufacturing methods such as laser-cutting, plastic injection molding, and 3D-printing could contribute to cost reductions.

Increased learning from manufacturing experience will help to de-risk system design and utilize finer margins (e.g., lower material requirements) to optimize cost, efficiency, and lifetimes. Costs of capital and building were identified by Mayyas as being large contributors to low-volume-production of stack elements such as the catalyst-coated membrane, bipolar plates, and porous transport layer (for PEMEL) and could be all but eliminated at large manufacturing volumes (of over 2,000 units per year) [27].

3.3.2 Design Improvements

There are several technical improvements that may increase efficiency or reduce the specific cost per unit of capacity and are specific to each electrolyzer technology type. For AELs, increasing current density from 0.2 to 0.4 up to 0.6 A/cm^2 via better mixed metal oxide catalysts may be achievable [51]. A higher temperature operation would enable increased efficiency with more stable electrodes and electrolytes, and zero-gap designs which remove the distance between electrodes would decrease the resistance associated with electrolyte and bubble formation [52, 53],.

For PEMELs, higher current densities can be achieved, from 0.6 to 2 A/cm^2 up to more than 3 A/cm^2 via improved electrode design, catalyst coating, and thinner membranes. Reducing the use of iridium and platinum with thinner coatings may reduce cost, as well as a replacement of titanium in bipolar plates and porous transport layers with high-conductivity/stable coatings on low-cost materials such as steel. The rectifier, which converts AC current to DC, represents a large proportion of CAPEX which could be reduced if a DC supply was used and required only a DC/DC converter.

For SOEC, CAPEX reductions are achievable via reducing operating temperatures to approximately 450 °C from reducing electrode polarization resistance. This would help to avoid the requirement for high-temperature exotic materials and enable the use of lower cost materials such as stainless steel. So far, SOECs are still at an early stage of development and there is a need to prove lifetimes and improve cell and stack designs.

Illustrating the combined potential cost reductions associated with design improvements, increased plant size, and manufacturing scale-up, Mayyas and Mann model the cost reduction for PEMEL systems [27]. They find the largest improvements are made from manufacturing economies of scale increasing production from 10 to 100 units per year, but total costs may be reduced from about 560–270 $/kW.

3.3.3 Shifting Production Centers

Another source of anticipated cost reductions is the shift of production from the west (primarily Europe and America) to China. This mirrors the experience seen with other low-carbon technologies; for example, the price of solar PV panels fell rapidly when production shifted from Germany and USA to China [54].

BNEF cite three reasons for lower production costs in China: lower costs for raw materials and labor, higher utilization rates for factories, and lower spending on R&D and marketing [55]. Others suggest that production quality is a factor, in particular lower durability and reliability [56]. BNEF announced that Chinese-made AELs were sold for 200 $/kW in 2019, 83% less than Western-made systems at the time [55].

The BNEF estimates were more bullish than other sources, as according to the IEA alkaline water electrolyzers cost rather about 500 $/kW. BNEF assumes that costs from Western manufacturers could converge with those from Chinese manufacturers over the coming decade [55]. Failing to become more competitive on cost could result in a declining market share for these manufacturers, and ultimately bankruptcy. Agora Energiewende propose that EU-wide innovation support is key to the success of electrolysis manufacturing in Europe [56].

4 Levelized Cost of Hydrogen Production

While capital costs are important, the total cost of construction and operation—and thus the cost of hydrogen produced—also depends strongly on the cost of electricity purchased or generated, and on technical parameters such as the electrolyzer cell efficiency and lifetime.

Just as renewable and conventional power stations can be summarized by their levelized cost of energy (LCOE), the total cost of electrolysis can be summarized by the levelized cost of hydrogen (LCOH), also known as the levelized cost of gas (LCOG). This quantifies the total cost of production discounted over the system's lifetime, per unit of hydrogen generated (e.g., $/kg or $/MWh).

The LCOH provides a fair comparison by factoring in all technical and economic parameters; capital cost, operating costs, production efficiency, system lifetime, performance degradation, and the cost of energy used. This concept can be used to explore important trade-offs, for example the use of better materials to increase the durability or efficiency of the system. This will likely increase the capital cost but reduce operating costs due to less maintenance required or less electricity needing to be purchased.

Calculation of LCOH The levelized cost of hydrogen (LCOH) can be described as the total lifetime cost of the investment in a hydrogen production technology divided by its cumulative delivered hydrogen. Its value reveals the average price that hydrogen must be sold for to make the system break-even financially [19]. Both costs and hydrogen production are discounted according to the investment's cost of capital (also known as the discount rate), to reflect the time-value of money. Costs incurred many years into the future, or the value of hydrogen that is sold far into the future will have less importance to the viability of the investment decision made today.

As with the levelized cost of storage (LCOS), there are various definitions employed which may include or exclude relevant parameters such as end-of-life disposal of the system, electrolyzer stack replacement, or capacity degradation over the lifetime [57].

The levelized cost of hydrogen [57] is given by:

$$
\text{COH} = \frac{\text{Investment cost}_n + \sum_n^N \frac{O\&M \text{ cost}_n}{(1+r)^n} + \sum_n^N \frac{\text{Energy cost}_n}{(1+r)^n} + \frac{\text{Endoflife cost}}{(1+r)^{N+1}}}{\sum_n^N \frac{\text{Hydrogen produced}_n}{(1+r)^n}}
\tag{2}
$$

Summing up all cost categories in each year (n) up to the system's lifetime (N), and discounting each by the project's discount rate (r).

4.1 The Importance of Electricity Costs

The total cost of hydrogen production from electricity chiefly comprises the electrolyzer CAPEX and the cost of electricity used as input to the electrolyzer. The IEA notes that with increasing utilization, CAPEX has a decreasing impact on hydrogen costs, whereas electricity purchase becomes the main cost component for water electrolysis [7].

The latter is governed by the applied technology and the regional environment. A key distinction is whether electricity is purchased from a region's power grid or directly from a renewable generation source. Wholesale electricity market prices vary around the world due to differences in generation mix and the fuels used, emissions prices, and taxation; but a primary driver in most markets is the global or regional price of fossil fuels [58, 59]. Electricity prices also see substantial short-term and long-term volatility, varying diurnally with demand and availability of renewable energy, and seasonally with fluctuating fossil-fuel prices [60, 61].

Many studies [7, 11, 23, 57] consider electricity prices in the range of 40–60 $/MWh, as this broadly reflects the long-term average seen across Europe and North America, or 20 $/MWh as a sensitivity to reflect the trend of power prices falling as the share of renewable energy increases [62]. Figure 13 shows the impact of electricity price on the cost of delivered hydrogen.

Given the role of water electrolysis in decarbonizing energy systems, there is a key focus on "green hydrogen" produced solely from renewable electricity. The cost of electricity generation from solar PV has fallen by a factor of 7 between 2010 and 2020, and for wind, it has halved over the same period [30]. This is primarily due to falling capital costs which are experienced worldwide, but there are also strong regional variations due to the underlying productivity of wind and solar farms [63, 64].

Every region has different solar and wind generation characteristics which affects hydrogen production and costs (Fig. 14). For regions with high capacity factors, the cost for electricity generation is low, reducing the cost of hydrogen production.

If green hydrogen is produced from hard-linking wind or solar PV with an electrolyzer, the lowest cost hydrogen production requires consideration of the trade-off

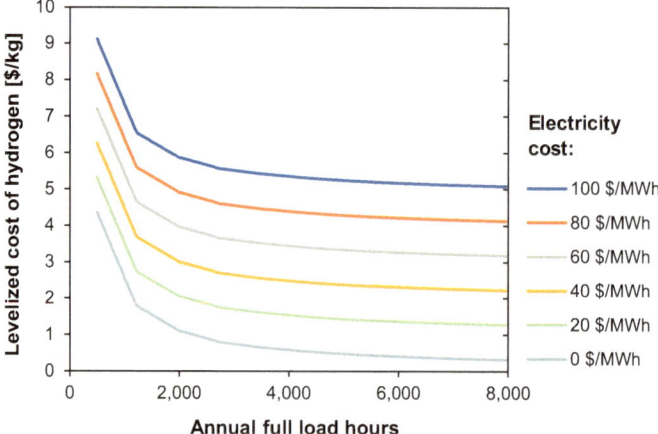

Fig. 13 Hypothetical future levelized cost of hydrogen (LCOH) production from electrolyzers as a function of electricity cost. Calculations assume a discount rate of 8% and efficiency of 69% (LHV) (data from IEA [7])

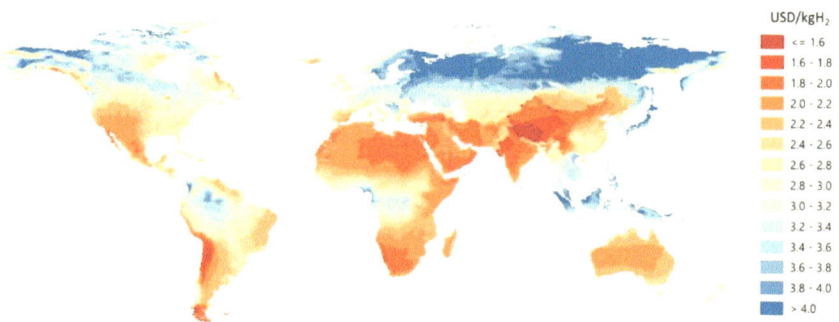

Fig. 14 Modeled cost of hydrogen production using solar PV or wind as electricity source (reproduced from IEA [7], CC-BY 4.0)

between installed solar/wind capacity and installed electrolysis capacity in combination with an energy storage technology (e.g., battery storage): this governs the average utilization rate of the electrolyzer.

For example, for a 1 MW electrolyzer system, 1 MW of installed wind capacity would supply an average of 400 kW with an average capacity factor of 40%. The utilization rate of the electrolyzer would be the same as the capacity factor of the wind (40%). To achieve higher electrolyzer utilization and to decrease the levelized electrolyzer CAPEX, higher quantities of wind must be installed. The increase in utilization will be governed by the wind output curve, and installing extra capacity will yield an oversupply of electricity at some points during the year. This oversupply could be exported if there is an available connection or used on-site; otherwise, it would have to be curtailed. Consequently, there is a trade-off between lowering cost from increased electrolyzer utilization and increasing cost from curtailed wind capacity.

Green hydrogen projects in some jurisdictions have the option to use electricity purchased from the grid, provided that the hydrogen production is time-matched to production of electricity from renewables. For example, the UK Low Carbon Hydrogen Standard allows for this, [65] and at the time of writing, the European Union and USA were also moving toward 24/7 matching being a requirement. [66] This allows for greater flexibility and could lower the cost of hydrogen production, but care must be taken about the whole-system consequential impacts. For example, if hydrogen is produced from existing renewables, this electricity is no longer available for other uses, thus raising the carbon intensity of grid electricity. If green hydrogen is produced from additional renewables, installed specifically for the purpose of producing hydrogen, but not necessarily at the same location, then these negative consequential impacts can be avoided.

4.2 Hydrogen Production Cost Estimates

Studies have converged around a cost of around 5 $/kg for electrolytic hydrogen produced today. This can be converted to 150 $/MWh via the energy content of hydrogen (33.3 kWh/kg at lower heating value (LHV)) [67] to give easier comparison with electricity prices.

IRENA projects that the levelized cost of hydrogen could fall from around 5 $/kg today (2.7–6.0 $/kg range depending on conditions) to 1 $/kg in the future [23]. Most of this saving comes from two key interventions: an 80% reduction in electrolyzer CAPEX (from 750 to 150 $/kW) which saves 1.80 $/kg; and a halving of electricity input cost (from 53–20 $/MWh) which saves 1.40 $/kg [23]. Similarly, ETC models hydrogen costs in Europe at being 5.70 $/kg today (assuming 780 $/kW capital costs for electrolyzers) [11]. This could fall to 4.00 $/kg in future with 500 TWh (10 Mt) annual demand for hydrogen, and further to 1.90 $/kg with 1,100 TWh (22 Mt) annual hydrogen demand. Again, the main savings come from reducing capital costs (1.30 $/kg) and abundant cheap renewable electricity (1.10 $/kg) [11].

Agora Energiewende is more optimistic, suggesting hydrogen could cost 2.60 $/kg today when using photovoltaics (PV) in North Africa as the electricity source [56]. This cost could fall to 1.90 to 2.20 $/kg in 2025, and further to 1.30 $/kg in 2030 if there is convergence toward Chinese manufacturing costs (115 $/kW), or to 1.90 $/kW with IEA's assumption for minimum CAPEX [56].

Academic studies similarly estimate hydrogen production costs of 3.50 to 4.50 $/kg [25, 68], when using offshore wind, and 5.00 to 6.10 $/kg when using other forms of renewable electricity in locations with favorable capacity factors [69]. Provided that recent market trends continue the hydrogen production costs that are assumed to reduce to 2.70 $/kg [25]. These scenarios also agree with current industry announcements. Areva H2Gen report a cost of 3.90 $/kg from a fully-utilized 1 MW PEMEL system (8,000 operating hours per year) at an electricity price of 55 $/MWh [70]. Enapter wishes to reduce the cost of hydrogen from their household-scale (2.4 kW) AEMELs from 10.60 $/kg in 2020 to 4.60 $/kg in 2030 (inclusive of 3 $/kg for electricity consumed) [70].

5 Conclusion

While electrolysis supplies only 7 million tons of hydrogen per annum today, the current hydrogen renaissance anticipates electrolysis-based hydrogen production to grow by 75% per annum over the next decade. Hydrogen is now seen as a critical vector in decarbonizing hard-to-reach areas of heavy industry (e.g., steel production, fertilizer) and heavy goods transport (e.g., shipping), where there are cost and technical limitations associated with electrification.

The capital cost of electrolyzers is a key factor in governing system feasibility, and costs have experienced a steep decline over the past decade. Current estimates range from 200 to 2,500 $/kW capacity across technologies and regional manufacturing hubs. Costs are likely to continue to decline as manufacturing increases, with learning rates associated with both material efficiency gains and increasing economies of scale. Cost reductions are particularly steep for less mature PEMEL and SOEC technologies, projected to be 10 to 20% per doubling of capacity. The shift of electrolyzer production from the USA and EU to China is also expected to yield substantial cost reductions, as previously experienced with solar PV.

More broadly, the levelized cost of hydrogen (LCOH) is a useful metric to judge total cost, particularly accounting for electricity price as the other critical cost of production. Green hydrogen requires the use of a renewable energy source, where the cost of electricity production from solar PV and wind has fallen dramatically between 2010 and 2020: by a factor of 7 for solar PV and 2 for wind. LCOH of green hydrogen today is estimated to be 2.5 to 5 $/kg, typically around 50% above higher-carbon sources of production, but these costs are projected to converge up to 2050. However, producing hydrogen solely from intermittent renewables limits the utilization of the electrolyzer, representing a design challenge to size the system for minimum cost while ensuring a low-carbon supply from a life cycle perspective. Connecting to grid electricity will help to maximize electrolyzer utilization, but care must be taken to ensure that supply is low carbon by time-matching grid off-take with known renewable supply.

References

1. Cembalest M (2020) J.P. Morgan tenth annual energy paper. https://tinyurl.com/jpmorgan-10aep

2. Deloitte (2020) Investing in hydrogen: ready, set, net-zero. https://tinyurl.com/deloitte-invest ing-in-hydrogen
3. HSBC Centre of Sustainable Finance (2020) Hydrogen for the future: delivering zero-carbon in heavy industry. https://tinyurl.com/hsbc-hydrogen-for-the-future
4. European Commission (2020) A hydrogen strategy for a climate-neutral Europe. https://ec.eur opa.eu/energy/sites/ener/files/hydrogen_strategy.pdf
5. Collins L (2021) Green hydrogen: ITM power's new gigafactory will cut costs of electrolysers by almost 40%. World-Energy. https://www.world-energy.org/article/15430.html
6. Bullard N (2021) A gigafactory for hydrogen could be a game-changer. Bloomberg. https://www.bloomberg.com/news/articles/2021-07-01/a-gigafactory-for-hydrogen-could-be-a-game-changer
7. IEA (2019) The future of hydrogen: seizing today's opportunities. International Energy Agency
8. Aurora Energy Research (2021) Hydrogen market attractiveness report (HyMAR)
9. Buttler A, Spliethoff H (2018) Renew Sustain Energy Rev 82:2440–2454
10. IEA (2020) Hydrogen projects database
11. ETC (2021) Making the hydrogen economy possible: accelerating clean hydrogen in an electrified economy. Energy Transitions Commission
12. Staffell I, Scamman D, Velazquez Abad A, Balcombe P, Dodds PE, Ekins P, Shah N, Ward KR (2019) Energy Environ Sci 12:463–491
13. Gregory DP, Ng DYC, Long GM (1972) The hydrogen economy. In: Bockris JO (1972) Electrochemistry of cleaner environments. Springer
14. Momirlan M, Veziroglu TN (2002) Renew Sustain Energy Rev 6:141–179
15. Sanderson H (2020) Fuel-cell producers jump on new hydrogen 'hype cycle'. Financial Times
16. Bakker S (2010) Energy Policy 38:6540–6544
17. Pollet BG, Kocha SS, Staffell I (2019) Curr Opin Electrochem 16:90–95
18. Dodds PE, Staffell I, Hawkes AD, Li F, Grünewald P, McDowall W, Ekins P (2014) Int J Hydrogen Energy 40:2065–2083
19. IEA (2021) Global EV outlook
20. Staffell I, Brett D, Brandon N, Hawkes A (2012) Energy Environ Sci 5:9291–9306
21. Liebreich M (2023) Clean hydrogen ladder, version 5.0. Concept credit: Adrian Hiel, energy cities. Liebreich Associates. https://www.linkedin.com/pulse/hydrogen-ladder-version-50-mic hael-liebreich/
22. Fuel Cells Bulletin (2019) 10:10
23. IRENA (2020) Green hydrogen cost reduction: scaling up electrolysers to meet the 1.5 °C climate goal. International Renewable Energy Agency
24. Smolinka T, Wiebe N, Thomassen M (2017) Cost break down and cost reduction strategies for PEM water electrolysis systems. In: 6th EUROPEAN PEFC & electrolyser forum
25. Glenk G, Reichelstein S (2019) Nat Energy 4:216–222
26. Bertuccioli L et al (2014) Fuel cells and hydrogen joint undertaking development of water electrolysis in the European Union. E4Tech/ Element Energy
27. Mayyas A, Mann M (2019) Procedia Manufacturing 33:508515
28. Böhm H, Goers S, Zauner A (2019) Int J Hydrogen Energy 44:30789–30805
29. James J (2019) Towards sustainable ammonia production. TNO
30. IRENA (2021) Renewable power generation costs in 2020. International Renewable Energy Agency
31. Jansen M, Staffell I, Kitzing L, Quoilin S, Wiggelinkhuizen E, Bulder B, Riepin I, Müsgens F (2020) Nat Energy 5:614–622
32. Schmidt O, Hawkes A, Gambhir A, Staffell I (2017) Nat Energy 2:17110
33. Staffell I, Green R (2013) Int J Hydrogen Energy 38:1088–1102
34. BNEF (2020) Hydrogen economy outlook
35. Schmidt O, Staffell I (2023) Monetizing energy storage: a toolkit to assess future cost and value. Oxford University Press, Oxford
36. BCG (1970) Perspectives on experience. Boston Consulting Group

37. Junginger M, van Sark W, Faaij A (2010) Technological learning in the energy sector: lessons for policy, industry and science. Edward Elgar Publishing
38. Wright TP (1936) J Aeronaut Sci 3:122–128
39. ITRPV (2021) Maturity report, 11th edn
40. McDonald A, Schrattenholzer L (2001) Energy Policy 29:255–261
41. Neij L (2008) Energy Policy 36:2200–2211
42. Malhotra A, Schmidt TS (2020) Joule 4:2259–2267
43. Saba SM, Müller M, Robinius M, Stolten D (2018) Int J Hydrogen Energy 43:1209–1223
44. Staffell I, Green R (2009) Int J Hydrogen Energy 34:5617–5628
45. Schoots K, Kramer GJ, van der Zwaan BCC (2010) Energy Policy 38:2887–2897
46. Wei M, Smith SJ, Sohn MD (2017) Appl Energy 191:346357
47. McDowall W (2012) Endogenous technology learning for hydrogen and fuel cell technology. In: UKSHEC II: literature review, research questions and data
48. Thomas CE, Kuhn IF Jr (1995) Electrolytic hydrogen production infrastructure options evaluation. NREL. https://doi.org/10.2172/125028
49. Schoots K (2008) Int J Hydrogen Energy 33:2630–2645
50. Hydrogen Council (202) Path to hydrogen competitiveness: a cost perspective
51. Schmidt O, Gambhir A, Staffell I, Hawkes A, Nelson J, Few S (2017) Int J Hydrogen Energy 42:30470–30492
52. Grigoriev SA, Fateev VN, Bessarabov DG, Millet P (2020) Int J Hydrogen Energy 45:26036–26058
53. Phillips R, Dunnill CW (2016) RSC Adv 6:100643–100651
54. Staffell I, Brett DJL, Brandon NP, Hawkes AD (2015) Domestic microgeneration: renewable and distributed energy technologies, policies and economics. Routledge, London
55. BloombergNEF (2020) Hydrogen economy outlook
56. Deutsch M, Graf A (2019) EU-wide innovation support is key to the success of electrolysis manufacturing in Europe. Agora Energiewende
57. Schmidt O, Melchior S, Hawkes A, Staffell I (2019) Joule 3:81–100
58. Girish GP, Vijayalakshmi S (2013) Int J Bus Manage 8:70–75
59. Zakeri B, Staffell I, Dodds PE, Grubb M, Ekins P, Jääskeläinen J, Cross S, Helin K, Castagneto Gissey G (2023) Energy Reports 10:2778–2792
60. Staffell I (2017) Energy Policy 102:463–475
61. Ward KR, Green R, Staffell I (2019) Energy Policy 129:1190–1206
62. Ketterer JC (2014) Energy Econ 44:270–280
63. Staffell I, Pfenninger S (2016) Energy 114:1224–1239
64. Pfenninger S, Staffell I (2016) Energy 114:1251–1265
65. HM Government (2022) Net zero hydrogen fund strands 1 and 2: successful applicants. https://www.gov.uk/government/publications/net-zero-hydrogen-fund-strands-1-and-2-successful-applicants
66. Peacock B (2022) 'Not actually green': why time matching is key to becoming a hydrogen superpower. PV Magazine Australia
67. Staffell I (2011) The energy and fuel data sheet. https://tinyurl.com/energy-fuel-data-sheet
68. McDonagh S, Ahmed S, Desmond C, Murphy JD (2020) Appl Energy 265:114732
69. Guo X, Li X, Xu Z, He G, Miao P (2020) Energy Storage Sci Technol 9:688–695
70. Lichner C (2020) Electrolyzer overview: lowering the cost of hydrogen and distributing its production. PV Magazine

Iridium and Platinum Availability for Electrolyser Production up to 2030

Maria Greitzer

Abstract To tackle anthropogenic climate change, the large-scale production and use of green hydrogen stand out as a key solution for replacing technologies dependent on fossil fuels. Achieving this, however, necessitates substantial amounts of electricity from renewable energy sources and the rapid expansion of electrolyser capacities. Notably, the production of green hydrogen using polymer electrolyte membrane (PEM) electrolysers relies heavily on the catalytic materials iridium and platinum. Due to the Earth's limited supply of these metals and their extensive use in other industries, scaling up electrolysis operations faces significant resource constraints. Addressing these challenges is essential for enabling a sustainable and cost-effective growth of green hydrogen production.

Keywords Iridium · Platinum · Ressource Availability · PEM · Electrolysis

1 Introduction

In addressing anthropogenic climate change, the widespread production and utilisation of green hydrogen emerges as a solution to replace various technologies reliant on fossil fuels. This requires large amounts of electricity produced by renewable sources of energy and the rapid deployment of electrolyser capacities. Yet, the production of green hydrogen through polymer electrolyte membrane (PEM) electrolysers also heavily relies on catalytic materials iridium and platinum. Given the limited resources of these metals on Earth and their wide-ranging applications in other industries, there is a pronounced resource criticality associated with scaling up electrolysis operations. Addressing these challenges is crucial for ensuring a sustainable and economically viable expansion of green hydrogen production.

On the demand side, Europe is regarded as a green hydrogen consumer region [1]. However, the know-how in PEM electrolysis, electrolyser production capacity

M. Greitzer (✉)
Fraunhofer Research Institution for Energy Infrastructures and Geothermal Systems IEG
(Cottbus), International University of Applied Sciences, Düsseldorf, Germany
e-mail: maria.greitzer@ieg.fraunhofer.de

and the willingness of European states to back up new technologies with incentives are not to be underestimated. Currently, from a total of 51 electrolyser manufacturers identified [2], most are located in Germany (18), France (6), the UK (3), Switzerland (2), and Norway, Sweden, Netherlands, and Denmark (each one).

The USA and China—due to their vast domestic potential and the ownership of the know-how (eight electrolyser producers in the USA and four in China)—serve both functions—as a technology exporter and a hydrogen utilisation region. The USA also has remarkable reserves of platinum group metals, estimated to be about 900 t [3]. Nevertheless, the USA relies fully on the iridium import. Similarly, China imported almost double the amount of iridium and ruthenium than necessary for covering the country's demand in 2021 [4], probably anticipating its scarcity in years to come.[1]

South Africa and Russia are rich in metals currently necessary for the production of electrolysers as their reserves are estimated to be about 63,000 t and 5,500 t, respectively [3]. The region of Middle East has the potential of low-cost PV electricity and the role of India, with one producer of alkaline electrolysers [2], remains to be clarified.

In general, an increased production of hydrogen in Australia, the Middle East, Africa as well as Central and South America (especially Chile and Colombia) is expected. This is true due to an immense solar potential or solar and wind potential (Chile, Argentina) for electricity production [1, 5]. However, countries deploying big hydrogen production projects or owning know-how for electrolysers' production do not match with countries with optimal conditions for producing hydrogen. For the years to come, the first MW scale PEM electrolyser projects are located in high-tech world regions with long-lasting engineering traditions. Only Australia is characterised by both qualities: high availability of solar energy and access to electrolyser technology. China also aspires to a unique position on the global market, the country leads globally in producing solar panels, batteries, and electric vehicles and is assured a strategic position for producing electrolysers, albeit mainly for export.

Given the aspects presented above, this chapter discusses material requirements[2] and availability for producing hydrogen in proton exchange membrane (PEM) electrolysers.[3] In this analysis, the physical scarcity of materials and the economic scarcity are distinguished as follows:

---

[1] At the same time, Chinese manufactures are trying to replace iridium with platinum for lithium tantalate crystal production. As a consequence, iridium was sold back to market [25].

[2] Studies on material limitations for renewable energy technologies started to appear around the beginning of the twenty-first century. As hydrogen production was considered too costly compared to then-low prices of fossil fuels, authors discussed photovoltaic (PV) panel and wind power plant production. For example, [5] assessed possible bottlenecks of lithium, cobalt and silver for PV, wind turbines, and batteries.

[3] Solid oxide electrolysis (SOE) has not been considered because this technology is not yet available to be scaled up to a MW scale [12]. Hydrogen production via bio-chemical and thermal conversion, as well as the production through conventional technology from natural gas (i.e. steam methane reforming (SMR)), are also out of the scope of this work.

- The physical scarcity of materials is defined as the lower amount of produced raw materials than the demanded amount from industry for a certain period, due to geological conditions.
- Economic scarcity refers to the delivery of lower amounts of materials from producing countries (e.g. South Africa, Russia) to countries of electrolysers' production (e.g. Germany, UK, the USA), due to socio-economic or geo-political reasons.

In this context, the analysis attempts to answer the question whether the physical scarcity of materials for PEM electrolysers could slow down the deployment of this technology on a global scale in its first scale-up phase until 2030. Therefore, in Sect. 2, the demand side characterised by the annual increase of electrolyser capacity is described on a global scale. Additionally, the focus on PEM electrolysers in this analysis is further explained. In Sect. 3, the supply side of two main platinum group metals, namely platinum and iridium,[4] is presented. The results of the analysis are presented in Sect. 4, and this article is concluded with an outlook and a conclusion in Sect. 5.

2   Iridium and Platinum Demands

Electrolysers, which are fundamental for green hydrogen production, are composed of hundreds of materials, each with its own supply and demand dynamics. As global efforts for carbon neutral energy solutions intensify, there is a risk that some of these materials could become scarce. Addressing potential bottlenecks and ensuring a steady supply of these materials through production and recycling enable scaling up hydrogen technology in the future. In this context, several studies and reports address material requirements for electrolysers, such as the following:

- Joint Research Centre [6]. In 2013, the report on critical materials for a hydrogen storage system of the Joint Research Centre listed aluminium and copper as scarce materials, since the increased cell area of electrolysers leads to the increased use of both materials. Aluminium and copper were followed by resin, iridium, platinum, polyvinyl chloride (PVC), and steel [6]. Their estimates for the global production of platinum and iridium in 2011 were 259 and 9 t/a, respectively. For the year 2020, their projections amounted to 337 and 11 t/a. However, unexpected events, the COVID-19 pandemic, decreased the actual production of these metals in 2020 down to 164 and 7.3 t/a (see Table 4).
- World Bank [7]. The World Bank Report on minerals for the deployment of technology using renewable sources did not include hydrogen technology although it was identified as "a key potential low-carbon technology" [7]. The reason in

---

[4] In literature, iridium and platinum have been often falsely characterised as rare earth materials. There are 17 rare earth elements (e.g. neodymium, samarium, terbium), some of them used in direct-drive magnetics in wind turbine generators.

2020 for excluding hydrogen production technology was the limitation of the technology due to high costs as well as infrastructure constraints [7]. Similarly, the European Commission (EC) in its 2021 report on "Critical Raw Materials for Strategic Technologies and Sectors in the EU" regarded the criticality of materials for fuel cells (e.g. platinum) but omitted the hydrogens' production side [8].[5] However, the latest, the fourth list of critical materials for the EU [8] contains platinum group metals and 29 other materials (e.g. heavy and light rare earth elements, lithium, and titanium). The main criteria for the EU to include material on the list are the economic importance of the material, corrected by the substitution index (SIEI)[6] and the supply risk (economic scarcity).

- IRENA [9]. In its publication "Geopolitics of the Energy Transformation" IRENA identifies platinum, cobalt, nickel,[7] iridium, tantalum, gadolinium, zirconium, lanthanum, cerium, and yttrium as critical materials. The last five mentioned are mined almost exclusively in China [10].

As shown above, some reports underscore the rising demand for metals like aluminium and copper, while others highlight the criticality of platinum group metals like iridium or platinum. Overall, different metals are considered essential across these studies. In this context, Table 1 sums up different material needs for three main types of electrolysers.

This work assesses whether the physical scarcity of materials for PEM electrolysers could slow down the deployment of this technology at a global scale. The criterion used is to select materials relevant to the core function of electrolysers, namely splitting water. In this context and within the model's boundary conditions, platinum and iridium have been chosen due to their catalyst function in membrane electrode assemblies (MEAs).

- Platinum (Pt). Platinum is used as a catalyst on both the anode and the cathode in PEM electrolysers to accelerate water splitting.
- Iridium (Ir). Iridium is used as a catalyst on the anode in PEM electrolysers to facilitate the oxygen evolution reaction (OER) due to its excellent corrosion-resistance property.

As explained above, platinum and iridium are central metals for PEM electrolysers. These electrolysers are especially relevant in the context of an electrolyser capacity scale-up, as compared to other electrolysis technologies, they offer rapid load switching times, making them advantageous for utilising electricity from renewable sources. Simultaneously, they exhibit relatively high efficiency and a compact

---

[5] The European Commission (EC) report states steam reforming of natural gas as the currently preferred option for hydrogen production.

[6] The "Substitution Index for Essentiality and Irreplaceability" (SIEI) for raw materials measures the replaceability of material in specific applications or industries to assess its criticality in product production or industrial processes. The availability of alternative metals reduces both the economic importance and the supply risk dimension.

[7] Availability of conventional catalysts such as nickel and cobalt for alkaline electrolysers is not examined in this chapter.

Table 1 Metals used for different electrolyser types (*AEM* Anion exchange membrane; *PEM* Polymer electrolyte membrane; less abundant materials are shown in italics)

Metals used	Electrolyser component	Rare metals (affecting electrolysers performance if reduced)	Alternatives
Alkaline electrolysers			
Nickel (Ni)	Bipolar plates, electrodes, diaphragm	−/−	Platinum-free and cobalt-free designs have been already developed
Titanium (Ti)	Diaphragm		
Potassium (K)	Diaphragm		
Iron (Fe)	Cathode		
PEM electrolysers[a]			
Platinum (Pt)	Membrane electrode assembly (MEA), corrosion-resistant coating	Due to poor stability of platinum-free catalysts	Ruthenium (Ru)
Iridium (Ir) *black*	Anode (MEA), coating for bipolar plates	Due to much slower oxygen evolution reaction (OER) kinetics of iridium alternatives [11, 12])	Iridium (Ir) oxides + Niobium (Nb), Molybdenum (Mo), Hafnium (Hf), Tantalum (Ta), Tin (Sn);[b] reengineering the electrode concept [9]
Titanium (Ti)	Anode, bipolar plates	Due to the decrease of the contact area and the catalytic efficiency of the reaction	
Ruthenium (Ru)	Anode (MEA)		
Platinum (Pt) *black Gold (Au)*	Coating for bipolar plates	Due to the lack of protection against titanium corrosion which results in the degradation of bipolar plates	Carbon-coated 316L stainless steel [13]
Platinum–palladium (Pt–Pd)	Cathode		
Graphite (C)/ *Rhodium* (Rh)	Cathode/Bipolar plates		
Copper (Cu) / Aluminium (Al)	Stack		
AEM electrolysers			

(continued)

Table 1 (continued)

Metals used	Electrolyser component	Rare metals (affecting electrolysers performance if reduced)	Alternatives
Nickel (Ni)	Cathode	No critical materials were used. Further research on materials to increase the chemical stability of cells	−/−
Nickel (Ni)/Iron (Fe)	Anode	No critical materials were used. Further research on materials to increase the chemical stability of cells	−/−

[a] Additional materials for PEM electrolysers, such as low and high alloyed steel, aluminium, copper, plastic, or concrete have not been considered; [b] for further alternatives see [12]

design. In this context, the subsequent sections of this analysis primarily address the demands and quantities required of platinum and iridium in relation to the development of PEM electrolyser capacities.

2.1 Methodology and Assumptions

In the following, the methodology to estimate the demand for iridium and platinum based on different assumptions on the development of PEM electrolysis capacities is outlined. Essentially, two fundamental methodological components can be distinguished: development projections of annual PEM electrolysis capacities and the subsequent derivation of respective iridium and platinum demands, based on specific metal requirements per unit of electrolyser capacity. This is explained in more detail below.

The installed capacity of all types of electrolysers estimated at around 286 MW[8] in 2020 [14], was chosen as the starting point. The IEA reported an increase of more than 200 MW in electrolyser capacity in 2021, and the Ningxia Solar Hydrogen project with alkaline electrolysers in China accounted for three-quarters of this increase [1].

Yearly electrolysers' capacity increase (ECI) is defined as the difference between the cumulative electrolyser capacity (CCI) of the following and the respective year (*t*) (Eq. 1). Accordingly, the cumulative electrolyser capacity is the sum of electrolysers' capacity in the reference year and of yearly increases of electrolyser capacity up to the forecasted year.

---

[8] International Energy Agency [14] reported the estimation of 176 MW for alkaline electrolysers, 89 MW for PEM electrolysers, and 21 MW reserved for unknown types in 2020.

$$ECI_{t+1} = CCI_{t+1} - CCI_t \qquad (1)$$

From the calculated annual development of electrolysis capacities, the associated demand for specific metals is derived, which in this context relates to capacities of PEM electrolysers. Therefore, the yearly demand $D_{(PEM)\,p,t}$ for chosen metals from the platinum group is calculated according to the following (Eq. 2).

$$D_{(PEM)p,t} = I_{(PEM)t} \cdot Mi_{(PEM)p,t} \qquad (2)$$

$I_{(PEM)\,t}$ represents the annual increase in PEM electrolyser capacity, while the metal intensity $Mi_{(PEM)\,p,t}$ stands for the specific amount of required metal from the platinum group (p) per unit of PEM electrolyser capacity in the respective year (t). For platinum (Pt) and iridium (Ir), $Mi_{(PEM)\,p,t}$ is set as follows:

- The platinum (Pt) loading at both the cathode and the anode is set to 0.4 kg_{Pt}/MW_{PEM} up to 2030 according to the average supplier data available in 2023.
- The anode iridium (Ir) loading is set to 0.67 kg_{Ir}/MW_{PEM} up to 2025 and to 0.35 kg_{Ir}/MW_{PEM} from 2026 onwards.

With the current state of technology, reducing iridium catalyst loading means increasing current density; both changes lead to an increase in cell voltage and thus to lower efficiency [11]. Table 2 depicts various platinum and iridium catalyst loading based on literature data.

To calculate the demands for iridium and platinum $(D_{(PEM)p,t})$ based on specific PEM electrolysis developments, the following assumptions are additionally applied:

- Iridium is the co-product of platinum. This means, the main economic interest in mining iridium will stay unchanged (i.e. remains primarily driven by the demand of platinum). Results are conservative as the value of platinum reserves remains unchanged (ca. 70,000 t worldwide), albeit historically reserves grew on average

Table 2 Iridium and platinum loading in electrolysers

Source	Year (reality or projection)	Iridium (Ir)				Platinum (Pt)		
		Current density (A/cm^2)	Loading (g/kW)	Anode catalyst (mg/cm^2)		Current density (A/cm^2)	Loading (g/kW)	Cathode catalyst (mg/cm^2)
[12, 15]	2017	2	0.67–0.75	2		2	0.333	0.5–1
[15]	2035	3	0.05	0.4		3	0.0375	0.3
[9]	Projection	5	0.4	0.2		−/−	0.1	0.05
[16]	2017	1.5	0.75	−/−		1.5	0.075	−/−
[16]	2030+	3	0.037	−/−		3	0.010	−/−
[17], innovative scenario	2035	3	0.05	0.4		−/−	−/−	−/−

of 138 t/a [18]. Recycling from autocatalysts, jewellery and industry accounts for ca. 30–40% of global primary supply but has not been considered as the interest lies in the by-production of iridium.

- Demand-driven platinum production will be influenced more by the expected deployment of hydrogen production (electrolysers) and hydrogen use (fuel cells) technologies than by macroeconomic factors such as global GDP growth. Unexpected events, especially in the mining industry of South Africa, may decrease platinum production in a short term only and thus have not been modelled.
- Primary production of metals is supposed to satisfy the demand for the first generation of scaled-up PEM electrolysers. In 2040, recovery of metals of at least 80% would have been included in the analysis (which falls out of the time scope of this analysis). Similarly, population growth influences the demand for platinum and iridium in the longer term (up to 2040).

2.2 Development of Electrolysis Capacities

In this section, first, a literature-based overview of various hydrogen and electrolysis development scenarios is provided (Sect. 2.2.1). Following that, the specific PEM electrolysis capacity development assumed to calculate the specific metal demands of iridium and platinum is detailed (Sect. 2.2.2).

2.2.1 Scenarios on Electrolysis Development

A number of PEM electrolyser projects is already in the construction phase worldwide. The IEA hydrogen project database lists PEM projects with a total capacity of ca. 55 MW under construction from Spain, the USA, Germany, Poland, Japan, China, Netherlands, Denmark, Norway, Italy, New Zealand, and Australia [19]. To put this number into relation, the Energy Transitions Commission set to add 30 to 70 GW of additional electrolysis capacity globally to reach the cost parity of green hydrogen with grey hydrogen by 2030 [20]. Development of manufacturing capacities lies behind the most progressive scenarios although Thyssenkrupp announced a manufacturing capacity of 1 GW/a in 2020, the ITM of 350 MW/a in 2020, and NEL of 2 GW/a in 2021 [9]. The total available and announced manufacturing capacity by 2025 was stated to be more than 20 GW/a [21] and is set as the upper boundary for electrolyser capacity increase from 2025 on.

Regarding the future development of electrolysis capacities, there is a number of different scenarios and developments predicted.

- The Energy Transitions Commission, the global alliance of energy-related organisations, stated in 2020: *"Hydrogen production will need to scale up from a 60 Mt/a capacity of carbon-intensive hydrogen production today to between 580 and 900 Mt/a capacity of low- or zero-carbon hydrogen by mid-century. If 60% of this is derived from electrolysis, total electrolyser capacity will need to be between*

1,700 and 5,800 GW depending on the load factor, versus less than 25 GW today."
[20]. The capacity of 25 GW refers to electrolysers used in the chemical industry.

- GlobalData [22], a consultancy and research company, assumes that the cumulative electrolyser capacity reaches 8.54 GW in 2026 globally and—under the assumption that 350 projects in planning worldwide will be realised—54 GW by 2030 for all types of electrolysers [22].

- The Sky Shell scenario begins with the present structure of the global economy and computes figures for a scenario in which net-zero emissions would have been achieved by 2070 [23]. To cover total final consumption of hydrogen in Shell Scenario for 2030, the total electrolysers' capacity in the range of 28.8–30.1 GW, depending on the full-load hours, shall be installed. In this scenario, hydrogen as a material energy carrier will be used after 2040, covering 10% of the global final energy consumption by the end of century.

- According to the IEA, if all hydrogen projects currently in the pipeline on a global scale materialise, global electrolysis capacity could reach 134–240 GW by 2030. Europe and Australia lead the scene, with about 30% of the overall installed electrolysis capacity each, followed by Latin America with more than 10% of the announced projects. The numbers above depict a significant increase in cumulative electrolyser capacity compared with 2021 when the projected total electrolysis capacity of the projects in the global project pipeline aiming to become operational by 2030 was 54–90 GW. However, to get on track with the Net Zero Scenario, in which over 700 GW of electrolysers is installed globally by 2030, the project pipeline needs to scale up even faster. The IEA assumes a global population increase from 7,753 to 8,505 million people in 2030 in its Net Zero Scenario [1]. In 2050, 60% of total hydrogen production comes from electrolysers in the Net Zero Scenario. To reach the goal of this scenario, the cumulative capacity of electrolysers needs to increase up to 3,585 GW by 2050 [1]. The lower value of 700 GW by 2030 corresponds to a hydrogen demand of 10 EJ (83.3 Mt) within the IEA scenario, while assuming an average of 4,000 annual full-load hours for the electrolysers.

Table 3 sums up the main predictions and scenarios of electrolyser capacity development from leading institutions (IEA; IRENA, Energy Transitions Commission) and companies (Shell Sky Scenario, Global Data).

2.2.2   Electrolysis Development Scenarios

For the purposes of this analysis, deriving the material requirements for PEM electrolysers capacity up to the year 2030, two scenarios from the Table 3 have been considered, namely the Shell Sky and Global Data scenarios.

- Shell Sky scenario. The Shell Sky scenario has been chosen as an example of scenario modelling to model the global energy system and meets the December 2015 Paris Agreement on climate change with significant carbon emissions sinks. As for electrolysers' capacity, target values for 2030 were 28.9 GW (Shell Sky

Table 3 Overview of electrolyser capacity by scenarios and predictions based on planned electrolyser projects

Scenarios	Unit	2020	2025	2026	2030	2035	2040[a]	2045	2050	2055	2060
Shell Sky scenario, hydrogen consumption [23]	EJ/a	0.04	0.14	–	0.43	1.04	2.18	4.37	8.74	15.2	23.52
Corresponding yearly capacity for covering hydrogen demand in Shell Sky Scenario (by the author)	GW	–	–	–	28.86	72.2	–	–	–	–	–
[9] Future (based on announcement and projects) electrolyser capacity	GW	< 1	6	16.5	25	–	–	–	–	–	–
IEA Net Zero Scenario, (2019–2030), PEM [1]	MW	93	–	–	–	–	–	–	–	–	–
IEA Net Zero Scenario, electrolysers—all types [1]	GW	–	–	–	700/ 850[d]	–	2,400	3,000[d]	3,585	–	–
[22] Electrolyser capacity	GW	0.027	4.61	8.52[c]	54[b]	–	–	–	–	–	–
[22] hydrogen potential production	TWh	–	–	–	–	–	–	–	2,250	–	–

Description. Total final consumption by hydrogen and capacity of electrolysers, various predictions. Assumptions: [a]The first generation of electrolysers will be taken from operation in 2040, and recycled materials are available for the construction of further electrolysers. [b]Under the assumption that 350 projects in planning worldwide will be realised. [c]Excluding feasibility-stage projects (2022) [d]IEA, Net Zero by 2050, A Roadmap for the Global Energy sector, estimated 60% of all hydrogen production will come from electrolysers by 2050. IEA does not distinguish among types of electrolysers in this report

Scenario), corresponding capacity to the hydrogen final consumption in the Shell Sky Scenario to be at 0.14 EJ considering 4,000 annual full-load hours. The cost-optimum range for electrolysers is at 4,000–6,500 annual full-load hours. It is assumed that the whole capacity increase will be covered by PEM electrolysers. This rather conservative scenario is comparable with IRENA data on summed installed capacity if all future projects based on announcements in 2020 come true (25 GW). In 2022, the global electrolyser capacity increased by 130 MW.

- Global Data Scenario. In the Global Data Scenario, it is assumed that 350 announced hydrogen projects with a total capacity of 54 GW will be realised by 2030. Up to 2025, it was computed with a 35% share of PEM electrolysers in the electrolyser market. From 2026, a 40% share has been considered. For comparison, in 2021, the market leader in the electrolyser market was an alkaline electrolyser (60%), PEM electrolyser share accounted for ca. 28% with the rest of the shares reserved for solid oxide electrolysers. As for electrolysers' capacity, target values for 2030 were 54 GW [22], this estimation assumes that all 350 hydrogen projects in planning worldwide will be realised and 21.4 of 54 GW of installed cumulated capacity will be of PEM type (ca. 40%). Iridium and platinum loading calculations relate to only the PEM part of the yearly capacity increase.

- Both scenarios are conservative and realistic, e.g. Wappler et al. [21] calculated with existing and announced capacities with an average annual growth rate of 10%/a and reached a cumulative capacity of ca. 150 GW in 2030.
- Furthermore, it is assumed that the yearly electrolyser capacity increase ($ECI_{(t+1)}$) does not grow linearly, instead a higher increase is expected during 2027 to 2030, because many final investment decisions take place in 2025 and iridium from the closed-loop recycling will be available after 2027.
- In the scenarios considered here, no values in all types of electrolyser capacity increase were higher up to 2027 than the maximum manufacturing capacity for all types of electrolysers, which was stated to be 6 GW/a by [21] and 8 GW/a by [24]. As stated above, for simplicity reasons, the PEM market share on the new capacity of electrolysers is set to 35% up to 2025 and 40% from 2026 on. Figure 1 depicts projections for yearly PEM electrolysers increase from 2023 to 2030 for both scenarios considered here.

Annual PEM capacity increases between 2020 and 2022 have been derived from statistical data on electrolysers being built until 2020 and from the list of completed

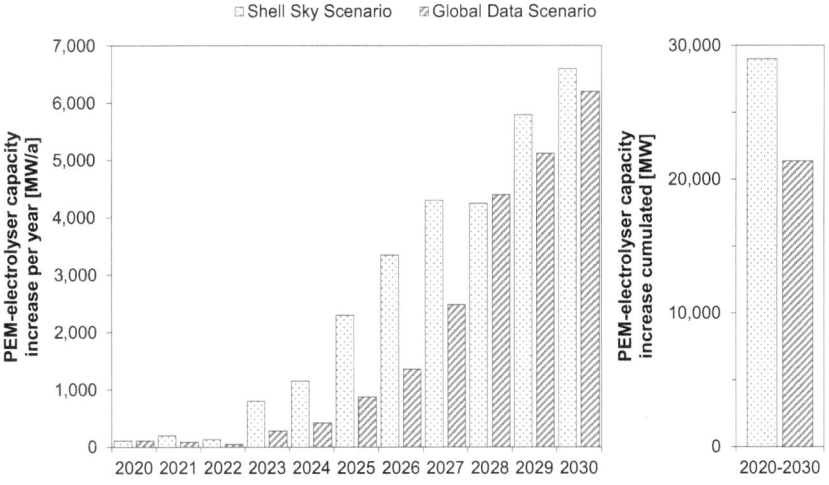

Fig. 1 Yearly and cumulated capacity increase of PEM electrolysers globally

projects. The year 2020 is considered as a starting point for initiating PEM electrolyser projects at a global scale to fulfil energy needs of industry, mobility, and energy sector.

2.3   Iridium and Platinum Demands

Both, iridium as the most corrosion-resistant metal and platinum due to its high chemical and thermal resistance found application in chemical sector, electrical sector, and electronics, automotive (catalysts), and jewellery to name a few. The first noticeable demand for iridium for PEM electrolysers took place in 2021, category electrochemical industry [4]. It is estimated that 283.5 kg of iridium was used for electrolysers' production in 2022 [25]. In this chapter, exclusively iridium and platinum demand derived from the two scenarios of electrolyser capacity increase described above (Fig. 1) are discussed. The results are shown in Fig. 2.

The above results can be interpreted as follows:

- First, loadings of iridium in electrolysers need to be addressed as a priority. Under ceteris paribus condition, any potential decline of platinum use would decrease incentives for platinum production; thus, iridium production would decline accordingly. In both scenarios, if platinum loadings were covered by platinum production only (no recycling), this would cause a 2% increase in global iridium production under the assumption of a 5% Ir/Pt ratio in South African mining fields (Table 5). Therefore, either automotive, electronics, glass

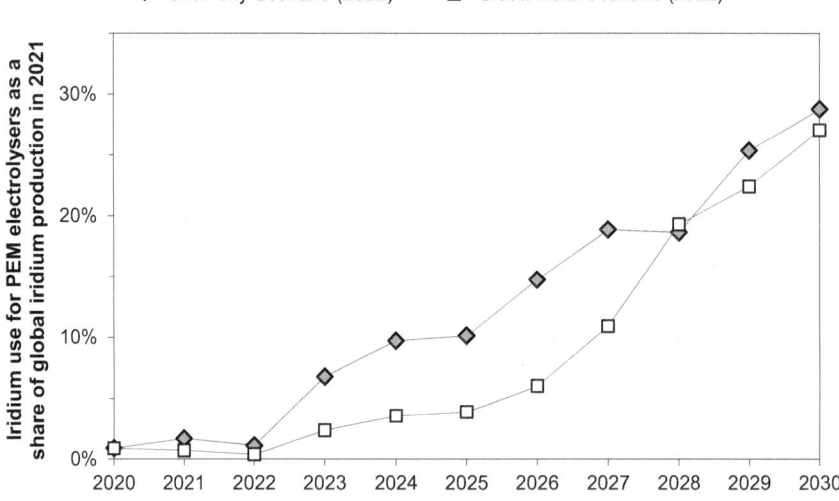

Fig. 2 Iridium use for PEM electrolysers (presented as a share of global iridium production in 2021 and based on different PEM electrolyser development scenarios as of Fig. 1)

industry, and jewellery can increase platinum demand worldwide, or new pricing mechanisms for iridium need to be developed.

- Both scenarios are realistic up to 2030, consuming a 27–29% share of annual iridium production in 2021 (Fig. 2). Comparably, Johnson Matthey's work with the assumption of a 20% share to be realistic; the IEA in its conservative Announced Pledges Scenario counts with annual primary iridium demand for electrolysers to be over 20% of global iridium production up to 2032 [26].

As a rule of thumb, it is sufficient to compare the co-production of iridium while mining platinum and the consumption of iridium (Ir) loading to platinum (Pt) loading to notice the discrepancy between the future demand and future supply. Results for the short term are in line with the IRENA statement: "*global Pt production supports 200 GW of PEM electrolysers per year but global iridium production theoretically enables only 3 till 7.5 GW/a*" [10]. After 2030, the analysis would have included recycling rates for platinum and iridium, as well as projections on the lifetime of the first MW scale generation of electrolysers. However, since this falls out of the time scope of calculations in this analysis, recycling has not been modelled.

3 Iridium and Platinum Availability

3.1 Methodology and Assumptions

Statistical data on all platinum group metals have been collected only recently; most studies evaluate data published by the US National Minerals Information Center [27] or data from the World Bank. Concerning the second source, South Africa exported 26.7 t in the category called "Metals, iridium, osmium, ruthenium, unwrought or in powder form" in 2020, mostly to Hong Kong, Japan, and the USA [28]. To put this figure in perspective, before the pandemic year of 2020, South Africa exported 33.6 t of metals in this category in 2017. Exports in the category "metals; platinum, unwrought or in powder form" highlight South Africa's dominant position on the market, with 65.2 and 69.5 t in 2016 and 2017, respectively. These metals were mostly exported to countries with fuel cell and electrolyser production capacity: Japan, Germany, the UK and the USA [28].

In the scenarios of this chapter, as stated in Sect. 2.3, the increased platinum production due to increasing yearly electrolyser capacity is considered for computing future production of iridium. Besides this assumption, no other assumptions for increasing platinum and thus iridium production as well as availability due to metal recycling have been considered up to 2030. However, recycling and thrifting (reducing material requirement per MW of installed electrolyser capacity) will be key factors for required deployment of electrolyser capacity up to 2050.

3.2   Iridium and Platinum Production

Iridium and platinum are two of six platinum group metals; the remaining four are ruthenium, rhodium, palladium, and osmium. These elements occur in mineral form in a few deposits with global relevance, namely on the territory of South Africa, Russia, and Zimbabwe. Data availability on the production of platinum group metals (PGM) is low. Table 4 depicts the most comprehensive data for previous years available in the literature.

For a better overview, the data presented in Table 4 is illustrated in Fig. 3. It shows the total production amounts (sum of production amounts per country).

As can be seen in Fig. 3, the annual production quantities of palladium and platinum are considerably higher than those of iridium. For the former metals and for the years shown, the quantities for palladium are approximately in the ranges of 200 to 230 t/a, and for platinum slightly below, in the ranges of 165–190 t/a. For both metals, a slight drop can be observed in the year 2020. The considerably lower iridium production amounts to about 7–8 t/a for the years shown.

The considered metals are primarily located in a limited number of deposits worldwide. This distribution is exemplified in Fig. 4, using platinum production amounts for the year 2021 as an example. From a global perspective, only a handful of countries exhibit noteworthy production volumes. Among these states, South Africa, and Russia stand out, contributing the most substantial quantities of platinum to the global market.

3.3   Iridium and Platinum Availability

The fact that iridium is produced as a by-product of platinum[9] makes iridium depend on the production rate for platinum. Figure 5 shows the share of the annual supply of the by-product iridium on the platinum annual supply for the most relevant producer countries: South Africa, Zimbabwe, Canada, and Russia from 2016 to 2020. For South Africa and Zimbabwe, this share ranges from 4 to 6%, except for Russia (around 1%).

The data are in line with the ratio of approximate platinum and iridium ore concentrations in known PGM locations (Table 5), ranging from 1.7% in East Bushveld to 7.6% in West Bushveld, both located in South Africa. In other words, iridium does not cover costs for mining in world-relevant locations. In Russia, platinum is also mined as a by-product of nickel.

As it can be seen in Table 5, due to negligible concentration levels for iridium, the metal is not reported separately. Iridium is also not included in the "4E" (estimates and data for combined platinum, palladium, rhodium, and gold) with platinum with more than 50% of the combined 4E value but in 6E, covering six metals.

---

[9] Iridium production is also associated with palladium, osmium, rhodium, gold, chromium, ruthenium, and copper.

Table 4 Global production of palladium, platinum, and iridium

Country/locality[3]	2017		2018		2019		2020		2021	
Palladium (Pd)	(kg)		(kg)		(kg)		(kg)		(kg)	
Australia[e,4]	600		420		380		410		350	
Canada[e]	19,000		21,000	r	23,000	r	14,000	r	15,000	
China	1,400		1,300		1,300		1,300	e	1,000	e
Finland	1,021		1,157		699		858		1,036	
Russia[e]	88,000		90,000		98,000		93,000		86,000	
Serbia	38		55		100	e	100	e	150	e
South Africa	80,713		80,629		80,684		66,264	r	84,336	
United States[4]	14,000	e	14,300		14,300		14,600		13,700	
Zimbabwe	11,822		12,094		11,640		12,890		12,398	
Total	217,000	r	221,000	r	230,000		203,000		214,000	
Platinum (Pt)	(kg)		(kg)		(kg)		(kg)		(kg)	
Australia[e,4]	170		120		110		110		100	
Canada[e]	7,600		7,600	r	8,500	r	5,300		6,000	
China	2,500		2,500		2,500		2,500	e	2,300	e
Colombia	567	r	270	r	178	r	414	r	400	e
Ethiopia[e]	4		4		2		2		4	
Finland	1,418		1,576		953		1,277		1,447	
Russia[e]	22,000		22,000		24,000		23,000		21,000	
Serbia	2		5		10	r,e	20	e	20	e
South Africa	132,500	r	137,053		132,989		111,993		141,626	
United States[4]	4,000	e	4,160		4,150		4,200		4,020	
Zimbabwe	14,257		14,703		13,857	r	15,005		14,732	
Total	185,000	r	190,000		187,000		164,000		192,000	
Iridium (Ir)	(kg)		(kg)		(kg)		(kg)		(kg)	
Canada[e]	200		400	r	300	r	0	r	0	
Russia[e]	300		200		300		250		230	
South Africa	6,057	r	6,357		6,464		6,186	r	7,006	

(continued)

Table 4 (continued)

Country/locality[3]	2017		2018		2019		2020		2021	
Zimbabwe	619		586		845	r	836		641	
Total	7,180		7,540		8,010	r	7,270		7,880	

[e] estimated; [r] revised; [1] the Table includes data available through June 6, 2022, totals may include estimated data; [2] mine production only, not including refinery production; [3] in addition to the countries and / or localities listed, Indonesia and the Philippines may have produced limited quantities of PGMs, but available information was inadequate to make reliable estimates of output; [4] by-product platinum and palladium produced from gold–copper and nickel ores are excluded (*Source* U.S. Geological Survey [27])

Table 5 Approximate concentrations for platinum (Pt) and iridium (Ir) (PGM: platinum group metals; ratio added by the author; data source: [29])

Country	Mining field	Approximate PMG ore concentrations		Ir/Pt ratio
		Platinum (g/t)	Iridium (g/t)	(%)
South Africa	Merensky Reef	2.7	0.05	1.85
	East Bushveld	2.32	0.04	1.72
	West Bushveld	3.60	0.14	3.89
	Platreef	1.90	0.04	2.11
	UG2	2.00	0.13	6.50
	East Bushveld	2.42	0.18	7.44
	West Bushveld	2.89	0.22	7.61
USA	Stillwater Complex	3.30	0.21	6.36
Canada	Sudbury Complex	0.30	0.01	3.33

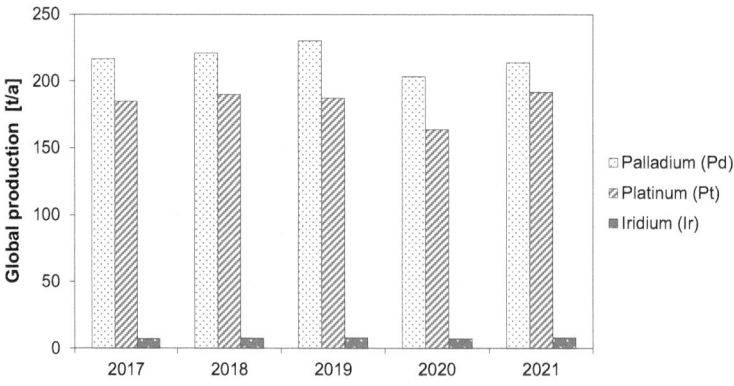

Fig. 3 Global production of palladium, platinum, and iridium (data based on Table 4)

Fig. 4 Global production of platinum by country in 2021 (data based on Table 4)

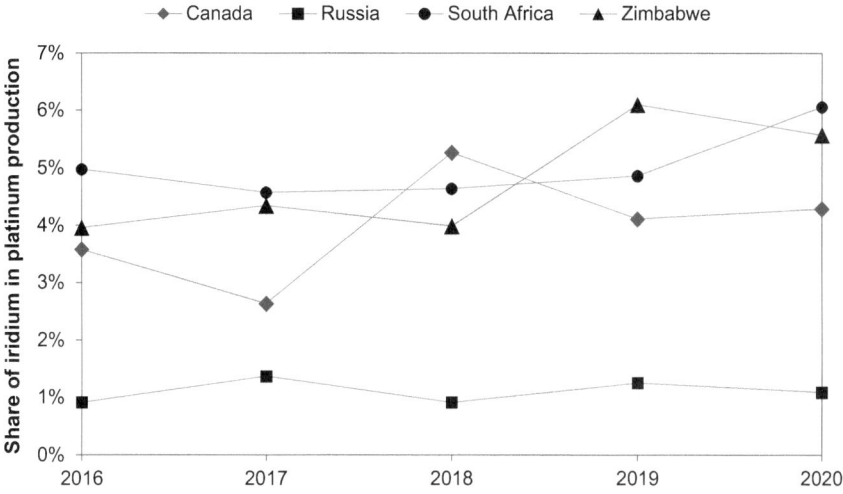

Fig. 5 Iridium production share on the platinum production for four main producing countries from 2016 to 2020

4 Further Demand and Availability Aspects

Physical availability, expressed in attributable reserves, sets the upper boundary for calculating the economic availability of metal production in years to come. The following parameters are considered for economic availability at the mining company level: forecasts on prices for main metals and by-products of their production, exchange rates, inflation indices, capital and operating expenditures, and production profiles.

The production of metals has already been affected by a power shortage (South Africa) or lower global demand due to COVID pandemic in 2020/2021. Figure 6 depicts a Sankey diagram for iridium production and use in the pandemic year 2020. Platinum group metals are known for having high values of companionality; i.e. being produced as a by-product of another material. With an economic contribution of more than 20%, the by-product is re-classified to a co-product. [30] determined companionality (the global percentage mined as a companion) based on data from 2008 with the following results for platinum (17), palladium (80), rhodium (48), ruthenium (95), and iridium (99). These values shall be understood as informed estimates as the data on companion metals are often subject to private contracts. Platinum is a principal host, i.e. a metal with the largest economic contribution for iridium, rhodium, ruthenium, and osmium [30].

Another factor influencing availability of iridium and platinum availability for electrolysers production is the competitive use for other sectors. Demand for iridium comes from chemical industry as well as from electronics. Table 6 depicts the use of chosen metals in the hydrogen economy and competitors' use of these metals in the industry. Platinum is still mainly used in the automotive industry (73.1 t in 2021) and in jewellery (45 t in 2021) [4].

As shown in previous parts, the demand and availability of iridium for electrolysers cannot be analysed separately from each other and other platinum group metals, especially platinum. Until now, considering the lack of statistical data, the annual supply of iridium has been assumed to be derived from global demand. Figure 6 illustrates this fact and the estimation of a recycling loop in the form of a Sankey Diagram. Alongside physical limitations of iridium production, the future iridium production will also be determined by the economic availability, which relates to world prices for metals and economic and social development in producing countries (e.g. strikes), by the demand for platinum and by anticipated trends in relevant industries. For example, it is expected that steadily decreasing platinum demand for autocatalysts in diesel engines (Table 6) will be offset by increasing platinum demand in, e.g.

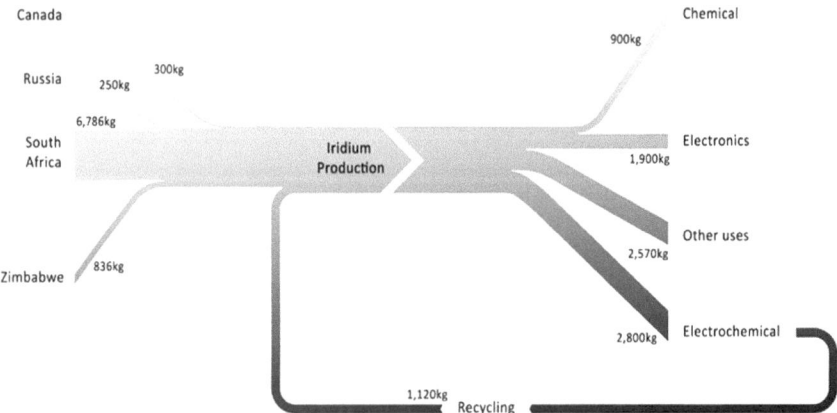

Fig. 6 Sankey diagram for iridium production, use, and recycling, reference year 2020

Table 6 Competitive use of metals

Metal	Use in hydrogen production plants and hydrogen use facilities	Competitive use
Cobalt	Catalyst (alkaline electrolyser)	Lithium-ion batteries
Nickel	Catalyst (alkaline electrolyser, steam reforming) [31]	Lithium-ion batteries, wind power plants (tower, nacelle)
Platinum	PEM fuel cell vehicles, electrolysers (proton exchange membrane), the phosphoric acid fuel cells (PAFC) [31]	Diesel vehicles (platinum as a catalytic converter, ca. 73 t/a), glass industry, jewellery, mobile phones, pacemakers, dental fillings
Titanium	Bipolar plates in PEM electrolysers, photoelectrolysis [31]	Photovoltaics [31]
Iridium	Electrolysers (PEM)	Polymer LED technology, 5G technology, radiation therapy in medicine, chemical treatment of methanol, crucibles, filters, defibrillators
Ruthenium	Electrolysers (membrane electrode assemblies) [32]	Crucibles, hard discs, resistors, ballast water treatment technology
Palladium	Hydrogen purification technologies due to selectivity of Pd for hydrogen and quick sorption kinetics [33], selective sensors, electrocatalyst in low-temperature fuel cells [34]	Capacitors, mobile phones, dental fillings Gasoline engine catalytic converters [3]
Carbon	Hydrogen storage [34], low-temperature fuel cells [33]	

the electrochemical sector; otherwise, willingness to invest in platinum production worldwide may diminish.

Second, iridium demand in other sectors and investment costs for electrolysers will determine the amount of iridium available for ramping up electrolyser capacity worldwide. Third, recycling mechanisms for iridium and platinum will start to impact the total amount of iridium needed when the first stacks are taken out of operation at the end of their lifetime in the second half of the 2020s. One more argument for recycling, alongside economic and availability aspects, it is the lower emission intensity, i.e. the life cycle of the functional unit expressed as 1 MW of electrolyser capacity with lower values of impact indicators compared to the life cycle with no recycling. To support this development, new stacks shall be designed so that metals are easily recycled at the end of their lifetime.

As for platinum use in electrolysers, geological constraints will not be faced; economic availability depends on the abovementioned factors.

5 Conclusion and Outlook

The present work investigates global platinum and iridium demand and availability under two chosen scenarios to ramp up of electrolyser capacity expressed in the yearly capacity increase. The short overview of listed critical materials puts PEM electrolysers and platinum and iridium loadings into the spotlight. As non-iridium catalysts are unlikely to replace iridium-based catalysts by 2030, iridium scarcity as the main challenge for PEM electrolyser capacity increase has been stated in almost every source of the reviewed literature. Reducing iridium loading in PEM electrolysers has limits at ca. 0.5 mg_{Ir}/cm^2 [12]. Based on literature studies, the second-order scarcity is automatically assumed for iridium and platinum, i.e. a scarcity to be overcome by "thrifting" lowering material use per unit (cm^2, kW). Although decreasing material use is a fact in basically all components becoming more market mature, it takes longer than previously anticipated.[10]

Global electrolyser production faces two main limitations: firstly, the relatively high platinum and iridium loading required for membrane electrode assemblies (MEAs), and secondly, the manufacturing limitations that need to be overcome by the automation of electrolyser production (e.g. assembling the electrolysis stack). However, electrolysers, fuel cells, and electric vehicles will benefit from the mutual deployment in their respective markets from a manufacturing point of view. Any delay in deployment can negatively affect efforts to decrease platinum group metals loading and would require alternative materials to replace iridium. All future materials should perform comparably well with iridium and platinum regarding current density, mechanical and chemical stability, and other factors such as catalyst degradation. Future policies may decrease the scarcity of materials by:

- Further investing in research for reducing PGM metals in electrolysers without compromising degradation grades.
- Creating financial incentives to increase the recycling rate of PGM metals of the first generation of electrolysers at a megawatt scale in operation. As a first step, novel products shall be designed and installed under the "design-to-recycle" concept. Given that the first generation of electrolysers at a megawatt scale with relatively high platinum and iridium loadings will be out of operation by 2040, increased recycling rates for iridium are expected to decrease primary production of the metal only after 2040. Platinum scrap is estimated to be over-proportionally higher due to catalytic converters, jewellery, and electronic equipment recycling cycles.

---

[10] For example, platinum loading in fuel cell vehicles (FCV) was assumed to decline from 60 g/ FCV in 2005 to 15 g/FCV in 2025 [35]. Nel [36] assumed a platinum loading for hydrogen FCV (Category 1) of ca. 10 g and for methanol-driven FCV of ca. 20 g (both by 2030). However, in 2020, the EC assumed 28 g of platinum for a FCV, referring to 0.25 g/kW of platinum for the 113 kW fuel cell engine of the Toyota Mirai [8].

Further work is needed to assess iridium and platinum demand after 2030 with more information on realised projects and projects in realisation in the second half of the 2020s.

6 Conversion Factors

Ounce, troy 31.103 g (g).
　Gram (g) 0.03215 oz, troy.

Acknowledgements The author thanks Pieter J Perold, the Principal Mineral Economist: Platinum & Silver of the Department of Mineral Resources and Energy of Republic of South Africa for advice on assumptions for the future production of platinum group metals. Furthermore, the author thanks Dr.-Ing. Nils Bullerdiek, the editor of *Powerfuels* and Research Associate of the Institute of Environmental Technology and Energy Economics (IUE), Hamburg University of Technology (TUHH), for fruitful discussions on chapter content and significant improvement of the final version of the chapter. Thank you.

References

1. IEA (2021b) Net Zero By 2050, IEA, Paris https://www.iea.org/reports/net-zero-by-2050, Licence: CC BY 4.0
2. Prior J, Bartelt M, Sinnemann J, Kuhlenkötter B (2022) Investigation of the automation capability of electrolyzers production. Procedia CIRP 107:718–723. https://doi.org/10.1016/j.procir.2022.05.051
3. U.S. Geological Survey (2022) Mineral commodity summaries. https://www.usgs.gov/centers/national-minerals-information-center/platinum-group-metals-statistics-and-information
4. Johnson Matthey (2022) PGM market report. Johnson Matthey. https://matthey.com/documents/161599/509428/PGM-market-report-May-2022.pdf/542bcada-f4ac-a673-5f95-ad1bbfca5106?t=1655877358676
5. Teske S (Hrsg.) (2019) Springer eBook collection. Achieving the Paris climate agreement goals: global and regional 100% renewable energy scenarios with non-energy GHG pathways for + 1.5 and +2 °C. Springer. https://link.springer.com/book/10.1007/978-3-030-05843-2. https://doi.org/10.1007/978-3-030-05843-2
6. Joint Research Centre (2013) Critical metals in the path towards the decarbonisation of the EU energy sector: assessing rare metals as supply-chain bottlenecks in low-carbon energy technologies. Joint Research Centre. Institute for Energy and Transport. https://core.ac.uk/download/pdf/38627063.pd. https://doi.org/10.2790/46338
7. World Bank Group (2020) The growing role of minerals and metals for a low carbon future. World Bank, Washington. https://openknowledge.worldbank.org/handle/10986/28312, https://doi.org/10.1596/28312
8. European Commission (2020) Critical raw materials for strategic technologies and sectors in the EU: a foresight study. Publications Office. https://doi.org/10.2873/58081
9. IRENA (2020) GreenHydrogen cost reduction: scaling up electrolysers to meet the 1.5 °C climate goal. International Renewable Energy Agency. Abu Dhabi
10. IRENA (2022) Geopolitics of the energy transformation: the hydrogen factor, International Renewable Energy Agency, Abu Dhabi

11. Bernt M, Siebel A, Gasteiger HA (2018) Analysis of voltage losses in PEM water electrolyzers with low platinum group metal loadings. J Electrochem Soc 165(5):F305–F314. https://doi.org/10.1149/2.0641805jes

12. Möckl M, Ernst MF, Kornherr M, Allebrod F, Bernt M, Byrknes J, Eickes C, Gebauer C, Moskovtseva A, Gasteiger HA (2022) Durability testing of low-iridium PEM water electrolysis membrane electrode assemblies. J Electrochem Soc 169(6):64505. https://doi.org/10.1149/1945-7111/ac6d14

13. Becker HF, Dickinson EJ, Lu X, Bexell U, Proch S, Moffatt C, Stenström M, Smith G, Hinds G (2022) Assessing potential profiles in water electrolysers to minimise titanium use. Energy & Environ Sci 15(6):2508–2518. https://doi.org/10.1039/D2EE00876A

14. IEA (2021) Global hydrogen review 2021, IEA, Paris https://www.iea.org/reports/global-hydrogen-review-2021, Licence: CC BY 4.0

15. Smolinka T, Wiebe N, Sterchele P, Palzer A, Lehner F, Jansen M, Kiemel S, Miehe R, Wahren S, Zimmermann F (2018). Studie IndWEDe - Industrialisierung der Wasserelektrolyse in Deutschland: Chancen und Herausforderungen für nachhaltigen Wasserstoff für Verkehr, Strom und Wärme. Nationale Organisation Wasserstoff- und Brennstoffzellentechnologie - NOW GmbH, Berlin. https://www.now-gmbh.de/wp-content/uploads/2020/09/181127_bro_a4_indwede-studie_kurzfassung_de_v03.pdf

16. Bareiß K (2020) An enhanced methodology for energy system modeling including life-cycle analysis. Dissertation, Technische Universität München, München. mediatum.ub.tum.de. https://mediatum.ub.tum.de/1547067

17. Minke C, Suermann M, Bensmann B, Hanke-Rauschenbach R (2021) Is iridium demand a potential bottleneck in the realization of large-scale PEM water electrolysis? Int J Hydrogen Energy 46(46):23581–23590. https://doi.org/10.1016/j.ijhydene.2021.04.174

18. Rasmussen KD, Wenzel H, Bangs C, Petavratzi E, Liu G (2019) Platinum demand and potential bottlenecks in the global green transition: a dynamic material flow analysis. Environ Sci Technol 53(19):11541–11551. https://doi.org/10.1021/acs.est.9b01912

19. IEA (2021a) Hydrogen projects database. https://www.iea.org/data-and-statistics/data-product/hydrogen-projects-database

20. Energy Transitions Commission (2020) Making mission possible: Delivering a net-zero economy. https://www.energy-transitions.org/publications/making-mission-possible

21. Wappler M, Unguder D, Lu X, Ohlmeyer H, Teschke H, Lueke W (2022) Building the green hydrogen market—Current state and outlook on green hydrogen demand and electrolyzer manufacturing. Int J Hydrogen Energy 47(79):33551–33570. https://doi.org/10.1016/j.ijhydene.2022.07.253

22. GlobalData (2022) Global electrolyzer capacity to reach 8.52 GW by 2026 led by the Asia-Pacific region. GlobalData. https://www.globaldata.com/media/power/global-electrolyzer-capacity-reach-8-52-gw-2026-led-asia-pacific-region-says-globaldata/

23. Shell International B.V (2018) Shell scenarios sky. Meeting the goals of the Paris agreement: the numbers behind Sky (xlsx). https://www.shell.com/energy-and-innovation/the-energy-future/scenarios/shell-scenario-sky.html

24. IEA (2022) Electrolysers. IEA, Paris. https://www.iea.org/reports/electrolysers

25. Johnson Matthey (2023) PGM market report. Johnson Matthey. https://matthey.com/documents/161599/404086/PGM+Market+Report+May23.pdf/2f048a72-74a8-8b23-f18e-c875000ed76b?t=1684144507321

26. IEA (2023) Tracking Clean Energy Progress 2023. IEA, Paris. https://www.iea.org/reports/tracking-clean-energy-progress-2023, Licence: CC BY 4.0

27. U.S. Geological Survey (2020) Platinum-group metals [advanced release-tables only]. Reston, VA, USA. https://www.usgs.gov/centers/national-minerals-information-center/platinum-group-metals-statistics-and-information

28. World Bank. World Integrated Trade Solution (2022) Metals; iridium, osmium, ruthenium, unwrought or in powder form exports by country in 2020. https://wits.worldbank.org/trade/comtrade/en/country/ALL/year/2020/tradeflow/Exports/partner/WLD/product/711041

29. Smith B, Graziano DJ, Riddle ME, Liu D, Sun P, Iloeje C et al (2022) Platinum group metal catalysts. Supply chain deep dive assessment. U.S. Department of Energy. https://doi.org/10.2172/1871583

30. Nassar NT, Graedel TE, Harper EM (2015) By-product metals are technologically essential but have problematic supply. Sci Adv 1(3):e1400180. https://doi.org/10.1126/sciadv.1400180

31. Rand D, Dell RM (2005) The hydrogen economy: a threat or an opportunity for lead–acid batteries? J Power Sources 144(2):568–578. https://doi.org/10.1016/j.jpowsour.2004.11.017

32. Clayton JA, Walton RI (2022) Development of new mixed-metal ruthenium and iridium oxides as electrocatalysts for oxygen evolution: Part I: survey of crystal structures and synthesis methods. Johnson Matthey Technol Rev 66(4):393–405. https://doi.org/10.1595/205651322X16529612227119

33. Abe JO, Popoola A, Ajenifuja E, Popoola OM (2019) Hydrogen energy, economy and storage: Review and recommendation. Int J Hydrogen Energy 44(29):15072–15086. https://doi.org/10.1016/j.ijhydene.2019.04.068

34. Adams BD, Chen A (2011) The role of palladium in a hydrogen economy. Mater Today 14(6):282–289. https://doi.org/10.1016/S1369-7021(11)70143-2

35. Anderson AF, Carlson EJ (2003) Platinum availability and economics for PEMFC commercialization: Report to US Department of Energy. https://www1.eere.energy.gov/hydrogenandfuelcells/pdfs/tiax_platinum.pdf

36. Nel WP (2004) The diffusion of fuel cell vehicles and its impact on the. In: International platinum conference 'platinum adding value'. https://www.saimm.co.za/Conferences/Pt2004/287_Nel.pdf

Hydrogen Transport and Storage Options

Lars Baetcke

Abstract In this article, different options for storage and transportation of hydrogen are discussed. First, different technologies, which are currently technically mature, are explained. Subsequently, these technologies are compared on the basis of different technical parameters. In addition, for the different technologies the variants and differences are explained. Also, it is discussed which storage technology is best suited for specific applications. Finally, an impression of the influence of the main cost driver for the different systems is given.

Keywords Pressurized Storage · Liquid Storage · Cryo-Compressed Storage · Metal Hydrides · LOHC

1 Introduction

To enable the transition to a climate-neutral energy system, hydrogen is a key factor for energy storage as well as the power fuels production. Therefore, the need to store and transport hydrogen in a future energy system increase drastically in comparison to nower days. This enables the necessary energy storage and flexibility to the energy system to replace strategic energy reserve and provide control power. Furthermore, hydrogen will be moved from a niche product for chemical industry, in which main use cases are ammonia production and refinery processes, to one of the most important globally traded energy carrier. Therefore, in this article the transportation and storage of hydrogen are discussed.

L. Baetcke (✉)
German Aerospace Center (DLR), Institute of Maritime Energy Systems, Geesthacht, Germany
e-mail: lars.baetcke@dlr.de

2   Hydrogen Storage Options

The storage of hydrogen is an important part of the overall value chain of power fuels. Hereby hydrogen storages are primarily used to couple a volatile production of hydrogen based on renewable energies with a hydrogen usage for the production of synthetic fuels. Besides stationary storage, various storage options are also used to transport hydrogen in trailers, containers or, if pressurized, also in pipelines. To enable an efficient hydrogen storage and transportation, it is always necessary to increase its density. Otherwise, the required storage volume would be far too large for any industrial application. This is why for all storage technologies the main focus is on increasing the volumetric energy density of hydrogen (3 kWh/Nm^3 at 273.15 K; 0.1013 MPa), since its gravimetric energy density, at 33 kWh/kg, is already the highest of all energy carriers. Furthermore, all storage systems result in a lower gravimetric energy density than hydrogen itself, since the storage system always adds weight. However, the systems always increase the volumetric energy density in comparison with hydrogen at standard conditions.

Overall, there are two different types of hydrogen storages, direct/pure hydrogen storage and material-based hydrogen storage (Fig. 1).

- The direct storages of hydrogen contain all forms of hydrogen storage where the hydrogen molecules are stored directly and only the pressure, temperature or a combination of both is changed. This also includes a phase change of hydrogen.
- The material-based storages include all storage systems where the hydrogen interacts with other materials to increase the storage density. This includes chemical reactions (e.g. metal hydrides, LOHC) and physical interactions (e.g. Metal–Organic Framework (MOF)).

However, this article is focused on hydrogen storage and transport technologies with a high technology readiness level (TRL). In this case, this means a TRL of

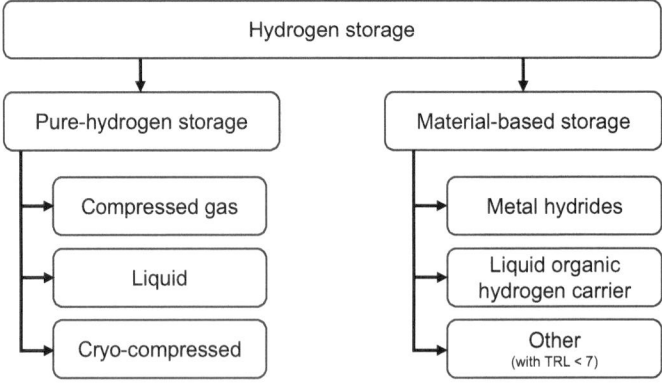

Fig. 1 Overview of different hydrogen storage options (TRL: technology readiness level)

seven or higher. By definition, this stands for a prototype that runs in a relevant environment. Therefore, the discussed storage options are pressurized hydrogen, cryo-compressed hydrogen, liquid hydrogen as well as liquid organic hydrogen carrier (LOHC) and metal hydrides. For these storage technologies demonstrators exist, which have proven that the concept works. The storage systems with a TRL below seven are mainly material-based systems, which are based on adsorption or absorption processes. One of the difficulties, for example, for materials with an adsorption effect is the necessary hydrogen condition for this storage system. To enable the adsorption of hydrogen to different materials, such as MOFs, it must be cooled to temperatures below 100 K, since the needed van der Waals forces have very low energy.

Besides the aspect of storage, most hydrogen storage technologies can also be used for hydrogen transport due to the increased energy density inside the storage systems. This either results in the usage as storage for mobile applications or just as a system for the transportation of hydrogen.

2.1 Pressurized Hydrogen Storage

The most common method to store and transport hydrogen is in the form of pressurized hydrogen. The reason therefore is that from a technical point of view, it is relatively simple to increase the pressure and it offers a wide variety of use cases. Therefore, many different pressure levels and tank sizes as well as different tank hulls are available for pressurized hydrogen storage. The pressure range starts at 30 bar and goes up to 1,000 bar. This leads to the fact that pressure vessels are the most common hydrogen storage option.

Storage systems for pressurized gases are very similar, regardless of the gas they store (e.g. hydrogen, methane and nitrogen). Their primary purpose is to create a stable and gas-tight hull around the volume that stores the gas, which can hold the stresses inside the materials resulting from the high pressure. In the hull, also the valves for the loading and unloading as well as sensors and safety features are integrated. The different kinds of pressurized hydrogen storage can be classified into four types of storage tanks. The main difference lies in the construction of the tank hull as well as the maximum pressure and the sealing for gas-tightness [1]:

- Type I. This type of storage tank is made of steel. Therefore, there is only one material to take the stresses and seal the hull. Due to the simple construction, this storage type is widely used in the industry at various pressure levels. The maximum pressure of type I storage is 300 bar, which is a common value for the used gas bottles to deliver industry gases. At the other end of the pressure range also large type I storage with pressure of 30–50 bar are very common as a stationary system in industrial applications.
- Type II. This type of storage tank is mostly used for stationary high-pressure storage and allows the highest pressure between the different storage types. The

tank hull of type II storage allows pressure levels up to 1000 bar. This pressure is possible due to the combination of a steel tank hull in combination with a layer of unidirectional carbon fibre-reinforced plastic (CFRP) around the steel cylinder. A very common use case for this kind of storage is the high pressure part of refuelling stations to refuel 700 bar tank systems [2].

- Type III. To reduce the weight and still allow high pressures, the type III storage tanks reduce the amount of metal inside the tank hull. Therefore, a hull made of glass fibre-reinforced plastic (GFRP) or CFRP handles the material stresses coming from the pressure. To seal the storage, a thin layer made of steel or aluminium is placed inside the tank hull. This layer is called a liner. Type III storage tanks are usually running with a pressure of 350 bar but can also handle higher pressure levels if needed.
- Type IV. For a further weight reduction compared to the other storage tank types, type IV storage tanks are developed. These storage vessels have a tank wall made of CFRP and have a polymer liner inside for the sealing. This allows to withstand high pressures of up to 700 bar and partly even higher, including a light-weight construction. This is the reason why most mobile applications today use type IV storage tanks [3].

As an example, a type IV tank for mobile application is shown in Fig. 2. This tank type can hold a pressure of 700 bar. The tank hull has three major components. From the inside, the first layer is a polymer liner for the sealing of the storage. The next layer is the thickest and is made out of CFRP. This layer takes all the stresses which result out of the high pressure. The third layer is also a composite layer, which is made out of GFRP and has the task to protect the storage from outside damages (e.g. rock chipping). This layer is not a standard for all type IV tanks. It is only used if the tank is mounted in a critical environment, such as below a vehicle where special safety measures are required. Two rubber protectors are placed at the outside of the tank to reduce the risk of cracking if the tank falls or gets hit by a foreign object. Two aluminium plates are placed at the top and the bottom of the cylinder. They are used to stabilize the dome. Also, different parts for the operation and safety of the tank are mounted into these plates. This includes the in- and outlet valve, the safety valve (e.g. bursting disc) and different sensors for temperature and pressure control of the storage [5–7].

Transportation of Pressurized Hydrogen There are various approaches for transporting pressurized hydrogen. For small amounts of hydrogen, the simplest option is to use gas bottles (type I storages) that can be transported by truck or train. This allows hydrogen to be handled in the same manner as other industrial gases. This transportation and delivery is done either in single bottles (at 200 bar, approx. 0.75 kg_{H2} or at 300 bar, approx. 1.05 kg_{H2}) or with cylinder racks (at 200 bar, approx. 9 kg_{H2} or at 300 bar, approx. 12.6 kg_{H2}). Another way to transport pressurized hydrogen is using special hydrogen trailers with storage pressure levels between 200 and 300 bar. These trailers can have a transport capacity of up to 500 kg_{H2}. These trailers are easily recognizable due to their very specific design. They have a pyramid out of long steel tubes (type I storages), which are permanently mounted on the trailer. Modern trailers

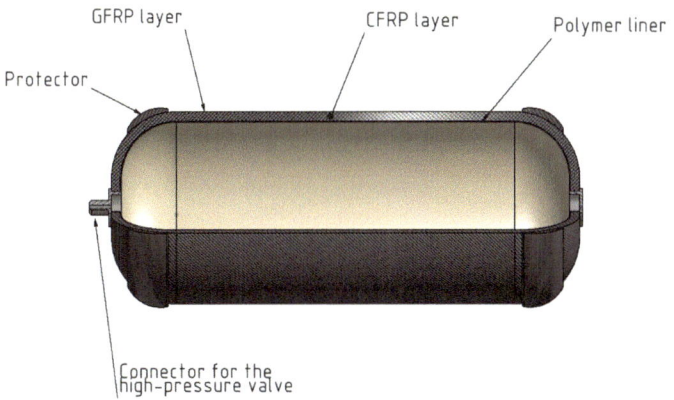

Fig. 2 Type IV storage tank for gaseous hydrogen up to 700 bar, based on [4] (CFRP: carbon fibre-reinforced plastic; GFRP: glass fibre-reinforced plastic)

with type III or IV storage vessels and a storage pressure of 500 bar can transport up to 1,000 kg of hydrogen. In this case, the storage cylinders are mounted either in a container or directly onto a standard truck trailer. A further difference to trailers with type I storages is the orientation of the cylinder. In this case, they are mounted vertically rather than horizontally. More hydrogen fits into a trailer with this kind of tank modification. The main reason is a better space utilization due to the lighter storage cylinder. Thus, for trailers with steel storage systems (type I storages), the maximum allowed weight for road transportation by truck (40 tons) is reached more quickly compared to trailers with type III or IV storages.

Another approach for gaseous hydrogen transportation is the usage of a pipeline. Today, pipelines offer the largest hydrogen transport capacities and have the benefit to enable a constant hydrogen supply for consumers. Some regional hydrogen pipelines already exiting at chemical industry parks. One example is the pipeline at the industrial park in the Ruhr area, where a 240-km-long dedicated hydrogen pipeline stretched from the Marl chemical park to Castrop-Rauxel and Leverkusen in Germany [8]. Additionally, there are concepts and plans underway for international hydrogen pipeline systems, such as the European Hydrogen Backbone [9], as well as newly built regional networks for hydrogen distribution. However, large industrial consumers are likely to be the primary customers of these pipeline networks, as their substantial hydrogen consumption makes a cost-efficient pipeline infrastructure feasible, enabling the development of such networks.

Underground Storage of Hydrogen A hydrogen storage technology that does not fit into the previously mentioned four categories is underground hydrogen storage. Two options exist here: salt caverns and porous rocks. They offer the largest storage capacities among all storage options and can be used for long-term energy storage, for example, as seasonal storage. In addition, they offer the possibility to replace portions of the strategic energy reserves.

- Salt Caverns. Salt caverns are already used today for natural gas storage and different projects are currently testing hydrogen handling and storage inside a salt cavern. Salt caverns can hold pressure up to 200 bar and store up to 140 million Nm^3 of hydrogen [10–12]. However, unfortunately the underground storage of hydrogen often comes along with some unwanted contamination. At salt caverns, the hydrogen is mostly contaminated by water, which is usually inside the cavern from the exploitation process.
- Porous Rocks. Another potential formation for hydrogen storage is porous rocks, a subject still under research. In these porous underground storage formations, hydrogen is pressed into a porous rock layer. For this purpose, a location is used where the cap rock forms a dome for the hydrogen to accumulate in this area. A gas-tight cap rock must always be present above the porous rock layer. Old gas or oil deposits could serve this purpose. Such porous storage can reach pressure up to 200 bar with the potential to store up to several billion Nm^3 of hydrogen [13, 14].

Another challenge of underground hydrogen storage, besides water contamination, is microorganisms that metabolize the hydrogen and produce other gases out of it like methane or hydrogen sulphide (H_2S). Therefore, gas purification is always required after storing hydrogen underground, regardless of the type of underground storage used [15].

Overall, looking at the size of different hydrogen storage and transport options, pressurized hydrogen operates on both ends of this scale. The smallest amounts of hydrogen are stored and transported in gaseous form using type I bottles, while also the largest amounts of hydrogen are stored and transported in gaseous form using pipelines and underground storage systems.

2.2   Liquid Hydrogen Storage

The second most common direct storage option for hydrogen is liquid hydrogen storage. The primary consumer of liquid hydrogen is the space industry, using hydrogen extensively as rocket fuel, in part due to its high gravimetric energy density. Additionally, the semiconductor industry utilizes large amounts of liquid hydrogen because of its high purity level. Furthermore, liquid hydrogen is supplied to various industrial applications with medium hydrogen demand using trailers.

Hydrogen is liquefied because handling a liquid is simpler than handling and transporting compressed gas. Moreover, with liquid hydrogen, higher volumetric and gravimetric energy densities can be achieved than with gaseous hydrogen. Due to the liquefaction, the density of the hydrogen rises from 0.09 kg/m^3 at standard conditions to 70.8 kg /m^3 as a liquid. The largest challenge of liquid hydrogen storage and transportation is its low boiling point of $-$ 253 °C (20 K), which is the second lowest boiling point of all elements after helium. Consequently, it requires a considerable effort to liquefy hydrogen. This results in an energy consumption of 10 to 30 kWh/kg$_{H2}$, depending on the liquefaction process and the size of the liquefaction facility. Today, three different types of hydrogen liquefiers are in use: Stirling compressors, Claude cycle and Bryton cycle. Hereby, Stirling compressors are used for liquefying smaller amounts of hydrogen (8 up to 500 kg$_{H2}$/day), while the Claude and the Bryton cycles are used for plants of larger sizes (starting from 1,000 kg$_{H2}$/day). The difference in the usual plant size is due to the occurring costs as well as the energy required for liquefaction. Stirling compressors have lower investment costs but need more energy for the liquefaction (approx. 30 kWh/kg$_{H2}$). Therefore, it is used for small liquefaction plants when the compressor only needs to run occasionally. In comparison, the Claude cycles are large systems. They have higher investment costs but through the high efficiency for the liquefaction (approx. 10 kWh/kg$_{H2}$) the operating costs are lower. Thus, liquefaction plants using a Claude cycle are suitable for high annual full-load hours.

To store liquid hydrogen, a storage system with a particularly effective isolation is necessary. This isolation mostly consists of different layers to create the best possible isolation. Usually, this is a combination of aluminium foil and vacuum chambers, to reduce the transfer of conductive heat, convective heat as well as radiation heat. Additionally, liquid hydrogen storages are built in forms with a small surface-volume ratio. Therefore, they are typically constructed as spheres or occasionally as cylinders. This design helps to reduce the heat impact on the stored liquid. However, due to the high temperature difference between the storage and its environment (about 270 K), some heat inevitably enters the storage, causing some of the hydrogen to vaporize, a phenomenon referred to as "boil-off". The hydrogen has its critical point at 33.18 K and 13 bar. Therefore, liquid hydrogen tanks are mostly designed to hold a maximum pressure of 10 bar and a temperature of 30 K. If this is reached, part of the stored hydrogen evaporates and has to be released through a boil-off valve. This is one of the most important safety systems for liquid hydrogen storage. Furthermore, heat sensors, valves, pumps and other auxiliary systems are required to operate this kind of storage. For the handling of the boil-off gas, different options exist. The most suitable option is to use the boil-off hydrogen directly for a gaseous hydrogen application. If that is not possible, it should either be used in a fuel cell to provide heat and power or be reliquefied. A common handling of boil-off gas is to release it into the atmosphere. But since hydrogen has a global warming potential (GWP$_{100}$) of about 11 ± 5 [16, 17], this should be avoided and only done in emergencies.

Currently, the primary bottleneck for storing or transporting liquid hydrogen is the limited number of hydrogen liquefaction plants available. Presently, only three such active plants with a combined daily capacity of 25 tons of liquid hydrogen exist in Europe [18]. However, new plants are under construction, including a 6 t$_{LH2}$/day facility in Norway [19].

Transportation of Liquid Hydrogen For the transportation of liquid hydrogen, several systems are already in existence: these are truck trailers, containers and one demonstrator ship. Truck trailers are the most common systems and can transport about 4 t of hydrogen per trailer. Primarily, these trailers are utilized to deliver liquid hydrogen to customers with high hydrogen demand who do not have their own hydrogen production unit or pipeline connection. The containerized storage is designed for the long-distance transportation of liquid hydrogen using standard container vessels. In addition to the storage, a small reliquefier is also included in the container, since no boil-off gas is permitted on container vessels and during the transportation routes. These containers are usually 40-feet containers and can transport 3 t of liquid hydrogen per container. The last currently existing technology for the transport of liquid hydrogen is a demonstrator ship; it is a ship for long-distance transportation of liquid hydrogen, the "Suiso Frontier". This ship has a hydrogen capacity of 77 t of liquid hydrogen [20]. In future, larger ships are planned capable of transporting up to 30,000 t of liquid hydrogen. There are also mobile applications under development, which use liquid hydrogen as energy storages. Currently, trucks with liquid hydrogen tanks are under development as well as ships and aeroplanes [21, 22].

2.3   Cryo-compressed Hydrogen Storage

A cryo-compressed storage of hydrogen is a combination of a compressed hydrogen storage and liquid hydrogen storage. In this case, hydrogen is stored at 350 bar and 30 K. This allows high energy storage densities, but also creates challenges for the tank hull. Inside the storage, the hydrogen is gaseous due to the high pressure, which is far above the critical point of hydrogen [23, 24]. Storage systems for cryo-compressed hydrogen are a combination of a type IV pressure vessel with an extreme effective isolation comparable to the isolation of liquid tanks. The combination of both options has the benefits of reduced boil-off rate and still reaching very high energy densities.

Due to the technical challenges associated with cryo-compressed hydrogen storage, these tanks are not being explored for hydrogen transportation. However, they are currently being developed for use within trucks. The objective here is to maximize the amount of hydrogen stored within the same space and to extend the duration of boil-off-free storage compared to mobile liquid hydrogen storage solutions. For the future integration into refuelling stations, two options are possible. On one hand, liquid hydrogen can be pressurized to 350 bar, and on the other hand, compressed gas can be cooled down to 30 K. The expected volumetric energy density is 1.2 kWh/L_{system}, and the gravimetric storage density is around 9 wt.-%$_{H2}$, both values considering the whole storage system [23]. When used as mobile storage, this allows for travel distances of over 1000 km per tank fill for hydrogen-powered cars and trucks [25, 26].

2.4 *Metal Hydride Storage*

Metal hydrides are a material, which can absorb hydrogen inside a metal lattice, comparable to a sponge absorbing water. Therefore, it is possible to store hydrogen in metal hydrides in a very compact way, without the need of high pressure or very low temperature. The absorption of hydrogen inside the metal hydride is an exothermic reaction, while the desorption out of the metal lattice is endothermic. Hence, the heat management of the storage controls the hydrogen flow. The general reaction is shown in Eq. (1).

$$H_2 + \text{metal} \leftrightarrow \text{metal hydride} + \Delta H_R; \quad \Delta H_R > 0. \tag{1}$$

During this reaction four steps are performed (Fig. 3). In the first step, the hydrogen molecule reaches the metallic surface (Fig. 3a). In the second step (Fig. 3b), the hydrogen molecule dissociates into the two hydrogen atoms and they diffuse into the metal lattice (Fig. 3c). The last step (Fig. 3d) is the reaction of the metal and the hydrogen to metal hydride under release of heat.

The pressure and the temperature level depend very much on the used material, but it can be chosen in a way, which fits to the overall process. In many cases, it is useful to select a metal hydride that absorbs hydrogen at the same pressure level at which the electrolyzer is operated to produce hydrogen. Within such a process layout neither an auxiliary heating device nor a hydrogen compression would be needed to integrate a metal hydride storage into the system. Besides the storage of hydrogen, metal hydrides can also be used for the thermal compression of hydrogen. In this process, hydrogen is stored at low pressure. Afterwards, the metal hydride is heated up and the hydrogen can be released at a higher pressure. This is especially beneficial when waste heat from other processes is available [27]. In the production of power fuels, exothermic processes within the overall production chain could provide the necessary heat for the storage control. Under those circumstances the metal hydride storage becomes very efficient (up to 98%).

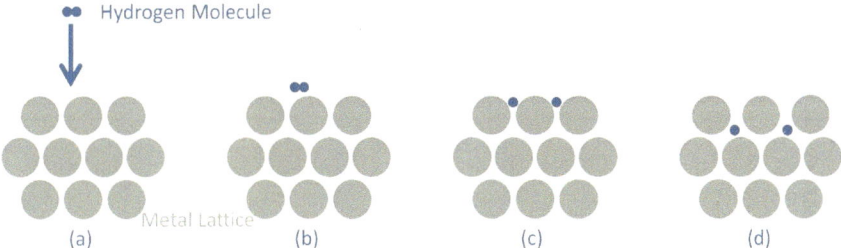

Fig. 3 Reaction steps of the adsorption reaction at a metal hydride storage (**a** hydrogen molecule reaches the surface; **b** dissociation of hydrogen at the surface; **c** the atomic hydrogen diffuses into the metal lattice; **d** reaction of hydrogen with the meta)

As an example of a material used for metal hydride hydrogen storages, iron–titanium alloys (FeTi-alloys) can absorb and desorb hydrogen within a temperature range of 0–40 °C and at pressure ranging from a few bars up to several tens of bar. Storage systems based on this material have a hydrogen storage density of around 60 kg_{H2}/m^3 [28–31].

Metal hydrides are mostly used as stationary hydrogen storage systems due to the high weight of the metal powder. They are employed either for intermediate storage, coupling hydrogen production with a process that requires hydrogen, or for long-term storage in combination with an emergency power system. There are some commercially available systems for storage capacities ranging from 10 to 250 kg of hydrogen [32, 33]. Additionally, metal hydride hydrogen storage is used in submarines. Here, the high volumetric energy density, as well as the high weight, is beneficial since space is extremely limited in submarines, and weight is needed as ballast [34]. Therefore, they have been used for this application for many years. Another possible mobile application is their use inside forklifts, as also in this case, both the weight and the small volume are beneficial [35, 36].

2.5 Liquid Organic Hydrogen Carrier (LOHC)

Another option for hydrogen storage in a material is using organic liquids, mostly various types of oils. These are referred to as liquid organic hydrogen carriers or LOHCs. Due to the chemical reaction and the liquid form of the material, handling, transportation, and storage are easy and comparable to the handling of liquid fossil energy carriers. Also, the hydrogen can be stored at ambient conditions for a long time since it is bound in a chemical reaction [37]. There are different organic fluids such as N-Ethylcarbazole, Dibenzyltoluol and Benzyltoluol, which can be used as an organic hydrogen carrier. The reaction to store hydrogen in this material is comparable to metal hydrides, since here also a hydrogenation of the material takes place. One of the main differences between LOHCs and metal hydrides is the aggregate state, and another is the nature of the chemical bonding. One example for the reaction of LOHC materials is shown in Eq. (2) [38].

$$+ 6\,H_2 \rightleftharpoons \qquad\qquad + \Delta H_R \quad \Delta H_R > 0 \quad (2)$$

(2)

For the absorption and desorption process, a reactor is needed to provide the necessary reaction temperature and pressures as well as the catalyst. The adsorption takes place at pressures between 20 and 50 bar and temperatures between 150 and 200 °C. The desorption takes place at a pressure close to the ambient pressure (1–3 bar) and at temperatures between 270 and 310 °C. Due to the high temperature, parts of the organic carrier are destroyed and have to be removed out of the fluid after several cycles [39]. Therefore, the losses of LOHC fluid must be replaced by new LOHC fluid frequently. Additional parts of the material evaporate and contaminate the released hydrogen stream. Therefore, hydrogen purification is needed for all applications that demand high-quality hydrogen. Furthermore, the high temperature for the different reactions can reduce the efficiency of this storage system drastically. Therefore, it should be coupled with some other processes, which can provide waste heat at the right temperature level to increase the storage efficiency [40].

LOHC systems are mainly discussed for transport of hydrogen over medium distances. The benefits of LOHC are the high hydrogen storage capacity and an easy handling of the storage material. However, the drawbacks are the necessary transportation and storage of the loaded and unloaded material. Furthermore, large reactors are needed at the place of hydrogen production as well as at the place of the hydrogen consumption. At the moment some hydrogen transport projects with LOHC for 5 up to 24 t_{H2}/day are announced and being built, for example, in Germany, in the USA and in Finland [41–43].

3   Comparison

For the comparison of the different hydrogen storage and transport options, the following aspects are discussed in the following: energy densities, storage conditions, hydrogen quality, the Technology Readiness Level (TRL) and cost.

Energy Densities In Fig. 4, the volumetric and gravimetric energy storage capacities are shown for the different hydrogen storage technologies presented above. This is one of the most important comparison parameters for many hydrogen applications, as it has a significant impact on the required space, weight for hydrogen storage and ultimately economic viability. For stationary applications, volume is typically the most critical criterion, while for mobile applications, both values are usually crucial (with exceptions for certain special applications such as forklifts, locomotives and pull tractors). For all hydrogen storage options described many parameters influence the energy density values. Therefore, the energy densities of different technologies can usually only be represented as a range, as seen in Fig. 4. Hereby, liquid hydrogen as well as cryo-compressed hydrogen shows the highest energy densities, whereas compressed gas storage has the lowest energy densities among all storage options. Between the direct storage options, but with especially high volumetric energy densities, metal hydrides and LOHC are located.

Storage Condition Another very interesting comparison especially for the long-term storage are the storage conditions. Ideally, these conditions should be close

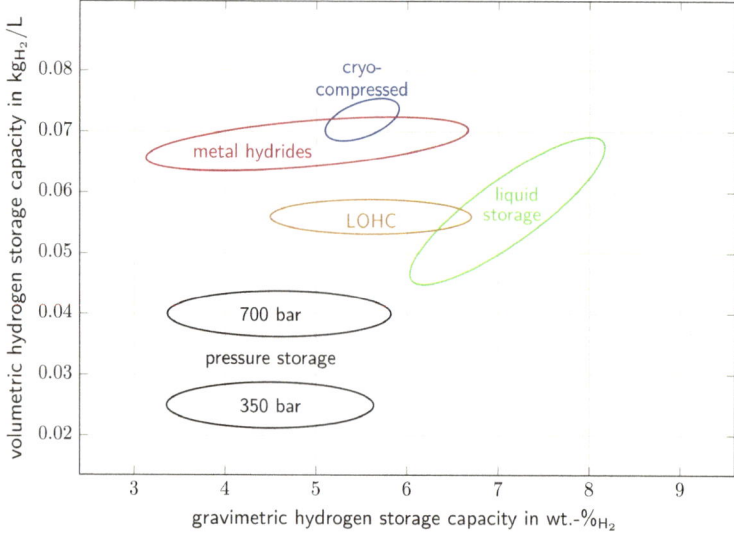

Fig. 4 Comparison of the volumetric and gravimetric hydrogen storage capacity for different hydrogen storage and transport technologies (LOHC: liquid organic hydrogen carrier)

to ambient conditions. Figure 5 illustrates the storage conditions of the various presented storage technologies. As can be seen, material-based storage technologies offer the best storage conditions, but unfortunately for the loading and unloading these storage systems need to be heated and therefore leave ambient conditions. Pressure vessels have a benefit over liquid storage, since it is easier to keep a high pressure over time in comparison to a low temperature. Due to the extreme low temperature, liquid hydrogen and cryo-compressed hydrogen storage systems are the most difficult systems for long-term storage with respect to the required storage conditions.

Hydrogen Quality The quality of the hydrogen provided from the different storage systems is very crucial since for some applications highly pure hydrogen is required. When comparing the hydrogen storage systems in terms of the hydrogen quality of the released hydrogen from the storage system, it becomes clear that impurities are not expected at the direct hydrogen storage methods (e.g. compressed or liquefied hydrogen storage). The only sources of potential contamination of the hydrogen stored in direct storage are compressors and cooling systems. These systems can cause potential contamination by lubricants in the compressors or other leaks in the hydrogen treatment systems. However, all these contaminations occur very rarely or can be avoided by technical measures. In case of material-based systems, it is possible for some parts of the storage material—in various forms—to pass into the stored hydrogen during hydrogen desorption. This problem sometimes occurs with metal hydrides when loose powder is used, i.e. small metal hydride particles can be carried out of the storage by the hydrogen flow. To avoid this, a sinter metal filter

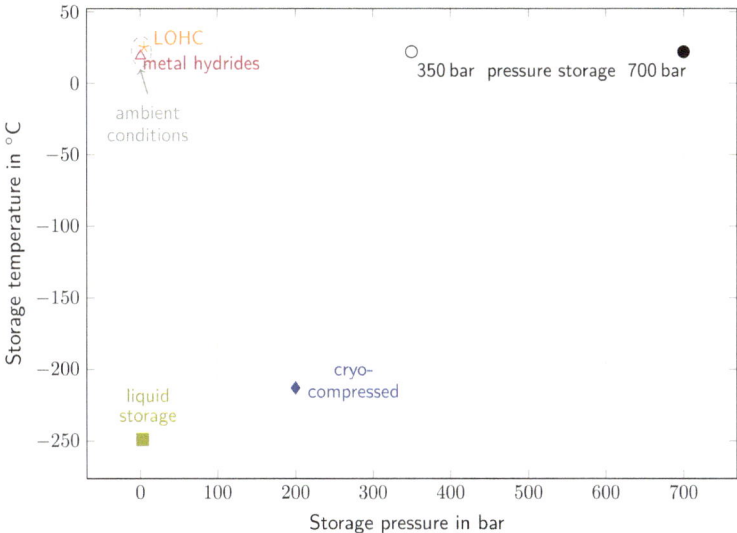

Fig. 5 Required conditions with regard to temperature and pressure during the storage of hydrogen for different technologies (LOHC: liquid organic hydrogen carrier)

is typically installed at the storage outlet. The biggest problem with contamination of the stored hydrogen occurs when liquid organic hydrogen carriers (LOHCs) are used. In this case, parts of the organic storage material can evaporate due to the high reaction temperature during dehydrogenation process and be released together with the hydrogen. Therefore, at LOHC storage systems, it is usually necessary to perform an appropriate gas purification after the hydrogen is released from the LOHC material.

Technology Readiness Level (TRL) At the comparison of the TRL, systems for pressure storage and liquid hydrogen storage are most advanced today. Both systems are already used commercially in various fields of application (TRL 9). In the case of metal hydride storage systems, there are first large-scale applications and one series application. In addition, metal hydrides are being used in a variety of demonstrators and in many experimental applications. Because of this overall picture, metal hydride storage currently has a TRL of seven to eight. For LOHC storage, the demonstration facilities are currently being built and the first large commercial applications are planned (TRL 7–8). The cryo-compressed storage is still at an earlier stage of research and development (TRL 7). Currently, this storage option is tested as tank system in a real truck application [26].

Cost The various hydrogen storage technologies have different associated costs. In addition to the cost of the storage systems themselves, auxiliary components such as compressors, liquefaction equipment, heat management systems and reactors must be taken into account. The cost of the storage systems depends on the complexity of the tank structure and the necessary hydrogen treatment processes. Furthermore, for the two material-based hydrogen storage options (metal hydrides and LOHCs), the choice of storage material itself significantly influences the cost of hydrogen storage.

Currently, the most cost-effective method is storing hydrogen in large low-pressure tanks, as this requires a relatively thin steel tank structure and only a few auxiliary components for storage. For pressurized storage, costs increase with the pressure level. Especially the high-pressure storage systems made of glass fibre-reinforced plastic (GFRP) or carbon fibre-reinforced plastic (CFRP) have high costs. Another long-established technology is the storage of liquid hydrogen, but this is generally more expensive overall due to the high energy demand for liquefaction, even though the storage tanks themselves may be cheaper than high-pressure tanks (Type III and IV). Cryo-compressed storage is expected to have even higher costs due to the combination of costly composite structures and expensive insulation.

The costs for material-based storage systems are primarily driven by the cost of the storage material itself, leading to a wide range of cost variations. This signifi-cant cost difference between the tank structure and the storage material is especially evident in the case of LOHC. Here, the main cost drivers are the material itself and the necessary loading and unloading reactors. The tanks themselves are extremely affordable, as they can be made from thin plastic. Similarly, in the case of metal hydrides, the main cost arises from the choice of storage material. The cost varies depending on the selected material, which can range from room temperature metals with lower costs and lower hydrogen storage capacities to complex metal hydrides with very high hydrogen storage capacities and higher costs. Furthermore, the tanks for metal hydrides are more expensive than LOHC tanks because they need to with-stand pressure (up to 100 bar, depending on the material) and need to have integrated heat management systems for loading and unloading.

4 Conclusion

At present, there are five technically mature options for hydrogen storage and trans-portation. The most widely used solution for hydrogen storage is pressurized gas. There are two main reasons for this: it is cost-efficient, and it is technically straight-forward and long known. Consequently, many use cases choose gaseous hydrogen. Liquid hydrogen is primarily used for the mobile transportation of large quanti-ties of hydrogen and for mobile applications that demand a low weight and high energy densities to for a longer range. The use of cryo-compressed hydrogen storage is gaining traction due to their high energy density; they are planned for different mobile applications. The two material-based storage options have two very different fields of application. While LOHC should mainly be used for the transportation of hydrogen from production to consumption side, since loading and unloading reactors are always needed as well as a lot of heat, it has the benefit that transportation in standard tankers, which are also used for other fluids, can be done. In comparison with that, metal hydrides are often used when volume is a critical criterion or when high pressures should be avoided. Besides that, metal hydrides are used either in combination with a fuel cell or other hydrogen application which offer waste heat, since the usage of waste heat makes the storage very efficient, but due to the weight, this storage type is mainly used for stationary applications.

References

1. Klell M (2010) Storage of hydrogen in the pure form, in handbook of hydrogen storage, pp 1–37
2. Rosen PA (2018) Beitrag zur Optimierung von Wasserstoffdruckbehältern: Thermische und geometrische Optimierung für die automobile Anwendung. Springer, Wiesbaden, p 172
3. Mori D, Hirose K (2009) Recent challenges of hydrogen storage technologies for fuel cell vehicles. Int J Hydrogen Energy 34(10):4569–4574
4. Nonobe Y (2017) Development of the fuel cell vehicle mirai. IEEJ Trans Electr Electron Eng 12(1):5–9
5. Simonovski I, Baraldi D, Melideo D, Acosta-Iborra B (2015) Thermal simulations of a hydrogen storage tank during fast filling. Int J Hydrogen Energy 40(36):12560–12571
6. Sdanghi G, Maranzana G, Celzard A, Fierro V (2019) Review of the current technologies and performances of hydrogen compression for stationary and automotive applications. Renew Sustain Energy Rev 102:150–170
7. Staffell I, Scamman, D, Velazquez Abad A, Balcombe P, Dodds PE, Ekins P, Shah N, Ward KR (2019) The role of hydrogen and fuel cells in the global energy system. Energy Environ Sci 12(2):463–491
8. TÜV Nord (2022). Hydrogen pipelines and hydrogen grids. Available from https://www.tuev-nord.de/en/company/energy/hydrogen/hydrogen-pipelines-grids/. Accessed 28 Oct 2022
9. European Hydrogen Backbone. The European hydrogen backbone (EHB) initiative. Available from https://ehb.eu/. Accessed 28 Oct 2022
10. EWE (2022). Forschungsprojekt HyCAVmobil - Speicherung von Wasserstoff. Available from https://www.ewe.com/de/zukunft-gestalten/wasserstoff/forschungsprojekt. Last accessed at 07 Nov 2022
11. H2CAST Etzel (2022) Wasserstoffspeicherung in Reichweite. Available from https://h2cast.com/de/projekt. . Last accessed at 07 Nov 2022
12. Foh S, Novil M, Rockar E, Randolph P (1979) Underground hydrogen storage. Final report. [Salt caverns, excavated caverns, aquifers and depleted fields]. United States. p. Medium: ED; size, p 268
13. HyStorPor. Hydrogen storage in Porous media. Available from https://blogs.ed.ac.uk/hystorpor/. Accessed 04 Nov 2022
14. Hystories. HYdrogen STORage in European subsurface. Available from https://hystories.eu/project-hystories/. Accessed 04 Nov 2022
15. Hassanpouryouzband A, Joonaki E, Edlmann K, Haszeldine RS (2021) Offshore geological storage of hydrogen: is this our best option to achieve net-zero? ACS Energy Lett 6(6):2181–2186
16. Warwick N, Griffiths P, Keeble J, Archibald A, Pyle J, Shine K (2022) Atmospheric implications of increased Hydrogen use. Available from https://assets.publishing.service.gov.uk/government/uploads/system/uploads/attachment_data/file/1067144/atmospheric-implications-of-increased-hydrogen-use.pdf. Accessed 25 Oct 2022
17. Ocko IB, Hamburg SP (2022) Climate consequences of hydrogen emissions. Atmos Chem Phys 22(14):9349–9368
18. CleanTech NM (2019) Norwegian future value chains for liquid hydrogen. Available from https://maritimecleantech.no/wp-content/uploads/2016/11/Report-liquid-hydrogen.pdf. Accessed 26 Oct 2022

19. Air Liquide (2022) Europe's first maritime value chain project for Liquid Hydrogen is short-listed as part of IPCEI. Available from https://no.airliquide.com/om-air-liquide/air-liquide-nyh eter/air-liquide-group-liquid-hydrogen-norway. Accessed 26 Oct 2022
20. Tekin N (2022) Hydrogen energy supply chain for decarbonization. In: H2-forum Hamburg. Kawasaki
21. Klell M, Eichlseder H, Trattner A (2018) Wasserstoff in der Fahrzeugtechnik, vol 4. Springer, XVIII, p 324
22. Airbus (2022) ZEROe - Towards the world's first zero-emission commercial aircraft. Available from https://www.airbus.com/en/innovation/zero-emission/hydrogen/zeroe. Accessed 26 Oct 2022
23. CRYOMOTIV (2021) Cryogas/Cryo-compressed hydrogen gas fueling. In: CEP—NOW heavy duty event
24. Brunner T, Kampitsch M, Kircher O (2016) Cryo-compressed hydrogen storage. Fuel cells. Wiley, Weinheim, pp 162–173
25. Aceves SM, Espinosa-Loza F, Ledesma-Orozco E, Ross TO, Weisberg AH, Brunner TC, Kircher O (2010) High-density automotive hydrogen storage with cryogenic capable pressure vessels. Int J Hydrogen Energy 35(3):1219–1226
26. Graß M (2022) CryoTRUCK-Konsortium startet die Entwicklung und Erprobung von „CRYO-GAS". Available from https://www.now-gmbh.de/aktuelles/pressemitteilungen/cryotruck-kon sortium-startet-die-entwicklung-und-erprobung-von-cryogas/. Accessed 26 Oct 2022
27. Dornheim M, Baetcke L, Akiba E, Ares J.-R, Autrey T, Barale J, Baricco M, Brooks K, Chalkiadakis N, Charbonnier V, Christensen S, Bellosta von Colbe J, Costamagna M, Dematteis EMM, Fernandez JF, Genett T, Grant DM, Heo TW, Hirscher M, Hurst K, Lototskyy MV, Metz O, Rizzi P, Sakaki K, Sartori S, Stamatakis E, Stuart A, Stubos A, Walker G, Webb CJ, Wood B, Yartys VA, Zoulias E (2022) Research and development of hydrogen carrier based solutions for hydrogen compression and storage. Progr Energy
28. Bellosta von Colbe J, Ares J-R, Barale J, Baricco M, Buckley C, Capurso G, Gallandat N, Grant DM, Guzik MN, Jacob I, Jensen EH, Jensen T, Jepsen J, Klassen T, Lototskyy MV, Manickam K, Montone A, Puszkiel J, Sartori S, Sheppard DA, Stuart A, Walker G, Webb CJ, Yang H, Yartys V, Züttel A, Dornheim M (2019) Application of hydrides in hydrogen storage and compression: achievements, outlook and perspectives. Int J Hydrogen Energy 44(15): 7780–7808
29. HyCARE (2020) Hydrogen carrier for renewable energy storage—Objectives. Available from https://hycare-project.eu/objectives/. Accessed 07 July 2020
30. Sujan GK, Pan Z, Li H, Liang D, Alam N (2019) An overview on TiFe intermetallic for solid-state hydrogen storage: microstructure, hydrogenation and fabrication processes. Crit Rev Solid State Mater Sci 1–18
31. Reilly JJ, Wiswall RH (1974) Formation and properties of iron titanium hydride. Inorg Chem 13(1):218–222
32. GKN Hydrogen. Green energy storage—HY2MEGA. Available from https://www.gknhyd rogen.com/wp-content/uploads/2022/07/GKN_HY2MEGA_ProductSheet.pdf. Accessed 26 Oct 2022
33. GKN Hydrogen. HY2 energy system. Available from https://www.gknhydrogen.com/. Accessed 26 Oct 2022
34. Stolten D, Emonts B (2016) Hydrogen science and engineering: Materials, processes, systems and technology. Wiley, Weinheim
35. Lototskyy MV, Tolj I, Davids MW, Klochko YV, Parsons A, Swanepoel D, Ehlers R, Louw G, Westhuizen B, van der Smith F, Pollet BG, Sita C, Linkov V (2016) Metal hydride hydrogen storage and supply systems for electric forklift with low-temperature proton exchange membrane fuel cell power module. Int J Hydrogen Energy 41:13831–13842
36. Lototskyy M, Tolj I, Klochko Y, Davids MW, Swanepoel D, Linkov V (2020) Metal hydride hydrogen storage tank for fuel cell utility vehicles. Int J Hydrogen Energy 45(14):7958–7967
37. Müller K, Völkl J, Arlt W (2013) Thermodynamic evaluation of potential organic hydrogen carriers. Energ Technol 1(1):20–24

38. Niermann M, Beckendorf A, Kaltschmitt M, Bonhoff K (2019) Liquid organic hydrogen carrier (LOHC)—Assessment based on chemical and economic properties. Int J Hydrogen Energy 44(13):6631–6654
39. Müller K, Aslam R, Fischer A, Stark K, Wasserscheid P, Arlt W (2016) Experimental assessment of the degree of hydrogen loading for the dibenzyl toluene based LOHC system. Int J Hydrogen Energy 41(47):22097–22103
40. Teichmann D, Arlt W, Wasserscheid P, Freymann R (2011) A future energy supply based on liquid organic hydrogen carriers (LOHC). Energy Environ Sci 4(8):2767
41. Niermann M, Drünert S, Kaltschmitt M, Bonhoff K (2019) Liquid organic hydrogen carriers (LOHCs)—Techno-economic analysis of LOHCs in a defined process chain. Energy Environ Sci 12:290–307
42. Chapman E, Sithole M (2016) Anglo American Platinum announces launch of Hydrogenious Technologies' first commercial hydrogen storage system and plans to enter the U.S. market: Hydrogenious Technologies to partner with United Hydrogen Group (UHG) to enter the U.S. market. Available from https://www.angloamericanplatinum.com/media/press-releases/2016/04-05-2016.aspx. Accessed 20 Dec 2021
43. EU-CORDIS (2022) Hydrogen supply and transportation using liquid organic hydrogen carriers. Available from https://cordis.europa.eu/project/id/779694/de. Accessed 28 Oct 2022

Hydrogen Supply Chains

Lucas Sens and Martin Kaltschmitt

Abstract Hydrogen is seen as a promising energy carrier due to its multifarious fields of possible application within climate change mitigation. Particularly, hard-to-abate sectors like aviation, shipping, steel production, and back up electricity supply can be promising applications for this energy carrier in its elementary form or as a derivative. In this context, different hydrogen supply chains are introduced, analyzed and discussed. While compressed gaseous hydrogen supply via pipeline networks offer cost advantages for shorter distances, seaborne transportation of liquid hydrogen can be a viable option for longer-distance transportation, leveraging regions with abundant renewable sources of energy. Material-based carriers like ammonia, methanol, and to some extend LOHCs are considered short-term solutions due to their established infrastructure, although they pose higher energy demands and costs for hydrogen release at the consumption site.

Keywords Green Hydrogen · Liquid Hydrogen · LOHC · Ammonia · Methanol · Supply Chains

1 Background

Hydrogen is seen as a promising energy carrier due to its multifarious fields of possible application within climate change mitigation. Particularly, hard-to-abate sectors like aviation, shipping, steel production, and back up electricity supply can be promising applications for this energy carrier in its elementary form or as a derivative. Additionally, due to the strong cost decline of wind power and especially solar (photovoltaics)-based electricity throughout the last decade [1–3], the production of green[1] hydrogen by water electrolysis becomes more and more attractive. As a result,

L. Sens · M. Kaltschmitt (✉)
Hamburg University of Technology (TUHH), Institute of Envirnmental Technology and Energy Economics (IUE), Hamburg, Germany
e-mail: kaltschmitt@tuhh.de

[1] Green hydrogen is defined as hydrogen produced by water electrolysis using electricity based on renewable sources of energy.

© The Author(s), under exclusive license to Springer Nature Switzerland AG 2025
N. Bullerdiek et al. (eds.), *Powerfuels*, Green Energy and Technology,
https://doi.org/10.1007/978-3-031-62411-7_12

a significantly growing hydrogen demand is expected globally being between 20 and 100 Mt_{H2} in 2030 and reaching 90–1,200 Mt_{H2} in 2050 [4–9].

For countries with a comparable high energy demand and limited options for a sufficient domestic energy supply based on renewable energies, the import of green hydrogen as a renewable energy carrier can help to cover the expected energy demand in a climate-neutral manner. In Germany, for example, being a potential net consumer of green hydrogen in the years to come, the overall hydrogen demand is projected to increase to an order of magnitude between 5 and 24 Mt_{H2} in 2050. Thereof, imported hydrogen is expected to contribute between 5 and 9 Mt_{H2} (2050) at the total hydrogen demand in Germany, which is a share of up to 50% and more [10–14]. Similar circumstances are expected for other industrialized countries with limitations on land availability and/or unfavorable conditions for an electricity generation based on renewable sources of energy.

Furthermore, since the hydrogen production costs, and therefore the electricity costs, determine the overall hydrogen supply costs, especially countries with a high solar radiation and/or high mean wind speeds—in combination with large unused and easily accessible areas—are typically promising for low-cost green hydrogen production. Figure 1 shows exemplarily cost estimations for a constant onsite hydrogen supply—including the costs for electricity generation (using photovoltaics and wind power), electrolysis, hydrogen compression, and storage—for (parts of) the European, Northern African, and Western Asian regions related to the year 2035.

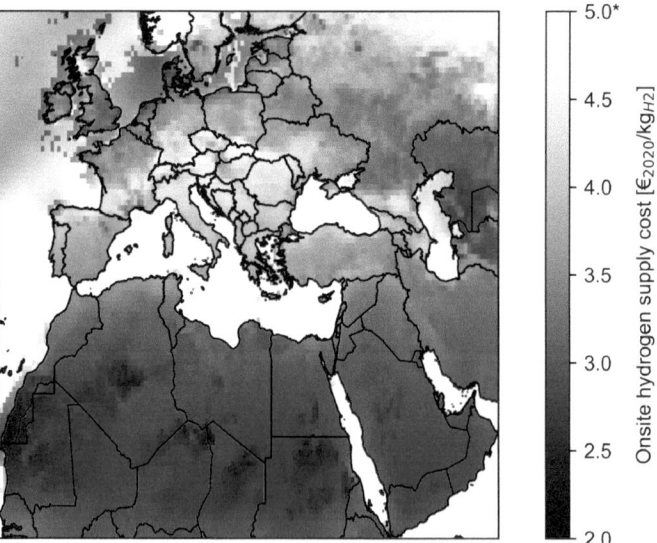

Fig. 1 Costs for a constant onsite hydrogen supply in 2035 (includes costs for electricity generation (photovoltaics and wind power), electrolysis, hydrogen compression and storage; *5.0 €$_{2020}$/kg$_{H2}$ and more; calculations based on [15])

Due to different solar radiation intensities and varying mean wind speeds, the costs for a constant onsite hydrogen supply in 2035 is about twice as high for locations in Western and Southern Europe as for the shown locations in Northern Africa. Having in mind that locations with a high hydrogen demand are often not located in regions where hydrogen can be produced at relatively low costs due to such favorable solar radiation/mean wind speed (and furthermore land availability is often limited at the hydrogen demand hubs), a need for regional, continental or even global hydrogen supply chains is likely to emerge in the future.

Due to the low volumetric energy density of hydrogen at ambient pressure and ambient temperatures (standard conditions), conditioning of hydrogen is needed with the goal to increase its volumetric energy density to enable (conditioned) hydrogen storage and transportation under economic constraints. So far, promising hydrogen conditioning options are compression (compressed gaseous hydrogen, CGH_2), liquefaction (liquid hydrogen, LH_2) and the chemical bond of hydrogen to a carrier molecule as liquid organic hydrogen carrier (LOHC), methanol (CH_3OH) or ammonia (NH_3). Against this background, this chapter discusses the characteristics of these conditioned hydrogen options, introduces different potential hydrogen supply chains, cursorily discusses the general assessment approach of hydrogen supply chains as well as evaluates such (selected) hydrogen supply chains at the example of a supply of hydrogen to Germany and concludes with a conclusion and outlook.

2 Hydrogen Conditioning

The conditioning of hydrogen to reach a higher volumetric energy density is crucial for hydrogen storage and transportation taking economic aspects into consideration. Such a hydrogen conditioning—here defined as the intermediate/reversible improvement of selected physical/chemical key figures of (the conditioned) hydrogen with the purpose to improve hydrogen storage and/or transportation characteristics—can be mainly distinguished in physical and material-based (also known as chemical) options.

- **Physical conditioning.** Hydrogen is compressed, cryo-compressed or liquefied and therefore still present in its elementary (molecular) form (exclusively as a single hydrogen molecule).
- **Material-based conditioning.** A carrier material, respectively, atom/molecule is needed, enabling a physically or chemically bound of hydrogen. Widely discussed options for a material-based conditioning of hydrogen are inter alia liquid organic hydrogen carrier, methanol, ammonia and to some extent methane as well as metal hydrides.

Each conditioned hydrogen option obtains different gravimetric and volumetric energy densities; Fig. 2 shows some examples. A selection of promising options will be introduced in the following.

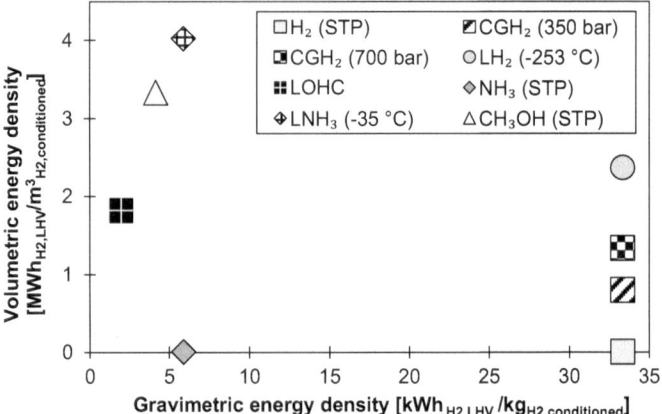

Fig. 2 Energy densities of different conditioned hydrogen options (values based on [16–18]) (CGH$_2$: compressed gaseous hydrogen; CH$_3$OH: methanol; H$_2$: gaseous hydrogen; NH$_3$: gaseous ammonia; LH$_2$: liquid hydrogen; LNH$_3$: liquid ammonia; LOHC: liquid organic hydrogen carrier (here: hydrogenated benzyltoluene); STP: standard temperature and pressure conditions)

2.1 Compressed Gaseous Hydrogen

The conditioning of hydrogen by compression is the simplest option realized since decades (e.g., within the (petro-)chemical industry). The (electrical) energy demand for (hydrogen) compression depends strongly on the inlet and desired outlet pressure and is often less than 3 kWh$_{el}$/kg$_{CGH2}$ (Fig. 3). Nevertheless, the volumetric energy density of below 1.5 MWh$_{H2,LHV}$/m$^3_{CGH2}$ achieved by such a compression (700 bar) is also the lowest of the options considered here. For hydrogen compression, reciprocating piston and ionic liquid compressors are the most common options.

- **Piston compressor**. Such compressors are technologically fully mature and obtain low specific investment costs. Possible contaminations of the compressed hydrogen due to lubrication oils (needed for greasing the piston) and a relatively low overall compression efficiency of around 65% are main drawbacks [19].
- **Ionic liquid compressor.** Ionic liquid compressors are a relatively new development with a comparable low market penetration so far (2024). Due to the limited market importance, this technology shows higher investment costs compared to piston compressors. However, the compression efficiency is relatively high with up to 90% and potential contamination due to the compression process is (in comparison with the piston compressor) negligible [19]. These advantages seem to make the use of ionic liquid compressors potentially favorable in the future.

Compressed gaseous hydrogen can be stored in pressure tanks or salt caverns. The storage pressure typically reaches levels of around 200 bar [23].

For the transportation of large quantities, hydrogen pipelines appear to be the most promising option in terms of specific transportation costs. In this case, the

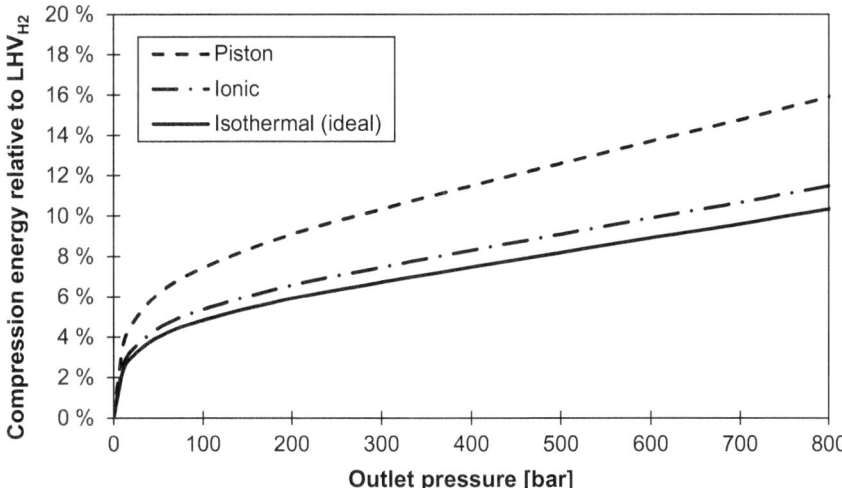

Fig. 3 Compression (electric) energy for hydrogen relative to its lower heating value (LHV) for an isothermal (ideal) compression, ionic liquid compressor and reciprocating piston compressor (own calculations based on [19–22])

hydrogen pressure level is in the range of 15 to 100 bar [24]. However, the impact on the hydrogen quality (e.g., ensuring 5.0^2) caused by, e.g., impurities due to abrasion needs to be investigated further. Additionally, inter alia the investment as well as the time frame required to build up a pipeline network is high, which typically increases the financial risk associated with such an investment (similar circumstance can be also valid for other infrastructure elements, e.g., salt caverns). Nevertheless, for large quantities and up to intermediate distances (up to 2,000–5,000 km) hydrogen transportation per pipeline is a very promising option in terms of specific hydrogen transportation costs [26–34].

For small quantities of hydrogen, short transportation distances and altering locations of hydrogen demand, compressed gaseous hydrogen transportation per truck via trailers (or per train) is often the most favorable option. For road transportation, hydrogen is typically stored at pressure levels around 500 bar, while higher pressure levels of up to 700 bar might be realized in the future. Impurities of hydrogen due to trailer transportation are not reported so far. A transportation of compressed gaseous hydrogen per ship, especially over longer distances, is considered less of a relevant option due to its low volumetric energy density and therefore potentially large transportation volumes and high specific hydrogen transportation costs.

Hydrogen losses during its compression and the subsequent storage and transportation as compressed hydrogen are from an economic point of view rather neglectable. Nevertheless, hydrogen being a colorless, odorless and tasteless gas is the molecule with the highest diffusion rate. It has a large flammability range

---

[2] Indicates a hydrogen content in the gas of 99.999 % [25]

between 4 and 78 vol.-% and low minimum ignition energy of 0.02 mJ. Therefore, specific safety precautions need to be met if hydrogen is transported in its elementary (molecular) form.

2.2 Liquid Hydrogen

The liquefaction of hydrogen is another physical conditioning option which increases the volumetric energy density of hydrogen to almost 2.4 $MWh_{H2,LHV}/m^3_{LH2}$. Since hydrogen reaches the liquid phase at -253 °C, liquefaction is characterized by a comparatively high energy demand. Therefore, the theoretical minimum energy demand for hydrogen liquefaction amounts to 3.92 kWh/kg_{LH2}. However, for currently existing and operated (small scale) hydrogen liquefaction units an (electrical) energy demand between 10 and 13 kWh_{el}/kg_{LH2} is typically stated. In the future, a decline of the (electrical) energy demand for hydrogen liquefaction down to approx. 6 kWh_{el}/kg_{LH2} is projected for further developed and optimized as well as steadily operated large-scale hydrogen liquefaction units [35].

The hydrogen liquefaction process itself represents a rather complex process. First hydrogen is compressed to at least 13 bar and cooled down to around 25 °C. Afterward, another cooling step is conducted where hydrogen is cooled down to around -193 °C by using liquid nitrogen. This is followed by a final cooling step—to reduce the temperature of the hydrogen further down to -228 to -251 °C—with (cryogenic) hydrogen or (cryogenic) helium and an integrated ortho-para[3] conversion. Finally, an adiabatic expansion takes place leading to a further cooling down to less than -253 °C and therefore a liquefaction of the hydrogen.

To prevent significant boil-off losses due to the high-temperature difference between the (cryogenic) liquid hydrogen and the ambient temperature, liquid hydrogen needs to be stored in highly insulated cryo-tanks at a pressure level between 1 and 10 bar.

Even though liquid hydrogen tanks are more material intensive compared to gaseous hydrogen pressure tanks, the specific hydrogen storage costs are typically lower as hydrogen can be stored in a liquid form at a notable higher volumetric energy density. However, boil-off losses are realistically often unavoidable (over time). For a fairly good insulated tank, boil-off losses are typically stated to amount to about 0.1 to 2%/d [36, 37]. But this order of magnitude strongly depends on the storage size (i.e., surface area to volume ratio), the filling level and the storage duration. Further losses, so-called flash rates, can occur during the transfer of liquid hydrogen to another (empty) storage tank with a higher temperature leading to an immediate evaporation of the liquid hydrogen during the filling/transfer process, reaching losses

---

[3] Ortho and para hydrogen are two configurations of the di-hydrogen molecule different in the symmetry of their nuclear spin configuration and their rotational status; even though they show the same chemical structure, the physical characteristics are partly different; hydrogen is typically a mixture of both configurations being defined by temperature.

up to 2%/filling and sometimes even significantly more (e.g., for the initial filling process) [38]. These flash rates can be inter alia reduced largely by reducing the vacancy time of the storage facilities and vessels used in a hydrogen supply chain and/or by a continuous cooling.

Due to its higher energy density, liquid hydrogen is typically favorable for longer distance road hydrogen transportation per truck as well as for overseas hydrogen transportation by ship. Nevertheless, transportation times of liquid hydrogen over several weeks should be avoided since the boil-off losses add up over time. The transportation of liquid hydrogen in cryo-pipelines is so far not discussed as a viable option for longer distances due to the high costs for the required insulation.

The reconditioning of liquid hydrogen back into a gaseous state—required prior to the respective use of hydrogen—is comparable simple as only a evaporation unit is needed typically using energy from ambient air or water (i.e., from the natural environment). Even if the required energy for regasification comes from the environment, the energy contained within the (cryogenic) liquid hydrogen is nowadays unused; however, concepts for the usage of this energy are under development. Furthermore, the purity of liquid hydrogen is the highest, and high-quality standards (e.g., 5.0^4 quality) can be easily complied with.

2.3 Liquid Organic Hydrogen Carrier

The term liquid organic hydrogen carrier (LOHC) is used for a material-based hydrogen conditioning option. They are typically a liquid aromatic or heteroaromatic compound able to bind hydrogen chemically during a catalytic reaction (hydrogenation) and to release the hydrogen again during a counterpart reaction, the dehydrogenation. Frequently discussed LOHCs are for example toluene, benzyltoluene and dibenzyltoluene [39, 40]. LOHCs enable storage and transportation of hydrogen (bound) as a liquid—under ambient conditions—with a higher volumetric energy density (typically around 1.5 to 2.0 $MWh_{H2,LHV}/m^3_{LOHC}$) compared to compressed gaseous hydrogen. The LOHC itself is circulated within the overall hydrogen supply chain involving also the return transportation of the dehydrogenated LOHC from the location of dehydrogenation back to the hydrogenation location. The basic LOHC reaction scheme is shown in Fig. 4.

The hydrogenation of LOHCs is an exothermic process taking place typically at temperatures of up to 300–400 °C and at elevated pressure levels. The occurring surplus thermal energy—due to the exothermic characteristics of the hydrogenation reaction—often sums up to around 20 to 30% [17, 30] of the bounded (and reversible releasable) hydrogen energy (lower heating value (LHV)). This "excess" thermal energy could potentially be used at the hydrogenation location for an external thermal energy supply (if there is any thermal energy demand). Due to the fact that a large-scale hydrogen production is expected to take place in less densely populated areas

---

[4] Indicates a hydrogen content in the gas of 99.999 % [25]

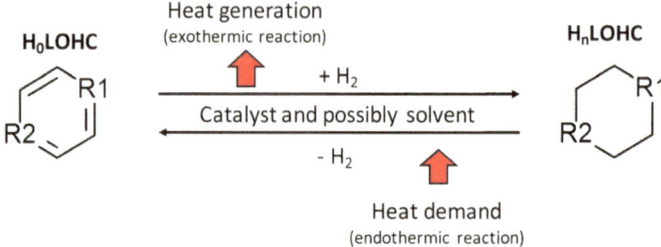

Fig. 4 LOHC reaction scheme [41] (H_0LOHC: dehydrogenated liquid organic hydrogen carrier; H_2: hydrogen; H_nLOHC: hydrogenated liquid organic hydrogen carrier; R1/R2: potential further (different) molecules)

(e.g., desert regions), the possibilities for a wide-ranging synergetic use of this surplus thermal energy are thus rather limited based on current knowledge. Most likely, this thermal energy has to be considered (mainly) as an energy loss.

The properties of the LOHC (hydrogenated and dehydrogenated) are similar to crude oil derivatives. Thus, its transportation characteristics are well known and transportation can be realized based on existing infrastructure (e.g., conventional crude oil ships, trains, trucks and pipelines). In addition, most LOHCs are not flammable or explosive if a 100% hydrogenation of the hydrogen can be assured, leading to fewer safety issues compared to the transportation/storage of, e.g., physical conditioned hydrogen (gaseous and liquid) or ammonia. However, the most promising LOHCs hold typically only around 6 wt% [17] of reversible stored hydrogen. This leads to large quantities of LOHC material needed for such a hydrogen storage and/or transportation. Consequently, the necessary LOHC material quantities are increasing with storage and transportation time (and therefore in general also transportation distance). Hence, similar to liquid hydrogen and its boil-off over time, hydrogen supply chains based on LOHCs are rather unsuitable for (very) long transportation distances and/ or long storage times since the amount of carrier molecules needed (the LOHC) increases notably and thus also the investment costs (i.e., the bound capital within the overall system) [29].

The dehydrogenation process is an endothermic process typically realized at ambient or slightly increased pressure levels. For most LOHCs, around 30–40% [17, 30] of the released hydrogen energy (lower heating value (LHV)) is required as a thermal energy input to release the hydrogen bound chemically within the (hydrogenated) LOHC molecule back into an elementary gaseous state. Hence, low-cost "green" thermal energy at a relatively high-temperature level (partly—depending on the respective LOHC carrier molecule—more than 300 °C) is needed at the dehydrogenation location. For these reasons, a promising use case for example would be a direct combination with a high-temperature fuel cell (solid oxide fuel cell, SOFC) characterized by waste heat at such a temperature level. Nevertheless, if LOHCs are "only" used as a carrier molecule for hydrogen imports, e.g., to Germany, and the goal is to feed the imported and released hydrogen into a hydrogen distribution pipeline network, "excess" thermal energy in the necessary order of magnitude on the needed temperature level is most likely not available within a defossilized energy system. Therefore, it can be expected that most likely the thermal energy has to be generated from the transported hydrogen itself, lowering the hydrogen supply efficiency and thus increasing necessarily the hydrogen supply costs.

After dehydrogenation of the LOHC, the released hydrogen potentially needs to be purified to meet the given quality standards, leading to additional hydrogen losses of 2–5% [38, 40]. However, these losses could be potentially used to cover parts of the dehydrogenation thermal energy demand.

2.4 *Methanol*

Methanol can not only be used as a fuel or bulk chemical itself but also as a material-based conditioned hydrogen carrier. For the conditioning of the hydrogen (here: the methanol synthesis), hydrogen (H_2) and carbon dioxide (CO_2) are converted to methanol (CH_3OH) and water (H_2O) within an exothermic synthesis reaction (Eq. 2). Vice versa for hydrogen release, methanol is cracked (by adding water) again to hydrogen and carbon dioxide within an endothermic reaction. Even though there are hydrogen losses within the methanol synthesis due to water formation, this lost hydrogen is recovered during the cracking process of methanol. Considering this, methanol shows an effective hydrogen loading capacity of 18.75 wt%.

$$3H_2 + CO_2 \rightleftharpoons CH_3OH + H_2O. \tag{1}$$

Similar to LOHCs methanol is liquid at ambient conditions, and handling of methanol (i.e., storage and transport) is carried out at large scale globally since decades. The most widely discussed concepts of methanol as a hydrogen carrier option require "green" (sustainable) carbon dioxide since—in comparison with LOHCs—the carrier molecule (carbon dioxide) is a gas and usually released into the environment after methanol cracking; means, so far most concept assume that the carbon dioxide is not returned to the methanol synthesis location. Since "green" (sustainable) carbon dioxide is expected to become a scarce commodity [42], the costs for this (carrier) molecule might have a significant impact on the overall hydrogen supply costs using methanol as a hydrogen carrier. Therefore, concepts are under discussion capturing the released carbon dioxide after the methanol cracking and consecutively transporting the carbon dioxide back to the methanol synthesis location.

The energy demand for the methanol cracking and subsequent hydrogen purification accounts to roughly 25% [30] of the released hydrogen energy (lower heating value (LHV)). A hydrogen purification is needed to ensure the requested hydrogen quality. This leads to hydrogen losses of roughly 15% [43] which can potentially be used to cover a share of the thermal energy input for the cracking.

2.5 Ammonia

Ammonia is another widely discussed material-based hydrogen conditioning option. Compared to LOHCs and methanol this option uses nitrogen as a carrier molecule (and is therefore carbon-free). The so-called Haber–Bosch-Synthesis is used to produce ammonia, where hydrogen (H_2) and nitrogen (N_2) (extracted from the air) react at temperatures between 300 and 500 °C and a pressure range between 150 and 350 bar to ammonia (NH_3) in an exothermic reaction (Eq. 2). Vice versa, the cracking of ammonia for the hydrogen release is an endothermic reaction releasing nitrogen and hydrogen.

$$3H_2 + N_2 \rightleftharpoons 2NH_3 \qquad (2)$$

Unlike LOHCs and methanol, ammonia is not liquid at ambient conditions. Nevertheless, only moderate temperatures below $- 33$ °C or pressure levels above 8.5 bar are needed to liquefy ammonia. Therefore, the energy demand for liquefaction of ammonia is comparatively low. Additionally, hydrogen loading capacity of ammonia is relatively high (17.6 wt%), leading to a high volumetric energy density of ca. 4 $MWh_{H2,LHV}/m^3_{LNH3}$ for liquid ammonia. The carrier molecule, here nitrogen, is released after the cracking back into the environment (i.e., the use cycle is closed via the atmosphere). The nitrogen required at the location of the ammonia synthesis can be supplied onsite by air separation units from ambient air being state of technology and usually associated with comparable low specific costs.

As ammonia is one of the most widely produced chemicals worldwide, international infrastructure for ammonia storage and transportation exists and the overall logistics are well known. However, since ammonia is not liquid at ambient conditions and highly toxic, ammonia storage and transportation involve more operational efforts and require higher (costly) safety standards compared to, e.g., LOHCs and methanol.

The energy demand for ammonia cracking and the necessary subsequent hydrogen purification is high and reported to sum up to roughly 35% [38, 44] of the released hydrogen energy (lower heating value (LHV)). Additionally, the hydrogen loss within the purification process is significant with roughly 20% according to current technology and current knowledge [45, 46]. This share might be lowered in the future to less than 10% by, e.g., the application of advanced membrane technologies. Furthermore, these hydrogen losses might be used to cover a large fraction of the thermal energy demand needed for the ammonia cracking.

3 Supply Chains

To use hydrogen in the energy system and to bridge potentially distances between the hydrogen production and demand side hydrogen supply chains are needed. Typically, these hydrogen supply chains can be separated into the following different sections.

- **Production**. For the production of green[5] hydrogen water electrolysis powered by electricity generated by the use of renewable sources of energy—most likely from wind, solar and if available (at reasonable costs) hydropower and geothermal energy—is compulsory. Since the electricity supply costs accounts typically to roughly two-thirds of the overall hydrogen production costs, locations with high solar radiation and/or wind speeds and a low seasonality of these are favorable from an economic point of view (i.e., low electricity generation costs). Furthermore, local salt formations, which can be potentially used as low-cost large-scale hydrogen storage options, might enhance economically promising hydrogen production configurations and therefore help to reduce the overall hydrogen production costs [15].
- **Conditioning**. A hydrogen conditioning process is needed to increase the volumetric energy density and thus allow an economically viable subsequent storage and/or transportation of hydrogen. The conditioning can vary strongly based on the chosen conditioned hydrogen option.
- **Storage**. Depending on the layout of the hydrogen supply chain, hydrogen storage systems are needed inter alia to enable cost-efficient hydrogen production system configurations (i.e., smoothing a fluctuating hydrogen production to ensure a specific hydrogen supply profile, for a (strategic) seasonal storage and/or for batch processes like a subsequent hydrogen transportation per vessel).
- **Transportation.** Hydrogen transportation is needed to supply low-cost hydrogen, if the hydrogen production is not onsite the demand centers. In general, it can be distinguished between hydrogen import via ships and pipelines (in general for large hydrogen quantities) and hydrogen distribution, e.g., per truck or pipeline network, directly to the consumer (in general for comparable smaller quantities).
- **Reconditioning.** Hydrogen reconditioning is necessary if the hydrogen is stored and transported in a physical state that differs from the physical state of hydrogen that is demanded by the consumer. As the conditioning, also the reconditioning can show very different characteristics depending on the conditioned hydrogen option.

Based on these sections, different (simplified) exemplary hydrogen supply chains can be set up (Fig. 5) depending on the respective frame conditions and local circumstances.

---

[5] Green hydrogen is defined as hydrogen produced by water electrolysis using electricity based on renewable sources of energiy. Nevertheless, other concepts (e.g., photocatalytic hydrogen production) might be also promising hydrogen production options in the future using renewable energies.

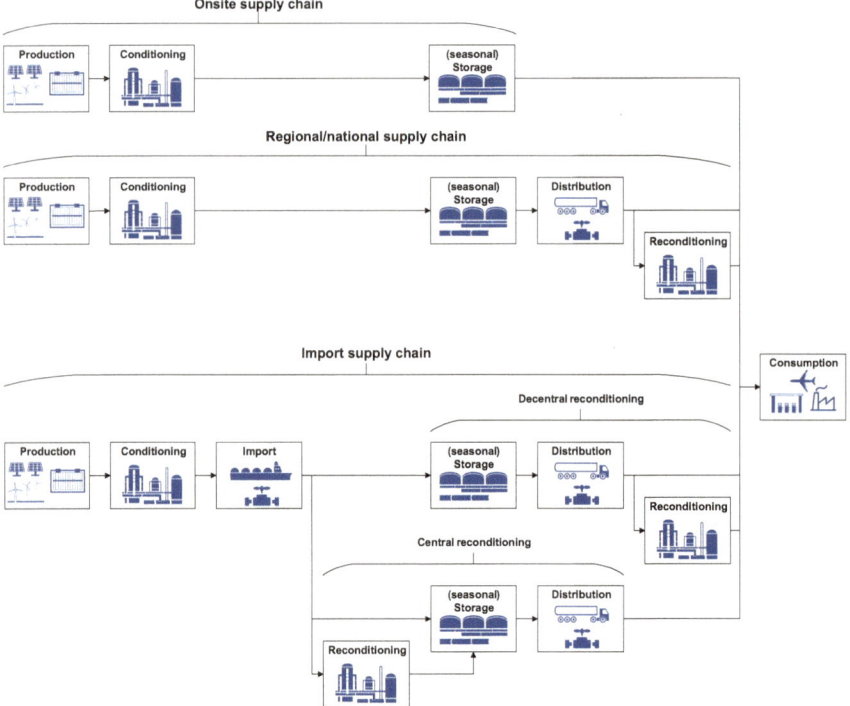

Fig. 5 Simplified hydrogen supply chain options

- **Onsite supply chain**. This is the simplest approach from a technical and logistical point of view since no (significant) hydrogen transportation infrastructure and only a (seasonal) hydrogen storage unit is needed. After the hydrogen production, the hydrogen can be directly conditioned as demanded by the consumer (compressed gaseous hydrogen or liquid hydrogen) and consequently (seasonal) stored to enable a hydrogen supply also in the case of insufficient electricity production based on renewable energies (e.g., dark doldrums). Such onsite hydrogen supply chains are promising options if low-cost electricity and sufficient "green" electricity/space (for the electricity generation technologies) are available at a location with a certain hydrogen demand. However, it is expected that most hydrogen demand centers do not obtain these characteristics.
- **Regional/national supply chain**. This represents hydrogen supply chains on a national/regional level. The hydrogen is produced here in regions with (relatively) low-cost electricity (based on systems using renewable sources of energy) and sufficient land availability (e.g., in the North Sea in the case of Germany). Subsequent to the hydrogen production, conditioning and seasonal storage in the extended proximity of the hydrogen production location is typically needed. Afterward, the conditioned hydrogen is transported per pipeline or via train or truck to the location, where the hydrogen demand is given and a potentially reconditioning

of the hydrogen (e.g., evaporation, dehydrogenation, cracking and liquefaction) to the desired state of matter (gaseous or liquid hydrogen) is needed.

- **Import supply chain**. This hydrogen supply chain describes the (export and) import of hydrogen. Thus, it involves a transportation of hydrogen—from countries/places with in general extraordinary high solar radiation and/or mean wind speeds—over longer distances (typically an intercontinental transport). Subsequently to hydrogen production and conditioning, the conditioned hydrogen is transported per ship or pipeline to the importing country. After the landing of the conditioned hydrogen—if a reconditioning of the hydrogen is needed—it can be distinguished between decentral and central reconditioning concepts (of the hydrogen).

  Decentral reconditioning. The imported hydrogen is stored and distributed in the same conditioning state as it was imported and reconditioned onsite at the consumption site (if another hydrogen conditioning state is demanded by the consumer).

  Central reconditioning. The imported conditioned hydrogen (e.g., liquid hydrogen per ship) is reconditioned (if needed; e.g., evaporation to gaseous hydrogen) stored and distributed (e.g., pipeline network) in its new and by the end-consumer demanded conditioned state.

4 Analysis and Assessment

Depending on the hydrogen production costs, transportation distances, seasonal (and strategic) storage capacities and demanded state of matter of the hydrogen by the consumer, different hydrogen supply chains and conditioned hydrogen options can be favorable. Therefore, an assessment of hydrogen supply chains can help to identify the most promising options. Hence, the general approach for the analysis and assessment of hydrogen supply chains will be introduced shortly, and afterward, an exemplary application, as well as the results, are shown.

4.1 Approach

For the analysis and assessment of hydrogen supply chains, the following steps need to be considered.

- **Definition**. First the hydrogen supply chain to be analyzed must be defined in detail. This is inter alia based on the given technical restriction and also includes the assumptions of specific framework conditions. In general, different sections in the hydrogen supply chain (e.g., boil-off losses of hydrogen during the transportation lead to higher hydrogen production quantities) can influence each other. Hence, if, e.g., a well-to-tank assessment should be realized it is crucial to consider (to a reasonable extend) all needed sections for the definition of the hydrogen supply chain in a way that the various parts of the chain fit together in an optimal/ sound way.

- **Modeling**. Based on the defined and coherent hydrogen supply chains, these chains need to be modeled. Depending on the desired assessment categories different technical, economic and environmental models can be applied. Nevertheless, independent from the assessment category, the basic laws of physics and chemistry (e.g., energy and mass balances) as well as technical/techno-economic restriction (e.g., plant start-up times) need to be considered, while simplifications during the development of the models (to reduce the computational efforts and thus to find a solid solution within a certain time frame) should be done carefully. Thereby, each section of the hydrogen supply chain should be modeled in the same modeling depth.

- **Data**. Since every model is only as good as its data input, data curation is from major importance. As most of the concepts for a hydrogen supply chain are not fully market mature yet and may be realized in the future, projections of the future development of the underlying values (e.g., capital expenditures and efficiencies of the considered technologies) are needed (e.g., by applying the experience curve theory, through historic analogies, by trend extrapolation or based on publication analyses). Furthermore, it is recommended to use the median instead of the arithmetic mean to have a robust data input (as long as sufficient quantities of data points can be curated).

- **Assessment**. For the final analysis and assessment of such hydrogen supply chains, adequate and target-oriented assessment parameters must be defined. Depending on the overall goal of the respective assessment, these assessment parameters are covering typically technical, economic and/or environmental aspects. As typically cost is a major decision parameter in industry, a commonly used and in general well-suited assessment parameter are the specific hydrogen supply costs (e.g., $€_{2023}/kg_{H2}$).

4.2   Application and Results

Based on the introduced approach, hydrogen supply chains can be set up, analyzed and assessed. Here an exemplary economic (specific hydrogen supply costs) well-to-tank analysis and assessment for a compressed gaseous (CGH_2, 700 bar) and liquid hydrogen (LH_2) supply of heavy-duty vehicles in Germany related to the year 2030 is introduced.

As hydrogen production locations/regions and (small-scale) onsite hydrogen production in Germany (OS), a (large-scale) hydrogen production within the German part of the North Sea (DE) as well as a (large-scale) hydrogen production in regions in Tunisia (TN) and Patagonia, Chile, (CL) is considered. The location/region-specific hydrogen production costs are determined by applying a linear optimization model [15] considering the location/region-specific (hourly) mean wind speed and solar radiation as a "primary" energy source for an hourly constant (onsite) hydrogen supply based on water electrolysis.

As hydrogen supply chains, respectively, conditioned hydrogen options compressed gaseous hydrogen (CGH_2), liquid hydrogen (LH_2), liquid organic hydrogen carriers (LOHCs), methanol (CH_3OH) and ammonia (NH_3) are considered for this example, but always with the goal to supply compressed gaseous and/or liquid hydrogen at the filling station for heavy-duty vehicles in the end. For the compressed gaseous hydrogen supply chain, the defined conditioned hydrogen import option is a pipeline (PI), while for all other conditioned hydrogen supply chain options, the conditioned hydrogen import is assumed by ship (SI). The conditioned hydrogen distribution to the filling stations is assumed to be done by truck (TD) for all conditioned hydrogen options and additionally per pipeline in the case of a compressed gaseous hydrogen supply chain (PD). The detailed definition, modeling, data input and assessment approach of the hydrogen supply chains are based on and can be found in detail in (Sens et al. [29]).

The results (of the specific hydrogen supply costs) of the well-to-tank analysis of the hydrogen supply chains can be found for the compressed gaseous hydrogen in Fig. 6 and for liquid hydrogen in Fig. 7 for filling of heavy-duty vehicles in Germany in the year 2030. The single sections of the hydrogen supply chains are defined as the following.

- **Production | excl losses**. Hydrogen production costs, exclusive additional hydrogen production costs due to hydrogen losses (e.g., hydrogen boil-off, purification losses) in the hydrogen supply chain.
- **Production | losses**. Additional hydrogen production costs due to hydrogen losses (e.g., hydrogen boil-off, hydrogen purification losses) occurring throughout the overall hydrogen supply chain.
- **Storage**. Sum of conditioned hydrogen storage costs (e.g., seasonal conditioned hydrogen storage, conditioned hydrogen storage at the terminal) given within the total hydrogen supply chain.
- **Conversion | excl carrier**. Costs for hydrogenation (LOHC)/synthesis (methanol, ammonia) excluding the carrier molecule (LOHC, carbon dioxide or nitrogen) costs.
- **Conversion | carrier**. Carrier (LOHC, carbon dioxide or nitrogen) molecule costs needed for hydrogenation (LOHC)/synthesis (methanol, ammonia).
- **Liquefaction | central**. Hydrogen liquefaction costs for a central (large-scale) liquefaction.
- **Transportation**. Conditioned hydrogen transportation costs (sum of potential hydrogen import and distribution).
- **Reconversion | excl th energy**. Costs for dehydrogenation (LOHC)/cracking (methanol, ammonia) excluding the thermal energy demand for the endothermic reaction.
- **Reconversion | th energy**. Thermal energy costs for dehydrogenation (LOHC)/ cracking (methanol, ammonia).

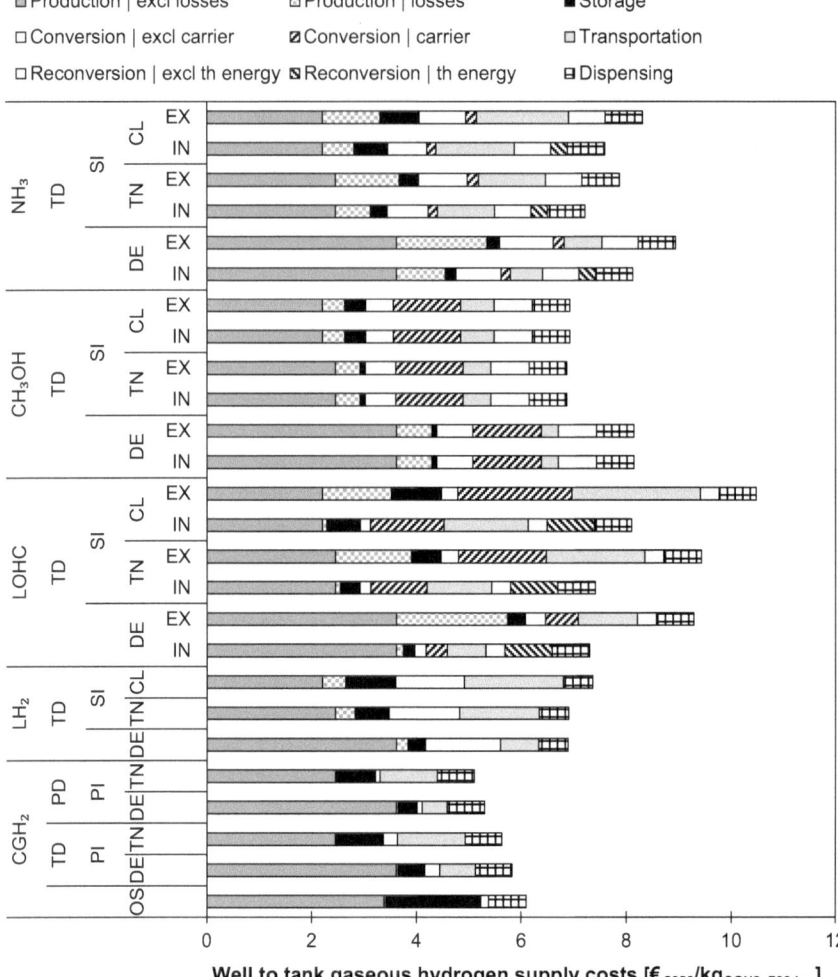

Fig. 6 Well-to-tank compressed gaseous hydrogen (700 bar) supply costs (filling of heavy-duty vehicles) in Germany in 2030 (CGH$_2$: compressed gaseous hydrogen-based hydrogen supply chain; CH$_3$OH: methanol-based hydrogen supply chain; CL: hydrogen production in Chile (Patagonia) (import supply chain); DE: centralized hydrogen production in the North Sea of Germany (regional/national supply chain); EX: reconversion thermal energy supply by external (low-cost) source; IN: reconversion thermal energy supply by internal hydrogen use; LH$_2$: liquid hydrogen-based hydrogen supply chain; LOHC: liquid organic hydrogen carrier-based hydrogen supply chain; NH$_3$: ammonia-based hydrogen supply chain; OS: onsite hydrogen production directly at the filling station in Central Germany (onsite supply chain); PD: hydrogen distribution to filling station per pipeline; PI: hydrogen import to Germany per pipeline; SI: hydrogen import to Germany per ship; TD: hydrogen distribution to filling station per truck; TN: hydrogen production in Tunisia) (results based on [29])

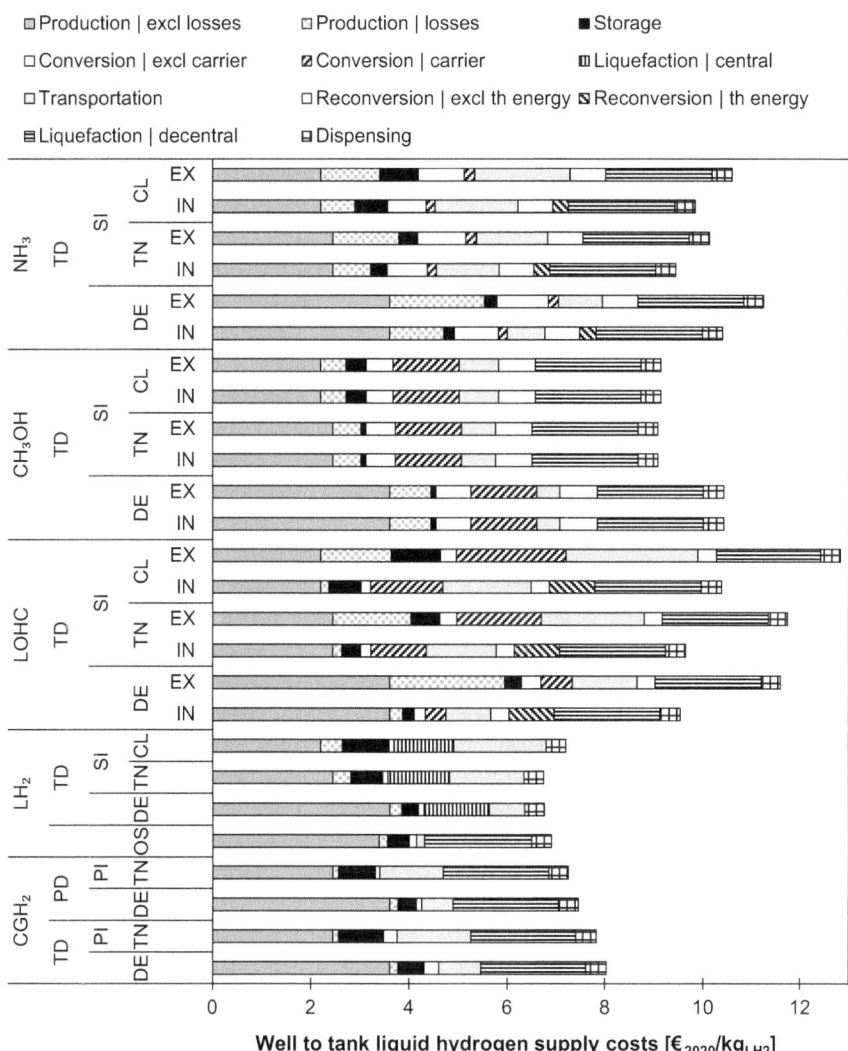

Fig. 7 Well-to-tank liquid hydrogen supply costs (filling of heavy-duty vehicles) in Germany in 2030 (CGH_2: compressed gaseous hydrogen-based hydrogen supply chain; CH_3OH: methanol-based hydrogen supply chain; CL: hydrogen production in Chile (Patagonia) (import supply chain); DE: centralized hydrogen production in the North Sea of Germany (regional/national supply chain); EX: reconversion thermal energy supply by external (low-cost) source; IN: reconversion thermal energy supply by internal hydrogen use; LH_2: liquid hydrogen-based hydrogen supply chain; LOHC: liquid organic hydrogen carrier-based hydrogen supply chain; NH_3: ammonia-based hydrogen supply chain; OS: onsite hydrogen production directly at the filling station in Central Germany (onsite supply chain); PD: hydrogen distribution to filling station per pipeline; PI: hydrogen import to Germany per pipeline; SI: hydrogen import to Germany per ship; TD: hydrogen distribution to filling station per truck; TN: hydrogen production in Tunisia) (results based on [29])

- **Liquefaction | decentral**. Hydrogen liquefaction costs for a small-scale liquefaction onsite the demand center.
- **Dispensing**. Costs of compressed gaseous (700 bar) respective liquid hydrogen dispensing for heavy-duty vehicles.

Subsequently the results for the different hydrogen supply chains can be summarized as the following.

- **Hydrogen supply using compressed gaseous hydrogen**. Due to the comparably low energy demand for the hydrogen conditioning (here compression), the low-cost hydrogen transportation in pipelines and the avoidance of hydrogen reconditioning efforts, compressed gaseous hydrogen supply chains obtain the lowest gaseous hydrogen supply costs (here: ca. 5 $€_{2020}$/kg$_{CGH2, 700 bar}$). However, the implementation of low-cost gaseous hydrogen storage options (e.g., salt caverns) and a compressed gaseous hydrogen transportation network per pipeline are crucial for achieving these favorable hydrogen supply costs. In addition, also in the case of a liquid hydrogen demand compressed gaseous hydrogen supply chains (here: around 7 to 8 $€_{2020}$/kg$_{LH2}$) appear to be a viable solution next to liquid hydrogen supply chains (here: approximately 7 $€_{2020}$/kg$_{LH2}$). However, a (compressed gaseous) hydrogen infrastructure based on a pipeline network and salt cavern storages comes with high investments and thus also high financial risks. Hence, the ramp-up is challenging and might need time to emerge. Overall, compressed gaseous hydrogen supply chains seem to be the most promising option for national hydrogen distribution and short hydrogen import distances (typically: less than 3,000 km).
- **Hydrogen supply using liquid hydrogen**. Liquid hydrogen is characterized by a relatively high volumetric energy density, resulting in comparably low hydrogen transportation and storage costs. Analogously to gaseous hydrogen supply chains, the purity of the hydrogen is very high and the reconditioning efforts (evaporation) are typically very low. However, boil-off losses during the handling of liquid hydrogen make the hydrogen transportation infrastructure rather challenging and demanding. Furthermore, these boil-off losses result in higher hydrogen production demands in comparison with gaseous hydrogen supply chains if the same amount of hydrogen needs to be supplied. In addition, hydrogen conditioning (liquefaction) causes high costs due to a high investment and high specific electricity demand. Also, a global liquid hydrogen infrastructure has not emerged so far (2024). This leads to the fact that liquid hydrogen supply chains are characterized by higher costs (here: around 7 $€_{2020}$/kg$_{CGH2, 700 bar}$) compared to gaseous hydrogen supply chains, if hydrogen is required in a gaseous state by the consumer. In the case of a liquid hydrogen demand (here: for heavy-duty vehicle filling), liquid hydrogen supply chains show the lowest overall hydrogen supply costs (here: slightly below 7 $€_{2020}$/kg$_{LH2}$), since hydrogen liquefaction takes place centralized at the hydrogen production location benefiting from scale effects and potentially low electricity supply costs. Nevertheless, in case of long-distance liquid hydrogen transportation (e.g., from South America to Europe) according

to current knowledge the hydrogen boil-off losses are most likely in a very high order. Hence, a seaborne liquid hydrogen transportation is more likely suitable for rather short to medium transportation distances (e.g., from Northern Africa or Middle East to Europe).

- **Hydrogen supply using liquid organic hydrogen carrier**. Liquid organic hydrogen carrier (LOHC) could easily be handled based on already existing transport infrastructure as the material properties are (more or less) similar to crude oil. Nevertheless, a significant amount of (thermal) energy is required for the hydrogen to be released from the carrier molecule at the hydrogen consumption site. In addition, the costs for the carrier material is of importance for long LOHC transportation distances and storage times since quantities of the required carrier materials are higher. Hence, even if LOHCs are operationally easy to handle within the existing infrastructure, in general it is not preferable to transport them over long distances or use them as a seasonal storage option from an economic point of view (here: at least 7 $\text{€}_{2020}/\text{kg}_{\text{CGH2, 700 bar}}$ and 9 $\text{€}_{2020}/\text{kg}_{\text{LH2}}$). In the end, a hydrogen supply based on LOHCs leads to partly significantly higher hydrogen supply costs compared to a gaseous or liquid hydrogen supply chain. This is mainly due to the (thermal) energy demand for dehydrogenation, the relatively low volumetric energy density of the bounded hydrogen and the high costs of the carrier material. If high-temperature thermal energy is available at the reconditioning site to low costs (e.g., industrial waste heat) or the generated thermal energy at the conditioning location has a monetary value, the hydrogen supply costs could be lowered [28]. However, this cost potential is not limited to LOHCs only, as this is also true for methanol and ammonia.

- **Hydrogen supply using methanol**. Methanol obtains very low transportation and storage costs due to its high volumetric energy density of the bound hydrogen and its easy handling within the existing infrastructure making it favorable for long-distance transports. Furthermore, in the future a significant demand for "green" methanol as a bulk chemical is expected ("green" chemistry) leading to the implementation of a no regret infrastructure. However, also for methanol a high energy demand is needed for the hydrogen reconditioning (methanol cracking) and the occurring hydrogen purification losses increase the hydrogen supply costs. In addition, the costs for the carrier molecule ("green" carbon dioxide) as well as its general availability have a clear impact on the overall supply costs. This leads to the fact that the costs for an elementary (molecular) hydrogen supply using methanol (here: around 7 $\text{€}_{2020}/\text{kg}_{\text{CGH2, 700 bar}}$ and 9 $\text{€}_{2020}/\text{kg}_{\text{LH2}}$) are higher than for gaseous or liquid hydrogen supply chains. Furthermore, it is more likely to use methanol directly as a (drop in) fuel or bulk chemical than cracking it back to elementary (molecular) hydrogen and releasing the carbon dioxide to the air with no further benefits.

- **Hydrogen supply using ammonia**. Similar to methanol, the handling of ammonia within the existing infrastructure is already realized at scale. Also, the demand of "green" ammonia will most likely emerge in the future leading to the build-up of a no regret ammonia infrastructure, which could thus also be used for a transportation of hydrogen in the form of ammonia. Nevertheless, due to the toxicity of ammonia the handling is challenging and demanding, while the carrier molecule (nitrogen) is rather cheap. Again, the hydrogen reconditioning (ammonia cracking) energy demand as well as the very high hydrogen purification losses weights on the overall hydrogen supply costs. Especially the hydrogen losses lead not only to higher hydrogen production quantities but also ammonia synthesis, storage, transportation and cracking costs, and therefore, the advantage of the high volumetric energy density of the bound hydrogen in liquefied ammonia is reduced. Overall, hydrogen supply costs by using ammonia as a carrier (here: at least 7 €$_{2020}$/kg$_{CGH2,\,700\,bar}$ and 9 €$_{2020}$/kg$_{LH2}$) are higher than for gaseous and liquid hydrogen supply chains as well as for methanol-based hydrogen supply chains. A cost reduction of the hydrogen supply using ammonia could be expected if the hydrogen purification losses after the ammonia cracking could be lowered significantly, e.g., due to the use of advanced membrane technology.

5   Conclusion and Outlook

The setup of an onsite, regional, national and international hydrogen supply infrastructure depends strongly on the development of the consumer's hydrogen demands. The final use of hydrogen is expected to dictate the conditioned hydrogen storage and transportation options too. If carbon-based fuels or ammonia are the demanded (final) product/energy carrier it is economically more reasonable that these products will be synthesized locally at the hydrogen production site and subsequently stored and transported as the final product. However, it is possible that elementary (molecular) hydrogen will be demanded in the future in larger quantities by sectors like steel production, aviation, heavy road transportation or back up electricity generation. Therefore, the production of low-cost hydrogen in regions like the Middle East and Northern Africa or in countries such as Australia, Namibia and Chile and consecutive conditioned hydrogen export to hydrogen demand regions such as in the USA, the European Union and selected Asian countries can be expected in the future.

Since the supply of gaseous hydrogen per pipeline transporting compressed gaseous hydrogen show in general the lowest specific (compressed) gaseous hydrogen supply costs for distances up to 3,000 km [28, 30–34], it seems most likely that compressed gaseous hydrogen pipeline networks will emerge in the future (if there is a significant elementary hydrogen end use demand in the future). Today, all compressed gaseous hydrogen pipelines combined cover in sum only about 5,000 km, mostly for the local compressed gaseous hydrogen transportation of (so-called gray) hydrogen at industrial sites like refineries and chemical plants. The European hydrogen backbone studies [33, 47] sketches that until the year 2030 the first transnational compressed gaseous hydrogen pipeline network will exist in the Netherlands, Belgium and the Western part of Germany leading to a compressed gaseous hydrogen pipeline backbone within the overall European Union in the year 2040 covering in total 28 countries. To have a reliable and economically transnational green compressed gaseous hydrogen pipeline network low-cost compressed gaseous hydrogen storage options like salt caverns are crucial to achieve this [15]. So far, a few compressed gaseous hydrogen salt caverns are in operation, which are located in Great Britain and the USA, while several projects are currently under development in the Netherlands, in Germany, in Sweden, in France and in Great Britain.

In addition to compressed gaseous hydrogen import and supply per pipelines, the seaborne conditioned hydrogen import is a further possibility to supply conditioned hydrogen over longer distances with a higher flexibility. An import of conditioned hydrogen is expected in the case of significantly lower renewable energy costs in the export countries and limited land availability in the import country. Liquid hydrogen seems to be the most promising option in the long term for a seaborne-based conditioned hydrogen import. In comparison with the material-based conditioned hydrogen options like LOHCs, methanol and ammonia, the main conditioning effort (electricity demand for liquefaction) occurs at the hydrogen production site/in the hydrogen production region with potentially low energy/electricity costs while in the case of the material-based conditioned hydrogen options the main conditioning energy demand for the hydrogen release occurs in the import country (with in general higher energy costs). The first international liquid hydrogen transportation per ship was demonstrated in the year 2022 between Australia and Japan [48]. Several more projects for liquid hydrogen transportation per ship are discussed nowadays but have not been realized so far.

In terms of specific well-to-tank hydrogen supply costs as an assessment parameter, hydrogen should always be transported and stored in the physical (molecular) state along the supply chain as it is demanded by the consumer in the end. Meaning, if (compressed) gaseous hydrogen is needed, compressed gaseous hydrogen supply chains obtain the lowest compressed gaseous hydrogen supply costs and likewise for liquid hydrogen supply chains if liquid hydrogen is needed. Furthermore, the concept of the use of material-based carriers (LOHCs, methanol or ammonia) for an elementary (molecular) hydrogen supply appears contra productive. Hydrogen is seen as a renewable energy carrier which should be produced at locations/in regions with low-cost electricity based on renewable energies and be transported to locations/regions, where the cost of electricity from renewable sources of energy is rather (relatively) high. In the case of the chemical bond to carrier molecules, there is always a very high (thermal) energy demand for the reconditioning (typically greater than 30% of the bound hydrogen energy), and this high energy demand occurs at/in the hydrogen consumption location/region with potentially high energy costs where additionally most likely no high-temperature waste heat is available within a widely defossilized industrial environment.

Even though the physical conditioned hydrogen supply chains are favorable in terms of costs in the long-term, material-based conditioned hydrogen supply chains (e.g., using methanol or ammonia) can be promising in the short term for a ramp-up. Due to the high investments into a compressed gaseous and liquid hydrogen infrastructure and the uncertainty of the development of the elementary (molecular) hydrogen demand in the future also the (investment) risk is high. Therefore, a no regret infrastructure is important to lower the financial risk. In the case of ammonia and methanol used as hydrogen carriers, the risk can be reduced significantly. For the future, the projected demand for "green" ammonia and "green" methanol needed as a bulk chemical substituting a fossil fuel-based use of ammonia and methanol is high. Hence, even if no elementary (molecular) hydrogen is needed in the future by the (end) consumer, ammonia and methanol (if no carbon dioxide recycle loop is considered) could be used as bulk chemicals anyway. Furthermore, a corresponding methanol and ammonia transportation infrastructure is already existing. Because of these facts, several projects for the export of ammonia [49] and methanol [50] have been announced recently (end of the year 2023) and are partly close to the start of construction.

In the case of LOHCs, although the existing infrastructure can be used for the LOCH transportation, the carrier (i.e., the unloaded LOHC) needs to be used again at the hydrogen production site. Consequently, it is only possible to supply elementary (molecular) hydrogen with the concept of LOHC and no further derivatives as in the case of ammonia and methanol (if no carbon dioxide recycle loop is considered). The first international seaborne transportation and hydrogen release using LOHCs took place in the year 2020 from Brunei to Japan [51]. Several further hydrogen supply projects using LOHCs are currently ongoing in Europe. However, the costs for an elementary (molecular) hydrogen supply is in most scenarios higher for the LOHC concept than for compressed gaseous and liquid hydrogen supply chains in the long term and also higher in comparison with using ammonia and methanol as a hydrogen carrier. Furthermore, in comparison with ammonia and methanol a no regret infrastructure (regarding the hydrogenation unit) is not given. Thus, a global emerging hydrogen supply chain based on LOHCs seems to be unlikely from today's (2024) perspective.

References

1. Kaltschmitt M, Özdirik B, Reimers B, Schlüter M, Schulz D, Sens L (2020) Stromerzeugung aus Windenergie. In: Kaltschmitt M, Streicher W, Wiese A (eds) Erneuerbare Energien. Springer Berlin, Heidelberg, pp 461–582
2. Bründlinger R, Christ D, Fechner H, Kaltschmitt M, Müller J, Peharz G et al. (2020) Photovoltaische Stromerzeugung. In: Kaltschmitt M, Streicher W, Wiese A (eds) Erneuerbare Energien. Springer, Berlin, Heidelberg, pp 339–460
3. Sens L, Neuling U, Kaltschmitt M (2022) Capital expenditure and levelized cost of electricity of photovoltaic plants and wind turbines—Development by 2050. Renew Energy 185:525–537. https://doi.org/10.1016/j.renene.2021.12.042
4. IEA (2020) Global hydrogen demand by sector in the Sustainable Development Scenario, 2019–2070, Paris
5. Deloitte (2019) Australian and global hydrogen demand growth scenario analysis Australian and global hydrogen demand growth scenario analysis
6. Fraunhofer ISI (2019) Study on the opportunities of "Power-to-X" in Morocco. Karlsruhe
7. IRENA (2021) World energy transitions outlook: 1.5 °C pathway. Abu Dhabi
8. BloombergNEF (2020) Hydrogen economy outlook
9. Energy Transition Commission (2021) Making the hydrogen economy possible: accelerating clean hydrogen in an electrified economy

10. Robinius M (2020) Wege für die Energiewende: Kosteneffiziente und klimagerechte Transformationsstrategien für das deutsche Energiesystem bis zum Jahr 2050. Forschungszentrum Jülich GmbH Zentralbibliothek, Verlag, Jülich
11. Dena (2018) Integrated energy transition
12. Becker Büttner Held (2018) Rechtliche Rahmenbedingungen für ein integriertes Energiekonzept 2050 und die Einbindung von EE-Kraftstoffen
13. Prognos (2021) Klimaneutrales Deutschland: In drei Schritten zu null Treibhausgasen bis 2050 über ein Zwischenziel von -65% im Jahr 2030 als Teil des EU-Green-Deals
14. Fraunhofer ISI. Eine Wasserstoff-Roadmap für Deutschland. Karlsruhe; 2019.
15. Sens L, Piguel Y, Neuling U, Timmerberg S, Wilbrand K, Kaltschmitt M (2022) Cost minimized hydrogen from solar and wind—Production and supply in the European catchment area. Energy Convers Manage 265:115742. https://doi.org/10.1016/j.enconman.2022.115742
16. NIST (2022) NIST chemistry webbook, SRD 69: thermophysical properties of fluid systems. Available from https://webbook.nist.gov/chemistry/fluid/. Accessed 21 Aug 2022
17. Niermann M, Beckendorff A, Kaltschmitt M, Bonhoff K (2019) Liquid Organic Hydrogen Carrier (LOHC)—Assessment based on chemical and economic properties. Int J Hydrogen Energy 44(13):6631–6654. https://doi.org/10.1016/j.ijhydene.2019.01.199
18. Tremel A, Wasserscheid P, Baldauf M, Hammer T (2015) Techno-economic analysis for the synthesis of liquid and gaseous fuels based on hydrogen production via electrolysis. Int J Hydrogen Energy 40(35):11457–11464. https://doi.org/10.1016/j.ijhydene.2015.01.097
19. Sdanghi G, Maranzana G, Celzard A, Fierro V (2019) Review of the current technologies and performances of hydrogen compression for stationary and automotive applications. Renew Sustain Energy Rev 102:150–170. https://doi.org/10.1016/j.rser.2018.11.028
20. Reuß M, Grube T, Robinius M, Preuster P, Wasserscheid P, Stolten D (2017) Seasonal storage and alternative carriers: a flexible hydrogen supply chain model. Appl Energy 200:290–302. https://doi.org/10.1016/j.apenergy.2017.05.050
21. Parks G (2014) Hydrogen station compression, storage, and dispensing technical status and costs independent review. Unpublished
22. Nexant (2008) H2A hydrogen delivery infrastructure analysis models and conventional pathway options analysis results
23. VAKO GmbH & Co. KG (2023) Wasserstoffbehälter, Druckbehälter—VAKO GmbH & Co. KG. Available from https://www.vako.net/produkte-behaelter/. Accessed 14 Apr 2023
24. Krieg D (2012) Konzept und Kosten eines Pipelinesystems zur Versorgung des deutschen Straßenverkehrs mit Wasserstoff. Forschungszentrum, Zentralbibliothek, Jülich
25. ZSW (2023) Wasserstoff: Qualität. Available from https://www.zsw-bw.de/forschung/wasserstoff-efuels/themen/wasserstoff-qualitaet.html. Accessed 05 June 2023
26. Adolf J, Warnecke W, Karzel P, Kolbeck A, van der Made A, Müller-Belau J et al (2020) On route to CO_2-free fuels. Hydrogen—Latest developments in its supply chain and applications in transport. Unpublished
27. Flis G, Deutsch M (2021) 12 insights on hydrogen
28. Sens L, Neuling U, Wilbrand K, Kaltschmitt M (2021) Green" hydrogen for ground-based heavy-duty longdistance transportation—A techno-economic analysis of various supply chains. In: Liebl J, Beidl C, Maus W (eds) Internationaler Motorenkongress 2021. Springer, Wiesbaden, pp 283–299
29. Sens L, Neuling U, Wilbrand K, Kaltschmitt M (2022) Conditioned hydrogen for a green hydrogen supply for heavy duty-vehicles in 2030 and 2050—A techno-economic well-to-tank assessment of various supply chains. Int J Hydrogen Energy. https://doi.org/10.1016/j.ijhydene.2022.07.113
30. Niermann M, Timmerberg S, Drünert S, Kaltschmitt M (2021) Liquid organic hydrogen carriers and alternatives for international transport of renewable hydrogen. Renew Sustain Energy Rev 135:110171. https://doi.org/10.1016/j.rser.2020.110171
31. Hampp J, Düren M, Brown T (2021) Import options for chemical energy carriers from renewable sources to Germany

32. IEA (2022) Global hydrogen review 2021—analysis. IEA. Available from https://www.iea. org/reports/global-hydrogen-review-2021. Accessed 01 Feb 2022
33. Wang A, Jens J, Mavins D, Moultak M, Schimmel M, van der Leun K et al (2021) European hydrogen backbone: analysing future demand, supply, and transport of hydrogen
34. Ortiz-Cebolla R, Dolci F, Weidner E (2021) Assessment of hydrogen delivery options
35. Stolzenburg K, Mubbala R (2013) Integrated design for demonstration of efficient liquefaction of hydrogen (IDEALHY)
36. DNV (2020) Study on the import of liquid renewable energy: technology cost assessment
37. IRENA (2022) Global hydrogen trade to meet the 1.5 °C climate goal: Part 2. Technology review of hydrogen carriers. Abu Dhabi
38. IEA (2021) The future of hydrogen. Available from https://www.iea.org/reports/the-future-of-hydrogen. Accessed 07 July 2021
39. Hydrogenious LOHC Technologies (2022) Wasserstoff sicher und effizient transportieren: Bilfinger und Hydrogenious werden Partner - Hydrogenious LOHC Technologies. Available from https://www.hydrogenious.net/index.php/de/2021/10/13/bilfinger_de/. Accessed 07 June 2022
40. Reuß M, Grube T, Robinius M, Stolten D (2019) A hydrogen supply chain with spatial resolution: comparative analysis of infrastructure technologies in Germany. Appl Energy 247:438–453. https://doi.org/10.1016/j.apenergy.2019.04.064
41. Niermann M, Drünert S, Kaltschmitt M, Bonhoff K (2019) Liquid organic hydrogen carriers (LOHCs)—techno-economic analysis of LOHCs in a defined process chain. Energy Environ Sci 12(1):290–307. https://doi.org/10.1039/C8EE02700E
42. Dögnitz N, Costa G, Hauschild S, Meisel K, Etzold H, Nieß S et al (2022) Monitoring erneuerbarer Energien im Verkehr, 1st edn.
43. Sircar S, Golden TC (2000) Purification of hydrogen by pressure swing adsorption. Sep Sci Technol 35(5):667–687. https://doi.org/10.1081/SS-100100183
44. Andersson J, Grönkvist S (2019) Large-scale storage of hydrogen. Int J Hydrogen Energy 44(23):11901–11919. https://doi.org/10.1016/j.ijhydene.2019.03.063
45. Giddey S, Badwal SPS, Munnings C, Dolan M (2017) Ammonia as a renewable energy transportation media. ACS Sustain Chem Eng 5(11):10231–10239. https://doi.org/10.1021/acssuschemeng.7b02219
46. Du Z, Liu C, Zhai J, Guo X, Xiong Y, Su W et al (2021) A review of hydrogen purification technologies for fuel cell vehicles. Catalysts 11(3):393. https://doi.org/10.3390/catal11030393
47. Rossum R, Jens J, Guardia G, Wang A, Kühnen L, Overgaag M (2022) European hydrogen backbone
48. Pekic S (2022) Suiso Frontier brings world's 1st LH2 shipment to Japan. Offshore Energy 2022, 25 February 2022; Available from https://www.offshore-energy.biz/suiso-frontier-brings-worlds-1st-lh2-shipment-to-japan/. Accessed 02 May 02 2022
49. Saudi Arabia to start construction of $5 bln green hydrogen plant in Neom. Al Arabiya English 2022, 17 March 2022; Available from https://english.alarabiya.net/business/energy/2022/03/17/Saudi-Arabia-to-start-construction-of-5-bln-green-hydrogen-plant-in-Neom. Accessed 02 May 2022
50. TUN-OL. Sonnenenergie in Tunesien zu Methanol. Available from https://tun-ol.com/. . Accessed 02 May 2022
51. 'World's first international hydrogen supply chain' realised between Brunei and Japan. Recharge. Available from https://www.rechargenews.com/transition/-world-s-first-international-hydrogen-supply-chain-realised-between-brunei-and-japan/2-1-798398. . Accessed 02 May 2022

Worldwide Hydrogen Production Potential

Joshua Fragoso Garcia

Abstract Green hydrogen offers a potential for broad global availability, unlike fossil fuels concentrated in a few countries. It can be produced in regions with ample wind and solar resources, and desalinated water can support production in areas lacking freshwater. Long viewed as a clean fuel, hydrogen combustion produces no CO_2, with applications spanning the chemical and steel industries, transportation (e.g., fuel cell vehicles), energy storage, and heating. This chapter outlines various hydrogen production methods and examines pathways to low-carbon hydrogen, focusing on "blue" and "green" hydrogen. It discusses requirements and potential capacities for each type and assesses global and regional projections for green hydrogen by 2050, concluding with key insights into future production potential.

Keywords Hydrogen Potential · Blue Hydrogen · Green Hydrogen · Economic Potential

1 Introduction

Unlike fossil fuels, which have major deposits concentrated in a few countries, green hydrogen has the potential for a broader availability. It can be produced by many nations around the world, especially in those where wind and solar resources or their combinations are abundant. In areas where freshwater is scarce, water supply through desalination can be procured for hydrogen production.

For decades, hydrogen has been touted as a key fuel for the future, as its combustion does not lead to any CO_2 emissions. Its future demand may include multiple sectors: as a feedstock for the chemical and steel industries, as an energy source in the transport sector for fuel cell vehicles or in the form of its derivatives in the maritime and aviation sectors, as an energy storage medium for electricity generation, and even in the heating sector where it can be integrated into heating networks via combined heat and power units [1, 2].

J. Fragoso Garcia (✉)
Fraunhofer Institute for Systems and Innovation Research (Fraunhofer ISI), Karlsruhe, Germany
e-mail: joshua.fragoso.garcia@isi.fraunhofer.de

© The Author(s), under exclusive license to Springer Nature Switzerland AG 2025   323
N. Bullerdiek et al. (eds.), *Powerfuels*, Green Energy and Technology,
https://doi.org/10.1007/978-3-031-62411-7_13

Therefore, in this chapter, the diverse approaches to producing hydrogen in order to meet both current and future needs are described and the potential for producing low-carbon hydrogen is explored. First, a brief overview of the different hydrogen production pathways is given (Sect. 2). This is followed by a detailed examination of the two most commonly discussed types of hydrogen, namely "blue" and "green" hydrogen (Sect. 3). In Sect. 4, the requirements for the production of blue hydrogen are detailed, and its potential production capacities are discussed. In Sect. 5, the requirements for the production of green hydrogen are presented. Section 6 outlines the methodology used to assess green hydrogen potentials, while Sect. 6 provides a comparative assessment of the existing literature on global and regional green hydrogen potentials projected for 2050. The chapter concludes with a summary of the main findings.

2 Hydrogen Color Spectrum

Hydrogen can be produced based on a variety of different feedstocks and processes. This includes, for example steam methane reforming (SMR) from natural gas, water splitting through electrolysis or biomass respectively coal gasification [3]. To cover the global hydrogen demand (about 94 Mt in 2021), hydrogen is currently mostly produced through steam methane reforming (SMR) of natural gas or coal/coke gasification [4]. This hydrogen provision resulted in the emission of about 900 Mt/a of carbon dioxide (CO_2) into the atmosphere [4]. However, other low-carbon hydrogen production methods, such as electrolysis or the combination of steam methane reforming (SMR) of natural gas and carbon capture and sequestration or utilization (CCSU), are beginning not only to attract interest but to gain momentum worldwide. This is also reflected by the increasing number of countries starting low-carbon hydrogen production projects or establishing national strategies for its development [5–7].

The different options for hydrogen production have been commonly classified according to their feedstock and employed processes following a color scale. A selection of this hydrogen color spectrum is presented in Fig. 1 and described in the following.

- Gray Hydrogen. Gray hydrogen is produced based on natural gas through steam methane reforming (SMR). It is the most common process for hydrogen production today accounting for 70% of the global production [3]. The process consists of mainly two chemical reactions, a reforming reaction where methane (CH_4) is combined with water (H_2O) at high temperatures (700–1000 °C) to produce carbon monoxide (CO) and hydrogen (H_2) (Eq. 1) [8].

$$CH_4 + H_2O + heat \rightarrow CO + 3H_2 \qquad (1)$$

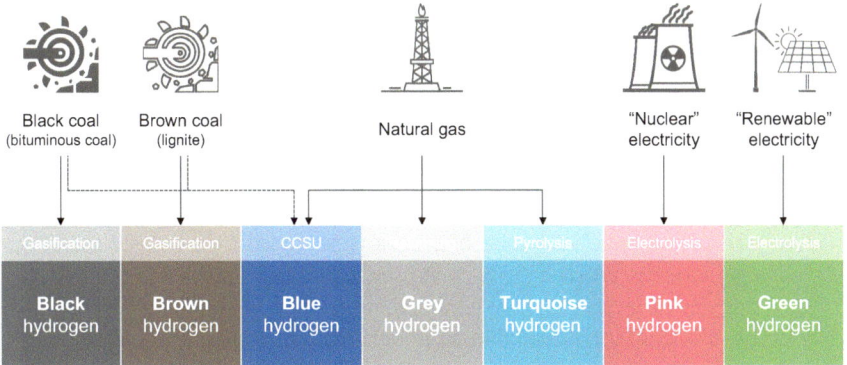

Fig. 1 Hydrogen color spectrum (CCSU: carbon capture and sequestration or utilization; image with courtesy of Bullerdiek/TUHH)

This is followed by a water-gas shift (WGS) reaction. Within this chemical reaction, the carbon monoxide (CO) is further oxidized with oxygen from water to CO_2 to achieve a higher hydrogen content (Eq. 2).

$$CO + H_2O \rightarrow CO_2 + H_2 \tag{2}$$

Additionally, other hydrogen colors are mentioned in the literature; this is true, for example, for white hydrogen (natural reservoirs). However, the colors explained above cover most of the current production (gray and black) and future expected hydrogen production (blue and green). As the future energy carriers need to be either CO_2-free or low-carbon, the main focus of current hydrogen research has been on the future use of blue hydrogen, green hydrogen, or a combination of both that may decrease CO_2 emissions. In the following section, these topics will be further elaborated on.

The produced CO_2 is typically released into the atmosphere during the production of such gray hydrogen from natural gas. The provision of 1 kg of gray hydrogen generates at least 10 kg of CO_2. Fugitive emissions, processing, and transport may increase the produced emissions [3].

- Black/Brown Hydrogen. Black or brown hydrogen is produced through the gasification of black coal, coke, or lignite (brown coal). Therefore, the coal/coke is converted at high temperatures together with water as a gasification agent to produce a syngas composed of hydrogen, carbon monoxide (CO) and carbon dioxide (CO_2). A subsequent water-gas shift reaction (Eq. 2) follows, producing additional hydrogen and CO_2. The production of 1 kg of black hydrogen, for instance, generates approx. 19 kg of CO_2 [9]. Similar to the case of gray hydrogen, other emission sources in the whole process chain like transport, process, or fugitive emissions may increase the carbon footprint [3].

- Blue Hydrogen. Blue hydrogen is produced using the same processes as gray, black, or brown hydrogen. However, a CCSU unit is added at the end of the value chain to capture CO_2 and prevent its release into the atmosphere. Between 70 and 95% of the CO_2 emissions can be prevented through the use of CCSU [10]. Nevertheless, carbon sequestration and/or use is still within its infancy. Blue hydrogen may be considered low-carbon according to the European hydrogen policy [11].
- Turquoise Hydrogen. Turquoise hydrogen is produced from natural gas through a process called methane pyrolysis. It uses high-temperature heat to separate the methane molecule into the two components hydrogen and solid carbon [10, 12] Because the carbon is available as a solid product (carbon black), no CO_2 is released into the atmosphere; nevertheless, this solid carbon need to be dumped safely to avoid oxidation and the subsequent CO_2 release over time. This process is still within the research phase and is expected to be commercially available around 2030–2035 [13].
- Green Hydrogen. Green hydrogen is produced through electrochemical water splitting (water electrolysis). In electrolysis, electricity is used to decompose water into hydrogen and oxygen. If the required electricity originates from renewable energy sources like solar radiation, wind, or geothermal energy, the production of green hydrogen produces no additional CO_2 emissions [10].
- Pink Hydrogen. Pink hydrogen is produced through electrochemical water splitting (water electrolysis) using electricity generated by nuclear power. Nuclear power reactors do not produce CO_2 emissions directly during their operation. Hence, the production of hydrogen using nuclear electricity would be considered low-carbon [14, 15].

3   Blue and Green Hydrogen

There is an ongoing debate on whether blue or green hydrogen will cover future hydrogen demand. One study considers a mixture of hydrogen sources in 2050 with 27.5% from fossil fuels with CCSU (blue), 25.5% from grid-connected electrolysis, as well as 17.5 and 13% from solar- and wind-based electrolysis (green), respectively. The remaining 1% comes from nuclear-based electrolysis (pink). The supply mix for the grid-connected electrolysis is considered low-carbon. Additionally, according to this study, by 2050 still 15% of black and gray hydrogen will be produced [16]. Another investigation estimates a global production of 530 Mt/a of hydrogen in the Net Zero Emissions Scenario in 2050. Hydrogen that is produced based on electricity has the largest share with 62%, followed by production based on fossil fuel resources and CCSU with 38% [17]. A third study elaborates on three different scenarios, blue, green, and mixed. It concludes that the mixture of both technologies may achieve a lower average hydrogen cost than either of the boundary scenarios [18].

Several countries consider both sources, blue and green hydrogen, in their plans to cover their future hydrogen demand. Canada contemplates different low-carbon hydrogen sources depending on their availability in each province. On the one hand, the Northwest Territories are expected to produce their hydrogen from wind energy. On the other hand, the provinces of Alberta and Saskatchewan may produce hydrogen based on fossil fuels with CCSU (blue) [12]. Similarly, the USA approach includes all their locally available resources. They consider the use of low-carbon options, renewable energy (green), nuclear (pink), and natural gas with CCSU (blue) to produce hydrogen [19]. Australia plans to use their potential for blue hydrogen, based on its vast fossil resources, and green hydrogen using its solar and wind resources [20] Green hydrogen is identified as the main source of hydrogen production for countries like Chile, Portugal, and Spain. Japan, South Korea, and the Netherlands consider the use of blue hydrogen as a bridge technology in the mid-term and a switch to green hydrogen afterward [7]. Current oil-producing countries have shown interest in becoming future hydrogen producers. Saudi Arabia contemplates the production of blue and green hydrogen. It has announced plans to build a green hydrogen plant that will produce 650 tons daily (about 237,000 tons a year). Additionally, it has also completed the first shipment of low-carbon ammonia, produced from blue hydrogen [21]. The United Arab Emirates (UAE) has recently signed a partnership to develop its national energy strategy [22]. The UAE has vast gas resources for the production of blue hydrogen and low-cost solar potential to produce green hydrogen [23].

In summary, both blue and green hydrogen are expected to play significant roles in a future decarbonized world, as suggested by the above studies and the strategies of several countries. Furthermore, the importance of blue hydrogen as a bridge technology cannot be overstated.

- Firstly, the use of green hydrogen requires a large escalation of the global renewable installed capacity.
- Secondly, the newly installed renewable capacity may be better used to decarbonize the power generation system itself first. The decarbonization factor of substituting one generation unit of electricity based on fossil fuel resources by electricity based on renewable sources of energy may be higher than the production of green hydrogen [24].
- Thirdly, the use of blue hydrogen may take advantage of the current natural gas infrastructure where the storage capacities for CCSU are available [10]. It may also enable the construction of further hydrogen infrastructure that may be used for green hydrogen later.

In summary, blue hydrogen serves as a bridge technology due to the need for expanded renewable capacity, the priority of power generation decarbonization, and the advantage of using existing gas infrastructure, which can later accommodate green hydrogen.

4   Blue Hydrogen

Blue hydrogen is produced by combining, e.g., steam methane reforming (SMR) technology from gray hydrogen with a CCSU unit. Thus, blue hydrogen has the advantage that it allows for further use of the currently existing gas infrastructure and represents an opportunity to lower climate change impacts related to current (fossil-based) hydrogen production. To fulfill this challenge, the integration of the CCSU units is crucial. Even though there are several blue hydrogen projects currently operational (the oldest one was installed in 1982 [25]), the CCSU technology is still developing and considered not fully mature [26]. The combination of SMR together with CCSU has achieved a Technology Readiness Level (TRL) of 8 [10].

Another critical factor of blue hydrogen is the long-term storage of the associated CO_2. The chosen storage option has to be safe, environmentally sustainable, and cost-effective [27]. Geological formations are considered for this purpose.

- Saline formations. Saline formations are formed of deep, porous rock that might contain trapped fluid in it, e.g., high salt/mineral content water. The CO_2 will be stored in the water, eventually dissolving and becoming part of the rock. Saline formations are found in various locations around the world [28].
- Oil and natural gas reservoirs. For decades, CO_2 has been pressed into crude oil and natural gas reservoirs to enhance the recovery rate; this is also known as enhanced oil recovery (EOR). It increases the pressure of the oil/gas field allowing an increase in its extraction [29]. Based on these experiences, the approach is to store CO_2 within existing crude oil and natural gas reservoirs.
- No longer productive coal seams. In this case, CO_2 may be injected into the coal seam where it would be absorbed by the coal, storing it permanently unless the coal is mined. The injection of CO_2 may also increase the production of methane from the coal seams [28].

In addition to the unsolved problem of long-term CO_2 storage, the production of blue hydrogen requires a significant amount of water. This water requirement can be divided into different types of demand [30]. Firstly, there is a stoichiometric water requirement, which amounts to 4.5 L_{H_2O}/kg_{H_2}. However, in real processes, there can be a slightly super-stoichiometric water requirement. For instance, a case study found that up to 5.9 L_{H_2O}/kg_{H_2} are required. Another source reports a value of 6 L_{H_2O}/kg_{H_2} [25]. In addition to the water demand for the actual reaction/conversion, further water requirements can arise, depending on the specific process design, e.g., for cooling to keep the system running efficiently. Water demands of 38 L_{H_2O}/kg_{H_2} are reported for system cooling.

Other sources report different water consumptions depending on the water that may be recovered during the process. One study identifies a total water consumption of 24 L_{H_2O}/kg_{H_2} [31]. A different source estimates a water consumption between 18 and 44 L_{H_2O}/kg_{H_2} [32].

The potential availability of blue hydrogen mainly depends on the availability of two resources:

- fossil fuels,
- storage options for CO_2 within a CCS concept.

The main fossil fuel currently used to produce hydrogen is natural gas (gray hydrogen). The potential and the costs of using this resource for the production of blue hydrogen will be analyzed in the following section.

4.1 Blue Hydrogen from Natural Gas

The global natural gas reserves in 2020 are estimated to amount to approximately 188 trillion m^3N [33]. The top ten countries with the highest natural gas resources are Russia, Iran, Qatar, Turkmenistan, the USA, China, Venezuela, Saudi Arabia, the United Arab Emirates, and Nigeria. Their cumulated natural gas resources amount to more than 83% of the known global reserves, totaling 155 trillion m^3N. Hence, they could play a major role in blue hydrogen production. Considering a net factor conversion of natural gas to hydrogen between 59 and 76%, the worldwide blue hydrogen potential would be between 1,573 and 1,221 PWh_{H_2}, if all the natural gas reserves were converted into blue hydrogen [34, 35]. The top ten countries show a total blue hydrogen potential between 1,304 and 1,012 PWh_{H_2}. Figure 2 exhibits the overall potential for the top 10 countries for both high and low efficiencies.

The second factor to be considered is the CO_2 storage potential. The worldwide porous geological storage is estimated between 6,880 and 29,511 billion t of CO_2 [25]. It is distributed among 19 countries and two world regions (Europe and the

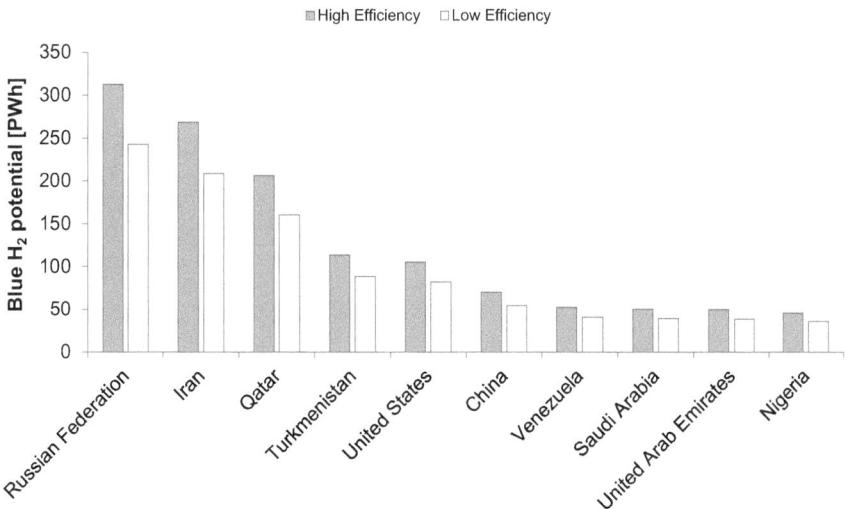

Fig. 2 Blue hydrogen potential for the top ten countries with the largest natural gas reserves. A conversion factor between 59 and 76% was used to calculate the hydrogen potential

Table 1 Blue hydrogen cost prognosis from different sources

Source	Blue hydrogen cost [€/kg$_{H_2}$]
The future of hydrogen IEA [3]	1.22–2.10
Frederick et al. [37]	2.60
Noussan et al. [10]	1.14–2.47
Oni et al. [37]	2.36
IRENA 2020 [38]	1.0–2.00

North Sea). The USA possesses the highest storage potential which is estimated to be in the range between 2000 and 21,000 billion t of CO_2.

Although the potential of geological storage does not match geographically with many of the potential blue hydrogen producers, the natural gas producer may use enhanced oil recovery (EOR) to store the produced emissions. The potential to store CO_2 through EOR is estimated to be between 60 and 360 billion tons of CO_2 for the next 50 years [36]. The largest storage potential lies in Russia, Saudi Arabia, United Arab Emirates, Iraq, Iran, Libya, Algeria, Kazakhstan, and Azerbaijan.

The upper bound of the CO_2 storage requirement, taking into account a production of 1,573 PWh of blue hydrogen by converting all of the remaining global natural gas reserves, can be determined by assuming a capture rate of 100% and a production rate of 10 kg of CO_2 per kg of hydrogen. Using these parameters, the total CO_2 storage requirement is about 471 billion t of CO_2.[1] Therefore, the need for CO_2 storage does not appear to be a limiting factor for the blue hydrogen production potential.

4.2 Blue Hydrogen Cost

Natural gas is the main feedstock for the production of blue hydrogen. The cost of natural gas amounts to a share between 45 and 75% of the blue hydrogen production cost. The levelized cost of hydrogen (LCOH) for blue hydrogen has a good consensus in the literature, with a lower limit of 1 €/kg$_{H_2}$ and an upper boundary of 2.6 €/kg$_{H_2}$. A selection of blue hydrogen cost estimations found in different references is shown in Table 1.

5 Green Hydrogen

Green hydrogen is produced by an electrolysis process that uses electricity from renewable energies. Thus, the sources and technologies to provide electricity to be used for the production of green hydrogen are exemplarily outlined below.

---

[1] A value of 33.33 TWh per Mt of hydrogen is considered.

- solar photovoltaic (PV),
- concentrated solar power (CSP),
- wind onshore and wind offshore,
- hydropower,
- geothermal.

One of the inherent requirements of electrolysis is water. The stoichiometric demand is 9 L_{H_2O}/kg_{H_2}. However, further water may be required to convert tap water into deionized water [32]. Nevertheless, the electrolysis itself is not the only process that demands water. Renewable energy sources, like PV, require water for cleaning. CSP needs water for cleaning, for the steam cycle and cooling purposes. Table 2 provides a range for the water requirements of the different processes within a CSP plant [39].

Water requirements of CSP vary greatly depending on the type of technology used. On the one hand, if a dry cooling system is implemented, no water is required for the cooling system. On the other hand, the use of a wet cooling system requires more than 3,400 L_{H_2O}/MWh_{el}. However, it shall be mentioned that water may be partially recovered and reused. A total of 3,785 L_{H_2O}/MWh_{el} is required for the whole process. Other sources consider a requirement of around 4,300 L_{H_2O}/MWh_{el} [40, 41].

As PV has no cooling or steam requirements, the water it consumes is mainly for cleaning purposes. The amount of cleaning depends on the dust and dirt accumulation on the PV modules. The accumulation of soiling is location and installation-specific, depending on meteorological conditions, tilt angle, and proximity to soiling sources, like factories or deserts. Therefore, the amount of water required for cleaning varies from site to site [25]. An average value might be about 22 L_{H_2O}/MWh_{el} [42].

Hence, considering an electrolyzer efficiency of 65%, the amount of additional water required per kg of hydrogen using a CSP plant will be between 9.7 and 193 L_{H_2O}/kg_{H_2}. For PV, the value is minimum as it will just add 1.1 L_{H_2O}/kg_{H_2} to the electrolysis needs.

Table 2 Concentrated solar power (CSP) water requirements per process [39]

Process	Minimum water needed [L_{H_2O}/MWh_{el}]	Maximum water needed [L_{H_2O}/MWh_{el}]
Cleaning	75.7	151.4
Steam cycle	113.6	227.1
Cooling system	0.0	3,406.5
Total	**189.3**	**3,785**

5.1 Green Hydrogen Production Potential Calculation

The potential to produce green hydrogen depends on different factors, including—among others—renewable energy availability and its cost, water availability, and the type of electrolyzer. Green electricity is the key factor for the production of green hydrogen. Hence, the green hydrogen production potential deeply depends on the potential and availability of renewable sources of energy. Green hydrogen production potentials may be categorized in a similar way as the renewable energy potential. The categories are as follows [43–45]:

1. Theoretical potential. This refers to the total energy content of wind or solar energy available in a specific region or globally. It serves as the maximum conceivable energy that can be harvested from these sources and sets the upper boundary for possible energy production, without considering technical or economic constraints.
2. Geographical potential. This pertains to the areas of land that are suitable and accessible for setting up wind or solar installations, for example. It considers the spatial dimensions of a region and identifies locations where infrastructure for harnessing renewable energy can be established, without yet considering technical limitations or environmental constraints.
3. Technical potential. This refers to the power that can be generated within the accessible areas, taking into account the limitations posed by (current) technology. It assesses how the state-of-the-art equipment and methods would perform in the identified locations, factoring in, for example, efficiencies and other technical parameters to determine an energy output.
4. Economic potential. This refers to the portion of the technical potential that can be realized in a cost-effective manner. In other words, it represents the amount of energy that can be produced at a competitive levelized cost of hydrogen (LCOH). It factors in financial considerations and investment returns to determine the feasibility of energy production from an economic perspective.
5. Market potential. This potential represents the fraction of the economic potential for which there is an actual demand or off-taker/buyer in the market. It considers the balance and opportunity costs between selling green electricity directly or converting and selling it as green hydrogen. Not all economically viable hydrogen production will necessarily find a buyer, so this potential provides an estimation of real-world adoption and consumption rates [45].

So far, there are just a handful of global hydrogen potentials which are described below. Africa is a continent that is deeply studied. Therefore, African hydrogen potentials are considered as well [46–49]. A study for Europe is added as the global studies partially exclude Europe, as it has been identified as a hydrogen demand center [50, 51]. The following subchapters review the methodologies of the aforementioned global and regional hydrogen potential calculations across different studies.

5.1.1 Land Utilization (Geographical Potential)

The available land to install renewable energies heavily influences the total generation potential [52]. Several studies restrict the use of natural protected areas and wetlands for renewable energy sources. A study from IRENA excludes additionally the use of forest land for both PV and onshore wind. Besides this, croplands are excluded only for PV, however, croplands natural being a mixture of croplands and herbaceous vegetation is excluded by just 60%. A terrain slope limitation is also considered, 20% for wind and 5% for PV. Urban areas (including rooftop PV) are completely excluded. Offshore wind is limited to a distance of 5 km from the coast and 40 m depth. The use of urban, water bodies, and wetlands is restricted as well [45]. Lux et al. [49, 50] utilize a factor, representing the percentage of land that is allocated for renewable energy utilization per technology [49, 50].

Heuser et al. [53] used a GIS-based model that restricts the available land based on technical, environmental, and social availability criteria. Similar to other studies, the use of water, natural protected, and urban areas are restricted. Besides this, they limit their study to locations with high energy resources, wind and irradiation, and countries/regions with coastal access to supply the necessary water [53].

Pfennig et al. [54] use a Boolean superposition of several exclusion criteria. The criteria may be classified as general, economic, and PTX-specific. In general criteria, land use like forests, urban (> 50 inhabitants/km^2), cropland, water bodies as well as snow and ice are completely excluded. On the economic part, all regions with a higher LCOE than 40 €/MWh for wind and 30 €/MWh for PV are excluded. Finally, the PTX-specific exclusion considers the distance to ports, pipelines, and inland water sources. Additionally, the region has to have a low water stress level to be considered [42]. Mukelabai et al. [48] use a similar set of factors as IRENA as their study is based on one of their previous studies [47].

5.1.2 Renewable Energy Sources (Technical Potential)

Wind onshore and offshore, photovoltaic, and the mixture of wind and photovoltaic (PV) are the predominant technologies considered to provide electricity for the production of green hydrogen. However, different studies consider other technologies like concentrated solar power (CSP), tidal energy, bioenergy, hydropower or geothermal energy depending on the availability at each location. Two different paths are followed—modeling and literature review.

- IRENA [45] considers wind onshore, wind offshore, and PV. They utilize weather data with an hourly resolution of 31×31 km^2 and make a characterization for each technology via hourly capacity factors.
- Pfenning et al. [54] calculate hourly time series for wind, PV, and hybrid systems. They assume four different wind configurations with two different hub heights (180 and 130 m) and two specific area power values (250 and 300 W/m^2). For the photovoltaic systems, optimal tilt and orientation angles are used.

- Heuser et al. [53] calculate hourly time series for both wind and PV. A basic wind turbine design with a hub height of 120 m, rotor diameter of 136 m, and rated power of 4.2 MW (289 W/m^2) is selected. The synthetic power curve is calculated for this wind turbine and modified to consider different hub heights according to the site characteristics and wind class. A photovoltaic module efficiency of 24% in 2050 is assumed.
- Fasihi et al. [55] determine hourly time series for both fixed tilted and single-axis tracking photovoltaic. A 3 MW wind turbine with a hub height of 150 m is assumed.
- Lux et al. [49, 50] calculate hourly resolution profiles. The wind turbine is selected through an optimization process for each location. The PV systems are calculated using optimal tilt and orientation angle. A 19% efficiency of PV is assumed in 2050.

5.1.3 Economic Potential

The main costs for the production of hydrogen are the electricity needed for the electrolyzer and the costs for the electrolyzer itself. Current literature, that has calculated hydrogen production potentials, global, and regional, focuses mainly on three different technologies for water electrolysis; alkaline, proton exchange membrane (PEM), and solid oxide electrolyzer (SOEC). The current costs of such electrolyzers range between 657 and 1,534 €/kW$_{el}$ [56]. However, a cost reduction is expected in the medium and long terms. Several technological improvements are assumed for the year 2050 that would affect the development of electrolyzer technologies. Some of them are outlined below [38].

- Economies of scale. The escalation of the projects to a gigawatt (GW) scale. Currently, the largest installed project of PEM technology has a 20 MW rated power in Canada. However, several GW projects are currently in the pipeline.
- Stack size increase. A large part of the costs comes from the balance of plant and not from the electrolyzer stack. Therefore, the increase in the stack size would further reduce costs.
- Material efficiency. The reduction of rare earth material used within the electrolyzers, especially iridium, would lead to a price reduction. Similarly, the substitution or redesign of the platinum plates would decrease the cost and electricity consumption.

Figure 3 shows the expected cost range taken from the literature for the three electrolyzer technologies in 2050. PEM has the lowest expected costs in an optimistic scenario (175 €/kW$_{el}$); however, in a pessimistic scenario, it has higher costs than alkaline technologies. The SOEC electrolyzers show the highest cost range of the three electrolyzer technologies, with the highest cost of 877 €/kW$_{el}$. Alkaline technologies exhibit the lowest average expected cost at 538 €/kW$_{el}$.

The technology improvements will not only be reflected in the costs; the efficiency is expected to increase in 2050 for all technologies as well. All three technologies

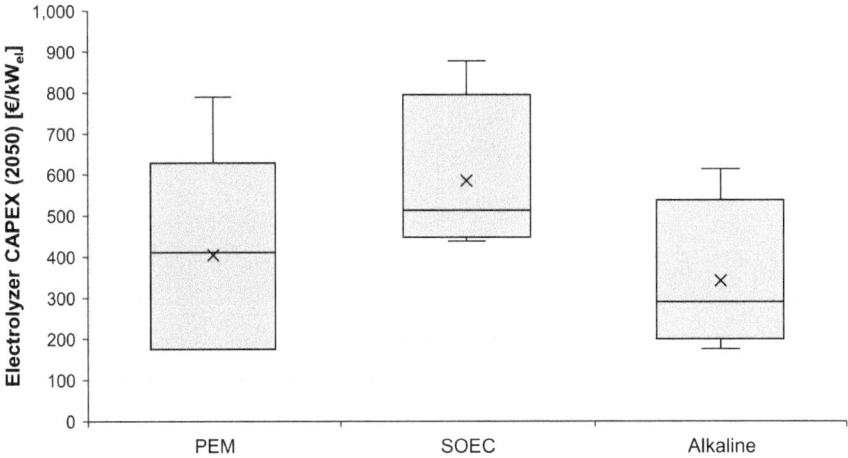

Fig. 3 Expected electrolyzer capital expenditures (CAPEX) in 2050 (self-elaboration with information from [3, 49, 53, 54, 57]; PEM: proton exchange membrane; SOEC: solid oxide electrolyzer cell)

exhibit efficiencies above 66%, which is the floor for PEM. SOEC exhibits the highest expected efficiency with up to 90%. Figure 4 shows the expected efficiencies in the year 2050. The higher efficiency of the electrolyzers would naturally lead to a higher production and therefore a lower levelized cost of hydrogen (LCOH).

Finally, the weighted average cost of capital (WACC) is used to determine the risks. Several studies used a single WACC for their calculation. IRENA uses a WACC

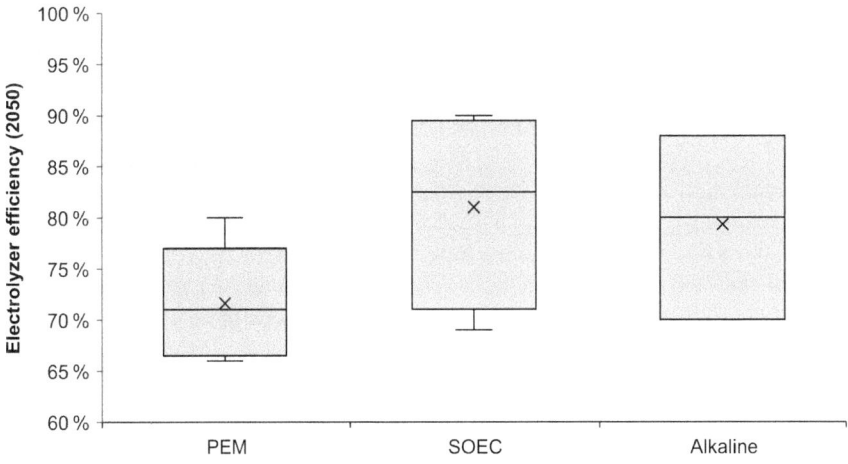

Fig. 4 Expected electrolyzer efficiencies in 2050 (self-elaboration with information from [1, 36, 42, 43, 48]; PEM: proton exchange membrane; SOEC: solid oxide electrolyzer)

per studied region [41]. Other investigations used a uniform WACC for all studied regions. Table 3 summarizes the different factors used in the literature.

The different assumptions used in the aforementioned studies lead to different results in the global or regional hydrogen production potentials. In the following subchapter, the calculated potentials are compared.

5.2   Green Hydrogen Production Potentials in 2050

The different methodologies lead to distinctive hydrogen production potentials. Hydrogen production potentials are explored on a global scale and/or regional scale. The levelized cost of hydrogen (LCOH) is a decisive factor. The LCOH provides a limit to quantify the hydrogen potential for example 30 TWh under 3 €/kg_{H_2}.

5.2.1   Green Hydrogen Production—Global Potentials

Figure 5 illustrates total green hydrogen production potentials in 2050 considering PEM electrolyzer and gaseous hydrogen.

Among the green hydrogen production potentials (techno-economic) presented in Fig. 5, IRENA [45] offers the highest estimates. Their optimistic scenario suggests a total potential of about 1,542 PWh, limited to those with a levelized cost of hydrogen (LCOH) below 1.75 €/kg_{H_2} (or 2.00 USD/kg_{H_2}). For its pessimistic scenario, IRENA factors in a higher weighted average cost of capital (WACC) for both technologies and countries, leading to a reduced estimate of 624 PWh. In contrast, two other studies depicted in the same graph provide considerably lower green hydrogen potentials, with values of about 116 PWh and 52.5 PWh, respectively [53, 54].

5.2.2   Green Hydrogen Production—Regional Potentials

A regional split is necessary to obtain a more complete depiction of the global green hydrogen potential. Figure 6 illustrates the ranges for the hydrogen production potential (techno-economic) for the different world continents.

Africa has the highest potential for green hydrogen with an average of 182 PWh. It's closely followed by the Americas with 99 PWh and Oceania with 93 PWh. Asia has an average potential of 67 PWh, while Europe has the lowest potential at 23 PWh. However, an IRENA report cites a remarkably high value for Europe of 95 PWh, which effectively raises its average [45].

Going deeper, Africa's potential peaks at 611 PWh [45]—a significant difference from its minimum of 23 PWh [54]. This large difference may be due to the inclusion of water restrictions and land exclusion criteria. Asia's potential ranges from a high of 164 PWh [45] to a low of 5 PWh. The latter could be due to the strict selection of only the best renewable potentials.

Table 3 Assumptions summary from different studies of hydrogen potential (WACC: weighted average cost of capital)

Source	Renewable energy sources (RES) considered	Renewable energy sources (RES) technical factors	Land utilization	Electrolyzer characteristics and WACC	Water restrictions
IRENA: Global hydrogen to meet the 1.5 °C Climate Goal [45]	• Wind onshore • Wind offshore • PV	• Year: 2030, 2050 • Hourly resolution and 31 × 31 km^2 spatial resolution; • Weather data ERA5 2018 • Characterization via hourly capacity factor profiles; • Power density of 45 MW/km^2 for PV,5 MW/km^2 for onshore wind and 7.43 MW/km^2 for offshore wind	• Excluded for PV: crops, forest and urban, croplands natural 60% • Excluded for wind: forest • Terrain slope: 20% for wind and 5% for PV • Wind offshore: 5 km from the coast and 40 m depth	• Individual WACCs per country and technology	• Water-stressed areas are considered and excluded

(continued)

Table 3 (continued)

Source	Renewable energy sources (RES) considered	Renewable energy sources (RES) technical factors	Land utilization	Electrolyzer characteristics and WACC	Water restrictions
Global GIS-based potential analysis and cost assessment of Power-to-X fuels in 2050—Pfennig et al. [54]	• Wind • PV • Hybrid (wind and PV)	• Year:2050 • Hourly resolution • Weather data from ERA5 using five historical weather years • Wind considers two different hub heights (180 m and 130 m) and specific area power (250 W/m² and 300 W/m²). A synthetic power curve is generated for each variation • Optimal tilt, orientation angles, and temperature are considered for PV	• Three different exclusion criteria • General: forests, urban, cropland, water bodies, and snow and ice are excluded • Economic: regions with higher LCOE than 30 Euros/MWh and 40 Euros/MWh for PV and wind respectively are excluded • PTX-specific: distance to ports, pipelines, and inland water sources • Wind offshore till 25 m is considered	• PEM and SOEC are considered	• Both fresh and desalinated water are considered • Regions need to have a low water stress level to be considered

(continued)

Table 3 (continued)

Source	Renewable energy sources (RES) considered	Renewable energy sources (RES) technical factors	Land utilization	Electrolyzer characteristics and WACC	Water restrictions
Worldwide Hydrogen Provision Scheme Based on Renewable Energy- Heuser et al. [53]	• Wind • PV	• Year:2050 • Hourly resolution for both technologies • MERRA as weather data together with the data from the wind atlas and the global solar atlas • Selected a basic wind turbine design, with a hub height of 120 m, rotor diameter of 136 m, and rated power of 4.2 MW (289 W/m^2) • Calculated a synthetic power curve to perform the simulations • Choice of a single PV module for all the simulations. Assuming a 24% efficiency for 2050 • Optimal tilt, orientation angles, and temperature are considered for PV	• GIS-based system to define restrictions • Water, natural protected areas, and urban are restricted • Locations with high energy available and close to the coast	NA	• Desalinated water is considered

(continued)

Table 3 (continued)

Source	Renewable energy sources (RES) considered	Renewable energy sources (RES) technical factors	Land utilization	Electrolyzer characteristics and WACC	Water restrictions
Long-term hydrocarbon trade options for the Maghreb region and Europe [46]—Fasihi et al	• Hybrid PV and wind plants	• 2030 and 2040 • Hourly resolution • Fixed tilted or single-axis tracking photovoltaic • Weather data from NASA • 3 MW wind turbine with a hub height of 150 m is considered	NA	• WACC: 7%	• Desalinated water is considered

(continued)

Table 3 (continued)

Source	Renewable energy sources (RES) considered	Renewable energy sources (RES) technical factors	Land utilization	Electrolyzer characteristics and WACC	Water restrictions
Renewable hydrogen economy outlook in Africa [47]—Mukelabai et al	• PV • Wind • CSP • Bioenergy • Hydroenergy • Geoenergy • Tidal	• Year:2050 • Based on literature, mainly IRENA report on Africa's renewable energy potentials • Assume different capacity factors for each technology • Considers the electricity necessary for the electrification in Africa	Factors taken from IRENA [47]	NA	• Water availability for potentials is explored
Supply curves of electricity-based gaseous fuels in the MENA region [48]	• PV • Wind onshore • Wind offshore • CSP	• Year:2050 • Real weather data • Hourly resolution • 6.5 × 6.5 km2 tile resolution • Wind turbine selection is optimized through a huge variety • 19% efficiency PV in 2050	• Land use allocated in 6.5 × 6.5 km^2 tiles • A use factor is given to each land use per technology	• WACC: 7% and 12% • PEM and SOEC are considered	• Desalinated water is considered

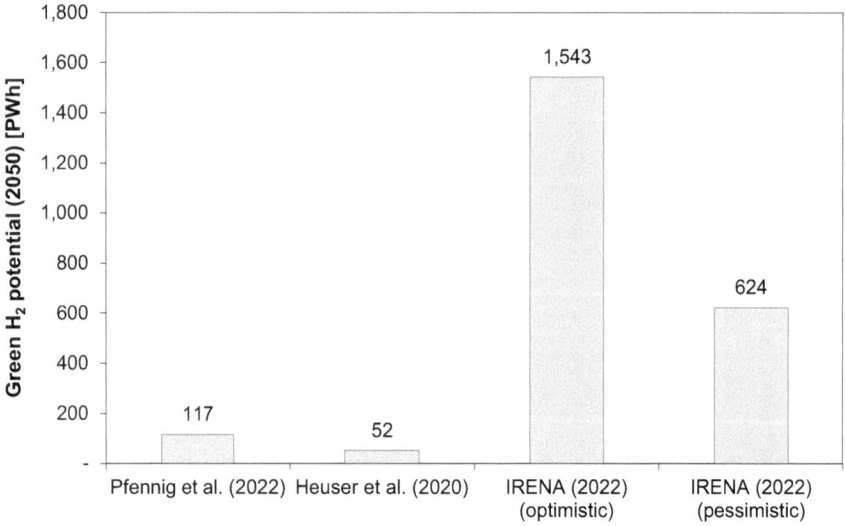

Fig. 5 Global green hydrogen potentials (techno-economic) in 2050 from different studies

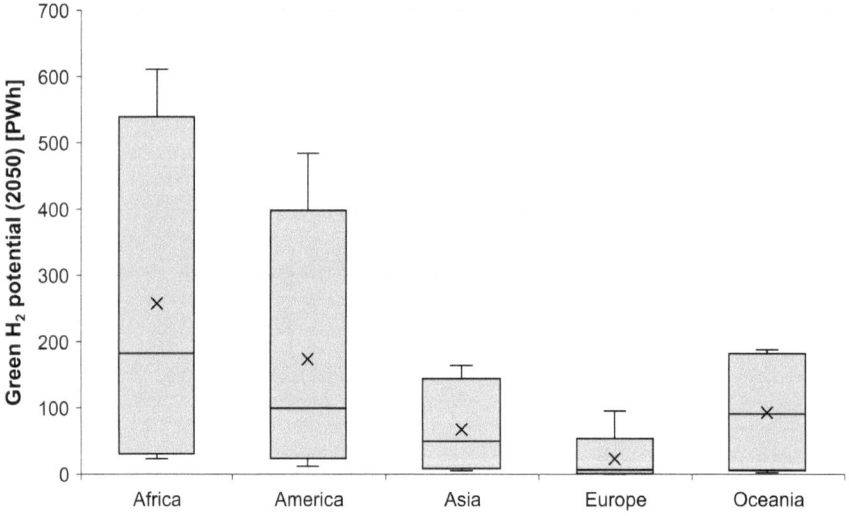

Fig. 6 Green hydrogen potential (techno-economic) per continent in 2050 (self-elaboration with information from [45, 48–50, 53, 54])

Europe studies report a wide range. On the one hand, the lowest reported potential is 0.03 PWh, considering only Belarus and Ukraine [54]. On the other hand, the potential of 95 PWh reported by IRENA includes the whole of Europe. This could even include Russia, which would significantly increase the potential [45].

America, Africa, and Oceania are the three regions with the most significant green hydrogen potentials, giving them a distinct advantage for production. However, studies in Asia have been concentrated on sites with the utmost potential or where water is accessible [53, 54]. It's important to examine not only the quantity of green hydrogen but also its correlation to the LCOH. In the next section, a focus on the potential per cost step and region is presented.

5.2.3 Green Hydrogen Regional Economic Potential

Figure 7 illustrates Africa's green hydrogen potential as a function of cost. Africa shows significant potential even at more affordable thresholds such as 2.00 €/kg$_{H_2}$ [54]. There's a consistent increase in potential across all studies, culminating in values below 5.00 €/kg$_{H_2}$. IRENA's projections for the optimistic and pessimistic scenarios are 611 PWh and 227 PWh, respectively, at costs below 1.75 €/kg$_{H_2}$ (see Fig. 12). Furthermore, IRENA's LCOH estimates for 2050 are 0.60 €/kg$_{H_2}$ and 1.25 €/kg$_{H_2}$ for Morocco and 0.86 €/kg$_{H_2}$ and 1.3 €/kg$_{H_2}$ for sub-Saharan Africa in the optimistic and pessimistic scenarios, respectively [45].

Figure 8 presents the green hydrogen potential of the continent of America (North and South America) in terms of cost. There is a notable difference between the two studies shown. The first study shows a significant potential of 18 PWh at a cost of less than 2 €/kg$_{H_2}$ [53, 54] which rises to more than three times this amount at a cost of less than 3 €/kg$_{H_2}$. In contrast, the second study reports a lower potential, especially at higher costs [53]. IRENA's data show potentials of 484 and 138 PWh for the optimistic and pessimistic scenarios, respectively, both below 1.75 €/kg$_{H_2}$

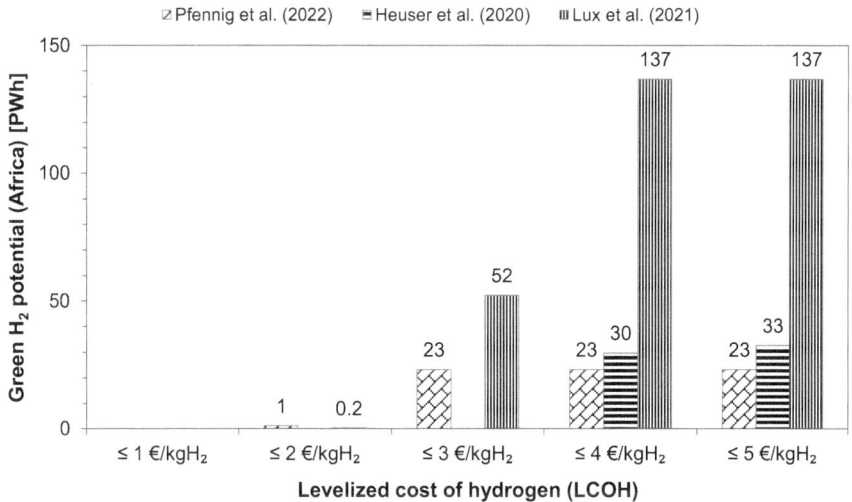

Fig. 7 Green hydrogen potentials per LCOH for Africa (LCOH: levelized cost of hydrogen)

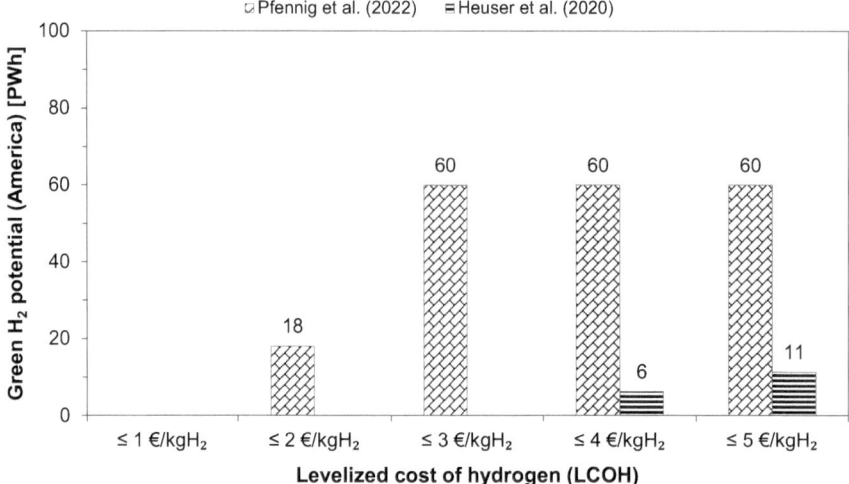

Fig. 8 Green hydrogen potentials per LCOH for the continent of America (North and South America; LCOH: levelized cost of hydrogen)

(see Fig. 12). Furthermore, IRENA's LCOH projections for Chile in 2050 are 0.59 $€/kg_{H_2}$ in the optimistic scenario and 1.04 $€/kg_{H_2}$ in the pessimistic scenario [45].

Figure 9 shows the green hydrogen potential in Asia. As with the American continent case, there's a marked difference between the two studies presented. One study shows a substantial potential of 16.8 PWh at costs as low as 3.00 $€/kg_{H_2}$ [54], while the other shows a more modest potential of 5.3 PWh under 5.00 $€/kg_{H_2}$ [53]. IRENA's projections for the optimistic and pessimistic scenarios are 164 PWh and 82 PWh, respectively, both at costs below 1.75 $€/kg_{H_2}$ (see Fig. 12). For China, IRENA projects a 2050 LCOH of 0.56 $€/kg_{H_2}$ in the optimistic scenario and 0.92 $€/kg_{H_2}$ in the pessimistic scenario, which is the lowest LCOH reported by IRENA for that year [45].

The potential for European green hydrogen is shown in Fig. 10. Three studies are shown that estimated the potential for Europe. One of these studies projected the highest potential across all cost levels peaking at 6.5 PWh under 5 $€/kg_{H_2}$ [49]. Almost 1.00 PWh potential was calculated under 3.00 $€/kg_{H_2}$. The other two studies showed lower values, with one peaking at less than 0.1 PWh and the other at 1.1 PWh [53, 54]. IRENA presents values of 95 and 12 PWh for its optimistic and pessimistic scenarios respectively, both below 1.75 $€/kg_{H_2}$ (see Fig. 12). In addition, Spain reports an LCOH of 0.73 $€/kg_{H_2}$ for the optimistic scenario and 1.47 $€/kg_{H_2}$ for the pessimistic scenario [45].

Similar to the cases of America and Asia, just two studies report potentials for Oceania (Fig. 11). They have a sharp contrast with one of them reporting potentials of 16.7 PWh under 3 $€/kg_{H_2}$ and the other having a value below 2.0 PWh under 5 $€/kg_{H_2}$. IRENA reports values of 188 and 164 PWh values for its optimistic and pessimistic scenarios under 1.75 $€/kg_{H_2}$ (see Fig. 12). It shall be pointed out that the

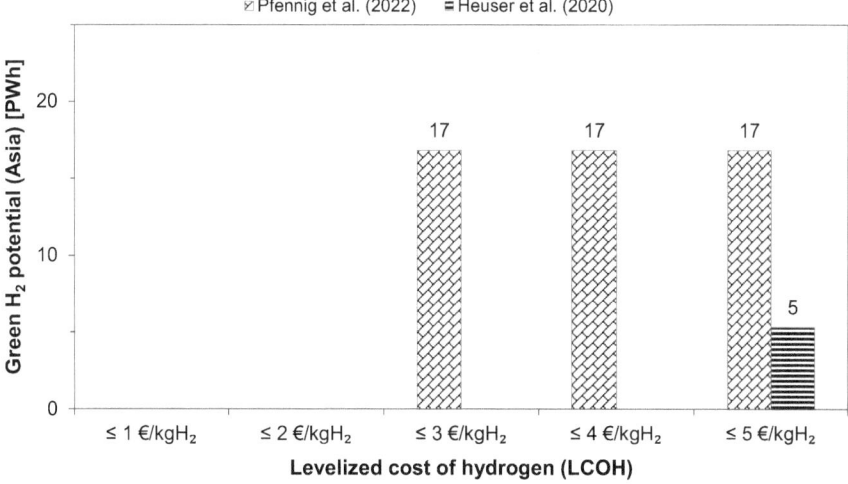

Fig. 9 Green hydrogen potentials per LCOH for Asia (LCOH: levelized cost of hydrogen)

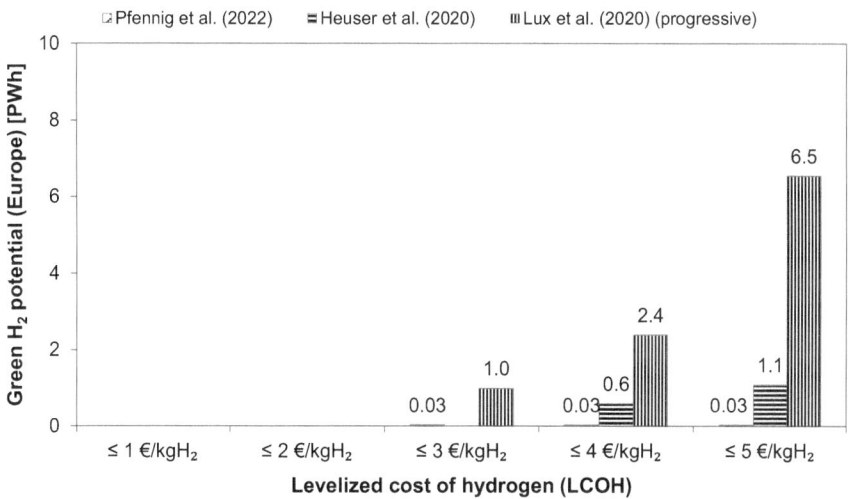

Fig. 10 Green hydrogen potentials per LCOH for Europe (LCOH: levelized cost of hydrogen)

difference between optimistic and pessimistic values is smaller for this region than it is for others. A LCOH value of 0.66 €/kg$_{H_2}$ and 1.31 €/kg$_{H_2}$ is reported in Oceania for the optimistic and pessimistic scenarios [45].

From a global perspective, the lowest LCOH values, excluding IRENA figures, are found for America and Africa. For America, a significant potential is estimated as depicted in Fig. 8. The expected potential for green hydrogen in Africa varies considerably. This variation can be attributed to two main factors: the different

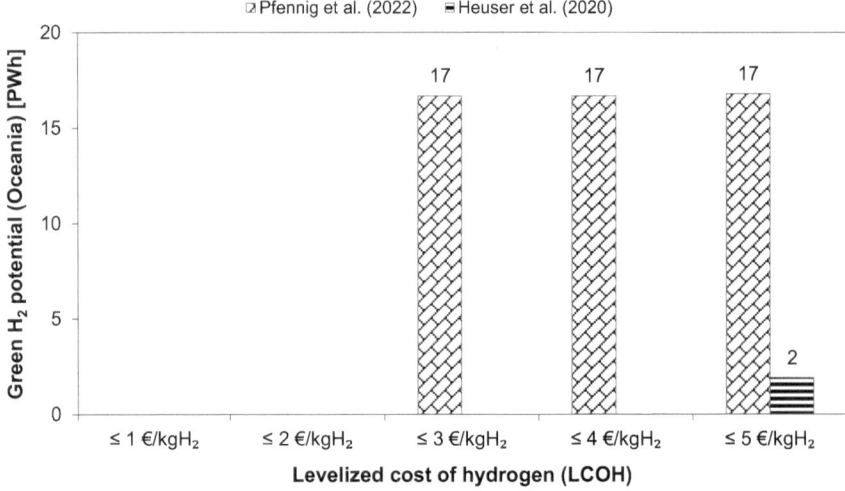

Fig. 11 Green hydrogen potentials per LCOH for Oceania (LCOH: levelized cost of hydrogen)

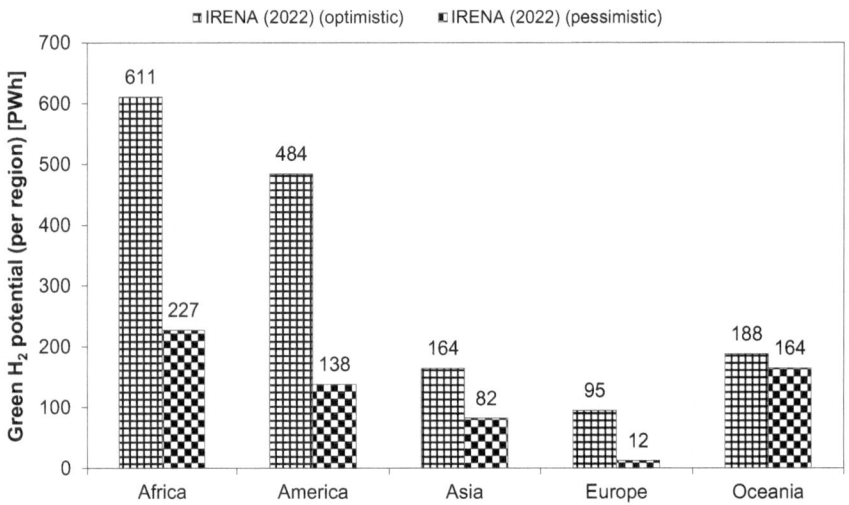

Fig. 12 Green hydrogen potentials per region estimated by IRENA

techno-economic assumptions, which affect the costs, the allocated land for RE, and the availability of water, which determines the amount of green hydrogen produced. IRENA highlights this in its water scarcity scenario, noting that in several regions in Africa and the Middle East, water scarcity reduces the potential by up to 94% [45]. Europe, on the other hand, has one of the lower potentials at 6.54 PWh (Fig. 10).

6 Conclusion

In this chapter, different pathways for hydrogen production have been described. As of now, hydrogen is mostly produced through steam methane reforming and coal gasification. However, in a future decarbonized world, most likely mainly blue and green hydrogen will be used which are considered low-carbon hydrogen options. Pink hydrogen, i.e., based on nuclear power, is considered low-carbon hydrogen as well. Nevertheless, its use is controversial due to the respective nuclear waste. Turquoise hydrogen, i.e., based on methane pyrolysis, may come into the market in the future, offering an interesting alternative to blue hydrogen, as the carbon is captured as a solid where no leakages are possible making it a cleaner alternative.

Both blue and green hydrogen potentials are sufficient to fulfill the global hydrogen demand in 2050. However, several aspects need to be considered, especially regarding the increased water demand for their production. This is particularly important for green hydrogen, as the best photovoltaic generation sites are usually in arid regions where water is scarce. The impact of water is reflected in the study of IRENA, where water availability severely limits the green hydrogen potential. Blue hydrogen water consumption is similar to the one of green hydrogen and shall not be underestimated.

The future role of blue hydrogen is important. Several countries, like Canada and the USA, consider it in their respective hydrogen strategies. Current natural gas producers like the United Arab Emirates (UAE) and Saudi Arabia consider it as a low-carbon alternative to their current exports. Blue hydrogen may be a bridge technology before a future transit to green hydrogen. The use of blue hydrogen will decrease the demand for green hydrogen which is the main component of Power-to-X (PtX) products.

The geopolitical difference between blue and green hydrogen is interesting. The top ten blue hydrogen potential producers hoard about 83% of its total global potential. In contrast, green hydrogen is available on a broader basis as renewable energy sources are available everywhere. Besides the production potential and its related cost, the transportation costs would be crucial in determining which countries would lead the green hydrogen market. The green hydrogen potentials show an ample range among the studies. The difference lies in two main factors: the land available for the installation of renewable energy sources and the water considerations for the production of green hydrogen. Land is already a scarce resource in places with high population densities. The "not in my backyard" is a common reference to the acceptance of different technologies, including renewable energy sources. Therefore, policies focused on increasing renewable energy sources (RES) acceptance and as a result allocating more land to renewable energy sources will increase green hydrogen availability.

Africa and America are the two continents with the largest green hydrogen potential. The North African region has a large hydrogen potential due to its high solar potential. In the case of America, the wind potential in Canada and the solar potential in some regions in Latin America (Chile and Argentina) are suitable for low-cost hydrogen production. Oceania has a high potential for the production of green hydrogen thanks to its vast solar resources. Europe is the region with the smallest green hydrogen potential.

Regional green hydrogen potential is sufficient to cover the domestic hydrogen demand. Especially the optimistic scenario of IRENA exhibits high potential in all of the world's regions. Even regions like Europe, which exhibits the lowest potential, possess enough potential to satisfy their needs in the future. However, the region may still import hydrogen as some other countries could be capable of producing hydrogen at a lower cost. One option is the MENA region due to the high solar potential.

Most of the studies consider just island hydrogen systems, meaning that the renewable energy systems would be allocated exclusively for green hydrogen production and are separated from the country's energy system. Therefore, further studies considering the local demand and energy system are necessary. However, some studies are exceptions and they considered a system overview [49, 50].

Solar technologies, especially PV, and wind onshore/offshore are the technologies mostly considered in these studies. Further studies that consider other technologies, e.g., concentrated solar power (CSP) and tidal or geothermal energy, might be necessary to calculate the total potential more accurately.

References

1. Riemer M, Zheng L, Eckstein J, Wietschel M, Pieton N, Kunze R (2022) Future hydrogen demand: a cross-sectoral, global meta-analysis: HYPAT Working Paper 04/2022. Fraunhofer Institut für System- und Innovationsforschung (Fraunhofer ISI); Fraunhofer-Institut für Solare Energiesysteme ISE (Fraunhofer ISE); Fraunhofer-Einrichtung für Energieinfrastrukturen und Geothermie IEG (Fraunhofer IEG); Ruhr Universität Bochum (RUB); Energy Systems Analysis Associates (SA2); Deutsches Institut für Entwicklungspolitik (die); German Develpement Instiute; IASS Potsdam; Deutsche Gesellschaft für Internationale Zusammenarbeit GmbH (GIZ), Deutsche Energie-Agentur (dena), September. Available from: URL: https://www.hypat.de/hypat-wAssets/docs/new/publications/HYPAT_Working_Paper_04_2022_Future_hydrogen_demand.pdf
2. Fleiter T, Rehfeldt M, Neuwirth M, Hersbst A (2021) Langfristszenarien für die Transformation des Energiesystems in Deutschland 3: Treibhausgasneutrale Hauptszenarien; Modul Industrie. Fraunhofer ISI. Available from: URL: https://www.langfristszenarien.de/enertile-explorer-wAssets/docs/Modul_TN_Hauptszenarien_Industrie.pdf.
3. IEA–International Energy Agency (2019) The future of hydrogen. Paris [cited 2022 Sep 12]. Available from: URL: https://iea.blob.core.windows.net/assets/9e3a3493-b9a6-4b7d-b499-7ca48e357561/The_Future_of_Hydrogen.pdf
4. IEA—International Energy Agency (2022) Global hydrogen review 2022. Available from: URL: https://iea.blob.core.windows.net/assets/c5bc75b1-9e4d-460d-9056-6e8e626a11c4/GlobalHydrogenReview2022.pdf
5. IEA—International Energy Agency (2022) Hydrogen projects database [cited 2022 Sep 12]. Available from: URL: https://www.iea.org/data-and-statistics/data-product/hydrogen-projects-database
6. World Energy Council (2022) Regional insights into low-carbon hydrogen scale up [cited 2022 Sep 12]. Available from: URL: https://www.worldenergy.org/assets/downloads/World_Energy_Insights_Working_Paper_Regional_insights_into_low-carbon_hydrogen_scale_up.pdf?v=1654526979

7. World Energy Council (2021) Working paper—national hydrogen strategies [cited 2021 Sep 12]. Available from: URL: https://www.worldenergy.org/assets/downloads/Working_Paper_-_National_Hydrogen_Strategies_-_September_2021.pdf

8. Rapier R (2020) Estimating the carbon footprint of hydrogen production. Forbes [cited 2022 Sep 12]. Available from: URL: https://www.forbes.com/sites/rrapier/2020/06/06/estimating-the-carbon-footprint-of-hydrogen-production/?sh=35b8154024bd

9. Coal Age (2021) Hydrogen from coal [cited 2021 Oct 11]. Available from: URL: https://www.coalage.com/features/hydrogen-from-coal/#:~:text=The%20estimated%20CO2%20emissions,from%20SMR%20of%20natural%20gas

10. Noussan M, Raimondi PP, Scita R, Manfred H (2020) The role of green and blue hydrogen in the energy transition—a technological and geopolitical perspective. Sustainability [cited 2022 Sep 19]; 13:298. Available from: URL: https://www.mdpi.com/2071-1050/13/1/298/htm

11. Erbach G, Liselotte J (2021) EU hydrogen policy: Hydrogen as an energy carrier for a climate-neutral economy. European Parliament, April [cited 2022 Sep 19]. Available from: URL: https://www.europarl.europa.eu/RegData/etudes/BRIE/2021/689332/EPRS_BRI(2021)689332_EN.pdf

12. Government of Canada (2020) Hydrogen strategy for Canada: seizing the opportunities for hydrogen [cited 19092022]. Available from: URL: https://www.nrcan.gc.ca/sites/nrcan/files/environment/hydrogen/NRCan_Hydrogen-Strategy-Canada-na-en-v3.pdf

13. Frens W (2022) Methane pyrolysis: producing hydrogen without carbon emissions. TNO [cited 2022 Sep 19] Available from: URL: https://www.tno.nl/en/technology-science/technologies/methane-pyrolysis/

14. U.S. Energy Information Administration (2021) Nuclear explained: nuclear power and the environment [cited 2022 Sep 19]. Available from: URL: https://www.eia.gov/energyexplained/nuclear/nuclear-power-and-the-environment.php#:~:text=Nuclear%20power%20reactors%20do%20not,or%20carbon%20dioxide%20while%20operating

15. Taylor K (2020) Hydrogen produced from nuclear will be considered 'low-carbon', EU official says: Euractiv [cited 2022 Sep 19]. Available from: URL: https://www.euractiv.com/section/energy-environment/news/hydrogen-produced-from-nuclear-will-be-considered-low-carbon-eu-official-says/

16. DNV (2022) Hydrogen forecast to 2050: energy transition outlook 2022. Oslo [cited 2022 Sep 19]. Available from: URL: https://www.dnv.com/focus-areas/hydrogen/forecast-to-2050.html

17. IEA, International Energy Agency (2021) Net zero by 2050—a roadmap for the global energy sector. Paris. Available from: URL: https://iea.blob.core.windows.net/assets/deebef5d-0c34-4539-9d0c-10b13d840027/NetZeroby2050-ARoadmapfortheGlobalEnergySector_CORR.pdf

18. Hydrogen Council (2021) Hydrogen decarbonization pathways: potential supply scenarios [cited 2022 Sep 19]. Available from: URL: https://hydrogencouncil.com/wp-content/uploads/2021/01/Hydrogen-Council-Report_Decarbonization-Pathways_Part-2_Supply-Scenarios.pdf

19. U.S. Department of Energy (2020) Hydrogen strategy: enabling a low-carbon economy. Washington, DC [cited 2022 Oct 11]. Available from: URL: https://www.energy.gov/sites/prod/files/2020/07/f76/USDOE_FE_Hydrogen_Strategy_July2020.pdf

20. COAG Energy Council Hydrogen Working Group (2019) Australia's national hydrogen strategy [cited 2022 Sep 19]. Available from: URL: https://www.industry.gov.au/sites/default/files/2019-11/australias-national-hydrogen-strategy.pdf

21. Hasan S, Shabaneh R (2021) The economics and resource potential of hydrogen production in Saudi Arabia. King Abdullah Petroleum Studies and Research Center (KAPSARC)

22. Kamel D (2022) UAE in deal to develop national hydrogen strategy. The National News [cited 2022 Mar 11]. Available from: URL: https://www.thenationalnews.com/business/energy/2022/09/26/uae-in-deal-to-develop-national-hydrogen-strategy/

23. Arvind S, Mackay J (2022) Harnessing the power of hydrogen in the UAE: KPMG [cited 2022 Mar 11]. Available from: URL: https://home.kpmg/ae/en/home/insights/2022/08/harnessing-the-power-of-hydrogen-in-the-uae.html

24. Dickel R (2020) Blue hydrogen as an enabler of green hydrogen: the case of Germany. The Oxford Institute for Energy Studies, Oxford (OIES paper NG; vol 159). Available from: URL: https://www.oxfordenergy.org/wpcms/wp-content/uploads/2020/06/Blue-hydrogen-as-an-enabler-of-green-hydrogen-the-case-of-Germany-NG-159.pdf
25. Global CCS Institute (2021) Blue hydrogen: circular carbon economy [cited 2022 Sep 19]. Available from: URL: https://www.globalccsinstitute.com/wp-content/uploads/2021/04/Circular-Carbon-Economy-series-Blue-Hydrogen.pdf
26. Yu M, Wang K, Vredenburg H (2021) Insights into low-carbon hydrogen production methods: Green, blue and aqua hydrogen. Int J Hydrogen Energy 46(41):21261–21273
27. U.S. Department of Energy (2022) National energy technology laboratory. Carbon storage FAQs [cited 2022 Sep 19]. Available from: URL: https://netl.doe.gov/carbon-management/carbon-storage/faqs/carbon-storage-faqs
28. CO$_2$ Capture Project (2008) Rock formations suitable for CCS [cited 2022 Sep 19]. Available from: URL: https://www.co2captureproject.org/pdfs/rock_formations_suitable_for_ccs.pdf
29. Gaurina-Međimurec N, Novak Mavar K (2019) CO$_2$ sequestration: carbon capture and storage (CCS): geological sequestration of CO$_2$. Available from: URL: https://www.intechopen.com/chapters/66365
30. Saulnier R, Minnich K, Sturgess K (2020) Water for the hydrogen economy. WaterSMART Solutions Ltd. [cited 2022 Sep 19]. Available from: URL: https://watersmartsolutions.ca/wp-content/uploads/2020/12/Water-for-the-Hydrogen-Economy_WaterSMART-Whitepaper_November-2020.pdf
31. Al-Qahtani A, Parkinson B, Hellgardt K, Shah N, Guillen-Gosalbez G (2021) Uncovering the true cost of hydrogen production routes using life cycle monetisation. Appl Energy 281:115958. Available from: URL: https://www.sciencedirect.com/science/article/pii/S0306261920314136
32. Coertzen R, Potts K, Brannock M, Dagg B (2022) Water for hydrogen: GHD [cited 2022 Sep 19]. Available from: URL: https://www.ghd.com/en/perspectives/water-for-hydrogen.aspx
33. bp. bp Statistical Review of World Energy 2021. London [cited 2022 Sep 19]. Available from: URL: https://www.bp.com/content/dam/bp/business-sites/en/global/corporate/pdfs/energy-economics/statistical-review/bp-stats-review-2021-full-report.pdf
34. Bauer C, Treyer K, Antonini C, Bergerson J, Gazzani M, Gencer E et al (2021) On the climate impacts of blue hydrogen production. Sustain Energy Fuels 6(1):66–75 [cited 2022 Sep 19]. Available from: URL: https://pubs.rsc.org/en/content/articlepdf/2022/se/d1se01508g
35. Spath PL, Mann MK (2000) Life cycle assessment of hydrogen production via natural gas steam reforming. NREL [cited 2022 Sep 19]. Available from: URL: https://www.nrel.gov/docs/fy01osti/27637.pdf
36. IEA—International Energy Agency (2015) Storing CO$_2$ through enhanced oil recovery: combining EOR with CO$_2$ storage (EOR+) for profit. Paris [cited 2022 Sep 19]. Available from: URL: https://iea.blob.core.windows.net/assets/bf99f0f1-f4e2-43d8-b123-309c1af66555/Storing_CO2_through_Enhanced_Oil_Recovery.pdf
37. Oni AO, Anaya K, Giwa T, Di Lullo G, Kumar A (2022) Comparative assessment of blue hydrogen from steam methane reforming, autothermal reforming, and natural gas decomposition technologies for natural gas-producing regions. Energy Conv Manage 254:115245 [cited 2022 Sep 19]. Available from: URL: https://www.sciencedirect.com/science/article/pii/S0196890422000413
38. International Renewable Energy Agency (2020) Green hydrogen cost reduction: scaling up electrolysers to meet the 1.5 °C climate goal [cited 2022 Sep 22]. Available from: URL: https://irena.org/-/media/Files/IRENA/Agency/Publication/2020/Dec/IRENA_Green_hydrogen_cost_2020.pdf
39. Bracken N, Macknick J, Tovar-Hastings A, Komor P, Gerritsen M, Mehta S (2015) Concentrating solar power and water issues in the U.S. Southwest. Joint Institute for Strategic Analysis [cited 2022 Sep 19]. Available from: URL: https://www.nrel.gov/docs/fy15osti/61376.pdf
40. Bruch A, Patchigolla K, Asfand f, Douard S (eds) (2019) SOLARPACES 2018. In: International Conference on Concentrating Solar Power and Chemical Energy Systems. AIP Publishing (AIP Conference Proceedings)

41. Zelt O, Kobiela G, Ortiz W, Scholz A, Monnerie N, Rosenstiel A et al (2021) Multikri-
    terielle Bewertung von Bereitstellungstechnologien synthetischer Kraftstoffe. Wuppertal [cited
    2022 Sep 19]. Available from: URL: https://wupperinst.org/fa/redaktion/downloads/projects/
    MENA-Fuels_Teilbericht3_D2-1_Technologiebewertung.pdf
42. Thomann J, Marscheider-Weidemann F, Stamm A, Lorych L, Hank C, Weise F et al (2022)
    Background paper on sustainable green hydrogen and synthesis products. HYPAT [cited 2022
    Sep 19]. Available from: URL: https://www.hypat.de/hypat-wAssets/docs/new/publications/
    HYPAT-Working-Paper-01-2022_background-paper-sustainable-green-hydrogen-synthesis-
    products.pdf
43. Franke K, Sensfuß F, Bernath C, Lux B (2021) Carbon-neutral energy systems and the
    importance of flexibility options: a case study in China. Comput Ind Eng 162:107712
44. Hoogwijk M, de Vries B, Turkenburg W (2004) Assessment of the global and regional
    geographical, technical and economic potential of onshore wind energy. Energy Econ
    26(5):889–919
45. International Renewable Energy Agency (2022) Global hydrogen trade to meet the 1.5 °C
    climate goal: Part III—Green hydrogen cost and potential. Abu Dhabi [cited 2022 Sep 19].
    Available from: URL: https://www.irena.org/-/media/Files/IRENA/Agency/Publication/2022/
    May/IRENA_Global_Hydrogen_Trade_Costs_2022.pdf
46. Nwabueze S (2022) H2 Atlas Africa: atlas of green hydrogen generation potentials in Africa:
    Forschungszentrum Jülich. Available from: URL: https://www.h2atlas.de/
47. International Renewable Energy Agency (2014) Estimating the renewable energy
    potential in Africa: a GIS-based approach. Abu Dhabi [cited 20/19/2022]. Avail-
    able from: URL: https://www.irena.org/publications/2014/Aug/Estimating-the-Renewable-
    Energy-Potential-in-Africa-A-GIS-based-approach
48. Mukelabai MD, Wijayantha UK, Blanchard RE (2022) Renewable hydrogen economy outlook
    in Africa. Renew Sustain Energy Rev 167:112705
49. Lux B, Gegenheimer J, Franke K, Sensfuß F, Pfluger B (2021) Supply curves of electricity-based
    gaseous fuels in the MENA region. Comput Ind Eng 162:107647
50. Lux B, Pfluger B (2020) A supply curve of electricity-based hydrogen in a decarbonized
    European energy system in 2050. Appl Energy 269:115011. Available from: URL: https://
    www.sciencedirect.com/science/article/abs/pii/S0306261920305237
51. International Renewable Energy Agency (2022) Global hydrogen trade to meet the 1.5 °C
    climate goal: trade outlook for 2050 and way forward. Abu Dhabi. Available from: URL:
    https://www.irena.org/publications/2022/Jul/Global-Hydrogen-Trade-Outlook
52. Franke K, Sensfuß F, Deac G, Kleinschmitt C, Ragwitz M (2021) Factors affecting the calcula-
    tion of wind power potentials: a case study of China. Renew Sustain Energy Rev 149:111351.
    Available from: URL: https://www.sciencedirect.com/science/article/pii/S1364032121006377
53. Heuser P-M, Grube T, Heinrichs H, Robinius M, Stolten D (2020) Worldwide hydrogen provi-
    sion sheme based on renewable energy. Preprints [cited 2022 Sep 19]. Available from: URL:
    https://www.preprints.org/manuscript/202002.0100/v1
54. Pfennig M, Böttger D, Häckner B, Geiger D, Zink C, Bisevic A et al (2022) Global GIS-
    based potential analysis and cost assessment of Power-to-X fuels in 2050 [cited 2022 Sep 19].
    Available from: URL: https://arxiv.org/ftp/arxiv/papers/2208/2208.14887.pdf
55. Fasihi M, Bogdanov D, Breyer C (2017) Long-term hydrocarbon trade options for the Maghreb
    Region and Europe—renewable energy based synthetic fuels for a net zero emissions world.
    Sustain 9(2):306
56. IEA—International Energy Agency (2021) Global hydrogen review. Paris [cited 2022 Sep 19].
    Available from: URL: https://iea.blob.core.windows.net/assets/e57fd1ee-aac7-494d-a351-f2a
    4024909b4/GlobalHydrogenReview2021.pdf
57. George JF, Müller VP, Winkler J, Ragwitz M (2022) Is blue hydrogen a bridging technology?—
    the limits of a CO_2 price and the role of state-induced price components for green hydrogen
    production in Germany. Energy Pol 167:113072 [cited 2022 Sep 19]. Available from: URL:
    https://www.sciencedirect.com/science/article/pii/S030142152200297X

Carbon as a Feedstock

Carbon—Classification, Sources, and Potentials

Tjerk Zitscher, Nils Bullerdiek, and Martin Kaltschmitt

Abstract Carbon is a vital element in the biosphere due to its abundance in living organisms. It exists in both pure and chemically bound forms and has the capacity to form a wide range of complex compounds, making it integral to industrial society. Since the Industrial Revolution, carbon-based fuels (e.g., coal, oil, natural gas) have been the primary sources of energy, but their combustion releases CO_2, contributing significantly to anthropogenic climate change. Despite the need to limit fossil fuels, hydrocarbons will remain crucial for sectors like chemicals and transportation, where alternative energy sources are challenging to implement. Sustainable synthetic or biologically produced hydrocarbons may offer alternatives for hard-to-decarbonize sectors. This chapter examines carbon availability, focusing on CO_2 provision options, categorizes carbon sources (fossil, biogenic, mixed), and assesses their potential in transitioning toward a defossilised energy and economic system.

Keywords Carbon sources · Carbon potentials · Carbon cycle · Carbon capture · Utilization · Carbon classification · CO_2 · Alternative carbon

1 Introduction

Carbon is an important element within the biosphere since it is the second most abundant element (by weight) in living organisms. In contrast, carbon geologically plays a subordinate role, since with a weight share of about 0.02%, it is scarce compared to oxygen (ca. 49.2%) or silicon (ca. 25.7%) [1]. It is found both in the native/pure form mainly within the underground and in a chemically bound form primarily within the atmosphere and biosphere. Due to its electron configuration, carbon can form a large number of—partly even very complex—chemical compounds. Furthermore, carbon and especially its various compounds are an integral part of our modern civilization and our highly industrialized society. Since the beginning of the industrial

T. Zitscher (✉) · N. Bullerdiek · M. Kaltschmitt
Hamburg University of Technology (TUHH), Institute of Environmental Technology and Energy Economics (IUE), Hamburg, Germany
e-mail: tjerk.zitscher@tuhh.de

revolution, carbon has been the most essential element for the provision of energy in (modern) energy systems—initially mainly in the form of hard coal and lignite, and successively with over time increasing shares in the form of crude oil and lately also natural gas. Besides the use for energy provision carbon is also essential as a material component for various products (e.g., plastics).

The technical use of carbon or hydrocarbons might result in the emission of carbon dioxide (CO_2) as this reflects the highest level of oxidation of a carbon atom. This is true if a carbon atom is used for the provision of energy but also at the end of the life cycle of carbon-containing products, such as plastics. CO_2 emissions resulting from the oxidation of carbon of fossil origin (i.e., from fossil fuels like hard coal, lignite, crude oil, natural gas) into CO_2 are largely responsible for the so-called anthropogenic climate change and its far-reaching consequences. Thus, science agree in unison that the amount of CO_2 within the atmosphere must be stabilized on the longer term to reduce the further rise in temperature and all the resulting secondary effects like a rise in the global sea level, like desertification of large, so far fertile areas, and like increasing shares of extremely hot days throughout the year. The logical consequence is a far-reaching abandonment of the use of fossil carbon (reduction of greenhouse gas (GHG) emissions by 43% by 2030 on a global scale) [2].

However, despite these effects, according to current knowledge, carbon or hydrocarbons will continue to play an important role in our highly industrialized society [3–5]; this is true e.g., for the chemical industry as well as other industrial sectors because a significant share of industrial products provided today is based on hydrocarbons and—according to current knowledge—this will be the case also in the future. In 2020, for example, the demand for naphtha and petroleum derivatives being typical hydrocarbon-based bulk chemicals to be used as a raw material within the German chemical industry was about 14.3 Mt/a [6]; and most likely, the demand will grow in the time to come.

Additionally, the transportation sector will most likely need a certain amount of hydrocarbons within some specific transport sectors hardly to be transformed toward a direct use of electricity or to energy/fuel options that do not contain carbon (e.g., hydrogen, ammonia). For example, according to current knowledge in aviation and marine transportation, a clear demand for liquid hydrocarbon-based fuels with a high volumetric and/or gravimetric energy density will be given also in the future. Synthetically or biologically produced hydrocarbons can be a suitable sustainable alternative to, e.g., hydrogen or ammonia to satisfy such markets/sectors that are hard to be decarbonized in the years to come [7]. From a GHG reduction perspective, it is essential that the carbon to be used in such a more GHG-neutral society/production system is obtained from renewable as well as sustainable sources and thus from the earth's recent (fast) carbon cycle.

Against this background, availability aspects of carbon as a resource are analyzed, focusing on provision options for carbon dioxide (CO_2). Therefore, first, general aspects of carbon as a chemical element and its role within earth's carbon cycles are outlined. Subsequently, three classifications of carbon sources/origins are defined (fossil, biogenic, mixed), and their implications for future carbon use assessed from a climate protection perspective. Provision options for each of these carbon source classifications are then presented in more detail and potentially available amounts are derived. These availability estimates are then compared to each other, and the respective carbon sources discussed in relation to their role in transitioning toward a defossilised energy and economic system.

2   The Chemical Element Carbon

In the universe, after hydrogen, helium, and oxygen, carbon holds the fourth position in terms of abundance by mass. However, it makes up (only) about 0.025% of the earth's crust, ranking it as the 15th most prevalent element on earth. The vast array of organic compounds that carbon can form, combined with its capacity to create polymers at typical ambient temperatures occurring on our planet, makes it fundamental to (known) life. All known life forms are based on carbon-containing compounds such as proteins, carbohydrates, and lipids, and potentially a skeleton rich in calcium carbonate [8–10].

2.1   Carbon Element

Chemically, carbon is represented by the symbol "C" and has an atomic number of 6. It belongs to the fourth main group of the periodic table (or the 14th IUPAC[1] group) and is located in the second period.[2] Thus, the nucleus of a carbon atom contains six protons, in addition to neutrons, resulting in a total of six positive charges. Surrounding the nucleus, the atomic shell comprises six electrons, equating to six negative charges, ensuring the atom's overall electrical neutrality. According to the atomic shell model, the innermost shell (K-shell) holds two electrons, while the outer shell (L-shell) contains four electrons. These four electrons in the L-shell, known as valence electrons, make carbon tetravalent, allowing it to form four covalent chemical bonds (Fig. 1) [9, 10].

Elementary carbon is non-metallic and occurs in several allotropic forms, with main ones being diamond, graphite, and fullerene [11]. The structure of these carbon allotropes is schematically shown in Fig. 2.

- Diamond is a crystalline form of carbon, where carbon is three-dimensionally bonded through four covalent bonds, oriented toward the corners of a tetrahedral shape [10]. Carbon in the form of a diamond is transparent, usually colorless, with a high refractive index, and is a typical insulator. In the diamond lattice, each carbon atom is at the center of a tetrahedron formed by other carbon atoms

---

[1] International Union of Pure and Applied Chemistry.

[2] The atomic number represents an element's position in the periodic table and equals the number of protons in its nucleus. The final digit of the main group number indicates the number of (valence) electrons in the outermost shell of the element ([8] S 33).

Fig. 1 Shell model of the
carbon atom (based on [9])

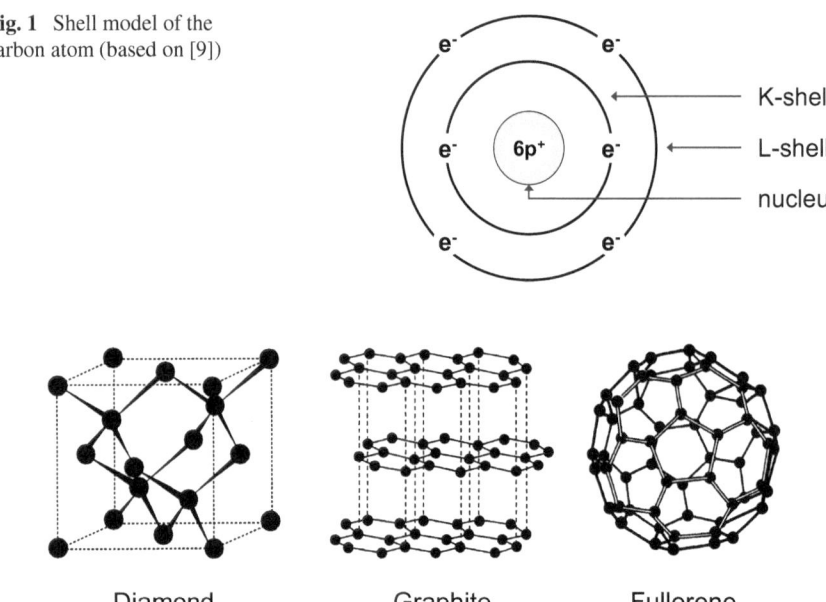

Fig. 2 Schematic structure of diamond, graphite, and fullerene carbon allotropes [11]

(atomic lattice). This accounts for diamond's extreme hardness; it is the hardest known naturally occurring material, scoring 10 on the Mohs scale of hardness, and is used technically as an abrasive [11].

- Graphite has a crystal lattice consisting of flat, two-dimensional layers of carbon atoms in a hexagonal arrangement linked on all sides. These layers are stacked in such a way that every third layer aligns identically with the starting layer. To form such hexagonal layers, only three electrons from each carbon atom are needed, leaving an extra electron for each carbon atom. These surplus electrons can move almost freely between the layers. This gives graphite its conductivity along the layers and its characteristic black color. Thus, graphite is used as an electrode, but also, for example, as a high-temperature sealing material, for lubricants, or for pencil leads [10, 11].

- Fullerenes, also known as "buckyballs," are molecules made of pure carbon that exist in spherical structures, being hollow inside. Their shell is formed from pentagons and hexagons. They are more stable the larger they are. Depending on their combination with other atoms, they become electrical conductors, insulators, or even superconductors [10, 11].

However, besides carbon in its pure form, its unique electron configuration (a half-filled outer L-shell, i.e., with four of eight electrons) enables carbon to form a variety of additional complex molecules. Unique characteristics of carbon include its ability to form chains and rings with itself and other elements, as well as double and triple bonds. Due to its moderate electronegativity, it exhibits good bonding capacity with both electropositive and electronegative elements. All oxidation states from − IV to +IV are found in nature, in either inorganic or organic compounds [11].

Thus, in nature, carbon is found not only in its pure form but also in various chemically bonded forms. It displays the widest range of chemical compounds of any element, second only to hydrogen (hydrogen takes the first position because most carbon compounds also contain hydrogen). A significant example is the group of hydrocarbons. The multiple bonding capabilities of carbon atoms with each other— unbranched and branched chains, rings, and single, double, and triple bonds—indicate that even with hydrocarbons, which consist only of carbon and hydrogen, a vast number of different compounds are possible (e.g., alkanes, alkenes, alkynes, aromatic compounds, alcohols, phenols, ethers, carboxylic acids, lipids). Considering that carbon atoms can bond not only with hydrogen but also with other elements, an immense number of organic compounds can be formed. These compounds contain other elements such as oxygen, nitrogen, sulfur, phosphorus, halogens, or metals [9, 11].

2.2 Carbon Dioxide Compound

One of the many possible compounds of carbon is carbon dioxide (CO_2). It consists of one carbon atom in the center of the molecule, covalently double bonded to two oxygen atoms (structural formula: $O = C = O$). Both oxygen atoms have two free electron pairs. The carbon dioxide molecule has a linear structure, with all three atoms lying on a straight line (Fig. 3) [11].

The molar mass of carbon dioxide is 44.01 g/mol and is gaseous under standard conditions. It is colorless, odorless, and tasteless. Carbon dioxide gas is soluble in water and heavier than air. Although it is non-flammable and non-toxic, in higher concentrations it can have a narcotic effect and act as an asphyxiant. Carbon dioxide cannot be detected by human sensory organs; when inhaled in concentrations of 10% or higher, it can lead to (spontaneous) unconsciousness and cause death within minutes. Under pressure, carbon dioxide can be condensed into a colorless liquid. When liquid carbon dioxide rapidly evaporates, it cools down sufficiently to freeze into a solid, also known as solid carbonic acid, "carbonic acid snow" or, when compressed, "dry ice" [8, 11].

Carbon dioxide is present as a natural component of the ambient air (approx. 420 ppm or 0.042%), in seawater, in mineral springs ("carbonated springs"), and is found bound, for instance, in carbonates. It is produced, for example, during respiration, fermentation, decomposition, or the combustion of carbon-containing substances (e.g., hydrocarbon fuels) [11].

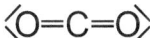

Fig. 3 Structural formula of carbon dioxide (C: carbon atom; O: oxygen atom)

3   Carbon Cycles

The carbon cycle is one of the various biogeochemical cycles on earth involving the movement and transformation of carbon through various (slow) processes within geological time periods. Over long periods of time (a few 100,000 years), the earth's balance of carbon remains approximately steady because of this carbon cycle, resulting in a nearly stable global temperature regime. In the magnitude of millions to tens of millions of years—and thus over clearly long time periods—the balance of carbon accessible within the biosphere can vary greatly due to tectonic plate movements, to changes within the land surfaces as well as to, e.g., volcanic eruptions, which could lead to extreme warm climate (Cretaceous) and glacial climate (Pleistocene). Depending on the time required for the respective processes to absorb and release carbon and its compounds, the global carbon cycle can be subdivided into the slow and the fast carbon cycle.

- Slow carbon cycle. The slow carbon cycle describes the carbon embedded in geological formations and released back into the atmosphere/biosphere within geological time intervals (several millions of years).
- Fast carbon cycle. The fast carbon cycle consists of the carbon removed by plants from the atmosphere during their growth and released again into the atmosphere during the degradation of the biomass through processes such as decomposition and respiration (less than a year and up to 1,000 years).

Understanding the dynamics of these carbon cycles is crucial in comprehending the impact of carbon use on climate and the earth's systems. Different forces drive the slow and fast carbon cycles over distinct periods. Emissions of anthropogenic CO_2 and their impact and effects on the carbon balance of the earth, and thus also on the global climate, can be explained by the respective carbon cycle [12]. Thus, below, a brief overview of the mechanisms of the slow and fast carbon cycle is given.

3.1   Slow Carbon Cycle

The slow carbon cycle refers mainly to the natural process of transfer and transformation of carbon between rocks, oceans and the atmosphere through tectonics, chemical weathering, sedimentation, and volcanism. Within this natural so-called slow carbon cycle, about 10–100 Mt/a of carbon is moved globally completely outside the influence of humans. The slow carbon cycle is more or less self-adjusting over time meaning that an increase of CO_2 in the atmosphere leads to a temperature rise resulting in more rain thus dissolves more rocks. The time to rebalance this carbon cycle through weathering typically takes several millions of years [12]. A schematic overview of the slow carbon cycle, highlighting the various processes and carbon fluxes allocated to this cycle, is given in Fig. 4.

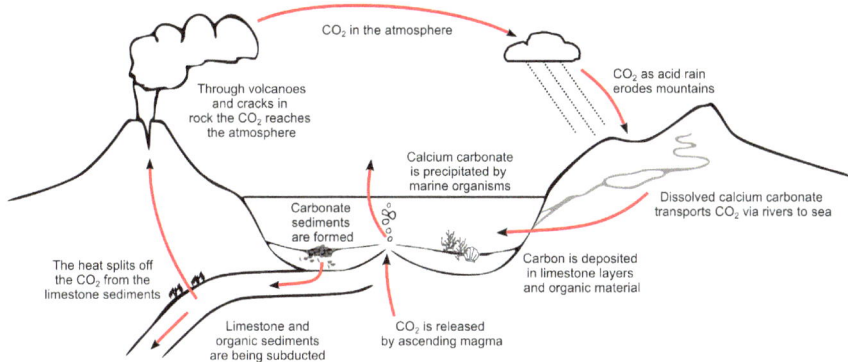

Fig. 4 Scheme of the slow carbon cycle (based on [13])

The first step of the slow carbon cycle begins with the transfer of carbon from the atmosphere to the lithosphere (i.e., the rigid outer shell of the planet). Carbon in form of CO_2 dissolved in rain (carbonic acid) falls onto the earth's surface as slightly acidic rain dissolving rocks (limestone and silicate rocks). This results in the formation of ions of various alkali and alkaline earth metals transported via rivers into the oceans. There, the ions form carbonates by the reaction with hydrogen carbonate from ocean water, resulting in, e.g., calcium carbonate. Organisms, that use the precipitation of calcite to build their skeletons, e.g., shells, plankton, and corals, also perform this process. When these organisms die, they sink on the seafloor over time resulting in the formation of limestone.

Another process of the slow carbon cycle is the formation of rocks via organic carbon from organisms (e.g., shale). Organic material grown on (wet-)land and/or within water either by plants or by animals sinks on the ground of oceans/lakes/wetlands; under these circumstances, oxidation of the biomass is avoided due to a lack of oxygen (i.e., the organic carbon is not released back into the atmosphere). Embedded within these sediments, the organic material could be oxidized only to a very limited extent over time; typically, only a chemical re-formulation of the organic molecules takes place due to typically increasing temperature and pressure over time. In parallel, inorganic material might pile up and embed the organic material into the sediment, respectively, in geological formations. This process is a substantial part of the formation process of fossil energy carriers, such as coal (hard coal and lignite), crude oil, and/or natural gas, if certain conditions such as high pressure and high temperatures are given over time. This embedded carbon might be released back to the atmosphere due to tectonic and other geological activities (e.g., volcanic activities) closing the slow carbon cycle within the very long time periods outlined above.

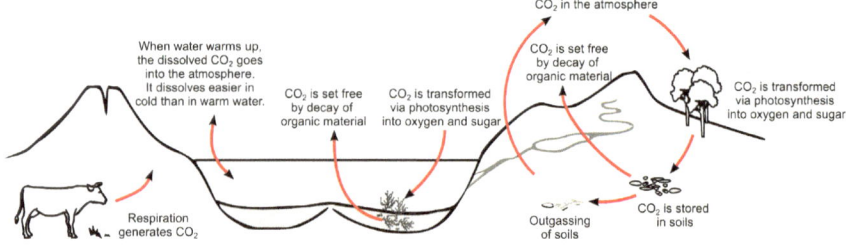

Fig. 5 Scheme of the fast carbon cycle (based on [13])

3.2 Fast Carbon Cycle

The fast carbon cycle mainly refers to the natural process of transfer and transformation of carbon between the atmosphere, oceans, and living organisms. The driving force behind is photosynthesis on the one side and natural degradation/oxidation of organic matter (biomass) on the other. The amount of carbon moved by the fast carbon cycle is around 1–1,000 Gt/a (1,000–1,000,000 Mt/a). The time period of this cycle is less than a year to roughly 1,000 years. A schematic overview of the fast carbon cycle, highlighting the various processes and carbon fluxes, is depicted in Fig. 5.

Land- and water-based plants and micro-organism (e.g., plankton) use the energy of sunlight to form sugar as the basic building block of organic material out of carbon dioxide and water derived from the environment (Eq. 1).

$$6\,CO_2 + 12\,H_2O + energy \rightarrow C_6H_{12}O_6 + 6\,O_2 + 6\,H_2O \qquad (1)$$

When plants and plankton decay, biocatalysts use the energy to power their life from the respective biomass by oxidizing it resulting in the release of water and carbon dioxide (CO_2) back into the environment (Eq. 2).

$$C_6H_{12}O_6 + 6\,O_2 \rightarrow 6\,CO_2 + 6\,H_2O + energy \qquad (2)$$

Thus, the fast carbon cycle moves carbon between the atmosphere and biosphere comprising the ocean and its surface sediments as well as the soils on land. The key processes that influence the fast carbon cycle besides photosynthesis and decay/decomposition are respiration, digestion, and combustion of plants and animals.

3.3 Comparison

In previous chapters, the earth's (slow and fast) carbon cycles, along with associated earth system compartments (e.g., atmosphere, oceans, and soils) are outlined. In the

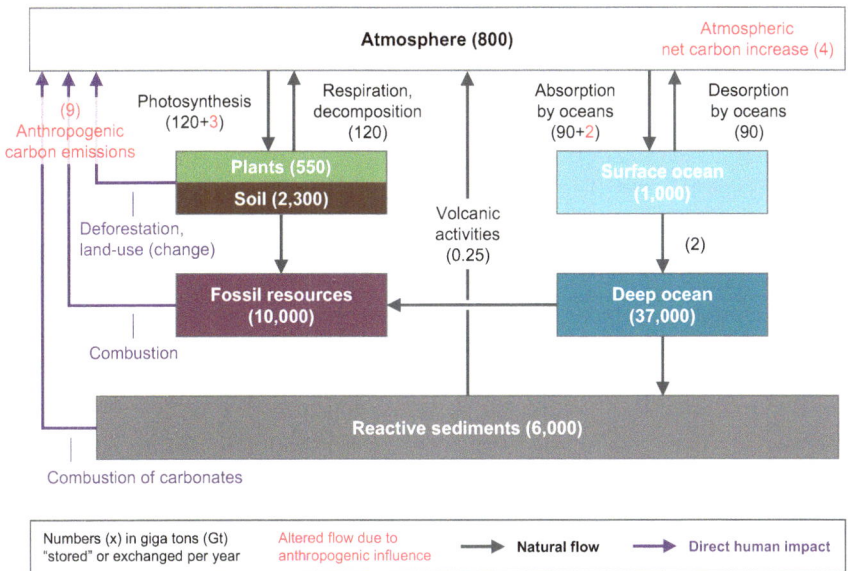

Fig. 6 Scheme of the carbon cycles (based on [12])

following, these cycles and compartments are discussed in an interconnected way with a focus on the amounts of stored carbon and annual carbon exchanges. To illustrate this, Fig. 6 displays the amount of carbon circulated within such a carbon cycle and the carbon stored in the various reservoirs within earth's compartments. This is intended to provide a general, approximated, and quantified overview; actual values may deviate from these approximations.

As Fig. 6 shows, the stored amounts of carbon in the individual reservoirs and the amounts of carbon exchanged annually in (slow and fast) carbon cycles vary considerably. The approximated total amount of stored carbon amounts to about 57,650 Gt. By far the largest portion of this is stored within the deep ocean; in the shown data this amounts to about two-thirds (ca. 64%) of the total carbon amount. The next largest amounts of stored carbon on earth are fossil resources (i.e., crude oil, natural gas, hard coal, lignite) and reactive sediments (e.g., limestone, clay), which account for about a fifth (ca. 17%) and a tenth (ca. 10%) of the total carbon stored. An even smaller portion of the earth's stored carbon is in plants and soils (combined about 5%) and in the surface ocean (ca. 2%). The proportion of carbon that is present in the atmosphere is only a fraction of the total carbon available on earth, amounting to an order of magnitude of 1–2%.

Regarding the annual exchange or transfer of carbon between these reservoirs, the aspects and proportions differ from those of stored carbon; only a fraction of the stored carbon in the reservoirs is actually exchanged. The largest exchange occurs primarily between the atmosphere and other reservoirs. Here, the largest carbon amounts (ca. 120 Gt/a) are exchanged through the fast carbon cycle between the atmosphere and plants during their growth (photosynthesis) and are released back into the atmosphere during degradation of the grown organic matter. Significant carbon exchanges also occur between the atmosphere and the oceans via their surface, amounting here to about 90 Gt/a. Overall, as depicted in Fig. 6, deep oceans act as a significant carbon sink.

Figure 6 also shows that the utilization of fossil resources significantly contributes to the carbon cycle—or disrupts the natural carbon cycles outlined above. Carbon stored within geological time frames within fossil fuel reservoirs is added to the atmosphere's carbon inventory, primarily in the form of carbon dioxide (CO_2), through the combustion of fossil fuels. These carbon dioxide emissions contribute to human-induced global warming. According to the data shown in Fig. 6, this amount of carbon is approximately 9 Gt/a. However, not all of this carbon ultimately remains in the atmosphere fostering global warming. A portion is absorbed by plants and some soils. Another significant portion is absorbed by the upper layers of the oceans. Consequently, since these systems act as buffers, only roughly half of the carbon released due to anthropogenic use of fossil resources ends up in the atmosphere, contributing to global warming. The carbon absorption characteristics of these two systems will be discussed in greater detail below.

- Carbon absorption by plants (and soils). Land plants absorb about 25% of the CO_2 released by human activities. Over the past decades, increased levels of atmospheric CO_2 have increased photosynthesis, leading to greater plant growth (known as "carbon fertilization"). However, while elevated CO_2 levels can enhance plant growth, an increased atmospheric CO_2 level can also contribute to global warming; reduced rainfall and rising temperatures as a direct consequence of climate change can slow down plant growth or even result in their mortality, thereby releasing stored carbon back into the atmosphere. Additionally, stressed plants become more susceptible to various threats, including fires. In this respect, it is important to note that naturally the conversion of the vegetation toward plants adapted to the changed climatic conditions cannot follow the climate change with the same speed as it takes place at the moment. In the event of forest fires (e.g., tropical forests, which might partly act as carbon sinks), significant amounts of stored carbon are released into the atmosphere. As for soils, another concern is the warming or thawing of permafrost, which contains large amounts of carbon accumulated over millennia. When it thaws, decaying organic matter releases methane and CO_2, further increasing the atmospheric carbon inventory [12].
- Carbon absorption by (surface) oceans. Approximately, 30% of the CO_2 released into the atmosphere is absorbed by the oceans through direct chemical interactions. When CO_2 dissolves in seawater, it forms carbonic acid, leading to increased water acidity. While subtle, this shift has (significant) implications for marine life. Firstly, carbonic acid reacts with water's carbonate ions to produce bicarbonate. These carbonate ions are essential for marine organisms like corals to form calcium carbonate shells. With fewer carbonate ions available, these organisms expend more energy, resulting in the development of thinner, more fragile shells. Secondly, increased water acidity enhances the dissolution of calcium carbonate.

Over time, this process can enhance the oceans' ability to absorb excess CO_2, as heightened acidity leads to greater rock dissolution, releasing more carbonate ions. However, in the short term, this increase in acidity can weaken marine organisms' carbonate shells, making them fragile as described. Besides these two effects, increasing oceans' temperatures, as a consequence of global warming, might reduce phytoplankton populations, which thrive in cooler, nutrient-rich waters. This could reduce the oceans' ability to absorb atmospheric carbon through the fast carbon cycle. Conversely, while CO_2 is vital for plant and phytoplankton growth, only a few species benefit from increased CO_2 levels, as most do not directly absorb it from the water [12].

The explanations above demonstrate that plants, soils, and oceans absorb a portion of the carbon released by human activities (due to fossil resource utilization). While adding more carbon to the atmosphere can enhance certain processes and their ability to absorb carbon, other direct and indirect effects may be induced that reduce this capacity and (significantly) increase the risk that stored carbon is released back into the atmosphere.

4 Carbon Source Classification

The explanations above reveal that carbon on earth is continuously transferred by the slow and fast carbon cycle and stored in different compartments of the earth (e.g., atmosphere, oceans, soils, fossil resources) in distinct quantities. Based thereupon, a distinction between fossil and biogenic carbon as well as carbon from ambient air can be made.

- Fossil carbon can be defined as carbon that circulates within the slow carbon cycle; i.e., mainly from rocks or minerals as well as crude oil, natural gas and coal (hard coal and lignite).
- Biogenic carbon can be defined as carbon that circulates within the fast carbon cycle; i.e., this type of carbon is primarily bound in biomass.
- Carbon contained within ambient air (mainly in the form of carbon dioxide (CO_2)) is, according to this definition, a mixture of fossil and biogenic carbon. The same applies for CO_2 from the incineration of mixed carbon stocks, e.g., municipal waste incineration.

Thus, there are various sources of renewable and sustainable carbon that can be utilized. These include recent carbon derived from organic matter of biological origin (i.e., biomass, e.g., from forest wood, from food waste, from agricultural residues), and carbon obtained from the atmosphere through technologies like direct-air-capture (DAC). Thus, recent carbon can be recycled from the atmosphere with the help of physical and technological systems created by humans [14–16].

Each of these sources comes with its own set of advantages and disadvantages. Biogenic capture, which involves capturing CO_2 through plant photosynthesis, is limited by the growth rates of plants and land availability [17, 18]. This limited land availability leads to competition for land usage, particularly between food production—especially due to a still strongly increasing global population—and other demands such as biomass provision for the subsequent use as a raw material [19, 20]. Direct-air-capture (DAC) is another source for the provision of renewable carbon if "green" energy is utilized [21, 22]. Such a direct-air-capture competes with the CO_2 capture from gas streams with an increased percentage of CO_2 in comparison to the percentage of CO_2 contained within the atmosphere (ca. 420 ppm). Typical examples are flue gas streams, such as those derived from heating plants using solid biofuels or from biogas plants based on animal manure.

In the discussion of different possible sources of carbon supply, categorized as either of fossil or biogenic origin—based on their impact on the carbon cycle—and their potential use as future carbon sources, it is important to consider criteria to avoid disrupting the global average temperature regime associated with the fast carbon cycle. These criteria include:

- Stabilizing the amount of carbon or CO_2 in the atmosphere, respectively, which means avoiding additional carbon emissions, is crucial.
- From a physico-chemical perspective, there is no difference between the impact that CO_2 emissions from energy-related processes have on the atmosphere and those from the production of goods, such as cement production.
- The atmosphere "perceives" the overall sum of CO_2 released, regardless of its source, since emitted CO_2 disperses evenly across the globe over time. Therefore, the removal and use of CO_2 from ambient air through direct-air-capture (DAC) is equivalent to using CO_2 from biomass, as long as the biomass is supplied sustainably and maintains a stable carbon stock over several years, particularly in the case of plant-based biomass.

Based on the aforementioned aspects and definitions, the specific carbon source classifications are described in more detail in the following chapters.

4.1 Fossil Carbon

Fossil carbon can be classified into gaseous, liquid, and solid carbon carriers, which can further be subdivided into (a) primary, (b) secondary, and (c) tertiary sources. Primary sources (a) refer to naturally occurring carbon carriers or raw materials, secondary sources (b) refer to products derived from these raw materials, and tertiary sources (c) represent the end-of-use stage with the highest oxidation state, typically as CO_2. Examples of these different carbon carriers and their respective classification are presented in Fig. 7.

A detailed discussion on the most relevant fossil carbon carriers in terms of further use is provided below. These include mainly tertiary CO_2 emissions from various sources and secondary solid carbon carriers.

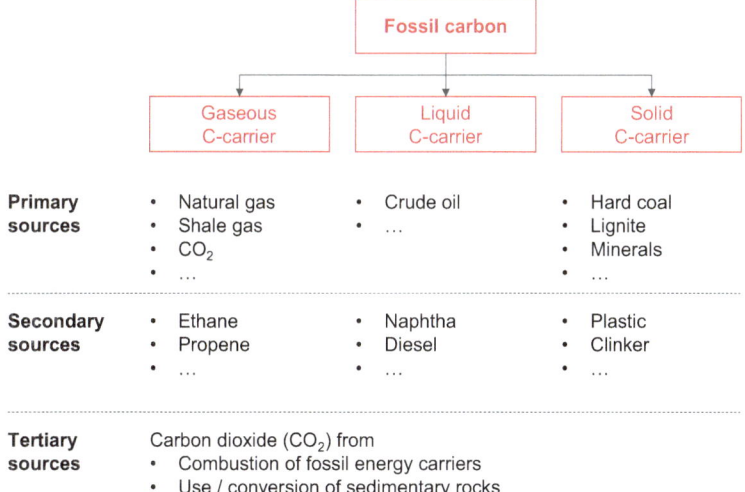

Fig. 7 Classification of fossil carbon sources

Fossil carbon from tertiary gaseous carbon carriers. The primary source of anthropogenic fossil CO_2 emissions is the combustion of fossil energy carriers, including hard coal, lignite, natural gas, crude oil, and its derivatives, for the purpose of generating electricity, producing heat, and for transportation purposes. As a result, these tertiary CO_2 emissions are primarily associated with energy-related activities [23]. Additional sources of CO_2 emissions include the utilization of fossil carbon resources as raw materials in specific industrial applications, as well as CO_2 emissions from the deacidification of minerals containing carbonates. These emissions are classified as process-related, as they result from specific industrial conversion processes rather than energy production [23].

The use of energy-related fossil CO_2 emissions from electricity and heat provision as a carbon source for subsequent industrial applications, such as for the production of synthetic hydrocarbons, would inevitably result in an increase in greenhouse gas (GHG) emissions within the biosphere and thereby contributing to further global warming. Moreover, the long-term availability of CO_2 from electricity generation is uncertain, as there is a growing shift toward electricity generation based on renewable sources of energy to meet the valid GHG reduction goals. Alternative technologies, e.g., the usage of wind and solar power, are already available on a large scale [24].

The utilization of process-related fossil CO_2 is a topic of widespread discussion, as it could potentially enable further cascade uses as a precursor for the production of synthetic carbon-containing compounds. This is because even in a highly defossilized energy and production system, some (small) amounts of fossil CO_2 emissions may still occur [25]. The mineral industry, including the production of cement clinker, quicklime, glass, and ceramics, is a primary source of process-related CO_2 emissions of fossil origin. These emissions are necessarily released during the production of these products, as CO_2 is driven out from carbonate compounds used as raw materials due to a chemical transformation. The resulting properties of the burnt or sintered products are crucial for their intended use [26]. Despite the lack of available alternatives for many of these processes so far, it is important to note that these CO_2 emissions from the mineral industry contribute significantly to the overall greenhouse gas (GHG) emissions in the atmosphere.

Fossil carbon from secondary solid carbon carriers. Carbon of fossil origin can also be obtained from other, secondary sources. A widely discussed example is plastic recycling or—in more general terms—the recycling of carbon-containing waste streams. Direct use of carbon via material recycling is only feasible for certain applications such as the production of new plastics from granulate obtained from plastic waste (mechanical recycling) in some specific cases [27]. A distinction between direct use of carbon and the intermediate state of CO_2 is important in the context of carbon recycling. Processes such as combustion, pyrolysis, or gasification can convert carbon-containing waste streams into CO_2 or syngas, which can then be used as feedstock for the production of new carbon-containing products and fuels.

However, the use of CO_2 or syngas from such fossil-derived sources based on these processes may result in near-term increase of greenhouse gas (GHG) emissions of fossil origin, as the carbon in these feedstocks so far fully originated from fossil sources. This is true if after a subsequent "new" life cycle of the carbon-containing product the respective product is treated as a waste and e.g., burnt within a waste incineration plant. A contribution to anthropogenic global warming is therefore the case because eventually the carbon will be released into the atmosphere. Nevertheless, secondary use of carbon from fossil sources, such as through material recycling, helps to keep the carbon within the industrial use cycle for a certain period of time.

4.2 Biogenic Carbon

Plants perform photosynthesis to capture carbon in the form of carbon dioxide (CO_2) from the atmosphere, incorporating it into carbohydrates to form complex macromolecular building blocks such as cellulose, hemicellulose, and lignin. Thus, the overall available biomass can basically be considered as a carbon carrier. Based on the availability of this carbon, the following biogenic carbon sources can be distinguished.

- Gaseous biogenic carbon carrier. Biogas, which consists of methane and CO_2, is a common example of gaseous biogenic carbon carrier. Other examples include CO_2 generated from bioethanol production, composting, and biomass combustion/thermal conversion. A further distinction can be made between energy-related processes (e.g., heat provision from solid biomass) and process-related sources (e.g., alcoholic fermentation).
- Liquid biogenic carbon carrier. Plant-based biomass can also serve as a source of liquid carbon carrier. Typical examples include plant oils, different types of

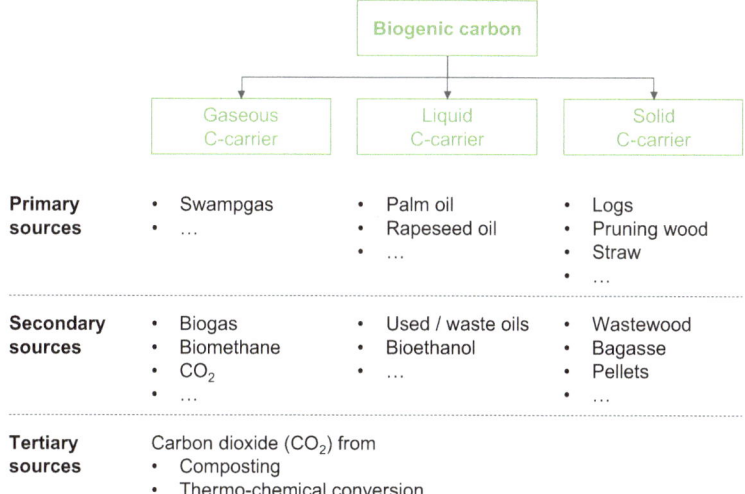

Fig. 8 Classification of biogenic carbon sources

alcohols and liquids released from plants (e.g., sugar containing fluids, rubber containing fluids).

- Solid biogenic carbon carrier. Lignocellulose, such as wood, also serves as a solid carbon carrier. Therefore, various forms of solid biomass, including products, by-products, waste-products, and residues, could potentially be used for this purpose.

Similar to the classification of fossil carbon, the carbon carriers can also be further divided into three categories: (a) primary or raw material, (b) secondary or processed products, and (c) tertiary or end-of-use phase with the highest oxidation state typically being CO_2 (Fig. 8).

The CO_2 emissions from sustainably produced biomass being part of the fast carbon cycle do not contribute to an increase of the atmospheric carbon inventory, as they are naturally balanced within the mentioned (fast) carbon cycle. However, the use of biogenic CO_2 released from sustainably produced cash crops in the current industrial system may result in a slight increase of fossil greenhouse gas (GHG) emissions. This is because the cultivation, harvesting, and transport of cultivated biomass for industrial use is usually associated with a use of fossil energy carriers, such as fertilizers and transportation fuels, leading to a release of additional GHG emissions [28]. The extent to which GHG emissions from a certain biomass supply chain contribute to climate change depends on the level of implementation of a defossilized system and sustainable cultivation methods. As the global energy system becomes increasingly defossilized, the additional GHG emissions from the supply chain of biogenic carbon, or biomass, are expected to decrease, making biogenic carbon more and more climate neutral.

In order for biogenic carbon to be used as a substitute for fossil fuel-based carbon, certain preconditions must be fulfilled.

- Biomass must be provided in a sustainable manner, meaning that organic carbon from sources such as rainforests converted into grasslands does not count as sustainable, as this reduces the carbon stock on a particular piece of land.
- Biomass waste streams, organic residues, and by-products from agricultural production, food processing, and final use can potentially be used, as long as food production is carried out in a sustainable manner.
- Similarly, wood from forests can be used in a figurative sense if the forests are managed according to the globally valid sustainability criteria.

4.3 Mixed Carbon Sources

Beyond the previously mentioned carbon sources, mixed carbon sources primarily consist of a combination of fossil and biogenic carbon. Within this category, two distinct sub-classifications can be considered; carbon from ambient air and carbon from waste.

Carbon from ambient air. The current atmospheric concentration of CO_2 amounts to approximately 422 ppm and is steadily increasing, due to the combustion of fossil energy carriers. This CO_2 can be extracted from the air and utilized as a carbon source. The utilization of carbon from ambient air is not expected to have negative effects on the fast carbon cycle and global warming, as it does not disrupt the natural carbon cycle. Therefore, it is anticipated that the technical use of atmospheric CO_2 will not result in any adverse consequences with regard to climate change.

Carbon from waste. One potential option for utilizing CO_2 emissions generated from electricity and heat production is through thermal waste treatment. However, when municipal solid waste is used as a feedstock, the resulting CO_2 emissions are typically a combination of carbon from both biogenic sources, such as food and yard waste, and fossil-based sources, such as plastics or other waste of non-biological origin [29]. Thus, the climate impact depends on the proportion of fossil and biogenic carbon within the respective waste streams.

5   Fossil Carbon Sources

In the following sections, a detailed description of fossil carbon sources and their respective potentials is given. First, the different possible sources are characterized followed by the determination of their respective potential. Based on the previous conclusion that carbon (respectively CO_2) resulting, e.g., from the provision of electrical or thermal energy based on fossil fuel energy should be excluded as a possible source for carbon, the following analysis of various fossil carbon sources focuses particularly on those sources that emit process-related CO_2. Therefore, a detailed description of the respective processes is given in order to determine the share of process- and energy-related carbon, respectively CO_2. The production processes and their descriptions are mainly related to the conditions of the industrial location Germany, as representative processes of the state of the art.

5.1 Mineral Industry

5.1.1 Cement

Cement is a complex mixture of various pure materials, with the specific composition of each type of cement tailored to meet the desired properties for different applications, such as durability or sulfate resistance. Cement clinker, which makes up a significant portion (ranging from 5 to 100% as per DIN EN 197-1), is a common component in almost all types of cement. Other materials, such as granulated slag, pozzolana, fly ash, silica dust, plaster, and burnt oil shale, may also be present in cement compositions.

The production of cement involves three main steps. Firstly, raw materials such as clay and limestone are extracted from the earth's surface and crushed. Additional components, such as quartz sand and iron ore, if needed, may be added, and the mixture is then homogenized, dried, ground, and burned in kilns. The thermally treated material is subsequently homogenized, ground again, and mixed with additives to meet required standards.

The process of clinker production is responsible for the largest amount of CO_2 emissions in the overall cement production process. These emissions can be categorized as energy-related and process-related emissions. Process-related emissions occur during the burning of input materials at high temperatures (gas temperature up to 2,000 °C) in a rotary or shaft kiln. The chemical transformation that occurs during this thermal treatment leads to the release of CO_2 from the carbonate, a process known as "deacidification" (Eq. 3).

$$CaCO_3 + heat \rightarrow CaO + CO_2 \tag{3}$$

The chemical reaction that results in the release of CO_2 from the carbonate occurs in the presence of SiO_2, Al_2O_3, and Fe_2O_3 at temperatures above 550 °C, with a significant increase in reaction rate observed above 900 °C. The provision of heat for the kiln, which is necessary for the cement production process, is responsible for the majority of energy-related CO_2 emissions released into the atmosphere by the cement industry. In addition to thermal energy, electricity is also required as auxiliary energy for electrically driven devices, such as grinders, conveyer belts, and exhaust gas treatment systems.

The CO_2 emissions from the production of cement, thus, originate from burning fuel for heat provision, electricity demand for auxiliary energy supply and process-related emissions from processing the material itself. The share of the process- and energy-related CO_2 emissions from the kilns and furnaces varies depending on the mixture of the input material (fuels, clay, limestone, etc.). Nevertheless, the variations are relatively small. Therefore, it is assumed that the share of the process- and energy-related CO_2 is similar for each cement production site. The share of process-related emission is about 60% of the overall emissions [19].

5.1.2 Limestone and Quicklime

The majority of limestone consists of over 90% $CaCO_3$ with a small percentage of $MgCO_3$, while dolomite is a mixture of approximately 55% $CaCO_3$ and 45% $MgCO_3$. Despite this composition, the lime industry in Germany is dominated by limestone processing [22]. A special case within this context is the production of refined sugar, which involves the processing of lime into quicklime, resulting in both energy-related and process-related emissions.

The processing of limestone usually begins with the extraction of raw material from the earth's crust typically realized in open-pit mines located in close proximity to the lime plant. The majority of limestone is obtained through quarrying using explosives. Dolomite as an alternative is mined in a few quarries and processed at various locations. After blasting, the extracted material is crushed, washed, and screened. It is then fed into kilns for calcination, or processed without burning for other purposes such as aggregates for the cement or steel industry (e.g., for lime putty).

During the calcination process in kilns, process-related CO_2 emissions are released from limestone or dolomite. This occurs at temperatures ranging from 900 to 1,200 °C for limestone and 500–750 °C for dolomite. At these temperatures, calcium and magnesium carbonates are thermally decomposed, resulting in the emission of CO_2 originally bound within these minerals. This can be described by Eq. (4) for calcium carbonate and Eq. (5) for magnesium carbonate.

$$CaCO_3 + heat \rightarrow CaO + CO_2 \tag{4}$$

$$CaCO_3 \cdot MgCO_3 + heat \rightarrow CaO \cdot MgO + 2\,CO_2 \tag{5}$$

Once the calcination process in the kiln is complete, the resulting quicklime is stored and can be hydrated to form calcium and magnesium hydroxide. These hydrated products are commonly used in applications such as in mortar. Additionally, quicklime can be directly utilized for various purposes, such as desulphurization of raw steel or as a fertilizer [23].

The production of refined sugar also relies on limestone and quicklime used for the purification of raw sugar juice after the extraction process. As a result, the sugar industry often operates its own lime kilns, with many of them being located on-site at sugar refineries [23].

The emissions from the limestone industry are attributed to both, fuel burning and the calcination process. It is estimated that approximately 35% of the emissions are energy-related, while 65% are process-related CO_2 emissions [24].

5.1.3 Glass and Ceramics

While there are various types of glass and ceramics with different compositions and properties, the production processes are generally similar in nature. Thus, the following analysis takes into account both production processes for glass and ceramics.

The production of glass requires raw materials such as silica sand, limestone, caustic soda, and various additives for purification, dyeing, or melting point reduction. The initial step in the production process involves batch preparation and mixture with recycled materials, which are then fed into the melting process. The resulting melt is then formed, conditioned, cured, and dried. The final steps in the production process include cutting, milling, and packaging. During the melting process, the majority of process-related and energy-related CO_2 emissions are released [17]. The process-related emissions in glass production arise from the carbonates used, such as $CaCO_3$, Na_2CO_3, $MgCO_3$, and $BaCO_3$. During the melting process, CO_2 is released from these carbonates, similar to the production of clinker and quicklime [14]. For the time being natural gas is the primary source of process-related energy used in kilns during glass production in Germany.

The production process for ceramics is similar to that of glass, involving the extraction and preparation of silicates and oxidic raw materials, shaping, drying, surface finishing, and burning. The drying and burning process is energy-intensive and accounts for the majority of the overall energy demand. Process-related emissions, which include both carbonate and fossil components of the raw materials, occur during the burning process. Additionally, porosification agents like sawdust can contribute to the overall emissions. Further finishing treatments may be applied as the final processing step, if needed [14].

Since the glass and ceramics industry produces a variety of goods (e.g., container glass, flat glass, bricks, tiles, technical ceramics) showing different shares of specific process-related CO_2 emission [17], an average proportion is assumed (based on [23]). The average share of process-related CO_2 from the overall CO_2 assigned to glass facilities is 15% and that of the ceramics industry is 55%.

5.2 Iron and Steel Industry

In the following section, the production of iron and steel using the oxygen steel and electric arc furnace routes are discussed. In Germany, these two main processes for crude steel production are currently in use. Approximately, one third of the production is based on the electric arc furnace route, producing electrical steel, while the remaining two-thirds are produced using the oxygen steel process in integrated mills [15]. Typically, oxygen steel is primarily made from iron ore and scrap metal as raw materials.

5.2.1 Steel Mills

The production of crude steel in blast furnaces involves several steps, including the extraction of ore and its processing. Once the ore is extracted, it undergoes a reduction process in blast furnaces to produce crude steel. The resulting crude steel is then further processed into intermediate and final products in rolling mills. This processing typically involves hot forming, which can result in emissions of CO_2 from the burning of gas or natural gas used during the heating process.

At integrated mills, the production of crude steel starts with grinded iron ore fines, which are obtained from ore dressing to remove gangue from the ore. These fines are then mixed with aggregates such as coke breeze, as well as a mixture of shales and lime as slack formers in the blast furnace. The mixture is then sintered at temperatures ranging from 900 to 1,200 °C, resulting in caking of the materials. Afterward, the sintered iron ore is further processed in blast furnaces to remove oxygen and remaining gangue.

In the blast furnace process for pig iron production, reduction agents are needed to reduce the iron oxide. The main reduction agent used is coke, and during the process, the carbon contained within the coke is partially oxidized, resulting in the discharge of CO and CO_2, along with other gases known as blast furnace gas. The reduction process occurs through indirect reduction via CO at temperatures ranging from 450 to 1,000 °C (Eq. 6), as well as direct reduction via carbon at temperatures above 1,000 °C (Eq. 7).

$$Fe_3O_4 + 4\,CO \rightarrow 4\,CO_2 + 3\,Fe \tag{6}$$

$$2\,Fe_2O_3 + 3\,C \rightarrow 3\,CO_2 + 4\,Fe \tag{7}$$

The pig iron obtained from the blast furnace is then processed in the oxygen steel converter. In this step, the iron is first desulfurized and then decarburized by blowing in oxygen into the converter, along with steel scrap and other additives. This process further reduces the carbon content, resulting in crude steel with an average carbon content of around 0.15%. The gaseous emissions, which contain CO and CO_2 among other gases escaping from the converter, are known as converter gas.

This converter gas, along with blast furnace gas, is used as an energy carrier. These gases are utilized as a fuel in various applications, such as in the hot blast stoves of the blast furnace plant, coke oven under-grate firing, process firing in sinter and rolled steel production, and electricity generation in industrial power plants.

The main source of process-related CO_2 emissions in integrated steel mills is the use of reduction agents, primarily coke, and limestone. Blast furnace gas consisting mostly of CO and CO_2 is utilized for heat supply in various sections of the steel mill as well as for electricity production in attached industrial power plants (since 2016, no gas is used in public power plants anymore). Since all these processes are located within the integrated steel mill or nearby, it is assumed that all CO_2 emissions are emitted at the location of the respective steel mill, as they originate from the reduction agent.

The use of coke as a reduction agent results in an emission factor of approximately 1.67 t CO_2 per ton of pig iron. Limestone is used in sintering plants and blast furnaces, and burnt lime is used as a slag generator in electric steel production and in the converter of oxygen steelworks. Quicklime is mainly obtained from the lime industry; i.e., the emissions associated with limestone are attributed to the lime industry. The arithmetic mean of these emissions leads to a specific emission factor of 76 kg CO_2 per ton of pig iron. The combined emission factor used to determine process-related CO_2 emissions is approximately 1.75 t CO_2 per ton of pig iron (see [23]).

5.2.2 Electro-Steel Plants

Electric arc furnaces can process various input materials. In Germany, scrap is the primary input material used in all electro-steel plants, with the exception of one plant in Hamburg utilizing also direct reduced iron or iron sponge as an input [31]. Other potential input materials include liquid or solid pig iron, quicklime, and foaming coal as slag formers and reduction agents. Iron sponge is produced in shaft furnaces using a mixture of H_2 and CO as reduction gas to extract oxygen from the iron ore (Eqs. 8 and 9). The input material is melted by electric energy applied through the electrodes gradually consumed during the process (Eq. 10). Once the required temperature and chemical composition are achieved, the hot melt is transferred to a steel ladle for secondary metallurgical treatment; this is very much similar to integrated steel mills. This treatment may involve processes such as decarburization, alloying, and homogenization. Finally, the melt is casted and solidified using continuous or block casting methods.

$$Fe_2O_3 + 3\,CO \rightarrow 3\,CO_2 + 2\,Fe \qquad (8)$$

$$Fe_2O_3 + 3\,H_2 \rightarrow 3\,H_2O + 2\,Fe \qquad (9)$$

$$C + O_2 \rightarrow CO_2 \qquad (10)$$

The process-related CO_2 emissions resulting from this process are caused by the combustion of the graphite electrodes and the oxidation reaction of the foaming coal. The electrodes consist of graphite burned with a ratio of 2.06 kg/t steel during the process. The specific emission factor for electric steel production is 7.1 kg CO_2/t steel [14]. The usage of injection/foaming coal leads to CO_2 emissions of approximately

23 kg CO_2/t crude steel [31] resulting in an overall emissions factor of 0.03 t CO_2/t steel.

5.3 Non-ferrous Metal Industry

Below, the production of primary aluminum, zinc, lead, and magnesium is described in detail.

5.3.1 Primary Aluminum

The production of aluminum involves three main steps. First, bauxite being the raw material for aluminum production is extracted from open-pit mines located in various regions around the world; major producers are Australia, China, and Guinea [33]. Next, aluminum oxide (also known as alumina) is extracted from bauxite using the Bayer process, which involves dissolving aluminum hydroxide in sodium hydroxide to separate iron-containing minerals. The resulting aluminum hydroxide is then calcined in kilns to obtain aluminum oxide. This process does not result in process-related CO_2 emissions as per definition provided earlier. However, the high energy demand associated with mining and the Bayer process leads to energy-related and indirect CO_2 emissions, primarily associated with electricity production.

The extracted aluminum oxide is further refined using the Hall-Héroult process, wherein the aluminum oxide is dissolved in molten cryolite to lower the melting point of the electrolyte. The electrolyte is contained in a steel pot lined with graphite serving as the cathode. Graphite blocks, acting as anodes in the form of prebaked electrode cells, are immersed from above into the electrolyte. A direct current (DC) voltage and high ampere current are applied to the anode and cathode, causing the dissociated aluminum cations to migrate to the cathode at the bottom of the cell and the oxygen-containing ions to migrate to the anode on top.

The overall chemical reaction is shown in Eqs. (11) and (12). The process-related CO_2 emissions from primary aluminum production mainly result from the combustion of the carbon anodes during the electrolysis of the aluminum oxide.

$$Al_2O_3 + 3\,C\ \rightarrow 2\,Al + 3\,CO \tag{11}$$

$$Al_2O_3 + \frac{3}{2}\,C\ \rightarrow 2\,Al + \frac{3}{2}\,CO_2 \tag{12}$$

The main compound formed at the anode during the electrolysis of aluminum is CO_2. Carbon from the anode is continuously consumed during this process and needs to be replaced periodically. The liquid elemental aluminum, which is denser than molten cryolite, is collected and discharged at the bottom of the cell via a siphon [34].

Some aluminum electrolyzing cells are equipped with prebaked anodes, also known as prebaked cells. The manufacturing of these prebaked anodes is another source of CO_2 emissions, in addition to their consumption during the electrolysis process [35]. To manufacture prebaked anodes, a mixture of petroleum coke, coal tar pitch, and crushed spent anodes is pressed into pits and baked for about 22–32 h at temperatures around 1,200–1,300 °C in oil- or gas-fired furnaces. After cooling, steel stubs are inserted into stub holes on top of the anode block and fixed with poured cast iron. These stubs are then connected to an aluminum or copper rod to facilitate the flow of electricity into the anode [34].

The carbon from the anodes reacts with the oxygen separated from alumina to form CO_2 during the electrolysis of aluminum. On average, 430 kg of anodes are consumed per ton of aluminum produced, assuming an average carbon content of 857 kg per ton of anode and stoichiometric conversion to CO_2. This results in a specific emission factor of 1,367 kg CO_2 per ton of aluminum [14]. It is also assumed that the CO formed during the electrolysis is further oxidized to CO_2 upon contact with ambient air and is included in the aforementioned emission factor [37].

CO_2 emissions from anode baking primarily result from three combustion processes: fuel combustion for firing the furnace, combustion of volatile matter, and combustion of baking furnace packing material [37]. Half of the total emissions from anode manufacturing are emitted by burning fuel, while the other 50% are process-related emissions from the combustion of pitch volatiles and packing coke. Total emissions from prebaked anode production vary between 320 and 575 kg CO_2 per ton of baked anode produced [38].

5.3.2 Zinc, Lead, and Magnesium

Zinc. Zinc is primarily extracted from its ore, sphalerite, which is a mineral composed of zinc and sulfur. The ore is mined using conventional mining methods, and then crushed and ground to liberate the zinc-bearing minerals. The crushed ore is then concentrated using a series of flotation and gravity separation techniques, which separate the zinc minerals from the other minerals in the ore. Primary zinc is produced using a hydrometallurgical process. In this process, ore concentrates, primarily zinc blende, are desulfurized in a roasting process. Carbon carriers are added, and zinc oxide and sulfuric acid are formed at around 900 °C. Sulfuric acid is removed as a by-product, and zinc oxide is then fed to the electrolysis, where pure zinc accumulates at the cathode. The anode is typically made of aluminum, which dissolves in the solution to release electrons that are used to reduce the zinc ions to metallic zinc. Once the zinc is deposited on the cathode, it can be cast into different shapes and forms, such as ingots, billets, and sheets. These shapes can then be further processed using various techniques, such as rolling, extrusion, and drawing, to produce finished products, such as pipes, wires, and galvanized steel. CO_2 emissions can occur at several stages of the process. One major source of CO_2 emissions is during the roasting process, which requires the use of fossil fuels, such as coal, oil, or natural gas, to generate the high temperatures required to convert the zinc sulfide minerals to zinc oxide.

In secondary zinc production, secondary materials containing zinc are oxidized to zinc oxide in rotary kilns by adding petroleum coke, which is fed into the production process of leaching and electrolysis. Process heat is obtained from using mainly crude oil and /or natural gas [41]. The recycling process also generates CO_2 emissions, primarily due to the energy required to melt the scrap and the use of fossil fuels (petroleum coke) in the melting process. However, the CO_2 emissions from the recycling process are generally lower than those from the primary production process, as the energy required to melt the scrap is usually lower than the energy required to roast and leach the ore.

Lead. Lead is primarily extracted from its ore, galena, which is a mineral composed of lead and sulfur. The ore is mined using conventional mining methods, and then crushed and ground to liberate the lead-bearing minerals. The crushed ore is then concentrated using a series of flotation and gravity separation techniques, which separate the lead minerals from the other minerals in the ore. The concentrate is then smelted in a furnace, where it is heated to high temperatures along with coke and other additives. The smelting process separates the lead from the other minerals, and produces lead bullion, which has a purity of about 99%. The lead bullion produced through smelting contains impurities, such as copper, silver, and antimony. Therefore, it needs to be refined to meet the required purity standards. This is typically done through a process called desilverization, in which the lead bullion is melted and treated with zinc. The zinc reacts with the silver and other impurities, forming a slag that can be skimmed off. The resulting lead is then further refined using various techniques, such as electrolysis, to remove any remaining impurities. Once the lead is refined, it can be cast into different shapes and forms, such as ingots, billets, and sheets. These shapes can then be further processed using various techniques, such as rolling, extrusion, and drawing, to produce finished products, such as pipes, wires, and batteries.

The production of primary and secondary lead emits process-related CO_2 due to the usage of fossil reduction agents such as hard coal and petroleum coke in the direct smelting process in bath smelting furnaces (Isasmelt-Ausmelt) or QSL-Reactors (primary). Short rotary furnaces or shaft furnaces processes are used for secondary lead production.

Magnesium. Magnesium is not found in its pure form in nature, but is widely distributed in the form of minerals, such as dolomite and magnesite. These minerals are mined using conventional mining methods, and then crushed and refined to extract magnesium. The extracted magnesium is typically in the form of magnesium chloride or magnesium oxide. The next production step involves converting the extracted magnesium into a form that can be used to make alloys and other products. This is usually done through electrolysis. In this process, magnesium chloride or magnesium oxide is dissolved in molten salt, and then an electric current is passed through the solution, which causes the magnesium to separate out. Then it can be collected at the bottom of the container. The magnesium produced through electrolysis is not pure and contains impurities, such as iron and aluminum. Therefore, it also needs to be purified to meet the required purity standards. This is typically done through vacuum distillation, in which the magnesium is heated in a vacuum chamber, and the impurities evaporate. After the magnesium is purified, it can be cast into different shapes and forms, such as ingots, billets, and slabs. These shapes can then be further processed using various techniques, such as rolling, forging, and extrusion, to produce finished products, such as sheets, plates, bars, and wires.

During the extraction process, CO_2 emissions can occur due to the use of fossil fuels, such as hard coal, crude oil, or natural gas, to generate the high temperatures required for thermal reduction or vacuum distillation. The refining and casting stages also require energy and can result in CO_2 emissions. The refining process typically involves heating the magnesium to high temperatures and subjecting it to a vacuum or other purification techniques, which can require the use of fossil fuels. The casting process can also require the use of fossil fuels to melt the magnesium and produce the desired shapes. Process-related emission result from the carbonates being part of the minerals used as input materials (e.g., dolomite and magnesite). During the production process, these carbonates emit in the form of CO_2, similar to the process which occurs during the burning of lime and cement clinker.

5.4 Chemical Industry

The production of soda, carbon black, and ammonia, as well as ethylene oxide is described in detail below.

5.4.1 Soda

Soda is produced based on the Solvay process. The first step is the purification of brine being a concentrated solution of salt (sodium chloride) in water by removing impurities such as calcium and magnesium ions. These impurities might interfere with the chemical reactions in the subsequent process steps. The purified brine is then mixed with ammonia (NH_3) and carbon dioxide (CO_2) gas. This causes the formation of sodium bicarbonate ($NaHCO_3$) and ammonium chloride (NH_4Cl), as shown in Eq. (13).

$$NH_3 + CO_2 + NaCl + H_2O \rightarrow NaHCO_3 + NH_4Cl \qquad (13)$$

The sodium bicarbonate is then heated to high temperatures, causing it to break down into soda ash (Na_2CO_3), water, and carbon dioxide (Eq. 14).

$$2NaHCO_3 \rightarrow Na_2CO_3 + H_2O + CO_2 \qquad (14)$$

The ammonium chloride that was formed in the previous step is recovered by treating it with lime (CaO) to produce ammonia gas and calcium chloride ($CaCl_2$), as shown in Eq. (15).

$$2NH_4Cl + CaO \rightarrow 2NH_3 + CaCl_2 + H_2O \tag{15}$$

Soda is used in great amounts during glass production where it is used to reduce the melting temperature of sand. Other major applications of soda are the production of detergents and cleaners as well as the chemical industry. The synthetic production of soda emits process-related CO_2 during the Solvay process. The CO_2 occurring from the input material calcium carbonate is partly embedded into the product Na_2CO_3, as described above. However, the process runs under CO_2 excess, resulting in overall emissions of 200–300 kg CO_2/t Na_2CO_3 [30]. Besides calcium carbonate, anthracite and coke are the main sources for CO_2 emissions. A more specific emission factor for the overall process is given for one soda ash plant located in Europe, which is 0.277 t CO_2/t Na_2CO_3 [31]. A determination between process-related and energy-related share within the stated emission factors is not given.

5.4.2 Carbon Black

Carbon black is produced in lamp black or gas black facilities and furnace black process. The CO_2 emissions from this process result from the combustion of hydrocarbons for heat supply in the reactor and incomplete combustion of, e.g., cracker or distillation residues as product feedstock; the latter is classified as process-related.

The dominant process for the global production of carbon black is the furnace process. This process involves the incomplete combustion of heavy petroleum products, such as fluid catalytic cracking (FCC) tar, coal tar, or ethylene cracking tar, in a furnace with a controlled amount of air. The resulting mixture of gases and fine particles is then cooled rapidly, causing the carbon black to condense and form a black powder. The powder is then collected, cleaned, and processed to produce different grades of carbon black with varying properties. This process shows an emissions factor of 2.26 t CO_2/t carbon black [32].

5.4.3 Ammonia

Ammonia production, known as the Haber–Bosch process, involves the synthesis of ammonia (NH_3) from hydrogen and nitrogen gas. Nitrogen is obtained from air through air separation units. As of today, the required hydrogen is mostly obtained from hydrocarbon sources, typically natural gas, through steam methane reforming. Steam reforming is an endothermic process in which desulfurized hydrocarbons are mixed with steam and heated up to a temperature level of 700–900 °C, forming CO and hydrogen (H_2). The CO is then mixed with steam again at 250–300 °C in the water–gas-shift-reaction (WGS-reaction), which produces CO_2 and increases the yield of H_2 in parallel. The hydrogen obtained from this reforming process is synthesized together with nitrogen (obtained from air separation) in the Haber–Bosch process to ammonia. A total of 1.8 t CO_2/t ammonia is emitted. Approximately, 1.3 t CO_2 of these emissions are process-related due to the reforming of natural gas and 0.5 t CO_2 result from the generation of steam (heat provision) [16].

5.4.4 Ethylene Oxide

Ethylene is typically used as the feedstock for ethylene oxide production. Ethylene is obtained through the cracking of hydrocarbons, such as natural gas or crude oil, in the presence of heat and/or catalysts. The ethylene feedstock is typically purified to remove impurities, such as water, oxygen, and other hydrocarbons, to ensure high-quality ethylene for the subsequent steps. Ethylene is then reacted with pure oxygen or air in the presence of a catalyst to undergo an oxidation reaction. This reaction is typically carried out in a fixed-bed reactor or a fluidized-bed reactor at elevated temperatures (usually between 200 and 300 °C) and pressures (usually between 1 and 3 MPa). The catalyst used in the oxidation reaction is typically silver-based, such as silver supported on alumina or silica, which helps to promote the selective oxidation of ethylene to ethylene oxide.

After the oxidation reaction, the reaction mixture is cooled and the resulting product stream contains a mixture of ethylene oxide, unreacted ethylene, carbon dioxide, water, and other by-products. The product stream is then subjected to a series of separation and purification steps to separate and remove these components. This typically involves processes such as distillation, absorption, and scrubbing, which help to separate ethylene oxide from other components in the mixture and purify it to a high degree. Once ethylene oxide has been separated and purified, it is typically stored in suitable containers or tanks for further handling and distribution. The by-products generated during the ethylene oxide production process, such as unreacted ethylene, carbon dioxide, and water, are typically recovered and recycled back into the process to minimize waste and optimize resource utilization.

The specific emission factors of ethylene production are 0.863 t CO_2/t C_2H_4O and 0.663 t CO_2/t C_2H_4O for the air process and oxygen process, respectively [33].

5.5 Fossil Carbon Potentials

The amount of CO_2 emissions from the industrial sectors and applications described above are presented in the following. This relates to global CO_2 emissions and only direct energy- and process-related emissions are considered. CO_2 emissions that might result from, e.g., the provision of electricity are not considered. In part, no consistent quantities are available over the same time series. Therefore, in the following, the CO_2 emissions for the year 2019 will be determined and reported as a basis. The quantities given in the corresponding summaries are given as annual quantities in Table 1 although it is recognized that the CO_2 emissions may vary depending on the production output or economic situation. Where possible, total sector CO_2 emissions are reported, as well as process emissions.

The CO_2 emissions from the industrial sectors and applications detailed in Table 1 are graphically represented in Fig. 9 for easier comparison. The left side of the figure displays the annual CO_2 emissions for each sector, as outlined in the table, while the right side aggregates the CO_2 emissions by the considered industries.

The total of CO_2 emissions from the mineral industry amounts to about 2,705 Mt CO_2/a, of which about 1,595 Mt CO_2/a (ca. 59%) are process-related emissions. Within the mineral industry, the greatest proportion of CO_2 emissions is released by the cement clinker production amounting to approximately 2,300 Mt CO_2/a in 2019 [34]. The countries with the greatest production capacity of cement clinker, thus resulting in the greatest amount of CO_2 emissions, are China, India, Vietnam, Turkey, and the USA. These countries emit about 68% of the total CO_2 emissions from global cement production. In 2019, the global emissions from lime production are estimated to about 291 Mt CO_2/a [35]. The global glass production emitted 95 Mt CO_2/a [36]. At present, CO_2 emissions from global ceramics production are not explicitly reported. The CO_2 emissions of the EU ceramic production amounted to 19 Mt CO_2/a in 2019 [37]. The total amount of CO_2 emissions from iron and steel production is about 2,600 Mt CO_2/a, of which about 300 Mt CO_2/a (ca. 12%) are process-related emissions [38].

Table 1 Fossil CO_2 emissions from different industries (globally for the year 2019)

Industry	Sector	Total CO_2 emissions [Mt CO_2/a]	Process-related CO_2 emissions[a] [Mt CO_2/a]
Mineral	Cement	2,300	1,380
	Lime/quicklime	291	189
	Glass	95	14
	Ceramics	19[b]	10
Iron and steel	Iron and steel	2,600	300
Non-ferrous metals	Primary aluminum	230	111[c]
	Zinc	–	23
	Lead	–	6
	Magnesium	–	3
Chemical	Soda	12	–
	Carbon black	25	–
	Ammonia	450	324
	Ethylene oxide	20	–
Total		**6,042**	**2,360**

[a] As defined in the respective process description given in the previous Sects. 5.1–5.4; [b] lack of data for global CO_2 emissions, therefore CO_2 emissions within Europe are given; [c] including anode production

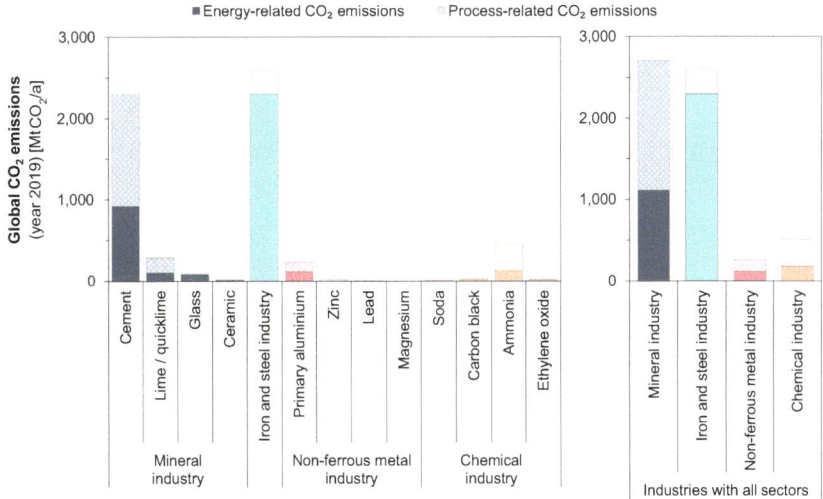

Fig. 9 Fossil CO_2 emissions from different industries (globally for the year 2019; process-related CO_2 emissions are defined as in the respective process description given in the Sects. 5.1–5.4)

The primary aluminum production as part of the non-ferrous metal industry leads to 230 Mt CO_2/a and about 111 Mt CO_2/a (ca. 48%) are process-related [39]. The CO_2 emissions from zinc production were calculated based on the default emission factor (1.72 t CO_2/t zinc) [40] and the total amount of zinc produced worldwide (2019) [41], resulting in a total amount of about 23 Mt CO_2/a. The same method is used for the overall CO_2 emissions from lead production amounting to about 6 Mt CO_2/a in 2019 [42]. These CO_2 emissions only represent the process-related emissions resulting from electrode and reductant use. The global primary magnesium production was about 1 Mt in 2019 [43]. The specific CO_2 emission factors are 5.13 t CO_2/t Mg (Dolomite) and 2.83 t CO_2/t Mg (Magnesite) depending on the used input mineral [40]. Since about 90% of magnesium is obtained by processing Magnesite [44] the overall CO_2 emissions amount to about 3 Mt CO_2/a.

In the considered sectors of the chemical industry, the global production of synthetic soda ash in 2019 was about 42 Mt [45], which results in overall CO_2 emissions of around 12 Mt CO_2/a applying an emission factor of 0.277 t CO_2/t Na_2CO_3. Up to date production as well as total CO_2 emission numbers for carbon black are not publicly available. The latest stated production is given for 2012 with a total amount of about 11 Mt [46]. Since the furnace black process is the primary production process for global carbon black production, the overall emissions by carbon black production are calculated using the specific emission factor of the furnace black process. This amounts to about 25 Mt CO_2/a. Ammonia production resulted in global CO_2 emissions of 450 Mt CO_2/a, of which 324 Mt CO_2/a (ca. 72%) are considered as process-related emissions [47]. The global production of ethylene oxide in 2019 was about 30 Mt, which led to CO_2 emissions of about 20 Mt CO_2/a, using the emission factor for the catalytic oxidation process.

6 Biogenic Carbon Sources

6.1 Biogas and Biomethane

Biogas is a mixture of methane (CH_4) and carbon dioxide (CO_2), along with trace amounts of other gases such as nitrogen (N_2), hydrogen (H_2), and hydrogen sulfide (H_2S), depending on the composition of the substrate and the specific conditions of the anaerobic digester. Biogas can be produced from a variety of organic materials such as agricultural residues, animal manure, food waste, sewage sludge, or energy crops. The selection of appropriate feedstocks depends on factors such as availability, composition, and energy content. The feedstock is typically chopped or shredded to increase its surface area and thus facilitate the microbial decomposition. This step can also involve pretreatment methods such as heating, grinding, or acid/enzyme hydrolysis to enhance the biodegradability of the substrate.

The substrate is then mixed with water and introduced into an anaerobic digester, which is a sealed container where the biogas production occurs. The absence of oxygen creates an environment suitable for anaerobic microorganisms, such as bacteria, archaea, and fungi, to thrive and carry out the process of anaerobic digestion. During this degradation process, the microorganisms break down the complex organic molecules in the substrate into simpler compounds through a series of biochemical reactions. This process involves four main stages: hydrolysis, acidogenesis, acetogenesis, and methanogenesis. As the microbial activity progresses, biogas is generated as a by-product. The biogas is collected from the anaerobic digester and stored in a separate container. The collection methods can vary depending on the size and type of the digester. The collected biogas can be utilized as a renewable energy source for various purposes; e.g., it can be burned directly for heat or electricity generation, used as a cooking fuel, or upgraded to biomethane for a subsequent injection into the natural gas grid or as a transportation fuel. The remaining material after biogas production, known as digestate, is a nutrient-rich liquid or solid residue that can be used as a fertilizer for agricultural fields, or processed further for nutrient recovery or solid–liquid separation.

In general, the production of biomethane is equal to the production of biogas, when fermentation of biomass is used. The difference between these two products is the gas composition. Biomethane is obtained from biogas by separating mainly CO_2 as well as other gases, e.g., hydrogen sulfide (H_2S) or water vapor from the methane to achieve a higher energy content and to fulfill the requirements to substitute natural gas by fulfilling the given standardized demands to be used subsequently, e.g., as a transportation fuel or for injection into the gas network.

6.2   Bioethanol

Bioethanol is a renewable biofuel that is produced through fermentation. It can be produced from various feedstocks, such as sugary crops (e.g., sugarcane, sugar beets) or starch-containing crops (e.g., corn, wheat). The selection of appropriate feedstocks depends on factors such as availability, composition, and sugar/starch content. The selected feedstock is typically subjected to pretreatment methods to make the sugars or starch more accessible for fermentation. Pretreatment can involve physical, chemical, or biochemical (enzymatic) processes, such as grinding, heating, acid/enzyme hydrolysis, or steam explosion, to break down the complex carbohydrates into simpler molecules being more easily degradable.

The pretreated feedstock is then mixed with water and introduced into a fermenter, where the fermentation process occurs. This fermentation is a biochemical process involving the conversion of sugars or starch into ethanol and carbon dioxide (CO_2) by the action of microorganisms, typically yeasts or bacteria. During fermentation, the microorganisms utilize the sugars or starch as a source of energy and produce ethanol as a metabolic by-product. The fermentation process can be carried out under anaerobic (oxygen-free) or aerobic (oxygen-rich) conditions, depending on the type of microorganisms used and the desired ethanol concentration. As the fermentation progresses, ethanol accumulates within the liquid phase, while CO_2 is released as a gas.

The amount of CO_2 emitted during the production of one metric ton of bioethanol can vary depending on the specific production process, feedstock used, and other factors. However, as a general estimate, the production of one metric ton of bioethanol typically produces around 0.96 t of CO_2 as a by-product. This estimate takes into account the stoichiometric ratio of ethanol production, which involves one molecule of ethanol being produced from one molecule of glucose (or other sugar) and yielding one molecule of carbon dioxide as a by-product. The ethanol concentration in the fermenter increases until it reaches a desired level, typically around 10–15% v/v, followed by distillation to separate ethanol from water and impurities. Dehydration using molecular sieves or other adsorbents can further increase ethanol concentration to up to 99% v/v.

6.3   Thermochemical Biomass Conversion

Industrial thermochemical biomass conversion refers to the process of converting solid biomass, such as wood, agricultural residues (e.g., straw), or dedicated energy crops (e.g., willow) into various forms of energy and value-added products using thermal and chemical reactions. Below the different processes of thermochemical biomass conversion are briefly described.

Pyrolysis. Pyrolysis involves heating biomass in the absence of oxygen to break it down into solid char, liquid bio-oil, and gaseous products like syngas. The solid char can be used as a fuel, while the bio-oil can be upgraded into transportation fuels or used for heat and power generation.

Gasification. Gasification converts biomass into a product gas that can then be cleaned and conditioned into a proper syngas, a mixture of carbon monoxide, hydrogen, and other gases. Syngas can be used as a fuel for power generation, heating, or converted into valuable chemicals, such as methanol or synthetic natural gas.

Combustion. Biomass combustion involves the direct burning of biomass to produce heat or steam, which can be used for industrial processes, heating, or electricity generation. Combustion is widely applied in biomass-fired power plants, district heating systems, and industrial boilers.

Torrefaction. Torrefaction is a thermal process that involves heating biomass at a moderate temperature. It leads to the partial decomposition of the solid biomass, resulting in a more energy-dense solid called torrefied biomass. Torrefied biomass can be used as a solid fuel in power generation characterized by improved fuel characteristics compared to virgin solid biomass.

6.4 Biogenic Carbon Potentials

The indirect estimation of CO_2 emissions from bioethanol and biomethane production can provide insights into global potentials. The distribution and respective amount of these CO_2 emissions are given in Fig. 10. However, due to a lack of data, the global potential to provide carbon (or CO_2 respectively) from biogas production was not calculated. Meant is the potential of biogas, where it is directly produced and deployed (e.g., for electricity and heat generation). Biogas facilities that separate biomethane—such as to substitute natural gas or for use in transportation—are shown Fig. 10. Therefore, this database serves as a basic approximation and a baseline indicator, providing a general overview of such biogenic carbon potential where CO_2 is already available in its molecular form.

Figure 11 illustrates the country-specific distribution of CO_2 emissions from bioethanol and biomethane production, based on the data from Fig. 10. On the left side of the figure, the annual CO_2 emissions from bioethanol production are presented by country, while the right side depicts the CO_2 emissions from biomethane production by country.

For bioethanol production, the three largest producers in 2019 were the USA, Brazil, and China. However, the USA from maize and sugar cane from Brazil are by far the biggest producers. The respective CO_2 emissions for these countries are approximately 46.3 Mt CO_2/a in case of the USA and 24.5 Mt CO_2/a for Brazil. Among these, the USA is the only country with a high potential for stationary CO_2 emissions from both biomethane and bioethanol production. The largest biomethane producers worldwide are Germany (accounting for approximately half of the global biomethane production), the USA, and UK, followed by other EU countries. The

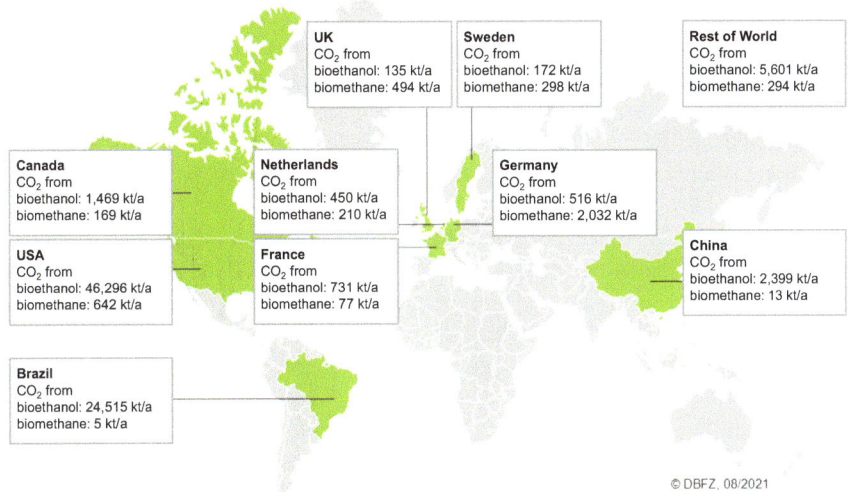

Fig. 10 Global biogenic CO_2 emissions [48]

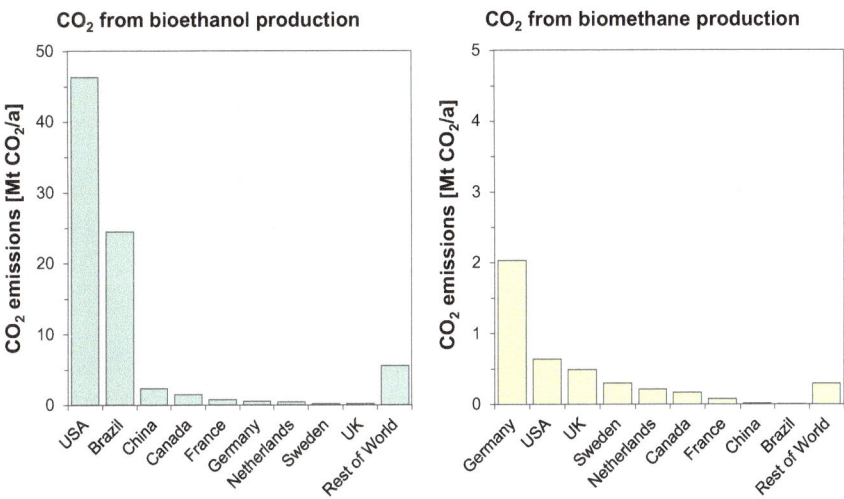

Fig. 11 Comparison of global biogenic CO_2 emissions by country [48]

combined global CO_2 emissions from stationary CO_2 sources for biomethane and bioethanol production, based on this database, amount to about 4.3 and 82.3 Mt CO_2/a, respectively. In this database, the biogenic CO_2 emissions amount to almost 87 Mt CO_2/a, with the vast majority globally attributed to the production of bioethanol (ca. 95%) and thus only a minor share to the biomethane production.

Carbon from biomass. Besides these biogenic CO_2 sources and quantities, where CO_2 is produced or available in its molecular form from processes utilizing biomass, biogenic carbon can also be sourced directly from the available biomass itself. The total biogenic carbon content within the overall biomass composition of the biosphere is about 550 Gt. The majority of carbon is embedded in plants (ca. 450 Gt of carbon), which are mainly terrestrial. The carbon content in crops cultivated by humans is about 10 Gt [49]. These values are to be regarded as theoretical maximum carbon potentials. A precise determination of the global carbon potential of biogenic origin that can be mobilized/recovered is difficult, as it depends on many factors, assumptions, and/or definition respective in- or exclusion of biogenic main products (e.g., energy crops, palm oil), by-products (e.g., residual timber, straw), as well as waste and residual materials (e.g., used cooking oil, organic fraction of municipal solid waste, manure).

There are various studies analyzing the biomass potential. In general, these studies focus on the energy content of the respective biomass rather than carbon content. The global biogenic resource potential for the period 2025–2035 varies between 45 and 375 EJ/a (mean 135 EJ/a) [48].

7 Mixed Carbon Sources

7.1 *Waste Incineration*

Municipal waste incineration is a process that involves the burning of solid waste from households, businesses, and other sources in a controlled combustion chamber. Municipal solid waste (MSW) is collected from various sources and transported to the incineration facility. The waste may undergo initial sorting and separation to remove recyclable materials such as metal, glass, and plastics.

Waste reception and preparation. At the incineration facility, the MSW is received and prepared for combustion. This may involve shredding or grinding the waste to reduce its size and improve combustion efficiency. The prepared waste is then fed into a combustion chamber, where it is burned at high temperatures (usually around 800–1,000 °C) in the presence of oxygen. This combustion process generates heat energy, which can be used to produce steam that drives a turbine to generate electricity. CO_2 is emitted from the combustion process as a by-product. The amount of emitted CO_2 depends on various factors, such as the composition of the waste being incinerated, the efficiency of the combustion process, and the air pollution control systems in place. The amount of CO_2 emitted by the incineration of municipal solid waste typically ranges between 0.7 and 1.7 t CO_2/t MSW [50].

7.2 Atmosphere

In addition to CO_2 emissions from biogenic and industrial sources, CO_2 can also be obtained by capturing it directly from the atmosphere using various technologies known as direct-air-capture (DAC). Unlike other CO_2 sources, the concentration of CO_2 in the atmosphere is relatively uniform worldwide, with minor fluctuations due to factors like vegetation, time of day, and time of the year (i.e., growing season). However, capturing CO_2 from the atmosphere using DAC technologies requires more technical effort due to the low atmospheric CO_2 concentration (ca. 420 ppm) compared to industrial or biogenic sources, and it requires a significant amount of thermal energy for sorbent regeneration. However, direct-air-capture (DAC) offers the advantage of being able to supply CO_2 on-site, independent of nearby stationary sources, which can be advantageous for applications such as renewable electricity production for synthetic fuel. To meet the thermal energy requirements, integration with synthesis processes or alternative renewable energy sources like solar thermal applications or deep geothermal energy may be considered.

7.3 Potentials

The amount of municipal waste which is incinerated in the OECD member states was about 146 Mt (2018) [51]. Assuming the range of 0.7–1.7 t CO_2/t MSW, the corresponding total amount of CO_2 emissions from waste incineration was between 102 and 248 Mt CO_2/a. Within the EU, CO_2 emissions from the combustion of municipal waste resulted in over 40 Mt CO_2/a of fossil CO_2 (2017) [52]. The total CO_2 emissions of global waste incineration could not be determined. However, it can be assumed that the emissions from OECD member states represent a great proportion of the global CO_2 from waste incineration since waste incineration is usually more widely realized in highly industrialized countries compared to less industrialized countries.

The amount of carbon that can be made available by direct-air-capture (DAC) units in the form of CO_2 is currently and most likely also in the future basically unlimited, if it is assumed that a reduction of the current CO_2 concentration from over 420 ppm to about 280 ppm, as it was in pre-industrial times (before about the year 1850), is technically possible [53]. This corresponds roughly to a cumulative amount of 1,500 Gt of CO_2 [54]. Since it is foreseeable that more CO_2 emissions will continue to be emitted in the coming years because most countries have defined the achievement of the net zero CO_2 emissions target by 2050 or even beyond, it can be assumed that the amount of carbon in the atmosphere will still increase in the future (and thus even more CO_2 can be removed from the atmosphere assuming that direct-air-capture systems become available on a large scale and are ecologically and economically feasible).

8 Summary and Conclusions

Carbon is a crucial element, essential not only for every living organism on our planet but also for its diverse applications in today's global energy and economic systems. Fossil energy carriers provide the greatest share of carbon so far, mainly as pure carbon or hydrocarbons (e.g., C from hard coal, CH_4 from natural gas). Global production of natural gas in 2020 was approximately 4 trillion m^3. The oil industry produced 4.16 Gt of crude oil and the mining of lignite and hard coal amounted to about 7.6 Gt in 2020 [55]. Though these numbers do not represent the amount of pure carbon used worldwide, they indicate well the order of magnitude of the global carbon usage. Since, e.g., crude oil contains about 82–85% carbon by mass, this means that the amount of carbon used or processed in the form of crude oil is approximately 3.5 Gt. The greatest amount is used by industry, followed by heat generation in buildings and transportation, respectively [56]. Carbon used within chemicals and derived products was about 450 Mt (average of 2015–2020) [57]. The greatest share (85%) is provided by fossil carbon sources (crude oil, natural gas, hard coal and lignite), 10% was obtained from biogenic sources (bioethanol, natural rubber, plant oil, starch/sugar, other types of biomass). The remaining 5% is obtained from recycling.

The above explanations show that carbon is used in various forms, such as for electricity and heat generation, transportation, and even as a material to produce specific products (e.g., plastics). However, currently, also CO_2 is already being utilized as a feedstock in a range of processes and applications for diverse purposes in specialized markets within the industrial sector, i.e., in its pure molecular form. Examples of these multiple uses are provided in Table 2.

Globally, the demand for CO_2 was approximately 236 Mt in 2022. The largest share of CO_2 usage was for the production of food and beverages, which accounted for 87 Mt in 2022, followed by medical applications, which utilized about 50 Mt in 2015 [58]. As also shown in Table 2, the direct utilization of CO_2 reflects a possible or prospective option for the provision of carbon in specific applications, which use carbon of fossil origin today, e.g., in the form of methane (from natural gas), methanol (based on natural gas or coal), or gasoline/naphtha (based on crude oil).

As presented in the previous chapters, there are different options to source carbon required by industry, which are categorized into the three classifications of carbon sources here: fossil, biogenic, and mixed.

- Fossil carbon can be defined as carbon that circulates within the slow carbon cycle; i.e., mainly from rocks or minerals as well as crude oil, natural gas, and coal (hard coal and lignite).
- Biogenic carbon can be defined as carbon that circulates within the fast carbon cycle; i.e., this type of carbon is primarily bound in organic matter/biomass.
- Carbon contained within ambient air (mainly in the form of carbon dioxide (CO_2)) is, according to this definition, a mixture of fossil and biogenic carbon. The same applies for CO_2 from the incineration of mixed carbon stocks (e.g., municipal waste incineration).

Table 2 Example application of CO_2 in different industries

Industry	Rather established applications and processes	Rather prospective applications and processes
Food industry	Dry ice for blast freezing and cooling	
	Inert gas for packaging	
	Carbonic acid in beverages	
	Solvents for supercritical extraction	
Agriculture	Fumigation in greenhouses	
Chemical industry	Urea production	Chemical intermediates (methanol, methane)
		Polymers
Oil/petro industry	Enhanced oil recovery	Methane
		Methanol
		Gasoline, diesel, kerosene
Other industrial applications	Refrigerant/heating agent	
Construction industry	Inert gas for welding	Cement
		Concrete

The potential availabilities for these carbon source classifications, as previously derived, are summarized and compared below. Therefore, the total CO_2 emissions from different sources are summarized in Table 3.

The total global CO_2 emissions amount to about 6,200–6,400 Mt CO_2/a, of which fossil CO_2 emission have by far the greatest share with more than 90%. About a third of the overall CO_2 emissions are considered to be process-related emissions. Mixed sources, without considering the atmosphere, show the second greatest amount with 102–248 Mt CO_2/a. Biogenic CO_2 emissions only amount to about 87 Mt CO_2/a. Including the potential from anthropogenically induced CO_2 in the atmosphere adds three orders of magnitude more CO_2 compared to the total annual CO_2 emission from all sectors analyzed.

For a better overview, the amounts for fossil, biogenic, and mixed CO_2 emissions as of Table 3 are graphically presented in Fig. 12. For the fossil CO_2 sources, these are summarized for the considered industries, with process-related CO_2 emissions depicted separately. For the mixed CO_2 sources from waste incineration, both the derived minimum and maximum amounts are shown.

Figure 12 shows that among the fossil CO_2 emissions, the mineral as well as iron and steel industries in particular emit the largest quantities of CO_2 worldwide. Within the mineral industry, cement production dominates, followed by quicklime (Sect. 5.5). In comparison, the non-ferrous metals and chemical industries play a minor role in the potential supply of carbon in the form of CO_2. Within the non-ferrous metals, primary aluminum production has the largest CO_2 quantities (Sect. 5.5). For the emissions from zinc, lead, and magnesium production, only the process-related

Table 3 Summary data of global CO_2 emissions

	Industry	Sector	Total CO_2 emissions [Mt CO_2/a]	Process-related CO_2 emissions[a] [Mt CO_2/a]
Fossil	Mineral	Cement	2,300	1,380
		Lime/quicklime	291	189
		Glass	95	14
		Ceramics	19[b]	10
	Iron and steel	Iron and steel	2,600	300
	Non-ferrous metal	Primary aluminum	230	111[c]
		Zinc	–	23
		Lead	–	6
		Magnesium	–	3
	Chemical	Soda	12	–
		Carbon black	25	–
		Ammonia	450	324
		Ethylene oxide	20	–
Biogenic	Biomethane		4	–
	Bioethanol		82	–
Mixed	Waste incineration		102–248	–
	Atmosphere		1,500,000	–
Total[d]			**6,231–6,377**	**2,360**
Thereof	Fossil		6,042	2,360
	Biogenic		87	–
	Mixed[d]		102–248	–

[a] As defined in the respective process description given in the previous Sects. 5.1–5.4; [b] lack of data for global CO_2 emissions, therefore CO_2 emissions within Europe are given; [c] including anode production; [d] without including the atmosphere

emissions could be reported. Within the chemical industry, ammonia production shows the largest amount, whereas soda, carbon black and ethylene oxide production are on a similar level, which is negligible compared to ammonia production in this industry (Sect. 5.5).

Process emissions are mainly emitted in the mineral industry, followed by the iron and steel as well as the non-ferrous metal industry. Process emissions are released in all sectors of the mineral and non-ferrous metal industries. The largest share of process emissions in relation to the total emissions of the respective sectors is recorded in cement production, quicklime production, and ammonia production, where the share accounts for more than half of the total emissions.

In case of biogenic CO_2, the global potentials from bioethanol production are significantly larger than those from biomethane production, although these represent only a fraction of the total potential compared to the fossil potentials. This is also

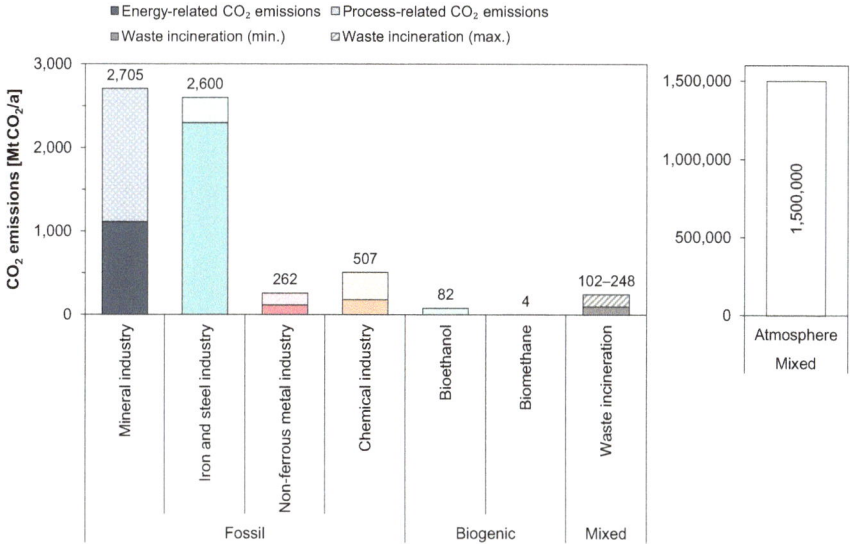

Fig. 12 Summary data of available CO_2 emissions

true for the CO_2 emissions from waste incineration. However, they are still about two to three times higher than the mentioned biogenic CO_2 availabilities. All these sources are negligible compared to the mixed CO_2 emissions that could potentially be captured from the atmosphere.

Overall, in terms of quantity, the greatest potential for carbon is by far in the atmosphere and, in principle, in the biosphere, even if this potential has so far been relatively little utilized on the basis of the results relating to CO_2 emissions from biomethane and bioethanol production. In comparison, the potentials from the various fossil sources currently dominate.

For a sustainable technical use, it can be stated that biogenic carbon or CO_2 sources should be the first choice for carbon procurement, as they do not short-circuit the fast carbon cycle. This also requires that the energy needed for the provision of this carbon is sustainably provided. From the perspective of the carbon cycle, CO_2 from mixed sources, i.e., the atmosphere, and, e.g., waste incineration, should be mentioned in second place. For a truly "green" technical use, fossil energy-related CO_2 emissions should be excluded as a carbon source, as should the direct use of fossil carbon. The process-related CO_2 emissions, that result especially from the mineral industry, also have a negative climate impact. However, these CO_2 emissions are most likely not substitutable at least in the medium term, which is why they could be utilized in a cascading manner (i.e., in the form of a hierarchical and sequential utilization for different applications).

A fact becoming evident from the examination is the low current availability of easily accessible carbon sources. Even the entire CO_2 emissions analyzed here, along with the carbon contained therein (about 2.4 Gt), would not be sufficient to substitute for the carbon used in the current scale of crude oil extraction (3.5 Gt), let alone when natural gas and coal are added to the equation.

References

1. Encyclopaedia Britannica, Carbon: compounds [Online]. Available: https://www.britannica. com/science/carbon-chemical-element/Compounds (accessed: 5.5.23)
2. Pörtner H-O, Roberts DC, Tignor M, Poloczanska ES, Mintenbeck K, Alegría A, Craig M, Langsdorf S, Löschke S, Möller V, Okem A, Rama B (eds) (2022) IPCC, 2022: climate change 2022: impacts, adaptation, and vulnerability: contribution of working group II to the sixth assessment report of the intergovernmental panel on climate change. Cambridge University Press, Cambridge, UK and New York, NY, USA
3. Purr K, Garvens H-J, Bernicke M, Brieschke J, Kaliske J, Kessler H, Malsch D, Plickert S, Proske C, Rothe B (2021) Contribution to the discussion on the evaluation of carbon capture and utilization. Dessau-Roßlau, Background
4. Gabrielli P, Gazzani M, Mazzotti M (2020) The role of carbon capture and utilization, carbon capture and storage, and biomass to enable a net-zero-CO_2 emissions chemical industry. Ind Eng Chem Res 59(15):7033–7045. https://doi.org/10.1021/acs.iecr.9b06579
5. Galimova T et al (2022) Global demand analysis for carbon dioxide as raw material from key industrial sources and direct air capture to produce renewable electricity-based fuels and chemicals. J Clean Prod 373:133920. https://doi.org/10.1016/j.jclepro.2022.133920
6. Verband der Chemischen Industrie e.V, Daten und Fakten zu Energie und Rohstoffen: Energiestatistik im Überblick [Online]. Available: https://www.vci.de/die-branche/zahlen-berichte/ vci-statistik-grafiken-energie-klima-rohstoffe-chemie.jsp (accessed: 27.03.23)
7. Stolz B, Held M, Georges G, Boulouchos K (2022) Techno-economic analysis of renewable fuels for ships carrying bulk cargo in Europe. Nat Energy 7(2):203–212. https://doi.org/10.1038/s41560-021-00957-9
8. Felixberger JK (2017) Chemie für Einsteiger. Berlin, Heidelberg: Springer Spektrum
9. Wollrab A (2014) Organische Chemie: Eine Einführung für Lehramts-und Nebenfachstudenten, 4th edn. Springer Spektrum, Berlin, Heidelberg
10. Gupta T (2018) Carbon: the black, the gray and the transparent. Springer International Publishing, Cham
11. Latscha HP, Mutz M (2011) Chemie der Elemente: Chemie-Basiswissen IV. Springer-Verlag, Berlin, Heidelberg
12. Riebeek H, The carbon cycle [Online]. Available: https://www.earthobservatory.nasa.gov/fea tures/CarbonCycle (accessed: Jan. 24 2023)
13. Wissenschaftszentrum Umwelt Universität Augsburg, CO_2-Ein Stoff und seine Geschicht: Die Kohlenstoffkreisläufe und der Mensch [Online]. Available: http://www.co2-story.de/aus stellung/die-kohlenstoffkreislaeufe-und-der-mensch.html (accessed: 10.07.23)
14. Schmid C, Hahn A (2021) Potential CO_2 utilisation in Germany: an analysis of theoretical CO_2 demand by 2030. J CO_2 Utili 50:101580. https://doi.org/10.1016/j.jcou.2021.101580.
15. Fischedick M, Görner K, Thomeczek M (2015) CO_2: Abtrennung, Speicherung, Nutzung. Springer Berlin Heidelberg, Berlin, Heidelberg
16. Bringezu S, Kaiser S, Turnau S (2020) Zukünftige Nutzung von CO_2 als Rohstoffbasis der deutschen Chemie-und Kunststoffindustrie
17. Saygin D, Gielen DJ, Draeck M, Worrell E, Patel MK (2014) Assessment of the technical and economic potentials of biomass use for the production of steam, chemicals and polymers. Renew Sustain Energy Rev 40:1153–1167. https://doi.org/10.1016/j.rser.2014.07.114
18. Zhu K, Zhang J, Niu S, Chu C, Luo Y (2018) Limits to growth of forest biomass carbon sink under climate change. Nat Commun 9(1):2709. https://doi.org/10.1038/s41467-018-05132-5
19. Gaseous carbon waste streams utilization, National Academies Press, Washington, DC

20. Tripathi AD, Mishra R, Maurya KK, Singh RB, Wilson DW (2019) Estimates for world population and global food availability for global health. In: The role of functional food security in global health. Elsevier, pp 3–24

21. Viebahn P, Horst J, Scholz A, Zelt O (2018) Technologiebericht 4.4 Verfahren der CO_2-Abtrennung aus Faulgasen und Umgebungsluft," Bundesministerium für Wirtschaft und Energie (BMWi). Wuppertal, Karlsruhe, Saarbrücken

22. Socolow R et al (2011) Direct air capture of CO_2 with chemicals: a technology assessment for the APS panel on public affairs [Online]. Available: http://www.cchem.berkeley.edu/molsim/articles/2011/direct-air-capture-of-co2.html (accessed: 22.12.2022)

23. Zitscher T, Neuling U, Habersetzer A, Kaltschmitt M (2020) Analysis of the German industry to determine the resource potential of CO_2 emissions for PtX applications in 2017 and 2050. Resources 9(12):149. https://doi.org/10.3390/resources9120149

24. Kost C, Shammugam A, Fluri V, Peper D, da Memar AV, Schlegl T (2021) Studie: Stromgestehungskosten erneuerbare Energien

25. International Energy Agency (IEA) (2019) Putting CO_2 to use: creating value from emissions

26. Leeson D, Dowell NM, Shah N, Petit C, Fennell PS (2017) A Techno-economic analysis and systematic review of carbon capture and storage (CCS) applied to the iron and steel, cement, oil refining and pulp and paper industries, as well as other high purity sources. Int J Greenhouse Gas Cont 61:71–84. https://doi.org/10.1016/j.ijggc.2017.03.020

27. Kähler F, Carus M, Porc O, vom Berg C (2021) Turning off the tap for fossil carbon: future prospects for a global chemical and derived material sector based on renewable carbon

28. Huang K, Peng X, Kong L, Wu W, Chen Y, Maravelias CT (2021) Greenhouse gas emission mitigation potential of chemicals produced from biomass. ACS Sustain Chem Eng 9(43):14480–14487. https://doi.org/10.1021/acssuschemeng.1c04836

29. Mohn J et al (2008) Determination of biogenic and fossil CO(2) emitted by waste incineration based on (14)CO(2) and mass balances. Biores Technol 99(14):6471–6479. https://doi.org/10.1016/j.biortech.2007.11.042

30. European Environmental Agency (EEA) (ed) (2019) EMEP/EEA air pollutant emission inventory guidebook 2019: 2.B.7 Soda ash production [Online]. Available: https://www.eea.europa.eu/publications/emep-eea-guidebook-2019 (accessed: 22.05.23)

31. Cichosz M et al (2022) Changes in synthetic soda ash production and its consequences for the environment. Materials (Basel, Switzerland) 15(14). https://doi.org/10.3390/ma15144828

32. Intergovernmental Panel on Climate Change, Emission factor database [Online]. Available: https://www.ipcc-nggip.iges.or.jp/EFDB/find_ef.php?ipcc_code=2.B.8.f&ipcc_level=3

33. Eggleston HS (ed) (2006) 2006 IPCC guidelines for national greenhouse gas inventories: Chapter 3: chemical industry emissions, Institute for Global Environmental Strategies, Hayama, Japan

34. Hasanbeigi A, Global cement industry's GHG emissions [Online]. Available: https://www.globalefficiencyintel.com/new-blog/2021/global-cement-industry-ghg-emissions

35. Bing L, Ma M, Liu L, Wang J, Niu L, Xi F (2022) An investigation of the global uptake of CO_2 by lime from 1963 to 2020. Earth Syst Sci Data. https://doi.org/10.5194/essd-2022-327

36. European Climate, Infrastructure and Environment Executive Agency, How LIFE is reducing emissions from glass production [Online]. Available: https://cinea.ec.europa.eu/news-events/news/how-life-reducing-emissions-glass-production-2022-03-16_en (accessed: 10.10.23)

37. Del Rio DDF et al (2022) Decarbonizing the ceramics industry: a systematic and critical review of policy options, developments and sociotechnical systems. Renew Sustain Energy Rev 157:112081. https://doi.org/10.1016/j.rser.2022.112081

38. International Energy Agency (IEA) (ed) (2020) Iron and steel technology roadmap: towards more sustainable steelmaking. Paris, France [Online]. Available: https://www.iea.org/reports/iron-and-steel-technology-roadmap (accessed: 10.10.23)

39. International Aluminium Institute, Statistics—greenhouse gas emissions—aluminium sector [Online]. Available: https://international-aluminium.org/statistics/greenhouse-gas-emissions-aluminium-sector/

40. Eggleston HS (ed) (2006) 2006 IPCC guidelines for national greenhouse gas inventories: chapter 4: metal industry emissions. Institute for Global Environmental Strategies, Hayama, Japan
41. British Geological Survey, World mineral statistics data: production of zinc, slab [Online]. Available: https://www2.bgs.ac.uk/mineralsUK/statistics/wms.cfc?method=listResults&dataType=Production&commodity=154&dateFrom=2015&dateTo=2021&country=&agreeToTsAndCs=agreed
42. British Geological Survey, World mineral statistics data: production of lead, refined. [Online]. Available: https://www2.bgs.ac.uk/mineralsUK/statistics/wms.cfc?method=listResults&dataType=Production&commodity=81&dateFrom=2019&dateTo=2020&country=&agreeToTsAndCs=agreed (accessed: 10.10.23)
43. British Geological Survey, World mineral statistics data: production of magnesium metal, primary [Online]. Available: https://www2.bgs.ac.uk/mineralsUK/statistics/wms.cfc?method=listResults&dataType=Production&commodity=163&dateFrom=2019&dateTo=2020&country=&agreeToTsAndCs=agreed (accessed: 10.10.23)
44. Czerwinski F (ed) (2014) Magnesium alloys: properties in solid and liquid states. Erscheinungsort nicht ermittelbar: IntechOpen [Online]. Available: https://directory.doabooks.org/handle/20.500.12854/66667
45. Bolen WP (2020) Mineral commodity summaries: SODA ASH [Online]. Available: https://pubs.usgs.gov/periodicals/mcs2020/mcs2020-soda-ash.pdf (accessed: 08.09.23)
46. International Carbon Black Association (ed) (2016) Carbon black user's guide: safety, health, & environmental information [Online]. Available: https://static1.squarespace.com/static/5fd161c5b1bc2872873bd5ee/t/5fdcecfb8f882041f1b03849/1608314109006/2016-ICBA-Carbon-Black-User-Guide_english.pdf (accessed: 15.10.23)
47. International Energy Agency (IEA) (ed) (2021) Ammonia technology roadmap: towards more sustainable nitrogen fertiliser production. Paris, France [Online]. Available: https://iea.blob.core.windows.net/assets/6ee41bb9-8e81-4b64-8701-2acc064ff6e4/AmmoniaTechnologyRoadmap.pdf (accessed: 10.10.23)
48. Schröder J, Naumann K (2022) Monitoring erneuerbarer Energien im Verkehr: DBFZ Deutsches Biomasseforschungszentrum gemeinnützige GmbH
49. Bar-On YM, Phillips R, Milo R (2018) The biomass distribution on Earth. Proc Natl Acad Sci USA 115(25):6506–6511. https://doi.org/10.1073/pnas.1711842115
50. Environment Agency (ed) (2020) Pollution inventory reporting—incineration activities guidance note: Environmental Permitting (England and Wales) Regulations 2016 Regulation 61(1), Bristol, UK [Online]. Available: https://assets.publishing.service.gov.uk/media/5f74a90ee90e0740d5228fcb/Pollution-inventory-reporting-incineration-activities-guidance-note.pdf
51. Organisation for Economic Co-operation and Development (OECD), Environment database—municipal waste, generation and treatment [Online]. Available: https://stats.oecd.org/index.aspx?DataSetCode=MUNW (accessed: 5.8.23)
52. Vähk J (2019) The impact of Waste-to-Energy incineration on climate: policy briefing [Online]. Available: https://zerowasteeurope.eu/wp-content/uploads/edd/2019/09/ZWE_Policy-briefing_The-impact-of-Waste-to-Energy-incineration-on-Climate.pdf
53. National Oceanic and Atmospheric Administration, Carbon dioxide now more than 50% higher than pre-industrial levels [Online]. Available: https://www.noaa.gov/news-release/carbon-dioxide-now-more-than-50-higher-than-pre-industrial-levels
54. Friedlingstein P et al (2022) Global carbon budget 2022. Earth Syst Sci Data 14(11):4811–4900. https://doi.org/10.5194/essd-14-4811-2022
55. Franke D et al (2022) BGR Energiestudie 2021—Daten und Entwicklungen der deutschen und globalen Energieversorgung
56. bp Energy Outlook 2022 (2022) [Online]. Available: https://www.bp.com/content/dam/bp/business-sites/en/global/corporate/pdfs/energy-economics/energy-outlook/bp-energy-outlook-2022.pdf
57. Levi PG, Cullen JM (2018) Mapping global flows of chemicals: from fossil fuel feedstocks to chemical products. Environ Sci Technol 52(4):1725–1734. https://doi.org/10.1021/acs.est.7b04573

58. ChemAnalyst, Decode the future of Carbon Dioxide (CO_2) [Online]. Available: https://www.chemanalyst.com/industry-report/carbon-dioxide-market-630 (accessed: 10.10.23)

Provision of Pure Carbon Dioxide Streams – Possibilities and Constraints

Wolfram Georg Tuschewitzki and Martin Kaltschmitt

Abstract Carbon, for example in the form of carbon-based fuels, plays a vital role in numerous industries, particularly in the energy sector. This results in the release of significant amounts of carbon dioxide (CO_2), which is a major contributor to global warming. To mitigate climate change, CO_2 emissions must be reduced and sustainable carbon sources must be used. Carbon capture can play a key role in this. This paper provides an overview of different technological approaches to capture and purify CO_2 from point sources. It discusses the main strategies for carbon capture: primary measures, which focus on modifying industrial processes to pre-concentrate CO_2, and secondary measures, which involve extracting and purifying CO_2. Pre-concentration techniques such as pre-combustion, post-combustion, and oxyfuel combustion are analyzed, and their efficiencies, energy requirements, and costs are compared. In addition, various CO_2 extraction methods (secondary measures), including ab- and adsorption, membrane, and cryogenic separation, will be evaluated for their energy consumption, CO_2 purity, and maturity. Optimal integration of primary and secondary measures, tailored to specific industrial processes, is critical to reducing energy requirements and costs. Ultimately, carbon capture technology plays a pivotal role in achieving the global climate goals set by the Paris Agreement, enabling industries with limited decarbonization options, such as cement and steel, to achieve carbon neutrality and contribute to a circular economy.

Keywords Carbon capture · CO_2 capture · CO_2 separation · CO_2 extraction · Capture technologies · Pre-combustion · Oxyfuel-combustion · Post-combustion · Absorption · Adsorption · Membrane separation

W. G. Tuschewitzki (✉) · M. Kaltschmitt
Hamburg University of Technology (TUHH), Institute of Environmental Technology and Energy Economics (IUE), Hamburg, Germany
e-mail: wolfram.tuschewitzki@tuhh.de

N. Bullerdiek et al. (eds.), *Powerfuels*, Green Energy and Technology,
https://doi.org/10.1007/978-3-031-62411-7_15

399

1 Introduction

Carbon is a vital element on Earth, forming the basis of all known life and serving as a building block for countless molecules in organic chemistry [1]. It is one of the chemical elements most extensively used in various fields of the global economic system and thus plays a crucial role in numerous products. The significance of carbon usage is particularly true in the energy industry, where naturally provided hydrocarbons, like crude oil, hard coal, lignite, and natural gas, are oxidized to cover the respective given energy demands. The resulting product of such oxidization, carbon dioxide (CO_2), is released into the atmosphere, which serves as a carbon sink for these waste streams. Thus, so far, anthropogenic CO_2 emissions originate mainly from the aforementioned fossil fuel sources [2]. Over time, this CO_2 dumped into the atmosphere accumulates and increasingly contributes to anthropogenic climate change or global warming, respectively [3]. To mitigate global warming, it is crucial to aim for net-zero additional CO_2 emissions, which means substantially reducing the release of CO_2 from non-recent sources [2].

In face of climate change and mankind's responsibility to limit global warming [2], the international community has resolved to phase out technologies leading to avoidable CO_2 emissions. Additionally, commitments to reduce both CO_2 emissions as well as emissions of further greenhouse gases have been laid down in legally binding contracts, legislations or regulations, such as the Kyoto Protocol and the Paris Climate Agreement [2–4]. As an example, Germany has committed to phase out the use of coal for power generation by the year 2038 [5]. In alignment with this overarching goal, there is the tendency to increasingly substitute the energy supply based on fossil fuel energy with carbon-free energy provision options, for example based on photovoltaics and wind turbines [3].

However, as previously mentioned, carbon is a substantial chemical component in numerous industrial and everyday products (e.g., plastics, fine chemicals). A decarbonization of these products is logically impossible as it would necessitate changing their very nature. Thus, most likely also in the years to come, at least in the foreseeable future, carbon will likely remain necessary for such products as long as no sufficient substitute materials emerge. Additionally, certain other sectors, such as aviation, maritime, or heavy-duty road transportation, will probably also continue to rely on carbon-based energy carriers to significant shares. Therefore, alternative sustainable carbon sources are required to realize the desired phase-out of the use of fossil carbon within the overall economy [6–11].

Against this background, the overarching goal of this paper is to provide an overview of various carbon sources as well as the technological options to use them for the provision of pure CO_2 streams, not increasing the overall carbon inventory within the biosphere on Earth. In particular, the fundamentals of carbon capture from point sources are explained. The following explanations are organized into several key sections.

- **Basics:** This part provides insights into CO_2 point sources, the varying concentrations of CO_2 in different processes, the energy requirements for obtaining pure CO_2, and the overarching challenges associated with pure CO_2 provision.
- **CO_2 pre-concentration:** Here, pre-concentration techniques for CO_2 on the basis of combustion-based concepts are discussed. Pre-combustion carbon capture, oxyfuel-combustion carbon capture, and post-combustion carbon capture are presented and compared.
- **CO_2 extraction:** This part compares different CO_2 extraction and separation methods. Sorption processes, including absorption, adsorption, and heterogeneous gas-solid reactions, are discussed, as well as membrane separation and low-temperature "cryogenic" separation.

Finally, a synthesis of the key concepts discussed throughout the chapter is given within a summary and for a better understanding of the overarching role of carbon capture and utilization, some conclusions are drawn.

2 Basics

Various sources of renewable and sustainable carbon can be utilized in theory. These include recent carbon derived from organic matter of biological origin (i.e., biomass like, e.g., from forest wood, from food waste, and from agricultural residues) and carbon obtained from the atmosphere. Additionally, through recycling, carbon bound in products exceeding their technical life time can be reused with the help of physical and/or chemical approaches created by humans [12–14].

Each of these sources comes with its own set of advantages and disadvantages. Biogenic capture, which involves capturing CO_2 through plants based on the photosynthesis process, is limited by the growth rate of plants and the availability of fertile land [15–17]. This a priori limited land availability leads to competition for land usage, particularly between food production – especially due to a still strongly increasing global population – and other demands such as biomass provision for the subsequent use as a raw material [18, 19].

Direct removal from ambient air – often called direct air capture (DAC) – is another source for the provision of renewable carbon if "green" energy is utilized [20, 21]. Such a direct air capture competes with the CO_2 capture from gas streams with an increased percentage of CO_2 in comparison with the share of CO_2 contained within the atmosphere. Typical examples are flue gas streams, such as, e.g., those derived from heating plants using solid biofuels or from biogas plants based on animal manure.

Historically, the carbon capture technology was developed by the energy industry while upgrading fossil resources for the market. This has been true for the process of capturing CO_2 to separate it from natural methane resources ("natural gas sweetening") for an enhanced heating value of the provided gas in the 1920s [22]. Later, in the 1970s, the energy industry started utilizing carbon capture for enhanced oil recovery (EOR) [23, 24] by injecting CO_2 into oil reservoirs showing a decreasing oil production. Since then, a lot of work has been carried out in the field of removal of CO_2 from the flue gas stream of large-scale combustors using fossil fuel energy (e.g., lignite power plant) with the goal to allow for a carbon neutral coal usage. The basic idea was to separate the CO_2 from the flue gas released from the power plant into the atmosphere so far and to dump it into the underground (so-called carbon capture and storage (CCS) approach). So far, due to economic constraints no commercial large-scale application of such a concept is known.

Today, carbon capture is seen as a very important mitigation option within selected industry branches such as the cement industry, the iron and steel production, or the waste incineration sector, where it can help to reduce CO_2 emissions. The importance of this technology option has been emphasized by the Intergovernmental Panel on Climate Change (IPCC) stating that limiting human-caused global warming requires comprehensive reductions mainly in CO_2 as well as net negative CO_2 emissions achieved through CO_2 removal (CDR) and carbon capture [2]. Whereas energy-related CO_2 emissions can be avoided, process-related CO_2 emissions remain (e.g., from cement production). Therefore, despite the fact that the CO_2 storage within the underground is not considered a proven technology so far and most likely the possibilities to store CO_2 within the underground in a largely safe manner is limited, this concept is seen as an important option to contribute to achieve net-zero emissions in the years to come [13].

The above-mentioned carbon capture from gas streams with a share of CO_2 in the upper one-digit percentage and above is referred to as carbon capture from point sources. It involves separating CO_2 from other gas components/other gases to obtain a "pure" CO_2 gas stream. There is a wide variety of possible point sources of gas streams with strongly varying CO_2 concentrations available for such a CO_2 separation (Sect. 2.1).

Additionally, multiple processes have been developed in recent years for capturing and/or extracting CO_2 from point sources/from flue gases released by various processes. The underlying physical/chemical methods include absorption, adsorption, heterogeneous gas-solid reactions, membrane separation, and low-temperature "cryogenic" separation. Each option has its specific advantages and limitations, depending on the source of CO_2 and the desired outcome. The selection of the most suitable combination of process and method depends on various factors, such as the CO_2 concentration in the flue gas, energy requirements, economic considerations, and system scalability. There is no universally valid blue print.

The most important factor determining the decision for a specific approach is the energy demand, coupled with the capture costs. The lower the energy demand for the provision of a pure CO_2 stream from a certain source, the more likely this technology will be implemented because typically the operation expenditures are determining the overall CO_2 provision costs. The operation costs are determined by the energy, respectively, the electricity and heat demand.

2.1 CO_2 Point Sources and CO_2 Concentrations

Various point sources are available containing varying shares of CO_2 within the respective gas stream. The CO_2 content within the respective point source either results from the combustion of carbon-based solid, liquid and/or gaseous fuels or are released from a certain production process with a respective material transformation (e.g., cement production). Furthermore, the CO_2 can be categorized by origin as biogenic, fossil, atmospheric, or other (mixed) sources.

- Combustion-based sources are a major category of CO_2 point sources. These sources primarily arise from the combustion of carbon-based fuels such as, e.g., hard coal, crude oil, natural gas, or biomass in industrial production processes, for electricity generation, and from transportation vehicles [2]. The resulting CO_2 concentration within the gas stream released by such combustion-based sources depend mainly on the type and composition of the fuel used as well as the combustion agent (e.g., ambient air, pure oxygen).
- Process-based sources are another category of CO_2 point sources. Unlike combustion-based sources, these emissions are not linked to the combustion of carbon-based fuels. Instead, process-based sources arise from industrial/chemical transformation processes where a certain carbon-containing material is transformed into another chemical bound under release of CO_2 (and maybe other gases). Cement production, the synthesis of selected fine chemicals, and the iron ore reduction within a blast furnace are only some examples of numerous others characterized by the release of CO_2 [13]. The CO_2 concentration in the released gas streams is determined by the chemistry of the underlying processes, the respective process conditions, and the feedstock involved.

In Table 1, typical CO_2 concentrations for different sources are listed. Figure 1 shows a respective bar chart with an additional classification according to the origin into biogenic, fossil, atmospheric, and other. These data show that the concentrations span a vast array with a range from about 100 vol.-% in the case of methane steam reforming to concentrations of 0.04 vol.-% in ambient air. Besides ambient air, all listed sources can be considered to be point sources.

Regarding the CO_2 concentrations, several "groups" can be observed.

- Air-based combustion processes typically yield a CO_2 concentration of about 3–15 %.
- Industrial process-based emissions from iron, steel, pulp, paper, and cement production processes are in the range of 7–35 % CO_2 concentration within the gas stream released into the atmosphere.
- Biogas from biogas plants, wastewater treatment, or landfills commonly have CO_2 concentrations of 20 up to 50/60 % (and sometimes even slightly more).
- Oxygen-based combustion processes show CO_2 concentrations of about 35–80 %.
- Nearly pure CO_2 streams result from steam reforming and bioethanol production.

Table 1 Sources of CO_2 and typical CO_2 concentrations (IGCC: integrated gasification combined cycle; based on [13, 25–32])

CO_2-source	CO_2 concentration [vol.-%]	References
Ambient air	0.04	[25]
Bioethanol (dry gas stream)	> 99	[26, 27]
Biogas (upgrading)	20–45	[28]
Biogas (wastewater treatment)	25–50	[29]
Biomass combustion	3–8	[27]
Cement plant	14–35	[13]
Hard coal	15.1	[13]
Hard coal (oxyfuel-combustion)	80.3	[13]
Heating oil	13.0–13.5	[13]
IGCC power plant (coal)	35–40	[32]
Iron and steel production	17–35	[30]
Landfill gas (phase IV: stable methane phase)	41–46	[31]
Lignite	12.3–14.8	[13]
Lignite (oxyfuel-combustion)	54.4–75.8	[13]
Natural gas	8.7–9.0	[13]
Pulp and paper production	7–20	[30]
Refineries	3–13	[30]
Steam reforming	Up to 100	[30]

Due to the high energy demand of our highly industrialized society covered mainly by fossil fuel energy, CO_2 emissions emitted by point sources are dominated by energy-related emissions today. With the phase-out of fossil fuels to fulfill the commitments defined within the Paris Agreement and the fundamental limitations of the availability of biomass on a global scale, most likely many of the CO_2 point sources enumerated and discussed above will be reduced in their availability and maybe even disappear in the years to come if the decarbonization and defossilization of our overall economy will get momentum [28]. However, some processes like iron, steel, and cement production will most likely remain relevant as CO_2 sources, as well as the sources based on biomass, respectively, bioenergy.

2.2 Energy Demand for Pure CO_2 Provision

In theory, the energy required to separate different gases depends mainly on the mole fraction of the gas to be extracted within the respective gas mixture. Molar fraction and partial pressure in a gas mixture are directly related. This is shown in Eqs. 1 and 2 describing the partial pressure and the mole fraction within an ideal gas. p is the total pressure of the gas mixture, p_i is the partial pressure of component i, x_i is the

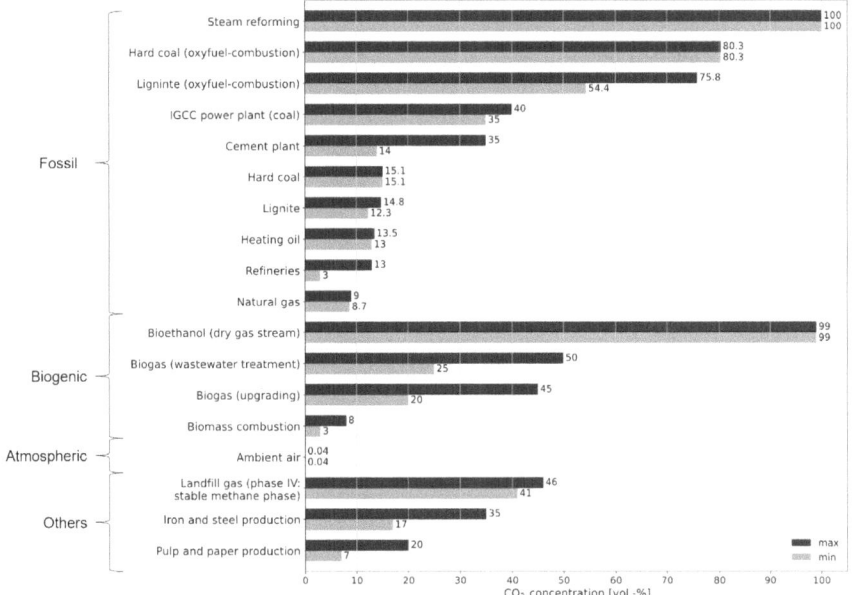

Fig. 1 Sources of CO_2 and typical CO_2 concentrations (IGCC: integrated gasification combined cycle)

mole fraction of component i, n is the total moles in the gas mixture, and n_i is the moles of component i.

$$p = p_1 + \ldots + p_n \tag{1}$$

$$x_i = \frac{p_i}{p} = \frac{n_i}{n} \tag{2}$$

Furthermore, gas concentration and energy demand for separation depend on each other. The lower the concentration – and thus the partial pressure – the more energy is required to separate a specific quantity (mass) of, e.g., CO_2 from a gas mixture according to the basic rules of thermodynamics. Separating one kilogram of CO_2 from ambient air (0.04 vol.-% CO_2) theoretically requires at least three times the energy compared to the separation of the same amount of CO_2 from a point source with 15 vol.-% CO_2 contained within a gas stream.

The theoretical minimum work w_{min} for the separation of CO_2 from a gas stream can be calculated based on Eq. 3 describing the theoretical minimum thermodynamic work required per kilogram of separated CO_2 for its complete removal from ambient air. The initial CO_2 mole fraction is described by x_{CO_2} [21]. R is the universal gas constant (8.314 J/(mol K)), T is the absolute temperature in K, x_{CO_2} is the initial mole fraction of CO_2 in the air, and M_{CO_2} is the molar mass of CO_2.

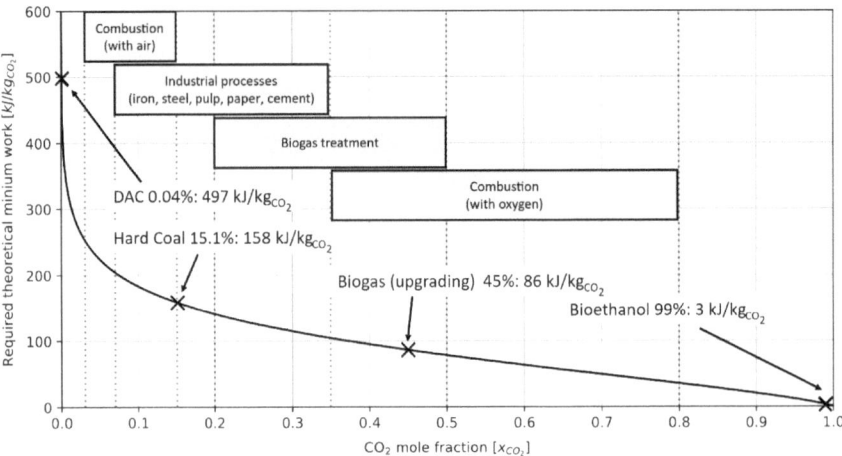

Fig. 2 Minimum separation work for 100 % CO_2 separation from ambient air ($T = 298.15$ K). Marked areas describe typical CO_2 concentrations of point source groups

$$w_{min} = -\frac{RT}{x_{CO_2}M_{CO_2}}\left(x_{CO_2}\ln(x_{CO_2}) + (1 - x_{CO_2})\ln(1 - x_{CO_2})\right) \tag{3}$$

The curve for the minimum work required per kilogram of CO_2 with different initial CO_2 mole fractions resulting from Eq. 3 is displayed in Fig. 2. Additionally, this graphic shows typical concentration spans for different technologies, as described in Sect. 2.1. Marked are the theoretical minimum separation work requirements:

- 497 kJ/kg CO_2 for a direct air capturing device
- 158 kJ/kg CO_2 for a flue gas stream derived from a combustion unit powered by a carbon-based solid energy carrier (e.g., charcoal, hard coal)
- 86 kJ/kg CO_2 for a typical biogas stream
- 3 kJ/kg CO_2 for a very high concentration point source like, e.g., from a bioethanol production facility (from the alcoholic fermentation vessel)

Thus, the separation energy requirement for CO_2 removal from a gas stream strongly depends on the respective source. Typically, the higher the CO_2 concentration is within the original gas stream, the less energy per kilogram of captured CO_2 is needed. Therefore, it could make sense to increase the share of CO_2 within the respective point source based on other measures (e.g., combustion with pure oxygen). This is the reason why for the provision of a pure CO_2 stream two approaches are distinguished:

(1) separation of CO_2 from the respective gas stream (which is also possible from ambient air) and
(2) removal of other gases to ensure that a point source just releases a pure CO_2 stream.

2.3 Challenges for Pure CO$_2$ Provision

Descriptions of carbon capture systems from point sources found in literature mostly focus on the modification of "classical" combustion processes to provide a flue gas rich in CO$_2$ (with the categories pre-combustion, oxyfuel-combustion, and post-combustion capture), while CO$_2$ sources besides that "classical" combustion field are disregarded. Alternative descriptions of CO$_2$ capture concepts distinguish between two fundamental concepts: process-related modifications that allow CO$_2$ pre-concentration within the gas stream (hereby referred to as "primary measures") and subsequent procedures for CO$_2$ extraction from the gas stream to produce a nearly pure CO$_2$ stream (referred to as "secondary measures") (Fig. 3). This viewpoint provides a more thorough understanding of the difficulties associated with supplying a source of pure CO$_2$, with each set of measures posing specific difficulties and trade-offs.

The first step in the carbon capture process is the CO$_2$ pre-concentration. It involves altering the underlying industrial processes so that the final gas composition contains a higher concentration of CO$_2$ (Fig. 3, left). Such changes to modify the composition of the resulting gas mixture must be made at the process level since they are inherent to the process itself. Within such processes, typically a raw material or an energy carrier containing carbon is introduced into a transformation process with optional inputs, and processed into a product. A gas mixture with a certain concentration of CO$_2$ is a by-product. Examples of such primary measures for "classical" combustion processes are oxyfuel and pre-combustion concepts. For instance, oxyfuel-combustion involves substituting ambient air with oxygen for combustion, which effectively lowers the amount of nitrogen in the gas mixture and thus increases the concentration of CO$_2$ in the resulting gas mixture.

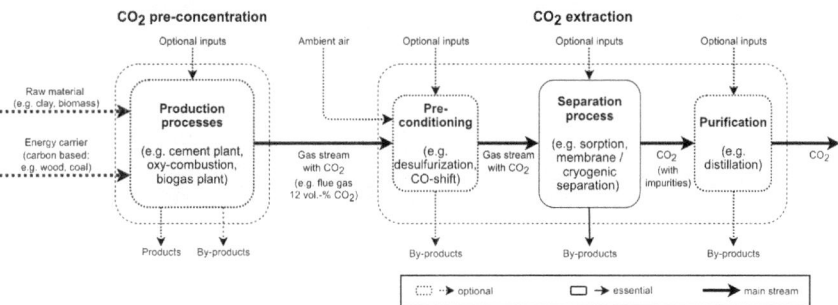

Fig. 3 Carbon capture process: Carbon-based energy source or raw material is processed in the CO$_2$ pre-concentration step (left) to create a gas stream with an elevated CO$_2$ concentration. The gas mixture is fed into the CO$_2$ extraction (right), where a pure CO$_2$ stream is extracted after passing optional pre-conditioning and purification steps. As an alternative input stream, ambient air can be fed into the CO$_2$ extraction process instead of a gas stream from a primary process. The ambient air stream represents the direct air capture (DAC) feed stream

Although primary measures typically do not yield a pure CO_2 stream, they are instrumental in achieving CO_2 pre-concentration. However, this enhancement comes at the cost of energy and resources required (e.g., for oxygen production in the example mentioned above). Therefore, determining the optimal configuration of primary measures involves carefully evaluating the trade-offs between increased CO_2 concentration and energy input, costs, resource availability, and other factors in relation to the necessary secondary measures.

In contrast, secondary measures are dedicated to extracting CO_2 from the gas stream, ultimately yielding a nearly pure CO_2 gas product. In Fig. 3, it is shown that a gas stream with CO_2 is fed in, and after an optional pre-processing step – like dedusting, desulfurization, or denitrification [33] – the separation of CO_2 takes place. Typical technological options are membrane separation or sorption processes; they are discussed in detail below (Sect. 4). The result – with an optional purification step – is a nearly pure CO_2 stream. These measures are particularly pertinent in post-combustion scenarios or end-of-pipe applications. Following the primary process, these methods target the extraction and separation of CO_2 from gas mixtures. A special case is direct air capture, where ambient air is directly supplied to the separation instead of a gas stream that results from an industrial process.

While secondary measures directly yield high-purity CO_2, they introduce their own challenges, including energy-intensive separation technologies and the selection of efficient adsorption or absorption materials. The final separation is bound to the thermodynamics of gas separation discussed in Sect. 2.2. The higher the gas concentration of the supplied gas mixture, the lower the required energy for separation (Eq. 3).

The key to an effective carbon capture system is strategically integrating primary and secondary measures. Balancing these approaches in an optimal way is essential to minimize energy consumption, reduce costs, and manage spatial requirements, as well as reduce the overall climate footprint. There is no universal solution; primary and secondary measures depend on the specific industrial processes and resource availability.

In conclusion, providing pure CO_2 through carbon capture technologies is a multifaceted challenge that benefits from a consideration that distinguishes between primary and secondary measures. While primary measures focus on altering processes to pre-concentrate CO_2, secondary measures extract and enrich the captured CO_2 to a pure stream. Primary and secondary measures need to be balanced for an optimal overall capture process.

3 CO_2 Pre-Concentration

Well-known approaches exist for providing an enriched CO_2 stream from point sources – especially from combustion units using carbon-based fuels. This chapter delves into more detail, providing examples of combustion-based processes to achieve a pre-concentration of CO_2 in the resulting gas stream. Specifically,

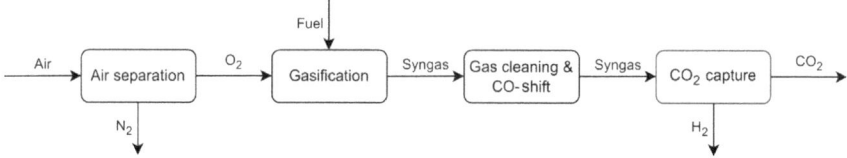

Fig. 4 Simplified schematic of a pre-combustion carbon capture concept (based on [35])

it explores three key concepts: pre-combustion, oxyfuel-combustion, and post-combustion approaches. While these concepts are discussed in the context of carbon-based fuel combustion, similar principles of modifying the conversion, production, upgrading, or synthesis process to pre-concentrate CO_2 in exhaust gas streams can be applied across various industries. After the pre-concentration of CO_2 as a primary measure, the extraction of CO_2 is done in an additional step as a secondary measure. This concept has been applied widely in other areas for decades; a typical example is the NO_x removal from power plants where usually a combination of primary and secondary (end-of-pipe) measures are realized to fulfill the legal emissions threshold cost-efficiently.

3.1 Pre-Combustion Carbon Capture

In a pre-combustion concept, air is separated in an air separation unit as a first step to yield a high-purity oxygen feed gas stream for subsequent fuel gasification (Fig. 4). Gasification, an under-stoichiometric combustion, of the carbon-based fuel is then realized under temperatures of approx. 1,000–1,700 °C and pressures of about 30 bar [14, 34, 35]. A gas mixture consisting – depending on the chemical composition of the carbon-based fuel used – mainly of hydrogen (H_2) and carbon monoxide (CO), is produced within this thermo-chemical gasification process. This gas mixture is then cleaned by means of a gas scrubber to remove harmful or toxic substances. Based on a subsequent water-gas-shift (WGS) reaction, the CO contained within the syngas reacts with water (H_2O) by being converted into CO_2 under release of H_2. The resulting gas stream is then separated into a gas stream rich in CO_2 and another one rich in H_2. The latter is then burned for power generation under the release of basically only water vapor [36]. The CO_2-rich stream is available for subsequent use [37].

3.2 Oxyfuel-Combustion Carbon Capture

The basic idea of oxyfuel-combustion of carbon-based fuels is to obtain a high-purity CO_2 output stream mainly undiluted by nitrogen. This is achieved by burning the

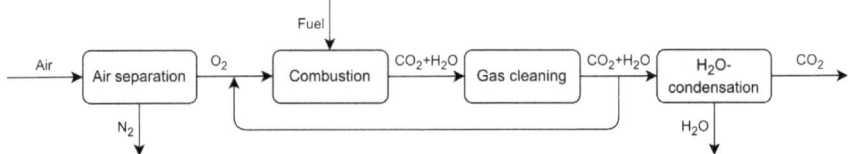

Fig. 5 Simplified schematic of an oxyfuel-combustion carbon capture concept (based on [35])

fuel with a high-purity oxygen stream, typically around 96 vol.-% purity [38, 39]. Like in pre-combustion carbon capture, oxygen is provided by an air separation unit (Fig. 5). The combustion of the carbon-based fuel results in a CO_2 flue gas stream saturated with water vapor. A subsequent condensation below the dew point of water removes the water in the liquid phase and obtains a purified CO_2 stream. The largely pure CO_2 is then available for further use [13, 40].

The combustion of the carbon-based fuel with pure oxygen (O_2) happens at a high-temperature level of over 2,000 °C [35] since fewer gases that are not contributing to the fuel oxidation (like nitrogen) need to be heated (as is the case within a "classical" combustion unit using ambient air as an oxidation agent) [40]. Due to material challenges at such high temperatures, a part of the flue gas is recirculated to keep combustion temperatures within a manageable range. The recirculation has the beneficial side effect of reducing the remaining O_2 concentration further within the flue gas stream. Thus, the output gas stream shows a relatively high CO_2 purity [13, 41].

3.3   Post-Combustion Carbon Capture

Carbon capture with a post-combustion concept is most widely discussed for the time being. Here, the first steps are identical to a typical ("classical") combustion unit using carbon-based fuels; i.e., the fuel is burned with ambient air within a combustor. The resulting flue gas is then cleaned by removing impurities. Instead of being released directly into the atmosphere, the gas stream is subsequently moved to the CO_2 capture unit (Fig. 6). Depending on the carbon-based fuel used within the combustion unit, a further pretreatment of the flue gas stream might be necessary (e.g., SO_2 removal, dust removal) [35]. However, in typical post-combustion carbon capture, no extra effort is put into the pre-concentration of CO_2; this results in relatively low CO_2 concentrations in the flue gas. Thus, a lot of energy has to be put into the final CO_2 extraction step.

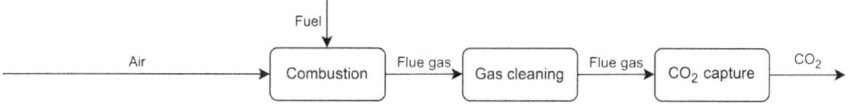

Fig. 6 Simplified schematic of a post-combustion carbon capture process (based on [35])

3.4 Comparison

The three approaches for integrating carbon capture units into a combustion concept using carbon-based fuels show specific characteristics, advantages, and disadvantages. These are compared below and summarized in Table 2.

- **Pre-combustion:** Pre-combustion carbon capture is more efficient than post-combustion capture because the CO_2 mole fraction in the syngas is generally higher than in the flue gas of post-combustion capture [32, 42]. However, it shows the disadvantage of relatively high investment costs. The required energy demand is high if oxygen (O_2) must be obtained from a specific air separation unit [40].

Table 2 Advantages and disadvantages of different carbon capture concepts (based on [21, 32, 34, 37, 40, 42, 43])

Concept	Advantages	Disadvantages
Pre-combustion	• High concentration of CO_2 in the syngas (~40 %) resulting in decreased energy need for separation • High pressure of product gas	• Relatively high investment costs • High energy penalty for O_2 production • Retrofitting not possible • Only suitable for energy carriers, allowing for the provision of syngas • Need for high-temperature stable materials
Oxyfuel-combustion	• High concentration of CO_2 in resulting flue gas (> 54 %) resulting in decreased energy need for separation	• High energy penalty for O_2 production • Prone to corrosion due to increased water content • Need for high-temperature stable materials • Demanding retrofitting • Challenging combustion with pure oxygen
Post-combustion	• Well-suited for retro fitting • Most mature combustion-based route (typically: low capital costs) • Proven technology • Extra removal of NO_x and SO_x	• Low concentration of CO_2 in the syngas (< 15 %) resulting in increased energy need for separation • High operational costs for auxiliary equipment

However, the resulting CO_2-rich flue gas stream is available at a high pressure. The pre-combustion concept cannot be retrofitted in typical large-scale combustion units because the entire combustion unit must be adapted [32, 34, 37, 40].

- **Oxyfuel-combustion:** Due to the combustion of the carbon-based fuel with oxygen instead of air in oxyfuel carbon capture, the flue gas stream contains a high concentration of CO_2 [42]; this is similar to the pre-combustion carbon capture and beneficial because an elevated CO_2 concentration means low energy input for capture per weight fraction [21]. However, the high energy input for oxygen generation is disadvantageous compared to post-combustion carbon capture [40]. Oxyfuel-combustion is not as limited in potential applications as pre-combustion carbon capture, but retrofitting of existing large-scale combustion units requires significant technical effort [13], especially since combustion of a fuel in pure oxygen might be challenging [32] since high-temperature stable materials are needed [40, 44]. Additionally, a lot of water vapor must be removed, making corrosion mitigation important.

- **Post-combustion:** Since the combustion unit is not changed for the pre-concentration of CO_2 in post-combustion capture, the concept is well-suited for retrofitting. The technology is the most mature of the three combustion-based routes, resulting in relatively low capital costs compared to the other two concepts. A disadvantage is the low CO_2 partial pressure (low concentration of CO_2) within the flue gas stream, i.e., large amounts of flue gas must be processed, resulting in high energy requirements for CO_2 sequestration and auxiliary equipment such as pumps and ventilation. Furthermore, a flue gas treatment is typically needed before CO_2 capture [32, 34].

Each carbon capture method has its advantages and disadvantages. Pre-combustion capture offers high capture efficiency but involves high equipment costs and yields gas streams with relatively high CO_2 percentage. In contrast, oxyfuel-combustion capture is less expensive but shows a lower capture efficiency but still a relatively high CO_2 pre-concentration. Post-combustion capture is the most flexible and adaptable technology but requires significant amounts of energy in the subsequent CO_2 extraction due to no extra effort in CO_2 pre-concentration, resulting in high operating costs. Ultimately, the choice of carbon capture route depends on the specific context of the combustion unit as well as the used carbon-based fuel. If the overarching goal is to provide a pure CO_2 stream, additionally, a compromise has to be found between the provision of a CO_2 stream due to measures at the combustion unit and the end-of-pipe removal/CO_2 extraction and purification.

4 CO_2 Extraction

Based on an existing gas stream containing CO_2 or a CO_2-enriched gas stream due to adequate upstream measures, some options are described to provide a "pure" CO_2 stream. Depending on the concentration of the feed gas, different physical and/or chemical options are possible, and the various resulting technical methods for such an extraction and purification process are discussed below (Fig. 7), i.e., to separate CO_2 from gas streams.

The spectrum of possible CO_2 extraction options spans from sorption over cryogenic processes toward membranes to temporarily binding CO_2 in construction materials or biomass. Most separation techniques require a gas stream with low contaminants, i.e., a cleaning step for the gas is usually required (e.g., desulfurization, dedusting, and denitrification). This is, for example, true for absorbent- or adsorbent-based processes, where the impurities might react with the sorbents and deactivate them or form unwanted substances [35].

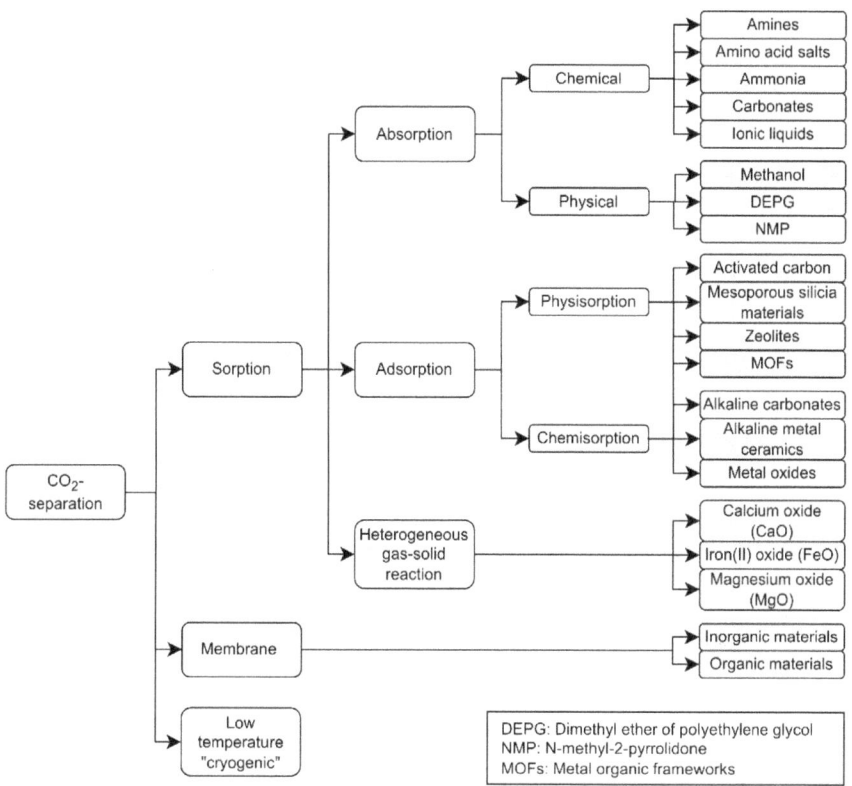

Fig. 7 Methods for CO_2 separation with associated subgroups, sorbents, and materials (based on [13, 32, 41, 45, 46])

4.1 Sorption Processes

Sorption processes are defined as processes by which a substance (sorbate) is sorbed on or in another substance (sorbent). The sorbate can be either absorbed or adsorbed. The reverse process for both is called desorption; the third form of sorption is ion exchange. Sorption can be either physical or chemical [47]. Various absorption and adsorption processes can be utilized for the extraction of CO_2 from a gas mixture. This is discussed below.

4.1.1 Absorption

Absorption is a term used in both chemistry and physics. In chemistry, absorption is defined as a process where a material (absorbate), typically a gas, is dissolved in another material (absorbent), usually a liquid or a solid [47]. Absorption is a bulk phenomenon; this is in contrast to adsorption being a surface phenomenon. The bound component is then separated from the absorbent within a subsequent desorption process (Fig. 8). Absorption is typically exothermal, whereas desorption is, in most cases, endothermic [37, 42].

There are two different types of absorption.

- Physical absorption: No chemical reactions occur in the absorption process. Physical absorption typically involves high pressure for absorption and low pressure for desorption [49]. Most common sorbents besides water are methanol, dimethyl ether of polyethylene glycol (DEPG) and N-methly-2-pyrrolidone (NMP) [46, 50].
- Chemical absorption: A third component, typically solved in a scrubbing agent, functions as a chemical sink that chemically binds with the absorbed substance [35]. There is a vast number of such sorbents. Commonly used fluid sorbents are amines, ammonia, or potassium carbonate, and widely used solid sorbents are lime or cement raw meal [34, 43, 51].

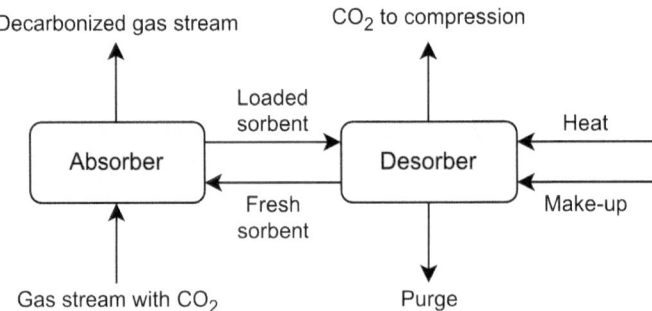

Fig. 8 Schematic of absorption-based CO_2 separation (based on [48])

Typically, the main challenges for an industrial process involving absorption and desorption are the (significant) energy requirement for desorption and sorbent deactivation and disposal due to impurities [18, 45].

A huge amount of possible absorption methods exists. They are used for a multitude of applications. Especially for CO_2 separation by absorption, amine scrubbing – a chemical absorption method – seems to be a promising option [51]. Further, suitable chemical absorption methods include amino acid salts, ammonia, carbonates, or ionic liquids. Typical absorbents for physical absorption of CO_2 from a gas stream are methanol, dimethyl ether of polyethylene glycol (DEPG), and N-methly-2-pyrrolidone (NMP) [34, 37, 46].

Amine scrubbing to remove CO_2 from gas mixtures involves two process steps in which an amine solution is circulated (absorber and desorber) and a third for the subsequent conditioning of the gas stream (Fig. 9). First, the gas mixture is passed through an absorber column containing a solution of amines (lean amine solution). The amines react with the CO_2 in the gas by forming chemical bonds binding the CO_2. The result is an amine solution loaded with CO_2 (rich/loaded amine solution). Second, the loaded amine solution is transported to the desorber, where it is heated to release the CO_2. This step is called the regeneration of the sorbent. The lean amine solution can now be passed back to the absorber to absorb further CO_2 [18, 42, 43, 52].

The absorption in amine scrubbing is exothermic and elevated pressure and low temperatures promote it. Therefore, cooling the absorber is essential. The desorption, in contrast, is endothermic, and elevated temperatures and reduced pressures promote the desorption process of CO_2 from the amine solution [18]. The CO_2-rich gas stream (purities > 99 vol.-% are reachable [43, 53]) produced in the desorber is then directed to a compressor, where it is pressurized for downstream use.

The major energy demand for the process derives from the endothermic desorption process, which requires a significant amount of energy in the form of heat (usually provided by low-temperature steam) for desorption. Additionally to the energy requirement, the atmosphere in the desorber is vital as oxygen and sulfur in the chamber accelerate amine degradation [18, 52]. Amine that is degraded must

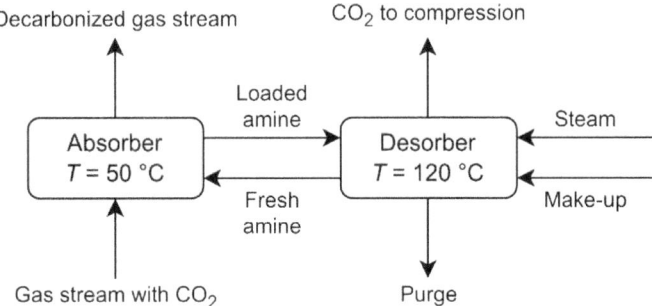

Fig. 9 Schematic of CO_2 separation by amine scrubbing with the temperature as T (based on [48])

be replaced with make-up amines. Therefore, additional desulfurization is required before the flue gas is fed into the absorption unit to minimize the need for make-up.

A variety of possible amines can be used in this process. The most commonly used amine is monoethanolamine (MEA), an aqueous ethanolamine [54]. This is due to its high reactivity with CO_2, low price, fast kinetics, and high mass transfer rates [39, 54]. However, MEA usage leads to increased energy consumption due to the needed regeneration energy as well as trouble with the degradation of the solvent through oxidation or thermal dissociation [21, 54]. Due to these reasons, the replacement of MEA by other amines is widely discussed [24, 42, 43, 54].

4.1.2 Adsorption

In adsorption, a substance (adsorbate) is attached to the surface of a liquid or solid (adsorbent). This is a surface phenomenon between two phases, driven by adhesive surface forces [47]. There is always a dynamic equilibrium between adsorption and the reverse process, desorption, influenced by pressure, the concentration of the adsorbate, temperature, material properties, and size of the interface between adsorbate and adsorbent [43]. Desorption can be controlled, e.g., by lowering pressure, raising the temperature, or both.

Adsorption phenomena can be further differentiated into chemisorption and physisorption.

- Classified as physisorption is adsorption where only intermolecular forces (van der Waals forces) are involved. Therefore, it is sometimes referred to as van der Waals adsorption. Physical adsorbents include zeolites, mesoporous silica materials, or metal-organic frameworks (MOFs) [45].
- Chemisorption is adsorption, which results from chemical bonds between adsorbate and adsorbent [47]. Chemical adsorbents include alkaline metal ceramics, alkaline carbonates, metal oxides, and potentially MOFs [45, 55]. Since chemisorption requires the formation and breaking of chemical bonds, the necessary activation energy is much more significant compared to physical adsorption [45].

There are different types of adsorption methods used for carbon capture, e.g., pressure swing adsorption (PSA), temperature swing adsorption (TSA), and electrical swing adsorption (ESA) [32, 42, 45].

- PSA uses high-pressure gas to induce the adsorption of CO_2 in the adsorbent, while low-pressure release desorbs the gas from the adsorbent.
- TSA employs temperature changes to stimulate the chemical bonding of CO_2 in the adsorbent.
- ESA is a type of TSA where the desorption is supported by a supplied electrical current that heats the material to speed up desorption [32, 42].

Carbon capture through adsorption is a versatile capture method due to the variety of classes of adsorbents, ranging from inexpensive natural zeolites to advanced MOFs with high specific surface area. Unlike other carbon capture methods, such as chemical absorption, adsorption does not require additional chemicals, and the energy needed for regeneration is low and suitable for CO_2 capture from gas streams with low CO_2 concentrations. However, despite the advantages of adsorption, some disadvantages still limit its utility for large-scale carbon capture. One of these limitations is that the adsorbents are prone to degradation over time, requiring frequent maintenance. Moisture and pollutants like SO_x, NO_x, and fly ash can be especially challenging [56]. Additionally, the adsorption capture capacity is lower in comparison with chemical absorption. Thus, an important parameter for an adsorbent is the capacity for adsorption per mass of adsorbents. So far, adsorption is at a lab-scale level but has not yet been implemented at a commercial state [42, 43, 51, 55–57].

4.1.3   Heterogeneous Gas-Solid Reactions

Besides pure adsorption and absorption methods, a combination of both can be used, where in the first step, the CO_2 is adsorbed at the surface of a medium and then absorbed into the material. A typical carbon capture process where ab- and adsorption are used is heterogeneous gas-solid reactions [39, 47]. Typically, CO_2 capture by heterogeneous gas-solid reactions is done with alkaline earth metals. The metals are carbonated with CO_2 in the carbonator, and the CO_2 is separated from the carbonates in a gaseous form in a second chamber, the calciner. The metals are transported back to the carbonator, where the cycle begins again [43, 45] (Fig. 10).

One typical example of heterogeneous gas-solid reactions is the so-called carbonate looping, which is based on an equilibrium reaction of a metal oxide, its carbonate, and CO_2 [58]. The process is explained by an example of the most common type, calcium looping (CaL), also known as the regenerative calcium cycle [59]. In calcium looping, a CO_2-containing gas mixture is supplied to the carbonator, where calcium oxide (CaO) is carbonated to calcium carbonate ($CaCO_3$) in an exothermic reaction at about 650 °C (Eq. 4; total carbonation reaction). The calcium

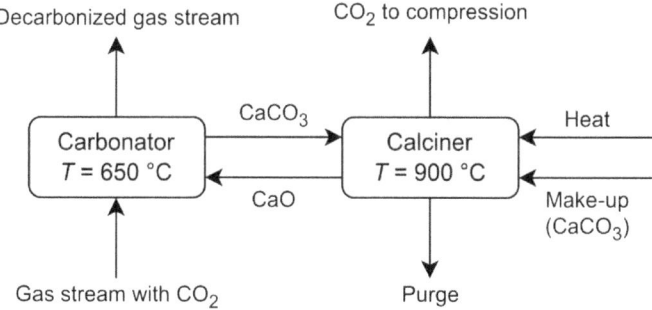

Fig. 10 Schematic of CO_2 separation with calcium looping with the temperature as T (based on [43, 48])

carbonate is then passed on to the calciner, where, in an endothermic reaction, at around 900 °C, the CO_2 is scrubbed from the metal and removed in gaseous form. The calcium carbonate is thus reduced to calcium oxide again, which can be cycled back to the carbonator to close the loop [13, 58].

$$CaO(s) + CO_2(g) \leftrightarrow CaCO_3(s) \left[+178.2 \frac{kJ}{mol}, \quad T \approx 650 \text{ °C} \right] \tag{4}$$

The energy for the reaction in the calciner can be directly supplied by the heat of internal combustion. The combustion has to be performed with pure oxygen to prevent nitrogen from taking part in the chemical reactions. The downside of this method is that air separation for oxygen generation is energy-intensive. Alternatively, the calciner can be heated indirectly with an external heat supply through heat exchangers [43, 60].

A typical side reaction of carbonate looping is the sulfation of calcium oxide to gypsum (Eq. 5 (indirect sulfation (irreversible in the temperature range)) and Eq. 6 (direct sulfation)). Due to this reaction, some of the sorbent (CaO) is bound and deactivated for the further uptake of CO_2. The sulfation is irreversible at the given temperatures inside the carbonator and calciner. To keep the uptake of CO_2 at a reasonable level, make-up material of CaO has to be supplied. However, the sulfation takes place at an about two orders of magnitude lower scale than the CO_2 uptake [61]. Furthermore, due to the high temperatures inside the calciner material, failures may occur, and the surface of the CaO tends to sinter, steadily lowering the binding capacity for CO_2 until a steady state is met after a couple of cycles [43, 45].

$$CaO(s) + SO_2(g) + \frac{1}{2}O_2(g) \rightarrow CaSO_4(s) \tag{5}$$

$$CaCO_3(s) + SO_2(g) + \frac{1}{2}O_2(g) \rightarrow CaSO_4(s) + CO_2(g) \tag{6}$$

The advantages of CaO as a sorbent are its abundance, low cost compared to other sorbents of CO_2, especially for calcium (crushed limestone [62]), and high CO_2 uptake capacity. Furthermore, calcium looping can be well-integrated with cement production [62, 63]. The technology has been tested on an MW scale and reaches capture rates of up to 94 % [60, 64, 65].

4.2 Membrane Separation

In principle, in membrane separation, two chambers (or parallel pipes) filled with gas are separated by a membrane. On the one side, the gas mixture containing CO_2 is supplied as a feed. It travels along the membrane surface area. Based on such a membrane, gases are separated due to permeability differences of the gases. That

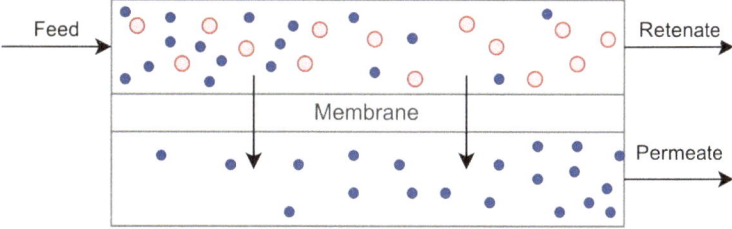

Fig. 11 Schematic of membrane separation (based on [32])

means some molecules and atoms in the feed gas stream can travel through the membrane to the other side (i.e., the second chamber) more readily than others; some are even hindered from passing over completely [32, 43] (Fig. 11).

A disequilibrium of chemical potential drives the transport through the membrane, often enhanced by pressure gradients. In the second chamber, the gas permeated through the membrane – the so-called permeate – can be retrieved. The remainder of the feed stream is called retentate [43]. There is a trade-off between selectivity (i.e., the purity of the permeate) and the flux through the membrane. A higher surface area for the same purity is needed if there is less flux to compensate for the lower permeability. High selectivity generally comes with lower permeability; thus, for high-purity CO_2 streams, immense surface areas are required [32, 66].

Membrane techniques can be categorized into organic membranes (primarily polymers) and inorganic membranes [42, 66]. Organic membranes are widely used in industries for CO_2 separation since they are relatively cheap. The downsides are their durability and low selectivity, which makes the technique unsuitable for large-scale CO_2 separation and purification. Inorganic membranes show better stability, corrosion resistance, and higher selectivity; but at present, they are characterized by high costs [56, 66].

Overall, membrane technologies are promising separation techniques once a sufficient selectivity and permeability can be ensured [34, 40, 42, 43, 56]. However, in most cases, the separation efficiency is low compared to other techniques, and reaching high-purity CO_2 is demanding. Furthermore, membranes may encounter issues with sulfur poisoning, high temperatures, and pressure stability, as well as water condensation [40]. The benefits of membrane technologies are their low energy consumption (e.g., no thermal energy is required since it is a purely physical process) as well as low costs, and the simple approach has no secondary pollutants [32, 40, 67].

4.3 Low-Temperature "Cryogenic" Separation

Different components in a gas mixture have individual condensation and sublimation temperatures. Therefore, gas mixtures can be separated by cooling it down. One example is the "cryogenic" separation of CO_2 from a gas mixture – cryogenic in science typically refers to temperatures of less than 120 K [68], which is not necessary for CO_2 separation; however, the term "cryogenic" is hereby generally used for low-temperature separation. If the condensation and sublimation temperature of CO_2 exceeds that of the other components inside the gas mixture, CO_2 will then undergo a phase change to liquid or, depending on the pressure level, solid form during cooling down and can then be removed from the gas mixture as a liquid or solid. This way, CO_2 is physically separated by condensation, sublimation, or distillation [43, 51].

The process works at low temperatures around the critical temperature of CO_2 (critical point CO_2: 304.13 K, 7.38 MPa). With pressure adjustments, the operational temperature can be changed. Cryogenic distillation is commercially used in the food industry [40]. However, for CO_2 capture of diluted streams, cryogenic carbon capture is considered to be in the early stages of development [51]. It is usually applied to gas mixtures with a relatively high purity (> 90 vol.-% CO_2) [32, 40, 43]. Furthermore, it is well-suited for processes with high pressure, like pre-combustion capture or oxyfuel-combustion [36, 51]. Purity values as high as 99.99 % CO_2 purity have been published [51], but more commonly, CO_2 purities of about 90 % are reported [21, 51]. A drawback is that the flue gas stream must be undiluted since NO_x, SO_x, and water are problematic [43, 51]. Furthermore, cooling and pressurization are energy-intensive, which makes the process rather expensive [42]. However, the benefits of the procedure besides high purity are that high CO_2 recovery rates can be reached, and no toxic chemicals need to be used. Overall, "cryogenic" carbon capture still has a low technology readiness level [51].

4.4 Comparison

The previous sections describe the most prominent carbon capture techniques, absorption, adsorption, heterogeneous gas-solid reactions, membrane separation, and low-temperature "cryogenic" separation, as well as their advantages and disadvantages. In this section, the techniques are compared qualitatively to each other regarding their energy consumption, purity, state of development, and costs.

In summary, the following analysis shows that each of the five discussed carbon capture techniques has its strengths and weaknesses (Table 3). Absorption and heterogeneous gas-solid reactions are the most mature ones, tested large scale or on an industrial scale. However, they are still energy-intensive. With improvements in membrane separation and adsorption processes, these low-energy techniques might prove themselves in the future. Cryogenic capture can be used where very high-purity CO_2 is required.

Energy Consumption: The energy consumption of CO_2 extraction techniques varies considerably depending on the technique used. Absorption and low-temperature "cryogenic" separation are very energy-intensive techniques, with the former requiring mainly thermal energy for regeneration and the latter requiring great amounts of energy for cooling and compression [40, 51]. Heterogeneous gas-solid reactions and adsorption have lower energy requirements. The former requires energy mainly for the high pressures and temperatures, and the latter mainly needs energy for the pressurization and cooling for adsorption, whereas the desorption of an adsorbent is an exothermic process [45, 60]. Membrane separation has relatively little energy requirements [43, 66].

Purity: All the carbon capture techniques discussed can achieve a high level of CO_2 purity. Nonetheless, the range of purity varies depending on the approach utilized. Low-temperature "cryogenic" separation attains the highest purity levels above 99.9 % [51]. Absorption and heterogeneous gas-solid reactions can achieve high purities well above 90 % [43]. Adsorption and membrane separation techniques exhibit medium to high purities and can be coupled with other capture strategies or can be deployed in multiple stages to enhance purity [32, 56, 69, 70]. Generally, higher purities can be reached with more effort, but the processes get increasingly complex/demanding—and thus more and more expensive—with increasing purity requirements.

State of Development: The carbon capture techniques explained above show very diverse states of development. The current state of technology ranges from mature and industry-implemented absorption in the form of amine scrubbing with many large-scale industrial applications already in operation to cryogenic carbon capture, which is in a very early stage of development. Adsorption is also widely tested with many different sorbents in the lab but has not been implemented commercially yet. Carbon capture via heterogeneous gas-solid reactions has been tested in the MW scale, and polymeric membranes have already been implemented at a commercial level. However, for all the different capture methods, there are novel approaches on the lab scale, like ionic liquids for absorption or dense inorganic membranes. Carbon capture techniques still have much potential for development, with some techniques being already viable for industrial-scale installation [40, 51, 60].

Costs: Carbon capture costs are challenging to establish, as the costs differ highly in literature, and many papers do not provide costs at all (a study of 250 selected papers showed that only 10 % of the reviewed papers contained cost information [71]). Generally, a higher CO_2 concentration in the gas mixture stream from pre-concentration leads to cheaper separation as long as the pollution of the gas stream remains the same. Absorption is the most common method and has so far the lowest equipment costs on an industrial scale. However, it shows high energy costs due to the energy penalty for sorbent regeneration and operational issues with corrosion [43, 66]. Heterogeneous gas-solid reactions have high material replacement needs and material requirements for high temperatures but lower energy needs for thermal cycling and sorbent regeneration than absorption processes [43, 60, 61]. The operational costs can be considered to be lower than for absorption. Adsorption and membrane separation have even lower operational costs due to lower energy requirements, but material costs vary widely from low-cost zeolites to advanced, hard-to-produce metal-organic frameworks (MOFs) for adsorption and for membranes, the material costs can range from low-cost polymeric membranes to high-cost inorganic membranes [45, 51, 66]. Low-temperature carbon capture has high capital and operational costs due to large cooling equipment requirements, material costs, and energy needed for solvent regeneration [51].

5 Summary and Conclusion

Carbon is a vital element, essential to all life on Earth and foundational in organic chemistry. It is pivotal across numerous industries and products. Particularly in the energy and transportation sector, carbon is used in the form of carbon-based fuels for energy generation and mobility. Currently, much of the carbon in today's products originates from fossil resources. However, burning fuel resources also leads to the emission of CO_2 and other greenhouse gases, which contributes to global warming. To mitigate the underlying greenhouse effects and achieve net-zero anthropogenic CO_2 emissions in the years to come, particularly in areas where decarbonization is impossible, renewable and sustainable carbon sources must be increasingly adopted to replace fossil carbon-based fuels. Various sources of renewable carbon exist, including biogenic sources, extraction from the atmosphere, and recycling. Often, carbon can be captured from point sources like bioethanol plants, cement production, biogas plants, and many more, with the profit of having an increased CO_2 concentration in the gas mixture to have a reduced energy toll for the extraction of "pure" CO_2.

Against this background in this paper an overview of different carbon sources and the technological approaches to supply pure CO_2 streams is analyzed. In this context, two supplementary concepts for the most efficient provision of pure CO_2 streams are considered: primary measures and secondary measures.

- Primary measures involve modifying industrial processes to pre-concentrate CO_2 in the gas stream, while
- Secondary measures focus on extracting and enriching CO_2 to produce a "pure" (clean) CO_2 stream.

Balancing these approaches depending on the specific circumstances is crucial to minimize technical effort, energy consumption, costs, and environmental impacts. Ultimately, the effectiveness of a carbon capture system depends on integrating primary and secondary measures to address specific industrial processes and resource availability.

The pre-concentration step is discussed exemplarily for three combustion-based routes (pre-combustion, oxyfuel-combustion, and post-combustion) showing that each process has advantages and disadvantages, including capture efficiency, energy requirements, capital costs, and retrofitting suitability. Ultimately, the choice of the respective process depends on the specific context. However, while these concepts are primarily discussed in the context of carbon-based fuel combustion for energy generation in power plants, similar strategies could be adopted across other sectors

Table 3 Advantages and disadvantages of carbon capture methods (based on [18, 32, 42, 43, 51, 56, 60, 61, 66])

Method	Advantages	Disadvantages
Absorption	• Proven, well-established technique • Low-cost sorbents • Most solvents suitable for high-temperature applications • Stable operation • High absorption efficiency and capacity • Excellent CO_2 selectivity	• High energy penalty for solvent regeneration • High investment costs for plant equipment size and high-performance sorbents • High maintenance costs due to corrosion of amines and the need of make-up material due to the formation of side products and evaporation • Mostly toxic solvents
Adsorption	• Low-cost reusable sorbents • Low energy consumption for regeneration • Almost non-corrosive • No use of extra chemicals • Relatively simple operation • Suitable for capture at a low CO_2 concentration	• High energy penalty for pressurization and temperature adjustment • Need of a replacement sorbent due to degradation issues • Smaller capture capacity per mass than for chemical absorption • Flue gas treatment necessary to regulate moisture content and contamination by SO_x and NO_x • Low selectivity • Not yet implemented at commercial scale
Heterogeneous gas-solid reactions	• Tested at large scale (> 1.7 MW) with capture rates of up to 94 % • Small operational costs due to limited energy penalty • Very cheap sorbents (e.g., crushed limestone for CaL) • Easy to integrate into cement manufacturing	• Material challenges due to high-temperature operation • Need for large amounts of make-up material due to the formation of gypsum and due to CO_2 capture capacity loss after multiple cycles

(continued)

Table 3 (continued)

Method	Advantages	Disadvantages
Membrane separation	• Limited energy penalty • Simple, modular system with a short start-up time • High CO_2 purity and recovery • Commercially available membranes	• Low selectivity for CO_2 capture, thus need for large membrane area for high capture rates • Membranes not suitable for high-temperature operation conditions; potential need for energy-intensive cooling • Susceptible to moisture and plugging by impurities • Low activity for feed gas streams with CO_2 concentration below 20 %
Low-temperature "cryogenic" separation	• CO_2 can be recovered at high purity of up to 99.9 % • Operation possible at atmospheric pressure • No toxic chemicals • Practical for transport or storage of CO_2 as a liquid	• Uneconomical for diluted CO_2 feed streams; stripping of water from feed gas to prevent ice plugging • High energy penalty for cooling and compression • High CO_2 concentration needed (restricted range of application)

by modifying the conversion, production, upgrading, or synthesis process to pre-concentrate CO_2 in exhaust gas streams, for example, in the chemical industry.

In addition to such primary measures, then the functionalities of various CO_2 extraction methods are presented as well as their advantages and disadvantages, especially in terms of energy consumption, purity, state of development, costs, and CO_2 recovery rate. The selection of the most appropriate capture system should thus be based on these factors and the concentration of CO_2 in the feed gas mixture (depending on the pre-concentration concept) and its pressure, in addition to the carbon-containing source used that influences the composition and level of pollutants.

Absorption and low-temperature "cryogenic" separation are energy-intensive methods, while adsorption, heterogeneous gas-solid reactions, and membrane separation have lower energy requirements. Low-temperature "cryogenic" separation attains the highest purity levels, while all the other methods can achieve medium to high levels of CO_2 purity. The state of development of each method varies, ranging from mature absorption methods already being implemented on an industrial scale to early-stage "cryogenic" carbon capture methods. Overall, some of these techniques still have much potential for development, and several other techniques are already viable for application today.

Reducing the energy need and thus the costs of carbon capture technologies remains a crucial goal. Therefore, an optimal combination between primary and secondary measures needs to be identified. Furthermore, scaling and learning effects can result in cost reduction when the technology is brought from lab scale to industrial scale. Novel techniques and refining of chemical parameters of materials can lead to lower energy demands. For absorption, research and development of new, different amines with improved thermodynamic properties and ionic liquids have the potential to mitigate high energy consumption [39, 56]. In the field of adsorption, research is currently focused on reducing costs and increasing the thermal stability of the materials used, improving capture capacity, and developing catalysts and adsorbents while creating better production processes [40]. Plasma catalysis and electrochemical CO_2 reduction are among the techniques under discussion [45].

In conclusion, carbon capture technology is a critical technology in the scenarios for achieving the goal of minimizing global warming to 1.5 °C above pre-industrial levels as set by the Paris Agreement. Capturing process-bound emissions is crucial to reach net-zero CO_2 emissions. The technology allows industry to achieve carbon-neutrality and tackle CO_2 emissions in sectors with limited alternative options like cement, steel, chemicals, waste incineration, and the production of synthetic fuels for long-distance transport. Implementing such carbon capture technologies helps to make the industry carbon-neutral and close carbon loops in a circular economy. An optimum between primary measures of CO_2 pre-concentration and secondary measures of CO_2 extraction/purification has to be established on the process level.

References

1. Kitadai N, Maruyama S (2018) Origins of building blocks of life: a review. Geosci Front 9(4):1117–1153. https://doi.org/10.1016/j.gsf.2017.07.007
2. IPCC (2023) Climate change 2023: synthesis report: a report of the intergovernmental panel on climate change. Contribution of working groups I, II and III to the sixth assessment report of the intergovernmental panel on climate change [core writing team, H. Lee and J. Romero (eds.)]. Geneva, Switzerland
3. IPCC (ed) (2022) Global warming of 1.5 °C: IPCC special report on impacts of global warming of 1.5 °C above pre-industrial levels in context of strengthening response to climate change, sustainable development, and efforts to eradicate. Cambridge University Press
4. UN (2015) Paris agreement: report of the conference of the parties to the United Nations framework convention on climate change. Paris
5. Deutscher Bundestag (2020) Gesetz zur Reduzierung und zur Beendigung der Kohleverstromung (Kohleverstromungsbeendigungsgesetz-KVBG): KVBG
6. Allen MR, Dube OP, Solecki W, Aragón-Durand F, Cramer W, Humphreys S, Kainuma M, Kala J, Mahowald N, Mulugetta Y, Perez R, Wairiu M, Zickfeld K (2018) Framing and Context. In: Masson-Delmotte V, Zhai P, Pörtner H-O, Roberts D, Skea J, Shukla PR, Pirani A, Moufouma-Okia W, Péan C, Pidcock R, Connors S, Matthews JBR, Chen Y, Zhou X, Gomis MI, Lonnoy E, Maycock T, Tignor M, Waterfield T (eds.) Global Warming of 1.5°C. An IPCC Special Report on the impacts of global warming of 1.5°C above pre-industrial levels and related global greenhouse gas emission pathways, in the context of strengthening the global response to the threat of climate change, sustainable development, and efforts to eradicate poverty. Cambridge University Press, UK and New York, NY, USA, pp 49–92. https://doi.org/10.1017/978100915 7940.003
7. Burrell AL, Evans JP, de Kauwe MG (2020) Anthropogenic climate change has driven over 5 million km^2 of drylands towards desertification. Nat Commun 11(1):3853. https://doi.org/10.1038/s41467-020-17710-7

8. van Oldenborgh GJ, Krikken F, Lewis S, Leach NJ, Lehner F, Saunders KR et al (2021) Attribution of the Australian bushfire risk to anthropogenic climate change. Nat Hazards Earth Syst Sci 21(3):941–960. https://doi.org/10.5194/nhess-21-941-2021

9. Williams AP, Abatzoglou JT, Gershunov A, Guzman-Morales J, Bishop DA, Balch JK et al (2019) Observed impacts of anthropogenic climate change on wildfire in California. Earth's Future 7(8):892–910. https://doi.org/10.1029/2019EF001210

10. Storlazzi CD, Gingerich SB, van Dongeren A, Cheriton OM, Swarzenski PW, Quataert E et al (2018) Most atolls will be uninhabitable by the mid-21st century because of sea-level rise exacerbating wave-driven flooding. Sci Adv 4(4):eaap9741. https://doi.org/10.1126/sciadv.aap 9741

11. Purr K, Garvens HJ (2021) Diskussionsbeitrag zur Bewertung von Carbon Capture and Utilization. Dessau-Roßlau

12. Schmid C, Hahn A (2030) Potential CO_2 utilisation in Germany: an analysis of theoretical CO_2 demand by 2030. J CO_2 Utili 50:101580. https://doi.org/10.1016/j.jcou.2021.101580

13. Fischedick M, Görner K, Thomeczek M (eds) (2015) CO_2: Abtrennung, Speicherung, Nutzung. Springer, Berlin, Heidelberg

14. Bringezu S, Kaiser S, Turnau S (2020) Zukünftige Nutzung von CO_2 als Rohstoffbasis der deutschen Chemie- und Kunststoffindustrie. Universität Kassel

15. Saygin D, Gielen DJ, Draeck M, Worrell E, Patel MK (2014) Assessment of the technical and economic potentials of biomass use for the production of steam, chemicals and polymers. Renew Sustain Energy Rev 40:1153–1167. https://doi.org/10.1016/j.rser.2014.07.114

16. Zhu K, Zhang J, Niu S, Chu C, Luo Y (2018) Limits to growth of forest biomass carbon sink under climate change. Nat Commun 9(1):2709. https://doi.org/10.1038/s41467-018-05132-5

17. Zhu X-G, Long SP, Ort DR (2008) What is the maximum efficiency with which photosynthesis can convert solar energy into biomass? Curr Opin Biotechnol 19(2):153–159. https://doi.org/10.1016/j.copbio.2008.02.004

18. National Academies of Sciences, Engineering, and Medicine, Division on Earth and Life Studies, Board on Chemical Sciences and Technology, Committee on Developing a Research Agenda for Utilization of Gaseous Carbon Waste Streams (eds) (2019) Gaseous carbon waste streams utilization: status and research needs. National Academies Press, Washington, DC

19. Tripathi AD, Mishra R, Maurya KK, Singh RB, Wilson DW (2019) Estimates for world population and global food availability for global health. In: The role of functional food security in global health. Elsevier, pp 3–24

20. Viebahn P, Horst J, Scholz A, Zelt O (2018) Technologiebericht 4.4 Verfahren der CO_2-Abtrennung aus Faulgasen und Umgebungsluft. Wuppertal, Karlsruhe, Saarbrücken

21. Socolow R, Desmond M, Aines R, Blackstock J, Bolland O, Kaarsberg T et al (2011) Direct air capture of CO_2 with chemicals: a technology assessment for the APS panel on public affairs

22. Cherepovitsyn A, Chvileva T, Fedoseev S (2020) Popularization of carbon capture and storage technology in society: principles and methods. Int J Environ Res Public Health 17(22). https://doi.org/10.3390/ijerph17228368

23. IEA (2016) 20 years of carbon capture and storage: accelerating future deployment. Paris

24. Mohamadi-Baghmolaei M, Hajizadeh A, Zahedizadeh P, Azin R, Zendehboudi S (2021) Evaluation of hybridized performance of amine scrubbing plant based on exergy, energy, environmental, and economic prospects: a gas sweetening plant case study. Energy 214:118715. https://doi.org/10.1016/j.energy.2020.118715

25. Lan X, Tans P, Thoning K, NOAA Global Monitoring Laboratory (2023) Trends in globally-averaged CO_2 determined from NOAA global monitoring laboratory measurements. NOAA GML

26. Gollakota S, McDonald S (2012) CO_2 capture from ethanol production and storage into the Mt Simon Sandstone. Greenhouse Gas Sci Technol 2(5):346–351. https://doi.org/10.1002/ghg.1305

27. Rodin V, Lindorfer J, Böhm H, Vieira L (2020) Assessing the potential of carbon dioxide valorisation in Europe with focus on biogenic CO_2. J CO_2 Utili 41:101219. https://doi.org/10.1016/j.jcou.2020.101219

28. Fröhlich T, Blömer S, Münter D, Brischke L-A (2019) CO_2-Quellen für die PtX-Herstellung in Deutschland — Technologien, Umweltwirkung. Verfügbarkeit, Heidelberg
29. Oles J, Büßelberg F, Brockmann M (2015) Sicherheit und Umweltrelevanz. In: Rosenwinkel K-H, Kroiss H, Dichtl N, Seyfried C-F, Weiland P (eds) Anaerobtechnik: Abwasser-, Schlamm- und Reststoffbehandlung, Biogasgewinnung, 3rd edn. Springer Vieweg, Berlin, Heidelberg, pp 741–799
30. von der Assen N, Müller LJ, Steingrube A, Voll P, Bardow A (2016) Selecting CO_2 sources for CO_2 utilization by environmental-merit-order curves. Environ Sci Technol 50(3):1093–1101. https://doi.org/10.1021/acs.est.5b03474
31. Ritzkowski M (2016) Biogaserzeugung und -nutzung - Exkurs: Deponiegas. In: Kaltschmitt M, Hartmann H, Hofbauer H (eds) Energie aus Biomasse: Grundlagen, Techniken und Verfahren, 3rd edn. Springer Vieweg, Berlin, Heidelberg, pp 1736–1755
32. Olajire AA (2010) CO_2 capture and separation technologies for end-of-pipe applications—a review. Energy 35(6):2610–2628. https://doi.org/10.1016/j.energy.2010.02.030
33. Görner K (2015) Umwandlungsprozesse bei fossilen Energieträgern. In: Fischedick M, Görner K, Thomeczek M (eds) CO_2: Abtrennung, Speicherung, Nutzung. Springer, Berlin, Heidelberg, pp 153–254
34. Bongartz R, Markewitz P, Biß K (2015) CO_2-Abscheidung. In: Wietschel M, Ullrich S, Markewitz P, Schulte F, Genoese F (eds) Energietechnologien der Zukunft. Springer Fachmedien Wiesbaden, Wiesbaden, pp 77–92
35. Görner K (2015) CCS-Prozesskette. In: Fischedick M, Görner K, Thomeczek M (eds) CO_2: Abtrennung, Speicherung, Nutzung. Springer, Berlin, Heidelberg, pp 255–482
36. Figueroa JD, Fout T, Plasynski S, McIlvried H, Srivastava RD (2008) Advances in CO_2 capture technology—The U.S. Department of Energy's Carbon Sequestration Program. Int J Greenhouse Gas Cont 2(1):9–20. https://doi.org/10.1016/S1750-5836(07)00094-1
37. Jansen D, Gazzani M, Manzolini G, van Dijk E, Carbo M (2015) Pre-combustion CO_2 capture. Int J Greenhouse Gas Control 40:167–187. https://doi.org/10.1016/j.ijggc.2015.05.028
38. Span R (2015) Stoffeigenschaften von Kohlendioxid. In: Fischedick M, Görner K, Thomeczek M (eds) CO_2: Abtrennung, Speicherung, Nutzung. Springer, Berlin, Heidelberg, pp 69–92
39. Pant D, Kumar Nadda A, Pant KK, Agarwal AK (eds) (2021) Advances in carbon capture and utilization, 1st edn. Springer, Singapore. Imprint Springer
40. Osman AI, Hefny M, Abdel Maksoud MIA, Elgarahy AM, Rooney DW (2021) Recent advances in carbon capture storage and utilisation technologies: a review. Environ Chem Lett 19(2):797–849. https://doi.org/10.1007/s10311-020-01133-3
41. Wietschel M, Ullrich S, Markewitz P, Schulte F, Genoese F (eds) (2015) Energietechnologien der Zukunft. Springer Fachmedien Wiesbaden, Wiesbaden
42. Fu L, Ren Z, Si W, Ma Q, Huang W, Liao K et al (2022) Research progress on CO_2 capture and utilization technology. J CO_2 Utili 66:102260. https://doi.org/10.1016/j.jcou.2022.102260
43. Görner K, Hübner K, Behr P, Heischkamp E, Rehfeldt S, Bergins C et al (2015) CCS-Prozesskette. In: Fischedick M, Görner K, Thomeczek M (eds) CO_2: Abtrennung, Speicherung, Nutzung. Springer, Berlin, Heidelberg, pp 255–482
44. Görner K, Oeljeklaus G, Bockhorn H, Pfeifer H, Hoenig V, Hoppe H et al (2015) Umwandlungsprozesse bei fossilen Energieträgern. In: Fischedick M, Görner K, Thomeczek M (eds) CO_2: Abtrennung, Speicherung, Nutzung. Springer, Berlin, Heidelberg, pp 153–254
45. Buckingham J, Reina TR, Duyar MS (2022) Recent advances in carbon dioxide capture for process intensification. Carbon Capture Sci Tech 2:100031. https://doi.org/10.1016/j.ccst.2022.100031
46. Burr B, Lyddon L (2008) A comparison of physical solvents for acid gas removal
47. International Union of Pure and Applied Chemistry (ed) (2014) Compendium of chemical terminology: gold book, 2nd edn. International Union of Pure and Applied Chemistry (IUPAC), Research Triangle Park, NC
48. Greco-Coppi M (2022) CO_2 capture from flue gases. YouTube
49. Borhani TN, Wang M (2019) Role of solvents in CO_2 capture processes: The review of selection and design methods. Renew Sustain Energy Rev 114:109299. https://doi.org/10.1016/j.rser.2019.109299

50. Zhang Z, Borhani TN, Olabi AG (2020) Status and perspective of CO_2 absorption process. Energy 205:118057. https://doi.org/10.1016/j.energy.2020.118057
51. Font-Palma C, Cann D, Udemu C (2021) Review of cryogenic carbon capture innovations and their potential applications 7(3):58. https://doi.org/10.3390/c7030058
52. Rochelle GT (2009) Amine scrubbing for CO_2 capture. Science 325(5948):1652–1654. https://doi.org/10.1126/science.1176731
53. Panja P, McPherson B, Deo M (2022) Techno-economic analysis of amine-based CO_2 capture technology: hunter plant case study. Carbon Capt Sci Tech 3:100041. https://doi.org/10.1016/j.ccst.2022.100041
54. Krótki A, Więcław Solny L, Stec M, Spietz T, Wilk A, Chwoła T et al (2020) Experimental results of advanced technological modifications for a CO_2 capture process using amine scrubbing. Int J Greenhouse Gas Control 96:103014. https://doi.org/10.1016/j.ijggc.2020.103014
55. Lee S-Y, Park S-J (2015) A review on solid adsorbents for carbon dioxide capture. J Ind Eng Chem 23:1–11. https://doi.org/10.1016/j.jiec.2014.09.001
56. Al-Mamoori A, Krishnamurthy A, Rownaghi AA, Rezaei F (2017) Carbon capture and utilization update. Energy Technol 5(6):834–849. https://doi.org/10.1002/ente.201600747
57. Cui J (2022) Recent advances in post-combustion CO_2 capture via adsorption methods. HSET 6:172–181. https://doi.org/10.54097/hset.v6i.959
58. Fedunik-Hofman L, Bayon A, Donne SW (2019) Kinetics of solid-gas reactions and their application to carbonate looping systems. Energies 12(15):2981. https://doi.org/10.3390/en12152981
59. Balfe MC, Augustsson O, Tahoces-soto R, Bjerge L-MH (2014) Alstom's regenerative calcium cycle—Norcem Derisking study: risk mitigation in the development of a 2nd generation CCS technology. Energy Proced 63:6440–6454. https://doi.org/10.1016/j.egypro.2014.11.679
60. Helbig M, Hilz J, Haaf M, Daikeler A, Ströhle J, Epple B (2017) Long-term carbonate looping testing in a 1 MWth pilot plant with hard coal and lignite. Energy Proced 114:179–190. https://doi.org/10.1016/j.egypro.2017.03.1160
61. Sun P, Grace JR, Lim CJ, Anthony EJ (2007) Removal of CO_2 by calcium-based sorbents in the presence of SO_2. Energy Fuels 21(1):163–170. https://doi.org/10.1021/ef060329r
62. Blamey J, Anthony EJ, Wang J, Fennell PS (2010) The calcium looping cycle for large-scale CO_2 capture. Prog Energy Combust Sci 36(2):260–279. https://doi.org/10.1016/j.pecs.2009.10.001
63. MacDowell N, Florin N, Buchard A, Hallett J, Galindo A, Jackson G et al (2010) An overview of CO_2 capture technologies. Energy Environ Sci 3(11):1645. https://doi.org/10.1039/c004106h
64. Arias B, Diego ME, Abanades JC, Lorenzo M, Díaz LAV, Martinez D et al (2013) Demonstration of steady state CO_2 capture in a 1.7 MWth Calcium looping pilot. Int J Greenhouse Gas Cont. https://doi.org/10.1016/j.ijggc.2013.07.014
65. Haaf M, Hilz J, Peters J, Unger A, Ströhle J, Epple B (2020) Operation of a 1 MWth calcium looping pilot plant firing waste-derived fuels in the calciner. Powder Technol 372:267–274. https://doi.org/10.1016/j.powtec.2020.05.074
66. Godin J, Liu W, Ren S, Xu CC (2021) Advances in recovery and utilization of carbon dioxide: a brief review. J Environ Chem Eng 9(4):105644. https://doi.org/10.1016/j.jece.2021.105644
67. Wilcox J, Haghpanah R, Rupp EC, He J, Lee K (2014) Advancing adsorption and membrane separation processes for the gigaton carbon capture challenge. Annu Rev Chem Biomol Eng 5:479–505. https://doi.org/10.1146/annurev-chembioeng-060713-040100
68. Timmerhaus KD, Reed RP (eds) (2007) Cryogenic engineering: fifty years of progress. Springer, New York
69. Liu Z, Grande CA, Li P, Yu J, Rodrigues AE (2011) Multi-bed vacuum pressure swing adsorption for carbon dioxide capture from flue gas. Sep Purif Technol 81(3):307–317. https://doi.org/10.1016/j.seppur.2011.07.037
70. Li G, Xiao P, Webley PA, Zhang J, Singh R (2009) Competition of CO_2/H_2O in adsorption based CO_2 capture. Energy Proced 1(1):1123–1130. https://doi.org/10.1016/j.egypro.2009.01.148

71. Leeson D, Mac Dowell N, Shah N, Petit C, Fennell PS (2017) A Techno-economic analysis and systematic review of carbon capture and storage (CCS) applied to the iron and steel, cement, oil refining and pulp and paper industries, as well as other high purity sources. Int J Greenhouse Gas Cont 61:71–84. https://doi.org/10.1016/j.ijggc.2017.03.020

Direct Air Capture

Kathrin Ebner, Lily Koops, Leonard Moser, and Andreas Sizmann

Abstract The production of carbohydrates via Power-to-X processes generally, and the synthesis of Power-to-Liquid fuels more specifically, requires a carbon feedstock. In order to close the carbon cycle and yield low- or zero-emission fuels, the carbon source needs to be regenerative. In that context, direct capture of carbon dioxide (CO_2) from ambient air represents an ubiquitous and scalable option. As of today, however, this process is associated with a low technological maturity and significant techno-economic uncertainty. Against this background, the following chapter aims to introduce the general principle, provide an overview on different technological approaches and current research directions, and briefly discuss cost perspectives.

Keywords Direct air capture · Carbon feedstock · Carbon source · Renewable hydrocarbon fuel · Low-emission fuel · Sustainable fuel

1 Introduction

The term direct air capture (DAC) refers to the direct removal of carbon dioxide (CO_2) from the atmosphere. In most cases, this is achieved through a sorption-based separation process. Thereby, air is passed through a capture unit in which CO_2 is bound to a sorbent material via physical or chemical interaction. As indicated in Fig. 1, this step is often termed "loading." To "unload," i.e. release the captured CO_2, the ambient conditions need to be adapted, for example by decreasing the pressure and/or enhancing the temperature. Concentrated CO_2 will then be collected and either permanently stored or reused. The sorbent is cycled between loading and unloading conditions, which is practically realized spatially or temporally, depending on the DAC approach. Several technology options employing different temperature ranges

K. Ebner · L. Koops · L. Moser · A. Sizmann (✉)
Bauhaus Luftfahrt e.V., Taufkirchen, Germany
e-mail: andreas.sizmann@bauhaus-luftfahrt.net

L. Koops
e-mail: lily.koops@bauhaus-luftfahrt.net

© The Author(s), under exclusive license to Springer Nature Switzerland AG 2025   431
N. Bullerdiek et al. (eds.), *Powerfuels*, Green Energy and Technology,
https://doi.org/10.1007/978-3-031-62411-7_16

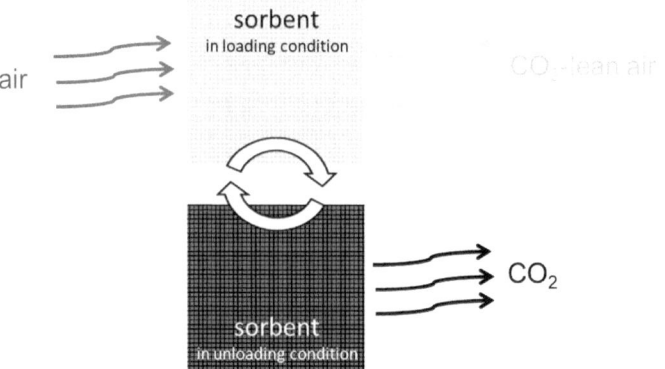

Fig. 1 Schematic illustration of the general principle of sorption-based direct air capture (DAC)

for the desorption step exist, as further discussed in Sect. 4. As the useable output is an increased CO_2 concentration by several orders of magnitude, the fundamental thermodynamics of this process in any technical realization demand significant amounts of thermal and/or electrical energy.

Sorption-based separation processes are well-established in chemical industry and often used for purification applications. In recent years, sorption-based carbon capture has mainly been employed to reduce CO_2 emissions in the form of concentrated waste streams, such as flue gas and exhaust from cement or steel industry [1]. However, the availability of accessible CO_2-rich streams as a feedstock, e.g. for fuel production, is limited [2]. In contrast to capturing CO_2 from those more concentrated sources, DAC is scalable and—in terms of the availability of CO_2—not location-dependent. Importantly, a decoupling of the carbon source from fossil infrastructures is enabled, thereby avoiding associated lock-in effects of emission-intense point sources.

Despite these benefits, significant challenges remain when it comes to an operation of DAC units based on renewable energies and under economically viable conditions. This is due to the comparatively low concentration of the feed involved, i.e. ambient air, which is 2–3 orders of magnitude lower than for CO_2 point sources. While the continuous upwards trend of the global atmospheric CO_2 content is undoubtedly associated with severe climate impact [3] and the concentration is rapidly approaching record values of 420 ppm at the time of writing [4], this value is still extremely small when looking at it from a chemical process engineering perspective. As a consequence, extracting CO_2 from ambient air via DAC processes, results in massive volumes of air needing to be processed. Even if the CO_2 removal was 100% effective, it would require processing approx. 1.4 million m^3 of ambient air to capture 1 tonne of CO_2 [5]. This translates into high energy consumption, which needs to be met with renewable energies in order to allow for a sustainable provision of CO_2 with minimized life cycle greenhouse gas emissions. When it comes to DAC's performance, first published life cycle assessments indeed confirm that the main penalties for several environmental indicators are associated with its energy demand [6, 7]. Note that meeting this demand with low-carbon intensity sources is a key requirement: Frankly, if the environmental impact of the energy required for DAC exceeds the one associated with the amount of greenhouse gas captured, there is no point of DAC.

To that end, it is of interest to take a look at the thermodynamically minimal energy demand, which equals approx. 22 kJ/mol for DAC and approx. 7 kJ/mol for point sources assuming a CO_2 concentration of 13% in the latter case, which is within the commonly observed range for flue gas [8–10]. It becomes evident that capture from more concentrated sources is thermodynamically favourable—though the energy demand differs only by a factor of about 3, due to the non-linear quasi-logarithmic dependence on concentration. Furthermore, higher concentrations also enhance the kinetics of the reaction.

So where is the growing interest for DAC—as a negative emission technology (NET) permanently storing the CO_2 after capture, on the one hand, and for carbon capture and use, i.e. the provision of a carbon feedstock, on the other hand—coming from? In addition to the above-mentioned advantages of avoiding lock-in effects and flexibility, scalability beyond what suitable point sources can provide, is a major incentive, and will be further addressed in the next sections. When it comes to the use of DAC as a NET, the removal of greenhouse gas emissions to compensate for emission overshoot—expectable even if the most ambitious climate strategies were followed [11–14]—is a cornerstone of reaching climate targets [3]. In that context, the economic performance of DAC might not be the central consideration and subsidizing early deployment may be motivated solely from a climate protection perspective. The use of DAC as a NET is, however, not in the focus of this chapter (for more in-depth considerations on this matter see [15] and [16]).

Furthermore, as in comparison with other sub-technologies within a Power-to-Liquid (PtL) conversion pathway DAC is still a comparatively immature technology, it represents the main bottleneck in terms of overall technological PtL maturity [17]. In this respect, research efforts have only increased significantly over the last decade; pronounced further development is expected. The large discrepancy of reported energy demands for approaches in early commercialization and the minimum separation energy (around factor 20–30 lower, when the reported total energy demand is considered) is a further indicator of a significant improvement potential, even under the further constraint of irreversible (far-from-equilibrium) operating conditions for practical CO_2 separation flow rates.

2 Direct Air Capture in the Context of Powerfuels

From the perspective of powerfuel production, DAC-sourced CO_2 can be considered a scalable feedstock for various fuel production pathways. Beyond advantages in scalability, DAC would theoretically allow closing the carbon cycle as schematically

Fig. 2 Graphical representation of the concept of recycling CO_2 for powerfuel production

illustrated in Fig. 2 and, consequently, beneficially influence the overall environmental footprint of the fuel. With that, DAC could act as an enabler for high life cycle greenhouse gas emission reduction potentials as are increasingly in demand for improved sustainability and adhering to policies designed to ensure compliance with climate targets (e.g. the European Union Renewable Energy Directive II). However, processes can never be ideally realized and, therefore, recycling the CO_2 released upon fuel combustion comes with penalties, which ought to be minimized. Again, the high energy demand of capturing CO_2 represents the most significant drawback in this context. For the production of kerosene via different Power-to-Liquid (PtL) pathways, the difference in total electrical efficiency has been estimated to be about 9–12% lower over the entire fuel production pathway as compared to CO_2 point sources [17]. Additionally, depending on the DAC approach, significant amounts of heat are required. As mentioned above for a favourable life cycle balance, it is absolutely essential to meet the overall energy demand (i.e. thermal and electrical) renewably.

An interesting aspect of powerfuel production is that certain steps in the conversion of the reactants to hydrocarbon fuels are exothermic [18], i.e. heat is released in the process. Depending on the plant design and process configuration, this thermal energy could be repurposed for (partially) meeting DAC's demand. Some authors have shown full heat integration when low-quality heat is required by DAC, while DAC approaches relying on high-quality heat represent a bigger challenge and typically need to be fed from other sources, such as electrical or methane-fuelled heating systems [19, 20]. Once again, the avoidance of fossil resources to operate these heating systems is of key importance to minimize process-related emissions and achieve an adequate carbon balance.

Beyond the impact of the associated energy requirements, also other indicators of environmental impact are often under vivid debate in the context of air capturing the carbon feedstock. One example is the use of land for DAC farms: Depending on the DAC approach and annual production capacity of the plant, estimates of specific land use requirements for DAC farms range over orders of magnitude from below 0.001 to more than 1 km^2/Mt a [21]. Main sources of uncertainty stem from the limited understanding of scaling properties, for instance, related with the ideal spacing and arrangement of individual units to avoid interference effects similar to wind farms [5]. In any case, it should be noted that the actual land coverage, i.e. the footprint of the capture units themselves in comparison with the total area, represents only a small share of the total land use and as is suggested for onshore wind farms, the uncovered area could still be utilized in a "dual use" manner, e.g. for farming purposes. To put things in perspective for the case of powerfuel production, the land use associated with

DAC itself, while non-negligible, will be significantly smaller than the area required for power generation based on renewable energies and, therefore, most likely not be the major lever in this impact category. Most importantly, overall land use of first generation (1G) and second generation (2G) biofuels—often considered alternatives to powerfuels—way exceeds land use for DAC-based powerfuel production as shown for synthetic kerosene previously [17].

Other relevant environmental aspects in the context of DAC at scale include resource depletion and water use among many others: It is evident that similar to other chemical plants significant impact is associated with the initial construction of equipment and the site itself (requiring concrete, steel and metal, plastic housing, etc.). And, while only a very limited amount of comprehensive quantitative life cycle assessments for different technological approaches of DAC can be found in scientific literature, it has been demonstrated that key components needing regular replacement and maintenance (most notably the sorbent), cause a prominent contribution to the normalized environmental impact of CO_2 captured via DAC approaches [6]. Beyond the system boundary of the DAC plant itself, also the generation of renewable electricity required for its operation does not come without an associated environmental footprint and contributes notably to material and metal depletion and other impact categories [6]. This in turns implies that minimizing energy demand represents a significant lever for reducing life cycle impact of DAC further motivating optimization to that effect, e.g. via heat integration as described above.

3   Development and Deployment Level

A larger-scale deployment of DAC-based CO_2 supply as a feedstock for the production of powerfuels is currently hampered by its high cost. Furthermore, DAC is associated with significant techno-economic uncertainties due to its limited development level and data availability, which in turn is a consequence of the aforementioned limited deployment level. At the time of writing, there is only a small number of commercial players in the field of DAC technologies [21, 22]. The CO_2 cumulatively captured in all plants operative in 2022 amounts to 10,000 t(CO_2)/a or 0.01 Mt(CO_2)/a respectively, which is several orders of magnitude below the total negative CO_2 emissions required for scenarios aiming at net-zero at 2050 [23]. As illustrated in Fig. 3, to a lesser extent, this still holds true when looking at the scale required to satisfy the feedstock demand with respect to CO_2 of only one hard-to-abate sector such as aviation [24, 25].

While deployment and maturity clearly represent a bottleneck, interest in furthering DAC is growing—not only in the scientific, but also in the private sector. Moreover, carbon capture (and especially DAC with its potential as a negative emission technology (NET)) has been in the focus of several recently implemented policies [26] and is a cornerstone of strategies aiming at avoiding overshooting globally remaining carbon budgets [3]. Whether DAC will undergo an unconventionally large

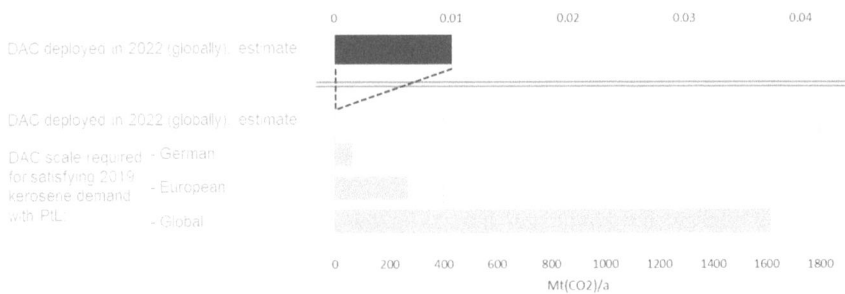

Fig. 3 Total current DAC deployment [23] versus estimated deployment required if aviation's total kerosene demand were fed with PtL fuels from DAC (theoretical scenario). Figure adapted from [24]

growth rate similar to certain energy technologies [27, 28] as argued by some authors (e.g. [29]) remains to be seen.

Research and development efforts in the field of DAC currently focus on a number of critical key performance indicators, which are briefly introduced below:

- capacity of the sorbent—*How much CO_2 can be taken up per amount of sorbent per cycle?*
- selectivity—*Does the sorbent have high affinity only for CO_2 or are there other relevant competitive species?*
- kinetics—*How fast does the sorption/desorption process proceed?*
- overall effectiveness—*What share of CO_2 can be removed from the air flow intake?*

The latter indicator is directly related to the former ones and strongly influenced by the design of the contactor (contactor design—*How can the air be best brought in contact with the sorbent?*). Overall effectiveness is critical as it determines the amount of air needing to be processed for a given amount of CO_2, affecting—among others—the overall energy demand. Furthermore, for smooth, low-maintenance operation, low overall cost, and good scalability, the following criteria are of high relevance:

- Sorbent stability—*Resistance to degradation caused by other species.*
- Sorbent cyclability—*Reversibility and resistance to degradation caused by the change from loading to unloading conditions.*
- Suitability for local climatic conditions (e.g. ambient temperature, humidity).
- Use of materials (inexpensive, abundant, non-toxic).
- Energy consumption (low).
- Infrastructure requirements (e.g. for thermal and electrical energy provision).

The limited development and deployment status is at the root of the techno-economic uncertainties associated with DAC. Data availability is low, especially when it comes to the envisaged large-scale DAC farms that can capture more than 1 Mt(CO_2)/a. To the best of the authors' knowledge, operating demonstration plants currently exist up to an annual capture capacity of 4,000 t(CO_2)/a [30] and for process-model-based estimations for larger-scale DAC plants a relatively wide range of energy and material demands is reported [31–33], which reflects the current uncertainties that accompany the scaling of DAC systems. The latter, alongside the variance in technological maturity, impede in-depth quantitative comparative technology assessments between different DAC approaches. For that reason, the following section is aimed at providing a mainly qualitative overview of the different technologies proposed, illustrating the capture principle and reporting the main challenges and opportunities associated with each.

4 Direct Air Capture Technologies

In the following chapter, the main technological approaches to DAC are described in more detail.

4.1 Solid Sorption

Solid sorption DAC refers to a process, in which the sorbent is either a self-standing solid itself or immobilized on a solid substrate. The majority of currently investigated solid sorbents for DAC are capturing via chemisorption, i.e. forming a chemical bond to CO_2, but also physisorption-based approaches, e.g. on metal organic frameworks, are proposed. To release the captured CO_2 after the loading step, the process conditions are temporarily adapted, which typically involves a change in pressure, temperature, and/or humidity. The reactor bed is cycled between loading and unloading conditions, as illustrated in Fig. 4.

A well-thought out design of the air contactor and a high surface area of the reactor bed are key factors for effectively exploiting the sorbent's CO_2-capacity. Research on various support materials (silica, ceramics, polymers, etc.) aims at sophisticated material and porous design to ensure optimum interaction of the CO_2 with the sorbent, while minimizing the resistance to air flow (i.e. pressure drop) through the reactor bed [22, 34]. The latter is of high relevance, as it directly influences the electrical energy demand for processing large volumes of air. Notably, to the best of the authors' knowledge, all currently prototyped DAC approaches require active flow—natural air flow is insufficient. Furthermore, effective heat transfer is targeted in order to avoid high thermal energy demand upon temperature cycling. With its impact on energy demand (and consequently operating cost) and its significant contribution to plant cost [5, 33, 35], it is unsurprising that contactor and reactor bed design generally represents one of the major research directions for solid sorption DAC today.

The second major focus of research efforts lies in the improvement of reaction kinetics and optimization of the sorbent-CO_2 interaction as a major lever on overall effectiveness and economic feasibility as theoretical assessments have demonstrated [33]. For that reason, a variety of sorbent options is currently under investigation. In the following, the two dominant ones—amines and metal organic frameworks

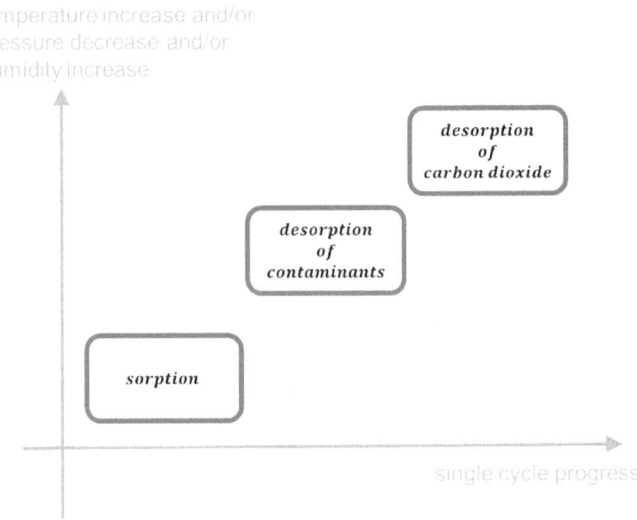

Fig. 4 Schematic illustration of an ideal single sorption–desorption cycle of a solid sorption unit

(MOFs)—will be briefly introduced (for a more in-depth review, it is referred to other works [36, 37]).

Amines. Amine-based solid sorption could be considered the most mature DAC approach at the time of writing, as the majority of operational plants as well as the DAC site with the currently highest capacity are all run by a company employing amines immobilized on cellulose [26, 30]. Also other players have demonstrated solid amine DAC at scale [21]. Despite these scaling and early deployment efforts, fundamental research on the sorbents and amine functionalization to further improve properties like CO_2 adsorption kinetics, selectivity, and heat transfer is still needed and vital for an energy-effective and low-cost DAC operation [33, 36, 38]. In that context, researchers strive for a better understanding of the impacts of the co-presence of water as a competitive species in the absorption and desorption process, on the one hand, that can also enhance required thermal input in the form of parasitic heat losses [38, 39]. On the other hand, it is considered as a possible lever for overall energy efficiency improvements for instance by co-optimizing adsorption kinetics of both species [38, 40]. What is more, amines are associated with some drawbacks when it comes to limited cyclability and degradation from contaminants gradually reducing capacity [22].

Scalability for larger-scale PtX production can be achieved by means of a modular approach, i.e. adding further contactor units. Due to the moderate temperature within the desorption step (usually $\leq$ 100 °C), heat integration with the fuel production process is possible and the thermal energy demand can (partially) be met with low-grade off-heat from the synthesis step [20, 25].

Metal organic frameworks (MOFs). Due to their "sponge-like" structure and their excellent chemical tunability, metal organic frameworks (MOFs) have attracted considerable attention in the context of sorption processes, among them DAC. Utilization for DAC processes, however, represents a rather challenging use case for MOFs as water and other contaminants present in ambient air can lead to severe degradation and limited cyclability of the sorbent [22]. MOF-based approaches typically bind CO_2 via physisorption, which comes with a trade-off with regards to sorption effectiveness and energy demand. In recent works, physisorption-based DAC has been associated with overall higher energy consumption for loading and lower effectiveness due to comparatively weaker sorbent-CO_2 interaction [37]. The desorption step, on the other hand, is less energy-intensive for the same reason and can be performed at only slightly elevated temperature enabling independence from external heat sources [41]. This may be of specific interest when no heat integration options are available. Besides obtaining a deeper understanding of this interplay and enhancing MOF durability for this application case, also inexpensively scalable synthesis approaches for suitable MOFs are in the focus of current research [42].

Finally, recent investigations have demonstrated the importance of optimal design and operation of solid sorbent DAC processes taking into account ambient conditions that vary by time and geographical location [43]. Since especially ambient temperature and humidity can have a significant influence on the energy consumption and productivity of DAC, research efforts go towards DAC processes allowing tunable operation variables and the development of corresponding control strategies (e.g. air pretreatment).

4.2 Liquid Sorption

Liquid sorption approaches entail the chemisorption of CO_2 on alkaline substances, however, in this case, the sorbent is not immobilized, but present in liquid form (typically in a solution). In practice, liquid sorption is realized by running the liquid over a high surface area support, whereby the latter is covered by a thin layer of the sorbent. That way, a large contact area is ensured. To optimize this wetting procedure and the interaction with the air flow while keeping energy consumption low, sophisticated flow designs have been proposed and their further development is still a subject of research [44]. A number of liquid sorption processes have been proposed mainly based on aqueous alkali hydroxides, amines, and ionic liquids (in order of technological maturity for DAC) [21, 44, 45]. These options will be briefly introduced in the following.

Aqueous alkali hydroxides. Within this type of DAC technology, dissolved alkali hydroxides (mainly potassium hydroxide (KOH) [33, 46]) react with CO_2 from air during the capture step to yield carbonates (Fig. 5). The latter can be precipitated from solution by adding calcium hydroxide ($Ca(OH)_2$) and exploiting the comparatively poor solubility of calcium carbonate ($CaCO_3$). As alkali hydroxides are strong bases, they strongly interact with CO_2, yielding stable bonds forming carbonates. This qualitatively explains not only the low energy demand of the capture step, but also a major drawback that comes with that approach; the subsequent release of CO_2 from $CaCO_3$ ("calcination") requires temperatures of around 900 °C and is therefore energy intensive (i.e. significant demand of high-quality heat) [47]. Finally, $Ca(OH)_2$ is regenerated from calcium oxide (CaO), also obtained from the calcination step, and can be reused in the process. In practice, capture, release of CO_2, and regeneration of the sorbent are spatially separated as illustrated in Fig. 5. Generally, this approach employs only widely available and inexpensive basic chemicals, enhancing the feasibility of rapid deployment of the already commercially available approach [48], from a material perspective. Disadvantages of the approach lie in the significant water use and the above-mentioned high thermal energy demand, which is currently met with fossil-fuelled calciners heavily impacting the overall life cycle balance of the aqueous alkali approach [7]. Research efforts towards all-electric non-fossil systems relying on the electrification of the calcination step aim at meeting the high-quality heat demand in a renewable way at the expense of higher electricity consumption [31, 32, 49]. Hereby, the presently very limited throughput of available electrical calciners (i.e. three to four orders of magnitude lower than what would be required for a commercial plant capturing 1 Mt of CO_2 annually [49]) is considered a relevant technology development gap.

Aqueous amines. Aqueous amines have been routinely commercially employed for CO_2 capture from point sources [50, 51]; however, their application for direct air capture is only under early investigation [36, 52, 53]. This is mainly due to the fact that amines are weak bases and, with that, CO_2 is not as strongly bound and as effectively removed from a low-concentrated feed as with their hydroxide counterparts. This results in large volumes of air and aqueous sorbent that need to be processed. In contrast to such drawbacks, this weaker bond is associated with advantages for the CO_2 release and sorbent regeneration, as it requires significantly lower temperatures (below 150 °C) as compared to aqueous hydroxides (ca. 900 °C), incentivizing studies exploring the use of aqueous amines also for CO_2 capture from low-concentrated feeds. The general process is illustrated in Fig. 6. Nevertheless, it should be noted

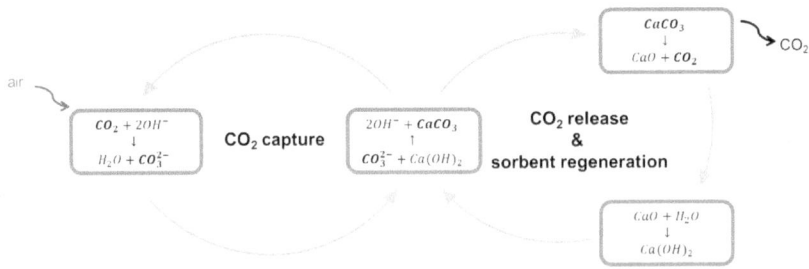

Fig. 5 Illustration of the DAC process employing alkali hydroxides as the sorbent. The spatial separation of the CO_2 capture, CO_2 release, and sorbent regeneration steps is indicated in the figure [46]

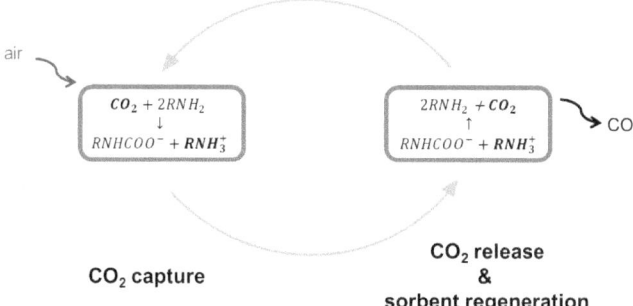

Fig. 6 Illustration of DAC process using aqueous amines as capture medium. "R" represents an organic side chain

that the substantial heat capacity of water as the solvent represents a limitation for minimizing thermal energy demand [33]. Furthermore, evaporation during regeneration can cause significant sorbent losses upon temperature cycling putting a toll on durability and safety, as commonly employed amines are toxic compounds. It has been found that immobilization aids cyclability [33] further explaining the early focus on solid sorption approaches using this group of sorbents as discussed above. To overcome the described disadvantages, research for aqueous amine-based DAC is largely focussed on improving kinetics and durability.

Ionic liquids. More recently, ionic liquids have been suggested to be utilized for DAC. Employing this material class opens up a vast design space to explore when it comes to fine-tuning the chemical and physical properties and, with that, optimizing their suitability for capturing very dilute species. Alongside excellent tunability comes the potential for attractive CO_2 selectivity, adequate stability, and low volatility [54]. Their typically high viscosity ranging all the way to gel-like states, however, presents an immense challenge to practical implementation and gas penetration. Polymeric or encapsulated types may be better suited and may allow adapting a more penetrable structure [45, 55]. What is more, the typically high cost associated with these sorbents have so far discouraged intensive investigation urgently required for further maturing this DAC approach. Novel bio-derived materials from "waste" amino acids with a priori low toxicity have recently been found to be effective sorbents for ready and reversible DAC within a first example of an ionic-liquid-enabled approach on a laboratory scale [56]. As ionic liquids are explored for a variety of applications at the moment, an enhanced understanding of ionic liquid design and increased production volumes may aid tackling some of the mentioned challenges in the future.

4.3 Electrochemical Approaches

The two main strategies for making use of electrochemistry for DAC could be summarized as follows:

- Replacing a single step of an already known DAC approach with an electrochemical reaction as an alternative (e.g. sorbent regeneration [10, 57]) or
- realizing DAC solely based on an electrochemical process, e.g. [8, 58–60].

The latter approach could again be subcategorized into direct electrochemical activation of the sorbent, i.e. the sorbent directly responds to the change in potential (the sorbent is "redox-active") or indirect activation referring to another species in the system being redox-active, thereby indirectly influencing the affinity of the sorbent to CO_2.

As the properties of the sorbent, most importantly its nucleophilicity, can be tuned via the electrochemical potential as illustrated in Fig. 7—and with that its interaction with CO_2—electrochemical approaches have the potential for very high selectivity. Furthermore, the required condition change between loading and unloading conditions can be achieved by the adaption of the electrochemical potential as well, eliminating the need for temperature or pressure swings and enhancing flexibility. Consequently, they hold the promise for energy-efficient capture and release of CO_2. Notably, no additional thermal energy is required as the mentioned electrochemical processes are typically conducted at ambient temperature.

With research efforts on electrochemical utilization of CO_2 continuously increasing [61–63], the perspective of combining electrochemical DAC and utilization in a single process further fosters the search for adequate electrochemical DAC approaches [64]. Benefits would include a decrease of process complexity and increase in efficiencies as well as full elimination of the need for intermediate CO_2 desorption and handling. As is the case with many other electrochemical processes,

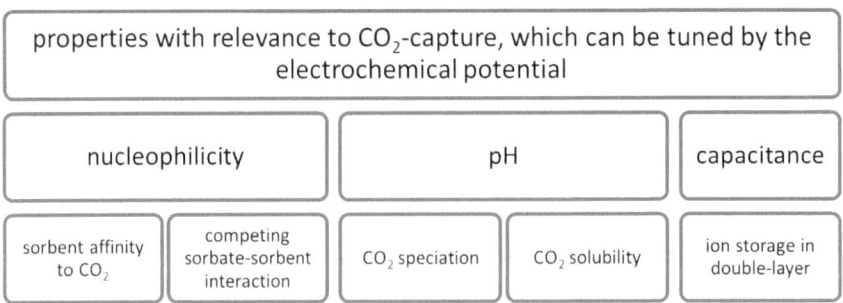

Fig. 7 Selected properties, which can be tuned by the electrochemical potential and have been exploited for CO_2 capture

however, fundamental molecular understanding not only of the sorption and desorption of CO_2 itself, but also of the interaction with impurities and of undesired side reactions (and more) is in many cases still limited. This aspect currently impairs fully leveraging the above-described potential. Despite these uncertainties and the limited technological maturity, many further developments have recently been reported on a conceptual or laboratory-scale level. A Massachusetts Institute of Technology spinoff is currently working on prototype implementations for capturing CO_2 by an electrochemical approach (respectively from air and from flue gases). The aim is to capture up to 100 kg of CO_2 per day, which corresponds to about 35 tonnes of CO_2 per year, thus considered a notable exception going way beyond lab-scale testing [65].

As most published electrochemical DAC approaches show a low technological readiness level (TRL), the energy demand is typically only reported for the electrochemical step. While this will likely represent the main contribution, (publically available) results on further evaluation and optimization of process-related energy flows are currently lacking for electrochemical DAC, despite being key factors for comparing to more conventional DAC approaches. Furthermore, for a commercially feasible operation on a larger scale, high current densities are of utmost importance, but again, most recent experiments have been focussing on demonstrating proof-of-concept, not commercially competitive performance. Another key requirement for practical realization of larger-scale electrochemical DAC is achieving a significantly higher affinity of the sorbent to CO_2 over oxygen (O_2), as has recently been pointed out and has previously not been a major research focus [65]. For these reasons, the possibility for direct comparison of electrochemical approaches with already commercialized liquid and solid sorption processes is still limited.

The range of electrochemical approaches is rather broad and an in-depth review is not aimed at herein, but some examples for the most relevant concepts are summarized in the following: Organic, redox-active sorbents (or competing sorbates) have been investigated quite thoroughly. Most works have utilized a variety of quinones to demonstrate electrochemical CO_2 capture; an in-depth review is provided, e.g. in reference [66]. DAC approaches based on dissolved quinones range from batch methods to continuous flow systems, whereby decoupling electrochemical and chemical reactions leaves more room for optimizing each component of the system [67]. As the dissolution of this compound class in organic solvents limits current density, more recent advancements focus on solvent-free processes immobilizing the quinone at the electrode [58, 68]. In summary, trade-offs between effective gas-sorbent interaction and energy-efficient cycling are associated with capturing only at the electrode surface or in a liquid electrolyte, similar to what is observed for solid compared to liquid sorption approaches as described above.

Alternative redox-active sorbents are, for example, transition metal complexes, e.g. based on nickel (Ni) or copper (Cu), which have been proposed in context of the two main strategies mentioned above (i.e. for directly (mediating) the capture or solely for the regeneration step of the sorbent) [10, 69]. Exploiting potential-induced pH swings is another path of incorporating electrochemical approaches. As is well known, CO_2 forms different species in aqueous media depending on the pH (carbonic acid, CO_2 (aq) bicarbonates, carbonates), which affects its solubility in water and can, therefore, be used as a driver for CO_2 capture and release [10, 57, 66]. Also, electrodialysis, which is a well-established process for water deionization, has been proposed for DAC, whereby one again relies on the principle of potential-induced pH change at the electrodes or over a (bipolar) membrane [8, 9].

Bipolar membrane dialysis can also be exploited for the replacement of the thermal regeneration step in the aqueous alkali hydroxide approach as suggested in reference [57], for instance. While the regeneration process was found to be less energy-intensive when employing this electrochemical approach, economic competitiveness was shown to suffer from currently high costs of the membrane materials.

4.4 Membrane-Based Approaches

Fully membrane-based CO_2 capture (i.e. no potential applied, unlike the above-described electrodialysis approach) follows one of two main ideas, which are illustrated in Fig. 8. Either the membrane allows solely CO_2 to pass through (part (A) in Fig. 8) or it prevents only CO_2 from passing (part (B) in Fig. 8). This results in a (theoretically) pure stream of CO_2.

While membrane-based approaches are a routinely implemented gas separation technique and have already been successfully demonstrated at scale for removing CO_2 from flue gas [1], applying it for the direct capture of CO_2 from ambient air is challenging due to the extremely low concentration of CO_2 in this feed. For membrane-based DAC approaches to achieve high concentration in the CO_2 product stream (i.e. > 50%) would require at least two separation stages, even when highly selective membranes are employed [70]. Current membranes suffer from low fluxes and would consequently ask for pressurized air flow and large membrane areas, resulting in an excessive energy demand that would exceed the demands of current solid and liquid sorption approaches already at early commercialization stage by far [70–72]. Adding to this are currently high costs of membranes that challenge economic competitiveness [57]. Unsurprisingly, development efforts are currently seeking to drastically improve separation efficiency and selectivity in an effort to exploit the theoretically high productivity achievable with this type of gas separation process, but are characterized by comparatively low technological maturity [22, 70].

Fig. 8 Schematic illustration of membrane-based CO_2 separation

4.5 Further Methods

A number of further methods for DAC have been suggested, but are currently not in the focus for the provision of CO_2 as a carbon feedstock for Power-to-X (PtX) applications. This is due to a variety of reasons, such as very limited technological readiness, comparatively excessive energy demand or challenges when it comes to CO_2 release. Nevertheless, some of them are briefly mentioned in the following:

- The first example is cryogenic distillation—a process currently well-established in industry for separating air into its constituents, yielding high-purity products. While this process is effective for species present in significant quantities (nitrogen, oxygen, etc.) and side products with high market value (e.g. noble gases), an enhanced deployment for the sole purpose of DAC would be associated with a prohibitive energy demand due to the low concentration of CO_2 in ambient air [73, 74].
- Another case would be passive DAC, where natural convection of air is exploited instead of relying on energy-intensive active flow [22]. It has been proposed to integrate suitable sorbents into existing infrastructure (e.g. [75]) to optimize land use and capital expenditures (CAPEX). However, for the provision of a scalable carbon source, effectivity and logistics will likely be a show-stopper. To enhance the airflow while still avoiding the need for external energy, integration of DAC into a solar updraft tower has been conceptualized [76].

5 A Note on Direct Air Capture Cost

In the following paragraph, a brief summary on the current perspectives on future DAC cost is provided, shedding a bit of light on this missing puzzle piece. Taking the techno-economic uncertainties, limited data availability and (for some of the approaches described above) low technological maturity into account, it does not come as a surprise that cost estimates for DAC at scale vary drastically depending on the underlying assumptions, e.g. with regard to the technical process layout (e.g. heat integration), scaling properties, the geographical location of a DAC site, or production and learning rates. More specifically, estimates for the cost for the provision of CO_2 by DAC range from more than 50 \$/t(CO_2) all the way to up to above 600 \$/t(CO_2) for the year 2050. In this context, the approaches for cost assessments range from detailed chemical process modelling to purely market-based assessments [5, 32, 46, 49, 77]. Sabatino et al. well-illustrated the dependencies of liquid alkali, liquid amine, and solid sorbent capture cost on key parameters such as reaction kinetics, reactor bed cost, and electricity cost [33]. Such analysis aids understanding the development required and the most relevant levers in reaching targeted capture costs. What both expert surveys and model-based assessments universally agree on, is that the cost of CO_2 supply by DAC will significantly decrease in the next decades, especially when an enhanced deployment is endorsed and capital expenditures (CAPEX) can be reduced [5, 29, 49, 77, 78]. A value of 100 \$/t(CO_2) has been adopted as a future target value for the economic competitiveness of DAC technologies for carbon feedstock provision [29]. For the purpose of Power-to-Liquid (PtL) fuel production, this target equates to a cost contribution of about 25–30 ct/L. For other purposes, however, higher cost for CO_2 captured from ambient air by DAC approaches may be tolerable.

6 Summary

Direct air capture (DAC) refers to a range of processes to capture the greenhouse gas carbon dioxide (CO_2) from ambient air. The CO_2 can subsequently be used as a carbon resource, e.g. for the production of powerfuels. Alternatively, it could be permanently stored. While economic considerations differ for these two general pathways, there is a common interest of furthering DAC for providing sustainable carbon feedstock, on the one hand, and supporting the fulfilment of climate targets by achieving negative CO_2 emissions, on the other. Deployment of DAC is still in an early phase with currently (at the time of writing) only 18 operational plants globally [26] that capture a cumulative CO_2 amount in negligible orders of magnitude compared to what would be required for a production of powerfuels at scale. Continued technological progress is required with a long way to go towards considering DAC technologies a fully energy-optimized, easily scalable, and readily deployable solution. Nevertheless, important advancements have been made and significant research efforts are devoted towards improving more established technologies and conceptualizing and proofing the feasibility of technologically less mature, but promising alternatives. While DAC is currently still associated with significant uncertainties regarding cost and performance at scale, further research and accelerating commercialization activities will help closing these knowledge gaps.

References

1. Bui M, Adjiman CS, Bardow A, Anthony EJ, Boston A, Brown S, Fennell PS, Fuss S, Galindo A, Hackett LA et al (2018) Carbon capture and storage (CCS): the way forward. Energy Environ Sci 11:1062–1176. https://doi.org/10.1039/C7EE02342A
2. Zitscher T, Neuling U, Habersetzer A, Kaltschmitt M (2020) Analysis of the German industry to determine the resource potential of CO_2 emissions for PtX applications in 2017 and 2050. Resources 9:149. https://doi.org/10.3390/resources9120149
3. IPCC (2021) Climate change 2021: the physical science basis. Contribution of Working Group I to the Sixth Assessment Report of the Intergovernmental Panel on Climate Change: Summary for Policy Makers
4. NOAA GML, Trends in atmospheric carbon dioxide. Available online: https://gml.noaa.gov/ccgg/trends/global.html (accessed on 25 September 2022)
5. Herzog H (2022) Direct air capture. In: Bui M, MacDowell N (eds) Greenhouse gas removal technologies. Royal Society of Chemistry, Cambridge, UK. ISBN 1839165251

6. Deutz S, Bardow A (2021) Life-cycle assessment of an industrial direct air capture process based on temperature–vacuum swing adsorption. Nat Energy 6:203–213. https://doi.org/10.1038/s41560-020-00771-9

7. de Jonge MMJ, Daemen J, Loriaux JM, Steinmann ZJN, Huijbregts MAJ (2019) Life cycle carbon efficiency of direct air capture systems with strong hydroxide sorbents. Int J Greenhouse Gas Cont 80:25–31. https://doi.org/10.1016/j.ijggc.2018.11.011

8. Sharifian R, Wagterveld RM, Digdaya IA, Xiang C, Vermaas DA (2021) Electrochemical carbon dioxide capture to close the carbon cycle. Energy Environ Sci 59:7007. https://doi.org/10.1039/D0EE03382K

9. Muroyama AP, Pătru A, Gubler L (2020) Review—CO_2 separation and transport via electrochemical methods. J Electrochem Soc 167:133504. https://doi.org/10.1149/1945-7111/abbbb9

10. Renfrew SE, Starr DE, Strasser P (2020) Electrochemical approaches toward CO_2 capture and concentration. ACS Catal 10:13058–13074. https://doi.org/10.1021/acscatal.0c03639

11. Rogelj J, Luderer G, Pietzcker RC, Kriegler E, Schaeffer M, Krey V, Riahi K (2015) Energy system transformations for limiting end-of-century warming to below 1.5 °C. Nature Clim Change 5:519–527. https://doi.org/10.1038/nclimate2572

12. Rogelj J, Popp A, Calvin KV, Luderer G, Emmerling J, Gernaat D, Fujimori S, Strefler J, Hasegawa T, Marangoni G et al (2018) Scenarios towards limiting global mean temperature increase below 1.5 °C. Nat Clim Chang 8:325–332. https://doi.org/10.1038/s41558-018-0091-3

13. Lackner KS, Brennan S, Matter JM, Park A-HA, Wright A, van der Zwaan B (2012) The urgency of the development of CO_2 capture from ambient air. Proc Natl Acad Sci USA 109:13156–13162. https://doi.org/10.1073/pnas.1108765109

14. Zeman FS, Lackner KS (2004) Capturing carbon dioxide directly from the atmosphere. World Res Rev

15. Minx JC, Lamb WF, Callaghan MW, Fuss S, Hilaire J, Creutzig F, Amann T, Beringer T, de Garcia WO, Hartmann J et al (2018) Negative emissions—Part 1: research landscape and synthesis. Environ Res Lett 13:63001. https://doi.org/10.1088/1748-9326/aabf9b

16. Fuss S, Lamb WF, Callaghan MW, Hilaire J, Creutzig F, Amann T, Beringer T, de Garcia WO, Hartmann J, Khanna T et al (2018) Negative emissions—Part 2: costs, potentials and side effects. Environ Res Lett 13:63002. https://doi.org/10.1088/1748-9326/aabf9f

17. Batteiger V, Ebner K, Habersetzer A; Moser L, Schmidt P, Weindorf W, Rakscha T (2022) Power-to-liquids—a scalable and sustainable fuel supply perspective for aviation; Background, Dessau-Roßlau. Available online: https://www.umweltbundesamt.de/publikationen/power-to-liquids

18. Dieterich V, Buttler A, Hanel A, Spliethoff H, Fendt S (2020) Power-to-liquid via synthesis of methanol, DME or Fischer–Tropsch-fuels: a review. Energy Environ Sci 13:3207–3252. https://doi.org/10.1039/D0EE01187H

19. Marchese M, Buffo G, Santarelli M, Lanzini A (2021) CO_2 from direct air capture as carbon feedstock for Fischer-Tropsch chemicals and fuels: energy and economic analysis. J CO_2 Util 46:101487, https://doi.org/10.1016/j.jcou.2021.101487

20. Heß D, Klumpp M, Dittmeyer R (2020) Nutzung von CO_2 aus Luft als Rohstoff für synthetische Kraftstoffe und Chemikalien: Studie im Auftrag des Ministeriums für Verkehr Baden-Württemberg

21. Viebahn P, Scholz A, Zelt O (2019) The potential role of direct air capture in the German energy research program—results of a multi-dimensional analysis. Energies 12:3443. https://doi.org/10.3390/en12183443

22. Erans M, Sanz-Pérez ES, Hanak DP, Clulow Z, Reiner DM, Mutch GA (2022) Direct air capture: process technology, techno-economic and socio-political challenges. Energy Environ Sci 15:1360–1405. https://doi.org/10.1039/D1EE03523A

23. IEA, Direct air capture. Available online: https://www.iea.org/reports/direct-air-capture (accessed on 28 September 2022)

24. Lufahrt B (2021) Jahrbuch 2021. Taufkirchen, Germany. Available online: https://www.bauhaus-luftfahrt.net/

25. Ebner K, Moser L, Koops L, Batteiger V (2022) Power-to-liquids: shedding light on levers and uncertainties in the process chain. In: Proceedings of the 9th European Conference for Aerospace Sciences. Lille, France, 27 June–1 July. https://doi.org/10.13009/EUCASS2022-7340

26. IEA, Direct air capture. Available online (accessed on 28 September 2022): https://www.iea.org/reports/direct-air-capture

27. Odenweller A, Ueckerdt F, Nemet GF, Jensterle M, Luderer G (2022) Probabilistic feasibility space of scaling up green hydrogen supply. Nat Energy 7:854–865. https://doi.org/10.1038/s41560-022-01097-4

28. Our World in Data, Installed solar energy capacity. Available online: https://ourworldindata.org/grapher/installed-solar-pv-capacity (accessed on 28 September 2022)

29. Lackner KS, Azarabadi H (2021) Buying down the cost of direct air capture. Ind Eng Chem Res 60:8196–8208. https://doi.org/10.1021/acs.iecr.0c04839

30. Climeworks, Climeworks begins operations of Orca, the world's largest direct air capture and CO_2 storage plant. Available online: https://climeworks.com/news/climeworks-launches-orca (accessed on 28 September 2022)

31. Hanna R, Abdulla A, Xu Y, Victor DG (2021) Emergency deployment of direct air capture as a response to the climate crisis. Nat Commun 12:368. https://doi.org/10.1038/s41467-020-20437-0

32. Fasihi M, Efimova O, Breyer C (2019) Techno-economic assessment of CO_2 direct air capture plants. J Clean Prod 224:957–980. https://doi.org/10.1016/j.jclepro.2019.03.086

33. Sabatino F, Grimm A, Gallucci F, van Sint Annaland M, Kramer GJ, Gazzani M (2021) A comparative energy and costs assessment and optimization for direct air capture technologies. Joule 13:63002. https://doi.org/10.1016/j.joule.2021.05.023

34. Zhu X, Ge T, Yang F, Lyu M, Chen C, O'Hare D, Wang R (2020) Efficient CO_2 capture from ambient air with amine-functionalized Mg–Al mixed metal oxides. J Mater Chem A 8:16421–16428. https://doi.org/10.1039/d0ta05079b

35. McQueen N, Psarras P, Pilorgé H, Liguori S, He J, Yuan M, Woodall CM, Kian K, Pierpoint L, Jurewicz J et al (2020) Cost analysis of direct air capture and sequestration coupled to low-carbon thermal energy in the United States. Environ Sci Technol 54:7542–7551. https://doi.org/10.1021/acs.est.0c00476

36. Barzagli F, Giorgi C, Mani F, Peruzzini M (2020) Screening study of different amine-based solutions as sorbents for direct CO_2 capture from air. ACS Sustain Chem Eng 8:14013–14021. https://doi.org/10.1021/acssuschemeng.0c03800

37. Leonzio G, Fennell PS, Shah N (2022) A comparative study of different sorbents in the context of direct air capture (DAC): evaluation of key performance indicators and comparisons. Appl Sci 12:2618. https://doi.org/10.3390/app12052618

38. Young J, Mcilwaine F, Smit B, Garcia S, van der Spek M (2022) Process-informed adsorbent design for direct air capture. ChemRxiv. https://doi.org/10.26434/chemrxiv-2022-gn7r2

39. Young J, García-Díez E, Garcia S, van der Spek M (2021) The impact of binary water—CO_2 isotherm models on the optimal performance of sorbent-based direct air capture processes. Energy Environ Sci 13:63001. https://doi.org/10.1039/D1EE01272J

40. Wurzbacher JA, Gebald C, Piatkowski N, Steinfeld A (2012) Concurrent separation of CO_2 and H_2O from air by a temperature-vacuum swing adsorption/desorption cycle. Environ Sci Technol 46:9191–9198. https://doi.org/10.1021/es301953k

41. AspiraDAC Pty Ltd., AspiraDAC website. Available online: https://www.aspiradac.com/ (accessed on 18 November 2022)

42. Olajire AA (2018) Synthesis chemistry of metal-organic frameworks for CO_2 capture and conversion for sustainable energy future. Renew Sustain Energy Rev 92:570–607. https://doi.org/10.1016/j.rser.2018.04.073

43. Wiegner JF, Grimm A, Weimann L, Gazzani M (2022) Optimal design and operation of solid sorbent direct air capture processes at varying ambient conditions. Ind Eng Chem Res 61:12649–12667. https://doi.org/10.1021/acs.iecr.2c00681

44. McQueen N, Gomes KV, McCormick C, Blumanthal K, Pisciotta M, Wilcox J (2021) A review of direct air capture (DAC): scaling up commercial technologies and innovating for the future. Prog Energy 3:32001. https://doi.org/10.1088/2516-1083/abf1ce

45. Yang H, Singh M, Schaefer J (2018) Humidity-swing mechanism for CO_2 capture from ambient air. Chem Commun (Camb) 54:4915–4918. https://doi.org/10.1039/C8CC02109K

46. Keith DW, Holmes G, St. Angelo D, Heidel K (2018) A process for capturing CO_2 from the atmosphere. Joule 2:1573–1594. https://doi.org/10.1016/j.joule.2018.05.006

47. Socolow R, Desmond M, Aines R, Blackstock J, Bolland O, Kaarsberg T, Lewis N, Mazzotti M, Pfeffer A, Sawyer K, Siirola J, Smit B, Wilcox J (2011) Direct air capture of CO_2 with chemicals: a technology assessment for the APS Panel on Public Affairs

48. Carbon Engineering Ltd., Carbon engineering website. Available online: https://carbonengineering.com/ (accessed on 11 November 2022)

49. McQueen N, Desmond MJ, Socolow RH, Psarras P, Wilcox J (2021) Natural gas vs. electricity for solvent-based direct air capture. Front Clim 2:186. https://doi.org/10.3389/fclim.2020.618644

50. Bui M, MacDowell N (eds) Greenhouse gas removal technologies. Royal Society of Chemistry, Cambridge, UK. ISBN 1839165251

51. Rochelle GT (2009) Amine scrubbing for CO_2 capture. Science 325:1652–1654. https://doi.org/10.1126/science.1176731

52. Kiani A, Jiang K, Feron P (2020) Techno-economic assessment for CO_2 capture from air using a conventional liquid-based absorption process. Front Energy Res 8:475. https://doi.org/10.3389/fenrg.2020.00092

53. Kothandaraman J, Goeppert A, Czaun M, Olah GA, Prakash GKS (2016) Conversion of CO_2 from air into methanol using a polyamine and a homogeneous ruthenium catalyst. J Am Chem Soc 138:778–781. https://doi.org/10.1021/jacs.5b12354

54. Yang Z, Dai S (2021) Challenges in engineering the structure of ionic liquids towards direct air capture of CO_2. Green Chem Eng 2:342–345. https://doi.org/10.1016/j.gce.2021.08.003

55. Lee Y-Y, Edgehouse K, Klemm A, Mao H, Pentzer E, Gurkan B (2020) Capsules of reactive ionic liquids for selective capture of carbon dioxide at low concentrations. ACS Appl Mater Interfaces 12:19184–19193. https://doi.org/10.1021/acsami.0c01622

56. Recker EA, Green M, Soltani M, Paull DH, McManus GJ, Davis JH, Mirjafari A (2022) Direct air capture of CO_2 via ionic liquids derived from "waste" amino acids. ACS Sustain Chem Eng 10:11885–11890. https://doi.org/10.1021/acssuschemeng.2c02883

57. Sabatino F, Mehta M, Grimm A, Gazzani M, Gallucci F, Kramer GJ, van Sint Annaland M (2020) Evaluation of a direct air capture process combining wet scrubbing and bipolar membrane electrodialysis. Ind Eng Chem Res 59:7007–7020. https://doi.org/10.1021/acs.iecr.9b05641

58. Diederichsen KM, Liu Y, Ozbek N, Seo H, Hatton TA (2021) Toward solvent-free continuous-flow electrochemically mediated carbon capture with high-concentration liquid quinone chemistry. Joule 13:8. https://doi.org/10.1016/j.joule.2021.12.001

59. Liu Y, Ye H-Z, Diederichsen KM, van Voorhis T, Hatton TA (2020) Electrochemically mediated carbon dioxide separation with quinone chemistry in salt-concentrated aqueous media. Nat Commun 11:2278. https://doi.org/10.1038/s41467-020-16150-7

60. Kang JS, Kim S, Hatton TA (2021) Redox-responsive sorbents and mediators for electrochemically based CO_2 capture. Current Opinion in Green Sustain Chem 31:100504. https://doi.org/10.1016/j.cogsc.2021.100504

61. Zhu P, Wang H (2021) High-purity and high-concentration liquid fuels through CO_2 electroreduction. Nat Catal 4:943–951. https://doi.org/10.1038/s41929-021-00694-y

62. Casebolt R, Levine K, Suntivich J, Hanrath T (2021) Pulse check: potential opportunities in pulsed electrochemical CO_2 reduction. Joule 20:1. https://doi.org/10.1016/j.joule.2021.05.014

63. Küngas R (2020) Review—electrochemical CO_2 reduction for CO production: comparison of low- and high-temperature electrolysis technologies. J Electrochem Soc 167:44508. https://doi.org/10.1149/1945-7111/ab7099

64. Pérez-Gallent E, Vankani C, Sánchez-Martínez C, Anastasopol A, Goetheer E (2021) Integrating CO_2 capture with electrochemical conversion using amine-based capture solvents as electrolytes: PREPRINT. Ind Eng Chem Res 60:4269–4278. https://doi.org/10.1021/acs.iecr.0c05848

65. Rathi A (2022) Bill Gates invests in carbon capture startup after tech breakthrough. Bloomberg [Online], February 2. Available online: https://www.bloomberg.com/news/articles/2022-02-02/new-tech-could-cut-carbon-capture-energy-use-by-70

66. Rheinhardt JH, Singh P, Tarakeshwar P, Buttry DA (2017) Electrochemical capture and release of carbon dioxide. ACS Energy Lett 2:454–461. https://doi.org/10.1021/acsenergylett.6b00608

67. Liu Y, Lucas É, Sullivan I, Li X, Xiang C (2022) Challenges and opportunities in continuous flow processes for electrochemically mediated carbon capture. iScience 25:105153. https://doi.org/10.1016/j.isci.2022.105153

68. Voskian S, Hatton TA (2019) Faradaic electro-swing reactive adsorption for CO_2 capture. Energy Environ Sci 12:3530–3547. https://doi.org/10.1039/C9EE02412C

69. Wang M, Rahimi M, Kumar A, Hariharan S, Choi W, Hatton TA, Flue gas CO_2 capture via electrochemically mediated amine regeneration: system design and performance

70. Castel C, Bounaceur R, Favre E (2021) Membrane processes for direct carbon dioxide capture from air: possibilities and limitations. Front Chem Eng 3:1047. https://doi.org/10.3389/fceng.2021.668867

71. Fujikawa S, Selyanchyn R, Kunitake T (2021) A new strategy for membrane-based direct air capture. Polym J 53:111–119. https://doi.org/10.1038/s41428-020-00429-z

72. Keith DW, Heidel K, Cherry R (2009) Capturing CO_2 from the atmosphere: rationale and process design considerations. In: Launder BE, Thompson JMT (eds) Geoengineering climate change: environmental necessity or Pandora's box? Cambridge University Press, pp 107–126

73. Song C, Liu Q, Deng S, Li H, Kitamura Y (2019) Cryogenic-based CO_2 capture technologies: state-of-the-art developments and current challenges. Renew Sustain Energy Rev 101:265–278. https://doi.org/10.1016/j.rser.2018.11.018

74. Lockley A, von Hippel T (2021) The carbon dioxide removal potential of liquid air energy storage: a high-level technical and economic appraisal. Front Eng Manag 8:456–464. https://doi.org/10.1007/s42524-020-0102-8

75. Carbon Collect, Carbon collect website. Available online: https://carboncollect.com/ (accessed on 10 June 2023)

76. Brady C, Davis ME, Xu B (2019) Integration of thermochemical water splitting with CO_2 direct air capture. Proc Natl Acad Sci USA 116:25001–25007. https://doi.org/10.1073/pnas.1915951116.77

77. Breyer C, Fasihi M, Bajamundi C, Creutzig F (2019) Direct air capture of CO_2: a key technology for ambitious climate change mitigation. Joule 3:2053–2057. https://doi.org/10.1016/j.joule.2019.08.010

78. Shayegh S, Bosetti V, Tavoni M (2021) Future prospects of direct air capture technologies: insights from an expert elicitation survey. Front Clim 3:1. https://doi.org/10.3389/fclim.2021.630893

CO$_2$ Transport and Storage Options

Larissa Fink, Michael Schulthoff, Nils Bullerdiek, and Martin Kaltschmitt

Abstract Since the onset of industrialization, humanity has been releasing increasing amounts of carbon dioxide (CO$_2$) into the atmosphere due to the use of fossil fuels. CO$_2$ is a greenhouse gas that significantly impacts the global climate when concentrated in the atmosphere. To limit global warming to below 2°C compared to pre-industrial levels, as outlined in the 2015 Paris Climate Agreement, greenhouse gas emissions—especially CO$_2$ from fossil fuels—must be drastically reduced. In addition to replacing fossil fuels with renewable energy sources, strategies like carbon capture and storage (CCS) or carbon capture and utilization (CCU), collectively known as CCUS, offer ways to mitigate CO$_2$ emissions. As new applications for CO$_2$ are expected to emerge, and as the volume and transport distance of CO$_2$ will likely increase in the future, there is an urgent need to further develop the global CO$_2$ infrastructure. A suitable infrastructure must be designed to handle the specific characteristics of CO$_2$, including transportation methods and facilities for conditioning and intermediate storage. Therefore, first the key properties of CO$_2$ are presented and then the current state and future advancements in CO$_2$ conditioning, storage, and transportation are reviewed. Finally, scenarios are created to compare different CO$_2$ logistics chains, evaluated from a techno-economic perspective, followed by an outlook on future developments and concluding remarks.

Keywords CO$_2$ transport · CO$_2$ storage · CO$_2$ · CO$_2$ logistics · CO$_2$ infrastructure

1 Introduction

Since the beginning of industrialization, humanity has emitted increasing amounts of carbon dioxide into the atmosphere due to fossil fuel energy usage. However, carbon dioxide (CO$_2$) is a greenhouse gas that impacts global climate if it concentrates within the atmosphere [1]. To limit the resulting global warming to below 2 °C

L. Fink (✉) · M. Schulthoff · N. Bullerdiek · M. Kaltschmitt
Hamburg University of Technology (TUHH), Institute of Environmental Technology and Energy Economics (IUE), Hamburg, Germany
e-mail: Larissa.fink@tuhh.de

© The Author(s), under exclusive license to Springer Nature Switzerland AG 2025
N. Bullerdiek et al. (eds.), *Powerfuels*, Green Energy and Technology,
https://doi.org/10.1007/978-3-031-62411-7_17

compared to pre-industrial levels according to the Paris Climate Agreement from 2015, greenhouse gas (GHG) emissions—and here primarily carbon dioxide from fossil fuel resources—need to be reduced significantly. Besides the substitution of fossil fuel energy by energy based on renewable sources of energy (often called "green" energy), approaches like carbon capture and storage (CCS) or carbon capture and utilization (CCU)—summarized as "CCUS"—do also allow for mitigation of carbon dioxide emissions.

The overall process chains of CCS and CCU applications consist of three steps, shown in Fig. 1. As a first step, CO_2 is captured, e.g., from coal-fired power plants (i.e., fossil fuel-based CO_2), from bioethanol production plants (i.e., biomass-based CO_2), or ambient air (i.e., atmospheric CO_2). Subsequently, being typically the second step, the "collected" CO_2 is transported, as the locations of CO_2 capturing and the subsequent storage or use of this CO_2 usually differ; for transportation of this CO_2, different options can be used. As a third step, the CO_2 is permanently stored (CCS), for example, injected into the underground, e.g., into exploited natural gas deposits (at both onshore and offshore locations), or used (CCU), for example, as the carbon feedstock for the production of synthetic hydrocarbons.

Nowadays, CO_2 is already used for different processes, applications, and products. The fertilizer industry uses most of the CO_2 utilized on an industrial scale to produce urea. The oil industry also has a CO_2 demand for enhanced oil recovery, mainly in North America. Smaller CO_2 demanders are, for example, the food and beverage sector (e.g., carbonated beverages), as well as industries like metalworking. CO_2, which is used nowadays, is mainly a by-product from other industrial processes (e.g., ammonia production and biomass fermentation) [3].

If climate protection is taken seriously and the GHG reduction goals agreed upon within the Paris Agreement are met, CO_2 will be used to a much wider extent and thus for a clearly broader spectrum of other products and processes in the future. Additionally, using CO_2 emissions can help close carbon loops by utilizing a "waste stream" within the context of a circular economy. Besides this, carbon dioxide can

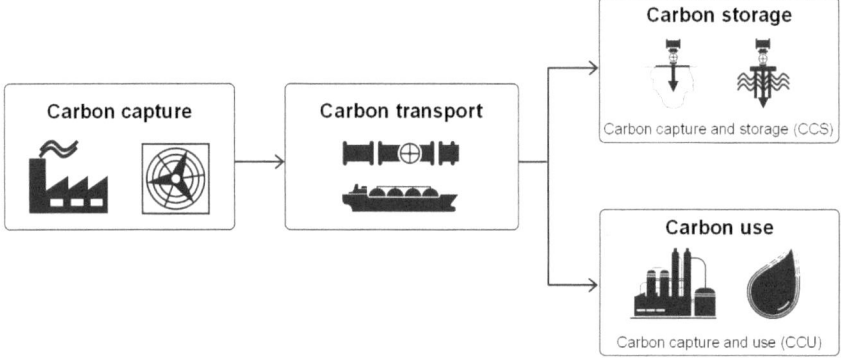

Fig. 1 Generic process chains for carbon capture and storage (CCS) as well as carbon capture and utilization (CCU) (according to [2])

be bound into a product used over a long period (e.g., as a building material) and thus be removed from the atmosphere. This allows greenhouse gas (GHG) emissions to be stored over more extended periods, contributing partly during this storage period to achieve valid GHG reduction targets [3].

Technological solutions related to CCUS have been investigated and tested for quite some time. However, rising interest in the respective technical solutions has significantly increased in more recent years due to an increasing awareness of the negative consequences associated with climate change accelerating over time. In 2022, 35 commercial CCUS facilities were in operation, with an annual CO$_2$ capture capacity of approximately 45 Mt CO$_2$. Another 12 sites to capture CO$_2$ are currently under construction. More than 200 further projects are being considered for implementation in the near future. If these projects materialize, these plants will have additional capacities to capture CO$_2$ of more than 220 Mt/a CO$_2$ by 2030 [4].

Since it is expected that, on the one hand, new applications for CO$_2$ will arise and emerge in the future and, on the other hand, the amount of CO$_2$ to be transported and thus also the transport distance will increase, the existing CO$_2$ infrastructure in the year 2022 must urgently be further developed. An adequate infrastructure for the logistics of CO$_2$ via short and long distances has to be adapted to the specific properties of CO$_2$. It implies means of transport as well as facilities for conditioning and intermediate storage of CO$_2$.

Against this background, based on the CO$_2$ characteristics presented first, the current state of the art and future developments in the fields of conditioning, intermediate storage, and transport of CO$_2$ are given below. Subsequently, scenarios are created to compare different CO$_2$ chains and evaluated techno-economically. Finally, an outlook on expected developments is given, and a conclusion is made.

2 Carbon Dioxide (CO$_2$) Characteristics

At ambient temperature and pressure (standard condition), carbon dioxide (CO$_2$) is a non-flammable, non-toxic, and odorless gas heavier than ambient air. By increasing the pressure, CO$_2$ can be condensed/liquefied at ambient temperature; i.e., in this respect, it shows similar characteristics to propane or butane. However, due to the position and length of its phase boundaries (Fig. 2), CO$_2$ reacts differently under pressure at ambient temperature than other condensable gases.

Since the critical temperature of CO$_2$, 31.1 °C, is close to the typical ambient temperature level, compression and heat transfer processes often pass through the extended critical region. Here, the equations of state for describing the changes of thermo-physical state variables are challenging to apply because the properties of supercritical fluids are composed of the properties of the liquid and the gaseous phase. For example, the density and viscosity of supercritical CO$_2$ correspond to a liquid's density and a gas's viscosity [6]. However, for the design of processes and technical apparatuses, the state variables of CO$_2$ must be determined. This is possible via published tables with the respective material data or equations of state;

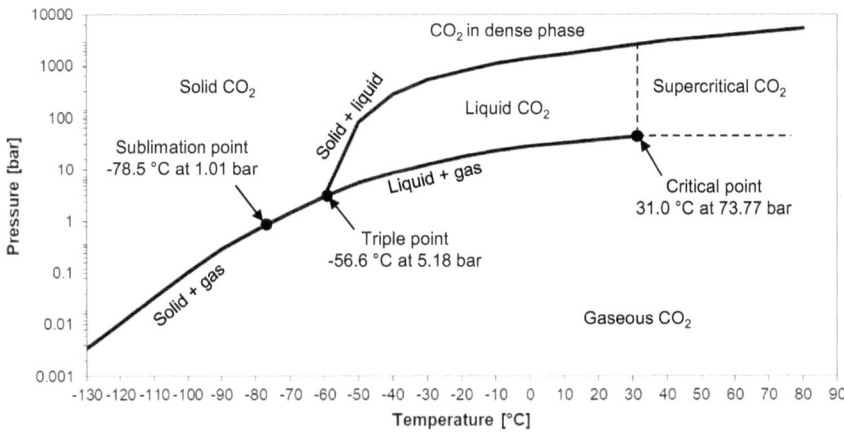

Fig. 2 Phase diagram (*p-T*-diagram) of carbon dioxide (according to [5])

for temperatures up to 827 °C and pressures up to 8000 bar, the equation of state by Span and Wagner can be used [7, 8].

The pressure of CO_2 at the triple point (Fig. 2) is above ambient pressure. Therefore, CO_2 has a normal sublimation point instead of a normal boiling point. The consequence is that at ambient pressure, solid CO_2 (dry ice) passes directly into the gas phase (i.e., it sublimates). Relevant substance properties of CO_2 are given in Table 1 [7, 9].

Table 1 Properties of carbon dioxide [5–7]

Property	Value	Unit
Molecular weight	44.01	kg/kmol
Density at standard conditions (temperature and pressure)	1.98	kg/m^3
Temperature at critical point	30.98	°C
Pressure at critical point	73.77	bar
Density at critical point	467.60	kg/m^3
Temperature at triple point	−56.56	°C
Pressure at triple point	5.18	bar
Density of saturated vapor at triple point	13.76	kg/m^3
Density of boiling liquid at triple point	1,178.53	kg/m^3
Density of dry ice at triple point	1,545.00	kg/m^3

2.1 Density

Density is an important factor determining the transport and storage of a specific substance, as transport and storage capacities are usually limited by mass and/or volume. For gases and liquids, like CO$_2$, the transport capacity and storage capacity are limited by volume. The density, defined as the ratio of mass to volume, thus determines the amount of CO$_2$ that can be transported or stored within a given volume. To minimize costs, typically, the overarching aim is to transport or store the largest possible mass at a limited volume, achieved by maximizing the density as far as possible, considering techno-economic aspects.

The density of CO$_2$ can be calculated as a function of temperature and pressure [7, 8]. The respective results show that solid CO$_2$ is characterized by the highest density (1,545 kg/m^3) (Table 1). However, providing solid CO$_2$ requires a comparatively large amount of energy. Therefore, this aggregate state for transportation is generally only used when solid CO$_2$ is actually required for the final application [10]. The density of liquid CO$_2$ (1,179 kg/m^3) is lower but still within a comparable order of magnitude to solid CO$_2$ and is also significantly higher compared to the density of CO$_2$ at the critical point (468 kg/m^3). Therefore, except for pipeline transport, CO$_2$ is preferably transported and stored in a liquid form when intermediate above-ground storage is required.

However, the density of liquid CO$_2$ is a function of temperature and pressure (Fig. 3). For liquid CO$_2$, the density increases with increasing pressure and is higher at higher temperatures than at lower temperatures. For the latter, the density increases only slightly with increasing pressure. As the temperature decreases, the density of CO$_2$ increases.

Since the pressure of CO$_2$ at its triple point is above ambient pressure, CO$_2$ needs to be compressed in order to be liquefied (Fig. 2). Thus, to keep CO$_2$ in a liquid state, respective tanks for the transportation and storage of liquid CO$_2$ can either be semi-pressurized/semi-refrigerated (higher pressure and lower temperature than ambient)

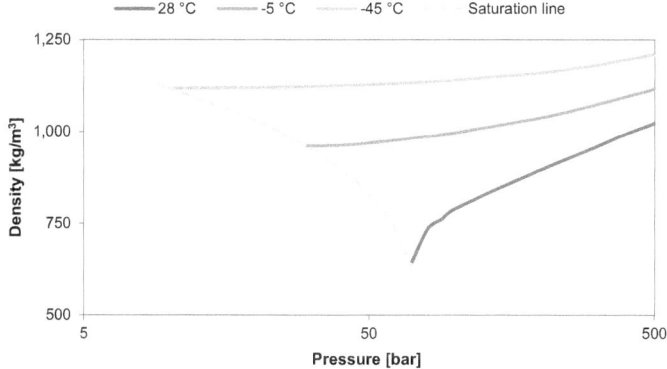

Fig. 3 Density of liquid CO$_2$ as a function of temperature and pressure (according to [11])

Table 2 Density of liquefied CO_2 depending on the pressure and temperature [15]

Pressure (bar)	Temperature (°C)	Density (kg/m^3)
6	−52.3	1,159
15	−27.7	1,064
25	−11.4	991
35	0.5	927
45	10.2	864
55	18.3	785
65	25.4	649

or fully pressurized (higher pressure than ambient and ambient temperature) [12]. Table 2 shows the liquefaction parameters with regard to pressure and temperature, as well as the associated density. With increasing pressure and temperature, the density of liquid CO_2 decreases significantly. Therefore, especially in the context of shipping, low pressures and low temperatures seem to be the preferred condition of CO_2 within an overall logistic chain [12–14]. To increase the density after liquefaction (i.e., to improve its transportation characteristics further), the CO_2 can be additionally compressed or subcooled.

The liquefaction parameters in Table 2 can be separated into three possible main pressure and temperature conditions for liquid CO_2 [16, 17].

- low pressure and temperature: around 7 to 9 bar and −55 to −45 °C,
- medium pressure and temperature: around 15 to 20 bar and −30 to −20 °C,
- high pressure and temperature: around 45 to 70 bar and 10 to 30 °C.

These three conditions have, in each case, advantages and disadvantages beyond the density aspects discussed above, which need to be considered to identify the techno-economic optimum of different supply chains. While CO_2 is mainly transported and stored at medium pressures and temperatures, tanks for CO_2 are more easily scalable at lower pressures and temperatures while also being lighter and, therefore, less expensive. However, CO_2 transported and stored at lower pressures and temperatures is closer to the triple point, resulting in different safety concerns and risks. CO_2 within the higher temperature and pressure range leads to higher tank weights and, thus, higher costs. However, the lower requirements for the materials and the lower costs for conditioning are advantageous. Table 3 summarizes the advantages and disadvantages of the three pressure and temperature ranges.

2.2 Impurities

The physical properties mentioned so far apply to pure CO_2. However, volumes of CO_2 often contain impurities when captured from flue gases of diverse industrial production facilities. Depending on the impurities contained within the CO_2 and

Table 3 Advantages and disadvantages of the three main pressure conditions of liquid CO_2 [16–18]

Pressure condition	Advantages	Disadvantages
Low pressure and temperature	• High density • Already existing know-how from LPG supply chains • Easily scalable tank sizes • Lower capital costs due to larger storage tanks and lower wall thickness • Lower storage weight	• Near triple point: safety concerns and risk of freezing • Higher conditioning costs due to higher energy demand for liquefaction • Higher insulation and material requirements due to lower temperatures can lead to higher material costs • Stricter requirements for water content limit to avoid the formation of hydrates
Medium pressure and temperature	• Already existing know-how from the food and brewery industries	• Relatively high volume of steel in the tank system • Technically challenging tank structure
High pressure and temperature	• Lower conditioning costs due to lower energy demand • Lower material costs due to lower material requirements • Operation is far from triple point: Well-known risks and safety requirements • Lowest energy demand for • Direct injection from ship in a CCS chain	• Low density • Higher costs due to smaller tank volumes and higher wall thickness • Higher weight • Challenging piping • Risk for cold boiling liquid expanding vapor explosion (BLEVE)

their respective quantity, they might affect the properties of CO_2 with regard to, e.g., density, viscosity, or phase equilibrium. Additionally, the operational handling of CO_2, health and safety aspects, as well as the integrity of CO_2 storage or the means of transport, can be influenced. Some issues, as well as impurities that can trigger them, are assigned to the mentioned effects in Table 4.

The properties of CO_2 thus change not only with pressure and temperature but also with the impurities and their respective concentrations. Therefore, the definition of maximum limits for the corresponding impurities is recommended [17, 20, 21]. The basis for that is a comprehensive understanding of the influence of such impurities, which, however, has not been fully understood and analyzed in all possible cases [16, 20].

Since water has a significant influence on the transport and storage of CO_2 through corrosion and the formation of hydrates, Fig. 4 shows the solubility of water in CO_2 depending on pressure and temperature. If CO_2 is gaseous, the water solubility decreases with increasing pressure and decreasing temperature. If CO_2 is transferred to the liquid phase, the water solubility increases abruptly at constant temperature. Liquid CO_2 exhibits higher water solubilities with increasing temperatures, which also rise with pressure. At lower temperatures, the influence of pressure on water

Table 4 Influence of impurities on carbon dioxide [16, 17, 19, 20]

Affected category	Effects/issues	Exemplary impurities
Properties	Affect phase equilibria (a) Non-condensable: increase saturation pressure, elevate bubble-point pressure (b) Reduce bubble pressure	(a) Oxygen (O_2), Hydrogen (H_2), Nitrogen (N_2), Methane (CH_4), Argon (Ar) (b) Sulfur dioxide (SO_2)
	Affect hydrate stability zones (a) Shift them to higher temperatures (b) Shift them to lower temperatures	(a) N_2, O_2 (b) SO_2, Hydrogen sulfide (H_2S)
	Decrease density of liquid CO_2	O_2, H_2, N_2, CH_4, Ar
Operation	Formation of hydrates	Water (H_2O)
	Freezing	H_2O
	Compression (non-condensable): require additional compression work	N_2, O_2, Ar, CH_4, H_2
Health and safety	Toxicity	Sulfur oxides (SO_x), H_2S, Carbon monoxide (CO), Nitrogen oxide (NO_x)
Integrity	Corrosion	H_2O, O_2, NO_x, SO_x
	Water solubility (a) Reduce water solubility (b) Increase water solubility	(a) Nitrogen dioxide (NO_2), SO_2, CH_4, N_2, O_2 (b) H_2S
	Materials (a) Risk of embrittlement (b) Damage to seals and components (c) Ductile fracture potential	(a) H_2S (b) Glycol (c) N_2, CH_4, H_2

solubility decreases. At constant low temperatures, the water solubility of liquid CO_2 is significantly higher than the water solubility of gaseous CO_2 [21].

Water solubility within CO_2 is an important parameter for transporting and storing CO_2 since a higher water solubility increases the risk of hydrate formation. Hydrates are solids and consist of water molecules agglomerated in the presence of CO_2. Such hydrates can be formed in gaseous and liquid CO_2 and can result in blockages (e.g., in pipes). The respective hydrate-water equilibrium depends on pressure and temperature; i.e., the risk of such a hydrate formation has to be considered only in certain pressure/temperature areas, which are shown in Fig. 5. Above the solid line (vapor pressure curve), CO_2 is liquid, and below this line, it is gaseous. Figure 5 illustrates that, mainly at temperatures below 0 °C, hydrate formation must be considered for transportation and storage and can become, under these conditions, a risk in operation [16, 22]. To reduce the risk of hydrate formation, water can be separated from the CO_2, the CO_2 can be handled in non-hydrate zones by adjusting pressure and temperature accordingly, or chemical inhibitors can be used [16, 22].

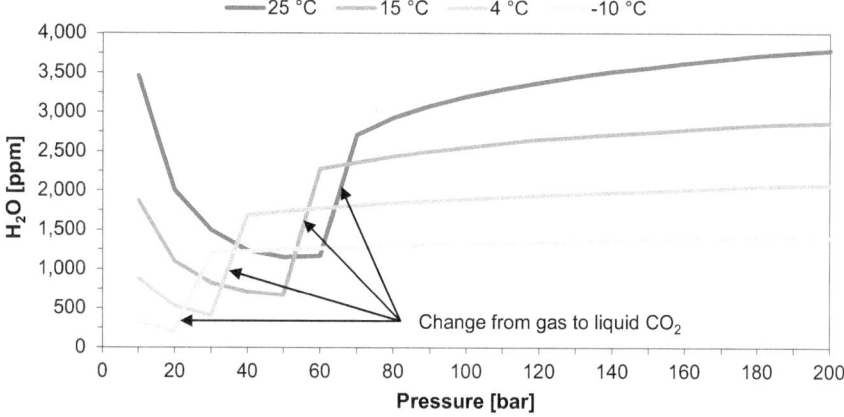

Fig. 4 Solubility of water in pure carbon dioxide depending on pressure and temperature [21]

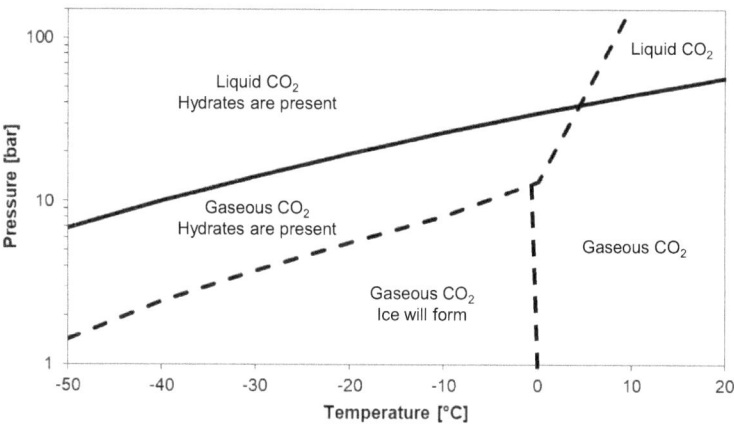

Fig. 5 Risk of hydrate formation depending on pressure and temperature in pure saturated CO_2 (solid line—vapor pressure curve of CO_2; dashed line—division into areas where hydrates are present or ice is formed) [22]

Since water is not the only impurity that can have an adverse effect (Table 4), removing other impurities may also be necessary. These processing steps can be seen as a part of CO_2 conditioning and are explained below in more detail.

2.3 Risks and Hazards

So far, CO_2 has been stored and transported mainly in smaller quantities. Therefore, experiences with large amounts of CO_2 handled during international/long-distance transportation and/or large-scale storage are missing.

CO_2 is a non-flammable gas, resulting in a fairly low fire risk. If CO_2 is stored or transported in liquid form at low temperatures and pressures (close to the triple point), dry ice may be formed. Additionally, the formation of hydrates is also possible (see above). Other risks and hazards may arise from material deterioration or damage, for example, corrosion, rapid embrittlement, or ductile fracture. These hazards mainly originate from or are aggravated by impurities [23]. If CO_2 is stored or transported in a liquid form under high pressure and temperature conditions, the risk of a boiling liquid expanding vapor explosion (BLEVE) might be considered. This effect (BLEVE) describes the high evaporation rate of a superheated liquid in the atmosphere due to a rapid pressure reduction, which might cause blast waves, material damage, and fragments flying around [9, 16].

Leakages should be avoided by granting structural integrity. In case of a leakage, gaseous or liquid CO_2 can be released into the environment. When liquid CO_2 escapes, the formation of dry ice and hydrates is most likely; i.e., due to the higher ambient temperature and the lower ambient pressure of the natural environment compared to the condition inside the storage facility, CO_2 cannot exist in the liquid phase and changes into a two-phase gas and solid mixture. Gaseous CO_2 is heavier than air and can pool around the leakage. In combination with wind, such a CO_2 pool might travel. An increased concentration of CO_2 within the air can, e.g., have a toxicological impact on the human body by the displacement of oxygen. Depending on the CO_2 concentration within the air and the inhalation exposure time, the symptoms vary from headaches, increased heart rate, and confusion to coma and death caused by asphyxiation. In addition, due to the low temperature of cryogenic liquids, humans face the risk of cryogenic burns during handling or in case of leakage [6, 16, 24].

2.4 Operational Aspects

For operation, appropriate materials for CO_2 need to be selected. Carbon steel can be used if CO_2 is transported or stored under high pressures and temperatures. However, higher water impurities can lead to corrosion. This risk can be reduced by selecting, e.g., stainless steel as a material for the tank. However, removing water may be less expensive than selecting materials that are less susceptible to corrosion.

With decreasing temperature and pressure, foam or vacuum insulation (or a combination) is required. If CO_2 is transported and stored under low pressures and temperatures, specialized low-temperature materials like carbon manganese steel or stainless steel are recommended to cope with thermal-induced stresses. Additionally, loading and unloading processes result in increased requirements for the used materials since temperature and pressure fluctuations occur; this need to be considered during the design phase [14, 16].

If CO_2 is transported and stored as a liquid at temperatures below ambient, boil-off gas results. This term describes the amount of liquid CO_2 that passes into the gas phase due to heat input from the ambient environment or sloshing (sloshing describes the movement of a liquid inside a tank caused, e.g., by the movements of a means of transport). The increasing amount of vapor within the tank increases pressure if no gas is removed from the tank. Thus, tanks are designed to be operated up to a defined maximum pressure level. The boil-off gas must be released from the tank to avoid structural damage or explosion-like effects if this maximum pressure level is reached. The actual amount of such necessarily generated boil-off gas depends on many factors, such as the ambient temperature, the temperature of stored CO_2, the insulation material and its thickness, the tank capacity, the CO_2 storage level in the tank, the level of impurities, the storage time, or the overall pressure design of the tank [16, 25]. Typically, two options are possible to handle the boil-off gas: (a) re-liquefaction of the boil-off gas (additional energy demand) or (b) release into the atmosphere (loss).

3 Conditioning

CO_2 must be conditioned according to the requirements for its mode and state of transportation, storage, or further use. Conditioning means, e.g., pressure and/or temperature adjustments, but also the removal of impurities. Since CO_2 is usually transported and stored as a gas or liquid, and the uses for CO_2 vary, the respective conditioning steps might differ.

3.1 Removal of Impurities

Impurities must be removed according to the recommended limits for impurities for the respective application and depending on the means of transport and the intended use of the CO_2. Impurities can be removed either by a distillation column or a flash separator. The former option is usually preferred since less CO_2 enters the volatile purge stream. A two-column system can be used to achieve a higher purity of the CO_2 stream while reducing the CO_2 losses within the purge stream. When designing the system, it has to be considered that the impurities interact with each other. Valuable impurities like argon or hydrogen may be recovered for sale [26].

Water is also an impurity whose removal may be necessary to prevent corrosion, freezing during liquefaction, or the formation of hydrates. However, there is no general rule for acceptable moisture levels. A maximum water content of less than 60% of the dew point is often suggested [5]. The share of water can be reduced, e.g., by vapor–liquid separator drums, which use gravity for separation to achieve a

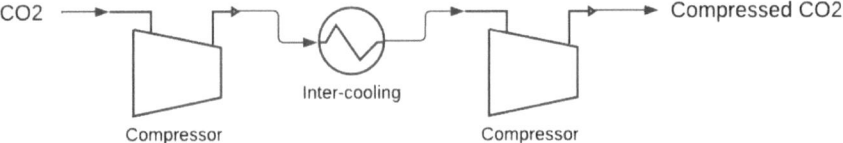

Fig. 6 Flow diagram of a two-stage compression with inter-cooling (according to [27])

water content of 400–500 ppm. If a lower water content is requested or necessary, regenerative adsorption columns can dry the CO_2 stream up to a water content in the single-digit ppm range. The regeneration of the column is done by heated CO_2. At least dehydration is typically conducted before the CO_2 stream enters the low-pressure compressor [16, 26, 27].

3.2 Compression

Compression can be divided into the low- and high-pressure range. In low-pressure compression, gaseous CO_2 is compressed up to 80 bar above ambient pressure, e.g., to be pre-conditioned for the inlet pressure of a liquefaction unit. High-pressure compression describes the increase of the CO_2 pressure from 80 to around 150 bar; the latter is a typical CO_2 pipeline pressure.

Different compressor types, like a screw compressor, a reciprocating compressor, or a centrifugal compressor, can be used to increase the pressure of a CO_2 stream. Screw compressors or reciprocating compressors are usually used for smaller volumes, while for larger volumes, centrifugal compressors seem to be more suitable; they are already used within the ammonia and urea synthesis industry.

With increasing pressure difference between a compressor's inlet and outlet pressure, several compression stages are realized. Additional inter-cooling exchangers between the compression stages increase the energy efficiency of the compressors. The flow diagram of a two-stage compression with inter-cooling exchangers is shown as an example in Fig. 6 [5].

After low-pressure compression, CO_2 is in the pressure range around the critical point. For further compression into the high-pressure range, pumps are used instead of compressors. Typically, CO_2 is condensed by a cooling medium, such as cooling water, between the low- and the high-pressure compression [5].

3.3 Liquefaction

For the liquefaction of CO_2, water has to be removed from the CO_2 stream to prevent freezing. Furthermore, gases that do not condense under the temperature conditions of liquid CO_2 (so-called non-condensable gases like nitrogen, argon, hydrogen, or methane) must be removed. Open- or closed-cycle refrigeration processes can be used for liquefaction depending on the type of refrigerant used.

- In open cycles, a refrigerant (e.g., water) is taken from the environment, used for cooling, and returned into the atmosphere.
- In closed-cycle processes, an external refrigerant (e.g., ammonia, propane) is circulated within a closed loop.

The difference between an open- and a closed-cycle liquefaction can be seen in Fig. 7. Since the refrigerant for open-cycle processes is taken from the environment, the plant layout and energy consumption might vary due to seasonal and site-specific changing environmental conditions [16, 27]. In the open cycle, the CO$_2$ needs to be expanded after the cooler (i.e., one or more multi-fluid heat exchangers) to achieve the liquefaction of the stream. This can be done by using either a control valve or a turbine.

The energy requirement for CO$_2$ liquefaction depends on several factors (e.g., quantity, process type). In addition, the specified pressure and temperature level of the liquid CO$_2$ significantly influences the liquefaction process. If the outlet pressure and temperature of CO$_2$ should be within the low range, some refrigerants like

Open-cycle process

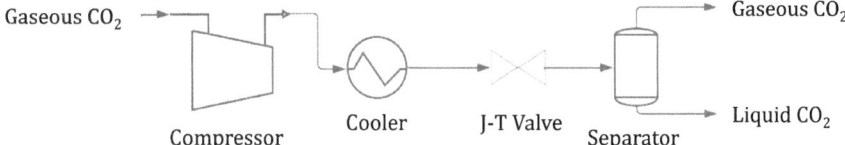

Closed-cycle process

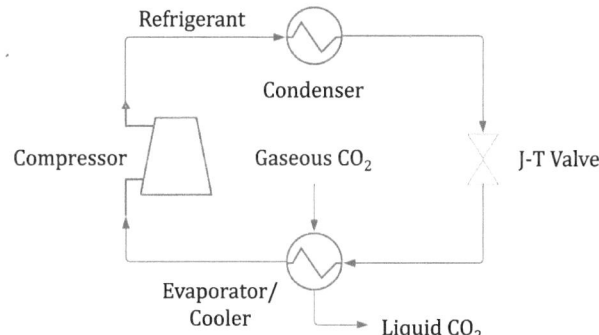

Fig. 7 Open- (top) and closed-cycle (bottom) refrigeration process (according to [16])

ammonia or propane might not be feasible anymore due to their very low operating pressures. Possible optimizations of the liquefaction process are the addition of several refrigeration stages in series for the closed-cycle refrigerant, a pre-cooling of the CO_2 stream, or energy recovery within the process. However, choosing a suitable liquefaction process cannot be made only based on energetic parameters but should also consider parameters of the whole chain and techno-economic constraints [16, 27].

4 CO_2 Interim Storage

In addition to the actual means of transport, intermediate storage tanks for CO_2 are often necessary, especially for loading and unloading operations or if it is collected and transported on an irregular basis. Furthermore, storage can provide flexibility if the aggregation state changes during transportation. Depending on the logistic cycle, the storage capacity within a CO_2 transportation chain is an important parameter. If the storage is before or after a means of transport, its capacity should be at least equal to the capacity of the means of transport [5, 17].

Storage infrastructure. In general, a distinction can be made between dedicated infrastructure and multi-substance infrastructure.

- Dedicated infrastructure can store and transport only one substance (e.g., CO_2) and is optimized with regard to the corresponding properties and requirements.
- Multi-substance infrastructure can store and transport defined substances with similar properties.

Since CO_2 can be liquefied with a combination of temperature and pressure, it has similar properties to liquefied petroleum gas (LPG), which is a mixture of propane and butane. Hence, as LPG is traded globally, CO_2 can be stored and transported within the existing LPG infrastructure if it comes to semi-refrigerated/semi-pressurized or fully pressurized storage and means of transport. Fully refrigerated LPG tanks (i.e., unpressurized) cannot be used for CO_2 because liquid CO_2 cannot exist at ambient pressure (the pressure within the tank has to be above the triple point of CO_2 at 5.18 bar). Furthermore, the densities of LPG and CO_2 differ, which might result in restrictions, e.g., on the maximum filling quantity [16]. The differences between fully refrigerated, semi-refrigerated/semi-pressurized, and fully pressurized are defined as follows:

- Fully refrigerated: The gas is liquefied by reducing its temperature at ambient pressure,
- Semi-refrigerated/semi-pressurized: The gas is liquefied by a combination of temperature and pressure,
- Fully pressurized: The gas is liquefied by increasing its pressure at ambient temperature.

While fully pressurized tanks and containers do not need insulation, semi-pressurized tanks and containers require necessary insulation, e.g., polyurethane foam [28, 29].

Status quo. The compressed gas industry, as well as the soft drinks industry, stores CO$_2$ typically cryogenically liquefied within medium pressure and temperature conditions. Typical tank capacities are up to roughly 350 t CO$_2$ [6]. In the following, fully pressurized and semi-pressurized/semi-refrigerated storage options for CO$_2$ are discussed.

4.1 Fully Pressurized Storage

Fully pressurized tanks store liquefied gases under pressure at ambient temperature (i.e., no thermal insulation is required). For larger tanks, insulation can protect against fire or solar heating. According to the withdrawal conditions, either gaseous or liquid CO$_2$ can be withdrawn from a fully pressurized tank [30].

Design. With increasing design pressure, the storage tank needs thicker walls. Moreover, the maximum allowable diameter of a single storage tank decreases with increasing design pressure due to wall thickness limitations (i.e., the maximum capacity of one storage tank decreases with increasing design pressure). Figure 8 shows the maximum storage capacity of a pressurized sphere depending on the design pressure. The pressure vessel design code, a standard for the design and construction of pressure vessels, can be used to determine the design pressure of a fully pressurized tank [30].

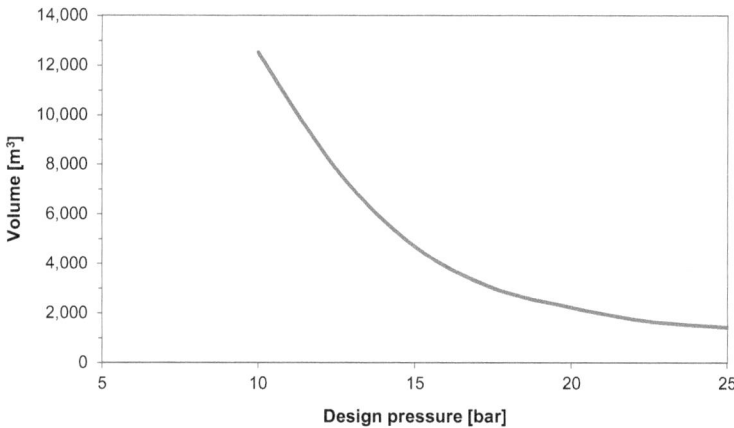

Fig. 8 Maximum storage capacity of a pressurized sphere depending on the design pressure (according to [30])

Due to the limitation of the maximum volume of a single tank and the decreasing density of liquid CO_2 with increasing pressure, fully pressurized tanks are more suitable for smaller to medium storage volumes. For very small quantities, compressed gas cylinders can be used to store CO_2. The capacity of such a small gas cylinder varies depending on the given application and can be up to 50 l, corresponding to a filling quantity of 37 kg of CO_2. Compressed gas cylinders can be combined into cylinder bundles to increase the storage quantity. The capacity of such a cylinder bundle is 600 l resp. 450 kg CO_2 [31].

Operation. To protect the tank, safety valves are required to prevent overpressure. Also, a minimum pressure within the tank must be retained to ensure the stored CO_2 remains in a liquid aggregate state. With a pressure-building vaporizer, the pressure can be automatically kept above a set value. This is also necessary if the operating pressure drops due to CO_2 vapor or liquid withdrawal. The pressure-building vaporizer increases the pressure in the tank by vaporizing a part of the liquid CO_2 and returning it to the tank. The heat required for this can be provided by electricity, steam, or hot water [28].

4.2 Semi-pressurized/Semi-refrigerated Storage

Cryogenically liquefied CO_2 is stored at a pressure above ambient pressure, as CO_2 only adopts this state of aggregation at pressures higher than the triple point pressure (5.18 bar). The storage tank requires insulation since cryogenically liquefied gases necessitate a temperature reduction below ambient temperature. Such tank insulation generally consists of materials with low thermal conductivity. The material chosen for the tank wall depends mainly on the design temperature and the impurities (i.e., the type of impurity and its proportion). For temperatures above -50 °C, carbon steel can be used. For lower temperatures, alloys like nickel steel or carbon steel with manganese seem appropriate [17]. Semi-pressurized/semi-refrigerated storage tanks for CO_2 with a capacity of up to 100 t CO_2 are already commercially available [32].

Design. The design pressure and temperature can be determined using the so-called Tebodin guidelines [5]. In this case, the design pressure is determined based on three steps.

1. Definition of the operating pressure as a pressure range.
2. Setting the maximum operating pressure to a value greater than or equal to the sum of the highest operating pressure derived and 1 bar (higher flexibility).
3. Adding 1.5 bar to the maximum operating pressure to account for unintended events.

The pressure value resulting from these three steps represents the minimum design pressure for the CO_2 storage tank. If CO_2 is stored close to the triple point, solidifying

CO$_2$ is a risk. Therefore, the minimum operating pressure of CO$_2$ has to be above the pressure at the triple point, taking unintended events under consideration [5].

Depending on the design pressure, the minimum operating temperature is equal to the temperature corresponding to the minimum operating pressure. To account for unintended events, the minimum design temperature equals the minimum operating temperature reduced by 5 K. However, e.g., due to leakages, a rapid depressurization to the ambient pressure level is possible. In this case, the temperature drops to the equilibrium temperature of solid CO$_2$, which is -78.5 °C, and could thus also be set as the minimum design temperature. Since these two possible minimum design temperatures differ greatly, the respective project authorities have to decide which to choose. The selected minimum design temperature determines the tank material and, thus, the costs of the tank [5].

Operation. The production of boil-off gas (partial evaporation of the liquid) and the filling of the tank leads to an increase in pressure, while the withdrawal of CO$_2$ results in a decrease in pressure. To ensure that the operating pressure is within a specified bandwidth, additional equipment needs to be installed to increase and decrease the tank pressure. A pressure-building vaporizer can be used to increase the tank pressure. In addition, CO$_2$ vapor can be fed into the tank. To lower the tank pressure, CO$_2$ vapor needs to be withdrawn from the tank and can be condensed again within a refrigeration unit. The normal heat input to the tank is used to design and size the refrigeration unit [28].

5   CO$_2$ Transport

In general, CO$_2$ may most likely be transported by truck, railway, ship, or pipeline. The most suitable transport option depends on various factors, such as transport distance, quantity, or general infrastructure conditions. If CO$_2$ is transported by truck, railway, or ship, the transport volume represents a limiting factor. Therefore, CO$_2$ is liquefied before transportation to increase the transported mass; different temperature and pressure conditions can be applied for such a liquefaction.

If CO$_2$ is transported by pipeline, volume is not a decisive factor. For such a pipeline transport, the diameter and maximum pressure level defined by the transport medium by the pipeline material determine the overall transport volume within a defined time period [16]. The following sections describe the mentioned means of transportation for CO$_2$ in more detail.

5.1   Pipeline Transport

CO$_2$ has been transported by pipelines since the 1970s and is predominantly used for enhanced oil recovery. In 2015, the overall length of global CO$_2$ pipelines amounted to more than 8,000 km, with most pipelines located in the USA (more than 7,200 km). Both onshore and offshore pipelines are used for this type of CO_2 transport.

The transport of CO_2 by pipeline systems can take place in the form of a gaseous, liquid, dense phase, or supercritical state of CO_2, which is determined by the pressure and the temperature within the pipeline. Typically, the pipeline transport of gaseous CO_2 has relatively low technical requirements due to ambient temperature and lower pressures. Furthermore, no insulation is required if the CO_2 is transported at ambient temperatures. However, pipeline transport of gaseous CO_2, like pipeline transport of liquid CO_2, is mainly suitable for lower capacities and shorter distances because it is less economically representative. In addition, when transporting liquid CO_2, pressure and temperature control are essential to keep the CO_2 in a liquid state. However, the lower friction and viscosity of the liquid CO_2 are advantageous. For larger transport volumes and longer transport distances, dense phase and supercritical CO_2 pipeline transport are usually more suitable from a cost perspective [23].

Design. Several parameters are important for the design of a pipeline: length, diameter, wall thickness, pressure, temperature, and the route. Pressure and temperature influence the condition of the CO_2. As the length of the pipeline increases, the pressure drop increases, and the temperature may change. Therefore, pressure and temperature need to be controlled.

In the year 2020, the length of existing CO_2 pipelines varies from relatively very short distances up to about 800 km. Since the specific pipeline costs per kilometer decrease with increasing capacity, longer pipelines often have higher capacities. However, the pressure drop increases with increasing length, making booster stations along the pipeline necessary [23].

The pipeline has to withstand the internal and external pressure acting on it. This defines the wall thickness, which increases with increasing design pressure, leading to higher investment costs. To calculate the wall thickness required, a stress analysis can be used depending on the chosen material (the higher the steel grade, the thinner the wall thickness can be, but the higher the material cost) [23, 33, 34, 35].

The course of the pipeline may depend on the conditions of construction. Depending on the area through which the pipeline passes, the installation may be more complex or simpler; e.g., in densely populated and built-up urban areas, the installation of the pipeline is more complex, and there is a higher operational risk. The installation is usually simpler if the pipelines are laid along (water) roads or railroad tracks. Additionally, offshore pipeline installation is more complex than onshore pipelines [23].

Operation. The transport of CO_2 by pipeline requires an increase in pressure to compensate for the pressure drop carried out with compressors or pumps, depending on the condition of the CO_2. In addition, a temperature change may occur if the temperature of the CO_2 is lower than the ambient temperature [23]. Process flow diagrams for the pipeline transport of gaseous, liquid, dense phase, and supercritical CO_2 are shown in Fig. 9.

Depending on the state of the CO_2, the pressure drop and the temperature drop vary, as well as their influence. If CO_2 is transported in a gaseous state, the pressure

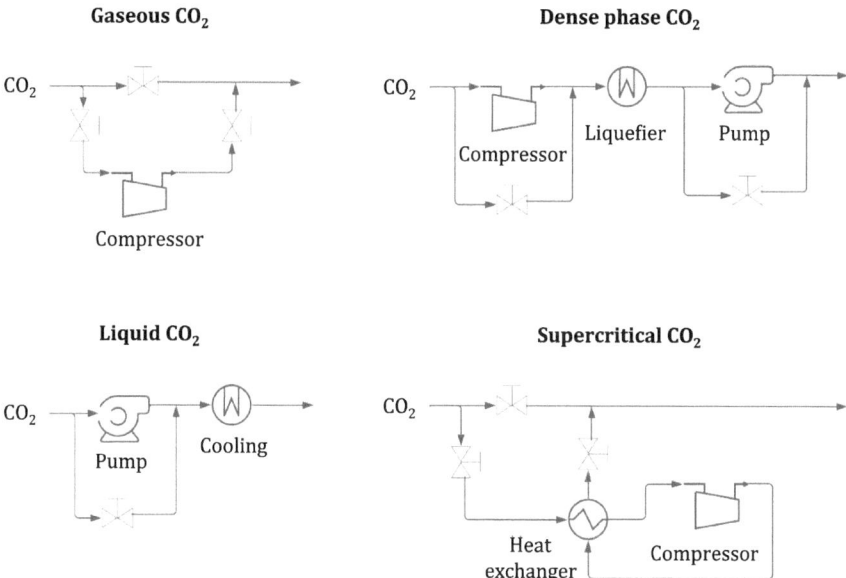

Fig. 9 Process flow diagrams for the transport of CO$_2$ corresponding to the different states of CO$_2$ (according to [36])

drop increases as the inlet temperature decreases [37]. If CO$_2$ is transported as a liquid, the pressure drop is lower than for CO$_2$ transport in a supercritical state but higher than for transport in a dense phase [38]. For the CO$_2$ transport in a dense phase, the pipeline inlet temperature substantially affects the temperature drop, but the effect on the pressure drop is comparatively small. However, the influence of the ambient temperature on the pressure drop increases with increasing distance. In the case of supercritical transport, the ambient temperature significantly influences the temperature drop. This may lead to a phase transition of CO$_2$ if no additional heating stations are implemented. Therefore, temperature and pressure control are important for the pipeline transport of CO$_2$ [23].

Also, impurities influence the pressure drop during the transport of CO$_2$. In addition, they can impact health, safety, and risk aspects. Therefore, the mole fraction of CO$_2$ should be higher than 95% in pipeline systems for CO$_2$, and specific limits are set for impurities (Table 5). However, these recommendations are not legally binding [23].

5.2 Ship Transport

Ship transport offers greater flexibility than pipeline transportation, as the individual transportation routes are adaptable. Furthermore, in the case of a ship transport, it

Table 5 Limits for impurities in CO_2 pipeline transport (according to [19, 20, 39])

Impurity	Impurity limit	Issue
Water	No free water $\leq 4.8 \cdot 10^{-4}$ kg/m^3 in gas phase	Corrosion
Sulfur (total)	≤ 35 ppm	Health and safety
Nitrogen	$\leq 4\%$ of the mole percentage	Minimum miscibility pressure concern
Hydrogen sulfide	≤ 20 ppm	Health and safety
Hydrocarbons	$\leq 5\%$ of the mole percentage dew point temperature ≤ -28.9 °C	Minimum miscibility pressure concern
Oxygen	≤ 10 ppm	Corrosion
Glycol	$\leq 4 \cdot 10^{-8}$ m^3/m^3	operations (damage to seals)

is possible to respond to a change in the amount of CO_2 to be transported by either using more ships or chartering ships with a larger transport volume. Ships for CO_2 transport can be categorized into seagoing vessels and inland vessels depending on the waters for which they are constructed [16, 40]. CO_2 is already transported by seagoing ships for the food and beverage industry as well as for use in horticulture, especially within greenhouses. Therefore, dedicated carriers with CO_2 tanks are used, characterized by a medium temperature and pressure level of around − 30 °C and 15–50 bar. The capacity of each of these transportation ships is currently comparatively low. One of the largest CO_2 traders uses four vessels for liquid CO_2 transportation converted from general cargo carriers, with capacities from 1,200 to 1,800 t CO_2 [17, 41]. In the future, larger ships and volumes will probably be needed for the transportation of CO_2 due to the increasing relevance of carbon capture and storage (CCS). Due to the volume restriction of ships, the state of high-density liquid CO_2 will be close to the triple point. In respect thereof, lower temperatures of about − 50 °C and lower pressures of around 7 bar are targeted. However, if CO_2 is injected directly from the ship to offshore storage in the context of CCS, higher pressure and temperature conditions of CO_2 on the vessel are the most appropriate conditions. Therefore, consideration of the entire chain may be necessary for selecting an appropriate pressure and temperature range of CO_2 [16, 17].

Design. CO_2 can be transported in multi-use ships, LPG or ethylene carriers, or dedicated CO_2 ships. LPG or ethylene carriers with 7–11 bar pressures over ambient pressure and temperatures between −140 and −43 °C can probably transport CO_2. However, the density of CO_2 (1,050–1,200 kg/m^3) is significantly higher than the density of LPG (550–700 kg/m^3), which leads to a restriction of usable tank capacity due to the difference in the ship's displacement. The usable tank capacity for CO_2 on LPG or ethylene carriers is assumed to be around 50–60% [17, 42].

Instead of a multi-use ship, dedicated CO_2 ships can either be a retrofit of an existing ship or a new build. The current CO_2 ships are retrofits of general cargo ships. Although a retrofit usually has a shorter construction time and lower capital expenditures (CAPEX) compared to newly built ships, retrofitting has some drawbacks. The usage of existing ships will normally lead to a shorter operational life after the retrofit compared to a dedicated new build ship since the total life of the vessel includes the use before and after the respective retrofit. Furthermore, the existing ship—i.e., the ship that is intended to be retrofitted—with its hull, geometry, and cargo holds is not optimized for CO$_2$ tanks, which can lead to an increased number of cargo tanks and a smaller cargo capacity compared to a dedicated new build CO$_2$ tanker of the same size. These drawbacks can lead to higher CAPEX and higher operational expenditures (OPEX) [17].

Because of the mentioned drawbacks of the multi-use and the retrofitting of existing ships, building new dedicated CO$_2$ ships can be favorable. As larger vessels and volumes for CO$_2$ are expected in the future, newly built ships can be designed for higher CO$_2$ transport capacities. In the year 2022, there is no experience in transporting large quantities of CO$_2$ by ship. However, experiences in handling LNG and LPG can serve as a basis for process safety and cargo handling procedures for CO$_2$. There are different ship designs for the transportation of larger volumes of CO$_2$. For a planned cargo capacity, the number of tanks and the tank size depend strongly on the assumed temperature and pressure. In general, larger tank diameters can be reached with lower pressures. Furthermore, tanks with larger diameters typically use the space on a ship more efficiently. Table 6 gives an overview of selected aspects of possible CO$_2$ ship designs depending on the desired pressure and temperature range.

First projects for the development, construction, and operation of ships for the transport of CO$_2$ have already been announced [43–46].

Table 6 Comparison of possible future CO$_2$ ship designs depending on pressure and temperature range [17, 18]

Parameter	Low pressure and temperature	Medium pressure and temperature	High pressure and temperature
Experience	Based on LPG ships	CO$_2$ transport of smaller capacities	Based on compressed natural gas (CNG) ships
Tank geometry	Horizontal cylinder (larger tank diameters compared to tanks within the medium pressure and temperature range)	Horizontal cylinder	Cylindrical bottles (smaller diameters, around 700–900 bottles)
Capacity range of the ships	20,000–30,000 t CO$_2$	Max. 10,000 t CO$_2$	~ 10,000 t CO$_2$
Cost	Lower cost is expected compared to ships of the same capacity but higher pressures and temperatures	Costs are scalable based on existing ships	Higher cost is expected compared to ships of the same capacity but lower pressures and temperatures

Operation. The loading capacity of multi-use ships to transport CO_2 will be smaller due to the higher density of CO_2. The resulting partial filling of the tanks leads to increased sloshing, describing the movement of the liquid inside the tanks due to weather, wave movements, and driving characteristics (e.g., the ship's speed). It has a comparatively low impact on the swim stability of a ship and the structural integrity of a tank in nearly full or empty tanks and a comparatively high impact in partially filled tanks. Furthermore, the tank geometry influences the effect of sloshing. Thus, sloshing results in higher impacts for cylindrical tanks and lower impacts for spherical tanks. Depending on the tank geometry, permitted tank levels are specified to reduce the influence of sloshing and, thus, to ensure the ship's tank integrity and swim stability.

Besides restricting usable tank capacity, a multi-use ship is connected with a higher cleaning effort between the cargo changes. This means higher operational costs due to the longer laytime of the ship in the harbor and costs for cleaning. Therefore, a single conversion of a ship's cargo seems more likely in practice [17, 42].

For the transport of CO_2 by ship, suitable terminals are needed. These terminals consist of one or more intermediate CO_2 storage tanks, a loading arm system, and a pipeline system as a connection between the ship and storage tanks. The number of storage tanks required and their capacity depends on the quantity of CO_2 within the supply chain and the shipping schedule. Due to heat in leaks, boil-off gas is generated and needs to be handled, for example, by using a re-liquefaction system [42].

5.3 Rail and Road Transport

The tanks for transporting CO_2 in bulk by rail or road are usually designed for liquefied gases and have a design temperature of -40 to $+50$ °C [29]. CO_2 is usually transported at around -30 °C and 15 bar (i.e., cryogenically liquefied). Therefore, depending on the transport temperature and pressure, the formation of dry ice can pose a risk [47]. Due to the temperature, the tanks need insulation. The tank material is usually carbon steel [28, 29]. Tank capacities can vary up to around 60 m^3 [29].

Design. Rail tank cars can be used for rail-bound CO_2 transport. Figure 10 shows an example of an insulated and pressurized rail tank car, which can be used to transport liquefied gases. Even though CO_2 is currently commercially transported by rail, the majority of CO_2 is transported by other means of transport [48].

For road-based CO_2 transport, tank trucks, semi-trailers, or transportable tanks mounted on flatbed trailers can be used. Figure 11 shows an example of a semi-trailer, which can be used to transport liquefied gases. Trucks are mainly used commercially to transport smaller quantities of CO_2 and shorter distances.

Operation. If CO_2 is transported in cryogenically liquefied form, heat enters the tank due to the typically higher ambient temperature. The evaporated boil-off gas leads to a pressure increase in the tank. The transport tank or the container for CO_2 has to

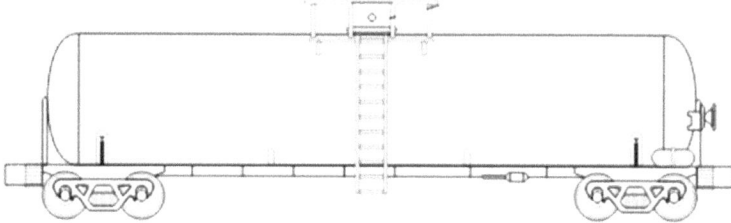

Fig. 10 Example of an insulated and pressurized rail tank car for the rail transport of liquefied gases (according to [49])

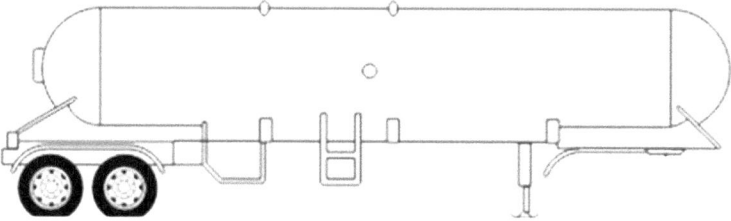

Fig. 11 Example of a semi-trailer for the road transport of liquefied gases (according to [49])

be designed to either withstand the pressure rise in the tank for a sufficient period or be equipped with appropriate equipment for pressure regulation [50].

6 Comparison

The conditioning, storage, and transport of CO_2 are generally not separate steps but are an integral part of an overall CO_2 provision chain. The quantity to be transported and the transport distance have a major influence on selecting a suitable means of transport. This also determines whether storage is needed and how the CO_2 will be conditioned.

In this section, different CO_2 chains related to techno-economic criteria are defined and evaluated in order to compare the means of transport, including intermediate storage and conditioning. For this comparison, the chains to be defined consist of the processing steps (Fig. 12), briefly explained below.

- **Start condition**. Ambient pressure and temperature are assumed as the start conditions for the CO_2 chain; i.e., the CO_2 is present as a gas. For example, this is the case when CO_2 is extracted from biomass conversion processes (e.g., biogas production and bioethanol production).
- **Conditioning**. During the conditioning of the CO_2, the pressure and/or temperature of the CO_2 is changed. The pressure can, e.g., be increased with compressors/

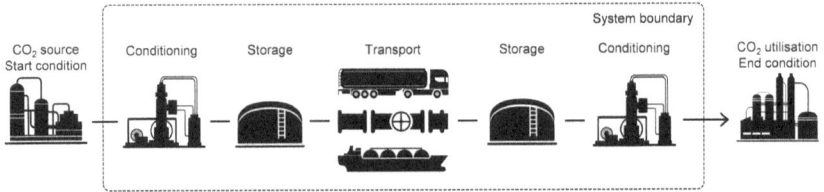

Fig. 12 Steps of a CO_2 chain for comparison

pumps depending on the aggregation state or reduced with valves. An adjustment of the temperature is necessary for a phase change. In this process, liquefaction converts the gas into a liquid, while evaporation converts the liquid into a gas.

- **Storage**. An intermediate storage is needed in the chain when CO_2 is transported batch-wise. Thus, storage is required for the chains in which CO_2 is transported by truck or ship. The storage capacity must be at least equal to the capacity of the means of transport. Furthermore, it is assumed that the storage must be able to store at least the amount of CO_2 produced in a continuous process over two days. The intermediate storage before and after transport by truck or ship is assumed to be identical.

- **Transport**. Various means of transport, such as trucks, pipelines, or ships, can be used to transport CO_2. Depending on the transport volume, a suitable means of transport is selected. Dimensioning is based on the pipeline's volume flow and typical transport capacities for the truck and ship, so several trucks and ships may be required.

- **End condition**. The assumed final state of CO_2 is ambient temperature and 20 bar, so CO_2 is again present as a gas. The CO_2 can then be used, e.g., for a Power-to-Liquid (PtL) process.

The comparison of the respective CO_2 chains to be defined below based on the various system elements within the outlined system boundaries is based on techno-economic evaluation parameters. These are defined and explained below.

- **Technical evaluation**. For a technical comparison, the specific energy demand (kWh/t_{CO2}) is determined. For this purpose, the energy requirements (i.e., electrical, thermal, and fuel requirements) are added up for each step of the CO_2 chain and divided by the annual amount of CO_2. The specific CO_2 energy demand is calculated for a fixed transport distance.

- **Economic evaluation**. For the economic comparison, the costs in each chain step, divided into CAPEX and OPEX, are determined. The annual CAPEX are calculated using the annuity method. For this, the weighted average cost of capital (WACC) at each step is assumed to be 6.4%, while the depreciation period may vary. OPEX consist of fixed OPEX, calculated as a percentage of annual CAPEX, and variable OPEX. The variable OPEX are calculated based on the energy demand of each step. The economic comparison is based on the following parameters.

- **Specific CO₂ cost**. To calculate the specific CO_2 costs [€/t_{CO2}], the annual costs of each step of the chain are added and divided by the annual amount of CO_2. The specific CO_2 cost is calculated for a fixed transport distance.
- **Annual costs as a function of transport distance**. The transport distance greatly influences the costs. CAPEX increases with increasing transport distance because more means of transport, or a longer pipeline, are required. Also, both fixed and variable OPEX increase. Therefore, the total annual costs, CAPEX and OPEX, for a fixed yearly CO_2 amount are presented and compared depending on the transport distance.

The specific energy requirements and costs of the various steps within such a CO_2 chain depend strongly on the amount of CO_2 to be transported. Typically, the efficiency can be increased with larger quantities due to efficiency gains; i.e., the specific energy requirement decreases with increasing quantity. Furthermore, economies of scale can be assumed for large quantities; i.e., the specific costs decrease with an increasing transport quantity. In addition to the CO_2 quantity, the transport distance in the realized transport step influences the specific energy requirement and the specific costs due to fuel consumption or electricity demand. Therefore, for a comparison of different CO_2 chains, three scenarios with varying CO_2 quantity and transport distance are set up. The CO_2 quantities and transport distances for these three scenarios are given below.

- **Small-scale scenario**. Initial PtL demonstration projects are smaller plants that require a small amount of CO_2 annually. Such small quantities can be sourced locally and regionally, allowing for relatively short transport distances. For comparison, an amount of 50,000 t/a of CO_2 and a transport distance of up to 1,000 km is assumed. The specific energy demand and the specific costs are compared for a distance of 500 km.
- **Medium-scale scenario**. Larger planned PtL units require a larger amount of CO_2 compared to the demonstration plants under discussion. Therefore, larger transport distances may be necessary to supply these quantities. For this scenario, an amount of 500,000 t/a of CO_2 and a transport distance of up to 7,000 km is assumed. The specific energy demand and specific costs are compared for a transport distance of 2,500 km.
- **Large-scale scenario**. As the demand for PtL fuels will most likely increase over time, it is expected that such fully commercial large-scale PtL plants will be clearly larger than in the year 2023 announced plants. For the supply of such large quantities of CO_2, global transport may become necessary, and thus long transport distances. For comparison, a quantity of 2 Mio. t/a of CO_2 and a transport distance of up to 15,000 km is assumed. The specific energy demand and the specific costs are compared for a distance of 7,500 km.

For each of the three scenarios, two CO_2 chains are set up for comparison. The CO_2 chains within the three scenarios are explained in detail below, and the results of the evaluation parameters are discussed.

6.1 Small-Scale Scenario

Within this scenario, a CO_2 quantity of 50,000 t/a is transported over a transport distance of up to 1,000 km. For comparison, two different provision chains, shown in Fig. 13, are set up. In one chain, the CO_2 is transported by pipeline, and in the other by truck. Below is a brief description of these provision chains, including the assumed pressure and temperature range of the transported CO_2.

- **Pipeline**. Smaller quantities of CO_2 can be transported in a gaseous form within a pipeline [23]. Since such a pipeline is a continuous means of transport, no storage facilities are required. The advantage of such a gaseous transport of CO_2 is that compression requires less energy than liquefaction. In addition, the CAPEX for a pipeline for transporting gaseous CO_2 is lower than for transporting liquid CO_2 due to lower material requirements [23]. In this chain, the CO_2 is transported in an onshore pipeline at ambient temperature at a pressure level of 70 bar. It is assumed that the pipeline will only be used to transport the defined amount of CO_2 resulting in a small, calculated pipeline diameter of 6.5 mm [15, 23, 51].
- **Truck**. Using a truck for CO_2 transportation is typically suitable for small quantities and short transport distances. In order to make maximum use of the limited truck volume, the CO_2 is liquefied. In the year 2022, truck transports of liquefied CO_2 were mainly realized at a medium temperature and pressure range; thus, − 30 °C and 15 bar are assumed. The truck shows a transport capacity of 24.6 t CO_2 [47]. Since the truck capacity is less than a CO_2 storage capacity of two days for a continuous process, the latter is considered for the storage design capacity, resulting in 274 t.

From the chains defined above, the specific energy demand and the specific costs are presented below. The parameters of the techno-economic analysis are shown in Table 7.Then, the annual costs are discussed as a function of the transport distance.

Figure 14 shows the specific energy demand on the left and the specific costs on the right, separated into the steps conditioning, storage, and transport for a CO_2 amount of 50,000 t/a at a transport distance of 500 km (according to the parameters shown in Table 7). Subsequently, the results of the two chains are explained and discussed.

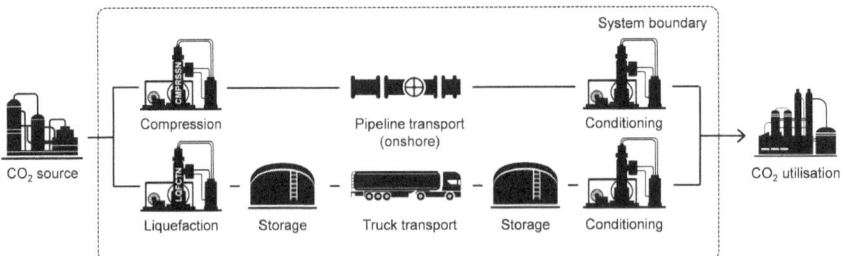

Fig. 13 Representation of the two chains investigated within the small-scale scenario

Table 7 Techno-economic parameters of the small-scale scenario

Parameter	Unit	Chain	Conditioning	Storage	Transport (500 km)	Sources
CAPEX	(Mio. €)	Pipeline	9.9	/	148.6	[52, 53]
		Truck	14.5	1.2	7.1	[15, 53, 54]
Depreciation	(a)	Pipeline	15	/	40	[53, 55]
		Truck	20	30	8 (TU) 12 (ST)	[53, 56, 57]
WACC	(%)	Pipeline	6.4			[58]
		Truck				
Fixed OPEX (share of annual CAPEX)	(%)	Pipeline	5	/	5	[55]
		Truck	4	2	12 (TU) 2 (ST)	[53, 59, 60, 61]
Specific energy demand	(kWh/t)	Pipeline	80	/	42.7	[52, 55, 62, 63]
		Truck	116.5	0	116.9	[15, 61, 64]
Energy cost	(€/kWh)	Pipeline	0.08	/	0.08	[65]
		Truck		0	0.13	[57, 65]

ST: semi-trailer; TU: tractor unit

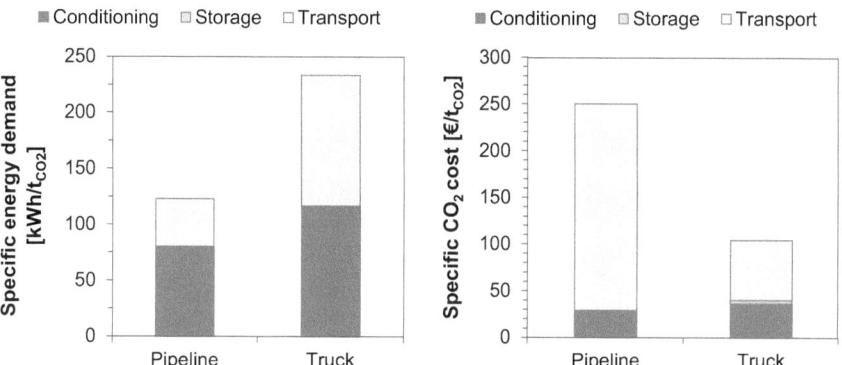

Fig. 14 Specific energy demand (left) and costs (right) for a CO$_2$ chain with a CO$_2$ quantity of 50,000 t/a and a transport distance of 500 km

- **Pipeline**. The pipeline transport chain is characterized by a lower specific energy requirement compared to the truck transport chain. Compression necessary for pipeline transport requires less energy than liquefaction. Although pipeline transport is more efficient than truck transport, these lower energy requirements are only partially reflected in the costs. Nevertheless, the specific costs of compression are lower than those of liquefaction because cooling is not required; i.e., CAPEX and OPEX are lower. However, the specific transport costs by pipeline are around 3.5 times higher than the truck's transport costs [53]. This is because

even for pipelines that transport small quantities, high CAPEX is incurred for the
pipeline and, above all, its installation [15, 51, 66].

- **Truck**. The CO_2 transport chain by truck shows higher specific energy requirements due to liquefaction and due to transport by truck. For liquefaction, energy
is needed for both compression and cooling, resulting in a higher specific energy
demand. Transportation by truck requires a higher energy demand because the
resistances (e.g., air and rolling resistance) to be overcome are comparatively
high. However, the high specific energy demand is not fully reflected within the
specific costs. While the liquefaction costs are higher and relatively low specific
costs for storage are added, the specific costs of transport are significantly lower.
This can be explained by the fact that the CAPEX of the truck, like those of the
pipeline, are high, but account for a smaller share of the specific transportation
costs for a distance of 500 km [48, 53, 66].

Figure 15 shows the annual cost of the CO_2 chain plotted against the transport
distance, which has a significant influence on the specific costs of transport. Subsequently, the influence of the transport distance on the costs of the two chains is
explained and discussed.

- **Pipeline**. Figure 14 indicates that the transport costs of the pipeline make up the
majority of the specific CO_2 transportation costs of this chain. The CAPEX responsible for this, the material and installation of the pipeline, depends on the distance;
i.e., the annual costs of the CO_2 chain for pipeline transport increase sharply with
the transport distance. Thus, pipeline transport of the assumed amount of CO_2
is only cheaper compared to truck transport over shorter distances (here: around
100 km). However, if the transport volume increases (e.g., due to an increased
demand or the use of a shared pipeline), the CO_2 chain with pipeline transport
can be cheaper compared to the truck-based CO_2 chain, even for longer transport
routes.

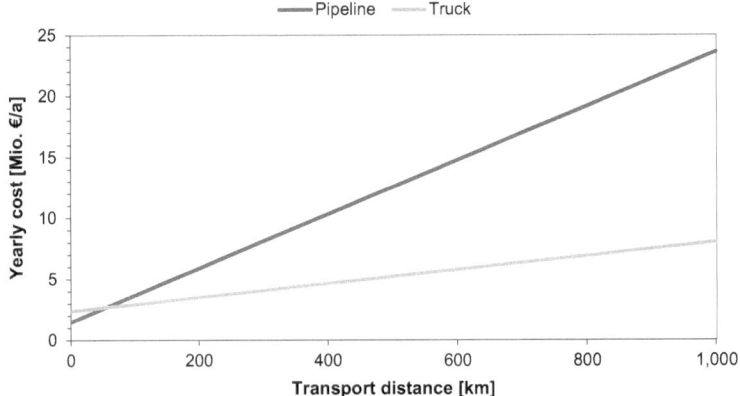

Fig. 15 Annual costs for a CO_2 chain depending on transport distance with a CO_2 amount of
50,000 t/a

- **Truck**. The CO$_2$ chain with truck transport causes higher annual costs over short distances than pipeline transport due to the high CAPEX for trucks. However, due to the small amount of transported CO$_2$ assumed here, the truck transport costs are more independent of the distance than is the case for the costs of pipeline transport (i.e., the slope of the truck graph is lower than the slope of the pipeline graph). This is because the CAPEX of the truck has a smaller share related to the total costs as the distance increases. Also, the fuel costs are so low in comparison that, despite the increasing proportion due to the increasing distance, they do not lead to significantly higher annual costs compared to the CO$_2$ chain with pipeline transport. In addition to the lower annual costs, transportation by truck is more flexible than transportation by pipeline. Not only can alternating start and end points be served, but the transport route can also vary. Furthermore, the number of trucks can be varied in response to changes in the amount of CO$_2$ to be transported.

6.2 Medium-Scale Scenario

In the medium-scale scenario, a CO$_2$ quantity of 500,000 t/a is transported over a transport distance of up to 7000 km. Here, the two different chains presented in Fig. 16 are defined. In one chain, the CO$_2$ is transported by offshore pipeline, and in the other, by ship. Below, these chains are briefly described, including the pressure and temperature range of the transported CO$_2$.

- **Pipeline**. Larger amounts of CO$_2$ can be transported as a liquid via pipeline. Compared to the pipeline transport of gaseous CO$_2$, the pipeline transport of liquid CO$_2$ has higher CAPEX but lower OPEX; the latter costs are lower because lower pump capacities than compressor capacities have to be installed to compensate for the pressure loss (i.e., the energy requirement is lower) [23]. Due to the long transport distance, an offshore pipeline is assumed. Compared to onshore pipelines, such offshore pipelines show higher CAPEX because, among other aspects, installing the pipeline is more complex [39]. The CO$_2$ is transported at 4 °C and 140 bar. The pipeline is only used to transport the assumed amount of CO$_2$, resulting in a calculated pipeline diameter of 50 mm.

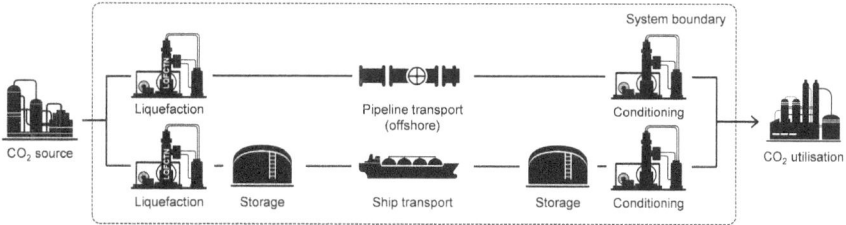

Fig. 16 Representation of the two chains analyzed within the medium-scale scenario

- **Ship**. Larger amounts of CO_2 can also be transported by ship. Since the transport volume of a ship is a priori limited, CO_2 needs to be liquefied. In 2022, existing ships transported CO_2 at medium temperatures (here: $-30\,°C$) and pressure ranges (here: 15 bar) [17]. The ship is assumed to have a transport capacity of 10,000 t CO_2, corresponding to the capacity of the intermediate storage facility.

Table 8 shows the techno-economic parameters of the two chains defined above. Figure 17 shows the specific energy demand and the specific costs for transporting 0.5 Mio. t_{CO2}/a at a transport distance of 2,500 km (according to the parameters shown in Table 8). Subsequently, the results of the two chains are explained and discussed below.

Table 8 Techno-economic parameters of medium-scale scenario

Parameter	Unit	Chain	Conditioning	Storage	Transport (2500 km)	Sources
CAPEX	(Mio. €)	Pipeline	26.4	/	610.6	[52–54]
		Ship	27.9	10.5	77.7	[15, 17, 53, 54]
Depreciation	(a)	Pipeline	20	/	40	[53, 55]
		Ship		30	25	[53, 56, 57, 67]
WACC	(%)	Pipeline	6.4			[58]
		Ship				
Fixed OPEX (share of annual CAPEX)	(%)	Pipeline	4	/	5	[53, 68]
		Ship		2		[17, 53, 59, 60, 61, 68]
Specific energy demand	(kWh/t)	Pipeline	102.7	/	148.4	[15, 52, 55, 62, 63]
		Ship	110.33	0	268.5	[15, 17, 61, 64]
Energy cost	(€/ kWh)	Pipeline	0.08	/	0.08	[65]
		Ship		0	0.04	[57, 65, 69]

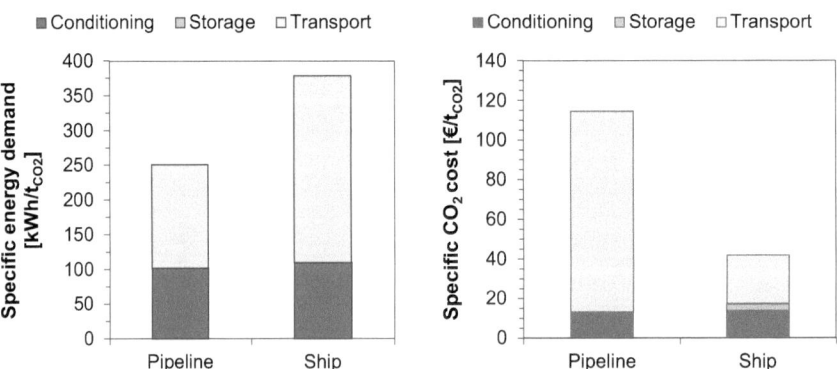

Fig. 17 Specific energy demand (left) and costs (right) for a CO_2 chain with a CO_2 quantity of 0.5 Mio. t/a and a transport distance of 2,500 km

- **Pipeline**. The pipeline transport is characterized by a lower specific energy requirement compared to the ship transport. Liquefaction to a higher temperature level requires less energy (pipeline 4 °C, ship −30 °C), and pipeline transport is more efficient than ship transport. Besides this, pipeline transport of liquid CO_2 has lower friction than pipeline transport of gaseous CO_2 [23]. However, this lower energy requirement is only reflected in the costs to a limited extent. While liquefaction to a higher temperature level also has lower CAPEX and OPEX, a pipeline has higher CAPEX than a ship for the assumed distance of 2,500 km [15, 17, 53]. The high transport share in the specific costs shows that the CAPEX of the pipeline contributes significantly to the specific transport costs.
- **Ship**. The specific energy requirement for the CO_2 transport by ship is higher compared to the respective pipeline option, as both liquefaction to a lower temperature level and transportation by ship have higher energy requirements [15, 51, 66]. Due to scaling effects, larger ships show lower specific energy requirements than smaller ships, and ships have significantly lower specific energy requirements compared to trucks. Within the CO_2 chain with ship transport, the specific conditioning costs are higher compared to the chain with pipeline transport due to the higher specific energy requirements for achieving the lower temperature level [15]. In addition, there are costs for intermediate storage, which, however, account only for a small proportion of the chain. Also, the specific transport costs of the ship are significantly lower than the specific transport costs of pipeline transport [17]. Like in the small-scale scenario, the ratio of CAPEX and OPEX depends on the transport distance.

Figure 18 shows the annual cost of the CO_2 chain depending on the transport distance, as the distance significantly influences the specific transport costs. Below, this influence of the transport distance on the costs of the two chains is discussed.

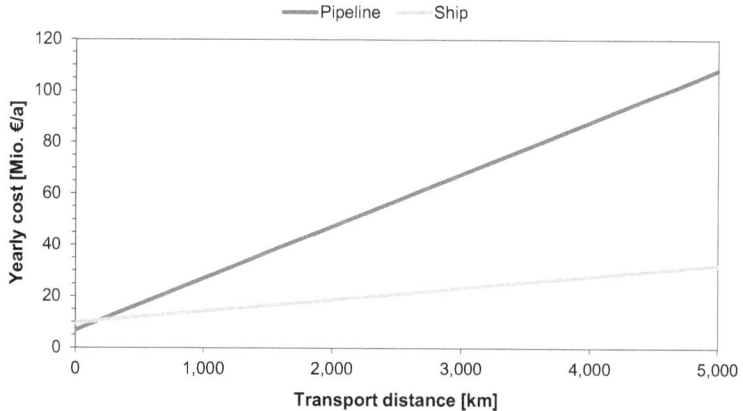

Fig. 18 Annual costs for a CO_2 chain depending on transport distance with a CO_2 amount of 0.5 Mio. t/a

- **Pipeline**. Figure 17 already indicates that the transport costs of the pipeline make up a significant proportion of the specific CO_2 costs. The CAPEX responsible for this depends on the distance; i.e., the annual costs of the CO_2 chain with pipeline transport increase (nearly) proportional to the transport distance increases. This means that the pipeline is only cheaper than ship transport for shorter transport distances (here: up to 250 km). Depending on the start and end point of the transport, the transport distance of a pipeline might be shorter than the transport distance of a ship since a pipeline can be installed both on- and offshore; the CAPEX of the former option is typically clearly lower compared to the latter variant. Thus, transport by pipeline can be cheaper than transport by ship—even over longer distances. Furthermore, if the transport volume is increased, for example, due to increased demand, the CO_2 chain with pipeline transport might be cheaper compared to the CO_2 ship transport—even for longer routes.
- **Ship**. Due to the high CAPEX for ships, the CO_2 ship chain over short distances incurs higher annual costs than the chain with pipeline transport. However, the CAPEX of a ship is more independent of the transport distance than the CAPEX of a pipeline (i.e., the slope of the ship graph is lower than the slope of the pipeline graph). The fuel costs of a ship, variable OPEX, are so low that, despite increased specific energy requirements, transport by ship is cheaper for longer distances. This makes ships particularly suitable for longer distances compared to pipelines. In addition, transportation by ship is also more flexible than transportation by pipeline. Not only can alternating start and end points be served, but the transport route can also be varied. Furthermore, both the number of ships can be varied and ships with different transport capacities can be used in response to changes in the amount of CO_2 to be transported.

6.3 Large-Scale Scenario

In the large-scale scenario, a CO_2 quantity of 2 Mio. t/a is transported over a transport distance of up to 15,000 km. Therefore, two different chains (Fig. 19) are defined. In both cases, CO_2 is transported by ship, but the temperature and pressure range of the liquid CO_2 differs. Below, the respective chains (Fig. 19), including the assumed pressure and temperature range of the transported CO_2, are outlined.

- **Ship—15 bar**. Analogous to the medium-scale scenario, the CO_2 is transported at a medium temperature and pressure level similar to the existing CO_2 transport ships in the year 2022. In this chain, this corresponds to $-30\,°C$ and 15 bar. However, the large-scale scenario assumes a larger capacity of around 32,100 t of the ship because the CO_2 to be transported is higher compared to the medium-scale scenario. The intermediate storages show the same capacity as the ship.
- **Ship—8 bar**. Here, CO_2 is transported at a lower temperature and pressure range ($-46\,°C$ and 8 bar). Assuming the same ship volume as within the variant above, more CO_2, in this case, around 34,200 t, can be transported at lower pressures

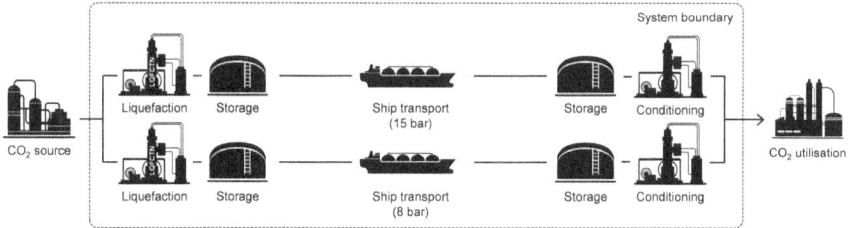

Fig. 19 Representation of two chains assessed within large-scale scenario

[15]. The intermediate storages show the same capacity like the ship. The storage costs are lower at lower pressure ranges because thinner walls and, therefore, less material are required. However, as the pressure range decreases, both the CAPEX and OPEX of liquefaction increase due to the increased cooling requirements [54].

Table 9 shows the techno-economic parameters of the two chains defined above.

Table 9 Techno-economic parameters of large-scale scenario

Parameter	Unit	Chain	Conditioning	Storage	Transport (7500 km)	Sources
CAPEX	(Mio. €)	Ship (15 bar)	72.5	21.2	454.1	[15, 17, 53, 54]
		Ship (8 bar)	81.1	15.8	317.0	
Depreciation	(a)	Ship (15 bar)	20	30	25	[53, 56, 57, 67]
		Ship (8 bar)				
WACC	(%)	Ship (15 bar)	6.4			[58]
		Ship (8 bar)				
Fixed OPEX (share of annual CAPEX)	(%)	Ship (15 bar)	4	2	5	[17, 53, 59, 60, 61, 68]
		Ship (8 bar)				
Specific energy demand	(kWh/t)	Ship (15 bar)	99.8	0	723.5	[15, 17, 61, 64]
		Ship (8 bar)	119.8		723.5	
Energy cost	(€/ kWh)	Ship (15 bar)	0.08	0	0.04	[57, 65, 69]
		Ship (8 bar)				

Now, the specific energy demand and the specific costs are presented (Fig. 20; according to the parameters shown in Table 9), and then, the annual costs are discussed as a function of the transport distance (Fig. 21).

- **Ship—15 bar**. The chain with the ship transport at 15 bar and −30 °C has a slightly lower specific energy requirement than the chain with the ship transport at 8 bar and −46 °C due to the lower specific energy requirement for liquefaction at higher pressures and temperatures [54]. However, when considering the specific CO_2 costs, the chain with the ship transport at 15 bar is slightly more expensive due to the more expensive specific transport costs. Due to the higher pressure, both the transport tanks on the ship and the storage tanks in the CO_2 chain have

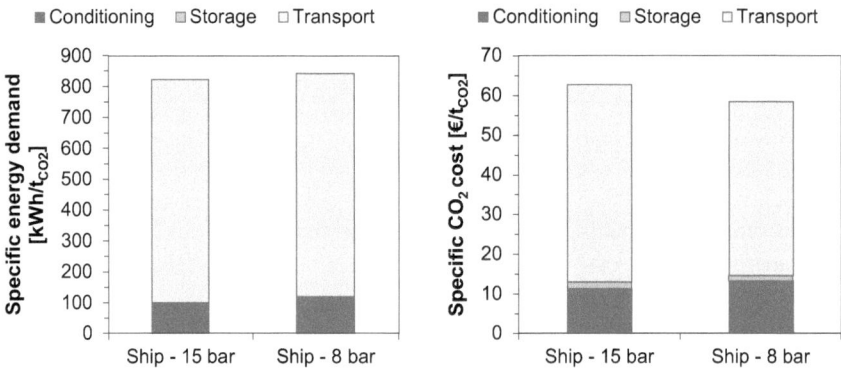

Fig. 20 Specific energy demand (left) and costs (right) for a CO_2 chain with a CO_2 quantity of 2 Mio. t/a and a transport distance of 7,500 km

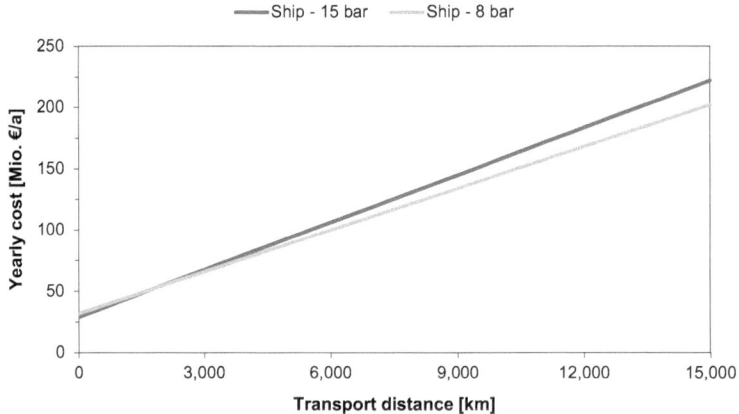

Fig. 21 Annual costs for a CO_2 chain depending on transport distance with a CO_2 amount of 2 Mio. t/a

to be manufactured with larger wall thicknesses; i.e., more material is required, resulting in higher costs [17].

- **Ship—8 bar**. The chain with the ship transport at 8 bar and -46 °C not only has the advantage of smaller wall thicknesses and the associated lower CAPEX due to the lower pressures but also that CO$_2$ shows a higher density under these conditions, i.e., more CO$_2$ can be transported or stored within the same volume resulting in reduced CAPEX. However, this advantage is limited due to the slight pressure difference. In addition, the costs and energy requirements of liquefaction are higher, so the CO$_2$ chain at 8 bar can also have higher costs than the chain at 15 bar, depending on the transport distance (ratio of conditioning costs to storage and transport costs).

Figure 21 shows the annual cost of the CO$_2$ chains depending on the transport distance. Subsequently, the influence of the transport distance on the costs of the two chains is discussed.

- **Ship—15 bar**. The annual costs of the CO$_2$ ship transport at 15 bar are similar to the annual cost of the CO$_2$ ship transport at 8 bar (Fig. 21). From a distance of around 1,500 km onwards, the annual costs of the ship transport at 15 bar are higher, and the difference between the two options grow as the distance increases; i.e., up to a distance of 1,500 km, the influence of cheaper liquefaction predominates and from this distance onwards the influence of the higher CAPEX due to the higher pressure predominates.

- **Ship—8 bar**. Due to the fact that in the year 2022, no ships for transporting CO$_2$ with the transport capacity assumed here nor ships that can transport CO$_2$ at 8 bar and -46 °C are available on the market, the costs assumed here are merely forecasts from literature; i.e., both, the intersection of the two graphs and the difference between the two graphs may differ. Nevertheless, the CO$_2$ chain has advantages at a pressure of 8 bar and -46 °C, especially for long transport routes and large transport quantities. On the one hand, a larger mass of CO$_2$ can be transported with the same ship volume, so the annual cost difference will continue to increase for longer transport routes than assumed here. On the other hand, lower pressures are particularly suitable for larger transport quantities, as larger tanks can be manufactured due to the tank wall restrictions, and less material is required for the same tank volume.

7 Final Consideration

The emission of greenhouse gases, mainly carbon dioxide (CO$_2$), increases global warming. With the goal of limiting global warming, the political focus moves more toward concepts for carbon capture and storage (CCS). For the implementation of such approaches, CO$_2$ needs to be transported and temporarily stored (in larger quantities). Therefore, this chapter aims to assess the characteristics of CO$_2$ influencing conditioning, intermediate storage, and transport.

Transport and interim storage are mainly limited by volume for gaseous substances at ambient conditions. Thus, transport and storage are often realized in a liquid aggregate state due to the higher density. CO_2 is liquid above ambient pressure and/or below ambient temperature. Typically, the following three ranges are subdivided.

- Low pressure and temperature range—7 to 9 bar and -55 to -45 °C,
- Medium pressure and temperature range—15 to 20 bar and -30 to -20 °C,
- High pressure and temperature range—45 to 70 bar and 10 to 30 °C.

Impurities, such as water, methane, or hydrogen sulfide, can also affect the behavior of CO_2 and may lead to operational challenges and safety risks. Therefore, depending on the impurities within the CO_2 as well as the selected aggregate state and pressure and temperature level or the selected transport and storage option, appropriate conditioning must be carried out.

During conditioning, impurities can be removed, but most importantly, the pressure and temperature of the CO_2 are adjusted by compression or liquefaction. Liquefaction has a slightly higher energy demand than compression because CO_2 is first compressed and then cooled. This leads also to marginally higher costs.

Two different storage concepts can be used for the intermediate storage of liquid CO_2. In fully pressurized storages, liquefied CO_2 is stored under pressure at ambient temperature (high pressure and temperature range). In semi-pressurized/semi-refrigerated storages, CO_2 is stored above ambient pressure and below ambient temperature (low/medium pressure and temperature range). The choice of the respective storage concept is mainly determined by the quantity to be stored. Smaller quantities are typically stored fully pressurized, and larger quantities are semi-pressurized/semi-refrigerated.

In 2022, CO_2 is transported and stored in smaller quantities by pipeline, ship, train, and truck, mainly in the medium pressure and temperature range. For the use of CO_2 in enhanced oil recovery, larger quantities are also transported by pipeline; i.e., the technical readiness level (TRL) of the means of transport for the transportation of CO_2 is already high. Only the transport of large quantities by ship has not yet been demonstrated. This requires new ship concepts with generally lower pressures and temperatures as well as the corresponding infrastructure.

The techno-economic comparison of different CO_2 transportation chains has shown that the transport of small CO_2 amounts over short distances by truck is by far the cheapest option. Compared to that, pipelines have lower specific energy requirements and are therefore preferable from an energy efficiency point of view. From a cost perspective, they are particularly suitable for medium to large quantities and shorter transport distances. Transport by ship is most cost-effective for smaller to larger amounts of CO_2 and longer transport routes. While trucks and pipelines, both for small and large amounts of CO_2, are already used commercially to transport CO_2, ships and the corresponding port infrastructure for the transport of larger amounts still need to be developed.

If large amounts of CO_2 have to be transported in the future, appropriate means of transport, probably pipelines and ships, will be necessary. This would make it likely that large CO$_2$ transport ships and the corresponding infrastructure need to be developed. This is also true for the reasons outlined below [14, 16, 70, 71, 72].

- In some regions, such as Japan, frequent earthquakes make it difficult to build (offshore) pipelines.
- Ships can realize an easy ramp-up by increasing the number of vessels.
- Ships are flexible since they can change their routes as well as their start and end points.

In addition to technical developments, economic developments can also take place. The latter developments, like technical advances, become more likely as the market implementation of a technology increases. Therefore, expected developments in the field of CO$_2$ transport and storage depend significantly on the global development of the CO$_2$ market (quantity and distance).

References

1. Fischedick M, Esken A, Pastowski A, Schüwer D, Supersberger N, Nitsch J, Viebahn P, Bandi A, Zuberbühler U, Edenhofer O (2007) RECCS: Strukturell-ökonomisch-ökologischer Vergleich regenerativer Energietechnologien (RE) mit Carbon Capture and Storage (CSS). Wuppertal
2. International Energy Agency (IEA) (2020) Energy technology perspectives 2020: special report on carbon capture utilisation and storage
3. International Energy Agency (IEA) (2019) Putting CO$_2$ to use: creating value from emissions
4. Budinis S, Fajardy M, Greenfield C, Moore R (2022) Carbon capture, utilisation and storage. https://www.iea.org/reports/carbon-capture-utilisation-and-storage-2. Accessed 19 January 2023
5. Vermeulen TN (2011) Knowledge sharing report: CO$_2$ liquid logistics shipping concepts (LLSC) overall supply chain optimization. Den Haag
6. Harper P, Wilday J, Bilio M (2011) Assessment of the major hazard potential of carbon dioxide (CO$_2$)
7. Fischedick M, Görner K, Thomeczek M (eds) (2015) CO$_2$: Abtrennung, Speicherung, Nutzung. Springer Berlin Heidelberg, Berlin, Heidelberg
8. Span R, Wagner W (1996) J Phys Chem Refer Data 25
9. Bjerketvedt D, Egeberg K, Ke W, Gaathaug A, Vaagsaether K (2011) Energy Proced 4
10. Ausfelder F, Bazzanella A (2008) Diskussionspapier - Verwertung und Speicherung von CO$_2$
11. Al-Siyabi I (2013) Effect of impurities on CO$_2$ stream properties. Edinburgh
12. Aspelund A, Mølnvik MJ, de Koeijer G (2006) Chem Eng Res Design 84
13. Neele F, Haugen HA, Skagestad R (2014) Energy Proced 63
14. Durusut E, Joos M (2018) Shipping CO$_2$—UK cost estimation study. Cambridge
15. Seo Y, Huh C, Lee S, Chang D (2016) Int J Greenhouse Gas Cont 52
16. Al Baroudi H, Awoyomi A, Patchigolla K, Jonnalagadda K, Anthony EJ (2021) Appl Energy 287
17. IEAGHG (2020) The status and challenges of CO$_2$ shipping infrastructures: IEAGHG Technical Report 2020-10. Cheltenham
18. Ministry of Petroleum and Energy, Feasibility study for full-scale CCS in Norway
19. Oosterkamp A, Ramsen J (2008) State-of-the-art overview of CO$_2$ pipeline transport with relevance to offshore pipelines. Haugesund
20. National Energy Technology Laboratory (2012) CO$_2$ impurity design parameters: quality guidelines for energy system studies

21. de Visser E, Hendriks C, de Koeijer G, Liljemark S, Barrio M, Austegard A, Brown A (2007) Towards hydrogen and electricity production with carbon dioxide capture and storage
22. Energy Institute (2013) Hazard analysis for offshore carbon capture platforms and offshore pipelines. London
23. Lu H, Ma X, Huang K, Fu L, Azimi M (2020) J Clean Prod 266
24. Strobl M (2022) USA will CO_2-Pipelines bauen, doch deren Sicherheit sorgt für Bedenken (2022). https://www.trendingtopics.eu/usa-will-co2-pipelines-bauen-doch-deren-sicherheit-sorgt-fuer-bedenken/. Accessed 1 December 2022
25. Chorowski M, Duda P, Polinski J, Skrzypacz J (2015) IOP Conf Series Mat Sci Eng 101
26. Aspelund A, Jordal K (2007) Int J Greenhouse Gas Cont 1
27. Gong W, Remiezowicz E, Fosbol PL, von Solms N (2022) Energies 15
28. Asia Industrial Gases Association (AIGA), Carbon dioxide. Singapore
29. VTG Rail Europe GmbH, Flüssiggas-Kesselwagen für Tiefkalte Gase: Kohlendioxid. Hamburg
30. Dharmadhikari S, Heck G (1991) The chemical engineer
31. Aquaspender, Kohlendioxid (CO_2). https://www.aquaspender.de/Downloads/Kohlendioxid-Daten.pdf. Accessed 5 January 2022.
32. ASCO KOHLENSÄURE AG (2018) Produktkatalog: Die komplette CO_2-Lösung. Romanshorn
33. Kang K, Seo Y, Chang D, Kang S-G, Huh C (2015) Energies 8
34. Knoope MMJ, Guijt W, Ramirez A, Faaij A (2014) Int J Greenhouse Gas Cont 22
35. Lazic T, Oko E, Wang M (2013) J Process Mech Eng 228
36. Zheng J, Shi J, Liu Z, Jiang S, Liu C (2018) Sino-global energy 6
37. Wang D (2017) Research on liquefaction, transportation and storage technology of CO_2. Jingzhou, Wuhan
38. Yu X, Li Z, Pan X, Li Y, Zheng X, Wang Y (2009) Natural gas industry 29
39. Metz B, Davidson O, de Coninck HC, Loos M, Meyer LA (2005) IPCC Special report on carbon dioxide capture and storage. United Kingdom and New York, NY, USA, Cambridge
40. Deutsche Flagge and Bundesministerium für Digitales und Verkehr, Frachtschiffe. https://www.deutsche-flagge.de/de/flagge/schiffsarten/frachtschiffe-1/frachtschiffe#:~:text=Ein%20Binnenschiff%20ist%20ein%20Schiff,Navigationsinstrumente%20und%20Rettungsger%C3%A4te%20zu%20stellen. Accessed 5 December 2022
41. Yara International ASA (2015) New liquid CO_2 ship for Yara. https://www.yara.com/news-and-media/news/archive/2015/new-liquid-co2-ship-for-yara/. Accessed 22 November 2022
42. Zahid U, An J, Lee C-J, Han C (2015) Indust Eng Chem Res 54
43. DNV (2022) DNV supports innovations in CO_2 carrier design. https://www.dnv.com/expert-story/maritime-impact/DNV-supports-innovations-in-CO2-carrier-design.html. Accessed 14 April 2023
44. MAN Energy Solutions (2022) ME-GI engines to power liquid-CO_2 carriers in groundbreaking carbon-transport-and-storage project. https://www.man-es.com/company/press-releases/press-details/2022/03/09/me-gi-engines-to-power-liquid-co2-carriers-in-groundbre aking-carbon-transport-and-storage-project. Accessed 14 April 2023
45. Ship Technology (2021) Posco, others collaborate to develop liquid CO_2 carrier tanker. https://www.ship-technology.com/news/posco-co2-carrier-tanker/. Accessed 14 April 2023
46. Vaughn Entwistle (2023) Mitsubishi shipbuilding launches liquid CO_2 carrier. https://www.gasworld.com/story/mitsubishi-shipbuilding-launches-liquid-co2-carrier/. Accessed 14 April 2023
47. Eckle P (2019) CO_2-transport. Olten
48. Stolaroff JK, Pange SH, Li W, Kirkendal WG, Goldstein HM, Aines RD, Baker SE (2021) Transport cost for carbon removal projects with biomass and CO_2 storage
49. SafeRack, Chlorine handling design, loading, and installation. https://www.saferack.com/bulk-chemical/chlorine-loading-platforms/. Accessed 5 January 2023
50. Air Products (2014) Safetygram 18: carbon dioxide. Allentown
51. Jackson S, Brodal E (2019) Energies 12

52. McCollum DL, Ogden JM (2006) Techno-economic models for carbon dioxide compression, transport, and storage correlations for estimating carbon dioxide density and viscosity. Davis
53. Reuß ME (2019) Techno-ökonomische Analyse alternativer Wasserstoffinfrastruktur. Aachen
54. Lee JH (2011) CCS chain design and optimization. Daejeon
55. Krieg D (2012) Energie & Umwelt: Konzept und Kosten eines Pipelinesystems zur Versorgung des deutschen Straßenverkehrs mit Wasserstoff. Jülich
56. Tremel A, Wasserscheid P, Baldauf M, Hammer T (2015) Int J Hydrogen Energy 40
57. Reuß ME, Dimos P, Leon A, Grube T, Robinius M, Stolten D (2021) Energies 14
58. European Central Bank (2020) Econ Bull
59. International Energy Agency (IEA) (2019) The future of hydrogen
60. DNV GL (2020) Study on the import of liquid renewable energy: technology cost assessment
61. Niermann M, Timmerberg S, Drünert S, Kaltschmitt M (2021) Renew Sustain Energy Rev 135
62. DeSantis D, James BD, Houchins C, Saur G, Lyubovsky M (2021) iScience 24
63. Bartels JR (2008) A feasibility study of implementing an ammonia economy
64. Powertrust GmbH (2021) Energiebilanzen von Diesel-und Elektro Pkw. https://www.powert rust.de/news/news-detail/energiebilanz-diesel-pkw. Accessed 20 November 2023
65. Fasihi M, Breyer C (2020) J Clean Product 243
66. Garcia-Alvarez A, Perez-Martinez PJ, Gonzales-Franco I (2012) J Intell Transp Syst Techn Plan Operat 17
67. Teichmann D, Arlt W, Wasserscheid P (2012) Int J Hydrogen Energy 37
68. Stolzenburg K, Mubbala R (2013) Integrated design for demonstration of efficient liquefaction of hydrogen (IDEALHY): fuel cells and hydrogen joint undertaking (FCH JU)
69. Ramboll Deutschland GmbH (2022) Kraftstoffanalyse in der Schifffahrt nach Segmenten
70. Alabi O, Andreas J, Brain N, Del Granado PC, d'Elloy M, Heffron R, Molnvik M, Skar C, Soothill C, Turner K (2018) Role of CCUS in a below 2 degrees scenario
71. Nam H, Lee T, Lee J, Chung H (2013) Int J Greenhouse Gas Control 12
72. Ozaki M, Ohsumi T (2011) Energy Proced 4

Conversion Processes and Technologies

Classification of Power-to-Gas (PtG) and Power-to-Liquid (PtL) Processes

Ulf Neuling and Nils Bullerdiek

Abstract The fight against climate change to limit global warming is one of the biggest challenges of mankind. Therefore, strict decarbonization efforts within all areas of our society are necessary. This is also seen by international governments, resulting in strict national and international climate protection policies. Besides the energy sector, the transport sector is responsible for the second largest share of global emissions. Today, transport accounts for around 30% of the annual greenhouse gas emissions in the EU-27, of which road transport accounts for more than two-thirds (about 72%). A vast consensus within politics, science, and industry sees direct electrification as the most efficient and cheapest solution to reduce emissions. However, in hard-to-abate sectors like aviation or maritime transport, where direct decarbonization is difficult, reducing greenhouse gas emissions is particularly challenging. Partly, this is due to the reliance on energy-dense fuels and long operational lifetimes of aircraft and vessels. For these transport modes, a defossilization based on renewable electricity is only possible indirectly via synthetic fuels, so-called Power-to-X (PtX) fuels. Against this background, feasible areas of application as well as general technical aspects of PtX fuels are described in this article. A subsequent classification of Power-to-X products based on the hydrogen rainbow and potential carbon sources is presented. According to this, the combination of biogenic or atmospheric CO_2 with green hydrogen shows the highest climate mitigation potential. However, other CO_2 and low-carbon hydrogen sources may be a short-term option that is required for a timely ramp-up of PtX production capacities if strict sustainability and environmental protection measures are taken.

Keywords Powerfuels · Efficiency · Sustainability · Carbon Sources · Hydrogen Sources

---

U. Neuling (✉) · N. Bullerdiek
Hamburg University of Technology (TUHH), Institute of Environmental Technology and Energy Economics (IUE), Hamburg, Germany
e-mail: ulf.neuling@tuhh.de

© The Author(s), under exclusive license to Springer Nature Switzerland AG 2025
N. Bullerdiek et al. (eds.), *Powerfuels*, Green Energy and Technology,
https://doi.org/10.1007/978-3-031-62411-7_18

493

1 Introduction

Based on the reports of the Intergovernmental Panel on Climate Change (IPCC), global warming of the earth's climate system is unequivocal. To prevent further climate deterioration, there is a consensus among scientists and research institutes charged with measures to mitigate climate change in order to keep global average temperature warming well below 2 °C above preindustrial levels—and if possible, even below 1.5 °C [1]. In the EU-27, the transport sector accounted for almost 30% of total CO_2 emissions in 2019. Of this, road transport was responsible for the major portion, contributing more than two-thirds (roughly 72%) [2]. To reduce net greenhouse gas (GHG) emissions within the so-called Fit for 55 package, the EU has set targets to reduce GHG emissions by 55% until 2030 compared to 1990 levels. This also includes the transport sector. One way to achieve this target is using powerfuels, also referred to as PtX fuels, synthetic fuels, electrofuels, or e-fuels. They can be supplied almost GHG neutral if produced based on renewable energy sources. The processes utilized to produce such fuels are referred to as Power-to-X (PtX) processes, where "X" refers to the final product, be it gas in Power-to-Gas (PtG) processes (e.g., to produce hydrogen or methane) or liquids in Power-to-Liquid (PtL) processes (e.g., to produce hydrocarbons or methanol).

The term PtX in general covers a broad variety of technical process combinations. All these processes involve the utilization of (renewable) electricity to produce hydrogen and a (sustainably provided) carrier molecule, usually CO_2 or nitrogen, to produce gaseous or liquid fuels. Based on this general characterization, the technical process layout can be distinguished into different subsequent process steps, i.e., the supply of hydrogen, the supply of nitrogen or CO_2, and the production of synthesis gas (syngas) followed by the actual synthesis step. Depending on the specific process and desired end product, further upgrading or refining processes of the synthesis products might be necessary. A schematic overview of the process flow and available technologies for the different steps to produce hydrogen and further PtX fuels is shown in Fig. 1.

The production of such PtX products needs to fulfill high sustainability standards to contribute to climate change mitigation and enable an environmentally sound transition of the transport sector. Hence, the electricity as well as the carrier molecule have to be supplied from sustainable sources. Therefore, electricity generation from renewable energies, such as solar power (solar PV as well as thermal systems), wind power, geothermal, and hydro power, is seen to be suitable. In terms of sustainable or "green" CO_2, different sources are currently discussed to fulfill the respective sustainability criteria. In general, one can distinguish between CO_2 from point sources or diffuse sources. The first comprises (waste) gas streams with an already high share of CO_2, resulting from industrial processes (e.g., cement or steel production) or conversion processes based on biomass (e.g., biogas or bioethanol plants, biomass combustion power plants). The latter in general refers to the atmosphere, which, compared to CO_2 point sources, contains a much lower CO_2 concentration of about

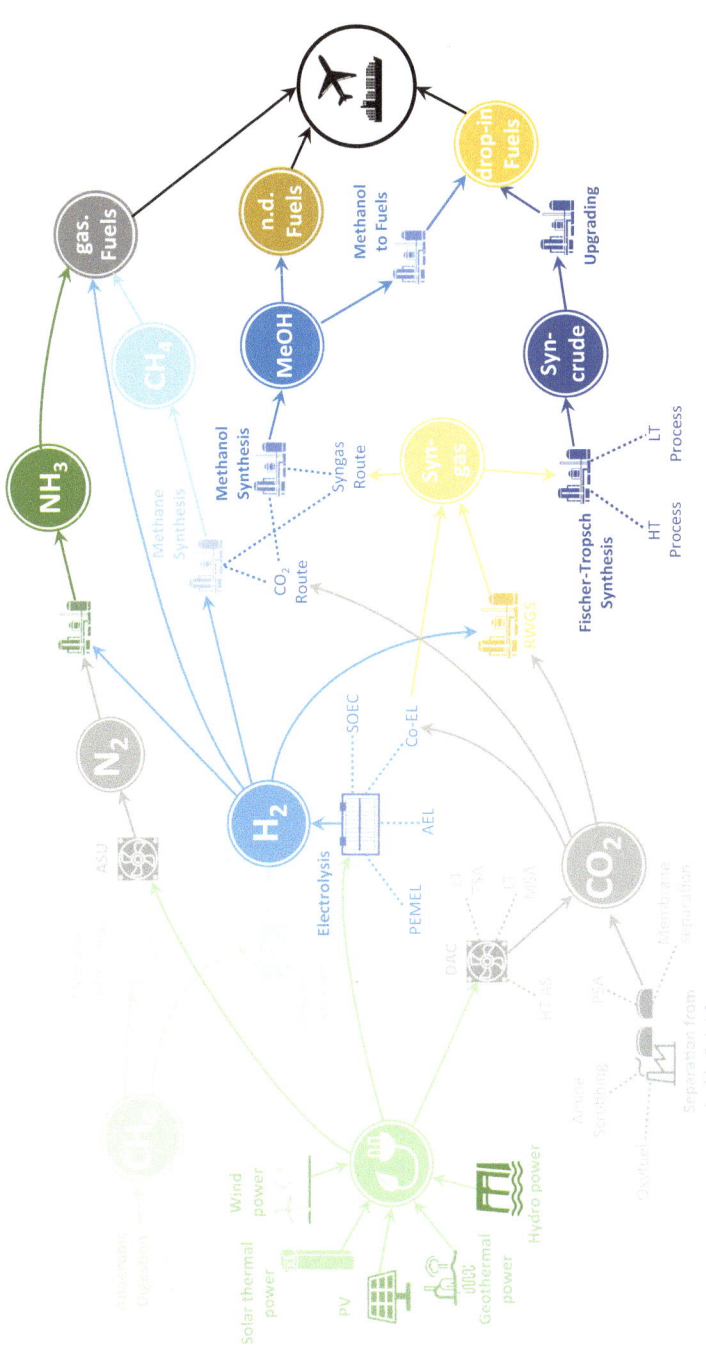

Fig. 1 Overview of carbon-based PtX conversion chains and possible technologies (AEL = alkaline electrolysis, AS = aqueous solution, ASU = Air separation unit, Co-EL = Co-electrolysis, DAC = direct air capture, HT = high temperature, LT = low temperature, MSA = moisture swing adsorption, n.d. = non-drop-in, PEMEL = proton exchange membrane electrolysis, PSA = pressure swing adsorption, PV = photovoltaic, RWGS = reverse water–gas shift, SOEC = solid oxide electrolysis cell, TSA = temperature swing adsorption)

420 ppm or 0.042%. The technological separation of CO_2 from ambient air is generally referred to as direct air capture (DAC). Whether CO_2 can be considered sustainable predominantly depends on the feedstock (origin). In this respect, especially the availability as well as the sustainability of CO_2 from industrial point sources (such as waste incineration plants, cement plants, and steel mills) is discussed critically [3, 4]. Therefore, CO_2 from ambient air provided by DAC technologies as well as biogenic point sources are considered to be the only long-term viable sources for sustainable, "green" CO_2. In contrast to CO_2, the nitrogen concentration in ambient air is significantly higher (about 78%), making its separation from ambient air considerably easier compared to CO_2. Thus, ambient air is generally seen the preferred source to provide sustainable nitrogen as reactant for the respective PtX production routes.

2 Application Areas of Powerfuels

In addition to the sustainability aspects related to the production of powerfuels mentioned above, the most efficient use of energy—and thereby the reduction of primary energy consumption—is very important. This is particularly important for the transport sector, where a large proportion of the energy content of fuels today is lost as waste heat in internal combustion engines. This is even more important when talking about powerfuels, as their production is already associated with high energy losses. Also, as renewable energies are needed to decarbonize the global economy as much as possible, among others heating and cooling of buildings and the industrial sector, renewable electricity will be scarce for the next decades. Therefore, also in the transport sector a so-called efficiency first approach has to be adapted. As the production of powerfuels is usually characterized by several conversion steps, the provision of energy in the form of powerfuels requires (significantly) more renewable electricity compared to the direct utilization of electricity, e.g., in battery electric vehicles. This is exemplarily illustrated in Fig. 2 for different drive train options powered by renewable electricity.

Thus, powerfuels should only be considered for application areas that face a severe lack of other feasible fuel or energy options to substitute fossil-based energy carriers in the coming decades—so-called hard-to-abate applications. Such applications mainly include long-haul flights and oceanborne maritime transportation (Table 1). In addition, some niche applications of land-based transport, such as heavy special-purpose vehicles, will most likely also be relying on powerfuels in the future. As far as possible, direct electrification should be implemented for all other modes of transport, such as cars, light-duty vehicles, trucks as well as small ships, and short-haul airplanes.

As the general requirements for different modes of transport vary, the same applies to suitable powerfuels and their respective fuel properties. Among these, civil aviation is characterized by the most stringent requirements to be fulfilled by powerfuel options introduced to substitute current kerosene-type aviation fuels of fossil origin. Each synthetic fuel option that is supposed to be used as an aviation fuel eventually has

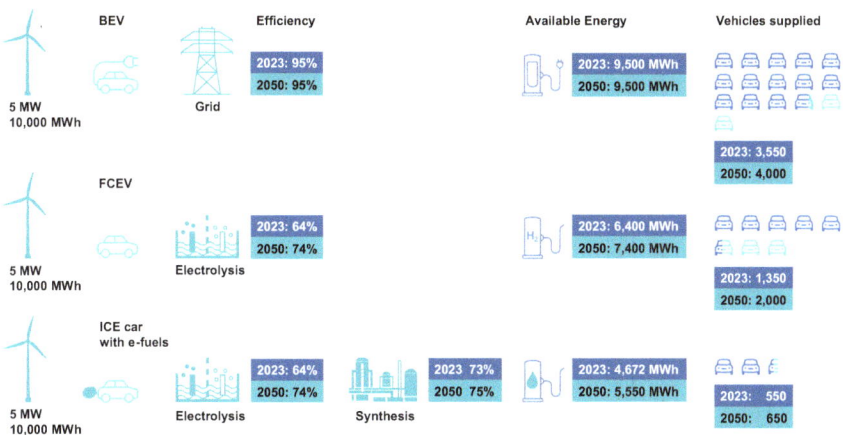

Fig. 2 Comparison of efficiency of different drive train options for use of renewable electricity in road transport. Above figures are for a compact car that is driven 14,000 km/year. Future efficiency improvements result from expected efficiency gains in electrolysis and synthesis processes. Assumed average full load hours for a wind turbine in Germany: 2000 h/year. (BEV = battery electric vehicle, FECV = fuel cell electric vehicle, ICE = internal combustion engine). According to [5]

to be approved by defined standardization committees. This standardization process is meant to guarantee a safe and seamless utilization of such alternative aviation fuels, i.e., different (renewable) fuels. However, alternative sustainable fuels have to fulfill predefined properties and characteristics similar to the currently used conventional (fossil) aviation fuel. This leads to the fact that also powerfuels have to fulfill these criteria's if they are considered to be used in aviation.

Maritime transportation also has classification and standardization procedures and bodies. However, these are mainly related to the ship class and propulsion systems. The fuel properties are not standardized to the same extend as it is true for aviation. This is due to different characteristics of marine applications, differing safety standards as well as other more general aspects of fuel storage and use, such as space and weight restrictions. This is why currently different fuel types are discussed as renewable and power-based options to replace fossil-based marine fuels, such as marine fuel oil and marine diesel. This includes a whole bandwidth of different fuel options, ranging from gaseous to liquid fuels, such as hydrogen, ammonia, methanol, or synthetic diesel (such as Fischer–Tropsch diesel).

Regarding the reasonable application of powerfuels in land-based transportation—specifically from an energy efficiency perspective, two approaches can be distinguished. For vehicles with high performance characteristics, e.g., high range requirements, high power demand, and high transport weight, direct electrification often presents technical and economic challenges. For these applications, hydrogen could be a feasible fuel option to decarbonize, which however would require the build-up of a completely new supply infrastructure. Secondly, for some niche applications, synthetic power-based gasoline or diesel may be needed to also defossilize

Table 1 Overview of sustainably feasible application areas for powerfuels in transportation and required changes on overall supply infrastructure to use these fuels

Fuel option	Reactant/feedstock		Application mode of transport	Application in transport mode
(Green) hydrogen	Electricity from renewable sources of energy	Water	Aviation (short and medium haul flights, currently under development) Shipping (small and medium ships, currently under development)	Fuel cell Combustion engine or turbine (currently under development)
Ammonia	Green hydrogen	Nitrogen	Shipping	Potentially in existing combustion engines (engine modifications necessary), fuel cells (under development)
Synthetic methane (liquid)	Green hydrogen	Green CO_2	Shipping	Combustion engine (similar to LNG)
Methanol	Green hydrogen	Green CO_2	Shipping	Potentially in existing combustion engines (engine modifications necessary), fuel cells (under development)
Methanol-to-X fuels (kerosene, diesel)	Green methanol		Aviation and shipping	Drop-in fuels, existing combustion engines and turbines
Fischer–Tropsch fuels (kerosene, diesel)	Green hydrogen	Green CO_2	Aviation and shipping	Drop-in fuels, existing combustion engines and turbines

LNG liquefied natural gas

these areas of transportation (e.g., heavy goods transports, emergency or off-road vehicles).

3 Power-to-Gas (PtG) and Power-to-Liquid (PtL) Processes

Extensive research has been conducted in recent years on various Power-to-X processes, which have seen further development. However, the production of power-fuels on an industrial scale has not yet been achieved. Even though many of the technologies that are used along the entire PtX process chain have been known and optimized for decades, the combination of these technologies together with a production of green hydrogen based on (fluctuating) renewable electricity used for this purpose poses new technical challenges [6–9].

An overview of common PtX processes is given in Fig. 3.

Although the basic procedural flow of these processes is similar, individual process steps, operating conditions, and thus products can differ significantly. All processes have in common that green hydrogen is required at first—i.e., hydrogen produced via water electrolysis with the aid of renewable electricity as initial feedstock. The hydrogen is then further processed with another reactant (CO_2 or nitrogen) in a synthesis process, resulting in synthetic energy carriers. These can be used as fuels either directly, as blend-stock or after further treatment or processing steps.

The following sections discuss the technologies required to supply the initial feedstock and to carry out the subsequent synthesis steps. Since the availability of electricity generation from renewable sources can be intermittent—especially in the case of solar and wind power, PtX plants often need to incorporate storage facilities to ensure a continuous fuel production. These storage options may include batteries or hydrogen storage, as well as CO_2 or nitrogen storage to bridge supply shortages. However, storage technologies are not discussed in detail in this chapter.

3.1   Feedstock Supply

Various technical options are available for supplying the feedstocks hydrogen, nitrogen, and CO_2 required for the production of PtX fuels.

Water: seawater desalination and treatment. The production of large quantities of green hydrogen requires correspondingly large quantities of water—about 10 kg of water per kilogram of hydrogen produced via electrolysis (on a stoichiometric basis). Most electrolysis processes require highly pure water (so-called ultrapure water). Even drinking water must be further purified to separate the salts and minerals dissolved in it. This purification usually takes place in a water treatment unit directly upstream of the electrolyzer.

Seawater desalination plants are generally seen as the most suitable option to provide the required high water demand without adverse effects on groundwater. This is particularly relevant for regions that have optimal conditions for renewable power generation, which often coincide with arid environments. Such plants are already used worldwide for drinking and industrial water supply and can thus be considered an established technology [13]. The most widespread technology is the so-called multi-stage flash evaporation, in which the seawater is evaporated by adding heat (often waste heat from nearby power plants) through several condensation stages, leaving the salt on the condensation bottoms. Due to the high energy requirements of this process, so-called reverse osmosis plants are also increasingly used. In this process, seawater is forced through a membrane under high pressure, allowing salt and other impurities to be separated from the water. However, to avoid this step research is also being conducted on electrolysis processes that can directly use seawater. However, these technologies are still at the laboratory scale [14].

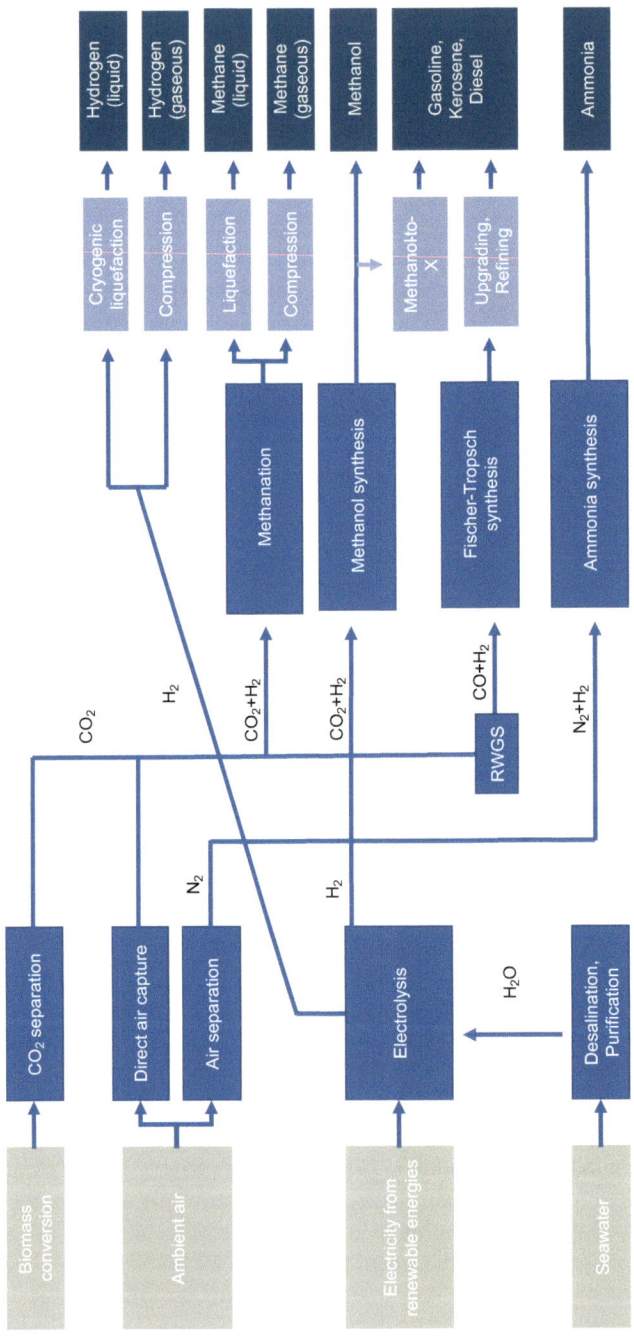

Fig. 3 Overview of Power-to-X (PtX) production processes (RWGS = reverse water gas shift)

Hydrogen: Electrolysis. In the context of green hydrogen production, electrolysis is generally understood to be the splitting of water (H_2O) into hydrogen (H_2) and oxygen (O_2) through the use of renewable electricity. Depending on the temperature level at which this process takes place, a distinction can be made between so-called low- and high-temperature electrolysis. Currently, the most widely used technology is alkaline electrolysis (AEL), in which an alkaline solution is used as the electrolyte. Especially in connection with the use of fluctuating renewable electricity, polymer electrolyte membrane electrolysis (PEMEL) has become established as another low-temperature electrolysis in recent years due to its higher load flexibility. Solid oxide electrolysis (SOEL), often referred to as high-temperature electrolysis, operates at significantly higher temperatures and usually requires water vapor as input, but is characterized by higher efficiencies than the low-temperature processes. A variant of SOEL is the so-called co-electrolysis, in which CO_2 is used as a feedstock in addition to water vapor, directly producing a synthesis gas consisting of hydrogen and carbon monoxide (CO). Compared to the low-temperature electrolyzers now established on the market, the high-temperature processes have so far only been realized in smaller demonstration plants [10].

CO_2: Biomass Conversion and Direct Air Capture. There are basically two different options available for capturing CO_2 from the atmosphere: direct capture via so-called direct air capture (DAC) processes or indirect capture via biomass. In the latter case, CO_2 is absorbed from the air by photosynthesis during plant growth and released again in subsequent conversion processes. For example, this conversion can be biomass combustion, biogas or bioethanol production. In biogas and bioethanol production, CO_2 is produced as a by-product of the biological conversion processes in relatively high concentrations and only needs to be purified in order to use it for PtX processes. As a by-product of combustion processes, CO_2 is present in the exhaust gas in considerably lower concentrations and must be separated and purified with considerably greater effort [11].

On the contrary, the DAC process is used to capture CO_2 directly from the atmosphere. These processes usually follow three basic steps: first, ambient air is directed onto a sorbent by means of fans which binds the CO_2 to itself. Secondly, CO_2 can be separated from the other substances present in the air. At last, the CO_2 is then separated from the sorbent again, usually under the addition of thermal energy, so that pure CO_2 is available at the end of the process chain. This can then either be injected into the ground (Direct Air Capture with Carbon Storage, DACCS) or used for Power-to-X applications (Direct Air Capture with Carbon Usage, DACCU) [12].

Nitrogen: Air Separation. In case of ammonia, in addition to hydrogen, pure nitrogen (N_2) is required as a feedstock for its production. The provision of nitrogen can be realized by separating it directly from the atmosphere in an air separation plant using the so-called Linde process. In this process, air is gradually liquefied and broken down into its main components, primarily nitrogen and oxygen. This process was developed at the end of the nineteenth century and has since been used on an industrial scale worldwide. Due to the significantly higher nitrogen content in the atmosphere compared to CO_2 (about 78% by volume compared to about 0.04% by volume), this process is characterized by significantly lower specific energy consumption, thus leading to significantly lower costs than CO_2 separation from the atmosphere using DAC processes.

3.2   Power-to-Gas Processes

The generic term Power-to-Gas (PtG) essentially covers processes for the production of synthetic methane or ammonia.

Methane synthesis (methanation). For the production of synthetic methane, CO_2 is required as a feedstock in addition to hydrogen. The two substances are then converted into synthetic methane in a methanation reactor at elevated pressure and temperature (also known as a Sabatier process); water is produced as a by-product of the reaction. This methane can then be used, after liquefaction, as eLNG in suitably equipped ships and thus substitute fossil-based liquefied natural gas (LNG) [15].

The production of synthetic methane has already been successfully tested in Germany for several years in various demonstration plants (e.g., the "Audi E-Gas" plant in Werlte in operation from 2013 to 2021 [16]). Even though there is still potential for optimizing the efficiency of the overall process and reducing costs, the technology is basically mature for industrial use. Large-scale plants for importing e-methane to Germany are currently under development. For example, the company TES plans to import 15 TWh of e-methane per year via Wilhelmshaven by 2030 [17].

Ammonia synthesis. Ammonia synthesis (usually done via the Haber–Bosch process) is a process that has been known since the beginning of the twentieth century and is now used on a large industrial scale in all regions of the world. In addition to hydrogen, ammonia synthesis requires nitrogen, which is usually separated from ambient air by processes that are also available on an industrial scale. Green hydrogen is also needed to produce green ammonia—current ammonia production plants today almost exclusively use gray hydrogen derived from natural gas. In addition, ammonia has so far not been used as a fuel, but as a bulk commodity for the chemical industry—however, the maritime industry in particular sees great (especially economic) potential for ammonia as a fuel for ocean-going shipping [18].

3.3   Power-to-Liquid Processes

The generic term Power-to-Liquid (PtL) covers various synthesis processes for the production of liquid, synthetic fuels. This essentially includes Fischer–Tropsch synthesis for the production of synthetic hydrocarbons (e-kerosene, e-diesel), methanol synthesis, and processes for further processing of methanol into synthetic hydrocarbons, so-called Methanol-to-X processes. Green hydrogen and sustainable CO_2 are again required as feedstocks for all of these processes.

Fischer–Tropsch synthesis. Fischer–Tropsch synthesis has been known since the beginning of the twentieth century and was originally developed for the production of synthetic diesel from coal gasification. Fischer–Tropsch synthesis first requires a synthesis gas consisting mainly of hydrogen and carbon monoxide (CO). To produce e-fuels, this synthesis gas is obtained from the reduction of CO_2 in combination with green hydrogen. This is done via the so-called reverse water–gas shift (RWGS) reaction. The technical implementation of the RWGS reaction currently represents the greatest technical challenge in e-fuel production. The product of the subsequent Fischer–Tropsch synthesis is a mixture of different hydrocarbons, often referred to as synthetic crude oil or syncrude for short.

The syncrude can be further processed in conventional refinery processes (including cracking, isomerization, and distillation) to produce chemical feedstocks such as naphtha (crude gasoline) or common fuels such as kerosene and diesel. This further processing can take place either directly at the syncrude plant or at existing fossil feedstock (i.e., primarily crude oil) processing refineries. If the syncrude is added to crude oil-based products, this is referred to as co-processing. In principle, existing refineries can also be used exclusively for the further processing of syncrude after minor technical modifications to the process control or catalysts used have been made. E-kerosene produced via Fischer–Tropsch synthesis is approved for use in civil aviation and may currently be blended with fossil kerosene in a proportion of up to 50% by volume. In Germany, e-diesel may currently be blended with fossil diesel, but not sold in pure form at service stations. However, the corresponding regulation (10. BImSchV) is currently being revised to also allow the use in its pure form (100%).

Today, large-scale Fischer–Tropsch plants exist, but so far they have been exclusively operated based on fossil feedstocks such as coal (Coal-to-Liquids, CtL) or natural gas (Gas-to-Liquids, GtL). The world's two largest GtL plants, QatarEnergy and Shell's Pearl GtL plant and QatarEnergy and Sasol's ORYX GtL plant, both in Ras Laffan, Qatar, have a combined production capacity of 8.5 million tons per year. Work has also been underway for several years to use biomass (Biomass-to-Liquids, BtL) or waste materials (Waste-to-Liquids, WtL) as feedstock for Fischer–Tropsch synthesis. The first such industrial plants are currently being built or have recently gone into operation in the USA and elsewhere [19]. They are significantly smaller than the aforementioned plants for processing fossil feedstocks (a few 100,000 metric tons of production per year). The first demonstration plants for the production of e-fuels via Fischer–Tropsch synthesis, which are also much smaller (production capacity of about 350 metric tons per year), are being commissioned in Germany [20]. Further larger PtL plants are currently being planned.

Methanol synthesis. Methanol is one of the world's most widely produced organic chemical feedstocks. There is an installed production capacity of more than 100 million tons of methanol per year and large-scale methanol production (up to 10,000 tons per day) is a well-known modern process. These conventional methanol production plants typically use natural gas or coal as feedstock.

Renewable methanol can be produced by providing a non-fossil synthesis gas (similar to Fischer–Tropsch synthesis) without significant changes to the established synthesis process. Again, there are several commercial biomass or municipal waste conversion plants to produce methanol currently planned or under construction. In addition to the use of synthesis gas, however, there are also synthesis processes that can convert hydrogen and CO_2 directly, i.e., without prior reduction of CO_2 to carbon monoxide (CO) as feedstocks. In addition to smaller demonstration plants for the production of e-methanol, for example in Iceland (production capacity about 4,000 metric tons per year), much larger plants are currently being planned [21, 22].

Methanol-to-Fuels. In addition to direct use as a fuel, methanol can be further processed into hydrocarbon drop-in fuels such as gasoline or kerosene via various process variants. Here, too, the processes are often named after the target product: e.g., Methanol-to-Gasoline (MtG) for the production of gasoline and Methanol-to-Jet (MtJ) for the production of kerosene. These processes are similar in terms of their basic process technology and structure but are currently at different levels of technical maturity. For example, e-kerosene via the MtJ route has not yet been approved for use in aviation; a corresponding certification process was initiated at the beginning of 2023 [23].

Plants to synthesize gasoline from methanol have been operating for the past century, and a first demonstration plant (with a production capacity of about 600 tons per year) for the production of e-fuels came on stream in Chile at the end of 2022 [24]. In contrast, the process for producing e-kerosene from methanol is still at the research stage and only laboratory-scale plants exist to date.

3.4   Demarcation from Biofuels

In addition to e-fuels, advanced biofuels represent another viable—at least transitional—and potentially sustainable option for reducing transport-related GHG emissions. Compared to conventional biofuels such as biodiesel or bioethanol produced from food or fodder crops, they are produced from residual and waste materials such as waste cooking oil, manure, straw, waste and residual wood, and similar feedstock.

In principle, different technical processes are available for the production of advanced biofuels. The main options can be distinguished as follows:

- Production of advanced biogas or biomethane from manure or straw in biogas plants with subsequent purification and conditioning into bio-CNG or bio-LNG, respectively. Although already a fuel itself, advanced biogas or biomethane can be further processed into synthesis gas by reforming steps, which can then be further converted into various advanced biofuels, such as methanol, kerosene, or diesel, through additional conversion processes.
- Production of advanced bioethanol from lignocellulosic feedstock such as straw. This bioethanol can again be directly used as fuel or be further converted into other advanced biofuels, including kerosene and diesel.

- Conversion of used cooking oils or animal fats to hydrogenated vegetable oil (HVO) for road transport or hydrotreated esters and fatty acids (HEFA) for air transport, respectively.
- Gasification of materials such as straw, waste/residual wood, or municipal solid waste (MSW) followed by fuel synthesis via so-called Biomass (BtL) or Waste-to-Liquid (WtL) processes, which are comparable to methanol or Fischer–Tropsch synthesis for the production of e-fuels.

Due to their technical complexity, the respective conversion technologies show a different degree of technological maturity. HVO or HEFA production is already state of the art and widely applied on a large scale globally. Biogas from manure is also produced on a commercial scale, e.g., in Germany. Yet, only a few industrial plants exist for the conversion of straw. Commercial BtL or WtL plants are not yet in operation in Germany and Europe, even though there have been many research and development projects and up-scaling projects in this area in recent years. However, first commercial projects are in operation, e.g., in the US [19, 25].

However, providing the necessary biomass feedstock, which is typically produced decentralized, can be quite costly and—comparably to PtX fuel production—the subsequent conversion into fuels is often both technically complex and very energy intensive. In addition, the availability of biomass from wastes and residues, for which land competitions do not occur, is expected to be too limited in order to supply fuel for transportation purposes in sufficient quantities. Additionally, such feedstock usually accrue very locally, which results in additional efforts regarding the feedstock supply chain, such as very decentralized collection systems [26–28].

Additionally, so-called hybrid processes, which integrate at least a share of renewable electricity or green hydrogen into the final energy content of the product, are also discussed. In this regard, two different approaches can be distinguished:

- Substituting (gas) fired heaters, e.g., in steam methane reformers, by directly electrified processes.
- Integration of additional green hydrogen into biofuel conversion processes, such as biomass gasification to enhance the overall carbon efficiency, among other benefits. This approach is often referred to as Power-and-Biomass-to-Liquid (PBtL) processes.

The first option is mainly suitable with respect to the production of hydrogen or syngas, since they are mainly produced via thermo-chemical conversion pathways, where the needed reaction heat is provided by gas fired heaters. This may be realized by reforming processes supplying external energy via (renewable) electricity [29, 30] or methane pyrolysis (i.e., the decomposition of methane into carbon and hydrogen) [31]. If biomethane is used as a feedstock the produced gas stream may account for as renewable.

Contrary to this, PBtL processes combine the conversion of biomass to biofuels with green hydrogen produced via electrolysis utilizing green electricity. In doing so, common biofuel production processes are combined with the key process step of

PtX processes—water electrolysis. On the one hand, this can be achieved by substituting hydrogen derived from natural gas (known as gray hydrogen) in downstream refining processes, such as hydroprocessing, with green hydrogen. This is seen as a promising option to significantly reduce the overall lifecycle GHG emissions of biofuels produced from biogenic oils and fats. On the other hand, by introducing green hydrogen to the synthesis gas, e.g., from biomass or waste gasification processes, the hydrogen-to-carbon ratio of synthesis gas can be enhanced. This improvement enhances the overall carbon efficiency of such processes and can optimize the overall process.

Although both options for such hybrid processes are seen to be promising pathways to efficiently produce renewable fuels with a comparably low GHG footprint, they are not realized in industrial scales so far. This results at least to some extend from still unclear regulatory frameworks, such as quota obligations for advanced biofuels or e-fuels, not especially considering fuels from such hybrid processes. Strictly speaking, these processes do neither fall into the category of PtX processes nor biofuels. In general legislation (e.g., the European renewable energy directive, EU RED), such hybrid processes are not especially covered until today.

4   Classification of Power-to-X Products

From a technical point of view, PtX processes can be build up on any available hydrogen and CO_2 source. However, to contribute to international climate mitigation goals and thereby limit global warming, these feedstocks also have to be supplied from sustainable sources. Therefore, discussed sources of hydrogen (i.e., the so-called hydrogen rainbow) as well as potential CO_2 sources are categorized and discussed in the following. Afterwards, sustainable feedstock combinations for a carbon neutral PtX production are derived. As nitrogen for ammonia synthesis is already extracted from the air, it is not further discussed within this section.

4.1   The Hydrogen Rainbow

As the importance of hydrogen for a successful energy transition has grown within recent years, a color code to categorize different hydrogen sources has evolved, the so-called hydrogen rainbow. Although no globally standardized classification has been agreed on and minor differences exist within the international excursus, the following colors of hydrogen are commonly referred to.

Green hydrogen. As mentioned before, green hydrogen is produced via water electrolysis, utilizing electricity from renewable energy sources. Mainly solar and wind power are discussed for large-scale applications; however, also waterpower generation may be used, especially within the market ramp-up phase. As the lifecycle GHG emissions of hydrogen from electrolysis is dominated by the emission factor of the utilized source of electricity, green hydrogen can be seen as nearly GHG neutral [32].

Red hydrogen. If electricity generated by nuclear power is used for water electrolysis, the so-produced hydrogen is usually referred to as red hydrogen (sometimes pink hydrogen). Since nuclear power can be seen as carbon neutral in a wider sense, red hydrogen also is mostly carbon neutral. However, all adverse environmental aspects of nuclear power generation have to be attributed to the red hydrogen as well, whereas it is generally not seen as a genuinely sustainable option.

Orange hydrogen. Hydrogen can also be produced from biogenic sources via thermo-chemical (e.g., gasification) or biological processes (e.g., fermentation). In general, orange hydrogen can also be seen as carbon neutral, but the overall lifecycle impact is strongly linked to the biomass used as feedstock. Additionally, biomass always also contains large carbon shares; therefore, it is highly questionable if hydrogen production from biomass is a sustainable option, especially if the hydrogen is combined with CO_2 from other sources later within a PtX process. However, the carbon might be sequestrated (bioenergy with carbon capture and storage, BECCS), leading to negative CO_2 emissions.

Grey hydrogen. The most common way to produce hydrogen for industrial applications today is via steam methane reforming (SMR) of natural gas. The so-produced hydrogen is referred to as grey hydrogen. A by-product of the reforming process is CO_2, which is usually released into the atmosphere. Additionally, the main component of natural gas, i.e., methane, is a quite potent greenhouse gas (depending on the timeframe its global warming potential (GWP) is at least 30 times higher than that of CO_2). Due to leakages during natural gas extraction and methane supply, its overall lifecycle impact is usually underestimated [33]. Due to these facts, grey hydrogen is characterized by high GHG emissions.

Blue hydrogen. If the CO_2 released during the SMR process is captured and stored, the so-produced hydrogen is called blue. Therefore, lower lifecycle GHG emissions compared to gray hydrogen can be achieved. However, as natural gas is still used as feedstock, blue hydrogen may be associated with a high GWP due to the aforementioned effects and therefore cannot be seen as climate neutral [33].

Turquoise hydrogen. Another technical option to produce hydrogen from methane is the so-called methane pyrolysis. During this high-temperature process, methane is split up into hydrogen and carbon. The hydrogen is referred to as turquoise hydrogen, the carbon can be used as a feedstock for different applications or sequestrated to produce nearly carbon neutral hydrogen. If biomethane is used as feedstock (instead of natural gas), sequestration can even lead to negative CO_2 emissions (i.e., it can act as carbon sink). However, due to the GWP of methane discussed before and the high energy demand of methane pyrolysis, overall lifecycle emissions are comparably high until today [31].

Due to the discussed drawbacks of most hydrogen production options, it is widely agreed on that green hydrogen is the most sustainable option to supply sufficiently large amounts of hydrogen within a future, climate-neutral energy system.

4.2   Carbon Sources for PtX Applications

Similar to the hydrogen supply options discussed before, different CO_2 options are currently discussed as potential carbon sources for PtX applications. However, although discussions about the sustainability of these sources have gained more importance in recent years, no color-code classification similar to that of hydrogen has been established to date. Therefore, the different carbon sources are usually classified based on their origin.

Atmospheric CO_2. CO_2 can be directly extracted from ambient air via direct air capture (DAC) technologies. Although the concentration of CO_2 in the atmosphere is constantly increasing—thereby accelerating global warming—it is still very low (roughly 420 ppm) in terms of technical separation processes. Therefore, DAC technologies are characterized by a high specific energy demand, leading to high specific costs. If this energy is supplied from renewable sources, the supplied CO_2 is characterized by very low lifecycle GHG emissions. If used for PtX applications, a closed carbon loop can be achieved by utilizing atmospheric CO_2. Therefore, it is seen as the preferable carbon source in a long-term perspective.

Biogenic CO_2. Since biogenic matter is build up by converting CO_2 into carbon matter via photosynthesis, biomass is another option to utilize atmospheric CO_2. The carbon bound within the biomass can then again be converted into so-called biogenic CO_2 by different processes. These can be combustion processes (e.g., within power or heat plants) or biogenic processes (e.g., biogas or bioethanol production). Although the CO_2 is a by-product and depending on the attribution of CO_2 emissions are usually not associated with any upstream emissions, a sustainable biomass supply is very important for the biogenic CO_2 to be really considered as carbon neutral. However, similar to atmospheric CO_2, a closed carbon loop and thereby carbon neutral PtX products can be achieved by utilizing biogenic CO_2.

Industrial point sources. Some industrial production processes, such as the mineral industry (e.g., cement production) or the pulp and paper industry, are characterized by unavoidable CO_2 emissions, as these are inherent to the actual production processes. The emissions of such processes will most likely be significantly reduced within the next decades due to efficiency gains or innovative production process; however, some process inherent CO_2 emissions will still remain. The CO_2 emissions of other production processes currently responsible for high process-related emissions, such as iron and steel production or the chemical industry, however, will most likely be completely mitigated within the next decades due to global decarbonization efforts [3]. Although utilizing such CO_2 sources in PtX pathways results in additional CO_2

being released into the atmosphere, the cascade utilization of this CO_2 still contributes to climate mitigation. This is due to the substitution of fossil energy carriers by PtX products in such cases. However, given the limited availability of these unavoidable industrial CO_2 point sources, they might rather serve as a short-term option during the initial ramp-up of a PtX market.

CO_2 from fossil point sources. CO_2 from the combustion of fossil energy carriers, such as coal, natural gas, or crude oil-based fuels, within power plants, is usually referred to as fossil CO_2 point sources. Although such power plants usually emit large amounts of CO_2 which could be utilized in an economically feasible way, their utilization within PtX applications would still lead to additional fossil CO_2 emissions. In contrast to the industrial point sources discussed before, fossil point sources themselves have to be substituted by renewable sources of energy within the next decades. Therefore, they are not seen to be a long-time feasible CO_2 source for PtX applications. To avoid lock-in effects on fossil energy infrastructures, they are however not seen as a sustainable sound CO_2 source.

Overall, while some industrial CO_2 point sources might be feasible during the PtX market ramp-up, a truly sustainable and predominantly climate-neutral PtX production requires a closed carbon loop. This can only be achieved by utilizing atmospheric CO_2—ideally through DAC technologies or, to a lesser extend due to feedstock limitations, from biogenic sources. Consequently, these are seen as the aspirational carbon sources for future PtX production.

4.3   Sustainable Carbon-Based PtX Combinations

The environmental impact of PtX products, and consequently their potential for climate mitigation, is closely interlinked to the utilized feedstock, namely hydrogen and CO_2. Thus, by combining the hydrogen rainbow as well as the classification of sustainable carbon sources defined above, feedstock combinations with the highest suitability to mitigate further global warming can be derived. The resulting matrix of sustainable combinations of hydrogen and carbon sources for PtX applications is visualized in Fig. 4. However, it is crucial to note that this is a basic qualitative assessment only and the actual sustainability of a PtX project might differ from this classification.

Within a stricter view, only the utilization of green and red hydrogen to produce synthetic fuels qualifies the produced fuels as PtX fuels. Process combinations using non-electrolytic hydrogen (i.e., blue, orange, gray, or turquoise), the feedstock already contains both hydrogen and carbon. Producing hydrocarbon fuels from such feedstock makes it impractical to separate the carbon and supply only hydrogen (increased process complexity and energy demand). Rather in the context of a broader hydrogen economy and supply with a developed distribution infrastructure, there could be cases where a dedicated supply of hydrogen from such feedstocks

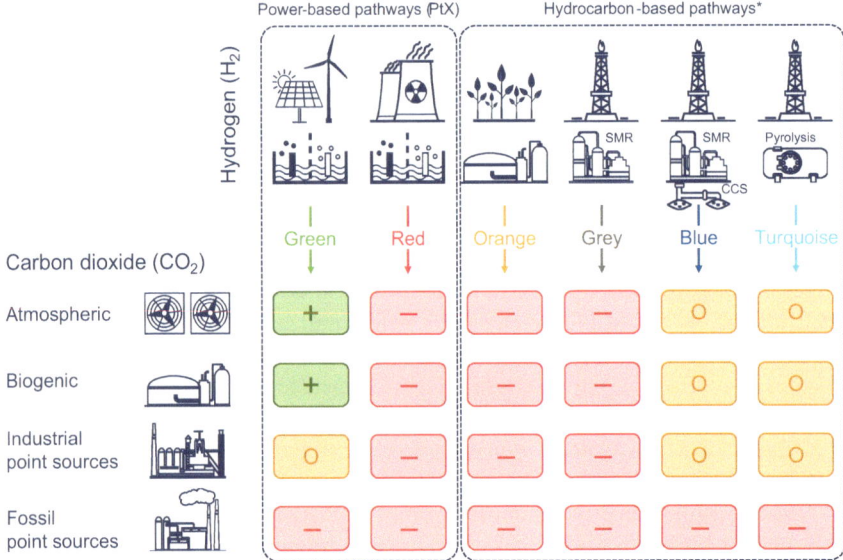

Fig. 4 Potential combination of hydrogen and CO_2 sources for PtX production and their classification regarding highest climate mitigation effect. *In these pathways, feedstocks already contain both hydrogen and carbon. Producing hydrocarbon fuels from such feedstocks makes it impractical to separate carbon and supply only hydrogen (increased process complexity and energy demand). Rather in context of a broader hydrogen economy and supply with a developed distribution infrastructure, there could be cases where a dedicated supply of hydrogen from such feedstocks is feasible (e.g., if it is sourced from various origins through a common distribution system like a pipeline). (green/+: preferable combination under sustainability aspects; yellow/o: potential application for a PtX market ramp-up; red/–: no preferable combination under sustainability aspects. Overview of Power-to-X (PtX) production processes; SMR = steam methane reforming; CCS = carbon capture and storage)

is feasible (e.g., if it is sourced from various origins through a common distribution system like a pipeline).

According to this classification, the combination of green hydrogen with biogenic or atmospheric CO_2 is the most sustainable option and should be the long-term goal for a future large-scale PtX economy. However, as mentioned before, sustainable biogenic CO_2 sources are scarce and DAC systems are far from market maturity today. Therefore, the near-term availability of PtX products from these combinations will be limited to biogenic CO_2 point sources or be even more expensive than other combinations if small scale DAC systems are utilized. Affordable DAC systems (i.e., CO_2 supply costs in the order of 100 €/t$_{CO_2}$) in sufficient scales ($\geq$ 100,000 t$_{CO_2}$/a) and thereby the silver bullet of CO_2 supply for PtX production are rather expected to be available in the second half of this century.

However, to still realize a swift PtX market ramp-up in order to defossilize sectors that most likely will continue to rely on high density energy carrier in the future, it might be necessary to accept hydrogen–carbon combinations with some environmental drawbacks as a compromise. In this regard, especially the utilization of industrial CO_2 point sources in combination with green hydrogen are discussed intensely. Such point sources pledge high amounts of low-cost CO_2 at defined locations and therefore can lead to comparably low PtX production costs, especially at the beginning of the international market ramp-up. However, due to their adverse environmental effects (i.e., mainly the release of additional fossil CO_2 into the atmosphere), they can only be pursued for a limited period of time. Additionally, it must be avoided that the decarbonization of distinct industrial branches (such as steel production or the chemical industry) is delayed due to so-called lock-in effects.

Much like the anticipated availability of sustainable CO_2 sources within the next decade, the global expansion of electrolyzer capacities and corresponding electricity production from renewable sources of energy for green hydrogen production may face delays. These can, for instance, arise from regulatory uncertainties and resulting concerns about financing risk. Moreover, a global competition for the limited availability of green hydrogen might impede the ramp-up of PtX production capacities—at least in the near future. Therefore, the supply of blue and turquoise hydrogen may be considered as an option to expedite a PtX market ramp in the near term, while it is crucial to impose temporal limitations for its use in the medium to long term at the same time. Particularly when combined with biogenic or atmospheric CO_2 or (to a much lower extent) industrial point sources, this might be an option with a comparably good climate mitigation potential. However, besides the inherent drawbacks of these combinations, it is essential to minimize methane leakages along the natural gas supply chain to ensure its viability from a climate perspective.

5 Conclusion

Within this chapter, general technical aspects of PtX and especially Power-to-Gas (PtG) and Power-to-Liquid (PtL) processes have been described. Additionally, they have briefly been delimited from biofuels as further renewable fuel options. Adding up to this, a classification of Power-to-X products based on the hydrogen rainbow and potential carbon sources is presented. According to this, the combination of biogenic or atmospheric CO_2 with green hydrogen results into the highest potential to mitigate further climate change. However, other CO_2 and low-carbon hydrogen sources might be viable short-term options required for a timely ramp-up of PtX production capacities, provided that strict sustainability and environmental protection measures are taken. Yet, it is important to note that this is purely a qualitative indication, and the actual climate impact or the corresponding GHG emissions of a specific PtX production layout depends on various different aspects, potentially leading to different conclusions.

References

1. United Nations (2015) Paris Agreement
2. European Parliament (2019) CO_2-emissions from cars: facts and figures, 22 März [Online]. Available: https://www.europarl.europa.eu/news/de/headlines/society/20190313S TO31218/co2-emissionen-von-autos-zahlen-und-fakten-infografik [Zugriff am 28 Dezember 2022]
3. Zitscher T, Neuling U, Habersetzer A, Kaltschmitt M (2020) Analysis of the German industry to determine the resource potential of CO_2 emissions for PtX applications in 2017 and 2050. Resources, Bd. 9, Nr. 12
4. Billig E, Decker M, Benzinger W, Ketelsen F, Pfeifer P, Peters R, Stolten D, Thrän D (2019) Non-fossil CO_2 recycling—the technical potential for the present and future utilization for fuels in Germany. J CO_2 Utilization 30:130–141
5. Agora Verkehrswende (2024) E-Fuels : Separating the substance from the hype - How electricity-based synthetic fuels can contribute to the energy transition in transport. Berlin
6. DLR, TUHH, JBV (2021) Konzeptionelle und technische Ausgestaltung einer Entwicklungsplattform für Power-to-Liquid-Kraftstoffe - Abschlussbericht," Im Auftrag des Bundesministeriums für Verkehr und digitale Infrastruktur (BMVI)
7. DECHEMA (2021) 3. Roadmap des Kopernikus-Projektes „P2X": Phase II - Technischer Anhang
8. The Federal Government (2021) PtL roadmap—Sustainable aviation fuel from renewable energy sources for aviation in Germany
9. Brynolf S, Hansson J, Anderson JE, Skov IR, Wallington TJ, Grahn M, Korberg AD, Malmgren E, Taljegård M (2022) Review of electrofuel feasibility—prospects for road, ocean, and air transport. Prog Energy 4:12
10. IEA (2022) Global hydrogen review
11. Fröhlich T, Blömer S, Münter D, Brischke L-A (2019) CO_2-Quellen für die PtX-Herstellung in Deutschland - Technologien, Umweltwirkung, Verfügbarkeit, IFEU, Heidelberg
12. Viebahn P, Scholz A, Zelt O (2019) The potential role of direct air capture in the German energy research program—results from a multi-dimensional analysis. Energies 12:18
13. Jones E, Qadir M, van Vliet MT, Smakhtin V, Kang S-M (2019) The state of desalination and brine production: a global outlook. Sci Total Environ 657:1343–1356
14. Asghari E, Abdullah MI, Foroughi F, Lamb JJ, Pollet BG (2021) Advances, opportunities, and challenges of hydrogen and oxygen production from seawater electrolysis: an electrocatalysis perspective. Current Opinion Electrochem
15. Comer B, O'Malley J, Osipova L, Pavlenko N (2022) Comparing the future demand for, supply of, and life-cycle emissions from bio, synthetic, and fossil LNG marine fuels in the European Union. In: ICCT. Washington, DC
16. Audi AG, Audi e-gas [Online]. Available: https://www.audi-technology-portal.de//en/mobility-for-the-future/audi-future-lab-mobility_en/audi-future-energies_en/audi-e-gas_en [Zugriff am 12 Oktober 2023]
17. Tree Energy Solutions (TES), Green energy, naturally [Online]. Available: https://tes-h2.com/ [Zugriff am 12 Oktober 2023]
18. ABS; CE-Delft; Arcsilea (2022) Potential of ammonia as fuel in shipping. EMSA
19. Cision PR Newswire, Fulcrum bioEnergy successfully produces first ever low-carbon fuel from landfill waste at its Sierra BioFuels Plant [Online]. Available: https://www.prnewswire.com/news-releases/fulcrum-bioenergy-successfully-produces-first-ever-low-carbon-fuel-from-lan dfill-waste-at-its-sierra-biofuels-plant-301707331.html [Zugriff am 12 Oktober 2023]
20. Atmosfair, Anlage [Online]. Available: https://fairfuel.atmosfair.de/anlage-techn-details/ [Zugriff am 31 März 2023]
21. Climeworks, Orca: the first large-scale plant [Online]. Available: https://climeworks.com/roa dmap/orca [Zugriff am 12 Oktober 2023]
22. Climeworks, Mammoth: our newest facility [Online]. Available: https://climeworks.com/plant-mammoth [Zugriff am 12 Oktober 2023]

23. Biofuels Central, Nacero is developing a world scale methanol to jet fuel complex with TOPSOE support using their MTJet technology for sustainable aviation fuel [Online]. Available: https://biofuelscentral.com/nacero-developing-world-scale-methanol-to-jet-fuel-com plex-with-topsoe-support-using-their-mtjet-technology-for-sustainable-aviation-fuel/ [Zugriff am 13 April 2023]
24. HIF, HIF global and its partners celebrate the first liters of synthetic fuels from Haru Oni, Chile [Online]. Available: https://hifglobal.com/wp-content/uploads/2022/12/HIFGlobal-and-its-partners-celebrate-the-first-litersof-synthetic-fuels-from-Haru-Oni-Chile.pdf [Zugriff am 22 August 2023]
25. Enerkem, Facilities & projects [Online]. Available: https://enerkem.com/company/facilities-projects/ [Zugriff am 12 Oktober 2023]
26. ICCTcct (2018) Bioenergy can solve some of our climate problems, but not all of them at once
27. IFEU, IZES, Öko-Institut (2019) BioRest: Verfügbarkeit und Nutzungsoptionen biogener Abfall- und Reststoffe im Energiesystem (Strom-, Wärme- und Verkehrssektor), Umweltbundesamt, Dessau-Roßlau
28. NRW.Energy4Climate GmbH (2023) Nachhaltiger Einsatz von Biomasse - Die Rolle von Biomasse in der Energiewende und in einer klimaneutralen Industrie, Düsseldorf
29. BASF(2021) BASF, SABIC and Linde join forces to realize the world's first electrically heated steam cracker furnace [Online]. Available: https://www.basf.com/global/en/who-we-are/sus tainability/whats-new/sustainability-news/2021/basf-sabic-and-linde-join-forces-to-realize-wolds-first-electrically-heated-steam-cracker-furnace.html [Zugriff am 28 Dezember 2022]
30. SYPOX (2020)Technology [Online]. Available: https://www.sypox.eu/technology [Zugriff am 28 Dezember 2022]
31. Timmerberg S, Kaltschmitt M, Finkbeiner M (2020) Hydrogen and hydrogen-derived fuels through methane decomposition of natural gas—GHG emissions and costs. Energy Conv Manage: X 7:100043
32. Weidner T, Tulus V, Guillén-Gosálbez G (2023) Environmental sustainability assessment of large-scale hydrogen production using prospective life cycle analysis. Int J Hydrogen Energy 48(22):8310–8327
33. Bauer C, Treyer K, Antonini C, Bergerson J, Gazzani M, Gencer E, Gibbins J, Mazzotti M, McCoy ST, McKenna R, Pietzcker R, Ravikumar AP, Romano MC, Ueckerdt F, Vente J (2022) On the climate impacts of blue hydrogen production. Sustain Energy Fuels 6:66–75

Reverse Water–Gas Shift for Synthesis Gas Provision—A Core Technology for Powerfuel Production

Steffen Voß, Stefan Bube, Nils Bullerdiek, and Martin Kaltschmitt

Abstract The supply of synthesis gas (syngas) plays a crucial role in the synthesis-based production of hydrocarbon-based liquid and gaseous energy carriers. The production of sustainable carbon-based products requires sustainable carbon feedstocks, such as CO_2 from biogenic sources or the atmosphere. Due to the thermodynamic stability of CO_2, its activation for chemical conversion is complex and potentially costly. In the reverse water–gas shift reaction (RWGS), CO is reduced to CO using hydrogen, whereby synthesis gas can be provided for downstream synthesis. This article examines the fundamentals of RWGS, reactor concepts, and integration into synthesis processes. The provision of thermal energy for adequate CO conversion and minimal methanation side reactions is of crucial importance. While fired RWGS reactors are used industrially, electrically heated reactors are still in the pilot stage.

Keywords Reverse Water–Gas Shift (RWGS) · Synthesis gas · CO_2 activation · Synthesis

1 Introduction

The provision of synthesis gas (syngas) plays a decisive role in the synthesis-based production of liquid and gaseous hydrocarbon-based energy carriers and/or bulk chemicals. Synthesis gas, which is already provided today by conventional technologies for the production of hydrocarbons with specific characteristics, alcohols, ethers, and other organic hydrocarbon-based chemical compounds, is mainly a mixture of carbon monoxide (CO) and hydrogen (H_2). The specific ratio between these two gases typically varies depending on the requirements of the particular downstream synthesis process. In addition to these main components needed primarily for the subsequent synthesis, it may also include other gas components, such as carbon

S. Voß (✉) · S. Bube · N. Bullerdiek · M. Kaltschmitt
Hamburg University of Technology (TUHH), Institute of Environmental Technology and Energy Economics (IUE), Hamburg, Germany
e-mail: steffen.voss@tuhh.de

© The Author(s), under exclusive license to Springer Nature Switzerland AG 2025   515
N. Bullerdiek et al. (eds.), *Powerfuels*, Green Energy and Technology,
https://doi.org/10.1007/978-3-031-62411-7_19

dioxide (CO_2), methane (CH_4), and water (H_2O); these additional components can be further converted within the downstream synthesis, may act inertly and/or, can affect subsequent processes.

Today, synthesis gas is mainly produced based on natural gas, hard coal, and residual streams of crude oil processing (i.e., petrol coke) [1]. However, the use of these fossil, carbon-containing energy carriers, which serve as both an energetic as well as a material basis of the provision of synthesis gas, is necessarily linked to CO_2 emissions contributing to an additional increase in atmospheric carbon content (atmospheric CO_2 concentration) due to their fossil origin. Thus, if the climate change issue is taken seriously, these emissions need to be avoided.

Depending on the feedstock and process technology, the chemical composition of the produced (raw) synthesis gas can vary widely. Adjustments to the respective composition can be made downstream, for example, via the water–gas shift reaction or by adding hydrogen (e.g., from electrolysis processes). Currently, it is common to utilize part of the fossil energy-based feedstock to supply additionally the process energy, primarily in order to heat the reactors (i.e., to provide the reaction enthalpy of the endothermic reactions) [1].

In light of the drive toward defossilization within the energy and various other industrial sectors, a continued use of these fossil fuel-based resources for synthesis gas production cannot be continued, necessitating the exploration and exploitation of alternative and especially sustainable resources. Among such alternative options, synthesis gas provision based on sustainably provided biogenic resources (biomass), such as by gasification of lignocellulosic biomass, presents a viable alternative pathway. In addition, carbon dioxide (CO_2), e.g., from biogenic point sources (e.g., bioethanol plants, biogas plants) or from direct air capture (DAC), itself serves as a carbon-containing resource for such synthesis reactions. But, CO_2 is thermodynamically very stable and thus inactive in many synthesis reactions; i.e., there is a need that CO_2 is "activated" upstream of the synthesis reaction [2]. Therefore, using CO_2 as a carbon material feedstock for the provision of synthesis gas presents decisive differences compared to using fossil-based carbon resources. Specifically, using CO_2 as a raw material requires both the separate provision of hydrogen as well as the supply of reaction enthalpy. Also, for the provision of the hydrogen needed for synthesis gas production, additional upstream infrastructure is required. On a process level, the activation of CO_2 with the addition of hydrogen and the resulting reduction to CO with the formation of water as a by-product is possible via the so-called reverse water–gas shift (RWGS) reaction [3]. The main product of such a reverse water–gas shift (RWGS) reaction being CO is then available for synthesis reactions with the optional addition of further hydrogen depending on the desired spectrum of the respective synthesis products. Against this background, Fig. 1 shows a schematic and general overview of the RWGS technology in the context of a power-based provision of synthetic fuels. It becomes clear that this technology forms a core component of the overall process chain starting from natural resources like water and "renewable" electricity as well as sustainable CO_2 on the educt side and synthetic provided hydrocarbons on the product side.

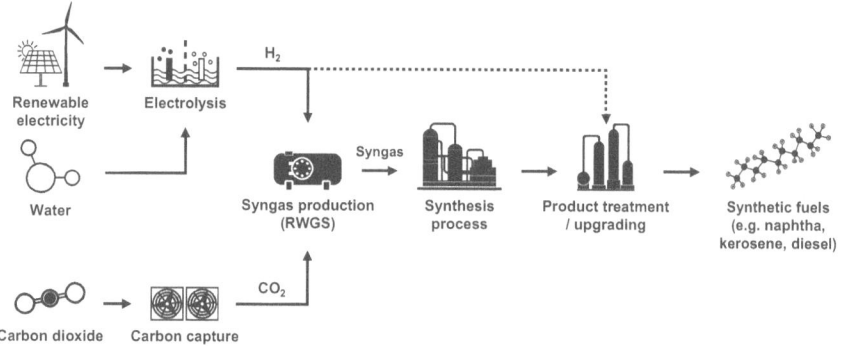

Fig. 1 Schematic overview of RWGS technology in production of power-based synthetic fuels (RWGS: reverse water–gas shift)

Within this context, this article provides an overall introduction to the reverse water–gas shift (RWGS) reaction and presents approaches for its technological integration into power-based fuel production. Therefore, first potential feedstock options are presented, followed by the requirements and objectives of synthesis gas as the product. Subsequently, basics of the reverse water–gas shift (RWGS) reaction and its technical implementation are described in more detail, followed by an illustration of its integration into a Power-to-Liquid process system. The paper concludes by summarizing the results and by giving an outlook on further technical developments.

2 Feedstock

In the following, the provision and material-related properties of the feedstocks (or main input streams) for the RWGS are briefly described; a breakdown of the provision of carbon dioxide and hydrogen is given.

2.1 Carbon Dioxide (CO₂)

Carbon dioxide (CO_2) can be obtained from a wide range of different sources or origins. In the following, various options are categorized into concentrated and highly diluted as well as stationary and diffuse sources, respectively.

Concentrated, stationary CO_2 sources. These CO_2 sources include industrial or energy conversion plants that emit CO_2 (e.g., bioethanol plant, biogas plant, biomass cogeneration plant). Therefore, the provision of a pure, highly concentrated CO_2 steams based on such sources requires a technical treatment through carbon capture

(CC) involving the separation of impurities and dilution components. Due to thermodynamic constraints, lower CO_2 concentrations in the feed initially result in higher technical efforts to separate and isolate the CO_2, which in turn leads to a higher energy demand for CO_2 provision [4].

Within such highly concentrated CO_2 sources, bioethanol plants typically reach the highest direct CO_2 concentration, reaching up to 100 vol.-% in a by-product stream; but, typically, this CO_2 stream is polluted at least by water vapor. Biogas plants produce a product mixture consisting of approximately 40–50 vol.-% CO_2 and the remaining portion primarily composed of methane; within existing biomethane plant, this CO_2 is removed anyway (and released so far into the atmosphere) to fulfill the demands for a feed into the public natural gas grid. Another biogenic, highly concentrated point source of CO_2 is produced during biomass combustion, whose exhaust gases contain CO_2 concentrations of 10–15 vol.-%.

The carbon material input for all these facilities is derived from biogenic resources (biomass). Thus, in these cases, CO_2 has been absorbed from the atmosphere through photosynthesis. Consequently, the utilization of the respective CO_2 emissions and the further downstream processing of the contained carbon into fuels, along with their subsequent use or combustion, do not contribute to a long-term or overall increase in CO_2 concentration in the atmosphere as long as the biomass is produced in a sustainable way (i.e., no clearing of forests).

During the biochemical or thermochemical conversion of biogenic resources, gaseous by-products, such as hydrogen sulfide (H_2S) or ammonia (NH_3), may be produced. These substances can act as catalyst poisons in the downstream RWGS and synthesis processes [5]. Hence, utilizing CO_2 from the mentioned processes/ sources can also require an in-depth purification of the CO_2 before it can be used as a feedstock for a subsequent synthesis.

Diffuse, highly diluted CO_2 sources. In the case of highly dilute CO_2 sources, such as the atmosphere—with a concentration of approximately 420 ppm [6]—a substantial volume of ambient air must fed through a separation process to capture and obtain the desired amount of CO_2. Due to thermodynamic reasons, the energy demand for the provision of a pure CO_2 stream from ambient air is quite high; nevertheless, technology wise a provision of CO_2 from air is possible and has been demonstrated already.

2.2 Hydrogen (H_2)

In addition to carbon dioxide (CO_2), hydrogen (H_2) is required as a feed for the thermochemical reduction of CO_2 to CO and further downstream synthesis. Today, this hydrogen is mainly produced from the reforming of hydrocarbon components (e.g., natural gas, crude oil fractions) and is used, among others, during various processing steps to transform crude oil into market ready fuels and within numerous chemical syntheses like, e.g., ammonia or methanol production being globally the most important examples [7]. For the provision of sustainable climate-neutral hydrocarbon products, also the supply of hydrogen, as outlined already for the supply of CO_2, must be achieved while largely avoiding the use of fossil fuel resources.

Hydrogen of non-fossil origin can be provided either from biogenic resources used as a feedstock and their subsequent biochemical and/or thermochemical conversion into hydrogen or hydrogen-containing carbon derivates (e.g., biomethane). Alternatively, hydrogen can be provided from water split into hydrogen (and oxygen) via an electrochemical reaction; to ensure that the hydrogen in GHG-neutral the respective technical process (i.e., electrolyzer) need to be powered by electricity from renewable sources of energy.

If biomass (biogenic resources like lignocellulosic material) is the sole source of hydrogen, there is a deficit in hydrogen (and an excess in oxygen) for the synthesis of the desired end product hydrocarbons. Thus, for a complete utilization of the biogenic carbon (being a valuable respectively expensive component) contained within the biomass, it is necessary to provide hydrogen from additional/extern sources (i.e., hydrogen beyond what is contained within the biogenic resource) in order to fully utilize the carbon content of the biogenic feedstock [8]. Such an external hydrogen source can be hydrogen from an electrolyzer powered by "green" energy.

During the electrolysis of water, the water molecule (H_2O) is split into hydrogen (H_2) and oxygen (O_2) (Eq. 1). Depending on the specific electrolysis process, a portion of the required process energy can be supplied in the form heat, while another part must be provided through electrical current [9]. The electrochemical reaction of splitting water occurs in two compartments (electrodes), separated by a membrane (separation layer). In the anode compartment, oxygen is produced, while hydrogen is generated in the cathode compartment. As a result, highly pure H_2 gas is provided during electrolysis, which can be directly used in further downstream processes, depending on the individual application.

$$2 \, H_2O \rightarrow 2 \, H_2 + O_2 \tag{1}$$

Currently, three different electrolysis processes are mainly discussed.

- Alkaline Electrolysis (AEL). The AEL technology is the electrolysis technology with the highest level of technological maturity. In both compartments of the electrolyzer, which are separated by a diaphragm, a KOH solution acting as an electrolyte is present. Based on the lower heating value of H_2, this electrolysis achieves efficiencies of around 63 to $71\%_{LHV}$ [9, 10].
- Polymer Electrolyte Membrane Electrolysis (PEMEL). The PEMEL achieves efficiencies between 60 and $68\%_{LHV}$ being similar to the AEL (see above). A proton exchange membrane separates the anode from the cathode compartment, and the protons can directly move from the anode compartment to the cathode compartment, where H_2 formation takes place [11]. This allows for higher overall H_2 purities, as there is less cross-permeation [9]. In addition, the PEMEL technology enables a higher load-flexibility and can thus respond to fluctuating renewable electricity production [12].

- Solid oxide electrolysis (SOEL). The SOEL can achieve efficiencies greater than $100\%_{LHV}$ based on the LHV of hydrogen and calculated only based on the electrical power input. The reason for that is that the electrolysis operates at high temperatures (700–900 °C) and steam is used as an additional energy input stream beside the electrical energy [9]. If a high-temperature (waste) heat source is available at the production site, the integration of this thermal energy stream can lead to efficiency advantages for this electrolysis technology compared to the other two options.

3   Synthesis Gas

In the following, the desired properties of synthesis gas within synthesis gas production processes are presented and discussed in more detail. Attaining these targeted, adequate properties is crucial for RWGS-based fuel production, thus representing an overall significant objective for RWGS technology.

3.1   Properties

Synthesis gas, often shortened to syngas, is understood to be a gas mixture of different compositions intended to be used as a feedstock for a subsequent chemical synthesis. This gas mixture contains non-oxidized and/or partially oxidized (and in some cases additionally fully oxidized) gas components [13]. These various gas components serve as the feedstock for the downstream located chemical synthesis realized within corresponding synthesis processes.

Depending on the specific synthesis reaction being pursued, this synthesis gas (syngas) may need to contain different components (reactants) in various concentrations or mixing ratios characterized by a predefined level of purity. For instance, the synthesis of hydrocarbon-based fuels and corresponding bulk chemicals primarily requires the reactive main components H_2 and CO, and in some of these syntheses routes currently under discussion, also CO_2. In the context of other syntheses targeting to molecules being not pure hydrocarbons, such as ammonia synthesis, the synthesis gas may also contain other main components, for example, nitrogen (N_2) and H_2.

However, in the following, the term synthesis gas (or syngas) refers to a gas mixture that is mainly based on H_2 and CO, as the key chemical components for various synthesis processes considered within this article.

To evaluate and further compare different types of synthesis gases and their chemical compositions, various H_2:CO ratios can be defined. In addition to the specific synthesis gas ratio (SGR) with regard to H_2:CO (Eq. 2), CO_2 can also be a synthesis gas component. The stoichiometric number (SN, Eq. 3) serves as a characteristic figure for synthesis processes where CO_2 participates within the chemical synthesis

Table 1 Synthesis gas requirements by different synthesis process (SGR synthesis gas ratio; SN stoichiometric number)

Synthesis	SGR	SN
Fischer–Tropsch synthesis (low temperature)	2.0–2.2	–
Fischer–Tropsch synthesis (high temperature)	1–2	
Methanol synthesis	–	2.05–2.15
Dimethyl ether (DME) synthesis	2	–
Methane synthesis	3	–
Mixed alcohol synthesis (MAS)	2	–

reaction, which thus leads to a demand in (additional) hydrogen to directly reduce this CO_2 inside the reactor being an integral part of the respective synthesis process. If this is the case, the amount of CO_2 in relation to the total carbon content, the so-called carbon oxide ratio (COR), can also be considered (Eq. 4). x describes the respective share of the respective gas component.

$$SGR = \frac{x_{H_2}}{x_{CO}} \tag{2}$$

$$SN = \frac{x_{H_2} - x_{CO_2}}{x_{CO} + x_{CO_2}} \tag{3}$$

$$COR = \frac{x_{CO_2}}{x_{CO} + x_{CO_2}} \tag{4}$$

Table 1 shows the synthesis gas ratios (SGR) required for various stoichiometric syntheses. In practically implemented plant concepts, it can happen that a oversto-ichiometric mixture is used due to conversion and selectivity reasons. This table presents a selection of different syntheses that require a stoichiometric synthesis gas composition.

3.2 Modification of Composition

If the synthesis gas, which can come from the reverse water–gas shift reaction or other (conventional) synthesis gas provision pathways (e.g., Methane steam reforming, Petrol Coke gasification), does not have the required composition for the respective downstream synthesis, various process steps can be realized to change its composition. These processing include reforming, CO_2 separation, and the water–gas shift reaction being briefly outlined below.

- **Reforming**. Reforming is used when hydrocarbon components are present in a gas stream to be used as a synthesis gas. For this purpose, an oxidizing agent, such as H_2O (steam reforming), CO_2 (dry reforming), or O_2 (partial oxidation), is added to this gas stream. Then, the hydrocarbon components are reacting with the oxidizing agent under the given process conditions to form primarily CO and H_2. The choice of the respective oxidizing agent thus determines the achievable synthesis gas ratio on the one hand and the required external energy input on the other hand. In case of an endothermic steam reforming, where H_2O is used as the oxidizing agent, synthesis gas ratios (H_2: CO ratios) of 3 and higher can be achieved [14]. In dry reforming processes, where CO_2 is used as the oxidizing agent, a high external energy input is necessary to activate the CO_2 for the intended chemical reaction. The synthesis gas ratios achieved are relatively low (~ 1) as additional hydrogen components are not added if only CO_2 serves as the oxidizing agent [15]. Alternatively, pure oxygen can be used as an oxidizing agent in partial oxidation. The respective exothermic reaction also leads to lower synthesis gas ratios (< 2) but can be operated without external heating. In addition to their individual application, a combination of the described reforming reactions is also conceivable, such as in autothermal reforming, where an initial partial oxidation provides the heat for a subsequent steam reforming [16].

- **CO_2 Separation**. Should the produced raw synthesis gas contain an excess of CO_2 that is undesirable for the downstream synthesis process, separating CO_2 to increase the partial pressure of the reactants might be advisable and thus necessary. Such a CO_2 separation from gas streams can be carried out by different separation processes, such as absorption, adsorption, gas–solid reactions, cryogenic processes, or membrane processes [4]. In large-scale CO_2 separation processes, the Rectisol and Selexol processes being based on a physical absorption of the gaseous CO_2 at a liquid are well-developed technologies widely established on a global scale. In both processes, the gas stream passes through a liquid phase, the so-called washing medium, in which the CO_2 dissolves. In case of the Rectisol process, methanol is used as the washing medium, while a mixture of dimethyl ether (DME) and polyethylene glycol (PEG) serves as the washing medium for the Selexol process. After the CO_2 is separated from the gas phase, it is released at higher temperatures of the respective washing media again, resulting in a pure gas stream of CO_2 for recycling or further processing.

- **Water–Gas Shift**. If it is not possible to add pure (external) hydrogen to the raw synthesis gas stream (e.g., due to system or infrastructure circumstances of a specific process plant), the hydrogen content of the syngas can be adjusted by the addition of water and a subsequent water–gas shift (WGS) reaction. Within this equilibrium reaction, the water–gas shift (WGS) reaction is the forward reaction of the reverse water–gas shift reaction (Chap. 4); here, CO is oxidized with water to CO_2 by forming/releasing hydrogen (Eq. 5). This exothermic reaction takes place at temperatures between 200 and 550 °C and pressures of up to 30 bar. By using iron-chromium (Fe_2O_3/CrO), copper-zinc (Cu/ZnO), or cobalt-molybdenum (CoMo) catalysts, the activation energy is minimized and the reaction rate is accelerated [17, 18].

$$CO + H_2O \rightarrow CO_2 + H_2 \quad \Delta H_R = -41 \, kJ/mol \tag{5}$$

The water–gas shift (WGS) reaction can also be used to produce pure hydrogen out of a synthesis gas stream or gas streams that are rich in CO. To transfer this chemical reaction into a process, two reactor stages are typically used to achieve a high CO conversion [17]. In the first reactor, the water–gas shift reaction is realized at high temperatures of up to 550 °C, whereas in the second reactor the temperature is reduced to 100–300 °C in order to shift the equilibrium to the product side, whereby the reaction rate decreases.

Inert and trace gases. In addition to the main components mentioned above, synthesis gas may also contain other components that can affect further downstream synthesis processes.

One the one hand, these include inert gases, such as nitrogen, methane, helium, and depending on the synthesis, CO_2. Such inert gases do not directly participate in the downstream synthesis, but they still play a role in the quality of the synthesis gas. Such inert gases may reduce the conversion capacity of the downstream processes because they consume volume without contributing to the desired reactions. Additionally, an increased proportion of inert gas directly leads to a reduction in the partial pressure of the reactants, which can thus influence the reaction behavior in the reactor. In addition, the inert gases are separated along with unreacted gas after the synthesis. In this enriched inert gas stream, a portion of the total gas flow must be purged or additionally purified to avoid a up concentration of the inert gas within the synthesis loop.

On the other hand, trace gases, which can come from the reactant of the synthesis gas production, can be present in the synthesis gas. Components like tar, ammonia, or hydrogen sulfide act as catalyst poisons and impair the catalytic conversion through various deactivation mechanisms [19]. Additionally, such trace gases may contribute to blocking of pipes and gluing of valves at selected parts of the downstream processes.

4 Reverse Water–Gas Shift

Below, the reverse water–gas shift (RWGS) reaction is described as a process for producing synthesis gas, utilizing CO_2 and H_2 as feedstocks to generate synthesis gas for various applications and syntheses. Initially, the fundamentals of the RWGS reaction are addressed, followed by a focus of its technical implementation, focusing on resulting demands on the reactor respective the corresponding reactor design. Finally, the integration of this synthesis gas production process into an overall synthesis process is discussed.

4.1   Thermodynamic Basics

Within the RWGS reaction (Eq. 6), CO_2 is chemically activated by its reduction to CO with the aid of H_2 by forming H_2O. This reaction is the reverse reaction of the WGS reaction (Eq. 5). The rate of both reaction directions is equilibrium-dependent; i.e., the chemical conversion is limited by the respective equilibrium composition. The endothermic RWGS reaction is characterized by a specific heat requirement of 41 kJ/mol being equivalent to the heat released in the WGS reaction under standard conditions [20].

Stoichiometrically, the same amount of H_2 as CO_2 is required for such a chemical reaction. Given the fact that the product mixture of a RWGS reactor is typically directly processed further, and a proper synthesis gas ratio tailored for the subsequent synthesis (e.g., Fischer–Tropsch or methanol synthesis) must be set (e.g., *SGR* ~ 2; Table 1), the hydrogen required for the synthesis can already be added to the RWGS reactors in excess. This excess hydrogen can be used to favor a high rate of CO_2 conversion within the RWGS [21, 22].

$$CO_2 + H_2 \rightarrow CO + H_2O \; \Delta H_R = 41 \, \text{kJ/mol} \tag{6}$$

In this reaction system on the basis of CO_2 and H_2, various side reactions can occur in the presence of catalysts. The main side reactions are the Sabatier reaction (Eq. 7) and the CO-methanation (Eq. 8), where methane is formed from CO_2 or CO and H_2. The methane formed acts as an inert gas in the subsequent synthesis and is therefore undesirable unless methane is the intended final synthesis product (e.g., within a methane synthesis). These (undesired) chemical side reaction increases with an increasing pressure and a growing hydrogen content within the reaction vessel [21].

$$CO_2 + 4 \, H_2 \rightarrow CH_4 + 2 \, H_2O \; \Delta H_R = -165 \, \text{kJ/mol} \tag{7}$$

$$CO + 3 \, H_2 \rightarrow CH_4 + H_2O \; \Delta H_R = -206 \, \text{kJ/mol} \tag{8}$$

The location of the thermochemical equilibrium of the main components shown in the equations outlined above is visualized as a function of temperature and pressure in Fig. 2 [22]. According to Le Chatelier's principle, high temperatures favor RWGS over WGS and methanation. This is due to the endothermic RWGS reaction, in comparison with the exothermic side reactions or equilibrium shift toward the WGS reaction. Within this graphic, it can be seen that at low pressures, temperatures of more than 750 °C, and a feed composition of H_2:CO_2 of 3, no large quantities of methane are present in the product. However, there is no complete CO_2 conversion; i.e., a CO_2 content of 5–10% in the synthesis gas is possible even at higher temperatures. Additionally, there is no pressure dependence of the chemical equilibrium, as the RWGS reaction neither reduces nor increases the molar number. Compared to this,

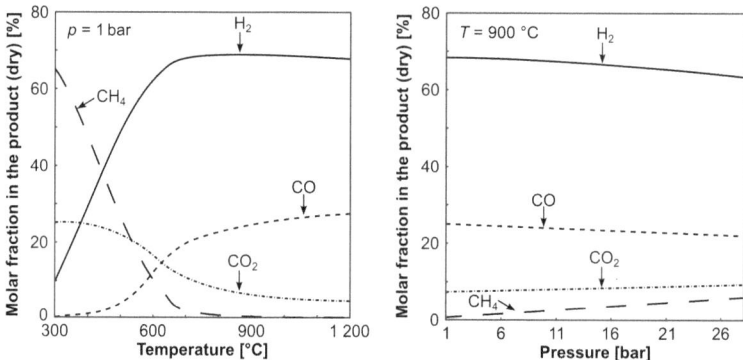

Fig. 2 Equilibrium of reverse water–gas shift (RWGS) reaction including methane formation as a function of temperature T (left) and pressure p (right) in dry synthesis gas ($H_2/CO_2 = 3$) [22]

methanation is a molar number-reducing chemical reaction; this is the reason why the equilibrium concentration increases at high pressures.

4.2 Catalysts

A catalyst is required to control and enhance the RWGS reactions by reducing the activation energy and accelerating the reaction kinetics. Here, the dissociation of CO_2 on the catalyst surface is the rate-determining step within the RWGS. In this respect, metals and, respectively, their metallic surfaces are the active site of the catalyst.

Several RWGS selective catalysts based on Pt, Pd, Ru, and Au have already been tested [21, 23]. There are selective catalysts promoting the RWGS reaction and catalysts that also catalytically support the methanation reaction. Selective RWGS catalysts are based on metal oxides or highly dispersed metals applied on supports or oxides. The catalyst performance is increased by the reducible oxides, such as CeO_2 or CrO_2. The key function of the catalyst is to provide oxygen vacancies that are able to dissociate CO_2 molecules. This incorporation of oxygen vacancies can also be achieved by the incorporating of transition metals or rare earths, such as Zr, La, and Y, and the fine dispersion of the metal particles onto the support [21]. Bimetallic catalysts tested as 100% CO-selective are, for example, M/La-ZrO_2 (M = Fe, Cu). Ni/La-ZrO_2 also shows a CO selectivity of 96.5%, but methane is also produced, as nickel also supports methanation catalytically [3, 21, 23].

Several tested RWGS catalysts, such as Ni, Rh, and Ru, are selective for the methanation reaction and are thus nonselective RWGS catalysts [24]. Although these catalysts support the formation of methane, at high reaction temperatures the reverse reaction (Eq. 8), the steam reforming of methane can take place [25]. Figure 3 illustrated the experimental synthesis gas compositions achievable across different temperature levels. A Ni catalyst was used, which also supports methanation. At low

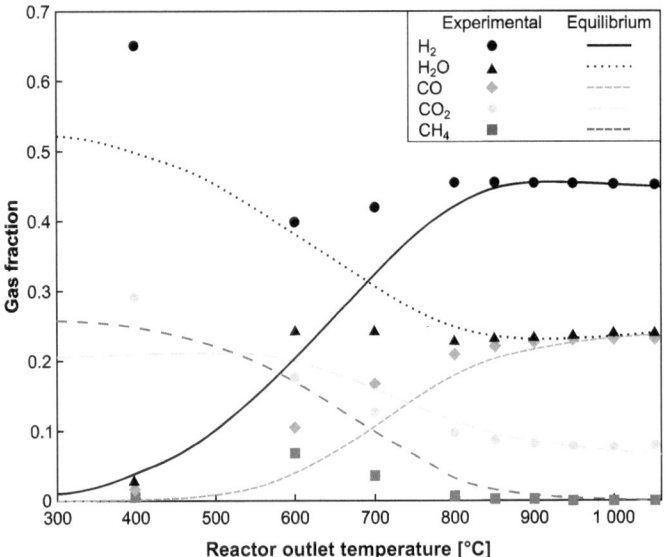

Fig.3 Experimental and equilibrium concentrations with a nonselective RWGS catalyst; molar gas composition over reactor outlet temperature (10 bar; H_2:CO_2 = 2.25, 200 °C; Gas hourly space velocity 12,000 h^{-1}) [25]

temperatures around 400 °C, no change to the feedstock composition is observed, as no reactions occur. At higher temperatures of around 600 °C, larger amounts of methane are present in the product gas, though still below the equilibrium composition, whereas the CO proportion exceeds the equilibrium composition. This indicates a sequential reaction where the RWGS reaction occurs more rapidly the methanation reaction. At temperatures above 800 °C, the experimental data converge toward the equilibrium values, and the methane fraction is further reduced [25].

The catalysts within the RWGS can be deactivated due to metal sintering, carbon deposits, and sulfur impurities. Carbon deposits (caused by the Boudouard reaction, among others) occur more frequently with non-noble and nonselective RWGS catalysts, such as Cu and Fe, at elevated temperatures. Noble metals and also nickel show lower carbon deposits at higher temperatures above 500 °C [26]. The dispersion of the metal particles on the support material can also reduce the deposition of carbon at the active sites of the catalyst [21]. In general, sulfur attaches to the active metal centers of the catalysts deactivate them. For example, NiAl catalysts in which a Fe or Cr phase was added show increased sulfur resistance, which is related to H_2S adsorption on the promoter phase [27]. However, as sulfur also has a poisoning effect in the downstream synthesis catalysts, it is advisable to remove it before the RWGS.

4.3 Reactor Concepts

Below, possible reactor concepts required for a technical realization of the described RWGS reaction are described. The basic principles already require certain framework conditions that must be ensured by the various concepts. For example, thermodynamics dictate that heat transfer must take place in the reaction system in order to supply the endothermic reaction with heat. This heat input should be introduced into the reactor at a high temperature level of more than 700 °C in order to achieve acceptable conversion rates and low by-product formation [28]. Furthermore, considering the integration into a synthesis process, an over-stoichiometric feed gas can be fed into the reactor in order to achieve the correct synthesis gas composition at the reactor outlet. The reactor and process design's currently under discussion primarily differ in terms of heat supply further described below. Conceptually, an overall distinction can be made between externally and internally heated RWGS reactors. While in externally heated reactors the heat is provided through external sources, such as electric heaters or combustion systems, in internally heated reactors, the heat is provided by the exothermic reactions of the reactants themselves.

In contrast to the WGS reaction, which is already realized on a large industrial scale for hydrogen production or hydrogen enrichment, the RWGS reaction has not yet been demonstrated at an industrial scale on a commercial basis for the activation of CO_2 for the time being.

4.3.1 Externally Heated Reactors

One of the main characteristics of externally heated reactors is that only the reactants and a catalyst are present in the reaction chamber, and the external heating takes place in a materially separate section allowing the heat to enter the reaction chamber by heat conduction and radiation. The heat source can be either a flame (fired) or a direct electrical source (electric); both options are discussed below.

Fuel heat reactors. An externally fired tubular reactor is one of the state-of-the-art technologies for steam cracking of methane, among other applications. Due to the similarity of the processes and reactions, such a reactor system is conceivable for the RWGS reaction. In this case, a fuel gas is fed into a furnace chamber from below or from the side of the respective reactor, in which pipes run along the edge or in the middle of the furnace (Fig. 4). These pipes contain the catalyst with the reactant material flow. A fuel gas is therefore required for this reactor system. This gas may come, for example, from light hydrocarbon purge gases from the synthesis plant itself or from further processing plants for the synthesis products. However, a use of these gases, especially if they originate from the plant's own production, results in an overall reduction in carbon and hydrogen efficiency since these components can then no longer be incorporated into the final synthesis products (end products) [29]. Typical reaction temperatures for steam reforming are between 800 and 900 °C [30, 31].

Fig. 4 Side-fired reactor
concept [32]

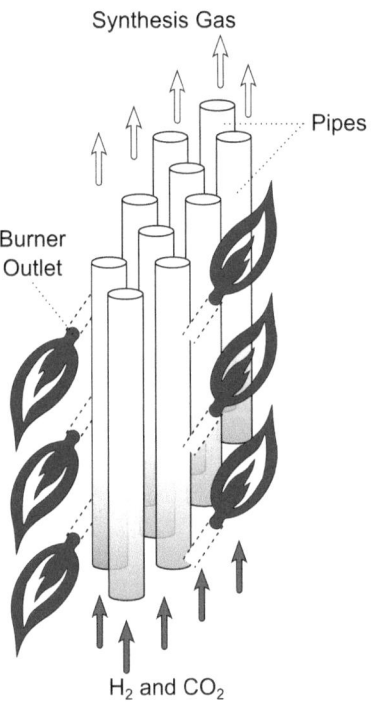

Electrical heated reactors. A fuel gas for firing—as in the process above—must be produced in the process plant itself or must come from external sources. However, in order to ensure a CO_2—respectively, GHG-neutral process, externally generated gas must also originate from a sustainable source and thus potentially an electricity- or a biomass-based production. To avoid such an expensive production of a fuel gas, a RWGS reactor can also be heated directly by electrical energy. Such reactor concepts are not yet applied at large scales, but are being demonstrated in pilot plants [33].

The electric power can be transformed into heat via induction heating, microwave energy, or electrical resistance heating [33–35]. When installing an electrical resistor, the electrical power is applied directly to the tube, which acts as the resistor and thus heats up. The catalyst can be applied directly to the inside of the tube so that there is direct heat conduction to the catalyst. The direct electrical heat transfer provides the advantage that higher heat transfer coefficients can be realized compared to externally fired heaters and therefore higher temperatures at the catalyst, thus shifting the thermodynamic equilibrium toward the target products CO and H_2O. In addition to that, smaller reactor volumes can be realized, as no combustion chamber and space for the heat transfer is needed [33, 36]. Additionally, no material components are used, so the carbon and hydrogen efficiencies remain unaffected by the heating concept (Fig. 5).

Fig. 5 Electrically heated
resistor reactor concept [36]

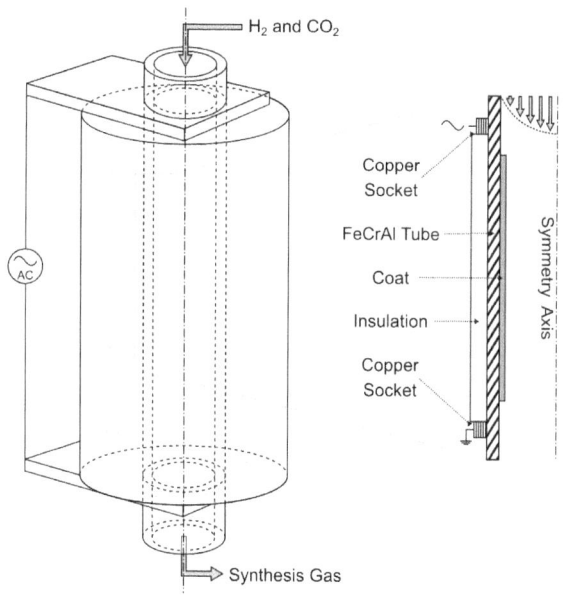

4.3.2 Internal Heat Generation

In contrast to reactor concepts with external heating, internally heated reactor concepts require adjusted feed compositions as part of the reactants are used for internal heat generation. In an autothermal RWGS reactor, additional hydrogen is added to the feed for heat provision. This feed mixture is partially burned with oxygen in the autothermal reactor, resulting in high temperatures in the flame. This gas mixture then enters a catalyst bed, where the further reaction is catalyzed (Fig. 6). Depending on the particular plant concept and hydrogen supply, the oxygen required for this process may be generated on-site in a Power-to-Liquid (PtL) process via the oxygen formed during hydrogen (H_2) production through water electrolysis. For such a reactor concept, reaction temperatures of over 1,000 °C can be achieved; under these circumstances, high equilibrium conversions and low methane contents in the product are to be expected (Fig. 3). Light hydrocarbon gases from downstream processes can also be reformed [14, 16]. This reactor concept, in which a partial flow of the H_2 is then oxidized, has not yet been implemented at larger industrial scales, but is already applied for methane reforming. The output gas steam of a first pre-RWGS stage, where a methanation of the CO_2 and H_2 is carried out at lower temperatures, can then be led to an autothermal reactor, in which the methane is partially oxidized [29].

Fig. 6 Autothermal reactor
with partial combustion and
downstream catalytic
conversion [16]

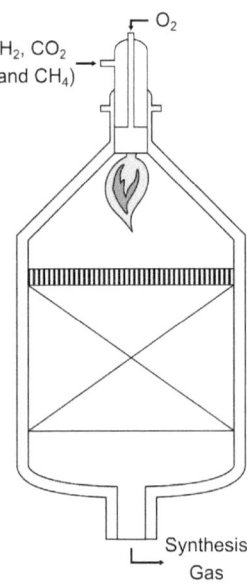

4.4 Process Concepts

In a complete power-based process and plant system for the production of synthetic
fuels or products, it is essential to integrate the fundamental RWGS principles with
various reactor technologies and downstream syntheses in order to create an efficient overall system that is tailored to the specific framework conditions. Against
this background, three different process concepts for the production of synthetic
(renewable) fuels based on different RWGS technologies are presented below, using
the example of a subsequent Fischer–Tropsch (FT) synthesis. Beyond these process
concepts, which mainly differ in their specific reactor concepts, further variations
can be considered in other concepts. These include optimizing the pressure level of
the RWGS reaction and determining whether the compression takes place before
or after the RWGS reaction. In addition, CO_2 separation between the RWGS and a
synthesis unit is conceivable in order to increase the partial pressure of the reactive
components and to be able to recycle the CO_2 directly.

4.4.1 Fired RWGS

Figure 7 shows a fired RWGS reactor integrated into an overall process pathway with
a Fischer–Tropsch synthesis and a hydrotreatment unit. In this case, the tail gas from
the Fischer–Tropsch synthesis is used to generate heat within the RWGS combustion
chamber. If the heat requirement is already met by a partial flow of the tail gas, the
rest of the tail gas can be fed into the reaction chamber of the RWGS, possibly

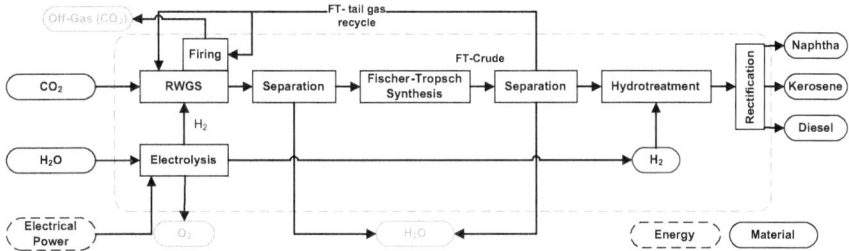

Fig. 7 Fired RWGS concept with downstream Fischer–Tropsch (FT) synthesis

integrating pretreatment, in order to be able to recycle a larger share of carbon and hydrogen. The synthesis gas is cooled down, and the heat is used internally and for steam production. During this process, the water formed in the RWGS is separated to avoid feeding large amounts of water into the Fischer–Tropsch synthesis, as this would lower the partial pressure of the reactants. The only larger energetic input flow within the system boundary remains electric power to run electrolysis. The carbon and hydrogen losses resulting from the combustion of the light gases (Fischer–Tropsch tail gases) must be compensated for by increasing the supply (process input streams) of CO_2 and H_2O as well as electrical power in order to provide a specified quantity of a target product (increased input–output ratio).

4.4.2 Electrified RWGS

In a process concept of the electrified RWGS (eRWGS), the entire tail gas of the Fischer–Tropsch synthesis can be fed back into the reaction chamber of the RWGS, with possible prior pretreatment. Heat is provided by the direct use of electric power (Fig. 8). This material-efficient approach allows for the carbon that enters the system boundary to be fully processed into the target synthetic products. Moreover, it required the production of only as much hydrogen as needed for the processes and reaction equations, eliminating the need for additional hydrogen to fire the RWGS reactor. Compared to fired RWGS concepts, this allows for lower electrolysis capacities (reduced electrolysis dimensions) and, depending on the electrical heating efficiency, also to reduce the absolute amount of electrical power required to produce a specified quantity of target product. So far, the integration of such electrified RWGS reactors has been announced by industrial market participants, and respective patents have been secured [37, 38].

4.4.3 Autothermal RWGS

Autothermal RWGS represents another material- and thus carbon-efficient variant among possible RWGS concepts. Figure 9 shows a two-stage autothermal RWGS

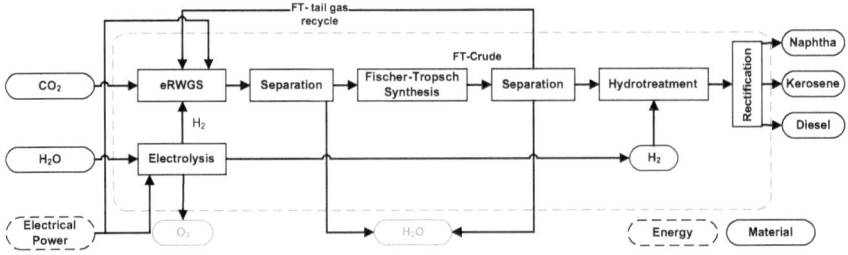

Fig. 8 Electrified RWGS concept with downstream Fischer–Tropsch (FT) synthesis

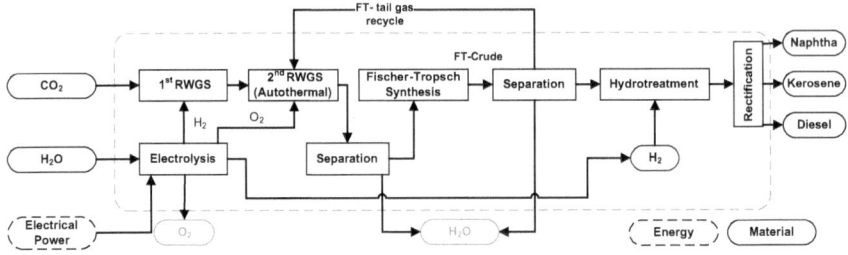

Fig. 9 Two-stage autothermal RWGS concept with downstream Fischer–Tropsch (FT) synthesis

where, in the first catalytic stage, CO_2 and H_2 are already converted to methane. However, complete conversion of CO_2 is not attainable in this first stage. Yet, an advantage of this stage is the ability to use technologically mature and commercially available autothermal reactors for the autothermal reforming of methane into the required synthesis gas in the second RWGS stage. In this concept, pure oxygen from electrolysis is added to the methane, CO_2, and H_2 mixture in the second RWGS stage in an under-stoichiometric quantity. This facilitates partial combustion of the mixture and achieving high reaction temperatures of more than 1,000 °C. The complete conversion into synthesis gas then takes place in another catalytic bed in the autothermal reactor. The heat supply is achieved by the partial combustion of hydrogen, which requires higher electricity demand for electrolysis and larger electrolysis capacities to supply the additional hydrogen required. The tail gas recycle from the Fischer–Tropsch synthesis can also be added to the material mixture from the first RWGS stage in order to close the carbon recycle within the process. Such a process concept is being presented by industrial providers of RWGS technology [29].

4.4.4 Concept Comparison

Following the above presentation of the various RWGS process concepts, a qualitative comparison among them is provided below. Table 2 sums up this qualitative comparison with regard to four basic criteria, namely carbon efficiency, hydrogen efficiency, reaction temperature, and technological maturity.

Following the comparison summary provided in Table 2, a concise overall examination of each of the four comparison criteria is outlined below.

- **Carbon efficiency**. The carbon efficiency indicates the proportion of the corresponding carbon that is provided (generated by electrolysis) and incorporated in the final synthesis products. With regard to carbon, every carbon atom can be incorporated into the valuable product, except in the option in which hydrocarbons are used to provide the heat required by the RWGS. In this process option, further CO_2 capture would be required to recycle the CO_2 back into the process. In both electrified and autothermal RWGS, there are no major carbon losses, although the carbon cycle cannot be completely closed due to some losses of CO_2 in the wastewater.

- **Hydrogen efficiency**. The hydrogen efficiency indicates the proportion of the corresponding hydrogen molecules that is provided (e.g., generated by water electrolysis) and incorporated in the final synthesis products. In terms of hydrogen efficiency, the autothermal RWGS also shows disadvantages compared to the electrified RWGS, as hydrogen is produced, which is used to provide heat and leaves the system boundary as water. This leads to higher electrolyzer capacities in this process option. The fired RWGS shows a slightly lower hydrogen efficiency, as an efficient heat transfer within the fired RWGS still has to be ensured and therefore potentially more heat has to be provided by combustion than by direct partial combustion in the reaction chamber, as is the case with the autothermal reactor.

- **Reaction temperature**. This can also be seen in the achievable reaction temperature. In the autothermal reactor, the highest reaction temperatures can be realized in the existing reactor concepts through the direct partial combustion of the hydrogen. With the increased temperatures, the methane content in the synthesis gas can be reduced and the CO_2 conversion increased. At lower temperatures and thus lower CO_2 conversions and higher methane content, there is a reduction in the partial pressure of the reactants in the downstream synthesis. These light gases can be recycled in an electrified RWGS, whereby the process conversion remains high. The respective equipment within the synthesis cycle only has to be designed accordingly larger. In addition, the reactor maturity can be evaluated.

Table 2 Qualitative comparison of different RWGS process concepts (+ more advantageous|o neutral|– more disadvantageous)

	Reactor/process concept		
	Fired RWGS	Electrified RWGS	Autothermal RWGS
Carbon efficiency	–	+	+
Hydrogen efficiency	–	+	O
Reaction temperature	–	o	+
Technological maturity	+	–	O

- **Technological maturity**. Fired reactors are state-of-the-art technologies, whereas electrically heated reactors are not yet used on an industrial scale. Although autothermal reactors are used industrially for the partial oxidation of methane, reactor-side adjustments may be necessary for the combustion of a hydrogen-rich feed gas. Electrically heated reactors are not only demonstrated for the RWGS reaction at high temperatures, but are also evaluated for other endothermic thermo-chemical reactions, such as naphtha steam cracking for olefin production, which may influence the development of such reactor systems.

5   Final Consideration

The production of sustainable carbon-containing products requires access to a sustainable carbon-containing feedstock. In addition to carbon-containing feedstocks and molecules of biogenic origin (biomass), such feedstock can also include carbon dioxide (CO_2) itself; this CO_2 can come from so-called biogenic point sources or directly from the atmosphere. However, CO_2 in particular is thermodynamically very stable, so that activating it in order to convert it chemically is technically complex and consequently expensive. The reverse water–gas shift (RWGS) reaction, outlined in this paper, is one technological approach aimed to make the carbon contained in a CO_2 molecule available for further downstream synthesis technologies. In general, this can enable the production of carbon-containing synthetic end products, such as bulk chemicals or transportation fuels, of non-fossil origin and in addition a use of sustainable biomass feedstocks.

In the RWGS process, the CO_2 is reduced to carbon monoxide (CO) using hydrogen (H_2) in an endothermic reaction. The resulting gas mixture is considered as synthesis gas, which is already used today—on the basis of feedstocks consisting of fossil fuels—to synthesize various hydrocarbon products (e.g., Fischer–Tropsch-Naphtha, Fischer–Tropsch-Waxes, Methanol). Against this background, the down-stream synthesis processes are a state-of-the-art technology, although the method of synthesis gas production based on CO_2 and H_2 feedstocks is a fairly novel approach. In this article, in addition to the fundamentals of RWGS for synthesis gas supply, different reactor concepts and the integration of those into an overall synthesis process are described.

In particular, the provision of thermal energy at a temperature level at which adequate CO_2 conversions can be achieved and the methanation side reaction only takes place to a minor extent defines the RWGS reactor design and the process-side integration. Heat can be provided by burning waste gases, such as light hydrocarbons from the synthesis section, or by direct combustion of hydrogen out of the elec-trolyzer. In addition, combustion can take place in a separate combustion chamber or within in the reaction chamber. For these RWGS concepts, the respective combustion fuel must first be produced. This is avoided in direct electric reactor heating, where the H_2 produced and the CO_2 provided for the overall process can be incorporated into the final synthetic products without (significant) material losses. However, unlike the fired RWGS reactor concepts, electrically heated RWGS reactor concepts are not yet used on an industrial scale. Various pilot and demonstration plants and smaller commercial plants have been announced for the use of RWGS reactors in processes, which why a demonstration of the various concepts is expected on a larger scale in the years to come [39–41].

References

1. Hiller H, Reimert R, Marschner F, Renner H.-J, Boll W, Supp E, Brejc M, Liebner W, Schaub G, Hochgesand G, Higman C, Kalteier P, Müller W.-D, Kriebel M, Schlichting H, Tanz H, Stönner H.-M, Klein H, Hilsebein W, Gronemann V, Zwiefelhofer U, Albrecht J, Cowper CJ, Driesen HE (2000) Gas production. In: Ullmann's encyclopedia of industrial chemistry. Wiley-VCH Verlag GmbH & Co. KGaA, Weinheim, Germany
2. Lang X-D, He X, Li Z-M, He L-N (2017) New routes for CO_2 activation and subsequent conversion. Curr Opin Green Sustain Chem 7:31–38. https://doi.org/10.1016/j.cogsc.2017.07.001
3. Daza YA, Kuhn JN (2016) CO_2 conversion by reverse water gas shift catalysis: comparison of catalysts, mechanisms and their consequences for CO_2 conversion to liquid fuels. RSC Adv 6:49675–49691. https://doi.org/10.1039/C6RA05414E
4. Fischedick M, Görner K, Thomeczek M (2015) CO_2: Abtrennung, Speicherung, Nutzung. Springer Berlin Heidelberg, Berlin, Heidelberg
5. Kaltschmitt M, Hartmann H, Hofbauer H (2016) Energie aus Biomasse. Springer, Berlin Heidelberg, Berlin, Heidelberg
6. Change IPOC (2021) Climate change 2021. The physical science basis: summary for policy-makers: working group I contribution to the sixth assessment report of the intergovernmental panel on climate change. IPCC, Geneva, Switzerland
7. Kaiwen L, Bin Y, Tao Z (2018) Economic analysis of hydrogen production from steam reforming process: a literature review. Energy Sources Part B: Econ Plan Policy. https://doi.org/10.1080/15567249.2017.1387619
8. Bube S, Sens L, Drawer C, Kaltschmitt M (2024) Power and biogas to methanol—a techno-economic analysis of carbon-maximized green methanol production via two reforming approaches. Energy Convers Manage 304:118220. https://doi.org/10.1016/j.enconman.2024.118220
9. Buttler A, Spliethoff H (2018) Current status of water electrolysis for energy storage, grid balancing and sector coupling via power-to-gas and power-to-liquids: a review. Renew Sustain Energy Rev 82:2440–2454. https://doi.org/10.1016/j.rser.2017.09.003
10. Roy A, Watson S, Infield D (2006) Comparison of electrical energy efficiency of atmospheric and high-pressure electrolysers. Int J Hydrogen Energy 31:1964–1979. https://doi.org/10.1016/j.ijhydene.2006.01.018
11. Bernt M, Gasteiger HA (2016) Influence of ionomer content in IrO_2/TiO_2 electrodes on PEM water electrolyzer performance. J Electrochem Soc 163:3179. https://doi.org/10.1149/2.0231611jes
12. Lange H, Klose A, Lippmann W, Urbas L (2023) Technical evaluation of the flexibility of water electrolysis systems to increase energy flexibility: a review. Int J Hydrogen Energy 42:15771–15783. https://doi.org/10.1016/j.ijhydene.2023.01.044
13. Franck H-G (1979) Kohleveredlung. Chemie und Technologie. Hochschultext. Springer, Berlin, Heidelberg
14. de Klerk A (2011) Fischer-Tropsch refining. Wiley-VCH, Weinheim

15. Chein RY, Chen YC, Yu CT, Chung JN (2015) Thermodynamic analysis of dry reforming of CH_4 with CO_2 at high pressures. J Nat Gas Sci Eng 26:617–629. https://doi.org/10.1016/j.jngse.2015.07.001

16. Aasberg-Petersen K, Christensen TS, Stub Nielsen C, Dybkjær I (2003) Recent developments in autothermal reforming and pre-reforming for synthesis gas production in GTL applications. Fuel Process Technol 83:253–261. https://doi.org/10.1016/S0378-3820(03)00073-0

17. Hamelinck CN, Faaij AP (2002) Future prospects for production of methanol and hydrogen from biomass. J Power Sources 111:1–22. https://doi.org/10.1016/S0378-7753(02)00220-3

18. Platon A, Wang Y (2009) Water-gas shift technologies. In: Liu K, Song C, Subramani V (eds) Hydrogen and syngas production and purification technologies. Wiley, pp 311–328

19. Kaltschmitt M, Hofbauer H, Lenz V (eds) (2023) Energie aus Biomasse. Thermo-chemische Konversion, 4th edn. Energie aus Biomasse. Springer Fachmedien Wiesbaden GmbH; Springer Vieweg, Wiesbaden

20. Bown RM, Joyce M, Zhang Q, Reina TR, Duyar MS (2021) Identifying commercial opportunities for the reverse water gas shift reaction. Energy Tech 9:210054. https://doi.org/10.1002/ente.202100554

21. González-Castaño M, Dorneanu B, Arellano-García H (2021) The reverse water gas shift reaction: a process systems engineering perspective. React Chem Eng 6:654–976. https://doi.org/10.1039/D0RE00478B

22. König DH (2016) Techno-ökonomische Prozessbewertung der Herstellung synthetischen Flugturbinentreibstoffes aus CO_2 und H_2. Dissertationsschrift, Universität Stuttgart

23. Oshima K, Shinagawa T, Nogami Y, Manabe R, Ogo S, Sekine Y (2014) Low temperature catalytic reverse water gas shift reaction assisted by an electric field. Catal Today 232:27–32. https://doi.org/10.1016/j.cattod.2013.11.035

24. Zhu M, Ge Q, Zhu X (2020) Catalytic reduction of CO_2 to CO via reverse water gas shift reaction: recent advances in the design of active and selective supported metal catalysts (trans: Tianjin Univ). https://doi.org/10.1007/s12209-020-00246-8

25. Thor Wismann S, Larsen K-E, Mølgaard Mortensen P (2022) Electrical reverse shift: sustainable CO_2 valorization for industrial scale. Angewandte Chemie (International ed. in English). https://doi.org/10.1002/anie.202109696

26. Panaritis C, Edake M, Couillard M, Einakchi R, Baranova EA (2018) Insight towards the role of ceria-based supports for reverse water gas shift reaction over RuFe nanoparticles. J CO_2 Util. https://doi.org/10.1016/j.jcou.2018.05.024

27. Wolf M, Wong LH, Schüler C, Hinrichsen O (2020) CO_2 methanation on transition-metal-promoted Ni-Al catalysts: sulfur poisoning and the role of CO_2 adsorption capacity for catalyst activity. J CO_2 Util. https://doi.org/10.1016/j.jcou.2019.10.014

28. Rezaei E, Dzuryk S (2019) Techno-economic comparison of reverse water gas shift reaction to steam and dry methane reforming reactions for syngas production. Chem Eng Res Des 144:354–369. https://doi.org/10.1016/j.cherd.2019.02.005

29. Christensen TS, Aasberg-Petersen K, Mortensen PM (2021) eFuels technology for converting CO_2 and renewable electricity to renewable synthetic fuels. Topsoe reverse water gas shift (RWGS) technologies enable production of eFuels and eChemicals from green hydrogen and captured CO_2. Haldor Topsoe. https://info.topsoe.com/en/erwgs-wp-dlp-0. Accessed 28 Feb 2024

30. Barelli L, Bidini G, Gallorinil F, Servili S (2008) Hydrogen production through sorption-enhanced steam methane reforming and membrane technology: a review. Energy. https://doi.org/10.1016/j.energy.2007.10.018

31. Simpson A, Lutz A (2007) Exergy analysis of hydrogen production via steam methane reforming. Int J Hydrogen Energy. https://doi.org/10.1016/j.ijhydene.2007.08.025

32. Engel S, Liesche G, Sundmacher K, Janiga G, Thévenin D (2020) Optimal tube bundle arrangements in side-fired methane steam reforming furnaces. Front Energy Res 6:583346. https://doi.org/10.3389/fenrg.2020.583346

33. From TN, Partoon B, Rautenbach M, Østberg M, Bentien A, Aasberg-Petersen K, Mortensen PM (2024) Electrified steam methane reforming of biogas for sustainable syngas manufacturing

and next-generation of plant design: a pilot plant study. Chem Eng J 479:147205. https://doi.org/10.1016/j.cej.2023.147205

34. Almind MR, Vendelbo SB, Hansen MF, Vinum MG, Frandsen C, Mortensen PM, Engbæk JS (2020) Improving performance of induction-heated steam methane reforming. Catal Today 342:13–20. https://doi.org/10.1016/j.cattod.2019.05.005

35. Meloni E, Saraceno E, Martino M, Corrado A, Iervolino G, Palma V (2023) SiC-based structured catalysts for a high-efficiency electrified dry reforming of methane. Renew Energy 211:336–346. https://doi.org/10.1016/j.renene.2023.04.082

36. Wismann ST, Engbæk JS, Vendelbo SB, Bendixen FB, Eriksen WL, Aasberg-Petersen K, Frandsen C, Chorkendorff I, Mortensen PM (2019) Electrified methane reforming: a compact approach to greener industrial hydrogen production. Science (New York, N.Y.). https://doi.org/10.1126/science.aaw8775

37. Topsoe (2024) eREACT™ fuels: new technology essential for electrofuels production. https://www.topsoe.com/our-resources/knowledge/our-products/equipment/e-react-fuels. Accessed 5 Mar 2024

38. Van der Ploeg GGP (2020) Electrically heated reactor and a process for gas conversions using said reactor Patent WO 2020/002326 A1, 2 January 2020

39. Deutsches Zentrum für Luft- und Raumfahrt (2023) Technologieplattform PtL—Blick in die Zukunft. https://www.dlr.de/de/aktuelles/nachrichten/2023/04/technologieplattform-ptl-blick-in-die-zukunft. Accessed 6 Mar 2024

40. Matthey J (2023) JM and bp chosen by EDL to support production of SAF at HyKero plant in Germany. https://matthey.com/media/2023/edl-hykero. Accessed 6 Mar 2024

41. norsk e-fuel: Alpha Plant, Mosjøen/Nesbruket, Vefsn Municipality, Norway. https://www.norsk-e-fuel.com/projects. Accessed 6 Mar 2024

Co-electrolysis

Erik Reichelt, Gregor Herz, and Matthias Jahn

Abstract In the context of power-to-liquid (PtL) processes for the production of synthetic fuels the generation of syngas is a key step. Co-electrolysis in high-temperature electrolyzers is a promising option for single-step generation of syngas from H_2O and CO_2. In this chapter an overview on co-electrolysis is given, covering the thermodynamic and technological basics as well as the applied approach for scale-up. Experimental results from a small-scale demonstration unit allow for an assessment of the importance of heat integration within the system. The advantage of increased PtL energetic efficiency by thermal coupling of co-electrolysis and synthesis is discussed, highlighting the future potential of co-electrolysis in the context of synthetic fuel production.

Keywords Co-electrolysis · Solid oxide electrolysis · High-temperature electrolysis · Syngas generation · Efficiency

1 Introduction

To meet global climate objectives, many sectors are transitioning toward direct electrification or the use of hydrogen as an energy carrier or fuel. However, certain "harder-to-abate" sectors, like aviation, are anticipated to predominantly depend on carbon-containing fuels with high energy density like kerosene for the foreseeable future. When producing such fuels from renewable feedstocks and energies, the respective synthesis technologies (e.g., Fischer-Tropsch synthesis) typically employ a syngas made up of hydrogen (H_2) and carbon monoxide (CO). Starting point of power-to-liquid (PtL) processes are water (H_2O) and carbon dioxide (CO_2). Thus, a conversion step into syngas is necessary.

E. Reichelt (✉) · G. Herz · M. Jahn
Fraunhofer Institute for Ceramic Technologies and Systems (Fraunhofer IKTS), Dresden, Germany
e-mail: erik.reichelt@ikts.fraunhofer.de

To produce the required syngas from H_2O and CO_2, low-temperature electrolysis technologies, like alkaline electrolysis (AEL) and proton-exchange membrane electrolysis (PEMEL), require an extra process step in addition to the production of H_2 from H_2O. In this step, H_2 and CO_2 are combined and converted via reverse water–gas shift reaction (RWGS). On the other hand, co-electrolysis provides an option to convert H_2O and CO_2 directly into syngas, enhancing the overall electrolysis efficiency. Particularly as part of PtL pathways, this technology might hold significant promise.

Therefore, this chapter intends to give a deeper insight into the co-electrolysis technology. The theoretical thermodynamic background of co-electrolysis is addressed first. This is followed by a detailed examination of the technical implementation, illustrating the various steps and stages from an initial electrolysis cell to a full-scale electrolysis system. Using data from a small-scale plant, various technical and economic parameters are then presented to offer a more detailed categorization of this technology. At times, the performance of co-electrolysis is compared with electrolysis-based processes that employ reverse water–gas shift reaction (RWGS) for syngas production. The overall aim of this chapter is to provide a comprehensive insight into the co-electrolysis technology.

2   Co-electrolysis Fundamentals

2.1   Basic Pathway Classification

Syngas is the basis for the production of different synthetic fuels. In the case of hydrocarbon-based fuels it consists mainly of hydrogen (H_2) and carbon monoxide (CO). The most relevant products in that context are methane, methanol, and Fischer-Tropsch products like kerosene or diesel. Although methane and methanol can also be directly produced from hydrogen (H_2) and carbon dioxide (CO_2) via corresponding synthesis processes [1–3], the Fischer-Tropsch synthesis, particularly when aiming for products suitable as synthetic fuels, requires a syngas rich in carbon monoxide (CO). The corresponding ratio of hydrogen to carbon monoxide is approximately two (H_2:CO $\approx$ 2). Production of such a syngas on the basis of electrolysis can be realized via two pathways, which are depicted in Fig. 1.

As shown in Fig. 1, the first pathway for the production of syngas involves initially the separate production of H_2 and its partial subsequent conversion with CO_2 via reverse water–gas shift reaction (RWGS) to obtain the desired synthesis gas (Eq. 1). This pathway is the option of choice when applying low-temperature electrolyzers.

$$H_2 + CO_2 \rightleftarrows H_2O + CO \quad \Delta h_R^0 = 41.2 \text{ kJ/mol} \quad (1)$$

While H_2 generation in this process concept could generally also be based on high-temperature electrolysis, it is more convenient in that case to produce syngas

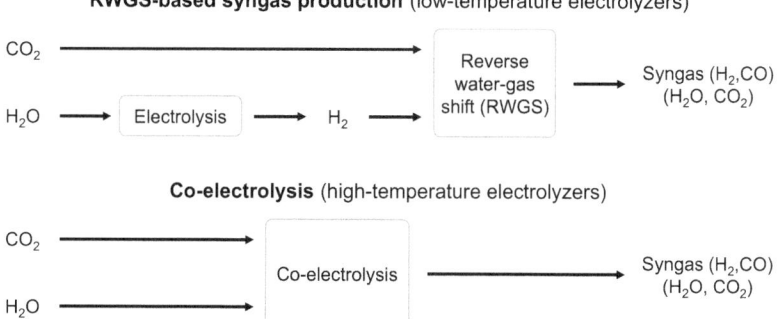

Fig. 1 Electrolysis-based pathways for syngas generation

via so-called co-electrolysis, which is presented as the second route in Fig. 1. Co-electrolysis, which can be conducted in high-temperature electrolyzers operated at temperatures in the range of $\vartheta \approx 700$–$900\ °C$, allows for a single-step generation of syngas from H_2O and CO_2. The underlying reactions are described in detail in the following section.

As stated above, syngas is mainly compost of H_2 and CO. However, the production processes depicted in Fig. 1—due to incomplete conversion—yield a mixture of these gases and the feed gases H_2O and CO_2. Water can be easily separated via condensation, but CO_2 can only be separated via more complex process steps like amine scrubbing, if necessary. In general, the CO_2 fraction in the produced syngas is smaller for the co-electrolysis-based route.

2.2 Thermodynamics

Co-electrolysis is no independent technology, but rather an operation mode of high-temperature electrolysis (HTEL)—also called solid oxide electrolysis (SOEL). However, as the ability to perform co-electrolysis within a high-temperature electrolyzer as well as the resulting performance is directly related to the thermodynamic fundamentals of this technology, the necessary background will be presented in the following.

2.2.1 Reaction Enthalpies

In Fig. 2, the reaction enthalpy Δh_R and free reaction enthalpy Δg_R for water (H_2O) reduction as well as for carbon dioxide (CO_2) reduction are plotted in the range of $\vartheta = 0$–$1,000\ °C$.

The respective reactions for the water and carbon dioxide reduction are presented in Eqs. (2) and (3).

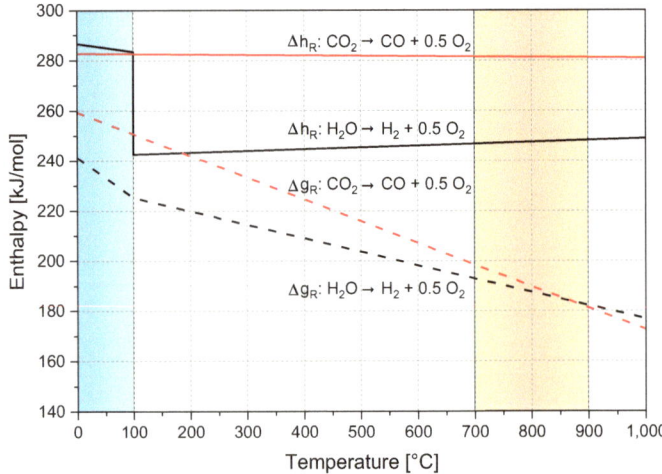

Fig. 2 Temperature dependency of reaction enthalpy Δh_R and free reaction enthalpy Δg_R for the electrochemical reduction of H_2O and CO_2

$$H_2O \rightarrow H_2 + 0.5O_2 \tag{2}$$

$$CO_2 \rightarrow CO + 0.5O_2 \tag{3}$$

From the reaction enthalpy Δh_R, the amount of energy necessary to convert water (H_2O) to hydrogen (H_2) and carbon dioxide (CO_2) to carbon monoxide (CO) can be derived. The step at $\vartheta = 100\,°C$ for water reduction is due to the phase transition from liquid water to steam and corresponds to the evaporation enthalpy Δh_{ev}. The resulting decrease in Δh_R for temperatures $\vartheta > 100\,°C$ is the reason for the considerably increased energetic efficiency of high-temperature electrolyzers in comparison to low-temperature water electrolysis—if steam can be generated applying waste heat of associated processes instead of electric energy. For the reduction of carbon dioxide, the reaction enthalpy that needs to be provided is higher in comparison to the water reduction reaction for temperatures $\vartheta > 100\,°C$. The difference between both values corresponds to the reaction enthalpy of the reverse water–gas shift (RWGS) reaction (Eq. 1).

While the reaction enthalpy Δh_R gives an information about the overall energy demand, the amount of energy that needs to be supplied in form of electric energy can be derived from the free reaction enthalpy Δg_R. The difference between the two enthalpies corresponds to the amount of energy that could be supplied in form of heat. As high-temperature waste heat is generally already utilized in industrial processes, the heat that is externally supplied to the high-temperature electrolyzer is most often limited to temperature levels that allow for the evaporation of the water fraction in the supplied feed.

2.2.2 Basic Reactions

A scheme depicting the occurring reactions in a high-temperature co-electrolysis cell is given in Fig. 3. Water (H_2O) and carbon dioxide (CO_2) are fed to the cathode side of the cell where the reduction reactions occur.

For the conversion of carbon dioxide within the cell, generally two pathways are possible—a direct and an indirect conversion. For the first pathway, the direct conversion according to Eq. (3), the resulting cathode-side reaction would occur according to Eq. (4):

$$CO_2 + 2e^- \rightarrow CO + O^{2-} \tag{4}$$

For the second pathway, the indirect conversion via the reverse water–gas shift (RWGS) reaction (Eq. 1), hydrogen (H_2) would be formed via the following half-cell reaction:

$$H_2O + 2e^- \rightarrow H_2 + O^{2-} \tag{5}$$

The hydrogen is subsequently converted according to Eq. (1). Even though both pathways (direct and indirect conversion) are generally possible, CO_2 conversion via RWGS has often been found to be prevailing in co-electrolysis operation [4–7].

Independent of the prevailing cathode-side reaction pathway, the other transport and reaction steps within the cell do not differ from the ones within an SOEL for hydrogen generation. The oxygen ions are transported through the solid oxide electrolyte via oxygen vacancies. On the anode side, oxygen molecules are formed from the oxygen ions:

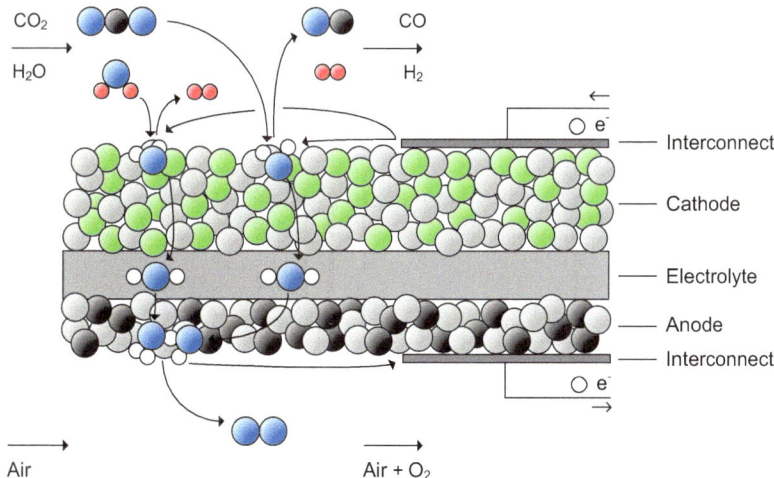

Fig. 3 Scheme of a high-temperature co-electrolysis cell

$$O^{2-} \rightarrow 0.5O_2 + 2e^- \tag{6}$$

The composition of the syngas to be produced via co-electrolysis depends on the demand of the downstream synthesis process. For the Fischer-Tropsch synthesis, for instance, a H_2:CO ratio of approximately two is required [8]. The H_2:CO ratio produced in a co-electrolysis cell mainly depends on the ratio of the H_2O and CO_2 feed. However, as the applied cathode materials are active for the RWGS reaction (Eq. 1), the resulting H_2:CO ratio is also dependent on the state of equilibrium of this reaction. Thus, the composition of the syngas produced can be calculated for a given conversion within the cell on the basis of feed composition and temperature by a minimization of the free enthalpy with good accuracy [9].

2.2.3   Modes of Operation

The operation of a high-temperature co-electrolyzer can be performed under different modes. While low-temperature electrolyzers can only be operated—for industrially relevant current densities—under exothermic conditions [10], leading to the necessity of cooling during operation, high-temperature electrolyzers can be operated in three modes: exothermal, endothermal, and thermoneutral. A schematic depiction of the three different operation modes described above is given in Fig. 4.

The so-called thermoneutral voltage U_{tn} can be calculated from the reaction enthalpy Δh_R, the number of transferred electrons z and the Faraday constant F:

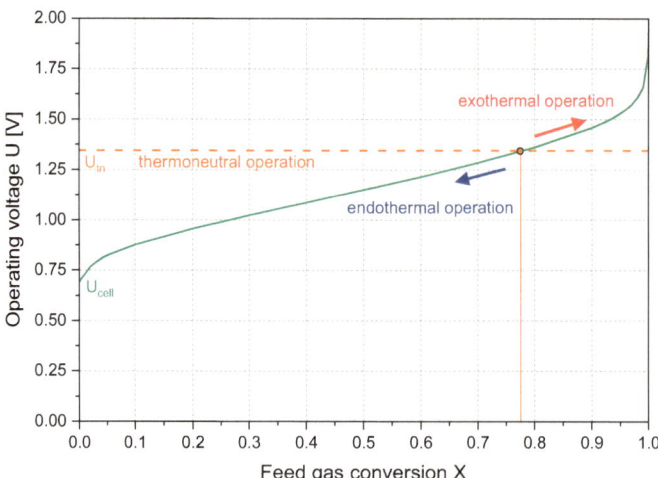

Fig. 4 Thermal operation modes of a co-electrolyzer in dependence of feed gas conversion X and operating voltage U for an exemplary cell

$$U_{tn} = \frac{\Delta h_R}{zF} \tag{7}$$

At this operating voltage, all necessary energy for driving the electrolysis reaction is provided in form of electric energy. The thermal energy demand of the reaction is covered by the ohmic losses of the cell.

If an operating voltage higher than U_{tn} is chosen, more energy than needed is integrated into the system. This leads to an exothermal operation with a subsequent increase in operation temperature or the need for cooling of the electrolyzer. If the electrolyzer is operated at voltages lower than U_{tn}, the amount of energy provided is not sufficient to drive the reaction. Thus, the temperature of the electrolysis cell decreases, if no additional heat is supplied to the electrolyzer, e.g., by overheating the feed gases to the cell.

In general, the co-electrolyzer will often be operated close to thermoneutral operation, as it decreases the demand for cooling or integration of additional heat from external heat sources or via electric pre-heating of the feed gases.

As Δh_R differs for the conversion of H_2O and CO_2 (Fig. 2), U_{tn} is a function of the feed gas composition. The resulting thermoneutral voltage U_{tn} in dependence of $H_2O:CO_2$ ratio at the inlet of the electrolyzer is given in Fig. 5.

With decreasing $H_2O:CO_2$ ratio U_{tn} increases. This corresponds to the higher energy demand for the conversion of CO_2 that can be derived from Fig. 2. Thus, high-temperature electrolyzers running in co-electrolysis mode are generally operated at higher voltages than electrolyzers in steam electrolysis mode. The cell voltage of a co-electrolyzer can be calculated on the basis of the ideal cell voltage:

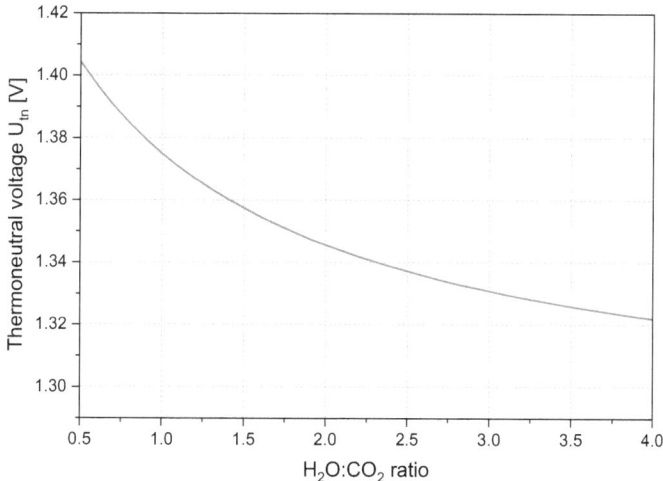

Fig. 5 Thermoneutral voltage U_{tn} in dependence of $H_2O:CO_2$ ratio at the inlet of the electrolysis cell for a cell temperature of $\vartheta = 800\ °C$

$$U_{ideal} = U_{rev} - \frac{RT}{zF} \ln \left(\frac{x_{H_2O}}{x_{H_2} x_{O_2}^{\frac{1}{2}}} \right) \tag{8}$$

with R as the universal gas constant, T as the absolute temperature, and x as the molar fraction of the reactants. As Eqs. (2) and (3) are coupled via Eq. (1), the ideal cell voltage within a co-electrolysis cell can be calculated on the basis of Eq. (2) solely. The reversible voltage (U_{rev}) can be calculated by:

$$U_{rev} = \frac{\Delta g_R}{zF} \tag{9}$$

The voltage of a real cell U_{cell} is additionally influenced by different overpotentials:

$$U_{cell} = U_{ideal} + \eta_{act, an} + \eta_{act, ca} + \eta_{ohm} + \eta_{conc, an} + \eta_{conc, ca} \tag{10}$$

Activation overpotentials result from the charge transfer in the electrodes—i.e., at the anode ($\eta_{act, an}$) and cathode ($\eta_{act, ca}$). The gas molecules must be adsorbed or desorbed at the so-called triple-phase boundary, where the dissociation or formation of the molecules takes place, as discussed below. Depending on the applied materials as well as on the operating conditions, these processes take place via different elementary reaction steps with different sub-steps [11].

Overpotentials due to ohmic resistances (η_{ohm}) are determined by the ionic conductivity of the electrolyte, the electrical conductivity of the anode and cathode as well as by the thickness of the respective layers. Ohmic losses are linearly dependent on the applied current [12].

The concentration overpotentials ($\eta_{conc, an}$ and $\eta_{conc, ca}$), which are also referred to as diffusion overpotentials, are caused by a deviating gas composition at the triple-phase boundary compared to the bulk gas composition. This occurs if the mass transfer to the triple-phase boundary is slower than the surface reactions. Such mass transfer limitations occur especially at high current densities and play a minor role in case of thin electrodes.

The intrinsic kinetics determining the activation overpotential as well as the ionic conductivity of the electrolyte increase with temperature, thus leading to decreasing overpotentials. As also U_{ideal} follows the same trend for higher temperatures, according to Eq. (8), the cell voltage of high-temperature co-electrolyzers generally decreases with temperature.

3   Technical Realization

The explanations shown in the previous section primarily concentrate on the electrolyzer cell level. However, to achieve larger electrolysis capacities at the system level, that allow for the large-scale production of synthetic fuels, upscaling from a single electrolyzer cell to a complete electrolysis system is necessary. An example for such a scale-up and the different scaling steps involved is given in Fig. 6.

Starting from cells with rather small nominal power input and, thus, syngas production capacity, stacks are prepared. The high operating temperature of the technology necessitates additional apparatuses like heat exchangers and electric heaters as well as thermal insulation. Within a module, these apparatuses are combined with a stack unit comprising several stacks. In order to finally reach sufficient syngas production capacities, several modules are arranged in parallel. The nominal power values given in Fig. 6 can be regarded as a representative example for the current state of the technology. However, increasing overall production capacity via upscaling on every of the given levels is focus of current R&D efforts. A deeper look on the different scales is given in the following sections.

3.1   *Cells*

In general, the cell concept for co-electrolyzers does not differ from the one for high-temperature steam electrolyzers. Also, the main components of a solid oxide electrolysis cell (SOEC) are already given in Fig. 3. On the cathode side, mainly nickel-based cermets (e.g., $Ni/Zr_{1-x}Y_xO_{2-x/2}$ or $Ni/Ce_{1-x}Gd_xO_{2-x}$) are applied as material as they combine electrochemical activity and durability at acceptable cost. Nickel as a catalyst for the occurring cathode-side reactions described above needs to be combined

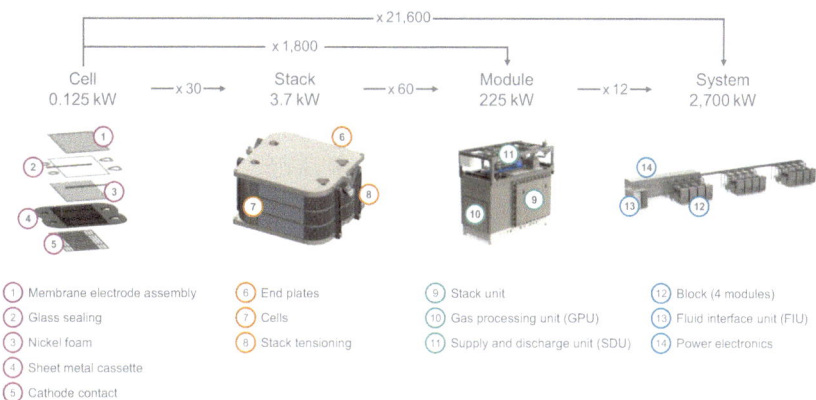

Fig. 6 Examples of different scales of co-electrolysis development. © Sunfire; adapted version

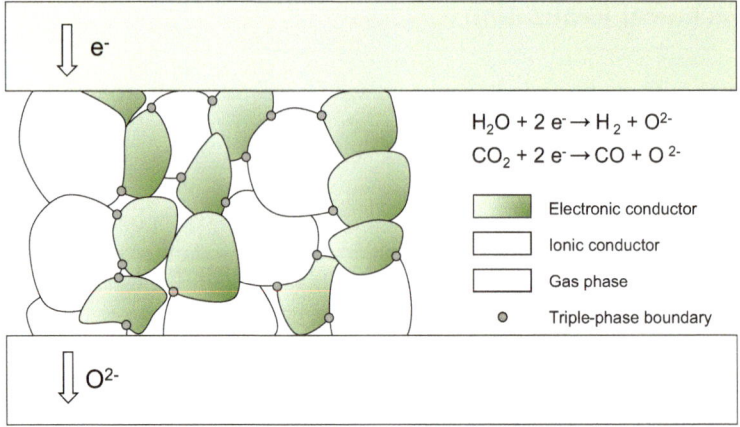

$$H_2O + 2\,e^- \rightarrow H_2 + O^{2-}$$
$$CO_2 + 2\,e^- \rightarrow CO + O^{2-}$$

Electronic conductor

Ionic conductor

Gas phase

Triple-phase boundary

Fig. 7 Schematical drawing of the electrode microstructure on the cathode side of the cell showing triple-phase boundaries between gas phase and the applied electronic and ionic conductor

with an ionic conductor creating the above-mentioned triple-phase boundaries. As ionic conductor most often 8YSZ (8 mol-% yttria-stabilized zirconia) is applied. But also other materials are applied or currently studied [13–15]. A depiction of the triple-phase boundaries resulting from the combination of nickel with the ionic conductor is given in Fig. 7. The scheme indicates the challenges in development of cathode materials as high triple-phase boundary length, high ionic as well as electronic conductivity, good accessibility for the gas phase, and long-term stability have to be achieved.

The standard material for the solid oxide electrolyte is YSZ ($Zr_{1-x}Y_xO_{2-x/2}$) as it has good ionic and low electronic conductivity as well as sufficient mechanical strength with low creep up to very high temperatures. In the last two decades also, other materials like ScSZ (scandium-doped zirconia, $Zr_{1-x}Sc_xO_{2-x/2}$) or GDC (gadolinium-doped ceria, $Ce_{1-x}Gd_xO_{2-x}$) have been successfully introduced [15]. The electrolyte transports oxygen ions via oxygen vacancies but blocking transport of electrons. The ability to sustain the thermal gradients and stability under reducing atmosphere (low oxygen partial pressure) are additionally required material properties for the application as electrolyte.

On the anode side, perovskite-type oxygen electrodes with the general structure ABO_3 and their composites with YSZ and GDC are applied. Here, different perovskite materials can be utilized as catalysts, e.g., $La_{1-x}Sr_xMnO_3$ (LSM), $La_{0.8}Sr_{0.2}CoO_3$ (LSC), or $La_{0.8}Sr_{0.2}Co_{0.2}Fe_{0.8}O_3$ (LSCF) [13].

During the long development history of solid oxide cells, planar cell concepts have been recognized to have the highest potential for power density and affordable costs [16]. As the planar cell is basically a ceramic component with limited mechanical stability, it needs to be supported. In general, there are three different types of planar solid oxide cells: electrolyte-supported cells, electrode-supported cells, and metal-supported cells (Fig. 8).

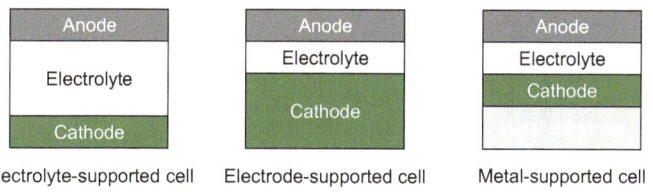

Fig. 8 Different types of cells for high-temperature electrolyzers

For the first two cell types, a specific layer of the cell is increased in thickness to provide the necessary mechanical stability. In case of metal-supported cells an additional porous metal substrate is introduced as supporting structure. These concepts come with different advantages and disadvantages, as specific overpotentials are increased like higher ohmic overpotential for electrolyte-supported cells, higher concentration overpotential for electrode-supported cell, and higher ohmic losses due to oxide scale formation in porous metal substrates in steam. A comprehensive overview on cell design and materials can be found, e.g., in reference [17].

Due to the high operating temperature and resulting thermal stresses in the ceramic cell materials coming from the stack environment due to differences in the coefficients of thermal expansion and thermal gradients, especially during heating and cooling, the lateral dimensions of the cells are limited to values of up to approximately 25×25 cm^2 [10, 18, 19]. The current density for long-term stable is in the range of 0.5–1 A/cm^2. Therefore, the resulting nominal power input of a single cell is currently in the range of less than 650 W.

3.2 Stacks

As syngas production capacity of a single cell is limited, several cells need to be combined to stacks. A scheme of a solid oxide electrolysis stack is given in Fig. 9. Several cells—generally a number of less than 50—are stacked with interconnects separating the anode and cathode compartments of different cells but allowing for electric in-series connection along the stack. The interconnects are manufactured from high-chromium stainless steels or from chromium alloys produced via powder metallurgical methods [15]. The interconnects contain gas channels allowing for a distribution of the cathode and anode gases and are coated by protective layers to reduce the oxide scale growth due to oxidation at high temperature. In order to ensure good contacting between the cells and interconnects, additional contact layers are used. Often also nickel meshes are applied on the cathode side as current collectors in order to improve the electron transport throughout the stack [20–22].

In order to reach a gas-tight connection, glass sealants are applied. After assembly of all components, the stacks are joined at high temperatures, generally in the range of $\vartheta > 900$ °C. The joining step is an additional process step in comparison to the manufacturing process of low-temperature electrolyzers, leading to the stack being

Fig. 9 Exemplary scheme of a cross-flow solid oxide electrolysis stack

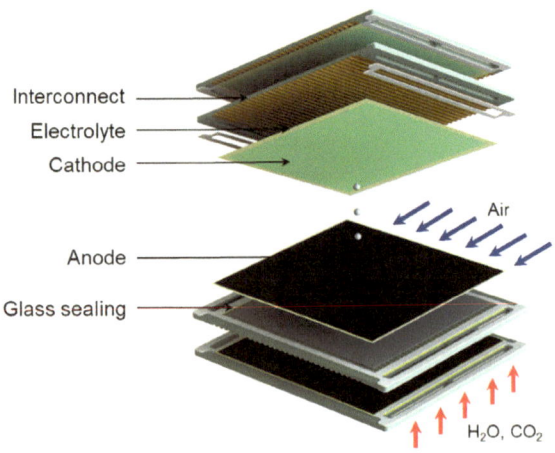

one single gas-tight unit. Therefore, in comparison to low-temperature electrolyzers, single cells within a stack cannot be replaced.

As the sealing between the gas volume within the stack and the environment is based on glass, high-temperature co-electrolysis stacks can often only be operated at atmospheric pressure. Pressurized operation of co-electrolyzers currently necessitates pressure chambers, ensuring a small pressure difference between the inside of the stack and its environment. Such concepts have been demonstrated [23, 24] but were found to be economically unfeasible [25]. The atmospheric operation of solid oxide electrolyzers is currently a drawback in comparison to low-temperature electrolyzers like alkaline or polymer electrolyte membrane (PEM) electrolyzers that can be operated under pressure. In case of co-electrolysis, the complete syngas needs to be compressed to the pressure level of the downstream synthesis step. For low-temperature electrolyzers, liquid water can be compressed with negligible energy demand upstream of the electrolysis unit and only carbon dioxide (CO_2) has to be compressed as a gas.

Another disadvantage in comparison to low-temperature electrolysis-based syngas production routes is related to oxygen generation. The generation of pure oxygen is challenging not only due to the anode material [26–28] but at the high temperatures also due to the metallic interconnect material [29]. Therefore, production of pure oxygen is often not focused, and the anode is flushed with air (Fig. 9) under operation to reduce the oxygen partial pressure.

For the current stack technology, the nominal power input to a co-electrolysis stack ($P_{el, DC}$) is generally less than 10 kW. Therefore, scaling-up the co-electrolysis stack technology is one of the directions of current R&D efforts. As the size of the applied cells is limited, as discussed above, other approaches are discussed to increase the footprint of the stacks. One potential pathway is a window-frame design, where several cells are aligned on one interconnect [30, 31]. However, as tolerances of the stack components add up during stacking—increasing the risk of failures like leakages—also the height of the stacks is limited. Therefore, in order to reach higher syngas production capacities, several stacks are combined in modules (as described in the next section).

3.3   Modules

The manufacturing of modules can follow two approaches that are most often combined. The first approach is connecting several stacks in series, like it is shown in Fig. 10. The second approach is a connection in parallel, as several of these "towers" can be arranged and electrically connected in parallel. For the example given in Fig. 6, 60 stacks are arranged in one unit following the described two approaches. The gases entering and leaving the cathode and anode compartment of the stacks need to be distributed evenly. The necessary gas manifolds are therefore also part of the stack unit. Due to the operation temperature, the unit comprising the electrolysis stacks needs to be insulated in order to reduce heat losses to the surrounding and, thus, a decrease in efficiency of the electrolyzer.

Efficient operation at temperatures of approx. 800 °C necessitates additional components like heat exchangers and electric heaters. To reduce heat losses, these components should be placed in close proximity to the stack unit. Therefore, these peripheral components are often also counted to the so-called stack module, as shown in Fig. 6. In the given scheme, heat exchangers and heaters are summarized under the term gas processing unit. As the cost share for these peripheral components decrease with increasing stack module size due to economies of scale, it is of advantage to increase the module size as far as technically possible. Additionally, such an increase reduces the surface-to-volume ratio and, therefore, the heat losses to the environment. The module size given in Fig. 6 is representative for the current state of the art.

Fig. 10 Stacks within a small-scale stack unit ($P_{el, DC} = 12$ kW) [32]. Reprinted with permission of Wiley

3.4 Systems

As the size of a single module is generally too small to meet the syngas demand of a downstream synthesis process, several modules are run in parallel to reach an acceptable system or synthesis plant scale. The general process scheme of such a co-electrolysis system is given in Fig. 11.

Carbon dioxide (CO_2) and water (H_2O) are fed to the system (1, 2). The necessary steam can be produced from waste heat of an associated process or via electric heating (3). As outlined above, the option utilizing waste heat offers a considerable reduction in electric energy demand. A scheme depicting the integration of a co-electrolysis system with an exothermal methanol synthesis as an example is given in Fig. 12.

As shown in Fig. 11, downstream of the evaporation step, the cathode feed gases have to be distributed to the m modules (4–6) of the electrolysis system. Within a specific module, the cathode feed gases are heated against the produced syngas in a counter-current heat exchanger (4). Due to heat losses and limited exchanger efficiency the inlet gases cannot be heated up to the operating temperature of the n stacks. To compensate these losses, an electric pre-heater (5) is often applied downstream of the heat exchanger to heat up the gases to the temperature of the electrolysis stacks or above, in case of endothermal operation. The electric heater must not be operated in case of exothermal operation of the stacks. However, it is always necessary during startup, in order to heat up the system to operation temperature.

After conversion of the feed gases on the cathode of the cells within the stack unit (6), the produced syngas is cooled down via the counter-current heat exchanger. The conversion within the cells is limited. This is due to a considerably increasing voltage at high conversion (Fig. 4). Additionally, at the resulting, high current densities degradation rate increases considerably [26]. Thus, the conversion within the electrolyzer is generally limited to values in the range of approximately 80%. Unconverted water is condensed in the syngas cooling step (7, 8) prior to compression of the produced

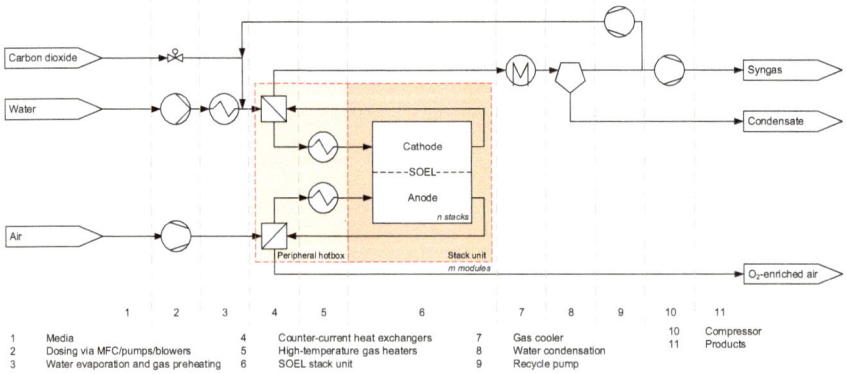

Fig. 11 General process scheme of a high-temperature co-electrolysis system (MFC: mass flow controller; SOEL: solid oxide electrolysis)

Fig. 12 Example for waste
heat integration of a
co-electrolysis system with a
subsequent methanol
synthesis based on CO-rich
syngas [10] (SOEC: solid
oxide electrolysis cell;
reprinted with permission of
The American Association
for the Advancement of
Science)

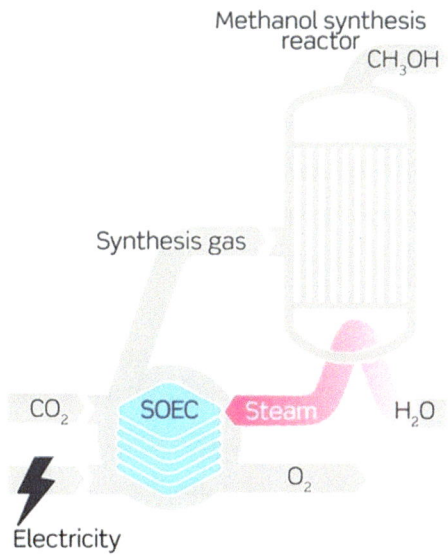

syngas (10) to the desired operating pressure of the downstream synthesis process. The condensed water can be recycled to increase the utilization of water.

High-temperature electrolyzers in general need a certain amount of reducing gases at the inlet of the cathode to prevent oxidation of the electrode material [4, 33, 34]. Therefore, part of the produced syngas is recycled (9) to achieve a fraction of reducing gases (H_2, CO) at the inlet of the cathode, generally in the range of approximately 5%.

As discussed above, the generation of pure oxygen (O_2) in high-temperature electrolyzers is challenging. Thus, the anode is often purged with a gaseous medium, most often air. The purge medium is also heated up in a counter-current heat exchanger (4) and an electric pre-heater (5). On the anode side the air flow is enriched with O_2 via the half-cell reaction given in Eq. (6).

The necessity to isolate the stack unit was already mentioned above. For the same reason, the peripheral components also operated at elevated temperatures (heat exchangers, electric pre-heaters) need to be isolated. For convenient manufacturing and maintenance, these components are often installed in a separate hotbox closely connected to the stack unit in order to decrease heat losses.

4   Operation and Performance Evaluation

In order to provide some insights into the operation of a co-electrolysis system, different technical parameters, namely cell voltage, energy flows, and energetic efficiency, are examined on the basis of experimental and modeling results. This is followed by an overview on the equipment costs associated with co-electrolysis, compared to other electrolysis technologies.

4.1   Technical Parameters

In the following, the technical parameters of cell voltage, energy flows, and energetic efficiency are presented for co-electrolysis systems.

Cell voltages. The general behavior of cell voltages in co-electrolysis mode is discussed on the basis of experimental results from a small-scale co-electrolyzer. According to the definition applied in Fig. 11, the studied system comprised a single module ($m = 1$). The nominal power input to the stack unit of $n = 4$ stacks with 30 cells each (Fig. 10) was $P_{el,DC} = 12$ kW. A comparison to the state of the art given in Fig. 6 shows that the size is rather small. However, the system contained all components given in Fig. 11, except for the syngas compression step. A detailed description of the system and the derived experimental results is given in reference [32].

   In Fig. 13a, results of the measured cell voltages $U_{cell, meas}$ in steam and co-electrolysis in dependence of the applied current density are given. A comparison of the cell voltages with the calculated thermoneutral voltage U_{tn} shows that the electrolyzer was operated in endothermal operation mode. With decreasing load or current density, overpotentials are decreasing which leads to less heat generation within the cells. Thus, temperature of the stacks—here, defined via the anode outlet temperature—decreases, as can be seen in Fig. 13b. In order to hold the stack on an appropriate operation temperature, additional heat needs to be supplied by the electric heaters upstream of the stacks. Generally, endothermicity of co-electrolysis is higher than in case of steam electrolysis (Fig. 2). Accordingly, the presented results given in Fig. 13b show a lower operating temperature of the stacks. Thus, cell voltages increase (Fig. 13a) due to the inverse temperature dependency of the overpotentials, as described above.

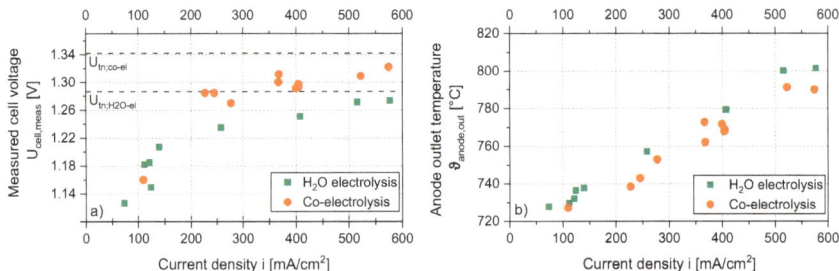

Fig. 13 Measured cell voltages $U_{cell, meas}$ (**a**) and anode outlet temperatures $\vartheta_{anode, out}$ (**b**) in dependence of applied current density i for a small-scale co-electrolysis system. Thermoneutral voltage U_{tn} is indicated by a dotted line (data from reference [32])

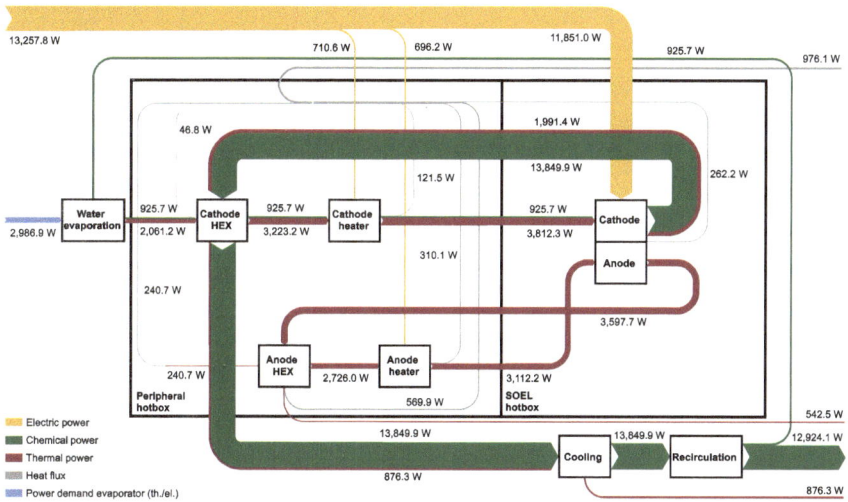

Fig. 14 Sankey diagram on the energy balance of a small-scale co-electrolysis system [32]. Chemical enthalpy flows are calculated based on the higher heating value (HHV) (HEX: heat exchanger; SOEL: solid oxide electrolysis; reprinted with permission of Wiley)

Energy balance. A Sankey diagram depicting the energy flows within the small-scale co-electrolyzer is given in Fig. 14. Due to the high operating temperature, large parts of the energy intrinsic to the streams are thermal energy. In order to reach a high energetic efficiency, this energy must be kept within the system. This is achieved by the heat exchangers on cathode and anode side, as outlined above. Figure 14 shows that considerable amounts of heat must be transferred in these two heat exchangers. Thus, heat exchanger design affects the efficiency of the system, but depending on the chosen size also the capital expenditures of a co-electrolysis system [25].

Another important aspect of high-temperature co-electrolyzers are heat losses. The data in Fig. 14 shows that for such a small system, as it was considered here for demonstration purpose, heat losses are not negligible. This is due to the unfavorable surface-to-volume ratio of small modules. The influence of heat losses decreases with increasing module size, which is one advantage of upscaling the technology. But even with the occurring heat losses and for the small scale, the high efficiency of co-electrolysis can be demonstrated by a comparison between the energy input into the system and the output of chemical energy stored within the produced syngas. A more detailed look on energetic efficiency is given in the following section.

Energetic efficiency. The energetic efficiency of syngas generation can be calculated according to Eq. (11):

$$\eta_{en, syngas} = \frac{P_{ch, syngas}}{P_{el, electrolyzer} + P_{el, pre\text{-}heating} + P_{el, compression} + P_{ch, purge\ gas}\left(+P_{el, evaporation}\right)} \quad (11)$$

The chemical power of the produced syngas can be calculated based on the lower heating value (LHV). In general, different forms of energy can be input streams to the system. The electric energy demand of the electrolyzer and the mentioned electric gas-heaters are the most obvious. If the syngas is to be compressed, the energy demand for this purpose can be included in the calculation. Especially in the context of power-to-liquid (PtL) processes, a gas stream has to be purged from the system. The chemical power of this purge gas can be converted to thermal power and utilized, e.g., for gas pre-heating. As the energy would else have to be provided in form of electric energy, it has to be accounted for in the calculation of energetic efficiency of syngas generation. The most important energy input for the assessment of high-temperature co-electrolysis is the necessary heat for evaporation of feed water. If waste heat is applied for this purpose, the energy demand is not considered as an input for the calculation of energetic efficiency. If it is done by electric energy, it has to be included.

For the small-scale demonstration plant discussed above, the given equation simplifies, as no compression and no purge-gas utilization are considered. From the energy balance given in Fig. 14, an energetic efficiency of 76.2% results in case of electric evaporation, as it was applied in the plant. If the energy for evaporation would have been supplied by waste heat, an efficiency of 88.1% would have resulted. These are considerably high values. However, it has to be kept in mind, that for the presented results no syngas compression—which is unavoidable for coupling with a downstream synthesis—and no AC/DC conversion efficiency have been considered.

A more detailed evaluation of the performance of co-electrolysis in a system context is given in the following paragraphs. As shown previously in Fig. 1, there are basically two pathways for the production of syngas based on electrolysis. Besides co-electrolysis, the combination of water electrolysis and a reverse water–gas shift (RWGS) step can be applied. This approach is the one to be chosen if low-temperature electrolysis is to be applied for syngas generation. In the following, both pathways are compared in the context of a Fischer-Tropsch-based power-to-liquid (PtL) process, as described in reference [25]. For the low-temperature electrolysis approach, a polymer electrolyte membrane electrolysis (PEMEL) is chosen. In order to assess both tech-nological options, the energetic efficiency of the syngas generation step within the power-to-liquid (PtL) process was calculated according to Eq. (11). In contrast to the results discussed above, for the electric energy demand of the electrolyzer losses due to AC/DC conversion were included. Additional assumptions can be derived from references [9, 25]. As syngas generation is most often directly coupled with a synthesis step, it is beneficial to integrate both processes as effectively as possible. For the exemplary case presented in Figs. 15 and 16, combustion of the purge gas of the considered power-to-liquid (PtL) process was integrated into the syngas generation step. The chemical power of the purge gas, that is converted into thermal power by combustion, was integrated into the calculation of efficiency according to Eq. (11), as discussed above.

For the Fischer-Tropsch synthesis, an operating pressure (p_{FT}) of 20 bar was assumed. Thus, for the PEMEL-based syngas generation pressurized operation was assumed, which leads to a negligible energy demand for water compression and only

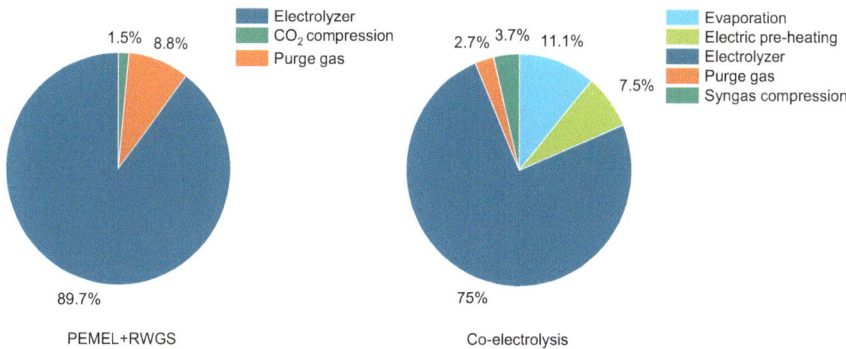

Fig. 15 Comparison of energy input of PEM electrolysis and subsequent reverse water–gas shift reaction with co-electrolysis for syngas generation within a FT-based PtL process (FT: Fischer-Tropsch; PEMEL: polymer electrolyte membrane electrolysis; PtL: power-to-liquid; RWGS: reverse water–gas shift)

Fig. 16 Comparison of energetic efficiency of PEM electrolysis and subsequent reverse water–gas shift reaction with co-electrolysis for syngas generation within a FT-based PtL process (FT: Fischer-Tropsch; PEMEL: polymer electrolyte membrane electrolysis; PtL: Power-to-Liquid; RWGS: reverse water–gas shift)

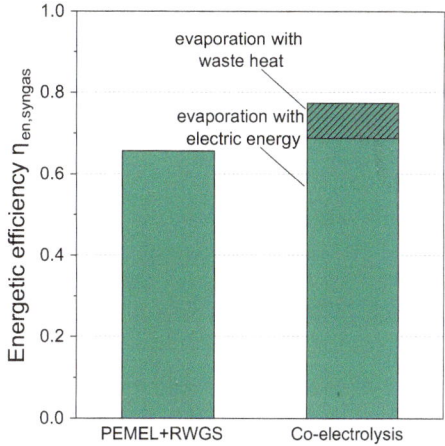

a comparably small energy demand for CO_2 compression. In case of co-electrolysis, the produced syngas has to be compressed, which leads to a higher energy demand for that purpose, as can be seen in Fig. 15.

However, co-electrolysis benefits from the fact that in high-temperature electrolysis the energy demand of evaporation can be decoupled from the electrolysis step, as discussed above. Thus, if waste heat at a temperature level sufficient for steam generation is available, the electric energy demand can be significantly decreased, and the energetic efficiency of syngas production can be increased according to Eq. (11). As especially Fischer-Tropsch synthesis is highly exothermal and is generally cooled with boiling water [35, 36], steam is frequently available in such a power-to-liquid (PtL) process.

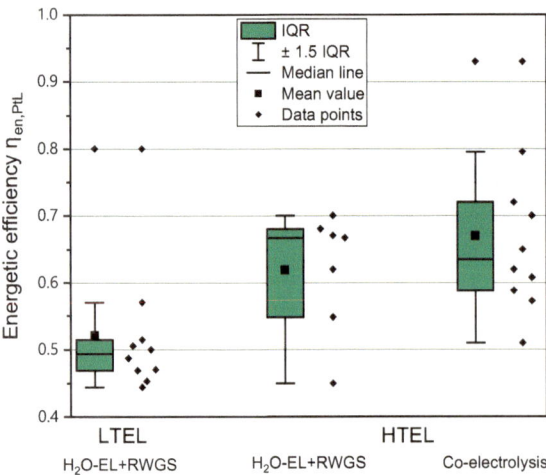

Fig. 17 Literature data on the energetic efficiency of PtL processes based on different syngas generation schemes [37–61] (IQR: interquartile range; HT-EL: high-temperature electrolysis; LT-EL: low-temperature electrolysis; PtL: power-to-liquid; RWGS: reverse water–gas shift; SOEL: solid oxide electrolysis)

Figure 16 illustrates that when electric energy is used for feed water evaporation, both pathways—PEMEL + RWGS and co-electrolysis—can achieve comparable energetic efficiencies of approximately 66 and 69%, respectively, for syngas generation. However, if waste heat from the synthesis reaction is utilized, energetic efficiency ($\eta_{en, syngas}$) of the co-electrolysis approach can be increased to values close to 80%.

The case of a Fischer-Tropsch-based power-to-liquid (PtL) process, as discussed above, is only one potential application example. However, the advantages of co-electrolysis were confirmed in several literature studies on the assessment of power-to-liquid (PtL) processes. In Fig. 17, literature data on the overall energetic efficiency of the process are compared depending on the applied syngas generation technology. The energetic efficiency applied for this comparison is based on the chemical power of the produced products, e.g., methanol or Fischer-Tropsch hydrocarbons, and the electric energy input to the overall process, also including the synthesis step.

The results illustrate that the difference in syngas production efficiency also transfers to overall process efficiency. The results in Fig. 17 also show that this increase in efficiency is mainly due to the waste heat integration for steam generation, as concepts based on the two-step approach applying high-temperature water electrolysis and an RWGS step also yield nearly comparable process efficiencies as co-electrolysis-based concepts allow for. However, when it comes to the operational implementation applying co-electrolysis allows to omit one costly process step in the overall process—the RWGS reactor.

Fig. 18 Expected cost degradation for the different electrolysis technologies (AEL: alkaline electrolysis; PEMEL: polymer electrolyte membrane electrolysis; SOEL: solid oxide electrolysis) (adapted from [62])

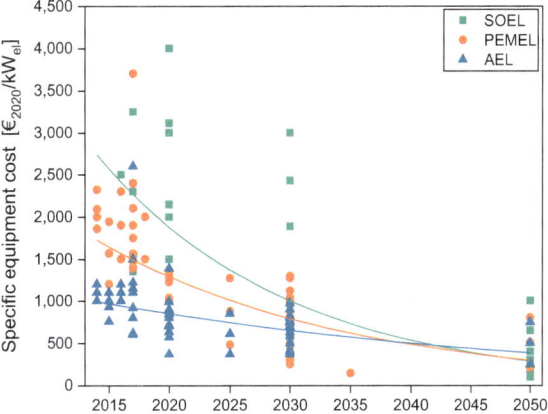

4.2 Economic Parameters

The advantages of co-electrolysis with regards to efficiency are obvious—provided that the required waste heat can be provided. However, besides technical aspects, for implementation and utilization at industrial scale also economics are of major importance. Currently, cost for high-temperature electrolyzers are considerably higher than for low-temperature electrolyzers (Fig. 18).

Figure 18 shows that so far (2015–2020), alkaline and PEM electrolysis technologies have exhibited lower specific equipment costs, whereas SOEL show higher costs due to the still lower maturity of the technology. Therefore, the effect of decreased operational expenditures in form of electric energy—due to the high efficiency of SOEL—is currently offset by higher capital expenditures, when it comes to an overall economic assessment of the technology [25].

For AEL, current costs are approximately in the range of 500–1,000 €$_{(2020)}$/kW$_{el}$, and for PEM, they are more in the range of about 1,500–2,500 €$_{(2020)}$/kW$_{el}$. According to the data shown and in correspondence with the already reached maturity of the technologies, less cost reduction is expected for AEL in the future, while higher cost reduction is anticipated for PEM. As a result, by 2030, these technologies are expected to have a largely comparable range of potential specific equipment costs. This range for AEL and PEM in 2030 is expected to be around 500–1,000 €$_{(2020)}$/kW$_{el}$.

The currently higher costs for SOEL are expected to decrease in future with increasing maturity of the technology. Improvements in cell technology, increased stack footprints, and especially more automation in the manufacturing process are expected to drive the cost degradation. As a result, in the longer term (2030–2050) converging specific costs in the range of 500 €$_{(2020)}$/kW$_{el}$ for all technologies are expected. Overall, according to the data shown in the present and foreseeable future, SOEL may still bear higher equipment costs than other electrolysis technologies. However, there is a long-term potential for substantial decrease in specific equipment costs, making it possible for SOEL to align with the costs of other electrolysis techniques, namely AEL and PEMEL.

5 Conclusion

Co-electrolysis is a promising technology for the efficient production of syngas for downstream conversion into hydrocarbon-based synthetic fuels and other chemical products. High-temperature or solid oxide electrolyzers allow for the direct conversion of CO_2 and H_2O into syngas due to the activity of the applied cathode material. However, the advantages of this technology are not only limited to this direct conversion but also to a considerable decrease in electric energy demand if waste heat is available for steam generation. In the context of power-to-liquid (PtL) processes, the synthesis step offers a convenient heat source.

The advantages in terms of energetic efficiency cannot only be calculated on a theoretical basis but have also been proven on demonstration scale. However, in comparison to low-temperature electrolysis, high-temperature electrolysis and, therefore, co-electrolysis currently exhibits a lower technological maturity. Especially the scale of co-electrolysis systems is currently rather small in comparison to the scales reached for low-temperature water electrolysis systems. It is expected that the obvious advantages of high-temperature electrolysis in general and specifically the ones of co-electrolysis drive significant development activities in future, enabling widespread industrial application in the production of synthetic fuels.

References

1. Ghaib K, Nitz K, Ben-Fares F-Z (2016) Chemical methanation of CO_2 a review. ChemBioEng Rev 3(6):266–275. https://doi.org/10.1002/cben.201600022
2. Haag S, Drosdzol C, Williams B, Peña V, Palauschek N, Wilken M et al (2022) Recent developments in methanol technology by air liquid for CO_2 reduction and CO_2 usage. Chem Ing Tec 94(11):1655–1666. https://doi.org/10.1002/cite.202200061
3. Roode-Gutzmer QI, Kaiser D, Bertau M (2019) Renewable methanol synthesis. ChemBioEng Rev 6(6):209–236. https://doi.org/10.1002/cben.201900012
4. Diethelm S, van herle J, Montinaro D, Bucheli O (2013) Electrolysis and co-electrolysis performance of SOE short stacks. Fuel Cells 13(4):631–637. https://doi.org/10.1002/fuce.201 200178
5. Foit SR, Vinke IC, de Haart LGJ, Eichel R-A (2017) Power-to-syngas: an enabling technology for the transition of the energy system? Angew Chem Int Ed Engl 56(20):5402–5411. https:// doi.org/10.1002/anie.201607552
6. Kim S-W, Kim H, Yoon KJ, Lee J-H, Kim B-K, Choi W et al (2015) Reactions and mass transport in high temperature co-electrolysis of steam/CO_2 mixtures for syngas production. J Power Sources 280:630–639. https://doi.org/10.1016/j.jpowsour.2015.01.083
7. Ebbesen SD, Knibbe R, Mogensen M (2012) Co-electrolysis of steam and carbon dioxide in solid oxide cells. J Electrochem Soc 159(8):F482–F489. https://doi.org/10.1149/2.076208jes

8. van der Laan GP, Beenackers AACM (1999) Kinetics and selectivity of the Fischer-Tropsch synthesis: a literature review. Catal Rev 41(3–4):255–318. https://doi.org/10.1081/CR-100 101170

9. Herz G, Müller N, Adam P, Megel S, Reichelt E, Jahn M (2020) High temperature co-electrolysis as a key technology for CO_2 emission mitigation—a model-based assessment of CDA and CCU. Chem Ing Tec 92(8):1044–1058. https://doi.org/10.1002/cite.202000012

10. Hauch A, Küngas R, Blennow P, Hansen AB, Hansen JB, Mathiesen BV et al (2020) Recent advances in solid oxide cell technology for electrolysis. Science 370(6513). https://doi.org/10.1126/science.aba6118

11. Godula-Jopek A (2015) Hydrogen production. Wiley-VCH Verlag GmbH & Co. KGaA, Weinheim, Germany

12. Aicart J, Laurencin J, Petitjean M, Dessemond L (2014) Experimental validation of two-dimensional H_2O and CO_2 co-electrolysis modeling. Fuel Cells 14(3):430–447. https://doi.org/10.1002/fuce.201300214

13. Zheng Y, Chen Z, Zhang J (2021) Solid oxide electrolysis of H_2O and CO_2 to produce hydrogen and low-carbon fuels. Electrochem Energ Rev 4(3):508–517. https://doi.org/10.1007/s41918-021-00097-4

14. Biswas S, Kaur G, Paul G, Giddey S (2023) A critical review on cathode materials for steam electrolysis in solid oxide electrolysis. Int J Hydrogen Energy 48(34):12541–12570. https://doi.org/10.1016/j.ijhydene.2022.11.307

15. Nechache A, Hody S (2021) Alternative and innovative solid oxide electrolysis cell materials: a short review. Renew Sustain Energy Rev 149:111322. https://doi.org/10.1016/j.rser.2021.111322

16. Udomsilp D, Lenser C, Guillon O, Menzler NH (2021) Performance benchmark of planar solid oxide cells based on material development and designs. Energy Tech 9(4):2001062. https://doi.org/10.1002/ente.202001062

17. Wolf SE, Winterhalder FE, Vibhu V, de Haart LGJ, Guillon O, Eichel R-A et al (2023) Solid oxide electrolysis cells—current material development and industrial application. J Mater Chem A 11(34):17977–18028. https://doi.org/10.1039/d3ta02161k

18. Mogensen MB (2020) Materials for reversible solid oxide cells. Curr Opin Electrochem 21:265–273. https://doi.org/10.1016/j.coelec.2020.03.014

19. Mogensen MB, Chen M, Frandsen HL, Graves C, Hansen JB, Hansen KV et al (2019) Reversible solid-oxide cells for clean and sustainable energy. Clean Energy 3(3):175–201. https://doi.org/10.1093/ce/zkz023

20. Kusnezoff M, Trofimenko N, Müller M, Michaelis A (2016) Influence of electrode design and contacting layers on performance of electrolyte supported SOFC/SOEC single cells. Materials (Basel) 9(11). https://doi.org/10.3390/ma9110906

21. Ioannidou E, Neophytides SG, Niakolas DK (2019) Distinguishing the CO_2 electro-catalytic reduction pathway on modified Ni/GDC electrodes for the SOEC H_2O/CO_2 co-electrolysis process. ECS Trans 91(1):2687–2696. https://doi.org/10.1149/09101.2687ecst

22. Sarner S, Schreiber A, Menzler NH, Guillon O (2022) Recycling strategies for solid oxide cells. Adv Energy Mater 12(35):2201805. https://doi.org/10.1002/aenm.202201805

23. Jensen SH, Sun X, Ebbesen SD, Knibbe R, Mogensen M (2010) Hydrogen and synthetic fuel production using pressurized solid oxide electrolysis cells. Int J Hydrogen Energy 35(18):9544–9549. https://doi.org/10.1016/j.ijhydene.2010.06.065

24. Riedel M, Heddrich MP, Ansar A, Fang Q, Blum L, Friedrich KA (2020) Pressurized operation of solid oxide electrolysis stacks: an experimental comparison of the performance of 10-layer stacks with fuel electrode and electrolyte supported cell concepts. J Power Sources 475:228682. https://doi.org/10.1016/j.jpowsour.2020.228682

25. Herz G, Rix C, Jacobasch E, Müller N, Reichelt E, Jahn M et al (2021) Economic assessment of power-to-liquid processes—influence of electrolysis technology and operating conditions. Appl Energy 292:116655. https://doi.org/10.1016/j.apenergy.2021.116655

26. Wang Y, Li W, Ma L, Li W, Liu X (2020) Degradation of solid oxide electrolysis cells: phenomena, mechanisms, and emerging mitigation strategies—a review. J Mater Sci Technol 55:35–55. https://doi.org/10.1016/j.jmst.2019.07.026

27. Park B-K, Zhang Q, Voorhees PW, Barnett SA (2019) Conditions for stable operation of solid oxide electrolysis cells: oxygen electrode effects. Energy Environ Sci 12(10):3053–3062. https://doi.org/10.1039/c9ee01664c

28. Khan MS, Xu X, Knibbe R, Zhu Z (2021) Air electrodes and related degradation mechanisms in solid oxide electrolysis and reversible solid oxide cells. Renew Sustain Energy Rev 143:110918. https://doi.org/10.1016/j.rser.2021.110918

29. Mikkola J, Couturier K, Talic B, Frangini S, Giacometti N, Pelissier N et al (2022) Protective coatings for ferritic stainless steel interconnect materials in high temperature solid oxide electrolyser atmospheres. Energies 15(3):1168. https://doi.org/10.3390/en15031168

30. Peters R, Frank M, Tiedemann W, Hoven I, Deja R, Kruse N et al (2021) Long-term experience with a 5/15kW-class reversible solid oxide cell system. J Electrochem Soc 168(1):14508. https://doi.org/10.1149/1945-7111/abdc79

31. Blum L, Fang Q, de Haart L, Malzbender J, Margaritis N, Menzler NH et al (2019) Forschungszentrum Jülich—progress in SOC development. ECS Trans 91(1):2443–2453. https://doi.org/10.1149/09101.2443ecst

32. Reichelt E, Adam P, Näke R, Herz G, Megel S (2023) Small-scale demonstration of a high-temperature coelectrolyzer—experimental results and model validation. Energy Tech. https://doi.org/10.1002/ente.202300086

33. Cai Q, Luna-Ortiz E, Adjiman CS, Brandon NP (2010) The effects of operating conditions on the performance of a solid oxide steam electrolyser: a model-based study. Fuel Cells 10(6):1114–1128. https://doi.org/10.1002/fuce.200900211

34. Cinti G, Discepoli G, Bidini G, Lanzini A, Santarelli M (2016) Co-electrolysis of water and CO_2 in a solid oxide electrolyzer (SOE) stack. Int J Energy Res 40(2):207–215. https://doi.org/10.1002/er.3450

35. Iglesias Gonzalez M, Kraushaar-Czarnetzki B, Schaub G (2011) Process comparison of biomass-to-liquid (BtL) routes Fischer-Tropsch synthesis and methanol to gasoline. Biomass Conv Bioref 1(4):229–243. https://doi.org/10.1007/s13399-011-0022-2

36. Konarova M, Aslam W, Perkins G (2022) Fischer-Tropsch synthesis to hydrocarbon biofuels: present status and challenges involved. In: Hydrocarbon biorefinery. Elsevier, pp 77–96

37. Albrecht FG, Nguyen T-V (2020) Prospects of electrofuels to defossilize transportation in Denmark—a techno-economic and ecological analysis. Energy 192:116511. https://doi.org/10.1016/j.energy.2019.116511

38. Bos MJ, Kersten S, Brilman D (2020) Wind power to methanol: Renewable methanol production using electricity, electrolysis of water and CO_2 air capture. Appl Energy 264:114672. https://doi.org/10.1016/j.apenergy.2020.114672

39. Botta G, Solimeo M, Leone P, Aravind PV (2015) Thermodynamic analysis of coupling a SOEC in Co-electrolysis mode with the dimethyl ether synthesis. Fuel Cells 15(5):669–681. https://doi.org/10.1002/fuce.201500016

40. Fu Q, Dailly J, Brisse A, Zahid M (2011) High-temperature CO_2 and H_2O electrolysis with an electrolyte-supported solid oxide cell. ECS Trans 35(1):2949–2956. https://doi.org/10.1149/1.3570294

41. Graves C, Ebbesen SD, Mogensen M, Lackner KS (2011) Sustainable hydrocarbon fuels by recycling CO_2 and H_2O with renewable or nuclear energy. Renew Sustain Energy Rev 15(1):1–23. https://doi.org/10.1016/j.rser.2010.07.014

42. Hansen JB, Christiansen N, Nielsen JU (2011) Production of sustainable fuels by means of solid oxide electrolysis. ECS Trans 35(1):2941–2948. https://doi.org/10.1149/1.3570293

43. Jess A, Kaiser P, Kern C, Unde RB, von Olshausen C (2011) Considerations concerning the energy demand and energy mix for global welfare and stable ecosystems. Chem Ing Tec 83(11):1777–1791. https://doi.org/10.1002/cite.201100066

44. Marchese M, Giglio E, Santarelli M, Lanzini A (2020) Energy performance of power-to-Liquid applications integrating biogas upgrading, reverse water gas shift, solid oxide electrolysis and Fischer-Tropsch technologies. Energy Conver Manag X 6:100041. https://doi.org/10.1016/j.ecmx.2020.100041

45. Mignard D, Pritchard C (2006) Processes for the synthesis of liquid fuels from CO_2 and marine energy. Chem Eng Res Des 84(9):828–836. https://doi.org/10.1205/cherd.05204
46. Michailos S, McCord S, Sick V, Stokes G, Styring P (2019) Dimethyl ether synthesis via captured CO_2 hydrogenation within the power to liquids concept: a techno-economic assessment. Energy Convers Manage 184:262–276. https://doi.org/10.1016/j.enconman.2019.01.046
47. Parigi D, Giglio E, Soto A, Santarelli M (2019) Power-to-fuels through carbon dioxide re-utilization and high-temperature electrolysis: a technical and economical comparison between synthetic methanol and methane. J Cleaner Prod 226:679–691. https://doi.org/10.1016/j.jclepro.2019.04.087
48. Rivera-Tinoco R, Farran M, Bouallou C, Auprêtre F, Valentin S, Millet P et al (2016) Investigation of power-to-methanol processes coupling electrolytic hydrogen production and catalytic CO_2 reduction. Int J Hydrogen Energy 41(8):4546–4559. https://doi.org/10.1016/j.ijhydene.2016.01.059
49. Schemme S, Breuer JL, Köller M, Meschede S, Walman F, Samsun RC et al (2020) H2-based synthetic fuels: a techno-economic comparison of alcohol, ether and hydrocarbon production. Int J Hydrogen Energy 45(8):5395–5414. https://doi.org/10.1016/j.ijhydene.2019.05.028
50. Tarutin AP, Filonova EA, Ricote S, Medvedev DA, Shao Z (2023) Chemical design of oxygen electrodes for solid oxide electrochemical cells: a guide. Sustain Energy Technol Assess 57:103185. https://doi.org/10.1016/j.seta.2023.103185
51. Tremel A, Wasserscheid P, Baldauf M, Hammer T (2015) Techno-economic analysis for the synthesis of liquid and gaseous fuels based on hydrogen production via electrolysis. Int J Hydrogen Energy 40(35):11457–11464. https://doi.org/10.1016/j.ijhydene.2015.01.097
52. Wang L, Chen M, Küngas R, Lin T-E, Diethelm S, Maréchal F et al (2019) Power-to-fuels via solid-oxide electrolyzer: operating window and techno-economics. Renew Sustain Energy Rev 110:174–187. https://doi.org/10.1016/j.rser.2019.04.071
53. Zhang H, Desideri U (2020) Techno-economic optimization of power-to-methanol with co-electrolysis of CO_2 and H_2O in solid-oxide electrolyzers. Energy 199:117498. https://doi.org/10.1016/j.energy.2020.117498
54. Zhang H, Wang L, van herle J, Maréchal F, Desideri U (2019) Techno-economic optimization of CO_2-to-methanol with solid-oxide electrolyzer. Energies 12(19):3742. https://doi.org/10.3390/en12193742
55. Albrecht FG, König DH, Baucks N, Dietrich R-U (2017) A standardized methodology for the techno-economic evaluation of alternative fuels—a case study. Fuel 194:511–526. https://doi.org/10.1016/j.fuel.2016.12.003
56. König DH (2016) Techno-ökonomische Prozessbewertung der Herstellung synthetischen Flugturbinentreibstoffes aus CO_2 und H_2
57. Becker WL, Braun RJ, Penev M, Melaina M (2012) Production of Fischer-Tropsch liquid fuels from high temperature solid oxide co-electrolysis units. Energy 47(1):99–115. https://doi.org/10.1016/j.energy.2012.08.047
58. König DH, Freiberg M, Dietrich R-U, Wörner A (2015) Techno-economic study of the storage of fluctuating renewable energy in liquid hydrocarbons. Fuel 159:289–297. https://doi.org/10.1016/j.fuel.2015.06.085
59. Stempien JP, Ni M, Sun Q, Chan SH (2015) Thermodynamic analysis of combined solid oxide electrolyzer and Fischer-Tropsch processes. Energy 81:682–690. https://doi.org/10.1016/j.energy.2015.01.013
60. Cinti G, Baldinelli A, Di Michele A, Desideri U (2016) Integration of solid oxide electrolyzer and Fischer-Tropsch: a sustainable pathway for synthetic fuel. Appl Energy 162:308–320. https://doi.org/10.1016/j.apenergy.2015.10.053

61. Vázquez FV, Koponen J, Ruuskanen V, Bajamundi C, Kosonen A, Simell P et al (2018) Power-to-X technology using renewable electricity and carbon dioxide from ambient air: SOLETAIR proof-of-concept and improved process concept. J CO_2 Util 28:235–46. https://doi.org/10.1016/j.jcou.2018.09.026
62. Jacobasch E, Herz G, Rix C, Müller N, Reichelt E, Jahn M et al (2021) Economic evaluation of low-carbon steelmaking via coupling of electrolysis and direct reduction. J Clean Prod 328:129502. https://doi.org/10.1016/j.jclepro.2021.129502

Methanation

T. J. Schildhauer and A. Gantenbein

Abstract The future energy system demands for a higher flexibility and higher versatility of energy carriers and conversion methods to incorporate the growing amount of renewable electricity generation methods. Renewable chemical energy carriers such as hydrogen, bio- or synthetic methane as well as further synthetic hydrocarbons have limited application nowadays, but they will most likely play an increasingly important role in the future. In this context, power-to-gas (PtG) conversion routes will be crucial for the provision of gaseous synthetic energy carriers. A power-to-gas (PtG) approach based on methanation presents a possible option to couple electricity, gas, and heat distribution networks. Integrating the respective conversion and storage technologies allows to offset fluctuations within these interconnected grids. In this context, in the chapter, the fundamentals of the methanation reaction system, the thermodynamic limitations, applied catalysts, and reactor types will be highlighted.

Keywords Power-to-Gas · Catalysts · Reactors · Synthetic Methane

1 Introduction

The future energy system demands for a higher flexibility and higher versatility of energy carriers and conversion methods to incorporate the growing amount of renewable electricity generation methods. Renewable chemical energy carriers such as hydrogen, bio- or synthetic methane as well as further synthetic hydrocarbons have limited application nowadays, but they will most likely play an increasingly important role in the future. In this context, power-to-gas (PtG) conversion routes will be crucial for the provision of gaseous synthetic energy carriers. The first step in such a PtG process is represented by the electrolysis of water where renewable electrical energy is used to produce hydrogen. The direct application of hydrogen in various processes is well established, and therefore, an efficient use of the stored

T. J. Schildhauer (✉) · A. Gantenbein
Paul Scherrer Institut (PSI), Villigen, Switzerland
e-mail: tilman.schildhauer@psi.ch

© The Author(s), under exclusive license to Springer Nature Switzerland AG 2025
N. Bullerdiek et al. (eds.), *Powerfuels*, Green Energy and Technology,
https://doi.org/10.1007/978-3-031-62411-7_21

energy is possible. Nevertheless, the handling and storage of hydrogen in current infrastructure is limited without major adaptations. It can be therefore advantageous to further convert hydrogen to methane, even more, if longer-term energy storage is required. This has among others the following positive implications:

- Use of existing infrastructure. Europe, for example, has a well-established grid of natural gas pipelines operating at various pressure levels. Due to material properties and existing downstream applications, its tolerance for high hydrogen concentrations is limited [1, 2]. Injection of methane from a PtG process allows using the existing, low-cost transport infrastructure to supply customers. Furthermore, applications currently operating on natural gas (industrial high temperature processes, such as steel, glass, and brick industry, residential heating, electricity generation, storage, mobility, and especially heavy transport) can be further operated without modification until better alternatives are found. To fully exploit the possibilities of the gas distribution system, the product gas of a PtG system has to be injected at the maximum possible pressure level, or additional compression stations have to be built to re-inject gas to large transport pipelines.
- Gas storage. Along the existing gas grid, gas storages exist at suitable pressure levels. They allow for long-term storage of energy. Since methane has an around 3.5-fold higher volumetric energy density than hydrogen, larger amounts of energy can be stored at about same compression cost. Additionally, due to the vast system of pipelines and distribution grid and their ability to operate at changing pressure levels, a large amount of energy can be buffered in this existing infrastructure.
- Addition of flexibility. Power-to-hydrogen in the form of an electrolyser and a hydrogen tank is highly flexible and is therefore suitable for short-term energy storage and absorbing short peaks of electricity production for grid stabilisation. The stored hydrogen can be re-electrified by fuel cells or conventional electricity generation methods. Longer-term storage of chemical energy requires higher energy density to be technically and economically feasible. Methanation therefore adds another degree of freedom to the energy system: it is possible to absorb unused energy into a system, which allows for long-term storage (Fig. 1). Furthermore, it enables the use of this stored energy in other, conventional sectors, such as mobility, residential or industrial heating, and chemical industry. Each involved conversion step emits off-heat, which can be fed to heat distribution grids. In the case of catalytic methanation, the temperature is even suitable for industrial heating processes. Methanation, together with combined heat and power (CHP) is a key part in linking the electricity grid with the gas and heat distribution grids. This interconnection of distribution grids allows for a higher operational flexibility of the systems and therefore a higher resilience of the overall energy systems. Additionally to hydrogen production, methanation can play a key role in the integration of renewable energies in the future energy system and helps to close the carbon cycle.

As exemplary shown in Fig. 1, a power-to-gas (PtG) approach based on methanation presents a possible option to couple electricity, gas, and heat distribution networks. Integrating the respective conversion and storage technologies allows

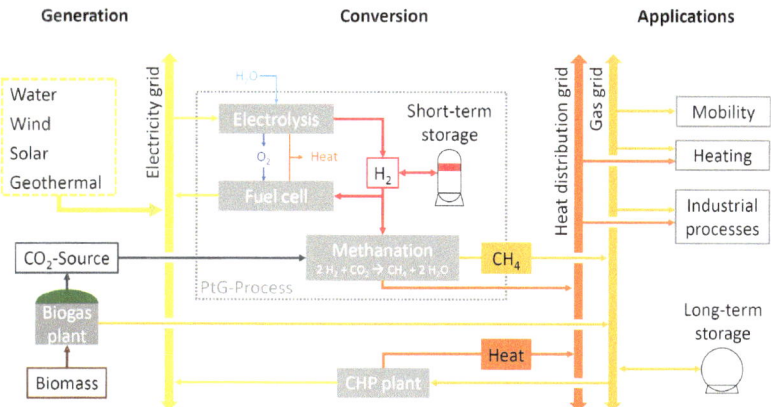

Fig. 1 Methanation-based power-to-gas (PtG) approach to couple electricity, gas, and heat distribution grids (CHP: combined heat and power; source [3])

to offset fluctuations within these interconnected grids. In the illustrated example, biogas derived from a digester is utilised as the source of carbon dioxide (CO_2) for the methanation process.

Against this background, in the following sections, the fundamentals of the methanation reaction system, the thermodynamic limitations, applied catalysts, and reactor types will be highlighted. Further, Sect. 3 discusses the efforts towards a higher flexibility of methanation processes, while Sect. 4 reports typical process chains, long-duration tests, and recent progress in commercialisation. Section 5 concludes this chapter with a comparison and an outlook.

2 Fundamentals of Methanation

The methanation reaction was discovered by Sabatier and Senderens in 1902. It allows to convert hydrogen (H_2) with carbon monoxide (CO) or carbon dioxide (CO_2), Eqs. (1) and (2), into methane (CH_4) [4]. This reaction is highly exothermic and limited by thermodynamic equilibrium:

$$3H_2 + CO \rightleftharpoons CH_4 + H_2O \text{ with } \Delta H_r^0 = -206 \text{ kJ/mol} \tag{1}$$

$$4H_2 + CO_2 \rightleftharpoons CH_4 + 2\,H_2O \text{ with } \Delta H_r^0 = -165 \text{ kJ/mol} \tag{2}$$

These two reactions are connected by the reversed homogeneous water–gas shift reaction:

$$H_2 + CO_2 \rightleftharpoons CO + H_2O \text{ with } \Delta H_r^0 = 41 \text{ kJ/mol} \tag{3}$$

It is a specific advantage of methanation that the thermodynamics limits the conversion of carbon oxides to methane (see next section), but that the selectivity to methane is strongly favoured. This allows converting many different gas mixtures containing carbon oxides without the necessity to adjust the stoichiometric ratios before. Therefore, besides mixtures of pure CO and/or CO_2 with hydrogen, also cleaned biogas (a mixture of methane and CO_2 stemming from anaerobic digestion) and, in case of catalytic methanation, product gas from gasification and pyrolysis (containing CO, CO_2, CH_4, C_2H_x and higher hydrocarbons) can be brought in contact with hydrogen and a suited catalyst to produce renewable methane.

Sufficient addition of hydrogen or steam and in some cases downstream separation of CO_2 (and recycle of unreacted hydrogen) enable very high concentration levels of methane (> 96% after drying). The reason is that typical methanation catalysts also support the (Reverse) water–gas shift reaction as well as hydrogenolysis of higher hydrocarbons (C_2H_x and larger molecules) to methane. Research to find suitable conditions for converting these species by biological methanation is ongoing. In a later section, typical process chains will be discussed in detail.

2.1 Thermodynamic Limitations of Methanation

Apart from the conversion of pure CO_2 or syngas streams, methanation is often applied in biomass conversion pathways. In such a case, the feed gas may originate from a digester or gasifier. Although such gases mostly contain CO_2, CO, H_2, CH_4, and water, they may also contain C_2-species and higher hydrocarbons in low percentages. In the case of a gasification process, also aromatic compounds occur. Such hydrocarbons, and especially CO, are prone to form carbon deposits on the catalyst surface. In the case of CO, the Boudouard reaction plays an important role Eq. (4):

$$2\,CO \rightleftharpoons C(s) + CO_2 \text{ with } \Delta H_r^0 = -172.54\,kJ/mol \tag{4}$$

The carbon deposits on the catalysts surface formed by this reaction may further polymerise and lead to catalyst deactivation. On the other hand, by applying steam or H_2, the deposits can be re-gasified by the heterogeneous water–gas shift reaction:

$$2\,H_2 + C(s) \rightleftharpoons CH_4 \tag{5}$$

$$H_2O\,(g) + C(s) \rightleftharpoons CO + H_2 \tag{6}$$

All involved reactions are reversible, and an equilibrium concentration is formed between the species. To what extent the equilibria are formed and what gas compositions occur is highly dependent on the applied gas composition, the process conditions, the catalysts involved, and the reactor design used.

The reversibility of the involved reactions allows to some extent to tune the product gas composition by adjusting the temperature and the pressure of the system. A variety of software tools are available to predict the molecular distribution. Databases—such as the DIPPR Project 801 database [5]—provide thermodynamic data to predict the enthalpies of the involved reactions based on the heat of formation. The temperature dependency of the standard equilibrium constant can be described by the van't Hoff equation Eq. (7):

$$\frac{\partial \ln(K_p)}{\partial T} = \frac{\Delta H_R^0}{R \cdot T^2} \tag{7}$$

The reaction enthalpy has to be calculated for each temperature, relative to the reference temperature (298.15 K) at standard conditions. The value is obtained by integrating the change of heat capacities Eq. (8):

$$\Delta H_R(T) = \Delta H_R^0(T_0) + \int_{T_0}^{T} \nu_i \cdot c_{p,i} \cdot dT \tag{8}$$

2.1.1 Gas Composition at Equilibrium

In the following, three common cases of methanation are presented: stoichiometric methanation of pure carbon monoxide (CO), stoichiometric methanation of pure carbon dioxide (CO_2), and direct methanation of biogas, represented by a mixture of equal parts of methane (CH_4) and carbon dioxide (CO_2). The first two cases are shown in Fig. 2, while the latter case is shown in Fig. 3.

As indicated previously by the reaction equations, a stoichiometric reaction of CO requires three equivalents of H_2 for full conversion Eq. (1). In the case of CO_2,

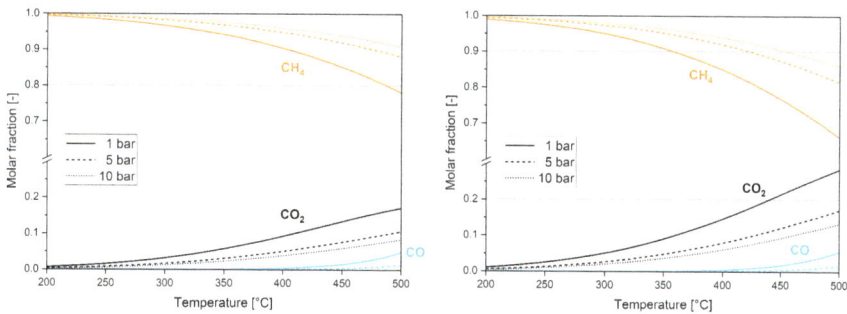

Fig. 2 Distribution of carbon atoms among the main species at thermodynamic equilibrium. The distribution was predicted in a temperature range of 200–500 °C and for the system pressures 1, 5, and 10 bar (HSC® [6]). The left figure shows the distribution for a stoichiometric mixture of H_2 and CO (H_2/CO = 3); the right figure represents stoichiometric methanation of CO_2 (H_2/CO_2 = 4)

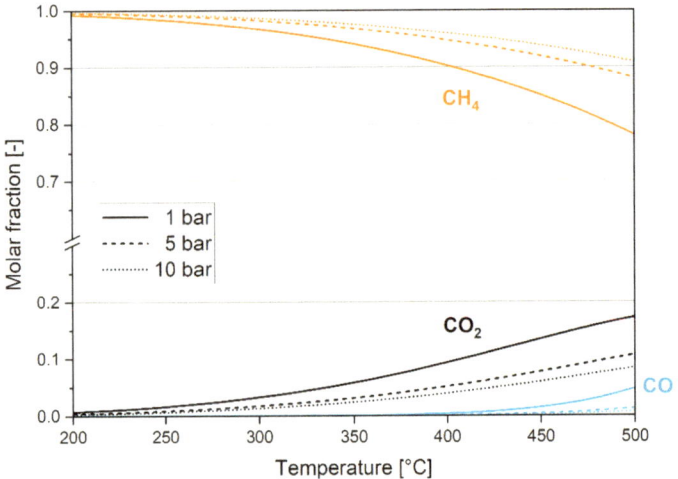

Fig. 3 Distribution of carbon atoms among the main species at thermodynamic equilibrium. The distribution was predicted in a temperature range of 200–500 °C and for the system pressures 1, 5, and 10 bar (HSC® [6]). Shown is the methanation of biogas: 50% of the biogas is assumed to consist of methane and the remaining 50% of CO_2. Analogously to CO_2 methanation, the ratio H_2/CO_2 equals 4

four equivalents of H_2 are required, respectively, Eq. (2). Nevertheless, due to the reversibility of the reactions and their exothermicity, thermodynamic equilibrium is reached and full conversion in not possible at working temperatures of common catalysts ($T > 250$ °C).

In all cases (Figs. 2 and 3), a high conversion of the carbon monoxide can be observed, as the concentrations are low. Nevertheless, due to the coupled water–gas shift reaction, a certain percentage of CO_2 is present, which reduces the maximum possible methane (CH_4) concentration in the product gas.

Since the methanation reaction is exothermic, the CO_2 fraction in the product gas increases towards higher temperatures. Since the coupled homogeneous water–gas shift reaction Eq. (3) is endothermic, the thermodynamic equilibrium is shifted towards the product side at higher temperatures. Therefore, a higher amount of CO_2 and CO is present at the upper temperature range and the overall equilibrium is shifted towards lower CH_4 concentrations. In the lowest temperature range, the carbon distribution is very similar in all three gas mixtures, as low temperature strongly favours full conversion towards methane. At elevated temperatures above 300 °C, methanation of pure CO tends to show higher methane concentrations in the product gas as CO methanation.

Due to the volume-contracting nature of the methanation reactions, increased pressure has a positive influence on conversion. In Figs. 2 and 3, the molecular distribution for 1, 5, and 10 bar system pressure is shown. When starting from CO_2 as feed gas, two water molecules are formed along the reaction pathway towards methane, whereas in the case of CO as feed gas, only a single molecule of water is produced. This results in a higher volume contraction in the CO methanation case than in CO_2 methanation, leading to a higher influence of the system pressure. This effect is clearly visible in the molecular distribution, as with higher pressures, the increase of methane concentration is higher in the CO methanation case than with CO_2.

In the example for direct methanation of biogas, the initial gas mixture already contains 50% of methane. Adding a stoichiometric amount of hydrogen results in an initial mixture of 16.7% of CO_2 and CH_4, and 66.7% hydrogen. Due to the high exothermicity of the methanation reaction and the equilibrium being shifted towards the product side, only a very small amount is converted to CO_2 and H_2. Therefore, the higher initial amount of CH_4 in the feed gas leads subsequently to a higher methane content in the product gas. Since CO_2 is converted to additional methane, high methane contents can be reached by direct methanation of biogas.

2.1.2   Alternative Feed Gases

Gasification of biomass is an alternative conversion pathway. Depending on the composition of the biomass feed, a range of compositions can be expected. Prior to methanation, the producer gas needs to be purified, i.e. sulphur-containing contaminants need to be removed, and the tar content needs to be lowered to a level suitable for the downstream methanation technology. The ratio of hydrogen to carbon oxides is below the stoichiometric values presented above. Furthermore, C_2-species are commonly found, such as ethane, ethene, and ethyne, as well as benzene, and other aromatic compounds. Hydrogenation and hydrogenolysis of these species to methane requires additional hydrogen. Due to the high entropy increase, the reactions are thermodynamically highly favourable. Thus, in a thermodynamic equilibrium state, the concentration of these species is negligible.

Raw producer gas also contains up to 40% of steam. Depending on the gas cleaning technology applied, it is passed on to the methanation reactor. In the case of autothermal gasification, high CO_2 contents can be expected from a partial combustion of biomass. Allothermal gasifiers are built with a separate combustion chamber for heat provision. Combustion products are discharged through a separate flue gas pathway, which results in a significantly lower steam and CO_2 content in the producer gas.

Compared to product gas from stoichiometric methanation of carbon oxides, methanation of producer gas results in a relatively high amount of CO_2 at thermodynamic equilibrium. The CO_2 content in the product gas is even slightly higher than methane, resulting in a concentration of both species in the range of 45–55%, depending on the elemental composition of the producer gas. The carbon monoxide content in the temperature range below 350 °C is marginal but increases sharply at higher temperatures. Compared to the stoichiometric gas mixtures, the CO_2 content is increased by the already existing CO_2 fraction in the producer gas. Furthermore, the under-stoichiometric hydrogen content and large amount of steam in the producer gas favour the water–gas shift reaction towards higher amounts of CO_2.

Similar to the molecular distribution in the stoichiometric mixtures in the previous section, high pressures and low temperatures favour a shift of the equilibrium towards methane formation. Due to the lower hydrogen content, the water–gas shift reaction is more pronounced, increasing the temperature dependency of the equilibrium.

In a process supposed to produce grid-ready synthetic natural gas (SNG), high CO_2 contents require downstream upgrading. Optionally, hydrogen can be added to the producer gas, enabling power-to-gas operation using producer gas as feedstock. This leads to a shift of the equilibrium towards very high methane concentrations, as all carbon species are converted through the methanation reaction and the reverse water–gas shift. In such a case, the equilibria are very similar to the stoichiometric cases.

2.2 Catalysts for Methanation

Although the methanation reaction is thermodynamically favoured at lower temperatures, it does not proceed without help due to the activation barrier of the involved molecules. At temperatures below 75 °C, microorganisms can take this role; chemical catalysts (usually metals) become active above 200–250 °C.

2.2.1 Catalytic Methanation

Many metals from the group eight to ten are able to catalyse the reaction [7]. The most suitable for methanation are ruthenium, nickel (Ni), cobalt (Co), iron (Fe), and molybdenum (Mo) [7, 8]. Most often, nickel-based catalysts are used, as the metal shows good selectivity towards CH_4, relatively high activity and a low raw material price [7, 9]. Ruthenium shows the highest activity for the methanation of CO and CO_2, or mixtures thereof, but its cost is significantly higher than that of nickel [7]. Although exhibiting a similar activity as nickel, cobalt is also more expensive and is therefore not as widely used as nickel [7]. Iron and molybdenum show a lower selectivity towards methane and are more suitable for the synthesis of higher hydrocarbons [7].

Commercial systems typically operate in the temperature range of 200–550 °C [7, 9]. In industrial scale, the methanation reaction is performed at elevated pressures up to around 30 bar, as reported for Haldor Topsøe's methanation technology [10, 11].

Several kinetic approaches and mechanisms are under discussion for catalytic methanation [12, 13]. The proposed reaction pathways mainly differ in the surface intermediates and subsequent rate-determining steps. Kopyscinski et al. [14] list a series of kinetic models proposed for nickel-based catalysts, ranging from simple power law approaches to Langmuir–Hinshelwood-type (L–H) models.

Kopyscinski et al. [14] performed methanation experiments on a plate reactor. By comparing the space-resolved gas concentration measurements to 32 different L–H-type models, it could be shown that the rate-determining step RDS can be assumed to be the reaction of the surface carbon species (i.e. C*, CH*, or COH*) with adsorbed hydrogen (H*). In the kinetic models, also the water–gas shift (WGS) reaction is considered, which is especially relevant for CO_2 methanation. It was later shown that by combination of CO methanation and the reverse water–gas shift (rWGS), the CO_2-methanation in an isothermal reactor could be sufficiently described after adding the term to describe the equilibrium limitation [15]. Koschany et al. [16] give an overview over different kinetic models used for simulation of cooled fixed bed reactors and compare them with own experimental results. Similarly Parlikkad et al. [17] adapted the classical kinetics determined by Xu and Froment [18] to correctly predict experimental results of adiabatic fixed bed methanation.

The methanation catalyst may suffer from one or a combination of the following deactivation phenomena described by Bartholomew et al. [19]:

- Fouling and pore blockage. Fouling and pore blockage can result from polymerisation of carbon species ("gum") or coke formation, especially in the presence of unsaturated hydrocarbons such as ethylene, acetylene, or benzene or at high CO concentrations [20–22]. The resulting carbon depositions cover the active sites and can block the physical access to complete domains of the catalyst particle. Another potential cause of fouling are siloxanes (organic compounds containing silicon atoms) that can be found in biogas from wastewater treatment plants and can form amorphous silica depositions in catalysts [23, 24].
- Poisoning or reaction with the active sites. The most common reason for poisoning of nickel-based catalyst is sulphur poisoning where relatively small concentrations even in the ppm range of any sulphur-containing species can cause complete loss of catalyst activity within one year of operation by formation of stable sulphides. Regeneration of the catalyst is difficult and practically impossible in existing reactors due to the high temperatures necessary [19]. Another potential challenge is the oxidation of the active nickel metal sites by oxygen or water molecules. Oxygen can also be considered a catalyst poison in biological methanation, since the microorganisms require anaerobic process conditions [25].
- Loss of active surface (e.g. by sintering of nickel crystallites). Sintering of nickel crystallites may happen by two different ways: thermal sintering due to increased mobility of nickel at high temperatures or chemical sintering. The latter may occur at higher CO concentrations and low temperatures in the reactor due to the formation of the volatile and highly toxic compound nickel tetra-carbonyl. Both lead to the formation of large nickel crystallites with low specific surface area, i.e. low dispersion and in consequence decreasing number of active sites which makes the catalyst even more vulnerable to poisoning [26, 27].
- Loss of active phase. Under certain operation conditions, carbon atoms may dissolve in the nickel crystallite and form carbon whiskers (nanofibers) between the crystallite and the ceramic support. Their growth separates the crystallite from the support and may thus lead to transport of the fibres and the crystallites out of the reactor. More often, the growth of carbon fibres in the pores of the catalyst particles leads to the physical destruction of the latter and subsequent blockage of the reactor or at least to very high-pressured drop.

To avoid these different ways of catalyst deactivation, suitable temperature control, and appropriate ratio of hydrogen, water and gaseous carbon species in the reactor are of utmost importance. Further removal of catalyst poisons such as sulphur species or siloxanes in a suited gas cleaning section are needed. Catalytic fixed bed reactors also necessitate the upstream removal or conversion of unsaturated hydrocarbons, while reactors with moving catalyst particles, e.g. gas solid fluidised beds, can convert these species to methane without undergoing catalyst deactivation [28, 29].

2.2.2 Biological Methanation

Alternatively to the catalytic methanation, the methanation reaction can also be performed by specialised microorganisms in a mesophilic (35–40 °C) or thermophilic (55–65 °C) temperature range [25].

The processes leading to biological methanation are linked to the processes for methane formation in anaerobic digesters. In these units, complex organic matter (carbohydrates, proteins, fats, and polysaccharides) are broken down to simple volatile fatty acids (VFA) by hydrolysis and fermenting bacteria (acidogenesis) [25]. These VFAs are further converted to CO_2 and CH_4 by processes known as acetogenesis and acetotrophic methanogenesis [25]. A secondary pathway comes into play when high partial pressures of H_2 are present in the reactor: CH_4 is produced by hydrogenotrophic methanogenesis. Nevertheless, it is known that high H_2 partial pressures inhibit the degradation of VFAs leading to a potential breakdown of the system [30]. It is therefore recommendable to physically separate the two processes to maintain efficient operation. This *ex-situ* methanation separates the anaerobic digestion process and the methanation step in two vessels which maintain specific, individually optimised process conditions.

In biological methanation, a water phase is required, which serves as living medium for the microorganisms catalysing the methanation reaction. In this water phase, also nutrients needed for the growth are dissolved and provided to the organisms. A variety of different growth media exist, depending on the organism used in the reactor [25]. They consist of salts, mainly containing phosphates, sulphates, carbonates and chlorides of potassium, ammonium, sodium, calcium, and magnesium. Further trace minerals containing zinc (Zn), nickel (Ni), molybdenum (Mo), boron (B), manganese (Mn), and iron (Fe) in various concentrations are also added [25, 31–34]. Schill et al. [35] report the addition of Na_2S-solution as sulphur source. It was added pulse-wise to reduce stripping of H_2S from the liquid phase.

After all, Rusmanis et al. [25] state that the elemental composition of growth media used in biological methanation are not very different from those used in anaerobic digestion [36, 37]. Therefore, concepts exist, where effluent from an anaerobic digestion unit is directly used as medium [31, 36, 38].

In pilot tests, the biological methanation showed good performance in untreated raw biogas [39]. It was shown that sulphur contaminants present in the raw gas stream did not affect the performance of the methanation reactor, which shows that biological methanation is much more tolerant against sulphur contaminants than catalytic methanation.

In anaerobic methanation, the reaction is catalysed by microorganisms, which metabolise H_2 and CO_2 and produce CH_4 and water as side products. The two reactant gases are only available to the organisms when they are dissolved in the liquid phase. A big challenge for biological methanation is therefore to increase the gas-to-liquid mass transfer rate of H_2, which is the main limiting factor for the methane conversion rate [40].

2.3   Reactors for Methanation

Due to its strong exothermicity and the equilibrium limitation, the methanation reaction requires reactor concepts, which either handle the developing hot spot by applying series of adiabatic reactors with intermittent and recirculation cooling or maximise methane conversion by good heat removal to prevent the hot-spot formation. In the case of direct methanation of biogas, the feed gas to the reactor already contains a large share of methane gas. The heat formation per volume flow is therefore dampened and hot-spot formation in catalytic reactors reduced [41]. The same holds to lower extent for methanation of gasification derived producer gas that can contain up to 10% methane and 25% CO_2. Many different reactor types exist (Fig. 4), which will be discussed in the following sections.

2.3.1   Adiabatic, Cooled Fixed Bed, and Structured Reactors

Adiabatic fixed bed reactors. In adiabatic fixed bed reactors, the catalyst is dumped in the reactor vessel as irregular bed of particles in the size of few millimetres. Such reactors can reach several meters in diameter and heights of ten meters and more. Depending on the exact gas composition, temperatures up to 700 °C can be

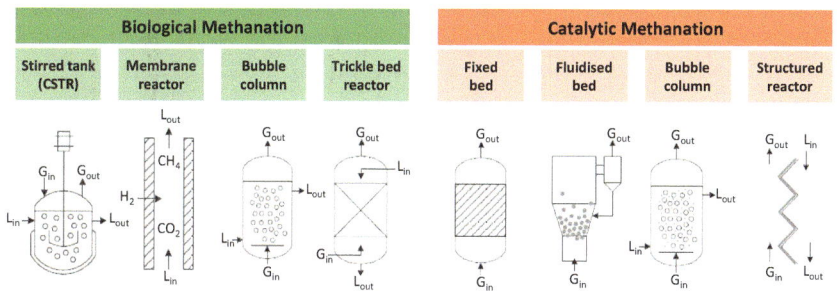

Fig. 4   Reactor types for methanation. Adapted from Ref. [42]; CSTR: continuously stirred tank reactor

reached as result of the exothermic reaction and the equilibrium limitation that avoids even higher hot-spot temperatures. Such high temperatures require the catalyst to withstand cracking and sintering effects under such conditions [9, 19]. This reactor type is the standard in the production of synthetic natural gas (SNG) from coal by gasification and methanation; examples are operated in the US and today mainly in China. Usually these plants are in the scale of 100 s of MW; therefore, the steam parameters in the cooling system are of high economic importance to maximise the electricity production in the steam turbine connected to the cooling system [20]. In such systems, cascades of adiabatic fixed bed reactors with intermediate cooling and product gas recycling are applied [43].

Power-to-gas (PtG) plants are of significantly smaller size so far (< 10 MW$_{CH4}$); therefore, the energy recuperation by steam turbine is not applied. As a result, the reactors can be cooled actively which leads to higher conversion in the reactor and thus to significantly lower number of reactors in series to reach sufficiently high methane concentration of $> 96\%$ [10]. Often, two reactors with intermediate cooling and water condensation are applied [44].

In fixed bed reactors, the cooling can be achieved most easily by filling the catalyst particles into tubes (diameter of one inch or lower) that are surrounded by cooling medium. In the first large-scale PtG plant in Werlte (3 MW$_{CH4}$ output [45]), molten salt was used as cooling medium. Alternatively, direct evaporation of water to raise steam can be applied [11]. Recently, the potentially more economic concept is tested to fill the catalyst in the reactors vessels, while cooling plates with running water are placed into the catalyst bed [46], which is a similar concept as the Linde reactor in the 1970s [10]. With the catalyst on the shell side instead inside the tubes, the challenge is sufficient flow distribution over the cross section to avoid preferential paths and dead zones which could decrease the heat removal capacity of the reactor.

Despite the cooling, a strong temperature rise has to be expected due to the very high reaction rates at the reactor inlet and the limited available heat transfer area in this region [11, 20] (Fig. 5).

At the hot spot, no further reaction progress can be seen; further conversion downstream of the hot spot depends on the cooling performance, i.e. the reactor is fully heat transfer limited in most parts. To improve this situation, reactors with staged reactant addition (usually CO_2) and/or different cooling zones have been tested. A strongly integrated reactor concept was developed by Moioli et al. [47]. The underlying concept is to divide the reactor into three zones for optimal reactor performance: an nearly adiabatic zone to start the reaction from low entrance temperatures, a strongly cooled zone where the bulk conversion takes place and a final isothermal zone at low temperatures to ensure high conversion. Further, the gas flows leaving the reactor are used to heat up the entering gas streams.

Structured reactors. Another option to improve the heat transfer is to introduce metal structures with high thermal conductivity and significant cross section [48]. In power-to-gas systems, this concept has been realised by Atmostat (80 kW plant in Troia within the European Union project Store&Go [49]) and by Ineratec [41], demonstrated at a wastewater treatment plant in Spain. In these structured reactors,

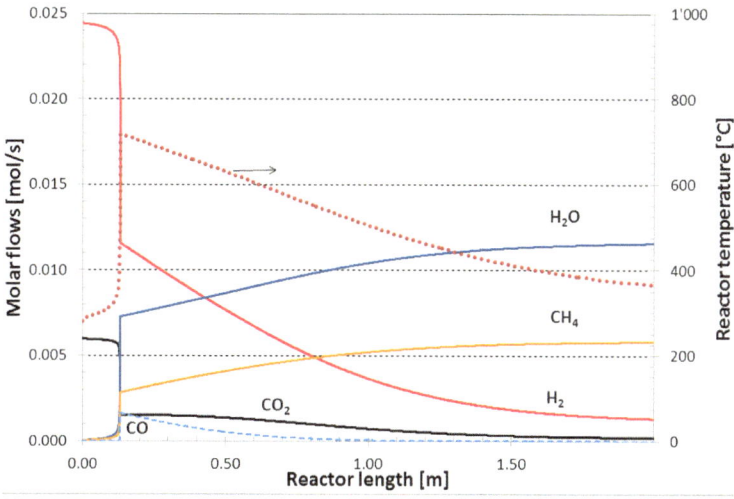

Fig. 5 Simulated results for a cooled tubular fixed bed reactor (inner tube diameter 25 mm) with slightly over-stoichiometric (4.1–1) mixture of H_2 and CO_2 at 10 bara with inlet temperature 280 °C and cooling temperature 340 °C: molar flows (full and dashed lines), temperature (dotted line)

reactor channels of a few millimetres diameter are present in the metal plates which are connected by high temperature diffusion bonding to improve the conduction. In one layer, the channels are filled or coated with catalyst (particles), while in the adjacent layers, cooling medium (water or air) is flowing [50]. In these reactors, the temperature hot spots are significantly lower. The disadvantage is the higher volume and weight and the fact that scaling up happens by numbering up, i.e. the specific costs do not decrease as much as with other reactor concepts. The numbering up for larger plants offers, however, the option for manufacturing many of standardised reactors with significantly lower CAPEX from the beginning.

Alternatively, catalyst can be coated on metal monoliths consisting of corrugated sheets and straight layers that are rolled up and hold together by a thin-wall metal tube. While the axial and radial heat transport by thermal conduction is limited due to relatively thin metal layers, the radiative heat exchange is facilitated by the high voidage that also leads to very low pressure drop. The challenge is however to operate in parallel many of these monolith-filled tubes and to ensure the equal distribution of flow over the complete cross section. So far, this lead to relatively large reactor volumes [51].

The cooled fixed bed reactors mentioned so far usually do not reach full conversion, i.e. the necessary grid injection specification of > 96% methane needed in many countries so far. Therefore, an intermediate condensation step is used to remove the reaction product water [11, 52] and to allow thus for higher conversion in the second reactor, which then is of similar size as the first reactor.

To overcome this limitation, sorption enhanced methanation reactors are developed. The idea is to adsorb the water produced in the reactor by solid sorbents such as zeolites which then changes the thermodynamic boundary conditions in the next catalyst section downstream. As long as sorbent capacity is available, very high conversions can be reached [53–55]. Here, the specific challenge is to find the optimum ratio between catalyst and sorbent volumes and to optimally align the thermal fronts and the moving area of remaining sorbent capacity. Further, the heat of reaction in one compartment is needed to regenerate the sorbent in another compartment. Therefore, this heat is not available anymore for other purposes such as regenerating of amine solution in amine scrubbers or to raise steam for energy integration. Very recently, higher H_2 contents [1, 2] are accepted in many gas grids, which decreases the necessity of very high conversions in one step.

The steam available from catalytic methanation without sorption enhancement could be used in a high temperature electrolyser [11, 56–58]. As in these devices, steam is split rather than liquid water, very high efficiencies can be reached in the electrolysis, which therefore is the best energy recuperation from a total plant perspective. Overall efficiencies of up to 81% from electricity to the higher heating value (HHV) of methane can be reached. [56, 58, 59].

2.3.2 Reactors with Catalyst Fluidisation

While all fixed bed methanation reactors show significant temperature gradients within the reactor, moving the catalyst particles allows for near-isothermal operation [60]. When internal cooling systems are in place, the high turbulence and particle movement in such reactors allows an efficient distribution of the reaction heat. The particle movement also facilitates a uniform distribution of the heat production, as the reaction front is more distributed than in fixed bed reactors.

Fluidised bed technology has been investigated and further developed since decades [10]. The more often applied concept is the gas–solid bubbling fluidised bed (BFB) where catalyst particles in the sub-millimetre range are brought in contact with an up-flow of the reactant gases such that the particles are lifted and form a fluid-type mixture of gas and particles. Excess of gas flow leads to the formation of nearly catalyst-free voids in the bed that behave to a large extent like bubbles in a liquid, i.e. they rise, coalesce and cause a constant re-mixing of the fluidised catalyst-particle rich phase. As mentioned above, this constant moving and mixing of the catalyst particles increases both, the local heat transfer and the axial dispersion of the heat production and by this, the usable heat transfer area in such reactors. As a result, the temperature can be kept permanently at values below 400 or 450 °C, which enables the use of more common types of steels for the pressurised equipment and thus lower capital costs. Although the movement of the catalyst particles leads to higher mechanical stress than in fixed bed reactors, the choice of proper catalyst materials, and suitable, relatively mild operation conditions in bubbling fluidised beds allow to operate such reactors without significant problems caused by particle attrition [28, 61, 62].

While the first methanation tests with fluidised bed reactors were conducted in the 1960s [10, 28], the concept was scaled up to 20 MW within the Comflux process in Germany [61, 62] in the late 1970s and early 1980s, at this time with synthesis gas from coal gasification. The reactor concept was picked up again in Switzerland after the turn of the millennium with focus on decentral production of renewable methane, both via gasification of wood [28] and power-to-gas, especially with biogas [15]. Meanwhile, the technology is investigated also by Korean [63], Chinese [64], Austrian [65], and French groups or consortia, and plants up to TRL 7 are built [66] or in operation [67, 68].

A second option is to use non-reactive liquids (e.g. large aromatic molecules) that contain catalyst particles in the range of 0.1 mm or smaller. The liquid hold-up improves the heat transfer to cooling coils and guarantees isothermal operation. Like in the gas solid fluidised bed, the gas is supplied from bottom and rises in bubbles up through the reactor. The reactants have to dissolve into the liquid that covers the surface and the pores of the catalyst particles. This causes a significant limitation to mass transfer that is counteracted by higher pressures. Development of such bubble column based methanation reactors was ceased in the 1970ies [20] and restarted by German research groups in the second decade of this century where meanwhile TRL 5–6 was reached [69].

2.3.3   Reactors for Biological Methanation

As mentioned above, biological methanation suffers from the fact that the gaseous reactants, H_2 and CO_2 have to dissolve in water to be converted by the microorganisms. Especially at elevated pressures, CO_2 readily dissolves in water and is therefore less of a problem than hydrogen whose gas-to-liquid mass transfer rate is the main limiting factor for the methane conversion rate [40]. Various reactor concepts are under investigation to overcome this problem. The most dominant for biogas upgrading are continuously stirred tank reactor (CSTR)-based bubble columns and counter-current trickle bed reactors [25, 32, 38, 70].

In the bubble column reactor, vigorous stirring increases the gas–liquid interface area. The energy required for stirring amounts to around 0.7–2.5% of the electricity used in the electrolyser, stating an efficiency loss of the same magnitude to the overall system [71, 72]. Often, these systems are operated at elevated pressures around 10 bar to facilitate the gas-to-liquid mass transfer. This necessitates the use of suitable stainless-steel equipment to withstand the pressure and the corrosion by CO_2 dissolved in water. Due to the water content and the lower productivity of the microorganisms compared to catalytic systems at elevated temperatures, the reactors for biological methanation are at least one order of magnitude larger than typical catalytic reactors [59]. From the main suppliers, electrochaea focused on dedicated type of microorganisms with comparatively high productivity and stirred bubble columns with high aspect ratio (height-diameter ratio > 5) [49], while HZI/microbenergy works with microorganisms available at the biogas sites and designs reactors with an aspect ratio between 2 and 4 [73].

For trickle bed reactors, additional energy input is required, as the liquid phase has to be circulated in order to provide a liquid film and nutrients to the biofilm present in the bed. The recirculation of the liquid can either be done continuously or pulsated [32, 36]. On the other hand, the compression costs for the gaseous reactants are lower due to the decreased pressure drop. Ullrich et al. [74] showed for trickle bed reactors, that increased system pressures indicate higher gas-to-liquid mass transfer and therefore result in a higher methane formation rates. All published results so far indicate that the productivity per volume of biological trickle bed reactors is even lower than that of the stirred bubble columns, which translates into higher volumes and costs of the reactors. One strategy by certain suppliers [75, 76] is to operate at atmospheric pressure such that no steel equipment is needed. Due to the lower mass transfer rates at low pressures, this then turns into very large equipment and footprint, but low costs per volume of reactor.

3 Efforts Towards a Higher Flexibility of Methanation Processes

The operation of power-to-gas reactors strongly depends on the availability of hydrogen and of a carbon source. The hydrogen production is usually based on electrolysis with renewable electricity, which especially in the case of photovoltaics (PV) and wind is not constant, but faces both, diurnal and seasonal variation. With the help of even relatively small, about half day-capacity sized hydrogen tanks, the variations in electricity supply and electrolysis operation can completely be separated from the methanation reactor such that the latter does not undergo changes in operation more often than once a day. Still, to keep the hydrogen tank small and to avoid complete shutdown of the methanation, it is helpful to operate it with part load if necessary. Therefore, it is of interest to understand to which extent methanation reactors are flexible and how fast they can start-up, shut down, or change their load between full- and part-load operation.

3.1 Simulation Studies

Gantenbein et al. conducted process chain simulation to investigate the systemic effects of flexible power-to-gas process concepts. While it is intuitively clear that a larger hydrogen tank decouples the variations in renewable electricity supply and hydrogen production from its consumption in the downstream methanation, the investigation of part-load operation led to some surprises. Part-load operation allows longer operation times between two shutdowns when a certain hydrogen tank volume is available, but only if part load is applied during emptying the tank, but not when it must be refilled. When the tank is empty, it first should be fully filled before starting the methanation to avoid too short operation of the methanation in case of only short hydrogen production phases.

Matthischke et al. [77] developed a one-dimensional fixed bed reactor model for CO_2 methanation. They tested the start-up behaviour and load range of reactors operated under adiabatic conditions or with external cooling. They showed that product gas recirculation is highly beneficial for the operation of cooled fixed bed reactors.

Cooled fixed bed reactors were found to start up considerably faster than adiabatic reactors, as in the latter, more heat is required to increase the reactor temperature to a steady-state level, which corresponds to the hot-spot temperature. In the cooled system, the temperature is considerably lower and approaches the temperature of the cooling medium asymptotically. In this study [77], the load range was set by adjustment of the feed flow rate. In the adiabatic reactor, increased product gas recirculation leads to a decreased hot-spot formation due to the cooling effect of the recirculated gas. At lower recycle rates, the conversion to methane is limited by equilibrium. With increasing recycle rates, the conversion rises due to a decreased hot-spot temperature. At high recycle rates, the reaction becomes kinetically limited, as a temperature profile forms along the reactor forming a hot spot at the outlet of the reactor. Therefore, in adiabatic fixed beds, an optimal recycle rate exists for a maximum conversion. In cooled reactors, recirculation of product gas also has a cooling effect, leading to a lower reaction rate and therefore less conversion. Nevertheless, this dampening effect has a positive impact on the stability of the reactor, when load changes are applied, as hot-spot formation and subsequent thermal deactivation is decreased. It furthermore increases the load range of the system.

In a prior study of Matthischke et al. [78], effects of the recycle stream and load ramps on an adiabatic fixed bed reactor were experimentally investigated. In addition, the benefits of product gas recirculation on system stability could be shown. Experiments with load ramps (by feed flow rate) showed that the temperature response of the reactor is slower than the concentration response, due to the heat capacity of the catalyst bed and the reactor vessel. A recycle loop in the system allows to manipulate the reactor temperature and therefore the conversion of CO_2.

Giglio et al. [79] present a one-dimensional dynamic model for a multi-tubular, fixed bed reactor. It was used for the dynamic simulation of a cascade of three boiling water-cooled CO_2 methanation reactors. The system allowed inter-stage feeding of CO_2 and condensation of water after the second and third reactor. The system was sized to a H_2 feed stream of 40 mol/s, which corresponds to an electrolyser size of 10–15 MW_{el}. Giglio et al. tested the hot start-up and part-load operation of the system. They found, that after a hot stand-by, their system can reach grid-quality product gas within 130 s and requires a sequential start-up of the reactors to avoid hot-spot formation. In full-load operation, 39% of the CO_2 is fed to the first reactor, and the remainder is mixed to the feed gas of the second reactor. In part-load operation, the share of gas fed to the first stage decreases. At lower part loads, an increasing amount of CO_2 is fed to the feed of the third reactor. The inter-stage feeding was required, as the lower gas velocity in the reactors caused hot-spot formation.

In order to optimise the start-up control of a tubular, fixed bed reactor, Bremer et al. [80] used a dynamic two-dimensional reactor model. The considered reactor tube is controlled via the cooling of the reactor jacket. The goal of this study is to avoid hot-spot formation and to reduce the start-up time to reach a steady state. They report an optimised temperature trajectory for the cooling medium, which initially serves as heating to reach reaction temperature in short time, but then is reduced in order to prevent hot-spot formations in steady state.

Zimmermann et al. [81] present a simulation study, which investigates the dilution of the active material in a fixed bed reactor and the introduction of catalyst particles with an active core and an inert shell. Sequential load changes were applied to the model and the step response analysed for several reactor designs. The reactors with core–shell catalysts revealed a significantly higher space–time yield, compared to diluted fixed beds, leading to a reduction of tube length. This effect originates in the reduction of the effective reaction rate at high temperatures, as the mass transport through the inert shell becomes rate-determining and the effective reaction rate becomes almost independent on temperature. In a diluted fixed bed reactor, the effective reaction rate is equally reduced over the whole temperature range. The study reveals that fixed bed reactors filled with core–shell catalysts are beneficial for load-flexible methanation, as no moving fronts were observed, and steady state was reached quickly.

For the biological methanation, Inkeri et al. [71] developed a dynamic model of a tall, continuously stirred tank reactor (CSTR)-based bubble column. The response of a 1 MW_{el} reactor to load changes was simulated, including an intermediate shut-down. The load profile applied two-part load levels at 40 and 80%, which were set by reducing the feed-gas flow. A single operation point was kept for 10 min. After each positive load change, spikes in the output gas flow were observed, which may be explained by the inability of the reaction to adapt fast enough to the new conditions, limiting the H_2 consumption. It could be seen, that after the load changes, the CH_4 concentration dropped below the desired 95%-level, which demands for additional stabilisation mechanisms (e.g. storage, control logics). Nevertheless, the model showed that the reactor is able to regain steady state in less than 10 min.

3.2   Experimental Studies

As shown above, several simulation studies show the options to operate methanation plants in part load and under dynamically changing loads. The experimental demonstration needs significantly more effort because plants at Technical Readiness Level (TRL) of at least 4 are needed to obtain convincing and meaningful results, as in this scale, different from laboratory setups, limiting effects of heat/mass transfer or fluid-dynamic patterns become visible.

Catalytic fixed bed. Specht et al. [45] demonstrated the ability of fixed bed methanation for changing to 70% part load in their 250 kW plant (TRL 7). In this plant, two different reactor types were used in parallel to test both concepts: (a) the more classical tube-bundle reactor where the catalyst is inside the tubes and the cooling is realised by a cooling agent (in this case molten salt) on the shell side; (b) a plate reactor where the catalyst is filled in between cooling plates which are flushed with evaporating water. While the hot spot in these reactors reached 600–650 °C, both systems allowed to maintain the gas quality (L-Gas, 90–93% methane) without problems during the load changes of 10 min.

Catalytic bubble column. The dynamics of a three-phase methanation reactor were experimentally investigated by Lefebvre et al. [82]. The system consists of an unstirred slurry bubble column (DN25 and DN50) filled with a heat transfer liquid. Therein, a commercial Ni/Al_2O_3 catalyst is suspended. Similar to biological methanation systems, this system tends to be limited by the gas-to-liquid mass transfer rate. In steady-state experiments, they investigated the preferred conditions for maximum CO_2 conversion. They show that high partial pressures of H_2 favour the gas-to-liquid transfer and are beneficial for the methane formation rate. The partial pressure of CO_2 shows a lower influence, due to the higher solubility of CO_2 in the heat transfer liquid. In general, higher system pressure increases the CO_2 conversion. They emphasise that an optimum reaction temperature has to be found, which ensures high reaction rates and therefore high CO_2 conversions, but also ensures stability of the suspension, which is prone to decomposition at elevated temperatures. Increasing feed-gas velocities resulted in an increase of the methane formation rate (MFR), but a decrease in CO_2 conversion, as larger bubbles are formed, which foster a larger bypassing of feed gas.

To assess the dynamic operation of the three-phase reactor, experiments with gas velocity steps were performed. It was found that the reactor time constant is only a function of the gas residence time. Therefore, the reaction kinetics are not limiting factors which influence the reactor time constant.

Biological methanation. During the European Union project Store&Go [72], a stirred bubble column by electrochaea was operated with CO_2 separated from a wastewater treatment plant in Solothurn/Switzerland. The maximum capacity of the plant was 700 kW, but it could be operated without any problem in the product gas composition in part load down to 60% with the part-load change happening within few minutes. Generally, the suppliers indicate that the possible part-load ranges from 0 to 100%. Given that the microorganisms can survive for quite some time without any feed, this appears to be realistic.

Similarly, it was even possible to completely turn off the hydrogen supply and restart the plant after hours within minutes. The only limitation to this flexibility is, however, the necessity to avoid freezing of the liquids during operation interruption in wintertime (when electricity and accordingly hydrogen shortages are more probable). Further, after long periods of operation interruption, the hold-up of microorganisms has to be slowly adapted.

Catalytic fluidised bed. To show that flexibility of fluidised bed methanation is technically possible, the 20 kW methanation COSYMA of Paul Scherrer Institut was connected to a commercial biogas plant in Inwil/Switzerland [83]. A suited hydrogen supply simulated synchronically the H_2 production by a system consisting of a photovoltaics (PV) installation, a battery, an electrolyser, and a H_2 tank to which it was electronically connected by means of data exchange within the framework of the ReMaP project (Renewable Management and Real-Time Control Platform, ReMaP [84]). The PV profile required multiple changes of the methanation operation between 40 and 100%, while maintaining the injected gas quality (> 96% methane), see Fig. 6. This was possible by means of a separation membrane downstream of the reactor that separated and recycled hydrogen. By systematic tests, it was found that the part load of the reactor itself is possible from 20 to 100%. For the low part loads, also low pressures are needed in the reactor to maintain the gas velocity inside the reactor and thus sufficient fluidisation. In consequence, the lower operation pressures of the reactor necessitated a larger membrane area than built in due to the lower pressure difference. Therefore, the input to the complete system including the actual separation membrane and hydrogen recycle was only varied from 40 to 100%, a value that was also reached without any problems in the 1 MW TRL 7 plant built in Güssing/Austria during the EU project BioSNG [28].

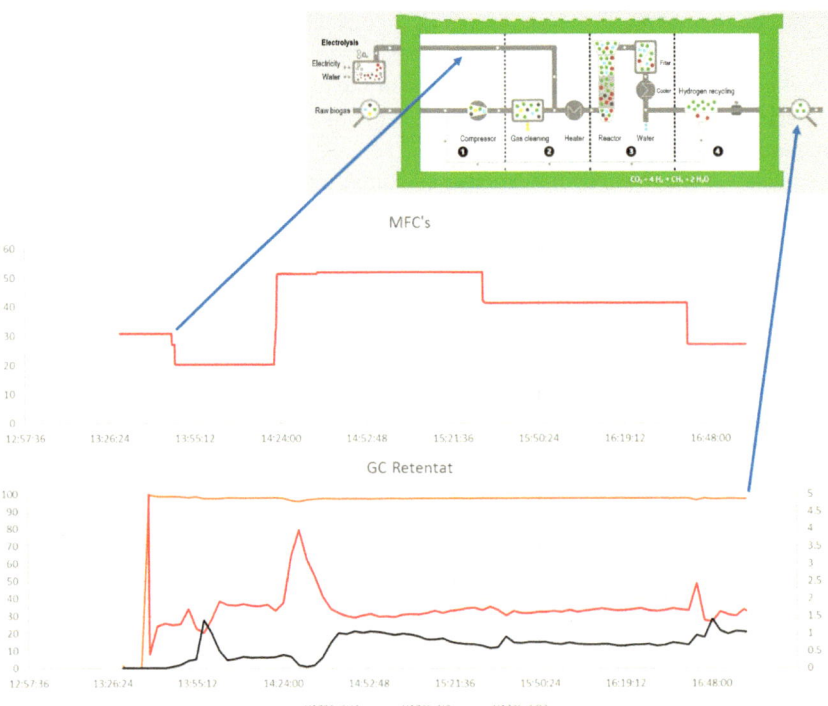

Fig. 6 Part-load study of TRL5 fluidised bed methanation plant (with integrated gas cleaning, upgrading membrane and recycle-loop), connected to the biogas plant in Inwil. Variation of the hydrogen feed flow rate (upper diagram, between 45 and 100% of the nominal capacity) and resulting concentrations of methane (VA726_CH4, left y-axis), hydrogen (VA726_H2, right y-axis), and carbon dioxide (VA726_CO2, right y-axis) in the injection line to the gas grid

4 Typical Process Chains, Long-Duration Tests, and Commercialisation

In both main technologies—catalytic and biological methanation—commercial operation or long-duration tests were performed at TRLs of around 5. Such tests are important to prove the stability of the process and show the feasibility of the respective methanation process technology [44].

In the following sections, long-duration tests at TRL 5 or higher and commercial operation of methanation plants will be discussed for different main pathways to supply the carbon atom for the power-to-gas process. In general, this carbon could be supplied as CO_2 from combustion processes, anaerobic digestion, alcoholic fermentation, or even from the atmosphere by direct air capture (DAC) technologies. Another option is gas mixtures from pyrolysis and gasification processes that usually contain besides CO_2 also CO and unsaturated hydrocarbons such as ethylene or benzene (see Fig. 7).

4.1 Supply of Pure CO_2 for Power-To-Gas

To obtain pure CO_2, a separation step is needed. Upgrading of biogas, a mixture of methane and 35–50% CO_2, is conducted by chemical scrubbers with amine solutions, water scrubbers, pressure swing adsorption (PSA), or membrane processes [85]. All these separation processes produce a relatively clean CO_2 flow, as usually the raw biogas has to be desulphurised before the upgrading, and because the methane in the CO_2 released to the atmosphere should be well below 1%. In future, the limit for this component with strong global warming potential may become even more stringent,

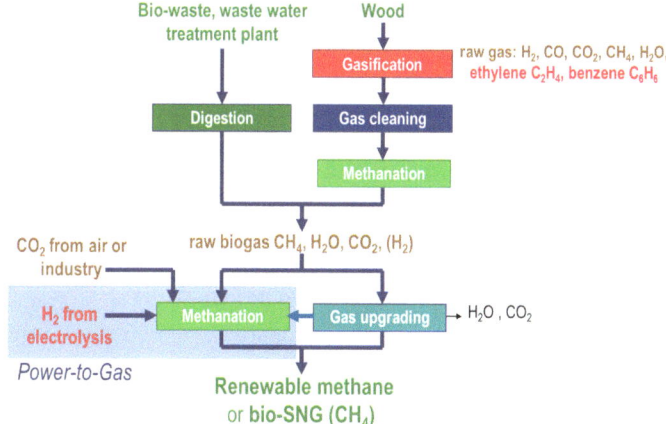

Fig. 7 Potential process chains to produce renewable methane

which might become a challenge especially for PSA and water scrubbing, but maybe also for membrane processes. In such situation, chemical scrubbing or conversion by power-to-gas become more favourable; however, they both have specific energy requirements.

For power-to-gas processes, it is mainly the electricity consumption of the hydrogen production, and to minor extent for compressors and further equipment. The energy input to separation processes can be steam in case of amine scrubbers to regenerate the washing liquid, or electricity in case of the other processes to operate pumps and compressors. In the chemical adsorption processes to separate CO_2 from air, usually a combination of partial vacuum and temperature increase to around 90 °C is used, i.e. pressure and temperature are changed at the same time, but to less expensive levels.

Like biogas upgrading, also most gasification or pyrolysis-based processes from biomass to methane, methanol (and its derivatives) or Fischer–Tropsch-type of products comprise inherently a CO_2 separation, often as amine scrubbing due to the relatively simple heat integration. The reason is the oxygen input within the biomass and depending on the process also the oxygen of the gasification agent which has to be separated as CO_2 from the oxygen-free product stream. It can be expected that the application of gasification-based processes will increase in the next decades due to its potential to use low grade biomass and waste streams to produce biogenic hydrocarbons. These hydrocarbons help to defossilise industry sectors that are hard to electrify (e.g. aviation, shipping or heavy-duty road transportation), provide high temperature process heat, and represent a feedstock for the chemical industry.

For large-scale combustion processes, e.g. cement plants with 10–20% CO_2 content or waste incineration plants with around 10% CO_2, both amine scrubbing or chilled ammonia scrubbing could be applied and were tested successfully at the CO_2 sequestration test site Mongsted in Norway [86].

Generally, the cost of the CO_2 supply are lowest at biogas upgrading plants, fermenters, and gasification-based synthesis processes, as the CO_2 separation is inherent to the process and already considered in the efficiencies and the economic calculation. More effort with respect to cost and energy input is needed to separate CO_2 from oxygen containing flue gases. The effort decreases with increasing CO_2 content, which ranges from below 10% for normal combustion conditions to 10–20% for cement plants (due to the geogenic CO_2 originating from the conversion of calcium carbonate to calcium oxide).

In consequence, costs are highest for separating CO_2 from air due to the (though too high) concentration of around 0.04% (400 ppm). These costs could be justified at locations that are far from any CO_2 source and very close to sequestration sites and/or areas where renewable electricity and thus hydrogen can be generated at low costs and in large amounts. While there is important ongoing work to increase the efficiency and to lower the costs of direct air capture (DAC), it is in direct competition with the most probably soon starting CO_2 logistics by pipelines, ships, and railway transport.

4.2 Operation of Power-To-Gas Plants with Pure CO_2

Specht et al. [45] reported successful operation of 25 and 250 kW_{el} methanation plants. The systems were mainly based on tubular fixed bed reactors with molten salt (250 kW_{el} system) and oil (25 kW_{el} system) cooling. The larger system had a plate reactor with water cooling installed. The systems were operated as two-stage reactors with intermediate gas cooling and condensation of product water. A part of the product gas was recycled to the feed to limit hot-spot formation. The duration of the test runs is not known, nevertheless, the results lead to the implementation of the technology in a 6 MW_{el}-scale plant in Werlte, Germany. For this commercial plant, CO_2 from a 10 MW biogas plant is separated by amine scrubbing. Interestingly, the heat for raising the steam to regenerate the amine scrubbing liquid is taken from the methanation reactor which is a good option for energy integration as otherwise other feedstock or even biogas needed to be burned for raising the steam. Moreover, the off-heat streams are large enough to also conduct the hygienisation, a process step to heat up critical feedstock to the fermenter to 70 °C by steam to control pathogens.

Within the European Union funded project Store & Go, methanation of pure CO_2 by several further technologies was tested in TRL 7 plants.

In Falkenhagen/Germany, CO_2 from an industrial fermentation process was converted in 525 kW_{LHV_CH4} scale plant. The oil-cooled reactor is equipped with 181 parallel catalyst coated 200 cpsi metal monoliths of 8.1 cm diameter and 40 cm length, resulting in an reactor of > 2 m diameter and few metres height that was delivered by Thyssen-Krupp. The applied pressure and temperature were 13–15 bar and 220 °C (with hot spot of 550–600 °C), respectively. The hydrogen was taken from an alkaline electrolyser of 2 MW_{el} input. The off-heat of the methanation was used in a nearby veneer mill, improving the overall efficiency. During a long-durations tests (1,186 h), renewable methane with grid quality could be produced due to intermediate condensation and a second fixed bed methanation reactor, and the gas was injected into the transport gas grid at 50 bar [51].

In Zuchwil, close to Solothurn in Switzerland, a stirred bubble column of 70 cm diameter and 9 m length was operated at 10 bar and 62 °C with specialised microorganisms (archaeae) obtained from Electrochaea. Up to 300 kW_{LHV_CH4} were produced from hydrogen and CO_2 for more than 1,200 h. While the hydrogen was produced on-site by PEM-electrolysis, the CO_2 was separated from the anaerobic digester of a wastewater treatment plant in a few kilometres distance and transported by a small pipeline.

In Troia, Italy, 2–8 Nm^3/h of CO_2 from direct air capture (from Climeworks) were converted at up to 11 bar in four parallel "milli-structured" reactors which consist of a metal block (24 L volume) with catalyst-filled channels in millimetre scale and was cooled by heat transfer oil. A membrane unit removed unreacted H_2 and CO_2, before a liquefaction unit produced renewable LNG.

Recently, the three-phase methanation in a bubble column was scaled up to 100 kW_{LHV_CH4} at KIT [69]. The reactor of 26 cm diameter and 2.5 m length is filled with dibenzo-toluene (DBT) as solvent and nickel-based catalyst with particles of <

100 μm size. CO_2 and stoichiometric amounts of hydrogen are fed from the bottom through a sparger. Cooling is achieved by welded cooling coils on the outer side of the vessel that are flushed with heat transfer oil (DBT). The reactor is operated at 6–20 bar and 260–330 °C and reached around 90% conversion of the 10 Nm^3/h of CO_2 flow, which necessitates a second reactor or membrane upgrading to reach gas grid injection specification. Further, hydrogenation and evaporation of the solvent DBT are reported [69].

4.3   Power-To-Gas with Biogas

Biogas is as mentioned above a mixture of 50–65% methane and CO_2 where the CO_2 content depends on the oxygen content of the biogenic feedstock. Further, biogas can contain some traces of oxygen or nitrogen and impurities such as ammonia, sulphur species (H_2S, mercaptans/thio-alcohols, thio-ethers, etc.), and sometimes siloxanes. The feedstock ranges from sewage sludge, manure, energy crops, agricultural wastes, wastes from food production, etc.

The number of biogas plants is still growing in many countries due to separate collection of green wastes as well as better use of manure and residuals from food production and agriculture. Until recent years, the business case of biogas plants was to produce heat and electricity by combustion of biogas in gas engines; meanwhile, the price of renewable electricity decreases due to photovoltaic (PV) and wind turbines which prompts the operators of biogas plants to seek for new markets such as the upgrading to biomethane and subsequent injection into the gas grid. Besides upgrading by separation of CO_2 (by means of scrubbers, pressure swing adsorption, or membranes), conversion of the CO_2 in the biogas with hydrogen by means of methanation is an interesting option.

While biological methanation reactors are relatively tolerant to the impurities, especially sulphur species (but not to oxygen), catalytic methanation reactors need upstream a gas cleaning section that removes reliably siloxanes and sulphur species from the biogas stream. The interactions of the other impurities with the available sorbents (e.g. active carbons, active carbons with metal impregnation, and metal oxides such as zinc oxide and or copper oxide) for sulphur removal typically also necessitate removing humidity and organic species such as terpenes [44].

In Avedøre/Denmark, long-duration tests were performed by the company Electrochaea using the stirred bubble column biologic methanation [44, 70]. While the reactor is approximately as large as the one in Zuchwil used for pure CO_2 methanation (see previous section), in the TRL 7 plant in Denmark biogas from a wastewater treatment plant was used as feedstock. About 50 Nm^3, equivalent to 550 kW_{HHV_CH4} additional methane were produced at 10 bar leading to a biomethane with 97% purity [44]. Similarly, the company MicrobEnergy (now a subsidiary of Hitachi Zosen INOVA) has a combined experience of more than a several 1,000 h for its stirred bubble column technology. Both companies work with thermophilic (60–65 °C) microorganisms. Electrochaea uses a single strain of archaea, whereas the technology of MicrobEnergy relies on a mixed set of organisms, which adapts itself to the conditions given by the digester effluent used as growth medium [38, 87]. Based on the technology of MicrobEnergy, that was tested for several 1,000 h with their 5 m^3 size reactor of 165 kW$_{HHV\_CH4}$ capacity in Germany, Limeco AG has built and commissioned a TRL 8 methanation plant in Switzerland [73]. The demonstration plant produces biomethane for grid injection, based on raw biogas from anaerobic digestion of sewage sludge. The electricity for the PEM electrolysis is generated on-site in the steam-cycle of a waste incineration plant close by Ref. [87]. The stirred tank reactor of about 50 m^3 volume converts at 10 bar the CO_2 in the biogas to additional 1.2 MW$_{HHV\_CH4}$ which makes it to the second-largest power-to-gas plant (with respect to methane production) after the Werlte plant.

Even larger by volume are the trickle bed reactors for biological methanation. Gicon GmbH in Germany offers, based on a simulation study [88], a 180 m^3 large reactor for 583 kW$_{HHV\_CH4}$ methane production from biogas at 55 °C. Similarly low volumetric productivity was found in TRL 7 trickle bed reactor experiments by Danish universities [89], in the TRL 4–5 plants in Bavaria [34] and in Austria [32]. Also the Q power technology from Finland, a bioreactor with filamentous/sponge type support, needs six reactor modules of 20' container size and a 40' container for the control unit to produce 750 kW$_{HHV\_CH4}$ methane from biogas and hydrogen [75, 90, 91]. The very large reactor size of trickle beds is accepted because the plant can be operated at close to atmospheric pressure which simplifies the reactor construction and limits capital costs [76]. Biological stirred bubble column methanation reactors operated at 10 bar (see previous paragraph) need stainless-steel pressurised vessels to avoid corrosion by CO_2 and water. Therefore, the material costs per volume are as high as the ones for catalytic reactors which leads to significantly higher capital costs of the reactor compared to the more compact catalytic methanation reactors [59].

Catalytic direct methanation of biogas was tested for 1,000 h by Dannesboe et al. [11] in collaboration with Haldor Topsøe A/S. A two-stage fixed bed methanation was used with intermediate cooling and condensation of reaction water, which allowed the production of grid-ready biomethane without further upgrading. The system was designed as boiling water reactor, producing steam at 280 °C (65 bar). The methanation was operated at 20 bar, and a commercial Ni/Al$_2$O$_3$ catalyst was used. As no axial movement of the hot spot could be observed during the whole test, no catalyst deactivation was occurring [11]. An important feature of the plant is the high temperature electrolysis (solid oxide electrolysis, SOE) which converts the steam produced by cooling of the methanation reactors at 700 °C to hydrogen. This leads to an HHV-based efficiency from electricity to methane of around 80%, considerably higher than any system with PEM or alkaline water electrolysis [56, 59]. Therefore, the thermal integration of a catalytic methanation with the SOE is the optimal heat recuperation in PtG processes.

In Spain, two microstructured heat exchange reactors in series with intermediate condensation (supplied by Ineratec GmbH) were tested with biogas from sewage sludge digestion at around 5 bar. The space between the heat conducting plates was filled with diluted nickel catalyst. Both reactors had a volume of around 15 L and were cooled by water and air, respectively, leading to moderate temperature gradients of 475–375 °C (first reactor) and 375–275 °C (second reactor). The conversion of CO_2 in the biogas with hydrogen lead to additional 1.4 Nm^3/h of biomethane, equal to 15.4 kW_{HHV_CH4}. With 72.66% conversion in the first and 93.48% after the second reactor, the methanation reached the Spanish grid injection specification which allow up to 5% H_2 [41].

Direct methanation of biogas in a catalytic fluidised bed reactor (nickel catalyst, up to 6 bar) was tested during a long-duration test at the wastewater treatment plant of Zurich, Switzerland. The experiments were performed using the TRL 5 methanation setup COSYMA (10 kW), which allowed to use a split-stream of the raw biogas produced in an anaerobic digester for sewage sludge. A part of the biogas also originated in an anaerobic digester fed with municipal green waste. The tests showed stable operation of the methanation plant for more than 1,100 h [15]. To protect the methanation catalyst, two serial sorbent beds were included in the plant, which consisted of combinations of commercially available materials. During the long-duration test, a slow activity loss of the catalyst could be observed due to sulphur poisoning. Several commonly occurring sulphur contaminants were regularly monitored during the field tests and the gas cleaning measures adapted accordingly [92]. The improvements in the gas cleaning reduced the deactivation rate to a level, which would allow the operation of the plant for a full year without catalyst exchange [15].

The results from this long-duration test was compared to the kinetics used in a simulation model for fluidised bed methanation developed in our group. The reaction temperature was varied between 300 and 360 °C during the experiments, leading to changing methane concentration in the product gas. The model was able to correctly predict the concentration maximum as a function of reaction temperature, which indicates that the kinetics applied describe CO_2 methanation appropriately [15].

4.4   Product Gas from Gasification as Carbon Source

Low temperature gasification (around 850 °C) of carbonaceous feedstock produces a gas mixture containing at least hydrogen and CO, further steam, CO_2, methane and, besides impurities such as sulphur, also particles and unsaturated hydrocarbons (ethylene, acetylene, benzene, and other aromatics). Besides wood residues, also other dry biomass streams as well as municipal solid waste can be gasified. After gas cleaning (tar removal, at least partial steam condensation and desulphurisation), product gas from gasification can be converted in methanation reactors where the CO reacts with H_2 to methane. Usually, the hydrogen to CO ratio is below two. Therefore, part of the CO has to be shifted by steam addition, this way internally producing the necessary hydrogen to convert the remaining CO. Due to this internal water–gas shift reaction (that is catalysed by nickel as well), the gas mixture leaving the methanation reactor consists of about 50% each methane and CO_2. To obtain pure renewable methane (also referred to as bio-SNG, bio-synthetic natural gas) according to grid injection specifications, remaining hydrogen has to be recycled and the CO_2 has to be separated (allowing for negative CO_2 emissions by sequestration). Alternatively, the CO_2 could be converted to additional methane by sufficiently high hydrogen addition to the methanation reactor (whose design needs to be adapted accordingly). By this, the plant can be converted into a power-to-gas (PtG) plant, which at least doubles the methane output from the same feedstock.

Although product gas from gasification and similar sources has not yet been used as carbon source for power-to-gas (PtG) beyond laboratory scale [93] or process simulations [94, 95], it can be expected to become more significant in the future for several reasons:

- Biogas plants are usually relatively small and represent only a small share of the CO_2 from stationary sources in a country. Product gas from gasification is a similarly suitable carbon source for power-to-gas processes (or in this case power- and biomass-to-gas), and the additional feedstock potential is large.
- As the addition of hydrogen within gasification based power-to-gas processes is not inherently necessary during the whole year, the flexibility of such a power-to-gas plant enables to avoid phases with high hydrogen prices due to expensive electricity.
- In a recently started European Union project HyFuelUp [66], a TRL 7 pilot plant with biomass gasification and flexible hydrogen addition is planned and will be built in Portugal.

Therefore, in the following, some of the so far built and operated gasification-based SNG plants are presented, also highlighting the gas cleaning steps, the methanation reactors, and the gas upgrading.

The largest wood-to-methane plant built and operated so far is the Gothenburg Biogas plant (GoBiGas project). A 32 MW_{th} wood input is gasified in an allothermal dual fluidised bed gasifier where the feedstock is converted with steam at 850 °C in a bubbling fluidised bed while the bed material and unconverted char are flowing through a loop seal to a riser. In the riser reactor, char and bed material are transported upwards by air. This initiates the combustion of the char and brings the bed material to about 950 °C. The hot bed material is separated by a cyclone and flows back into the gasification reactor where its sensible heat drives the endothermic gasification [96]. After the gasifier, a relatively complex gas cleaning procedure is used to remove impurities from the product gas that otherwise could damage the catalyst. Besides heavy tars and sulphur species, also ethylene and light aromatics such as benzene have to be removed or converted. The heavy tars are partly removed after condensation on fine mineral powder in the particle filter, and partly in an oil scrubber that also condensates water. Organic sulphur species are converted by a hydrodesulphurisation catalyst and a COS hydrolysis catalyst and then removed by a liquid scrubber and a guard bed. Benzene is removed by a large active carbon bed (with regeneration), while ethylene and acetylene are converted on the HDS catalyst and a special pre-reformer. This is necessary to protect the downstream series of four methanation reactors, as the unsaturated hydrocarbons lead to heavy carbon deposition and catalyst deactivation in the applied fixed bed reactors. A water–gas shift reactor and a CO_2 scrubber (amine based) allow to reach the desired mixture of H_2 and CO upstream of the four methanation reactors to obtain bio-SNG (20 MW) according to the grid injection specifications. While the gasifier is atmospheric, a compressor increases the pressure before the catalytic reactors to 16 bar leading to an electricity consumption of 3 MW_{el}. The plant was commissioned and operated but is today moth-balled due to the present economically unfavourable situation. This allows to restart the plant as soon as rising prices for bio-SNG make operation attractive. Like the ESME process developed by TNO-ECN in the Netherlands [29, 97, 98], the process demonstrated by the GoBiGas plant is meant to be built in scale of 100 MW and larger. Such large plants take advantage from the economy of scale and thus allow the use of the fixed bed reactors which necessitate the complex gas cleaning section.

Applying fluidised bed reactors for the methanation allows to significantly simplify the gas cleaning section between gasifiers and the methanation reactor. While the desulphurisation to below 1 ppm total sulphur is also here of utmost importance, the removal or conversion of unsaturated hydrocarbons such as benzene, ethylene, and acetylene is not necessary. This can be explained with the internal "regeneration" of the catalyst due to the constant movement of the particles: at the bottom of the reactor, where the reactants enter, the catalyst surface is covered with species that tend to cause carbon deposition (CO, unsaturated hydrocarbons). Due to the bubbling character of the fluidised bed reactor, the catalyst particles are moved constantly upwards to the top of the reactor. As the strongly carbon depositing compounds are converted before reaching the upper part of the reactor, hydrogen and steam that are present in the gas stream can react with the carbon atoms on the surface before they cause catalyst deactivation. Such "carbon management" allows to limit successfully the carbon deposition and therefore to convert unsaturated hydrocarbons to methane [60, 99]. As a result, smaller active carbon beds can be used, and no pre-reformer or olefin conversion step is needed leading to lower CAPEX [29]. The process from wood via dual fluidised bed gasifier and fluidised bed methanation (nickel/alumina catalyst, 2–5 bar, 300–400 °C) was demonstrated in 2009 by a 1 MW PDU in Güssing/Ausria within the EU project BioSNG (see Fig. 8) [28]. The gas cleaning section consisted of a cold oil scrubber, impregnated active carbon bed and zinc oxide bed, while the raw SNG was upgraded for a CNG fuelling station by amine-based CO_2 scrubbing, glycole-based drying, and a polymer membrane for hydrogen recycle. In this plant, bio-SNG in H-Gas quality (Wobbe index = 14.0, HHV = 10.67 kWh/Nm^3) was reached with an LHV-based efficiency from wood to bio-SNG of 61%. Within the EU project HyFuelUp [66], this process concept will be further developed by integrating gasification of low-grade biomass and flexible addition of hydrogen.

The proven suitability of fluidised bed methanation prompted recently a French industry consortium to test the combination of dual fluidised bed gasification and fluidised bed methanation at 500 kW scale for a large variety of feedstocks within the GAYA project [100]. As was already shown in the Comflux process (coal to SNG) in the early 1980s and by the EU project BioSNG, the fluidised bed reactor is nearly isothermal, adapts very fast to changing loads, and allows high conversion in one step due to the excellent temperature control.

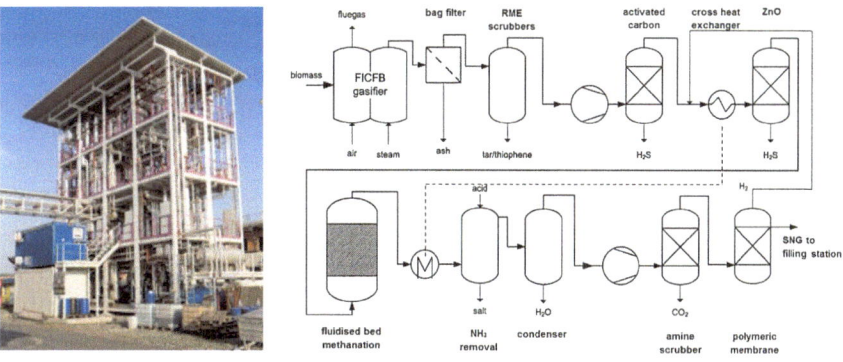

Fig. 8 Photograph and scheme of the 1 MW pilot plant to produce renewable methane from wood gasification gas; erected within the European Union project BioSNG. Adapted from Ref. [10]

The presence of unsaturated hydrocarbons such as benzene and ethylene (which cause problems for fixed beds, but not for fluidised bed reactors) is connected to the production of significant amounts of methane (up to 10%) in low temperature gasifiers (850 °C). The higher amount of methane in the product gas is an advantage for the overall process chain compared to full conversion of methane to syngas. The reason is that no energy has to be used in the gasifier at 850 °C to decompose methane (which is strongly endothermic), and no heat of reaction from the exothermic methanation has to be removed in the methane synthesis reactor. Both effects increase the cold gas efficiency. Still, to avoid the unsaturated hydrocarbons and the connected challenges in as cleaning, process chains with higher temperature in the gasification section have been tested. These lead to a syngas with methane contents below 2%, but free of unsaturated hydrocarbons, tars, and organic sulphur compounds [101, 102] which allow the use of fixed bed methanation reactors with less complex gas cleaning. In Sweden, the wood roll gasifier concept combines a low temperature pyrolysis with high temperature steam gasification of the pyrolysis char. The necessary heat for the endothermic gasification is supplied by the combustion of the volatile pyrolysis products inside tubes who transfer their heat by radiation and convection into the surrounding gasification section. This gasifier was connected to a 50 kW scale methanation system with catalytically coated metal honeycombs as reactors [103]. In the UK, within the GoGreenGas project, a steam/oxygen blown fluid bed gasifier (700–800 °C) is connected to a plasma reformer at 1100–1200 °C leading to a similar tar-free and methane-poor raw gas. A combination of dry (filtration with sorbent addition) and wet (one acidic and one caustic scrubber) gas cleaning steps allows to convert the syngas from waste and residue gasification in a series of fixed bed reactors at 16 bar (one water gas shift and four adiabatic methanation steps) according to the Vestas process. In this methanation process, the late separation of CO_2 and steam and the distributed addition of feed gas allow suitable temperature control to avoid catalyst deactivation. After first commissioning tests, this TRL 7 was moth-balled due to unfavourable financial support situation.

5   Comparison, Conclusion, and Outlook

The relatively large number of different technologies presented in the previous sections is a clear sign of a not yet mature market where many projects rely on support schemes and research grants. Accordingly, the necessary number of demo plants has not yet been built and the operation experience is missing, which would be needed to identify the best performing technologies. Table 1 gives an overview over the published experimental results so far, especially with respect to reactor volume (= capital costs), operation experience and feed-gas flexibility (Tables 2 and 3).

Still, some first conclusions can be drawn. While adiabatic fixed bed reactors seem to be limited to large scale (coal based) applications, cooled fixed bed reactors are of interest in the smaller scale power-to-gas systems due to their simplicity. It has, however, not yet turned out which type of cooled fixed bed reactors has the best overall performance. Further, according to the few detailed and comparative studies, catalytic fluidised bed reactors promise to have lower capital costs due to improved heat transfer performance and allow for simpler gas cleaning in case of gasification gas as feedstock [29, 106, 107]. However, more experience in at least pilot-scale has to be added to the encouraging results of the Comflux process in the 1980s [61, 62] to prove this.

Similarly, compared to biological reactors, catalytic reactors seem to have the potential to have lower overall costs due to their more compact reactors, lower foot print, and the options for heat integration [59]. This might, however, be balanced out by easier operation and synergies with existing infrastructure, especially at biogas plants.

It can be expected that these open questions will be answered in the next years, as the need for non-fossil methane is increasing to enable a defossilation of industrial processes and hard-to-abate sectors while decreasing the dependency from imports of energy carriers. This will allow the development of a more mature market where those technologies will compete that have now already reached a certain technical

Table 1 Overview over the published experimental results in the field of methanation for power-to-gas (technologies TRL ≥ 5) (1/3)

	Haldor Topsøe A/S	Solarfuel/etogas (today: Hitachi Zosen Inova/Hitz)	Microbenergy (today: Hitachi Zosen Inova)	Paul Scherrer Institut / AlphaSYNT
Capacity in CH_4 output and location of pilot plant	I: 20 MW Gothenburg, Sweden II: 45 kW Foulum, Denmark	I: 3 MW Werlte, Germany II: 1.2 MW: Japan III: 210 kW Gabersdorf Austria	I: 1.2 MW Dietikon, Switzerland II: 165 kW Allendorf, Germany	I: 1 MW Güssing, Austria II: 200 kW Villigen, Switzerland III:20 kW Werdhölzli, Switzerland

(continued)

Table 1 (continued)

	Haldor Topsøe A/S	Solarfuel/etogas (today: Hitachi Zosen Inova/Hitz)	Microbenergy (today: Hitachi Zosen Inova)	Paul Scherrer Institut / AlphaSYNT
Raw gas source	I: wood gasification II: anaerobic digestion of agricultural waste and manure	I: CO$_2$ separated from commercial anaerobic digestion II: CO$_2$ III: anaerobic digestion	I–II: commercial anaerobic digestion	I: wood gasification II: gas bottles III: commercial anaerobic digestion of sewage sludge and green waste
Reactor type	I: adiabatic fixed bed reactors II: 2 cooled tubular fixed bed in series with intermediate condensation	I: 2 cooled tubular fixed bed with intermediate condensation II, III: 2 fixed beds with immersed heat exchanger plates with intermediate condensation	Stirred bubble column	Bubbling fluidised bed (gas—solid)
Technology readiness level	I: 8 II: 7	7–8	7–8	I: 7 II: 6 III: 5
Catalyst	Nickel/alumina	Nickel/alumina	Mixed microorganisms	Nickel/alumina
Cooling system	II: Boiling water	I: molten salt (shell side) II, III: water in heat exchanger plates	Cooling water	I, II: thermos-oil III: cold air
Reactor temperature	280–680 °C	250–550 °C	60–70 °C	320–360 °C
Reactor pressure	I: 16 bar II: 20 bar	6–7 bar	5–10 bar	3–10 bar
Volume main reactor	I: pre-methanator plus > 4 reactors in series undisclosed II: length 2.3 m, diameter not disclosed	I, III: undisclosed II: 2 reactors, length > 5 m, diameter > 60 cm	I: 50 m^3 II: 5 m^3	I: 1–2 m^3 II: < 120 L III: < 15 L
Duration of operation	I: II: about 1,000 h	I: since several years II, III: since 2022/ 2023	Several 1,000 h experience	I: < 1,000 h II. Since 2022 III: 1,150 h
Additional information			Electricity consumption stirrer 2—2.5% of electrolyser-input	
Reference(s)	I: [29, 96] I: [11, 56]	[45, 46]	[46, 104]	[15, 28, 92]

Table 2 Overview over the published experimental results in the field of methanation for power-to-gas (technologies TRL ≥ 5) (2/3)

	Electrochaea	KIT/Thyssen (reactor)	Atmostat	KIT
Capacity in CH_4 output and location of pilot plant	I: 500 kW Avedøre, Denmark II: 300 kW Zuchwil, Switzerland III: 150 kW Golden, USA	525 kW Falkenhagen, Germany	80 kW Troia, Italy	100 kW Karlsruhe, Germany
Raw gas source	I: Commercial anaerobic digestion of sewage sludge II: CO_2 from anaerobic digestion III: CO_2	CO_2 from fermentation	CO_2 from direct air capture	CO_2
Reactor type	Stirred bubble column	1st reactor: catalyst-coated metal monoliths 2nd reactor fixed bed	4 milli-structured heat exchange reactors in parallel	Slurry bubble column
Technology readiness level	6–7	6–7	7	6
Catalyst	Specific archaea type	Nickel/alumina	Unknown catalyst as powder in millimetre-scale channels	Nickel/alumina
Cooling system	Cooling water	Thermo-oil	Thermo-oil	Thermo-oil
Reactor temperature	60–65 °C	220–600 °C	> 300 °C	260–330 °C
Reactor pressure	I, II: 10 bar III: 18 bar	13–16 bar	3–11 bar	5–20 bar
Volume main reactor	I, II: Ca. 7 m^3 (3,600 L liquid)	> 6 m^3	96 L	Length 2.5 m, 26 cm diameter; > 120 L hold-up
Duration of operation	I: > 500 h II: > 1,200 h III: since 2019	1186 h	761 h	Since 2022
Additional information	Electricity consumption stirrer W/l = 1.3—2.5% of electrolyser-input			
Reference(s)	[72, 105]	[105]	[105]	[69]

Table 3 Overview over the published experimental results in the field of methanation for power-to-gas (technologies TRL $\geq$ 5) (3/3)

	Gicon	Ineratec GmbH	Q Power
Capacity in CH_4 output and location of pilot plant	30 kW Cottbus, Germany	15 kW Sabadell, Spain	Vantaa, Finland
Raw gas source	Anaerobic digestion of agricultural waste and manure	Commercial anaerobic digestion of sewage sludge	Anaerobic digestion of agricultural waste and manure
Reactor type	Atmospheric trickle bed reactor	2 microstructured heat exchange reactors in series with intermediate condensation	Similar to atmospheric trickle bed with filamentous/sponge type support
Technology readiness level	5	5	5
Catalyst	Microorganisms (mixed archaea)	105 g of Ni/CeO_2/alumina, 400—500 μm, diluted with SiC	Mixed microorganisms (hydrotrophic methanogens)
Cooling system	Cooling water	Water evaporation 1st stage, air cooling 2nd stage	Cooling water
Reactor temperature	60–70 °C	475–375 °C (1st), 375–275 °C (2nd)	50–70 °C
Reactor pressure	Atmospheric	2.5–8 bar, mainly 5 bar	Atmospheric
Volume main reactor	10 m^3	2 Reactors of 29.5 × 15 × 33.5 cm = 2 × 14.8 L	Design for TRL 7–8 plant: 6 containers, each with 29 reactors of 2 m height and about 50 cm diameter, in total about 50 m^3 for 700 kW CH_4
Duration of operation	Unknown	Unknown	> 3 months operation of TRL 5 plant
Additional information			
Reference(s)	[76]	[41]	[75, 91]

readiness level (TRL 6–7). This inherently means that the focus in methanation will shift from academic research to more industrial development.

References

1. Deutscher Verein des Gas- und Wasserfaches (DVGW) (2013) Technische Regel Arbeisblatt G260
2. Schweizerischer Verein des Gas- und Wasserfaches (SVGW) (2016) Richtlinie für die Einspeisung von erneuerbaren Gasen—G13/G18
3. Gantenbein A (2022) Flexibilisation of biogas-based power-to-gas processes: a techno-economic and experimental assessment. EPFL. https://doi.org/10.5075/epfl-thesis-10032
4. Sabatier JB, Senderes P (1902) Nouvelle synthèse du méthane. C R Acad Sci 514–516
5. Design Institute for Physical Properties (n.d.) DIPPR project 801 database. https://www.aiche.org/dippr
6. Metso Outotec Finland Oy (n.d.) HSC® Chemistry 7.0
7. Rönsch S, Schneider J, Matthischke S, Schlüter M, Götz M, Lefebvre J et al (2016) Review on methanation—from fundamentals to current projects. Fuel 166:276–296. https://doi.org/10.1016/j.fuel.2015.10.111
8. Mills GA, Steffgen FW (1974) Catalytic methanation. Catal Rev 8:159–210. https://doi.org/10.1080/01614947408071860
9. Götz M, Lefebvre J, Mörs F, McDaniel Koch A, Graf F, Bajohr S et al (2016) Renewable power-to-gas: a technological and economic review. Renew Energy 85:1371–1390. https://doi.org/10.1016/j.renene.2015.07.066
10. Kopyscinski J, Schildhauer TJ, Biollaz SMA (2010) Production of synthetic natural gas (SNG) from coal and dry biomass—a technology review from 1950 to 2009. Fuel 89:1763–1783. https://doi.org/10.1016/j.fuel.2010.01.027
11. Dannesboe C, Hansen JB, Johannsen I (2020) Catalytic methanation of CO_2 in biogas: experimental results from a reactor at full scale. React Chem Eng 5:183–189. https://doi.org/10.1039/C9RE00351G
12. Schmider D, Maier L, Deutschmann O (2021) Reaction kinetics of CO and CO_2 methanation over nickel. Ind Eng Chem Res 60:5792–5805. https://doi.org/10.1021/acs.iecr.1c00389
13. Hernandez Lalinde JA, Roongruangsree P, Ilsemann J, Bäumer M, Kopyscinski J (2020) CO_2 methanation and reverse water gas shift reaction. Kinetic study based on in situ spatially-resolved measurements. Chem Eng J 390:124629. https://doi.org/10.1016/j.cej.2020.124629
14. Kopyscinski J, Schildhauer TJ, Vogel F, Biollaz SMA, Wokaun A (2010) Applying spatially resolved concentration and temperature measurements in a catalytic plate reactor for the kinetic study of CO methanation. J Catal 271:262–279. https://doi.org/10.1016/j.jcat.2010.02.008
15. Witte J, Calbry-Muzyka A, Wieseler T, Hottinger P, Biollaz SMA, Schildhauer TJ (2019) Demonstrating direct methanation of real biogas in a fluidised bed reactor. Appl Energy 240:359–371. https://doi.org/10.1016/j.apenergy.2019.01.230
16. Koschany F, Schlereth D, Hinrichsen O (2016) On the kinetics of the methanation of carbon dioxide on coprecipitated NiAl(O)x. Appl Catal B Environ 181:504–516. https://doi.org/10.1016/j.apcatb.2015.07.026
17. Parlikkad NR, Chambrey S, Fongarland P, Fatah N, Khodakov A, Capela S et al (2013) Modeling of fixed bed methanation reactor for syngas production: operating window and performance characteristics. Fuel 107:254–260. https://doi.org/10.1016/j.fuel.2013.01.024
18. Xu J, Froment GF (1989) Methane steam reforming, methanation and water-gas shift: I. Intrinsic kinetics. AIChE J 35:88–96. https://doi.org/10.1002/aic.690350109
19. Bartholomew CH (2001) Mechanisms of catalyst deactivation. Appl Catal A Gen 212:17–60. https://doi.org/10.1016/S0926-860X(00)00843-7
20. Schildhauer TJ (2016) Methanation for synthetic natural gas production—chemical reaction engineering aspects. In: Schildhauer TJ, Biollaz SMA (eds) Synthetic natural gas from coal, dry biomass, power-to-gas application. John Wiley & Sons, Inc., Hoboken, NJ, USA, pp 77–159. https://doi.org/10.1002/9781119191339.ch4

21. Czekaj I, Loviat F, Raimondi F, Wambach J, Biollaz S, Wokaun A (2007) Characterization of surface processes at the Ni-based catalyst during the methanation of biomass-derived synthesis gas: X-ray photoelectron spectroscopy (XPS). Appl Catal A Gen 329:68–78. https://doi.org/10.1016/j.apcata.2007.06.027

22. McCarty JG, Wise H (1979) Hydrogenation of surface carbon on alumina-supported nickel. J Catal 57:406–416. https://doi.org/10.1016/0021-9517(79)90007-1

23. Foppiano D, Tarik M, Schneebeli J, Calbry-Muzyka A, Biollaz S, Ludwig C (2020) Siloxane compounds in biogas from manure and mixed organic waste: method development and speciation analysis with GC-ICP-MS. Talanta 208:120398. https://doi.org/10.1016/j.talanta.2019.120398

24. Calbry-Muzyka A, Tarik M, Gandiglio M, Li J, Foppiano D, de Krom I et al (2021) Sampling, on-line and off-line measurement of organic silicon compounds at an industrial biogas-fed 175-kWe SOFC plant. Renew Energy 177:61–71. https://doi.org/10.1016/j.renene.2021.05.047

25. Rusmanis D, O'Shea R, Wall DM, Murphy JD (2019) Biological hydrogen methanation systems—an overview of design and efficiency. Bioengineered 10:604–634. https://doi.org/10.1080/21655979.2019.1684607

26. Rostrup-Nielsen JR, Sehested J, Nørskov JK (2002) Hydrogen and synthesis gas by steam-and CO$_2$ reforming. Adv Catal 47:65–139. https://doi.org/10.1016/S0360-0564(02)47006-X

27. Ross JRH (1985) Metal catalysed methanation and steam reforming. Catalysis 7:1–45

28. Schildhauer TJ, Biollaz SMA (2016) Fluidised bed methanation for SNG production—process development at the Paul Scherrer Institut. In: Synthetic natural gas from coal, dry biomass, power-to-gas application. John Wiley & Sons, Inc., Hoboken, NJ, USA, pp 221–230. https://doi.org/10.1002/9781119191339.ch8

29. Schildhauer TJ (2019) Biosynthetic natural gas (Bio-SNG). In: Kaltschmitt M (ed) Energy from organization material. Springer New York, New York, NY, pp 1065–1080. https://doi.org/10.1007/978-1-4939-7813-7_996

30. Fukuzaki S, Nishio N, Shobayashi M, Nagai S (1990) Inhibition of the fermentation of propionate to methane by hydrogen, acetate, and propionate. Appl Environ Microbiol 56:719–723. https://doi.org/10.1128/aem.56.3.719-723.1990

31. Strübing D, Huber B, Lebuhn M, Drewes JE, Koch K (2017) High performance biological methanation in a thermophilic anaerobic trickle bed reactor. Bioresour Technol 245:1176–1183. https://doi.org/10.1016/j.biortech.2017.08.088

32. Rachbauer L, Voitl G, Bochmann G, Fuchs W (2016) Biological biogas upgrading capacity of a hydrogenotrophic community in a trickle-bed reactor. Appl Energy 180:483–490. https://doi.org/10.1016/j.apenergy.2016.07.109

33. Peillex J-P, Fardeau M-L, Boussand R, Navarro J-M, Belaich J-P (1988) Growth of methanococcus thermolithotrophicus in batch and continuous culture on H$_2$ and CO$_2$: influence of agitation. Appl Microbiol Biotechnol 29:560–564. https://doi.org/10.1007/BF00260985

34. Thema M, Weidlich T, Kaul A, Böllmann A, Huber H, Bellack A et al (2021) Optimized biological CO$_2$-methanation with a pure culture of thermophilic methanogenic archaea in a trickle-bed reactor. Bioresour Technol 333:125135. https://doi.org/10.1016/j.biortech.2021.125135

35. Schill N, van Gulik WM, Voisard D, von Stockar U (1996) Continuous cultures limited by a gaseous substrate: development of a simple, unstructured mathematical model and experimental verification with Methanobacterium thermoautotrophicum. Biotechnol Bioeng 51:645–658. https://doi.org/10.1002/(SICI)1097-0290(19960920)51:6%3c645::AID-BIT4%3e3.0.CO;2-H

36. Ullrich T, Lemmer A (2019) Performance enhancement of biological methanation with trickle bed reactors by liquid flow modulation. GCB Bioenergy 11:63–71. https://doi.org/10.1111/gcbb.12547

37. Vintiloiu A, Lemmer A, Oechsner H, Jungbluth T (2012) Mineral substances and macronutrients in the anaerobic conversion of biomass: an impact evaluation. Eng Life Sci 12:287–294. https://doi.org/10.1002/elsc.201100159

38. microbEnergy GmbH (2019) Web page microbEnergy 2019. https://www.microbenergy.de/. Accessed 19 Jul 2022
39. Graf F, Krajete A, Schmack U (2014) Abschlussbericht: Techno-ökonomische Studie zur biologischen Methanisierung bei Power-to-Gas-Konzepten
40. Seifert AH, Rittmann S, Herwig C (2014) Analysis of process related factors to increase volumetric productivity and quality of biomethane with Methanothermobacter marburgensis. Appl Energy 132:155–162. https://doi.org/10.1016/j.apenergy.2014.07.002
41. Guilera J, Andreu T, Basset N, Boeltken T, Timm F, Mallol I et al (2020) Synthetic natural gas production from biogas in a waste water treatment plant. Renew Energy 146:1301–1308. https://doi.org/10.1016/j.renene.2019.07.044
42. Bajohr S (2020) Power-to-gas technologies, oral presentation at the store and go conference. Processing Store&Go Conf., Karlsruhe
43. Schaaf T, Grünig J, Schuster MR, Rothenfluh T, Orth A (2014) Methanation of CO_2—storage of renewable energy in a gas distribution system. Energy Sustain Soc 4:2. https://doi.org/10.1186/s13705-014-0029-1
44. Calbry-Muzyka AS, Schildhauer TJ (2020) Direct methanation of biogas—technical challenges and recent progress. Front Energy Res 8. https://doi.org/10.3389/fenrg.2020.570887
45. Specht M, Brellochs J, Frick V, Stürmer B, Zuberbühler U (2016) The power to gas process. In: Schildhauer TJ, Biollaz SMA (eds) Synthetic natural gas from coal, dry biomass, power-to-gas application. John Wiley & Sons, Inc., Hoboken, NJ, USA, pp 191–220. https://doi.org/10.1002/9781119191339.ch7
46. Moioli E (2023) HZI methanation technologies. In: Proceeding 6th nuremb. work. Methanation 2nd generation fuels
47. Moioli E, Mutschler R, Borsay A, Calizzi M, Züttel A (2020) Synthesis of grid compliant substitute natural gas from a representative biogas mixture in a hybrid Ni/Ru catalysed reactor. Chem Eng Sci X 8:100078. https://doi.org/10.1016/j.cesx.2020.100078
48. Groppi G, Tronconi E (2005) Honeycomb supports with high thermal conductivity for gas/solid chemical processes. Catal Today 105:297–304. https://doi.org/10.1016/j.cattod.2005.06.041
49. Store&Go Project (2016). https://www.storeandgo.info/. Accessed 11 Feb 2020
50. Dittmeyer R, Boeltken T, Piermartini P, Selinsek M, Loewert M, Dallmann F et al (2017) Micro and micro membrane reactors for advanced applications in chemical energy conversion. Curr Opin Chem Eng 17:108–125. https://doi.org/10.1016/j.coche.2017.08.001
51. Schirrmeister S, von Morstein O, Föcker H (2019) Innovative large-scale energy storage technologies and power-to-gas concepts after optimisation. Demonstration plant Falkenhagen commissioned/commissioning report (D2.3)
52. Neubert M, Hauser A, Pourhossein B, Dillig M, Karl J (2018) Experimental evaluation of a heat pipe cooled structured reactor as part of a two-stage catalytic methanation process in power-to-gas applications. Appl Energy 229:289–298. https://doi.org/10.1016/j.apenergy.2018.08.002
53. Borgschulte A, Gallandat N, Probst B, Suter R, Callini E, Ferri D et al (2013) Sorption enhanced CO_2 methanation. Phys Chem Chem Phys 15:9620. https://doi.org/10.1039/c3cp51408k
54. Kiefer F, Nikolic M, Borgschulte A, Dimopoulos EP (2022) Sorption-enhanced methane synthesis in fixed-bed reactors. Chem Eng J 449:137872. https://doi.org/10.1016/j.cej.2022.137872
55. Delmelle R, Duarte RB, Franken T, Burnat D, Holzer L, Borgschulte A et al (2016) Development of improved nickel catalysts for sorption enhanced CO_2 methanation. Int J Hydrogen Energy 41:20185–20191. https://doi.org/10.1016/j.ijhydene.2016.09.045
56. Dannesboe C (2019) Catalytic upgrading of CO_2 in Biogas. Aarhus University
57. Karlsruher Institut für Technologie (KIT). Helmeth Project (2023). http://helmeth.eu/. Accessed 19 Jul 2023

58. Gruber M, Weinbrecht P, Biffar L, Harth S, Trimis D, Brabandt J et al (2018) Power-to-Gas through thermal integration of high-temperature steam electrolysis and carbon dioxide methanation—experimental results. Fuel Process Technol 181:61–74. https://doi.org/10.1016/j.fuproc.2018.09.003

59. Gantenbein A, Kröcher O, Biollaz SMA, Schildhauer TJ (2022) Techno-economic evaluation of biological and fluidised-bed based methanation process chains for grid-ready biomethane production. Front Energy Res 9. https://doi.org/10.3389/fenrg.2021.775259

60. Seemann MC, Schildhauer TJ, Biollaz SMA (2010) Fluidized bed methanation of wood-derived producer gas for the production of synthetic natural gas. Ind Eng Chem Res 49:11119–11119. https://doi.org/10.1021/ie101898w

61. Hedden K, Anderlohr A, Becker J, Zeeb H-P, Cheng Y-H (1986) Gleichzeitige Konvertierung und Methanisierung CO-reicher Gase. Eggenstein-Leopoldshafen: Fachinformationszentrum Energie, Physik, Mathematik Karlsruhe. https://worldcat.org/title/636620241

62. Friedrichs G, Proplesch P, Wismann G, Lommerzheim W (1985) Methanisierung von Kohlenvergasungsgasen im Wirbelbett, Pilot Entwicklungsstufe, prepared by Thyssengas GmbH for Bundesministerium für Forschung und Technologie, Forschungsbericht T 85–106

63. Nam H, Kim JH, Kim H, Kim MJ, Jeon S-G, Jin G-T et al (2021) CO_2 methanation in a bench-scale bubbling fluidized bed reactor using Ni-based catalyst and its exothermic heat transfer analysis. Energy 214:118895. https://doi.org/10.1016/j.energy.2020.118895

64. Liu J, Cui D, Yao C, Yu J, Su F, Xu G (2016) Syngas methanation in fluidized bed for an advanced two-stage process of SNG production. Fuel Process Technol 141:130–137. https://doi.org/10.1016/j.fuproc.2015.03.016

65. Bartik A, Fuchs J, Pacholik G, Föttinger K, Hofbauer H, Müller S et al (2022) Experimental investigation on the methanation of hydrogen-rich syngas in a bubbling fluidized bed reactor utilizing an optimized catalyst. Fuel Process Technol 237:1–12. https://doi.org/10.1016/j.fuproc.2022.107402

66. EU Project HYFUELUP (2023). https://hyfuelup.eu/

67. Hervy M, Maistrello J, Brito L, Rizand M, Basset E, Kara Y et al (2021) Power-to-gas: CO_2 methanation in a catalytic fluidized bed reactor at demonstration scale, experimental results and simulation. J CO_2 Util 50:101610. https://doi.org/10.1016/j.jcou.2021.101610

68. Riechmann P, Schildhauer TJ (2023) Applying bubbling fluidized-bed reactors for strongly exothermic reactions: focus on methanation. KONA Powder Part J 2024009. https://doi.org/10.14356/kona.2024009

69. Sauerschell S, Bajohr S, Kolb T (2022) Methanation pilot plant with a slurry bubble column reactor: setup and first experimental results. Energy Fuels 36:7166–7176. https://doi.org/10.1021/acs.energyfuels.2c00655

70. Electrochaea.dk ApS. BioCatProject (2014). http://biocat-project.com/. Accessed 11 Feb 2020

71. Inkeri E, Tynjälä T, Laari A, Hyppänen T (2018) Dynamic one-dimensional model for biological methanation in a stirred tank reactor. Appl Energy 209:95–107. https://doi.org/10.1016/j.apenergy.2017.10.073

72. Hafenbradl D (2019) Oral Presentation: Biological methanation

73. Limeco (2020) Project web page: power-to-gas plant Dietikon (Switzerland). https://www.powertogas.ch/. Accessed 19 Jul 2022

74. Ullrich T, Lindner J, Bär K, Mörs F, Graf F, Lemmer A (2018) Influence of operating pressure on the biological hydrogen methanation in trickle-bed reactors. Bioresour Technol 247:7–13. https://doi.org/10.1016/j.biortech.2017.09.069

75. Q Power (2023) Company brochure 2023

76. Burkhardt M (2022) Biological methanation by GICON® trickle bed process—an upgrade for conventional biogas plants. Proc. Regatec, Malmö

77. Matthischke S, Roensch S, Güttel R (2018) Start-up time and load range for the methanation of carbon dioxide in a fixed-bed recycle reactor. Ind Eng Chem Res 57:6391–6400. https://doi.org/10.1021/acs.iecr.8b00755

78. Matthischke S, Krüger R, Rönsch S, Güttel R (2016) Unsteady-state methanation of carbon dioxide in a fixed-bed recycle reactor—experimental results for transient flow rate ramps. Fuel Process Technol 153:87–93. https://doi.org/10.1016/j.fuproc.2016.07.021

79. Giglio E, Pirone R, Bensaid S (2021) Dynamic modelling of methanation reactors during start-up and regulation in intermittent power-to-gas applications. Renew Energy 170:1040–1051. https://doi.org/10.1016/j.renene.2021.01.153

80. Bremer J, Rätze KHG, Sundmacher K (2017) CO_2 methanation: optimal start-up control of a fixed-bed reactor for power-to-gas applications. AIChE J 63:23–31. https://doi.org/10.1002/aic.15496

81. Zimmermann RT, Bremer J, Sundmacher K (2022) Load-flexible fixed-bed reactors by multi-period design optimization. Chem Eng J 428:130771. https://doi.org/10.1016/j.cej.2021.130771

82. Lefebvre J, Götz M, Bajohr S, Reimert R, Kolb T (2015) Improvement of three-phase methanation reactor performance for steady-state and transient operation. Fuel Process Technol 132:83–90. https://doi.org/10.1016/j.fuproc.2014.10.040

83. Gantenbein A, Schildhauer TJ (2022) Challenges in part load operation of biogas-based power-to-gas processes (part I). Front Energy Res 10:1–14. https://doi.org/10.3389/fenrg.2022.1049687

84. Schildhauer T, Madi H, Peter C, Wan Y, Densing M, Hofer M (2022) Renewable management and real-time control platform (ReMaP): final report of experiment T3.10 Electrochem Energy Storage

85. Angelidaki I, Treu L, Tsapekos P, Luo G, Campanaro S, Wenzel H et al (2018) Biogas upgrading and utilization: current status and perspectives. Biotechnol Adv 36:452–466. https://doi.org/10.1016/j.biotechadv.2018.01.011

86. Bui M, Flø NE, de Cazenove T, Mac DN (2020) Demonstrating flexible operation of the technology centre mongstad (TCM) CO_2 capture plant. Int J Greenh Gas Control 93:102879. https://doi.org/10.1016/j.ijggc.2019.102879

87. Peyer T, Flückiger J, Di Lorenzo T, Gündel N (2018) Das Stadtgas der Zukunft. Aqua Gas 34–9

88. GICON Advanced Environmental Technologies GmbH (2023) Company information

89. Jønson BD, Tsapekos P, Tahir Ashraf M, Jeppesen M, Ejbye Schmidt J, Bastidas-Oyanedel J-R (2022) Pilot-scale study of biomethanation in biological trickle bed reactors converting impure CO_2 from a full-scale biogas plant. Bioresour Technol 365:128160. https://doi.org/10.1016/j.biortech.2022.128160

90. Alitalo A, Aura E (2013) Means and methods for methane production. WO2013167806A1

91. Q Power Company Web Page (2023). https://qpower.fi/

92. Calbry-Muzyka AS, Gantenbein A, Schneebeli J, Frei A, Knorpp AJ, Schildhauer TJ et al (2019) Deep removal of sulfur and trace organic compounds from biogas to protect a catalytic methanation reactor. Chem Eng J 360:577–590. https://doi.org/10.1016/j.cej.2018.12.012

93. Leimert JM, Neubert M, Treiber P, Dillig M, Karl J (2018) Combining the Heatpipe Reformer technology with hydrogen-intensified methanation for production of synthetic natural gas. Appl Energy 217:37–46. https://doi.org/10.1016/j.apenergy.2018.02.127

94. Teske LS (2014) Integrating rate based models into a multi-objective process design and optimisation framework using surrogate models. https://doi.org/10.5075/epfl-thesis-6302

95. Gassner M, Maréchal F (2008) Thermo-economic optimisation of the integration of electrolysis in synthetic natural gas production from wood. Energy 33:189–198. https://doi.org/10.1016/j.energy.2007.09.010

96. Held J (2016) SNG from wood—the GoBiGas project. In: Synthetic natural gas from coal, dry biomass, power-to-gas application. John Wiley & Sons, Inc., Hoboken, NJ, USA, pp 181–90. https://doi.org/10.1002/9781119191339.ch6

97. Rabou LPLM, Bos L (2012) High efficiency production of substitute natural gas from biomass. Appl Catal B Environ 111–112:456–460. https://doi.org/10.1016/j.apcatb.2011.10.034

98. Rabou LPLM, Zwart RWR, Vreugdenhil BJ, Bos L (2009) Tar in biomass producer gas, the energy research centre of The Netherlands (ECN) experience: an enduring challenge. Energy Fuels 23:6189–6198. https://doi.org/10.1021/ef9007032

99. Kopyscinski J, Seemann MC, Moergeli R, Biollaz SMA, Schildhauer TJ (2013) Synthetic natural gas from wood: reactions of ethylene in fluidised bed methanation. Appl Catal A Gen 462–463:150–156. https://doi.org/10.1016/j.apcata.2013.04.038
100. Maheut M (2022) GAYA: production of SNG from dry biomass and waste pyrogasification. IEA Bioenergy Work
101. Materazzi M, Lettieri P, Mazzei L, Taylor R, Chapman C (2015) Reforming of tars and organic sulphur compounds in a plasma-assisted process for waste gasification. Fuel Process Technol 137:259–268. https://doi.org/10.1016/j.fuproc.2015.03.007
102. Heyne S, Seemann M, Schildhauer TJ (2016) Coal and biomass gasification for SNG production. Synthetic natural gas from coal, dry biomass, power-to-gas application. John Wiley & Sons, Inc., Hoboken, NJ, USA, pp 5–40. https://doi.org/10.1002/9781119191339.ch2
103. Bajohr S, Schollenberger D, Buchholz D, Weinfurtner T, Götz M (2014) Kopplung der PtG-Technologie mit thermochemischer Biomassevergasung: Das KIC-Projekt "DemoSNG." GWF/Gas, Erdgas 155:470–475
104. Heller T (2016) BioPower2Gas—power-to-gas with biological methanation. In: Presentation at of the biomass for Swiss energy future conference, Brugg, Switzerland
105. Gorre J (2020) Oral presentation at the Store&Go conference. Proceeding Store&Go conference, Karlsruhe
106. Witte J, Settino J, Biollaz SMA, Schildhauer TJ (2018) Direct catalytic methanation of biogas—part I: new insights into biomethane production using rate-based modelling and detailed process analysis. Energy Convers Manag 171:750–768. https://doi.org/10.1016/j.enconman.2018.05.056
107. Witte J, Kunz A, Biollaz SMA, Schildhauer TJ (2018) Direct catalytic methanation of biogas—part II: techno-economic process assessment and feasibility reflections. Energy Convers Manag 178:26–43. https://doi.org/10.1016/j.enconman.2018.09.079

Fischer-Tropsch Synthesis

Lucas Brübach, Moritz Wolf, and Peter Pfeifer

Abstract For the production and utilization of powerfuels, particularly hydrocarbon-based power-to-liquid (PtL) fuels like kerosene and diesel, the Fischer-Tropsch (FT) synthesis is seen as an essential technology and pathway in recent and upcoming years. Starting from carbon dioxide (CO_2) as a primary source for producing the fuels in a carbon neutral pathway, FT has the advantage to eliminate the oxygen molecules directly in the fuels synthesis by producing water as byproduct. The latter could be recycled for generating hydrogen through electrolysis. Also, fuel composition matches quite well to the current standards and regulations such as ASTM 7566-1 and DIN EN 15940 exist for blending hydro-processed FT syncrude product to kerosene and diesel fuel into existing turbine and motor fleet, respectively. The chapter on FT is focussing on the recent developments of reactor technology for process intensification and integration regarding CO_2 activation with specific attention to volatility of the hydrogen source when hydrogen is produced from renewable electricity.

Keywords Fischer-Tropsch synthesis · Process intensification · Reactor technology · Process integration · CO_2 activation · Fluctuating feed/Dynamic operation

L. Brübach
Karlsruhe Institute of Technology (KIT), Institute for Micro Process Engineering (IMVT), Karlsruhe, Germany

M. Wolf
Karlsruhe Institute of Technology (KIT), Institute of Catalysis Research and Technology (IKFT) and Engler-Bunte-Institute (EBI), Karlsruhe, Germany

P. Pfeifer (✉)
INERATEC GmbH, Karlsruhe, Germany
e-mail: peter.pfeifer@ineratec.de

© The Author(s), under exclusive license to Springer Nature Switzerland AG 2025
N. Bullerdiek et al. (eds.), *Powerfuels*, Green Energy and Technology,
https://doi.org/10.1007/978-3-031-62411-7_22

1 Introduction

For the production and utilization of powerfuels, particularly hydrocarbon-based power-to-liquid (PtL) fuels like kerosene and diesel, the Fischer-Tropsch (FT) synthesis is seen as an essential technology and pathway in recent and upcoming years. Starting from carbon dioxide (CO_2) as a primary source for producing the fuels in a carbon neutral pathway, FT has the advantage to eliminate the oxygen molecules directly in the fuels synthesis by producing water as byproduct. The latter could be recycled for generating hydrogen through electrolysis. Also, fuel composition matches quite well to the current standards and regulations such as ASTM 7566-1 and DIN EN 15940 exist for blending hydro-processed FT syncrude product to kerosene and diesel fuel into existing turbine and motor fleet, respectively.

Nevertheless, the application and development of this technology date back to the last century already. Since its discovery in the early nineteenth century, the Fischer-Tropsch synthesis (FTS) has always received huge interest whenever crude oil was scarce due to geographical, economical, or political reasons.[1] The Fischer-Tropsch process was first commercialized and intensively used by the Nazi regime in Germany in the 1930s with the goal of industrial autarky preceding World War II. Synthetic fuels were produced from coal in a coal-to-liquid (CTL) process including FTS, which was adapted by the South African government during apartheid due to sanctions. The oil crisis in the 1970s and the price hike of crude oil in recent decades again put FTS in the spotlight for alternative production of fuels from other fossil resources than oil, while in particular natural gas became the primary feedstock for commercial applications of FTS via a gas-to-liquid (GTL) approach. However, the strong link of the profitability in the production of FTS products to the crude oil price always rendered investments into new large-scale FTS plants risky.

Eventually, the urgent need in seeking sustainable alternatives to fossil energy sources due to climate change revived the FTS technology once again, at first focusing on biomass as a renewable feedstock (biomass-to-liquid, BTL) and later using CO_2 and green electricity via the power-to-liquid (PtL) route. In this context, i.e., for the deployment of FTS technologies for PtL fuel production in the upcoming years, an in-depth understanding and assessment of precise process conditions and specifications is crucial. This includes key areas such as

- process combinations to activate CO_2,
- understanding catalyst optimization,
- reactor technologies to improve process efficiency, and
- challenges of fluctuating renewable energy on reactor and process level.

However, in this field, comparative studies are somewhat limited. Thus, this chapter aims to provide a broader overview of the specifics in FTS processes by consolidating research results from respective individual studies.

2 Traditional Fischer-Tropsch process

The overall classical Fischer-Tropsch process consists of four major steps: production of synthesis gas, cleaning/purification of synthesis gas, the actual FT synthesis step, and product conditioning (Fig. 1). All carbonaceous sources, such as coal, natural gas, biomass, or organic wastes may be reformed or gasified yielding synthesis gas. However, the particular processes and process conditions strongly vary for different raw materials. For instance, due to high ash contents in biomass and coal, the purification of synthesis gas from biomass or coal gasification is much more elaborate than for synthesis gas from steam reforming of natural gas [1, 2]. In regard to the FTS, a high temperature (320–350 °C) and a low temperature (220–250 °C) pathway are commercially established [3], while a medium temperature Fischer-Tropsch (MTFT) process (270–290 °C) has recently been commercialized by Synfuels China [4, 5].

2.1 Heterogeneous Catalysts

Franz Fischer and Hans Tropsch reported that group VIII metals were capable of activating carbon monoxide (CO) hydrogenation, while only iron, cobalt, and nickel were identified as viable catalytic material [7]. Iron showed a superior performance at 350–500 °C, while a decreased temperature range of 250–270 °C was observed for cobalt. For nickel, only a reduction of the reaction temperature to 160 °C resulted in a reasonable methane selectivity [7]. Combinations of these active metals with an equal or three times higher amount of aluminum, beryllium, chromium, magnesium, manganese, uranium, or zinc in their oxidic phase, but also the addition of silicic acid, rare earth (RE) metal oxides, activated carbon, or amorphous carbon was tested for a promotional effect accordingly. Especially the metal oxides had a beneficial effect on the catalytic behavior, while the addition of palladium, copper, and iron oxide to nickel and cobalt enhanced their performance [7].

These early findings are still in line with nickel being the metal of choice for industrial methanation and an exclusive commercial application of iron-based catalysts

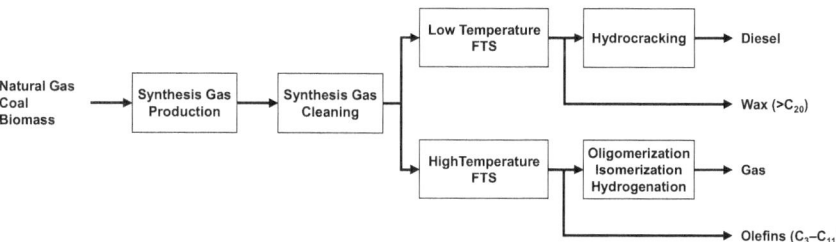

Fig. 1 Scheme of the overall process for the production of synthetic liquid fuels and other hydrocarbons by the Fischer-Tropsch synthesis. Reproduced from Guettel et al. [6]

in the high-temperature Fischer-Tropsch synthesis (HTFTS) [8, 9]. The catalysts are typically chemically promoted by K_2O and contain structural promoters such as magnesium oxide (MgO) or aluminum oxide (Al_2O_3). However, iron catalysts have been significantly improved and are nowadays, together with cobalt-based catalysts, also applied in the low-temperature Fischer-Tropsch synthesis (LTFTS) [8, 9]. In case of Sasol for instance, a major global company in FTS technology, a typical iron-based catalyst for LTFTS contains Fe, SiO_2, Cu, and K_2O with a 20:5:1:1 mass ratio [10]. One major characteristic of iron-based catalysts is their water–gas shift activity allowing for the use of hydrogen (H_2)-poor synthesis gas feeds as obtained from the gasification of coal. The conversion of CO to the favor of H_2 via the water–gas shift reaction is typically thermodynamically controlled under FT conditions [11, 12, 13]. In contrast, this reaction path is mostly negligible for cobalt-based FT catalysts, but a pronounced formation of methane is observed over cobalt-based catalysts, especially at increased temperatures [11, 12, 13]. The first commercial cobalt-based catalysts in the German coal-to-liquid (CTL) plants were promoted by thorium, zirconium, or magnesium and mixed with Kieselguhr, a siliceous sedimentary rock, at a 1:1 mass ratio [14]. Nowadays, cobalt is typically supported on (mostly modified) large surface area metal oxide supports such as aluminum oxide (Al_2O_3), silicon dioxide (SiO_2), and titanium dioxide (TiO_2), while platinum (Pt), rhenium (Re), or ruthenium (Ru) are applied as reduction promoters. Lanthana, zirconia, or alkali oxides represent common structural oxidic promoters [15, 16, 17].

Cobalt-based catalysts are more expensive and active than iron-based catalyst [18]. An increased mass-specific activity over the lifetime of approximately two magnitudes is required and achieved via utilization of supported nanoparticles, i.e., by increasing the specific surface area of the active cobalt phase drastically. In contrast, fused or precipitated, rather bulky iron-based catalysts are typically applied as a commercial catalyst due to the low price of iron-based ores [8, 18].

2.2   Reactions and Product Distribution

The broad product spectrum is one of the major characteristics of the FTS and includes hydrocarbon chain lengths of more than 100. Paraffins, olefins, oxygenates, and aromatics are all primary organic products. The reaction proceeds via in situ produced monomers in a polymerization-type mechanism. Hence, the reaction pathway includes the generation of a chain initiator, which is followed by chain growth or propagation. Subsequent chain growth termination or simple desorption results in the (intermediate) products. A representative overall reaction may be described as the hydrogenation of CO forming a (CH_2) hydrocarbon chain segment and water (H_2O) (Eq. 1) [10, 12]. A general reaction equation for the formation of paraffins (Eq. 2) and olefins (Eq. 3) can be formed as well.

$$CO + 2H_2 \rightarrow (CH_2) + H_2O \tag{1}$$

$$nCO + (2n + 1)H_2- \rightarrow C_nH_{(2n+2)} + nH_2O \tag{2}$$

$$nCO + 2nH_2 \rightarrow C_nH_{2n} + nH_2O \tag{3}$$

The required stoichiometric composition of the reactants H_2 and CO in the synthesis gas feed decreases with the average chain length in the final product mixture [19, 20]. Methanation (Eq. 4) and the water–gas shift (WGS) reaction (Eq. 5) are the most important side reactions in the FTS [19, 20]. Thermodynamically, methane is the most stable hydrocarbon under FT conditions. However, the FT reaction is kinetically controlled allowing for the formation of other hydrocarbons as well. Therefore, the main reactions of the FTS are far away from the thermodynamic equilibrium [12, 21].

$$CO + 3H_2 \leftrightarrow CH_4 + H_2O \tag{4}$$

$$CO + H_2O \leftrightarrow CO_2 + H_2 \tag{5}$$

The exact FT mechanisms are one of the most discussed topics in heterogeneous catalysis with various proposed pathways. It is generally understood that a number of parallel pathways exist; i.e., a combination of proposed mechanisms may describe the FTS best [9, 19, 20, 22–26]. The four most popular mechanisms are the alkyl mechanism, the alkenyl mechanism, the enol mechanism, and the CO-insertion mechanism [22]. They all propose a step-wise chain growth as the product distribution can be predicted by a chain growth probability [10, 14].

The chain growth probability depends on the reaction conditions such as the overall pressure, the temperature, or the composition of the gas phase, and strongly on the catalysts. For the industrial LTFTS, a probability exceeding 90% is desired to mostly produce long-chained hydrocarbons, while the probability in the HTFTS is typically in the range of 70–75%.[10] Idealized, statistical models to calculate the product distribution for reactions like polymerizations have been developed. The Anderson-Schulz-Flory model (ASF) has been widely accepted as descriptor of the FT product spectrum (Fig. 2).

In the ASF model, a chain growth probability (p_G) is determined by the rates of chain growth (r_{cg}) and chain growth termination (r_{cgt}; Eq. 6). Aside from the actual chain length (N_C), this probability is the only parameter in the calculation of the selectivity of the particular chain lengths ($S_{N,C}$; Eq. 7) [22, 27].

$$p_G = \frac{r_{cg}}{r_{cg} + r_{cgt}} \tag{6}$$

$$S_{N,C} = (1 - p_G)p_G^{N_C-1} \tag{7}$$

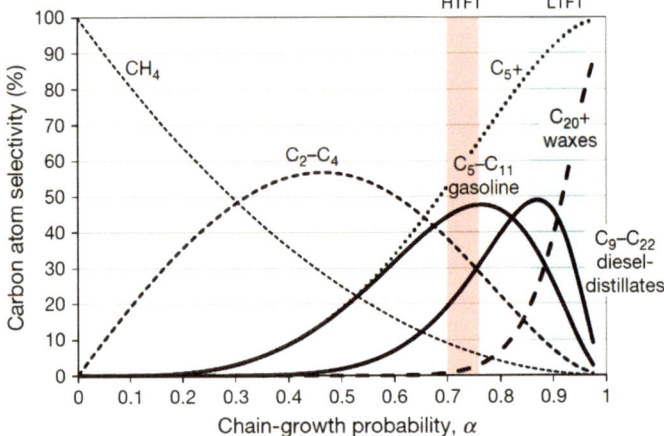

Fig. 2 Hydrocarbon product spectrum according to the Anderson-Schulz-Flory (ASF) model with regimes for the chain growth probability for the commercial HTFT and LTFT. Reproduced from van de Loosdrecht et al. [16]

2.3 Traditional Reactor Solutions

Commercial FT reactors are typically operated with heterogeneous catalysts. The most important reactor designs (Fig. 3) for the high-temperature Fischer-Tropsch synthesis (HTFTS) are the circulating and the fixed fluidized bed reactors. Multi-tubular fixed bed reactors and slurry phase bubbling reactors are preferred for the low-temperature Fischer-Tropsch synthesis (LTFTS) [3]. The released energy of the highly exothermic FT reaction is equal to around 25% of the heat of combustion of the input synthesis gas [12]. Removing this heat of reaction from the FT reactor is of utmost importance in order to avoid thermal degradation of the catalyst by overheating, as well as associated undesired selectivity changes, such as an increased methane yield. The selectivity is strongly temperature-dependent, and hence, an ideal isothermal reactor is mandatory to control the product distribution [3].

3 Fischer-Tropsch Processes for Powerfuels

The overall process layout for producing powerfuels from carbon dioxide can heavily determine the approach in which Fischer-Tropsch is conducted. Currently, the following process step combinations, as shown in Table 1 and Fig. 4, are being discussed.

For most process combinations, examples can be found in recent research. Moreover, the Fischer-Tropsch process seems advantageous with cobalt as active catalyst

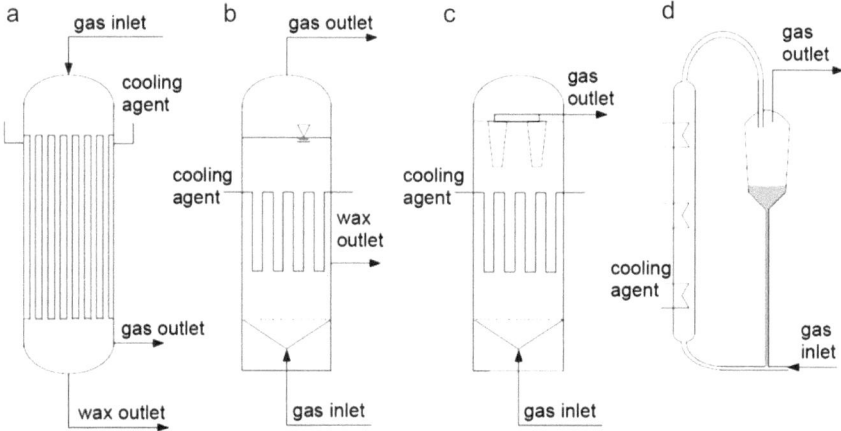

Fig. 3 Schematics of major large-scale Fischer-Tropsch reactors: **a** multi-tubular fixed bed reactor, **b** slurry phase bubbling reactor, **c** fixed fluidized bed reactor, and **d** circulating fluidized bed reactor

Table 1 Overview of main Fischer-Tropsch (FT) process step combinations

	FT process variant	Description
1	Water electrolysis, rWGS, and FT	Water electrolysis in proton exchange membrane (PEM) or alkaline (AE) cells, summarized as water electrolysis, for hydrogen production combined with reverse water–gas shift (rWGS) and FT
2	Direct activation of CO_2	Water electrolysis in PEM or AE and simultaneous activation of CO_2 and FT synthesis in a single reactor (often called CO_2-FT or direct activation of CO_2)
3	SOEC steam electrolysis, rWGS, and FT	Steam electrolysis in solid oxide electrolyzer cells (SOEC) for hydrogen production combined with rWGS and FT, whereas steam is exported from FT to feed the electrolysis
4	SOEC co-electrolysis of CO_2 and steam and FT	Co-electrolysis of CO_2 and steam in SOEC for synthesis gas production combined with cobalt-based FT, whereas steam is exported from FT to feed the electrolysis
5	Separate CO_2 and H_2O electrolysis plus FT	Electrolysis of CO_2 in PEM or SOEC to generate CO for co-feeding to hydrogen from other electrolysis for FT
6	CO_2 plasma splitting and electrolysis plus FT	Plasma splitting of CO_2 to generate CO for co-feeding to hydrogen from other electrolysis for FT
7	CO_2 plasma splitting, WGS, and plus FT	Plasma splitting of CO_2 to generate CO and the water–gas shift (WGS) to form additional hydrogen from parts of the CO for FT

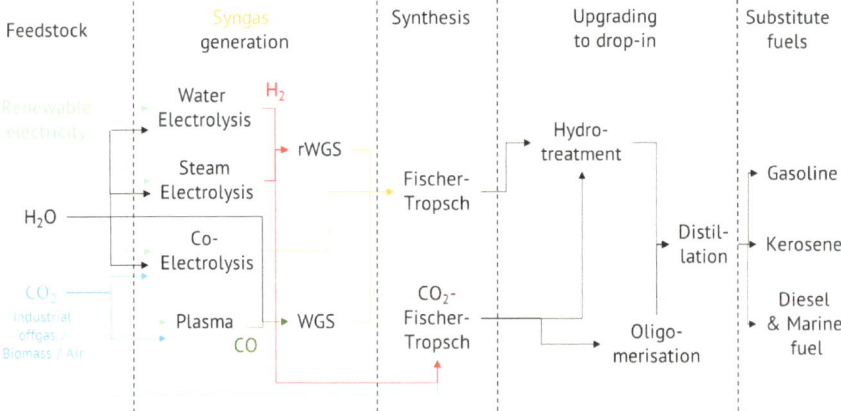

Fig. 4 Process variants for conduction of FT from renewable power and CO_2 with simplified view on the syncrude upgrading to fuels

for all these combinations because different but commonly above or near the thermodynamic CO concentration at temperature up to 300 °C may be present from the preceding processes in the feed gas of the FT. As such, water–gas shift, a possible side reaction on iron catalyst, would often reduce the CO level than increasing it. A recycle over the FT reactor may be required if no process intensification is performed. In such case, low conversions of CO are required to prevent hot-spots or an overall high steam partial pressure in the reactor. Both would lead to accelerated catalyst deactivation. Such necessary recycle would be additional to any required recycling of unreacted CO_2 in the preceding processes of CO generation for high carbon utilization. To avoid the FT recycle, process intensification can be applied.

The following sections will elaborate on the current trends in process intensification for Fischer-Tropsch processes, as presented in Table 1, with focus on cobalt catalysts and reactors. Nevertheless, the following list will introduce the overall processes and provide some general aspects.

- Process variant #1 is the most common one. As water electrolysis is state-of-the-art, different companies like INERATEC GmbH or Haldor Topsoe A/S apply reverse water–gas shift to obtain syngas with H_2/CO of 2:1 from a H_2/CO_2 molar mixture 3:1. This process variant is also quite frequently investigated in research. It has been explored, e.g., in the BMWi-funded PowerFuel project [28] under the aspects of transient operation along the process chain containing DAC, PEM electrolysis, rWGS, and FT together with upgrading of the synthetic crude product to kerosene. The aspect of transient operation in the FT synthesis will be discussed in the subsection "The Aspect of Transient Operation in PtL" (Sect. 4). The project partners Siemens Energy, Karlsruhe Institute of Technology (KIT), INERATEC GmbH and Climeworks AG utilized KIT's Energy Lab 2.0 infrastructure [29] to investigate the process chain, which also currently serves for the Kopernikus Power-to-X project.

- A specialty from the topic of process intensification may be process variant #2; there, process integration is the key factor. The direct route has recently gained attention through the intended avoidance of the high-temperature rWGS process for CO_2 activation and will be discussed in the section "Process Integration of rWGS and FT" (Sect. 3.4). Nevertheless, the upgrading could be more complex than in standard cobalt-based FTS due to the obtained rather short hydrocarbon chain length.

- Process variant #3 is frequently discussed to improve the overall process efficiency through the higher Faradaic efficiency of steam electrolysis in solid oxide electrolyzer cells (SOEC) versus water electrolysis. Heat from the exothermic FT process can further be utilized to generate the steam. Thus, beneficial combination of individual steps in the overall process is possible. At Energy Lab 2.0 [29] at Karlsruhe Institute of Technology (KIT) such combination is established and could be investigated further.

- A further example for research on the above process combinations is the BMBF-funded Kopernikus Power-to-X project [30] assembling the worlds' first complete process of CO_2 direct air capture (DAC) with Co-SOEC and FT system. Within the project, the partners KIT, INERATEC GmbH, Climeworks, and Sunfire GmbH showed that there is evidence of much higher overall efficiencies through coupling these process steps. Around 50 L of liquid already hydro-processed FT crude has been produced with this process combination. Heat, effectively recovered in microstructured FT reactors, could be delivered for steam generation of the SOEC system and residual heat to CO_2 desorption, a special benefit of process variant #4.

- Process variant #5 separates H_2 and CO supply to the FT process. For generation of both molecules, different types of electrolysis may be applied, i.e., low- or high-temperature type. It has been suggested that CO could be produced in similar arrangements like water electrolysis or SOEC cells in different projects. Separating the two feed streams could be beneficial in terms of complexity over co-electrolysis as CO could be sequestrated from unconverted CO_2 after the electrolysis. Nevertheless, this would not allow recycling of gaseous species from FT into a preceding process and thus reduce carbon efficiency for C_{5+} species. To our knowledge, no example for this process variant is known.

- Plasma splitting is also frequently under debate to serve CO for downstream processes like FT. Process variant #6 is also a separate supply approach for CO and H_2 while using plasma splitting of CO_2 for CO production. However, from literature only process variant #7 is known. In the EU project Kerogreen, the world's first assembly has been evaluated following this process variant. Figure 5 shows the assembly together with the project partners' contributions. CO and O_2 are produced through plasma splitting, while oxygen is separated in an electrochemical membrane system [31]. Unreacted CO_2 is recovered in a pressure swing adsorption. To generate synthesis gas, two-thirds of the CO is converted by a sorption-enhanced WGS (SE-WGS) with steam into CO_2 and H_2 for obtaining a H_2/CO ratio of 2 while simultaneously adsorbing and recovering CO_2 [32]. Further FT is used to produce syncrude, which is hydrocracked in situ to a kerosene

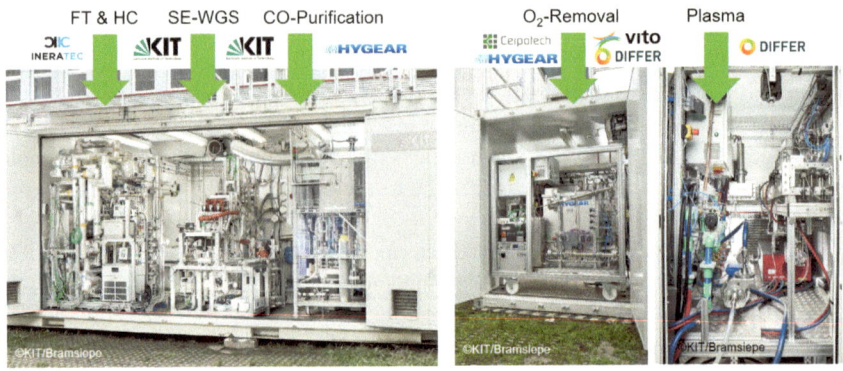

Fig. 5 Process variant 7 as demonstrated by Kerogreen's project partners [34]

pre-product [33]. Since the electrochemical cell for oxygen removal was not in the desired size, the overall process could not be fully evaluated. However, an upstream and downstream process testing was performed successfully.

In the following subsections, different aspects and approaches for process intensification and integration will be elaborated under the aspect of PtL technology. After a short section about catalyst insights and improvements, several sections will be devoted to different reactor designs and approaches with the FT core process.

3.1 Catalyst Improvement Through Operando Spectroscopy

Catalyst improvement is also a key factor for commercial viability of methods for process intensification as higher catalyst activity and selectivity are reducing capital expenditures for advanced reactor design through minimizing reactor size. Reduced holdup using more active and selective catalysts enables also reduced costs for operation and feedstock. This chapter does intend to detail recent advances around FT catalysts, since there are already elaborated studies on several aspects available, e.g., by Wolf et al. [35] or Suo et al. [36, 37]. Nevertheless, operando spectroscopy is shortly introduced as it can help to further understand and intensify FT for PtL application.

Ravenhorst et al. [38] used operando spectroscopic techniques for Co/TiO_2 catalysts, which account for nearly half of the world's transportation fuels produced by Fischer-Tropsch catalysis. Being able to obtain a spatial resolution of approximately 50 nm, they showed the interdependence of formed hydrocarbon species on the catalyst valence state. Their observed trends showed furthermore intra- and interparticular heterogeneities previously believed not to occur in particles below 200 µm. Capturing the genesis of an active FTS particle over its propagation to steady-state operation, they conclude that hydrocarbon deposits (differing in their chemical composition) form a film on the surface of the catalyst particle, with inter- and intraparticle heterogeneities. These heterogeneities can be correlated to slight changes in the local Co oxidation/valence states also influenced by the H_2/CO ratio.

These observations are in line with operando deactivation studies by Loewert et al. [39] at a synchrotron radiation facility. They investigated a commercial Co/ alumina catalyst over 300 h under harsh FT conditions (250 °C, 30 bar) including a detailed analysis of the resulting liquid products. They found that within the first hours on stream, the initial state of reduced Co remained without detecting any pronounced structural changes on the catalyst. With knowledge of a high C_{5+} selectivity, they conclude that the formation of a liquid film on the catalyst particles leads to decreasing catalytic activity due to mass transport limitations. In the following 8– 80 h of operation, near-edge XANES spectra evidenced the formation of $CoAl_2O_4$. Furthermore, the ongoing increase in methane selectivity accompanied by the decline in C_{5+} selectivity was attributed from the authors to the filling of the catalyst pores by liquid products, which was still ongoing, leading them to the conclusion that the deactivation originated from a combination of changes in mass transport and solid-state reactions in this activity regime. Finally, they identified in the last hours of operation a third regime, in which no change of the Co state nor selectivity change occurred. Finding ideal graphitic carbon depositions on the catalyst by Raman spectroscopy after the long-term experiments leads them to the conclusion that long-term deactivation originates from a carburization process.

Fischer and Claeys [40] wrapped up some recent studies on operando spectroscopy and also highlighted the importance of such techniques to study catalyst deactivation [35, 41, 42]. Their conclusion might ideally summarize what can be expected from such kinds of studies. Well-designed laboratory-based in situ characterization techniques, such as XRD, magnetometry, and XAS, with representative hydrodynamics provide fully relevant kinetic data. Further, these methods could play an important role in the development of understanding of the FT process.

Through the availability of operando technology, extensive research is enabled and could provide a solid ground for further improvement on catalysts. Being able to observe the deactivation mechanism, one can expect that material science can help to reduce the need for catalyst exchange. Long catalyst life is a further key for costs associated with catalyst exchange and constant product quality. Reducing the requirement for increasing the temperature to cope for catalyst deactivation typically reduces the C_{5+} selectivity and thus the total output of a PtL plant. Avoiding "intrinsic" carbon formation on the Co catalyst is only one aspect. The role of specific sites for improved C_{5+} selectivity leading to a higher wax yield and thus a greater kerosene output after hydrocracking is another aspect, which is investigated in current projects like BMBF-funded CARE-O-SENE [43]. Further, the importance of the water byproduct on the oxidation or sintering of the catalyst, which is nowadays largely suppressed by applying mono-modal Co crystallite size slightly above 8 nm in industrial catalysts [41, 44, 45], could potentially be solved differently. This also applies for oxidation via solid-state reaction with the metal oxide support forming mixed cobalt oxides, which may be hindered via (surface) modification of the support [35].

3.2 Reactor Technologies

3.2.1 Microstructured or Microchannel Reactors

Microstructured devices were already introduced in the 1990s as potential technology to enhance heat and mass transfer [46]. Following the principle of process intensification, introduced by the 1st International Conference on Process Intensification for the Chemical Industry [47, 48], the microstructured technology takes advantage of the short heat and mass transfer distance within the tiny structures and, moreover, the large surface to volume ratio. It has been explored largely for enhancing chemical processes from batch to continuous processes [36]. Microchannel reactors have been proposed as a tool for enabling greener processes [49], for being applied for commercial size [50], and are more recently reviewed under the aspect of process intensification in the catalytic conversion of natural gas to fuels and chemicals [51].

One of the first published comparisons of a microstructured reactor with a highly diluted fixed bed by Myrstad et al. [52] clearly shows the advantage of the microsystem. Packed with undiluted catalyst of near industrial recipe, the selectivity did not differ from the highly diluted fixed bed, meaning that the catalyst is operated in near isothermal regime and is not influenced by any wall effects, neither side reactions on the large surface area nor wall slip effects were measurable. They also showed that the pressure drop in such catalyst-packed systems is tolerable. Piermartini et al. [53] intensified this study and varied the microchannel size with catalyst packing. They conclude that microstructures have two major advantages over a slurry bubble column: no backmixing suppressing catalyst deactivation through water vapor and a high selectivity by being able to work at high single pass conversion leading to an intrinsic high chain growth probability at a H_2/CO ratio of 1.8. Even at larger slit sizes of 1.5 mm the superior heat transfer leads to a high chain growth probability.

Knochen et al. [54] provided some design aspects for the millistructure-like systems under the aspect of process intensification. Alongside with the experimental verification, they used a modeling of the system and concluded that a particle size range of 100–350 μm and channel width of 1.5–3.0 mm are interesting options for commercial FT applications. In their study, the reactor productivity is the largest with 1.5 mm slit size. The catalyst productivity is larger for a 3.0 mm slit size, mainly due to slight over-temperatures which are of course on expense of the selectivity. Chambrey et al. [55] came to similar conclusions when comparing a standard fixed bed with a milli-fixed bed. Moreover, they highlighted that an uncontrolled temperature increase during the reactor startup seems to be a common problem of FT synthesis in the standard fixed bed reactors. This temperature increase during the reactor startup can lead to catalyst deactivation. Interestingly, Chambrey et al. [55] also found a similar conclusion on suppressed long-term catalyst deactivation in the milli-type reactor as Piermartini et al. [53] but compared to a stirred tank reactor. Nevertheless, they explained the difference by the cobalt reduction procedure and not via the water vapor backmixing effect on the catalyst.

Catalyst incorporation by coating for FT synthesis is proposed in several studies. For example, the study of Chin et al. [56] proposed aligned carbon nanotube arrays for microchannel systems. In their study, they produced carbon nanotubes grown on FeCrAlY foam structures with Al_2O_3 coating. Since the catalyst preparation involved combined methods of metal-organic chemical vapor deposition (MOCVD), catalytic nanotube growth, and dip-coating of Co catalyst components, this technology seems rather complex. Although they were able to demonstrate an increase in Fischer-Tropsch synthesis activity by a factor of four when compared to an engineered catalyst structure without the carbon nanotube arrays, the catalyst stability might limit commercial viability. Again, this study showed that improved mass and heat transfer allow for operation of the Fischer-Tropsch synthesis at higher temperatures without selectivity runaway favoring methane formation.

Almeida et al. [57, 58] also studied the coating but with a simple standard wash-coating procedure for cobalt catalysts. They demonstrated that with different catalyst coating thickness, a good selectivity for C_{5+} can be achieved during FT synthesis due to the excellent temperature control of the microchannel technology. Nevertheless, they also found that the coating thickness can influence the selectivity for both C_{5+} and olefin/paraffin ratio. Thus, it can be concluded from their study that industrial relevant thicknesses to reach a high reactor productivity [59] discourages the use of coatings because they tend to get inefficient. Ying et al. [60, 61] also conducted a study on washcoating of a microchannel structure with cobalt catalyst. They obtained a 0.8 g g^{-1} h^{-1} C_{5+} catalyst productivity. However, this is also not converted to reactor productivity, which would be required to detail this regarding commercial viability. Sun et al. [62] performed research on coated iron catalyst for determining the kinetics. Even though they were able to show a methane selectivity as low as 6.5%, the application of coatings seems quite far from application.

Generally, a broad range of geometric configurations within the microstructure are proposed to enable higher conversion per pass and higher selectivity to C_{5+} applying packed microsystems. For example, a fractal geometry was suggested by Zhang et al. [63] following the structure design of Myrstad et al. [52] and Piermartini et al. [53]. They compared a mini-fixed bed reactor, a parallel straight microchannel reactor, and honeycomb fractal microreactor in the framework of olefin production in a HTFT process and concluded that the fractal configuration is better than the other concepts. Straight channels have been applied, e.g., by Cao et al. [64]. At a high gas hourly space velocity of 60,000 h^{-1}, they found a catalyst productivity of 2.1 g g^{-1} h^{-1} C_{2+} while still maintaining a methane selectivity less than 10%.

Loewert et al. [65] showed in their study that microstructured reactors from INER-ATEC GmbH can reach at least 2.1 g g^{-1} h^{-1} C_{5+} catalyst productivity and a record value of reactor space time yield of 1785 kg m^{-3} h^{-1} at isothermal conditions all over the reactor in pilot scale. These reactors are further equipped with a more complex heat removal microstructure allowing for optimized water evaporation with high efficiency. Catalyst deactivation was plotted in this study over 1000 h in a lab-scale experiment with near stable conversion in the end of the period. The run was performed at a constant temperature as high as 240 °C to show that the catalyst

Fig. 6 a) Fin-strip-like structure of a Velocys reactor [67] and b) commercial scale 175 bpd reactor [68], https://futurefuels.blog/in-der-praxis/kerosin-aus-abfaellen/ (last accessed 08.09.2024)

can be operated for long time without pronounced deactivation under these harsh conditions.

In commercial scale straight channels are often reported, e.g., from the company Velocys Inc. (Fig. 6a)—these channels are adapted from fin-strip heat exchangers regarding costs' aspects. Water is evaporated in standard cross-flow to keep the temperature of the system under control. In the study by Steynberg et al. [66], the conversion drop over 60 days at around 70% per pass conversion of the commercial unit is shown. The average reactor temperature was manipulated to increase in the run from 205 to 210 °C with an accompanied methane selectivity increase by roughly 1% (absolute).

Regarding manufacture of microchannel systems, 3D printing has recently been proposed as well, e.g., from Mohammad et al. [69]. Nevertheless, this study is still quite preliminary, as they used a self-prepared catalyst made from silica with a methane selectivity of 14%. Under these conditions, they showed that single and parallel reactor arrangements yield the same product. From industrial point of view, production of the reactor with 3D printing still requires improvement regarding achievable reactor size. In addition, it needs to be considered that the pressure requirement of FT synthesis is an issue of wall thickness for 3D printing.

The developments toward commercial application of microchannel or microstructure technology are more and more increasing. Fischer-Tropsch microchannel reactors are discussed in the framework of BtL from Velocys with their scale-up of individual reactor units from 125 up to 175 barrels per day (bpd) (Fig. 6b). Jung et al. [68] and Na et al. [70] with the Korea Gas Corporation recently demonstrated an up-scale to a 1 bpd unit. Their approach is similar to the Velocys reactor concept (Fig. 7). They could show that the operation of the microreactor prevented thermal runaway in contrast to a fixed bed, which was more complex to operate [70].

INERATEC has also reported about the commercial scale reactors in a recent study [71]. Following the pilot scale results [65], they showed 2 bpd reactors (Fig. 8) and their current scale-up to 1.25 MW units (based on electrical power for hydrogen production capacity). These units come along with also rWGS reactors, which are

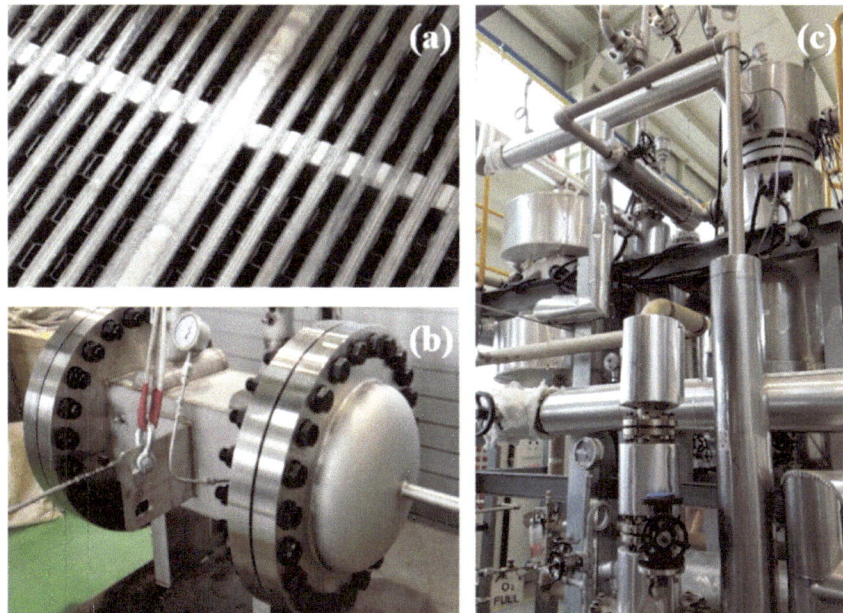

Fig. 7 Reactor module of 1 bpd dimension [68]: **a** internal structure with fin-strip-like structure, **b** reactor, and **c** setup for testing (bpd: barrel per day)

also made from microstructure technology, and a complete process starting from CO_2 and hydrogen for PtL applications.

3.2.2 Monolithic and Loop Reactors

Monolithic reactors have been proposed frequently as possible solution in the form of a loop reactor. One of the first publications on the use of monoliths for FT is from Mesheryakov et al. [72]. They showed that thin catalyst sheets permit to avoid intradiffusional resistance. They concluded that the small geometry-equivalent sizes of the gas bubbles and the high specific surface of the catalyst packing provide the intense gas-liquid-solid mass transfer. Further advantages were discussed from the point of low-pressure drop. Deugd et al. [73] have investigated this concept (Fig. 9) in depth from modeling point of view including the external recycle with heat removal which is necessary to achieve industrial relevant conversions per pass.

They compared the reactor volume to a study of the required size of a slurry bubble column from Maretto and Krishna [74] and came up with the conclusion that the monolithic reactor is smaller for an equal production capacity of 5,000 tons/day of middle distillates. The required slurry reactor volume is 4,410 m^3 divided into three vessels of 30 m height and 7 m diameter.

Fig. 8 Approved 2 bpd reactor module from INERATEC GmbH inside the skid of a 1 MW(el) PtL plant [71] (bpd: barrel per day)

Deugd et al. estimated a significantly smaller monolith reactor volume. Even though the internal heat removal and a lower temperature rise in case of the slurry bubble column, they conclude that plug flow behavior in the channels of the monolith reactor is a clear advantage enabling a higher productivity compared to the slurry bubble column. They also propose that the advantage of the monolith system is the fixed catalyst eliminating problems of catalyst attrition and separation. Attrition is, however, to some extent negligible over deactivation due to poisonous trace components which could occur in the context of PtL. As such, operation requires the exchange of the monoliths, which could be a considerable cost disadvantage.

Figure 10 shows the coated monolith from the work of Güttel [75]. He also pointed out that, although the preparation is reproducible, differences in the distribution of the active material were observed against the normal catalyst powder preparation. This is a general issue frequently observed for coatings in microstructured reactors as well. Further, catalyst accumulation in the corners of the quadratic channels needs to be avoided since it affects the diffusion distance. If the coatings are thick, the selectivity can be manipulated as reported also from larger particles [76].

During catalytic measurements in the catalytic setup, the group found that internal and external mass transfer affect the reaction rate and methane selectivity. External mass transfer constrains the reaction rate when the catalyst is coated. It is explained by the small relative velocity between catalyst particles and surrounding liquid, since the catalyst particles are entrained in the liquid. However, it was confirmed that the

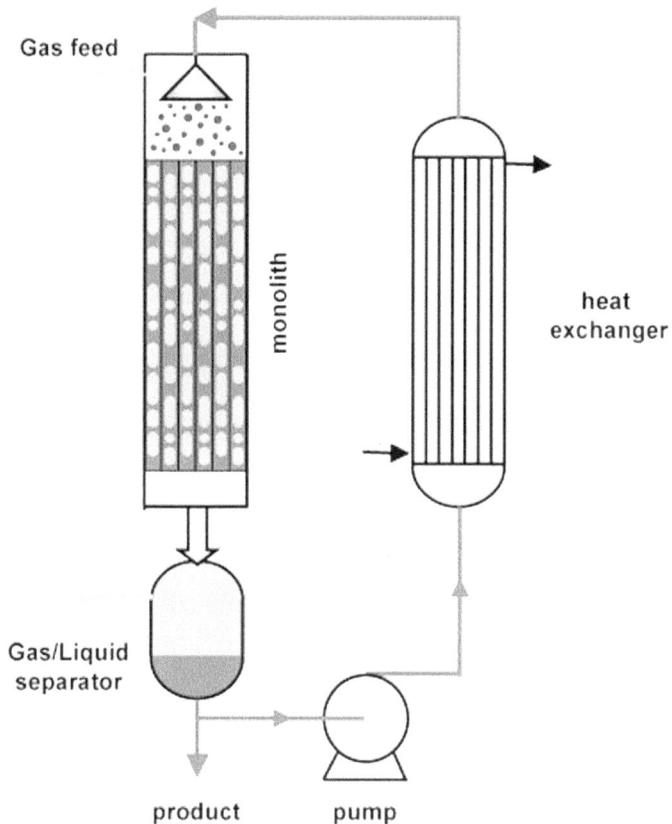

Fig. 9 Schematic drawing of the monolithic loop reactor with liquid recycle [73]

option of catalyst-filled channels is a preferred over coated channels by comparing the latter approach with a standard fixed bed. The enhancement in reaction rate is most probably caused by the advantageous external mass transfer characteristics of the monolithic catalyst in slug flow regime.

Merino et al. [77] investigated the effect of catalyst layer macroporosity in high-thermal-conductive monolithic Fischer-Tropsch catalysts with cobalt (Co) as active site. They applied coatings with one pore distribution centered at ~ 6 nm and another consisting of a meso-macroporous support with a wider pore size distribution centered at mesopores of ~10 nm and a significant macropore ($d_{pore} = 50$–$1,000$ nm) contribution. At iso-conversion and isothermal conditions with negligible differences of cobalt loading, it was concluded that the reduced methane selectivity and higher C_{5+} selectivity in the microporous coating arises from less diffusion resistance even though the coating was thicker due to lower density. This finding is obviously like the study of layer thickness in microchannels [57, 58] and special care must be taken when designing the catalyst layer.

Fig. 10 Photograph of an uncoated (left) and coated (right) honeycomb monolith [75]

Visconti et al. [78] studied the importance of the thermal conductivity of the applied monoliths. They found by modeling that heat conduction in an aluminum support of the catalyst can be exploited to effectively remove the heat generated by the strongly exothermic FT reaction. Flat axial and radial temperature profiles along the catalytic bed, according to their conclusion, can guarantee an excellent temperature control without the need of recycling a fraction of the liquid reaction products, neither during reactor startup nor in continuous operation. Thus, monoliths may not only be used in the framework of loop reactors. In a later study [79], they tested 3D-printed highly conductive periodic open cellular structures (POCS). Arguing with the requirement of scaling down the conventional packed-bed multi-tubular reactors to modular compact units in the framework of biomass-to-liquid (BtL) and power-to-liquid (PtL) applications, their results seem promising. They indicated that the adoption of a conductive POCS enables to operate an FT reactor in isothermal conditions even under very severe conditions of high CO conversions corresponding to high volumetric heat duties. Comparing their results with open cell foams (Fig. 11), they found improved heat transfer coefficients for the ordered structure of the POCS over foams. Further, they highlighted that packing of the POCS structure with catalyst particles is also boosting the overall space–time yield compared to coated POCS or foams. These results are indicative for other approaches to intensify standard fixed bed or slurry bed reactors as discussed in the following section.

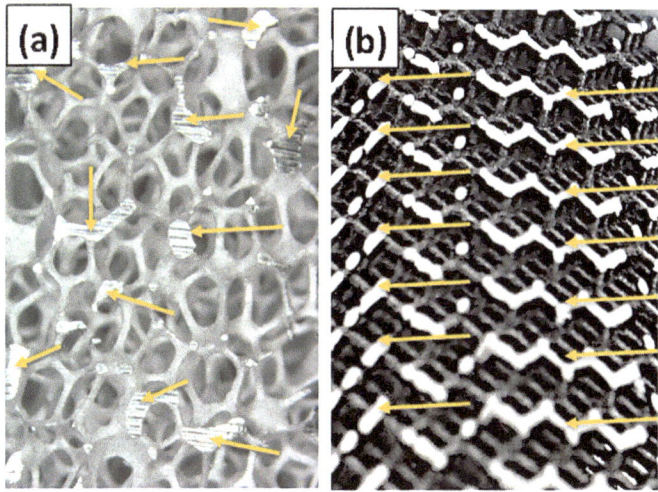

Fig. 11 **a** Open cell foam and **b** periodic open cellular structure applied in the study by Fratalocchi et al. [79]

3.2.3 Intensification in Traditional Reactors

Fixed bed bundle-type reactors are often used for low-temperature FT synthesis. As indicated in the previous section there are some attempts to intensify this type of reactor further. Johnson Matthey and BP recently introduced their CANS catalyst carrier system for Co-based FT synthesis [80]. The system works according to the following description:

1. Syngas from the catalyst carrier body above is guided downwards a porous central channel A (see Fig. 12).
2. The syngas then flows radially through the catalyst bed where heat is evolved (Fig. 12, arrows B).
3. The gas exits via a porous outer wall, flowing toward the top inner side of the catalyst carrier body (Fig. 12, arrows C).
4. Cooling occurs as the gas flows down the narrow annulus between the body and inside wall of the tube, through the transference of heat to boiling water on the shell side (Fig. 12, arrows D).
5. A seal prevents gas bypassing of the next catalyst carrier body, and the gas then enters the body below with the process then repeated (Fig. 12, arrows E).

According to the authors, a wider tube with improved heat transfer through the annular gas is feasible. In addition, higher gas velocities are said to be possible due to the short path through the catalyst bed. Smaller catalyst particles are also envisaged to be used through the short catalyst bed length. Nevertheless, from the point of flow directions, this concept seems to introduce some hurdles for Co-based FT synthesis as the liquid formed is entrapped in the core of the system until it is flooded.

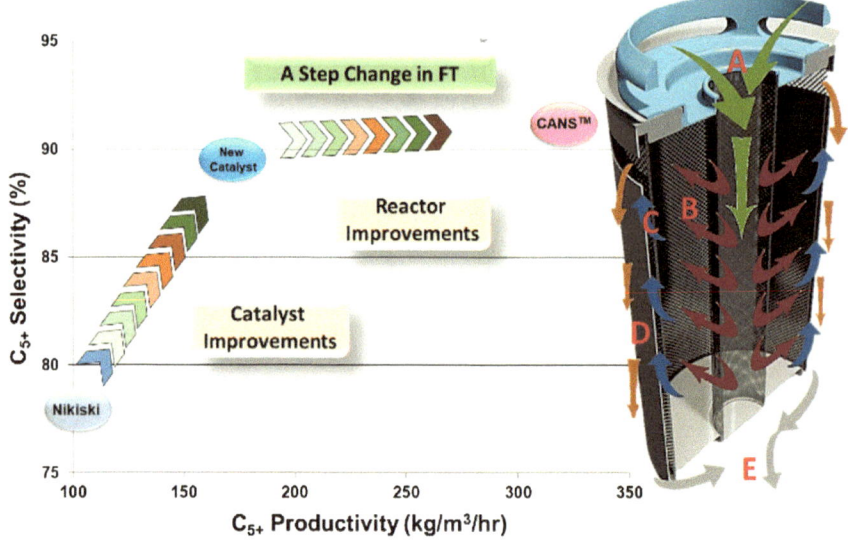

Fig. 12 CANS catalyst carrier system from Johnson Matthey and BP [80]

Another approach is based on improved heat conductivity on the catalyst level. Asalieva et al. [81] tested thermally conductive additive (Al, Cu, or Zn metal powder) for cobalt-based catalyst.

Pangarkar et al. [82] did an early evaluation of the concept of implementation of packings into the fixed bed tube. Structured packings such as open cell flow structures (OCFS) and closed cell flow structures (CCFS), but also knitted wires and open cell aluminum foams were investigated. They found that the heat transfer coefficient in regular flow structures is superior to that of randomly ordered structures, which can be explained by the shorter heat transfer path. The randomly ordered catalyst packing is even worse in heat transfer due to the almost missing solid heat conduction.

Among studies on integrating foams [83], a recent study on packed-POCS with skin was published by Fratalocchi et al. [84]. The so-called POCS with skin (Fig. 13) is characterized by a surrounding tube thermally connected with the internal diamond structure of the POCS through the printing process. According to the authors, this resulted in an increase of the contact to the reactor tube, which enabled heat transfer from the POCS to skin and the wall to be governed by heat conduction rather than convective heat transfer. Thus, a further improvement in this technology is envisaged.

Hooshyar et al. [85] modeled the improvement of structured packings in fixed bed arrangements as well as in slurry bubble columns. They found out that the insert is eligible for both reactor types. In the slurry bubble column backmixing is also reduced, which could lead to an improvement in reactor productivity of 20% in the slurry bubble column and up to 40% in the fixed bed reactor.

Frost et al. [86] showed the development of a compact Fischer-Tropsch reactor based on heat conductive internals shown in Fig. 14a and the pilot scale system in

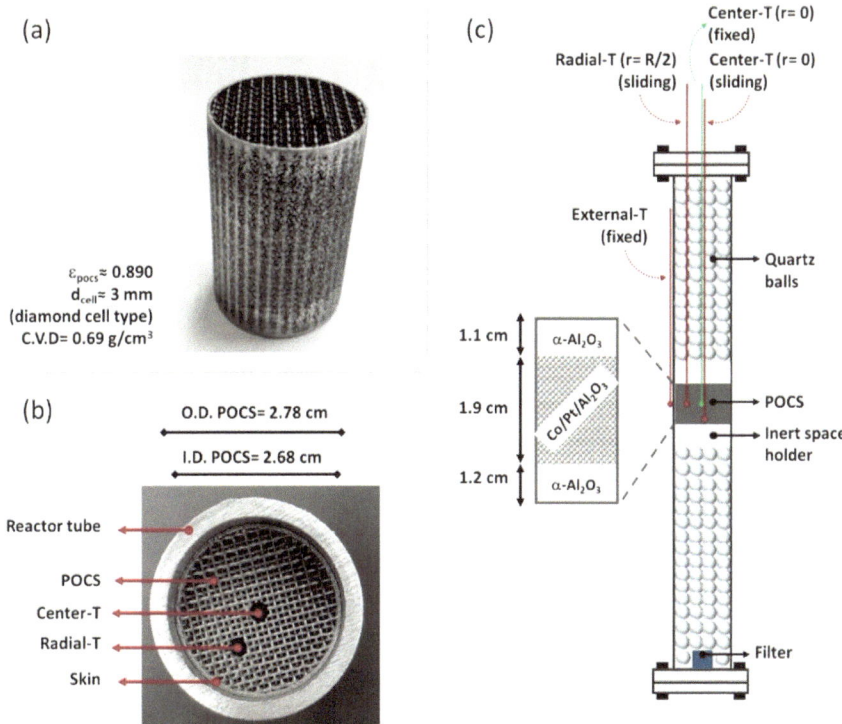

Fig. 13 (a) and (b) side and top view images of POCS and (**c**) schematic diagram of a reactor tube with a single POCS insert packed with catalyst pellets and α-Al₂O₃ [84] (POCS: periodic open cellular structures)

Fig. 14b. Unfortunately, no later reports are available of the Ceramatec Inc. system. Nevertheless, recently a study on a methodological study on the reactor internals by Barrera et al. [87] with coauthors of Frost was published.

Fig. 14 **a** Internal reactor insert for thermal transfer and **b** pilot scale FT reactor system [86]

Kapteijn and Moulijn [88] also gave a recent more general perspective for the structured catalysts and reactors from the viewpoint of heat conductive inserts, which provides a more detailed overview on the aforementioned studies.

A completely other approach to improve the multi-tubular fixed bed reactor is the use of phase change materials (PCM) to manage temperature control. Odunsi et al. [89] suggested that the PCM material should be located at the outer perimeter of the individual tubes. The delay in temperature rise due to phase transition by the PCM, according to their calculations, could keep the catalyst bed in a narrow temperature window. This technology is, however, not being exploited commercially, maybe because of the rough environmental conditions for the PCM material in the multi-component mixture of the FT synthesis.

Apart from the suggestion of cell structures for intensifying the slurry phase reactor [85], An et al. [90] also modeled an increasing reactor productivity with increasing number of internal tubes in the slurry phase. However, there is no experimental proof for these simulative results. Geng et al. [90, 91] wrapped up the approaches of agitated slurry reactors in a general perspective. These approaches could also be transferred to FT synthesis regarding process intensification. Agitation can help to improve hydrodynamics and mass transfer, i.e., superficial velocity increase between gas, liquid, and solid phase as well as less backmixing, higher gas holdup and associated higher interfacial surface area.

3.3 Other Means of Intensification

3.3.1 Supercritical FT

Supercritical operation conditions are often discussed as method to avoid mass transport limitation. A review from 2009 indicated that these conditions are promising although implementation of supercritical fluid in FT was still at laboratory scale at that time [92]. Supercritical condition for FT synthesis has been discussed by Mogalicherla et al. [93], but they found that supercritical conditions are not able to eliminate the diffusion effects in the pores of the catalyst. Larger particles will still be prone to a H_2/CO ratio shift due to different diffusion coefficients of H_2 and CO in the catalyst pores filled with hydrocarbons. This leads to enhanced chain termination.

In a follow-up simulation by Abusrafa et al. [94], the supercritical conditions were demonstrated to provide exceptional reduction in hot-spot formation with a maximum radial bed temperature variation < 15 K for a 4 inch internal diameter (ID) reactor bed as opposed to 800 K in an equivalent standard FT reactor bed. They concluded that process intensification by supercritical conditions can lead to up to 16-fold reduction in the number of tubes required to achieve a throughput equivalent to a conventional 1 inch ID reactor bed. Nevertheless, this technology has not been scaled yet to any pilot plant to our knowledge. This is most probably a matter of operational costs for implementation of the supercritical conditions and dominates over capital expenditure for the reactor size.

3.3.2 Integrated Upgrading

Tandem catalysis for FT and hydrocracking has also been proposed to enhance product selectivity and reaction rate by avoiding the accumulation of long-chain hydrocarbons in the catalyst bed [95, 96]. Addition of ZSM5 in a catalyst mixture [97], such as a combined coating on channel walls [98] or a core–shell structure, has often been applied, e.g., Refs. [99, 100].

Different schemes of integration of hydrocracking into the process of rWGS and FTS have been investigated by Kirsch et al. [101]. The authors suggest direct coupling of FTS and HC as otherwise intermediate product separation improves the liquid fuel yield just by a few percent based on modeling. Arguing that compact processes are particularly important for flexible small-scale plants for decentralized application regarding the energy transition, it is believed to be advantageous to directly convert the formed wax in situ. In a follow-up study, advanced process design of a special bifunctional catalyst bed is proposed (Fig. 15) [102]. While hydrocracking in the presence of CO can lead to secondary cracking and is largely prevented by a liquid film entrapping the hydrocracking catalyst, either a gradient with increasing hydrocracking catalyst along the catalyst bed or a sequential bed with a second stage having a low FT catalyst mass contribution is proposed. Nevertheless, Brosius [103] recently explored the possibility of ambient pressure hydrocracking on a Pt/HY catalyst (a catalyst where platinum (Pt) particles are dispersed or supported on a HY zeolite material), which allows to reach much higher yields of liquids. This may be more advantageous also in the field of smaller-scale plants.

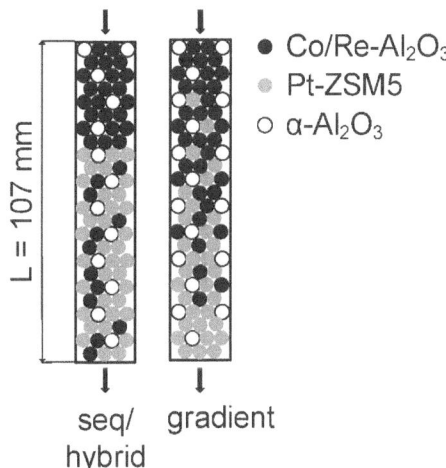

Fig. 15 Distribution of FT and hydrocracking catalyst over the catalyst bed for avoiding secondary cracking in the presence of CO: in the gradient configuration the amount of FT catalyst is linearly decreasing, whereas the amount of HC catalyst is linearly increasing. The catalysts are diluted with inert α—Al_2O_3 [102]

3.3.3 Sequential Reactors

A sequence of FT reactors with intermediate removal of products and the formed byproduct of water is frequently discussed in patents [104] and literature [105, 106]. Common sense is that the per pass conversion can be limited to a level where hot-spot formation is under control, inert gases do not accumulate, and water partial pressure increase is avoided. In this way, a critical H_2O/H_2 ratio for oxidation of the cobalt or iron catalyst is prevented. Pandey et al. [105, 106] recently investigated the optimal reaction trajectory for a three-stage process and found that an under-stoichiometric H_2/CO feed with maximum possible total CO conversion (93.6%) and operation at a conversion per pass of 60% is an optimal strategy for once-through three-stage FT synthesis, which is generally applicable for the FTS over a cobalt catalyst.

In the framework of power-to-liquid, however, the conversion in the FT stage may not be as relevant as in the gas-to-liquid or biomass-to-liquid application. Typically, low inert composition is intended via CO_2 and H_2 in the PtL case. Thus, a recycling of unreacted gas over the rWGS stage is feasible and recovers the already existent CO and H_2 in the process while methane can be converted via in situ reforming on the catalyst.

3.4 Process Integration of rWGS and FT

It has already been discovered in the 1930s during early Fischer-Tropsch research in Germany that carbon dioxide (CO_2) can, similarly to carbon monoxide (CO), be converted to higher hydrocarbons [107, 108] using alkalized iron, cobalt, or ruthenium catalysts. Presumably due to the lack of commercial relevance, a small number of studies or patents were published up to the mid-1990s [109, 110, 110–114]. Along with the global awareness for the effects of greenhouse gas (GHG) emissions, a growing interest in the utilization of CO_2 as a carbon source developed in the last decades [115, 116]. Especially worth mentioning are the early works at the Korea Research Institute of Chemical Technology (KRICT), which are summarized in a review of Sai Prasad et al. [117]. The current research is mainly focused on the development of suitable catalysts and is summarized in several detailed reviews [115, 116, 118, 119, 120, 121].

3.4.1 Reaction Network

As the current research mainly focuses on iron-based catalysts (see below), the focus on any reaction characteristics will be put on these catalysts. It seems to be consensus that the reaction proceeds via two consecutive reactions: CO_2 is first reduced to CO via the reverse water–gas shift (RWGS, Eq. 8), which is then converted to a hydrocarbon mixture through a conventional Fischer-Tropsch (FT) synthesis [122–125]. Iron-based catalysts are especially suitable for this process as they catalyze both reactions. Under reaction conditions, several iron oxide and carbide phases coexist [126]. In this mixture, the rWGS activity is mainly ascribed to Fe_3O_4 while the FTS activity is ascribed to χ-Fe_5C_2 (Hägg carbide) [124, 125]. A direct formation pathway of hydrocarbons from CO_2 may exist to some extent but is likely negligible compared to the indirect pathway [122].

$$CO_2 + H_2 \leftrightarrow CO + H_2O \quad \Delta_R H^0 = 41 \ kJ \, mol^{-1} \tag{8}$$

The overall balance is given by Eq. 9. One should note the large amount of water that is formed in the reaction, being the main product.

$$CO_2 + 2H_2 \rightarrow (CH_2) + 2H_2O \quad \Delta_R H^0 = -122 \, kJ \, mol^{-1} \tag{9}$$

The large amount of water that is formed in the reaction is a main challenge for a realization of the process as it does not only strongly inhibit the reaction [127, 128, 129], but may also lead to catalyst degradation via oxidation at high water partial pressures [128, 130, 131].

3.4.2 Proposed Catalysts

Iron-Based Catalysts

Iron-based catalysts have been widely applied for the CO_2-FTS (process variant #2, Table 1) due to their activity for both, rWGS and FTS, and their remarkable flexibility regarding the feed gas composition. Unlike cobalt, iron catalysts do not require high CO partial pressures to establish a chain growth regime and can yield a product spectrum similar to the traditional FTS when using H_2/CO_2 feed gas mixtures [132]. CO that is formed in situ via the rWGS is sufficient to sustain chain growth. The catalysts usually require a stronger alkali promotion [133]. Under typical operating conditions of 300 °C, the product spectrum does usually resemble a high-temperature FTS, with short-chain 1-alkenes being the main product [128].

Iron-based catalysts may be divided into three categories depending on the preparation pathway: precipitated, supported, and fused catalysts. Iron (Fe) bulk catalysts prepared via precipitation have been applied for decades in the low-temperature FTS and have been demonstrated to be also applicable for the CO_2-FTS [122, 126, 132, 134]. Supported iron catalysts have hardly been applied for the traditional FTS due to the poor reducibility and lowered promoter effectiveness [135, 136]. For the CO_2-FTS, however, potassium-promoted alumina-supported iron catalysts have been adopted in several studies, which demonstrated a remarkable activity and selectivity [137]. They still require a high reduction temperature (>400 °C) and high potassium content. For the high-temperature FTS fused iron catalysts are applied commercially by Sasol [135]. This type of catalyst has gained limited attention for the CO_2-FTS so far [138].

Among promoters, the importance of alkaline promoters (potassium (K) [133, 137, 139, 140, 141, 142] or sodium (Na) [143, 144, 145, 146, 147]) is usually high-lighted to tune the catalyst performance. Alkali metals improve the adsorption of CO_2 and CO, increase the selectivity to alkenes, and drive the product spectrum to long-chain hydrocarbons [116, 118]. An approach that has gained considerable attention in the last years is the combination of an iron-based catalyst with a zeolite (usually H-ZSM-5) to directly convert short-chain alkenes to aromatic components [123, 148, 149]. This route may be of particular interest for gasoline applications.

Cobalt and Ruthenium

Cobalt-based catalysts for the traditional FTS are usually supported catalysts using a porous carrier (alumina, silica, titania, etc.) with a noble metal promotion (platinum (Pt), rhodium (Rh), rhenium (Re)) [135]. These catalysts are not applicable for the CO_2-FTS as they require a high CO partial pressure to establish a chain growth regime and thus mainly act as methanation catalysts when exposed to CO_2-H_2 mixtures [124, 132]. However, alkali promotion was shown to suppress methane formation and enhance the selectivity to long-chain hydrocarbons [150, 151, 152]. Hence, it may also be possible to employ Co-based catalysts for the CO_2-FTS upon further development.

Ruthenium (Ru)-based catalysts are well known for their high activity and selec-tivity to long-chain hydrocarbons in the traditional FTS, but are not applied commer-cially due to the high costs and poor availability of the noble metal [135]. Unpromoted Ru catalysts mainly catalyze methane formation under the conditions of CO_2-FTS [107, 111]. Like for cobalt, however, alkali promotion showed to suppress methane formation and to improve the selectivity to higher hydrocarbons [107]. For the same reasons as for the CO-FTS, it is unlikely that Ru will become a relevant catalyst candidate for the CO_2-FTS.

3.4.3   Process Considerations

Unlike the two-stage rWGS/FTS process, there has been very limited work beyond lab scale for the CO_2-FTS. The main challenges have been identified and are outlined below.

Implications of Water Formation on Process Concept

A key part of a direct-FT scale-up concept is the effective removal of water vapor from the reaction mixture to overcome the limitation of a small conversion per pass. In a patent from 1954, Kölbel and Ackermann proposed two possible concepts: multiple reactors in series with intermediate condensation or a recycle reactor with continuous product condensation [109].

Both concepts have been adopted in more recent studies: Landau et al. achieved a CO_2 conversion of up to 89% in a bench-scale setup with three reactors in series with intermediate condensation and Guo et al. demonstrated a conversion of almost 70% in a two-stage reactor system. Choi et al. [153] realized the recycle reactor concept and achieved a CO_2 conversion of up to 88%. The only study at pilot scale has been reported by Willauer et al. [154] who achieved up to 69% CO_2 conversion under recycle conditions. Along with the conventional process concepts, Rohde et al. proposed a membrane reactor concept which allowed an in situ removal of water vapor from the reaction mixture [155]. This concept does still further development of membrane materials, though [156].

Reactor Design

So far, experimental studies have been reported almost exclusively in fixed bed reactors. However, the typically observed limited long-term stability (< 1 year) of iron-based FTS catalysts [135] may favor other reactor concepts for a large-scale application. Fluidized bed or slurry reactors are established reactor concepts for the conventional FTS and would allow for online catalyst replacement [157]. Kim et al. [158] studied different reactor concepts for the CO_2-FTS and demonstrated the applicability of fluidized bed or slurry reactors.

Brübach et al. [130, 159] studied the process on supported iron in two different scales of a tubular reactor. In the second part of their study, a large 1 inch tubular reactor scale was operated under recycling conditions. With a detailed kinetic model, they were able to predict the reactant consumption, as well as the hydrocarbon distribution, reliably within the experimental range studied (10 bar; 280–320 °C; 900–120,000 $mL_N h^{-1} g^{-1}$; and H_2/CO_2 molar inlet ratios of 2–4). High conversions up to 80% demonstrated the applicability of the process concept. Fractions of individual hydrocarbon classes (1-alkenes, n-alkanes, and iso-alkenes) were accounted in their model with chain-length-dependent kinetic parameters for branching and dissociative desorption. They found that hydrogenation is the main reaction involved while recycling short-chain olefins into the reactor. No chain growth occurred through re-adsorption. Further, hot-spots occurred in the larger reactor which required catalyst dilution, and thus, more advanced reactor concepts could be helpful here as well. Catalyst stability and potential coke formation could, nevertheless, be crucial.

Syncrude Refining

The refining of the synthetic crude (syncrude) is the final step for fuel applications. Due to many similarities to the cobalt-based high-temperature FTS, the refining may be conducted in a similar manner [128, 160]. When aiming for middle distillates (kerosene and diesel), oligomerization of alkenes would be a key refining step [161, 162]. Oligomerization on aluminosilicates like hydrogenated Zeolite Socony Mobil-5 (H-ZSM-5), amorphous silica-alumina (ASA), or solid phosphoric acids

(SPA) may be appropriate technologies [163]. The selectivity to oxygenate (mainly alcohols, carbonyls, and carboxylic acids) is neglected in many studies but presents a considerable part of the product fraction for cobalt-based catalysts and should be considered for the refining strategy [160, 163, 164]. Hydro-processing steps are especially susceptible to the presence of oxygenates and must be designed appropriately [164]. Lower molecular weight oxygenates are preferably dissolved in the aqueous phase. For a high carbon efficiency, recovery of hydrocarbons from the aqueous phase should be considered [165].

4 The Aspect of Transient Operation in PtL

In the framework of power-to-liquids, the production of the powerfuels based on renewable energies as well as carbon dioxide, e.g., from ambient air, biogenic sources, or unavoidable fossil point sources, is feasible. The entire supply chain is then CO_2 neutral or at least no additional CO_2 is emitted, respectively. The intermittency of renewable power, nevertheless, imposes a serious problem for continuously running a PtL plant. The peak power generation must be enlarged greatly or either certain/ large volumes of hydrogen storage are needed, or the synthesis plants must also run in transient modes.

Pfeifer et al. [166] have investigated the operation of INERATEC plants regarding electricity production from 80% photovoltaics and 20% wind energy for the German Federal State of Baden Württemberg on the basis of varying peak power versus electrolyzer and synthesis plant size in dependence on the load flexibility of the synthesis plant. They found that high renewable energy and synthesis plant utilization degree of above 80% can be reached with the turndown ratio of their plants as high as 83%; valid for a storage capacity of hydrogen of 10 h for the lowest load point of the synthesis plant. Figure 16 shows plots of time-related renewable power and load state of the synthesis. It shows that of course a certain larger peak power must be installed, while available power is transferred with above 80% utilization through small hydrogen storage capacity into fuel. Further, the hydrogen buffer volume can be reduced by operating the plants in transient mode feasible through modularity. Although no information on the capital expenditure saving is available, the concept should be followed for safety reasons.

Several authors have investigated the Fischer-Tropsch reaction under transient conditions [167, 168, 169, 170, 171, 172, 173] in the framework of varying load. Some research is also summarized in a review [174]. There is common sense that Fischer-Tropsch intrinsic reactions on cobalt, in contrast to iron where phase changes can occur even after days, are fast enough to be described by stationary kinetics [167, 168, 169, 170, 171, 172], even for highly transient conditions in the minute scale through monolithic or microstructured reactors [169, 171, 172].

De Swart and Krishna [175] enabled simulation of slurry bubble columns to better design and stabilize hydrodynamics of these reactor types in transient conditions. Nevertheless, Loewert et al. [171] demonstrated experimentally, that plants with

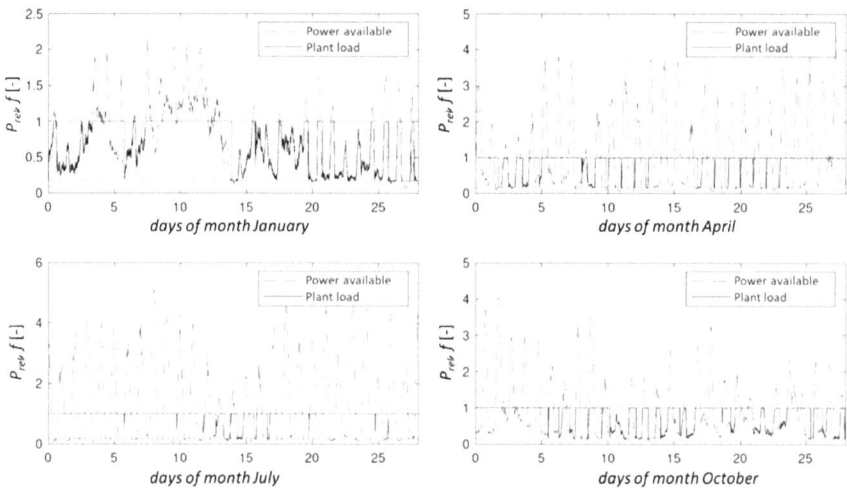

Fig. 16 Relative electric power and load of the fuel plant for selected months (for the German Federal State of Baden Württemberg) at a fuel plant capacity $F_{max} = 1$, a tank size for peak power $t_{peak} = 10$ h, and a wind/photovoltaic (PV) ratio of 20:80. Minimum tank capacity $t_{min} = 1.3$ h [166]

microstructured reactors can be easily manipulated regarding the process conditions during changing loads on day and even at single minute scale. They applied a real photovoltaics (PV) profile to a reactor assuming instantaneous conversion via electrolysis. They could show that manipulation of the reactor temperature by changing the water evaporation conditions on the cooling side to keep the conversion on a high level of 65% the volumetric content of formed methane from FT was less varying over the day profile and thus advantageous over keeping the reactor temperature constant (Fig. 17). Only near isothermal operation would be feasible in slurry due to the high-thermal masses of accompanied liquid.

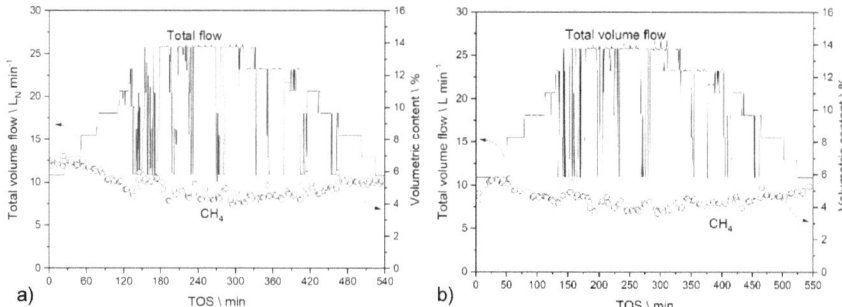

Fig. 17 Flow variation (constant H_2/CO ratio) according to the photovoltaic (PV) power and the resulting variation in methane volumetric content at **a** almost constant reactor temperature and **b** almost constant conversion of around 65% through reactor temperature manipulation [171]

In further studies by Pfeifer et al. [71], it is also shown that the microstructured reactors are controllable in serial and parallel operation also under transient conditions. They also concluded that start-stop can be performed with multiple reactors leading to high flexibilities regarding load so that, e.g., 83% of turndown ratio [166] seems feasible.

5 Comparison and Conclusion

In the context of integrating and utilizing powerfuels, particularly hydrocarbon Power-to-Liquid drop-in fuels like kerosene and diesel, the Fischer-Tropsch synthesis has emerged as a central technology and pathway in recent years. To rapidly deploy this technology at scale in the upcoming years and decades, besides the broader techno-economic and environmental aspects, understanding exact process conditions and specifications is equally vital. This includes in-depth analyses of aspects like catalyst utilization, reactor technologies, and approaches to effectively combine different sub-technologies and processes (e.g., electrolysis technologies, carbon dioxide activation, and FTS) as well as the aspect of intermittent feed and most suitable plant size at the point of use within an overall FTS pathway.

However, only a few comparative studies or reviews are available [51, 176, 177, 178] comparing reactor technologies and/or aspects in process intensification for Fischer-Tropsch synthesis. While Holmen et al. [178] compare different structured reactors and Venvik et al. [177] provide a review over microchannel reactors for FT, Saeidi et al. [176] also consider traditional reactors with internals. One of the most recent studies by Kee et al. [51] compares slurry, fixed bed, and microchannel reactors not only for FT but also for reforming and oxidative methane coupling. Therefore, this chapter aims to provide a concise overview of the diverse approaches for process intensification in FT, while also addressing the different aspects required for powerfuels on a process level. Such are, to name a few: CO_2 activation, space time yield, catalyst stability, and regeneration, as well as intermittent loads.

In light of the content and study results presented here, it seems that especially microstructured or monolithic and POCS are a good alternative to traditional FT reactor systems. While modularity is gaining importance for fast roll-out of the PtL technology and is best suited to address local availability of CO_2 and power, these technologies are just on the road. Giving an example, INERATEC's pre-assembled containerized or skid-mounted systems are already on the market as integrated rWGS and FT process in 1 MW industrial pilot and a scale of up to 10 MW (Fig. 18) shall start operation in 2024 at the Frankfurt Hoechst site [179].

Other attempts in this area to mention are, e.g., the strategic alliance of Sasol EcoFT and Haldor Topsoe for FT and rWGS, respectively. Combined rWGS and FT in a single reactor are followed from, e.g., AirCompany [180, 181], Sasol [182], to name just a few. Hence, the future will basically show if all these concepts will come alive.

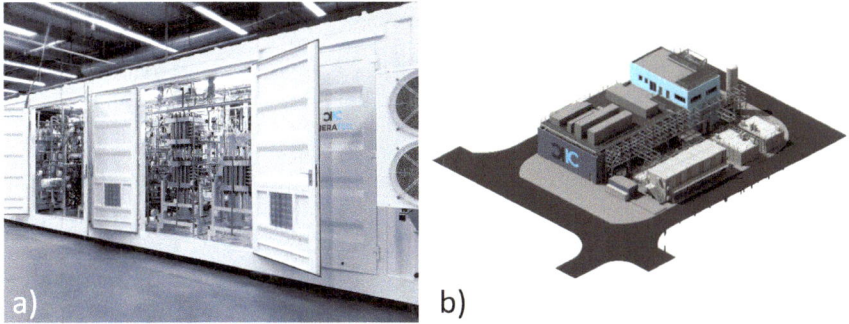

Fig. 18 **a** 1 MW industrial pilot and **b** concept of up to 10 MW plant for the Frankfurt Hoechst site [source: INERATEC GmbH]

In general, conventional cobalt-based as well as iron-based FT utilizing carbon monoxide are already approved for the production of synthetic kerosene from FT syncrude in the respective ASTM regulations for aviation turbine fuels (ASTM D7566). Looking at the future demand of CO_2 neutral aviation fuels, it is anticipated that a significant portion will be in the form of PtL kerosene, especially in the medium to long terms. Thus, FT could be a good option to fulfill this kerosene demand. Process-wise FT-based PtL kerosene, could be produced from all the FT process variants presented in this chapter while only direct CO_2-FT is not yet approved by ASTM D7566. Thus, for CO_2 originating from ambient air or biomass, the processes with upfront CO_2 reduction to CO (e.g., by rWGS) are ahead in time regarding today's regulation. Very good combustion properties, for example regarding reduced soot emissions, have repeatedly been demonstrated [183]. Climate effects from the formation of condensate are reduced additionally [184]. These better combustion properties arise from the absence of aromatics.

Furthermore, ASTM regulations fit for blending of up to 50% of the hydrotreated FT product to specification-compliant conventional (fossil) jet fuel. For CO_2-based FT a respective ASTM certification is still required, although, this could be relatively easily solved due to similar product quality compared to the HTFT process. Furthermore, it can be concluded that even with a higher demand for upgrading steps which is possible only in large installations, this route could play a role in the future. Nevertheless, cobalt-based FT processes might be the choice for distributed energy availability also in the smaller scale, i.e., in remote energy supply.

References

1. Unruh D (2006) Fischer-Tropsch Synthese mit Synthesegasen aus Biomasse—Verbesserung der Kohlenstoffnutzung durch Anwendung eines Membranreaktors (Ph.D. thesis, Universität Frider-iciana Karlsruhe (TH))

2. Steynberg AP (2004) Introduction to Fischer-Tropsch technology. In: Steynberg AP, Dry ME (eds) Fischer-Tropsch Technology, vol 152. Elsevier, pp 1–63
3. Steynberg AP, Dry ME, Davis BH, Breman BB (2004) Fischer-Tropsch reactors. In: Steynberg AP, Dry ME (eds) Fischer-Tropsch technology, vol 152. Elsevier, pp 64–195
4. Xu J, Yang Y, Li Y-W (2013) Fischer-Tropsch synthesis process development: steps from fundamentals to industrial practices. Curr Opin Chem Eng 2:354–362
5. Xu J, Yang Y, Li Y-W (2015) Recent development in converting coal to clean fuels in China. Fuel 152:122–130
6. Guettel R, Kunz U, Turek T (2008) Reactors for Fischer-Tropsch synthesis. Chem Eng Technol 31:746–754
7. Fischer F, Tropsch H (1926) Über die direkte Synthese von Erdöl-Kohlenwasserstoffen bei gewöhnlichem Druck. (Zweite Mitteilung). Berichte der deutschen chemischen Gesellschaft 59:832–836
8. Dry ME (2004) FT catalysts. In: Steynberg AP, Dry ME (eds) Fischer-Tropsch technology, vol 152. Elsevier, pp 533–600
9. Schulz H (1999) Short history and present trends of Fischer-Tropsch synthesis. Appl Catal A Gen 186:3–12
10. Dry ME (1981) The Fischer-Tropsch synthesis. In: Anderson JR, Boudart M (eds) Catalysis: science and technology. Springer, pp 159–255
11. Dry ME (2004) Chemical concepts used for engineering purposes. In: Steynberg AP, Dry ME (eds) Fischer-Tropsch technology, vol 152. Elsevier, pp 196–257
12. Frohning CD et al (1977) Fischer-Tropsch synthesis. In: Falbe J (ed) Chemierohstoffe aus Kohle, Thieme, pp 219–299
13. Khodakov AY, Chu W, Fongarland P (2007) Advances in the development of novel cobalt Fischer-Tropsch catalysts for synthesis of long-chain hydrocarbons and clean fuels. Chem Rev 107:1692–1744
14. Anderson RB (1956) Catalysts for the Fischer-Tropsch synthesis. In: Emmett PH, Dixon JK (eds) Hydrocarbon synthesis, hydrogenation and cyclization, vol 4. Reinhold, pp 29–255
15. Oukaci R, Singleton AH, Goodwin JG Jr (1999) Comparison of patented Co F-T catalysts using fixed-bed and slurry bubble column reactors. Appl Catal A Gen 186:129–144
16. van de Loosdrecht J et al (2013) Fischer-Tropsch synthesis: catalysts and chemistry. In: Comprehensive inorganic chemistry II (second edition): from elements to applications, vol 7. Elsevier Ltd., pp 525–557
17. Vogel AP, van Dyk B, Saib AM (2015) GTL using efficient cobalt Fischer-Tropsch catalysts. Catal Today 259:323–330
18. van Steen E, Claeys M (2008) Fischer-Tropsch catalysts for the biomass-to-liquid (BTL)-process. Chem Eng Technol 31:655–666
19. Dry ME (2002) The Fischer-Tropsch process: 1950–2000. Catal Today 71:227–241
20. van der Laan GP, Beenackers AACM (1999) Kinetics and selectivity of the Fischer-Tropsch synthesis: a literature review. Catal Rev Sci Eng 41:255–318
21. Dry ME (1996) Practical and theoretical aspects of the catalytic Fischer-Tropsch process. Appl Catal A Gen 138:319–344
22. Claeys M, van Steen E (2004) Basic studies. In: Steynberg AP, Dry ME (eds) Fischer-Tropsch technology, vol 152. Elsevie, pp 601–680
23. Schulz H (2003) Major and minor reactions in Fischer-Tropsch synthesis on cobalt catalysts. Top Catal 26:73–85
24. Dry ME (1982) Catalytic aspects of industrial Fischer-Tropsch synthesis. J Mol Catal 17:133–144
25. Schulz H (2013) Principles of Fischer-Tropsch synthesis—constraints on essential reactions ruling FT-selectivity. Catal Today 214:140–151
26. Weststrate CJ, van Helden P, van de Loosdrecht J, Niemantsverdriet JW (2015) Elemen-tary steps in Fischer-Tropsch synthesis: CO bond scission, CO oxidation and surface carbiding on Co(0001). Surf Sci 648:1–7

27. Iglesias González M, Kraushaar-Czarnetzki B, Schaub G (2011) Process comparison of biomass-to-liquid (BtL) routes Fischer-Tropsch synthesis and methanol to gasoline. Biomass Conver Biorefin 1:229–243

28. Project funding website. https://www.energiesystem-forschung.de/forschen/projekte/powerfuel. Last accessed 31 Jul 2023

29. Project website. https://www.elab2.kit.edu/index.php. Last accessed 31 Jul 2023

30. Kopernikus Press release. https://www.kopernikus-projekte.de/aktuelles/news/p2x_erste_strom_zu_kraftstoff_anlage_in_betrieb. Last accessed 31 Jul 2023

31. Pandiyan A, Kyriakou V, Neagu D, Welzel S, Goede A, van de Sanden MCM, Tsampas MN (2022) CO_2 conversion via coupled plasma-electrolysis process. J CO_2 Utilization 57:101904. https://doi.org/10.1016/j.jcou.2022.101904

32. Stadler TJ, Bender LJ, Pfeifer P (2022) Dynamic simulation of a compact sorption-enhanced water-gas shift reactor. Front Chem Eng 4. https://doi.org/10.3389/fceng.2022.1000064

33. Stadler TJ, Bertin-Mente B, Dittmeyer R, Brübach L, Böltken T, Pfeifer P (2022) Influence of CO_2-rich syngas on the selectivity to C_{10}–C_{14} in a coupled Fischer-Tropsch/hydrocracking process. Chem Ing Tec 94:289–298. https://doi.org/10.1002/cite.202100172

34. Welzel S, Pfeifer P (2023) Highlights from KEROGREEN's plasma route towards e-Kerosene. In: Final project event 27th September 2022 at KIT Germany. https://www.kerogreen.eu/downloads/220927_Final_event_Peter_Stefan_wo_video.pdf. Last accessed 31 Jul 2023

35. Wolf M, Fischer N, Claeys M (2021) Formation of metal-support compounds in cobalt-based Fischer-Tropsch synthesis: a review. Chem Catalysis 1(5):1014–1041. https://doi.org/10.1016/j.checat.2021.08.002

36. Pöchlauer P, Bohn L, Kotthaus M, Kraut M, Vorbach M, Wenka A, Schubert KF (2006) Micro-structured devices for the chemical research, process development, and production-opportunities and limits. In: Micro-structured devices for chemical research. Process development and production—opportunities and limits, F. Hoffmann-La Roche Ltd., Basel

37. Suo Y, Yao Y, Zhang Y, Xing S, Yuan Z-Y (2022) Recent advances in cobalt-based Fischer-Tropsch synthesis catalysts. J Ind Eng Chem 115:92–119. https://doi.org/10.1016/j.jiec.2022.08.026

38. van Ravenhorst IK, Vogt C, Oosterbeek H, Bossers KW, Moya-Cancino JG, van Bavel AP, van der Eerden AMJ, Vine D, de Groot FMF, Meirer F, Weckhuysen BM (2018) Capturing the genesis of an active Fischer-Tropsch synthesis catalyst with operando X-ray nanospectroscopy. Angew Chem Int Ed 57:11957. https://doi.org/10.1002/anie.201806354

39. Loewert M, Serrer M-A, Carambia T, Stehle M, Zimina CA, Kalz KF, Lichtenberg H, Saraçi E, Pfeifer P, Grunwaldt J-D (2020) Bridging the gap between industry and synchrotron: an operando study at 30 bar over 300 h during Fischer–Tropsch synthesis. React Chem Eng 5:1071–1082. https://doi.org/10.1039/C9RE00493A

40. Fischer N, Claeys M (2020) In situ characterization of Fischer-Tropsch catalysts: a review. J Phys D: Appl Phys 53:293001. https://doi.org/10.1088/1361-6463/ab761c

41. Wolf M, Fischer N, Claeys M (2020) Water-induced deactivation of cobalt-based Fischer-Tropsch catalysts. Nat Catal 3:962–965. https://doi.org/10.1038/s41929-020-00534-5

42. Claeys M, Dry ME, van Steen E, du Plessis E, van Berge PJ, Saib AM, Moodley DJ (2014) In situ magnetometer study on the formation and stability of cobalt carbide in Fischer-Tropsch synthesis. J Catal 318:193–202. https://doi.org/10.1016/j.jcat.2014.08.002

43. https://care-o-sene.com/en/. Last access 29 Jul 2023

44. Rytter E, Holmen A (2015) Deactivation and regeneration of commercial type Fischer-Tropsch co-catalysts—a mini-review. Catalysts 5:478–499. https://doi.org/10.3390/catal5020478

45. Claeys M, van Steen E, Botha T, Crous R, Ferreira A, Harilal A (2021) Oxidation of Hägg carbide during high-temperature Fischer–Tropsch synthesis: size-dependent thermodynamics and in situ observations. ACS Catal 11(22):13866–13879. https://pubs.acs.org/doi/abs/10.1021/acscatal.1c03719

46. Bier W, Keller W, Linder G, Seidel D, Schubert K (1990) Manufacturing and testing of compact micro heat exchangers with high volumetric heat transfer coefficients. Microstruct Sens Actuators, DSC 19:189–197

47. Ramshaw C (ed) (1995) 1st international conference on process intensification for the chemical industry, Antwerp, Belgium (6–8 Dec 1995). BHR group conference series. Publication No. 18. Mechanical Engineering Publications Limited, London

48. Ramshaw C (1983) Chem Eng 389(2):13–14

49. Lerou JJ, Tonkovich AL, Silva L, Perry S, McDaniel J (2010) Microchannel reactor architecture enables greener processes. Chem Eng Sci 65:380–385

50. Tonkovich A, Kuhlmann D, Rogers A, McDaniel J, Fitzgerald S, Arora R, Yuschak T (2005) Microchannel technology scale-up to commercial capacity. Chem Eng Res Des 83(6):634–639. https://doi.org/10.1205/cherd.04354

51. Kee RJ, Karakaya C, Zhu H (2017) Process intensification in the catalytic conversion of natural gas to fuels and chemicals. Proc Combust Inst 36(1):51–76. https://doi.org/10.1016/j.proci.2016.06.014

52. Myrstad R, Eri S, Pfeifer P, Rytter E, Holmen A (2009) Fischer-Tropsch synthesis in a microstructured reactor. Catal Today 147:S301–S304. https://doi.org/10.1016/j.cattod.2009.07.011

53. Piermartini P, Böltken T, Selinsek M, Pfeifer P (2017) Influence of channel geometry on Fischer-Tropsch synthesis in microstructured reactors. Chem Eng J 313:328–335. https://doi.org/10.1016/j.cej.2016.12.076

54. Knochen J, Güttel R, Knobloch C, Turek T (2010) Fischer-Tropsch synthesis in milli-structured fixed-bed reactors: experimental study and scale-up considerations. Chem Eng Process 49(9):958–964. https://doi.org/10.1016/j.cep.2010.04.013

55. Chambrey S, Fongarland P, Karaca H, Piché S, Griboval-Constant A, Schweich D, Luck F, Savin S, Khodakov AY (2011) Fischer-Tropsch synthesis in milli-fixed bed reactor: comparison with centimetric fixed bed and slurry stirred tank reactors. Catal Today 171(1):201–206. https://doi.org/10.1016/j.cattod.2011.04.046

56. Chin Y-H, Hu J, Cao C, Gao Y, Wang Y (2005) Preparation of a novel structured catalyst based on aligned carbon nanotube arrays for a microchannel Fischer-Tropsch synthesis reactor. Catal Today 110(1–2):47–52. https://doi.org/10.1016/j.cattod.2005.09.007

57. Almeida LC, Sanz O, Merino D, Arzamendi G, Gandía LM, Montes M (2013) Kinetic analysis and microstructured reactors modeling for the Fischer-Tropsch synthesis over a Co–Re/Al2O3 catalyst. Catal Today 215:103–111. https://doi.org/10.1016/j.cattod.2013.04.021

58. Almeida LC, Sanz O, D'olhaberriague J, Yunes S, Montes M (2013) Microchannel reactor for Fischer-Tropsch synthesis: adaptation of a commercial unit for testing microchannel blocks. Fuel 110:171–177. https://doi.org/10.1016/j.fuel.2012.09.063

59. Klemm E, Döring H, Geißelmann A, Schirrmeister S (2007) Mikrostrukturreaktoren für die heterogene Katalyse. Chemie Ingenieur Technik 79:697–706. https://doi.org/10.1002/cite.200700052

60. Vogel AP, van Dyk B, Saib AM (2016) GTL using efficient cobalt Fischer-Tropsch catalysts. Catal Today 259(Part 2):323–330. https://doi.org/10.1016/j.cattod.2015.06.018

61. Ying X, Zhang L, Xu H, Ren Y-L, Luo Q, Zhu H-W, Qu H, Xuan J (2016) Efficient Fischer-Tropsch microreactor with innovative aluminizing pretreatment on stainless steel substrate for Co/Al2O3 catalyst coating. Fuel Process Technol 143:51–59. https://doi.org/10.1016/j.fuproc.2015.11.005

62. Sun Y, Jia Z, Yang G, Zhang L, Sun Z (2017) Fischer-Tropsch synthesis using iron based catalyst in a microchannel reactor: performance evaluation and kinetic modeling. Int J Hydrogen Energy 42:49, 29222–29235. https://doi.org/10.1016/j.ijhydene.2017.10.022

63. Zhang X, Zhong L, Zeng G, Gu Y, Peng C, Yu F, Tang Z, Sun Y (2018) Process intensification of honeycomb fractal micro-reactor for the direct production of lower olefins from syngas. Chem Eng J 351:12–21. https://doi.org/10.1016/j.cej.2018.06.078

64. Cao C, Hu J, Li S, Wilcox W, Wang Y (2009) Intensified Fischer-Tropsch synthesis process with microchannel catalytic reactors. Catal Today 140(3–4):149–156. https://doi.org/10.1016/j.cattod.2008.10.016

65. Loewert M, Hoffmann J, Piermartini P, Selinsek M, Dittmeyer R, Pfeifer P (2019) Microstructured Fischer-Tropsch reactor scale-up and opportunities for decentralized application. Chem Eng Technol 42:2202–2214. https://doi.org/10.1002/ceat.201900136

66. Steynberg AP, Deshmukh SR, Robota HJ (2018) Fischer-Tropsch catalyst deactivation in commercial microchannel reactor operation. Catal Today 299:10–13. https://doi.org/10.1016/j.cattod.2017.05.064

67. https://www.edwardtdodge.com/2014/10/31/velocys-small-scale-gas-to-liquids/. Last accessed 30 Jul 2023

68. Jung I, Na J, Park S, Jeon J, Mo Y-G, Yi J-Y, Chung J-T, Han C (2017) Optimal design of a large scale Fischer-Tropsch microchannel reactor module using a cell-coupling method. Fuel Process Technol 159:448–459. https://doi.org/10.1016/j.fuproc.2016.12.004

69. Mohammad N, Chukwudoro C, Bepari S, Basha O, Aravamudhan S, Kuila D (2022) Scale-up of high-pressure F-T synthesis in 3D printed stainless steel microchannel microreactors: experiments and modeling. Catal Today 397–399:182–196. https://doi.org/10.1016/j.cattod.2021.09.038

70. Na J, Kshetrimayum KS, Jung I, Park S, Lee Y, Kwon O, Mo Y, Chung J, Yi J, Lee U, Han C (2018) Optimal design and operation of Fischer-Tropsch microchannel reactor for pilot-scale compact gas-to-liquid process. Chem Eng Proc Proc Intensification 128:63–76. https://doi.org/10.1016/j.cep.2018.04.013

71. Pfeifer P, Schmidt S, Betzner F, Kollmann M, Loewert M, Böltken T, Piermartini P (2022) Scale-up of microstructured Fischer-Tropsch reactors—status and perspectives. Curr Opin Chem Eng 36:100776. https://doi.org/10.1016/j.coche.2021.100776

72. Mesheryakov D, Kirillov VA, Kuzin NA (1999) A multifunctional reactor with a regular catalyst packing for Fischer-Tropsch synthesis. Chem Eng Sci 54(10):1565–1570. https://doi.org/10.1016/S0009-2509(99)00066-4

73. de Deugd RM, Chougule RB, Kreutzer MT, Michiel Meeuse F, Grievink J, Kapteijn F, Moulijn JA (2003) Is a monolithic loop reactor a viable option for Fischer–Tropsch synthesis? Chem Eng Sci 58:3–6, 583–591. https://doi.org/10.1016/S0009-2509(02)00583-3

74. Maretto C, Krishna R (1999) Modelling of a bubble column slurry reactor for Fischer-Tropsch synthesis. Catal Today 52:279–289. https://doi.org/10.1016/S0920-5861(99)00082-6

75. Güttel R (2009) Monolith loop reactors for Fischer-Tropsch synthesis. Doctoral thesis, TU Clausthal

76. Vervloet D, Kapteijn F, Nijenhuis J, van Ommen JR (2012) Fischer-Tropsch reaction–diffusion in a cobalt catalyst particle: aspects of activity and selectivity for a variable chain growth probability. Catal Sci Technol 2:1221–1233. https://doi.org/10.1039/C2CY20060K

77. Merino D, Sanz O, Montes M (2017) Effect of catalyst layer macroporosity in high-thermal-conductivity monolithic Fischer-Tropsch catalysts. Fuel 210:49–57. https://doi.org/10.1016/j.fuel.2017.08.040

78. Visconti CG, Tronconi E, Groppi G, Lietti L, Iovane M, Rossini S, Zennaro R (2011) Monolithic catalysts with high thermal conductivity for the Fischer-Tropsch synthesis in tubular reactors. Chem Eng J 171(3):1294–1307. https://doi.org/10.1016/j.cej.2011.05.014

79. Fratalocchi L, Groppi G, Visconti CG, Lietti L, Tronconi E (2020) Adoption of 3D printed highly conductive periodic open cellular structures as an effective solution to enhance the heat transfer performances of compact Fischer-Tropsch fixed-bed reactors. Chem Eng J 386:123988. https://doi.org/10.1016/j.cej.2019.123988

80. Peacock M, Paterson J, Reed L, Davies S, Carter S, Coe A, Clarkson J (2020) Innovation in Fischer–Tropsch: developing fundamental understanding to support commercial opportunities. Top Catal 63:328–339. https://doi.org/10.1007/s11244-020-01239-6

81. Asalieva E, Gryaznov K, Kulchakovskaya E, Ermolaev I, Sineva L, Mordkovich V (2015) Fischer-Tropsch synthesis on cobalt-based catalysts with different thermally conductive additives. Appl Catal A 505:260–266. https://doi.org/10.1016/j.apcata.2015.08.006

82. Pangarkar K, Schildhauer TJ, van Ommen JR, Nijenhuis J, Moulijn JA, Kapteijn F (2009) Experimental and numerical comparison of structured packings with a randomly packed bed reactor for Fischer-Tropsch synthesis. Catal Today 147S:S2–S9. https://doi.org/10.1016/j.cattod.2009.07.035

83. Fratalocchi L, Visconti CG, Groppi G, Lietti L, Tronconi E (2018) Intensifying heat transfer in Fischer-Tropsch tubular reactors through the adoption of conductive packed foams. Chem Eng J 349:829–837. https://doi.org/10.1016/j.cej.2018.05.108

84. Fratalocchi L, Visconti CG, Groppi G, Lietti L, Tronconi E (2022) Packed-POCS with skin: a novel concept for the intensification of non-adiabatic catalytic processes demonstrated in the case of the Fischer-Tropsch synthesis. Catal Today 383:15–20. https://doi.org/10.1016/j.cattod.2020.12.031

85. Hooshyar N, Vervloet D, Kapteijn F, Hamersma PJ, Mudde RF, van Ommen JR (2012) Intensifying the Fischer-Tropsch synthesis by reactor structuring—a model study. Chem Eng J 207–208:865–870. https://doi.org/10.1016/j.cej.2012.07.105

86. Frost L, Hartvigsen J, Elangovan D (2011) Development of a compact Fischer Tropsch reactor. In: Conference paper from 2011 AIChE annual meeting, assessed 31.07.2022 via https://www.researchgate.net/profile/Lyman-Frost/publication/267313463_Development_of_a_Compact_Fischer_Tropsch_Reactor/links/55491e1b0cf2ebfd8e3ad82e/Development-of-a-Compact-Fischer-Tropsch-Reactor.pdf

87. Barrera JL, Hartvigsen JJ, Hollist M, Pike J, Yarosh A, Fullilove NP, Beck VA (2023) Design optimization of integrated cooling inserts in modular Fischer-Tropsch reactors. Chem Eng Sci 268:118423. https://doi.org/10.1016/j.ces.2022.118423

88. Kapteijn F, Moulijn JA (2022) Structured catalysts and reactors—perspectives for demanding applications. Catal Today 383:5–14. https://doi.org/10.1016/j.cattod.2020.09.026

89. Odunsi AO, O'Donovan TS, Reay DA (2016) Temperature stabilisation in Fischer-Tropsch reactors using phase change material (PCM). Appl Therm Eng 93:1377–1393. https://doi.org/10.1016/j.applthermaleng.2015.08.084

90. An M, Guan X, Yang N, Bu Y, Xu M, Men Z (2018) Effects of internals on fluid dynamics and reactions in pilot-scale slurry bubble column reactors: a CFD study for Fischer-Tropsch synthesis. Chem Eng Proc Proc Intensification 132:194–207. https://doi.org/10.1016/j.cep.2018.09.004

91. Geng S, Mao Z-S, Huang Q, Yang C (2021) Process intensification in pneumatically agitated slurry reactors. Engineering 7(3):304–325. https://doi.org/10.1016/j.eng.2021.03.002

92. Abbaslou RMM, Mohammadzadeh JSS, Dalai AK (2009) Review on Fischer-Tropsch synthesis in supercritical media. Fuel Proc Technol 90(7–8):849–856. https://doi.org/10.1016/j.fuproc.2009.03.018

93. Mogalicherla AK, Elmalik EE, Elbashir NO (2012) Enhancement in the intraparticle diffusion in the supercritical phase Fischer-Tropsch synthesis. Chem Eng Process 62:59–68. https://doi.org/10.1016/j.cep.2012.09.008

94. Abusrafa AE, Challiwala MS, Choudhury HA, Wilhite BA, Elbashir NO (2020) Experimental verification of 2-dimensional computational fluid dynamics modeling of supercritical fluids Fischer Tropsch reactor bed. Catal Today 343:165–175. https://doi.org/10.1016/j.cattod.2019.05.027

95. Sun C, Luo Z, Choudhary A, Pfeifer P, Dittmeyer R (2017) Influence of the condensable hydrocarbons on an integrated Fischer-Tropsch synthesis and hydrocracking process: simulation and experimental validation. Ind Eng Chem Res 56:13075–13085. https://doi.org/10.1021/acs.iecr.7b01326

96. Sun C, Pfeifer P, Dittmeyer R (2017) One-stage syngas-to-fuel in a micro-structured reactor: investigation of integration pattern and operating conditions on the selectivity and productivity of liquid fuels. Chem Eng J 326:37–46. https://doi.org/10.1016/j.cej.2017.05.133

97. Sun C, Zhan T, Pfeifer P, Dittmeyer R (2017) Influence of Fischer-Tropsch synthesis (FTS) and hydrocracking (HC) conditions on the product distribution of an integrated FTS-HC process. Chem Eng J 310:272–281. https://doi.org/10.1016/j.cej.2016.10.118

98. Sun C, Klumpp M, Binder JR, Pfeifer P, Dittmeyer R (2017) Einstufige Treibstoffsynthese mittels gedruckter Katalysatorschichten in mikrostrukturierten Reaktoren = one-stage syngas-to-fuel conversion with printed catalyst layers in microstructured reactors. Chemie - Ingenieur - Technik 89(7):894–902. https://doi.org/10.1002/cite.201600180

99. Zhu C, Gamliel DP, Valla JA, Bollas GM (2021) Fischer-Tropsch synthesis in monolith catalysts coated with hierarchical ZSM-5. Appl Catal B 284:119719. https://doi.org/10.1016/j.apcatb.2020.119719

100. Zhu C, Bollas GM (2018) Gasoline selective Fischer-Tropsch synthesis in structured bifunctional catalysts. Appl Catal B 235:92–102. https://doi.org/10.1016/j.apcatb.2018.04.063

101. Kirsch H, Sommer U, Pfeifer P, Dittmeyer R (2020) Power-to-fuel conversion based on reverse water-gas-shift, Fischer-Tropsch synthesis and hydrocracking: mathematical modeling and simulation in Matlab/Simulink. Chem Eng Sci 227:115930. https://doi.org/10.1016/j.ces.2020.115930

102. Kirsch H, Lochmahr N, Staudt C, Pfeifer P (2020) Dittmeyer, production of CO_2-neutral liquid fuels by integrating Fischer-Tropsch synthesis and hydrocracking in a single micro-structured reactor: Performance evaluation of different configurations by factorial design experiments. Chem. Eng J 393:124553. https://doi.org/10.1016/j.cej.2020.124553

103. Brosius R (2019) WO/2020/016845, Low pressure hydrocracking process for the production of a high yield of middle distillates from a high boiling hydrocarbon feedstock, 19 Jul 2019

104. Rytter E (2006) WO2008/062208, Multi stage process for producing hydrocarbons from syngas, 23 Nov 2006

105. Pandey U, Putta KR, Rout KR, Blekkan EA, Rytter E, Hillestad M (2022) Staging and path optimization of Fischer-Tropsch synthesis. Chem Eng Res Des 187:276–289. https://doi.org/10.1016/j.cherd.2022.08.033

106. Rafiee A, Hillestad M (2013) Staging of the Fischer-Tropsch reactor with a cobalt-based catalyst. Chem Eng Technol 36:1729–1738. https://doi.org/10.1002/ceat.201200700

107. Fischer F, Bahr T, Meusel A (1935) Über die katalytische Reduktion des Kohlendioxyds zu Methan und höheren Kohlenwasserstoffen bei gewöhnlichem Druck 16:466–469

108. Küster H (1936) Über die Reduktion der Kohlensäure zu höheren Kohlenwasserstoffen bei Atmosphärendruck an Katalysatoren der Eisengruppe. Brennst.-Chem. 17:221–228

109. Kölbel H, Ackermann P (1954) U.S. Patent 2,692,274

110. Russell WW, Miller GH (1950) Catalytic hydrogenation of carbon dioxide to higher hydrocarbons. J Am Chem Soc 72:2446–2454. https://doi.org/10.1021/ja01162a025

111. Weatherbee GD, Bartholomew CH (1984) Hydrogenation of CO_2 on group VIII metals: IV. Specific activities and selectivities of silica-supported Co, Fe, and Ru. J Catal 87:352–362. https://doi.org/10.1016/0021-9517(84)90196-9

112. Fiato RA, Soled SL, Rice GW, Miseo S (1992) U.S. Patent 5,140,049A

113. Lee M-D, Lee J-F, Chang C-S (1989) Hydrogenation of carbon dioxide on unpromoted and potassium-promoted iron catalysts. Bull Chem Soc Jpn 62:2756–2758. https://doi.org/10.1246/bcsj.62.2756

114. Lee J-F, Chern W-S, Lee M-D, Dong T-Y (1992) Hydrogenation of carbon dioxide on iron catalysts doubly promoted with manganese and potassium. Can J Chem Eng 70:511–515. https://doi.org/10.1002/cjce.5450700314

115. Dorner RW, Hardy DR, Williams FW, Willauer HD (2010) Heterogeneous catalytic CO_2 conversion to value-added hydrocarbons. Energy Environ Sci 3:884. https://doi.org/10.1039/c001514h

116. Wei J, Yao R, Han Y, Ge Q, Sun J (2021) Towards the development of the emerging process of CO_2 heterogenous hydrogenation into high-value unsaturated heavy hydrocarbons. Chem Soc Rev 50:10764–10805. https://doi.org/10.1039/D1CS00260K

117. Sai Prasad PS, Bae JW, Jun K-W, Lee K-W (2008) Fischer–Tropsch synthesis by carbon dioxide hydrogenation on Fe-based catalysts. Catal Surv Asia 12:170–183

118. Panzone C, Philippe R, Chappaz A, Fongarland P, Bengaouer A (2020) Power-to-liquid catalytic CO_2 valorization into fuels and chemicals: focus on the Fischer-Tropsch route. J CO_2 Util 38:314–347. https://doi.org/10.1016/j.jcou.2020.02.009

119. Wang D, Xie Z, Porosoff MD, Chen JG (2021) Recent advances in carbon dioxide hydrogenation to produce olefins and aromatics. Chem 7:2277–2311. https://doi.org/10.1016/j.chempr.2021.02.024

120. Ye R-P, Ding J, Gong W, Argyle MD, Zhong Q, Wang Y, Russell CK, Xu Z, Russell AG, Li Q, Fan M, Yao Y-G (2019) CO_2 hydrogenation to high-value products via heterogeneous catalysis. Nat Commun 10:5698. https://doi.org/10.1038/s41467-019-13638-9

121. Atsbha TA, Yoon T, Seongho P, Lee C-J (2021) A review on the catalytic conversion of CO_2 using H2 for synthesis of CO, methanol, and hydrocarbons. J CO_2 Util 44:101413. https://doi.org/10.1016/j.jcou.2020.101413

122. Riedel T, Schaub G, Jun K-W, Lee K-W (2001) Kinetics of CO_2 hydrogenation on a K-promoted Fe catalyst. Ind Eng Chem Res 40:1355–1363. https://doi.org/10.1021/ie000084k

123. Wei J, Ge Q, Yao R, Wen Z, Fang C, Guo L, Xu H, Sun J (2017) Directly converting CO_2 into a gasoline fuel. Nat Commun 8:15174. https://doi.org/10.1038/ncomms15174

124. Visconti CG, Martinelli M, Falbo L, Fratalocchi L, Lietti L (2016) CO_2 hydrogenation to hydrocarbons over Co and Fe-based Fischer-Tropsch catalysts. Catal Today 277:161–170. https://doi.org/10.1016/j.cattod.2016.04.010

125. Visconti CG, Martinelli M, Falbo L, Infantes-Molina A, Lietti L, Forzatti P, Iaquaniello G, Palo E, Picutti B, Brignoli F (2017) CO_2 hydrogenation to lower olefins on a high surface area K-promoted bulk Fe-catalyst. Appl Catal B 200:530–542. https://doi.org/10.1016/j.apcatb.2016.07.047

126. Riedel T, Schulz H, Schaub G, Jun K-W, Hwang J-S, Lee K-W (2003) Fischer-Tropsch on iron with H2/CO and H2/CO_2 as synthesis gases: the episodes of formation of the Fischer-Tropsch regime and construction of the catalyst. Top Catal 26:41–54. https://doi.org/10.1023/B:TOCA.0000012986.46680.28

127. Meiri N, Radus R, Herskowitz M (2017) Simulation of novel process of CO_2 conversion to liquid fuels. J CO_2 Util 17:284–289. https://doi.org/10.1016/j.jcou.2016.12.008

128. Landau MV, Vidruk R, Herskowitz M (2014) Sustainable production of green feed from carbon dioxide and hydrogen. Chemsuschem 7:785–794. https://doi.org/10.1002/cssc.201301181

129. Brübach L, Hodonj D, Pfeifer P (2022) Kinetic analysis of CO_2 hydrogenation to long-chain hydrocarbons on a supported iron catalyst. Ind Eng Chem Res 61:1644–1654. https://doi.org/10.1021/acs.iecr.1c04018

130. Brübach L, Hodonj D, Pfeifer P (2022) Kinetic analysis of CO_2 hydrogenation to long-chain hydrocarbons on a supported iron catalyst. Ind Eng Chem Res 61(4):1644–1654. https://pubs.acs.org/doi/10.1021/acs.iecr.1c04018

131. Iglesias MG, de Vries C, Claeys M, Schaub G (2015) Chemical energy storage in gaseous hydrocarbons via iron Fischer–Tropsch synthesis from H2/CO_2—kinetics, selectivity and process considerations. Catal Today 242:184–192. https://doi.org/10.1016/j.cattod.2014.05.020

132. Riedel T, Claeys M, Schulz H, Schaub G, Nam S-S, Jun K-W, Choi M-J, Kishan G, Lee K-W (1999) Comparative study of Fischer–Tropsch synthesis with H2/CO and H2/CO_2 syngas using Fe- and Co-based catalysts. Appl Catal A 186:201–213. https://doi.org/10.1016/S0926-860X(99)00173-8

133. Fischer N, Henkel R, Hettel B, Iglesias M, Schaub G, Claeys M (2016) Hydrocarbons via CO_2 hydrogenation over iron catalysts: the effect of potassium on structure and performance. Catal Lett 146:509–517. https://doi.org/10.1007/s10562-015-1670-9

134. Hong J-S, Hwang JS, Jun K-W, Sur JC, Lee K-W (2001) Deactivation study on a coprecipitated Fe-Cu-K-Al catalyst in CO_2 hydrogenation. Appl Catal A 218:53–59. https://doi.org/10.1016/S0926-860X(01)00617-2

135. Dry ME (2004) Chapter 7—FT catalysts. In: Steynberg A, Dry M (eds) Fischer-Tropsch technology, Reprinted, Elsevier, Amsterdam, pp 533–600

136. Bukur DB, Sivaraj C (2002) Supported iron catalysts for slurry phase Fischer–Tropsch synthesis. Appl Catal A 231:201–214. https://doi.org/10.1016/S0926-860X(02)00053-4

137. Choi PH, Jun K-W, Lee S-J, Choi M-J, Lee K-W (1996) Hydrogenation of carbon dioxide over alumina supported Fe-K catalysts. Catal Lett 40:115–118. https://doi.org/10.1007/BF00807467

138. Kuei C-K, Lee M-D (1991) Hydrogenation of carbon dioxide by hybrid catalysts, direct synthesis of aromatics from carbon dioxide and hydrogen. Can J Chem Eng 69:347–354. https://doi.org/10.1002/cjce.5450690142

139. Amoyal M, Vidruk-Nehemya R, Landau MV, Herskowitz M (2017) Effect of potassium on the active phases of Fe catalysts for carbon dioxide conversion to liquid fuels through hydrogenation. J Catal 348:29–39. https://doi.org/10.1016/j.jcat.2017.01.020

140. Martinelli M, Visconti CG, Lietti L, Forzatti P, Bassano C, Deiana P (2014) CO_2 reactivity on Fe–Zn–Cu–K Fischer-Tropsch synthesis catalysts with different K-loadings. Catal Today 228:77–88. https://doi.org/10.1016/j.cattod.2013.11.018

141. Rodemerck U, Holeňa M, Wagner E, Smejkal Q, Barkschat A, Baerns M (2013) Catalyst development for CO_2 hydrogenation to fuels. ChemCatChem 5:1948–1955. https://doi.org/10.1002/cctc.201200879

142. Satthawong R, Koizumi N, Song C, Prasassarakich P (2013) Bimetallic Fe–Co catalysts for CO_2 hydrogenation to higher hydrocarbons. J CO_2 Util 3–4:102–106. https://doi.org/10.1016/j.jcou.2013.10.002

143. Liang B, Duan H, Sun T, Ma J, Liu X, Xu J, Su X, Huang Y, Zhang T (2019) Effect of Na promoter on Fe-based catalyst for CO 2 hydrogenation to alkenes. ACS Sustain Chem Eng 7:925–932. https://doi.org/10.1021/acssuschemeng.8b04538

144. Wei J, Sun J, Wen Z, Fang C, Ge Q, Xu H (2016) New insights into the effect of sodium on Fe 3 O 4—based nanocatalysts for CO 2 hydrogenation to light olefins. Catal Sci Technol 6:4786–4793. https://doi.org/10.1039/C6CY00160B

145. Liu X, Zhang C, Tian P, Xu M, Cao C, Yang Z, Zhu M, Xu J (2021) Revealing the effect of sodium on iron-based catalysts for CO_2 hydrogenation: insights from calculation and experiment. J Phys Chem C 125:7637–7646. https://doi.org/10.1021/acs.jpcc.0c11123

146. Liu B, Geng S, Zheng J, Jia X, Jiang F, Liu X (2018) Unravelling the new roles of Na and Mn promoter in CO 2 hydrogenation over Fe 3 O 4 -based catalysts for enhanced selectivity to light α-olefins. ChemCatChem 10:4718–4732. https://doi.org/10.1002/cctc.201800782

147. Schmidt C, Kureti S (2022) CO 2 conversion by Fischer-Tropsch synthesis using na-modified Fe catalysts. Chem Ing Tech cite 202200067. https://doi.org/10.1002/cite.202200067

148. Wen C, Jiang J, Chiliu C, Tian Z, Xu X, Wu J, Wang C, Ma L (2020) Single-step selective conversion of carbon dioxide to aromatics over Na-Fe 3 O 4 /hierarchical HZSM-5 zeolite catalyst. Energy Fuels 34:11282–11289. https://doi.org/10.1021/acs.energyfuels.0c02120

149. Wang Y, Kazumi S, Gao W, Gao X, Li H, Guo X, Yoneyama Y, Yang G, Tsubaki N (2020) Direct conversion of CO_2 to aromatics with high yield via a modified Fischer-Tropsch synthesis pathway. Appl Catal B 269:118792. https://doi.org/10.1016/j.apcatb.2020.118792

150. Khangale PR, Meijboom R, Jalama K (2020) CO_2 hydrogenation to liquid hydrocarbons via modified Fischer–Tropsch over alumina-supported cobalt catalysts: effect of operating temperature, pressure and potassium loading. J CO_2 Util 41:101268. https://doi.org/10.1016/j.jcou.2020.101268

151. Jo H, Khan MK, Irshad M, Arshad MW, Kim SK, Kim J (2022) Unraveling the role of cobalt in the direct conversion of CO_2 to high-yield liquid fuels and lube base oil. Appl Catal B 305:121041. https://doi.org/10.1016/j.apcatb.2021.121041

152. Shi Z, Yang H, Gao P, Li X, Zhong L, Wang H, Liu H, Wei W, Sun Y (2018) Direct conversion of CO_2 to long-chain hydrocarbon fuels over K–promoted CoCu/TiO_2 catalysts. Catal Today 311:65–73. https://doi.org/10.1016/j.cattod.2017.09.053

153. Choi MJ, Kim J, Lee S, Lee W, Lee K (2003) Promotion of CO_2 hydrogenation in fixed bed recycle reactors. In: Gale J, Kaya Y (eds) Greenhouse gas control technologies: proceedings of the 6th international conference on greenhouse gas control technologies, 1–4 October 2002, Kyoto, Japan, Pergamon, Amsterdam, Oxford, pp 1491–1496

154. Willauer HD, Bradley MJ, Baldwin JW, Hartvigsen JJ, Frost L, Morse JR, DiMascio F, Hardy DR, Hasler DJ (2020) Evaluation of CO_2 hydrogenation in a modular fixed-bed reactor prototype. Catalysts 10:970. https://doi.org/10.3390/catal10090970

155. Rohde MP, Unruh D, Schaub G (2005) Membrane application in fischer−tropsch synthesis to enhance CO 2 Hydrogenation. Ind Eng Chem Res 44:9653–9658

156. Li Z, Deng Y, Dewangan N, Hu J, Wang Z, Tan X, Liu S, Kawi S (2021) High temperature water permeable membrane reactors for CO_2 utilization. Chem Eng J 420:129834. https://doi.org/10.1016/j.cej.2021.129834

157. Steynberg AP, Dry ME, Davis BH, Breman BB (2004) Chapter 2—Fischer-Tropsch reactors. In: Steynberg A, Dry M (eds),Fischer-Tropsch technology, Reprinted, Elsevier, Amsterdam, pp 64–195

158. Kim J-S, Lee S, Lee S-B, Choi M-J, Lee K-W (2006) Performance of catalytic reactors for the hydrogenation of CO_2 to hydrocarbons. Catal Today 115:228–234. https://doi.org/10.1016/j. cattod.2006.02.038

159. Brübach L, Hodonj D, Biffar L, Pfeifer P (2022) Detailed Kinetic modeling of CO_2-based Fischer-Tropsch synthesis. Catalysts 12:630. https://doi.org/10.3390/catal12060630

160. de Klerk A (2011) Fischer-Tropsch refining, 1st edn. Wiley-VCH, Weinheim

161. Willauer HD, Ananth R, Olsen MT, Drab DM, Hardy DR, Williams FW (2013) Modeling and kinetic analysis of CO_2 hydrogenation using a Mn and K-promoted Fe catalyst in a fixed-bed reactor. J CO_2 Util 3–4:56–64. https://doi.org/10.1016/j.jcou.2013.10.003

162. Landau MV, Meiri N, Utsis N, Vidruk Nehemya R, Herskowitz M (2017) Conversion of CO_2, CO, and H_2 in CO_2 hydrogenation to fungible liquid fuels on Fe-based catalysts. Ind Eng Chem Res 56:13334–13355. https://doi.org/10.1021/acs.iecr.7b01817

163. de Klerk A (2011) Fischer-Tropsch fuels refinery design. Energy Environ Sci 4:1177–1205

164. de Klerk A (2008) Hydroprocessing peculiarities of Fischer-Tropsch syncrude. Catal Today 130:439–445. https://doi.org/10.1016/j.cattod.2007.10.006

165. Nel RJJ, de Klerk A (2007) Fischer−Tropsch aqueous phase refining by catalytic alcohol dehydration. Ind Eng Chem Res 46:3558–3565. https://doi.org/10.1021/ie061555r

166. Pfeifer P, Biffar L, Timm F, Böltken T (2020) Influence of power-to-fuel plant flexibility towards power and plant utilization and intermediate hydrogen buffer size. Chem Ing Tec 92:1976–1982. https://doi.org/10.1002/cite.202000084

167. González MJ, Schaub G (2015) Fischer-Tropsch synthesis with H_2/CO_2—catalyst behavior under transient conditions. Chem Ing Tec 87:848–854. https://doi.org/10.1002/cite.201 400137

168. Bremaud M, Fongarland P, Anfray J, Jallais S, Schweich D, Khodakov AY (2005) Influence of syngas composition on the transient behavior of a Fischer-Tropsch continuous slurry reactor. Catal Today 106(1–4):137–142. https://doi.org/10.1016/j.cattod.2005.07.126

169. Loewert M, Pfeifer P (2020) Dynamically operated fischer-tropsch synthesis in PtL-Part 1: system response on intermittent feed. Chem Eng 4:2, 21. https://doi.org/10.3390/chemengin eering4020021

170. Eilers H, González MJ, Schaub G (2016) Lab-scale experimental studies of Fischer-Tropsch kinetics in a three-phase slurry reactor under transient reaction conditions. Catal Today 275:164–171. https://doi.org/10.1016/j.cattod.2015.11.011

171. Loewert M, Riedinger M, Pfeifer P (2020) Dynamically operated Fischer–Tropsch synthesis in PtL—Part 2: coping with real PV profiles. Chem Eng 4:2, 27. https://doi.org/10.3390/che mengineering4020027

172. Nikačević N, Todić B, Mandić M, Petkovska M, Bukur DB (2020) Optimization of forced periodic operations in milli-scale fixed bed reactor for Fischer-Tropsch synthesis. Catal Today 343:156–164. https://doi.org/10.1016/j.cattod.2018.12.032

173. Mandić M, Dikić V, Petkovska M, Todić B, Bukur DB, Nikačević NM (2018) Dynamic analysis of millimetre-scale fixed bed reactors for Fischer-Tropsch synthesis. Chem Eng Sci 192:434–447. https://doi.org/10.1016/j.ces.2018.07.052

174. Wentrup J, Pesch GR, Thöming J (2022) Dynamic operation of Fischer-Tropsch reactors for power-to-liquid concepts: a review. Renew Sustain Energy Rev 162:112454. https://doi.org/ 10.1016/j.rser.2022.112454

175. de Swart JWA, Krishna R (2002) Simulation of the transient and steady state behaviour of a bubble column slurry reactor for Fischer-Tropsch synthesis. Chem Eng Process 41(1):35–47. https://doi.org/10.1016/S0255-2701(00)00159-8

176. Saeidi S, Nikoo MK, Mirvakili A, Bahrani S, Amin NAS, Rahimpour MR (2015) Recent advances in reactors for low-temperature Fischer-Tropsch synthesis: process intensification perspective. Rev Chem Eng 31(3):209–238. https://doi.org/10.1515/revce-2014-0042

177. Venvik HJ, Yang J (2017) Catalysis in microstructured reactors: short review on small-scale syngas production and further conversion into methanol, DME and Fischer-Tropsch products. Catal Today 285:135–146. https://doi.org/10.1016/j.cattod.2017.02.014

178. Holmen A, Venvik HJ, Myrstad R, Zhu J, Chen D (2013) Monolithic, microchannel and carbon nanofibers/carbon felt reactors for syngas conversion by Fischer-Tropsch synthesis. Catal Today 216:150–157. https://doi.org/10.1016/j.cattod.2013.06.006

179. Press release InfraServ, 20.04.2023. https://www.infraserv.com/en/news/presse/press-rel eases/pressemeldung-nc_62848.html. Last accessed 04 Aug 2023

180. Sheehan SW, Chen C, US20230234037, Molybdenum-based catalysts for carbon dioxide conversion, 05 May 2021

181. Sheehan SW, Chen C, Garedew-Ballard M, Luthria N, Shah MR, Wu Q (2023) WO2023137002, Methods and catalysts for carbon dioxide conversion to long-chain hydrocarbons, 10 Jan 2023

182. Press release Sasol, 06.09.2021. https://www.sasol.com/media-centre/media-releases/sasol-and-uct-researchers-collaborate-use-commercial-iron-catalysts. Last accessed 04 Aug 2023

183. Jürgens S, Oßwald P, Selinsek M, Piermartini P, Schwab J, Pfeifer P, Bauder U, Ruoff S, Rauch B, Köhler M (2019) Assessment of combustion properties of non-hydroprocessed Fischer-Tropsch Fuels for aviation. Fuel Process Technol 193:232–243. https://doi.org/10.1016/j.fup roc.2019.05.015

184. Voigt C, Kleine J, Sauer D, Moore RH, Bräuer T, Le Clercq P, Kaufmann MS, Jurkat-Witschas T, Aigner M, Bauder U, Boose Y, Borrmann S, Crosbie E, Diskin GS, DiGangi J, Hahn V, Heckl C, Huber F, Nowak JB, Rapp M, Rauch B, Robinson C, Schripp T, Shook M, Winstead E, Ziemba L, Schlager H, Anderson BE (2021) Cleaner burning aviation fuels can reduce contrail cloudiness. Commun Earth Environ 2:114. https://doi.org/10.1038/s43247-021-001 74-y

Refining of Fischer–Tropsch Products

Dagmar Beiermann

Abstract The Fischer–Tropsch synthesis is a chemical process that converts carbon monoxide (CO) and hydrogen (H_2), collectively known as "syngas," into liquid hydrocarbons. Traditionally used for coal liquefaction and natural gas conversion, this process is also expected to be crucial in producing renewable fuels, particularly kerosene for aviation, as well as naphtha and diesel. The Fischer–Tropsch process yields synthetic crude oil (syncrude), encompassing a broad hydrocarbon range (C_1 to C_{100}), and requires further upgrading or refining to produce usable fuels. In this context, this chapter examines the differences between conventional and Fischer–Tropsch crude oil refining, the specific methods required to refine Fischer–Tropsch products, and anticipated advancements in this field.

Keywords Refinery Processes · Syncrude Refining · Hydrocracking · Isomerization · Co-Processing

1 Introduction

The Fischer–Tropsch synthesis is a well-known process that comprises a series of chemical reactions to convert a mixture of carbon monoxide (CO) and hydrogen (H_2), commonly referred to as "synthesis gas" or "syngas," into liquid hydrocarbons. While this method is central to both liquefaction of coal and the conversion of natural gas to produce liquid hydrocarbons, it is also anticipated to play a pivotal role in the future production of renewable fuel options, especially for kerosene, as the aviation sector will be one of the most difficult sectors to decarbonize, but also for naphtha and diesel.

The Fischer–Tropsch synthesis produces a synthetic crude oil (syncrude), a wide hydrocarbon product range from C_1 to C_{100}, and is not selective. As these Fischer–Tropsch intermediates cannot be used as fuel directly, further upgrading or refining is required. This is also needed to increase the product quality and to transform more

D. Beiermann (✉)
BP Europa SE, Lingen Refinery, Lingen, Germany
e-mail: dagmar.beiermann@bp.com

© The Author(s), under exclusive license to Springer Nature Switzerland AG 2025
N. Bullerdiek et al. (eds.), *Powerfuels*, Green Energy and Technology,
https://doi.org/10.1007/978-3-031-62411-7_23

647

of the synthetic crude oil into useful products. The difference between upgrading and refining is that upgrading produces intermediates that must still be refined to produce final fuels that meet the respective fuel specifications. These intermediates can also be used as blend component for final fuel production. On the other hand, refining directly produces final products. Without an upgrading or refining, a Fischer–Tropsch based process is clearly very inefficient [1].

Synthetic crude oil from Fischer–Tropsch synthesis is comparable but different to conventional crude oil [1]. In this context, the upgrading and refining of Fischer–Tropsch products are detailed in the following sections. Therefore, the basics of conventional crude oil refining are outlined first. Based on this, specific aspects on the options to refine Fischer–Tropsch products are explained. Finally, a brief overview of anticipated future developments in this area is provided.

2 Conventional Refining

A conventional refinery consists of several units producing mainly diverse fuels or petrochemical components from fossil crude oil. Petroleum crude oil is separated into fractions, and undesired components are removed. Heavy fractions are converted into lighter ones, and all streams are upgraded or refined and blended to final products like heating or transportation fuels, lubricants, or petrochemical components.

History. The ability to produce a kerosene fraction from crude oil via a distillation vessel was first demonstrated in 1854. The first crude oil refineries were relatively simple and consisted of single batch distillation vessels to boil crude oil so that the middle-boiling kerosene fraction (180–300 °C) could be collected and sold as lamp oil. This batch processing was used until 1910, only then continuous operating boilers appeared, and the first real distillation columns were built in the 1920s. In the 1870s, refiners were just interested in producing maximum amount of the kerosene fraction from the crude oil, throwing away all by-products. Around the same time, the internal combustion engine was invented, and it was realized that the naphtha fraction from crude oil distillation was a good fuel for automotive engines. These two distillation products, naphtha for gasoline and kerosene for lamp oil, were highly demanded from 1890 onward. Few years later, the diesel fraction was found to be suited for compression-ignition engines used for heavy vehicles and ships.

Following these initial advancements, refineries underwent various stages of evolution and transformation, leading to different generations of refineries, influenced by diverse factors. The rapid spread of internal combustion engines after the First World War led to a rapid further development of the processes used in a refinery and the requirement to construct numerous new refineries. Improvement in the engine design required a better fuel quality leading to the second generation of refineries. The second-generation refineries introduced conversion processes to modify the quality beyond simple distillation, incorporating the introduction of conversion processes like hydrotreating, isomerization, cracking, alkylation, and catalytic reforming, which were developed between 1920 and 1960. They were able to influence the quality of the crude oil products and also the boiling range. The evolution of the third-generation refineries was mainly driven by the oil crisis in the 1970s and the subsequent spike in crude oil prices. Thus, third-generation refineries are related to the inclusion of vacuum distillations and residue upgrading capacity to utilize the residue fraction of crude oil efficiently. The emerging of fourth-generation refineries was primarily a consequence of stricter vehicle emission control standards that were introduced due to poor air quality, especially in densely populated cities. Thus, fourth-generation refineries adapted, for example, to produce high-octane, unleaded gasoline products or diesel fuel with significantly reduced sulfur content and improved attributes such as the cetane number [1].

2.1   Fossil Crude Oil as Feedstock

Fossil crude oil is a liquid mixture of hydrocarbons and related compounds found in the earth's crust in various regions of the world. Several hundred crude oil grades are traded worldwide—for example, North America's West Texas Intermediate (WTI) or North Sea Brent Crude (Brent)—and each crude oil grade has a unique chemical composition that determines physical properties such as density and viscosity. Sometimes there is also a deviation over time in the composition of the crude, depending on the type of oil field exploration.

Crude oil consists of a large number of different compounds [2]. The most common are linear or branched alkanes, cycloalkanes, and aromatics. To a lesser extent, fossil crude oil contains hydrocarbon compounds containing heteroatoms like sulfur, nitrogen, or oxygen in many forms, such as the mercaptan, amine, or alcohol series. In addition, there are metal compounds such as iron, copper, vanadium, and nickel. The carbon content is usually between 83 and 87% and the hydrogen content between 10 and 14%. Other main group elements like sulfur or nitrogen are between 0.1 and 1.4%, and the content of metal compounds is less than 1,000 ppm. However, the latter is enough to cause deposits on refining catalysts and to reduce catalyst lifetime [3]. Metals are normally found in the heavier fractions and end up in the residue processing units of a refinery.

No valid classification system for crude oil exists, as the varied nature of crude oil makes generalization difficult. Grouping crude oils is often based on parameters that affect the refining process, such as the hydrocarbon class, the distillation profile, and the content of heteroatoms [1].

Hydrocarbon class. The main hydrocarbon class present in the crude oil can be paraffinic (alkanes), naphthenic (cycloalkanes), or aromatic. Olefins (alkenes) content is normally a minor hydrocarbon class in crude oil or is even not present at all.

Distillation profile. Fractional distillation quantifies the portions of crude oil that boils in several temperature ranges, leading to a specific distillation profile. When the amount of material that boils in the naphtha range is high, the crude oil is classified as "light". If most material boils in the middle distillate range, the crude is classified as "heavy". Also, the density of a crude oil gives an indication if a crude oil is rather light or heavy.

Heteroatom content. The main heteroatoms in crude oil are sulfur, nitrogen, and oxygen. These heteroatoms are disruptive factors in refining as they can cause corrosion, act as catalyst inhibitor, or when combusted, emit air pollutants that harm the environment.

- Low sulfur crude oils are called "sweet" (< 0.5 wt.-%), and high sulfur-containing crudes are called "sour" and can contain up to 6 wt.-% of sulfur or even more. Sulfur can be present in inorganic (elemental sulfur, carbonyl sulfide, or hydrogen sulfide) or organic form (sulfides, thiols, and thiophenes). Sulfur is not equally distributed in the crude oil, and the sulfur content increases with the boiling point. The sourer a crude (the more sulfur it contains), the more intensive the refining process to meet fuel specification, where a maximum sulfur content is defined.
- Nitrogen-containing compounds inhibit acid catalysis that is part of most catalytic processes in a refinery. Nitrogen-containing compounds are mostly in crude oil boiling fractions above 250 °C. The total nitrogen content can meet 2 wt.-% but is often less than 0.5 wt.-%. A typical nitrogen-containing compound class in crude oil is amines.
- Oxygenates are generally present in low concentrations up to 1.5 wt.-%. A high amount of oxygenates leads to a high total acid number (TAN) of the crude, which is an indicator for the corrosion potential of a crude. Crude oil with a TAN of more than 1 mg/KOH·g is called a "high TAN" crude oil. To what extent a high TAN crude oil can be processed in a refinery depends on the metallurgical limitations of the refinery equipment [1].

2.2   Refinery Processes

Refinery configurations usually vary significantly since there is no one-size-fits-all standard. Usually, no refinery is like another due to the diverse nature of crude oils, differing geographical locations as well as unique market factors that influence each refinery's design and operations. Nevertheless, most refineries possess a set of typical or common processing units, while their arrangement may differ. Figure 1 illustrates an example of a possible refinery configuration that includes different upgrading units for the crude oil streams. There are multiple configurations possible, as refineries are specific assets and complexity can differ widely.

The desalted crude oil is separated into different fractions by an atmospheric crude distillation unit. Residues are fed into a vacuum distillation unit to split the heavy fraction into further intermediate streams. After fractionation, a series of treatment and conversion/refining processes are applied to remove heteroatoms, reduce the boiling point, and improve the quality of the intermediate products. The final products such as motor gasoline, kerosene, diesel fuel, or fuel oil are then mixed from various intermediate product streams that are produced in the different refining processes. By making careful use of all process streams, a refinery can work residue-free and use its own fuel gas by-product to fire its furnaces. Several of the process steps are described in more detail below.

Desalter. To prevent corrosion in the refinery assets, the crude oil is desalted to a salt content of less than 10 ppm. This is done by adding water to produce an emulsion of crude oil and water. The salt dissolves in the aqueous phase of this emulsion. The emulsion is then separated in an electrostatic field. The salt containing water at the bottom is fed to the appropriate wastewater treatment plant, and the desalted crude oil is pumped to the distillation [4].

Fractionation. After desalting the crude oil is heated in heat exchangers by recovering heat from product streams that leave the unit. Additional heat is supplied by furnaces to bring the oil up to around 400 °C. The heated oil is separated into diverse fractions by rectification in a column that can be up to 50 m high. Often these columns are called distillation columns, but as crude oil is a multi-component, gas–liquid mixture which flows in countercurrent, the correct technical term is rectification. The crude oil enters the column in a 2-phase flow (gas/liquid). The temperature profile drops toward the top. Since the temperature is highest in the bottom and the light components do not condense there, the light components continue to rise in gaseous form to the top. At the top part of the rectification column, gases, and light hydrocarbons (naphtha) accumulate, further down the middle distillates accumulate—including the kerosene fraction, the diesel, and light heating oil fraction. Even further down, gasoil and, at the bottom of the column, the residue are collected.

The rectification takes place at atmospheric pressure and is therefore called atmospheric rectification. The residue is stripped again under vacuum in another column. At low pressure (vacuum), the boiling temperature of the liquids to be separated is reduced which enables vacuum distillation to separate hydrocarbons with high-boiling ranges and high carbon chain length that would tend to crack thermally in an atmospheric distillation at higher temperatures, which is not desired [4].

Hydrotreating. The fractions generated by rectification (naphtha, kerosene, diesel, etc.) still include heteroatom-containing components. The concentration of these heteroatoms must be reduced to the parts per million (ppm) range; otherwise, they could harm subsequent catalytic refining processes (like catalytic reforming) irreversibly or, when the heteroatom-containing fraction is combusted as fuel, form chemical compounds like hydrogen sulfide (H_2S) or nitrogen oxide (NO_x) that harm the environment. Hydrotreating mainly means the hydrodesulfurisation of sulfur-containing components, as sulfur is the most prevalent heteroatom. Next to desulfurization, also denitrification and deoxygenation occur. In addition, olefins are also hydrogenated.

The oil streams to be hydrotreated are mixed with hydrogen and are heated. The hot mixture enters a reactor filled with catalyst. At a temperature between 300 and 400 °C and pressure between 20 and 80 bar, the hydrogen reacts at the catalyst, for instance with the sulfur-containing compounds to form hydrogen sulfide and desulfurized hydrocarbons. Catalysts made of nickel, molybdenum, or cobalt on aluminum oxide carriers are often used. The hydrogen sulfide (H_2S) obtained is converted into pure sulfur in a Claus plant and can be sold as valuable product [4].

Catalytic reforming. The aim of catalytic reforming is to increase the octane number of the naphtha fraction and to produce aromatic hydrocarbons. Importantly, hydrogen is obtained as a valuable by-product, which is used in further refining processes like hydrotreating or hydrocracking. The reforming takes place at approximately 500 °C and 5–40 bar in a moving or swing bed reactor. Bifunctional catalysts are used (platinum-tin or platinum–rhenium on chlorinated aluminum oxide or zeolite acidic support). Typical reactions during reforming are ring closure, dehydration, and isomerization Eqs. (1–3).

$$\text{Ring closure} \quad \text{alkanes and alkenes} \rightarrow \text{cycloalkanes} + H_2 \tag{1}$$

$$\text{Dehydrogenation} \quad \text{cycloalkanes} \rightarrow \text{aromatics} + H_2 \tag{2}$$

$$\text{Isomerization} \quad n\text{ - alkanes} \rightarrow \text{iso - cycloalkanes} \tag{3}$$

The hydrogenation/dehydrogenation reactions preferably take place at the metal centers of the catalyst, while the acid centers catalyze isomerization and ring-closure reactions. An undesirable side reaction is coking through polymerization and dehydrogenation reactions. Coke deposits on the catalyst are removed by burning off the coke and then oxychlorinating the catalyst [4].

Isomerization. Isomerization changes the structure of a hydrocarbon, without changing its molecular mass in order to modify (improve) one or more properties of a fraction. Depending on the hydrocarbon chain length, the isomerization catalyst and technology used are different. For the gasoline fraction, the aim is to improve the octane number (i.e., the antiknock behavior in spark-ignition engines). This effect is achieved by increasing the degree of branching of the molecules. For example, the octane number of n-heptane is zero while its branched isomers are around 80. Isomerization of middle distillate fractions is applied for improving cold-flow properties of diesel fuel and to meet the stringent freezing point specification of jet fuel (−40 °C for Jet A; −47 °C for Jet A-1) [1]. Isomerization of middle distillate alkanes is often performed as beneficial side reaction during hydrocracking. The reaction is carried out at temperatures around 250 °C and—to prevent catalyst deactivation by coking—at moderate pressure of about 15 bar in the presence of hydrogen. Isomerization catalysts used are similar as in catalytic reforming. Due to the milder process conditions compared to catalytic reforming, cracking, and ring-closure reactions are largely suppressed.

Cracking. Cracking is the most important type of conversion for heavy end and residue upgrading. These conversion processes allow to decompose heavy distillates into lighter products and thus better meet the demand structure of the market. Three main types of cracking can be distinguished: thermal cracking, catalytic cracking, and hydrocracking [2]. As the cracking mechanism and feed streams are different, the yield structure is different for each process [1].

- Thermal cracking. Thermal cracking relies on high temperatures above 400–450 °C to deliver enough energy to decompose large hydrocarbons into small ones. Alkanes are easily cracked, whereas aromatics are quite resilient to heat. Visbreaking is a thermal cracking technology, where heavy residual oils are partly cracked in a furnace with moderate residence time at temperatures around 460 °C and 15 bar nominally to lower the viscosity of the feed hydrocarbons. The fractions are subsequently separated by distillation. There is still a high amount of residue oil remaining as one product fraction. Another thermal cracking technology is delayed coking, where heavy residual oils are cracked into to gas, naphtha, gas oil, and coke at temperatures around 500 °C. Sufficient residence time under cracking conditions must be provided resulting in coke as a product. For this purpose, several coke drums are needed into which the feedstock enters after heating in a high-temperature furnace. A coke drum is filled after 18–24 h, and the hot feedstock switches to the next coke drum. Once the coking is complete, the drums are then emptied sequentially and returned to service. The petroleum or "pet" coke with a low ash and sulfur content can be calcined (i.e., heated to high temperature) to make purified coke that is used to manufacture electrodes for industrial purposes. Delayed coking liquid products are chemically highly unsaturated and need to be hydrotreated to reduce alkene and aromatic content before they can be used as a fuel product. Unreacted material is recycled, so there is no residue fraction generated. No catalysts are used in thermal cracking, so it is not a very selective process, fuel fractions have lower quality, and therefore, a further upgrading of these streams is needed [2].
- Catalytic cracking. The development of fluid catalytic cracking (FCC) represents a milestone in refining history. Compared to thermal cracking, the fuel yield and quality significantly improved by using acid catalysts. The key equipment consists of a reactor and regenerator, followed by a fractionation. High molecular weight hydrocarbon feedstocks from distillation or thermal cracking are mixed prior to the fluidized bed reactor with the fine powdered catalyst. The catalyst cracks the long-chain hydrocarbons into lighter fractions. During cracking, some coke is formed on the catalyst. Coked catalyst is sent to the regenerator, where coke is removed via combustion to make heat to drive the endothermic cracking reactions when the catalyst is recycled back to the reactor. Products are mainly high-octane gasoline streams, which are separated by fractionation. All product streams of FCCs still contain heteroatoms (such as sulfur), leading to the development of another cracking technology, the hydrocracking [1].
- Hydrocracking. Hydrocracking is an acid-catalyzed cracking process coupled with a catalytic hydrogenation process. This is possible due to bifunctional catalysts that contain both acid and metal functionalities [1]. The process conditions are at temperatures between 350 and 450 °C and pressures above 100 bar [2].

In the presence of hydrogen, unsaturated hydrocarbons are immediately saturated after cracking. Furthermore, heteroatoms are removed by hydrodeoxygenation, hydrodesulfurization, and hydrodenitrogenation, greatly improving the fuel quality. Another fuel quality improvement is obtained through hydroisomerization, as a higher degree of product branching improves the cold-flow properties of the products [1]. Hydrocracking is very flexible, as product yields are determined by different catalysts and operation modes. Thus, depending on catalyst and process design, mainly gasoline or middle distillates can be produced. In general, high fuel yields can be achieved. Hydrocracking requires significant amounts of hydrogen up to 300 m^3 of hydrogen per m^3 feed (m$^3_{hydrogen}$/m$^3_{feed}$). This is far more than a catalytic reformer can deliver. A separate hydrogen production plant is therefore required to supply a hydrocracker with enough hydrogen [2]. Consequently, in many refineries today, the necessary hydrogen is typically produced from (fossil) natural gas using the steam reforming of methane in dedicated plants within the refinery complex.

Claus process. Through hydrotreating and hydrocracking a significant amount of hydrogen sulfide (H$_2$S) is produced in refineries. In Claus process units the H$_2$S is first oxidized to sulfur dioxide (SO$_2$), then this SO$_2$ reacts with further H$_2$S to produce elemental sulfur that can be sold as a valuable product.

Blending. Blending can be described as mixing of process streams and is one of the final processes in a refinery. Diverse treated process streams and intermediates are blended together in dedicated storage tanks to achieve a certain product specification.

2.3 Conventional Products

As crude oil is composed of a myriad of different components, no refinery produces only a single product. Instead, refineries are typically designed to produce a wide range of products, each tailored to supply a specific market. Main product types are heating and transportation fuels as well as petrochemicals, but also lubricants and intermediates are produced [1]. The final products can be gaseous, liquid, or solid. Respective examples are shown in Table 1 [4].

Heating and transportation fuels act as an energy carrier and the chemical energy is transformed to heat or kinetic energy when the fuel is combusted. Fuels are separated based on their boiling range. Therefore, fuels are always mixtures of hydrocarbons and not pure compounds. The fuel quality is mainly assessed on the performance of the fuel for the intended application. Depending on geography and corresponding fuel specification, fuels can have different qualities, even when they are used for the same application. Lubricants are also a mixture of compounds differentiated based on their viscosity. In contrast to fuels and lubricants, petrochemicals are relatively pure compounds and their quality depends on if there are impurities present [1].

The quantity of each fuel type is mainly determined by its boiling range, which are overlapping and not precisely defined (Fig. 2). Since the goal is to meet the desired

Table 1 Example refinery products by main product types

	Gaseous	Liquid	Solid
Heating/transportation fuels	Methane, ethane, propane, butane	Motor gasoline, kerosene, diesel fuel, light/heavy fuel oil	-/-
Petrochemicals	Propylene, butylene, ethylene	Benzene, toluene, xylene, cyclohexane, methanol	-/-
Lubricants	-/-	Motor oils, hydraulic fluids, turbine oils, gear oils	Greases
Intermediates/ by-products	-/-	Naphtha	Bitumen, sulfur, calcined coke, petroleum coke

fuel specification or application, boiling ranges can vary and are not fixed. This is one reason why operations of a refinery are very flexible and can be continuously optimized.

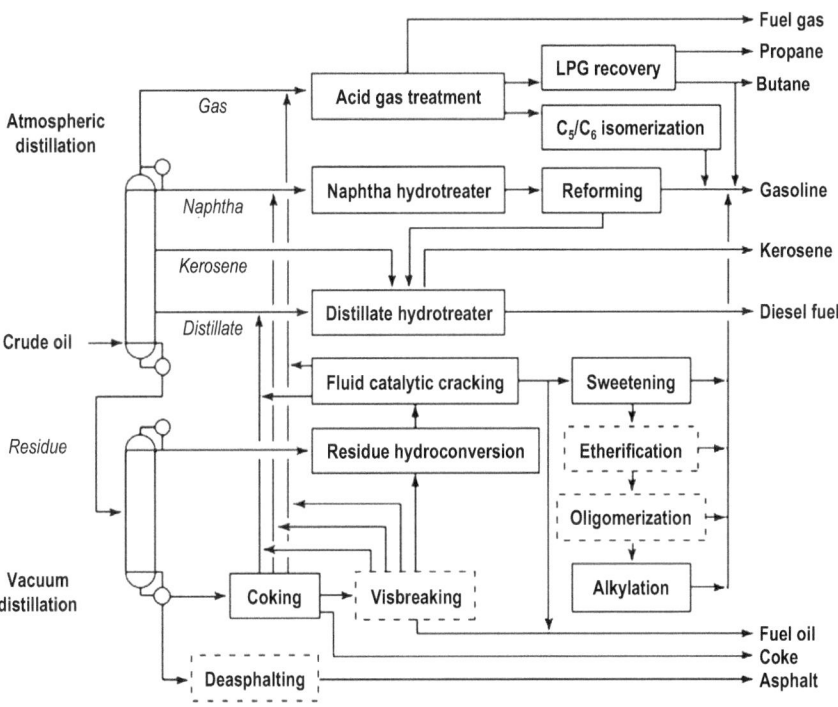

Fig. 1 Exemplary scheme of crude processing in a refinery (LPG: liquefied petroleum gas; Ref. [1])

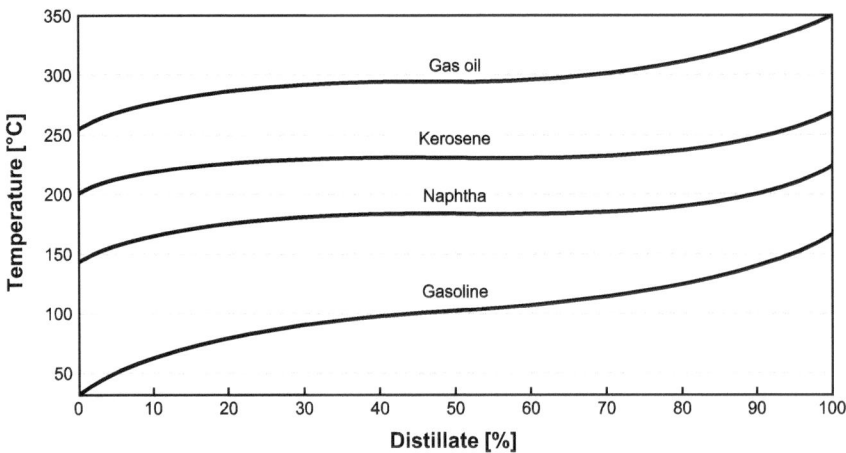

Fig. 2 Distillation curves of different refinery products [5]

There are some boundary conditions related to the boiling range that are fixed by fuel specification, e.g., the European norm EN 590 for diesel requires that at 360 °C at least 95 vol.-% of the diesel must have vaporized. In the end, the refiner must meet the fuel specification of each fuel type; otherwise the product cannot be marketed under that fuel's name [1]. Hence, fuel products are always blends of diverse naphtha or middle distillate streams that can vary depending on the crude oil feedstock and refinery operation modes. The blends are set up to meet all specification requirements. Besides boiling range, several further parameters need to be met, e.g., a maximum sulfur content or a specific density range. Additives are often added in the ppm range to meet specification criteria and to improve combustion. The main liquid transport fuel products are gasoline, jet fuel, and diesel:

- Gasoline. Gasoline, also called "petrol", is used for spark-ignition engines. As specified under EN 228, different grades exist, mainly determined by the octane number. The density of gasoline is specified to be between 720 and 775 kg/m^3. The hydrocarbon range is from approximate C_6 to C_{12} and the boiling range from 70 to 210 °C. Gasoline is blended from several naphtha streams from distillation or cracking and upgrading units.
- Jet fuel. Jet fuel mainly consists of the kerosene fraction and is used as aviation fuel for turbine engines. The boiling range is from 190 to 260 °C with a hydrocarbon chain length from approximately C_{10} to C_{15}. This relative flat boiling range results in a nearly complete and almost residue-free combustion with a well-defined ignition point. Jet fuel is blended from light middle distillate streams plus very few additives to meet the respective specification. Jet fuel specifications apply mostly international and are not limited to a country or a region. In international civil aviation Jet A-1 specification according to ASTM D1655 and DEF STAN 91–91 is used almost exclusively with a freezing point of −47 °C, flash point of about 38 °C, and a density from 775 to 840 kg/m^3.

- Diesel. Diesel fuel is used in compression-ignition engines. The hydrocarbon range is approximately between C_{11} and C_{22} with a boiling range from about 180–380 °C. Diesel consists of a mixture of middle distillate streams coming from desulfurization units or cracking units. Within the EU, the EN 590 standard describes the properties of diesel fuels and defines uniform conditions, like a density between 820 and 845 kg/m^3. For marine diesel, a separate specification exists which allows higher amounts of sulfur and a higher density [1].

3   Refining of Fischer–Tropsch Synthetic Crude

After delving into the production of various products from crude oil and highlighting its properties and refining processes, the refinement and processing of synthetic crude oil derived from the Fischer–Tropsch synthesis is described in more detail in the following. In this context, it is important to note that current Fischer–Tropsch facilities are much smaller than conventional refineries. The largest Fischer–Tropsch gas-to-liquids (GtL) facilities have the size of small conventional refineries with production capacities around 100,000 barrel per day (bbl/d). The first Fischer–Tropsch facilities producing syncrude based on "green" hydrogen and carbon dioxide (CO_2) are currently moving from pilot to demo size with capacities around 1,000 bbl/d. Also due to the limited numbers of industrial Fischer–Tropsch plants, there is no roadmap yet what kind of refining technology to use. It depends on local conditions, the available feed, its characteristics as well as the product requirements [6].

History. The Fischer–Tropsch synthesis was invented in Germany in 1925 by Franz Fischer and Hans Tropsch mainly due to increasing oil prices and geopolitical reasons based on coal as feedstock for the synthesis gas. In the following decades, this did not become economic compared to crude oil-based fuels. Political reasons led South Africa to the development of several Fischer–Tropsch coal-to-liquids (CtL) plants with a capacity of more than 100,000 bbl/d in total with chemical intermediates and liquid fuels as products. First gas-to-liquids (GtL) plants in industrial scale were erected in the nineties as it became economic in some areas to liquefy the stranded natural gas while exploring crude oil and the products are used mainly as fuel blend components [7]. The increasingly visible negative consequences of anthropogenic climate change led to the development of more sustainable synthetic fuels based on biomass, the biomass-to-liquids (BtL) fuels around turn of the millennium. Given the economic advantages of producing crude oil-based fuels and the complexity of the biomass-to-liquids (BtL) process, a significant large-scale industrial adoption of such plants and technologies has not emerged at this time. However, presently this technology is gaining relevance again, especially in the production of renewable kerosene. During the last years, "green" hydrogen and sustainable carbon sources like CO_2 from the atmosphere also gained importance as feedstocks for the production of synthetic power-to-liquid (PtL) fuels.

3.1 Fischer–Tropsch Syncrude as Feedstock

The Fischer–Tropsch synthesis is an established technology that involves various chemical reactions to convert a mixture of carbon monoxide (CO) and hydrogen (H_2), often called "synthesis gas" or "syngas", into a synthetic product that consists a wide range of hydrocarbons. This product is commonly referred to as "synthetic crude oil" or "syncrude". The syncrude is not produced as a single liquid phase; instead, it is a mixture of gas, liquid organics, waxes, and water. Prior to delivering the syncrude to refineries, some upgrading is required in advance: the gaseous fraction is separated, and the water fraction is removed from the liquid and waxy hydrocarbons fraction. If the gaseous fraction is large, it becomes inefficient not to recover or upgrade these products. Therefore, the focus is often to achieve a high carbon chain length probability within the Fischer–Tropsch synthesis to come out with a small share of gases and as many liquids and waxes as possible.

There are three types of Fischer–Tropsch syncrude produced commercially, differing mainly in terms of synthesis temperature and the catalyst used. These three syncrude types are produced by either iron-based high-temperature Fischer–Tropsch synthesis (Fe-HTFT), iron-based low-temperature Fischer–Tropsch synthesis (Fe-LTFT) or cobalt-based low-temperature Fischer–Tropsch synthesis (Co-LTFT). The main products from high-temperature and low-temperature Fischer–Tropsch synthesis are linear alkenes, linear alkanes, methane, and water. Some branched chain hydrocarbons and oxygenates are also produced. The oxygenates are mainly 1-alcohols, although aldehydes, ketones, and carboxylic acids are also included [6]. Table 2 summarizes the typical syncrude composition from the three main syncrude types. These values are illustrated in Fig. 3. As in the case of conventional crude oils, there is a deviation between those different Fischer–Tropsch syncrudes.

Syncrude can also contain dissolved or suspended metals or material from the Fischer–Tropsch synthesis, such as catalyst particles produced by attrition, that could find their way into the syncrude product, if they are not sufficiently separated [6]. Table 3 shows the comparison between the product composition of typical crude oil and syncrudes from Fe-HTFT and Co-LTFT.

There are similarities, but also fundamental differences between crude oil and syncrudes and also between the syncrudes themselves, depending on which production process they were obtained. Whereas water is the major by-product in Fischer–Tropsch synthesis, normally 0–2% water is found in crude oil. Also oxygenates are typical intermediates from the Fischer–Tropsch synthesis, but rarely found in crude oil. On the other hand, crude oil contains several heteroatomic compounds, which will not be found in intermediates from Fischer–Tropsch synthesis, as these contaminants are separated before synthesis to avoid catalyst deactivation. In low-temperature Fischer–Tropsch crude and in crude oil a high content of alkanes prevailed, whereas high-temperature Fischer–Tropsch crude contains less than 10% alkanes, but high amounts of alkenes.

In contrast to fossil crude oil, the composition and the fractions of syncrude can be strongly influenced by and during the Fischer–Tropsch synthesis itself via adjusting

Table 2 Typical syncrude composition from iron-based and cobalt-based Fischer–Tropsch syntheses [1] (Co: cobalt-based; Fe: iron-based; HTFT: high-temperature Fischer–Tropsch synthesis; LTFT: low-temperature Fischer–Tropsch synthesis)

Product fraction	Carbon range	Compound class	Syncrude composition[a] (wt.-%)		
			Fe-HTFT	Fe-LTFT	Co-LTFT
Tail gas	C_1	Alkane	12.7	4.3	5.6
	C_2	Alkene	5.6	1.0	0.1
		Alkane	4.5	1.0	1.0
LPG	C_3–C_4	Alkene	21.2	6.0	3.4
		Alkane	3.0	1.8	1.8
Naphtha	C_5–C_{10}	Alkene	25.8	7.7	7.8
		Alkane	4.3	3.3	12.0
		Aromatic	1.7	0.0	0.0
		Oxygenate	1.6	1.3	0.2
Distillate	C_{11}–C_{22}	Alkene	4.8	5.7	1.1
		Alkane	0.9	13.5	20.8
		Aromatic	0.8	0.0	0.0
		Oxygenate	0.5	0.3	0.0
Residue/wax	C_{22+}	Alkene	1.6	0.7	0.0
		Alkane	0.4	49.2	44.6
		Aromatic	0.7	0.0	0.0
		Oxygenate	0.2	0.0	0.0
Aqueous product	C_1–C_5	Alcohol	4.5	3.9	1.4
		Carbonyl	3.9	0.0	0.0
		Carboxylic acid	1.3	0.3	0.2

[a]The syncrude composition is expressed as the total mass of product from Fischer–Tropsch synthesis, excluding inert gases (N_2 and Ar) and water gas shift products (H_2O, CO, CO_2, and H_2). Zero indicates low concentration and not necessarily a total absence of such compounds

process parameters like temperature and pressure as well as downstream steps like cooling and separation. Nevertheless, similar to fossil crude oil, syncrude must be upgraded or refined in order to produce useful products [6].

3.2 Refining Process Options

The ability to process Fischer–Tropsch syncrude in an existing conventional crude oil refinery depends on the composition of the Fischer–Tropsch syncrude. There can be significant differences between low-temperature and high-temperature Fischer–Tropsch syncrudes, and the technology setup of a refinery plays a crucial role for what kind of products can be efficiently produced. In addition to carbon number

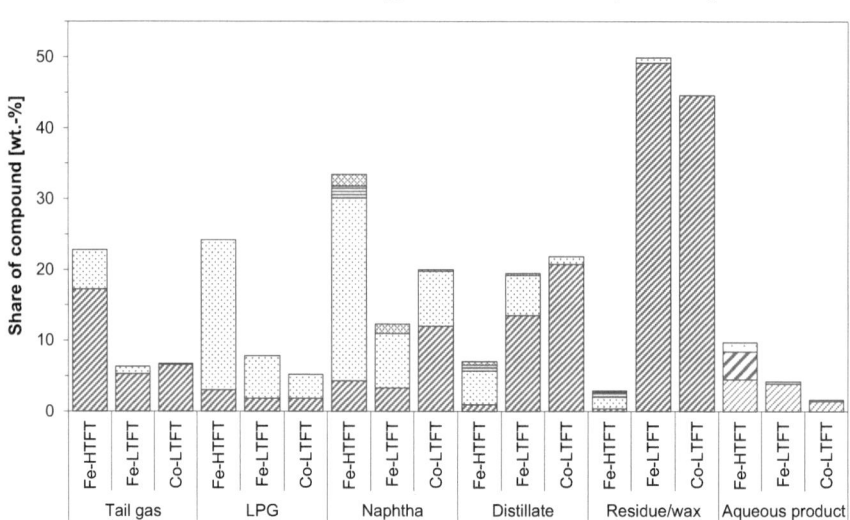

Fig. 3 Typical syncrude composition from iron-based and cobalt-based Fischer–Tropsch syntheses based on Table 2 (Co: cobalt-based; Fe: iron-based; HTFT: high-temperature Fischer–Tropsch synthesis; LTFT: low-temperature Fischer–Tropsch synthesis)

Table 3 Comparison between conventional crude oil and Fischer–Tropsch syncrudes [1] (HTFT: high-temperature Fischer–Tropsch synthesis; LTFT: low-temperature Fischer–Tropsch synthesis)

Compound class	HTFT	LTFT	Crude oil[a]
Alkanes (paraffins)	> 10%	Major product	Major product
Cycloalkanes (naphthenes)	< 1%	< 1%	Major product
Alkenes (olefins)	Major product	> 10%	None
Aromatics	5–10%	< 1%	Major product
Oxygenates	5–15%	5–15%	< 1% O (heavy)
Sulfur compounds	None	None	0.1–5% sulfur
Nitrogen compounds	None	None	< 1% nitrogen
Organometallics	Carboxylates	Carboxylates	Phorphyrines
Water	Major by-product	Major by-product	0–2%

[a] There is considerable variation between different crude oil types, and some crude oils may fall outside the boundaries indicated

distribution, the concentrations of various classes of compounds are important for an appropriate refining. It must also be considered that processing syncrude could be affected by dissolved or suspended material from the Fischer–Tropsch synthesis, and hence, the downstream equipment must be prepared for those contaminants; otherwise, fouling or plugging could occur, leading to lower efficiencies and reduced run lengths. Fischer–Tropsch syncrude has similarities but also differences compared to crude oil what necessitates a different approach to the refining process [6]. From a refining point of view, the differences between Fischer–Tropsch syncrude and crude oil have a significant impact on refinery technology selection, as the syncrudes contain multiple product phases, metal carboxylates, high amounts of oxygenates, high amounts of alkenes, and mostly linear products, but little cyclic compounds [1].

Water needs to be removed prior to refining. Oxygenates should be converted into hydrocarbons before distillation to reduce the corrosivity of the syncrude. As Fischer–Tropsch syncrudes are free of sulfur and heteroatom contaminants as well as free of aromatics, certain upgrading processes are necessary while others are not required at all.

In principle, different process options exists to convert Fischer–Tropsch syncrudes into more valuable products. Fischer–Tropsch syncrude can be fed into existing refinery units as single and pure feedstock. As amounts of available syncrudes are still very limited, a co-processing operation mode can make sense, where syncrude is processed together with fossil crude oil feeds to some percentage. From an economic point of view and regarding capabilities, it is highly advantageous to use existing assets with the suitable infrastructure and utilities setup or to integrate a dedicated upgrading or refining unit into an existing refinery system. When refining syncrudes in an existing refinery, also regulation and certification aspects need to be considered as well as guidelines that provide direction on how such an integration should be carried out.

For the processing of Fischer–Tropsch syncrude, two main approaches are described in more detail in the following. First, the dedicated "stand-alone" processing of syncrude is discussed, followed by its combined processing with crude oil in conventional refinery processes.

3.2.1   Stand-Alone Processing

A dedicated Fischer–Tropsch refinery will look different to a conventional crude oil refinery [8]. Conventional crude oil refineries can be used with Fischer–Tropsch syncrude, but this requires in most cases a pretreatment or modifications to the plant; otherwise, operations are sub-optimal. However, some refinery configurations with appropriate conversion processes can perform very well with syncrude [1]. Of high importance for assessing if a conventional refinery has a suitable setup is the type of Fischer–Tropsch technology used, namely high-temperature Fischer–Tropsch (HTFT) or low-temperature Fischer–Tropsch (LTFT) synthesis, as the hydrocarbon chain length and the composition of such crudes are different.

There are many ways to refine syncrude. For Fischer–Tropsch syncrudes deoxygenation and atmospheric distillation are always required, but no vacuum distillation or delayed coker is needed. A fluid catalytic cracking (FCC) is only required in the case of low-temperature Fischer–Tropsch syncrudes for olefin production to meet gasoline fuel specifications if that represents the desired product [6].

(Wax) Hydrocracking and isomerization. Hydrocracking and isomerization are very important especially for the produced synthetic low-temperature Fischer–Tropsch wax fraction that contains long-chain paraffinic hydrocarbons. Heavy syncrude fractions like Fischer–Tropsch waxes from low-temperature Fischer–Tropsch synthesis with hydrocarbon chain lengths longer than C_{22} can be quite selectively converted to middle distillate components in order to increase the yield of transportation fuels via hydrocracking and improve product quality—especially cold-flow properties—by isomerization. Hydrocracking is the key technology for Fischer–Tropsch syncrude upgrading as it is a combination of cracking, hydrogenation, and often isomerization using bifunctional catalysts. Compared to conventional hydrocracking, there are some differences. Hydrocracking of Fischer–Tropsch waxes requires milder conditions than crude oil hydrocracking as hydrocarbon chains are mainly linear paraffinic, and also, less hydrogen is needed due to the minimal heteroatom content [1, 9, 10]. These differences are outlined in Table 4.

Figure 4 illustrates the hydrocarbon product distribution prior and after hydrocracking, based on low-temperature Fischer–Tropsch (LTFT) syncrude.

Table 4 Process conditions for conventional crude oil hydrocracking, mild hydrocracking and LTFT wax hydrocracking [1] (LTFT: low-temperature Fischer–Tropsch synthesis)

Description	Unit	Conventional hydrocracking	Mild hydrocracking	LTFT wax hydrocracking
Operating conditions				
Temperature	[°C]	350–430	380–440	325–375
Pressure	[MPa]	10–20	5–8	3.5–7
Space velocity	[h^{-1}]	0.2–2	0.2–2	0.5–3
H_2: feed ratio	[$m^3 \cdot m^{-3}$]	800–2,000	400–800	500–1,800
Conversion properties				
Cracking conversion	[%]	70–100	20–40	20–100
H_2 consumption	[wt.-%$_{feed}$]	1.4–4	0.5–1	< 1
Heat release	–	Exothermic	Exothermic	Near isothermal

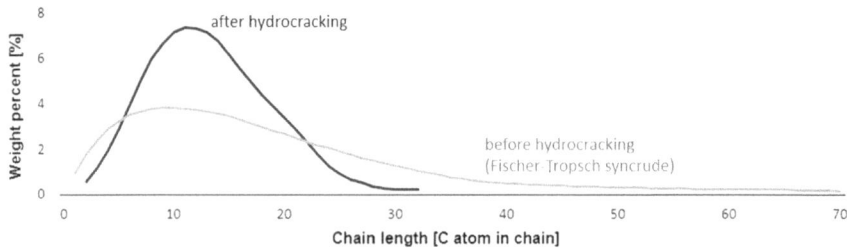

Fig. 4 Product distribution prior and after hydrocracking according to [9]

Via hydrocracking of Fischer–Tropsch syncrudes, the yield of the fuel fractions can be maximized. In trials, yields of 80% middle distillate (kerosene and diesel), 15% naphtha, and 5% light off-gases were generated [11, 12]. Afterward a distillation is required to separate the fuel fractions.

3.2.2   Co-processing

When the amount of Fischer–Tropsch syncrude is limited, a co-processing operation mode in an existing refinery represents an appropriate approach. Via co-processing, the syncrude is processed together with the fossil crude oil feedstock. Prior to the conversion unit, syncrude is mixed with the fossil feed-stream. Instead of running the syncrude through the whole refinery beginning with the distillation, rather feeding it to a dedicated unit like a hydrocracker can make sense. The syncrude can also be fractioned and the fractions can be sent to different units, for example, the naphtha fraction can be directed to isomerization or steam reformer and the wax fraction to the hydrocracker [1].

The reaction scheme in a co-processing mode is similar to a dedicated processing of syncrudes. Disadvantage of such co-processing modes is that the conversion optima from crude oil and its intermediates and syncrude are different, as the composition of both feeds will be slightly different. Thus, a compromise in process conditions needs to be defined. In contrast to a dedicated processing of either Fischer–Tropsch syncrude or conventional crude oil, this results in not ideal yields or quality from the fossil fraction, the syncrude, or both. When the product of co-processing should be a jet fuel, ASTM D1655 needs to be considered, which currently allows for a maximum of 5 vol.-% of co-processed material in the feed to the co-processing unit and also a maximum of 5 vol.-% of co-processed material in the final jet fuel product. However, it is advantageous that co-processing can directly take place as far as logistics and a storage infrastructure for syncrude and a tie-in to the relevant unit are available. This eliminates the need for significant capital expenditure to refine the syncrude to final products.

3.3   Synthetic Fischer–Tropsch Products

It is important to know the properties that are required for the final products as well as the possible syncrude composition. When syncrudes and products are well matched on a molecular level, the refining process becomes more efficient [1]. Both, the synthesis itself and the subsequent upgrading can influence the properties and yields of synthetic fuels, and therefore, these products are also referred to as taylor-made fuels in some cases. However, it is important to note that not all desired products can be obtained from each type of syncrude. Therefore, it is required to assess what is technically feasible, as well as viable from an environmental and economic perspective. The refinery design determines types and quality of the fuels [6].

- Synthetic natural gas (SNG). Synthetic natural gas is mainly methane. Its composition is close to that of conventional (fossil) natural gas. Methane is always produced during Fischer–Tropsch synthesis. Often Fischer–Tropsch facilities include steam methane reformer to turn the methane back to syngas with the aim of increasing the yield of long-chain hydrocarbons. It could also be used as fuel gas, but from an economic and environmental point of view this is relatively inefficient [6].
- Liquefied petroleum gas (LPG). Liquefied petroleum gas refers to an alkane mixture of C_2–C_4 components. It requires moderate pressures to remain in a liquid state, which is an enormous advantage, and can be directly used as fuel for road transportation in combination with appropriate LPG engines and powertrain systems in the vehicles. The composition depends on the region where the LPG is used; the higher the prevailing temperature (hotter climate conditions), the higher is the content of butane (C_4H_{10}) as it is the component with the highest boiling point in that mixture. Furthermore, C_2 to C_4 alkenes could have value as petrochemical components [6].
- Gasoline. Gasoline is one of the most refining-intensive fuels to produce as it requires a lot of molecular transformation, e.g., to achieve a high octane number, the straight-chain alkanes need to be transformed partly into branched molecules. The European specification to meet is EN 228, and it is easier to achieve it from high than from low-temperature Fischer–Tropsch syncrudes. As it is very hard to achieve all specification parameters like the minimum research octane number (RON) of 95, a density between 720 and 775 kg/m^3, a vapor pressure of a maximum of 60 kPa, and a maximum alkene content of 15 wt.-% [6] from the Fischer–Tropsch naphtha fraction, this distillate fraction is often used as blend component or petrochemical intermediate.
- Jet fuel. Jet fuel, also often referred to as "kerosene" or "aviation turbine fuel", is relatively easy to refine from Fischer–Tropsch syncrude as the properties of the refined Fischer–Tropsch derived synthetic jet fuel are within the range of jet fuel from conventional crude oil refining. Here it is easier to produce jet fuel in high yields from low- than from high-temperature Fischer–Tropsch synthesis [6]. Synthetic jet fuel produced from Fischer–Tropsch processes has undergone intensive and successful testing and continues to be the subject of various research activities. Diverse specifications can be met like Jet A/A-1 or military jet fuel specifications. When producing refined Fischer–Tropsch derived synthetic jet fuel in a stand-alone unit, the respective requirements of the ASTM D7566 (Annex 1) need to be met. It cannot be used as final fuel according to ASTM norms and must be blended up to 50% with conventional fossil jet fuel that is certified under ASTM D1655, and the final mixture needs to meet ASTM D1655 as well. If the Fischer–Tropsch syncrude-based jet fuel is used in a co-processing approach, the respective requirements of ASTM D1655 for the final jet fuel apply. According to that ASTM norm currently up to 5 vol.-% of synthetic kerosene feedstock are allowed.

- Diesel fuel. There is a difference in produced distillates from high and low-temperature Fischer–Tropsch syncrude refining. Distillates from high-temperature Fischer–Tropsch syncrudes are partly aromatic and similar to crude oil-based diesel fuel, whereas distillates from low-temperature Fischer–Tropsch syncrudes are low in aromatics but highly paraffinic with a high amount of linear alkanes, which leads to a very high cetane number and a low density around 770–780 kg/m^3. In regions where the minimum density of diesel fuel is regulated (e.g., in Europe), a low-temperature Fischer–Tropsch derived distillate cannot qualify for sale as a diesel fuel and can just be used as blending component [6] to meet in the end the specification EN 590 with a required minimum density of 820 kg/m^3.

Based on Fischer–Tropsch syncrude also lubricants and petrochemicals can be produced.

4 Developments in the Years to Come

There are still very limited amounts of sustainable Fischer–Tropsch syncrudes available, which are produced based on "green" hydrogen and a corresponding, alternative carbon source, mostly in the form of carbon dioxide (CO_2). For sure, there is the need to bring more demo plants and also industrial plants online and to integrate the syncrude processing into appropriate refineries. Clear and stable long-term policies and regulations play a crucial role to facilitate this. When processing units in existing refineries are too large for a pure processing of (smaller and limited) amounts of Fischer–Tropsch syncrudes, dedicated new small units or co-processing represent suitable approaches to refine the syncrude into valuable market products.

Regarding synthetic jet fuel there is the need for increasing the allowed co-processing percentage above 5 vol.-% by adjusting the ASTM D1655—which is to be expected in the years to come. In addition, the common processing of several alternative syncrudes and biomaterials to produce sustainable jet fuel is still not allowed under the current ASTM specification but would help to create a market ramp-up during the next years. Certification processes for such synthetic products are still underway and need to be clearly and pragmatically defined.

References

1. de Klerk A (2011) Fischer-Tropsch refining. Wiley-VCH-Verlag
2. Harms KP (1989) Das Buch vom Erdöl. Verlag Reuter und Klöckner
3. Speight JG (2014) The chemistry and technology of petroleum. CRC Press Inc.
4. Lumitos AG (2023) Erdölraffinerie. Online posting, internet: www.chemie.de/lexikon/Erdöll raffinerie.html
5. Francis W, Peters MC (1980) Petroleum oils—preliminary treatment and distillation. In: Fuels and fuel technology. Elsevier. Article online posting, Internet: www.sciencedirect.com/topics/engineering/fractional-distillation

6. Maitlis PM, de Klerk A (2013) Greener Fischer-Tropsch processes for fuels and feedstocks. Wiley-VCH Verlag
7. Schulz H (1999) Short history and present trends of Fischer-Tropsch synthesis. In: Applied catalysis a: general, vol 186. Elsevier, pp 3–12
8. de Klerk A (2008) Fischer-Tropsch refining: technology selection to match molecules. In: Green chemistry, cutting-edge research for a greener sustainable future, vol 10, no 12. RSC Publishing
9. Eilers J, Posthuma SA, Sie ST (1990) The shell middle distillate synthesis process (SMDS). Catal Lett 7
10. Lappas AA, Voutetakis SS, Drakaki N, Vasalos IA (2004) Production of transportation fuels from biomass. Published by Chemical Process Engineering Research Institute and Center for Research and Technology Hellas, Thessaloniki
11. Kaneko T, Derbyshire F, Makino E, Gray D, Tamura M (2005) Coal liquefaction. Ullmann´s encyclopedia of industrial chemistry. Wiley-VCH Verlag. Article online posting, Internet: www. mrw.interscience.wiley.com
12. Dry ME (2001) High quality diesel via the Fischer-Tropsch process—a review. J Chem Technol Biotechnol 77:43–50

Electricity-Based Methanol

Status and Prospects of Methanol as a Power-To-Liquid Product

Stefan Bube, Steffen Voß, Nils Bullerdiek, Ulf Neuling, and Martin Kaltschmitt

Abstract Methanol is one of the most widely used basic chemicals, with increasing demand driven by olefin production, especially in Asia. Traditionally produced from fossil fuels, particularly natural gas and coal, methanol production contributes significantly to global CO_2 emissions. Transitioning to renewable feedstocks, such as electricity-based methanol (e-methanol) via power-to-methanol (PtM) pathways, is critical for achieving carbon neutrality. This chapter explores current market developments, technical state-of-the-art, as well as technological and economic implications of a transition toward power-based methanol production. Despite mainly economic challenges, PtM offers a pathway to defossilize the chemical industry and transport sector, potentially increasing methanol demand in the future.

Keywords Methanol · Power-to-methanol (PtM) · E-methanol · Methanol marke · Synthesis · Techno-economics · Mixed alcohol synthesis

1 Introduction

Methanol is a widely discussed molecule, especially in the context of "green" power-to-liquid products, to mitigate anthropogenic climate change. Methanol (CH_3OH) is the simplest aliphatic alcohol and, despite its small molecular size, liquid under ambient conditions. This alcohol is already produced commercially on a large scale but almost exclusively using fossil fuel resources [1]. Today methanol plays a central role as a basic chemical for numerous applications within the industrial sector. Beside the use as a raw material, methanol contributes more and more as a fuel within the transportation sector throughout the last decades.

To transform the global economy toward sustainability—i.e., particularly concerning GHG emission reduction to achieve GHG neutrality in the years to

S. Bube (✉) · S. Voß · N. Bullerdiek · U. Neuling · M. Kaltschmitt
Hamburg University of Technology (TUHH), Institute of Environmental Technology and Energy Economics (IUE), Hamburg, Germany
e-mail: stefan.bube@tuhh.de

come—the production of methanol must be shifted from the use of fossil resources toward a provision based on renewable sources of energy. Thus, such a "renewable" or "green" methanol is assumed to extend its cross-sectoral role as a basic chemical for almost all carbon-based products and applications.

The necessary transformation of conventional, fossil fuel-based methanol production to renewable methanol provision for GHG reduction reasons can be performed based on different feedstock options and process routes. Depending on the used feedstock, a distinction can be made between biomass-based methanol (bio-methanol) and electricity-based methanol (e-methanol). Due to the a priori limited potential of sustainable biomass, future supply capabilities of bio-methanol will most likely be restricted. Thus, a considerable part of the renewable methanol required to reduce GHG emissions to net zero needs probably to be provided from electricity-based methanol produced within power-to-methanol (PtM) processes. The predominant technical, economic, and environmental differences between electricity-based and conventional, fossil fuel-based methanol production lie in the origin of the synthesis gas fed into the methanol synthesis. While conventional synthesis gas production is mainly based on the utilization of carbon from natural gas and coal, electricity-based synthesis gas production predominantly consists of individual provision of hydrogen from "green" electricity and "sustainable" carbon, e.g., from recent biomass or from ambient air.

Furthermore, the transformation of the synthesis gas supply might lead to technical and economic changes related to the methanol synthesis and subsequent purification as well as the overall production concept. These changes can mainly be attributed to adaptions in synthesis gas composition, production capacities, modes of operation, and overall process integration as well as the resulting change in production cost. Besides such production level aspects, large-scale electricity-based methanol production also impacts the systemic level. As an easily transportable molecule with a globally existing logistics infrastructure, methanol can contribute to make energy from regions with a promising energy supply from renewable sources of energy (e.g., high wind speeds, high solar radiation intensity, availability of geothermal resources or hydro power) available worldwide and relieve energy systems based on volatile energies with system-serving PtM processes. However, the conversion of conventional methanol production requires significant amounts of renewable energy and "sustainable" carbon to be provided in line with energy system transformation.

Given the previously outlined backgrounds, this paper elaborates on the technical, economic, and environmental aspects of electricity-based methanol production. To contextualize the evolved economic role of methanol, a brief description of the historical development of methanol production is provided. This is followed by an overview of fundamental properties of methanol and current methanol market aspects as well as resulting demands and prices. Then, a technical description of electricity-based methanol production is given, serving as a basis for the subsequent discussion on its technical, economic, and environmental implications. After a summarization and concluding remarks an excursus on mixed alcohol synthesis is given, which is additionally described due to its technical proximity to methanol synthesis.

2   Historical Development

Methanol as a pure substance was first obtained in 1661 by R. Boyle and named "adiaphorus spiritus lignorum" [2]. Boyle produced methanol by rectifying raw wood vinegar (primarily acetic acid) over lime milk. More than 150 years later, in 1835, based on the independent determination of the molecular composition of methanol from J. von Liebig and J. B. A. Dumas, the term "methyl" was introduced into chemistry [3]. However, until the end of the nineteenth century, the chemical industry had been dominated by inorganic chemistry and the chemistry of aromatics based on coal tar. In contrast, the chemistry of aliphatic components (i.e., non-aromatic hydrocarbons), mainly based on methanol as a feedstock, only played a minor role [4]. During this time, methanol was obtained almost exclusively by dry distillation of wood, which is why it was also called "wood alcohol". The so-produced methanol was used mainly for lighting, cooking, and heating purposes [5].

In 1913, as part of development work on ammonia synthesis, the first synthesis of oxygenated organic compounds, including methanol, was shown by A. Mittasch. The synthesis reaction was obtained from carbon monoxide and hydrogen on an iron oxide catalyst. By improvements of the catalysts especially in the 1920s, primarily by enabling a higher sulfur-resistance, coke-based synthetic methanol production was further developed up to its first commercial implementation in 1923. The reaction was realized at 320–450 °C and 250–350 bar and heterogeneously catalyzed on zinc oxide-chromium oxide ($ZnO-Cr_2O_3$) [3, 6]. This was the first time that the synthesis of a fuel directly from coal was achieved on an industrial scale and the first technical process of synthesis gas chemistry. In the following years, the production capacity was increased to 100,000 t/a and reached 200,000 t/a during World War II [4]. Among other things, through the new production pathways for methanol and the associated coke-based C1-chemistry, wood chemistry lost its importance and was ruled out by petrochemistry.

In the following decades, ongoing research into synthesis gas generation and catalyst technology enabled novel methanol synthesis concepts to reduce the energetic and constructional effort. In the 1960s, a methanol synthesis process has been developed to be realized at much milder reaction conditions of 200–300 °C and 50–100 bar [3]. The process was fed with sulfur-free synthesis gas with a high proportion of carbon dioxide and supported by a novel, highly selective copper, zinc, and chromium catalyst. Based on this low-pressure methanol process concept, the focus of the ongoing further development was predominantly put on the design of the reactor concepts to increase production capacity and yields.

The increasing global demand for methanol, especially during the last two decades drove its production continuously toward larger plant concepts. This led to a reduction in specific capital expenditure (economy of scale). However, the resulting large-scale production of methanol in plants with annual production capacities in the million ton scale (more than 5,000 t/d), built mainly in the past 15 years, is still based on the concept developed in the 1960s [3]. Today, the global demand for methanol exceeds 100 Mt/a and is expected to increase further [1]. Table 1 summarizes the main historical milestones of methanol.

Due to the finite nature of the fossil hydrocarbons currently serving as the feedstock for methanol production, and particularly due to the need to reduce fossil fuel-based CO_2 emissions throughout the methanol life cycle, current research has shifted its focus toward "renewable" methanol production und utilization. This essentially includes the defossilization of methanol production on the one hand and its cross-sectoral application as a potentially renewable and sustainable "green" carbon and energy carrier on the other hand. As part of these developments, the first plant for producing renewable, non-biogenic methanol was successfully commissioned in Iceland in 2011 [4]. From a production perspective, besides biomass-based production, methanol shows preferable properties as a power-to-liquid (PtL) product, which might become one of the predominant production pathways in the future.

On the application side, initial thoughts on methanol as a cross-sectoral and potentially renewable economic key component were developed as early as the 1960s in the so-called "Methanol Economy" [4]. While coal was the predominant energy and raw material source at the beginning of the nineteenth century, oil and gas gained importance in the second half of the century due to their simpler handling and use. Given the limited availability of fossil raw materials, especially crude oil and natural gas, and the wasteful use of chemical-grade carbon chains from oil to provide energy rather than for use in chemistry, F. Asinger—a former professor and pioneer in petrochemistry—was the first to present his thoughts on a methanol-driven economy. In the 1960s, he proposed that after the exhaust of oil and gas, methanol from coal, and later based on biomass and carbon dioxide, could build the basis for the economy by replacing crude oil as a chemical building block, as a fuel and as an energy carrier [4]. These thoughts and concepts were subsequently taken up and developed further by G. A. Olah—a professor of chemistry and Nobel Prize winner in 1994—who

Table 1 Historical milestones of methanol

Time	Milestone
Approximate 1000 B.C	First known use of methanol for embalming by the ancient Egyptians in a mixture of substances obtained from the pyrolysis of wood
1661	First-time extraction of pure methanol from wood distillation
1835	Determination of the molecular composition of methanols molecular composition and introduction of the term "methyl" into chemical nomenclature
1913	First synthesis of oxygenated organic compounds, including methanol from carbon monoxide and hydrogen
1923	Start of commercial methanol production from synthesis gas
1963	First patenting of a low-pressure methanol (LPM) process
2011	Completion of the "George Olah Plant" for the world's first production of renewable, non-biological methanol
2020	Total global demand for methanol exceeds 100 Mt/a

published the first edition of the book "Beyond Oil and Gas: The Methanol Economy" in 1986 [7, 8]. Despite the prevailing reliance on crude oil and natural gas within the global economy, the renewable "green" methanol economy is gaining attention as a solution for the urgent need to defossilize global economy and to fulfill the goals of the Paris Agreement. In this context, electricity-based methanol production via power-to-liquids (PtL) processes is a promising option for storing and transporting renewable energies on the one hand and for defossilizing the existing methanol market on the other.

3 Properties and Handling

Methanol (CH_3OH) is the smallest and first representative of the homologous series of alcohols, consisting of one carbon atom, three hydrogen atoms, and a hydroxyl group (OH-group). Methanol, also named methyl alcohol, methyl hydroxide, carbinol, or wood alcohol, is a colorless, pleasant to penetrative smelling, neutral, and polar liquid. Due to its flammability (flash point 12.2 °C, ignition temperature 470 °C) and toxicity, methanol requires appropriate handling to avoid safety problems. Therefore, international guidelines exist for safe handling, storage, and explosion protection [9, 10]. One of the most significant properties of a methanol molecule is that it exists as a liquid under standard conditions. Figure 1 shows the structural chemical formula of methanol (Fig. 5, left) and its three-dimensional configuration (Fig. 5, right).

In the following, an overview of methanol's physical and toxic properties is given. This is followed by a brief overview of current handling standards and the logistics of methanol.

3.1 Properties

Due the fact that methanol has a long history, the physical and toxic properties are widely known today.

Physical properties: A methanol molecule has a mass of 32.04 g/mol and thus consists of 50wt.% oxygen. The hydroxyl group is the functional group of the

Fig. 1 Molecular structure of methanol

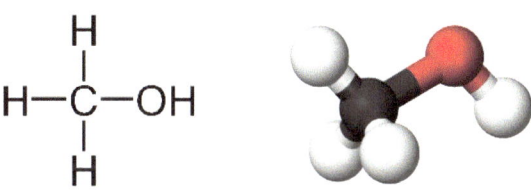

molecule. This OH-group is mainly responsible for the polarity and versatile reactive properties. Due to the polarity of methanol, it can be mixed as a protic solvent with water, most polar solvents, and to a limited extent with oils and greases. The hydrogen bonds formed by the hydroxyl group are decisive for the liquid aggregate state of methanol under standard/ambient conditions. Therefore, methanol is liquid over a wide temperature range from -97.6 (freezing point) to $64.6\,°C$ (boiling point). The critical point lies at around 81 bar and $240\,°C$, where the critical density is about $276\ \text{kg/m}^3$.

Methanol has a gravimetric energy density of $19.9\ \text{MJ}_{\text{LHV}}/\text{kg}$ and $22.7\ \text{MJ}_{\text{HHV}}/\text{kg}$ related to the lower (LHV) and the higher heating value (HHV), respectively. Considering the density of $786.7\ \text{kg/m}^3$ at $25\,°C$, the volumetric energy density is $15.7\ \text{GJ}_{\text{LHV}}/\text{m}^3$ and $17.9\ \text{GJ}_{\text{HHV}}/\text{m}^3$. A comparison of the gravimetric and volumetric energy densities of methanol against other energy carriers is shown in Fig. 2.

Methanol can be used directly as fuel in fuel cells, turbines, and stoves. Due to its high octane number (RON 109), methanol can also be used in internal combustion engines as an additive or even as a gasoline substitute. The use in modified diesel engines is associated with a significant adaptation effort but can also be realized in theory (the cetane number of methanol is < 5 and of "classical" diesel fuel ca. 51) [11, 12].

A methanol molecule contains inherent oxygen, reducing the gravimetric energy density. Compared to gasoline and diesel, methanol has a more than three times higher latent heat of vaporization, providing an extra cooling effect. This contributes to a higher efficiency in methanol-fueled engines compared to gasoline or diesel-fueled engines. However, in terms of range, this effect can only compensate for a small part of the increased fuel demand caused by the significantly lower heating values compared to gasoline and diesel. Therefore, the use of pure methanol requires adjusted tank sizes if similar ranges are to be achieved. The poor lubricity and the

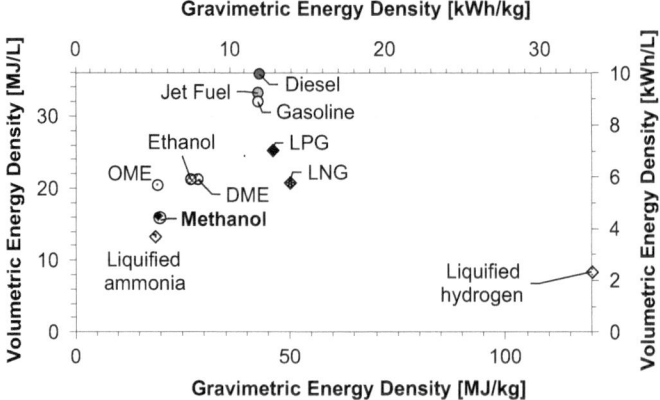

Fig. 2 Volumetric and gravimetric energy density of different energy carriers (LNG: liquefied natural gas, LPG: liquefied petroleum gas, OME: oxymethylene ethers, DME: dimethyl ether)

lower viscosity of methanol relative to conventional fossil fuel-based liquid fuels also require further engine adjustments and/or fuel additives.

Methanol can also be used in cookstoves since it burns without soot, fumes, or odor production [1]. In addition, there are various ways of further processing methanol to alternative and conventional fuels like dimethyl ether (DME) or oxymethylene ethers (OME) and gasoline or diesel.

Toxicology: Methanol can be formed by biological processes in nature and thus it is present within our natural environment. Here, it mainly occurs in plants like cotton, grasses, and fruits as well as a by-product of alcoholic fermentation (taking place also in nature, e.g., in overripe fruits). Therefore, alcoholic beverages, predominantly fruit spirits, may contain methanol to a certain extent. Due to the negative effect on organisms described below, the permissible limit is far below 2vol.% in beverage alcohol based on globally accepted standards. Methanol's environmental toxicology presents a lower risk compared to other commonly used fossil carbon-carriers or fuels such as gasoline or diesel. Methanol is completely miscible with water and shows high mobility when it enters an aqueous environment. It is readily biodegradable and shows no bioaccumulation. Diffused methanol in the atmosphere has an estimated lifetime of about 12 days before oxidizing to carbon dioxide and water. Therefore, it does not persist in the environment. Even though methanol maybe present in very low concentrations in the environment and within selected food materials, and thus also within the human body, it can have severe negative effects on living organisms, even in slightly excessive quantities. Regarding the toxicology for humans, methanol can be absorbed by oral ingestion, inhalation, and skin or eye contact. The lethal dosage for human adults is between 20 and 100 mL, while 5–15 mL can already lead to irreversible blindness. Metabolism of methanol in humans, which leads to poisoning, occurs via the degradation into formaldehyde and formic acid, leading to blindness and even death. Unlike most petrochemical fuels, methanol does not cause cancer from long-term exposure [4, 13].

The previously described and other properties of methanol are summarized in Table 2 [4, 13].

3.2 Handling

Due to its physical and toxicological properties, methanol is characterized by specific handling and logistic requirements. However, these requirements are similar to those of other frequently used (liquid) raw materials and fuels, i.e., it can be handled almost like any other conventional liquid fuel (e.g., gasoline) [16]. Since it has been produced, stored, and transported on a large scale for decades, there are comprehensive regulations allowing for safe handling [13]. A purchase and use of methanol by private end-users is severely restricted within the EU and many other industrialized countries. Methanol production and global trade are mainly carried out on

Table 2 Compilation of important methanol properties (based on [3, 4, 14, 15])

General properties	Value	Unit
Molecular weight	32.042	kg/kmol
Density		
At 20 °C	791.5	kg/m^3
At 25 °C	786.7	kg/m^3
Boiling point (1 bar)	64.6; 337.8	°C; K
Vapor pressure		
At 20 °C	0.128	bar
At 25 °C	0.170	bar
Freezing point	−97.6; 175.5	°C; K
Critical pressure	81 ± 1	bar
Critical temperature	240; 513 ± 1	°C; K
Critical volume	0.116 ± 2	m^3/kmol
Critical density	276 ± 5	kg/m^3
Enthalpy of evaporation		
At 20 °C	37.4	kJ/mol
At 64.6 °C	35.2	kJ/mol
Enthalpy of combustion (gas)	763.7	kJ/mol
Enthalpy of formation (gas)	−205	kJ/mol
Constant pressure heat capacity, cp		
Gas	44.1	kJ/(kmol K)
Liquid at 25 °C	81.1	kJ/(kmol K)
Fuel properties		
Lower heating value (LHV)	19.9	MJ/kg
Higher heating value (HHV)	22.7	MJ/kg
Cetane number (CN)	< 5	–
Research octane number (RON)	109	–
(Auto) ignition temperature	470.1; 743.2	°C; K
Heat of vaporization (1 bar)	1,089	kJ/kg
Stoichiometric air/fuel ratio	6.5	–
Peak flame temperature (1 bar)	1,890; 1362	°C; K
Explosion limits	6.7–36.5	vol%
Flash point	288.8; 15.6	K; °C
Viscosity (liquid, 25 °C)	0.544	mPas
Explosion group	II B, T1	
Toxicity properties		
Lethal dosage (adult human)	20–100	mL

(continued)

Table 2 (continued)

General properties	Value	Unit
Odor threshold	2,000	ppm
MAC [a]	200	ppm

[a] MAC: maximum permissible average concentration of an agent in the form of gas, vapor, or suspended matter in the air at the workplace

an industrial scale and thus based on a specialized/adapted infrastructure by trained personnel.

Since methanol is a liquid, it is basically easy to store and transport. This contributes to the fact that about 80% of the methanol traded is transported transcontinental between producer and user, mainly by ship. Therefore, a distinctive infrastructure for large-scale chemical transport already exists today. Methanol is received and stored in marine terminals and transshipped onto a truck, rail, or barge for subsequent distribution to production plants for further processing and bulk distributors, where it is stored in industrial tank farms. Typically, for short to medium distances and small to large capacities, pipeline grids are the preferred option, while trucks and trains are suitable for medium to long distances and small to medium capacities. Tankers are commonly utilized for large-capacity transportation over long distances, especially for oversea transport. Standard materials that are appropriate for methanol contact are carbon or stainless steel or, especially for mobile applications, aluminum alloys or fiberglass. Above-ground tanks equipped with floating roofs are often used. Tanks must be grounded to avoid static discharge hazards. Overlaying with inert gas pads is common. Tanker trucks and trailers complete the distribution network, delivering methanol to a wide range of final users in the methanol value chain [13].

Particular caution is required when handling methanol, mainly due to its high flammability and toxicity. Methanol is a flammable liquid that can burn in the air and become explosive within a wide concentration range. In case of ignition, methanol can be extinguished using water. However, it is completely miscible with water retaining its flammability even at high water concentrations exceeding 75 vol.-%. Therefore, precautionary measures such as implementing explosion protection zones or employing inert gases for blanketing methanol tanks are necessary to prevent ignition or explosion. Additional measures, such as extra ventilation, must be taken when storing methanol in sealed areas. This is because the molecular weight of methanol vapor (32 g/mol) is only slightly greater than that of air (28 g/mol). If methanol is released in enclosed spaces, such as aboard ships or in warehouses, methanol vapor can migrate close to the ground and accumulate in confined and low-lying areas.

4 Methanol Markets

The following section details the global production and demand development of methanol as well as the associated areas of application. This is followed by a regional breakdown of methanol production and demand aspects, concluding with a brief overview of recent methanol price developments.

4.1 *Global Production and Use*

Today, methanol is one of the most important basic chemicals globally. The major share of this methanol is used within the chemical industry (approximately 70%). Additionally, its importance as a fuel is growing, above all in China [1, 3]. The annual global usage volumes of methanol, which basically also represent the production volumes to a large extent, are shown in Fig. 3. In addition, the production capacity installed worldwide and the annual load factor of the plants are shown. The latter results from the ratio of annual production to production capacity. According to Fig. 3, the annual global production volumes of methanol were relatively constant in the early 2000s. A sharp rise began in the mid-2000s; since then (i.e., in the last 15 years), methanol production increased worldwide by a factor of 2.5. Production capacity increased even more strongly during this period reducing the overall average plant utilization. While the annual load factor was still over 90% at the beginning of the 2000s, today's production plants are operated only at an annual load factor of around 65% [1, 17]. Possible reasons for the resulting overcapacity include high expectations of the future market, economy of scale effects and avoidance of market share losses by individual manufacturers.

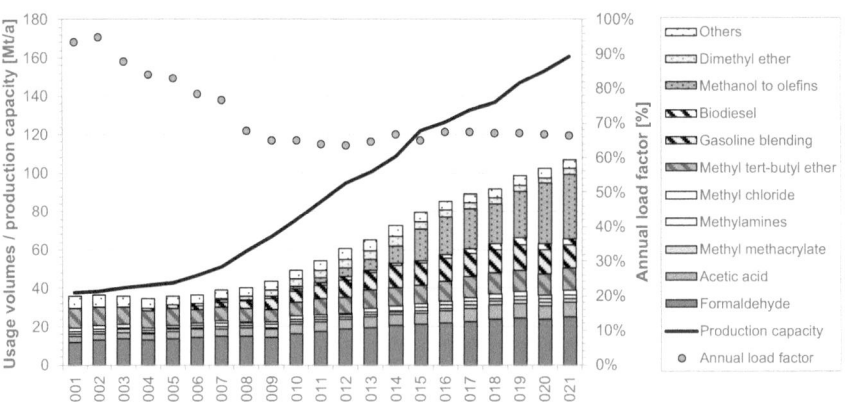

Fig. 3 Global methanol usage volumes and production capacity (based on [1, 17])

With regard to the development of methanol utilization applications, the usage volumes were nearly constant from the years 2001–2007. Methanol was mainly used as a basic chemical for producing formaldehyde and methyl chemicals, especially methyl tert-butyl ether (MTBE) used within the fuel market as an additive. While the use of methanol for these products has remained relatively constant or only increased slowly until today, the intensified use of methanol as a fuel component or as a feedstock for fuel production began shortly before the 2010s. This development was mainly driven by China's will to reduce dependence on crude oil imports by using methanol produced from domestic hard coal resources [1, 18]. The same can be seen in the conversion of methanol via methanol-to-olefin (MTO) processes for the production of ethene and propene. This use increased sharply in the last years and still continues until today.

The methanol use can be classified by its use in different economic sectors, as it is visualized by pattern in Fig. 3. In the chemical sector (blank), methanol is used as a molecular building block for further processing into higher-value products. Within the transportation sector (striped), methanol can be used directly as a fuel (so-called "M100" or "M85" in China or "M3" in Europe)[1] or for the production of fuels and additives.[2] However, this classification cannot be unambiguously made for all purposes. For example, olefins from methanol-to-olefin (MTO) processes can be used as a feedstock for the chemical sector (e.g., to produce plastics) or for further processing into fuels. The same applies to dimethyl ether (DME), which can also be used as a diesel substitute and, among others, for dimethyl sulfate production in the chemical industry. These uses, which cannot be clearly assigned, are shown in the diagram as dotted areas.

The future trends and specifics of methanol production and use in the global economic context are inherently unpredictable. Yet, the following two aspects could influence this development significantly.

- The growth of conventional methanol production is likely to continue depending among others on the overall development of the global economy and the prices of fossil feedstock—especially the price ratio of natural gas to hard coal on the production site and future crude oil prices on the demand side This growth will likely primarily rely on the existing production and distribution infrastructure for methanol.
- To meet global climate targets, a significant defossilization across all economic sectors is necessary. This is likely to also impact the current methanol production, steering a shift from conventional production to an increased production of "green"

---

[1] M100/85 means 100/85% methanol share (approved in Chinese gasoline), M3 means 3% methanol share (approved in European gasoline).

[2] For the production of fatty acid methyl esters (FAME) via transesterification, conversion into gasoline via methanol-to-gasoline (MtG) or the production of the additive methyl tert-butyl ether (MTBE).

methanol and expanded use in various sectors[3]; climate-friendly "green" methanol has the potential to support defossilization efforts beyond current applications and can be widely adopted as a fuel, energy carrier, or medium for energy storage.

Therefore, if a global adherence to the agreed-upon climate targets is assumed, also a rising demand for sustainable "green" methanol can be anticipated in the coming years.

4.2   Demand and Production by Region

Due to the simple logistical handling of methanol, the globally well-developed logistic infrastructure, and worldwide production overcapacity, the methanol market is very flexible and price-driven [13]. This implies that regional production is particularly dependent on site-specific feedstock prices and only to a certain extent on the local demand. In addition, global efforts to achieve a climate-neutral economy will most likely also have a certain impact on the global demand for methanol in the coming decades and its local origin. Against this background, the following section first analyzes current demands and then production volumes by region over time.

Demand: Currently, in most cases, the geographical locations of methanol demand and production are far apart. As the logistics of methanol (i.e., intercontinental transportation and intermediate storage) is rather simple, the availability of inexpensive feedstock options and/or governmental support is the decisive factor in the choice of the respective production location, rather than the transport distance to be overcome (transport of a liquid bulk good is typically relatively cheap).

Figure 4 shows the development of the overall methanol demand by region from 2001 to 2020. The largest consumers of methanol are located in regions with high industrial activities, such as Asia (China, Japan, Taiwan, and South Korea), Western Europe, and North America. Beyond that, methanol is consumed in smaller quantities almost all over the world. As depicted in Fig. 4, the substantial increasing demand in methanol primarily comes from Northeast Asia—and here especially from China. In this region, methanol consumption now significantly outweighs consumption in the rest of the world, which remained relatively stable in the years shown [18].

Production: Methanol is mainly produced in regions with large deposits of inexpensive feedstock (such as natural gas and hard coal) or in industrialized areas that can be supplied with these feedstocks at low costs. The easily accessible deposits of fossil fuel resources are often located in less industrialized regions with resulting relatively low local methanol demands; this is most likely the reason why approximately 80%

---

[3] In this context, the increased use of methanol in China is already an example of replacing crude oil-based products by products based on methanol. However, in this case the production of methanol is still based on fossil feedstock.

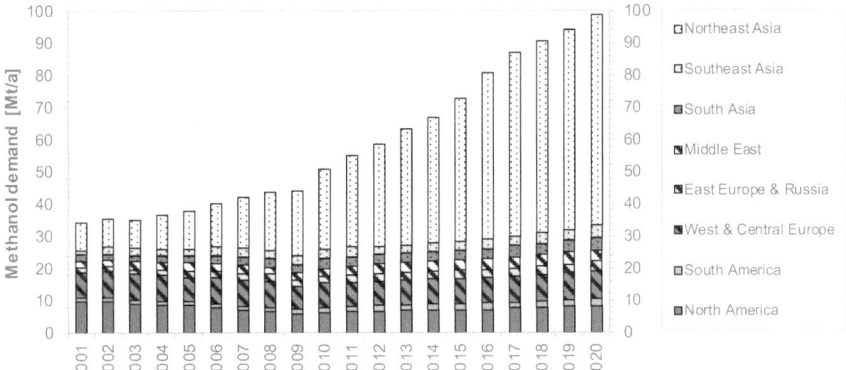

Fig. 4 Development of methanol demand by region (based on [18]; E: expected)

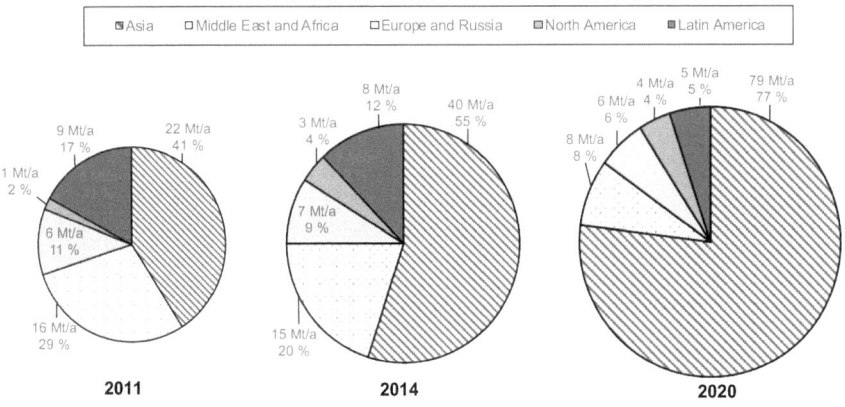

Fig. 5 Methanol production by region for the years 2011, 2014, and 2020 (according to [4] for 2011, [19] for 2014, [20] for 2020)

of the overall volumes of traded methanol[4] are transported between continents by trans-ocean shipping.

Figure 5 indicates that methanol production shifted more and more to Asia probably due to two main reasons: the availability of inexpensive hard coal as a feedstock and a substantial (and still growing) national/regional demand. The decrease in production shares in other regions can be attributed to the significant increase in production volumes in China rather than a decline in production output in those regions. As a result, in 2020, 37% of the methanol produced worldwide was produced

---

[4] It should be noted that the amount of methanol produced globally does not correspond to the total amount of methanol traded, as integrated plants such as coal-to-olefins (CTO) or gas-to-olefins (GTO) plants do not have an impact on the methanol market. The methanol produced in such concepts is neither bought nor sold, as it is merely an intermediate stage in the overall process [18].

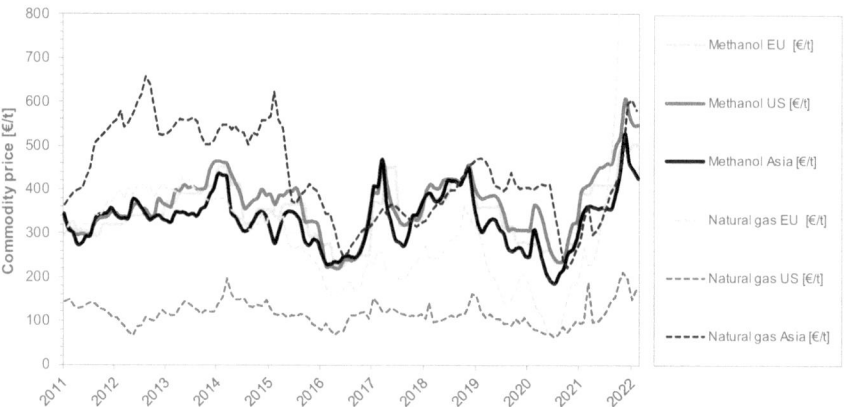

Fig. 6 Methanol and natural gas prices from 2011 to 2022 (according to [21, 22])

in China alone. Since coal-based methanol production is generally more complex and, therefore, more expensive than natural gas-based methanol production, regional load rates and production volumes can shift significantly depending on the price of natural gas. When natural gas prices are low, coal-based methanol of Chinese origin can be displaced by imports of natural gas-based methanol.

4.3 Market Prices

The market prices of methanol within the European Union (EU), the United States (US) of America, and Asia are shown in Fig. 6. It can be seen that prices differ only slightly between continental markets in most cases due to the overall well-developed international trade in methanol. Margins and transport surcharges result only in slightly more than a 5% average deviation between the markets, with an almost uniform price trend. Furthermore, methanol prices show a strong volatility being in line with the price for natural gas.

The exact pricing of methanol is very complex and depends on many factors. Decisive factors are the feedstock prices, the demand–supply situation, the crude oil price,[5] or the downstream sectors' market movement. However, since there is worldwide overcapacity in production and global logistics infrastructures for methanol—and thus a global methanol market—regional price formation effects contribute only slightly to the methanol price. Overall, there are large overcapacities for the production of methanol and strong competition among producers with presumably rather narrow profit margins. Thus, it can be assumed that market prices are strongly aligned

---

[5] Substitution of crude oil-based products by methanol-based products in case of higher crude oil prices (such as producing olefins from methanol conversion (MtO) instead of naphtha cracking).

with production costs, which mainly depend on costs for feedstock, energy, and investment costs [11].

Regarding renewable or "green" methanol its production is projected to have higher production costs compared to conventional, fossil fuel-based methanol in the foreseeable future. Also, this is likely to be coupled with anticipated higher revenues due to regulatory burdens and social incentives. These factors will likely result in higher overall market prices for "renewable" methanol compared to conventional, fossil fuel-based methanol in the future and thus also distinct markets for these products. Whether this will also affect the physical methanol trade depends on the future method of accounting for renewable methanol production.

5   Power-to-Methanol Processes

In order to produce methanol volumes that meet the current methanol demand of approximately 100 Mt/a, huge production capacities have been built globally in recent decades. Today, methanol is produced exclusively thermochemically via the heterogeneously catalyzed reaction of synthesis gas—i.e., a mixture of primarily carbon monoxide (CO), carbon dioxide (CO_2), and hydrogen (H_2). The formation of methanol can be described by the reaction Eqs. (1) and (2). Both reactions are coupled via the parallel water–gas shift (WGS) reaction Eq. (3). The methanol formation from CO and H_2 takes place stoichiometrically according to Eq. (1) without by-products. In contrast, reaction Eq. (2), starting from CO_2 and H_2 occurs with the formation of water (H_2O). The resulting water is partly consumed again by the WGS reaction Eq. (3), depending on the present CO content, which closes the reaction system.

$$CO + 2\,H_2 \rightleftharpoons CH_3OH \quad \Delta H_{300\,K} = -90.8 kJ/mol \qquad (1)$$

$$CO_2 + 3\,H_2 \rightleftharpoons CH_3OH + H_2O \quad \Delta H_{300\,K} = -49.2 kJ/mol \qquad (2)$$

$$CO + H_2O \rightleftharpoons CO_2 + H_2 \quad \Delta H_{300\,K} = -40.9 kJ/mol \qquad (3)$$

Today, the production of methanol is mainly based on CO-rich[6] synthesis gas produced almost exclusively from natural gas and hard coal. Historically, methanol synthesis was thus developed and optimized for the conversion of CO-rich synthesis gas. The use of these feedstocks of fossil origin is associated with the release of greenhouse gas (GHG) emitted during processing as well as during combustion of the methanol, or the final product. Therefore, the production of sustainable and climate-neutral methanol requires, above all, a switch to a synthesis gas derived from renewable energies and "sustainable" carbon. Furthermore, this altered supply

---

[6] "Rich" in this context refers to the ratio of CO to CO_2. Irrespective of this, the hydrogen content is set slightly higher than stoichiometrically required.

of synthesis gas can require significant adjustments in the subsequent processes, including the methanol synthesis section.

Power-to-methanol (PtM) is a form of power-to-liquid (PtL) and thus also power-to-x (PtX), where the liquid end-product of the process is electricity-based methanol. Similar to other PtX technologies, energy conversion from electrical to chemical energy is realized via hydrogen production by water electrolysis. The carbon source used within such a PtM process is mainly considered to be CO_2. This molecule is completely oxidized and thus does not bring any chemical energy input into the process. In addition to fully electricity-based methanol production, hybrid processes are also being considered. One such example is the co-feeding of reformed renewable hydrocarbons and hydrogen from electrolysis, allowing for a full carbon utilization. However, in the context of PtL production, the following consideration is limited to purely electricity-based production without using hydrocarbons or unoxidized carbon as feedstock.

The electricity-based production of methanol from CO_2 and H_2O can be carried out via two basic process routes. The utilization of CO_2 as a carbon source offers the advantage of directly converting CO_2 into the final synthesis product. Although this CO_2-converting methanol synthesis (Fig. 7, right) has already been demonstrated on a small industrial scale [23], it requires some synthesis adjustments compared with CO-rich methanol synthesis, mainly due to thermodynamic restrictions and the stoichiometric determined water formation during the reaction. A methanol synthesis operated on conventional CO-rich synthesis gas is also possible with CO_2 as the feedstock by using a reverse water–gas shift (RWGS) reaction before entering the methanol synthesis section (Fig. 7, left).

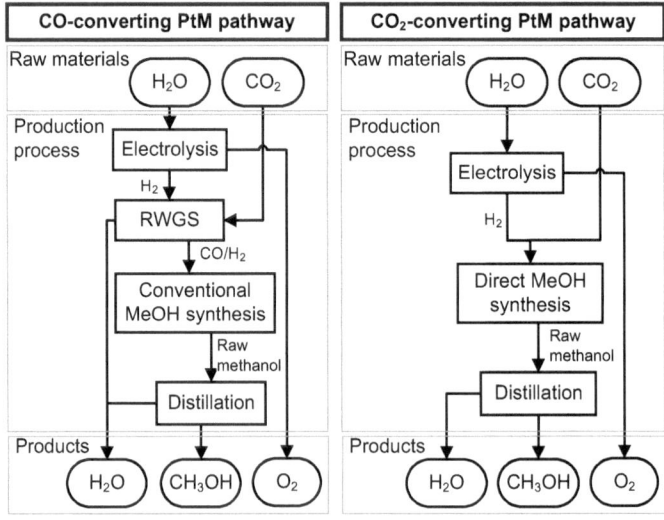

Fig. 7 Power-to-methanol production pathways (PtM: power-to-methanol)

The overall reaction equation for methanol production Eq. (4) from CO_2 and H_2O is pathway independent and shows the theoretical minimum energy requirement.

$$CO_2 + 2\,H_2O_{(l)} \rightarrow CH_3OH_{(l)} + 1.5\,O_2 \quad \Delta H_{300\,K} = 726.6 kJ/mol \quad (4)$$

- CO-converting PtM pathway: The process scheme shown on the left side of Fig. 7 is also known as the CAMERE process (carbon dioxide hydrogenation to form methanol via a reverse water–gas shift reaction). One advantage of the CO-rich route via the RWGS is that existing methanol synthesis plants can be converted to electricity-based production plants without adjustments to the synthesis loop and downstream processes. Only the reconstruction of the synthesis gas supply takes place, which, however, represents the predominant share of the installation costs in PtX processes. In addition, the RWGS is usually operated at temperatures significantly above 700 °C and therefore requires considerable amounts of high-temperature heat, which most likely would not be supplied via potential heat integration. The high temperature also places high demands on the materials used. Since the RWGS technology first gained technical relevance by using CO_2 as a carbon source for synthesis products, the process step has not yet been implemented beyond the pilot plant scale. Currently, there are some research and demonstration plants but no RWGS reactors is in commercial use. The technology readiness level (TRL) is estimated at 6–7 [24, 25]. Electrically heated reactors, which offer a significant efficiency benefit compared to renewable fueled fired systems, have an even lower TRL of 5–6. Challenges for developing and operating large scale, commercial RWGS reactors are preventing soot formation and guaranteeing feasible catalyst service lifetimes. In this route, the RWGS can be seen as a technological bottleneck. However, this PtM pathway could be an option, especially for retrofitting existing plants.
- CO_2-converting PtM pathway: Methanol synthesis already enables direct CO_2 conversion with hydrogen in a single synthesis step. This allows a shorter and more direct process route and the elimination of the technologically demanding RWGS reaction. In addition, high-temperature heat can be avoided, allowing for better heat integration and potentially higher process efficiencies. Due to several CO_2-converting demonstration plants and, since 2011, a small-scale commercial plant came into operation, the TRL can be classified as 8–9 [4]. However, the direct methanol synthesis based on CO_2 takes place with lower conversion rates compared to a CO-rich synthesis, which requires, among other things, an adjustment of the synthesis loop. In addition, the synthesis product has a significantly higher water content, which also requires catalysts' adjustments and modifications in downstream processing. Reconstruction of conventional methanol plants to direct CO_2 conversion plants is therefore associated with adjustments in all process stages. However, for new build PtM plants, the direct conversion route shows advantages in terms of complexity and potentially higher efficiency compared to the previously discussed route [23, 26].

Below the various process steps of these processes are discussed in more detail.

5.1 Synthesis Gas Provision

In the context of methanol production, synthesis gas (short form: syngas) is the gas mixture used as feedstock in the methanol synthesis reactor. The synthesis gas can contain different components in different concentrations and ratios depending on the starting material and synthesis technology. However, for the actual methanol synthesis, the desired reactive components are hydrogen (H_2), carbon monoxide (CO), and carbon dioxide (CO_2). The synthesis gas can be distinguished mainly by the ratio of these main components/reactants (synthesis gas composition) as well as by the proportions of inert gases and catalyst poisons (synthesis gas impurities).

The electricity-based production of synthesis gas comprises the provision of H_2 and CO_2 and the subsequent conditioning to synthesis gas via RWGS (CO-rich methanol synthesis) or mixing (CO_2-rich methanol synthesis). Hydrogen is supplied via electro-chemical water splitting, for which various electrolysis technologies are available. The technologies required for the supply of pure CO_2 depend on the CO_2 source used. The primary sustainable sources are biogenic point sources and direct air capture (DAC).

Synthesis gas composition: The theoretical "stoichiometric" required ratio of H_2/CO or H_2/CO_2 for the methanol reaction can be derived from the reaction Eqs. (1) and (2). Since the reactions are connected via the water–gas shift (WGS) equilibrium Eq. (3), the composition of synthesis gas for methanol synthesis is commonly described with the stoichiometric number (SN) defined in Eq. (5), considering the mole fractions of the reactants. Therefore, a SN of 2 is required for the ideal stoichiometric reaction to methanol [6]. Over- or under-stoichiometric synthesis gas ratios increase with increasing conversion, causing the SN to change along the length of the reactor. In practice, a lack of hydrogen leads to a drastic reduction in selectivity for methanol by favoring side reactions. However, a significant excess of hydrogen increases the amount of hydrogen recycling and with that the synthesis loop size and purge gas losses. Therefore, a synthesis gas composition with a stoichiometric number of slightly above 2 has been shown to be optimal for methanol synthesis. Commonly, SN is adjusted between 2.05 and 2.15 [3, 27]. The ratio of CO to CO_2, which defines the difference between CO– and CO_2-rich synthesis gas and thus also between the two pathways shown in Fig. 7, is expressed by the carbon oxide ratio (COR) according to Eq. (6). Since the water produced during the conversion of CO_2 to methanol has a deactivating effect on many methanol catalysts, the COR is kept as low as possible in conventional applications. A COR between 0.02 and 0.05 is the optimum ratio to allow maximum methanol production since a complete absence of CO_2 also hinders the conversion [27]. In addition, increasing COR leads to a decrease in by-product formation, which can probably be attributed to the lower heat release of CO_2 conversion and the associated flatter temperature profile at the catalyst. Conventional synthesis gas from gasification and steam methane reforming (SMR) usually results in *COR* from 0.05–0.3 and 0.2–0.4, respectively [27]. For CO_2-rich synthesis gas in PtM processes, the *COR* is accordingly 1.

$$SN = \frac{[H_2] - [CO_2]}{[CO] + [CO_2]} \tag{5}$$

$$COR = \frac{[CO_2]}{[CO] + [CO_2]} \tag{6}$$

Synthesis gas impurities: In addition to the reactants H_2, CO, and CO_2, further components may be present in the synthesis gas. A distinction can be made between inert gases, which are not directly involved in any reaction, and catalyst poisons, which negatively influence the reaction (e.g., the catalyst activity) even in low concentrations. Although inert components do not directly interact with the reaction system and therefore do not harm the catalyst, they do lead to a reduction in the partial pressure of the reactive components by diluting the synthesis gas. Furthermore, these inert gases can accumulate in the synthesis loop, meaning that parts of the unreacted gas stream have to be purged out of the process. In principle, the proportion of inert gases in the synthesis gas should be kept as low as possible to avoid energy and carbon losses. Typical inert gases in synthesis processes are nitrogen (N_2), methane (CH_4), argon (Ar), and helium (He) [27]. The proportion of inert gases in the synthesis gas depends primarily on the feedstock used and the process selected for synthesis gas production. While the inert fraction in synthesis gas from conventional steam methane reforming is up to 10% and from gasifiers up to 18%, only fractions below 2% of inert components are expected for CO_2-rich synthesis gas, depending on the CO_2 source and the applied separation process [11, 27].

In addition to inert gas concentrations that must not be exceeded, syntheses and synthesis gas production processes, or more precisely the catalysts used in these processes, have very high purity requirements concerning the content of catalyst poisons, i.e., catalyst-damaging impurities. This becomes particularly crucial when the processes rely on feedstock of fossil or biogenic origin for the synthesis gas supply, rather than being pure PtM processes; in addition to carbon, hydrogen, and oxygen, conventional fossil and biogenic feedstock used for synthesis gas provision contain further heteroatoms, such as sulfur or halogens, which might migrate into the raw synthesis gas during processing. The purity requirements for today's methanol catalysts are based on the resulting common interfering substances in today's synthesis gas. Particular attention is paid to sulfur (< 0.1 ppm) and chlorine (< 1 ppb) compounds, but also to particles (< 0.1 mg/m³$_N$), tar (< 1 mg/m³$_N$), alkalis (< 0.25 mg/m³$_N$), and ammonia (NH_3) (< 1 ppm) [4, 23, 28]. Since electricity-based synthesis gas production is built on other feedstock and processing methods, other interfering substances may become relevant.

In PtM processes, water electrolysis provides one part of the synthesis gas. Here, the main H_2 impurities are H_2O, N_2, and O_2. While H_2O and N_2 can be considered as unproblematic up to concentrations in the lower percentage range, O_2 can already be a catalyst poison in very low concentrations.[7] Therefore, many manufacturers already offer additional purification units in the electrolysis unit to achieve harmless concentrations in the lower ppm-range. Depending on the technology used, high to very high hydrogen purities (> 99.9–99.999%) are provided, while the main impurities are H_2O and N_2 (< 12 ppm) and O_2 (< 2 ppm) [29–31]. Accordingly, hydrogen production via electrolysis typically does not introduce any significant impurities into synthesis gas that needs to be considered.

By contrast, there are far more significant uncertainties about impurities with regard to the CO_2 supply. The concentration of the separated CO_2 stream and the type and proportion of impurities depend on the CO_2 source and the separation process used. While CO_2 streams from, e.g., fermentation processes (for example: ethanol production) are already present in high concentrations (99–100 vol.-%) and relatively free from impurities, flue gas streams contain only 3–15 vol.-% CO_2 but various impurities (e.g., sulfur components, halogens, heavy metals) [32]. In addition to the substances already contained in the CO_2 carrier stream, the capture process can also add further contaminants to the CO_2 stream. For example, amines or degradation products from the amine wash could be entrained and contaminate the synthesis gas. The interaction of the CO_2 carrier stream and separation process is still a relatively new issue concerning CO_2 quality since these processes have mainly been used to upgrade the carrier stream rather than provide pure CO_2. Therefore, additional and, if necessary, extensive cleaning steps must be carried out to ensure that potential impurities do not reach the catalyst [32].

5.2 Methanol Synthesis

Out of the previously described synthesis gas, methanol is produced in the methanol synthesis loop under specified operating conditions. In the following, the fundamentals of the methanol synthesis, as well as process-specific aspects of the synthesis section, are discussed. Potential differences between electricity-based and conventional production of methanol as well as CO-rich and CO_2-rich production are highlighted. Under optimal operating conditions, methanol synthesis can achieve a very high selectivity. However, deviating operating conditions can lead to increased formation of higher alcohols. This is used in the so-called "mixed alcohol synthesis" for the targeted synthetic production of higher alcohols. An excursus on the mixed alcohol synthesis can be found at the end of the paper.

Thermodynamics: Both methanol-building reactions Eqs. (1) and (2) are exothermic and accompanied by a reduction in volume. According to Le Chatelier's principle of moderation, methanol formation is thus favored by increasing pressure and decreasing temperature, where the theoretical maximum conversion is determined

---

[7] Oxygen has not played a significant role in the specification of synthesis gases up to now due to the usual processes for synthesis gas production, in which the starting materials are usually partially oxidized so that no unbound oxygen remains in the synthesis gas.

Fig. 8 Equilibrium of CO- and CO$_2$-rich methanol synthesis (according to [23])

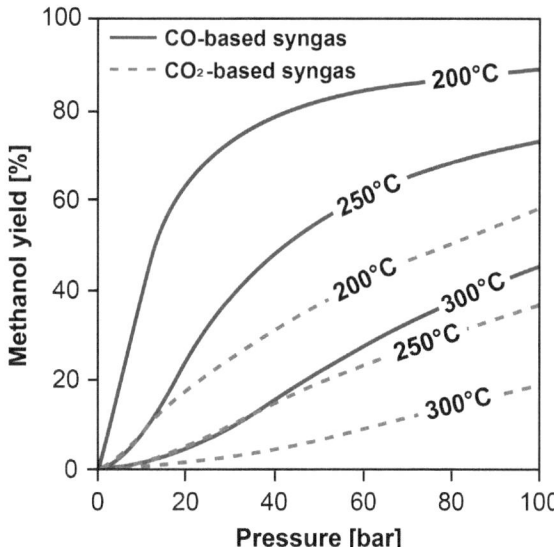

by the equilibrium composition [3, 23, 33]. Figure 8 shows the maximum methanol yield based on the reaction equilibrium as a function of pressure and temperature. The required minimum reaction temperature is about 200 °C. Below this temperature limit, common catalysts are not sufficiently active, and the reaction kinetics are significantly reduced. The curve plots in Fig. 8 show clearly lower equilibrium concentrations of methanol over the entire range from CO$_2$-rich synthesis gas compared to CO-rich synthesis gas. For CO$_2$-rich synthesis, the maximum conversion per reactor pass is therefore limited by the position of the thermodynamic equilibrium. In the operating range of the reactors or catalysts, this is well below 60%.

Despite many years of industrial use and intensive research, the detailed reaction mechanism from synthesis gas to methanol has not been clarified with final certainty. Among other things, it is discussed whether methanol is actually formed via the hydrogenation of CO or CO$_2$. However, most publications today assume a reaction starting from CO$_2$ adsorption. The CO to CO$_2$ ratio (*COR*) influence on the synthesis has not yet been clarified either, but maximum activity is assumed at a CO$_2$ content between 2 and 5 mol-%. Besides thermodynamic restrictions, the decreasing synthesis gas conversion with an increasing CO$_2$ content is mainly attributed to an oxygen enrichment at the catalyst surface. In addition, the resulting water vapor lowers the partial pressure of the reaction and accelerates the degradation of most catalysts [34]. The resulting challenges of CO$_2$-rich methanol synthesis (and therefore for most PtM concepts) primarily involve the development of suitable—i.e., highly active and water-resistant—catalysts and synthesis loops.

Reactors: Although the entire industrial methanol production is consistently based on the low-pressure methanol synthesis developed in the 1960s, different technology providers offer various application-specific synthesis concepts for which different reactor types are used. The technology selection is mainly determined by the synthesis gas composition, the plant capacity as well as the heat management at the site (availability of heating and cooling media and/or the integration into the heating/cooling network on-site). Apart from a few exceptions [35] methanol synthesis takes place as a heterogeneously catalyzed reaction of a gas phase on a solid catalyst without the formation of a liquid phase. Thus, a variety of relatively simple reactor concepts are feasible. Important design specifications are mainly a low catalyst volume, a suitable temperature profile (low outlet temperature for high conversion) and heat recovery at high temperature. Due to the strong exothermic nature of methanol formation and the large temperature influence on the reaction equilibrium and catalyst properties, a particular focus in reactor design is on thermal management. Firstly, temperature peaks in the catalyst bed must be limited in order to avoid catalyst damage (sintering). Secondly, a compromise between reaction kinetics and reaction equilibrium (line of maximum reaction rate) must be realized via the reactor's temperature profile. The heat released during the reaction inevitably leads to the local heating of the catalyst. Without heat dissipation, this adiabatic temperature increase would lead to lower yields, increased by-product formation, damage to the catalyst, and safety problems [4, 34].

Numerous reactor designs are available in the market today. Conventional reactors are mainly quasi-isothermal fixed-bed reactors, which are temperature-controlled utilizing steam generation. In addition, quench gas-cooled, or intercooler reactors can also be used individually or in combined reactor concepts. In the following, some standard reactor designs for methanol synthesis are briefly described and discussed as examples. Beyond that, various further variants are also conventionally established. However, those are fundamentally based on largely similar operating principles. In Fig. 9, a simplified illustration of the steam raising reactor (left) and the gas and water-cooled double-tube reactor (right) with approximate temperature profiles is shown.

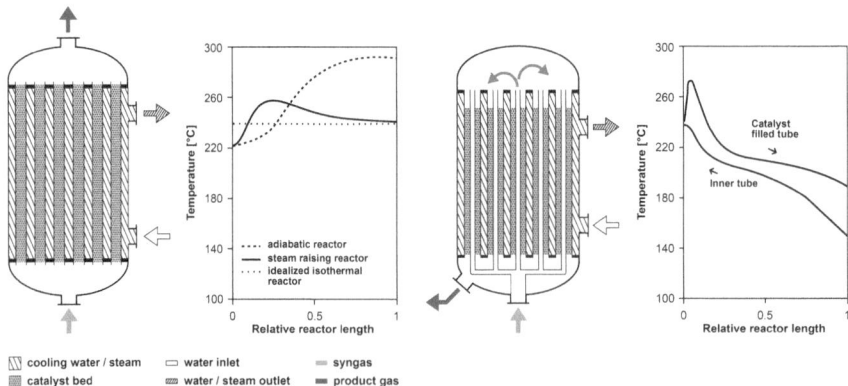

Fig. 9 Simplified illustration of steam raising reactor (left) and gas and water-cooled double-tube reactor (right) with approx. temperature profiles (according to [23])

Steam raising fixed-bed reactor: Figure 9 (left) shows a simple shell-and-tube reactor with a fixed catalyst bed, cooled on the shell side with boiling water (steam raising converter; SRC). Such cooling with the aid of boiling water enables quasi-isothermal operation as well as simple temperature setting and control through pressure and level control. The generated steam can be effectively used for heat integration. In the shown example, the synthesis gas flows axially/vertically through the catalyst-filled tubes. In the first section of the catalyst bed, the reaction occurs at a fast rate due to the high proportion of unreacted gases. Thereby large amounts of heat are released. Depending on the feed temperature, additional heat might be transferred from the cooling side, leading to a sharp temperature increase on the synthesis gas side in this reactor section. Due to unavoidable heat transport limitations, the temperature in the catalyst bed exceeds the temperature of the coolant. As the reaction rate decreases with the concentration of unreacted gas, the gas temperature also decreases over the remaining length of the reactor due to permanent cooling. The maximum capacity of today's steam raising converter is limited to approx. 2,200 t_{MeOH}/d. Against the temperature profile of the steam raising converter, an adiabatic reactor shows a wider preheating zone since no external heat is transferred to the gas stream. Along with the rising synthesis gas temperature, the catalyst activity and reaction kinetics rise, leading to a sharp temperature increase in the middle reactor section. High temperatures and gas compositions close to the equilibrium cause the reaction to slow down, which also causes the temperature increase to cease. In addition to a single steam raising converter, an adiabatic catalyst bed can be connected upstream of the reactor, which reduces necessary preheating power. In the idealized isothermal reactor, the reaction occurs at the selected temperature level over the entire reactor length. An ideal heat transfer without resistances is assumed in this theoretical case [23].

Gas and water-cooled double-tube reactor: To improve heat integration and the temperature profile in the methanol reactor, the "superconverter" was developed (Fig. 9, right). In this concept, the synthesis gas enters the reactor well below the reaction temperature. The "cold" gas flows through the inside of a double tube until it changes the flow direction at the tube outlet and enters the outer tube side, which is packed with the catalyst. As a result, the synthesis reaction only takes place in the outer tube. The waste heat from the exothermic synthesis reaction preheats the synthesis gas in the inner tube. In addition, the shell side of the outer tube is cooled with boiling water, as in the previously described shell-and-tube reactor, to remove excess energy as steam. The temperature profile is based on the line of the maximum reaction rate. The reactor concept combines a tubular gas cooling system with a steam-generating tubular reactor by using double tubes in a boiling water vessel.

Gas and water-cooled dual-stage reactor: In Fig. 10, a simplified illustration of a gas and water-cooled dual-stage reactor with approximate temperature profiles is depicted.

This reactor concept is commonly known as the "MegaMethanol" concept. It was developed to enable very large production capacities of up to 5,000 t/d. Here, the cold synthesis gas flows through the first reactor on the tube side while it heats up to the reaction temperature. The hot synthesis gas enters the catalyst bed of the second

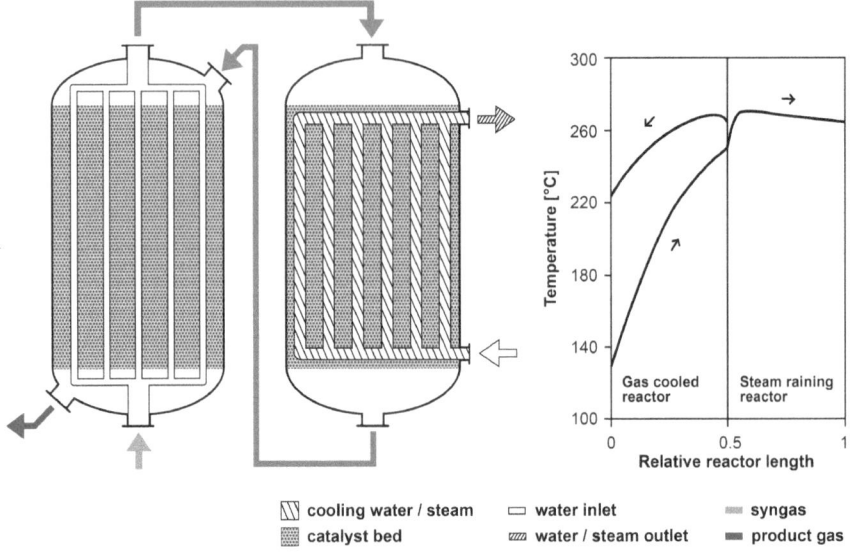

Fig. 10 Simplified illustration of a gas and water-cooled dual-stage reactor (MegaMethanol reactor concept) with the approx. temperature profiles (according to [23])

reactor on the shell side and begins to convert into methanol. The heat released in this process is dissipated through axially/vertically flowing cooling tubes based on isothermal temperature control and used to generate steam. Since the synthesis gas already enters the catalyst bed at reaction temperature and with maximum reactant concentration, the reaction takes place at a high rate and releases large proportions of reaction heat. The synthesis mixture then enters the catalyst bed of the first reactor on the shell side, which is cooled by the fresh synthesis gas (tube side) and reacts further in the direction of the equilibrium, which shifts to the methanol side with decreasing temperature. The concept has several advantages, including advanced heat integration, requiring only slight preheating of the fresh synthesis gas, and optimized temperature control, enabling high conversions to be achieved. In addition, the large cross-section of the shell-side catalyst beds enables low-pressure losses, which allow for large production capacities [4, 23].

The notable advancements in recent years have mainly focused on the design of increasingly larger reactor concepts. Apart from the heat removal challenges in this regard centered around a limitation of the pressure losses and constructional restrictions. Today, research efforts are focused on two main aspects: optimizing the transition from a CO-rich to a CO_2-rich synthesis gas and developing efficient, load-flexible, small-scale reactors. The latter is particularly important as non-fossil feedstock (e.g., biomass, renewable electricity) are often decentralized and subject to fluctuations over time.

Catalysts: The first industrial methanol production processes based on synthesis gas, contained considerable amounts of sulfur and chlorine components. The catalysts, therefore, had to be highly stable against such impurities. These catalyst systems were primarily based on zinc oxide (ZnO) and chromium oxide (Cr_2O_3) [3, 34]. However, such catalysts require high reaction temperatures (300–400 °C), whereby high pressures (250–350 bar) are necessary to achieve relevant conversions. The provision of a significantly purer synthesis gas enabled more active copper-based catalysts used since the 1960s. Today's commercial catalysts are all based on copper oxide (CuO) and ZnO enabling the low-pressure methanol process, operated at 200–300 °C and 40–100 bar [3, 34]. However, these systems are highly sensitive to impurities (Sect. 5.1). The active components are stabilized with Al_2O_3 or Cr_2O_3 and spiked with small amounts of promoters [23]. Common promoters are Zr, Cr, Mg, and rare earth metals influencing especially the Cu dispersion and particle size. The catalysts are produced in complex, multi-stage processes, aiming for an ideal distribution of the active components and the highest possible active surface area. In principle, these catalysts are used to convert both CO-rich and CO_2-rich synthesis gas. However, catalyst manufacturers are optimizing the components for the particular requirements of CO_2-rich synthesis, thus enabling higher activities and increasing resistance to high water contents.

The catalysts are mostly formed as porous cylindrical tablets with a diameter and height of 4–6 mm, resulting in a bulk density of around 1,000–1,300 kg/m^3 [3, 34]. Under the usual operating conditions and common residence times (gas hourly space velocity, GHSV) of 10,000 h^{-1}, space–time yields (STY) between 0.7 and 2.3 kg$_{MeOH}$/(L$_{Cat.}$ h) can be achieved for CO-rich synthesis gas [23]. However, for CO_2-rich synthesis gas, the resulting space–time yield is significantly lower ranging between 0.4 and 0.8 kg$_{MeOH}$/(L$_{Cat.}$ h) [23].

Before the catalyst can be used, activation with H_2, synthesis gas, or CO, diluted in inert gas at 150–250 °C, must occur, whereby the CuO is selectively reduced to metallic Cu [3, 34]. The high activity at the beginning of the catalyst life cycle decreases over the operating time due to various deactivation mechanisms. Some catalyst deactivation processes can be reversible (i.e., they can be reactivated), while others are irreversible and persist over the catalyst's lifetime. Typically, there is a substantial activity loss at first, stabilizing after a short time and only slowly decreasing further (Fig. 11). A loss of around 15 to more than 30% within the first 1,000 h of operation is not unusual, whereby the total running time of a catalyst can be well over 20,000 h [23, 34]. Mainly, deactivation effects can be attributed to unsuitable operating parameters or catalyst poisons. Excessively high operating temperatures, which can occur during operation and even during activation, lead to accelerated sintering, whereby the copper crystallite size increases, reducing the active surface area. Besides this, poisons can cause additional deactivation. Sulfur components like hydrogen sulfide (H_2S) block the active surface atoms and prevent further participation in reactions. Other components, like chlorides, accelerate sintering by forming copper (Cu) and zinc (Zn) compounds whose melting points are significantly lower than those of the respective oxides.

Other deactivation effects can occur due to iron and nickel carbonyls, which can be entrained from material detachments in upstream process parts. These compounds

Fig. 11 Characteristic catalyst deactivation curve during regular operation (according to [36])

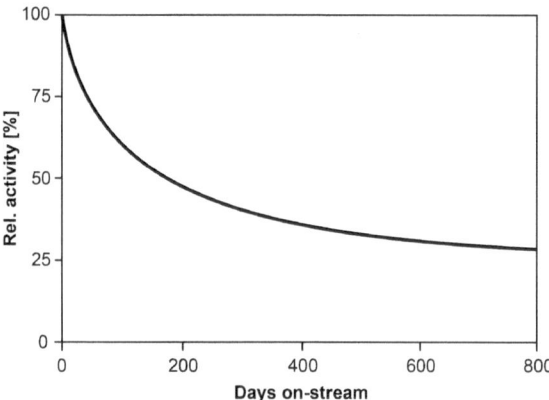

unintentionally catalyze the Fischer–Tropsch synthesis reactions, whereby the catalyst surface gets covered by waxes and thus becomes inactive [34].

With the use of CO_2-rich synthesis gas, the water content in the synthesis gas increases during methanol formation, Eq. (2). In addition to the previously discussed thermodynamic effects (i.e., equilibriums restrictions, decreasing partial pressure of reactants), several studies suggest that the higher water formation could also lead to accelerated catalyst degradation [11, 34].

The catalyst lifetime is usually specified as 3–5 years under optimal operation conditions [4, 23]. In order to maintain production performance despite decreasing catalyst activity, it is common to adjust the process conditions; i.e., reaction pressure and temperature are increased. The runtime or level of deactivation at which the catalyst is regenerated or replaced is not a fixed value, but rather depends on the specific process, maintenance practices, and market factors in each individual case. In general, however, shortened lifetimes due to improper operation should be avoided, as this would lead to a sharp increase in the operating costs of a methanol plant.

Synthesis Loop: To enable a high overall conversion despite only low to medium per-pass conversions, the synthesis reactor is commonly integrated into a synthesis loop. Compared to a synthesis with recirculation, once-through operation modes are less economic, especially in large-capacity plants or when the synthesis gas supply is costly. A typical synthesis loop is shown in Fig. 12.

The fresh synthesis gas is mixed with the recycle gas stream, heated if necessary, and fed to the synthesis reactors. After flowing through the reactor, the product gas stream is cooled or quenched. However, the hot gas stream can initially also be used for preheating the fresh gas. A gas separator then separates the condensed liquid phase from the non-condensed gases and feeds the raw methanol into the distillation columns. Since methanol synthesis is very selective, the remaining gas phase can be recycled without further reforming or conditioning. However, a certain amount must be purged to avoid the accumulation of inert gases and light by-products. The recirculated gas stream is compressed to overcome the pressure drop in the reactor,

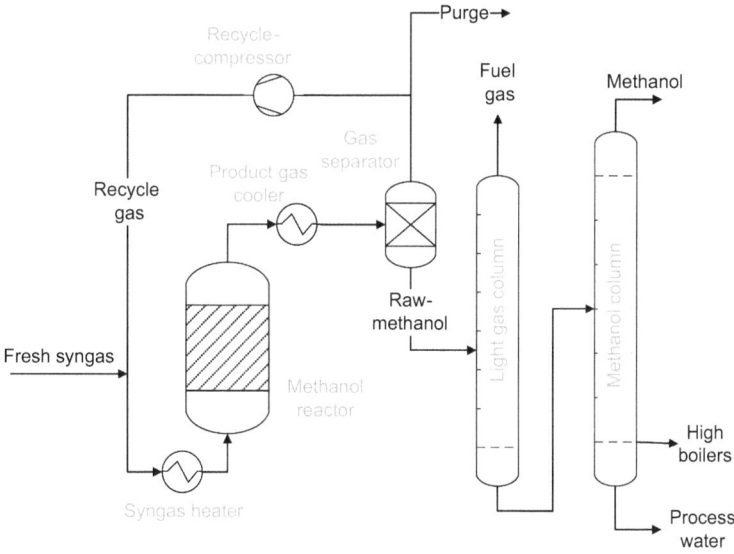

Fig. 12 Power-to-methanol synthesis loop with downstream processing (MeOH: methanol)

which is commonly between 2 and 5 bar. The recycling ratio[8] usually lies below 3 for CO-rich synthesis gases. Significantly higher ratios can be assumed for CO_2-rich synthesis due to the lower per-pass conversion [23]. The raw methanol is fed to the downstream processing, where it is separated from lighter and heavier boiling by-products.

Operating conditions: Only the low-pressure methanol synthesis process is commercially relevant today. Table 3 shows typical operating parameters and key figures of the reactor section.

The activity limit of the catalyst and the reaction kinetics determines the lower limit of the reaction temperature of about 200 °C. In contrast, the upper range of the temperature window is determined by the catalyst stability as well as by the reaction equilibrium that becomes less favorable with increasing temperature. The operating pressure is also limited to the lower limit by the reaction equilibrium and the associated conversions. The upper-pressure range is set by the design of the process components.

While the operating conditions of the synthesis reactor are almost equal for both CO- and CO_2-rich synthesis, the per-pass conversion and selectivity differ significantly. Accordingly, the higher selectivity of up to 99.96% can be noted as an advantage of CO_2-rich synthesis [23]. This is due to the less exothermic nature of the reaction and the associated flatter temperature profile in the catalyst bed as well as in the catalyst particle itself. However, a disadvantage is a significantly lower conversion resulting mainly from thermodynamic equilibrium and the increased water formation.

---

[8] Volume-based ratio of recycled gas to fresh gas.

Table 3 Operating parameters and key figures of the methanol synthesis (GHSV: gas hourly space velocity; based on [3, 23, 34])

Parameter/key figure	Unit	Methanol synthesis	
		CO-rich	CO_2-rich
Temperature	°C	200–300	
Pressure	Bar	40–100	
Stoichiometric number[a]	–	2.05–2.15	
GHSV[b]	10^3 h^{-1}	6–12	8–18
Per-pass conversion[c]	%	Up to 80	Below 45
Selectivity[d]	%	99.0–99.9	99.90–99.96

[a] according to Eq. (5); [b] space velocity (gas hourly space velocity) as the ratio of the feed volume flow under standard conditions to the catalyst volume; [c] based on converted carbon; [d] proportion of converted carbon in methanol

In terms of flexibility and load ranges, today's conventional plants are optimized for steady-state operation with start-up and shut-down operations being reduced to a minimum [37]. Nevertheless, wide load ranges can be achieved depending on the selected reactor types and storage volumes. Steam raising reactor concepts can enable a part-load operation with 10–15% and overload operation with up to 120% of the design load [23]. Between these operation constraints, relatively fast load change rates, which enable a switch from minimal to full load in a few minutes, are possible. In the context of PtX technologies, these dynamic properties become even more critical since the availability of electricity from renewable sources of energy is strongly volatile. Therefore, the flexibilization of methanol synthesis, through suitable reactor and control concepts, is part of current research projects [38]. For example, on a test scale, the direct coupling of a PtM concept with a wind power plant without storing intermediate products has already been demonstrated [39].

By-products: Despite the very high selectivity of the catalysts used (> 99%), the formation of by-products must be considered in the process design. Water is by far the largest by-product in terms of volume. Since the formation results stoichiometrically from the conversion of CO_2 to methanol (Eq. (2)), the amount of water correlates directly with the *COR* in the synthesis gas. Besides water, various compounds are formed by undesired side reactions. However, these occur to a low to very low extent. Typical by-products are higher alcohols (predominantly ethanol), esters (mainly methyl formate), ether (predominantly dimethyl ether), ketones (mainly acetone and butanone), and lower hydrocarbons [4, 25]. The low amount of these impurities is noteworthy as the formation is thermodynamically favored over methanol for most of these components. Various circumstances can favor by-product formation.

- Impurities in the synthesis gas or on the catalyst surface (e.g., alkali, iron, cobalt, nickel) can promote side reactions and the incorporation of heteroatoms. Furthermore, unsuitable process conditions, mainly high temperatures and low space velocities, lead to an increased formation of C_2 components since those reactions are controlled by kinetics rather than thermodynamic conditions. Additionally, the composition of the synthesis gas has a considerable influence on the course of the reaction and the formation of by-products (Sect. 5.1). On the one hand, under-stoichiometric hydrogen concentration in the synthesis gas ($SN < 2$) leads to a hydrogen deficit at the catalyst surface, inhibiting hydrogenation reactions, which significantly increases by-product formation.

• On the other hand, a high COR leads to higher methanol selectivity. The influences on selectivity described above can also be used in the opposite direction to deliberately produce a mixture of different alcohols on purpose (see excursus mixed alcohol synthesis).

5.3 Downstream Processing

Subsequently to the methanol synthesis, the downstream processing aims to purify methanol to a required specification purity. Thereby, downstream processing of raw methanol commonly includes mainly thermal separation operations. The reactor output is gaseous due to the given temperature and pressure conditions. The product mixture released by the reactor vessel contains unreacted synthesis gas, the desired and unavoidable reaction products methanol and water, and small amounts of lighter and heavier boiling by-products. The raw methanol quality and the purity requirements for the target product mainly define the type and extent of downstream processing.

The required purity of methanol usually depends on the applicable standard or specification. The most common ones for methanol are the IMPCA Methanol Reference Specification and the ASTM grades A and AA. The worldwide specifications for methanol that is used in the chemical sector normally correspond to ASTM grade AA. Table 4 lists concentration-related requirements for the respective specifications.

In addition to the use of methanol as a basic chemical, large quantities of methanol are also used directly as a fuel. Compared to purity requirements for its use in the chemical sector, the purity required for methanol as a fuel is significantly lower

Table 4 Typical methanol standard specifications

Criteria	Unit	ASTM grade A	ASTM grade AA	IMPCA
Methanol content	wt%	> 99.85	> 99.85	> 99.85
Water content	wt%	< 0.15	< 0.10	< 0.10
Ethanol content	wt-ppm	–	< 10	< 50
Acetone content	wt-ppm	< 30	< 20	< 30
Acidity as acetic acid	wt-ppm	< 30	< 30	< 30
Sulfur	wt-ppm	–	–	< 0.5
Chloride as Cl⁻	wt-ppm	–	–	< 0.5
Iron in solution	wt-ppm	–	–	< 0.1

and depends on the methanol concentration in the overall fuel mix and the combustion concept. So far, there is currently no internationally recognized standard for fuel-grade methanol. However, national and multinational standards for fuels with different methanol shares are already in place [1, 3].[9]

To achieve the respective purity standards, typically a downstream separation is necessary. First, the unreacted synthesis gas must be separated from the synthesis products for further use (recycling or thermal use). This is done by cooling and the resulting condensation of the raw methanol. If significant amounts of heavier boiling by-products are present in the product gas stream, these can be partially condensed and removed before the raw methanol is separated. Other than shown in Fig. 12, the methanol cooling can also be carried out via a quench. The condensed raw methanol is flashed from the reaction pressure to a lower pressure level of 5–10 bar to drive out dissolved gases, mainly CO_2 [4, 40]. Adding small amounts of aqueous caustic soda makes the methanol–water mixture slightly alkaline to neutralize lower carboxylic acids and partially hydrolyze esters [3]. Various concepts can be selected for the subsequent distillation-based methanol purification, with two to three rectification columns. One- to two-column designs can also be used for low to medium methanol concentrations, focusing on water separation. Other light and heavy boilers can also be removed to a certain extent via side cuts (Fig. 12). In order to meet high methanol purity requirements, two- to three-column concepts are selected, with different pressure levels in the columns enabling heat integration between the respective condensers and boilers. Since the by-products contained are only present in minor quantities, they are not extracted as individual pure substances but as substance fractions. These are then mostly only used energetically (e.g., to provide process heat). Common names for the lighter streams, which contain mainly dissolved CO_2, dimethyl ether (DME), methyl formate, and acetone, are fuel- or tail-gas and light ends. The heavier boiling fraction occurring beside the water is sometimes called fusel oil and contains predominantly ethanol, longer-chain hydrocarbons, higher ketones, and esters [3, 4].

For downstream processing in the context of PtM with high CO_2 contents in the synthesis gas (*COR* close to 1), significantly higher water contents are to be expected compared to conventional synthesis. Furthermore, the solubility of CO_2 in methanol is relatively high, leading to higher amounts of CO_2 in the distillation section. However, smaller amounts of other by-products are expected due to a higher methanol selectivity. In terms of plant flexibility, it should be noted that distillation columns require more continuous operation than the reactor section. Storage capacities for interim-storage before distillation of the raw methanol are therefore necessary.

---

[9] United States: ASTM D4814 (M2.7); Europe: EN 228:2012 + A1:2017(M3); China: GB/T 23,510–2009 (M100).

6 Techno-Economic and Environmental Aspects

The following section presents technical, economic, and environmental aspects of electricity-based methanol production, considering the overall production plant.

6.1 Technical Aspects

To evaluate methanol production from a technical point of view, it is common to define and determine key figures of the process (e.g., specific feedstock demand, energy efficiency). However, such material and energy-related assessment parameters depend strongly on the considered system boundaries and the input parameters. Since the production of methanol from conventional, fossil fuel-based feedstock differs fundamentally compared with production based on electricity—for example in terms of feedstock, energy supply, and process configuration—a comparison of both routes is challenging based on simple process indicators. Therefore, in the following, the key influencing parameters on energy and carbon efficiency[10] are discussed in a qualitative manner only.

Conventional methanol production: Today, the production of methanol is based on fossil fuel-based hydrocarbons. These feedstocks, almost exclusively natural gas and hard coal, are used to provide carbon and hydrogen on the one hand and process energy on the other.

The synthesis gas production processes must adjust the ratio of carbon to hydrogen provided by the feedstock according to the requirements of methanol synthesis. Product gas from hard coal gasification has a low hydrogen content to be enhanced/ adjusted by a water-gas shift (WGS) reaction step as well as a CO_2-separation resulting in significant carbon losses and the corresponding GHG emissions. These carbon losses are significantly lower for natural gas-based methanol production since the reforming product has a much higher hydrogen content already due to the much more favorable hydrogen-to-carbon ratio of natural gas compared to hard coal. Accordingly, the carbon efficiency is significantly lower for hard coal-based processes than for natural gas-based processes. The share of purged synthesis gas and, to a lower extent, the by-product formation mainly determine the carbon losses in the actual methanol synthesis section.

Since energy efficiency correlates directly with the specific raw material requirement, which is decisive for economic competitiveness, today's large-scale plants are highly integrated across the entire production concept, which means that they also operate efficiently in terms of energy. In natural gas-based production concepts, energy efficiencies of up to 67% and carbon efficiencies of up to 83% can be reached

---

[10] Carbon efficiency describes the proportion of carbon that ends up in the chemical bonds of the product compared to the carbon input from the overall feedstock. Energy efficiency in this paper describes the ratio of the energy transferred to the chemical energy of the product to the overall energy used for production.

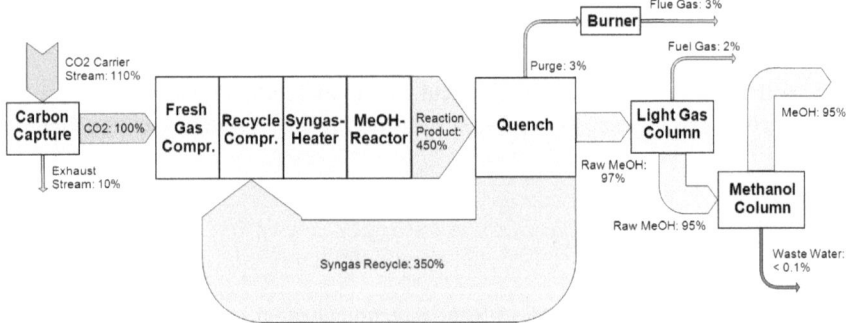

Fig. 13 Carbon streams of a CO_2-converting PtM concept (percentages indicate relative values in relation to the total carbon input (100%) in the form of CO_2. Carbon capture is considered via amine wash from CO_2 point source; MeOH: methanol)

[3]. In hard coal-based production concepts, the main energy consumption is required for the provision of reforming heat and compression power [3, 27]. The carbon efficiency of hard coal-based plants is usually below 35%.[11]

In terms of plant capacity, due to the almost unlimited local availability of the fossil raw materials used, ever-larger production concepts have been designed to exploit economy of scale effects. The current largest plant for methanol production was opened in China in 2020 in a coal-to-olefin (CTO) concept. The production capacity is approx. 2.2 Mt/a and 7,200 t/d [41]. Natural gas-based plants in similar orders of magnitude with regard to production capacity have recently been built in Iran and Turkmenistan [36, 41].

Electricity-to-Methanol production: For electricity-based methanol production, in contrast to production from hydrocarbons, the provision of material and energy resources takes place separately. The main material inputs are water (H_2O) and carbon dioxide (CO_2). Since both input material streams do not have a caloric value, the energy that is transferred into methanol as well as the necessary process energy is provided as electric energy. If technically possible, the provision of heat is also to be ensured via electricity or even as direct solar heat. In terms of carbon efficiency, decoupling carbon and hydrogen sources can enable a theoretically ideal carbon conversion. In real plants, however, small amounts of carbon are lost via purge gases and by-products. Nevertheless, carbon efficiencies of more than 90% can be expected [23]. Figure 13 shows an exemplary carbon flow diagram for a PtM concept with direct CO_2 conversion.

The carbon streams are given as a percentage of the total carbon feed in the form of CO_2. Due to relatively low per-pass conversion and associated high recycling quantities, impurities like inert gases and light by-products are only present in low concentrations. The purge gas, therefore, predominantly contains H_2 and CO_2. The required purge quantity depends mainly on the purity of the feedstock. The top stream

---

[11] Calculated from production-related emissions [1].

(fuel gas) separated in the light gas column (also called topping column) contains not only CO_2 but also low-boiling by-products such as DME and methyl formats; i.e., recirculation cannot take place without further separation of the components. Methanol losses in the methanol–water separation (methanol column) can be kept low by sufficient distillation. In the concept presented, a carbon efficiency of 95% is achieved, with purge streams set to keep the inert gas concentration below 5% in the synthesis loop.

Figure 14 shows an example of the energy flows for CO_2-converting methanol synthesis, using CO_2 from a point source and hydrogen generation through low-temperature electrolysis. The energy efficiency is predominantly determined by the efficiency of the electrolysis, which usually requires more than 90% of the electricity and more than 80% of the energy input. Further losses occur due to the exothermic nature of the methanol synthesis. This heat occurs at a relatively high-temperature level (ca. 200 °C) enabling an efficient heat integration. The main heat sinks in the process are distillation, synthesis gas heating, and, if located on the production side, also the carbon capture plant.

The PtM synthesis route via CO-based synthesis gas theoretically requires more energy since significant amounts of heat at high temperature levels (about 900 °C) are required for the RWGS reaction. Although the subsequent methanol synthesis also generates significantly more waste heat, which, however, cannot be used for the RWGS due to the lower temperature level. Thermodynamically, the thermal energy released under standard conditions is $14\%_{LHV}$ for CO-based methanol synthesis (CAMERE process) and only $8\%_{LHV}$ for CO_2-based methanol synthesis. However, for direct CO_2 conversion, higher energy demand for synthesis gas compression needs to be considered.

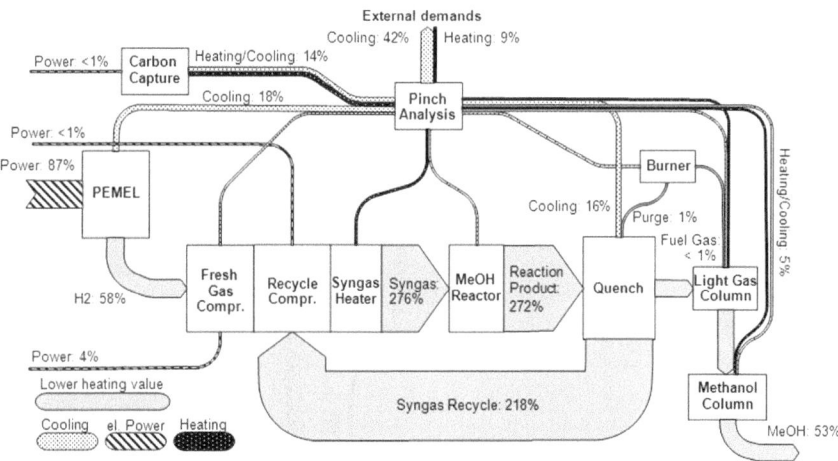

Fig. 14 Energy streams of a CO_2-converting PtM concept (percentages indicate relative values in relation to the total energy input; only main heat flows labeled. Carbon capture is considered via amine wash from CO_2 point source; PEMEL: proton exchange membrane electrolysis)

In summary, the energy efficiency of PtM concepts is predominantly determined by H_2-production, and to a lower extent, by CO_2 supply and synthesis technology. Improvements in overall efficiency could therefore be achieved primarily through advanced electrolysis technologies (such as high-temperature electrolysis), although these only have a positive effect if sufficient usable waste heat is available at the site. In ideal constellations, overall energy efficiency could rise to over 60% [42]. However, if sustainable CO_2 point sources are not available at the site and direct air capture (DAC) technologies must be used for CO_2 provision, the overall process energy efficiency could decrease to about 40% [23, 43].

From a technical perspective, the technically challenging RWGS is the biggest drawback in the CO-converting PtM pathway (CAMERE process). The main technical challenge of the direct CO_2-converting pathway can be allocated to the smaller per-pass conversion and the associated effects on the synthesis loop. From an energy point of view, this leads to increasing compression power and higher purge losses.

Regarding the capacity of PtM concepts, currently only pilot and small-scale industrial plants are in operation. However, from a technical point of view, there are no direct upscaling restrictions; therefore, much larger concepts could already be realized today. Compared to current conventional methanol plants, the high recycling flows expected for CO_2-based production could have a limiting effect on the capacity of individual reactors since these have already reached a capacity maximum for conventional plants (due to pressure losses and structural limitations). In the medium term, however, the capacity of future PtM plants will be much more limited by the availability of sustainable CO_2. Furthermore, since renewable energy is most likely not constantly available, lower full load hours (FLH) can be expected for PtM concepts, which is why the annual production output can be limited additionally to the maximum nominal load.

6.2 Economic Aspects

A decisive factor for the implementation of PtM concepts is the economic viability against conventional production and competing renewable energy carriers. Although electricity-based methanol is chemically identical to conventional methanol, it may have a higher market value due to its renewable origin and reduced climate impact. The actual monetary value, therefore, is influenced by the national and international regulatory framework and customers' willingness to pay more for a "green" product. A trade decoupled from conventional fossil methanol is to be expected, provided that political framework conditions aim to include the climate impact of products in pricing through a downstream burden. This could be achieved, for example, by setting mandates and/or effective CO_2 pricing systems. Another possibility could be the upstream burden of conventional methanol; under these circumstances, there would be less incentive to trade fossil and renewable methanol separately. However, as long as CO_2 emissions from fossil sources are not adequately incorporated into pricing, renewable methanol will face direct competition with conventionally produced methanol. In addition to economic competitiveness, the price difference between fossil and renewable methanol is also important in assessing potential cost changes for the long-term pricing of downstream methanol products.

Conventional methanol production: The cost of conventional methanol production depends mainly on the direct feedstock costs, the investment costs, and to a smaller amount on logistic- and infrastructure-related costs [3]. In principle, hard coal is less expensive as a feedstock compared to natural gas, but its specific demand is higher and conversion to synthesis gas is more complex, leading to significantly higher investment costs. Typically, synthesis gas generation accounts for the majority of capital expenditures (CAPEX); this is true for more than 50% of the total investment in natural gas-based plants and for coal-based plants this sums up to about 80% of the total investment [34]. Since most sub-processes in conventional methanol production plants are primarily scaled by increasing the size of the components rather than by numbering-up, the overall process can be assumed to have a nonlinear scaling with an exponent of approximately 0.7 [35, 44]. These economy-of-scale effects can significantly reduce the specific capital expenditure and their share of methanol production costs in large-scale plants. For example, compared to a 2,000 t/d plant, the investment costs of a 5,000 t/d plant and a 10,000 t/d plant are reduced by 30% and 40%, respectively [23]. The feedstock costs, on the other hand, account for over 90% of operational expenditures (OPEX), which makes the availability of low-cost feedstock the key for a competitive conventional methanol provision. The production costs, therefore, lie mainly in the range of 50–100 €/t_{MeOH} for countries with rather cheap feedstock (e.g., Iran, Turkmenistan) and 300–400 €/t_{MeOH} for locations with higher feedstock costs (e.g., Europe) [45].

Electricity-to-Methanol production: Methanol production costs from PtM plants are dominated by the cost of hydrogen production. These costs are mainly determined by the electricity costs and the electrolyzer costs. Electrolyzers are mainly scaled up by increasing the number of modules with only little economies of scale effects. However, costs for electrolyzers are expected to further decrease in the future due to the development of larger modules and a mass-market production. Compared to conventional methanol production, specific CAPEX are expected to be higher for PtM concepts since electrolyzer and, if required, direct air capture (DAC) plants are today still rather costly. Slightly increased costs can also be expected for the synthesis loop. An example of cost breakdown and cost sensitivity for a PtM concept is shown in Fig. 15.

The considered production plant for the data in Fig. 15 has a production capacity of approx. 40,000 t/a operated with 8,000 h/a (full load) (FLHs). The electricity price is assumed to be 30 €/$MWh_{el,}$ and the CO_2 is provided via capturing from a "green" point source. The OPEX resulting primarily from electricity costs are significantly higher than CAPEX.[12] The process-specific costs are predominantly based on the supply of hydrogen. The costs of the other process sections are evenly distributed,

---

[12] Interest rate of 6% and a depreciation period of 10 years for the electrolysis plant and 20 years for all other plant parts were assumed.

Fig. 15 PtM production cost
(left) and cost sensitivity
(right) (CAPEX: capital
expenditure, CCU: carbon
capture utilization, FLH: full
load hours, MeOH:
methanol, OPEX:
operational expenditure)

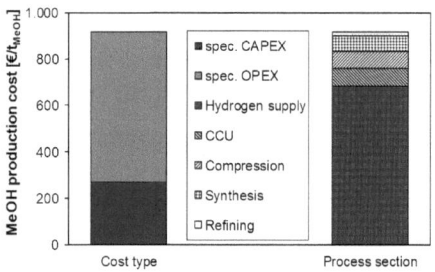

with the costs of refining being low due to extensive heat integration. Cost sensitivity shows a significant nonlinear increase in costs, especially for low annual full load hours and plants with small production capacities. Assuming an energetic efficiency of approx. 50%, the methanol production costs increase by approx. 11 €/t_{MeOH} per 1 €/MWh_{el}. The CO_2 supply costs are incorporated in the methanol production costs by a factor of 1.4, which means that, for example, CO_2 costs of 100 €/t_{CO2} result in a cost share of 140 €/t_{MeOH} for CO_2 supply.

Given the volatility and seasonal fluctuations of renewable energies, cost-optimized dimensioning of plant components becomes crucial, particularly in balancing the production of hydrogen and the methanol synthesis section. This may also result in different full load hours (FLH) for the individual components of the plant (e.g., the electrolyzer and the methanol synthesis section). As a consequence, this likely leads to higher costs compared to conventional methanol production plants; plant components may need to be oversized, so that significantly reduced annual full load hours or capacity utilization can be expected. Both of these factors contribute to increased costs. For plant concepts operated at high annual full load hours, methanol production costs are expected to be in the range of 300–900 €/t_{MeOH}, whereby costs below 700 €/t_{MeOH} only seem achievable for the most advantageous locations and in the medium term [23]. For current projects, the development costs are more likely to be in the range of 800–1,600 €/t_{MeOH} [1].

6.3   Environmental Aspects

The environmental aspects in the context of PtX technologies mainly focus on the climate impact and, thus the GHG emissions of the production and use. Currently, global methanol production and use are associated with GHG emissions of approximately 0.3 Gt_{CO2eq}/a [1]. This corresponds to about 1% of total global GHG emissions and confirms that methanol-related emissions play a major role in climate change mitigation. Figure 16 shows the specific GHG emissions from different methanol production pathways. The molecular carbon emissions correspond to the GHG emissions from methanol combustion. The carbon bound in the chemical composition of the methanol molecule is released in the end use (combustion of the fuel) or, for most

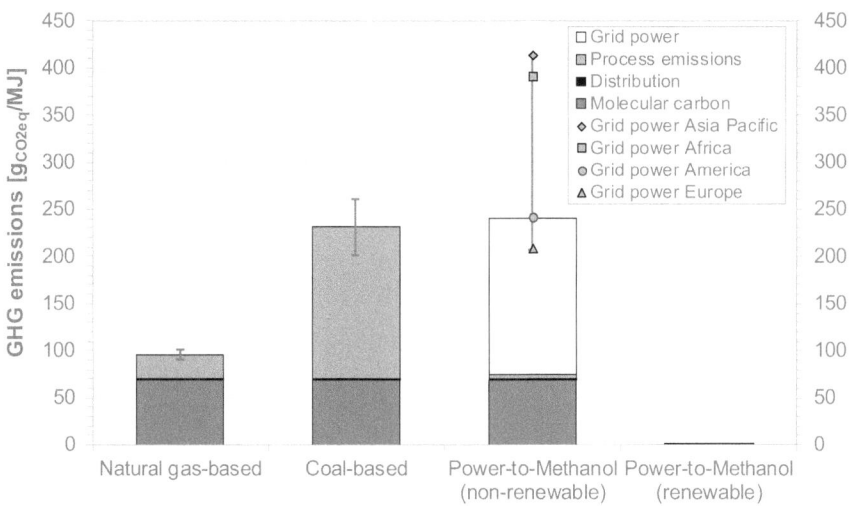

Fig. 16 Specific GHG emissions from methanol production pathways (electricity emissions from [46], transport and distribution from RED II standard value [47], process emissions calculated from [1, 48, 49])

other applications, in the end use of the downstream products (e.g., incineration of non-recyclable plastic). Other path-dependent emissions are described below.

Conventional methanol production: According to Fig. 16 the specific process emissions of natural gas-based methanol are significantly lower than those of hard coal-based methanol. This is mainly due to the higher hydrogen-to-carbon ratio in natural gas, i.e., less CO_2 has to be separated within the synthesis gas conditioning. In addition, the energy requirement of natural gas-based production is lower; i.e., less raw material is needed for heat generation [48–50]. Since these emissions vary depending on the specific raw material and plant characteristics, significant variations can occur [1, 51]. Because natural gas generally has little variation in composition, process-related emissions vary less than for hard coal, where different types of hard coal with significantly different molecular compositions occur.

Electricity-to-Methanol production: GHG emissions from electricity-based methanol production are predominantly driven by emissions resulting from electricity production as well as from methanol combustion emissions (molecular carbon). When non-renewable carbon dioxide is incorporated into the chemical composition of the alcohol, the molecular carbon emissions also apply to PtM-based methanol. For a renewable carbon feedstock, emissions from combustion do not increase climate-effective GHG concentration in the atmosphere, i.e., molecular carbon can be evaluated as carbon neutral. This also applies to process- or production-related emissions, although these are considerably lower than for conventional production (Sect. 6.1). The comparison between a fully renewable and a non-renewable PtM process in terms of GHG emissions also shows the considerable impact of electricity-related emissions. For non-renewable production, the average emission intensities of grid power from different continents are considered, while renewable electricity is assumed to have zero GHG emissions. It can be seen that PtM can result in significant emission reductions when only renewable sources are used.

7 Summary and Outlook

Today, methanol is one of the most widely used basic chemicals worldwide, and partly also used as a fuel additive or fuel, especially in Asia. In the past decade (2010–2020), the global demand for methanol has nearly doubled, primarily due to its increased utilization in olefin production. Also, in both production and consumption a regional shift toward China, where methanol is derived from domestic hard coal reserves, occurred in the past decade. In contrast, in almost all other parts of the world, methanol is primarily produced from natural gas, which offers process advantages due to the higher hydrogen content of methane compared to hard coal. However, given the CO_2 emissions associated with methanol production from such fossil fuel-based feedstocks, a transition toward more sustainable feedstocks from renewable energy sources is imperative to achieve carbon neutrality in the provision of methanol.

In this context, electricity-based production of methanol, so-called e-methanol, from CO_2 and H_2O via power-to-methanol (PtM) pathways presents a promising approach. From a technological perspective, the changes compared with conventional production processes are mainly related to the supply of synthesis gas and less to the methanol synthesis itself. Today, concepts and demonstration plants that ensure sufficient technical maturity of the PtM process already exist. If a PtM pathway is pursued, where CO_2 is directly converted into methanol in a single conversion step, adjustments to the synthesis section will be necessary compared to conventional methanol production. However, the direct conversion allows for higher methanol selectivity and a less complex overall process, i.e., the avoidance of a RWGS upstream and potentially simplified distillation downstream.

Regarding PtM-based production plants, they are expected to have significantly lower production capacities in most cases when compared to current world-scale methanol plants using natural gas or hard coal. This is partly due to the fact that renewable energy and CO_2 do not occur locally in the quantities of fossil resources, and also because the availability of renewable energy from wind and solar radiation fluctuates over time. Especially for methanol plants in first PtM projects, significantly lower production capacities are to be expected.

Differences in specific process technologies, feedstock supply characteristics, or plant scales compared to conventional world-scale methanol plants also result in economic differences. For instance, reduced plant utilization due to volatile energy provision increases specific annual capital costs. Additionally, if the overall availability of electricity or hydrogen is to be increased through additional storage capacities to allow for more continuous methanol synthesis, this will lead to higher investments for the upstream processes, thereby increasing overall methanol production costs. Thus, a key aspect for future PtM plant concepts is the (economic) optimization of electricity supply and the capacities of individual sub-processes throughout the complete production chain to achieve minimal methanol production costs. Given the significant cost disparities between PtM-based methanol and fossil fuel-based methanol, even at advantageous locations with high renewable energy availability, a crucial factor for competitive PtM production is effective regulatory support schemes that facilitate the purchase of more expensive "green" methanol.

From an environmental perspective considering GHG emissions, PtM processes are only beneficial if they predominantly rely on renewable energy and sustainably provided CO_2. In this case, such concepts can be used to produce "green" and potentially GHG-neutral methanol. The carbon that is chemically bound in methanol during production does not lead to GHG emissions during production, but is very likely to be released as CO_2 at the end of the respective methanol life cycle. Therefore, in this context, only the use of non-fossil CO_2 can be considered sufficiently sustainable.

While "green" methanol currently represents only a small fraction of the global methanol production, several factors indicate a possible substantial increase in its future demand. One such factor are the chemical properties of methanol; as a simple C_1-alcohol it can be processed in a wide range of conversion pathways, which, among others, enable to produce a variety of second-generation petrochemicals from methanol [4]. This would offer the possibility to replace crude oil as the carbon feedstock of today's petrochemical industry. As a fuel, methanol can be used either directly in spark-ignition combustion engines or even in fuel cells. In direct combustion, the alcohol shows good combustion properties, although its heating value is only about half that of conventional fuels (e.g., gasoline or diesel). Furthermore, the conversion of methanol into hydrocarbons is also possible to fulfill the given fuel standards for conventional fuels (e.g., gasoline, kerosene, diesel). Thus, as defossilization efforts advance, "green" methanol could have the potential not only to substitute significant shares of conventional methanol but also to increase the overall methanol demand, particularly in the chemical industry and the transport sector.

However, given the energy-intensive nature of synthesis gas provision and the demand for (non-fossil) sustainable carbon, which might become a scarce resource, e-methanol use should be prioritized for material use and applications where a defossilization by carbon-free technologies, such as direct electrification or hydrogen utilization, is not feasible. However, besides these efforts, overall methanol production via PtM concepts represents a carbon-efficient technology that allows to extend the options for defossilization where decarbonization is not feasible.

8 Excursus—Mixed Alcohol Synthesis

When process conditions deviate from the optimal conditions for methanol synthesis, other/higher alcohols are also formed in the product spectrum alongside methanol. This can be specifically exploited to synthesize mixtures of higher alcohols from synthesis gas, having a longer hydrocarbon chain (mainly C_2 to C_6) attached to the hydroxy (OH) group compared to methanol. In this context, particularly the production of ethanol, propanol, and butanol (C_2 to C_4 alcohols) is of special interest.

- Ethanol, with an average global production of around 100 Mt/a, is an important biotechnologically produced basic chemical and component in the food and beverage industry [52]. It is also used as a fuel admix component or even as a pure fuel for road transportation [53]. In the EU ethanol is added to gasoline in shares of 5–10 vol.-%, depending on legal requirements.
- Propanol and butanol are produced to a much more limited extent compared to ethanol (and also methanol) with global production volumes of about 3 Mt/a. Propanol is used, for instance, as a solvent in paints, varnishes or as an additive to antifreeze [3]. Butanol is mainly used as a solvent and basic chemical for the chemical industry. Similar to ethanol, both of these alcohols can also be added to gasoline as a fuel admix component.

Mixed alcohols can be produced by the thermo-catalytic/thermo-chemical and the biocatalytic/biochemical process route, each from partly different feedstock. In the following, the thermo-chemical mixed alcohol synthesis is discussed, in which a mixture of methanol to higher alcohols is produced from a synthesis gas. The biotechnological, fermentative conversion of synthesis gas is not described in detail here.

Thermodynamics: Besides methanol, longer-chain alcohols (i.e., ethanol, propanol and butanol) can also be synthesized from a hydrogen and carbon monoxide mixture in a catalyst-controlled manner. The formation reaction Eq. (7) generally describes the formation of alcohols starting from synthesis gas.

$$n\,CO + 2n\,H_2 \rightleftharpoons C_nH_{2n+1}OH + (n-1)H_2O \tag{7}$$

Depending on the catalyst, pressure and temperature as well as the respective reactor design, a broad product spectrum consisting primarily of methanol, higher unbranched alcohols (usually C_2 to C_8), alkanes, alkenes, dimethyl ether (DME), methyl esters, and higher branched alcohols is usually produced via further chemical secondary reactions. Based on the stoichiometry according to the aforementioned equation, a molar H_2 to CO ratio of 2:1 is ideal. In general, however, higher conversions and selectivities are achieved at a lower H_2 to CO ratio due to the higher C–C bond formation [54, 55].

Reactors: In order to increase the yield and selectivity toward the higher alcohols, a double-bed reactor system is discussed in addition to the "classic" fixed-bed reactor, which can lead to higher conversions and selectivities in comparison. The aim is to form methanol in the first catalyst bed at lower temperatures, which then reacts further in a second reactor bed at higher temperatures to form higher alcohols via the formation of C–C bonds. Due to the higher thermodynamic stability of iso-alcohols, this leads to their formation in particular and correspondingly to a lower formation of linear n-alcohols. Another reactor concept that allows the heat of the highly exothermic reaction to be safely removed is the slurry reactor. Compared to the fixed-bed reactor, this reactor concept offers increased heat and mass transfer and also allows catalyst regeneration during operation. In this reactor concept, by-product formation (e.g., CO_2 formation) is also suppressed [55].

Catalysts: The higher alcohols are initially formed via the synthesis of methanol and the subsequent insertion of a CO molecule into the previously formed synthesis product. Against this background, modified alkali, zinc oxide (ZnO), or chromium oxide (Cr_2O_3) catalysts from high-pressure methanol synthesis are particularly suitable for catalyzing this reaction. Furthermore, further developed catalysts from the low-pressure methanol synthesis (alkali, Cu, ZnO, Al_2O_3), modified Fischer–Tropsch catalysts (alkali, CoO, Al_2O_3) as well as alkali-doped sulfides (especially molybdenum disulfide (MoS_2)) are used [54].

The currently available catalysts show comparatively low conversions and selectivities. Thus, a CO conversion of approx. 5–30% per-pass and a selectivity to higher alcohols of 30–90% are achieved depending on temperature, pressure, and catalyst. Usually, a higher selectivity results in a lower conversion. Furthermore, the catalysts have a short lifetime, as they are deactivated by metal sintering and phase separation [56].

Synthesis loop: As mentioned above, only low per-pass conversions can be achieved in the synthesis reactor within a mixed alcohol synthesis. Thus, i.e., to enable high overall conversions along the synthesis loop, a recycle of unreacted synthesis gas is necessary. For this purpose, the alcohols produced are first condensed from the reaction product and separated. However, the remaining gas phase contains light by-products (such as short chain alkanes and alkenes) as well as unreacted synthesis gas components (H_2, CO). This is why additional reforming of this gas stream may be necessary before the unreacted synthesis gas is returned to the reactor. However, this requires a high temperature, which is associated with efficiency losses in the overall process. In order to minimize such losses, there is still a need for further research and development—especially on the catalyst side [57]. In addition, the implementation of purge streams is a common practice to prevent accumulation of inert gases in the synthesis loop.

The total synthesis route starting from the synthesis gas to the production of mixed alcohols is shown in Fig. 17.

The synthesis gas is first heated to the reaction conditions and fed into the reactor. The produced by-products can be separated in a gas separator and fed back to the reactor via a reforming step. Subsequently, the water can be separated from the alcohol mixture, for example by rectification and pressure swing adsorption. In the last step, the alcohol mixture is separated into its individual products via distillation/

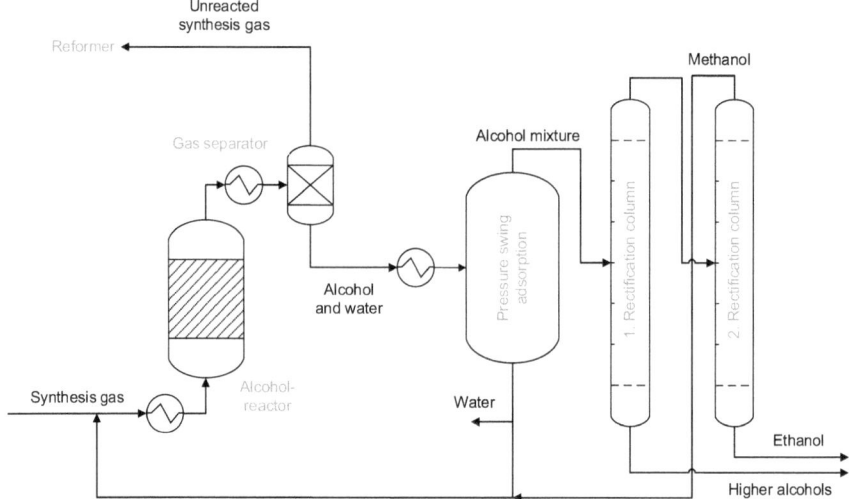

Fig. 17 Overall synthesis section of the mixed alcohol synthesis (according to [57])

rectification. Here, the methanol separation is of particular interest, as this can also be returned to the synthesis reactor, for example to increase the selectivity of the higher alcohols [57]. The methanol can also be recycled together with the separated water if the adjusted H_2 to CO ratio in the synthesis gas is too low. In this case, the water in the synthesis reactor can participate in the water–gas shift (WGS) reaction together with CO under the given process conditions Eq. (3) and thus directly increase the hydrogen concentration in the reactor [57].

Operating conditions: The operating conditions depend on which catalyst technology is used. If modified high-pressure methanol synthesis catalyst is used, pressures from 125 to 300 bar and temperatures from 300 to 425 °C are reached. If low-pressure methanol synthesis catalyst is used, pressures from 50 to 100 bar and temperatures from 275 to 310 °C are used. The other catalysts are also in a similar range of values like the low-pressure methanol catalysts [54].

By-products: As can be seen from Eq. (7), the main by-product of the reaction is water, which is increasingly produced as the chain length of the alcohol increases. However, there is also a high formation of methanol, which in this case is not the actual target product. In addition, CO_2 can be formed, as well as ether compounds and hydrocarbons, which have to be separated in the downstream process.

References

1. International Renewable Energy Agency (IRENA) and the Methanol Institute (2021) Innovation outlook: renewable methanol. Abu Dhabi

2. Sheldon D (2017) Methanol production—a technical history. Johnson Matthey Technol Rev 61(3):172–182. https://doi.org/10.1595/205651317X695622
3. Ullmann's encyclopedia of industrial chemistry: methanol. Wiley-VCH Verlag GmbH & Co. KgaA, Weinheim, Germany, 2012
4. Bertau M, Offermanns H, Plass L, Schmidt F, Wernicke H-J (2014) Methanol: the basic chemical and energy feedstock of the future. Springer Berlin Heidelberg, Berlin, Heidelberg
5. Bozzano G, Manenti F (2016) Efficient methanol synthesis: perspectives, technologies and optimization strategies. Prog Energy Combust Sci 56:71–105. https://doi.org/10.1016/j.pecs.2016.06.001
6. Methanol synthesis. In: Handbook of heterogeneous catalysis online
7. Goeppert A, Czaun M, Jones J-P, Surya Prakash GK, Olah GA (2014) Recycling of carbon dioxide to methanol and derived products—closing the loop. Chem Soc Rev 43(23):7995–8048. https://doi.org/10.1039/c4cs00122b
8. Olah GA, Goeppert A, Prakash GKS (2009) Beyond oil and gas: the methanol economy, 2nd edn. Wiley-VCH, Weinheim
9. Medina E, Methanol safe handling manual, 5th edn
10. Bechtold RL, Goodman MB, Timbario TA (2007) Use of methanol as a transportation fuel
11. Zukünftige Kraftstoffe MW (2019) Springer Berlin Heidelberg, Berlin, Heidelberg
12. Kowalewicz A (1993) Methanol as a fuel for spark ignition engines: a review and analysis. Proc Instn Mech Engrs, vol 207
13. Methanol Institut. Safe-Handling-Manual_5th-Edition_Final
14. Engineering ToolBox (2022) Methanol—thermophysical properties. Available from: https://www.engineeringtoolbox.com/methanol-methyl-alcohol-properties-CH3OH-d_2031.html
15. NIST Chemistry WebBook. Methyl alcohol. Available from: https://webbook.nist.gov/cgi/cbook.cgi?ID=67-56-1
16. Agarwal AK, Gautam A, Sharma N, Singh AP (2019) Methanol and the alternate fuel economy. Springer Singapore, Singapore
17. Methanol Market service Asia (2022) Methanol price and supply/demand. Available from: https://www.methanol.org/methanol-price-supply-demand/
18. Alvarado M (2022) The changing face of the global methanol industry. Available from: https://www.methanol.org/wp-content/uploads/2016/07/IHS-ChemicalBulletin-Issue3-Alvarado-Jun16.pdf
19. Harker J, Sanchez A, Gupta A, English A (2022) Process design overview for upgrading a gas-to-methanol facility. (28 Apr 2022). Available from: http://gasprocessingnews.com/features/201612/process-design-overview-for-upgrading-a-gas-to-methanol-facility.aspx
20. Methanol Market Services Asia (2022) Methanol markets. (28 Apr 2022). Available from: https://www.methanolmsa.com/methanol/
21. Methanex, Methanex monthly average regional posted contract price history. Available from: https://www.methanex.com/our-business/pricing
22. International Monetary Fund. Primary commodity prices. Available from: https://www.imf.org/en/Research/commodity-prices
23. Dieterich V, Buttler A, Hanel A, Spliethoff H, Fendt S (2020) Power-to-liquid via synthesis of methanol, DME or Fischer–Tropsch-fuels: a review. Energy Environ Sci 13(10):3207–3252. https://doi.org/10.1039/D0EE01187H
24. Arndt C, Neuling U, Vorsatz M, Prause J, Molzberger U, Le Clercq P, Frech G, Wollenweber H, Tauchnitz H, Jänisch T (2021) Konzeptionelle und technische Ausgestaltung einer Entwicklungsplattform für Power-to-Liquid-Kraftstoffe: Abschlussbericht
25. Schemme S (2020) Techno-ökonomische Bewertung von Verfahren zur Herstellung von Kraftstoffen aus H2 und CO2. Energie & Umwelt/Energy & Environment
26. Anicic B, Trop P, Goricanec D (2014) Comparison between two methods of methanol production from carbon dioxide. Energy 77:279–289. https://doi.org/10.1016/j.energy.2014.09.069
27. Nestler F, Krüger M, Full J, Hadrich MJ, White RJ, Schaadt A (2018) Methanol synthesis—industrial challenges within a changing raw material landscape. Chem Ing Tec 90(10):1409–1418. https://doi.org/10.1002/cite.201800026

28. Kaltschmitt M, Hartmann H, Hofbauer H (2009) Energie aus Biomasse. Springer Berlin Heidelberg, Berlin, Heidelberg

29. Buttler A, Spliethoff H (2018) Current status of water electrolysis for energy storage, grid balancing and sector coupling via power-to-gas and power-to-liquids: a review. Renew Sustain Energy Rev 82:2440–2454. https://doi.org/10.1016/j.rser.2017.09.003

30. Siemens Energy (2022) Silyzer 300: the next paradigm of PEM electrolysis. (01 May 2022). Available from: https://assets.siemens-energy.com/siemens/assets/api/uuid:a193b68f-7ab4-4536-abe2-c23e01d0b526/datasheet-silyzer300.pdf

31. Cummins Inc (2022) HyLYZER® water electrolyzers: electrolyzer brochure. (01 May 2022). Available from: https://www.cummins.com/new-power/applications/about-hydrogen

32. Rodin V, Lindorfer J, Böhm H, Vieira L (2020) Assessing the potential of carbon dioxide valorisation in Europe with focus on biogenic CO_2. J CO2 Utilization 41:101219. https://doi.org/10.1016/j.jcou.2020.101219

33. De Heer J (1957) The principle of Le Châtelier and Braun. University of Colorado. J Chem Educ 34:375–80

34. Hansen JB, Hjlund Nielsen PE (2008) Methanol synthesis. In: Ertl G, Knzinger H, Schth F, Weitkamp J (eds) Handbook of heterogeneous catalysis. Wiley-VCH Verlag GmbH & Co. KGaA, Weinheim, Germany

35. Biegler LT, Grossmann IE, Westerberg AW (1998) Solutions manual: systematic methods for chemical process desing, 5th edn. Prentice Hall, Upper Saddle River (NJ)

36. ParsToday (2022) World's largest methanol plant opened in southern Iran. Available from: https://parstoday.com/en/news/iran-i124964-world%E2%80%99s_largest_methanol_plant_opened_in_southern_iran

37. Seidel C, Nikolić D, Felischak M, Petkovska M, Seidel-Morgenstern A, Kienle A (2021) Optimization of methanol synthesis under forced periodic operation. Processes 9(5):872. https://doi.org/10.3390/pr9050872

38. Diermann R (2020) Fraunhofer ISE erforscht Herstellung von klimaneutralem Methanol auf Basis von Wasserstoff

39. Stralsund H (2020) Das IRES hat´s geschafft: Endlich flüssiger Strom

40. Butera G, Fendt S, Jensen SH, Ahrenfeldt J, Clausen LR (2020) Flexible methanol production units coupling solid oxide cells and thermochemical biomass conversion via different gasification technologies. Energy 208:118432. https://doi.org/10.1016/j.energy.2020.118432

41. Hydrocarbon Processing (2022) World's largest ATR-based methanol plant has been put into successful operation. Available from: https://www.hydrocarbonprocessing.com/news/2020/01/world-s-largest-atr-based-methanol-plant-has-been-put-into-successful-operation

42. Rivera-Tinoco R, Farran M, Bouallou C, Auprêtre F, Valentin S, Millet P et al (2016) Investigation of power-to-methanol processes coupling electrolytic hydrogen production and catalytic CO_2 reduction. Int J Hydrogen Energy 41(8):4546–4559. https://doi.org/10.1016/j.ijhydene.2016.01.059

43. Fasihi M, Efimova O, Breyer C (2019) Techno-economic assessment of CO_2 direct air capture plants. J Clean Prod 224:957–980. https://doi.org/10.1016/j.jclepro.2019.03.086

44. Peters MS, Timmerhaus KD, West RE (2006) Plant design and economics for chemical engineers, 5th edn. McGraw-Hill, Boston

45. Boulamanti A, Moya JA (2017) Production costs of the chemical industry in the EU and other countries: ammonia, methanol and light olefins. Renew Sustain Energy Rev 68:1205–1212. https://doi.org/10.1016/j.rser.2016.02.021

46. International Energy Agency (2022) Electricity market report

47. European Union (2018) DIRECTIVE (EU) 2018/2001 of the European parliament and of the council: red II

48. Svanberg M, Ellis J, Lundgren J, Landälv I (2018) Renewable methanol as a fuel for the shipping industry. Renew Sustain Energy Rev 94:1217–1228. https://doi.org/10.1016/j.rser.2018.06.058

49. Kajaste R, Hurme M, Oinas P (2018) Methanol-Managing greenhouse gas emissions in the production chain by optimizing the resource base. AIMS Energy 6(6):1074–1102. https://doi.org/10.3934/energy.2018.6.1074

50. Ellis J, Svanberg M, Expected benefits, strategies, and implementation of methanol as a marine fuel for the smaller vessel fleet. http://summeth.marinemethanol.com/. Deliverable D5.1
51. Matzen M, Alhajji M, Demirel Y (2015) Chemical storage of wind energy by renewable methanol production: feasibility analysis using a multi-criteria decision matrix. Energy 93:343–353. https://doi.org/10.1016/j.energy.2015.09.043
52. OECD-FAO Agricultural Outlook (2021) 2021–2030. OECD Publishing, Paris
53. Campos JN, Viglio JE (2022) Drivers of ethanol fuel development in Brazil: a sociotechnical review. MRS Energy Sustain 9(1):35–48. https://doi.org/10.1557/s43581-021-00016-6
54. Spath PL (2003) DDC. Preliminary screening—technical and economic assessment of synthesis gas to fuels and chemicals with emphasis on the potential for biomass-derived syngas. NREL/TP-510-34929
55. Fang K, Li D, Lin M, Xiang M, Wei W, Sun Y (2009) A short review of heterogeneous catalytic process for mixed alcohols synthesis via syngas. Catal Today 147(2):133–138. https://doi.org/10.1016/j.cattod.2009.01.038
56. Ao M, Pham GH, Sunarso J, Tade MO, Liu S (2018) Active centers of catalysts for higher alcohol synthesis from syngas: a review. ACS Catal 8(8):7025–7050. https://doi.org/10.1021/acscatal.8b01391
57. Phillips S, Aden A, Jechura J, Dayton D (2007) Thermochemical ethanol via indirect gasification and mixed alcohol synthesis of lignocellulosic biomass: technical report. NREL/TP-510-41168

From Conventional to Emerging Ammonia Production Technologies

Laura Collado, Alejandro Herrero, and Víctor A. de la Peña O'Shea

Abstract Chemicals are the foundation of modern society. Nowadays, most useful base chemicals (e.g., hydrogen, ammonia, methanol, olefins, etc.) that are building blocks of everyday products (i.e., plastics, fertilizers, clothes, or cosmetics) are still produced from fossil feedstocks, mainly oil and natural gas. Ammonia (NH_3) is one of the most critical chemicals in modern society, with an annual production of approximately 183 Mt, primarily used as a key component in fertilizers, supporting the global food supply. Beyond its role in agriculture, ammonia serves as a refrigerant and a chemical feedstock in various industries. Historically explored as a fuel, its application was limited by the availability of cheaper fossil resources. However, renewed interest positions ammonia as a zero-carbon fuel alternative for transportation and power generation, as well as an efficient hydrogen carrier. The transition to green ammonia production is essential to reduce the high carbon footprint of conventional methods, which rely on fossil feedstocks. Green ammonia offers a pathway to decarbonizing key sectors. Future success depends on fulfilling key criteria such as cost-effectiveness, safety, and carbon neutrality. As hydrogen is expected to meet 20% of global energy demand by 2050, the demand for ammonia is likely to rise, with industry players already planning green ammonia projects. If scaled effectively, ammonia could become a cornerstone of the low-carbon economy, replacing significant portions of fossil fuel consumption.

Keywords Green Ammonia · Haber-Bosch Process · Electrocatalysis · Photochemical · Ammonia Cracking

1 Ammonia—The Green Energy Carrier of the Future

Chemicals are the foundation of modern society. Nowadays, most useful base chemicals (e.g., hydrogen, ammonia, methanol, olefins, etc.) that are building blocks of everyday products (i.e., plastics, fertilizers, clothes, or cosmetics) are still produced

L. Collado (✉) · A. Herrero · V. A. de la Peña O'Shea
IMDEA Energy Institute, Photoactivated Processes Unit, Madrid, Spain
e-mail: laura.collado@imdea.org

© The Author(s), under exclusive license to Springer Nature Switzerland AG 2025
N. Bullerdiek et al. (eds.), *Powerfuels*, Green Energy and Technology,
https://doi.org/10.1007/978-3-031-62411-7_25

713

from fossil feedstocks, mainly oil and natural gas. Ammonia (NH_3) is the second most produced chemical worldwide with approximately 183 Mt of annual production [1], and it is one of the largest-volume industrial chemicals in terms of energy use and carbon footprint [2]. Ammonia is predominantly used as a key component of mineral fertilizers (Fig. 1). In the beginning of the twentieth century, estimations state that 45% of the fixed inorganic nitrogen were used in the fertilizer industry [3]. Nowadays, this sector consumes about 88% of the total NH_3 production [4], and supports the feeding of over 48% of the world's population [5], which is expected to rise 30% by 2050 with the consequent intensification in the use of fertilizers. Besides, ammonia is also used as a good industrial refrigerant (i.e., in breweries and warehouses) [6] and as chemical feedstock in the production of polyimides, nitric acid, nylon, pharmaceuticals, dyes, explosive materials, cleaning solutions, etc. [7, 8].

Besides current utilization, some attempts have been made since the end of the nineteenth century to power medium-sized locomotive systems with ammonia. However, its application as a fuel has been stopped due to the lower cost of fossil resources, low efficiencies, and poor public perception [9]. Nevertheless, a renewed interest has emerged in the twenty-first century proposing ammonia as a potential

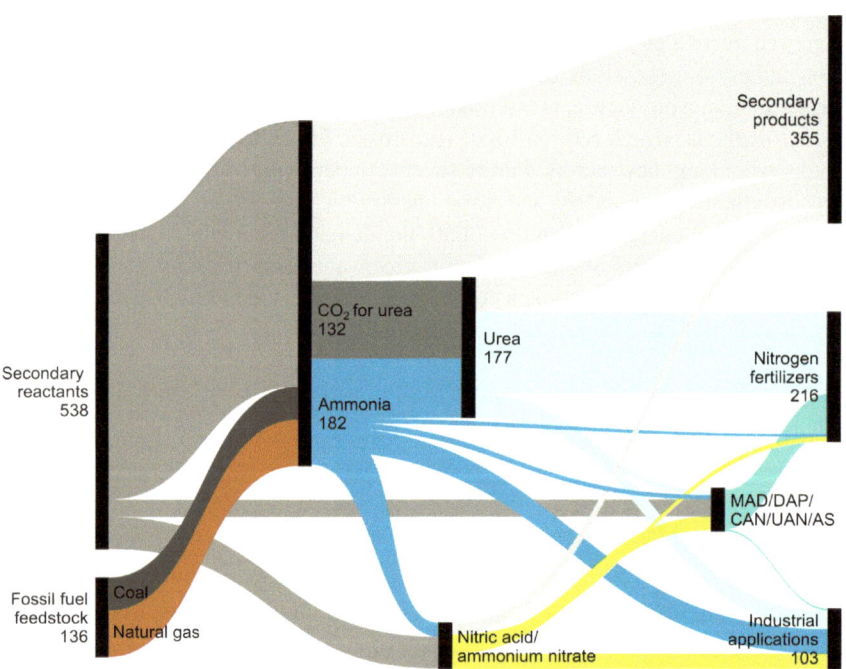

Fig. 1 Mass flows in the global ammonia supply chain in 2019 (AS: ammonium sulfate; CAN: calcium ammonium nitrate; DAP: diammonium phosphate; MAP: monoammonium phosphate; UAN: urea ammonium nitrate; numeric values in Mt/a; values do not represent process energy inputs; *Source* [1])

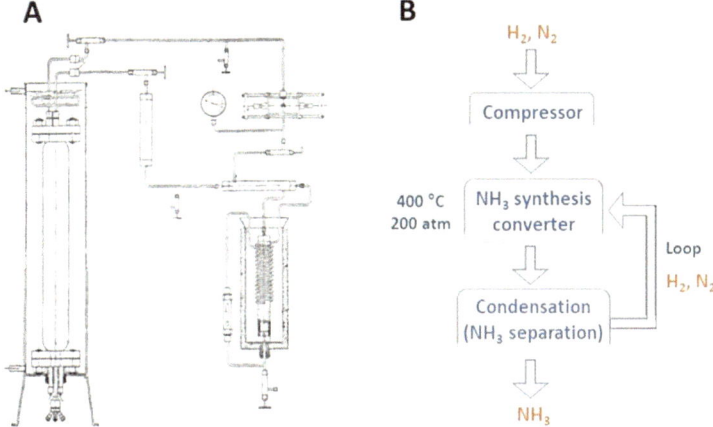

Fig. 2 **a** Apparatus used by Haber and Rossignol for the catalytic synthesis of ammonia (Copyright 1913 American Chemical Society. Reprinted with permission from reference [28]). **b** Simplified diagram of ammonia synthesis

zero-carbon fuel solution for various applications. Among its potential uses, ammonia stands out as an attractive candidate for the storage and delivery of hydrogen or as a viable liquid fuel replacement for many daily uses of fossil fuels, such as transportation or power generation. Indeed, the development of green fuels is essential to enable a transition to a low-carbon economy. Some estimations predict a 50% reduction in carbon emissions in the shipping sector by 2050 if ammonia is implemented as a fuel [9, 10]. Green routes to ammonia will succeed if they fulfill some key requirements such as low cost, flexibility, safety, reliability, and carbon neutrality. Although these routes still involve long-term renewable technologies, they present an attractive opportunity to mitigate the high carbon footprint of the conventional production of fertilizers and fuels.

Bearing in mind that ammonia is a highly effective vector for hydrogen transport, its demand will likely increase considering the growing trend in hydrogen use, which is foreseen to supply about 20% of the global energy demand by 2050 [11]. To meet this goal, many key players of the ammonia industry have announced plans to modify their existing plants or to develop demonstration projects for the production of green ammonia [12], as detailed in the following sections. If current and upcoming scale-up initiatives succeed, ammonia could replace a substantial fraction of daily fossil fuel consumption and take a step ahead as one of the most competitive renewable-sourced fuel of the future.

2 Conventional and Emerging Technologies for Ammonia Production

The subsequent section presents conventional methods for ammonia production as well as alternative production pathways based on renewable energies.

2.1 Conventional Ammonia Production Processes

Prior to the development of the thermal ammonia synthesis in the twentieth century, nitrogen was extracted from natural organic matter as manures, urine, guano (with uric acid), and saltpeter (Chilean nitrate or sodium nitrate and potassium nitrate). The development of new industrial processes for nitrogen fixation started during the second industrial revolution, when the exponential population growth demanded an increase in fertilizer production to prevent food shortages [13, 14]. Ammonium sulfate was the most important industrial source of nitrogen for ammonia (NH_3) production until World War I [15]. It was produced by treating the gas effluents of coal-based steel works (coal gas works or Mond Gas) and coke oven gases with sulfuric acid.

Besides, new nitrogen fixation processes started to emerge in that period. Several research groups investigated the use of electric arc, motivated by the experiences of Priestley and Cavendish with electric sparks. For instance, the process developed by Kristian Olaf Bernhard Birkeland (1867–1917) and Samuel Eyde (1866–1940) in Norway in 1903, known as the Birkeland–Eyde electric arc process, transformed nitrogen (N_2) and oxygen (O_2) from air with an electric arc in a plasma reactor to nitrogen oxides (NO and NO_2) with a high electric power consumption (60,000–70,000 kWh/t_N or 600 GJ/t_N) [16, 17]. This process consumed about 2.4 to 3.1 MJ/mol_N, producing 1 to 2 mol% NO and consuming 175 t_{air}/t_N [18]. NO_2 was then processed with an aqueous solution for the production of nitric acid (HNO_3), which could be processed to calcium nitrate.

Alternative nitrogen fixation processes developed during these years included the cyanide method [19], the Serpek aluminum nitride procedure [20], and the Frank-Caro or calcium cyanamide process, being the latter the most successful. This process was developed by Adolph Frank (1834–1916), Nikodem Caro (1871–1935), and Fritz Rothe using calcium carbide (CaC_2) with pure nitrogen at high temperature (1000–1600 °C) to produce calcium cyanamide ($CaCN_2$) (Eqs. 1 and 2).

$$CaO + 3C \leftrightarrow CaC_2 + CO \tag{1}$$

$$CaC_2 + N_2 \leftrightarrow CaCN_2 + C \tag{2}$$

In 1900, Caro found that cyanamide in contact with water vapor gave rise to the formation of ammonia, which could be absorbed in sulfuric acid to obtain ammonium sulfate [21]. Further, Frank demonstrated the release of ammonium when calcium cyanamide enter in contact with soil (Eq. 3) [22]:

$$CaCN_2 + 3H_2O \leftrightarrow CaCO_3 + 2NH_3 \tag{3}$$

The temperature of the Frank–Caro process was decreased to 700–800 °C by the addition of different catalysts, such as calcium chloride. This synthesis method still consumed large amounts of energy (ca. 190 GJ/t) [16]. However, this was the main industrial source of ammonia till World War I, together with ammonium sulfate. The global cyanamide production was around 220,000 tons by 1914 [22, 23]. Later on, new and more efficient thermal catalytic processes were developed for industrial ammonia synthesis [24, 25]. Despite this, the cyanamide production has continued until now, serving as a slow-release fertilizer [26].

New investigations of Nernst, Haber, and co-workers set the basis for the direct ammonia production by obtaining the equilibrium data of ammonia, hydrogen, and nitrogen. Fritz Haber and Robert Le Rossignol (1884–1976) patented a process for the industrial thermocatalytic synthesis of ammonia with hydrogen (H_2) and nitrogen (N_2) (Fig. 2) [27–29]. Haber previously demonstrated the problems associated with the catalytic decomposition of ammonia in its elements at high temperature [30]. Then, process improvements led to higher ammonia yields by increasing the pressure [31] and developing a recycling loop [32], where unreacted synthesis gas was recycled and ammonia was separated from the flow stream by condensation. This modification increased the ammonia conversion from 15 to 98% [33]. In 1908, Haber signed an agreement with BASF (Badische Anilin and Soda Fabrik), and the process was then industrially scaled up by Carl Bosch (1874–1940) with the collaboration of Alwin Mittasch (1869–1953). They tested a great number of catalysts and found out that promoted iron catalysts were the most interesting materials in terms of activity, stability, and costs [34].

The first commercial high pressure ammonia plant was built by BASF in Oppau (Germany) with a capacity of 30 t/d [35]. This plant used hydrogen and atmospheric nitrogen as starting compounds with a 75:25 percent ratio, which were usually extracted from electrolysis of brines and air liquefaction, respectively, but also from coke treatment with water or air [23, 36]. Ammonia synthesis was developed under intensive conditions of pressure (150–300 atm) and temperature (400–600 °C), and using an iron catalyst [37–39].

The production costs of ammonia and ammonium sulfate decreased during the following years due to the introduction of modifications and development of new processes. Among new developments, manufacturers obtained synthesis gas from coke or coal (lignite), cryogenic air separation or shift conversion. Besides, they introduced additional processes such as desulfurization, carbon dioxide (CO_2) removal by water scrubbing, carbon monoxide (CO) removal, ammonia condensation, and scrubbing. Hydrogen was also obtained from coke oven, water gas, and electrolysis using renewable energy from hydropower. In the 1940s, further developments enabled the use of other hydrocarbons, such as heavy oils, naphtha, or natural gas.

The latter became the main source of hydrogen by reforming during the last years of the century [16].

The durability of the catalytic converter was also improved by Bosch, who designed a double steel wall that avoided the decarbonization (reaction between hydrogen and carbon to form trapped methane [16]) and embrittlement of steel to form iron hydride [36] by hydrogen diffusion, which was a cause of explosions [40]. New alloys were introduced to avoid this problem, testing the stability of steels as a function of temperature and hydrogen partial pressure [16]. Additional difficulties arose from the embrittlement caused by the nitridation of iron, when it was in contact with ammonia at high temperatures.

Nowadays, most industrial ammonia plants use air, water, and fossil fuels (natural gas) as the raw materials for syngas production (Fig. 3). The steam reforming process is very sensitive to sulfur poisoning. Therefore, sulfur compounds are previously separated in a desulfuration unit that removes hydrogen sulfide by adsorption on zinc oxide (ZnO). Besides, primary natural gas reforming produces carbon monoxide (CO) that poisons nickel catalysts (Eq. 4).

$$CH_4 + H_2O \rightarrow CO + 3H_2 \quad \Delta H^0 = 206 \text{ kJ mol}^{-1} \tag{4}$$

During the secondary reforming, carbon monoxide (CO) is converted into carbon dioxide (CO_2) releasing hydrogen (H_2) in the process (Eq. 5).

$$CO + H_2O \rightarrow CO_2 + H_2 \quad \Delta H^0 = -41.2 \text{ kJ mol}^{-1} \tag{5}$$

Typical synthesis gas compositions contain low amounts of carbon monoxide (CO), carbon dioxide (CO_2), water (H_2O), and oxygen-containing compounds that are removed in a final purification step to avoid catalyst poisoning by methanation. Removal methods include dryers, cryogenic methods, physical and chemical solvents, among others [16]. Nowadays, carbon dioxide (CO_2) is employed in fertilizer plants to produce urea. This integration was suggested by Koatsu, Kellogg, or Snamprogetti [16].

The syngas stream containing nitrogen is compressed at high pressure (150–250 bar) using centrifugal compressors in the synthesis loop, leading to more than 20% of ammonia in each pass [16]. The catalytic converters can be classified into tube cooled and multi-bed converters patterns, either with axial, cross-flow, or radial gas flows [16]. The ammonia production units use space velocities between 10,000 and 40,000 mL/g·h, and usually obtain production yields higher than 100 mmol$_{NH3}$/ g·h [16, 41, 42].

Catalysts for thermal ammonia synthesis have been classified in those unable to form stable nitrides (Ru, Ir, or Os); those that form unstable nitrides under reaction conditions (Mn, Fe, Co, Ni, Tc, or Re); and metals that can be present as nitrides under operation conditions (Mn, V, or Mo) [16].

Among them, magnetite was considered the best performing catalyst during the first years of development of this technology. The synthesis consisted of an oxygen-melt process, carried out in a furnace under oxygen flow or in electric arc furnaces,

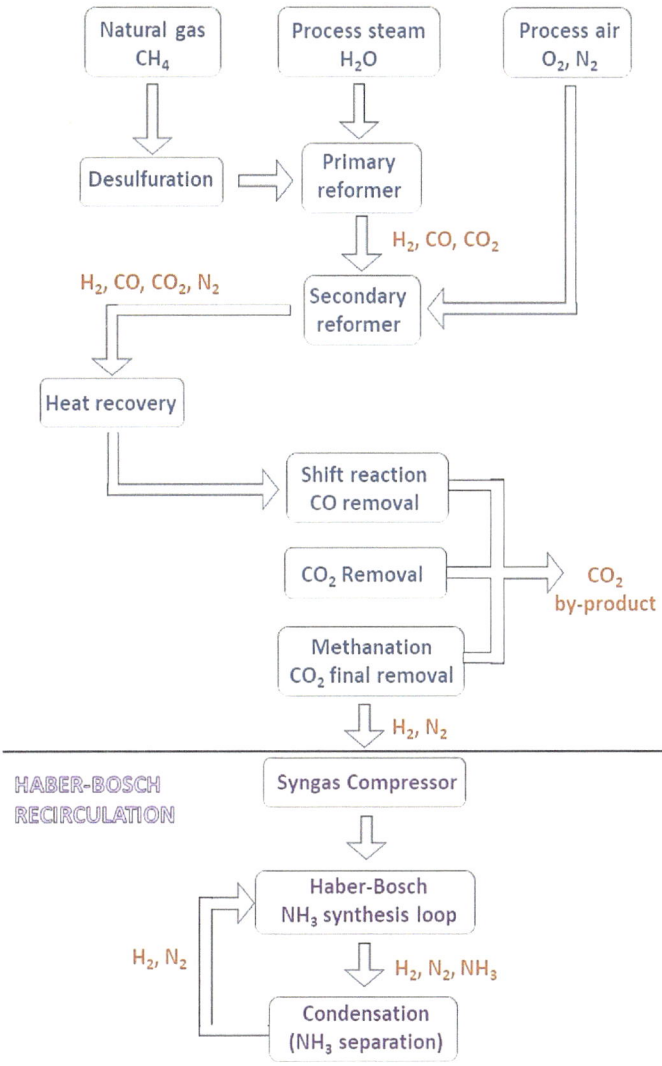

Fig. 3 Flow diagram for conventional Haber–Bosch ammonia production (based on Refs. [2, 43])

where iron and additives were burnt to magnetite (Fe_3O_4). Some studies found that the optimum proportion of Fe^{3+}/Fe^{2+} in the magnetite catalyst was 0.5 [44, 45].

Magnetite has been traditionally prepared using small amounts of electronic promoters, such as alkali oxides (e.g., K_2O), rare earth oxides or metallic oxides, such as cobalt. Promoters were used to enrich the electron density on the iron surface, aiming to enhance the nitrogen chemisorption and to improve the resistance to catalyst poisoning [46, 47]. The following elements and compounds were identified as catalyst poisoners: arsenic (As), sulfur (S), and phosphorus (P), from coal or heavy oils, as well as selenium (Se), tellurium (Te), boron (B), bismuth (Bi), lead (Pb), chlorine (Cl), among others. Oxygen compounds (e.g., H_2O, CO, or CO_2) at low concentrations were also identified as reversible poisoners [16]. Iron catalysts were also mixed with structural promoters (e.g., Al_2O_3, TiO_2, Cr_2O_3, MgO, MnO, CaO, or SiO_2) to increase the specific surface area and to improve the overall activity. Lantanide supports and promoters were used to mitigate hydrogen poisoning [48, 49].

As a result, the thermal ammonia synthesis was focused on improving the performance of promoted Fe_3O_4, FeO (wustite), $Co-Fe_3O_4$, CeO_2, Co-Mo-nitride, and Ru catalysts for over a century [41]. Ru catalysts, usually prepared by depositing Ru on graphite with metallic promoters, showed higher activity than iron catalysts at temperatures above 375 °C, and suffer less from oxygen poisoning. In contrast, they were affected by hydrogen poisoning at temperatures lower than 325 °C [50, 51]. Ru-based catalysts with alkali earth promoters showed higher activities due to a better nitrogen dissociation [52], usually identified as the rate-limiting step. The enhanced nitrogen cleavage was ascribed to the electrostatic attraction with Ru sites [53].

Over the past few years, investigations are focusing their efforts on the study of strong metal-support interactions, the development of electride supports, as well as the use of metal hydrides and active supports with nitrogen vacancies at low temperatures and pressures [54].

2.1.1 Haber–Bosch Process

Ammonia has been historically produced in large plants (1,000 to 1,500 t/d) [8] via the thermochemical Haber–Bosch (H-B) process, invented by the Nobel Prize in Chemistry Fritz Haber (1868–1934) in 1918. His patent on the synthesis of ammonia from its elements was then scaled up by Carl Bosch (Nobel Prize in Chemistry in 1931), resulting in the widely used Haber–Bosch process for industrial ammonia synthesis. This process enabled the massive production of synthetic fertilizers, avoiding the exhaustion of the natural nitrogen-containing resources and mitigating the fertilizer shortage in Germany caused by the blockage of Chilean nitrate imports (Chilean mineral and ammonium sulfate represented 90% of the nitrogen consumption in Germany before the First World War (WWI) [55]). During the following decades, the Haber–Bosch process enabled an exponential population growth through ammonia-based fertilization [56].

The Haber–Bosch process synthesizes ammonia using high purity (99.99%) hydrogen (H_2) and nitrogen (N_2) streams, which typically react together at high temperature and pressure (ca. 350 to 550 °C and 20 to 40 MPa) in the presence of an iron-based catalyst [2, 6] (Eq. 6).

$$\frac{1}{2}N_2(g) + \frac{3}{2}H_2(g) \leftrightarrow NH_3(g) \tag{6}$$

The synthesis of ammonia is exothermic ($\Delta H_{298K}° = -45.9$ kJ mol^{-1}, $\Delta G_{298K}° = -16.4$ kJ mol^{-1}, $K_{eq} = 750$) and reversible, and it is favored at low temperatures according to the equilibrium. However, the high chemical inertness of the nitrogen molecule requires high temperatures to break the strong nitrogen triple bond (N≡N; energy bond of 941 kJ mol^{-1}) at sufficient rates. The rate-determining step of the ammonia synthesis was demonstrated to be the chemisorption of the nitrogen molecule [47]. In fact, equilibrium calculations show that more than 99% of ammonia decomposes to nitrogen and hydrogen at temperatures higher than 400 °C at 0.1 MPa, and consequently, the Haber–Bosch process is operated under high pressures (20–40 MPa) with a suitable catalyst to shift the equilibrium to the formation of ammonia. The conversion of nitrogen and hydrogen to ammonia via Haber–Bosch in a single reaction is around 15%, and for that reason unreacted nitrogen and hydrogen are recirculated to achieve overall ammonia conversion of about 97% [2, 56].

Approximately 96% of the supplied hydrogen is derived from fossil fuels, mainly from steam reforming of methane (72%) and partial oxidation of coal (26%). The remaining 2% is generated from coal or natural gas electrical generation [6, 8]. Therefore, the Haber–Bosch process is currently one of the largest global energy consumers and greenhouse gas emitter process, with a consumption of 1–3% of the global electrical energy and about 5% of the world natural gas. Moreover, the strict operational conditions are responsible for about 1.8% of the global carbon dioxide (CO_2) emissions per year, surpassing 2.16 kg of carbon dioxide equivalent (CO_{2eq}) for each kg of ammonia (NH_3) produced [8, 12, 57]. Besides, this energy-intensive process also generates nitrogen oxides (NO_x) that need to be removed via selective non-catalytic reduction, which increases cost of operation [6].

In numbers, the Haber–Bosch dominates the production of ammonia, producing almost 180 Mt of fertilizers per year, mostly in form of anhydrous ammonia, which represent half of the fixed nitrogen nowadays [57]. Modern Haber–Bosch plants produce ammonia with an energy efficiency of only 65% and at an energy cost of at least 8 kWh/kg [58]. These energy costs have been substantially reduced over the past 100 years (ca. 110 kWh/kg) as a result of the increasing production scale and improvements in engineering [59]. In fact, the energy requirement for ammonia synthesis in the 1920s was about 100 GJ/t$_{NH3}$, and it decreased down to 27–32.6 GJ/t$_{NH3}$ (equivalent to 7.78–9.06 kWh/kg$_{NH3}$) in modern steam methane reforming (SMR) plants, compared to the theoretical minimum of 20.3 GJ/t$_{NH3}$ (equivalent to 5.64 kWh/kg) [42].

2.1.2   Other Conventional Processes

Several industrial processes for nitrogen fixation were developed over the last century, starting from the basic principles and conditions of the Haber–Bosch process, and introducing some changes related to operational conditions, catalysts, design of catalytic converters, and gas purification steps [60]. The source of hydrogen and nitrogen varied depending on the technology and the cost and availability of raw materials. These technologies were competitors of the process developed by BASF after World War I, and the main motivation for these changes was to guarantee food security [25, 61]. The following sections describe a chronological evolution of the industrial ammonia synthesis processes, highlighting the most relevant modifications starting from Haber–Bosch.

Claude Process

The Claude process was invented in France in 1921 by Georges Claude (1870–1960), a well-known chemist for his work on industrial liquefaction of air for the purification of nitrogen and other gases by fractional distillation, and the discovery of neon lights. Claude started to work on the synthesis of ammonia in 1917. The main difference between Claude process and Haber–Bosch is its higher temperature (500 °C) and extreme pressure (900–1,000 atm). Nitrogen was obtained from liquid–air machines, and hydrogen from coke oven gas or water electrolysis. This process used multiple reactors made of special alloyed steels with chrome, nickel, or molybdenum. They were equipped with condensers set in series without recirculation, which were smaller than those used in the Haber–Bosch process [62]. Claude incorporated an additional converter before the catalytic vessels to transform carbon monoxide (CO) and oxygen (O_2), which are by-products of the hydrogen (H_2) production from coke oven gas and methane reforming, but also catalyst poisons.

Claude process required a lower volume of catalyst than the Haber–Bosch process to obtain the same quantity of ammonia. The catalyst used was peroxidized iron, prepared by burning iron in oxygen with some promoters, and then reduced with hydrogen inside the catalytic converter.

Further, the use of higher pressures greatly improved the production of ammonia, according to the equilibrium. The increasing of pressure from 100 to 1,000 bar required half of the power needed for the compression from 1 to 100 bar. This fact made the reaction autothermic, without needing external heating once the reaction started [62].

Ammonia was recovered by cooling the gas stream after each stage, and heating again up to the reaction temperature. This approach reduced the heating time of reactors compared to that required for the big converters in Haber–Bosch. Residual ammonia was recovered from the gas stream by further cooling or upon absorption with sulfuric acid [62].

The Claude process produced higher ammonia yields in a single pass than the Haber-Bosch-based processes and did not require any recirculation of the gas mixture. In contrast, it used the residual gas for heating or H_2 production. The maximum ammonia yield achieved in the original Claude process was 40% in the first converter and 25% in the second for each single pass [63] (6 g_{NH3}/g_{cat}·h, compared to 0.5 g_{NH3}/g_{cat}·h in the BASF process [64]). Later on, new developments achieved more than 80% conversion after three or four reaction and condensation steps, although they found limitations associated with the dissipation of heat, the poisoning of catalysts and the corrosion caused by oxygen (O_2) or water (H_2O) [64, 65].

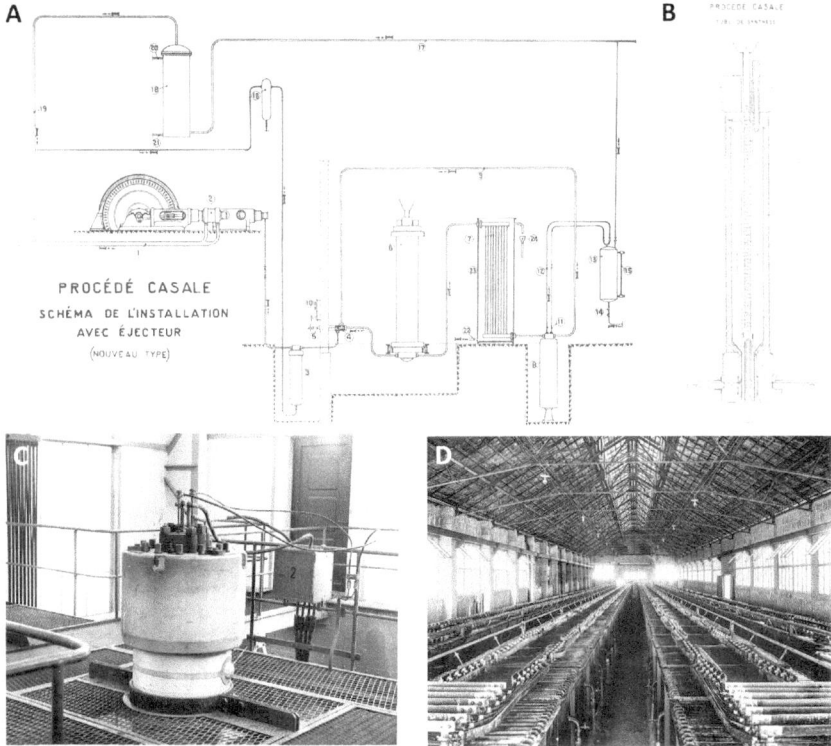

Fig. 4 Casale process scheme (**a**); Schematics of Casale reactor (**b**); Head of Casale reactor (**c**); Electrolyzers at Sabiñanigo (**d**). Images kindly provided by Casale S.A

Casale Process

Luigi Casale (1882–1927) was an Italian chemist who introduced a new competitor to BASF in Germany in the early 1900s, which was the sole ammonia producer using a catalytic thermal process (Fig. 4). In 1919, Casale invented a new ammonia synthesis process that was later licensed in 1921. The reactor was equipped with an axial flow converter containing concentric steel tubes [66]. The spaces between the concentric tubes were filled with the catalyst, used in lower amounts than in Haber–Bosch process. In general, this process used catalysts with spherical shape based on inexpensive iron oxide or iron alloys with manganese (Mn), chromium (Cr), molybdenum (Mo), carbon (C), or tungsten (W), which were prepared by calcination under pressurized oxygen [67]. These catalysts were less active and stable than those used in Haber–Bosch. Subsequent patents modified the synthesis methods with certain amounts of metal oxides not reducible under hydrogen [68], and some promoters such as calcium (Ca), carbon (C), magnesium (Mg), or aluminum (Al) [69, 70].

The Casale process was operated under more intense operation conditions (500 °C and 850 atm) than the Haber–Bosch process. A concentric heat exchanger, surrounding the catalyst bed chamber, heated the inlet cold compressed mixture of gases [63, 71]. This configuration reduced the temperature of the outer shell, thus allowing the use of conventional steel. The inner wall of the reactor shell was cooled down by the mixture of gases. In this regard, some produced ammonia was left in the recirculated gas stream to slow down the reaction rate in the catalyst chamber. The recirculation of gases was improved by installing a closed train loop and replacing the reciprocating compressors, which also simplified the process [63].

During the 1970s, the original reactor was changed to a three catalyst bed config-uration equipped with heat exchangers within the beds; a similar solution to that proposed in the Kellog process in 1950 [36]. The new catalyst beds harnessed an axial-radial technology, enabling the use of smaller size catalysts and achieving lower pressure drops than standard catalyst beds [4].

Another innovation of Casale process was the obtention of nitrogen through burning air [72] (with hydrogen) or by air liquefaction. Hydrogen was obtained from water electrolysis powered by hydroelectric stations, coke oven, or water gas [73]. In general, the process needed to feed lower gas volumes than Haber–Bosch to produce the same amount of ammonia [63].

Opposite to Haber–Bosch, the Casale process did not require expensive refrig-eration systems to produce liquid anhydrous ammonia, due to the high operating pressures (900–1,000 bar). This modification enabled the recovery of ammonia by direct condensation (cooling with water), rather than washing with water as in the Haber–Bosch. This innovation avoided the need for a refrigeration unit, considerably reducing the factory space [25, 63, 74].

The ammonia yields obtained in Casale process (15–18%) were higher than those of the Haber–Bosch process (5–8%), largely due to the recycling of the gas mixture in a closed train loop, shifting the equilibrium to the formation of ammonia [75].

Fauser Process

The process developed by Giacomo Fauser (1892–1971) in 1921 [76] was adopted by Montecatini corporation and restricted to Italy until 1926 [63]. Then, Fauser devel-oped other processes for the synthesis of nitric acid, ammonium sulfate, ammonium nitrate, or urea [63]. The process Fauser-Montecatini, also known as Fauser ammonia synthesis, worked at 500 °C and 250 to 300 atm. During the first years, it employed hydrogen from hydroelectric power stations, while nitrogen was obtained by oxida-tion of part of the hydrogen obtained by electrolysis [77]. Fauser addressed the removal of excess heating by introducing a heat exchanger between the catalytic layers, in which hydrogen (H_2) and nitrogen (N_2) passed in countercurrent before contacting the catalyst chamber [78, 79], thus shifting the reaction to ammonia formation. Catalysts were based on K-promoted iron materials. Casale and Fauser-Montecatini processes have been implemented in numerous plants currently under operation [80], including some improvements in the reactor chambers [78].

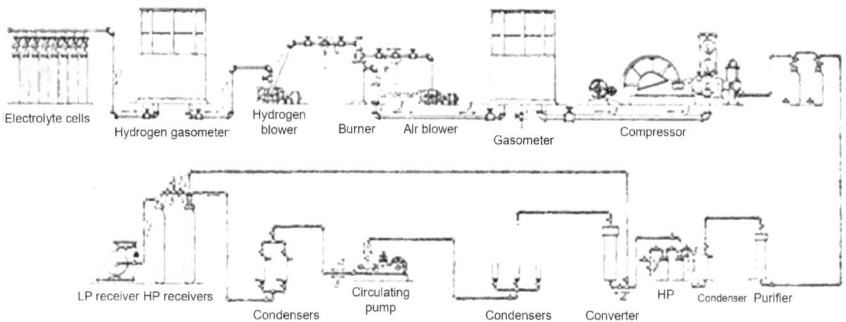

Electrolyte cells   Hydrogen gasometer   Hydrogen blower   Burner   Air blower   Gasometer   Compressor

LP receiver HP receivers   Condensers   Circulating pump   Condensers   Converter   HP   Condenser Purifier

Fig. 5 American process for direct synthesis of ammonia with electrolytic production of hydrogen (Copyright 1925 American Chemical Society. Reprinted with permission from reference [83])

Other Processes and Further Modifications

Similar technologies to the aforementioned processes were also developed in other countries to guarantee the ammonia production capacity and to reduce the production costs. The General Chemical/Allied Chemical process was developed in United States in 1921, employing 500 °C and 200 to 300 atm. In 1915, Frederick W. de Jahn et alii. developed a similar process to BASF, using an iron catalyst but incorporating sodamide [71].

The *American process* from the Fixed Nitrogen Research Laboratory (FNRL), also known as FNRL process or Nitrogen Engineering Corporation (NEC) process, was similar to other direct synthesis technologies and operated at 475 °C and 300 atm [81] (Fig. 5). The catalyst consisted in a novel iron oxide mixed with metal oxides and promoters, such as potassium oxide and alumina [82]. Hydrogen was produced by water electrolysis, generally employing alkaline electrolyzers, while nitrogen was obtained by burning oxygen with hydrogen, and removing water via condensation. The inlet cold gas entered in a pressure-sustaining vessel made of chrome-nickel-vanadium-steel alloy, which contained a heat exchanger to control the temperature of the steel as in other ammonia technologies. The American process achieved ammonia yields higher than 13%.

The Mont Cenis process was the second German process after Haber–Bosch process. It was invented in 1925–1926 in France. This process required 400 °C and lower pressures (100–300 atm) than Haber–Bosch and used an iron cyanide catalyst.

In 1928, the Japanese Showa Fertilizer Company developed the Showa Hiryo Fertilizer process for ammonia synthesis (Fig. 6), also known as TIEL process (Tokyo Industrial Experimental Laboratories). Until the late 1920s, the main industrial process for nitrogen fixation in Japan was the calcium cyanamide process and the coke treatment. During the 1930s, some processes as Claude, Casale, Fauser, NEC, and Haber–Bosch were introduced in this country. In 1931, the Showa process started to produce ammonium sulfate using hydrogen from water electrolysis. In the

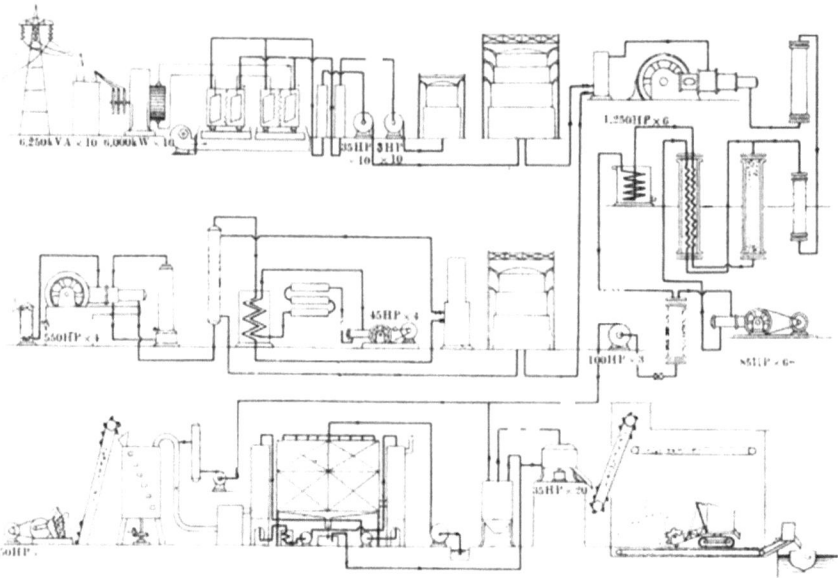

Fig. 6 Schematic diagram of the ammonium sulfate fertilizer process in the Kawasaki Factory of the Showa Hiryio K.K. [84]. Image kindly provided by Hitachi

1950s, the decrease in the cost of fossil fuel feedstocks caused a decline in the hydro-electric production, positioning natural gas as the main source of hydrogen. Besides, the development of new catalysts and the optimization of the energy consumption promoted the massive use of natural gas in that period [25]. In this regard, the Showa process was similar to the Haber–Bosch process except for the catalyst, which was a stable iron hydroxide patented by Shibata Hatsutaro and Buichi Yokohama in 1921 [17].

Some further modifications and adaptations of the ammonia synthesis process were introduced over the years by Du Pont in the so-called Du Pont process, which introduced some innovations from Casale and Claude processes for the production of hydrogen from coal, air, and water [85].

M.W. Kellog Co. used reactors operating at near isothermal conditions. Refrigeration was carried out with cooling tubes inside the catalyst beds, or flowing cooling gases between each layer in multi-bed converter configurations. In 1963, this process introduced a single-train process for steam reforming in plants with high ammonia production capacity [16]. This innovation together with the decrease in the operation pressures led to a 50% reduction in energy consumption and production costs [36].

In the following years, a renewed interest on Ru-promoted catalysts emerged. The use of Ru was first reported by Mittasch in 1917, although their use and related investigation was abandoned during decades until the 1970s. Then, several works proposed the use of Ru and metal promoters (Cs, K, Ba, or Ca) to increase the catalytic activity [86]. Integration of these catalysts in several plants during the 1990s enabled the operate at lower pressure and temperature but with higher catalyst costs (10 $/kg and over 500 $/kg for Fe and Ru catalysts, respectively [54]). These materials showed high performance but they suffered hydrogen poisoning or methanation in carbon-supported catalysts [48].

In 1992, a collaboration between British Petroleum and M.W. Kellog Co. led to a new successful synthesis process named Kellog Advanced Ammonia Process (KAAP). By operating at 370 to 400 °C and 50 to 100 atm, using Ru/C or Ru/oxides catalysts, this process obtained 40 to 50% ammonia yields, which were 10 to 20 times higher than that of traditional Fe catalysts [16, 33].

Imperial Chemical Industries (ICI) introduced the AMV process using a Fe-Co catalyst, and the Leading Ammonia Concept (LCA) process, which operated at lower pressures (70 atm) with 12 to 15% of ammonia yields (Table 1) [36].

Other processes with different configurations and small differences related to specific energy consumptions are: Haldor Topsøe, Krupp-Uhde, KBR, TVA, MMC, lead, Fluor, Lummus, Humphreys & Glasgow BYAS, Humphreys & Glasgow MDF, Montedison low pressure, Jacobs plus ammonia technology, KTI PARC, Texaco gasification, Texaco coal gasification, Koppers-Totzek, Linde (LAC), Lurgi, Kellogg's LEAP, Chiyoda, Kellogg Ammonia 2000, GIAP concept, Snamprogetti, or ANC [16].

In some of these processes, other intensively developed solutions were absorbent-enhanced and adsorbent-enhanced Haber–Bosch processes. These approaches improved the removal of liquid ammonia from the gas flow, enabling the operation of the synthesis loop at lower temperatures and pressures. Besides, pressure swing absorption (PSA) and temperature swing absorption have also been explored. Materials such as activated carbon, COFs, ionic liquids, MOFs, oxides, deep eutectic solvents, metal halides, polymers, or zeolites, among others, have been used as absorbents or adsorbents [42].

The evolution of ammonia plants until nowadays includes an integrated design to improve both efficiency and energy recovery. Among the most relevant improvements, the implementation of centrifugal compressors are highlighted (replaced reciprocating compressors that required higher maintenance efforts, leading to a 30% reduction of capital costs [36]), which promote the building of high production plants (more than 1,000 t/d); the evolution from multiple-train to single-train integrated plants; improvements in catalysts and in softer operation conditions; reduction in catalyst sizes (less than 3 mm); new supports and catalyst shapes for reforming, such as four hole (ICI), six-shooter (Topsøe), or wagon-wheel (United Catalyst) [16]; new gas purification technologies (C.F. Brawn process that improves the cryogenic purification of syngas, where methane and nitrogen are separated from the synthesis loop); revamping or modernizing of existing converters or the use of bigger ones with more efficient designs; the use of lower pressure systems and energy-saving designs (recovery of heat exchange from reformers) [26] such as those developed by KBR (Kellog, Brown and Root) [4]. Other companies such as Topsøe developed further innovations in the sector. For instance, they introduced an innovative radial-flow ammonia converter with a very low pressure drop, obtaining higher yields using

Table 1 Summary of the different thermocatalytic ammonia synthesis processes

Process	Pressure [atm]	Temperature [°C]	Catalyst	Yield in single pass [%]	Country	Year
Processes based on Haber–Bosch research						
Haber	200	550	Promoted Fe	7–8	Germany	1913
Modified Haber	250	550	Doubly promoted Fe	20	Germany	-
Fauser	250–300	500	Fe + K promoter	12–23	Italy	1921
FNRL/ American/ NEC	200–300	475	Doubly promoted Fe	20–22	USA	1921
General Chemical	200	500	Fe + sodamide	20–22	USA	1921
Mont Cenis	100	400–425	Fe cyanide	9–20	Germany	1925
TIEL	–	–	Fe hydroxide	–	Japan	1931
Uhde	300	500	–	18–22	Germany	–
ICI-AMV	70–100	400	Fe-Co	13	Great Britain	–
Topsøe	250	325	Fe	10	Denmark	–
Kellogg	300	530	Fe	20–30	USA	1960s–1990s
KAAP	50–100	370–400	Ru/C or Ru/ oxides + promoters	40–50	USA	1992
Processes working under hyper-pressures based on Casale and Claude research						
Casale	800–850	500	Fe with Mn, Cr, Mo, C or W + promoters	15–18	Italy	1921
Claude	900–1,000	500–650	Promoted Fe with Al and K	40–85[a]	France	1921
L'air liquide	900–1,000	500	Promoted Fe	–	France	–
Du Pont	1000	500	Promoted Fe	–	USA	–

Based on Refs. [63, 85, 87, 88]

[a] Without recycling; about 40% of ammonia (NH_3) yield obtained with a single converter and 85% yield after several reactions in several converters

catalyst with smaller particle sizes. Besides, the incorporation of a new heat exchange reformer considerably reduced the consumption of fuel and gas.

2.2 Emerging Green Ammonia Routes

The large amounts of energy required for the conventional synthesis of ammonia have triggered researchers to look for a step change in the ammonia (NH_3) production, aiming to fulfill three key criteria: energy efficiency, scalability/modularity, and CO_2-free emissions [89]. Therefore, new approaches have emerged powered by renewable energies to reduce the environmental impact of this process. As a result, and analogously to hydrogen (H_2), synthetic ammonia has been classified by colors depending on the production route (Fig. 7).

These color classifications for synthetic ammonia shown in Fig. 7 are described in the following.

- Brown/gray ammonia. Brown ammonia is produced using hydrogen from coal gasification, while gray ammonia uses hydrogen from natural gas reforming. Both categories correspond to conventional synthetic routes and emit between 2.5 and 3.8 t_{CO2}/t_{NH3} and 1.6 t_{CO2}/t_{NH3} for brown and gray ammonia, respectively.
- Blue ammonia. Blue ammonia uses the same feedstock as brown and gray ammonia but includes a carbon capture and storage (CCS) unit. This approach avoids more than 90% of the CO_2 emissions generated by conventional ammonia production routes and could be even entirely eliminated by supplying green hydrogen to the Haber–Bosch process.
- Green ammonia. The term "green" implies that the hydrogen is generated from water electrolysis, and the nitrogen is extracted from air through an air separation unit, with the process being completely based on renewable energy [10–12, 42]. Various electrolysis systems are currently available, such as alkaline electrolyzers

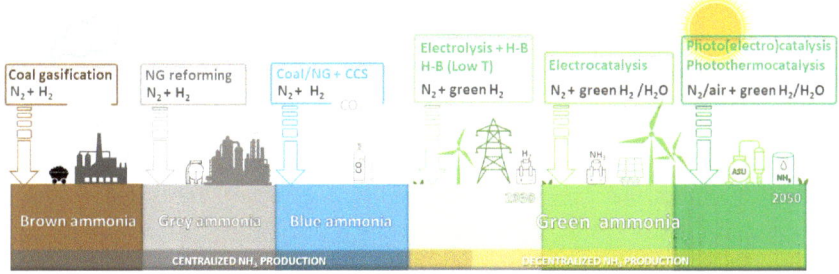

Fig. 7 Classification of synthetic ammonia categories based on the energy source and used technology (ASU: air separation unit; CCS: carbon capture and storage; H-B: Haber–Bosch; Low T: low temperature; NG: natural gas)

or proton exchange membrane (PEM) electrolyzers, while other technologies are still under development such as solid oxide electrolyzers, or the less developed anion exchange membranes (AEM) electrolyzers [42].

It should be borne in mind that every step in the production process of green ammonia must be renewably powered to achieve zero-carbon emissions along the entire production chain. This includes gas production (H_2, N_2), gas compression, heating, and separation of ammonia from unreacted nitrogen and hydrogen; even considering desalination of water to feed the process in areas where access to fresh water is limited [10]. Green ammonia can be produced both by established processes from conventional thermochemical ammonia production—Haber–Bosch process combined with renewable (green) hydrogen—as well as through newer, alternative processes like electrochemical and photochemical ammonia production, provided they are based on renewable energy sources. These approaches will be presented in the following sections.

2.2.1 Green Ammonia via Haber–Bosch Route

In the conventional Haber–Bosch approach for ammonia production, nitrogen (N_2) is extracted from ambient air using an air separation unit (ASU). The nitrogen is then combined with green hydrogen in the Haber–Bosch process under high pressures and temperatures. Using this approach, when (green) hydrogen is produced from water electrolysis using renewable energy instead of being derived from fossil fuels, and renewable energy powers the other processes as well (e.g., air separation), green ammonia can be produced. A number of different companies have developed innovative systems to produce green ammonia via electrolysis of water with renewable technologies (Fig. 8), consuming 9–11 kWh/kg_{NH3} [42].

Nitrogen purification technologies usually account for the 20 to 25% capital costs of the Haber–Bosch plant at large scales [90]. The catalytic combustion of air is usually employed in methane steam reforming, in which the oxygen contained in air reacts with hydrogen or methane. Other technologies include air separation units by cryogenic distillation (0.03–0.05 USD/kg_{N2}), pressure swing adsorption units (PSA) (0.05 USD/kg_{N2}), and membrane permeation (0.05–0.1 USD/kg_{N2}) [90]. Cryogenic separation is based on the different boiling points of nitrogen (N_2), oxygen (O_2), and argon (Ar) (77.4 K, 90.2 K, and 87.3 K).

In pressure swing adsorption (PSA), a continuously regenerated adsorbent is employed to separate nitrogen (N_2) and oxygen (O_2). A recent analysis of costs [91] shows that membrane permeation is preferred for small production units (< 1 MW), pressure swing for medium scale (1–100 MW), and cryogenic distillation for large productions (>100 MW). These technologies are used in combination with electrolyzers, obtaining different nitrogen purities and costs as detailed in Table 2.

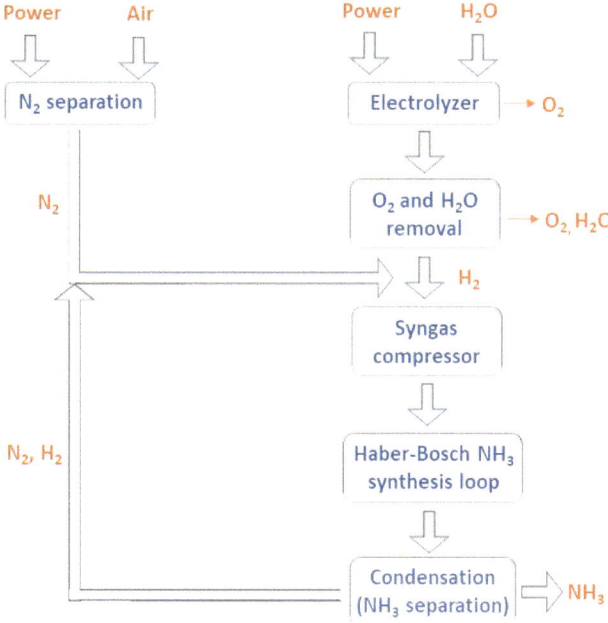

Fig. 8 Flow diagram for green ammonia production with electrolysis and renewable energy (based on reference [43])

Table 2 Nitrogen purification technologies

	Unit	Cryogenic distillation	Pressure swing adsorption	Membrane
Temperature	°C	−195 to −170	20–35	40–60
Pressure	bar	1–10	6–10	6–25
Purity	wt.%	99.999 (ppb)	99.8–99.9	99–99.5
Energy consumption	kWh/kg$_{N2}$	0.1	0.2–0.3	0.2–0.6
	GJ/t$_{NH3}$	0.3	0.7–1.0	0.7–2.0
Capacity range	Nm^3/h	200–400,000	5–5000	1–3,000
Load range	%	60–100	30–100	30–100
Investment cost	k€/t$_{NH3}$·d	<8	4–25	25–45
Start-up time	–	h	mins	min
TRL	–	9	9	8–9

TRL: technology readiness level; from Refs. [90, 92]

2.2.2 Other Green Ammonia Technologies

The so-called direct ammonia synthesis can be achieved via electrochemical or photochemical routes. Both approaches are still at an early stage of development at laboratory scale, since working under mild temperature and pressure requires highly active catalysts to overcome the thermodynamic challenge to activate nitrogen [10]. The deployment of ammonia synthesis units powered by solar energy in the near future could improve the production capacity in regions distant from production plants, where the transport of fertilizers considerably increases costs. Besides, the fluctuation of prices of fossil fuels during times of crisis dramatically impact the production costs of ammonia and other fertilizers and products. Thus, the introduction of decentralized small green ammonia modular plants, based on electrochemical or photochemical production systems powered with solar energy, could mitigate the increase in production costs during these periods, while reducing the dependence of importing countries.

Electrochemical Route

The electrochemical synthesis of ammonia proceeds on electrode surfaces by electrons supplied from an external circuit with simultaneous proton addition from hydrogen, which is obtained from the renewably powered electrolysis of water [10, 56]. This reaction can proceed via a direct electron and proton addition to the nitrogen (N_2) molecule, or indirectly using redox mediator, such as Li^+, which is first reduced to produce ammonia and then regenerated [58].

There are several electrochemical ammonia technologies that operate at different temperatures, such as high-temperature solid-state electrolyte reactors (400–750 °C), molten salt reactors (100–500 °C), liquid electrolyte reactors, and low-temperature solid-state electrolyte reactors (20–100 °C). In particular, the use of electrical energy to activate nitrogen using a low-temperature route has attracted interest in recent years as a milder alternative to the thermocatalytic Haber–Bosch process [56].

Active electrochemical systems are usually based on precious metal (Au, Pt, Pd, Rh, or Ru) or transition metal (Fe, Mo) electrocatalysts, and molten salt or polymer electrolytes. Most commonly used cell designs are similar to that for the well-established water splitting technology [10, 58]. Current systems have reached production rates lower than 10^{-6} mol h^{-1} cm^{-2} and Faradaic efficiencies (FE) lower than 30%, which are far from the target set by the US Department of Energy for a viable electrochemical ammonia production (i.e., 10^{-4} mol h^{-1} cm^{-2}, FE of 50%) to compete with Haber–Bosch [10]. The highest FE values reported to date are higher than 80% [93, 94]. The low activity and selectivity of currently available electrocatalysts are usually related to the predominance of the hydrogen evolution reaction (HER), in which the electrocatalyst surface is dominated by HER and no active sites are available for nitrogen reduction, thus reducing the Faradaic efficiencies [56]. Major improvements are needed in terms of materials and cells to increase the ammonia formation rates to reasonable levels for potential commercialization.

Photochemical Route

The photochemical approach is powered by sunlight and uses water and air to provide hydrogen and nitrogen, respectively. Unlike the electrochemical route, there is no need to supply electricity, which is very interesting for decentralized off-grid ammonia production. However, the photochemical synthesis is limited by the intermittency of solar energy [10]. This route requires the use of semiconductors with suitable bandgaps, which are able to absorb sunlight and generate charge carriers that can migrate to the catalyst surface to drive the redox reactions, namely the reduction of nitrogen and the oxidation of water.

The catalyst can be coated on a solid surface or suspended in water, with (humid) air bubbled into the reaction system. Active photocatalysts are mostly based on semiconductors modified with co-catalysts, doped with metals or modified with surface defects to improve the binding and activation of the nitrogen molecule. Most common photocatalysts include metal oxides (TiO_2, Fe_2O_3, $SrTiO_3$, $BaTiO_3$, etc.), metal sulfides (MoS_2, CdS), and carbon nitrides (g-C_3N_4) [95]. Current state-of-the-art systems have reached ammonia productions of 1 to 20 mmol/g_{cat}·h. In general, the reported ammonia production rates are still too low and far from scaled-up devices (Table 3). Such low productions are even in some cases the results of contamination or reduction of oxidized forms of nitrogen rather than pure nitrogen (N_2) [58]. Nevertheless, this route represents a promising long-term approach for energy production and storage at local level.

3 Techno-Economic Evaluations

The subsequent sections focus on technical–economic aspects of the processes outlined before. They address production costs (and market prices for conventional ammonia), and the related technical aspects. Therefore, conventional ammonia production via the Haber–Bosch synthesis is considered as well as the three previously described green ammonia production pathways, namely through Haber–Bosch (thermochemical route), electrocatalysis (electrochemical route), and direct solar technologies (photochemical route).

3.1 Conventional Ammonia (Haber–Bosch)

During the last 50 years, the ammonia (NH_3) synthesis technology has evolved to sophisticated and integrated plants with single-train designs, and capacities that can reach more than 3,000 t_{NH3}/d. The energy consumption has been reduced from 40 to 60 GJ/t_{NH3}, in the first natural gas reforming plants, to 28 to 32 GJ/t_{NH3} in current single-train plants with methane reforming [4]. In general, the higher energy consumption comes from reformers and CO_2 removal systems. Methane reforming

Table 3 Recent high performing catalysts tested in photo(thermo)catalytic N_2 fixation into ammonia (till the be-ginning of 2022)

Photocatalyst	T [°C]	Phase	Light source	Reagents	Hole scavenger	Calculated activity[a]	References
Porous graphdiyne (GDY) CoOx based quantum dots	20	Liquid	300 W xenon full wavelength	N_2, H_2O	Na_2SO_4	26.5 mmol/g·h	[96]
Bi_5O_7Br-40	Room	Liquid	300 W xenon lamp (200–800 nm)	N_2, H_2O	–	12.7 mmol/g·h	[97]
SV-1 T-MoS_2CdS and MoS_2	–	Liquid	AM 1.5G	N_2, H_2O	Methanol	8.2 mmol/L·h·g	[98]
Bi_2Te_3/BiOCl	Room	Liquid	UV and VIS	N_2, H_2O	Methanol	7.9 mmol/g·h	[99]
Cu/TiO_2	–	Liquid	300 W, AM 1.5 Filter, Xe lamp	N_2, H_2	Glycerol	6.8 mmol/g·h	[100]
Ru – Cs/MgO	60	Gas	broad spectrum white light	N_2, H_2	–	4.5 mmol/g·h	[101]
Ru-/$H_x MoO_{3-y}$	360	Gas	300 W Xe lamp	N_2	–	4.0 mmol/g·h	[102]
Au-Os/Cs_2O	60	Gas	200 mW/cm	H_2, N_2	–	2.7 mmol/g·h	[103]
Black phosphorus	Room	Liquid	LED lamp 420 nm	N_2	Na_2SO_3 and $Na_2S·9H_2O$	2.4 mmol/g·h	[104]

[a] Photocatalytic activity determined from experimental data from each reference

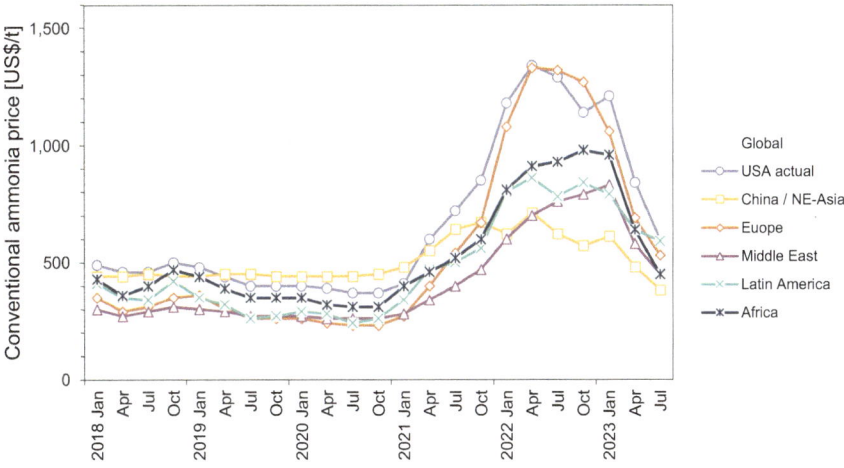

Fig. 9 Market prices of (conventional) ammonia from 2018 to July 2023. *Source* Reference [107]

consumes less energy and release less CO_2 emissions than other hydrocarbons. For that reason, natural gas is nowadays considered the most economic feedstock for ammonia synthesis compared to coal or heavy oils. The use of methane represents 80% of the total cost of ammonia production [36].

Large Haber–Bosch (H-B) plants can produce ammonia (>2,000 t_{NH3}/d on average) at low costs of about 160 $/$t_{NH3}$. The production costs of ammonia in small Haber–Bosch plants (91 t_{NH3}/d) double those of large Haber–Bosch plants [105], mainly due to higher operation and maintenance (O&M) costs and higher capital expenditures. This limits the implementation of small Haber–Bosch plants for decentralized ammonia production. For centralized ammonia production in large-scale plants, costs further increase due to transportation and storage, rising the market price of ammonia. Overall those market prices are thus higher and have been overall rather volatile in the past two years (Fig. 9) [106].

As shown in Fig. 9, from January 2018 to January 2020, ammonia prices remained relatively stable worldwide, fluctuating between approximately 300 to 500 US$/t, depending on the region. The prices then saw a rise in 2021, peaking around the middle of 2022. Regarding the global ammonia price shown, the prices 2022 were around 1,000 to 1,200 US$/t, with the majority of this increase being after March 2021. After these peak prices in 2022, the market prices then experienced a decline. As indicated, the global ammonia price for July 2023 was at roughly 500 US$/t.

This increase in ammonia prices—along with their overall trends in its price development—can be attributed to the fact that ammonia market prices are typically closely linked to natural gas prices (Fig. 10), its primary feedstock for production, and these natural gas prices escalated in this time period. Not only with regard to ammonia, but rather generally, there is a close correlation between prices for chemical commodities (like ammonia) and their feedstock (in this case natural gas).

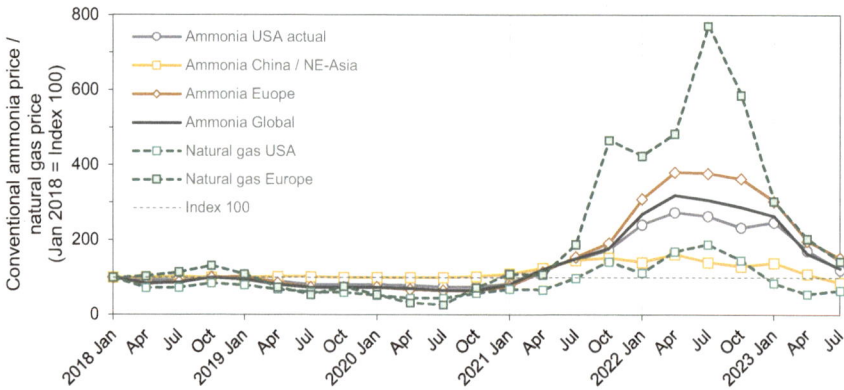

Fig. 10 Market prices of (conventional) ammonia and natural gas (as indices) from 2018 to July 2023 (source natural gas prices: [108]; source ammonia prices: [107])

3.2 Green Ammonia

The subsequent sections describe technical–economic aspects of green ammonia production, covering the production based on the Haber–Bosch synthesis (thermochemical route), electrocatalysis (electrochemical route), and direct solar technologies (photochemical route).

3.2.1 Green Ammonia by Haber–Bosch Synthesis

In this technology, hydrogen is no longer obtained from fossil fuels but from alternative processes such as biomass gasification (syngas), biomass reforming (biogas from anaerobic digestion), or water electrolysis based on renewable energy electricity (powered by solar, wind or hydropower) [87]. However, the latter is the main approach to provide hydrogen for green ammonia production. These three pathways have the same primary energy consumption, about 14 to 15 kWh/kg_{NH3}, but the electrocatalytic (electrolysis) approach avoids the CO_2 emissions released by the other processes (more than 3 kg_{CO2}/kg_{NH3}).

A recent study [109] estimated that ammonia production with biomass gasification has an average production costs of 488 to 1,500 \$/t_{NH3}, in which biomass price varies from 16 to 280 \$/odt (oven dry ton). The production scale of this technology varies between 10 and 1,200 t_{NH3}/d. Ammonia plants with production scales over 1,000t_{NH3}/d are considered competitive, generally with production costs close to 500 \$/t_{NH3}. CO_2 emissions were calculated to be 0.59 to 0.84 kg_{CO2eq}/kg_{NH3} compared to 2.16 kg_{CO2eq}/kg_{NH3} for conventional plants.

In other analyses, green decentralized or islanded ammonia production plants with electrocatalytic green hydrogen were assessed. Ammonia production in electrolysis-based Haber–Bosch processes at small- or medium-size modular plants, includes several advantages: competitive ammonia production costs with reduction of transportation costs, especially for remote locations where transportation expenses are prohibitive; non-fossil fuels consumption and pollution mitigation; and more environmental-friendly operation conditions.

The cost of ammonia production using electrolysis strongly depends on the levelized cost of electricity (LCOE). In this regard, the economic analysis of ruthenium (Ru)-based catalysts (270–415 €/kg_{cat}) and iron (Fe)-based catalysts (1.80 €/kg_{cat}) [65] showed that the use of Ru-based catalysts is advantageous in small-scale ammonia plants (<100 t_{NH3}/d) with lower costs, in which the cost associated with reactors and heat exchangers are dominant compared to the catalyst cost [110]. The use of non-noble metal catalysts at lower operation temperatures could reduce the production costs, increasing the feasibility of local ammonia production via electrolysis.

A recent techno-economic study [5] estimates that a levelized cost of ammonia (LCOA) of 473 $/$t_{NH3}$ is achievable today, while a LCOA below 310 $/$t_{NH3}$ is predicted to be possible by 2030. Other studies estimate a LCOA below 500 $/$t_{NH3}$ for an ammonia plant of 35,000 t_{NH3}/y powered by wind energy [111]. Other study estimated a cost of 798 $/$t_{NH3}$ for an all-electrical ammonia production facility (30 times smaller than large-scale conventional Haber–Bosch plants) with a capacity of 20,000 t_{NH3}/y. This facility used hydrogen from electrolysis and nitrogen from pressure swing adsorption, and was considered as cost-competitive since the LCOE was lower than 10 $/MWh [112].

Other studies estimated the costs for power-to-ammonia units. As an example, a system fed with hydrogen (H_2) and nitrogen (N_2), obtained from a battolyzer and a pressure swing adsorption (PSA), respectively, operating at milder conditions (275 °C and 8 bar) than Haber–Bosch alternatives, and equipped with a synthesis loop loaded with a $Ru/Ba-Ca(NH_2)_2$ catalyst, produced ammonia at a cost of 0.30 to 0.35 €/kWh. The energy consumption was 8.7 to 10.3 kWh/kg_{NH3}, which is comparable to that of a large-scale Haber–Bosch process with a low-temperature PEM electrolyzer (8.6–9.5 kWh/kg_{NH3}) [42]. Other studies estimated the cost of green ammonia around 600 to 1,000 $/$t_{NH3}$, when considering the cost of renewable energy input [7].

3.2.2 Green Ammonia by Electrocatalysis

Some techno-economic analyses developed during the past few years have demonstrated the technological feasibility and economic viability of ammonia synthesis via electrocatalysis for delocalized production. However, it should be noted that the implementation of direct electrochemical synthesis of ammonia using renewable energies would require a different supply chain (to avoid fossil fuels) and methods for leveling the electrical grid [105]. Feasibility evaluations of large- and small-scale electrochemical ammonia synthesis technologies show that LCOA is more sensitive to variations in operation factors than in capital costs. Particularly, the most important variables in LCOA are the cost of the electricity (0.03–0.06 $/kWh), the Faradaic efficiency (FE), and the current density, while the most relevant contributor to the capital cost are the electrochemical reactors and electrodes [113, 114]. In fact, LCOA as low as 500 $/t could be achievable by adjusting the Faradaic efficiency (FE) to 30% and current density to 10^{-7} A/cm^2. Moreover, these evaluations [113] highlight that electrochemical ammonia synthesis could be feasible for a fully decentralized ammonia production at lower production scales (<10 t/d), in locations where natural gas Haber–Bosch plants or electrolysis Haber–Bosch plants are not competitive.

As an example, a recent study demonstrated the feasibility of a small-scale decentralized ammonia production by solar electrochemistry (SECAM process). This process integrates nitrogen gas separation from air, electrocatalytic ammonia synthesis and separation, and recycling of excess hydrogen with an overall energy efficiency comparable to that of the Haber–Bosch process, ranging from 0.56 to 0.92 MJ/mol (at 85 or 20% FE, respectively) versus traditional gas-based Haber–Bosch with energy consumptions about 0.58 to 0.81 MJ/mol. The practical application of this system is awaiting more active electrocatalyst with higher current densities and higher Faradaic efficiencies (FE). Finally, techno-economic analysis and life cycle assessment have shown potentially viable pathways for the production of green ammonia via electrochemical synthesis, using hydrogen from electrolysis (32 t_{H2}/d) and cryogenic nitrogen (135 t_{N2}/d) with an energy consumption of 17 MWh/t_{NH3} [115].

3.2.3 Green Ammonia by Direct Solar Technologies

As highlighted before, one of the main drawbacks of many green ammonia technologies is their strong dependency on levelized cost of electricity (LCOE), which has been recently worsened due to increased costs for energy and rising electricity prices. However, an alternative direct synthesis of green ammonia can be carried out powered by solar photocatalysis. This approach avoids the use of electricity and offers the opportunity for decentralized production in remote locations. Moreover, liquid-phase photocatalytic systems offer the possibility to use electron donors like agricultural or industry side-products to increase the reactions rates, while reusing wasted compounds.

Besides, aqueous systems could operate using air as nitrogen source, the so-called "aerobic" photocatalytic nitrogen fixation, avoiding the need for high-purity nitrogen (like Haber–Bosch or electrocatalytic processes) and thus reducing the operation costs. Nevertheless, technical analyses of "aerobic" photocatalytic nitrogen fixation identified some drawbacks related to the presence of oxygen, such as its reaction with photogenerated carriers (holes and electrons) and/or already formed nitrogen species, or the lower availability of dissolved nitrogen that can impact the conversion efficiency [90, 116]. Notably, the techno-economic analysis [90] indicated that NH_4^+ could be obtained at competitive costs even considering the decrease in activity caused by dissolved oxygen (Table 4). Indeed, the suppression of the air separation unit would radically decrease the capital costs of decentralized ammonia photocatalytic plants. Estimations predict that production costs under 600 $/$t_{NH3}$ could be achievable with solar-to-ammonia (STA) efficiencies of 1, 2.5, and 10% using slurry

Table 4 Estimated production costs of ammonia ($/t_{NH3}$) depending on reactor configuration and nitrogen purity for a small size ammonia production plant assuming a solar-to-ammonia efficiency of 10%

Mode	Slurry reactor ($/t)	Panel ($/t)	Concentrated array ($/t)
With air	30	120	600
With membrane separation	110	200	–

From Ref. [90]

reactors, panels, or concentrator arrays, respectively (Table 4). On-going and future research on highly active catalysts, with high solar-to-ammonia (STA) efficiencies and resistant to oxygen, could facilitate the implementation of photocatalytic systems for decentralized ammonia production.

Further costs analyses of photocatalytic ammonia production systems highlight its use as "solar fertilizer" for the direct fertigation of soils [117]. This approach involves a more efficient fertilization that avoids fertilizer losses in soils by leaching/percolation, which cause nitrate accumulation in groundwaters. Besides, solar ammonia could be directly incorporated into high surface solid materials, such as biochars, as a low release fertilizer. In brief, photocatalytic ammonia synthesis could contribute to reduce pollution and promote a more rational use of nitrogen, following the *precision farming* concept. Besides, decentralized production would bring economic savings, avoiding transportation of fertilizers from the industry to the countryside or remote locations. This would also imply a more efficient fertilization processes in terms of time, effort, and energy consumption.

3.3 Comparison

As commented before, large (H-B) plants of more than 2,000 t/d can produce ammonia at low costs (about 160–600 $/t_{NH3}$) mainly depending on the feedstock prices (i.e., natural gas and other fossil fuels). These prices can be distorted during periods of crisis, and even increased with transportation costs. The production costs of green ammonia using hydrogen from biomass can reach values higher than 470 $/t_{NH3}$ for plants sizes bigger than 1,000 t/d. The techno-economic analyses of green ammonia obtained from electrolysis estimated production costs of more than 500$/t_{NH3}$ for plants with productions lower than 100 t/d.

Table 5 compares an estimation of costs for different energy scenarios, including (i) conventional Haber–Bosch, (ii) electrolysis coupled with Haber–Bosch (E/H-B) at different production capacities, and (iii) direct electrochemical nitrogen reduction (ENR) with nitrogen (N_2) and hydrogen (H_2), considering a benchmark electric power price of 50 $/MWh and a Faradaic efficiency (FE) of 62.2%. A substantial decrease in costs is foreseen when plants are operated at low electricity prices. Electrochemical nitrogen reduction (ENR) plants could achieve more competitive costs than E/H-B technology with higher FE values and overpotential [105].

Table 5 Estimated production costs for ammonia ($/t_{NH3}$) for different 2020 scenarios

Technology	Large H-B plant (ca. 2000 t/d)	Medium H-B plant (545 t/d)	Medium H-B plant (282 t/d)	Small H-B plant (91 t/d)
Scenario 2020				
Conventional H-B	159 $/t	229 $/t	–	325 $/t
Electrolysis coupled with H-B	629 $/t	669 $/t	–	725 $/t
Direct ENR	–	–	508 $/t	–
Scenario 2040				
Conventional H-B (with social cost)	201 $/t (588 $/t)	–	–	–
Electrolysis coupled with H-B	515 $/t	–	–	613 $/t
Direct ENR	–	–	423 $/t	–

From Ref. [105]; ENR: electrochemical nitrogen reduction; H-B: Haber–Bosch

Table 6 Comparison of ammonia with different combustion fuels

Properties	Units	Gasoline	Diesel	LPG	CNG	Gaseous hydrogen	Liquid hydrogen	Ammonia
Lower heating value	MJ/kg	44.5	43.5	45.7	47.5	120.1	120.1	18.8
Flammability limits, gas in air	vol.%	1.4–7.6	0.6–5.5	1.81–8.86	5.0–15.0	4–75	4–75	16.2
Flame speed	m/s	0.58	0.87	0.83	8.45	3.51	3.51	0.15
Flash point	°C	−42.7	73.8	−87.7	−184.4	N/A	N/A	−33.4
Octane	–	90–98	N/A	112	107	>130	>130	110
Fuel density	kg/m^3	698.3	838.8	1.9	187.2	17.5	71.1	602.8
Energy density	MJ/m^3	31,074	36,403	86,487	7,132	2,101	8,539	11,333
Storage method	–	Liquid	Liquid	Liquid	Gas	Gas	Liquid	Liquid
Storage temperature	°C	25	25	25	25	25	−253	25
Storage pressure	kPa	101.3	101.3	850	24,821	24,821	102	1,030
Cost	$/liter	0.58	0.65	0.72	0.57	0.14	0.18	0.24

CNG compressed natural gas; *LPG* liquefied petroleum gas; from Refs. [120, 121]

Regarding the production costs obtained with direct solar photocatalytic technologies, a substantial decrease in cost can be achieved through the optimization of the catalytic systems, photoreactors, and operational conditions (e.g., aerobic conditions) [90]. Besides, the most important advantage with respect to the previous technologies is the low dependence from electricity prices and transportation, which makes this technology invariable to electricity and fuel price fluctuations. The main drawback is the amount of land area required, which makes this technology suitable only for distributed production. For its implementation, it will be desirable that the area of the photoreactors occupy vacant lands or the lower amount of productive land, preferably less than 1 to 5%.

4 Application of Ammonia as a Green Energy Vector

Recently, ammonia has regained interest as an alternative carbon-free fuel for transportation, stationary energy, and industrial sectors [9]. Considering that around 80% of the world energy consumption is in the form of fuels, production of renewable fuels without net CO_2 emission is crucial for meeting the targets of the Paris Agreement [118]. The concept of ammonia fuel that is produced with zero-carbon emissions implies the use of green hydrogen and nitrogen production technologies, in which hydrogen is produced from electrolysis of water and nitrogen is extracted from air by air separation unit (ASU) with the use of renewable energy [43].

4.1 Basic Properties of Ammonia

The global goal for decarbonization of the energy industry could become a reality by converting electrical energy into chemical energy via renewable hydrogen, which is actually the foundation of most chemical energy vectors [11, 43]. Hydrogen (H_2) can be stored, used in a fuel cell to generate electricity, or combusted with air to generate heat, producing only water as a by-product. However, hydrogen has a very poor volumetric energy density compared to most liquid fuels (e.g., 2.4 kWh/L under cryogenic conditions compared to 9 kWh/L for gasoline) [11], and therefore, its chemical derivatives are considered a promising option as carbon-free energy vectors with a foreseen cost reduction in transportation of at least a factor of three [11].

Green ammonia is an appealing chemical derivative with application as fertilizer, hydrogen storage medium (17.3 wt.% hydrogen in liquid ammonia), chemical feedstock (e.g., production of polyimides, nitric acid, nylon, pharmaceuticals, dyes, explosive materials, refrigerants, etc.), and clean burning fuel for transportation and power generation sectors [2, 8, 11, 119]. In fact, green ammonia can be stored at relatively mild conditions (it can be liquefied at -33 °C and at atmospheric pressure, while liquid hydrogen has to be chilled below -253 °C), and it has a lower cost per unit of stored energy (0.54 $/kg_{H2}$ compared to 14.95 $/kg_{H2}$ of pure hydrogen storage) [6, 11]. Its wide potential application arises from its high energy density (22.5 MJ/kg at higher heating value (HHV)), comparable to that of fossil fuels like low-ranked coals (20 MJ/kg), natural gas (55 MJ/kg), liquefied natural gas (LNG) (54 MJ/kg), and hydrogen (142 MJ/kg) (Table 4) [6].

Among its competitors, liquid ammonia can store energy with higher energy density (LHV) (11.3 MJ/L) than liquid hydrogen (8.54 MJ/L at cryogenic temperature) and compressed hydrogen (2 MJ/L) [6, 122, 123]. The energy release from green ammonia can be achieved by clean combustion, releasing only water and nitrogen as by-products (Eq. 7). Then, ammonia can be cracked back into hydrogen to be used in a fuel cell.

$$4NH_3 + 3O_2 \rightarrow 2N_2 + 6H_2O \qquad (7)$$

Another major advantage is that there is an established and reliable infrastructure for ammonia storage and distribution worldwide. It can be pressurized or absorbed in metal halides (40 wt.%) or zeolites (60 wt.%) [42], and globally distributed by pipeline, rail, road, and ship. In fact, around 180 Mt of ammonia are produced and transported annually, reinforcing ammonia as a competitive option for zero-carbon energy carrier [6]. From an economical point of view, green ammonia from small-scale plants in the vicinity of the consumer is one of the most attractive options for a widespread utilization [43].

Despite these advantages, renewable green ammonia synthesis accounts today for just 0.01% of the global production [25]. Several production challenges exist such as tackling with the competing hydrogen evolution from water rather than hydrogen incorporation into ammonia, or the generation of nitrous oxides (NO_x) during combustion, especially if used in vehicle engines or gas turbines at high temperatures/pressures [10, 43]. Apart from that, the high cost is the largest barrier for its widespread utilization as energy vector. For instance, ammonia produced from fossil fuels needs 28 GJ/t, while renewable-based ammonia requires energy in the range of 32 to 54 GJ/t [124]. Nevertheless, costs reductions are foreseen at medium term through technical improvements in renewable electricity and transportation [2, 11].

4.2 Ammonia Cracking

Ammonia can be catalytically decomposed into hydrogen and nitrogen at high temperatures (>700 °C), using inexpensive materials such as iron- or nickel-based catalysts (Eq. 8).

$$2NH_3(g) \rightleftharpoons N_2(g) + 3H_2(g) \quad \Delta H^{\circ} = 92 kJ/mol \qquad (8)$$

Then, the obtained hydrogen can be used in fuel cells to generate electricity, mainly in PEM fuel cells, or be used as a fuel for combustion. At present, the main application of the ammonia decomposition reaction in industry is metal annealing and galvanizing [125]. However, extending the application of the ammonia cracking technology to fuel cell applications holds a great potential, based on the fact that liquefied ammonia contains 50% more hydrogen by volume than liquid hydrogen [12, 125].

Active research is now focused on the optimization of both catalyst and process to achieve an efficient and low-cost ammonia cracking. Lowering the decomposition temperature reduces energy costs but usually requires the use of rare and expensive metals such as ruthenium, although some examples can be found with inexpensive catalysts such as amide ($-NH_2$) [126] as well as imide ($-NH$) [127] materials, which operate at 450 to 500 °C and reach TRLs of 2 to 4 [12].

The U.S. Department of Energy report (2006) [128] concluded that the ammonia cracking technology was still too expensive for mobile applications, due to the high temperatures needed to obtain high-purity hydrogen gas streams, which must contain less than 1 ppm of ammonia [12]. A post-cracking purification step guarantees this level of purity that is crucial for PEM fuel cells applications, because they can be irreparably degraded in at low ammonia concentrations (ppm) [127]. The main technological solution to purify the hydrogen stream includes membranes and absorption-based systems (e.g., metal halide amines) [12]. These systems have proved to lower ammonia concentration < 0.1 ppm from inlet streams of about 10,000 ppm. The Australian Commonwealth Scientific and Industrial Research Organisation (CSIRO) has patented a metal-based selective H_2 permeable membrane combined with a pilot-scale NH_3 cracking reactor at high temperature (>400 °C) for the production of high-purity hydrogen from ammonia at a rate of 0.75 kg/d [129]. In 2018, CSIRO demonstrated that the obtained hydrogen was suitable for refueling PEM fuel cell vehicles, and now this technology is in commercial development.

On the other hand, alternative solutions to the purification step include the direct use of ammonia in alkaline fuel cells (AFC), less developed than PEM but able to operate in the presence of ammonia (up to 9%), or solid oxide fuel cells (SOFC) which are not ideal for transportation but are excellent for small-scale power generation [43, 126, 127]. In this regard, production of hydrogen by thermal cracking within a SOFC leads to an overall maximum amount of work of about 0.33 MJ/mol$_{NH3}$ or 0.22 MJ/mol of cracked hydrogen, which is close to other more conventional fuels (e.g., methane, n-octane, methanol, etc.) [6].

An alternative use includes the combustion of ammonia instead of its use in fuel cells. In this regard, ammonia alone is difficult to ignite due to its low burning velocity and high minimum ignition energy; however, 70/30 mixtures of ammonia and hydrogen by volume have been identified as a viable operating point for combustion [11, 126, 127]. Ammonia turbines are foreseen to be commercialized in the medium term, meanwhile a pilot scale with a 50 kW class micro gas-turbine system is already in operation at the National Institute of Advanced Industrial Science and Technology (AIST) in Japan [130].

4.3 Application Options as a Carbon-Free Energy Carrier

4.3.1 Transportation Sector

One of the most interesting applications of ammonia in the transportation sector is oriented to light-duty vehicles. Here, ammonia could be used as a monofuel or being cracked into hydrogen followed by injection into an internal combustion engine (ICE) or a fuel cell [43]. Ammonia fuel can also power heavier vehicles, such as heavy-duty vehicles (trucks), or even certain maritime vessels (ships). Regarding fuel cells, proton exchange membrane (PEM) fuel cells are the most typical devices due to their low operating temperature [10, 12, 43]. Direct ammonia fuel cells (DAFC) are another alternative for small-scale and mobile applications, in which ammonia is directly oxidized to nitrogen in an alkaline electrolyte and at moderate temperatures, simultaneously coupling the reduction of O_2 to OH^- [9, 58]. Moreover, solid oxide fuel cells (SOFCs) can be directly fed with ammonia without any external reformer or cracker reactors because they operate at high temperatures (600–800 °C) to promote the rapid decomposition of ammonia (NH_3) into nitrogen (N_2) and hydrogen (H_2) [9]. It should be noted that both DAFC- and SOFC-fuel cells can potentially generate nitrogen oxide emissions (NO_x emissions). In these cases, catalytic converters should be implemented in ammonia-fueled engines to control this issue [58].

The first work on power generation for vehicles with ammonia as a fuel dates back to 1933 [131]. Nowadays, there are some energy projects worldwide using ammonia for transportation from light- to heavy-duty vehicles. A prototype car [132] has been tested in South Korea, using a dual fuel containing 70% ammonia and 30% gasoline that powers a converted ICE. This fuel ratio results in a 70% reduction in CO_2 emissions, compared to pure gasoline. Recently, the world's first ammonia-powered tractor [133] successfully completed its first demonstration run at the Advanced Energy Center at the Stony Brook University (New York). This tractor can run for several hours on a 60-gallon tank of ammonia, which is cracked in a reactor to produce hydrogen that is then supply to a 100 kW hydrogen fuel cell.

Regarding long-distance transportation, the Australian company Aviation H_2 has recently announced the modification of their turbofan engines to use liquid ammonia to achieve carbon-free flights [134]. Of special significance is the application of ammonia in the shipping industry to lower the sulfur content of fuels and to meet the decarbonization goal of the sector by 2050, as required by the International Maritime Organization [58]. As an example, a 2 MW NH_3-SOFC is being tested by the shipping company Eidesvik in an offshore supply vessel (Viking Energy) [135] to demonstrate the relevance of carbon-free ammonia for long-distance transportation (expected results for 2024) [58]. MAN Energy Solutions, which is part of the Volkswagen Group, is currently developing an ammonia-fueled marine two-stroke engine predicted to be commercially available by 2024, following by a gradual rebuild of its existing maritime vessels by 2025 [136].

4.3.2 Stationary Power Plants

Ammonia (NH_3) is a solid candidate for electricity generation via stationary power plants using turbine engines and fuel cells. Turbine engines rely on ammonia combustion (900–1,100 °C), which is blended with other fuels (e.g., hydrogen, natural gas, gasoline, diesel, and kerosene) to tackle the low ignition point of ammonia. The main obstacle of this application is the emission of nitrogen oxides (NO_x) associated with the combustion of ammonia. For fuel cells, ammonia can be either directly fed or cracked into hydrogen and then be fed into a fuel cell. This technology includes alkaline fuel cells (room temperature), PEM fuel cells (60–80 °C), or solid oxide fuel cells (700–775 °C) [42].

An important advantage of ammonia for power generation is its safe storage in large quantities at relatively moderate pressures and temperatures, compared to hydrogen. In fact, ammonia can be safely stored without contamination for long periods of time (>10,000 h) [9]. Among relevant examples, IHI Corporation tested in 2018 a 1 kW NH_3-SOFC stack in Japan, and later on a co-combustion demonstrator with 2 MW ammonia gas turbines, achieving for the first time a 20% ammonia-coal co-firing [137]. The use of ammonia as a supplementary fuel in co-combustion coal-fired power generators (120 MW) has been demonstrated in Japan, using a 1% share of ammonia (in terms of total energy content). Higher blending shares of up to 20% ammonia have been successfully achieved in smaller furnaces with a capacity of 10 MW thermal. These blending values could be achievable in coal power plants with only minor technological adjustments [138]. On the other hand, researchers from the National Institute of Advanced Industrial Science and Technology (AIST) in Japan have demonstrated the operation of a 41.8 kW gas turbine power generator with a methane–ammonia gas mixture, which even run a mono-fuel combustion test burning 100% ammonia gas [139].

5 Ammonia in a Decarbonized Future

5.1 Scenarios for Ammonia Decarbonization

Over the last few years, the decarbonization of ammonia production has become imperative, considering the significant carbon footprint of conventional routes. Namely, fossil-based production pathways emit about 2.0 t_{CO2}/t_{NH3} on average. About two-third of the CO_2 is produced to generate hydrogen through hydrocarbons reforming, while a third originates from fuel combustion for the synthesis (about 7.2–9.0 GJ/t_{NH3}) [57].

Three possible strategies are currently being considered for decarbonization of ammonia production, which can be considered as three overlapping technology generations:

- Generation 1 (Gen 1) involves a conventional production with CO_2 sequestration (i.e., blue ammonia) to bring the net carbon impact of the ammonia production to zero.
- Generation 2 (Gen 2) refers to ammonia produced in a modified small-scale Haber–Bosch process using renewable-sourced (green) hydrogen from water electrolysis.
- Generation 3 (Gen 3) refers to the development of alternative synthetic routes using renewable electricity (electrochemical) or sunlight (photochemical), in which Haber–Bosch process is no longer required [10, 140].

Nowadays, ammonia is produced via Generation 1, considering the option of capturing CO_2 emissions to produce blue ammonia. However, carbon sequestration adds cost and plant complexity to the existing Haber–Bosch technology. For this reason, Generation 1 scenario appears as a transitional solution to establish a competitive ammonia market beyond the fertilizer and chemical industries.

In contrast, renewable ammonia from Generation 2 is predicted to be widespread by 2030 because the existing Haber–Bosch plants could be easily fed with renewable hydrogen without major modifications [58]. Nitrogen is obtained directly from air using an air separation unit, which accounts for 2 to 3% of the process energy used. A Generation 2 demonstration project is now in operation in the UK (i.e., Rutherford Appleton Laboratory). This plant, launched in 2018 by Siemens, is fully powered with renewable electricity from a 20 kW wind turbine, and uses a proton exchange membrane (PEM) water electrolyzer to produce hydrogen, which is fed to a Haber–Bosch unit that delivers 30 kg$_{NH3}$/d [141]. Other example is located in Japan (Fukushima), where JGC Group launched a 20 kg$_{NH3}$/d demonstrator powered by wind, and designed to test the cycle stability of different catalysts [12].

The main challenge in Generation 2 is to achieve a significant cost reduction, of which about 85% is electricity. Considering that in most parts of the world electricity is significantly more expensive than natural gas, this approach is likely to be implemented in areas with abundant renewable electricity, such as in North-Western Australia. Significant cost reductions could be achievable by improving the engineering of cells and stacks. ***For instance, increasing the cell areas or by developing cheaper bipolar flow field plates, membranes, and catalysts [58]. In this regard, megawatt scale proton exchange membrane (PEM) water electrolyzers are already commercially available [58], and the first gigawatt production of green hydrogen is scheduled for 2023 in Germany by Siemens [142].

Further pilot-scale green ammonia plants are in development in Australia, Denmark, Morocco, and the Netherlands. Undoubtedly, the achievement of cost-competitive and sustainable ammonia production will help to take a big step toward a hydrogen economy, and will contribute to reach the commercialization target of Generation 3 ammonia by 2050 [10].

5.2 Integration of Green Ammonia in a Low-Carbon Economy

The projected increase in population during the next decades and the consequent increase in cropland productivity require a constant increase in fertilizers production. Over the past years and decades, the synthesis of ammonia by thermal processes (160 to 180 Mt/year) have reached high levels of efficiency by decreasing the energy requirements close to minimum values. However, the historic episodes of economic crisis (i.e., oil crisis (1973), financial crisis (2008), and present energy crisis) have abruptly increased the prices of fossil fuels and electricity, consequently increasing the production costs of many industrial processes based on these energy sources. At the same time, the cost of fertilizers has also increased during the crisis periods, and this situation has caused elevated prices or even scarcity of food and ammonia-derived chemical products.

Today, the growing need for ammonia production not only responds to the increasing fertilizer demand, but also to the foreseen use of ammonia as an energy vector in a low-carbon economy system. In numbers, projections estimate that the ammonia demand (mostly for fertilizers synthesis) will be higher than 200 Mt_{NH3}/year by 2050 [143]. This implies that the industrial production capacity of ammonia should be multiplied several times to meet this future demand, especially for the use of ammonia for power generation. This assumption is clear considering the massive global primary energy production of 176,000 TWh in 2021 [143], from which 136,000 TWh was obtained from fossil fuels.

Among sectors, current electricity consumption accounts for 29,000 [144] to 30,000 TWh/year [145], while transportation represents around 28,000 TWh/year (ca. 20% of the global energy production from fossil fuels) [146]. In this regard, the IEA ammonia technology roadmap [147] includes some projections for *Sustainable Development* and *Net Zero Emissions* scenarios, which estimate that ammonia production would reach 400 to 600 Mt by 2050, considering existing uses, maritime fuel, and power generation. For both scenarios, maritime shipping is expected to use around 45 or 110% of the current ammonia production, while ammonia use for co-firing in coal power stations would climb to 60 or 85 Mt/year for *Sustainable Development* and *Net Zero Emissions* scenarios, respectively.

It is to be noted that ammonia projections will also need to meet the Paris Agreement goal of reaching zero emissions by 2050, which implies cutting 70% of CO_2 emissions from ammonia production by this year [147], considering current values of around 450 Mt_{CO2}/year.

In a low-carbon economy context, projections estimate that ammonia would be more easily implemented in the fertilizer sector, chemical industry, maritime transport, power generation tanks, distributed energy storage applications, and hydrogen commerce [7]. These applications will count with the current shipping and pipeline transfer infrastructure, which would be more easily and safely implemented for ammonia than for hydrogen [148].

Regarding production models (i.e., centralized vs decentralized), the historic centralized production of many chemicals and fuels, and even its agglomeration in specific regions, have produced a strong reliance of conventional Haber–Bosch on fossil fuels. This dependency generates vulnerability to face volatile global market prices (e.g., volatility in fertilizer production), and a complex intersection with geopolitical situations. In addition, it should be noted that current centralized ammonia production in large plants, usually located close to natural gas facilities or ports, is cost-effective when methane prices are low. In addition, a rise in fuel prices dramatically increases the transportation costs to remote locations.

During the last years, both secondary and tertiary economy sectors have decentralized their activities and production in developed countries. Following this trend, the production decentralization of strategic products, like ammonia, represents a potential solution. Among possible technologies that could be used for decentralization, we highlight that low temperature and pressure ammonia synthesis is one of the most interesting and energy-saving options. Other technologies such as the electrochemical ammonia synthesis with water and nitrogen arises as the most likely alternative to substitute the Haber–Bosch process in decentralized locations.

Recent works deal about the implementation of *solar ammonia refineries* in a near future. Wang et al. [2] studied several technologies currently available at laboratory scale, which could be scaled up although they would also be contingent on the electricity cost. Among them, conventional Haber–Bosch (30.5–61 MJ/kg$_{NH3}$) is compared in energy requirements terms to an electrolysis coupled Haber–Bosch process (264.9 MJ/kg$_{NH3}$), electrochemical cycles like lithium [149] or molybdenum (Mo) (339 MJ/kg$_{NH3}$), direct electrocatalytic or photocatalytic ammonia synthesis (190.5 and 208.3 MJ/kg$_{NH3}$, respectively), or aluminum redox cycle (299.2 MJ/kg$_{NH3}$). This work also evaluates how to store ammonia as an energy vector, contrasting two trends or "schools": on one hand, one that avoids the storage of electrical energy into chemical energy considering this pathway as an inefficient waste of energy; and on the other hand, the solar-to-chemical "school" that promotes solar harvesting, and its direct conversion to chemicals through photochemistry, photothermal chemistry, and solar thermal chemistry, avoiding also the inefficient or unnecessary use of electricity. The last strategy should be considered not only for ammonia production as an energy vector, but also to achieve the decentralization of *solar fertilizers* in remote locations for direct fertigation, or to incorporate ammonia into carbon-based adsorbents, avoiding nitrogen oxide (NO$_x$) emissions under certain conditions [150–152].

Following this last approach, Comer et al. [117] proposed a decentralized fertilizer production in three scenarios. A small remote farm-scale (70%) production in low-income countries could be assisted by inexpensive photocatalysis; larger farms (>100 ha that represent 2% of farm holdings and 45% of land area) with fertigation (aqueous fertilizer); or places with distribution infrastructure (multi-farm scale), in which the electrocatalytic ammonia production would avoid seasonality and would maintain a constant production. The excess of fertilizer could be stored on adsorbents, such as activated carbon or biochar, either for the fertilizer season or other uses.

A recent work by McFarlane et al. [58] discussed about the technology development and scale-up of an ammonia-based economy, describing three successive ammonia production scenarios for the gradual replacement of existing liquid fuels: Generation 1 corresponded to current technologies based on thermal processes combined with carbon sequestration (so-called *blue ammonia*); Generation 2 refers to immediate technologies to be used with hydrogen (produced in proton exchange membrane (PEM) or solid oxide electrolyzers) and powered with renewable electricity; and Generation 3 corresponds to electrocatalytic ammonia production without Haber–Bosch reactor. The last scenario is circumscribed to electrocatalysis due to its higher TRL, mainly because it is more resilient to intermittency, it operates at near ambient temperature, and it is less sensitive to nitrogen purity (lower poisoning effect). However, electrocatalytic processes still present some weaknesses such as low Faradaic efficiencies (FE), and the production of nitrogen oxides. In good agreement, Smith et al. [153] estimate that the success of a second ammonia revolution as an energy vector relies on increasing the energy efficiency, and developing small-scale distributed production in remote locations powered by renewables. This approach offers the opportunity to counter-balance the fluctuations in energy supply with ammonia, likely through the most mature technologies (i.e., electric-powered Haber–Bosch and electrochemical ammonia synthesis). Among other technologies, photocatalysis would require higher yields for practical implementation, and therefore, it is mostly foreseen for intensive agriculture applications in a longer timescale [58]. Additional alternatives such as biological ammonia synthesis, catalytic plasma, non-thermal synthesis, etc., already exist at laboratory scale but their TRL are still low [153].

The roadmap for ammonia economy proposed [58] describes a constant increase in ammonia production with a sharp rise by 2030 to 2040, corresponding to the introduction of ammonia as a large-scale energy vector, once Generation 2 and 3 technologies reach a Commercial Readiness Index (CRI) of 3. Further, it is worth noting the warning of the authors about the transformation of the nitrogen cycle during the transition from a low-carbon economy to a nitrogen economy. This change would originate an increase in nitrogen oxides (NO_x) and nitrate emissions due to inefficiencies and losses, which have an unknown impact on aquatic ecosystems that should be addressed.

On the other hand, the transition toward an ammonia economy may present some limitations and challenges, such as the improvement of renewable hydrogen production efficiency; the development of power generation systems; the development of hydrogen purification systems, or the elimination of nitrogen oxides (NO_x) from exhaust gases, among others [140, 153, 154].

Taking in consideration the current situation and the challenges for the incorporation of ammonia in a near-future low-carbon economy, the main policies should focus on R&D of ammonia clean technologies and dual centralized/decentralized production systems; implementation of these new synthesis technologies for fertilizers production, maritime transport or decentralized power generation; avoid the dependency from external fossil fuel resources; and impulse ammonia use by producers/consumers/governments.

6 Conclusion

Achieving a carbon-free energy sector and a greener chemical industry ask for a massive use of renewable-based technologies and energy carriers. For this transition, ammonia is likely to play a key role in the near future, aiding to replace a substantial fraction of fossil fuel consumption in multitude of applications, ranging from fertilizers production to transportation or decentralized power generation. Indeed, projections estimate an increasing trend in the ammonia demand, mostly for fertilizers synthesis and power generation, and this implies that the production capacity should be multiplied several times to meet this future demand.

The industrial production of ammonia has achieved important improvements over the past century, introducing operational changes and optimizing the design of reactors. However, conventional technologies are still highly energy-intensive processes, and this has triggered researchers to look for a step change in the ammonia production, aiming to fulfill three key criteria: energy efficiency, scalability/modularity, and CO_2-free emissions. The new synthetic approaches can be achieved via electrochemical or photochemical routes, and are powered by renewable energies to achieve zero-carbon emissions along the entire production chain. Both approaches are still at an early stage of development at laboratory scale, although they hold a great potential for the transition to a low-carbon economy. Techno-economic analysis and life-cycle assessment show the technological feasibility and economic viability of electrochemical synthesis for future decentralized ammonia production, although it would require a fossil-free supply chain and methods for leveling the electrical grid. Further cost analyses show the great potential of low temperature and pressure photochemical synthesis for a decentralized solar ammonia production, contributing to the development of a future circular and decarbonized economy.

Acknowledgements Laura Collado acknowledges the national project ARMONIA (PID2020–119125RJ-I00) funded by MCIN/AEI/https://doi.org/10.13039/501100011033. The authors would like to express their gratitude to Hita-chi and Casale for the images provided to illustrate the industrial processes in this chapter. The authors also acknowledge the project Maetzu Unit of Excellence award (MCIN/AEI/https://doi.org/10.13039/501100011033, CEX2019-000931-M).

References

1. International Energy Agency (IEA) (2021) Ammonia technology roadmap. https://doi.org/10.1787/f6daa4a0-en
2. Wang L, Xia M, Wang H, Huang K, Qian C, Maravelias CT, Ozin GA (2018) Greening ammonia toward the solar ammonia refinery. Joule 2:1055–1074. https://doi.org/10.1016/j.joule.2018.04.017
3. Scott EK (1923) Nitrates and ammonia from atmospheric nitrogen. Lecture I, J R Soc Arts 71:859–876
4. Pattabathula V, Richardson J (2016) Introduction to ammonia production. Chem Eng Prog 112:69–75

5. Nayak-Luke RM, Bañares-Alcántara R (2020) Techno-economic viability of islanded green ammonia as a carbon-free energy vector and as a substitute for conventional production. Energy Environ Sci 13:2957–2966. https://doi.org/10.1039/d0ee01707h
6. Valera-Medina A, Xiao H, Owen-Jones M, David WIF, Bowen PJ (2018) Ammonia for power. Prog Energy Combust Sci 69:63–102. https://doi.org/10.1016/j.pecs.2018.07.001
7. Yüzbaşıoğlu AE, Avşar C, Gezerman AO (2022) The current situation in the use of ammonia as a sustainable energy source and its industrial potential. Curr Res Green Sustain Chem 5:100307. https://doi.org/10.1016/j.crgsc.2022.100307
8. Ghavam S, Vahdati M, Wilson IAG, Styring P (2021) Sustainable ammonia production processes. Front Energy Res 9:1–19. https://doi.org/10.3389/fenrg.2021.580808
9. Elishav O, Mosevitzky Lis B, Miller EM, Arent DJ, Valera-Medina A, Grinberg Dana A, Shter GE, Grader GS (2020) Progress and prospective of nitrogen-based alternative fuels. Chem Rev 120(12):5352–5436. https://doi.org/10.1021/acs.chemrev.9b00538
10. Smart K (2021) Review of recent progress in green ammonia synthesis. Johnson Matthey Technol Rev 66:230–244. https://doi.org/10.1595/205651322x16334238659301
11. Salmon N, Bañares-Alcántara R (2021) Green ammonia as a spatial energy vector: a review. Sustain Energy Fuels 5:2814–2839. https://doi.org/10.1039/d1se00345c
12. The Royal Society (2020) Ammonia: zero-carbon fertiliser, fuel and energy store. Policy Briefing. https://royalsociety.org/topics-policy/projects/low-carbon-energy-progra mme/green-ammonia/
13. Malthus TR (1798) An essay on the principle of population
14. Crookes W (1899) The wheat problem based on remarks made in the presidential address to the British association at Bristol in 1898, revised, with an answer to various critics, 1st edn, Lomdon
15. Valera-Medina A, Roldan A (2020) Ammonia from steelworks. In: Sustainable ammonia production, pp 69–80
16. Appl M (1999) Process steps of ammonia production. In: Ammon Princ Ind Pract, Wiley Online Books, pp 65–176. https://doi.org/10.1002/9783527613885.ch04
17. Khosla A (2004) Entry barriers and the structure of the Japanese ammonium sulphate industry in the interwar period. Discussion Paper Series A458, Inst. Econ. Res. Hitotsubashi Univ
18. Rouwenhorst KHR, Jardali F, Bogaerts A, Lefferts L (2021) From the Birkeland-Eyde process towards energy-efficient plasma-based NO: Xsynthesis: A techno-economic analysis. Energy Environ Sci 14:2520–2534. https://doi.org/10.1039/d0ee03763j
19. Bartell FE (1922) Production of ammonia by the sodium cyanide method. J Ind Eng Chem 14:516–520. https://doi.org/10.1021/ie50150a017
20. Richards JW (1913) The Serpek process for the manufacture of aluminium nitride. J Ind Eng Chem 5:335–337. https://doi.org/10.1021/ie50052a024
21. Travis AS (1998) High pressure industrial chemistry: the first steps, 1909–1913, and the Impact BT—determinants in the evolution of the European chemical industry, 1900–1939: new technologies, political frameworks, markets and companies. In: Travis AS, Schröter HG, Homburg E, Morris PJT (eds) Springer Netherlands, Dordrecht, pp 3–21. https://doi.org/10.1007/978-94-017-1233-0_1
22. Travis AS (2018) Electricity and the chemical industry BT—nitrogen capture: the growth of an international industry (1900–1940). In: Travis AS (ed) Springer International Publishing, Cham, pp 49–92. https://doi.org/10.1007/978-3-319-68963-0_5
23. Travis AS (2015) Occasional papers no7 Nitrogen, novel high-pressure chemistry, and the German War Effort (1900–1918)
24. Patil BS, Hessel V, Seefeldt LC, Dean DR, Hoffman BM, Cook BJ, Murray LJ (2017) Nitrogen fixation. Ullmann's Encycl Ind Chem, pp 1–21. https://doi.org/10.1002/14356007.a17_471.pub2
25. Rouwenhorst KHR, Travis AS, Lefferts L (2022) 1921–2021: a century of renewable ammonia synthesis. Sustain Chem 3:149–171
26. Travis AS (2018) High-pressure synthesis and later developments BT—nitrogen capture: the growth of an international industry (1900–1940). In: Travis AS (ed) Springer International Publishing, Cham, pp 347–360. https://doi.org/10.1007/978-3-319-68963-0_16

27. Haber RF (1916) Le Rossignol, Production of ammonia, 1202995
28. Haber F, Le Rossignol R (1913) The production of synthetic ammonia. J Ind Eng Chem 5:328–331. https://doi.org/10.1021/ie50052a022
29. Haber F, Le Rossignol R (1913) Über die technische Darstellung von Ammoniak aus den Elementen, Zeitschrift Für Elektrochemie Und Angew. Phys Chemie 19:53–72. https://doi.org/10.1002/bbpc.19130190201
30. Haber F (1905) Thermodynamik technischer Gasreaktionen. Oldenbourg Wissenschaftsverlag
31. Haber F (1909) Verfahren zur Darstellung von Ammoniak aus den Elementen durch Katalyses unter Druck bei erhöhter Temperatur, 238450
32. BASF (1908) Verfahren zur synthetischen Darstellung von Ammoniak aus den Elementen., 235421
33. Puspitasari P, Yahya N (2011) Development of ammonia synthesis. In: 2011 National Postgraduate Conference, pp 1–4. https://doi.org/10.1109/NatPC.2011.6136449
34. Liu H (2014) Ammonia synthesis catalyst 100 years: practice, enlightenment and challenge, Cuihua Xuebao/Chinese. J Catal 35:1619–1640. https://doi.org/10.1016/S1872-2067(14)60118-2
35. Modak JM (2002) Haber process for ammonia synthesis. Resonance 7:69–77. https://doi.org/10.1007/BF02836187
36. Travis AS (2022) Casale: the first 100 years, Casale S.A.
37. Mittasch A, Frankenburg W (1950) Early studies of multicomponent catalysts. In: Frankenburg WG, Komarewsky VI, EKBT-A in Rideal C (eds) Academic Press, pp 81–104. https://doi.org/10.1016/S0360-0564(08)60375-2
38. Farber E (1966) From chemistry to philosophy: the way of Alwin Mittasch (1869–1953). Chymia 11:157–178. https://doi.org/10.2307/27757266
39. Ertl G (1980) Surface science and catalysis—studies on the mechanism of ammonia synthesis: the P. H. Emmett Award address. Catal Rev 21:201–223. https://doi.org/10.1080/03602458008067533
40. Travis AS (2018) The direct synthesis of ammonia BT—nitrogen capture: the growth of an international industry (1900–1940). In: Travis AS (ed) Springer International Publishing, Cham, pp 93–127. https://doi.org/10.1007/978-3-319-68963-0_6
41. Humphreys J, Lan R, Tao S (2021) Development and recent progress on ammonia synthesis catalysts for Haber-Bosch process. Adv Energy Sustain Res 2:2000043. https://doi.org/10.1002/aesr.202000043
42. Rouwenhorst KHR, Van der Ham AGJ, Mul G, Kersten SRA (2019) Islanded ammonia power systems: technology review & conceptual process design. Renew Sustain Energy Rev 114. https://doi.org/10.1016/j.rser.2019.109339
43. Chehade G, Dincer I (2021) Progress in green ammonia production as potential carbon-free fuel. Fuel 299:120845–120879. https://doi.org/10.1016/j.fuel.2021.120845
44. Almquist JA, Crittenden ED (1926) A study of pure-iron and promoted-iron catalysts for ammonia synthesis 1. Ind Eng Chem 18:1307–1309. https://doi.org/10.1021/ie50204a036
45. Bridger GL, Pole GR, Beinlich AW, Thompson HL (1947) Production and performance of ammonia-synthesis catalyst. Chem Eng Prog 43:291
46. Ertl G, Weiss M, Lee SB (2013) Reprint of: the role of potassium in the catalytic synthesis of ammonia. Chem Phys Lett 589:18–20. https://doi.org/10.1016/j.cplett.2013.08.060
47. Ozaki A (1981) Development of alkali-promoted ruthenium as a novel catalyst for ammonia synthesis. Acc Chem Res 14:16–21. https://doi.org/10.1021/ar00061a003
48. Siporin SE, Davis RJ (2004) Use of kinetic models to explore the role of base promoters on Ru/MgO ammonia synthesis catalysts. J Catal 225:359–368. https://doi.org/10.1016/j.jcat.2004.03.046
49. Rosowski F, Hornung A, Hinrichsen O, Herein D, Muhler M, Ertl G (1997) Ruthenium catalysts for ammonia synthesis at high pressures: preparation, characterization, and power-law kinetics. Appl Catal A Gen 151:443–460. https://doi.org/10.1016/S0926-860X(96)00304-3

50. Nielsen A, Kjaer J, Hansen B (1964) Rate equation and mechanism of ammonia synthesis at industrial conditions. J Catal 3:68–79. https://doi.org/10.1016/0021-9517(64)90094-6

51. Kitano M, Inoue Y, Sasase M, Kishida K, Kobayashi Y, Nishiyama K, Tada T, Kawamura S, Yokoyama T, Hara M, Hosono H (2018) Self-organized Ruthenium–Barium core–shell nanoparticles on a mesoporous calcium amide matrix for efficient low-temperature ammonia synthesis. Angew Chemie Int Ed 57:2648–2652. https://doi.org/10.1002/anie.201712398

52. Aika K (2017) Role of alkali promoter in ammonia synthesis over ruthenium catalysts—Effect on reaction mechanism. Catal Today 286:14–20. https://doi.org/10.1016/j.cattod.2016.08.012

53. Mortensen JJ, Hammer B, Nørskov JK (1998) Alkali promotion of N_2 dissociation over Ru(0001). Phys Rev Lett 80:4333–4336. https://doi.org/10.1103/PhysRevLett.80.4333

54. Ravi M, Makepeace JW (2022) Facilitating green ammonia manufacture under milder conditions: what do heterogeneous catalyst formulations have to offer? Chem Sci 13:890–908. https://doi.org/10.1039/D1SC04734E

55. Travis AS (2018) A time of guns and grain BT—nitrogen capture: the growth of an international industry (1900–1940). In: Travis AS (ed) Springer International Publishing, Cham, pp 129–149. https://doi.org/10.1007/978-3-319-68963-0_7

56. Qing G, Ghazfar R, Jackowski ST, Habibzadeh F, Ashtiani MM, Chen CP, Smith MR, Hamann TW (2020) Recent advances and challenges of electrocatalytic N_2 reduction to ammonia. Chem Rev 120:5437–5516. https://doi.org/10.1021/acs.chemrev.9b00659

57. Rouwenhorst KHR, Krzywda PM, Benes NE, Mul G, Lefferts L (2021) Ammonia production technologies. In: Techno-economic challenges green Ammon. As an energy vector. Elsevier Inc., pp 41–83. https://doi.org/10.1016/b978-0-12-820560-0.00004-7

58. MacFarlane DR, Cherepanov PV, Choi J, Suryanto BHR, Hodgetts RY, Bakker JM, Ferrero Vallana FM, Simonov AN (2020) A roadmap to the ammonia economy. Joule 4:1186–1205. https://doi.org/10.1016/j.joule.2020.04.004

59. Yüzbasıoglu AE, Tatarhan AH, Gezerman AO (2021) Decarbonization in ammonia production, new technological methods in industrial scale ammonia production and critical evaluations. Heliyon 7:e08257. https://doi.org/10.1016/j.heliyon.2021.e08257

60. Travis AS (2018) Catching up: mainly Italy, Japan, and the Soviet Union BT—nitrogen capture: the growth of an international industry (1900–1940). In: Travis AS (ed) Springer International Publishing, Cham, pp 379–381. https://doi.org/10.1007/978-3-319-68963-0_19

61. van Rooij A (2005) Engineering contractors in the chemical industry. The development of ammonia processes, 1910–1940. Hist Technol 21:345–366. https://doi.org/10.1080/073415 10500268215

62. West JH (1921) The Claude synthetic ammonia process and plant. J Soc Chem Ind 40:R420–R424. https://doi.org/10.1002/jctb.5000402202

63. Travis AS (2018) Non-BASF ammonia technologies BT—nitrogen capture: the growth of an international industry (1900–1940). In: Travis AS (ed) Springer International Publishing, Cham, pp 225–264. https://doi.org/10.1007/978-3-319-68963-0_10

64. The Claude Process for Ammonia Synthesis, Nature. 107 (1921) 765. https://doi.org/10.1038/107765a0.

65. Rouwenhorst KHR, Van der Ham AGJ, Lefferts L (2021) Beyond Haber-Bosch: the renaissance of the Claude process. Int J Hydrogen Energy 46:21566–21579. https://doi.org/10.1016/j.ijhydene.2021.04.014

66. Casale L (n.d.) Catalytic apparatus for the synthesis of ammonia. US1478549A

67. Casale L (1922) Improvements in catalysts more particularly for the synthesis of ammonia, GB197199A

68. Casale L (1923) Improvements in catalysts for the synthesis of ammonia, GB218237A

69. Travis AS (2022) Luigi Casale's enterprise: pioneer of global catalytic high-pressure industrial chemistry. Catal Today 387:4–8. https://doi.org/10.1016/j.cattod.2021.11.003

70. Casale L (1923) Improvements in the production of catalysts for the synthesis of ammonia, GB227491A

71. Travis AS (2021) First steps: synthetic ammonia in the United States. Substantia 5:55–77. https://doi.org/10.36253/Substantia-1181

72. Casale L (1921) Process for generating from air and hydrogen either nitrogen or hydrogen or nitrogen, 1384428
73. Casale L (1922) Apparatus for the electrolysis of water, 1547362
74. Filippi E, Pizzolitto C (2022) The past and the future of catalysis and technology in industry: a perspective from Casale SA point of view. Catal Today 387:9–11. https://doi.org/10.1016/j.cattod.2021.11.005
75. Casale L (1921) An improved process for the synthetic production of ammonia
76. Fauser G (1921) Process and plant for the production of synthetic ammonia. IT102751X
77. Scaglione P (2009), Il proceso Fauser per la sinesi di NH3: La nascita di un'empresa, Chim. e Ind, pp 118–121
78. Rosa L (1968) Evolution of Fauser-Montecatini process for ammonia synthesis and for methanol synthesis with heat recovery. In: Fauser GBT-CF (ed) Pergamon, pp 65–72. https://doi.org/10.1016/B978-0-08-003605-2.50012-4
79. Fauser G (1924) Apparatus for the production of synthetic ammonia, IT240436X
80. Zecchina A, Califano S (2017) From the onset to the first large-scale industrial processes. Dev Catal, pp 1–57. https://doi.org/10.1002/9781119181286.ch1
81. The "American Process" for Nitrogen fixation. Chem Metall Eng (1924)
82. Travis AS (2018) The United States BT—Nitrogen Capture: The Growth of an International Industry (1900–1940). In: Travis AS (ed) Springer International Publishing, Cham, pp 265–279. https://doi.org/10.1007/978-3-319-68963-0_11
83. Ernst FA, Reed FC, Edwards WL (1925) A direct synthetic ammonia plant. Ind Eng Chem 17:775–788. https://doi.org/10.1021/ie50188a002
84. Kagiwada R, Hazuka H (1931) Kawasaki factory of Showa Hiryo K.K., Hitachi Hyoron. 14
85. Shreve SN (1956) The chemical process industries, 2nd edn. McGraw-Hill Book Company, New York
86. Aika K, Hori H, Ozaki A (1972) Activation of nitrogen by alkali metal promoted transition metal I. Ammonia synthesis over ruthenium promoted by alkali metal. J Catal 27:424–431. https://doi.org/10.1016/0021-9517(72)90179-0
87. Frattini D, Cinti G, Bidini G, Desideri U, Cioffi R, Jannelli E (2016) A system approach in energy evaluation of different renewable energies sources integration in ammonia production plants. Renew Energy 99:472–482. https://doi.org/10.1016/j.renene.2016.07.040
88. Huazhang L (2013) Ammonia synthesis catalysts: innovation and practice. World Scientific Publishing Co Pte Ltd. https://doi.org/10.1142/8199
89. Greenlee LF, Renner JN, Foster SL (2018) The use of controls for consistent and accurate measurements of electrocatalytic ammonia synthesis from dinitrogen. ACS Catal 8:7820–7827. https://doi.org/10.1021/acscatal.8b02120
90. Liu YH, Fernández CA, Varanasi SA, Bui NN, Song L, Hatzell MC (2022) Prospects for aerobic photocatalytic nitrogen fixation. ACS Energy Lett 7:24–29. https://doi.org/10.1021/acsenergylett.1c02260
91. Sánchez A, Martín M (2018) Scale up and scale down issues of renewable ammonia plants: towards modular design. Sustain Prod Consum 16:176–192. https://doi.org/10.1016/j.spc.2018.08.001
92. Rouwenhorst KHR, Krzywda PM, Benes NE, Mul G, Lefferts L (2020) Ammonia, 4. Green ammonia production. Ullmann's Encycl Ind Chem, pp 1–20. https://doi.org/10.1002/14356007.w02_w02
93. George M, Michael S (1998) Ammonia synthesis at atmospheric pressure. Science (80-.) 282:98–100. https://doi.org/10.1126/science.282.5386.98
94. Du H-L, Chatti M, Hodgetts RY, Cherepanov PV, Nguyen CK, Matuszek K, MacFarlane DR, Simonov AN (2022) Electroreduction of nitrogen with almost 100% current-to-ammonia efficiency. Nature 609:722–727. https://doi.org/10.1038/s41586-022-05108-y
95. Ziegenbalg D, Zander J, Marschall R (2021) Photocatalytic nitrogen reduction: challenging materials with reaction engineering. ChemPhotoChem. 5:792–807. https://doi.org/10.1002/cptc.202100084

96. Liu Y, Xue Y, Hui L, Yu H, Fang Y, He F, Li Y (2021) Porous graphdiyne loading CoO_x quantum dots for fixation nitrogen reaction. Nano Energy 89:106333. https://doi.org/10.1016/j.nanoen.2021.106333

97. Li P, Zhou Z, Wang Q, Guo M, Chen S, Low J, Long R, Liu W, Ding P, Wu Y, Xiong Y (2020) Visible-light-driven nitrogen fixation catalyzed by Bi5O7Br nanostructures: enhanced performance by oxygen vacancies. J Am Chem Soc 142:12430–12439. https://doi.org/10.1021/jacs.0c05097

98. Sun B, Liang Z, Qian Y, Xu X, Han Y, Tian J (2020) Sulfur vacancy-rich o-doped 1t-mos2 nanosheets for exceptional photocatalytic nitrogen fixation over cds. ACS Appl Mater Interfaces 12:7257–7269. https://doi.org/10.1021/acsami.9b20767

99. Rong X, Mao Y, Xu J, Zhang X, Zhang L, Zhou X, Qiu F, Wu Z (2018) Bi2Te3 sheet contributing to the formation of flower-like BiOCl composite and its N_2 photofixation ability enhancement. Catal Commun 116:16–19. https://doi.org/10.1016/j.catcom.2018.07.018

100. Liu Y, Yu Z, Guo S, Yao L, Sun R, Huang X, Zhao W (2020) Photocatalytic nitrogen fixation on transition metal modified TiO_2 nanosheets under simulated sunlight. New J Chem 44:19924–19932. https://doi.org/10.1039/D0NJ04397D

101. Li X, Zhang X, Everitt HO, Liu J (2019) Light-induced thermal gradients in ruthenium catalysts significantly enhance ammonia production. Nano Lett 19:1706–1711. https://doi.org/10.1021/acs.nanolett.8b04706

102. Yin H, Chen Z, Peng Y, Xiong S, Li Y, Yamashita H, Li J (2022) Dual active centers bridged by oxygen vacancies of ruthenium single-atom hybrids supported on molybdenum oxide for photocatalytic ammonia synthesis. Angew Chemie Int Ed 61:e202114242. https://doi.org/10.1002/anie.202114242

103. Zeng H, Terazono S, Tanuma T (2015) A novel catalyst for ammonia synthesis at ambient temperature and pressure: visible light responsive photocatalyst using localized surface plasmon resonance. Catal Commun 59:40–44. https://doi.org/10.1016/j.catcom.2014.09.034

104. Bian S, Wen M, Wang J, Yang N, Chu PK, Yu XF (2020) Edge-rich black phosphorus for photocatalytic nitrogen fixation. J Phys Chem Lett 11:1052–1058. https://doi.org/10.1021/acs.jpclett.9b03507

105. Hochman G, Goldman AS, Felder FA, Mayer JM, Miller AJM, Holland PL, Goldman LA, Manocha P, Song Z, Aleti S (2020) Potential economic feasibility of direct electrochemical nitrogen reduction as a route to ammonia. ACS Sustain Chem Eng 8:8938–8948. https://doi.org/10.1021/acssuschemeng.0c01206

106. Schnitkey G (2018) Fertilizer prices higher for 2019 crop. Farmdoc Dly 8

107. Analytiq B (n.d.) Ammonia price index. https://businessanalytiq.com/procurementanalytics/index/ammonia-price-index

108. The WorldBank (n.d.) Commodity markets. Access 24 Sept 2023. https://www.worldbank.org/en/research/commodity-markets

109. Arora P, Hoadley AFA, Mahajani SM, Ganesh A (2016) Small-scale ammonia production from biomass: a techno-enviro-economic perspective. Ind Eng Chem Res 55:6422–6434. https://doi.org/10.1021/acs.iecr.5b04937

110. Yoshida M, Ogawa T, Imamura Y, Ishihara KN (2021) Economies of scale in ammonia synthesis loops embedded with iron- and ruthenium-based catalysts. Int J Hydrogen Energy 46:28840–28854. https://doi.org/10.1016/j.ijhydene.2020.12.081

111. Armijo J, Philibert C (2020) Flexible production of green hydrogen and ammonia from variable solar and wind energy: case study of Chile and Argentina. Int J Hydrogen Energy 45:1541–1558. https://doi.org/10.1016/j.ijhydene.2019.11.028

112. Lin B, Wiesner T, Malmali M (2020) Performance of a small-scale haber process: a techno-economic analysis. ACS Sustain Chem Eng 8:15517–15531. https://doi.org/10.1021/acssuschemeng.0c04313

113. Fernandez CA, Hatzell MC (2020) Editors' choice—economic considerations for low-temperature electrochemical ammonia production: achieving Haber-Bosch parity. J Electrochem Soc 167:143504. https://doi.org/10.1149/1945-7111/abc35b

114. Bicer Y, Dincer I (2019) Exergoeconomic analysis and optimization of a concentrated sunlight-driven integrated photoelectrochemical hydrogen and ammonia production system. Int J Hydrogen Energy 44:18875–18890. https://doi.org/10.1016/j.ijhydene.2018.10.074

115. Gomez JR, Baca J, Garzon F (20250) Techno-economic analysis and life cycle assessment for electrochemical ammonia production using proton conducting membrane, Int J Hydrogen Energy 45:721–737. https://doi.org/10.1016/j.ijhydene.2019.10.174

116. Hirakawa H, Hashimoto M, Shiraishi Y, Hirai T (2017) Photocatalytic conversion of nitrogen to ammonia with water on surface oxygen vacancies of titanium dioxide. J Am Chem Soc 139:10929–10936. https://doi.org/10.1021/jacs.7b06634

117. Comer BM, Fuentes P, Dimkpa CO, Liu YH, Fernandez CA, Arora P, Realff M, Singh U, Hatzell MC, Medford AJ (2019) Prospects and challenges for solar fertilizers. Joule. 3:1578–1605. https://doi.org/10.1016/j.joule.2019.05.001

118. Braun A, Bora DK, Lauterbach L, Lettau E, Wang H, Cramer SP, Yang F, Guo J (2022) From inert gas to fertilizer, fuel and fine chemicals: N_2 reduction and fixation. Catal Today 387:186–196. https://doi.org/10.1016/j.cattod.2021.04.020

119. Ithisuphalap K, Zhang H, Guo L, Yang Q, Yang H, Wu G (2019) Photocatalysis and photoelectrocatalysis methods of nitrogen reduction for sustainable ammonia synthesis. Small Methods 3:1–20. https://doi.org/10.1002/smtd.201800352

120. Erdemir D, Dincer I (2021) A perspective on the use of ammonia as a clean fuel: challenges and solutions. Int J Energy Res 45:4827–4834. https://doi.org/10.1002/er.6232

121. Khan MI, Yasmin T, Shakoor A (2015) Technical overview of compressed natural gas (CNG) as a transportation fuel. Renew Sustain Energy Rev 51:785–797. https://doi.org/10.1016/j.rser.2015.06.053

122. Koike M, Miyagawa H, Suzuoki T, Ogasawara K (2012) Ammonia as a hydrogen energy carrier and its application to internal combustion engines. In: Gaydon BT-SVT Warwickshire (eds). Woodhead Publishing, pp 61–70. https://doi.org/10.1533/9780857094575.2.61

123. Kenanoğlu R, Baltacioğlu E (2021) An experimental investigation on hydroxy (HHO) enriched ammonia as alternative fuel in gas turbine. Int J Hydrogen Energy 46:29638–29648. https://doi.org/10.1016/j.ijhydene.2020.11.189

124. Gezerman AO (2022) A critical assessment of green ammonia production and ammonia production technologies. Kem u Ind 71:57–66. https://doi.org/10.15255/kui.2021.013

125. Lucentini I, Garcia X, Vendrell X, Llorca J (2021) Review of the decomposition of ammonia to generate hydrogen. Ind Eng Chem Res 60:18560–18611. https://doi.org/10.1021/acs.iecr.1c00843

126. David WIF, Makepeace JW, Callear SK, Hunter HMA, Taylor JD, Wood TJ, Jones MO (2014) Hydrogen production from ammonia using sodium amide. J Am Chem Soc 136:13082–13085. https://doi.org/10.1021/ja5042836

127. Makepeace JW, Wood TJ, Hunter HMA, Jones MO, David WIF (2015) Ammonia decomposition catalysis using non-stoichiometric lithium imide. Chem Sci 6:3805–3815. https://doi.org/10.1039/c5sc00205b

128. Thomas G, Parks G (2006) Potential roles of ammonia in a hydrogen economy. U.S. Department of Energy

129. Lamb KE, Viano DM, Langley MJ, Hla SS, Dolan MD (2018) High-Purity H_2 produced from NH_3 via a ruthenium-based decomposition catalyst and vanadium-based membrane. Ind Eng Chem Res 57:7811–7816. https://doi.org/10.1021/acs.iecr.8b01476

130. Iki N, Kurata O, Inoue T, Takayuki M, TsujimuraT, Furutani H, Kawano M, Arai K, Kobayashi H, Hayakawa A, Okafor E (2019) Rich-lean combustor for a 50kW class micro gas turbine firing ammonia. In: Proceedings of global power propulsion societ.y ISSN-Nr 2504-4400

131. Mounaïm-Rousselle C, Bréquigny P, Medina AV, Boulet E, Emberson D, Løvås T (2022) Ammonia as fuel for transportation to mitigate zero carbon impact BT—engines and fuels for future transport. In: Kalghatgi G, Agarwal AK, Leach F, Senecal K (eds). Springer Singapore, Singapore, pp 257–279. https://doi.org/10.1007/978-981-16-8717-4_11

132. Ammonia Energy Association (n.d.) The AmVeh—an ammonia fueled car from South Korea. https://nh3fuelassociation.org/2013/06/20/the-amveh-an-ammonia-fueled-car-from-south-korea/

133. ars technical (n.d.) World's first ammonia-powered zero-emissions tractor starts testing. https://arstechnica.com/cars/2022/06/worlds-first-ammonia-powered-zero-emissions-tractor-starts-testing/
134. Green Car Congress (n.d.) Aviation H_2 selects liquid ammonia as carbon-free fuel of choice. https://www.greencarcongress.com/2022/04/20220428-aviationh2.html
135. Eidesvik (n.d.) Viking energy with ammonia-driven fuel cell. https://eidesvik.no/viking-energy-with-ammonia-driven-fuel-cell/
136. MAN Energy Solutions (n.d.) Unlocking ammonia's potential for shipping. https://www.man-es.com/discover/two-stroke-ammonia-engine
137. Ammonia Energy Association (n.d.) IHI corporation's recent technology demonstrations. https://www.ammoniaenergy.org/articles/ihi-corporation-pushes-its-ammonia-combustion-technologies-closer-to-commercialization/
138. Ammonia Energy Association (n.d.) The evolving context of ammonia-coal co-firing. https://www.ammoniaenergy.org/articles/the-evolving-context-of-ammonia-coal-co-firing/
139. AIST (n.d.) Gas turbine power generation with a methane-ammonia gas mixture and 100% ammonia. https://www.aist.go.jp/aist_e/list/latest_research/2016/20160412/en20160412.html
140. Ye L, Nayak-Luke R, Bañares-Alcántara R, Tsang E (2017) Reaction: "green" ammonia production. Chem 3:712–714. https://doi.org/10.1016/j.chempr.2017.10.016
141. Wilkinson I (2018) Siemens, Green ammonia—a significant contributor to a low- carbon energy system?
142. SIEMENS Energy press release—Siemens Energy to start production of hydrogen electrolyzers in Berlin (n.d.) https://press.siemens-energy.com/global/en/pressrelease/siemens-energy-start-production-hydrogen-electrolyzers-berlin
143. How much energy does the world consume? (2021)
144. Our World in Data (2021) Electricity production by source. World
145. Enerdata (2022) World energy & climate statistics-yearbook
146. Khalili S, Rantanen E, Bogdanov D, Breyer C (2019) Global transportation demand development with impacts on the energy demand and greenhouse gas emissions in a climate-constrained world. Energies 12. https://doi.org/10.3390/en12203870
147. IEA (2021) Ammonia technology roadmap towards more sustainable nitrogen fertiliser production
148. Barrels J (20008) A feasibility study of implementing an ammonia economy. Iowa State University. https://doi.org/10.31274/etd-180810-137.
149. McEnaney JM, Singh AR, Schwalbe JA, Kibsgaard J, Lin JC, Cargnello M, Jaramillo TF, Nørskov JK (2017) Ammonia synthesis from N_2 and H_2O using a lithium cycling electrification strategy at atmospheric pressure. Energy Environ Sci 10:1621–1630. https://doi.org/10.1039/c7ee01126a
150. Takaya CA, Parmar KR, Fletcher LA, Ross AB (2019) Biomass-derived carbonaceous adsorbents for trapping ammonia. Agric 9. https://doi.org/10.3390/agriculture9010016
151. Taghizadeh-Toosi A, Clough TJ, Sherlock RR, Condron LM (2012) Biochar adsorbed ammonia is bioavailable. Plant Soil 350:57–69. https://doi.org/10.1007/s11104-011-0870-3
152. Allohverdi T, Mohanty AK, Roy P, Misra M (2021) A review on current status of biochar uses in agriculture. Molecules 26. https://doi.org/10.3390/molecules26185584
153. Smith C, Hill AK, Torrente-Murciano L (2020) Current and future role of Haber-Bosch ammonia in a carbon-free energy landscape. Energy Environ Sci 13:331–344. https://doi.org/10.1039/c9ee02873k
154. Morlanés N, Katikaneni SP, Paglieri SN, Harale A, Solami B, Sarathy SM, Gascon J (2021) A technological roadmap to the ammonia energy economy: Current state and missing technologies. Chem Eng J 408:127310. https://doi.org/10.1016/j.cej.2020.127310

Alcohol to Hydrocarbons

Steffen Voß, Stefan Bube, Nils Bullerdiek, Ulf Neuling, and Martin Kaltschmitt

Abstract Alcohols such as methanol and ethanol are crucial for various applications, particularly in the chemical industry and as fuel components. Ethanol, which today is mainly produced from biomass, already contributes significantly to reducing greenhouse gas emissions (GHG) in the road-based transportation sector. However, the direct and pure use of renewable alcohols in various transportation sectors is associated with problems due to their hygroscopic properties and lower energy density. While blending with hydrocarbons is possible in road transport, it is not suitable for aviation. The conversion of these alcohols into long-chain hydrocarbons such as gasoline, kerosene and diesel offers a viable solution that fits seamlessly into the existing fuel infrastructure. This paper analyses the process routes from alcohol to hydrocarbon components and demonstrates their potential to defossilize the mobility sector. The alcohol-hydrocarbon route is not yet being implemented on a commercial scale. One of the main obstacles is the relatively higher cost of such alcohols compared to the price of fossil fuels and the possibility of using such alcohols directly for road-based transportation without further processing.

Keywords Alcohol-to-Jet · Hydrocarbon production · Dehydration · Oligomerization · Fuel production

1 Introduction

Alcohols, in this paper especially methanol and ethanol, play already today a significant role within the chemical industry as platform chemicals or for road transportation as fuel or fuel components. Particularly, ethanol is produced in large volumes worldwide (ca. 100 Mt/a) from plant material through biotechnological processes and already contributes significantly to the defossilization of the transportation sector. In light of the increasing challenges to defossilize the transportation and various further

S. Voß (✉) · S. Bube · N. Bullerdiek · U. Neuling · M. Kaltschmitt
Hamburg University of Technology (TUHH), Institute of Environmental Technology and Energy Economics (IUE), Hamburg, Germany
e-mail: steffen.voss@tuhh.de

© The Author(s), under exclusive license to Springer Nature Switzerland AG 2025
N. Bullerdiek et al. (eds.), *Powerfuels*, Green Energy and Technology,
https://doi.org/10.1007/978-3-031-62411-7_26

industrial sectors, and to further reduce greenhouse gas (GHG) emissions from fossil fuel energy in these areas, the sustainable provision of alcohols can play a crucial role.

As of today, alcohols are produced based on feedstocks of fossil origin (natural gas and coal in the case of methanol) or biogenic origin (sugar cane, corn, and wheat in the case of ethanol). However, alcohols can also be synthesized via different conversion routes based on power-to-liquid (PtL) pathways. These chemical processes enable the production of various alcohols, such as methanol, ethanol, and butanol, which can then serve as platform molecules for the downstream production of a wide variety of products and applications, such as for transportation (e.g., jet fuels) or industrial purposes (e.g., plastics). For specific transportation applications, such as aviation, the direct use of pure alcohols as a fuel or increasing their share as an admix component in fossil fuels presents challenges. One main issue with alcohols is their low energy density compared to pure hydrocarbons, as they contain at least one oxygen atom in the overall molecule that offers no energetic benefit. A lower heating value means that more fuel volume is needed to achieve a specific transportation performance. This becomes particularly problematic for transportation purposes, where both volume and mass are critical limiting factors, such as in air transportation. Additionally, the vaporization enthalpy and the high miscibility with water cause major changes in the previous fuel uses compared to the previously used fuel. However, instead of using alcohols directly, they can also be converted to longer-chain hydrocarbons, such as kerosene or diesel, via well-established process routes. These resulting hydrocarbons serve so-called drop-in fuels, meaning they can directly substitute currently used fossil fuels without requiring technological modifications to the existing fuel infrastructure, fuel storage systems, or propulsion systems of vehicle, vessel, or aircraft fleets [1].

Against this background, the primary objective of this paper is to outline the process pathway to convert different alcohols into pure longer-chain hydrocarbons that meet the given standards and specifications for chosen transportation fuels by removing the oxygen and adjusting the carbon chain length accordingly. The focus is put on the processing of alcohols and processes already implemented at a larger scale. In addition, the direct use of alcohols in various fields of application is briefly compared to the use of products from alcohol to hydrocarbon processes (e.g., jet fuel). This results in a full technical overview of the alcohol to hydrocarbon process routes for the production of fuel components. Therefore, the various alcohols and their existing and potential future production processes are first described, followed by a discussion of their use and market situation. Afterward, the alcohol to hydrocarbon process is explained in general and subsequently presented with regard to the individual conversion steps (i.e., dehydration, oligomerization, hydrogenation, and fractionation). Finally, the results are summarized, and key factors for alcohol processing are discussed.

2 Alcohols

Alcohols belong to the group of organic compounds in which a hydroxyl group ($-OH$) is attached to a hydrocarbon chain. This hydrocarbon chain might vary in its length and—depending on the overall amount of carbon atoms within the molecule—might be branched [2]. Alcohols that have exactly one hydroxyl group per molecule are called monohydric alcohols. The hydroxyl group can be located at the head of the chain, in the middle of the chain, or at a branch. The general chemical formulation of these simple alcohols is $C_nH_{2n+1}OH$. Alcohols can first be identified by their different aliphatic chain lengths, with methanol being the simplest alcohol with an attached methyl group. The longer aliphatic chain also leads to various possible isomers. Table 1 lists alcohols with a carbon chain length from C_1 to C_5, their main areas of application today, as well as characteristic physical properties.

Currently, methanol and ethanol in particular are already produced on a large industrial scale. The higher alcohols—in this case propanol, butanol, and pentanol—are used in particular in the chemical industry as solvents. Considering their current production volumes, they are of minor market importance compared to methanol and ethanol. However, with regard to their chemical properties, such as their lower heating value or enthalpy of vaporization, they could also serve as fuel components. In the following, the two most important industrial alcohols, methanol and ethanol, are described in more detail, along with butanol as the currently most prominent example of various production routes for higher alcohols to hydrocarbons.

Table 1 Description of the considered alcohols

Alcohol	Formula	Abbreviation	Application	Boiling Point (°C)	Freezing Point (°C)	LHV (MJ/kg)
Methanol	CH_3OH	MeOH	Basic chemical, fuel component	65.0	-96.0	19.9
Ethanol	C_2H_5OH	EtOH	Fuel component, food, and beverage	78.0	-114.0	27.2
Propanol	C_3H_7OH	PropOH	Disinfects, solvent	97.2	-126.1	30.6
Butanol	C_4H_9OH	ButOH	Basic chemical, solvent	117.7	-89.3	33.8
Pentanol	$C_5H_{11}OH$	PentOH	Solvent, extracting agent	137.8	-78.5	35.1

LHV: lower heating value; [3–8]

2.1 Methanol

Methanol is the simplest alcohol, with one carbon atom per molecule. Furthermore, methanol is the most basic organic molecule liquid under standard conditions. Methanol production is a synthesis-based process that enables the utilization of various feedstocks.

Production. Today, methanol is mainly produced in large-scale chemical plants. These have capacities of up to 10,000 t/d, which adds up to a global annual production capacity of about 180 Mt/a. Thereby, the largest annual production volume of methanol is concentrated in China [9]. The feedstock currently used in these methanol production plants is mainly natural gas and hard coal, with natural gas accounting for the majority of around 65%. In China, almost exclusively hard coal from domestic deposits is used as feedstock. The price range for fossil fuel based methanol is commonly observed to be between 200 and 400 €/t$_{MeOH}$; however, the prices can significantly differ due to market related dynamics in feedstock prices [10, 11]. In the currently operating production processes, hard coal is gasified or natural gas is reformed to produce synthesis gas (syngas), a mixture mainly composed of hydrogen (H_2), carbon monoxide (CO), and carbon dioxide (CO_2). Natural gas reforming, and especially hard coal gasification, are yielding a product gas/syngas with a hydrogen deficit related to the final desired product; i.e., a lower ratio of H_2 to CO and CO_2 than required for the methanol synthesis. Consequently, additional conditioning steps are necessary. One commonly employed method for further hydrogen production is the integration of the water–gas shift (WGS) reaction within the overall process scheme, with subsequent separation of CO_2 to adjust the syngas composition accordingly. Within the methanol synthesis, mainly CO, but also CO_2, are converted to methanol in an exothermic reaction, as shown in (Eq. 1) and (Eq. 2). Besides fossil feedstock, the syngas can also be of biogenic or non-biogenic renewable origin. The corresponding process routes are shown in Fig. 1.

The biomass-based process pathway can be based on the direct utilization of biogenic raw materials—e.g., agricultural and forestry residues, municipal solid waste, as well as biogenic intermediate products, mainly biogas. In addition to these pathways, a purely electricity-based process (PtL-pathway) for methanol production is possible. Here, hydrogen is obtained via water-electrolysis, and CO_2 is provided via direct air capture (DAC) or sustainable CO_2 point sources (e.g., bioethanol or biogas production). In order to conduct the conventional methanol synthesis based on an overall PtL-pathway (Eq. 1), the CO-rich syngas can be produced from H_2 and CO_2 via the reverse water–gas shift (RWGS) reaction. However, methanol synthesis also enables the direct conversion of CO_2 and H_2 (Eq. 2) by using adapted catalysts and an adjusted process design [12].

$$CO + 2\,H_2 \rightleftharpoons CH_3OH \;\; \Delta H_R = -91^{kJ}\!/_{mol} \qquad (1)$$

$$CO_2 + 3\,H_2 \rightleftharpoons CH_3OH + H_2O \;\; \Delta H_R = -50^{kJ}\!/_{mol} \qquad (2)$$

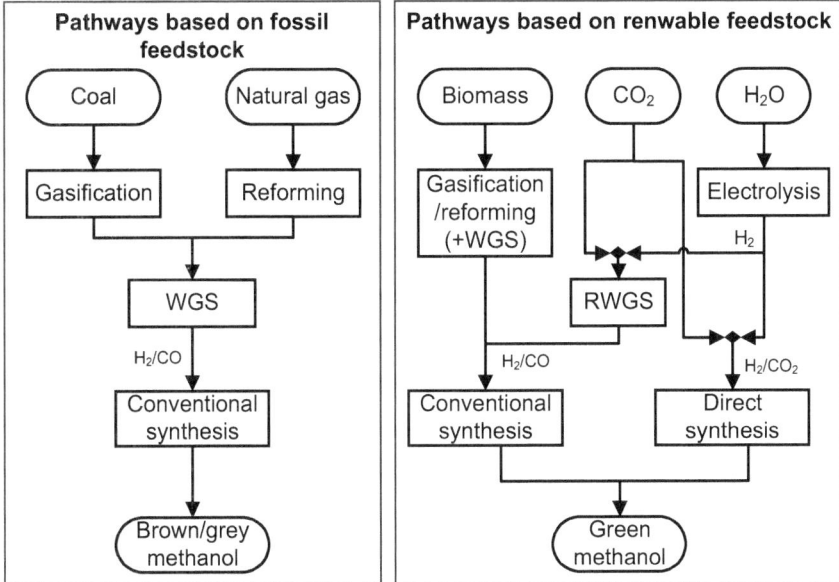

Fig. 1 Methanol production pathways (MeOH: Methanol; RWGS: Reverse water–gas shift; WGS: Water–gas shift)

Utilization. Methanol has a wide range of applications and is suited as a C_1 platform chemical or as a direct energy carrier currently and likely in future energy systems. It is already used as a substitute for crude oil-based products in China. Global methanol production has roughly doubled in recent decades, while some forecasts expect a further increase of up to 500 Mt/a up to the year 2050 [10, 11]. Today, methanol is mainly used as a primary chemical for the production of formaldehyde, formic acid, acetic acid, or plastics [13]. In addition, it can also be used directly as a (pure) fuel or indirectly as a fuel admix component. According to current EU fuel regulations, methanol can be mixed into gasoline fuel with up to 3% by volume, serving as a fuel admix component [14]; (DIN EN 228). Using pure methanol as a fuel in adjusted combustion engines is also possible—but so far only applied in China [15]. Methanol can also be converted to fuel additives like methyl tert-butyl ether (MTBE), increasing the knock resistance of the gasoline fuel mixture. Another possible future application of methanol could be its use as a hydrogen carrier. In this case, hydrogen is bound to carbon in the form of methanol, simplifying its transportation and overall logistical handling significantly compared to pure hydrogen logistics.

2.2 *Ethanol*

Ethanol is the second simplest alcohol after methanol, with two carbon atoms per molecule. From both the production and usage sides, ethanol seems promising as a platform chemical and fuel component.

Production. Today, ethanol is already being produced in large-scale production plants. Worldwide, about 100 Mt/a of ethanol are produced. Large portions of these production volumes are traded globally, with ethanol established as a global commodity. Consequently, a comprehensive ethanol transportation and trade infrastructure is already in place [16]. Ethanol is mainly produced by fermentation of sugars and starch-containing biomass. In the respective production process, sugars are directly fermented into ethanol using established biotechnological processes, while starch from feedstocks such as corn, wheat, or other grains is first saccharified before being fermented into ethanol [17]. In addition to the current bioethanol production from cultivated biomass, there is growing interest in converting/fermenting lignocellulosic biomass into ethanol to prevent competition in biomass utilization with food and fodder applications [18, 19]. Due to this food / fodder vs. fuel debate, there is also growing interest in ethanol production from synthesis gas. Similar to methanol production, this synthesis gas can be provided from a wide variety of biogenic and renewable non-biogenic sources. This diversifies the raw material supply situation for ethanol production and allows the use of feedstocks that do not conflict with food or fodder purposes. Ethanol production from synthesis gas can be achieved through either a thermochemical, catalytic process (mixed alcohol synthesis) or via a biochemical, fermentative process (syngas fermentation).

Mixed Alcohol Synthesis. The mixed alcohol synthesis is a thermochemical, catalytically controlled synthesis process producing ethanol, and other alcohols such as methanol, propanol, and butanol. The respective synthesis requires a CO-rich synthesis gas with an H_2 to CO ratio of 2. During the synthesis, methanol is formed first, which can grow into a longer carbon chain by adding further CO molecules. Based on the currently available catalysts, relatively low conversions of 5 to 30% and low selectivities for longer-chain alcohols of 30 to 90% are achieved [2]. A simultaneous high conversion and selectivity, and a high catalyst lifetime, which is currently being shortened by metal sintering and phase separation, continues to be a research focus [20].

Syngas Fermentation. Ethanol can also be obtained directly and as the main product in syngas fermentation and thus by a biochemical process. In this conversion process, CO_2 needs to be reduced, e.g., with H_2 serving as the reducing agent. The resulting synthesis gas is then fermented via the reductive acetyl-coenzyme A (Acetyl-CoA) pathway, also called the Wood–Ljungdahl pathway [2]. This metabolic pathway produces ethanol and acetic acid as the main products described by various overall reaction equations, shown in (Eq. 3) to (Eq. 6) [21].

$$2CO_2 + 4\,H_2 \rightarrow \ CH_3COOH + 2\,H_2O \ \Delta H_R = -75\,{}^{kJ}\!/_{mol} \qquad (3)$$

$$2CO_2 + 6\,H_2 \rightarrow C_2H_5OH + 3\,H_2O \; \Delta H_R = -97\,{}^{kJ}\!/_{mol} \qquad (4)$$

$$4CO + 2\,H_2O \rightarrow CH_3COOH + 2\,CO_2 \; \Delta H_R = -155\,{}^{kJ}\!/_{mol} \qquad (5)$$

$$6\,CO + 3\,H_2O \rightarrow C_2H_5OH + 4\,CO_2 \; \Delta H_R = -218\,{}^{kJ}\!/_{mol} \qquad (6)$$

The biochemical process pathway is shown in Fig. 2. The western, or so-called methyl branch, is shown on the left side of the figure, while the eastern, carbonyl branch is displayed on the right.

In the methyl branch of the metabolic pathway, carbon monoxide dehydrogenase converts CO to CO_2 and protons in the biological water–gas shift (WGS) reaction before the carbon dioxide is converted via formic acid to methyl corrinoid iron-sulfur protein. CO_2 can also be the direct substrate. In the carbonyl branch, CO is either directly processed further as a product, or CO is first obtained from CO_2. Together with the methyl corrinoid iron-sulfur protein from the western branch, the CO is converted in combination with coenzyme A to acetyl-CoA. The intermediate acetyl-CoA can be a starting product for various metabolic pathways, so different products can be formed from it. The pathway to acetate and ethanol is shown as an example in Fig. 2. During this process, the acetate can be further reduced to ethanol, forming the main product. Besides ethanol and acetic acid, butanol or butyric acid can also be formed as by-products [2].

In addition to the $CO:CO_2:H_2$ ratio in the feed, temperature and pH value during fermentation are important parameters for influencing the biocatalytic reaction. Due to the relatively low operating temperature of the fermenter, it is usually necessary to cool and clean the produced synthesis gas before it is fed into the synthesis gas fermenter. A temperature range of 30–40 °C is required for mesophilic microorganisms and a temperature range of 55–80 °C for thermophilic microorganisms [21]. In order to provide the optimal environment for the respective microorganisms, the aqueous fermentation solution must maintain a narrow pH range in addition and contain various trace elements [22]. The overall plant and process concept must ensure that the mass transfer from the gas phase to the liquid phase can be maximized as far as possible, as this is the only way to make the substrates available to the microorganisms for the metabolic pathway. In order to support gas–liquid mass transfer as far as possible, various reactor concepts are being discussed. These can include bubble column reactors, in which the synthesis gas rises through the fermentation broth in bubbles, or membrane-based reactors [22]. Furthermore, the product concentration in the fermenter solution after the reaction is low, leading to a considerable effort required for the separation of the pure product, especially with regard to the overall quantity. A recirculation of the fermentation solution as well as the microorganisms is necessary [2].

Utilization. As of today, ethanol is mainly used as a biomass-based blending component to gasoline fuel blends. Within the EU, blends of 5 and 10% by volume with

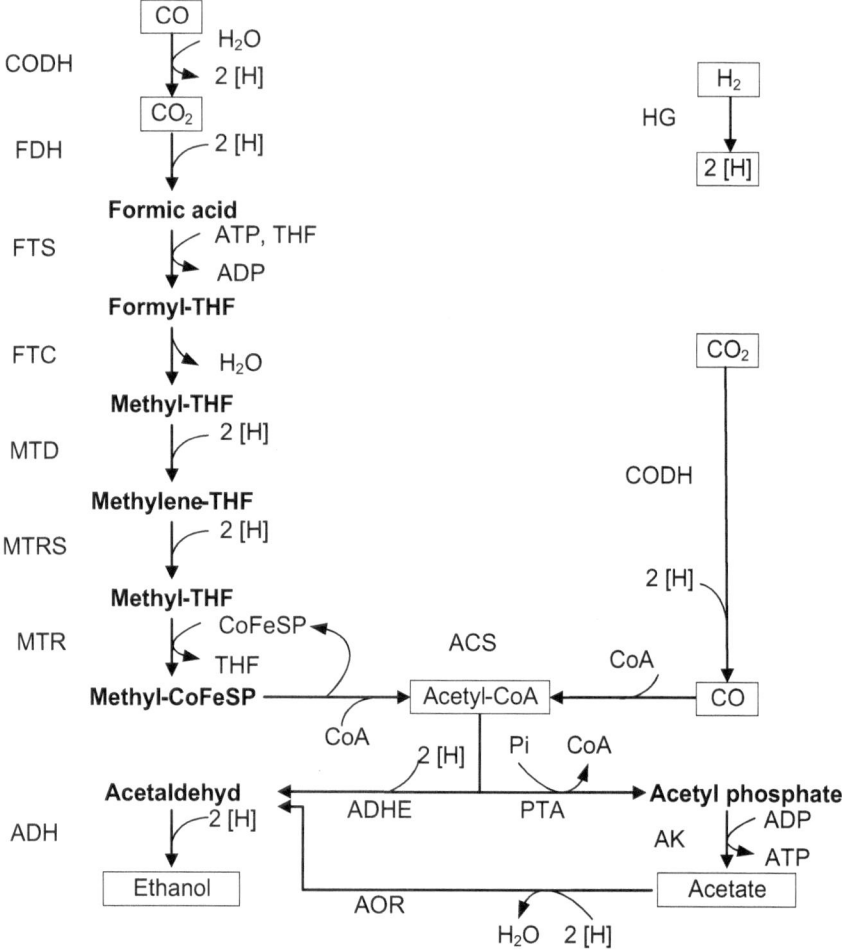

Fig. 2 Wood–Ljungdahl pathway (ACS: Acetyl-coenzyme A synthase; ADH: Alcohol dehydroge-
nase; ADHE: Aldehyde dehydrogenase; ADP: Adenosine diphosphate; AK: Acetate kinase; AOR:
Aldehyde-ferredoxin oxidoreductase; ATP: Adenosine triphosphate; COA: Coenzyme A; CODH:
Carbon monoxide dehydrogenase; CoFeSP: Corrinoid iron-sulfur protein; FDH: Formate dehy-
drogenase; FTC: Formyl-THF cyclohydrolase; FTS: Formyl-THF synthetase; HG: Hydrogenase;
MTD: Methylene-THF dehydrolase; MTR: Methyltransferase; MTRS: Methylene-THF reductase;
Pi: Inorganic phosphorus; PTA: Phosphotrans acetylase; THF: Tetrahydrofolate) [2]

gasoline are particularly common. In countries with high domestic ethanol production, such as Brazil, ethanol is also used as a pure fuel component, especially in
so-called flexible fuel vehicles (FFVs), which operate on both gasoline and any
blend of gasoline and ethanol. Furthermore, ethanol is also used as a solvent in the
chemical industry. In addition to those applications, ethanol is also used directly in
the food and beverage industry as a component of beverage alcohol [2].

2.3   Butanol

Butanol, as an alcohol with a hydrocarbon chain of four carbon atoms, has a total of four isomers, each of which is produced and used for different applications. In the context of implementing power-to-liquid (PtL) fuels, the production and use of n-butanol will be discussed below.

Production. Currently, the most industrially relevant isomer of butanol is n-butanol with an annual production of about 4.8 Mt/a [6]. N-butanol is mainly provided by hydroformylation of propene (Eq. 7). In such a chemical process, propene is converted together with a CO and H_2 containing synthesis gas with the help of a chemical catalysis to n-butanol or 2-methyl-1-propanol. A selectivity of about 95% with respect to n-butanol can be achieved. This process route relies on the use of a synthesis gas typically provided from natural gas or from coal combustion (but also a provision of sustainable synthesis gas based on a variety of process pathways and feedstock options is possible), and on the use of propene, which today comes to a large extent from naphtha steam cracking processes and thus from crude oil of fossil origin [6].

$$C_3H_6 + CO + H_2 \rightleftharpoons C_4H_7OH \tag{7}$$

Butanol can also be produced sustainably via biogenic and renewable non-biogenic process pathways in addition to this fossil-fuel-based process route. Historically, butanol was produced on an industrial scale at the beginning of the twentieth century via the fermentation of sugar and starch-containing biomasses using *Clostridium acetobutylicum* [23]. This bacterium was the basis for the so-called acetone-butanol-ethanol (ABE) fermentation. Due to the inherent toxicity of butanol to the microorganisms used for this fermentation, the maximum achievable product concentrations are typically limited to around 20 g/L [2]. Due to this low yield, the resulting high energy consumption for the product separation, and the incomplete utilization of all reaction products, this production pathway is no longer of industrial importance or used on a large scale [1].

As with ethanol, n-butanol can also be obtained directly from synthesis gas. On the one hand, mixed alcohol synthesis can be realized by forming a product mixture of different alcohols. However, the butanol selectivity is usually in the single-digit percentage range—and thus rather low [2, 24]. On the other hand, butanol can be produced from synthesis gas via a biotechnological route. Here, the reductive acetyl-CoA pathway is also used for the assimilation of carbon and to produce acetyl-CoA. Microorganisms such as *Clostridium carboxidivorans* can convert the produced acetyl-CoA via butyryl-CoA to butyrates and n-butanol. Subsequently, the provided butyrate can be further converted to butanol (Fig. 3). As with ABE fermentation, the product concentration in the fermentation broth is currently still very low at approximately 2 g/L. As a result, a high energy input is also required in this process concept in order to separate the product from the fermentation broth. Just like in the conversion of acetyl-CoA to ethanol, temperature, pH value, and substrate composition are important parameters for the adjustment of the process [25]. In butanol production, the gas–liquid mass transfer is the rate-limiting process step, which is why the mass transfer and subsequent reaction in various reactor concepts, such as bubble column or membrane reactors, are still being investigated [25].

Utilization. Today, n-butanol is mainly used as a solvent for coatings. Therefore, butanol can either be the solvent directly, or it is first processed into a derivative such as butyl acetate. As a butyl acrylate derivative, it is also used in a variety of applications as a plasticizer or process component in plastics production. The other three isomers can also be used as solvents. The 2-methyl-2-propanol as a butanol isomer can be added as an anti-knock agent and used as an additive to gasoline. However, this has been replaced by ethanol as an additive [6].

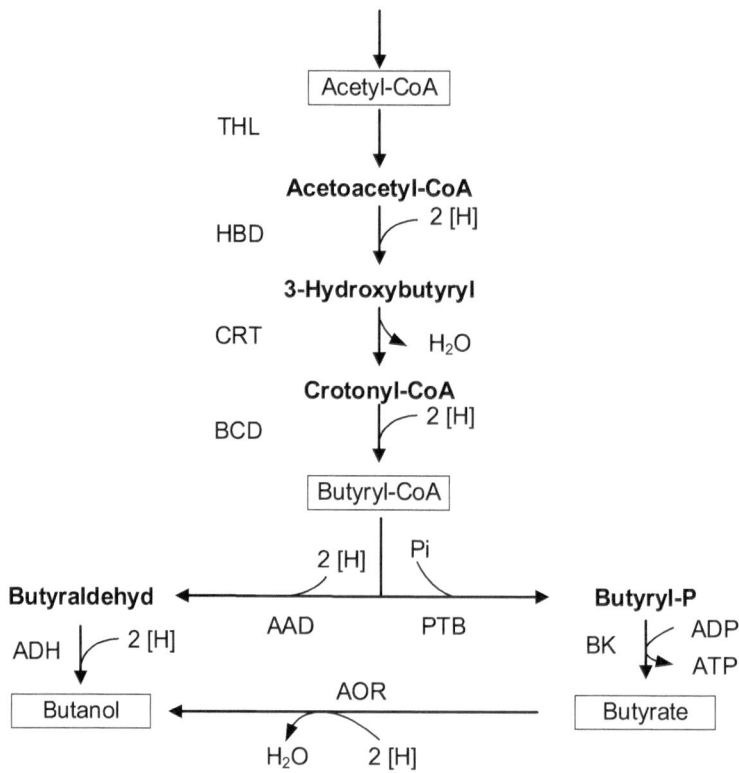

Fig. 3 Butanol production starting from acetyl-CoA (AAD: Alcohol/Aldehyde dehydrogenase; ADH: Alcohol dehydrogenase; ADP: Adenosine diphosphate; AOR: Aldehyde-ferredoxin oxidoreductase; ATP: Adenosine triphosphate; BCD: Butyryl-CoA dehydrogenase; CRT: Crotonase; HBD: 3-hydroxybutyryl-CoA dehydrogenase; Pi: Inoganic phosphor; PTB: Phosphotransbutyrylase; THL: Thiolase) [26]

3   Need for Alcohol to Hydrocarbon Processes

Besides the conversion of alcohols to hydrocarbon-based energy carriers, the direct use of alcohol as an energy carrier can be considered as well. This is because alcohols, as liquids, show similar properties to today's fossil hydrocarbon-based fuels, although their lower heating values may have disadvantages in specific applications. Therefore, this chapter briefly describes the extent to which alcohols can be used in mixed and pure form in today's gasoline-, kerosene- and diesel-based combustion processes and addresses the respective effects on the operation and usability of the application. Additionally, in all cases, the use of pure alcohol as a fuel would result in a change in the underlying infrastructure and handling of the respective liquids, which is not discussed here in detail.

3.1   Gasoline

Methanol can and already is partly used as a pure fuel in spark ignition engines, primarily for road transportation. In addition, mixtures of methanol and gasoline are typically used with a methanol content of 3, 15, 50, or 85% by volume; the latter is mainly realized in China for road transportation gasoline fuels. Below, some utilization aspects are addressed for the various alcohol options explained above.

- Methanol
  - An increase in the methanol content leads to an increased octane number of the fuel mixture, thus improving knock resistance.
  - Other materials and seals must be used for methanol contents greater than 3% by volume within the fuel mixture compared to the use of "classical" gasoline, as such a fuel mixture has a more corrosive effect on the materials currently used [27].
  - Using methanol and other alcohols in high mixing ratios, or as a pure fuel component, cold starting might be challenging. More evaporation energy must be provided due to the lower energy density of the alcohols. Thus, it might be necessary to modify the engine block or the injector by heating them directly electrically on the construction side [28].

- Ethanol
  - Both the octane number improvement and the higher corrosive characteristic on tank and engine materials are also properties that occur when ethanol is used in a mixture with gasoline components.
  - Ethanol can be added to the petrol mixture in concentrations of about 10 vol% which is commonly realized, e.g., within the EU [29].

- Butanol

- The use of butanol as a pure fuel component in gasoline engines is more beneficial compared to methanol and ethanol, due to higher calorific values.
- Butanol also shows reduced problems with cold starts, since the heat of vaporization of butanol is lower than that of the other alcohols.
- The disadvantage of the direct use of butanol in gasoline engines is the lower octane number compared to ethanol and methanol, which means that only smaller compression ratios and lower efficiencies can be achieved in the combustion engine [30].

3.2 Kerosene

For a direct use of alcohols in commercially operated aircrafts, the following aspects need to be considered.

- The entire aircraft fuel system (tank system, components in contact with fuel) would have to be designed to be adequately resistant to corrosion due to the corrosiveness of the alcohol.
- From a technical perspective, the combustion of the alcohols within existing jet turbine engine technology is possible, but would still need to be adapted for absolute operational functionality and optimal performance [31].
- Compared to kerosene, the alcohols considered have lower gravimetric and volumetric energy densities. Methanol has only about half the energy density of kerosene; the energy densities increase with increasingly higher alcohols due to a decreasing ratio of oxygen to hydrogen in the molecule. If methanol, ethanol, or butanol were used as pure fuel for aviation, the tank volumes of the aircraft would have to be increased due to the lower energy densities, as otherwise, the same ranges could not be achieved. If aircraft tank volumes remain unchanged, the combination of maximum achievable range and payload that can be carried would be reduced [31]. The reduction in range combined with the potential payload of the aircraft at this range through the use of different alcohols compared to kerosene is shown in Fig. 4 as an example of a short-haul aircraft [8]. According to Fig. 4, the lower energy densities of alcohols compared to kerosene would significantly impact on the respective payload-range characteristics. Using methanol as an example, a range of just over 1,000 km can be achieved with a full payload and an unchanged aircraft design, whereas a range of around 4,000 km can be achieved with the same payload of kerosene. The longer the carbon chain of a specific alcohol, the lower the range limitation, as the gravimetric and volumetric energy densities increase. Nevertheless, the range limitations for higher alcohols are still significant compared to the use of kerosene. The decrease in permissible payload to achieve an increased range is also significant [32]. This is due to the lower gravimetric energy density and the associated increase in the amount of fuel to be carried for a certain range compared to kerosene. These effects have an even greater influence on the range of long-haul aircraft [8].

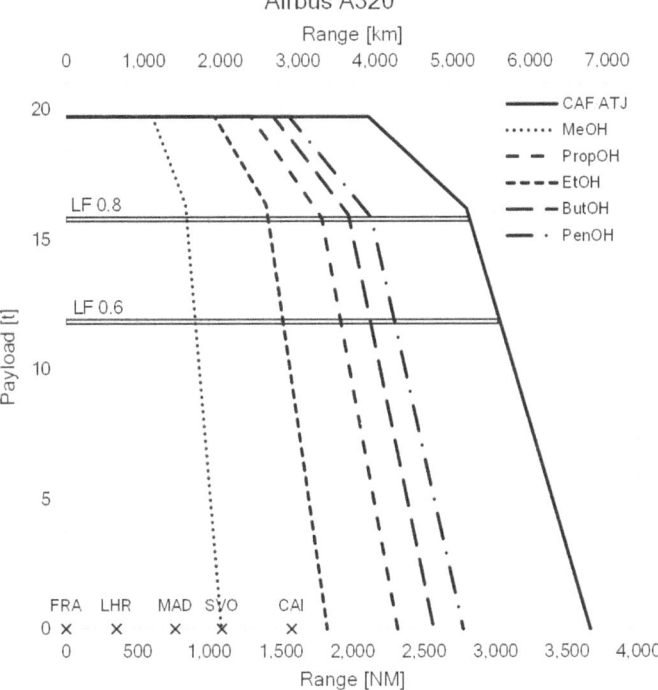

Fig. 4 Payload-range diagram for alcohols and short-/medium-range aircraft (ATJ: Alcohol-to-jet; ButOH: butanol; CAF: Conventional aviation fuel; CAI; Cairo; EtOH: Ethanol; FRA: Frankfurt; LF: Load-factor; LHR: London; MAD: Madrid; MeOH: Methanol; NM: Nautical miles (1.852 km); PentOH: Pentanol; PropOH: Propanol; SVO: Moscow) [8]

3.3 Diesel

Alcohols can also be used in diesel or compression ignition combustion engines of vehicles. Due to the specific characteristics of the diesel engine, the following constraints apply to the different alcohols.

- Methanol. The self-ignition of the fuel mixture within a diesel engine is achieved by a high compression of a fuel–air mixture. Thus, the cetane number describing the ignitability of a fuel mixture that is being compressed plays a decisive role. However, for methanol, the cetane number is notably low at 3, making its direct use in a diesel engine extremely challenging [33]. In order to allow for the use of methanol in diesel engines, the mixture of methanol with a high-cetane fuel component (e.g., diesel), which is fed into the combustion chamber together or separately, is being discussed [28]. The use of methanol in diesel engines would thus result in a major engine modification.
- Ethanol. The use of ethanol in diesel engines has the same disadvantages as the use of methanol. Ethanol has only a slightly better cetane number of 8, wich is

insufficient for pure use within this engine type. For this reason, high-cetane fuel components must also be added to ensure proper ignition [28, 30]. As a result, the so-called ED95 fuel has been developed, which consists of 95% ethanol and 5% ignition improver. This fuel mixture has so far been used only in a few test vehicles and is not yet commercially mature [28].

- Butanol. In contrast to diesel fuel, butanol also has a lower cetane number of around 25, which is much higher compared to the shorter-chain alcohols methanol and ethanol. Due to the favorable miscibility between diesel and butanol, a broad spectrum of mixing ratios is conceivable. The emissions of nitrogen oxides (NO_x) and carbon monoxide (CO) decrease with increasing butanol shares compared to the pure use of diesel fuel. However, with increasing butanol shares, a cetane improver is also required so that the mixture of butanol and diesel ignites properly [30].

4 Alcohol to Hydrocarbon

To obtain a hydrocarbon drop-in fuel like gasoline, kerosene, or diesel from alcohol, the alcohol molecule must be further processed. In most of today's fuel specifications for gasoline, kerosene, or diesel, the oxygen content in the fuel is capped with an upper limit. In the case of gasoline, for example, this upper limit is achieved by adding pure ethanol. In diesel for example, an oxygen component can be added in the form of fatty acid methyl esters (FAME). In both cases, the oxygen component is added to increase the proportion of biogenic fuel in the respective fuel mixture. In kerosene, on the other hand, no oxygen components are permitted in order to meet the respective ASTM D1655 aviation fuel specification [34]. Similarly, according to ASTM D7566, synthetic kerosene components, which could be produced from different alcohols, must be converted into an oxygen-free kerosene product similar to fossil-based kerosene before being blended up to a certain (and clearly defined) extend to fossil-based kerosene [35]. In addition to kerosene, other hydrocarbon fractions can be produced either as by-products or in a targeted manner. These products differ mainly in the hydrocarbon chain length [36], decisively determined by the specified boiling curve in the various fuel standards. Different key figures of the respective fuel standards are given in Table 2.

To ensure that a drop-in fuel is provided and these fuel products are fully compatible with the existing fuel provision infrastructure, compliance with the specific chain lengths / chain length distribution of the respective fuel fractions is necessary, among other things.

Against this background within this section, a process route is presented to produce gasoline, kerosene, and diesel components from different alcohols to adjust it as close as possible to today's specifications. For this purpose, the general process is described, which is already approved for the production of kerosene from alcohol according to ASTM D7566 [35]. Within such a process, not only kerosene components are produced, but also gasoline and diesel fractions.

Table 2 Properties of gasoline, kerosene, and diesel as hydrocarbon products [14, 35, 37, 38]

Hydrocarbon Product	Carbon Chain Length	10% Boiling Point [°C]	[a] Final/[b] 95% Boiling Point [°C]	[a] Freezing Point [°C]; [b] Cold Filter Plugging Point [°C]	[a] Oxygen Content [wt-%]; [b] FAME [vol-%]
Gasoline	C_4 to C_8	–	210 [b]	–	2.7 [a]
Kerosene	C_9 to C_{16}	205	300 [a]	−40/−47 [a]	75×10^{-6} [a]
Diesel	C_{11} to C_{22}	–	360 [b]	−5/−15 [b]	7 [b]

FAME: fatty acid methyl ester; depending on the fuel standard, different information is given for the same category (e.g., for the boiling end [a] the final or [b] 95% boiling point)

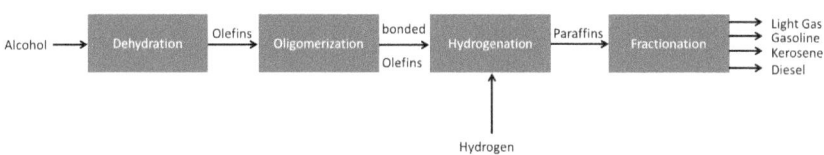

Fig. 5 Main alcohol to hydrocarbon process steps (in accordance with [39])

On the one hand, the choice of alcohol as the feed stream does not influence the overall process concept, as the four key process steps—dehydration, oligomerization, hydrogenation, and fractionation—for converting alcohols remain the same regardless of the specific alcohol used. The individual processes, on the other hand, differ greatly in terms of reaction conditions and separation techniques, depending on the alcohol as a feed stream. These four process steps for upgrading the different alcohols to the hydrocarbon fractions are shown in Fig. 5 and described in more detail below. Subsequently, currently industrially implemented process concepts are described and discussed.

4.1 Dehydration

The first step in alcohol processing is dehydration, with the main objective to remove oxygen from the alcohol molecule. Thus, the hydroxyl group bound in the alcohol is removed from the molecule in the form of water. The main product is a short-chain alkene, which serves as the raw material for the subsequent oligomerization step [29]. Depending on the alcohol used, different reaction conditions and purifications are applied in this process step, which are described in more detail below.

Methanol. The dehydration of methanol to alkenes represents a special case in the group of dehydration reactions described here. To form the smallest possible existing alkene, ethene, it is not sufficient to remove the hydroxyl group as water from the methanol molecule, but a new C–C bond must also be formed to create the alkene. This can be done one or more times during the dehydration of methanol, so that a product mixture of different short-chain alkenes is formed.

This overall process of dehydration of methanol has been known industrially since the 1980s as the so-called methanol-to-olefins (MTO) process [40]. The MTO process thus offers an alternative to steam cracking-based and, thus, crude oil-based production of alkenes. Ethene and propene are the main products of the dehydration process, although smaller amounts of higher alkenes are also formed. The propene-to-ethene ratio (P/E-ratio) can be variably adjusted, in contrast to the steam cracker process, and can follow possible market trends for selling these two bulk chemicals.

Alternatively, as it is the case for the processes discussed here, the alkenes can be converted into higher hydrocarbons by oligomerization (methanol to hydrocarbons; MTHC). Particularly in China, MTO process capacities have been expanded in the last decade, as this ensures alkene production based on cheap domestic hard coal [10].

The reaction equation of methanol dehydration is exemplified by ethene production shown in (Eq. 8). This reaction proceeds via the formation of dimethyl ether (DME) as an intermediate. In the process, a water molecule is first split off, and the bonding of two methyl groups is realized via the remaining oxygen atom. In the second step, this oxygen is also removed from the molecule by water. Through this reaction mechanism, additional hydrocarbon chain segments can be formed by eliminating a water molecule, resulting in a mixture of light alkenes [41].

$$2\,CH_3OH \;\rightarrow\; CH_3OCH_3 + H_2O \;\rightarrow\; C_2H_4 + 2\,H_2O \tag{8}$$

In this reaction, a maximum stoichiometric selectivity of 44 wt% toward ethene production is achievable. The applied process conditions, in particular the temperature, pressure, catalyst selection, and space–time velocity, can determine the yield structure, such as the P/E-ratio. Due to the formation of the new hydrocarbon bonds from at least two molecules, the reaction is strongly exothermic with an adiabatic temperature increase of 650 °C [40]. At reaction temperatures below 350 °C, the intermediate DME is produced in significant quantities. Above 350 °C and at pressures below 10 bar, an almost complete methanol conversion into the alkenes can be achieved. The applied zeolite catalyst is commonly operated with weight hourly space velocities (WHSV) of 0.8 to 4 1/h [42, 43]. In the described reaction mechanism, in addition to DME as an intermediate product at low temperatures, coke, short-chain alkanes, and CO_2 can also be formed as by-products. These can be separated from the alkenes depending on the further processing and use of the end products. Overall, selectivities of around 90% can be achieved with regard to the production of short-chain alkenes [1, 40, 42, 43].

This reaction can be carried out on a large scale in different plant concepts. To support the heterogeneously catalyzed, exothermic reaction, a fixed-bed reactor or a fluidized-bed reactor can be used, each of which is cooled to adjust the reaction temperature [40]. The fixed-bed reactor is characterized by its simple handling and the possibility of scale-up. In the fluidized-bed reactor, the catalyst can be regenerated within the process, leading to longer catalyst life times primarily due to the high coke formation on the catalyst surface. Furthermore, the mass and heat transport within the reactor is more pronounced in contrast to the fixed-bed reactor. Industrially, protonated ZSM-5 catalysts and SAPO-34 zeolites are used for the MTO process. Due to their structure, both catalysts exhibit a shape selectivity for the preferential formation of short-chain alkenes. Due to the smaller pore size of the SAPO-34 catalyst, its selectivity for ethene and propene formation is higher than that of the ZSM5. However, the SAPO-34 also leads to increased coke formation on the catalyst surface being the reason why a fluidized-bed reactor is often used in combination with the catalyst to regenerate the catalyst [40, 41, 44].

Since the MTO process was developed in the 1970s, various companies have created and improved processes for large-scale implementation [40, 42–46]. A summary of various process characteristics for these processes is given in Table 3.

One of the world's leading commercial technologies for the MTO process is the first technology concept in Table 3 [44]. The technology uses a proprietary SAPO-34 catalyst catalyzing the reaction in combination with a fluidized-bed reactor, including a catalyst regeneration. The process flow diagram shown in Fig. 6 represents this process, including alkene purification. In addition to the alkene separation, the by-products, such as CO_2 and alkanes, are removed from the product mixture. The methanol is almost completely converted to alkenes and water at the reaction temperature of about 400 °C and pressure between 1 and 6 bar. Selectivity with respect to alkene production can reach over 90%, while the propene to ethene ratio is highly adjustable by the process parameters [1, 40].

Ethanol. In the case of ethanol dehydration, various process concepts can be implemented. Common to all of them is the selective conversion of ethanol to ethene (Eq. 9), with a low formation of other by-products such as methane or CO_2.

$$C_2H_5OH \rightarrow C_2H_4 + H_2O \tag{9}$$

Table 3 Overview of major methanol-to-olefins (MTO) technologies [40, 44]

Technology	Reactor Type	Catalyst	T [°C]	P [bar]	Conversion [%]	Selectivity [%$_C$]^a	P/E [wt%]^b
MTO^c	Fluidized bed	SAPO-34	400–525	1–6	100	90	0.7–1.3
D-/S-MTO^c	Fluidized bed	SAPO-34^d	400–550	1–5	100	92	0.8–1.2/ 0.6–1.3^d
MTO/P^e	Fixed bed	Modified ZSM-5	430–450	1–4	99	–	–

^{a} based on carbon in alkene fraction; ^{b} mass-based propene (P) to ethene (E) ratio; ^{c} major licensing technologies in China; ^{d} SMTO with novel SAPO-34; ^{e} AirLiquide Lurgi—reaction via MeOH/ H2O/DME mixture

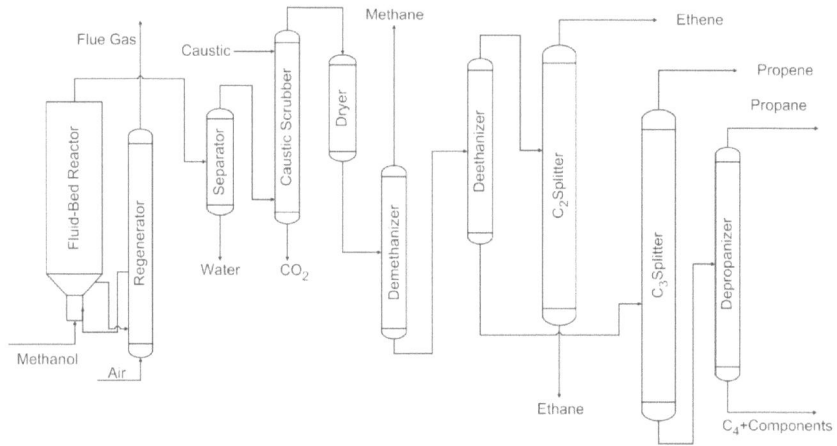

Fig. 6 UOP/Norsk Hydro-MTO process flowsheet [1]

The dehydration of ethanol is an endothermic process (1,632 J/g_{Ethene}); i.e., heat must be added externally for the reaction system to maintain a constant temperature during the process. In the context of reaction control, different reactor concepts can be useful, such as isothermal fixed-bed reactors, adiabatic fixed-bed reactors, or fluidized-bed reactors. In the adiabatic reactor, the heat supply is ensured by the installation of heated intermediate stages, so that the temperature is variable only in smaller ranges. Furthermore, a catalyst is required in the reactors; this can be an oxide catalyst, usually aluminum oxide, a molecular sieve catalyst, or a heteropoly acid catalyst [47–49].

Different technology providers offer various basic processes. For example, the Hummingbird technology uses a heteropoly acid in a fixed-bed reactor. The dehydration of ethanol takes place here at temperatures between 160 and 270 °C and at pressures up to 45 bar [50]. Due to the deactivation of the catalyst in the presence of water, only lower conversions are achieved here [48]. However, since unreacted ethanol can be returned to the reactor, selectivities of over 99% can be achieved.

Another dehydration technology is a multi-stage adiabatic reactor system with a syndol catalyst (Al_2O_3-MgO/SiO_2). A similar process concept has also already been implemented on a large scale with 200,000 t/a ethanol in Brazil for subsequent ethene polymerization. In this concept, four adiabatic reactors are operated with inlet temperatures of 425 °C and upstream pressures of the first reactor of 4.5 bar. The catalyst employed in this process is not deactivated by water, allowing for the utilization of an alcohol–water mixture, specifically below the azeotropic point. In addition, steam could be added to the reactor to lower the partial pressure of the reaction components and thus reduce coking reactions. Between the individual stages, the mixture is heated back up to the reaction temperature of 425 °C [47, 48, 51]. In this system, the individual reactors operate at a weight hourly space velocity (WHSV)

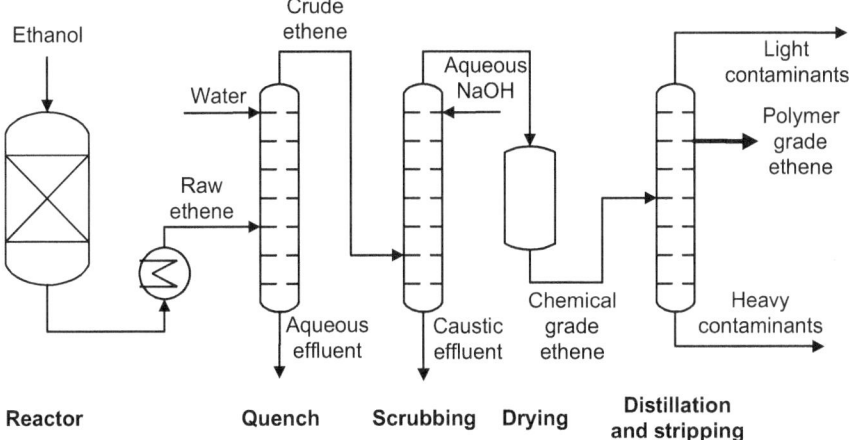

Fig. 7 Depiction of a generic ethanol dehydration process plant [52] (NaOH: Sodium hydroxide)

of 0.33 to 0.43 1/h and thus achieve conversions of over 95% and ethene selectivities of around 97% [47, 48].

Figure 7 shows the first purification of the ethene–water mixture coming out of the reactor. The mixture is quenched with water in a column, and the water is separated via the boiling point difference. The crude ethene is then washed with a sodium hydroxide (NaOH) solution, and the CO_2 by-product is removed in the form of sodium carbonate. Afterward, the remaining water is separated in molecular sieves, as the subsequent process steps may be water sensitive. If a higher purity of ethene is required, for example, for oligomerization or a sale of ethene in the polymer purity class, the technical ethene is fed to cryogenic rectification, which allows further heavier and lower boiling by-products to be separated [47, 52, 53].

Butanol. Both n-butanol and iso-butanol dehydration [54–56] proceed according to the following reaction equation (Eq. 10).

$$C_4H_9OH \rightarrow C_4H_8 + H_2O \tag{10}$$

Various catalysts can be used to control this chemical reaction, such as mildly acidic alumina catalysts, solid acids catalysts, zeolites, or other catalysts. The dehydration reaction can take place at a mild acid catalyst and low temperatures. In addition to dehydration, isomerization to iso-butenes can also occur on the same catalyst during this process step. However, higher temperatures of 300–500 °C and stronger acids must prevail to induce this reaction (isomerization) [55].

Today, butanol dehydration is not carried out on an industrial scale, since the alkene butene is obtained in particular from steam cracking of naphtha and thus provided based on a different feedstock and process chain [57]. This process route does not yet provide "sustainable" butene due to the fossil origin (crude oil) of the naphtha, whereby the provision of Fischer–Tropsch (FT) naphtha would be an indirect "sustainable" alternative.

4.2 Oligomerization

As shown in the previous section regarding the dehydration process step, different processes can be used to remove the oxygen from the different alcohols and produce an alkene or a mixture of alkenes. The oligomerization is the decisive process step within the process chain to adjust the final product distribution. Therefore, the overarching objective is to combine the short-chain alkenes in such a manner that the resulting mixture of oligomerized longer-chain alkenes already matches the carbon chain length range of the desired end product (gasoline, kerosene, and diesel) as closely as possible. This chain extension is achieved by the formation of C–C bonds between individual alkenes. Due to the fact that the dehydration of different alcohols yields different alkene products—methanol dehydration mainly produces a mixture of alkenes, ethanol dehydration mainly yields ethene, and butanol dehydration mainly yields butene—different process concepts can be employed for the subsequent oligomerization [1, 4, 29]. Thus, the various concepts depending on the specific alcohol used are described below.

Methanol (Alkene-Mixture-Based). While oligomerization from pure alkenes as well as from higher alkene blends is already applied at an industrial scale or at least comprehensively discussed in literature, oligomerization, in particular of ethene/propene blends to higher distillates in the kerosene and diesel range, is rather less implemented and not yet as extensively studied. From a technical point of view, the main challenge lies in the different operating conditions required for the conversion of the less reactive ethene on the one hand and the production of long-chain hydrocarbons on the other. While propene is already converted in higher quantities at moderate temperatures, ethene requires higher temperature levels at which the product distribution shifts from the distillate to the gasoline fraction. Nevertheless, there are process concepts that convert a methanol-based short-chain alkene mixture into a long-chain alkene mixture in one process step. In the processes to be mentioned, such as Mobil olefins to gasoline and distillate (MOGD) and methanol to synthetic fuels (MTSynfuel), an attempt is made to maximize the higher boiling fraction by recycling a gasoline-rich alkene fraction into the oligomerization reactor [58]. Both processes are well understood from a technical standpoint, but they are not yet optimized for the production of kerosene or diesel fuels.

The different starting materials (alkenes) in the process routes based on methanol, ethanol and butanol results in different product compositions. In the ethanol and butanol processes, primarily molecules with even carbon numbers (or multiples of 4) are formed. However, in the oligomerization of the alkene mixture derived from methanol, a continuous distribution of alkenes with both even and odd carbon numbers can be achieved. This results partly from the proportion of alkenes with odd carbon numbers and partly from cracking reactions, which occur more frequently at higher reaction temperatures [59, 60].

The optimal process conditions and the catalyst used are strongly dependent on the composition of the feedstock and the target product. Since the reaction is highly molecular reducing and exothermic, high pressures of 20 to 100 bar and temperatures of 100 to 450 °C (as a compromise of reaction equilibrium versus activity and kinetics) are generally used. Solid phosphoric acid and zeolite catalysts (H-ZSM-5/22/57) can be used for the conversion of alkenes into gasoline, kerosene, and diesel fuel. The weight hourly space velocity (WHSV) is given in wide ranges between 0.3 and 4.0 1/h [59–61]. Acidic zeolite catalysts can be operated under process conditions between 200 and 300 °C and allow controlled branching of the molecules. The product of oligomerization under these process conditions is a distribution of predominantly unsaturated, lightly to moderately branched hydrocarbons [1, 40, 60–63].

Ethanol (Ethene-Based). The general reaction equation for ethene oligomerization is shown in (Eq. 11), showing that during ethene oligomerization, mainly even-numbered alkenes are produced since one additional ethene molecule is bonded to each growing chain. Only in smaller numbers, longer alkene chains can also break and thus produce odd hydrocarbon chain lengths. However, these are distinctly outnumbered by the even-numbered hydrocarbon chains [29].

$$\mathrm{n\ C_2H_4 \rightarrow\ C_{2n}H_{4n}}(n = 2 \text{ to } 10) \tag{11}$$

For the oligomerization of ethene, different approaches exist [29, 59, 64–68] to be distinguished by the number of stages of oligomerization and the type of catalyst used. For the catalyst, a distinction can be made between homogeneous and heterogeneous catalysts and between acid-based and transition metal-catalyzed processes. In general, one- or two-stage concepts are discussed within as well, with oligomerization occurring in both stages. The advantage of a second stage is that the process conditions, in particular, can be adapted to the respective reactants. In the case of acid-based catalysts, high temperatures of over 300 °C are required for the activation of the ethene, as otherwise, the carbenium ion of the ethene is unstable. However, in the case of oligomerization of molecules with higher chain lengths (C_{4+}), there is the challenge that the higher temperatures lead to cracking of the alkene chains and thus prevent a product distribution of the chain lengths in kerosene and diesel length. Thus, in the first stage, the ethene can be synthesized into an alkene mixture of mainly C_4 to C_6 alkenes, which subsequently oligomerize to higher alkenes. The high temperatures required in the first oligomerization step also favor other side reactions. The use of a transition metal catalyst can reduce the temperature required there [29]. Ziegler–Natta catalysis is a respective approach using a homogeneous catalyst to produce longer-chain alkenes. Among others, the Shell higher olefin process (SHOP) uses this reaction scheme and thus achieves a Schulz–Flory distribution of the alkenes. Heterogeneous Ni-catalysts can also be used for the oligomerization of ethene. At 100 °C, the temperatures are lower than for acid-based catalysis [63, 64].

Another possible process technology for oligomerization from ethene is a two-step process, which is primarily optimized for the production of kerosene fractions [69]. This process concept uses a two-stage system in which a Ni-metal catalyst is used in the first stage. The product of the first stage is further oligomerized in the second stage with the use of an acid catalyst. In the first oligomerization stage, a pressure of 22 bar and a moderate temperature of about 85 °C are applied. This results in a total conversion of about 98% of the ethene to a mixture of C_4 to C_{10} alkenes, the majority of which is present as C_4 alkene with almost 70% selectivity. This alkene mixture is fed into the second oligomerization reactor, where longer-chain alkenes are formed at the same pressure conditions and a (higher) temperature of about 225 °C. After the second reaction stage, the unreacted, short alkenes are separated and are returned to the second oligomerization stage, thus increasing the yield of hydrocarbons in the fuel range. In a kerosene-optimized mode of operation, up to 70% of the carbon input can end up in hydrocarbons in the kerosene range, with hydrocarbons in the gasoline and diesel boiling ranges being the main by-products [64].

Butanol (Butene-Based). Like ethene oligomerization, butene oligomerization can be carried out using transition metal or acid catalysts. These can also be carried out as homogeneous and heterogenic catalysis [70]. The reaction equation is shown in (Eq. 12).

$$m\ C_4H_8 \rightarrow\ C_{4m}H_{8m}(m = 2\ \text{to}\ 5) \tag{12}$$

An example of an industrial process of butene oligomerization is the dimersol process, where either butenes or a mixture of propene and butene can be oligomerized. The main product of the butene-based dimersol process are iso-octanes. This high-octane molecule can either be added directly to the gasoline blend as a fuel component, or can be used in hydroformylation to produce plastic plasticizers [70]. During the oligomerization of butene into hydrocarbons in the kerosene and diesel boiling range, mainly iso-alkenes are formed, which are derived from integer multiples of butene. As a result, mainly C_{12} and C_{16} alkenes are present in the kerosene section [71].

In the case of butene, it can be directly utilized for hydrogenation without the need for an oligomerization step to produce longer-chain hydrocarbons. This is possible because there are already market applications for the hydrogenated product butane. For instance, it can be added in small quantities to the current gasoline blend. However, the amount of butane that can be added also depends on the time of year; in summer, the addition of very low-boiling components is more restricted in order to avoid off-gassing at high ambient temperatures.

4.3 Hydrogenation

The purpose of hydrogenation is to saturate the existing carbon double bonds with hydrogen in order to meet the various fuel specifications. These carbon double bonds are relatively unstable, which can cause problems, e.g., during storage. For the description of this process step, the origin of the alkene mixture produced within the oligomerization step is of secondary importance, which is why a feed-independent description of the hydrogenation is given.

Hydrogenation of alkenes is already a well-known process within the (petro-) chemistry. For this purpose, hydrogen is added within this reaction step in addition to the alkenes. The amount of hydrogen required corresponds to the molar amount of alkenes to be saturated in the hydrogenation process [72]. The respective reaction equation is shown below (Eq. 13).

$$C_{2n}H_{4n} + H_2 \rightarrow C_{2n}H_{4n+2}(n = 2 \text{ to } 10) \tag{13}$$

Hydrogenation of alkenes is normally controlled via a metal catalyst, such as platinum, nickel, or palladium. On an industrial scale, rather inexpensive Raney-nickel catalysts are often used. In this case, a pressure of 3 to 7 bar is required at a slightly elevated temperature (approx. 150 to 200 °C). When using the more noble platinum or palladium catalysts, the reaction can also be carried out at ambient pressure and temperature [4, 73]. To maximize the conversion (saturation) of the alkenes, the process is operated with an excess of hydrogen. The unused hydrogen can be separated downstream within a thermal separation and recycled to the reactor.

4.4 Fractionation

In the final fractionation stage, the produced hydrocarbon mixture is separated into different product fractions based on their chain length under fulfillment of the given fuel standards. This allows for the separation of gasoline, kerosene, and diesel fractions from each other. This separation is carried out separately according to the relative volatility of the substances within the mixture [74]. This is necessary in order to maintain the boiling curves of the products defined in the specific fuel specifications. In addition to these three main fractions, lighter hydrocarbons may also be dissolved in the hydrogenated process and separated as a light gas fraction as a by-product. Fractionation, like the hydrogenation, is a widely used process within the petroleum industry respective within "classical" crude oil refineries.

Figure 8 shows, as an example, a schematic concept for the separation of fractions in a single column. Various rectification concepts can be considered, such as the use of several columns or one column with the separation of the three fuel fractions gasoline, kerosene, and diesel on different trays [75]. In addition, side columns can

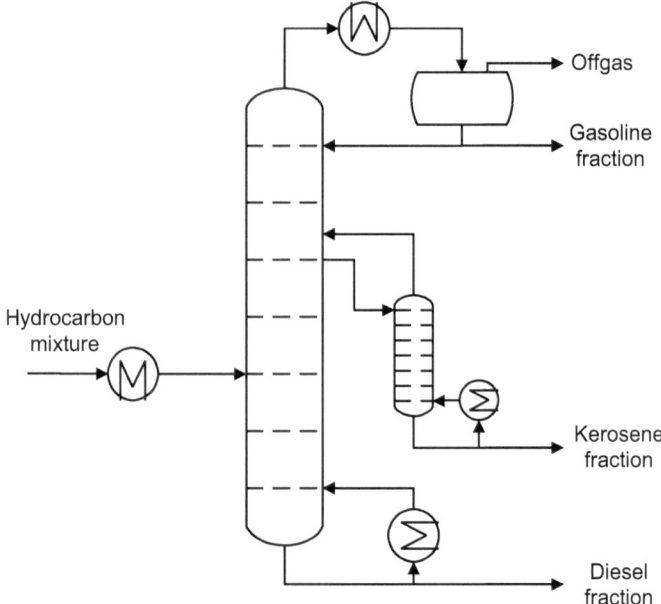

Fig. 8 Schematic rectification column

be used to further purify the withdrawn intermediate product and return its head product to the main column.

5 Overall Processes

In this section, overall processes already available and partly implemented in the industry for converting the various alcohols into hydrocarbons are briefly discussed.

5.1 *Methanol*

Methanol-based hydrocarbon production was developed in the 1970s by using zeolite catalysts (MTO process). Additional research enabled the successful testing of the further developed MOGD process in 1981. Combined with the upstream MTO process, MTO/MOGD produces long-chain hydrocarbons from methanol with distillate yields of more than 80% [27]. Figure 9 shows the flowsheet of the MOGD section of the process [76].

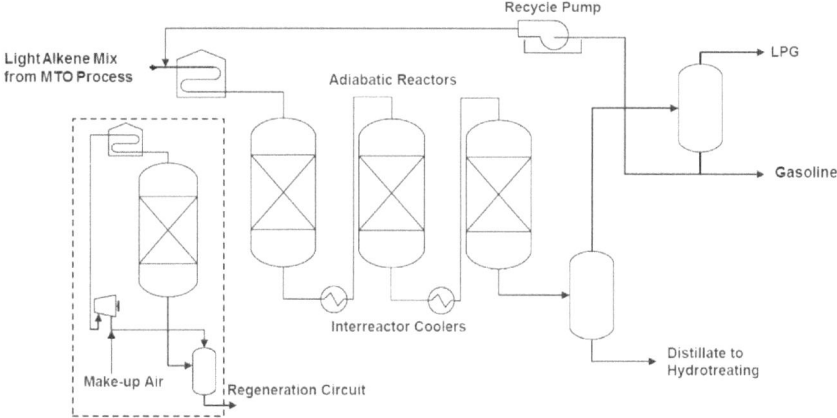

Fig. 9 Mobil's olefin to gasoline and distillate (MOGD) process (LPG: Liquefied petroleum gas; MTO: Methanol-to-olefins; according to [1, 68])

For the oligomerization of the alkene mixture, three fixed-bed reactors are operated in series with interstage cooling. While three reactors are in production operation, a fourth reactor is in cyclic regeneration. In production operation, the reactors are operated at about 40 bar and 200 °C, whereby several reactions occur simultaneously on the zeolite catalysts used—mainly oligomerization, isomerization, and cracking [1, 62]. When high distillate yields are desired, propene and higher alkenes are the preferred feedstock for this process. Therefore, lower temperatures in the upstream MTO process are preferable, as it leads mainly to C_3 to C_6 alkenes [1]. In addition, the distillate share can be increased by recycling considerable quantities of the gasoline fraction, which also has a stabilizing effect on process operations. Saturated hydrocarbons from the MTO process pass through the MOGD section as inert components. The product consists mainly of C_5 to C_{20} methylated iso-alkenes fractionated by downstream distillation. The distillate fraction must additionally be saturated entirely with hydrogen in further hydrogenation. In addition to the branched alkanes, there is also a considerable proportion of aromatic compounds in the final product. Since the plant concept has been technically tested, but no information on large-scale industrial implementation is known, the process can be estimated with a technical readiness level (TRL) of 8 [62, 76].

Another process for the production of long-chain hydrocarbons from methanol is the MtSynfuel process. The process has already been successfully tested in the 2000s on a pilot scale. It is based on the methanol-to-propene (MTP) process, which is already in operation on a large scale for propene production and has been extended in the MtSynfuel concept by an additional oligomerization section. The MTP process is carried out in fixed-bed reactors on zeolite catalyst, operated at 300 to 550 °C and a low pressure of 1 to 2 bar. The oligomerization also takes place on zeolite catalyst at temperatures of 150 to 350 °C and elevated pressure levels of around 35 to 85 bar. The concept offers high production flexibility. It allows the market-oriented production of DME, methanol, light olefins, gasoline, or diesel in optional ratios [40]

5.2 *Ethanol*

In ethanol-based hydrocarbon production, the first demonstration plants have been built since 2018, while the first commercial plants with an annual production of 30 kt/a are being planned [77, 78]. In this context, the current processes increasingly relate to the production of kerosene and thus follow the alcohol-to-jet (ATJ) process route. This process route based on ethanol is already approved for commercial use in aircraft by ASTM D7566 [35]. An industrial ATJ process, based on an ethanol feedstock, is shown in Fig. 10. This is also referred to as ethanol-to-jet (ETJ) pathway.

Here, the ethanol is dehydrated first, then oligomerized, and finally hydrogenated and fractionated into the desired products. In the dehydration step of ethanol, the Hummingbird technology is used, in which high ethene selectivities can be achieved through high pressures and moderate temperatures (Sect. 4.1) [77]. Two oligomerization stages are used for the subsequent oligomerization of ethene, where the first oligomerization stage produces mainly C_4 to C_6 alkenes, which are oligomerized to hydrocarbons in the kerosene or jet-fuel range in a second oligomerization stage. The yield of the overall oligomerization steps can be increased by internal recycling. A Ni-loaded aluminosilicate catalyst is used in the first stage, and a solid acid catalyst in the second stage [29, 64, 78]. In the subsequent hydrogenation, external hydrogen must then be added in order to carry out the hydrogenation. The main products of this process are then branched hydrocarbons, which are in the boiling range of diesel and kerosene.

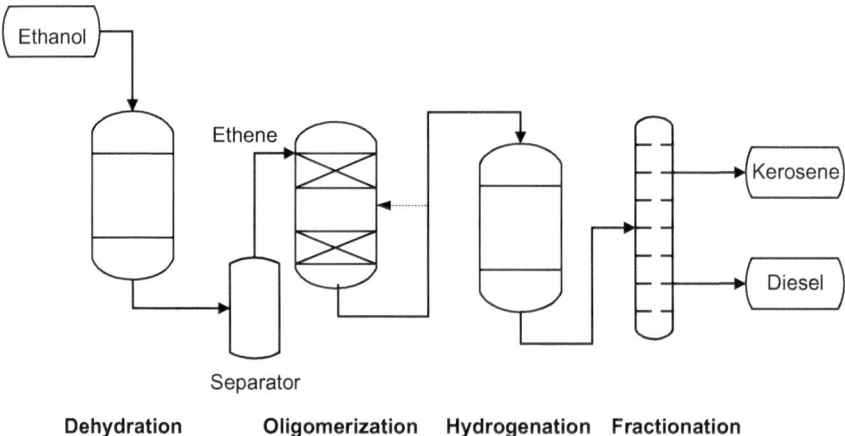

Fig. 10 Ethanol-to-jet process pathway (according to [77])

Within the overall ethanol-based ATJ process of another technology provider, the oligomerization products are first fractionated before the individual fractions are partially hydrogenated separately. The separated unreacted gas can be reacted in a hydrocarbon reformer to meet the hydrogen demand of the hydrogenation process [79].

5.3 *Butanol*

In addition to the ethanol-based route, the iso-butanol-based route for the production of kerosene or jet fuel in particular is already approved in ASTM D7566 [35]. The first pre-commercial process plants with a capacity of 300 t/a exist, and commercial process plants with larger capacities of 140 kt/a are planned and designed [80]. In this process, the iso-butanol is first dehydrated and subsequently converted to C_8, C_{12}, and C_{16} oligomers. The iso-octanes can be added to the gasoline fraction, among other things, to improve the knockability of the mixture [81]. In the final step, the long-chain alkene mixture is hydrogenated and can subsequently serve as a jet fuel component—mainly C_{12} and C_{16} alkenes within the kerosene boiling range result from the oligomerization of the butene mixture.

6 Final Consideration

Today, alcohols, notably methanol and ethanol, are significant for various applications, especially in the chemical industry and as fuel components in road transportation. Ethanol, produced mainly from biomass, already contributes substantially to the defossilization of transportation. As there is an increasing need to reduce greenhouse gas (GHG) emissions across sectors, the role of such alcohols and their provision based on renewable energies could become more crucial. In the transportation context, the direct use of such "renewable" alcohols can face challenges due to their hygroscopic nature, lower energy density, and poor self-ignition at high pressures. While "renewable" ethanol, as well as other "renewable" alcohols, can be blended directly with hydrocarbons for energy use in road transportation, this approach is not feasible for aviation. However, these alcohols can also be converted into long-chain hydrocarbons such as gasoline, kerosene, and diesel via the alcohol to hydrocarbon process steps. Depending on the specific application, this offers various advantages as they can then serve as drop-in fuels, fitting seamlessly into existing fuel infrastructure and fleets. Thus, to further defossilize the transportation sector, converting renewably produced alcohols into such fuel components may be of interest.

Against this background, the aim of this paper is to analyze the process routes from alcohol to hydrocarbon components and thus to present possibilities for a defossilization of the mobility sector based on alcohols. In addition to analyzing the current production and direct use of alcohols, the uses of hydrocarbons are also analyzed and compared.

This analysis is based on methanol, ethanol, and butanol as alcohol options. Today, ethanol is the only alcohol produced sustainably from biomass in relevant larger quantities. Methanol and butanol, in contrast, are mainly produced from fossil raw materials, meaning alcohols produced in this way cannot be used for defossilization. The production of all alcohols via a power-to-liquid (PtL)-based pathway can be an attractive alternative, since it does not require fossil feedstocks or cultivated biomass, and the process steps are already available in part on a large scale (e.g., methanol synthesis from synthesis gas).

In alcohol downstream processing to hydrocarbons, alcohols are usually processed in the same four steps: dehydration, oligomerization, hydrogenation, and fractionation. The dehydration and oligomerization depend on the used alcohol, as the process parameters and reaction characteristics change depending on this. Hydrogenation and fractionation are well-known chemical processes, which do not differ significantly in the case of an alkene mixture from an oligomerization.

The technical implementation of the process to convert alcohols into higher hydrocarbons is already taking place in China on the basis of fossil methanol in the so-called Mobil olefins to gasoline and distillate (MOGD) process. This process is used in particular to produce gasoline and diesel fuels for use in road transport. Current pilot and demonstration plants based on the ethanol and butanol process primarily focus on a kerosene-optimized approach. This is because the foreseeable future of the aviation sector's defossilization will largely depend on the use of kerosene-based fuels. Thus, the development and commercialization of processes to produce kerosene-based sustainable aviation fuels from alcohols like methanol, ethanol, or butanol, which can be used directly as drop-in fuels, is of great interest. This kerosene-optimized process is also known as alcohol-to-jet (ATJ).

Overall, a large-scale commercial implementation of the alcohol to hydrocarbon process step is not yet being carried out, but the first projects are being planned. The commercialization of the alcohol to hydrocarbon process route for fuel purposes faces certain challenges. One of the main obstacles is the relatively higher cost of such alcohols compared to existing fossil fuel prices as well as by the option to use such alcohols directly for road transportation purposes without further processing.

References

1. Kaltschmitt M, Neuling U (2018) Biokerosene. Springer Berlin Heidelberg, Berlin, Heidelberg
2. Kaltschmitt M, Hartmann H, Hofbauer H (2016) Energie aus Biomasse. Springer Berlin Heidelberg, Berlin, Heidelberg
3. Lappe P, Hofmann T (2000) Pentanols. In: Ullmann's encyclopedia of industrial chemistry. Wiley-VCH Verlag GmbH & Co. KGaA, Weinheim, Germany
4. Pechstein J, Neuling U, Kaltschmitt M (2015) Biokerosin aus Alkoholen. Substrate, Konzepte und Verfahren, Bewertung

5. Klabunde J, Bischoff C, Papa AJ (2000) Propanols. In: Ullmann's encyclopedia of industrial chemistry. Wiley-VCH Verlag GmbH & Co. KGaA, Weinheim, Germany, pp 1–14
6. Hahn H-D, Dämbkes G, Rupprich N, Bahl H, Frey GD (2000) Butanols. In: Ullmann's encyclopedia of industrial chemistry. Wiley-VCH Verlag GmbH & Co. KGaA, Weinheim, Germany
7. Tutak W, Jamrozik A, Pyrc M (2017) Co-combustion of biodiesel with oxygenated fuels in direct injection diesel engine. E3S Web Conf. https://doi.org/10.1051/e3sconf/20171402018
8. Kaltschmitt M, Neuling U, Bullerdiek N, Voß S (2022) Direct alcohol vs. alcohol-to-jet SPK utilisation in commercial aviation—an energetic-operational analysis. IJSA. https://doi.org/10.1504/IJSA.2022.10046511
9. Bazaluk O, Havrysh V, Nitsenko V, Baležentis T, Streimikiene D, Tarkhanova EA (2020) Assessment of green methanol production potential and related economic and environmental benefits: the case of China. Energies. https://doi.org/10.3390/en13123113
10. Kang S, Boshell F, Goeppert A, Prakash SG, Landälv I, Saygin D (2021) Innovation outlook. Renewable methanol. International Renewable Energy Agency, Abu Dhabi
11. Dieterich V, Buttler A, Hanel A, Spliethoff H, Fendt S (2020) Power-to-liquid via synthesis of methanol, DME or Fischer–Tropsch-fuels: a review. Energy Environ Sci. https://doi.org/10.1039/d0ee01187h
12. Anicic B, Trop P, Goricanec D (2014) Comparison between two methods of methanol production from carbon dioxide. Energy. https://doi.org/10.1016/j.energy.2014.09.069
13. Iaquaniello G, Centi G, Salladini A, Palo E, Perathoner S, Spadaccini L (2017) Waste-to-methanol: process and economics assessment. Biores Technol. https://doi.org/10.1016/j.biortech.2017.06.172
14. European Norm, Automotive fuels—Unleaded petrol—requirements and test methods, 1st edn (228)
15. Agarwal AK, Gautam A, Sharma N, Singh AP (2019) Methanol and the alternate fuel economy. Springer Singapore, Singapore
16. Michele P, OECD-FAO agricultural outlook 2021–2030. Biofuels 9:202–213
17. Kroumov AD, Módenes AN, Tait MCDA (2006) Development of new unstructured model for simultaneous saccharification and fermentation of starch to ethanol by recombinant strain. Biochem Eng J. https://doi.org/10.1016/j.bej.2005.11.008
18. Tang Y, Huang Y, Gan W, Xia A, Liao Q, Zhu X (2021) Ethanol production from gas fermentation: rapid enrichment and domestication of bacterial community with continuous CO/CO_2 gas. Renew Energy. https://doi.org/10.1016/j.renene.2021.04.134
19. Gonzalez-Contreras M, Lugo-Mendez H, Sales-Cruz M, Lopez-Arenas T (2021) Synthesis, design and evaluation of intensified lignocellulosic biorefineries—case study: ethanol production. Chem Eng Process Process Intensification. https://doi.org/10.1016/j.cep.2020.108220
20. Ao M, Pham GH, Sunarso J, Tade MO, Liu S (2018) Active centers of catalysts for higher alcohol synthesis from syngas: a review. ACS Catal. https://doi.org/10.1021/acscatal.8b01391
21. Min F, Kopke M, Dennis S (2013) Gas fermentation for commercial biofuels production. In: Fang Z (ed) Liquid, gaseous and solid biofuels—conversion techniques. InTech (2013)
22. Monir MU, Yousuf A, Aziz AA (2020) Syngas fermentation to bioethanol. In: Lignocellulosic biomass to liquid biofuels. Elsevier, pp 195–216
23. Veza I, Muhamad Said MF, Latiff ZA (2021) Recent advances in butanol production by acetone-butanol-ethanol (ABE) fermentation. Biomass Bioenerg. https://doi.org/10.1016/j.biombioe.2020.105919
24. Slaa J, van Ommen J, Ross J (1995) The synthesis of alcohols using $Cu/ZnO/Al_2O_3{}^+$ (Ce or Mn) catalysts. Top Catal 2:79–89
25. Fernández-Naveira Á, Veiga MC, Kennes C (2017) H-B-E (hexanol-butanol-ethanol) fermentation for the production of higher alcohols from syngas/waste gas. J Chem Technol Biotechnol. https://doi.org/10.1002/jctb.5194
26. Daniell J, Köpke M, Simpson S (2012) Commercial biomass syngas fermentation. Energies. https://doi.org/10.3390/en5125372
27. Maus W (2019) Zukünftige Kraftstoffe. Springer Berlin Heidelberg, Berlin, Heidelberg

28. Verhelst S, Turner JWG, Sileghem L, Vancoillie J (2019) Methanol as a fuel for internal combustion engines. Prog Energy Combust Sci. https://doi.org/10.1016/j.pecs.2018.10.001

29. Eagan NM, Kumbhalkar MD, Buchanan JS, Dumesic JA, Huber GW (2019) Chemistries and processes for the conversion of ethanol into middle-distillate fuels. Nat Rev Chem. https://doi.org/10.1038/s41570-019-0084-4

30. Da Trindade WRS, Santos RGD (2017) Review on the characteristics of butanol, its production and use as fuel in internal combustion engines. Renew Sustain Energy Rev. https://doi.org/10.1016/j.rser.2016.11.213

31. Bullerdiek N, Quante G, Bube S, Neuling U, Kaltschmitt M (2022) Non Drop-In Kraftstoffe im Luftverkehr. Technische Universität Hamburg, Ein gesamtsystemischer Vergleich von Nutzungs- und Einsatzmöglichkeiten

32. Escalante ESR, Ramos LS, Rodriguez Coronado CJ, de Carvalho Jr JA (2022) Evaluation of the potential feedstock for biojet fuel production: Focus in the Brazilian context. Renew Sustain Energy Rev. https://doi.org/10.1016/j.rser.2021.111716

33. Schröder J, Müller-Langer F, Winther K, Baumgarten W, Lindgren M (2020) Methanol as motor fuel summary report. In: A report from the advanced motor fuels technology collaboration

34. Edwards J (2020) Jet fuel properties, AFRL-RQ-WP-TR-2020-0017. Fuels & Energy Branch (AFRL/RQTF)

35. ASTM (2019) Specification for aviation turbine fuel containing synthesized hydrocarbons. ASTM International, West Conshohocken, PA (D7566)

36. de Klerk A (2011) Fischer-Tropsch refining. Wiley-VCH, Weinheim

37. ASTM (2013) Specification for aviation turbine fuels. D1655-13A. ASTM International, West Conshohocken, PA

38. European Norm, Automotive fuels—Diesel fuel—Requirements and test methods, 1st edn (590)

39. Neuling U (2018) Biokerosinherstellung. Dissertation, Technische Universität Hamburg; Verlag Dr. Kovač

40. Bertau M, Offermanns H, Plass L, Schmidt F, Wernicke H-J (2014) Methanol: the basic chemical and energy feedstock of the future. Springer Berlin Heidelberg, Berlin, Heidelberg

41. Keil F (1998) Methanol-to-hydrocarbon: process technology. Review. Microporous Mesoporous Mater 1999:49–66

42. Kianfar E (2019) Comparison and assessment of zeolite catalysts performance dimethyl ether and light olefins production through methanol: a review. Rev Inorg Chem. https://doi.org/10.1515/revic-2019-0001

43. Sun Q, Xie Z, Yu J (2018) The state-of-the-art synthetic strategies for SAPO-34 zeolite catalysts in methanol-to-olefin conversion. Natl Sci Rev. https://doi.org/10.1093/nsr/nwx103

44. Gogate MR (2019) Methanol-to-olefins process technology: current status and future prospects. Pet Sci Technol. https://doi.org/10.1080/10916466.2018.1555589

45. Samanta C, Das RK (2021) C3-based petrochemicals: recent advances in processes and catalysts. In: Pant KK, Gupta SK, Ahmad E (eds) Catalysis for clean energy and environmental sustainability. Springer International Publishing, Cham, pp 149–204

46. Ortiz-Espinoza AP, Noureldin MM, El-Halwagi MM, Jiménez-Gutiérrez A (2017) Design, simulation and techno-economic analysis of two processes for the conversion of shale gas to ethylene. Comput Chem Eng. https://doi.org/10.1016/j.compchemeng.2017.05.023

47. Mohsenzadeh A, Zamani A, Taherzadeh MJ (2017) Bioethylene production from ethanol: a review and techno-economical evaluation. ChemBioEng Rev. https://doi.org/10.1002/cben.201600025

48. Yakovleva IS, Banzaraktsaeva SP, Ovchinnikova EV, Chumachenko VA, Isupova LA (2016) Catalytic dehydration of bioethanol to ethylene. Catal Ind. https://doi.org/10.1134/S2070050416020148

49. Knözinger H (1968) Dehydration of alcohols on aluminum oxide. Angew Chem Int Ed Engl. https://doi.org/10.1002/anie.196807911

50. Bailey C, Bolton L, Gracey B, Lee M, Partington S, Process for producing ethylene. Patent 8426664 B2

51. Kagyrmanova AP, Chumachenko VA, Korotkikh VN, Kashkin VN, Noskov AS (2011) Catalytic dehydration of bioethanol to ethylene: Pilot-scale studies and process simulation. Chem Eng J. https://doi.org/10.1016/j.cej.2011.06.049
52. Morschbacker A (2009) Bio-ethanol based ethylene. Polym Rev. https://doi.org/10.1080/15583720902834791
53. Zhang M, Yu Y (2013) Dehydration of ethanol to ethylene. Ind Eng Chem Res. https://doi.org/10.1021/ie401157c
54. West RM, Braden DJ, Dumesic JA (2009) Dehydration of butanol to butene over solid acid catalysts in high water environments. J Catal. https://doi.org/10.1016/j.jcat.2008.12.009
55. Zhang D, Al-Hajri R, Barri SAI, Chadwick D (2010) One-step dehydration and isomerisation of n-butanol to iso-butene over zeolite catalysts. Chem Commun (Cambridge, England). https://doi.org/10.1039/c002240c
56. Taylor JD, Jenni MM, Peters MW (2010) Dehydration of fermented isobutanol for the production of renewable chemicals and fuels. Top Catal. https://doi.org/10.1007/s11244-010-9567-8
57. Busch H, Vogel H (2019) Dehydratisierung von Butanol in nah- und überkritischem Wasser. Chem Ing Tec. https://doi.org/10.1002/cite.201800070
58. Avidan AA (1988) Gasoline and distillate fuels from methanol. In: Methane conversion, proceedings of a symposium on the production of fuels and chemicals from natural gas, vol 36. Studies in Surface Science and Catalysis. Elsevier, pp 307–323
59. Saavedra Lopez J, Dagle RA, Dagle VL, Smith C, Albrecht KO (2019) Oligomerization of ethanol-derived propene and isobutene mixtures to transportation fuels: catalyst and process considerations. Catal Sci Technol. https://doi.org/10.1039/C8CY02297F
60. O'Connor CT, Kojima M (1990) Alkene oligomerization, pp 329–349
61. Bellussi G, Mizia F, Calemma V, Pollesel P, Millini R (2012) Oligomerization of olefins from light cracking naphtha over zeolite-based catalyst for the production of high quality diesel fuel. Microporous Mesoporous Mater. https://doi.org/10.1016/j.micromeso.2012.07.020
62. Ruokonen J, Nieminen H, Dahiru AR, Laari A, Koiranen T, Laaksonen P, Vuokila A, Huuhtanen M (2021) Modelling and cost estimation for conversion of green methanol to renewable liquid transport fuels via olefin oligomerisation. Processes. https://doi.org/10.3390/pr9061046
63. Nicholas CP (2017) Applications of light olefin oligomerization to the production of fuels and chemicals. Appl Catal A. https://doi.org/10.1016/j.apcata.2017.06.011
64. Lilga M, Hallen R, Albrecht KO, Cooper AR, Frye JG, Ramasamy KK, Systems and processes for conversion of ethylene feedstocks to hydrocarbon fuels. Patent 9771533
65. Heveling J, Nicolaides CP, Scurrell MS (1998) Catalysts and conditions for the highly efficient, selective and stable heterogeneous oligomerisation of ethylene. Appl Catal A. https://doi.org/10.1016/S0926-860X(98)00147-1
66. Nicholas C, Process for making diesel by oligomerization of gasoline. Patent 9,663.415 B2
67. Berard S, Process for the production of a fuel base from an ethylene feedstock implementing at least one oligomerization stage in the presence of a homogeneous catalytic system. Patent 8957270
68. Guillon E, Flexible process for transformation of ethanol into middle distillates. Patent US9475999B2
69. LanzaJet (2021) Lanzajet technology to be deployed across three different projects in the UK. https://www.lanzajet.com/lanzajet-technology-to-be-deployed-across-three-different-projects-in-the-uk-to-meet-growing-demand-for-sustainable-aviation-fuels/. Accessed 3 Jan 2022
70. Schmidt R, Griesbaum K, Behr A, Biedenkapp D, Voges H-W, Garbe D, Paetz C, Collin G, Mayer D, Höke H (eds) (2000) Ullmann's encyclopedia of industrial chemistry. Hydrocarbons. Wiley-VCH Verlag GmbH & Co. KGaA, Weinheim, Germany
71. Zschocke A, High biofuel belnds in aviation—final report
72. Geleynse S, Brandt K, Garcia-Perez M, Wolcott M, Zhang X (2018) The alcohol-to-jet conversion pathway for drop-in biofuels: techno-economic evaluation. Chemsuschem. https://doi.org/10.1002/cssc.201801690
73. Wollrab A (2009) Organische Chemie. Springer Berlin Heidelberg, Berlin, Heidelberg

74. Behr A, Agar DW, Jörissen J (2010) Einführung in die technische Chemie. Spektrum Akad. Verl, Heidelberg
75. Romero-Izquierdo AG, Gómez-Castro FI, Gutiérrez-Antonio C, Hernández S, Errico M (2021) Intensification of the alcohol-to-jet process to produce renewable aviation fuel. Chem Eng Process Process Intensification. https://doi.org/10.1016/j.cep.2020.108270
76. Tabak SA, Avidan AA, Krambeck FJ (1986) Production of synthetic gasoline and diesel fuel from nonpetroleum resources. Div Gas Fuel Chem
77. LanzaJet (2021) Technip Energies announces first Hummingbird® catalyst supply agreement for LanzaJet biorefinery. https://www.lanzajet.com/technip-energies-announces-first-hummin gbird-catalyst-supply-agreement-for-lanzajet-biorefinery/. Accessed 7 Apr 2022
78. Pacific Northwest National Laboratory (2018) PNNL and LanzaTech team to make new jet fuel. https://www.pnnl.gov/news/release.aspx?id=4527. Accessed 7 Apr 2022
79. Byogy (2021) Technology A 4-step process. https://byogy.com/technology/. Accessed 7 Apr 2022
80. Gevo Inc. (2022) About gevo. https://gevo.com/wp-content/uploads/2022/01/Gevo_LakePre ston_one-sheet_1.19.2223.pdf. Accessed 7 Apr 2022
81. Gevo Inc. (2022) Isobutanol in, isooctane and sustainable aviation fuel out. https://gevo.com/about-gevo/our-plants/silsbee/. Accessed 7 Apr 2022

Blending of Aviation Powerfuels

Alexander Zschocke, Gunnar Quante, Nils Bullerdiek, and Jan Pechstein

Abstract As the global energy landscape evolves in response to anthropogenic climate change, there is an increasing focus on using sustainable alternatives to conventional fossil-based fuels. In aviation such an alternative is provided by drop-in synthetic fuels which can blended with fossil kerosene to produce a blend that is virtually indistinguishable from conventional kerosene and is fully compatible with existing infrastructure. However, for most of these fuels blending is a crucial part of the process as the fuels are not fully drop-in on their own. This topic is discussed in the following chapter, with a focus on powerfuels.

Keywords Blending · Powerfuels · Power-to-kerosene

1 Introduction

As the global energy landscape evolves in response to anthropogenic climate change, there is an increasing focus on using sustainable alternatives to conventional fossil-based fuels. Yet, in sectors where end-use applications will continue to depend on liquid hydrocarbon fuels in the foreseeable future (e.g., in aviation and maritime), there will be a transition period during which synthetic (renewable) and fossil fuels will be consumed in parallel. During such a transition and in sectors where a standardized fuel specification is in use globally, conventional (fossil) fuels and synthetic (renewable) fuels will be distributed in the same infrastructure to simplify logistics. Prior to that blending may be necessary if required by the fuel specification. For

A. Zschocke (✉)
CENA-Hessen, Frankfurt, Germany
e-mail: alexander.zschocke@cena-hessen.de

G. Quante · N. Bullerdiek
Hamburg University of Technology (TUHH), Institute of Environmental Technology and Energy Economics (IUE), Hamburg, Germany

J. Pechstein
Deutsche Lufthansa AG, Frankfurt am Main, Germany

© The Author(s), under exclusive license to Springer Nature Switzerland AG 2025   791
N. Bullerdiek et al. (eds.), *Powerfuels*, Green Energy and Technology,
https://doi.org/10.1007/978-3-031-62411-7_27

this purpose, a thorough understanding of the implications of blending hydrocarbon fuels in general and powerfuels in particular is required. Thus, in this article, various aspects related to the blending of renewable and conventional fuels are assessed using the example of synthetic kerosene in aviation. The overarching intention is to highlight different fuel properties critical for blending and explain how these properties influence permissible blending ratios of conventional and synthetic fuel components.

Therefore, first an overview of specifications for both conventional (fossil) and synthetic kerosene is provided, along with aspects of blending these kerosene types. Subsequently, for conventional kerosene and synthetic kerosene of relevant powerfuel conversion pathways, characteristic physical and chemical properties are described, and their variations discussed. Based hereupon, conclusions toward maximum permissible blending shares are drawn from a present-day perspective. The chapter concludes with an outlook on potential future developments.

2   Kerosene Specifications and Blending

In this section, a concise overview of aviation (turbine) fuel specifications for both conventional and synthetic kerosene is presented, and overall aspects on blending of both options in light of existing specifications are outlined.

2.1   Kerosene Specifications

Aviation fuel characteristics are determined by fuel specifications that primarily define performance, material, and manufacturing properties. Given that the air transport system is an international market, some aviation fuel standards apply virtually globally, with minimal regional differences in fuel specifications (this uniformity is also the case in other sectors, such as maritime transport). The predominant kerosene specifications in the Western world are the Jet A/A-1 specification administrated by the organization ASTM. Based on these prevailing aviation fuels, kerosene can be categorized into conventional and synthetic types. The corresponding fuel specifications, or specific aspects thereof, are described in greater detail below.

Conventional kerosene (ASTM D1655 specification). For conventional kerosene, the central specification is known as ASTM D1655. The parameters and limits specified in ASTM D1655 are primarily based on historical reasons, stemming from experiences with kerosene derived from fossil crude oil, which has been the predominant feedstock. The properties of conventional kerosene vary significantly, depending on crude oil feedstock and refinery operation [1]. Even with the rigorous standardization of aviation fuels, ASTM D1655 does not prescribe an exact chemical composition for the fuel, such as specifying certain hydrocarbons. Rather, it sets out minimum and maximum requirements for performance, material, and manufacturing properties

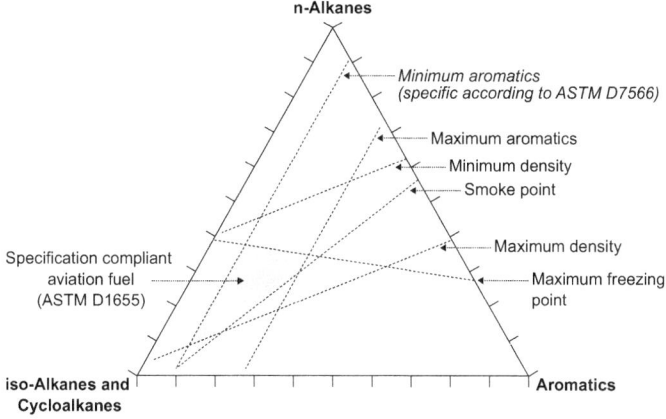

Fig. 1 Relationship between aviation fuel composition and the synthetic kerosene fuel specification domain. **Ar**: aromatics; **BC**: branched alkanes and cycloalkanes, **n**: n-Alkanes; based on [4]

[2]. This flexibility allows for a range of hydrocarbon compositions in the fuel, for example, with regard to variations in contents of n- and iso-alkanes, cycloalkanes, and aromatics (owing to the fact that the properties of a kerosene fraction can differ, depending on the particular crude oil processed and the refining technology). Despite these permissible variations, the fuel properties must remain within the specified compliant range of performance, material, and manufacturing properties, such as fuel density, freezing point, smoke point, or aromatic content (Fig. 1).

Synthetic Kerosene (ASTM D7566 specification). For synthetic kerosene, defined from a specification perspective as aviation fuel not produced from crude oil, ASTM is also the leading organization. The primary ASTM specification for synthetic kerosene is ASTM D7566 with its respective annexes [3].[1] Each annex specifies the production pathway and the required properties of a particular specific synthetic kerosene option. So far different respective production pathways have been approved by ASTM and are included in the ASTM D7566 specification. These are presented in the following Table 1, together with two co-processing pathways (specified in ASTM D1655), which involves the joint refining of certain feedstocks or intermediates alongside with conventional crude oil in refineries.

The first ASTM D7566 pathway for synthetic kerosene was already approved in 2009, based on the Fischer–Tropsch technology. In the subsequent years since then, various other pathways have been added. The maximum blending limit for all these options is currently at 50 vol.-%, with some pathways having significantly lower

---

[1] ASTM specifications are available at the ASTM Web site (www.astm.org/products-services/sta ndards-and-publications.html). Except where stated otherwise, references to ASTM D7566 in this paper refer to ASTM D7566-23a, which is the version in force at the time of writing. However, this specification is frequently updated, and at the time of printing, a more recent version may well be in force.

Table 1 Approved production pathways for synthetic kerosene

ASTM	Annex	Year of approval[a]	Process	Blending limit[b] (vol.-%)
D7566	1	2009	FT-SPK	50
	2	2011	HEFA-SPK	50
	3	2014	HFS-SIP	10
	4	2015	FT-SPK/A	50
	5	2016	AtJ-SPK	50
	6	2020	CH-SK	50
	7	2020	HC-HEFA-SPK	10
	8	2023	AtJ-SKA	50
D1655	–	2018	Co-Processing of biogenic lipids	5[c]
	–	2020	Co-Processing of FT-crude	5

At the time of writing; **A**: aromatics; **AtJ**: Alcohol-to-Jet; **CH**: Catalytic Hydrothermolysis; **FT**: Fischer–Tropsch; **HC**: Hydrocarbons; **HEFA**: Hydroprocessed Esters and Fatty Acids; **HFS**: Hydroprocessed Fermented Sugars; **SIP**: Synthetic Isoparaffins; **SK**: Synthetic Kerosene; **SKA**: Synthetic Kerosene with Aromatics; **SPK**: Synthetic Paraffinic Kerosene
[a] This refers to the initial approval. After the initial approval, approved processes were partially adapted, for example, by increasing the maximum blending limit or by expanding the raw materials that can be used
[b] Precautionary maximum blending limit of the synthetic kerosene component
[c] An increase of the blend ratio to 30 vol.-% is currently prepared for approval at ASTM. This increase will require that the co-processing includes a hydrocracking step

limits in the range of 5 to 10 vol.-%. The subsequent section describes these aspects and the corresponding blending procedure in more detail.

2.2 Blending of Synthetic Kerosene

In the following, first, a general overview of blending is provided. Thereafter, the blending of synthetic and conventional kerosene is described in the context of the previously presented specifications.

Blending in General. Blending, as used in oil industry parlance, generally refers to the intentional mixing of two or more substances in a controlled environment with the intention of creating a blend meeting a defined specification. It is different from incidental commingling of fluids meeting the same specification in an uncontrolled environment, which routinely happens when crude oils or products come together in pipelines or common tank storage:

- Blending of fluids. Blending may involve two different batches of a product meeting the same specification, as for example, if one parameter in one of the batches is marginal and needs to be improved by blending. But more typically, it involves blending of two or more different products. A well-known example is the blending of road gasoline with a certain percentage of bio-ethanol, where both the gasoline and the ethanol have to meet individual, but different, specifications.

• Commingling of fluids. Comingling refers to the random mixing of fuel batches of the same specification. In particular commodities (e.g., aviation or maritime fuels) are subject to comingling, since they are globally traded and thus undergo bulk transportation and storage. Especially the use of pipeline systems or large storage tanks cause comingling, which in turn can result in an averaging tendency of fuel properties. As in case of aviation only certified fuel is fed into the fuel distribution systems, commingling is common and considered uncritical. Therefore, it will not be addressed in this chapter in greater detail.

In blending, a complete mix of the blended fluids is aimed for, unlike in incidental commingling where mixing may be limited to boundary zones between two batches. Large-scale blending therefore typically involves the use of dedicated mixing (e.g., stirring) equipment. This, however, is not a legal requirement, and small-scale blending of synthetic and conventional kerosene in early trials has involved more improvised methods, like repeatedly pumping around in a single tank or pumping between several tanks.

Blending of Conventional and Synthetic Kerosene. From an operational perspective, synthetic kerosene options are primarily designed to serve as "drop-in" fuels. This designation necessitates their seamless integration into the existing aviation fuel supply infrastructure and compatibility with in-service aircraft types alongside other specification-compliant aviation fuels. Consequently, to achieve this, they must adhere to the general jet fuel requirements of ASTM D1655 (plus a few specific additions). Thus, they are not required to have a specific chemical composition (apart from saturated or unsaturated hydrocarbons), but similar to ASTM D1655, they need to meet a certain prescribed set of performance, material, and manufacturing properties, as illustrated in Fig. 1.

However, despite a multitude of nearly identical properties, (many of) the neat synthetic kerosene options produced in compliance with ASTM D7566 differ in some respects from typical ASTM D1655 Jet A/A-1 aviation fuel, for example, in terms of their aromatics content. Therefore, to use the synthetic kerosene as aviation fuel and ensure adherence to the properties defined by ASTM D1655, so far an additional process step, namely blending, is required. In this blending process, the synthetic kerosene (synthetic blending component) is mixed with conventional hydrocarbons or (conventional) ASTM D1655 aviation fuel. Therefore, ASTM D7566 not only contains the specification on synthetic kerosene in its respective annexes, but also the required properties of a blend between a conventional ASTM D1655 kerosene component and a synthetic blending component ("blend specifications"). This blending procedure and interplay of the aviation fuel specifications mentioned is schematically shown in Fig. 2.

The ASTM D7566 specification defines a precautionary maximum blending ratio (as shown in Table 1) and defines properties for both the neat synthetic component as well as the blended fuel ("blending requirements"). These blend specifications

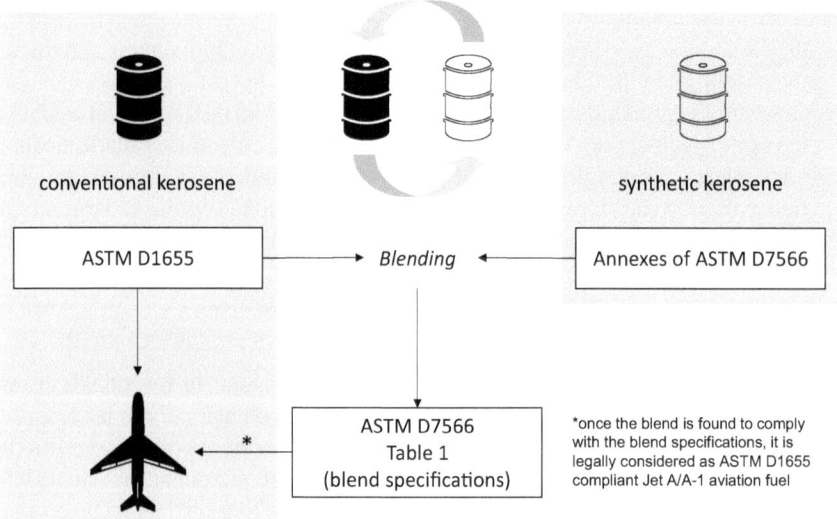

Fig. 2 Specification and certification of conventional and synthetic kerosene

cover the requirements of ASTM D1655 and some additional ones. The practically achievable blend ratio largely depends on the properties of the conventional blending component. Thus, a conventional blend component which is compliant with ASTM D1655 is not necessarily compliant with the ASTM D7566 blend specifications. If its properties are unfavorable (e.g., aromatic content), the allowable blending ratio must remain below the precautionary maximum. On the other hand, with particularly favorable conventional kerosene batches even higher blending ratios than currently approved (Table 1) would be achievable, if ASTM was to abolish the precautionary requirement. This is the reason, why also the properties of the conventional blending component need to be taken into account for fuel blending (Sect. 3).

3   Conventional Kerosene

The classical petroleum fuels—gasoline, kerosene, and diesel—are conventionally produced by fractionated distillation of crude oil. Basically they are defined by a certain boiling range. In case of kerosene, this is approximately 170–270 °C [1]. The properties of conventional kerosene vary to a great extent. Hence, they also influence permissible blend ratio significantly. The most important properties directly result from the chemical composition, which, however, is usually not analyzed itself. Therefore, in this chapter, a brief outline of the main constituents in kerosene with regard

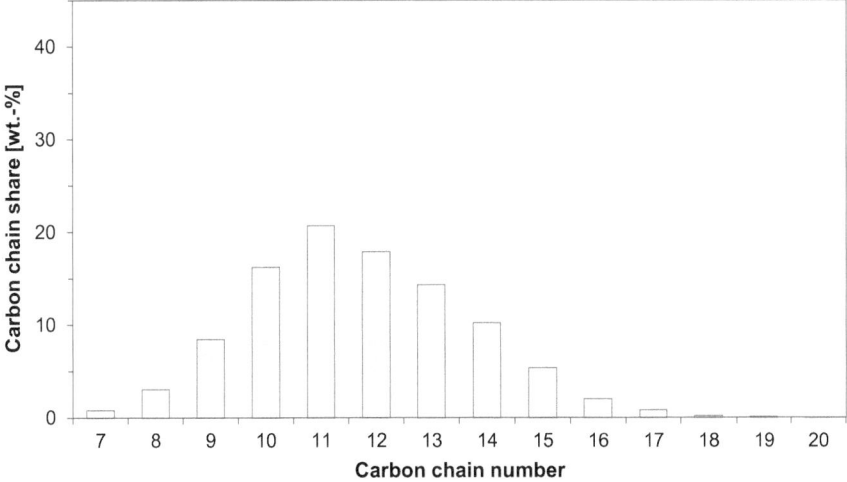

Fig. 3 Carbon chain distribution of a sample Jet A kerosene from fossil-based crude oil[2]

to the chemical composition in terms of hydrocarbon components, the aromatics content, the density as well as distillation curve characteristics is given.

In order to illustrate the variety of properties of conventional kerosene with regard to aromatics content, the density as well as distillation curve characteristics, a dataset of a study on German kerosene properties is used in the respective sections [5]. Other studies are available, e.g., the World Fuel Sampling Program and the Petroleum Quality Information Service (PQIS) in the UK [6–8]. However, to ensure comparability, the following section focusses on the dataset of the study on German kerosene properties only. It contains data from ca. 2,000 fuel certificates representing the fuel used at 14 major airports in Germany.

3.1 Chemical Composition

Kerosene derived from conventional sources typically consists of thousands of different molecules that can be categorized into a few main types, mainly hydrocarbons. Their carbon numbers are normally distributed as depicted in Fig. 3.

By far most molecules of conventional kerosene have a carbon number between C_8 and C_{16} [9]. However, compounds up to C_{19} may be detected [10]. The precise composition varies due to factors such as the type of crude oil used and the refining process. Overall, it is a complex mixture of various hydrocarbons. The main hydrocarbon species of aviation fuel include alkanes (paraffins) in the form of n-alkanes and iso-alkanes, cycloalkanes, and aromatics—as well as alkenes to very minor amounts:

---

[2] Data are for reference conventional kerosene in LanzaTech research report, Table A 3.1c.

- **Alkanes (paraffins).** Alkanes, or paraffins, are straight or branched chain molecules. The straight/linear alkanes are commonly referred to as n-alkanes while the branched alkanes are referred to as iso-alkanes. As a general rule, the freezing point of iso-alkanes is lower than their corresponding n-alkane with the same carbon number. Alkanes are the major constituent in conventional aviation fuel. Among all hydrocarbons, they offer the largest hydrogen-to-carbon ratio. Regarding the hydrocarbons found in aviation fuel, they burn cleanest and offer the highest net heat of combustion at the lowest density [7, 11]. Their composition conforms to Eq. (1):

$$C_n H_{2n+2} \tag{1}$$

- **Cycloalkanes.** Cycloalkanes are saturated ring molecules with a slightly lower hydrogen-to-carbon ratio than alkanes. This results in a reduced heating value and an increased density in comparison with alkanes [9]. Their composition conforms to Eq. (2):

$$C_n H_{2n} \tag{2}$$

- **Aromatics.** Aromatics also are ring-shaped molecules but contain carbon double binds. They have the lowest hydrogen-to-carbon ratio resulting in the lowest heating value and the greatest density. When burned, aromatics act as soot precursors [6, 11–13]. Short hydrocarbon chains may be attached to the ring structure [1, 9]. The chemical composition of aromatics is described by Eq. (3):

$$C_n H_{2n-6} \tag{3}$$

- **Alkenes (olefins).** Alkenes, or olefins, are unsaturated alkanes, i.e., at some point of the hydrocarbon chain is a double bound instead of a hydrogen atom. They are not chemically stable and therefore undesirable, and hence, their presence in kerosene is limited to minor amounts only. By far most of them originate from cracking processes within the refinery [3, 7]. Their composition conforms to Eq. (4):

$$C_nH_{2n} \qquad (4)$$

Saturated hydrocarbons (alkanes and cycloalkanes) account for approx. 70 to 90 vol.-% in aviation fuel [14]. Of these, cycloalkanes amount to approximately 30 vol.-% [7, 15]. When presuming that aromatics amount to ca. 18 vol.-% [5, 16] and alkenes do not exceed 1 vol.-% [3], a typical conventional aviation fuel may be approximated as shown in Fig. 4.

Besides those components described above, conventional aviation fuel also contains non-hydrocarbons. From this group, sulfur (sulfur compounds) is the most important representative. It is corrosive, and its presence is limited by specification to a maximum of 0.3 wt.-%. Usually 99% of the sulfur is bound in thiophenes [1, 7] belonging to the group of heteroorganic compounds. These are molecules containing sulfur, nitrogen, or oxygen. Even in very small amounts, they improve the lubricity

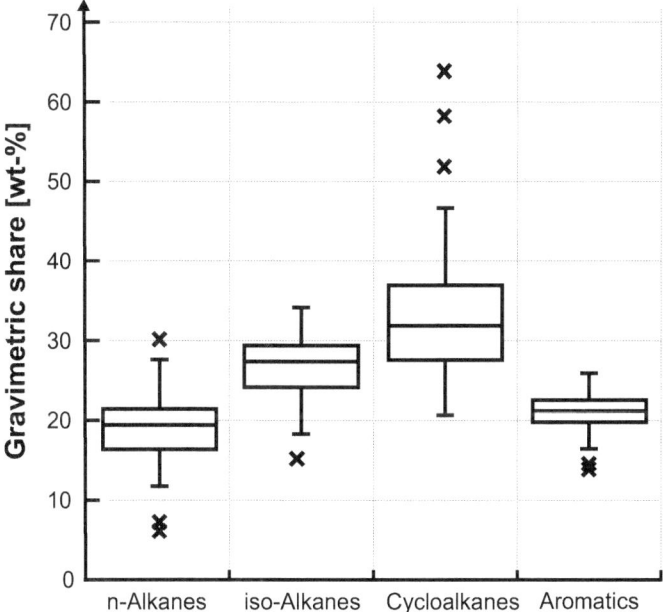

Fig. 4 World Fuel Survey distribution of hydrocarbon types in aviation fuel [7]

and inhibit oxidation. But they tend to form deposits and degrade the ability of water separation (due to polarity).

Furthermore, naphthenic acids may be present in conventional aviation fuel as their boiling range is similar to kerosene. They are aggressive and therefore undesired. When reacting with certain metals, they tend to form soaps. The described non-hydrocarbons can be removed by caustic washes or hydroprocessing in combination with clay filtering [6, 7]. The share of these compounds in conventional aviation fuel is variable. It may reflect the characteristics of the crude oil and is strongly influenced by the refinery technology used for kerosene production [3, 7]. As a rule of thumb, straight-run kerosene contains the highest amount of non-hydrocarbons. With increasing hydroprocessing severity, their number is successively reduced.

3.2 Aromatics Content

The specification limits for aromatics content of conventional kerosene are maximum 25 vol.-% if measured following ASTM D1319 and 26.5 vol.-% if measured in accordance with ASTM D6379. For the final aviation fuel blend, ASTM D7566 also requires a minimum aromatics content of 8 vol.-%. An exemplary aromatic content of aviation fuel batches when leaving the refinery is shown in the following Fig. 5, based on data from the mentioned study at 14 major airports in Germany [5].

The minimum aromatics content in the dataset of the study was 5.9 vol.-% and the highest 25.5 vol.-%. As shown in Fig. 5, the modal values range from 15 to 19 vol.-%. This covers roughly 60% of the corresponding samples in the dataset, while aviation

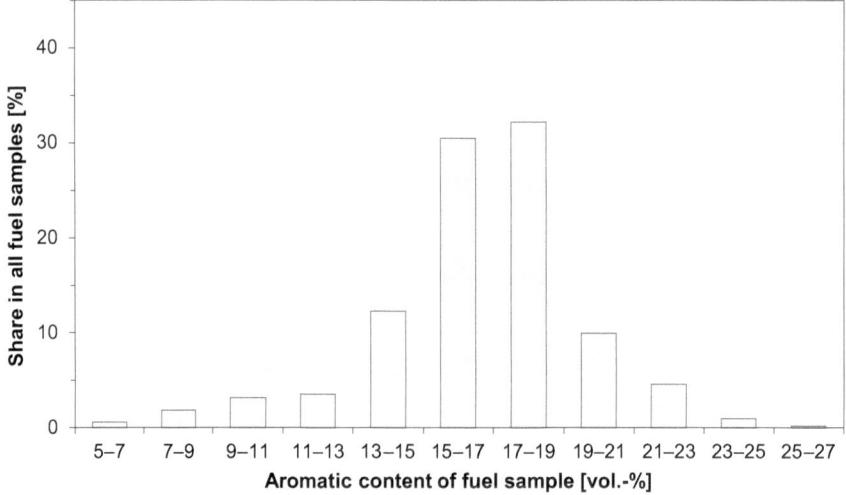

Fig. 5 Aromatic content of aviation fuel batches when leaving the refinery [5]

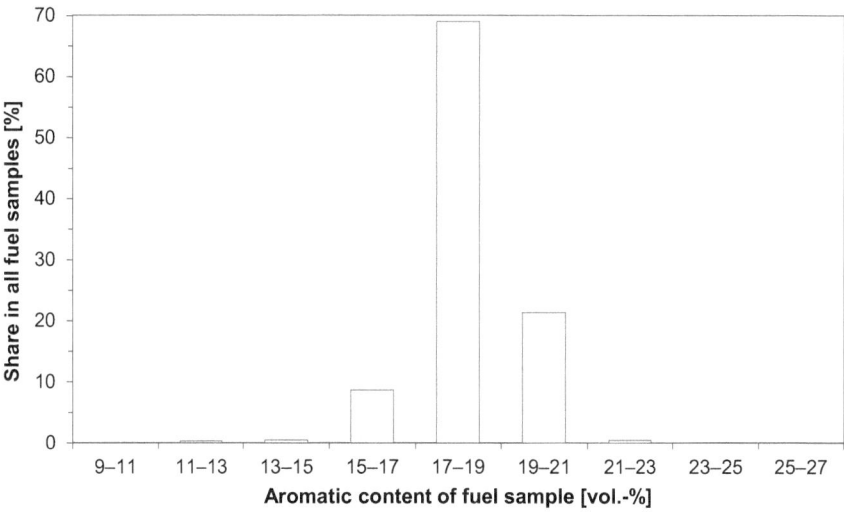

Fig. 6 Aromatic content of aviation fuel batches from tank farms and pipelines [5]

fuel batches with lower and higher aromatics contents account for less than 40%. Note that those values were obtained by ASTM D1319 and ASTM D6379. Both methods give similar results, but ASTM D6379 results in somewhat higher values.

When comparing the aromatic content of all samples with batches from tank farms and pipelines (Fig. 6), a clear averaging tendency can be observed. The comingling of fuel batches in tanks and pipelines yields fuel properties closer to the average values, thus reducing extreme values.

Bearing in mind the current requirement of a minimum aromatics content of 8 vol.-% for synthetic kerosene blends, it is noteworthy that 15 of the German batches had an aromatics content lower than 8 vol.-%, as conventional ASTM D1655 kerosene is not required to meet the 8 vol.-% minimum. These batches were all produced by one refinery over a course of two months and were delivered directly from the refinery to two airports, where this refinery was in one case the only and in the other case the major supplier. No adverse issues relating to the low aromatics content were observed at these airports while this low aromatics fuel was delivered and used.

3.3 Density

ASTM D1655 defines a density range from 775 to 840 kg/m^3. The minimum density found in the German dataset was 786.9 kg/m^3, and the maximum was 834.2 kg/m^3. As shown in Fig. 7, the majority of the samples falls in the range between 795 and 805 kg/m^3. These account for approximately 67% of the samples. In less than 10% of the fuel batches sampled, densities were discovered to be lower than 790 kg/

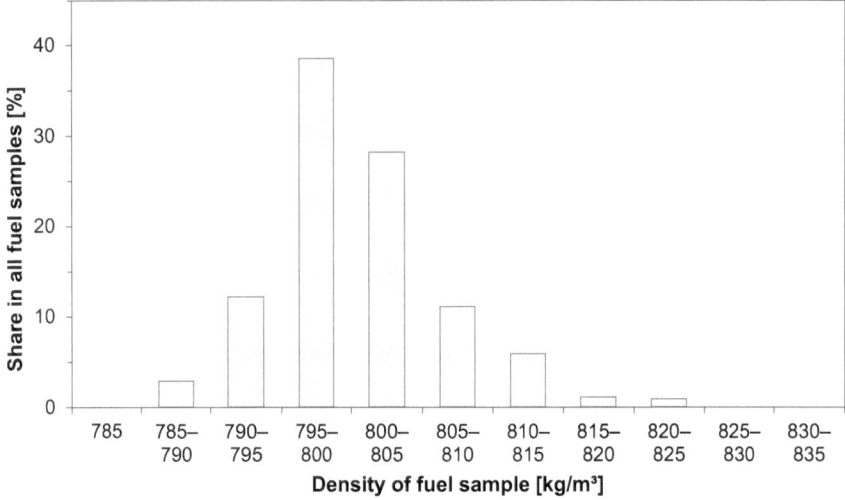

Fig. 7 Distribution of density in conventional kerosene in Germany [5]

m³ or higher than 815 kg/m³. This indicates that such extreme density values were relatively rare occurrences within the sampled batches.

All values are well within the specification range. ASTM D7566 uses the same density range for the blend as ASTM D1655 requires for the conventional kerosene component, and hence, the density distribution of the conventional blend component seems generally well suited for blending.

3.4 Distillation Curve

The variety of hydrocarbons (in terms of carbon chain lengths and molecular structure) of conventional kerosene (Fig. 3) results in the phenomenon that during distillation certain components evaporate earlier than others. Thus, a part of components in conventional kerosene evaporates at lower temperatures (e.g., 150–180 °C) while others need substantially higher temperatures for evaporation (ca. 240–260 °C). As per ASTM D1655, one temperature for the recovery of 10 vol.-% (205 °C) is defined to control the fuel's volatility and ease of evaporation. This temperature is referred to as "T10" and is important for engine start performance. In turn, the final boiling point (maximum 300 °C) prevents undesired heavy fractions (alkanes beyond C16, poly-aromatics) from being incorporated in the fuel [3]. Figure 8 illustrates that the distillation slopes of the conventional aviation fuel in Germany are very similar [17]. A share of 90% of the samples remain within a spread of less than 36 °C.

In order to replicate this behavior, ASTM D7566 requires the synthetic kerosene (blend) to show a similar behavior [3]. In addition to the maximum limits for T10

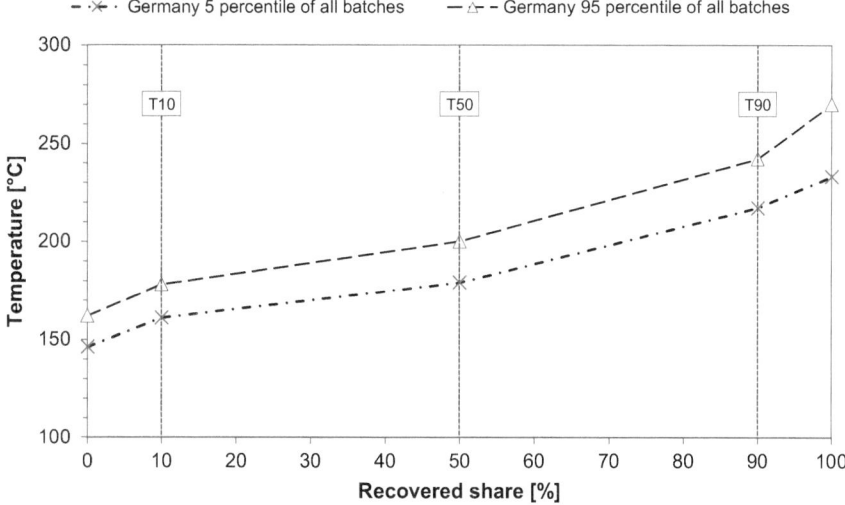

Fig. 8 Distillation curve gradient range for conventional kerosene in Germany [5]

and the final boiling point, a minimum temperature rise of 15 °C is required between the 10% recovery temperature and the 50% recovery temperature (T50-T10). From 50% recovery to 90% recovery (T90–T50), an increase in temperature of at least 40 °C is mandatory.

ASTM D1655 (conventional kerosene) does not specify a certain minimum gradient. Thus—as for aromatics—there may be conventional batches that do not conform with ASTM D7566 (blend specification). Actual requirements toward distillation gradients are under investigation [3].

4 Synthetic (Powerfuel) Kerosene

None of the ASTM D7566 annexes are specific to synthetic kerosene that can be produced by power-to-liquid (PtL) production pathways. Rather, the production pathways specified in the annexes typically start from an intermediate product and describe the conversion of the intermediate product to a kerosene blending component. For example, the alcohol-to-jet (AtJ) SPK pathway describes the conversion of either ethanol or iso-butanol to a synthetic alcohol-to-jet (AtJ) SPK blending component, regardless of how the alcohol itself has initially been produced. Such alcohol can possibly be the result of a power-to-liquid (PtL) process, but can equally result from biomass conversions or have been produced by using industrial off-gases. As long as the conversion pathway of the specific ASTM D7566 annex is followed, the resulting fuel will be a specification compliant, valid kerosene blending component, both legally and chemically.

There are basically three production pathways that start with an intermediate product, which can potentially be produced based on power-to-liquid (PtL) pathways, namely Fischer–Tropsch (FT), ethanol-to-jet (EtJ)[3], and AtJ-SKA. The approval process for methanol-to-jet (MtJ) is ongoing at the time of writing [18, 19]. The other pathways listed in Table 1 start from intermediate products based on vegetable oils, animal fats, or sugar, which currently cannot be replaced by power-to-liquid (PtL). A discussion of power-to-liquid (PtL) kerosene blending can therefore be limited to blending of these three pathways.

Within the alcohol-to-jet (AtJ) pathway, so far, ethanol and iso-butanol are the only applicable synthetic alcohol-to-jet (AtJ) kerosene feedstock options. However, the approval of methanol as an alcohol feedstock for the kerosene production is ongoing at the time of writing. Since a corresponding approval for the methanol-to-jet (MtJ) pathway is expected in the next years [18, 19] and since kerosene production based on methanol is already considered to play an important role in future, the methanol-to-jet (MtJ) pathway is also discussed here.

With regard to aviation fuels, most of the specification requirements only indirectly refer to the fuel composition (except for specific aspects like acidity, aromatics content, mercaptan sulfur content and total sulfur content) [20, 21]. Therefore, the connection between fuel composition and its' specification compliance is complex, and conclusions about specification compliance based on fuel composition are very challenging. A more pragmatic approach is to study kerosene blends empirically and derive critical properties for blending by testing the resulting blends for specification compliance. This has been done in the past, e.g., in the High Biofuel Blends in Aviation (HBBA) study [5]. This study examined a broad range of synthetic types, all certified by ASTM D7566. Fuel density, aromatics content, and the distillation curve gradient were found to be critical characteristics for some aviation fuel blends, partially resulting in permissible blending ratios well below the precautionary blending limit. In the following section, different powerfuel production pathways, namely the Fischer–Tropsch (FT) pathway and alcohol-to-jet (AtJ) pathway are described, the resulting fuel properties are discussed, and conclusions regarding permissible blending ratios and overall blending characteristics are drawn.

4.1 Fischer–Tropsch Kerosene

The Fischer–Tropsch (FT) pathway was the first alternative kerosene blending component pathway to be approved and is covered in Annex 1 of ASTM D7566. The approval process for this pathway was chiefly promoted by Sasol, with a focus

---

[3] ASTM D7566 refers to this process as alcohol-to-jet (AtJ) SPK. To allow for a more transparent distinction between the feedstock ethanol and methanol, the expressions ethanol-to-jet (EtJ) and methanol-to-jet (MtJ) are used. AtJ SPK may also be produced from iso-butanol or isobutene, but these feedstocks have so far not been discussed in connection with PtL.

on the FT conversion of coal to kerosene. However, the approved pathway is feed-stock independent, and any feedstock that can be converted into FT-synthesis gas (syngas) is suitable in principle. In this case, the syngas primarily consists of a mixture of hydrogen and carbon monoxide in a ratio of about 2 to 1. In addition to coal, feedstocks for the production of FT syngas can be, for example, natural gas, biomass, and municipal waste but also hydrogen and carbon dioxide that represent the two core components for power-to-liquid (PtL) processes.

Work on approval of the Fischer–Tropsch (FT) pathway took some ten years, from 1999 to 2009, not counting prior work by Sasol on an approval specific to their fuel [22]. This was partly due to this being the first ASTM approval of a production pathway for an alternative, synthetic kerosene blending component, but also because there is much potential flexibility in the design of the production process for FT liquids.

4.1.1 Chemical Composition

The direct result of the Fischer–Tropsch (FT) process is not a usable product, but a so-called syncrude consisting of hydrocarbon chains ranging from C_1 to well over C_{30}, i.e., from gases to waxes [2]. The distribution of the chain lengths of the syncrude is typically lognormal, but as Fig. 9 illustrates, both the range and the shape of the ultimately resulting kerosene chain lengths distribution vary depending on the kind of FT reaction (e.g., type of catalyst, type of reactor, and operating parameters). This FT syncrude must then be subject to a refining process to be converted into usable products, and again, there is scope to vary this refining process. As a result, there are potentially many ways to shape the kerosene resulting from the FT process, and much effort was spent to determine what was acceptable, and what was not.

In the end, it was found that a blend of conventional kerosene and FT kerosene meeting the requirements of the standard ASTM D1655 specification can be used for aviation purposes, except in those cases where the FT kerosene is extremely concentrated at a few carbon chain lengths, leading to a very spiky distribution of the blend. To control this factor, minimum limits were introduced for the difference between certain points of the distillation curve.

4.1.2 Aromatics Content

The Fischer–Tropsch (FT) pathway does not comprise an aromatization step, and hence, the resulting fuel does not contain aromatic components. ASTM D7566 Annex 1 is therefore known as FT synthetic paraffinic kerosene (SPK)—or FT-SPK—pathway annex, as the hydrocarbons in the fuel are all paraffins (alkanes). Aromatics burn worse than paraffinic components and hence increase undesirable soot emissions. However, they also cause the seals and valves in an aircraft's fuel system to swell and thus to remain tight. Aircraft trials with neat FT kerosene showed that fuel with no aromatics increases the risk of fuel leaks. Consequently, blends with

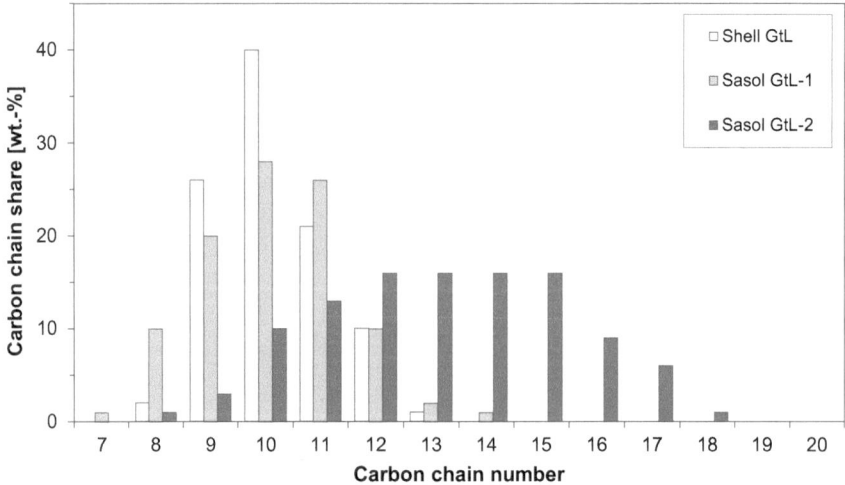

Fig. 9 Carbon chain distributions for various Fischer–Tropsch synthetic kerosene (**GtL:** Gas-to-liquid; based on [23])

alternative fuel components are required to have a minimum aromatics content of 8 vol.-%. This limit is another constraint, which may only permit blend ratios rather below 50 vol.-%.

4.1.3 Density

The practical aspects of blending conventional and Fischer–Tropsch (FT) kerosene were investigated in the HBBA study[5], which particularly looked at limiting factors. Although the HBBA study was about blending limits for biokerosene, the FT blending component used was provided by Sasol and produced from coal, as no biomass-based FT material was available. The lack of aromatics in Fischer–Tropsch (FT) SPK results in a lower density than conventional kerosene, in the case of the sample used in the HBBA study 761.2 kg/m^3. This is because aromatics are heavier than paraffins (alkanes), and their absence results in a fuel of a lower density. Density is therefore another constraining factor. However, the HBBA study found this constraint would only become relevant for FT kerosene at blend ratios beyond 50 vol.-%.

4.1.4 Distillation Curve

For distillation curve gradients, the most limiting factor identified in the HBBA study was T50-T10. The T50-T10 of the FT blendstock was only 8 °C. As a result, the T50-T10 of the conventional kerosene used for blending was a crucial factor for permissible blend ratios. For conventional kerosene itself having a very flat distillation curve, the HBBA study found that adding even a few percent of the FT blendstock resulted in the blend not meeting the specification requirements. Conventional kerosene with a steep distillation curve, on the other hand, permitted blending up to 50 vol.-%.

It needs to be emphasized, however, that to a certain extent these results only apply to the specific sample analyzed in the HBBA study. The sample was in line with other samples described in literature, but as discussed above there is a considerable variation for FT blendstock as allowed by the corresponding ASTM Annex, which potentially can be used to ease blending constraints.

4.2   Alcohol-to-Jet Kerosene

The alcohol-to-jet (AtJ) SPK pathway was first approved in 2016 and reflected in Annex 5 of ASTM D7566 [3]. This annex initially only covered synthetic kerosene produced from iso-butanol, a C_4 alcohol, and was produced from biomass. Later on, ethanol was also certified as potential alcoholic feedstock, and in 2023, iso-butene was approved as another feedstock. The ASTM certification of methanol as feedstock is currently ongoing. Ethanol and methanol can both be provided based on synthesis gas and thus represent potential powerfuel intermediate feedstock. In the case of ethanol, this would be possible via synthesis gas fermentation, while in the case of methanol, it is possible via thermochemical conversion.

4.2.1   Chemical Composition

Ethanol-based Alcohol-to-Jet (AtJ) Kerosene. In 2018, annex 5 was expanded to also permit alcohol-to-jet (AtJ) from ethanol, based on a production pathway developed by LanzaTech [24]. With the inclusion of ethanol as feedstock, the maximum blending limit in annex 5 was increased from 30 to 50 vol.-%, for both alcohol-to-jet (AtJ) from ethanol and from iso-butanol [3]. Ethanol is a C_2 alcohol, and the resulting carbon chain distribution is comparatively close to the one of conventional kerosene (Fig. 10). The distribution ranges from C_8 to C_{17} which is essentially the same as for conventional kerosene. In contrast to conventional kerosene, the carbon chain distribution of alcohol-to-jet (AtJ) SPK shows two distinct modal values at C_{10} and C_{12}. As expected, for alcohol-to-jet (AtJ) SKA from mixed alcohols only one modal values is recognizable, due to the range of alcohols used feedstock for the oligomerization.

Alcohol-to-jet (AtJ) SPK from ethanol was only certified after the HBBA study was finished and hence was not included in the study. Apparently, no other research on the blending of alcohol-to-jet (AtJ) SPK from ethanol has been performed, and hence, any statement on limiting factors needs to be based on the known properties of

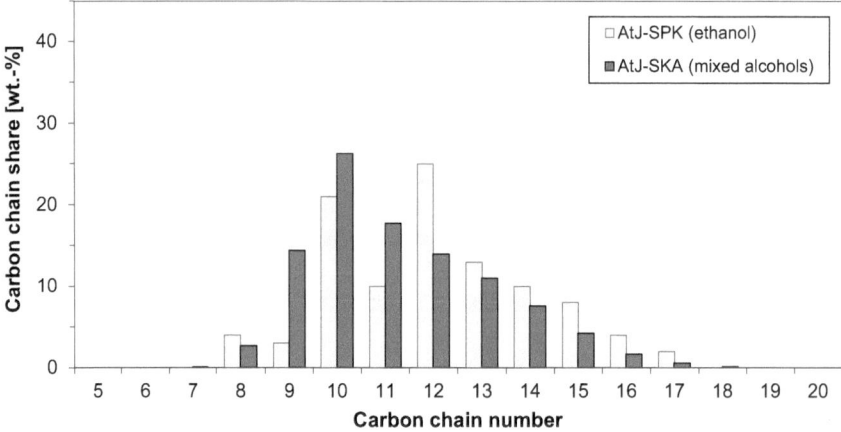

Fig. 10 Carbon chain distribution of alcohol-to-jet (AtJ) SPK from ethanol and alcohol-to-jet (AtJ) SKA from mixed alcohols (**SKA**: synthetic kerosene with aromatics; **SPK**: synthetic paraffinic kerosene; based on [24])

the fuel rather than any experimental verification. Aromatics content and fuel density are likely to play a similar role to what is the case for Fischer–Tropsch (FT) SPK (i.e., relatively low density of synthetic blending component and high aromatics content required in the conventional blending component). The reason is that both fuels are SPKs containing no aromatics. The boiling curve of alcohol-to-jet (AtJ) SPK from ethanol is essentially linear and sufficiently steep, which gives the fuel a sufficiently large T50-T10 and T90-T10 for these factors not to be a constraint [24].

Methanol-based Alcohol-to-Jet (AtJ) Kerosene. Several players have announced projects to produce alcohol-to-jet (AtJ) SPK (and potentially ATJ-SKA) based on methanol [18, 19], and an approval process is currently ongoing, with a task force established by ASTM in December 2022. Again, the AtJ-SPK process will produce negligible (if any) quantities of aromatics, meaning that again the same implications as for Fischer–Tropsch (FT) SPK can be assumed (i.e., relatively low density and high aromatics content required in the conventional blending component). If AtJ-SPK from methanol shows a similar, linear boiling curve like ethanol, T50-T10 and T90-T10 would most likely be large and probably not a constraint. Other challenges, specific to AtJ-SPK from methanol cannot be ruled out, as at the time of writing no study on the blending behavior of this fuel type is available.

4.2.2 Aromatics Content

By definition, the older synthetic kerosene production pathways do not contain an aromatization step, and the fuels therefore are also synthetic paraffinic kerosenes (SPKs), like Fischer–Tropsch (FT) blendstock. Therefore, both the requirement for a minimum aromatic content of 8 vol.-% and the fuel density are further constraining factors identified by the HBBA study. Still, in most cases these factors only become binding at blend ratios close to 40 vol.-% or above [5].

However, very recently the company Swedish Biofuels has received approval for an AtJ-SKA pathway—where "SKA" stands for synthetic kerosene with aromatics. The Swedish Biofuels kerosene is designed to correspond to conventional kerosene in every respect except burning cleaner, with almost no contaminants. Due to the similarity, the carbon chain distribution of this fuel therefore corresponds to that for conventional kerosene. This process is feedstock flexible, with all alcohols from C_2 to C_5 possible, either as single alcohols or as mixtures. Approval is for a blend ratio of 50 vol.-% at first, with approval as a 100 vol.-% fuel to be pursued as a second step. Blending with conventional kerosene is expected to be easiest for the Swedish Biofuels pathway, as the blending component is designed to correspond to conventional kerosene. As a rule, it should be possible always to blend up to the limit of 50 vol.-% [25].

4.2.3 Density

Like Fischer–Tropsch (FT) SPK, neat AtJ-SPK does not contain aromatics, and thus, its density is lower compared to conventional kerosene (in the HBBA study the density amounted to 757.1 kg/m^3, which is substantially below the specification minimum of 775 kg/m^3). The dependency of density on blend ratio is again linear for blends of conventional and synthetic kerosene, and therefore, the resulting density of the blend can be calculated from initial values of the blend components. For the fuels investigated in the HBBA study, the limits range from ca. 40 to 70 vol.-%. For AtJ-SKA, this will not be a problem.

4.2.4 Distillation Curve

In the HBBA study, T50-T10 and to a lesser extent T90-T50 was identified as a possible obstacle to higher blend ratios of 70 vol.-% or more, as the sample supplied had relatively low T50-T10 and T90-T10 values and did not on its own meet target values for the blend.[4] However, the HBBA study used AtJ-SPK from iso-butanol which has a different carbon chain distribution than AtJ-SPK from ethanol, for instance. Using iso-butanol (C_4) as feedstock results in high modal values for C_{12} and C_{16} molecules, as the oligomerization step primarily generates integer multiples of the feedstock. These modal values are likely to result in a rather flat distillation curve. In contrast, in the kerosene range (C_8–C_{16}) there are more integer multiples for ethanol (C_2). Accordingly, the carbon distribution in this case is most likely rather linear, and thus, the resulting distillation curve gradient could be closer to

---

[4] HBBA study, page 77.

the specification limits. Taking into account Fig. 10, most probably cracking reactions after oligomerization further improved the carbon number distribution in the ATJ-SPK positively influencing the distillation properties. The AtJ-SKA production pathway uses a process that is independent of the chain length distribution of the initial alcohol and therefore can produce a fuel with a carbon chain length distribution corresponding to that of fossil kerosene, independent of the feedstock alcohol [25]. Early versions of this fuel still had unsatisfactory values for T50-T10 [5], but in its current form, the distillation curve is not a limiting factor. The methanol-to-jet (MtJ) pathway involves a conversion into olefins of different chain lengths as an interim step [25], and hence, the resulting product should also be well-distributed [26].

5 Blending Properties

As described in Sect. 2, an unfavorable combination of a conventional and synthetic blending component might yield a fuel blend which does not comply with the ASTM D7566 blend specification or at least the maximum allowable blending ratio may not be achievable. Accordingly, synthetic and conventional fuel batches for blending—and thus also in case of the production of synthetic blending components by power-based or powerfuel production pathways—should be matched carefully in the future aviation fuel supply chain. From a technical point of view, even blending ratios beyond the current precautionary limit could be possible, if two favorable batches were chosen.

According to ASTM D7566, blending is considered manufacture of Jet A or Jet A-1, respectively. In Europe an explicit requirement demands that it shall be conducted upstream of the airport storage [27]. In the USA, such a requirement does not exist. However, in practice, general quality assurance standards can be expected to rule out blending at the airport [21]. Three analyses have to be performed before a fuel blend may be used in commercial aviation:

- an ASTM D1655 analysis of the conventional kerosene before blending,
- an ASTM D7566 analysis of the neat synthetic kerosene before blending,
- an analysis of the blend, which is described in ASTM D7566, but in practice, is an analysis of the ASTM D1655 parameters, plus some additional ones.

An entire ASTM D1655 analysis requires ca. 20 labor hours and the use of specialized, expensive equipment. These incur high cost, usually in the range of thousands of Euros. For the first two analyses, these costs are independent of the blend ratio and will usually be performed for large batches of thousands of tons, yielding comparatively low specific (per ton) costs in the range of a few Euros. This is not the case for the analysis of the blend, since the cost impact per ton of powerfuel is crucially dependent on the blend ratio. For single-digit blend ratios, the cost for the analysis will be incurred for selling only a few tons of powerfuel, leading to very high costs per ton. However, for high blend ratios these costs will not be an issue. Furthermore, using low blend ratios implies that large volumes of conventional kerosene have to be transported to the blending point, inducing complex and expensive logistics, and in particular potentially increase the transport distances and thus related emissions [5].

This is why the maximum allowable blend ratio holds significance for powerfuel producers and blenders, even in the early stages of the market. It will also be relevant for governments and communities involved in the planning of logistics and blending capacities [5] (Table 2).

Regarding the actual blending process, currently ASTM D7566 only permits binary blending. This means, a conventional component (meeting specification ASTM D1655) and an alternative, synthetic kerosene blending component as specified in any individual annex to ASTM D7566 may be blended. The resulting blend can be regarded acceptable as an aviation turbine fuel if it meets the requirements

Table 2 Comparison of properties relevant for the blending of conventional and synthetic kerosene types

Characteristic	Conventional	FT-SPK	AtJ-SPK
Carbon distribution	Normal	Lognormal	Linear
Composition	Alkanes (n- and iso-), cycloalkanes, aromatics, heteroatoms	Alkanes (n- and iso-)	Alkanes (n- and iso-)
Density	Reference	Usually reduced (lack of cyclic hydrocarbons)	Usually reduced (lack of cyclic hydrocarbons)
Aromatics content	Up to 25 vol.-%	Max 0.5 wt.-% (as per ASTM D7566)	Max 0.5 wt.-% (as per ASTM D7566)
Distillation curve	Reference	Strongly depends on feedstock and process conditions	Strongly depends on feedstock and process conditions
Lubricity	There is no general correlation between blend ratio and lubricity. In blends the component with the better lubricity usually positively influences this parameter. Usually straight-run kerosenes have a better lubricity and can be expected to impose less constraints in blending		
Freezing point	As long as both individual blend components meet the specification, also the freezing point of the blend was found to be compliant with the specification		
Smoke point	The smoke point primarily depends on the aromatics content of a fuel. Therefore, blends with aromatic-free synthetic fuels/blend components (FT-SPK, AtJ-SPK) show an improved smoke point		
Flash point	Flash point is mainly influenced by the presence of volatile fuel components. In blends this property strongly depends on the blend component with the lower flash point. For all blends investigated in the HBBA study, flash point was not an issue		

Properties critical for blending are marked **bold;** data based on [5]; **AtJ**: Alcohol-to-Jet; **FT**: Fischer–Tropsch; **SPK**: Synthetic Paraffinic Kerosene defined in the body of ASTM D7566 for a target blend. Such a blend may then again be blended with a blending component defined in an annex to ASTM D7566.[5] However, simultaneous blending of two or more alternative, synthetic kerosene blending components with conventional kerosene—so-called multiblends—is not covered by ASTM D7566 yet and hence not permitted. Against the background of increasing production volumes, a broader variety of synthetic kerosene types and the opportunity to achieve desirable fuel properties by multi-blending, this can become a practical problem in the future. An ASTM task force is working on also permitting the production of multiblends. However, as of now only binary blends are permitted.[6]

5.1 Aromatics Content

In general, a high aromatics content leads to soot and carbon deposit formation during combustion. For this reason, the maximum content of aromatics in kerosene is restricted to 25 vol.-% (or 26.5 vol.-%, depending on test method). Yet, fuels without any aromatics are neither desirable, as aromatics contribute to a swell and mass gain of certain elastomer seals and thus prevent leakage in the aircraft's fuel system [21]. This primarily affects nitrile butadiene rubber, a common elastomer in aviation. When in contact with aromatic-free fuel after having been exposed to highly aromatic fuel beforehand, the seal shrinks significantly [5]. Consequently, a minimum aromatic content of 8 vol.-% (8.4 vol.-%, depending on test method) has been introduced in ASTM D7566. However, this is only applicable to Jet A or Jet A-1 containing synthetic kerosene.

Figure 11 shows the relationship between aromatic content of the conventional kerosene and the resulting maximum share of synthetic kerosene, to remain within the lower limit for the aromatics content in the blend.

The gray area indicates a representative value range for aviation fuels leaving pipeline systems or airport tank farms in Germany. It shows that most likely for the majority of conventional fuel in use, blend ratios of at least 50 vol.-% of synthetic kerosene can be achieved. Still, at batch level, the variation of aromatics content is substantially larger (Fig. 5), and thus for blending individual batches, the permissible blend ratio might be substantially lower.

5.2 Density

Density describes the ratio of a fuel's mass per its volume. According to ASTM D1655 and ASTM D7566, the density of Jet A and Jet A-1 must be between 775 and 840 kg/m^3. Consequently, the density of the fuel and the tank volume of an

---

[5] See Sect. 1.2.2 of ASTM D7566-21 [3].

[6] For a discussion, see DEMO-SPK report, Sect. 6.1.4 and 6.4 [28].

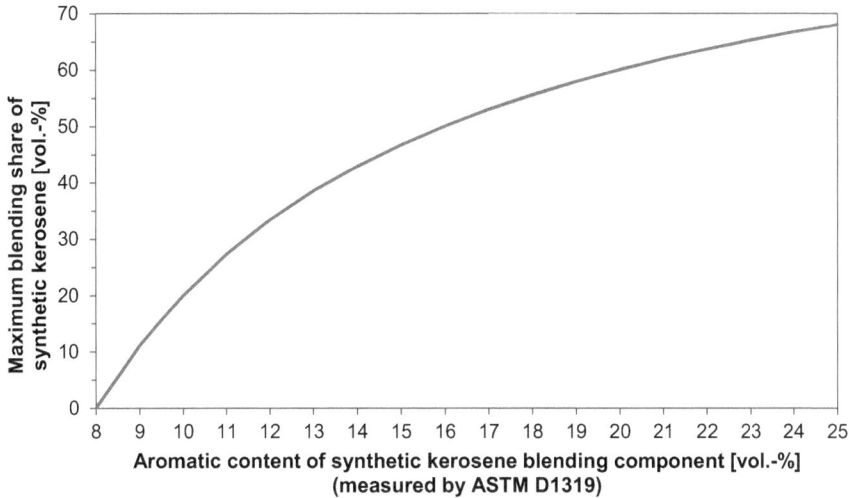

Fig. 11 Maximum share of synthetic kerosene as a function of aromatic content in conventional blending component (gray area indicates a representative value range for fuels leaving pipeline systems or airport tank farms in Germany; adopted from [21])

aircraft defines the maximum fuel mass available on board and hence its maximum range. In general, the lack of aromatics in the discussed aviation powerfuels results in a reduced fuel density. For example, the fuel densities for neat Fischer–Tropsch (FT) SPK and alcohol-to-jet (AtJ) SPK measured by the HBBA study were about 760 kg/m^3.

Figure 12 shows the maximum permissible blend ratio for different densities of the synthetic blend component (*x*-axis) and the conventional blend component (*y*-axis). Exemplarily, the density for both fuel types of the HBBA study is also indicated. Assuming a mean density of the kerosene in Germany of 800 kg/m^3 [5] blend ratios of about 60% seem feasible. However, this will strongly depend upon the individual conventional fuel batch.

There is a notable distinction between density as mass per volume and energy density as energy per mass or per volume. Even though the lack of aromatics reduces a fuel's density, the energy content increases due to the improved combustion behavior of straight-chain hydrocarbons compared with cyclic hydrocarbons. A higher energy density is generally preferable, as the same fuel mass leads to a higher energy stored on board of an aircraft, and consequently, the amount of fuel that an aircraft needs to carry for flight operations can be reduced, resulting in a decrease in the fuel consumption of the aircraft.

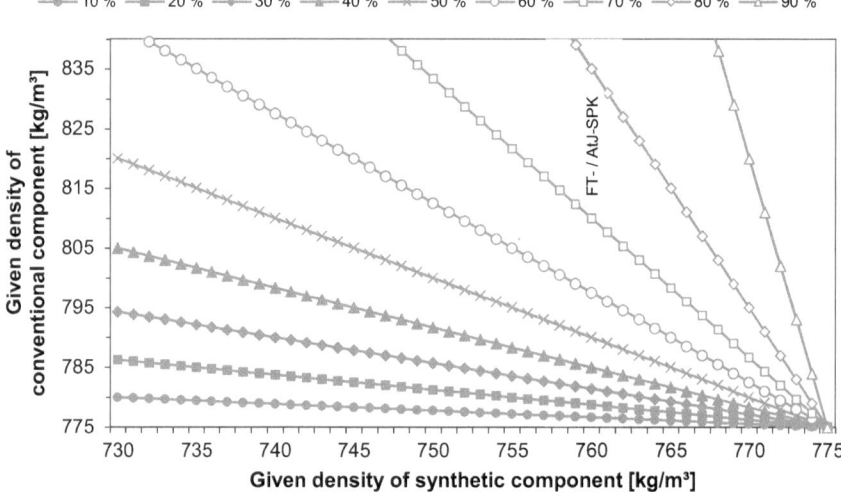

Fig. 12 Maximum share of synthetic kerosene as a function of aromatic content in conventional blending component. The gray bar indicates the density of Fischer–Tropsch (FT) SPK and AtJ-SPK measured by the HBBA study [5] (adopted from [21])

5.3 Distillation Curve

ASTM D7566 states that the volatility and ease of evaporation are controlled by the distillation requirement. The 10% recovery temperature guarantees engine start performance by ensuring that volatile compounds are present [21]. In turn, the final boiling point prevents undesired heavy fractions (alkanes beyond C16, poly-aromatics) from being incorporated [21]. However, the introduction of synthetic kerosene affects the aviation fuel's composition. As a precautionary measure minimum distillation gradients have been specified resembling the typical behavior of conventional aviation fuel. Between the 10% recovery temperature and the 50% recovery temperature (T50 to T10), a minimum temperature rise of 15 °C is required, while from 50% recovery to 90% recovery (T90 to T50) an increase in temperature of at least 40 °C is mandatory.

Figure 13 shows the distribution of distillation curve gradients for German kerosene batches and—as illustrative data—the distillation curve for a single hydro-carbon molecule ($C_{15}H_{32}$) and alcohol-to-jet (AtJ) SPK and alcohol-to-jet (AtJ) SKA as well as Fischer–Tropsch (FT) SPK. For conventional kerosene, ASTM D1655 only requires determining the 10% recovery temperature and the final boiling point. It does not specify a certain minimum gradient. Thus—as for aromatics—there may be conventional batches that do not conform to ASTM D7566. As conventional kerosene contains a broad range of different hydrocarbons with various chain lengths, the distil-lation curves commonly show a distinct slope. This would not be the case if the fuel would consist only of one molecule type. For example, pure farnesane ($C_{15}H_{32}$; gray

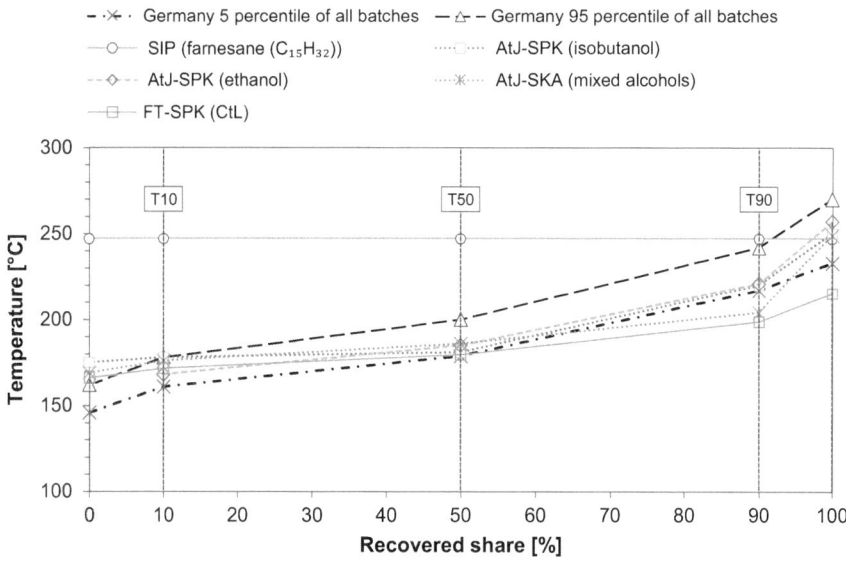

Fig. 13 Distillation curves for conventional kerosene (black lines) and indicative curves for a single hydrocarbon molecule ($C_{15}H_{32}$) and alcohol-to-jet (AtJ) SPK from iso-butanol and ethanol, alcohol-to-jet (AtJ) SKA from mixed alcohols and Fischer–Tropsch (FT)-SPK; (**SKA:** synthetic kerosene with aromatics; **SPK:** synthetic paraffinic kerosene; based on [5, 25]

line with circles) has a boiling point of 274 °C. Thus, it would entirely vaporize at this temperature, and its distillation curve would be a horizontal line (neglecting allowed contaminants in SIP like hexahydrofarnesol).

As it is composed primarily of three different alkanes, the distillation curve of AtJ-SPK from iso-butanol is extraordinarily flat (dashed gray line with boxes). This is particularly pronounced below the 50% recovery point. Special attention must therefore be paid to the conventional blending component. It must have a very steep gradient for compensation [5]. However, AtJ -SPK from ethanol shows a broader range of alkanes, and accordingly, their distillation curve gradient is slightly steeper (dashed gray line). A similar behavior can be observed for alcohol-to-jet (AtJ) SKA, which is produced from mixed alcohols.

A similar situation can be observed for Fischer–Tropsch (FT) SPK. The distillation curve of the sample investigated in the HBBA study was also fairly narrow (T50 to T10 = 8 °C). Consequently, the conventional blending component needs a substantially steeper slope in order to compensate for it [5]. However, this is not necessarily the case for all Fischer–Tropsch (FT) SPK batches, as reaction products are highly dependent on reaction conditions (Sect. 4.1) Thus, for blending Fischer–Tropsch (FT) SPK, the distillation characteristics require special attention; some conventional kerosene might not be suitable for blending at all [5].

6 Conclusion and Outlook

In sectors like aviation, where dependence on liquid hydrocarbon fuels will most likely persist in the foreseeable future, blending conventional and renewable fuels, such as power-to-liquid (PtL) hydrocarbons, is both a mandatory requirement per specification and an opportunity to leverage existing infrastructure. Thus, a thorough understanding of powerfuel blending requirements and implications is important. In this context, this article provides an overview of blending these fuels, using the example of aviation fuels and examining the respective physical and chemical properties of conventional and synthetic kerosene options for all presently approved aviation powerfuel production pathways.

In this context, it can be concluded that physical and chemical properties of synthetic and conventional kerosene options vary considerably. Attention needs to be paid to the specific properties of each blending component, to ensure that the resulting blend complies with fuel specifications. Fuel density, aromatics content, and the distillation curve gradient were found to be critical for some blends, potentially resulting in permissible blending ratios well below the precautionary blending limit laid out in the fuel specifications.

As synthetic paraffinic kerosene (SPK) fuels contain no aromatics, the conventional component needs to contain at least enough aromatics that the blend still achieves the minimum aromatics content of 8 vol.-%. The lack of aromatics also results in a rather low density, which again needs to be compensated for by the conventional kerosene component. In terms of distillation curve gradient, the results are less specific. As the distillation curve gradient of the neat synthetic kerosene strongly depends on process conditions (AtJ-SPK, FT-SPK) and the (intermediate) feedstock used (AtJ-SPK), a simple conclusion cannot be drawn. In general, a broader distribution of carbon chain lengths appears to be beneficial for this specification requirement.

The main goal of the blending procedure as discussed in this article is to produce a specification-compliant blend from a conventional and a synthetic kerosene component. However, in future, with increasing synthetic kerosene market shares, a more targeted approach toward blending might be beneficial, i.e., targeted blending might be used to achieve a high fuel energy content or low aromatics content. In the nearer term, permitting the use of 100% synthetic kerosene will permit the use of potentially fully greenhouse gas (GHG) neutral jet fuel, which is not possible with the current limitation to a 50% blend ratio. Similarly, a fully synthetic kerosene, in particular one with reduced aromatics content and thus improved combustion properties would be beneficial for other environmental effects besides GHG emissions, namely contrail climate impact and local air quality [29].

In the near future, inclusion of 100% synthetic fuels in ASTM D7566 can simplify the production process and potentially enable the use of fully GHG neutral kerosene while eliminating the requirement of blending completely. Such 100% synthetic fuels can either be fuels like AtJ-SKA, where the aromatics formation is a direct part of the fuel manufacture process, or blends of SPKs with synthetic aromatics. At the same time, research is required on the blend properties for powerfuel SPKs based on ethanol and/or methanol. In the longer term, the potential of "designing" fuels more specifically to utilize potentials toward improved local air quality and climate impact should be investigated. Still, if commingling must be excluded (e.g., due to restriction of the in-service fleet with 100% synthetic kerosenes), this automatically requires a designated supply infrastructure for such fuels.

References

1. Hsu CS, Robinson PR, Others (eds) (2017) Springer handbook of petroleum technology: With 754 figures and 282 Tables. Springer International Publishing, Cham
2. Rauch B (2018) Systematic accuracy assessment for alternative aviation fuel evaporation models: a thesis accepted by the Faculty of Aerospace Engineering and Geodesy of the University of Stuttgart in partial fulfillment of the requirements for the degree of Doctor of Engineering Sciences (Dr.-Ing.) [Ph.D.]. Stuttgart: University of Stuttgart
3. ASTM (2020) Standard specification for aviation turbine fuel containing synthesized hydrocarbons: D7566
4. de Klerk A (2011) Fischer-Tropsch refining. Wiley-VCH, Weinheim
5. Zschocke A, Scheuermann S, Ortner J (2017) High biofuel blends in aviation (HBBA): ENER/C2/2021/420-1 Final Report
6. Edwards T (2013) Reference jet fuels for combustion testing
7. Edwards J (2020) Jet fuel properties: interim report
8. PQIS (201) Petroleum quality information system
9. Chevron Corporation (2006) Aviation fuels: technical review
10. Rachner M (1998) Die Stoffeigenschaften von Kerosin Jet A-1
11. Schripp T, Anderson BE, Bauder U, Rauch B, Corbin JC, Smallwood GJ et al (2022) Aircraft engine particulate matter emissions from sustainable aviation fuels: Results from ground-based measurements during the NASA/DLR campaign ECLIF2/ND-MAX. Fuel 325:124764. https://doi.org/10.1016/j.fuel.2022.124764
12. Bräuer T, Voigt C, Sauer D, Kaufmann S, Hahn V, Scheibe M et al (2021) Airborne measurements of contrail ice properties—dependence on temperature and humidity. Geophys Res Lett 48(8). https://doi.org/10.1029/2020GL092166
13. Schripp T, Anderson B, Crosbie EC, Moore RH, Herrmann F, Oßwald P et al (2018) Impact of alternative jet fuels on engine exhaust composition during the 2015 ECLIF GRound-based measurements campaign. Environ Sci Technol 52(8):4969–4978. https://doi.org/10.1021/acs.est.7b06244
14. Petroleum HPV Testing Group (2010) Kerosene/jet fuel category assessment document: submitted to the US EPA
15. Graham J, Rahmes T, Kay M, Belieres J-P, Kinder J, Millett J et al (2011) Final report for alternative fuels task: impact of SPK fuels and fuel blends on non-metallic materials used in commercial aircraft fuel systems
16. Rickard D (2018) Quality of aviation fuel available in the United Kingdom: annual survey 2014
17. Zschocke A (2014) Abschlussbericht zu dem Vorhaben Projekt BurnFAIR
18. Zukunftstraum GK, „Mit grünem Treibstoff fliegen". Available from: https://www.uni-bremen.de/zukunftstraum-mit-gruenem-treibstoff-fliegen

19. ExxonMobil (2023) ExxonMobil methanol to jet technology to provide new route for sustainable aviation fuel production, 20 June 2023. Available from: https://www.exxonmobilchemical.com/en/resources/library/library-detail/101116/exxonmobil_sustainable_aviation_fuel_production_en?utm_source=google&utm_medium=cpc&utm_campaign=cl_downstream_none&ds_k=&gclsrc=aw.ds&&ppc_keyword=methanol%20to%20jet&gclid=Cj0KCQjwnMWkBhDLARIsAHBOftphOG4QcTIdIhvI5ZX1plfx_U2lrH7kd5PIs3dY8YExlFmuDH8krVUaAu5pEALw_wcB.

20. ASTM (2022) Standard specification for aviation turbine fuels: D1655

21. Pechstein J, Zschocke A, N N (2018) Blending of synthetic kerosene. In: Kaltschmitt M, Neuling U (eds) Biokerosene: status and prospects. Springer, 2018, pp 665–686

22. Aviation RM, Standards B, Approval A (2018). In: Kaltschmitt M, Neuling U (eds) Biokerosene: status and prospects. Springer, Berlin, pp 639–664

23. König DH (2016) Techno-ökonomische Prozessbewertung der Herstellung synthetischen Flugturbinentreibstoffes aus CO_2 und H_2. Dissertation

24. Harmon L, Hallen R, Lilga M, Heijstra B, Palou-Rivera I, Handler R (2017) A hybrid catalytic route to fuels from biomass syngas

25. Hull A, Hull A, Zschocke A (2023) Evaluation of alcohol-to-jet synthetic kerosene with aromatics (ATJ-SKA) fuels and blends: research report, Version 9

26. ASTM J Meeting (2023) Denver

27. UK Ministry of Defence (2022) Defence standard 91–91, turbine fuel, kerosene type, Jet A-1; NATO Code: F-35; Joint Service Designation: AVTUR

28. Bullerdiek N, Buse J, Dögnitz N, Feige A, Hallen A-M, Hauschild S et al (2019) Einsatz von Multiblend JET A-1 in der Praxis: Zusammenfassung der Ergebnisse aus dem Modellvorhaben der Mobilitäts- und Kraftstoffstrategie

29. Voigt C, Kleine J, Sauer D, Moore RH, Bräuer T, Le Clercq P et al (2021) Cleaner burning aviation fuels can reduce contrail cloudiness. Commun Earth Environ 2(1). https://doi.org/10.1038/s43247-021-00174-y

Application and Use

Powerfuels for Heavy-Duty Road Transportation

Till Gnann, Steffen Link, Daniel Speth, and Aline Scherrer

Abstract The transition towards European carbon neutrality by 2050 poses major challenges for heavy-duty road freight and necessitates technological changes. This study examines the current European truck market, including the legal framework aimed at enhancing sustainable transport via infrastructure deployment or CO_2 reduction targets. It evaluates the technological and economic prospects of various alternative technologies, and explores future pathways. While current political efforts predominantly focus on battery-electric and hydrogen-powered trucks, all alternative technologies can offer substantial and immediate CO_2 emission reduction potentials. However, battery-electric trucks are very likely the most promising pathway for achieving cost-effective operations, with also being at a more advanced technological readiness compared to fuel cell trucks, catenary electric systems, and renewable fuels. Apart from fast declining costs and technical feasibility in line with infrastructure or fuel availability, user acceptance will be a key to acceleration.

Keywords Truck · Battery-Electric · Fuel-cell · Alternative Infrastructure · TRL · Social Acceptance

1 European Road Freight Transport and Truck Market

Cutting greenhouse gas (GHG) emissions to comply with the Paris Agreement and eventually reaching climate neutrality by 2050 is challenging for a number of different sectors—as in the case of road freight transport that remains dominated by fossil fuels to date. In the European Union (EU), road freight transport accounts for nearly 160 billion kilometers per year and 13 billion tons of transported goods (EU-27). While there are around 406 million motor vehicles in use within Europe, there are only 11.8 million medium- and heavy-duty trucks (gross vehicle weight > 12 tons) while annual sales are around 230,000 to 280,000 trucks. Despite their minor significance in the total vehicle fleet, heavy-duty vehicles (HDVs) contributed about 8%

T. Gnann (✉) · S. Link · D. Speth · A. Scherrer
Fraunhofer Institute for Systems and Innovation Research (Fraunhofer ISI), Karlsruhe, Germany
e-mail: till.gnann@isi.fraunhofer.de

821

of the total EU GHG emissions in 2020, necessitating immediate action apart from drastic demand reduction. The same holds for other world regions [1]. All European truck manufacturers expressed their commitment toward low carbon solutions by announcing a 100% zero-emission truck sales target by 2040 [2] and partially high interim targets of up to 60% by 2030. This decarbonization—or rather defossiliza-tion—comprises two general technological pathways: direct usage of electric energy in battery electric trucks (BET) or the introduction of synthetic GHG emission-neutral fuels like hydrogen or synthetic liquid fuels that are produced based on renewable energy. Kindly check and verify that the author names are correctly recognized and presented in the correct sequence order, which is [given name, middle name/ initial, family name] for the author(s). In addition, please verify that the name(s) and respective affiliation(s) shown on the metadata page are valid and make any necessary amendments if required.

Hydrogen can be used in fuel cell trucks (FCET) and trucks with modified internal combustion engines (H_2-ICET). All these alternatives entail building a corresponding alternative refueling/charging infrastructure. In contrast, synthetic diesel may be seen as a drop-in replacement for current fossil diesel fuels, meaning that existing and new diesel trucks (D-ICE) may use them without modifications and that the existing filling station network can continue to be used [3]. One more electric solution may be catenary electric trucks (CET) that are charged on overhead lines. These options are at different stages of commercialization, and their significance for a future sustainable road transport system—from low-volume products aimed at specialist applications to the mass market backbone—is still under investigation. In this chapter, the following options will be analyzed:

- Battery electric trucks (BET),
- Catenary electric trucks (CET),
- Fuel Cell electric trucks (FCET),
- Trucks power with liquefied natural gas (LNG),
- Diesel trucks (ICE-D).

One key factor for the high uncertainty is the heterogeneous nature of the truck market. While trucks are commercial vehicles that are used primarily for transporta-tion of goods, there is a wide range of further use cases within different indus-tries (e.g., agriculture, timber industry, chemical industry, harbor logistics, food retail, mining, municipal, waste disposal, special purpose vehicles for firefighting, or airport services). Each industry and use case has its peculiarities regarding applica-tion profile, uptime, flexibility, payload, required range, required power, additional energy demand from accessories, or vehicle body type. As stated in ACEA [4], no truck is like another, nor is its application. This heterogeneity makes a "one tech-nology fits all" solution challenging and potentially unlikely. A typical high-level commercial vehicle market segmentation relies on the gross vehicle weight (GVW): light commercial vehicles (class N1, up to 3.5 t), medium-duty vehicles (class N2, 3.5 to 12 t), and heavy-duty (class N3, greater than 12 t). Vehicles over 40 t GVW require a special permit as heavy transport. A more detailed classification, based on the regulatory process for HDV CO_2 emissions in the EU, segments the market into

17 groups based on axle configuration (e.g., 4×2, 6×2, 6×4, or 8×4), chassis configuration (i.e., rigid or tractor), and GVW. Cabin type (i.e., day cap or sleeper cap) and engine power may be used to further split these main groups into subgroups [5].

This chapter will primarily focus on heavy-duty trucks (gross vehicle weight > 12 t) and explain the current legal framework and political goals in the following Sect. 2. Thereafter, a more detailed overview of technical possibilities to defossilize HDVs is presented in Sect. 3. Section 4 holds a techno-economical-environmental assessment of the technologies while Chap. 5 contains potential long-term pathways and their implications. Aspects on infrastructure requirements are given in Sect. 6, and an overview on user acceptance for alternative drive trains is presented in Sect. 7. This chapter is closed by a short summary in Sect. 8.

2 Legal Framework and Political Goals for Heavy-Duty Road Transport

Heavy-duty road freight transport in the European Union (EU) is affected by different legal and political framework conditions. As part of the "Fit for 55 Package", the EU aims to reduce greenhouse gas (GHG) emissions by 55% by 2030 compared to 1990 levels. Given this objective, also CO_2 emissions related to road freight transport need to be reduced significantly and as soon as possible. Different measures are designed on EU level to ensure this reduction [6]:

- firm-level emission performance standard for new vehicles,
- infrastructure mandates (e.g., the Alternative Fuel Infrastructure Regulation (AFIR)),
- CO_2 limitation for fuels (e.g., blending mandates for renewable fuels, cap-and-trade scheme),
- subsidies (e.g., regarding investments and research).

In the following, the measures mentioned above will be described in more detail.

Firm-level emission performance standard. The firm-level emission performance standard requires a reduction of 15% of CO_2 emissions of newly sold vehicles compared to 2019/2020 in 2025 and a reduction of 30% in 2030 [7]. The regulation covers vehicles of the classes N2 and N3 with a laden mass of more than 16 t and a 4×2 axle configuration as well as vehicles with a 6×2 axle configuration. Zero- and low-emission vehicles are credited with a multiplication factor, so that the average CO_2 emissions of the newly sold vehicles fleet can be additionally reduced by up to 3% by multiple crediting. In addition, credits and debts of no more than 5% of the manufacturer's specific CO_2 emissions target are possible within a reporting period (2019 to 2024; 2025 to 2029). If a manufacturer fails to meet his specific target, he will be subject to penalties. The penalties amount to 4,250 EUR/(g CO_2/

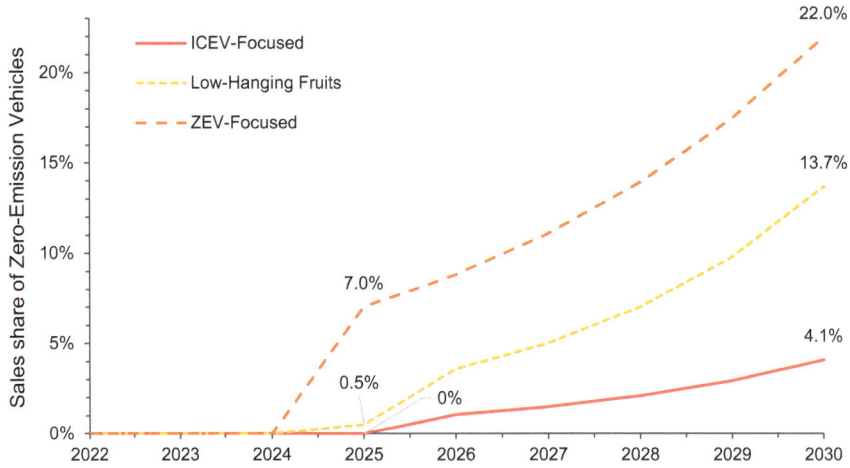

Fig. 1 Percentage of zero-emission vehicles sales share to meet the emission performance standard of new vehicles (-30% of CO_2 compared to 2020/2021; ICEV: internal combustion engine vehicle; ZEV: zero-emission vehicle; Breed et al. [8])

tkm) for excess CO_2 emissions from 2025 to 2029 and 6,800 EUR/(g CO_2/tkm) from 2030 onwards.

As shown in Fig. 1, newly registered vehicles will contain 4 to 22% zero-emission vehicles to achieve a CO_2 reduction of 30% compared to 2020/2021 [8]. The exact value depends on the efficiency improvements of the diesel vehicles. However, the manufacturers announced an average share of newly sold zero-emission vehicles of 7% in 2025 and 43% in 2030 [9]. A more recent manufacturer survey calculates a zero-emission vehicles share of 11% in 2025 and 63% in 2030 [10]. The ongoing revision of the firm-level emission performance standard therefore foresees a reduction of 45% in 2030, 65% in 2035, and 90% in 2040 [11].

Infrastructure mandates. A second aspect of the European regulation concerns infrastructure development. As part of the "Fit for 55 Package", there is a proposal from the European Commission to amend the Alternative Fuel Infrastructure Directive (AFID). The directive shall be enhanced and apply directly to all EU member states as Alternative Fuel Infrastructure Regulation (AFIR) [12]. The Commission's proposal outlines pathways to build an infrastructure for battery electric trucks (BET) as well as for fuel cell electric trucks (FCET) on the most important European highways (TEN-T network). In the TEN-T core network, charging infrastructure shall be provided at a maximum distance of 60 km. In the TEN-T comprehensive network, the distance is increased to 100 km. By 2025, a total of 1,400 kW of charging power and at least one charging point with 350 kW should be available at every station. By 2030, the capacity shall be increased to a total of 3,500 kW and at least two charging points with 350 kW. In addition, every "safe and secure parking area" will receive at least one 100 kW charging point by 2030. Each urban node will have a minimum charging capacity of 600 kW for trucks by 2025. The individual charging power per charging point amounts 150 kW. By 2030, the power per urban node will be doubled to 1,200 kW. However, the proposal does not mention the new Megawatt Charging System (MCS), which is expected to allow charging power above 1,000 kW per charging point. There is also no build-up pathway outlined for catenary overhead lines or other electric road systems (ERS) which are only mentioned in the annex. As part of the legislative process, the Commission and the Parliament agreed on the completion of the described charging infrastructure by 2030 [13].

Hydrogen is planned to be available on both the TEN-T core and the TEN-T comprehensive network with a maximum distance for refueling hydrogen stations of 150 km in 2030. The stations shall provide 2 t of hydrogen per day at 700 bar level. In addition, at least one hydrogen refueling station is to be located in each urban node in 2030. Liquid hydrogen for trucks should be made available at the TEN-T core network at intervals of no more than 450 km in 2030 [12].[1] If there is demand, a LNG charging infrastructure should be available at least on the TEN-T core network by 2025 according to European Commission [12]. However, the focus of the Commission's proposal with regard to heavy-duty vehicles is clearly on charging infrastructure for BET as well as hydrogen refueling stations.

Carbon dioxide (CO_2) limitation for fuels. A third pillar in the "Fit for 55" context consists of the direct reduction and limitation of CO_2 emissions. For this purpose, the greenhouse gas impact of transport fuels shall be reduced by 16% until 2030, and the share of renewable fuels of non-biological origin shall rise to at least 5.7% in 2030 [14, 15].[2] In addition, an Emissions Trading System (ETS) for the building and transport sector—ETS2—shall be introduced from 2027 onwards. This system will limit the CO_2 emissions from fuels put on the market. The amount of certificates (emission allowances) to bring fuels and their CO_2 emissions to market is to be reduced by slightly more than 5% per year, with 2024 as the base year [16]. The ETS will allow to determine the total amount of CO_2 emissions in the transport and building sector. Moreover, since the certificates are auctioned and traded, a CO_2 price is created in the transport and building sector. A predecessor of this system is the German national Emissions Trading System (nETS) for fuels. Companies that trade oil or gas products have to purchase pollution rights for their products. Starting with 25 €/t CO_2 in 2021 (~0.08 €/liter diesel), the price will increase to 55 to 65 €/t CO_2 in 2026.

Subsidies. As a fourth option, subsidies can accelerate both the development of new technologies and their diffusion into the vehicle fleet or the market, respectively. Various EU programs provide investment subsidies and research subsidies for the transport sector [6]. On national level in Germany, the KsNI program provides a good example in the context of heavy-duty vehicles. On the one hand, it subsidizes up to 80% of the additional investment for a vehicle with an electric powertrain (battery

---

[1] As part of the legislative procedure, the Commission's proposal was further adapted. The final legislation was still pending at the time of writing. Reference is therefore made to the Commissions proposal.

[2] At the time of writing, the Renewable Energy Directive (RED) III is in the final stage of approval.

electric or fuel cell electric) compared to a diesel vehicle. Related charging and refueling infrastructure can also be subsidized. On the other hand, feasibility studies are also funded to explore the potential use cases of alternative fueled vehicles in companies. As shown in Speth et al. [17] and Basma et al. [18], appropriate subsidies can accelerate the economic viability of alternative powertrains by several years.

3 Technological Assessment and Status Quo of Market Development

3.1 Status of Battery Electric Trucks

Introduction and market. Battery electric trucks (BETs) benefit from the experience and technical innovations—i.e., specific energy, volumetric energy density, power density, cycle life, and costs [19, 20]—of batteries from passenger cars. Plus, these synergies imply short-term commercial availability and rapid scalability. BETs seem to be set in the future product portfolios of all major manufacturers and all weight classes. The IEA model outlook shows that ranges from 200 to 450 km are commercially available, with higher ranges already announced [21]. This BET product portfolio already permits covering major parts of urban and regional truck distribution transport (TRL 9) [18]. Challenges remain for national and international long-haul truck transport as well as special operations (TRL 7).

Accordingly, scientific studies and reports increasingly confirm technical feasibility and cost-effective operations for BETs in urban and regional delivery with a daily mileage of less than 400 km [20]. Most recent studies even see heavy long-haul truck transport close to the threshold where BETs become technically feasible [19] insofar as possibilities for fast charging are available (e.g., in the mandatory European driving break after 4.5 h). Specifically, the megawatt charging system (MCS) standard—developed by the CharIN initiative and available from the mid-2020s—is the key enabling technology, allowing recharge with up to 1,000 kW peak in early specifications and up to 3,750 kW in maximum. From a feasibility perspective, market diffusion may happen for urban and regional transport first. In contrast, achieving cost parity with current diesel trucks and manufacturer compliance with the European CO_2 fleet targets may favor the market diffusion for long-haul trucks.

Technology and efficiency. The whole battery electric powertrain—including the high-voltage battery, battery management system, electric motors, power electronics, high-voltage system, and electrified accessories—benefits from the technical advancements achieved in the automotive passenger car market. However, higher continuous power requirements and thermal stress are more distinctive for heavy-duty trucks. Specifically for batteries, the achievable system specifications regarding achievable weight, required installation space, cyclic and calendar aging stability, and fast-charging capability are decisive.

Aiming for maximum flexibility and high truck ranges without recharging leads to increased battery capacities involving potential payload or volume losses. With current lithium-ion technologies and cell chemistries, this often requires trade-offs. While up to 600 kWh of gross battery capacity may be available today, heavy long-haul transport may require around 1 MWh of installed battery capacity, increasing the vehicle curb weight. Although the European Commission increased the maximum permitted GVW of certain zero-tailpipe-emission trucks by 2 tonnes, maximum permissible loads per axle and vehicle dimensions must not be exceeded. Studies indicate around 5 to 20% of payload losses with current technologies for long-haul trucking [18]. However, any disadvantage is expected to almost disappear within this decade with on-purpose vehicle platform architectures and other battery-specific advances [22].

The well-to-wheel (WTW) efficiency of BETs is typically around 70–75% [23, 24]. The battery-to-wheel (BTW) energy consumption varies from 120 to 150 kWh per 100 km for long-haul tractors. This high efficiency leads to advantages in operating costs versus other alternative technologies. While economic operations may require funding and subsidies as of today, any governmental support will likely become obsolete within the decade [18].

Infrastructure. A sufficient charging infrastructure network is pivotal for the market diffusion of BETs. This network involves private charging opportunities by the vehicle owner, semi-public charging locations at depots, stores, warehouses, or industry hubs (often called destination charging), and public infrastructure at urban nodes or along the highway network (often called opportunity charging) [25]. While private and semi-public charging may predominantly involve slow overnight charging (ONC) and only partially fast charging, both may be covered with onboard charging (22 to 44 kW AC charging) or the current combined charging system (CCS) standard (up to 350 kW). Public charging parks equipped with the newly developed MCS standard may facilitate heavy long-haul transport by fast recharging during driving breaks, while ONC opportunities must also be provided [26]. This additional electricity demand may vary regionally and require—in addition to the general power grid expansion—further local grid expansions, including new transformers, which may incur severe costs as well as long authorization and planning times. However, larger energy throughputs allow moving into higher electricity bandwidths and reduce the infrastructure cost share, with both lowering the price per consumed kWh of electricity. Large-scale empirical data on commercial truck charging tariffs are not yet available today.

Future challenges. Generally, improving energy efficiency may facilitate higher ranges with smaller batteries or higher payloads despite larger ones. These improvements may involve road-load reductions, including vehicle aerodynamics, rolling resistance, and lightweight construction, as well as minimizing further losses in the electric powertrain (mainly power electronics and the electric motor but also the battery efficiency under high C-rates). Consequently, up to around 100 kWh per 100 km may be reachable for long-haul BETs.

Specifically for batteries, achievable system specifications may be optimized. This includes increasing the specific energy and volumetric energy density to eliminate payload losses. Here, on-purpose vehicle platform architectures are also crucial to optimize battery placement and, thus, weight distribution. In parallel, truck batteries must have high cyclic stability and must withstand high-power loads, with fast charging at low temperatures [19] and effective cell cooling during fast-charging events posing particular challenges. Thus, advances in battery management systems (BMS) and thermal management systems (TMS) are crucial. Last, battery system costs must continue to decrease, for example, by using low-cost battery cell chemistries such as LF(M)P or optimizing cell-to-system costs.

Regarding power systems and infrastructure, accelerating the infrastructure deployment in parallel with an early regional grid expansion is pivotal for providing the right charging infrastructure, ready for drivers when and where it is needed.

3.2   Status of Fuel Cell Electric Trucks

Introduction and market. Fuel cell electric trucks (FCETs) may be considered the second pillar for sustainable road transport systems, in particular for long-haul and weight-constrained transport, highly flexible operation, and use cases where excessive energy is required to power additional accessories [27]. Manufacturer strategies are heterogeneous, so some explicitly target over 800 to 1,000 km per tank while others focus on lower ranges (200 to 500 km), typically associated with different onboard hydrogen storage solutions. These are gaseous hydrogen (cGH2) at either 350 or 700 bar, subcooled liquid hydrogen (sLH$_2$), and cryo-compressed hydrogen (CcH$_2$). While first fleet tests and series models with cGH2 are already available today (TRL 8–9), first sLH$_2$ prototypes are available (TRL 6–7) yet CcH$_2$ models are still at an early development stage (TRL 4–5) [28]. The general availability of FCET models is expected to expand around 2030 [21], although the importance of FCETs may vary widely across regions.

Technology and efficiency. The hydrogen system supplements a battery electric powertrain, where the battery is necessary to smooth load peaks, compensate for high continuous loads, and enable energy recuperation while braking. The actual hydrogen system consists of the fuel cell, its management and cooling system, hydrogen tanks, and the air intake system. High purity (hydrogen and air) and effective heat removal are important for long-life operations. The fuel cell dimensioning depends on the FCET concept, i.e., range extender concepts with larger batteries and smaller fuel cells or vice versa for fuel cell-dominated concepts. For mobile applications (high dynamics and high required power density) in general, the low-temperature polymer electrolyte membrane (LT-PEM) fuel cell system has been proven to be the most attractive technology [29].

Due to the significantly smaller batteries yet volume required to accommodate the hydrogen tanks (chassis integration or tower configuration behind the cabin), payload limitations are minor or even non-existing compared to diesel trucks. In contrast, the WTW efficiency is around 25%, which is only one-third of the BET efficiency [23, 24]. The tank-to-wheel (TTW) energy consumption varies from 200 to 300 kWh per 100 km (i.e., 6 to 9 kg H_2 per 100 km) for long-haul trucks [22]. However, higher shares of battery electric driving can significantly increase overall efficiency. The peak system efficiency of the LT-PEM is around 65%, while only 45 to 50% may be achieved during operations [29].

The different onboard hydrogen storage options complicate the development and standardization of both vehicle components and HRS infrastructure. While some manufacturers, such as Hyundai, rely on cGH_2 at 350 bar, other manufacturers, such as Toyota or Nikola Motors, rely on cGH_2 at 700 bar to increase energy density and reduce refueling times with future protocols. Since cGH_2 systems are typically used for passenger cars and buses, this is the most mature storage option, with many players already engaged. However, vehicle range and refueling times are potential pitfalls. In contrast, some manufacturers have announced their strategy toward sLH_2 at around $-245\,°C$ and 5 to 16 bar to realize higher energy densities and, thus, higher vehicle ranges. Plus refueling is more convenient and faster. Last, CcH_2 at around $-200\,°C$ and 300 bar may be conceivable to achieve the highest energy densities among all four technologies [30, 31], but it is also the least mature.

Infrastructure. Apart from implications on the vehicle design, those onboard hydrogen storage options also affect the HRS layout and its components, the whole hydrogen supply chain from production to pump (i.e., different distribution concepts), and synergies with other vehicles such as buses or cars. Due to the highly different HRS layouts and required components, a service station may offer multiple options from a technical perspective. However, it may be associated with higher costs, which questions economic viability.

According to H_2 Mobility, around 100 cGH_2 hydrogen refueling stations are available in Germany, Austria, and Switzerland, with additional facilities scheduled. This technology is currently limited to around 40 kg of H_2 (cGH_2 350 bar) per refueling event [30]. However, faster protocols with larger hydrogen volumes are being developed. cGH_2 stations are typically seen as the most flexible option for different supply options. While sLH_2 or CcH_2 stations may promise the fastest refueling times, there are no commercial stations in operation today, with many components for such HRS stations still under development and not yet commercially available [28].

Economic operations will be highly sensitive to the (green) hydrogen price. Typically, around 3 to 5 €/kg H_2 are perceived as threshold versus diesel trucks [28]. This price must include hydrogen production, import, local distribution and transport, potential liquefaction, HRS infrastructure, profits, as well as levies, taxes, and fees (excl. VAT as this refundable for commercial applications). H2 Mobility currently offers (mostly gray) cGH_2 at around 10.80 to 11.60 €/kg H_2 (excl. VAT). Given the limited supply and scarcity of green hydrogen and demand from other sectors, economic operations may require governmental support even beyond this decade.

Future challenges. There is little evidence on fuel cell aging (e.g., efficiency degradation due to catalyst degradation and impurities) and durability of large mobile systems [29] under real-world conditions. An adequate cooling architecture is crucial to remove the heat inside the fuel cell stack and, thus, enable long-life operations.

Apart from road-load reduction technologies and electric powertrain optimizations (see the previous chapter), high continuous efficiency, cold-start patterns [27], and the synergies between the fuel cell and the battery may be optimized to increase the overall efficiency and ensure high vehicle reliability [29]. Hydrogen boil-off—especially at high fuel levels—is a notable problem for some onboard storage technologies.

Providing sufficient infrastructure coverage relies on an accelerated HRS deployment, the standardization of required components, and potentially a technology decision by the HRS operators. In parallel, this requires establishing a whole hydrogen ecosystem to facilitate large-scale productions and, thus, both lower prices and larger availability.

3.3   Synopsis on Catenary Electric Trucks

Catenary trucks (CAT) may operate electrically on route sections that are equipped with an overhead line infrastructure. This concept may be conceivable for both BETs and FCETs that are additionally equipped with a pantograph and detection system. Plus, hybrid trucks may be conceivable where the conventional powertrain handles non-electrified route sections (typically off-highway) and a battery electric powertrain plus pantograph system handles electrified ones. Conceptually, one might distinguish between a patchwork and a full coverage concept [32]. Range limitations of current BETs may be avoided, the efficiency of FCET and D-ICE may be upgraded, and emissions from D-ICE may be significantly reduced. The overall efficiency of the CAT-BETs is expected to be comparable to pure BETs; however, no commercial models are available.

Overhead contact line systems have been successfully used in rail transport for decades and thus represent a proven and reliable technology. However, adaptations to the special features of road freight transport (e.g., irregularities in traffic flow and congestion) are necessary. However, the system is currently only used on demonstration and test sites and is not commercially available (TRL 6–7). Aging, maintenance, and degradation are highly uncertain (Jöhrens and Helms [33]). High upfront investments that are required for the infrastructure build-up and the low profitability in the early market phase (few CATs) are major hurdles. There are five relevant pilot projects with few electrified route sections up to 5 km, whereof three are located in Germany.

3.4 Status of ICE-Truck with Renewable Fuels

Introduction and market. Trucks with ICE certainly appeal by their "business as usual" aspect, which may be relevant for various reasons. For instance fleet operators may dislike any operational changes that are entailed with owning alternative trucks such as BET or FCET (i.e., charging at depots, longer and more frequent recharging stops on a lengthy journey, or scarce infrastructure availability) or depot electrification is hardly possible. Plus, this alternative may also be attractive for specialist applications with certain requirements on robustness and reliability, if vehicles are exposed to particular external stresses (like dirt, grime, or vibrations), or if excessive energy is required to power additional accessories. Apart from power-based synthetic fuels, different biofuel options can also be considered as non-fossil fuel options. However, given the biomass scarcity and demand from other harder-to-abate sectors where less alternatives are available, biomass-based fuel options have certain limitations that do not apply to power-based synthetic fuels—making the latter more attractive in terms of general availability. Three technological pathways may be distinguished for ICE-powered trucks. These are either diesel, methane, or hydrogen ICEs.

Technology and efficiency. Synthetic diesel fuels certainly have some merits. On the one hand, it may be considered as a drop-in replacement for current diesel fuels based on crude oil, meaning that existing and future diesel trucks could use them without or only minor modifications and that the existing filling station network could continue to be used. Additionally, current expertise in ICE technology, existing supply chains, and current vehicle architectures may be continued to be used. For synthetic diesel fuels, hydrogen—ideally produced from based on renewable energy—is used as raw material, while upstream processes capture the required carbon. Although the use of synthetic fuels does not allow for a fully climate-neutral WTW CO_2 emission balance, using synthetic fuels would tremendously reduce the climate-effective CO_2 emissions of ICEs. On the other hand, this transition to synthetic diesel fuel is not expected to be achieved fast enough for the mass market, as currently only prototypes and demonstrator systems are available (TRL 6–7). For instance, even the eFuel Alliance e.V. sees only a gradually increasing share of synthetic diesel fuels from 2030 onwards rather than an immediate swap. Given the scarcity of renewable energy, the low WTW efficiency—only around 15% and below [23]—is the largest pitfall. This results from large conversion losses, several process, and distribution steps, and only around 35 to 40% efficiency of the actual vehicle powertrain (i.e., ICE, transmission, shafts, and axle). Moreover, achievable synthetic diesel prices are probably very high, which questions economic viability without any governmental support even beyond 2040, in particular for high mileage.

The second pathway with methane-powered trucks barely has any market significance today, neither in registrations (at most between 1.8 to 3.6% in 2018 to 2021) nor in vehicle stock (about 0.5% in 2020). For methane powertrains typically CNG (i.e., compressed natural gas) and LNG (i.e., liquefied natural gas), powertrains are distinguished. This chapter is focused on the latter as being the more relevant option for heavy-duty trucks. For LNG powertrains one distinguishes between using gasoline engines (i.e., spark-ignition (SI) engines) and high-pressure direct injection (HPDI) engines. While SI engines are less expensive and benefit from simpler system operations as well as simpler emission after-treatment (i.e., 3-way catalytic converter and exhaust gas recirculation to reduce NO_x emissions), they provide less low-end torque given the same engine displacement, and their efficiency is clearly below current diesel engines (due to throttle losses and limited maximum compression). The additional energy consumption is equivalent to around 15 to 24% compared to diesel engines. HDPI engines, operating comparable to diesel engines, use dual-fuel technology, where a small amount of diesel fuel ignites the methane–air mixture, where the vaporized methane is directly injected into the resulting flame. On average, the additional energy consumption is equivalent to around 3 to 5% compared to diesel engines. However, the system is significantly more complex, which translates into higher costs. Also, emission after-treatment is more complex and similar to those of a modern diesel engines. The overall WTW efficiency for synthetic-based methane may be comparable to that from synthetic diesel, with minor differences regarding process steps or through the compression (CNG) or liquefaction (LNG).In contrast to diesel fuels, methane-powered trucks depend on the presence of a suitable refueling infrastructure (i.e., either CNG, LNG or both), whose availability is currently not comprehensive across Europe and considerably varies across regions.

Despite running hydrogen in FCET or via synthetic-based gaseous or liquid fuels, it can directly be used in modified ICEs. Modifications include the hydrogen injection system, compressors, application processes, and emission after-treatment. Both self-ignition and spark-ignition combustion processes are feasible. H_2-ICEs may reach peak efficiencies around 45% like current diesel engines, while best-point efficiencies of up to 50% are conceivable [29]. Thus, the TTW efficiency is around 5 to 10% lower than for FCET, and, in parallel, the WTW efficiency may be slightly lower than FCET (around 20%). However, the H_2-ICE is more robust than fuel cells, in particular concerning hydrogen purity or vibrations, and current expertise in ICE technology may be used further. In terms of particle mass, unburned hydrocarbons (UHC), and carbon monoxide (CO) emissions, H_2-ICEs are expected to meet current and future limits, although these emissions are not completely eliminated. However, nitrogen oxide (NO_x) emissions are an issue that must require sufficient after-treatment systems such as H_2-SCR (selective catalytic reduction), Urea-SCR, or diesel oxidation catalyst (DOC) [29]. Unlike synthetic diesel fuels and similar to FCETs, H_2-ICEs depend on the presence of an alternative refueling infrastructure matching the corresponding onboard hydrogen storage technology (see previous chapter).

In addition, ICE hybridization from mild hybrids to plug-in hybrids may increase efficiency, save operating costs yet increase the upfront investment, and reduce emissions.

Future challenges. Supplying sufficient quantities of synthetic fuels at reasonable costs will be pivotal for ICE trucks. While the vehicle technology is established (drop-in concept)—in particular for synthetic diesel fuels—only a few small-scale production plants have yet been announced and establishing a whole value chain and large-scale production upscaling will take years to decades. For others pathways, namely methane- or hydrogen-powered ICE trucks, additional infrastructure deployment will be required. Given recent natural gas price volatilities and limited availability due to the Ukraine war, the future of methane-powered trucks may rather be uncertain, both politically and strategically.

4 Economic and Environmental Assessment

The following section contains two perspectives. First, the total cost of ownership (TCO) are discussed from a private perspective that represent truck owners who are buying and using the vehicles for running their businesses. Second, different alternatives are evaluated from a federal perspective based on achievable reductions in CO_2 emissions.

Costs are a decisive factor for logistics companies [34, 35]. An exemplary calculation of the TCO for various powertrains in Germany from 2020 to 2050 can be found in Speth et al. [17]. A tractor-trailer with a typical annual mileage of 120,000 km is considered. For H_2-ICE, FCET, and BET, the vehicle range is varied between 100 km and 1,000 km and included in the results as a bandwidth. To reach climate-neutral transport in 2050, an increasing blend with synthetic diesel and gas is assumed. Simultaneously, existing subsidies—toll exemptions and purchase price subsidies—for FCET, BET, and PHET are fully reduced by 2030.

The main results of the TCO analysis are depicted in Fig. 2 and highlight three main findings across the years:

- In 2020, implemented massive subsidies—already included in Fig. 2—are needed for BET, FCET, and PHET powertrain options to reach cost-competitiveness versus advanced and mature D-ICE and NG-ICE, whose TCO are dominated by fuel costs. NG-ICE and PHET may be the most cost-effective solution.
- As of 2030, BET and PHET may reach TCO parity even without subsidies, whereas results are highly sensitive to battery capacity and range for BETs and electric driving share (EDS) for PHETs. The relevance of fuel costs decreases, and the share of acquisition costs increases. FCET and H2-ICE are significantly more expensive and highly sensitive toward hydrogen prices. NG-trucks are no longer competitive without toll exemption and reduced energy taxation.
- In 2050, all powertrain options are very close and clearly below the estimated cost level of D-ICE trucks. FCET, BET, and possibly PHET with high EDS may be the most cost-effective powertrain options. However, it is important to acknowledge the high uncertainty associated with such a long-term forecast. Nevertheless, the

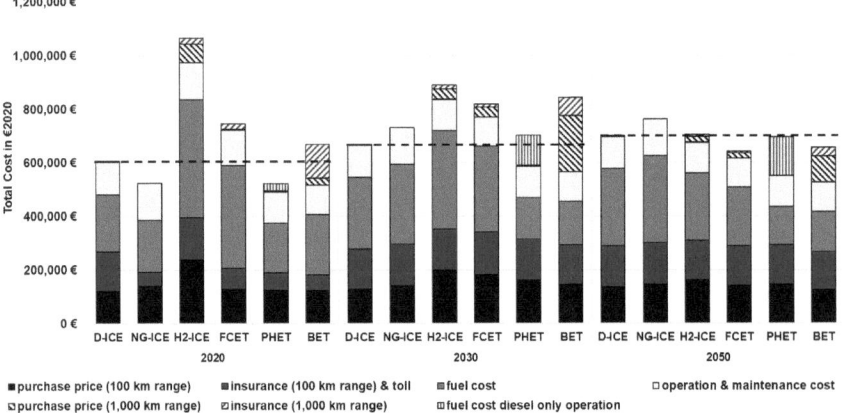

Fig. 2 Total Cost of Ownership calculation for five different HDT powertrains in Germany covering 2020 (left), 2030 (middle) and 2050 (right). Powertrain options comprise diesel trucks (D-ICE), methane trucks (NG-ICE), hydrogen ICE trucks (H2-ICE), fuel cell electric trucks (FCET), plug-in hybrid electric trucks (PHET), and battery electric trucks (BET). TCO for D-ICE are highlighted as dashed line. TCO in EUR2020 covers vehicle acquisition including federal subsidies, vehicle resale, energy costs, insurance, and maintenance. Results adopted from Speth et al. [17]

general competitiveness of alternative powertrains is also confirmed by other studies, for example, by Noll et al. [36].

In the environmental assessment, an artic lorry with 40 t GVW is compared for Germany and Sweden for 2020 and 2030. A lifetime mileage of 800,000 km is used, and a BEV battery capacity of 990 kWh in 2020 and 1,450 kWh in 2030 assumed. Electricity is produced with the country-specific electricity mix (with the respective share of renewable energies) while today's hydrogen production is completely based on natural gas (NG) steam reforming (100% in 2020) and contains a 10% share of green hydrogen from renewable energies in 2030. Figure 3 illustrates estimations in medium-term developments of CO_2 emissions per kilometer for a selection of different drive trains. It is worth noting that in the long term, all of these options have the potential to become climate-neutral (see the following section).

First of all, this comparison highlights the relatively small proportion of specific CO_2 emissions attributed to the vehicle production (indicated by the black bar with white dots) in comparison with the specific CO_2 emissions over its lifetime (all other bars). This stems from the high amounts of annual mileage and has also been the case in the TCO comparison. Second, all alternative fuel vehicles can reduce emissions today (2020) and in 2030 in Germany and Sweden, even with the current electricity mix. The reduction will be higher in future due to increasing renewable energy shares. Third, emissions of alternative fuels occur only during production while tailpipe CO_2 emissions are completely reduced. For conventional diesel fuels, about 20% of emissions occur while being produced (WTT). Fourth, maintenance and end-of-life emission only play a minor role for the specific CO_2 emissions.

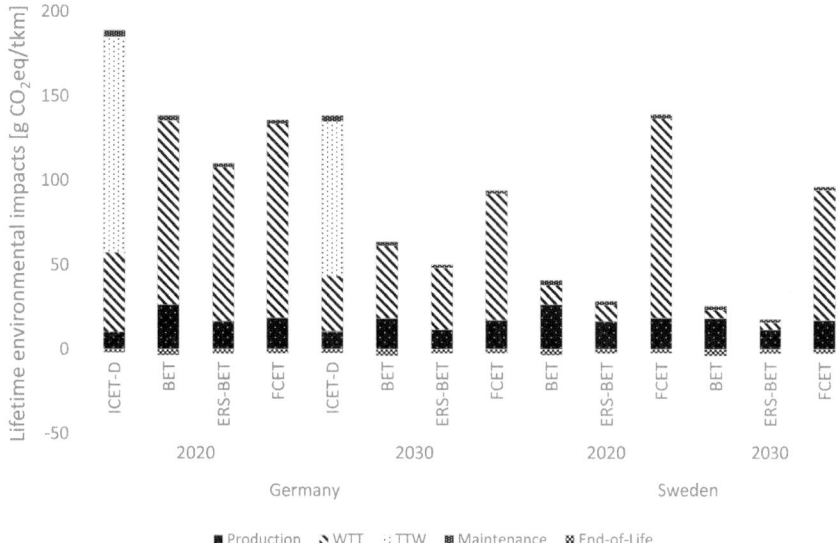

Fig. 3 Lifetime impacts by powertrain type[3] for Artic Lorry 40t GVW in Germany and Sweden 2020 and 2030. Lifetime mileage 800,000 km, BEV battery capacity of 990 kWh (2020) and 1,450 kWh (2030), hydrogen production mostly based on NG steam reforming (100% in 2020 and 90% in 2030). *Source* Results Viewer from (Hill et al. 2020). Abbreviations: WTT = Well-to-tank, TTW = Tank-to-wheel, ICET-D = Internal combustion engine truck with diesel, BET = Battery Electric Truck, ERS-BET = Electric road system battery electric truck; FCET = fuel cell electric truck; tkm = ton kilometer. Figure from Ref. [37]

Hence, while from an economic point of view, the introduction of alternative fuels is discussable and may only make sense for specific use cases, it has a positive environmental impact with regard to CO_2-emissions already with today's electricity mix.

5   Future Development Pathways

The comparison in the previous section showed an advantage in terms of cost and energy demand in a static observation without infrastructure. In this section, some development pathways for Germany are pointed out for trucks with a special focus on heavy-duty trucks.

In 2021 and 2022, two studies treating the long-term development of the German transport sector for the Federal Ministry of the Economy and Energy (BMWi) were conducted by Krail et al. [38], Wietschel et al. [32]. Both contained the analysis of the future development of trucks in Germany until 2050 with advantageous conditions for

---

[3] Due to the considerable deviation of the biofuel content in Swedish diesel fuel, ICETs are only displayed for Germany where the biofuel blend is similar to the average European situation.

electricity (here: focus electricity), hydrogen (focus hydrogen), and focus synfuels (focus synfuels) and methane (focus methane). Their results and assumptions are described in more detail in the respective studies. However, the core assumptions are described in the following:

- General assumptions: Germany should be climate-neutral by 2050. Thus, it is assumed that conventional, fossil fuels will be entirely replaced by a use of electricity, biofuels, or synthetic fuels based on renewable energy until 2050. The energy prices for trucks cover cost for energy and infrastructure. Alternative fuel trucks are exempted from road tolls until 2030.
- Scenario "focus electricity": A high-power charging infrastructure and an overhead line infrastructure for battery and catenary electric trucks for public charging is built over time. The electricity price is slightly lower than in the other scenarios in 2050 (2 €ct/kWh in 2050).
- Scenario "focus hydrogen": A hydrogen refueling infrastructure is built over time. The hydrogen price is slightly lower than in the other scenarios in 2050 (2 €ct/kWh in 2050).
- Scenario "focus methane": A hydrogen refueling infrastructure is built over time. The methane price is slightly lower than in the other scenarios in 2050 (2 €ct/kWh in 2050).
- Scenario "focus synfuels": No infrastructure for any alternative fuel is built over time. The synthetic fuel price is slightly lower than in the other scenarios in 2050 (2 €ct/kWh in 2050).

The development of the heavy-duty vehicle stock in the four scenarios is illustrated in Fig. 4. For further comparison, the development for light- and medium-duty vehicles is shown as well.

For light- and medium-duty vehicles, only results of the scenario focus electricity are shown in the upper left and center panel as all scenario results are fairly similar [39]. The future vehicle stock for both size classes will be dominated by electric vehicles. This is caused by low variable cost of the intensively used vehicles and a relatively low upfront infrastructure investment for the users. To date, a vehicle and infrastructure purchase incentive also helps to trigger market diffusion in these segments. The small share of fuel cell electric trucks (FCET) in medium size can only be fulfilled if there is a parallel diffusion of FCET in other size classes or countries; otherwise, the amount of vehicles is insufficient to create a supply for truck makers.

Results are different for heavy-duty trucks. In all scenarios, about 40 to 50% of battery electric trucks (BET) are projected to be in stock until 2050 that operate in the low- to medium-distance range. For long-haul trucks, results show the scenario-specific energy carriers as best option for heavy-duty vehicles. Thus, in focus electricity, long-distance trucking is performed by catenary electric trucks. Additional calculations show that these could also be replaced by BET with a sufficient high-power charging (HPC) infrastructure. Long-haul trucks could be operated with hydrogen (resp. methane) with a sufficiently low hydrogen (resp. methane) price and a good infrastructure development for hydrogen (resp. methane). In the scenario focus synfuels, the long-haul option remains diesel.

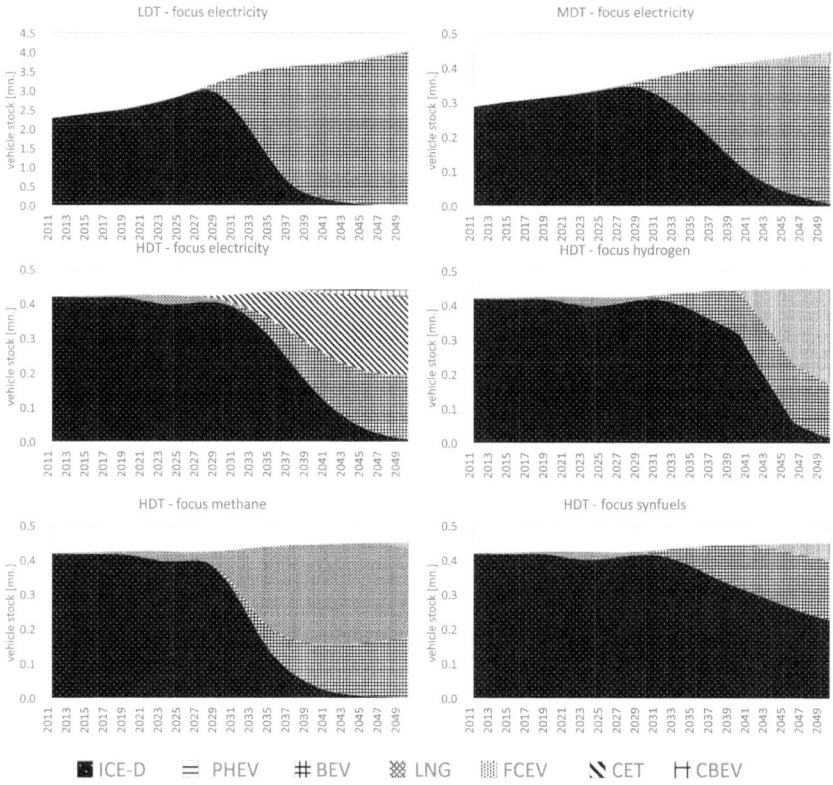

Fig. 4 Development pathways for trucks in Germany until 2050 differentiated by drive train. *Upper left panel*: Light duty trucks, *upper center panel*: medium-duty trucks, *upper right panel*: heavy-duty trucks (HDT) in focus electricity scenario, *lower left panel*: HDT in scenario focus hydrogen, *lower center panel*: HDT in scenario focus methane, *lower right panel*: HDT in scenario focus synfuels. ICE-D: internal combustion engine vehicles powered with diesel; PHEV: plug-in hybrid electric vehicles: BEV: battery electric vehicles; LNG: liquefied natural gas vehicles; FCEV: fuel cell electric vehicles; CBEV: catenary battery electric vehicles; CET: catenary electric vehicles

These results may seem obvious, but the large share of BET was not foreseen earlier, although many truck makers are focusing on offering BET now [40]. This could also be the case as there will surely be synergy effects with the light- and medium-duty vehicle sector. Furthermore, these vehicle developments require very different amounts of energy. The developments for the final energy consumption for heavy-duty trucks in the four scenarios are shown in Fig. 5.

In all scenarios, the amount of energy is reduced by nearly half in 2050. In the scenario focus electricity, the final energy demand is 90 TWh in 2050, while in focus synfuels about 110 TWh are needed then. These differences in the final energy consumption are small, yet the amount of electricity needed for hydrogen, methane, or synthetic fuels is two to three times higher. A back-of-the-envelope calculation

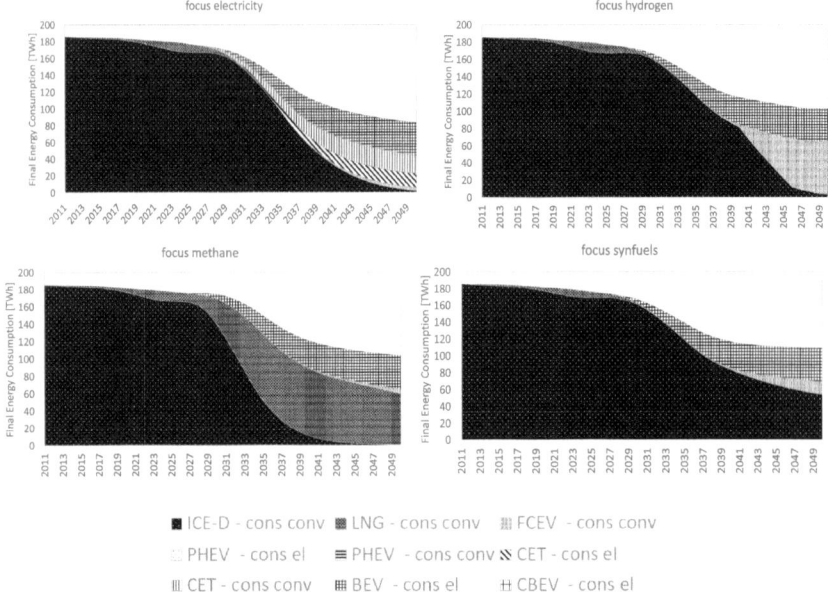

Fig. 5 Final energy consumption for development pathways for heavy-duty trucks in Germany until 2050 differentiated by drive train. *Upper left panel*: Heavy-duty trucks (HDT) in focus electricity scenario, *upper right panel*: HDT in scenario focus hydrogen, *lower left panel*: HDT in scenario focus methane, *lower right panel*: HDT in scenario focus synfuels. ICE-D: internal combustion engine vehicles powered with diesel; PHEV: plug-in hybrid electric vehicles: BEV: battery electric vehicles; LNG: liquefied natural gas vehicles; FCEV: fuel cell electric vehicles; CET: catenary electric vehicles; CBEV: catenary battery electric vehicles; cons el: electric consumption; cons conv: conventional consumption

with these figures returns an electricity demand of about 135 TWh in focus electricity, about 170 TWh in focus hydrogen and about 230 TWh in focus synfuels. As a result, the electricity demand is significantly higher in the latter scenarios due to conversion losses. Therefore, it is advisable to prioritize the use of electric vehicles.

This evaluation aligns with results and conclusions from comparable studies, where most studies assess that electric trucks are the most competitive option in climate protection scenarios [1].

6 Infrastructure Setup

To fulfill their transportation task, trucks with alternative powertrains are dependent on a corresponding refueling or charging infrastructure. In this context, two questions can be derived.

- The first question that arises is whether it is meaningful to determine the minimum infrastructure level that enables all vehicles to reach their destination. This is highly relevant for an initial infrastructure ramp-up.
- The second question that arises is how a network with a high level of user comfort could look like. Such an infrastructure network ensures that a charging or refueling station can be reached quickly at any time.

The smaller the range of the alternative powertrain the higher the need for a dense charging or refueling infrastructure network. Therefore, the proposal for the Alternative Fuel Infrastructure Regulation [12] foresees refueling options for electricity at intervals of 60 to 100 km, for compressed hydrogen at intervals of 150 km, and for liquid hydrogen at intervals of 450 km in 2030.

By using a capacity-constrained flow refueling location model, Rose et al. [41] calculated the minimum required number of hydrogen refueling stations for Germany. Assuming full conversion of the truck fleet to hydrogen and a maximum dispensing of 30 t of hydrogen per day per refueling station, they calculated a hydrogen refueling network of 142 stations. The optimizing approach chosen identifies the smallest possible number of highly utilized station locations. On the left panel of Fig. 6, it is evident that the determined number of hydrogen refueling stations more than halves the network of highway refueling stations available today for diesel fuel. Only few refueling stations are set up on less frequented highways, for example, in eastern Germany. The liquefied natural gas (LNG) stations network (right panel of Fig. 6), which is currently under construction, also shows a concentration on heavily traveled road sections, but to a lesser extent than in the optimized hydrogen refueling stations network.

To reduce the focus on heavily trafficked routes and to ensure a high level of coverage in the rest of the road network, it is reasonable to follow the approach of the European Commission. The Commission's AFIR proposal [12] foresees infrastructure at regular intervals along major transportation corridors. Depending on the local traffic volume, the individual locations can be individually dimensioned. Assuming that, for example in 2030, 15% of heavy road freight transport will be electrified and every second charging event will take place publically, Speth et al. [43] designed a public fast-charging network for trucks in Germany with an average distance of 100 km between stations. The network of megawatt chargers shown in Fig. 7 (left panel) thus has the same number of locations as the hydrogen refueling network in [41]. However, stations are much more spread out across the country, and scaling takes place in the local design. While small locations are equipped with two charging points, large locations contain up to 13 charging points. The same effect is shown in Speth et al. [44] in a European network, assuming 15% battery electric trucking, 25% public charging, and 100 km average distance between two charging locations. While stations with up to 18 charging points are positioned at central transshipment points, for example, ports, and along highly-trafficked roads in Central Europe, smaller stations are planned for Northern and Southern Europe.

The design of a future charging or refueling network probably will represent a compromise between the approaches of Speth et al. [43] and Rose et al. [41]. While

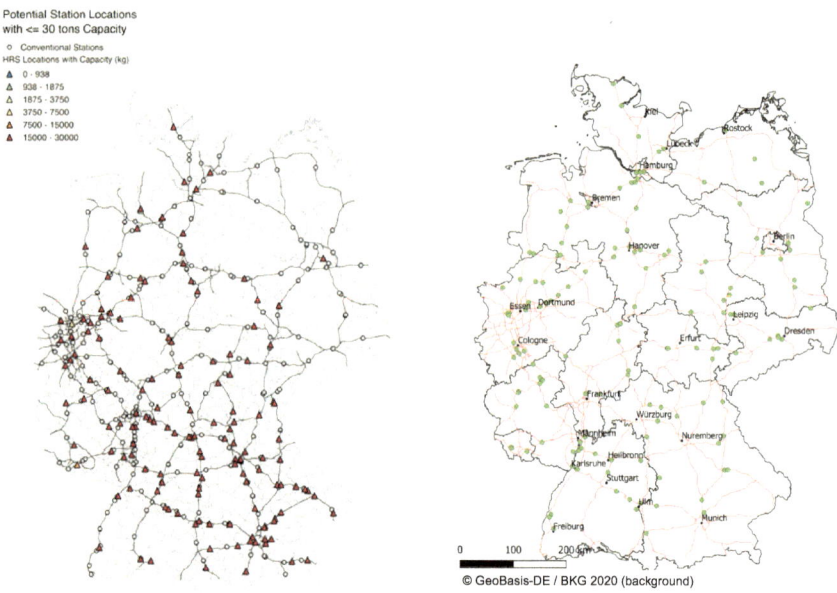

Fig. 6 *Left panel:* Regional distribution of 360 existing highway refueling stations (dots) and 142 potential hydrogen refueling stations with a maximum capacity of 30 t per day (triangles). Adopted from Rose et al. [41]. *Right panel:* Regional distribution of 150 existing liquefied natural gas (LNG) highway refueling stations (dots). Adopted from Deutsche Energie-Agentur [42]

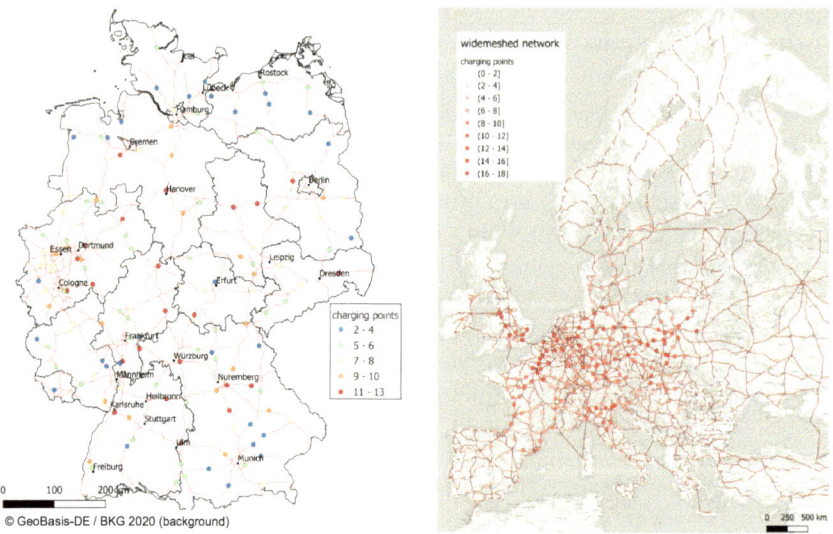

Fig. 7 *Left panel:* Regional distribution of 142 fast-charging locations with a mean distance of 100 km. Adopted from Speth et al. [43]. *Right panel:* Regional distribution of 660 fast-charging locations in Europe with a mean distance of 100 km. Adopted from Speth et al. [44]

Rose et al. [41] offer relevant insights for an initial minimal network, Speth et al. [43] show a network that is convenient for the users with a high level of resilience.

From today's (ex ante) perspective, infrastructure scenarios for road freight transport are inherently subject to uncertainty. Although electrification is a key factor in all scenarios (Fig. 4) and is probably a no-regret option, the role of each alternative powertrain remains still uncertain. Therefore, local infrastructure needs and the resulting local energy demand as well as infrastructure costs are also uncertain today. However, initial estimates show that the costs for an initial network for charging infrastructure, overhead catenary lines, and hydrogen refueling stations could be at least in the same order of magnitude. For example, Kühnel et al. [45] calculated infrastructure investments of 2.3 billion € for hydrogen refueling stations, 3.7 billion € for charging stations (megawatt charging and depot charging), or 5.1 billion € for overhead catenary lines. Each of the infrastructures in Kühnel et al. [45] can supply up to 40,000 vehicles. This corresponds to approximately 10% of the heavy-duty vehicles in Germany.

7   Social Acceptance of Alternative Fuels for Trucks

Acceptance research for low- and zero-emission heavy-duty road transport is still in the early stages, but first results are available for different actor groups. The social acceptance for alternative fuels can be captured based on three dimensions: local or community acceptance, sociopolitical acceptance, and market acceptance [46]. Local acceptance deals with the acceptance of individuals and groups, who live close to the implementation site of a new technology or project and are directly or potentially affected by it. It also includes the views of local authorities. Sociopolitical acceptance refers to the general public acceptance or societal climate toward a technology and includes views of the public as well as key stakeholders and policymakers. Finally, market acceptance captures the acceptance of individuals and firms on both the supply and demand side of the market that is most relevant for the technology.

Local acceptance. The local dimension and therefore acceptance of synthetic fuels for heavy-duty road transport depends on the specific characteristics of the respective fuel and on its supply chain. If the synthetic diesel fuels closely resemble the currently used diesel fuels, the local environment, which the local public is used to, will not change. Vehicles will basically look the same, and refueling stations and their infrastructure will predominantly remain unchanged.

If, for example, synthetic hydrogen is introduced as an option, changes in fueling infrastructure can be necessary, including high-pressure tanks and additional hydrogen grids. This could pose challenges for local acceptance both in terms of safety concerns and with regard to the build-up of new infrastructure on public land and in vicinity to communities [47, 48]. Local acceptance becomes increasingly complex and a potential barrier if parts of the supply chain of synthetic fuels are also envisioned to be carried out in Germany. Then additional renewable energy production, e.g., through wind turbines, and the large-scale build-up of electrolysers would be necessary. Visible new energy generation infrastructures, such as wind turbines, have posed challenges for local acceptance in the past already and need to be taken into account when envisioning local supply chains [49, 50].

Sociopolitical acceptance. Sociopolitical acceptance toward synthetic fuels can be analyzed based on the official standpoint of the German government and based on views of the general society as captured in household survey data and media. In its concept on climate-friendly heavy-duty vehicles, the German government focuses predominantly on battery electric, hydrogen fuel cell electric, and catenary electric vehicles [51]. However, the concept also states that a technology-open approach is still pursued and that power-to-liquid (PtL) diesel fuels (synthetic diesel fuels) are considered as "further fuels", which can contribute to defossilize road freight transportation next to these three main options (BMVI [51], p. 7).

A household survey in large German cities in 2020 found that attitudes toward synthetic fuels were neutral to slightly positive for both cars and trucks [52]. The public support for trucks powered by synthetic fuels was also slightly above medium. Scheer et al. [53] compare German stakeholder positions toward synthetic fuels and find a wide variety of narratives and positions. They derive three narratives across different stakeholders ranging from a narrative that consider synthetic fuels a key component for the mobility transition, to synthetic fuels only playing a niche role next to electrification or other approaches in the wider mobility transition, such as intermodal and interoperable transport. Finally, interviews about the usage of renewable methane for trucks with stakeholders in the heavy-duty road transport sector in 2021 reveal neutral to negative attitudes [32, 54]. On the one hand, synthetic methane is seen as a promising alternative for reducing CO_2 emissions in demanding applications, such as long-haul transport and for utilizing existing gas infrastructure and vehicles. On the other hand, challenges with costs, efficiency, and availability of both the fuel and the necessary production and refueling infrastructures are also mentioned. While there is agreement regarding the use of synthetic fuels in the maritime and aviation sector, opinions consequently differ when it comes to their use for heavy-duty road transportation.

Market acceptance. Market acceptance for renewable energy carriers on the supply side differs between truck manufacturers and depends on the application context. Current innovation strategies of truck manufacturers for new vehicles are clearly focused on electric drives, with room for hydrogen fuel cell vehicles for a few manufacturers. Biogas, hydrogen combustion, and synthetic diesel and gas fuels are discussed as supplementary options (Scherrer et al., forthcoming). For the current fleet, some manufacturers see synthetic diesel fuels as a potential defossilization option, while others have already announced their final diesel models and are no longer investing in R&D in this area. Statements and publications are, however, largely focused on the key markets for new vehicles in Europe, North America, and Japan, and applications for existing truck fleets and secondhand markets around the globe are possible.

On the demand side, interview results from research projects on renewable methane for trucks and on electric road systems for trucks show that logistics companies are generally technology open and willing to use alternative options which enable them to follow their previous business practices and fulfill their customers' needs [32, 54, 55]. However, logistics companies are also very price sensitive and mention total costs of ownership (TCO) as their main consideration for evaluating alternative fuels [56]. This can be an acceptance barrier for synthetic diesel fuels if other alternatives are more affordable in the long run. While other alternatives require larger upfront investments due to the necessary changes in vehicles and infrastructure, these could be captured by single-point subsidies, whereas high fuel costs would require continuous support throughout the usage phase of a vehicle. The challenge in acceptance increases for power-to-gas (PtG) fuels—such as hydrogen or methane—where both a modification of vehicles and a potentially more expensive fuel infrastructure are necessary [32, 54]. Overall, actors on the demand side are not opposed to synthetic fuels but remain observant of price developments and the future design of government subsidies.

8   Summary

Heavy-duty trucks are responsible for a relatively large share of CO_2 emissions in road transportation and are thus subject for a number of reduction efforts. For this reason, a number of policy measures and regulations have been adapted recently (e.g., the alternative fuels infrastructure regulation (AFIR) has been agreed on in the trilogue in March 2023) or will be updated soon (e.g., vehicle emission standards, infrastructure mandates, and CO_2 limitations for fuels and subsidies). A technical comparison reveals that the development of battery electric trucks (TRL 7–9) is much further than for fuel cell electric trucks (TRL 7), catenary electric trucks (TRL 6) and trucks powered by renewable fuels (TRL 6–7). An economic assessment does not show a clear advantage of alternative fuel trucks compared to conventional fuels in the future under the applied framework conditions and assumptions. However, from an environmental perspective, the use of these fuels enables a significant and immediate reduction of CO_2 emissions, highlighting their environmental benefits and advantageous application even today. The future transition toward climate neutrality in heavy-duty road transportation by 2050 will require varying amounts of renewable electricity, depending on the renewable fuel options and technologies employed. In this regard, electric options should be the preferred choice whenever feasible. Overall, this requires a large-scale infrastructure setup, yet at comparable investment between solutions (2.3 to 5.1 billion € for Germany). This has to go in hand with local user acceptance which is higher if local residents are not visually affected and potential users (truck drivers) can be convinced by the advantages. The current political efforts are mainly focused on direct electric or hydrogen electric trucks in the future.

Acknowledgements This comparison is part of the project "ewayBW" funded by the Federal Ministry for Economy and Energy (grant no. 16EM4017-1).

References

1. Kluschke P, Gnann T, Plötz P, Wietschel M (2019) Market diffusion of alternative fuels and powertrains in heavy-duty vehicles: a literature review. Energy Rep 5:1010–1024. https://doi.org/10.1016/j.egyr.2019.07.017
2. European Automobile Manufacturers' Association (2020) Road freight transport on the way to carbon neutrality. Policy Paper. European Automobile Manufacturers' Association. Available at https://acea.be/uploads/press_releases_files/ACEA_Policy_Paper-Road_freight_transport_on_the_way_to_carbon_neutrality.pdf
3. Plötz P, Jakob W, Gnann T, Neuner F, Speth D, Link S (2021) Net-zero-carbon transport in Europe until 2050—Targets, technologies and policies for a long-term EU strategy. Targets, technologies and policies for a long-term EU strategy. Fraunhofer ISI, Karlsruhe. Available at https://www.isi.fraunhofer.de/content/dam/isi/dokumente/cce/2021/EU_Transport_policybrief_long.pdf
4. European Automobile Manufacturers' Association (2016) Reducing CO_2 emissions from heavy-duty vehicles. Empowering customers, strengthening market forces and working in an integrated approach. Position Paper. European Automobile Manufacturers' Association
5. Ragon P-L, Rodriguez F (2021) CO_2 emissions from trucks in the EU: An analysis of the heavy-duty CO_2 standards baseline data. International Council on Clean Transportation, Berlin
6. Ovaere M, Proost S (2022) Cost-effective reduction of fossil energy use in the European transport sector: An assessment of the Fit for 55 Package. Energy Policy 168:113085. https://doi.org/10.1016/j.enpol.2022.113085
7. EU (ed) (2019) Regulation (EU) 2019/1242 of the European Parliament and of the Council of 20 June 2019 setting CO_2 emission performance standards for new heavy-duty vehicles and amending Regulations (EC) No 595/2009 and (EU) 2018/956 of the European Parliament and of the Council and Council Directive 96/53/EC
8. Breed AK, Speth D, Plötz P (2021) CO_2 fleet regulation and the future market diffusion of zero-emission trucks in Europe. Energy Policy 159:112640. https://doi.org/10.1016/j.enpol.2021.112640
9. Suzan S (2021) Easy ride: why the EU truck CO_2 targets are unfit for the 2020s. In: Transport & environment
10. NOW (2023) Market development of climate-friendly technologies in heavy-duty road freight transport in Germany and Europe. Evaluation of the 2022 cleanroom talks with truck manufacturers. Commissioned by the Federal Ministry for Digital and Transport (BMDV) in Germany. National Organisation hydrogen and Fuel Cell Technology
11. EC (2023b) Proposal for a regulation of the European parliament and of the council amending regulation (EU) 2019/1242 as regards strengthening the CO_2 emission performance standards for new heavy-duty vehicles and integrating reporting obligations, and repealing Regulation (EU) 2018/956. Strasbourg. Accessed 6 Apr 2023
12. European Commission (ed) (2021b) Proposal for a Regulation of the European Parliament and of the Council on the deployment of alternative fuels infrastructure, and repealing Directive 2014/94/EU of the European Parliament and of the Council
13. EC (2023a) European green deal: ambitious new law agreed to deploy sufficient alternative fuels infrastructure. Press release. European Commission, Brussels. Accessed 18 July 2023
14. European Commission (ed) (2021a) Proposal for a Directive of the European Parliament and of the Council amending Directive (EU) 2018/2001 of the European Parliament and of the Council, Regulation (EU) 2018/1999 of the European Parliament and of the Council and Directive 98/

70/EC of the European Parliament and of the Council as regards the promotion of energy from renewable sources, and repealing Council Directive (EU) 2015/652

15. European Parliament (ed) (2022) Renewable energy directive ***I. Amendments adopted by the European Parliament on 14 September 2022 on the proposal for a directive of the European Parliament and of the Council amending Directive (EU) 2018/2001 of the European Parliament and of the Council, Regulation (EU) 2018/1999 of the European Parliament and of the Council and Directive 98/70/EC of the European Parliament and of the Council as regards the promotion of energy from renewable sources, and repealing Council Directive (EU) 2015/652 (COM (2021)0557—C9–0329/2021—2021/0218(COD)) (1)

16. EU (ed) (2023) Directive (EU) 2023/959 of the European Parliament and of the Council of 10 May 2023 amending Directive 2003/87/EC establishing a system for greenhouse gas emission allowance trading within the Union and Decision (EU) 2015/1814 concerning the establishment and operation of a market stability reserve for the Union greenhouse gas emission trading system

17. Speth D, Kappler L, Link S, Keller M (2022a) Attractiveness of alternative fuel trucks with regard to current tax and incentive schemes in Germany: a total cost of ownership analysis. In: 35th International electric vehicle symposium and exhibition (EVS35)

18. Basma H, Saboori A, Rodriguez F (2021) Total cost of ownership for tractor-trailers in Europe: battery electric versus diesel. International Council on Clean Transportation, Berlin

19. Nykvist B, Olsson O (2021) The feasibility of heavy battery electric trucks. Joule 5(4):901–913. https://doi.org/10.1016/j.joule.2021.03.007

20. Phadke A, Khandekar A, Abhyankar N, Wooley D, Rajagopal D (2021) Why regional and long-haul trucks are primed for electrification now. Lawrence Berkeley National Laboratory. Available at https://escholarship.org/uc/item/3kj8s12f

21. International Energy Agency (2021) Global EV outlook 2021. Accelerating ambitions despite the pandemic. International Energy Agency. Available at https://iea.blob.core.windows.net/ass ets/ed5f4484-f556-4110-8c5c-4ede8bcba637/GlobalEVOutlook2021.pdf

22. Basma H, Rodriguez F (2022) Fuel cell electric tractor-trailers: technology overview and fuel economy. Berlin

23. Ash N, Davies A, Newton C (2020) Renewable electricity requirements to decarbonise transport in Europe with electric vehicles, hydrogen andelectrofuels. Investigating supply-side constraints to decarbonising the transport sector in the European Union to 2050. RicardoAEA Ltd; Transport & Environment, Brussels

24. Traton (ed) (2021) Fraunhofer analysis sees battery-electric trucks at an advantage over hydrogen trucks

25. Funke SÁ (2018) Techno-ökonomische Gesamtbewertung heterogener Maßnahmen zur Verlängerung der Tagesreichweite von batterieelektrischen Fahrzeugen. Dissertation, Universität Kassel (Ed.), Kassel

26. Muratori M, Borlaug B, Ledna C, Jadun P, Kailas A (2023) Road to zero: Research and industry perspectives on zero-emission commercial vehicles. iScience 26(5):106751. https://doi.org/10.1016/j.isci.2023.106751

27. Mauler L, Dahrendorf L, Duffner F, Winter M, Leker J (2022) Cost-effective technology choice in a decarbonized and diversified long-haul truck transportation sector: A U.S. case study. J Energy Storage 46:103891. https://doi.org/10.1016/j.est.2021.103891

28. Zerhusen J, Landinger H, Astono Y, Böhm M, Pagenkopf J, Heckert F (2023) H_2-Infrastruktur für Nutzfahrzeuge im Fernverkehr. Aktueller Entwicklungsstand und Perspektiven. Ludwig-Bölkow-Systemtechnik GmbH; Deutsches Zentrum für Luft- und Raumfahrt e.V., Stuttgart

29. AVL zsw (2021) Systemvergleich zwischen Wasserstoffverbrennungsmotor und Brennstoffzelleim schweren Nutzfahrzeug. Eine technische und ökonomische Analyse zweier Antriebskonzepte. AVL; zsw, Stuttgart

30. H2 Mobility (2021) Wasserstoffbetankung von Schwerlastfahrzeugen—die Optionen im Überblick. Available at https://h2-mobility.de/wp-content/uploads/sites/2/2021/10/H2M_Ueb erblick_BetankungsoptionenLNFSNF_TankRast_2021-10-21.pdf. Accessed 5 Oct 2022

31. Rivard, E.; Trudeau, M.; Zaghib, K. (2019): Hydrogen Storage for Mobility: A Review. In: Materials (Basel, Switzerland), 12 (12). https://doi.org/10.3390/ma12121973.

32. Wietschel M, Oberle S, Akca Subasi M, Speth D, Lux B, Scherrer A, Gnann T, Burghard U, Pfluger B, Kunze R, Steyer N, Erler R, Köppel W, Vayas L, Zubair A, Monsalve C, Pemsel J, Lozanovski A (2022) Systemanalytische Untersuchungen zur Evaluierung der Rolle von EE-Methan. Endbericht Verbund 6—MethSys. Studie im Auftrag des BMWi. Karlsruhe. Available at https://doi.org/10.24406/publica-199
33. Jöhrens J. Helms H (2020) Roadmap für die Einführung eines Oberleitungs-Lkw-Systems in Deutschland. ifeu GmbH. Heidelberg
34. Anderhofstadt B, Spinler S (2019) Factors affecting the purchasing decision and operation of alternative fuel-powered heavy-duty trucks in Germany—A Delphi study. Transp Res Part D: Transp Environ 73:87–107. https://doi.org/10.1016/j.trd.2019.06.003
35. Bae Y, Mitra SK, Rindt CR, Ritchie SG (2022) Factors influencing alternative fuel adoption decisions in heavy-duty vehicle fleets. Transp Res Part D Transp Environ 102:103150. https://doi.org/10.1016/j.trd.2021.103150
36. Noll B, Del Val S, Schmidt TS, Steffen B (2022) Analyzing the competitiveness of low-carbon drive-technologies in road-freight: A total cost of ownership analysis in Europe. Appl Energy 306:118079. https://doi.org/10.1016/j.apenergy.2021.118079
37. Widegren F, Helms H, Hacker F, Andersson M, Gnann T, Eriksson M, Plötz P (2022) Ready to go? Technology readiness and life-cycle emissions of electric road systems. A discussion paper from the CollERS2 project
38. Krail M, Speth D, Gnann T, Wietschel M (2021) Langfristszenarien für die Transformation des Energiesystems in Deutschland. Treibhausgasneutrale Hauptszenarien Modul Verkehr. Studie im Auftrag des BMWi. Fraunhofer ISI, Karlsruhe
39. Gnann T, Speth D, Krail M, Wietschel M, Oberle S (2022) Pathways to carbon-free transport in Germany until 2050. World Electr Veh J 13(8):136. https://doi.org/10.3390/wevj13080136
40. Plötz P (2022) Hydrogen technology is unlikely to play a major role in sustainable road transport. Nat Electron 5(1):8–10. https://doi.org/10.1038/s41928-021-00706-6
41. Rose PK, Nugroho R, Gnann T, Plötz P, Wietschel M, Reuter-Oppermann M (2020) Optimal development of alternative fuel station networks considering node capacity restrictions. Transp Res Part D Transp Environ 78:102189. https://doi.org/10.1016/j.trd.2019.11.018
42. Deutsche Energie-Agentur (2022) Öffentliche LNG-Tankstellen in Deutschland. Available at https://www.dena.de/themen-projekte/projekte/mobilitaet/lng-taskforce-und-initiative-erdgasmobilitaet/. Accessed 10 Oct 2022
43. Speth D, Plötz P, Funke S, Vallarella E (2022b) Public fast charging infrastructure for battery electric trucks—a model-based network for Germany. Environ Res Infrastruct Sustain 2(2):25004. https://doi.org/10.1088/2634-4505/ac6442
44. Speth D, Sauter V, Plötz P (2022c) Where to charge electric trucks in Europe—modelling a charging infrastructure network. World Electr Veh J 13(9):162. https://doi.org/10.3390/wevj13090162
45. Kühnel S, Hacker F, Görz W (2018) Oberleitungs-Lkw im Kontext weiterer Antriebs- und Energieversorgungsoptionen für den Straßengüterfernverkehr. Ein Technologie- und Wirtschaftlichkeitsvergleich. Öko-Institut, Berlin
46. Wüstenhagen R, Wolsink M, Bürer MJ (2007) Social acceptance of renewable energy innovation: an introduction to the concept. Energy Policy 35(5):2683–2691. https://doi.org/10.1016/j.enpol.2006.12.001
47. Emodi NV, Lovell H, Levitt C, Franklin E (2021) A systematic literature review of societal acceptance and stakeholders' perception of hydrogen technologies. Int J Hydrogen Energy 46(60):30669–30697. https://doi.org/10.1016/j.ijhydene.2021.06.212
48. Schönauer A-L, Glanz S (2022) Hydrogen in future energy systems: Social acceptance of the technology and its large-scale infrastructure. Int J Hydrogen Energy 47(24):12251–12263. https://doi.org/10.1016/j.ijhydene.2021.05.160
49. Devine-Wright P (2005) Beyond NIMBYism: towards an integrated framework for understanding public perceptions of wind energy. Wind Energy 8(2):125–139. https://doi.org/10.1002/we.124

50. Reusswig F, Braun F, Heger I, Ludewig T, Eichenauer E, Lass W (2016) Against the wind: local opposition to the German Energiewende. Utilities Policy 41:214–227. https://doi.org/10.1016/j.jup.2016.02.006
51. Bundesministerium für Verkehr und digitale Infrastruktur (2020) Gesamtkonzept klimafreundliche Nutzfahrzeuge. Mit alternativen Antrieben auf dem Weg zur Nullemissionslogistik auf der Straße. Berlin. Available at https://www.now-gmbh.de/wp-content/uploads/2021/06/Gesamtkonzept_klimafreundliche-nutzfahrzeuge_BMVI-1.pdf
52. Scherrer A (2023) How media coverage of technologies affects public opinion: Evidence from alternative fuel vehicles in Germany. Environ Innov Societ Trans 47:100727. https://doi.org/10.1016/j.eist.2023.100727
53. Scheer D, Schmieder L (2023) Stakeholder discourse on synthetic fuels: a positioning and narrative analysis. Fuels 4(3):264–278. https://doi.org/10.3390/fuels4030017
54. Burghard U, Scherrer A (2022) Are synthetic fuels a promising option for ships and trucks? An investigation of actors and acceptance of renewable methane in Germany. Hyères, France. Available at https://www.eceee.org/static/media/uploads/site-2/summerstudy2022/pdfs_docs/abstracts_2022_rev4june.pdf
55. Zembrot M, Böttiger H, Schollbach F, Flotho R, Wietschel M, Burghard U, Scherrer A, Gnann T, Speth D, Plötz P, Fritz J, Doll C, Brauer C, Waßmuth V, Köllermeier N, Schüller H, Lauer M, Burgert T, Berg LF (2021) Oberleitungs-LKW als ein Baustein für ein nachhaltiges Verkehrssystem: Das Projekt eWayBW in Baden-Württemberg. Fraunhofer ISI, Karlsruhe. Available at https://publica-rest.fraunhofer.de/server/api/core/bitstreams/8e46abf9-56cb-4205-b6b1-65d1cd805db2/content
56. Rose P (2020) Modeling a potential hydrogen refueling station network for fuel cell heavy-duty vehicles in Germany in 2050. https://doi.org/10.5445/IR/1000119521

Power-To-Liquid (PTL) Kerosene and Opportunities to Introduce Green Hydrogen in Aviation

Valentin Batteiger, Kathrin Ebner, Leonard Moser, Christina Penke, Benjamin Portner, and Andreas Sizmann

Abstract During the past years, power-derived fuels have moved into the center of the debate on a future renewable energy supply for the aviation sector, due to a continued need for liquid fuels in aviation and serious concerns about the sustainability of biofuel production at the scale of future fuel demand. Basic characteristics of kerosene synthesis via power-to-liquid pathways are reviewed with focus on the application of synthetic kerosenes in civil aviation. Furthermore, alternative use cases for green hydrogen in aviation are discussed.

Keywords Synthetic Kerosene · Fischer-Tropsch · Methanol-to-Jet · Hydrogen Aviation · Fully Synthetic Jet Fuel

1 Introduction

1.1 Kerosene-Type Aviation Fuels and the Need for Electricity-Based Energy Carriers

Throughout the history of commercial aviation, the aviation industry has relied on fuels derived from crude oil. Notable exceptions include aviation fuel production from coal and natural gas reserves at industrial scale [1]. Furthermore, an increasing number of smaller-scale demonstrations and first commercial plants for the production of renewable jet fuel via various conversion pathways exist. In 2021, approximately 125 million liters (ca. 100 kt) of renewable jet fuel were produced globally [2], mainly via the HEFA conversion pathway. Yet, this is less than 0.1% compared to a global jet fuel demand of about 300 Mt/a (pre-Covid level in 2019) [3].

These orders of magnitude are about to change. Several countries implement regulatory frameworks that will likely induce an uptake of renewable jet fuels at the 10 Mt/a level within this decade and on a global level. In the European Union (EU),

V. Batteiger (✉) · K. Ebner · L. Moser · C. Penke · B. Portner · A. Sizmann
Bauhaus Luftfahrt, Taufkirchen, Germany
e-mail: valentin.batteiger@bauhaus-luftfahrt.net

849

for instance, the ReFuelEU Aviation initiative foresees a mandate to foster the uptake of renewable jet fuels in the EU, targeting a minimum share of 6% in 2030. This includes a sub-target of 1.2% specifically for synthetic fuels that aims to scale-up Power-to-Liquid (PtL) jet fuels [4].

A transition to renewable fuels is widely acknowledged as necessity to reduce CO_2 emissions in aviation as most observers expect a future growth in jet fuel demand. During the past decades, annual fuel efficiency increases of about 1–1.5% [5] were outpaced by the annual increases in air transport demand of 3–6%. Projections of future jet fuel demand suggest a continued difference of air traffic growth and fuel efficiency improvements. One main driver of further fuel demand growth, despite a rising awareness about the climate impact of aviation, is an increasing share of the global population that can afford air travel. In this context, Fig. 1 shows a scenario in which annual fuel consumption increases toward 430–530 Mt/a by 2050. In such a scenario, replacing fossil fuels by renewable fuel options becomes the crucial option to significantly reduce CO_2 emission that would otherwise grow.

Furthermore, the aviation industry has set the target to operate net carbon–neutral by 2050, which requires an almost complete substitution of current jet fuel use by renewable and sustainable energy carriers within only three decades [6]. In contrast to road transportation, the potential for CO_2 emission reduction through battery-electrification is strongly limited for aviation purposes due to the relatively low volumetric and gravimetric energy density of batteries (over an order of magnitude lower than for kerosene). While electrifying urban and regional air traffic is certainly of interest [7], the majority of aviation CO_2 emissions result from mid- to long-range flights [8], for which batteries are not considered a suitable energy carrier. Instead, liquid fuels will be required to sustain air travel in a climate neutral world.

A certain share of future jet fuel demand will likely be met by renewable jet fuels based on biomass options. Renewable jet fuels derived from residual feedstocks

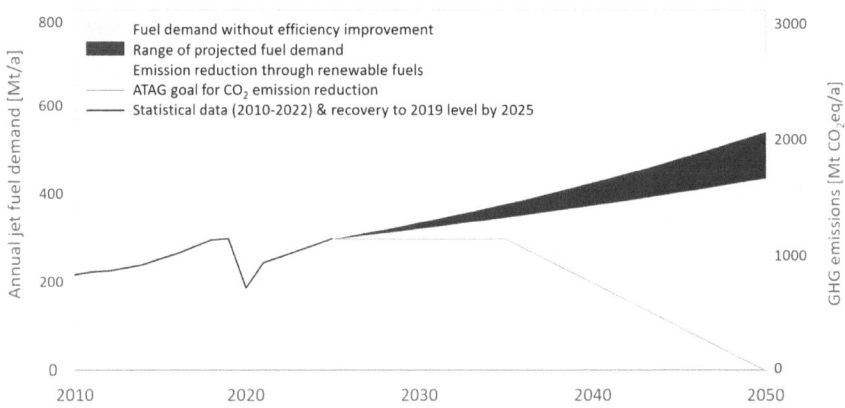

Fig. 1 Historic and potential future development of global jet fuel demand (this scenario assumes a recovery to 2019 fuel demand levels by 2025, and a projected fuel demand of 430–530 Mt by 2050)

such as used cooking oil (UCO) are expected to constitute a significant portion of the renewable jet fuel supply in the coming years as they are already produced based on technological mature production pathways. However, such fuels alone cannot meet future jet fuel demand due to strongly limited feedstock availability [9, 10]. Other alternative biomass-based options also face certain limitations and restrictions. For instance, an excessive expansion of the cultivation of traditional biogenic energy crops, such as vegetable oils, sugars or starch, should be avoided for sustainability reasons [11].

So-called next generation or advanced biofuel technologies, that can convert more abundant lignocellulosic feedstock into jet fuels are under development. However, a number of techno-economic evaluations suggest that electricity-based fuels can favorably compete with advanced biofuels from cultivated lignocellulosic feedstock [12, 13]. This estimation is essentially based on a continuing cost reduction of electricity generation from solar and wind energy. More recently, a corresponding development in cost reduction is observed for electrolysis technology. Accordingly, it is expected by many experts that electricity-based fuels from solar and wind energy become a corner stone of the future energy supply of aviation.

Against this background, various aspects on the supply and use of synthetic Power-to-Liquid (PtL) kerosene options are presented in this chapter. First, an overview of kerosene range hydrocarbons as aviation turbine fuel is given. The aspects on the supply and use of PtL-based synthetic kerosene options and on a use toward fully synthetic kerosene are presented in more detail. Lastly, different options to integrate hydrogen in the aviation sector, alongside the application of synthetic PtL jet fuel, are presented.

1.2 Clean Aviation Fuels and Their Role in Reducing Air Quality Emissions and Non-CO_2 Climate Impacts

The combustion of jet fuels in turbine engines is not only associated with CO_2 emissions, but also with the impact of further pollutant emissions such as nitrogen oxide (NO_x), sulfur oxides (SO_x), or soot. In addition to the climate impacts of CO_2 emissions, two important environmental impacts of civil aviation are air quality emissions and non-CO_2 climate impact. Fuel options, which allow for a "cleaner" combustion compared to conventional jet fuel, such as synthetic paraffinic kerosenes (SPK) resulting from Power-to-Liquid (PtL) pathways, are associated with significant reductions in air quality emissions and non-CO_2 climate impact at cruise altitudes. These effects and further environmental and technical benefits of synthetic jet fuels are discussed in the following.

Emissions affecting air quality. The air pollutant emissions from jet fuel combustion in current aircraft engines depend on fuel properties and combustion conditions. In turbine engines, jet fuel is sprayed through a nozzle into the combustion chamber where the fuel is continuously burned with pressurized air. Spray behavior, air/fuel mixing ratio, temperature and pressure levels are key levers to influence air pollutant emissions by combustion technology. The heated air mass drives a series of turbines to pressurize the incoming air flow and to provide shaft power to fans or rotors. The airflow, including the combustion products, is then exhausted to the environment; in particular, there is no filtering or other means of exhaust after treatment in the current fleet of aircraft. Removing components that affect pollution emissions from the fuel used is therefore an important option to reduce such emissions in aviation.

Pollutants such as sulfur oxides (SO_x) and particulate matter emissions are heavily linked to fuel composition. The correlation between sulfur content and sulfur oxides (SO_x) emissions is obvious. Synthetic jet fuels that are produced from water (H_2O) and carbon dioxide (CO_2) are essentially free of sulfur. Current specifications set the maximum sulfur content that may result from contamination of synthetic jet fuel to 2 ppm [14]. For comparison, conventional (fossil) jet fuel may contain up to 3000 ppm of sulfur [15]. Consequently, sulfur oxide (SO_x) emissions can be reduced—and ultimately eliminated—by synthetic fuels. It is also worth to note that regulations limit the sulfur content of road transportation fuels to much lower values of 10–15 ppm in most developed countries. Hydrodesulfurization is a well-established refinery process, consequently sulfur may also be removed from conventional (fossil) jet fuel in a similar way as it is currently removed from gasoline and diesel for road transportation.

Furthermore, synthetic paraffinic kerosenes (SPK) do practically not contain aromatics such as Power-to-Liquid (PtL) fuels in form of Fischer–Tropsch (FT) SPK, where the aromatic content is limited to less than 0.5 wt-%. In contrast, conventional jet fuel typically contains aromatics at a level of 14 to 20% [16]. The correlation between the aromatic content of the jet fuel and particle formation is well established [17]. Fuel blends with increasing shares in synthetic fuels can significantly reduce soot and ultra-fine particle emissions [18–20]. However, particle emissions from current aircraft engines are not completely suppressed. Recent in-flight measurements provide evidence that burning low aromatic synthetic fuel can result in a 50–70% reduction in soot concentrations [25].

In addition to the reduction of air pollution, which is of particular relevance in airport environments, reduced sulfur emissions and reduced sooting is also beneficial for the aircraft engines themselves; synthetic fuels that allow for a cleaner combustion can reduce wear and maintenance effort and may ultimately influence engine design in a positive way.

Measurements from the exhaust of aviation turbine engines indicate that further air pollutants, such as unburned hydrocarbons (UHC), carbon monoxide (CO), as well as nitrogen oxide (NO_x) are not measurably changed by replacing conventional jet fuel by blends with shares in synthetic kerosene [20]. A reduction of these pollutants must mainly come from adaptions of the combustion process. In this regard, synthetic fuel may provide an option for a future co-optimization of fuels and engines that may significantly reduce pollutant emissions in the longer-term future.

Non-CO_2 climate impact. The full climate impact of jet fuel combustion results from CO_2 emissions as well as the so-called non-CO_2 climate impact. Studies suggest that the climate impact of non-CO_2 emissions is of the same order of magnitude as the impact of aviation related CO_2 emissions. A much-cited study [21] estimated that climate impacts induced by non-CO_2 emissions may account for about two-thirds (2/3) of the total effective radiative forcing related to aviation.

The local short-term effect of CO_2 emissions from individual flights is marginal, instead CO_2 emissions from aviation add to the CO_2 concentration of the atmosphere in a cumulative manner. Regardless of the location of occurrence, i.e., during flight or at ground, a given amount of CO_2 will essentially lead to the same climate effect. In contrast, non-CO_2 climate impacts mainly act at cruise altitudes and change the radiative balance of the atmosphere on much shorter time-scales. At typical cruise altitudes of civil aircraft, the dominating contributions to non-CO_2 climate impacts stem from aviation induced cloudiness and effects induced by nitrogen oxide (NO_x) emissions. The latter emissions are strongly determined by the combustion process. Measurements with current aircraft engines indicate that a substitution of conventional (fossil) jet fuel by synthetic counterparts has little effect on nitrogen oxide (NO_x) emissions [20]. However, the composition of the fuel has a well-documented effect on the properties of contrails and induced cirrus clouds [22]. Various particles can act as condensation nuclei during contrail formation, but the nucleation of ice particles is currently dominated by particulate emissions that stem from the combustion of jet fuels [23]. A replacement of current jet fuel with a synthetic fuel that is less prone to particle formation will reduce the particle number. In consequence, smaller numbers of ice crystals are formed and their average mass tends to be larger. This reduces both the optical density and the lifetime of aviation induced clouds [24, 25]. Particle forming species may also be removed from conventional jet fuel via refinery processes such as hydrodearomatization, similar to the removal of sulfur via hydrodesulfurization. Such a "clean-up" of the conventional blend component may be the more effective way to reduce the non-CO_2 climate impact of aviation in the short-term [26]. Synthetic fuels that allow for a cleaner combustion reveal their potential once their production is scaled up and they are used at high shares. Non-CO_2 climate impacts strongly depend on ambient atmospheric conditions, therefore it can be beneficial to make targeted use of limited amounts of low aromatic fuels by supplying individual flights with high blending ratios. Such an approach aims to achieve the highest possible reduction in climate impact during a transition period where the overall availability of low aromatic fuels is still limited [27].

2 Kerosene Range Hydrocarbons as a Turbine Fuel for Civil Aviation

In the road transportation and maritime sector, several alternative fuel options, such as liquefied or compressed natural gas, liquefied petroleum gas (LPG), alcohols, gasoline, diesel and heavier fuel oils, are used in different vehicles and vessels within the overall fleet. In contrast, commercial aviation is almost exclusively using kerosene-type fuels. This reliance on a single type of fuel is a consequence of the success of gas turbine engines as main source of shaft power for almost all commercial transport aircraft as well as helicopters. In aviation, liquid hydrocarbons have been chosen due to their favorable energy density and handling properties, but also due to their availability at global scale and at viable price levels. At the time of gas turbine development, kerosene range hydrocarbons have been a widely available product from crude oil refining, with a boiling range in-between the main refining products gasoline and diesel (Fig. 2).

Some requirements of current specifications for jet fuels, such as the density requirements, are a direct consequence of the typical molecular composition of the kerosene fraction that results from crude oil refining. Further specifications reflect the high safety requirements in civil aviation. Special safety considerations have to be made due to the need of global operation under a wide range of climatic conditions as well as harsh ambient conditions at cruise altitudes at or above 10,000 m. Most prominently, flash point requirements (>38 °C for Jet A and Jet A-1) ensure that the fuel is not flammable in warm climates, while cold flow properties (e.g., with regard to a maximum freezing point of −40 °C for Jet A and −47 °C for Jet A-1) ensure safe operation at typical temperature levels at cruise altitudes. As shorter chained hydrocarbons are more volatile, flash point requirements define a boundary toward lighter refining products (naphtha, gasoline), while cold flow properties limit the composition of jet fuel toward longer-chained hydrocarbons (diesel).

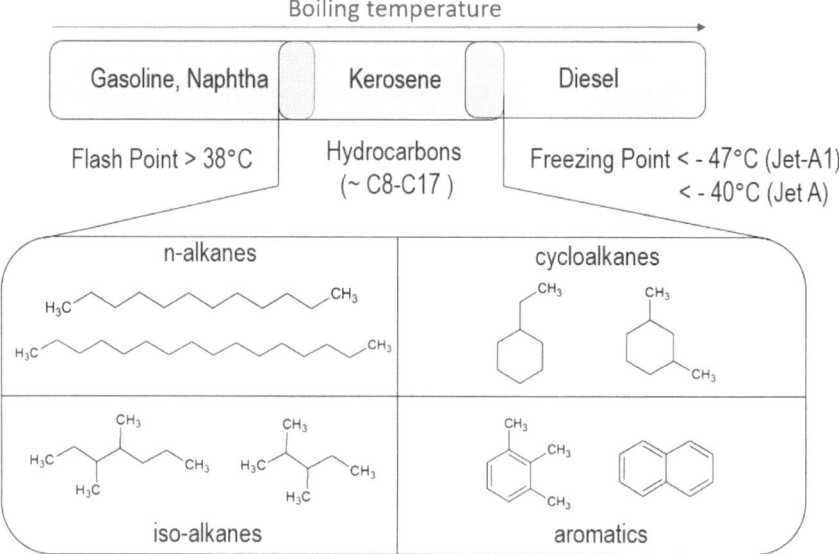

Fig. 2 Conventional jet fuel from crude oil refining contains a large variety of alkanes and aromatic hydrocarbons (only few molecules are shown for illustration). Synthetic paraffinic kerosenes, such as FT-SPK, typically contain only n-alkanes and iso-alkanes

Finally, the global character of the aviation industry led to a far-reaching harmonization of jet fuel specifications. Specifications for Jet A-1 as outlined in ASTM D1655 [15] define the most relevant international standard in civil aviation. Further national standard specifications for civil aviation fuels such as Jet A (USA), DEF STAN 91–91 (UK), N°3 jet fuel (China), or TS-1 (Russia) are kerosene-type jet fuels similar to Jet A-1.[1] The following discussion will focus on Jet A-1 specifications according to ASTM D1655. Specifications for synthetic jet fuels derived from Fischer–Tropsch (FT) synthesis are either defined as Annex to ASTM D1655, or refer to blends as defined in the ASTM D7566 specification [14] that again relate to ASTM D1655.

3 Power-to-Liquids—Synthetic Kerosene for Aviation

After the overview of the composition and use of hydrocarbons in aviation fuel, various aspects of synthetic Power-to-Liquid (PtL) kerosene are presented in this section. At the beginning, general characteristics of the PtL jet fuel production pathway are reviewed, followed by aspects on its current approval status. Differences between the chemical compositions of synthetic and conventional kerosene are then highlighted. Lastly, possibilities to supply methanol-based PtL kerosene are presented.

3.1 PtL Jet Fuel Production Pathway

In the following, a short review on basic features of production technologies regarding electricity-based jet fuels or Power-to-Liquid (PtL) jet fuels as well as their techno-economic and environmental performance is provided (for further explanations see [28, 29]). PtL fuels are produced from renewable electricity, water (H_2O) and a carbon source, usually in form of carbon dioxide (CO_2). A basic scheme for PtL jet fuel production via a Fischer–Tropsch (FT) pathway is given in Fig. 3.

As shown in Fig. 3, the main energy conversion steps are renewable electricity generation and hydrogen (H_2) production via water electrolysis. Hydrogen (H_2) is reacted with carbon dioxide (CO_2) to obtain a synthesis gas containing H_2 and CO at a molar ratio of about 2:1 for Fischer–Tropsch synthesis. The intermediate synthetic crude is refined to jet fuel and further by-products. The carbon cycle is closed by extracting the amount of CO_2 that results from fuel combustion from the atmosphere. The individual subsystems are separately discussed in the following paragraphs.

---

[1] Jet A has less stringent freezing point specifications (maximum of −40 °C). TS-1 allows for higher volatility (flash point > 28 °C) but has more stringent freezing point specifications (maximum of −50 °C) as it is typically operated in colder climates.

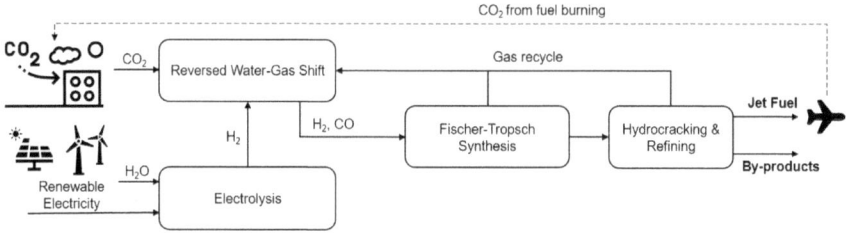

Fig. 3 Basic building blocks of a Fischer–Tropsch-based PtL pathway

Water provision. In techno-economic and environmental analyses of PtL fuel production, it turns out that the effort for water provision is almost negligible compared to the effort of water splitting via electrolysis [28]. If necessary, the water demand may be provided by seawater desalination and pipeline transport [30]. The overall low water demand, which is orders of magnitude lower compared to biofuel production, is a main benefit for PtL fuels, especially in arid areas [31]. The effort for water provision is also much lower than the effort for carbon provision.

Carbon provision. Carbon provision, usually in the form of CO_2, can become a bottleneck for PtL fuel production both at regional [32] and global scale [33]. Each PtL plant needs to secure an appropriate carbon supply over the lifetime of the plant to sustain fuel production. In addition, it is unlikely that CO_2 from jet fuel combustion will be captured on board of aircrafts. Therefore, a regenerative carbon source is required to close the carbon cycle in aviation. During early scale-up, PtL plants may be co-located or in close spatial proximity with facilities that represent suitable biogenic CO_2 sources (e.g., biomethane plants or bioethanol plants). In the long-term, it is expected that a capture of CO_2 from ambient air will be required for PtL jet fuel production at the scale of several hundreds of million tons per year. As a rule of thumb, a cost of 100 € for the provision of 1 metric ton of CO_2 translates into a cost contribution of about 25–30 ct per liter (ca. 310–375 €/t) of jet fuel product [33]. However, at the current state of art, the cost for the provision of CO_2 via direct air capture is a multiple of 100 €/t_{CO2}. Consequently, further technology scale-up and development effort is urgently needed to reduce the cost of direct air capture toward an economically tolerable regime. System analyses have shown that the energy, material and area demand of direct air capture systems is only a fraction of the energy, material and area demand of the corresponding green hydrogen production from solar or wind energy [33].

Electricity supply and area demand. The area-specific yield of PtL fuel production from solar and wind energy is much larger compared to the area-specific yield of biofuel production from cultivated energy crops [31]. Furthermore, solar and wind energy can also be harvested on land that is unsuitable for agriculture. In agricultural regions, wind parks are often co-located with farmland, as foundations and supply roads cover only a small fraction of total wind park area. Thus, PtL fuel production is largely complementary with agricultural land use. The potential availability of electricity based on solar and wind energy by far exceeds the requirements of global aviation fuel production [34]. From a pure energy supply perspective, a complete global substitution of current conventional (fossil) jet fuel consumption is plausible. However, it is important to acknowledge that the required area is still very large and the acceptance to construct solar and wind energy plants in populated areas or natural habitats may become a constraining factor.

The area demand for direct CO_2 air capture plants, electrolyzers and fuel synthesis is much lower than the area required to harvest solar and wind energy. Likewise, the cost and material demand associated with the upstream electricity generation from renewable energies is the dominating factor in techno-economic and live-cycle analyses of PtL fuel production [28, 35]. Consequently, the steeply falling cost for solar and wind electricity production [36] has fueled the interest into PtL over the past years. A continuation of this cost reduction of solar and wind electricity will be the decisive factor for the competitiveness of PtL compared to conventional jet fuel and biofuel options. The dominating role of electricity for fuel economics is the main argument why PtL is most suitable for areas with excellent resources of solar or wind energy.[2]

Electrolysis. Electrolysis is frequently the second largest cost component (after electricity generation) in techno-economic analyses of PtL fuel production, adding to the paramount role of green hydrogen production for the competitiveness of PtL. Again, excellent resources of wind or solar energy are favorable to increase the utilization of the electrolyzers. Significant cost reduction of electrolyzers may be expected from economies of scale. Alkaline electrolyzers are already a mature, cost-effective and scalable choice for PtL fuel production. Depending on future development, further technologies such as proton exchange membrane (PEM) and solid oxide electrolysis cell (SOEC) electrolyzers may offer additional benefits.

Hydrogen storage. The intermittent nature of solar and wind energy suggests hydrogen storage to utilize the downstream fuel synthesis at higher capacity factors. However, hydrogen storage comes at a cost and it may be beneficial to operate the downstream fuel synthesis in load following mode. This decision is part of the individual plant optimization and reaches beyond the scope of this book chapter.

Fuel synthesis and refining. The current baseline technology for PtL jet fuel production is Fischer–Tropsch (FT) synthesis using cobalt-based catalysts[3] followed by appropriate upgrading and refinement steps. Fischer–Tropsch (FT) synthesis requires a synthesis gas that consists of hydrogen (H_2) and carbon monoxide (CO) at a ratio of about 2:1. Consequently, it is necessary to reduce CO_2 to CO, e.g., by reacting H_2 with CO_2 (reversed water–gas shift) or by electrolytic schemes such as solid oxide

---

[2] Hydropower and high-enthalpy geothermal are further sources to supply low-cost electricity that can be used for PtL, similar to their current role in aluminium smelting, but their global potential is limited [34].

[3] Iron-based catalysts are an alternative and yield a similar product spectrum when operated at similar temperatures as cobalt-based catalysts. Iron-based catalysts can also be used for high-temperature FT synthesis; in this case an olefin rich product with shorter average chain length is obtained.

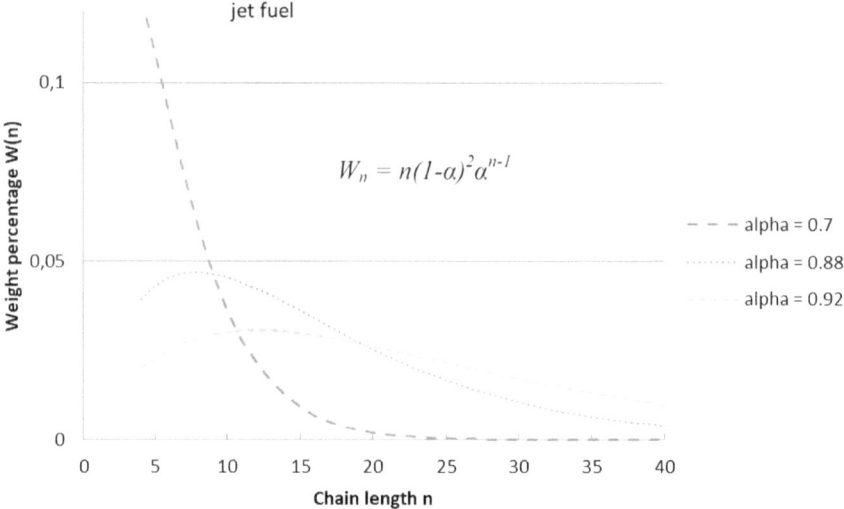

Fig. 4 Product distribution W_n of Fischer–Tropsch synthesis. Typically, a synthetic crude with long chain length is chosen and the yield within the jet fuel range (shaded area) is increased by hydrocracking

electrolysis cells (SOEC). Cobalt-based Fischer–Tropsch (FT) synthesis yields a synthetic crude with long average carbon chain length, and the majority of the raw product can be in a waxy phase (Fig. 4).

The intermediate Fischer–Tropsch (FT) synthetic crude contains a broad distribution of mainly unbranched hydrocarbons. This synthetic crude is not yet a suitable fuel product for aviation as it contains reactive unsaturated hydrocarbons (alkenes) and chemical compounds containing oxygen (oxygenates). Furthermore, the selectivity of the crude synthetic product toward kerosene range molecules is poor (Fig. 4) and the cold flow properties do not meet jet fuel requirements due to the high share of unbranched n-alkanes. Consequently, subsequent hydroprocessing steps are required to increase the selectivity toward the jet fuel product, to saturate carbon double bonds (transform the alkenes into alkanes), to remove oxygen heteroatoms and to meet jet fuel specification by producing a sufficient share of branched molecules (iso-alkanes).

Product spectrum of Fischer–Tropsch refineries. Uncomplicated refinery designs that are mainly based on hydrocracking can yield 60–65% selectivity to jet fuel [37]. More sophisticated conversion and refining schemes can move the jet fuel selectivity toward 70% and beyond. The specific PtL fuel refinery design of an individual plant can be optimized toward a variety of final product portfolios. Some basic statements on the final product basket may be made. The market potential of higher valued by-products such as waxes or lubricant base oils is limited, and the majority of PtL plants will therefore co-produce fuels or commodity chemicals.

Synthetic crude from cobalt-based Fischer–Tropsch (FT) synthesis is a favorable feedstock for diesel production due to the high cetane number of n-alkanes and cleaner combustion properties. Diesel is mainly used for heavy-duty road transportation.Battery-electric or fuel cell-electric drivetrains may significantly reduce the diesel demand in the future, while jet fuel demand is expected to grow. It is well-known from Fischer–Tropsch (FT) refining literature that jet fuel optimized plant designs do not co-produce appreciable amounts of diesel as longer-chained hydrocarbons get cracked to optimize jet fuel yield (if diesel is of lower market value) [1]. Nevertheless, diesel and jet fuel may be co-produced if the market value for the diesel product justifies such refinery designs. A jet fuel optimized Fischer–Tropsch (FT) refinery will produce a higher share of naphtha and gaseous hydrocarbons as a larger fraction of longer-chained product is cracked.

The naphtha range product can in principle be used in the gasoline blending pool, but additional processing steps are required to achieve a high octane number. It may be more favorable to use naphtha as a steam cracker feed to produce renewable commodity chemicals or as a general liquid fuel product for various purpose. The value of the gaseous hydrocarbons will critically depend on the location and the capacity of the individual plant. The C_3 and C_4 hydrocarbons can be liquefied at moderate pressure levels and may be marketed as (synthetic) liquefied petroleum gas (LPG). The separation of methane (C_1), ethane and ethene (C_2) requires more effort, but this effort may be justified depending on existing markets and infrastructures for these product gases.

Economic and environmental considerations. The concept of PtL jet fuel production was discussed for a long time [38]. Key technologies such as alkaline water electrolysis or Fischer–Tropsch (FT) synthesis were commercially available for decades. However, the availability of crude oil at more competitive price levels kept these technologies within niche applications in the past. Electricity generation is usually the largest cost-item in techno-economic analyses of PtL fuel production [35]. Historically, low-cost renewable electricity was only available in form of hydropower and in few regions also high-temperature geothermal power. A scalable generation of cost-effective electricity from solar and wind energy became broadly available during the last decade [36]. Steep cost reductions together with increasing societal pressure to reduce GHG emissions are main factors that led to the current interest in PtL fuels for aviation.

Further major cost items in techno-economic analyses of PtL fuel production are electrolysis, hydrogen storage, carbon dioxide provision and the capital cost of Fischer–Tropsch (FT) synthesis and refining. Each conversion step is associated with energy losses and cost, resulting in an overall fuel production cost that is not competitive with past and current prices for conventional (fossil) jet fuel [29]. In consequence, supporting schemes like specific mandates for PtL fuels will be required to ramp-up a production and market for PtL fuels in significant volumes. This market support needs to be justified by credible environmental benefits. Low water demand and higher area-specific yield are important consideration in the comparison to aviation fuels from cultivated bioenergy feedstock. The main motivation compared to conventional (fossil) jet fuel is based on a significant reduction in global warming potential (GWP).

However, the global warming potential of PtL can quickly exceed that of conventional (fossil) jet fuel if the PtL fuel production is based on electricity from fossil sources. A threshold value of 100–125 gCO_2eq/kWh_{el} can be derived from energy efficiency considerations [28]. Electricity with a higher global warming potential such as natural gas (>400 gCO_2eq/kWh_{el}) or coal (>800 gCO_2eq/kWh_{el}) would lead to a global warming potential for PtL fuels that is higher compared to conventional (fossil) jet fuel. Electricity that is generated based on solar energy (<20 gCO_2eq/kWh_{el} [39]) or wind energy (<10 gCO_2eq/kWh_{el} [40]) is well below the mentioned threshold value. Furthermore, as sustainable source of carbon needs to be employed [28]. Biogenic carbon sources may be the favorite choice for an early implementation of PtL fuels. Scaling PtL fuel production to several 100 million tons per year requires the development of scalable carbon sources such as by direct air capture of CO_2 from ambient air.

3.2 Approval Status of PtL Jet Fuel

Different from the initial situation in the middle of the twentieth century, large fleets of aircraft exist today and significant development effort was already made to ensure high performance at a high safety level, all based on experience with the existing types of petroleum derived fuels. For such reasons, the aviation industry has initially focused on the development of standard specifications for so-called drop-in fuels. A drop-in fuel is defined as a fuel that meets the requirements of ASTM D7566 and can in turn be regarded as an ASTM D1655 compliant jet fuel. Initial effort to supplement ASTM D1655 by Annex A1 *"Fuels from Non-Conventional Sources"* was successfully made by the South African company Sasol. Both Sasol semi-synthetic and Sasol fully synthetic fuel are recognized as meeting the requirements of ASTM D1655. However, this approval is specific for Fischer–Tropsch (FT) derived kerosene from coal produced by the company Sasol and it is not generally applicable for PtL jet fuel production.

For the time being, ASTM D7566 *Standard Specification for Aviation Turbine Fuel Containing Synthesized Hydrocarbons* provides the most relevant specification that covers the production of synthetic jet fuel. The majority of non-conventional jet fuels that are currently used are ASTM D7566 compliant fuels consisting of conventional and synthetic blending components. ASTM D7566 is a technical specification that makes no statement on the sustainability of a blend component, no differentiation is made, e.g., between Fischer–Tropsch (FT) fuel from coal and from renewable origin or between hydroprocessed lipids from used cooking oil or palm oil. Instead, only technical requirements are defined. In order to qualify as a drop-in fuel, detailed requirements need to be met:

- The final blend between conventional fuel and a synthetic blend component needs to fulfill a set of requirements that resembles the conventional jet fuel specifications as set in ASTM D1655 in most aspects.

- In addition, extended requirements are defined in ASTM D7566, such as a minimum aromatic content of 8 vol-% and additional volatility, lubricity and fluidity requirements. These extended requirements apply to the final blend.
- Individual Annexes to ASTM D7566 define further sets of detailed batch requirements. These batch requirements are specific to the typical product composition of the respective production pathways of non-conventional kerosene and apply only to the synthetic blend component.

At the moment, a total of eight synthetic kerosene production processes are approved as Annexes to ASTM D7566. After a specification compliant blending procedure, the final fuel blend can be regarded as Jet A/Jet A-1 according to ASTM D1655, if all batch requirements of the non-conventional blend component and all requirements for the final blend are met. In practice, this means that the final fuel can be handled as interchangeable substitute for Jet A/Jet A-1. In this way, full compatibility of the resulting fuel blend with the existing fleet of aircraft and fuel infrastructures is enabled. On the downside, the significant advantages of fleet and infrastructure-wide compatibility come at the cost of forcing synthetic fuel specifications into a rigid framework of requirements that were initially designed for crude oil derived products. The molecular composition of synthetic kerosene can be very different from conventional jet fuel (see above), consequently the individual batch requirements as detailed in the Annexes of ASTM D7566 differ significantly from ASTM D1655 requirements. Few general features of batch requirements as described in these Annexes may be summarized as follows:

- Basic properties that relate to safe operation such as distillation profile, flash point as well as thermal stability requirements largely resemble ASTM D1655 requirements (a notable exception is Annex A3—Synthetic Isoparaffins (SIP) from Hydroprocessed Fermented Sugars (HFS)).
- The freezing point requirement of a maximum of -40 °C (except for Annex A3—SIP) resembles the requirement of Jet A, and the stricter requirement of a maximum of −47 °C for the production of Jet A-1 applies only to the final blend.
- Fluidity and combustion properties are not listed in the batch requirements, and these properties need to be met for the final blend.
- Density requirements reflect the typical molecular composition of the blend material. Density requirements in the ASTM D7566 Annexes can differ significantly from the ASTM D1655 requirements, that need to be met only for the density of the final blend.
- Requirements relating to composition, such as acidity, aromatic content or non-hydrocarbon composition (sulfur, nitrogen, metals) tend to be much stricter for synthetic fuels than for conventional fuel.
- Most currently approved synthetic components, including FT-SPK, can be blended up to a ratio of 50 vol-%.[4]

---

[4] Exceptions: Two types of biofuels (SIP and HC-HEFA) have a lower blending limit of 10 vol-%.

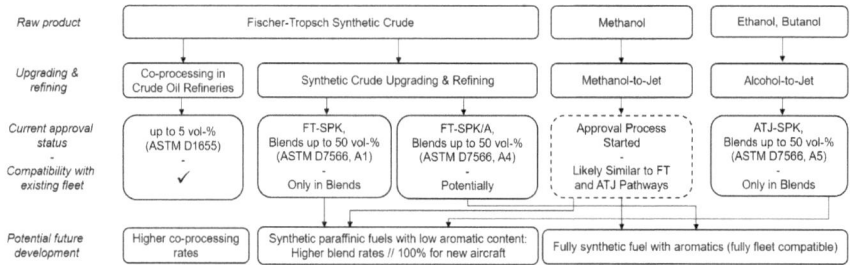

Fig. 5 Overview of the approval status of various types of PtL jet fuels. Both Fischer–Tropsch and methanol-based conversion schemes offer routes to fleet compatible drop-in fuels as well as to fully synthetic fuels with low aromatic content. Note added in proof: Fig. 5 shows the status at the time writing the manuscript, meanwhile ATJ-SKA is approved , preparing for fully fleet compatible synthetic fuels via alcohol based pathways

Annex A1 to ASTM D1655 foresees further options to introduce fuels from non-conventional sources, namely the co-processing of lipids (i.e., plant oils, fats and greases) and raw Fischer–Tropsch (FT) product (synthetic crude) in conventional refineries up to a fraction of 5 vol-%. This option is not only important for the aviation sector, as refinery processes are not (generally) optimized for the production of kerosene. With this respect, it also provides a market opportunity for refineries that co-process renewable streams to fulfill their renewable diesel quota and market the kerosene fraction produced for aviation purposes.[5] There are ongoing attempts to increase the fraction of raw products that can be co-processes in conventional refineries, which would facilitate a gradual conversion of crude oil refineries into renewable refineries. Figure 5 provides an overview of the current approval status of PtL kerosene and potential future developments.

3.3   Differences to Conventional Jet Fuel

Both synthetic jet fuels and conventional (fossil) jet fuels mainly consist of hydro-carbon mixtures in the kerosene range. The specific molecular composition, with the exception of the aromatic content and trace components, is not regulated in ASTM D1655. Instead, jet fuels need to fulfill a set of macroscopic requirements. In conse-quence, basic physical and chemical properties of synthetic jet fuels and conventional crude oil derived jet fuels are similar by definition. However, at molecular level, the specific composition of a synthetic jet fuel can be quite different from conventional jet fuel. These molecular differences can have a profound impact on important proper-ties, e.g., with regard to the emissions that result from the combustion of the respective fuels.

---

[5] Without this approval co-processing of such streams for diesel production would prohibit the use of the kerosene fraction as jet fuel for aviation.

Crude oil derived jet fuels for their part are complex mixtures of mainly hydrocarbon molecules with minor shares of contaminants and trace components. Conventional jet fuels also differ in their individual composition, not only because crude oils of various origin show different composition, but also due to different process designs of individual refineries and the chosen product spectrum of a given refinery. Conventional jet fuels contain large fractions of:

- n-alkanes (unbranched chains of saturated hydrocarbons),
- iso-alkanes (isomers of n-alkanes with various degree of branching),
- cyclo-alkanes (saturated alkanes with ring structure),
- aromatics (ring structured hydrocarbons with conjugated carbon double bonds).

The aromatic content of jet fuel is typically between 14 and 20% [16], and the remainder is made up from alkanes, where n-alkanes and iso-alkanes typically add up to a share of 55–60% [41]. The relative content of these hydrocarbon classes in conventional jet fuel is not regulated, only for aromatics, a maximum content of 25% is specified in ASTM D1655 [15]. In contrast to the specifications for synthetic fuels in ASTM D7566, no lower bound on aromatic content is given.

The molecular composition of synthetic jet fuel differs significantly from conventional jet fuel. In fact, no additional specifications for synthetic jet fuels would be required if their compositions were within the typical range for petroleum derived jet fuels. Synthetic jet fuel in form of FT-SPK, but also in form of HEFA-SPK and ATJ-SPK,[6] consists almost exclusively of n-alkanes and iso-alkanes. One direct consequence is a significantly lower density, which is reflected in a density requirement for FT-SPK of 730–770 kg/m^3 [14], which is outside the density specifications of conventional jet fuel of 775–840 kg/m^3 [15]. The lower density of FT-SPK goes in hand with a lower volumetric energy density, but 2–4% higher specific energy density [41]. From an operational perspective, lower volumetric energy density will reduce the design range of an aircraft, as less energy can be stored in a given tank volume. Higher specific energy density results in a reduction of fuel burn since the power requirement is proportional to the mass of the aircraft. Overall, a slight benefit may be expected for civil aviation due to the higher specific energy density of synthetic fuel.[7]

The most severe difference between synthetic and conventional jet fuel results from the absence of aromatics. Fuel leakage has occurred due to shrinking elastomers in the seals of aircraft fuel systems [42]. This problem could be linked to the absence of aromatics in SPK-type synthetic jet fuel. In consequence, a maximum blend rate of 50 vol-% was introduced and a minimum level of 8 vol-% aromatics is required in ASTM D7566 for all final blends between conventional jet fuel and a synthetic blend component. Meanwhile, aircraft and turbine manufacturers and their

---

[6] HEFA: Hydroprocessed esters and fatty acids, ATJ: Alcohol-to-Jet.

[7] At maximum take-off weight, fuel can account for about 40% of the weight of long-range aircraft. The advantage of higher specific energy density reduces over the mission as fuel is burned. The actual fuel saving of synthetic fuel vs. conventional fuel is heavily dependent on the mission. A typical level may be about 1% for long missions.

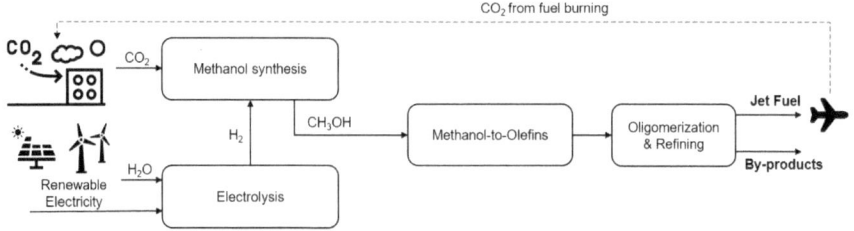

Fig. 6 Basic building blocks of jet fuel synthesis via methanol as an intermediate product. Methanol can be directly produced from H_2 and CO_2, furthermore methanol is a room temperature liquid that may be transported to centralized fuel synthesis plants, if favorable

component suppliers claim that this issue is solved by a replacement of the formerly used elastomer materials. Major aircraft manufacturers have announced the target that all new commercial aircraft will be compatible with 100% SPK-type synthetic fuel by 2030 [43–45].

3.4 Methanol-Derived PtL Kerosene

The existing approval of synthetic kerosene from Fischer–Tropsch (FT) pathways dates back to Coal-to-Liquid (CtL) and Gas-to-Liquid (GtL) conversion. In this way, Fischer–Tropsch (FT) fuels had already been approved before Power-to-Liquid (PtL) plants begun with the production of first fuel samples. The ground for methanol-based jet fuels has not been prepared in a similar way and a corresponding fuel approval is still pending. Figure 6 shows a sketch of a typical Methanol-to-Jet (MtJ) conversion pathway.

Methanol is a platform chemical, that can be produced with high selectivity from hydrogen (H_2) and carbon dioxide (CO_2) using technically mature processes. In particular, in case of a direct methanol synthesis from CO_2, a prior CO_2 to CO conversion is not needed and a reverse water gas shift (RWGS) step can be omitted.[8] Furthermore, methanol is a room temperature liquid that can be easily transported and stored. This opens the possibility to decouple methanol production and jet fuel synthesis. A Methanol-to-Olefin (MtO) conversion, which is the first step of Methanol-to-Jet (MtJ) conversion schemes, is commercial practice at large scale, especially for coal-based olefin production in China.[9] The olefins, with a high share of ethylene and propylene, are then oligomerized to jet fuel range hydrocarbons. The final fuel product is a synthetic paraffinic kerosene (SPK), which is rich in isoparaffins.

---

[8] Conventional methanol synthesis uses a synthesis gas that contains hydrogen (H_2), carbon monoxide (CO), and carbon dioxide (CO_2) [46].

[9] Methanol based olefins, such as ethylene and propylene, are important precursors for polymer production. Synergies of renewable polyethylene and polypropylene productions and Methanol-to-Jet (MtJ) are apparent.

Publicly available data on typical product compositions is still scarce, but the ASTM approval process for Methanol-to-Jet (MtJ) has been started. The underlying chemistry and apparent similarities to already approved Ethanol-to-Jet (EtJ) conversion strongly suggest that Methanol-to-Jet (MtJ) pathways will receive an approval as new Annex to ASTM D7566 within the next years. One likely outcome is that batch requirements of SPK from methanol will be similar to the existing batch requirements of FT-SPK and ATJ-SPK.

4 Toward Fully Synthetic Jet Fuel

It is clear that a transition to climate neutral aviation needs to be backed by specifications that no longer require a minimum share of 50 vol-% of hydrocarbons of fossil origin in jet fuels. During the upcoming years, new or adapted specifications need to be developed, which finally allow for a use of fully synthetic jet fuels—a 100% synthetic jet fuel use. A straightforward approach is to extend the maximum blending ratio of synthetic materials beyond 50 vol-%. There is no physics-based reason to restrict synthetic fuel use to 50 vol-% [41], in fact ASTM D1655 Annex 1 already approved one particular type of fully synthetic jet fuel. Multi-blends with hydrocarbons of renewable origin that fulfill all physical and chemical specifications of current Jet A/Jet A-1 should be as save to use as current fuels. However, mimicking the composition of jet fuel from conventional hydrocarbons has major disadvantages. First, it can pose an economic burden on synthetic jet fuels as additional production steps may be needed to meet the composition of crude oil-based jet fuels. Secondly, major benefits of synthetic jet fuel, especially in terms of cleaner combustion are not exploited to full extent. Consequently, two major approaches are followed for the development of fully synthetic fuels, namely fully synthetic drop-in and non-drop-in jet fuels, that will be further explained in the following (see also Fig. 7).

Fully synthetic drop-in jet fuels. This approach requires that a minimum set of specifications is met, such that fully synthetic fuels can be safely used in the existing fleet of aircraft. It was mentioned above that one specific version of a fully synthetic fuel is already approved in ASTM D1655 Annex 1.2.1.2 [15], which relates to the British Defense Standard (Def Stan) 91–91. SASOL fully synthetic fuel is a blend of up to five synthetic streams from the Secunda plant that is recognized as meeting the requirements for Jet A/A-1. The requirements for Sasol fully synthetic fuels include an aromatic content between 8 and 25 vol-%, as well as minimum slope requirements for the boiling point distributions. The 1:1 correspondence of these requirements to the extended requirements in ASTM D7566 illustrate the development from a specific Annex in ASTM D1655 to a dedicated standard specification in ASTM D7566. A potential road toward a general approval of fully synthetic drop-in fuel was prepared by the approval of *FT Synthesized Paraffinic Kerosine plus Aromatics* (SPK/A; Annex A4) as a blend component. SPK/A as a synthetic blend component may contain up

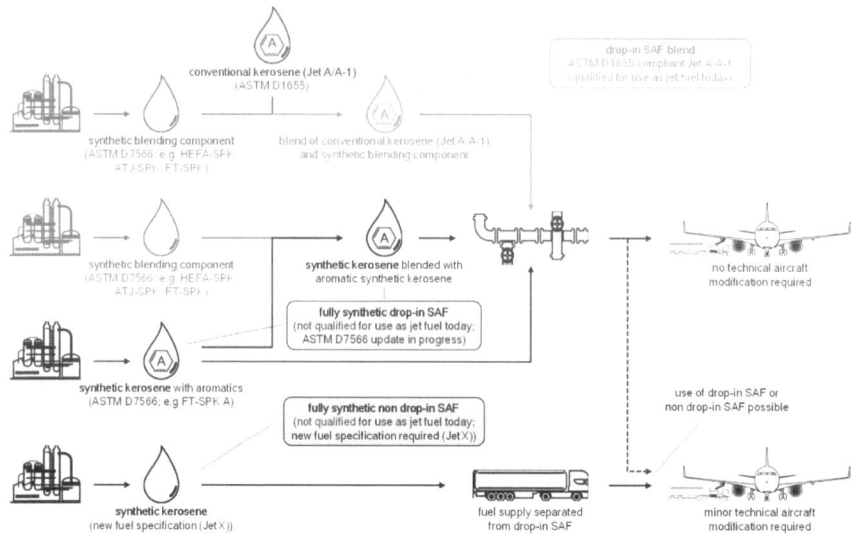

Fig. 7 Schematic representation of drop-in and non-drop-in pathways. Drop-in fuels can be used interchangeable with conventional jet fuel. Leveraging the benefits of fully synthetic non-drop-in fuel requires adapted aircraft and a separated fuel infrastructure (Image with courtesy of Bullerdiek, TUHH, Rauch, Enderle, DLR)

to 20 vol-% aromatics that can be produced by oligomerization and alkylation of lighter Fischer–Tropsch (FT) refining products [1].

So far, SPK/A use is limited to a blend ration of 50 vol-%, but this cautious restriction may be lifted once the synthetic jet fuel share comes anywhere near to the 50 vol-% blend wall. The clear advantage of fully synthetic drop-in fuels is that conventional (fossil) jet fuel use can be eliminated for the full fleet of legacy aircraft. In addition, major air quality benefits and a reduction of non-CO_2 climate impact may be achieved as SPK/A is sulfur free, and a lower content of aromatics will reduce soot emissions compared to conventional jet fuel. Prospectively, the threshold level for the minimum aromatic content for drop-in fuels could be further lowered, provided that investigations of material compatibility and sooting propensity could identify synthetic fuel compositions that preserve full fleet compatibility at much reduced particle emissions.

Fully synthetic non-drop-in jet fuels. This approach involves new specifications for synthetic fuels that require an adaption of aircraft but allow to use the benefits of synthetic fuels such as low aromaticity to a large extent. Current jet fuel specifications have evolved over time, taking into account not only safe operations of aircraft, but also economic considerations of crude oil refining. In the future, where synthetic jet fuels need to become the backbone of aviation's energy supply to reduce climate impacts, it is meaningful to develop specifications for fully synthetic fuels that reflect all relevant aspects of safe operation, but also economic and environmental considerations that relate to synthetic fuel use and production. One key argument for the development of such specifications is that synthetic fuels like FT-SPK, but also lipid-based HEFA-SPK and alcohol-based AtJ-SPK, have superior properties compared to conventional (fossil) jet fuel, mainly in terms of a cleaner combustion, but also specific energy density. On the downside, such fuels can no longer be used as "drop-in substitute" with regard to the ASTM D1655, due to a lack of aromaticity, but also due to a violation of density requirements. Most major aircraft manufacturers are already adapting the fuel systems of future aircraft to tolerate fuels with low aromatic content. Preparatory work for the definition of new specifications is also underway. The agreement on new specifications requires a multi-stakeholder approach. A few guidelines for this undertaking can be derived:

- It should be possible to meet the new physico-chemical requirements by a SPK-type jet fuel that contains almost exclusively n-alkanes and iso-alkanes.
- Fuels with low aromatic content (e.g., < 0.5 vol-%) should meet the requirements.
- The fuel should be sulfur free.

The detailed definition of new specifications is a complex undertaking, as there is a multitude of conversion pathways not only for PtL, but also for biofuels.

During a transition period, at least two types of fuels, a drop-in replacement that fulfills ASTM D7566 requirements and a non-drop-in SPK-type fuel with low aromatic content may co-exist. The drop-in fuels are fleet compatible and can be offered everywhere. The non-drop-in SPK-type synthetic fuel is only suitable for a subset of the fleet that can tolerate aromatic-free jet fuel. During the early phase of such a transition, compatibility of new aircraft with both types of fuel can greatly reduce operational challenges. In the long run, it may be beneficial to co-optimize synthetic fuels and aircraft engines at the cost of backward compatibility.

5 Using Green Hydrogen in Aviation

Power-to-Liquid (PtL) pathways are not the only opportunity to introduce green hydrogen in aviation. Figure 8 illustrates further options to use hydrogen for kerosene production or directly as aviation fuel.

A straightforward option is the utilization of green hydrogen for refinery processes in conventional and biofuel refining. A coupling of green hydrogen with advanced biofuel pathways enables a deeper integration, e.g., in form of Power-and-Biomass-to-Liquid (PBtL) pathways. In the future, hydrogen may also be used directly as an aviation fuel.[10] Each of these options is described in more detail in the following sections. First, Sect. 5.1 describes options for integrating green hydrogen into jet fuel production pathways, followed by Sect. 5.2, which discusses the direct use of hydrogen as aviation fuel.

---

[10] Further energy carriers, such as various hydrocarbons or ammonia can be produced via Power-to-X (PtX) pathways, but these fuels are currently not in the focus of aviation due to limitations in their physico-chemical properties [47].

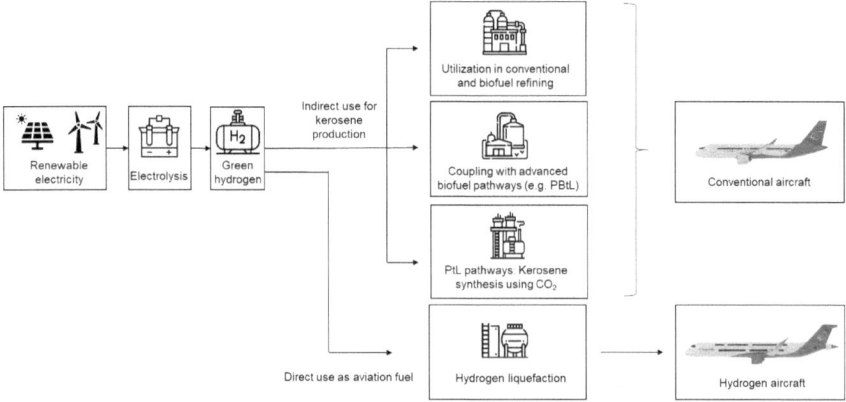

Fig. 8 Overview of various options to introduce green hydrogen to aviation

5.1 Integration to Jet Fuel Production Pathways

In the following, the mentioned options for integrating green hydrogen into jet fuel production pathways are described in more detail; (1) integration in conventional refining, (2) integration in biofuel refining, and (3) a deeper integration into advanced biofuel pathways (Power-and-Biomass-to-Liquid, PBtL). These serve as additional approaches to use green hydrogen in aviation to the previously described Power-to-Liquid (PtL) pathways.

Green hydrogen in conventional refining. The hydrogen demand in conventional refineries is usually met by steam reforming of natural gas or other fossil hydrocarbons that are available at the respective refinery site. Replacing this hydrogen by green hydrogen can reduce the life cycle greenhouse gas (GHG) balance of the produced fuels and stimulate an early uptake of green hydrogen technology [31]. The implementation thresholds for GHG savings by replacing fossil refinery hydrogen are comparably low as there are no major adaptions to current process steps required apart from the addition of renewable electricity generation and water electrolysis. Furthermore, conversion losses work in favor of green hydrogen when conventional hydrogen is substituted, while they work against green hydrogen in synthetic fuel production. Replacing refinery hydrogen is therefore a cost-effective and simple way of specific GHG emissions reductions. It should be noted, though, that the long-term impact and lever associated with this measure is limited as the main feedstock of the refinery, namely crude oil, is not substituted.

The amount of hydrogen that is consumed in conventional (fossil) jet fuel production depends on crude oil feedstock and refinery design. The majority of refinery hydrogen is usually required for refining steps that are necessary to meet the specifications for road transportation fuels. In contrast to that, the refining of jet fuel differs significantly. In simple refinery designs, jet fuel is derived from the kerosene fraction of the initial atmospheric distillation. A sweeting process removes specific organosulfur compounds (mercaptanes), but a deep reduction of sulfur content by hydrodesulfurization is usually not required, since the sulfur limit of Jet A/A-1 (3000 ppm) is well above the typical sulfur limits for road transportation fuels (10–15 ppm in most developed countries). In contrast to Fischer–Tropsch (FT) synthetic crude refining, a reconfiguration of molecules is also not required as straight run kerosene from crude oil distillation contains sufficient amounts of iso-alkanes, cyclo-alkanes and aromatics to meet fluidity and volatility requirements. Consequently, the overall hydrogen demand is comparatively low in these simple refinery approaches. However, the average share of jet fuel compared to other refinery products has steadily increased from about 4% in the early 1970s to more than 8% in 2019 to meet the growing demand [3]. This is why an increasing number of more complex refineries optimize their product spectrum toward higher jet fuel yield by processing heavier products resulting in a significantly enhanced consumption of refinery hydrogen for jet fuel production.

Green hydrogen in biomass-based jet fuel production. Similar to the substitution of hydrogen demand in conventional refineries, the hydrogen required for the refining of biofuels can also be met by green hydrogen instead of hydrogen that is provided based on fossil sources (e.g., via steam reforming of natural gas). Jet fuel obtained from hydroprocessing esters and fatty acids (HEFA) currently represents the most common type of aviation biokerosene. Processing of lipid feedstock to HEFA-SPK directly requires hydrogen to remove heteroatoms (mainly oxygen) from the molecules and to introduce a sufficient degree of isomerization, conceptually similar to the production of FT-SPK as described in Sect. 3.1. The various production pathways toward renewable jet fuel differ significantly in the specific hydrogen consumption (Table 1).

Direct liquefaction pathways based on pyrolysis or hydrothermal liquefaction (HTL) tend to have a particularly high demand for refinery hydrogen. This is a direct result of the high heteroatom content (O, N) of the intermediate pyrolysis oils or HTL biocrudes, whereby the removal of these species is again associated with high hydrogen consumption. Consequently, the utilization of green hydrogen for the

Table 1 Specific hydrogen demand for different renewable fuel production pathways

Conversion pathway	Hydrogen demand [$MJ_{H2}/MJ_{product}$]	References
Alcohol-to-Jet (AtJ)	0.06	[48]
Hydrotreated Vegetable Oil (HVO)/ Hydroprocessed Esters and Fatty Acids (HEFA)	0.12	[48]
Hydrothermal Liquefaction (HTL)	0.15	[49, 50]
Pyrolysis	0.38	[48]
Power-and-Biomass-to-Liquid (PBtL)	0.72	[51]
Power-to-Liquid (PtL)	1.51	[35]

hydroprocessing of pyrolysis oils or HTL biocrudes has a particular high impact on the GHG emissions of these pathways.

Power-and-Biomass-to-Liquid (PBtL). Other than employing green hydrogen for the final upgrading and refining steps in biokerosene production, electrolysis can also enable a significant boost of the carbon efficiency of many biokerosene pathways. The respective conversion pathways are usually referred to as Power-and-Biomass-to-Liquid (PBtL).

In the Biomass-to-Liquid (BtL) process, biomass is gasified to produce a synthesis gas that is subsequently fed to a Fischer–Tropsch (FT) reactor. The resulting synthesis gas composition depends mainly on the feedstock constitution and the selected gasification technology [52]. Steam gasification for example produces a synthesis gas with a relatively high hydrogen (H_2) to carbon monoxide (CO) ratio (H_2:CO ratio), while oxygen-based gasification systems are more energy-efficient, but result in a raw synthesis gas with a lower H_2:CO ratio and require an oxygen supply. Usually, the raw synthesis gas produced by gasification contains a lower H_2:CO ratio than required for Fischer–Tropsch (FT) synthesis—and thus a hydrogen deficiency. To adjust (increase) the H_2:CO ratio, a water–gas shift (WGS) reaction is typically employed. In this process step, steam (H_2O) is added to the synthesis gas, which allows a portion of the carbon monoxide (CO) to be oxidized into carbon dioxide (CO_2), while simultaneously releasing hydrogen (H_2). As a result, a portion of the carbon monoxide (CO) originally contained in the synthesis gas is converted to CO_2, leading to what is referred to as a "carbon loss" since CO_2 mainly acts as an inert gas in Fischer–Tropsch synthesis.

The addition of green hydrogen can resolve this deficiency without the separation of carbon in the form of carbon dioxide (CO_2) and, consequently, enable a more effective use of the carbon contained in the feedstock [12, 53]. The oxygen produced during electrolysis can be used as a gasification agent in biomass gasification, thus improving the quality of the raw synthesis gas [54]. Such pathways, that aim at synergies between Biomass-to-Liquid (BtL) and Power-to-Liquid (PtL) processes, are often called Power-and-Biomass-to-Liquid (PBtL) pathways. A schematic process diagram is shown in Fig. 9.

As described by the explanations above, one key advantage of integrating biomass gasification with water electrolysis is a significantly higher utilization of the carbon

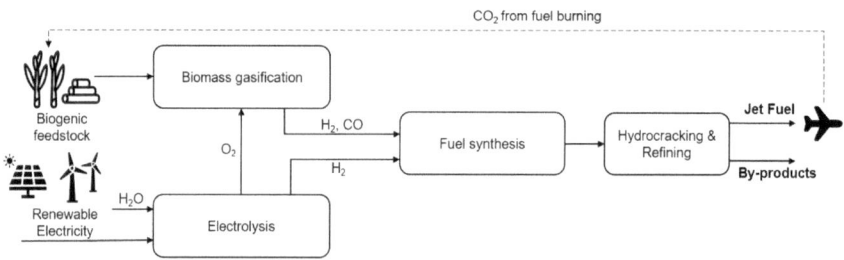

Fig. 9 Process scheme of a Power-and-Biomass-to-Liquid (PBtL) process

content in the biomass feedstock [55, 56]. At a simple level of Power-and-Biomass-to-Liquid (PBtL) integration, green hydrogen is mainly used to compensate for the hydrogen deficiency of the raw synthesis gas resulting from biomass gasification. A further boost in carbon efficiency is achieved when CO_2 streams that evolve along the conversion pathway, are ---fed back as a gasification agent or get combined with additional hydrogen for Power-to-Liquid (PtL) fuel synthesis. The latter option is also a lever to boost the carbon efficiency of other advanced biomass conversion schemes, such as hydrothermal liquefaction (HTL) or fast pyrolysis processes, that generate significant streams of CO_2 as a side product.

5.2 Hydrogen Aircraft

In the recent years, the perspective of directly using green hydrogen as a non-drop-in aviation fuel, instead of merely using it as a reactant for kerosene production, has drawn considerable attention as decarbonization efforts are increasing. Since hydrogen reacts with oxygen in an exothermal process yielding only water and heat as products, emissions are intrinsically carbon-free.

$$2 H_2 + O_2 \rightarrow 2 H_2O \tag{1}$$

Notably, its production from renewable electricity and water is independent of a carbon feedstock and the challenges and consequences related with it. Furthermore, green hydrogen represents a scalable fuel option and is associated with low life cycle global warming potential further motivating the direct use of this energy carrier for aircraft.

While all of this seems intriguing, hydrogen features drastically different properties than the tried and trusted kerosene and, therefore, demands radical developments not only when it comes to aircraft technology, but also in the context of handling, storage and transport infrastructure. This is in large parts due to its limited volumetric energy density: as a fixed volume of compressed hydrogen (e.g., at 350 bar) contains about an order of magnitude less energy than an equivalent volume of kerosene. Moreover, the required pressurized tank structures add a considerable weight penalty for aircraft operation. Consequently, the potential for using hydrogen in its gaseous form is limited to niche applications (i.e., small vehicles, short distances). Thus, the community is mainly focusing on liquefied hydrogen, a cryogenic liquid (boiling point of -253 °C) with a volumetric energy density of approximately 2.4 kWh/L, i.e., significantly higher than in compressed form and about one fourth of kerosene. Due to the ultracold temperatures required and the highly volatile characteristic of liquefied hydrogen, special well-insulated tank structures with optimized surface to volume ratios are required for onboard storage. Such tank structures add significant weight and also demand an integration in the fuselage due to their typically spherical or cylindrical shape—a fundamental difference to the conventional wing storage of kerosene resulting in additional drag and a reduced cabin space among others.

Overcoming these intrinsic drawbacks requires new aircraft designs and concepts and exploitation of opportunities and synergies that come along with the needed adjustments. One example in this context would be the drastically enlarged design space for the wing—where, unlike today's aircraft, no fuel would be held—enabling adapted wing shapes directly impacting the aerodynamic properties and the accomplishable lift. Significant research and development efforts are currently focused on maturing hydrogen aircraft concepts, from academic groups all the way to aircraft manufacturers.

Hydrogen as an onboard fuel also opens up new means of energy conversion: in addition to combustion-based gas turbine systems, electrochemical energy conversion in fuel cells can be used. Hereby, the half-reactions of the process described above, i.e., hydrogen oxidation and oxygen reduction reactions (Eq. 2) are spatially separated forcing the electrons involved to travel through an external circuit.

$$
\begin{aligned}
H_2 &\to 2H^+ + 2e^- \\
O_2 + 4H^+ + 4e^- &\to 2H_2O
\end{aligned}
\tag{2}
$$

The electrical power can then be used for propulsion. Fuel cells offer high conversion efficiency if operated in a suitable operating point, but come with a significantly lower system specific power (depending on system layout and operating point for both gas turbine and fuel cell about an order of magnitude difference). In addition to the mentioned tradeoff of conversion efficiency and specific power, also a number of other considerations such as operating temperature, cooling effort, balance of plant complexity, power management, maintenance and durability have to be taken into account when an electrochemical conversion-based propulsion system is envisaged. Furthermore, as aircraft span over orders of magnitude of power and energy demand (all the way from the regional to the long-range segment), it is crucial to match the propulsion train layout with the specific application to optimize overall system performance. A number of synergistic hybrid concepts have been proposed [57, 58], in order to leverage advantages of both energy conversion approaches—an in-depth discussion, however, beyond the scope of this chapter.

Importantly, to use liquid hydrogen as an aviation fuel alternative does not only depend on overcoming challenges in aircraft and propulsion design. In addition, infrastructures for the production of liquefied hydrogen and logistic chains tailored to the properties of the cryogenic liquid need to be established, as well as infrastructural adaptions on the airport side. This includes stationary storage facilities, but also extends to distribution and refueling infrastructure. The latter is especially critical and major developments are needed in that area, as high volumetric throughput is required to adhere to operational constraints and designated time windows for aircraft turnaround and refueling at airports.

Lastly, it should be noted that while it is evident that using green hydrogen as an aviation fuel would cut carbon-based emissions, there are large uncertainties associated with the full climate impact at high altitude [21]. More research is required to get a full understanding of the influencing factors and strategies to avoid any enhancement of radiative forcing (e.g., based on avoiding flying through atmospheric conditions favoring the formation of contrail cirrus or controlling the water droplet size forming). Literature studies expect a significant reduction of the climate impact over carbon-based synthetic kerosene [59].

To summarize, considering the expected emission reduction potential, the remaining technological hurdles to overcome and the long design cycles in aviation of usually more than 20 years, liquid hydrogen may represent a promising fuel option only for a more long-term perspective.

6 Summary and Outlook

During the past years, power-derived fuels have moved into the center of the debate on a future renewable energy supply for the aviation sector. Main arguments in favor of these fuel pathways include:

- The expectation of a continued need for liquid fuels in aviation due to a severe range limitation arising from the comparatively low gravimetric and volumetric energy density of alternatives such as batteries or gaseous hydrogen.
- Serious concerns about the sustainability of biofuel production at the scale of future aviation fuel demand.
- The high maturity level of key conversion technologies along Power-to-Liquid (PtL) fuel production chains.
- Steep technological improvements and cost reductions of electricity generation from solar and wind energy, and more recently also for water electrolysis.

Nevertheless, synthetic fuel production based on renewable electricity remains expensive compared to present conventional (fossil) jet fuel prices and a potential bottleneck with regard to their large-scale provision may arise from sustainable carbon provision at scale. In consequence, regulatory support is crucial, especially during the early scale-up phase of PtL jet fuels. Furthermore, continued development and deployment of technologies for direct air capture of carbon dioxide (CO_2) in anticipation of a mid-term scale-up of PtL jet fuel production toward hundreds of millions of tons per year is important. For the time being, PtL jet fuel production is continuously gaining momentum. For instance, Germany has established a mandate that requests a use of 2% of PtL jet fuel at German airports by 2030. A similar sub-mandate for PtL jet fuels of 1.2% by 2030 will likely be implemented at EU level.

From a fuel qualification perspective, PtL fuels from Fischer–Tropsch (FT) synthesis are already approved for civil aviation. FT-SPK can be used in the fleet of transport aircraft up to a blend ration of 50 vol-%, which may be considered as sufficient for many years to come. Nevertheless, various options for fully synthetic fueling are currently explored. The approval of FT-SPK/A prepared the field for a future specification for fully synthetic fuels that resembles all relevant physical and chemical properties of conventional (fossil) jet fuel and can be safely used in the existing fleet of aircraft. However, compatibility with the current fleet of aircraft still requires a certain share of aromatic hydrocarbons in the fuels. Aromatic hydrocarbons are associated with the formation of soot particles that cause air quality issues and contribute to the climate impact of aviation from contrail and cirrus formation. For these reasons, aircraft manufactures adapt the fuel systems, such that future aircraft will be compatible to cleaner burning SPK-type fuels with very low aromatic content. SPK-type fuels that consist almost exclusively of n-alkanes and iso-alkanes differ significantly from the molecular composition of crude oil derived jet fuel. Consequently, new fuel specifications will need to be established. Preparatory effort in this direction is already underway. Ultimately, a co-development of synthetic fuels and engine technology may drastically reduce exhaust emissions in the longer-term future.

The early approval of Fischer–Tropsch (FT) fuels dates back to Coal-to-Liquids (CtL) and Gas-to-Liquids (GtL) conversion pathways. So far, an approval of kerosene that is produced from methanol-based syntheses is pending. Based on the underlying reaction chemistry, it is expected that Methanol-to-Jet (MtJ) conversion yields a paraffinic kerosene that is rich in iso-alkanes. The fuel approval process has been started and it is likely that this process will result in similar batch requirements as for already approved FT-SPK and ATJ-SPK within few years. Both, methanol and FT-based synthesis pathways have specific advantages. However, basic techno-economic characteristics, including the need for a scalable CO_2 source, are not fundamentally changed.

Renewable electricity can also reduce emissions in the aviation sector via production of green hydrogen and utilization of hydrogen in crude oil or biofuel refining, coupling with advanced biofuels pathways such as Power-and-Biomass-to-Liquid (PBtL) processes or direct use of hydrogen as an aviation fuel. A substitution of refinery hydrogen involves rather low technological risks and the specific GHG emission reduction is large, since formerly fossil hydrogen generation is replaced and significant energy conversion losses only apply to the electrolysis step. However, the large specific effect per unit hydrogen goes in hand with limited overall GHG emission reduction potential as the main refinery feedstock remains unchanged. In contrast, utilization of large amounts of hydrogen for much deeper integration with advanced biomass conversion pathways can significantly improve the product yield of biofuel production. Examples include PBtL pathways and the utilization of CO_2 rich side streams from hydrothermal liquefaction (HTL) or pyrolysis pathways for additional synthetic fuel production. For such pathways, green hydrogen becomes a major contribution to the overall energy and cost balance. In consequence, excellent conditions for green hydrogen generation are required for the competitiveness of such approaches.

Finally, hydrogen may be directly used as an aviation fuel. This radically different approach, however, requires fundamental technological development on aircraft and infrastructure level. Beyond niche applications, the constrained design space of passenger aircraft necessitates storing hydrogen in its liquid, cryogenic form bringing along numerous challenges in fuel supply, handling and onboard use. Compared to PtL jet fuels, liquid hydrogen has the profound advantage of more direct and more energy-efficient fuel production from renewable electricity. Furthermore, air pollution emissions and non-CO_2 climate impact from fuel combustion may be significantly reduced. However, it remains to be shown if such advantages will finally outweigh the associated drawbacks. Furthermore, the development and fleet introduction of commercial hydrogen aircraft will require decades even if all technological hurdles can be overcome. Consequently, for the near- and mid-term, a transition from conventional (fossil) jet fuel to renewable kerosenes remains the only option for significantly reducing the carbon intensity of aviation's fuel demand.

Acknowledgements This research was funded by the German Federal Ministry of Economic Affairs and Climate Action under grant number 03EIV071C (Project PowerFuel).

References

1. Klerk A (2011) Fischer-Tropsch refining. Wiley-VCH Verlag & Co. KGaA, Weinheim
2. The International Air Transport Association (2022) Incentives needed to increase SAF production
3. The International Energy Agency. Key World Energy Statistics 2021; 2021.
4. The European Commission (2021) Proposal for a Regulation of the European Parliament and of the Council on ensuring a level playing field for sustainable air transport: com(2021)561
5. Kharina A, Rutherford D (2015) Fuel efficiency trends for new commercial jet aircraft: 1960 to 2014. Berlin
6. Air Transport Action Group (2021) Waypoint 2050, 2nd edn
7. Atanasov G, van Wensveen J, Peter F, Zill T (2019) Electric commuter transport concept enabled by combustion engine range extender. In: Deutsche Gesellschaft für Luft- und Raumfahrt (ed), vol 68. Deutscher Luft- und Raumfahrtkongress (DLRK)
8. Graver B, Zhang K, Rutherford D (2019) CO_2 emissions from commercial aviation, 2018
9. Staples MD, Malina R, Barrett SRH (2017) The limits of bioenergy for mitigating global life-cycle greenhouse gas emissions from fossil fuels. Nat Energy 2(2). https://doi.org/10.1038/nenergy.2016.202
10. van Grinsven A, van den Toorn E, van der Veen R, Kampman B (2020) Used cooking oil (UCO) as biofuel feedstock in the EU. Delft, Netherlands
11. Zhao X, Taheripour F, Malina R, Staples MD, Tyner WE (2021) Estimating induced land use change emissions for sustainable aviation biofuel pathways. Sci Total Environ 779:146238. https://doi.org/10.1016/j.scitotenv.2021.146238
12. Isaacs SA, Staples MD, Allroggen F, Mallapragada DS, Falter CP, Barrett SRH (2021) Environmental and economic performance of hybrid power-to-liquid and biomass-to-liquid fuel production in the United States. Environ Sci Technol 55(12):8247–8257. https://doi.org/10.1021/acs.est.0c07674
13. Dray LM, Schäfer AW, Grobler C, Falter C, Allroggen F, Stettler MEJ et al (2022) Cost and emissions pathways towards net-zero climate impacts in aviation. Nat Clim Chang 12(10):956–962. https://doi.org/10.1038/s41558-022-01485-4
14. ASTM International. Standard Specification for Aviation Turbine Fuel Containing Synthesized Hydrocarbons: D7566(D7566–20); 2020.
15. ASTM International. Standard Specification for Aviation Turbine Fuels: ASTM D1655(D1655-19) (2019)
16. Zschocke A, Scheuermann S, Ortner J (2017) High biofuel blends in aviation (HBBA): final report. ENER/C2/2012/420-1

17. Moore RH, Shook M, Beyersdorf A, Corr C, Herndon S, Knighton WB et al (2015) Influence of jet fuel composition on aircraft engine emissions: a synthesis of aerosol emissions data from the NASA APEX, AAFEX, and ACCESS missions. Energy Fuels 29(4):2591–2600. https://doi.org/10.1021/ef502618w

18. Corporan E, DeWitt MJ, Belovich V, Pawlik R, Lynch AC, Gord JR et al (2007) Emissions characteristics of a turbine engine and research combustor burning a Fischer−Tropsch jet fuel. Energy Fuels 21(5):2615–2626. https://doi.org/10.1021/ef070015j

19. Moore RH, Thornhill KL, Weinzierl B, Sauer D, D'Ascoli E, Kim J et al (2017) Biofuel blending reduces particle emissions from aircraft engines at cruise conditions. Nature 543(7645):411–415. https://doi.org/10.1038/nature21420

20. Schripp T, Anderson B, Crosbie EC, Moore RH, Herrmann F, Oßwald P et al (2018) Impact of alternative jet fuels on engine exhaust composition during the 2015 ECLIF ground-based measurements campaign. Environ Sci Technol 52(8):4969–4978. https://doi.org/10.1021/acs.est.7b06244

21. Lee DS, Fahey DW, Skowron A, Allen MR, Burkhardt U, Chen Q et al (2021) The contribution of global aviation to anthropogenic climate forcing for 2000 to 2018. Atmos Environ 244:1–29. https://doi.org/10.1016/j.atmosenv.2020.117834

22. Burkhardt U, Kärcher B (2011) Global radiative forcing from contrail cirrus. Nat Clim Chang 1(1):54–58. https://doi.org/10.1038/NCLIMATE1068

23. Kärcher B (2018) Formation and radiative forcing of contrail cirrus. Nat Commun 9(1):1824. https://doi.org/10.1038/s41467-018-04068-0

24. Burkhardt U, Bock L, Bier A (2018) Mitigating the contrail cirrus climate impact by reducing aircraft soot number emissions. npj Clim Atmos Sci 1(1):837. https://doi.org/10.1038/s41612-018-0046-4

25. Voigt C, Kleine J, Sauer D, Moore RH, Bräuer T, Le Clercq P et al (2021) Cleaner burning aviation fuels can reduce contrail cloudiness. Commun Earth Environ 2(1):465. https://doi.org/10.1038/s43247-021-00174-y

26. Faber J, Király J, Lee D, Owen B, O'Leary A (2022) Potential for reducing aviation non-CO_2 emissions through cleaner jet fuel. Delft

27. Teoh R, Schumann U, Voigt C, Schripp T, Shapiro M, Engberg Z et al (2022) Targeted use of sustainable aviation fuel to maximize climate benefits. Environ Sci Technol 56(23):17246–17255. https://doi.org/10.1021/acs.est.2c05781

28. Schmidt P, Batteiger V, Roth A, Weindorf W, Raksha T (2018) Power-to-liquids as renewable fuel option for aviation: a review. Chem Ing Tec 90(1–2):127–140. https://doi.org/10.1002/cite.201700129

29. Batteiger V, Ebner K, Antoine Habersetzer, Moser L, Schmidt P, Weindorf W et al (2022) Power-to-liquids—a scalable and sustainable fuel supply perspective for aviation. Dessau-Roßlau

30. Falter C, Pitz-Paal R (2017) Water footprint and land requirement of solar thermochemical jet-fuel production. Environ Sci Technol 51(21):12938–12947. https://doi.org/10.1021/acs.est.7b02633

31. Schmidt P, Weindorf W, Roth A, Batteiger V, Riegel F (2016) Power-to-liquids—potentials and perspectives for the future supply of renewable aviation fuel. Background Paper, UBA

32. Zitscher T, Neuling U, Habersetzer A, Kaltschmitt M (2020) Analysis of the German industry to determine the resource potential of CO_2 emissions for PtX applications in 2017 and 2050. Resources 9(12):149. https://doi.org/10.3390/resources9120149

33. Batteiger V, Falter C, Galvez JL, Dufour J, Iribarren D, Techno-economic and environmental analysis of Carbon dioxide provision from various sources.: Public report, SUN-to-LIQUID project. Available from: https://www.sun-to-liquid.eu/page/en/scientific-publications.php

34. Roth A, Riegel F, Batteiger V (eds) (2018) Potentials of biomass and renewable energy: the question of sustainable availability: in Biokerosene. Springer, Berlin, Heidelberg

35. König DH, Freiberg M, Dietrich R-U, Wörner A (2015) Techno-economic study of the storage of fluctuating renewable energy in liquid hydrocarbons. Fuel 159:289–297. https://doi.org/10.1016/j.fuel.2015.06.085

36. The International Renewable Energy Agency (2021) Renewable power generation costs 2020
37. Klerk A (2011) Fischer–Tropsch fuels refinery design. Energy Environ Sci 4(4):1177. https://doi.org/10.1039/c0ee00692k
38. Haije W, Geerlings H (2011) Efficient production of solar fuel using existing large scale production technologies. Environ Sci Technol 45(20):8609–8610. https://doi.org/10.1021/es203160k
39. Müller A, Friedrich L, Reichel C, Herceg S, Mittag M, Neuhaus DH (2021) A comparative life cycle assessment of silicon PV modules: impact of module design, manufacturing location and inventory. Sol Energy Mater Sol Cells 230:111277. https://doi.org/10.1016/j.solmat.2021.111277
40. Mali S, Garrett P (2022) Life cycle assessment of electricity production from an onshore V150–4.2 MW wind plant
41. Holladay J, Abdullah Z, Heyne J (2020) Sustainable aviation fuel: review of technical pathways
42. Chen K, Liu H, Xia Z (2013) The impacts of aromatic contents in aviation jet fuel on the volume swell of the aircraft fuel tank sealants. SAE Int J Aerosp 6(1):350–354. https://doi.org/10.4271/2013-01-9001
43. Boeing (2023) Boeing commits to deliver commercial airplanes ready to fly on 100% sustainable fuels, 30 Aug 2023. Available from: https://boeing.mediaroom.com/2021-01-22-Boeing-Commits-to-Deliver-Commercial-Airplanes-Ready-to-Fly-on-100-Sustainable-Fuels
44. Airbus (2023) Sustainable aviation fuel: towards 100% SAF capability by 2030, 30 Aug 2023. Available from: https://www.airbus.com/en/sustainability/respecting-the-planet/decarbonisation/sustainable-aviation-fuel
45. Embraer (2023) E195-E2 completes 100% SAF flight testing, 30 Aug 2023. Available from: https://www.embraercommercialaviation.com/news/e195-e2-completes-100-saf-flight-testing/
46. Nestler F, Krüger M, Full J, Hadrich MJ, White RJ, Schaadt A (2018) Methanol synthesis—industrial challenges within a changing raw material landscape. Chem Ing Tec 90(10):1409–1418. https://doi.org/10.1002/cite.201800026
47. Quante G, Bullerdiek N, Bube S, Neuling U, Kaltschmitt M (2023) Renewable fuel options for aviation—a system-wide comparison of drop-in and non drop-in fuel options. Fuel 333:126269. https://doi.org/10.1016/j.fuel.2022.126269
48. de Jong S, Antonissen K, Hoefnagels R, Lonza L, Wang M, Faaij A et al (2017) Life-cycle analysis of greenhouse gas emissions from renewable jet fuel production. Biotechnol Biofuels 10:64. https://doi.org/10.1186/s13068-017-0739-7
49. Haider MS, Castello D, Rosendahl LA (2020) Two-stage catalytic hydrotreatment of highly nitrogenous biocrude from continuous hydrothermal liquefaction: a rational design of the stabilization stage. Biomass Bioenerg 139:105658. https://doi.org/10.1016/j.biombioe.2020.105658
50. Haghighi S, Askari K, Hamidi S, Mahdi Rahimi M (2018) OPEM open source PEM cell simulation tool. JOSS 3(27):676. https://doi.org/10.21105/joss.00676
51. Ostadi M, Rytter E, Hillestad M (2019) Boosting carbon efficiency of the biomass to liquid process with hydrogen from power: the effect of H_2/CO ratio to the Fischer-Tropsch reactors on the production and power consumption. Biomass Bioenerg 127(12):105282. https://doi.org/10.1016/j.biombioe.2019.105282
52. Rauch R, Hrbek J, Hofbauer H (2014) Biomass gasification for synthesis gas production and applications of the syngas. WIREs Energy Environ 3(4):343–362. https://doi.org/10.1002/wene.97
53. Gruber H, Groß P, Rauch R, Reichhold A, Zweiler R, Aichernig C et al (2021) Fischer-Tropsch products from biomass-derived syngas and renewable hydrogen. Biomass Conv. Bioref. 11(6):2281–2292. https://doi.org/10.1007/s13399-019-00459-5
54. Dossow M, Dieterich V, Hanel A, Spliethoff H, Fendt S (2021) Improving carbon efficiency for an advanced Biomass-to-Liquid process using hydrogen and oxygen from electrolysis. Renew Sustain Energy Rev 152:111670. https://doi.org/10.1016/j.rser.2021.111670

55. Nielsen AS, Ostadi M, Austbø B, Hillestad M, del Alamo G, Burheim O (2022) Enhancing the efficiency of power- and biomass-to-liquid fuel processes using fuel-assisted solid oxide electrolysis cells. Fuel 321. https://doi.org/10.1016/j.fuel.2022.123987

56. Ostadi M, Paso KG, Rodriguez-Fabia S, Øi LE, Manenti F, Hillestad M (2020) Process integration of green hydrogen: decarbonization of chemical industries. Energies 13(18):4859. https://doi.org/10.3390/en13184859

57. Seitz A, Nickl M, Troeltsch F, Ebner K (2022) Initial assessment of a fuel cell—gas turbine hybrid propulsion concept. Aerospace 9(2):68. https://doi.org/10.3390/aerospace9020068

58. Fernandes MD, P. Andrade ST de, Bistritzki VN, Fonseca RM, Zacarias LG, Gonçalves H et al (2018) SOFC-APU systems for aircraft: a review. Int J Hydrogen Energy 43(33):16311–16333. https://doi.org/10.1016/j.ijhydene.2018.07.004

59. Ponater M, Pechtl S, Sausen R, Schuhmann U, Hüttig G (2006) Potential of the cryoplane technology to reduce aircraft climate impact: a state-of-the-art assessment. Atmos Environ 40(36):6928–6944. https://doi.org/10.1016/j.atmosenv.2006.06.036

Climate Impacts of Aviation and the Potential of Aviation Powerfuels Toward Their Mitigation

Gunnar Quante, Christiane Voigt, and Martin Kaltschmitt

Abstract The UN Conference of the Parties, representing nearly all UN member states, agreed in the Paris Agreement to "hold the increase in the global average temperature to well below 2 °C above pre-industrial levels and pursuing efforts to limit the temperature increase to 1.5 °C above pre-industrial levels". The International Civil Aviation Organization (ICAO) and the International Air Transport Association (IATA) have set a "Net-Zero Carbon Emissions" target by 2050. While the Paris Agreement focuses on limiting global temperature increase, the aviation industry targets primarily address CO_2 emissions. However, aviation's climate impact extends beyond CO_2 to contrails and indirect effects of nitrogen oxide (NO_x) emissions. Research shows that only one-third of aviation's climate impact (in terms of effective radiative forcing, ERF) is due to CO_2, with the remaining two-thirds from the other non-CO_2-related cliamte impacts. The largest climate impacts of aviation, by magnitude, are contrails, CO_2 emissions, and NO_x effects. Addressing only CO_2 emissions would overlook a significant portion of aviation's climate impact, making it essential to consider all related climate factors to align aviation industry targets with the Paris Agreement. This chapter first describes aviation's climate effects and then explores how increased use of powerfuels as example for renewably sourced kerosene might influence its overall climate impact.

Keywords Aviation Climate Impact · Non-CO_2 Effects · Contrails · Nitrogen Oxide · Aerosol Effects

G. Quante (✉) · M. Kaltschmitt
Hamburg University of Technology (TUHH), Institute of Environmental Technology and Energy Economics (IUE), Hamburg, Germany
e-mail: gunnar.quante@tuhh.de

M. Kaltschmitt
e-mail: kaltschmitt@tuhh.de

C. Voigt
German Aerospace Center (DLR), Institute of Atmospheric Physics, Oberpfaffenhofen-Wessling, Germany

© The Author(s), under exclusive license to Springer Nature Switzerland AG 2025
N. Bullerdiek et al. (eds.), *Powerfuels*, Green Energy and Technology,
https://doi.org/10.1007/978-3-031-62411-7_30

1 Introduction

Manifold human activities affect the earth's weather system in the short term and the climate system in the longer term. The primary cause of these substantial changes is the combustion of fossil fuel-based energy carriers, resulting mainly in carbon dioxide (CO_2) emissions. In addition, emissions of other climate-effective gases such as methane (CH_4), nitrous oxide and halogenated gases, aerosols and their direct and indirect effects on clouds (e.g., aviation contrails) contribute to anthropogenic climate change (Fig. 1).

The influence of human activities on weather and climate results not only in an increase of globally averaged surface temperature levels. Furthermore, global mean sea levels increase, glaciers retreat, extreme weather events such as concurrent heatwaves and heavy precipitation occur more frequently and climate zones are shifting poleward [1, 2]. These developments are more and more visible; e.g., basically all glaciers within the European Alps are rapidly disappearing since a couple of years.

Hence, the UN Conference of the Parties (COP), representing almost all United Nations (UN) member states, agreed within the so-called Paris Agreement to "hold the increase in the global average temperature to well below 2 °C above pre-industrial

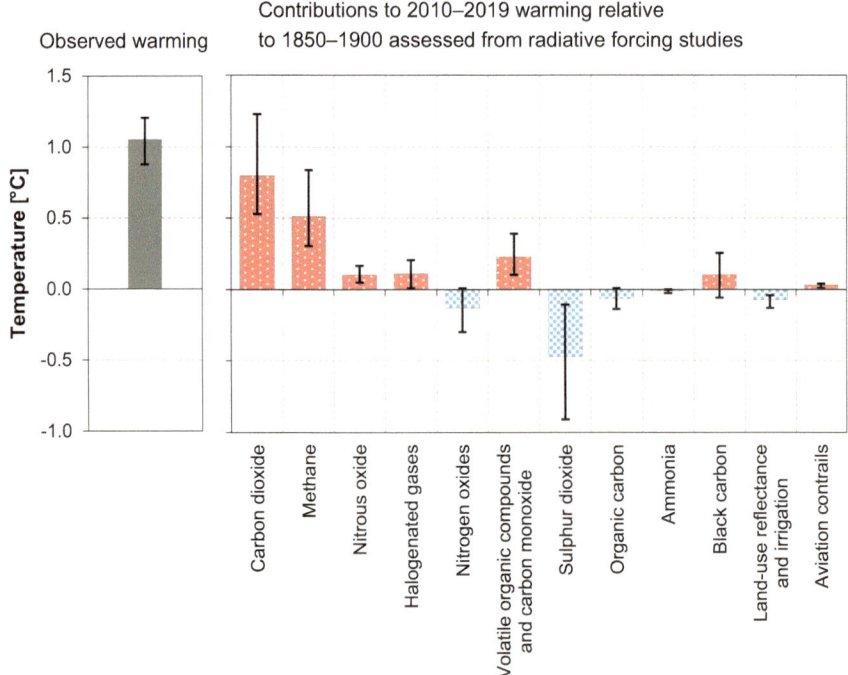

Fig. 1 Observed global warming (left) and associated drivers (right), adopted from [1]

levels and pursuing efforts to limit the temperature increase to 1.5 °C above pre-industrial levels" [3].

A central element of this Paris Agreement is the definition of nationally determined contributions, which require each party of the agreement to lay out mitigation measures to constrain anthropogenic climate change to the abovementioned global average temperature increase [3].

The national framework of the nationally determined contributions is difficult to apply to international transport modes, such as aviation and the maritime sector; e.g., the greenhouse gas (GHG) emissions for a flight/ship journey carrying passengers/freight of various nationalities from one country to another, potentially even with an intermediate stop in a third country, can be allocated in manifold ways.

The two international aviation organizations, the International Civil Aviation Organization (ICAO) and the International Air Transport Association (IATA) both defined a "Net-Zero Carbon Emissions" target by 2050 [4, 5]. Both targets focus on the abatement of CO_2 emissions, while the Paris Agreement defines a maximum in globally averaged temperature change. This difference is important because the climate impact of aviation does not only result from "classical" greenhouse gases like CO_2, but also from contrails and indirect effects of nitrogen oxide (NO_x) emissions (among others). The current state of research attributes around one-third of aviation's climate impact (in terms of effective radiative forcing (ERF)) to CO_2 emissions and the remaining two thirds to other, so-called non-CO_2 climate terms [6]. The largest climate effects of aviation are (by decreasing order of magnitude) contrails, CO_2 emissions and indirect effects from NO_x emissions. Mitigating only the climate impact of aviation's CO_2 emissions would omit a significant share of aviation's climate impact. Therefore, a consideration of all aviation-related climate terms is indispensable to align the aviation industries mitigation targets with the Paris Agreement.

The aim of this paper is to describe and characterize the climate effects of aviation and then to discuss the effects of a potentially increased use of powerfuels[1] onto aviation's overall climate impact. The paper is structured accordingly, while the discussion of the individual climate terms is sorted by their magnitude in terms of effective radiative forcing (ERF).

2 Climate Impacts of Aviation

The aviation sector consumed around 300 Mt of fuel (mainly fossil fuel-based kerosene) in the year 2018 resulting in CO_2 emissions of about 1 Gt (2018) [6, 7]. This corresponds to approximately 2.4% of the worldwide anthropogenic CO_2

---

[1] In the context of this chapter , the term "powerfuels" refers to fuels produced based on electrical power. "Aviation powerfuels" more specifically denotes powerfuels suitable for the use in aviation turbine engines.

emissions and ~0.9% of the anthropogenic effective radiative forcing due to emissions of greenhouse gases (GHG) [6, 8]. Contrails and indirect effects from NO_x emissions (among others) increase the overall climate impact beyond gaseous emissions. Keeping this in mind, an assessment of aviation's impact on climate available to date estimates a climate impact for the time period between 1940 and 2018 to 101 mW/m^2 in terms of effective radiative forcing (ERF) [6]. This corresponds to 3.5% of the overall anthropogenic climate forcing of 2,840 mW/m^2 (time span 1750 until 2019) [8].

Following Fig. 2, the most prominent effect of aviation on climate are contrails, in particular contrail cirrus cluster at night. Compared to that, CO_2 emissions and indirect effects of NO_x emissions contribute less to the effective radiative forcing (ERF). Beside this, water vapor emissions in the stratosphere, aerosol-radiation effects and aerosol-cloud interaction have a still relevant but smaller climate impact than the aforementioned climate terms [6]. Below, those climate terms and their cause-and-effect relationships are described in detail.

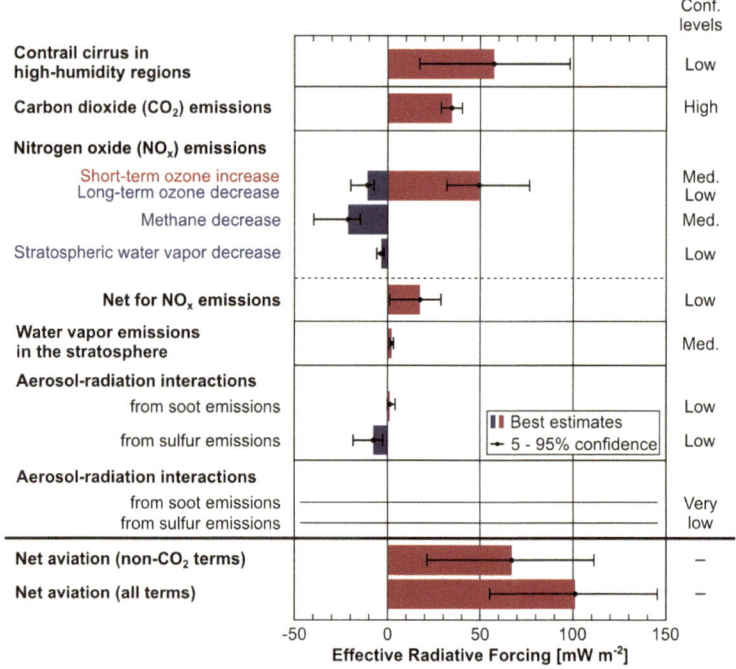

Fig. 2 Global Aviation Effective Radiative Forcing (ERF) Terms for the time 1940 to 2018, adopted from [6]

2.1 Contrails

The climate impact of contrails has been estimated as 57.4 (17 to 98) mW/m^2 in terms of effective radiative forcing (ERF). This corresponds to ~56% of the overall aviation-related effective radiative forcing (ERF) until 2018 [6].

The formation of contrails is triggered by particles (soot and aerosols) emitted from aircraft engines. These particles serve as condensation nuclei initiating the condensation of water. When during a flight at the typical cruising altitude of a commercial airplane, the hot engine exhaust mixes with the cool ambient air, rapidly cools down and can reach values above the saturation vapor pressure. In those cases, the water vapor will condense on the nuclei, freeze instantaneously and form a line-shaped condensation trail ("contrail") behind the respective aircraft engine/turbine [9]. These contrails affect the earth's radiative balance by scattering incoming solar radiation (cooling effect) and absorb outgoing terrestrial heat radiation (warming effect). Hence, the climate effect of an individual contrail strongly depends on the time of the day or more precisely the solar zenith angle. In general, around noon contrails tend to have a cooling climate impact, while they have a clearly warming impact at night, as they do not scatter any solar irradiation, but absorb the outgoing terrestrial heat radiation and reemit it at a colder temperature compared to the ground.

Figure 3 shows the cumulative density function for annual energy forcing (EF) by share of contrail forming flights for a global fleet dataset between 2019 and 2021 [10]. Thus, only ~20% of the flights produce contrails. Most of them are short and medium range flights during the day and are therefore characterized by a slightly cooling effect. Nevertheless, the impact of warming contrails clearly outweighs cooling contrails. As a result, the net impact of aviation contrails on global climate is warming. Simultaneously, the majority of the warming contrail climate impact is caused by less than 5% of all flights.

The contrail climate impact over the time of the day is shown for flights in the North Atlantic flight corridor for each day in the year 2019 in Fig. 4 [11]. The distinction between a mainly warming impact at night (red colors) and a predominantly cooling impact during day (green colors) can be seen by the horizontal change in contrail cirrus net radiative forcing (RF). As all investigated flights take place on the Northern hemisphere, a seasonal trend becomes visible by the vertical change in Fig. 4. On the Northern hemisphere, days are shorter in winter and accordingly, times with a warming contrail climate impact increase for winter days.

Contrail formation is initiated when the mixture of aircraft engine exhaust and ambient air becomes supersaturated with regard to water under the presence of condensation nuclei. The thermodynamic conditions (temperature, humidity) under which a supersaturation and thus contrail formation occurs, can be determined by the Schmidt–Appleman criterion (SAC) [12, 13]. They depend on the local meteorological situation (pressure and humidity of the surrounding air) and selected aircraft properties (e.g., propulsive efficiency, released combustion heat and water vapor emissions due to fuel combustion) [12, 14, 15].

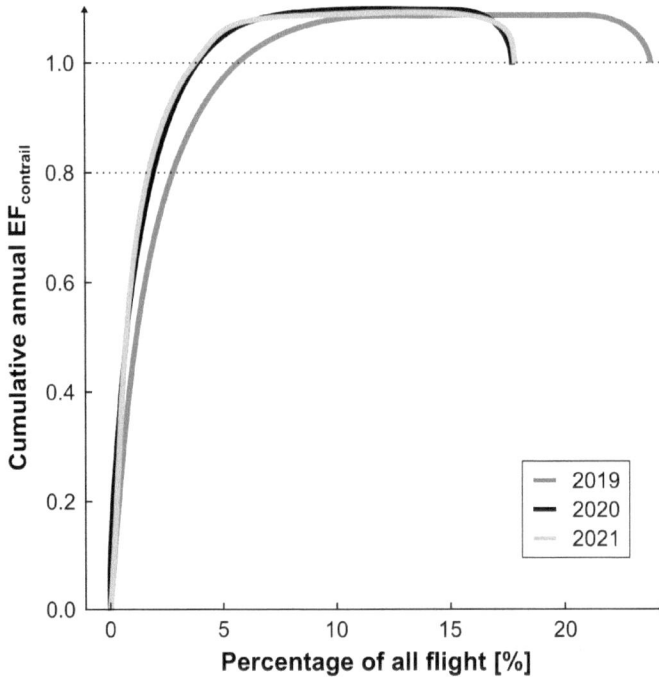

Fig. 3 Cumulative density function of the annual energy forcing (EF) of the percentage of contrail forming flights (for simplicity, warming contrails are described by EF values > 1 and cooling contrails by EF values < 1) adopted from [10]

Different particles can serve as condensation nuclei for aviation contrails. Aircraft engines emit soot and ultrafine aqueous particles. Additionally, the ambient air contains typically high numbers of aerosols. The amount of soot emissions for a given engine are strongly affected by fuel composition on the fuel side and design aspects as well as the maintenance situation on engine/turbine side. Aviation kerosene typically consists of different hydrocarbon species, such as straight chain alkanes, cycloalkanes and aromatics. Especially aromatics act as an initial soot precursor, due to their molecular structure, specifically the strong molecular bonds in the aromatic ring. Among different hydrocarbon components contained in jet fuel/kerosene, the soot formation tendency roughly decreases from poly- to monocyclic aromatics via cycloalkanes toward alkanes [16].

The variation of the molecular structure of the hydrocarbon species also affects the mass fraction of hydrogen contained. For example, the alkane Tridecan ($C_{13}H_{28}$) shows a hydrogen content of ~15.2 m-%, cycloalkanes have a hydrogen content of ~14.3 m-%, the mono-aromatic benzene is characterized by a hydrogen content of ~7.7 m-%, and the bicyclic naphthalene ($C_{10}H_8$) shows even a hydrogen content of ~6.3 m-%. Therefore, for currently used engines/turbines in commercial airplanes,

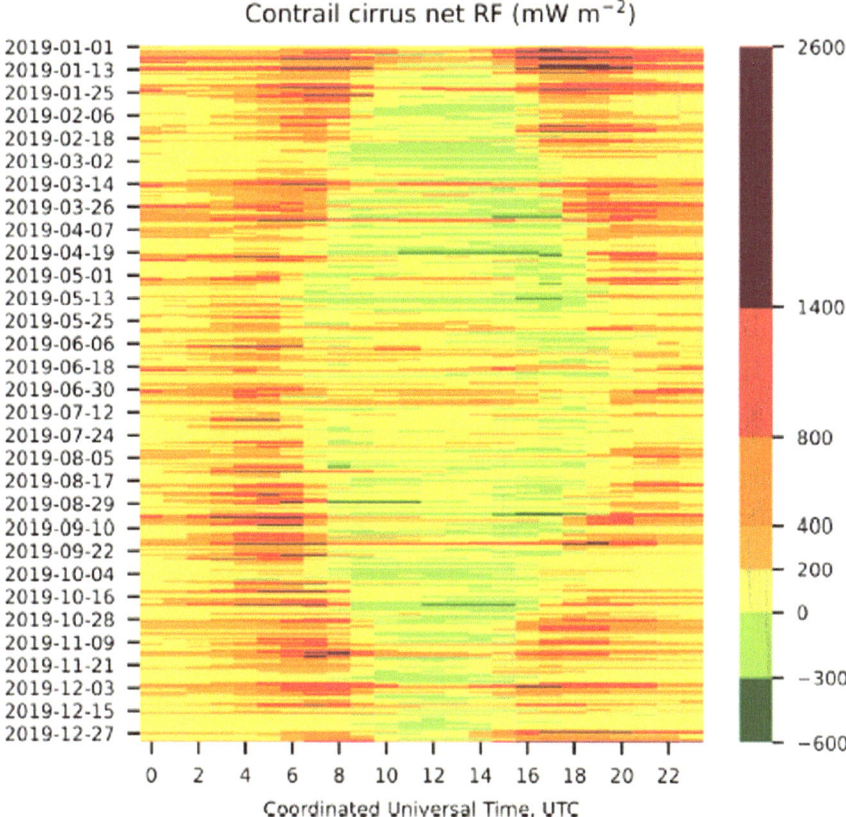

Fig. 4 Contrail cirrus net radiative forcing (RF) by time of day (x-axis) and calendar day (y-axis), adopted from [11]

the hydrogen content of a particular kerosene can serve as simplified estimate for its sooting tendency [16–20].

Particulate matter emissions of most modern aircraft engines/turbines range from 10^{14} to 10^{16} particles/kg—fuel (soot-rich regime, Fig. 5). For these engines/turbines, soot particles are the predominant condensation nuclei. Here, the number of ice crystals formed under the respective weather conditions within the cruising altitude is roughly proportionate to the number of soot particles emitted [18, 21–24].

Ambitions to further increase engine/turbine efficiency and to improve local air quality around airports have facilitated the development of engines/turbines with substantially reduced soot emissions. Assuming that modern or future engines/turbines can reduce soot emissions substantially below 10^{14} particles/kg of fuel burned, another process would become important for ice nucleation. In this range (soot-poor regime), ice nucleation is primarily initiated by ultrafine aqueous particles and aerosols from ambient air.

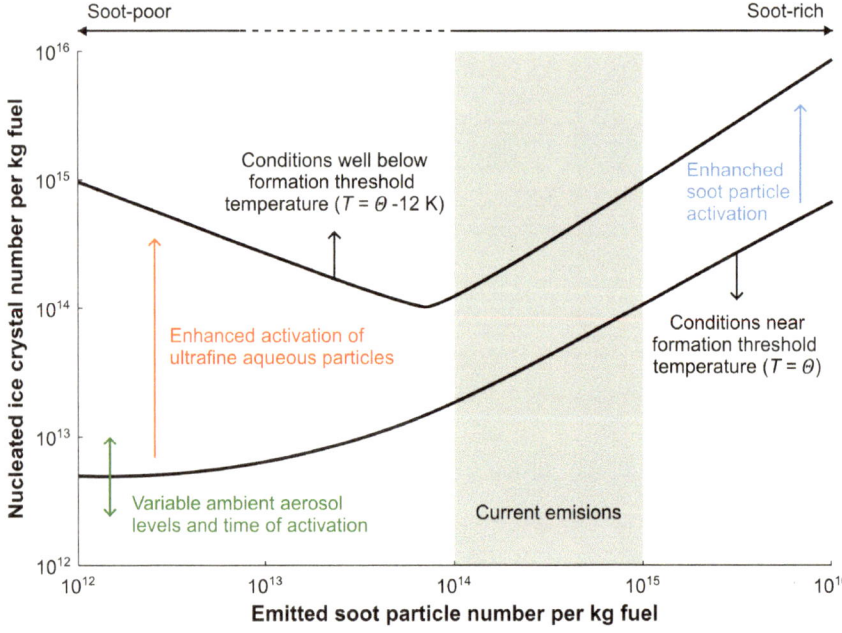

Fig. 5 Effect of soot particle emissions on nucleated ice crystal numbers, adopted from [9]

The lifetime of a contrail is mainly determined by processes taking place after ice nucleation. If further mixing and cooling of engine/turbine exhaust and ambient air results in sub-saturation, the contrail diminishes relatively fast and its climate impact is small. In some atmospheric regions, however, the relative humidity with respect to ice is greater than 100% (ice-super-saturated regions). If a contrail is formed under these circumstances, it grows by uptake of water vapor from the ambient air and can reach a contrail lifetime up to a few hours (so-called persistent contrail). Such persistent contrails typically exist for several hours but usually not longer than one day [25–27]. Thus, this type of contrails can cause a large climate impact, depending on the solar zenith angle and vertical wind shear [11, 28, 29]. Under the influence of vertical wind shear, persistent contrails can lose their linear shape, reach large horizontal extents and form so-called contrail cirrus or merge into contrail cirrus cluster. In these cases, the large surface area further increases the contrail's climate impact.

Concluding, the climate impact of contrails is severely influenced by atmospheric and aircraft parameters. The first group is comprised of the ice saturation of the ambient air, potential wind gradients and the solar zenith angle. The latter group depends on the aircraft's propulsive efficiency (a parameter of the Schmidt–Appleman criterion), soot and ultrafine aqueous particle emissions. Atmospheric parameters are externally defined by the prevailing weather situation, hence their impact can only be mitigated by avoiding regions with high contrail formation probability [30, 31]. High propulsive efficiencies are generally preferred for high fuel efficiencies and thus lower CO_2 emissions and operating cost. Another option in the high soot regime of current engine technologies would be to lower soot emissions, either by developing new engine/turbine technology [32] or by using fuels with a lower aromatics content/increased hydrogen content [9, 11, 18, 21]. Further studies are required to investigate the ice nucleation processes in the low soot regime, when the reduced abundance of soot might lead to the activation of volatile aerosol or background aerosol.

2.2 Carbon Dioxide (CO_2)

The climate impact of CO_2 has been estimated as 34.3 (28 to 40) mW/m^2 in terms of effective radiative forcing (ERF). This corresponds to ~34% of the overall aviation-related effective radiative forcing (ERF) until the year 2018 [6].

Virtually all aviation kerosene used today consists of hydrocarbons. For the short- and medium-range fleets, electric propulsion concepts, fuels cells and hydrogen combustion are currently investigated and tested, in order to advance their introduction into the global commercial airplane fleet. Still, especially for long-haul aircraft, due to the long development and use times for such airplanes and engines/turbines as well as their high development cost and associated risks it appears unlikely that a carbon-free fuel or battery-electric solutions can be introduced fleet-wide clearly before 2050 [33].

Despite differences in the exact composition of the various groups of hydrocarbons within aviation kerosene, for a hypothetical ideal combustion the products are always water (H_2O) and carbon dioxide (CO_2). An ideal combustion of fossil-based kerosene would yield ca. 74 g_{CO2}/MJ_{fuel} [34]. In reality, the formation of soot and other products of non-ideal combustion lowers this value slightly; nevertheless, due the fact that the unburned and partly burned fuel components oxidize over time within the atmosphere, it is advisable to stick to the value mentioned above.

Emissions from combustion are commonly referred to as "Tank-to-Wake (TtW)" emissions. Additionally, the provision of kerosene also requires energy and this causes necessarily further CO_{2e} emissions,[2] which have to be added to the emission budget calculations in terms of "Well-to-Wake (WtW)" emissions. But in basically all cases, the largest fraction of the "Well-to-Wake (WtW)" emissions are the "Tank-to-Wake (TtW)" emissions.

Crude oil extraction and processing in "classical" refineries and transportation contribute the majority of the so-called "Well-to-Tank (WtT)" GHG emissions. Figure 6 shows sources of CO_{2e} emissions from fossil-based kerosene provision and use in absolute and relative terms.

---

[2] In particular, the provision of fuels may cause the emissions of other greenhouse gases (e.g., CH_4). In the following, overall emissions of greenhouse gases are referred to as CO_{2e} emissions and emissions of CO_2 specifically are referred to as CO_2 emissions.

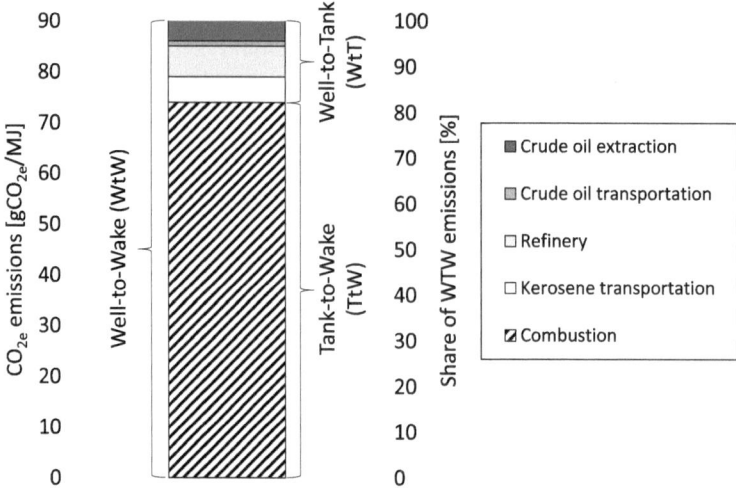

Fig. 6 CO_{2e} emissions of different steps in the kerosene value chain; based on [34–36] (values are indicative and will change for different crude oil types and supply chains)

The largest fraction of the "Well-to-Tank (WtT)" emissions is caused by crude oil refining (Fig. 6). The upgrading of crude oil toward specification compliant aviation fuels (kerosene type Jet A/A-1) requires hydrogen and heat. In present-day refineries, hydrogen is usually supplied by steam methane reforming (SMR) or petrol coke gasification [37]. Both processes incur GHG emissions, approximately 10 kg_{CO2}/kg_{H2} for steam methane reforming and about 20 kg_{CO2}/kg_{H2} for petrol coke gasification [38–41]. The provision of heat for the various upgrading processes and the catalyst regeneration incurs further CO_{2e} emissions release by conventional refineries to be allocated to kerosene [37]. As fossil resources become increasingly depleted, crude oils tend to become "heavier" (i.e., the density of the crude oil increases) and usually contains higher amounts of sulfur. As a result, upgrading toward specification compliant fuels requires larger amounts of hydrogen as well as thermal energy [42]. If this trend persists, emissions from refining of fossil-based kerosene can be expected to increase in the future [37].

Various modes of transport are available for the transportation of crude oil and aviation kerosene. In general, shorter transport distances and scale effects from larger transport volumes can reduce transport emissions; e.g., specific emissions of maritime transport are typically lower than for rail transport, which in most cases still shows lower specific GHG emissions than road transport.

The increasing depletion of fossil resources does not only affect emissions from refinery operations, but also from crude oil extraction. Enhanced oil recovery technologies require additional (thermal) energy usually resulting in increased CO_{2e} emissions. The second major influence on emissions from crude oil extraction are gas flaring practices [43]. If gas from a specific oil field is not economically saleable, it is either flared, vented or reinjected. Flaring refers to the burning of the crude oil gas released from the crude oil due to pressure relief during production (causing mainly CO_2 emissions), while venting refers to directly releasing the gas into the atmosphere. Gas from oil fields typically contains methane, which has a clearly higher mass-specific impact on global climate compared to CO_2. Thus, in most cases the climate impact of flaring is lower than the climate impact of venting. Once released, naturally occurring local and global atmospheric circulations distribute the emitted CO_2 virtually uniformly across the globe.

The atmospheric lifetime of a CO_2 pulse emission is regulated by the fast and the slow carbon cycle.

- The fast carbon cycle encloses land uptake of CO_2 by biomass growth and ocean uptake. It removes about a third up to half of the CO_2 pulse emission.
- The slow carbon cycle encloses the reaction of CO_2 with calcium carbonate and silicate weathering.

The land and ocean uptake of the CO_2 (fast cycle) takes place at timescales up to 100 years, and reactions with calcium carbonate have typical timescales of 1,000–10,000 years while silicate weathering rather takes place between 10,000 and millions of years (slow cycle) [44, 45]. The large durations of both cycles yield a thorough mixing of CO_2 across the entire atmosphere and a very slow removal of CO_2 from the atmosphere by natural processes.

In conclusion, CO_2 emissions from aviation are primarily caused by fossil-based kerosene combustion in the engines/turbines and additional emissions arise from the production/provision from crude oil. The removal of CO_2 from the atmosphere by natural processes takes decades to centuries and partly even longer. Thus, a pulse emission of CO_2—in contrast to an individual contrail—is characterized by a long-lasting climate impact. This is the reason CO_2 emissions partially accumulate in the atmosphere over time, provided that the emissions rate is greater than the removal rate, which is clearly the case for present-day CO_2 emission levels.

2.3 Nitrogen Oxide (NO_x)

The net climate impact of indirect effects from emissions of NO_x has been estimated at ~17.5 (0.6 to 29) mW/m^2 in terms of effective radiative forcing (ERF), corresponding to 18% of the overall aviation-related ERF until the year 2018 [6]. The term NO_x commonly refers to nitric oxide (NO) and nitrogen dioxide (NO_2), since most of the nitric oxide (NO) from combustion oxidizes to nitrogen dioxide (NO_2).

As the atmospheric lifetime of NO_x emissions is very short, its direct climate impact is negligibly small. However, NO_x emissions affect the climate by a short-term formation of ozone and a longer term reduction in atmospheric methane [46, 47]. The increase in atmospheric ozone (O_3) has a warming effect (~38 mW/m^2), while the methane reduction has a cooling effect (~- 21 mW/m^2). Accordingly, the resulting net effect of NO_x emissions from aviation is warming. The ozone production rate and lifetime depend on ambient NO_x and OH concentrations. For global NO_x emissions, the methane lifetime reduction is more pronounced and the ozone production rate is lower compared to the rather specific case of aviation NO_x emissions at aircraft cruise altitudes. Therefore, the net climate impact of global NO_x emissions is most likely cooling, while the net effect of aviation NO_x emissions is clearly warming (Fig. 1) [48].

NO_x are formed during fuel combustion, when molecular oxygen (O_2) is dissociated with atomic oxygen (O•) and reacts with molecular (N_2) or atomic (N•) nitrogen inside the hottest parts of the combustor of the turbine [49]. The formation of NO_x can be attributed to four different processes: Thermal NO formation, the N_2O mechanism, prompt NO formation and fuel nitric oxide formation 50. The first three processes describe different reactions for the oxidation of atmospheric N_2 into NO and N_2O, while the latter refers to the formation of NO_x from fuel-bound nitrogen. Aviation kerosene typically contains less than 2 m-% of nitrogen while the atmosphere contains ~75 m-% of nitrogen. Thus, the reaction of fuel-bound nitrogen is of lower importance for the overall NO_x formation [50]. Nevertheless, the contribution of each formation process largely depends on combustion temperature and equivalence ratio (actual vs. stoichiometric fuel/air ratio) [32, 50, 51].

In general, an equivalence ratio close to one yields the highest combustion temperature and pressure. This in turn results in increased NO_x emissions (Fig. 7). For high thrust settings and high fuel efficiency, however, higher combustion chamber temperatures would be desirable [32]. This causes a trade-off between engine/turbine efficiency and NO_x formation. To counteract this trade-off, the design of combustors has become increasingly advanced with the goal to facilitate both, low nitrogen oxide NO_x emissions as well as a high fuel efficiency (e.g., the "Rich-Quench-Lean" (RQL) combustor) [32, 50].

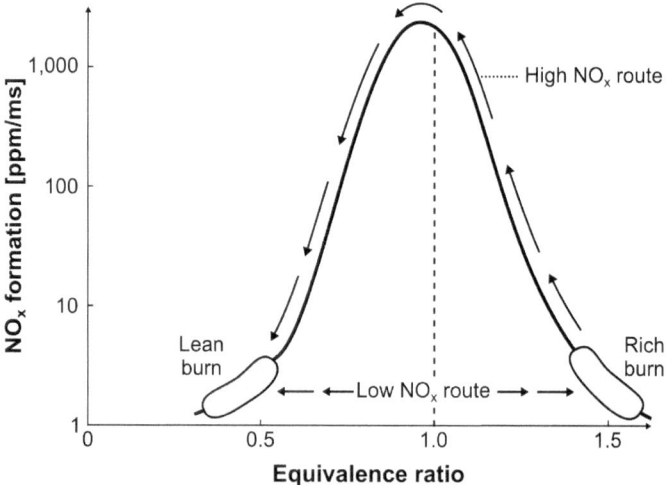

Fig. 7 Nitrogen oxide (NO_x) formation as function of fuel/air ratio ("equivalence ratio"), adopted from [50]

Once NO_x emissions are emitted from the engine/turbine exhaust, a combination of photochemical and catalytical processes alter the atmospheric composition and thus indirectly its radiative balance. Photochemical reactions with OH yield in an increase in atmospheric O_3 at the expense of NO_x and atmospheric CH_4. Catalytical reactions result in an O_3 decrease. These reactions strongly depend on the background concentration ratio of NO, OH and O_3. For low ratios the O_3 depletion is dominant, while higher ratios favor an O_3 increase [47]. As the NO/O_3 ratio changes geographically, also the effect of nitrogen oxide NO_x emissions on atmospheric ozone (O_3) varies regionally [52]. The CH_4 decrease is less regionally dependent, due to methane's CH_4 comparatively long average atmospheric lifetime of 8 to 9 a, allowing for thorough atmospheric mixing [53].

In conclusion, the climate impact of indirect NO_x effects is to a large extend determined by the engines/turbines thrust setting and combustor design, as well as the flight route. Aviation kerosene properties do not substantially influence overall NO_x emissions, since the amount of NO_x formed from fuel-bound nitrogen contribute only to a (very) small share to the overall NO_x emissions.

2.4 Water Vapor (H_2O) in the Stratosphere

The climate impact of water vapor within the stratosphere has been estimated as 2.0 (0.8 to 3.2) mW/m^2 in terms of effective radiative forcing (ERF). This corresponds to ~2% of the overall aviation-related effective radiative forcing (ERF) until 2018 [6].

Weather phenomena (e.g., wind, clouds and rain) occur in the troposphere, the lowest atmospheric layer. Within this layer, atmospheric circulations cause horizontal and vertical movements of air masses and thus wash out additional water emissions comparatively fast. The boundary between the troposphere and stratosphere (atmospheric layer above the troposphere) is called tropopause. Its height changes with latitude, reaching higher elevations above ground in equatorial (~16 km) than in polar regions (~8 km). Cruise altitudes of present-day aircraft vary between 9 and 13 km height and thus tend to take place in the stratosphere more frequently with increasing latitude [54]. Due to the atmospheric stability of the troposphere, the vertical exchange of air masses between troposphere and stratosphere is quite limited. Once injected into the stratosphere, the lifetime of water vapor increases as it follows the general circulation patterns. For this reason, the lifetime of water vapor emissions is substantially higher in the stratosphere compared to the troposphere. Hence, the climate impact of water vapor emissions becomes significant in the stratosphere. This is particularly relevant for potential supersonic flights, which would usually take place at even higher altitudes than most commercial flights today [55].

2.5 Aerosol Effects

Aerosol effects can be distinguished as direct influence of aerosols on radiation and indirect effects from aerosols altering the optical properties of clouds and thereby affecting the earth's radiative balance. The climate impact of aerosol-radiation effects has been estimated as 0.9 (0.1 to 4.0) mW/m^2 in terms of effective radiative forcing (ERF) for soot emissions and -7.4 (-19 to -2.6) mW/m^2 for sulfur aerosols. Estimates of aerosol–cloud interaction effects exhibit large uncertainties presently preventing the formulation of best estimate values [6].

2.5.1 Aerosol-Radiation Effects

Emissions of soot and sulfate aerosols have a direct forcing impact by scattering and absorbing radiation. Soot primarily absorbs incoming short-wave radiation and hence has a warming impact. Sulfate aerosols in turn mainly scatter incoming short-wave radiation, less solar radiation reaches the ground and thus they have a cooling impact [6, 56] (for the formation mechanism of soot see Sect. 2.1).

Sulfate aerosols are formed by the full oxidation of sulfur components contained within aviation kerosene. Lower levels of sulfur contained in aviation kerosene would directly reduce the emissions of sulfate aerosols as well as the activation of co-emitted soot particles and their associated climate impacts.

2.5.2 Aerosol–Cloud Interaction

In addition to contrail formation, both, supercooled aqueous solutions (e.g., sulfate aerosol particles) and insoluble ice nuclei particles (e.g., dust particles and soot) can trigger the formation of ice in cirrus clouds and they can modify existing cirrus and low level clouds [57]. They occur in the background atmosphere, and their concentration is affected by aircraft emitting soot and sulfate aerosols. As the processes governing the formation and the modification of clouds from those particles are not yet scientifically well understood, the magnitude and sign of the resulting impact on climate is still unclear. Some studies suggest that their absolute value may be even larger than contrail-related climate effects (warming or cooling) [58–64].

2.6 CO$_2$ and Non-CO$_2$ Effects

The comparison of aviation CO$_2$ and Non-CO$_2$ effects requires careful consideration of their spatiotemporal characteristics. This can be illustrated by an exemplary comparison of CO$_2$ emissions and contrails.

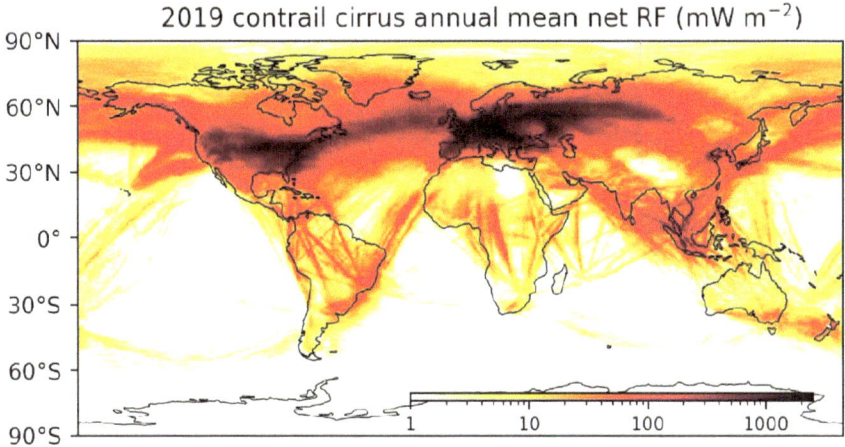

Fig. 8 Contrail cirrus annual mean net radiative forcing for 2019, adopted from [10]

In temporal terms, contrails exist not longer than one day, but a portion of a CO_2 pulse emission remains in the atmosphere for more than hundred years. For a given point in time, the climate impact of contrails is determined by all contrails existing in this particular moment. But for CO_2, not only current, but also historic emissions need to be considered, even for centuries. In other words, while the climate impact of contrails is primarily determined by the present situation, the climate impact of CO_2 (partially) accumulates over time.

In spatial terms, contrails affect mostly their coverage area resulting in a regionally constrained strong climate impact. Due to the large global variations in air traffic density, the contrail climate impact is regionally very inhomogeneous and concentrated within the flight corridors and in heavily flown areas, where they contribute significantly the anthropogenic radiative forcing (Fig. 8) (i.e., contrails have a large climate impact regionally). CO_2, however, mixes very well in the atmosphere and thus its regional differences are negligibly small.

Different climate metrics can be distinguished following the driver-response-impact chain (Fig. 9). Uncertainties increase from emissions/radiative forcing via temperature change estimates which require climate modeling toward socio-economic impacts. However, measures such as welfare loss or other indicators would be desirable from a socio-economic perspective, as these are common targets for political measures instead of physical state variables such as the earth's radiative balance.

These aspects complicate a generally valid comparison between different aviation climate terms. At first, the choice of a metric needs to find a balance between low uncertainty and relevance for policy. Secondly, the choice of time horizon has a strong influence on, e.g., global warming potentials (GWP) or absolute temperature change potential (ATP) used to compare different climate terms. The time horizon is decisive for the relative weight placed on effects taking place on different time scales.

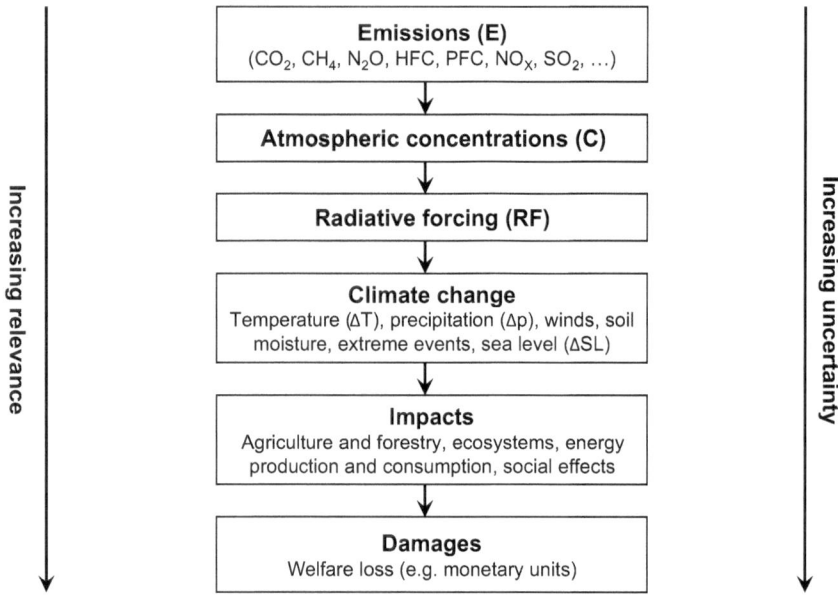

Fig. 9 Driver-response-impact chain of climate effects from greenhouse gas (GHG) emissions, adopted from [65]

Hence, their choice is rather a value judgment than a physically based selection [66]. Instead, the weighting of short- and long-term effects is rather a political than a scientific decision about how much emphasis should be placed on the situation today versus the mid- to long-term future.

3 Climate Impact Mitigation Potentials of Aviation Powerfuels

The average service duration for commercial aircraft is around 20 to 30 years. This results in long delays for the adoption of new technologies. In order to facilitate their timely introduction, renewably sourced kerosene is often required to be "drop-in capable". This means that they are approved for in-service aircraft and infrastructure without any modifications. To meet this requirement, renewably sourced kerosene and thus also aviation powerfuels have to fulfill the same specification as fossil-based kerosene plus some additional requirements. But in reality, some characteristic differences between aviation powerfuels and fossil-based kerosene remain.

There are virtually no aromatics in neat aviation powerfuels[3] yielding in a ~1 to 2 m-% higher hydrogen content, resulting in a slightly (<5%) increased gravimetric energy content and marginally increased water emissions from combustion. Another difference is that the carbon contained in aviation powerfuels is sourced from recent carbon sources (e.g., the atmosphere, biomass) instead of fossil resources. Aviation powerfuels also contain hardly any molecules contaminated with heteroatoms, such as sulfur or nitrogen.

Against this background, climate relevant properties of aviation powerfuels and their effect on aviation's climate impact are qualitatively discussed below. As there is hardly any effect of aviation kerosene type on aviation NO_x emissions [16, 68], a discussion of the impact of aviation powerfuels on this climate term is omitted. The following aspects are discussed in depth.

- The lack of aromatics in aviation powerfuels can reduce the contrail climate impact (Sect. 3.1).
- The use of carbon from renewable sources allows for large reductions in lifecycle CO_2 emissions (Sect. 3.2).
- Water vapor in the atmosphere, aerosol-radiation effects and aerosol–cloud interactions could be affected to some extent. Since their magnitude is most likely far smaller and some of their effects on weather and climate are not yet scientifically well understood, potential effects from using aviation powerfuels on these climate terms are briefly summarized (Sect. 3.3).

3.1 Contrails

Figure 10 shows the relationship between increasing the aviation kerosene hydrogen content by higher blend ratios of renewably sourced kerosene and the relative change in contrail climate impact relevant parameters. The lower amount of aromatic components in renewably sourced kerosene (including aviation powerfuels) results in a reduction of soot formation and their higher hydrogen content leads to a slight increase in water emissions from combustion.

Various in-flight measurement campaigns studied engines/turbines emitting in the soot-rich regime and several of them investigated the effect of a reduced content in aromatics and an increased content in hydrogen on the soot particle emission numbers and ice nucleation [15, 18, 23, 49, 69]. Evidently, ice crystal numbers decrease proportionally with soot particle emissions. Therefore, the optical thickness of the contrail is reduced compared to contrails from fossil-based kerosene. As the water available for ice particle formation remains roughly constant, it condenses on fewer ice particles growing under these circumstances to larger sizes [18, 21]. These heavier ice particles tend to sediment faster into warmer air masses/warmer parts of the atmosphere,

---

[3] Currently, alternative kerosene are only certified to be used as blend with fossil-based kerosene as per ASTM D7566. But, ASTM is currently developing a standard for unblended alternative aviation fuels (status 2023) [67].

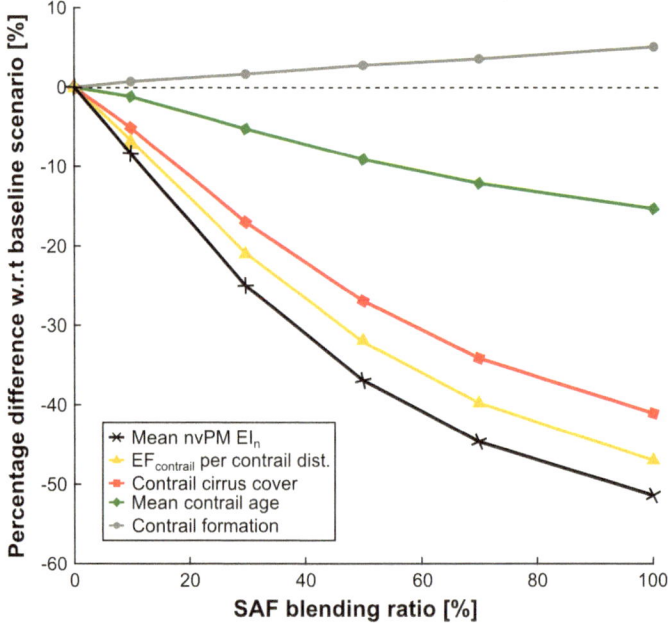

Fig. 10 Changes in non-volatile soot particle emissions, contrail formation potential, contrail lifetime, contrail cover, and contrail climate impact in terms of radiative forcing and energy forcing per contrail distance when increasing the hydrogen content of aviation kerosene (left), adopted from [11]

where they sublimate by reducing the lifetime of the contrails. Both effects—i.e., the faster sedimentation causing a reduced contrail lifetime and the reduced optical depth reducing the contrail cover—help to reduce the contrail climate impact [29]. For engines/turbines emitting in the soot-poor regime it is not yet clear to which extent this effect diminishes and how ice nucleation on ultrafine volatile particles and low levels of background aerosols affects the climate impact of contrails formed from those engines/turbines [23]. Aviation powerfuels, which typically contain hardly any sulfur, might reduce the ice nucleation on sulfate aerosols.

The increase in water emissions slightly enhances the occurrence of contrails (Fig. 10) [11]. The exhaust air contains more water from combustion, which increases the likelihood that its mixing with ambient air results (at least temporarily) in supersaturation and subsequent condensation. Also, the higher energy content of the fuel changes the range of atmospheric conditions under which the contrails can form. Even though the use of aviation powerfuels can slightly increase contrail occurrence, modeling studies suggest that this effect is not as pronounced as the reduction in contrail lifetime. Hence, a net reduction in contrail climate impact can be expected from using aviation powerfuels, especially for high blend shares or especially neat aviation powerfuels.

So far, only few studies on the effect of renewably sourced kerosene on the contrail climate impact exist. One uses a global model and finds that a 50% reduction in ice crystals lowers the contrail climate impact in terms of radiative forcing (RF) by almost 25% [29]. Another study of the North Atlantic flight corridor finds for a 50% reduction in soot emissions and ice crystals in contrails a related reduction in radiative forcing from contrails of ~ 42% [11].

Current market shares of renewably sourced aviation kerosene are below 1%, aviation powerfuels are not even produced at commercial scale yet [33]. As only a small fraction of all flights causes climate-relevant contrails, the targeted supply of renewably sourced aviation fuels could enable a faster contrail-climate impact reduction [11]. As of now, the implications of infrastructural changes and associated cost are unclear for a targeted use of renewably sourced kerosene. Hence, aviation powerfuels can substantially reduce the contrail-climate impact of aviation, but the exact extent of this reduction remains uncertain.

3.2 Carbon Dioxide (CO_2)

As shown by Fig. 6, the majority of CO_{2e} emissions from aviation kerosene stems from their combustion in aircraft engines/turbines. A smaller, but still relevant share of CO_{2e} emissions is created during fuel refining, transport and crude oil extraction.

The fundamental approach of renewably sourced kerosene (including aviation powerfuels) is to replace the carbon of fossil origin contained in fossil-based kerosene by carbon of renewable sources. The carbon of these sources stems more or less directly from atmospheric CO_2 and is bound by various biological and/or thermo-chemical processes within the powerfuel. In such a way, a closed carbon cycle is established preventing the net increase of atmospheric CO_2 levels. Since the "Tank-to-Wake (TtW)" emissions from combustion remain unaffected and only the net emissions of the fuel's entire lifecycle are reduced, the net effect is often referred to as "lifecycle CO_2 emissions".

However, the closed carbon cycle is a simplification, the fuel need to be produced and transported also resulting in so-called secondary emissions. These are caused by, e.g., energy provision or logistics. Literature indicates that the residual emissions of aviation powerfuels range from 1 to 27 gCO_{2e}/MJ [70–73]. This amounts to less than a third of the ICAO standard emissions value for fossil-based kerosene [34]. Additionally, it is most likely that over time these secondary emissions are more and more reduced because the overall energy system needs to be increasingly defossilized to fulfill the goals of the Paris Agreement and thus, e.g., fuel logistics should show less and less GHG emissions.

As Fig. 11 illustrates, hydrogen production and direct-air-capture (DAC) of CO_2 are the main contributors toward the residual CO_{2e} emissions of aviation powerfuels. This can be traced back to the provision of renewably sourced electricity in both cases. Hence, the key determinant for the CO_{2e} emissions of aviation powerfuels is the emission factor of the electricity source used. For both conversion routes, the

Fischer–Tropsch and the Methanol-to-Jet route, heat integration can cover most of the heat demands [74]. Hence, energy demands (and associated emissions) for the provision of heat can be neglected.

Another factor comes into play when comparing fossil-based kerosene and aviation powerfuels. Aviation powerfuels primarily consist of alkanes (Sect. 2.1), while fossil-based kerosene also contains various cyclic hydrocarbon components. This results in a slightly higher energy and hydrogen content of aviation power-fuels. In comparison with fossil-based kerosene, the increased hydrogen content yields slightly higher water and slightly lower CO_2 emissions assuming a fully stoichiometric combustion.

In terms of "Well-to-Tank (WtT)" emissions, aviation powerfuels provide a further mitigation potential. The provision of such synthetic fuels is realized by the Fischer–Tropsch or Methanol-to-Jet route and thus in theory powered fully by renewably sourced electricity; thus, the respective GHG emissions on the fossil fuel side can be avoided. The pendant to crude oil refining for aviation powerfuels is the upgrading to specification compliant aviation kerosene. If this step is also powered by renewably sourced electricity and renewably sourced hydrogen is used, CO_{2e} emissions of this step can also be mitigated to a large extent or even fully avoided. Due to the "drop-in" requirement, logistics for aviation powerfuels will most likely be similar to those of fossil-based kerosene. Emissions in this area might decrease, provided that the overall defossilation of the transport sector progresses and "green" fuels are used during the various transport processes.

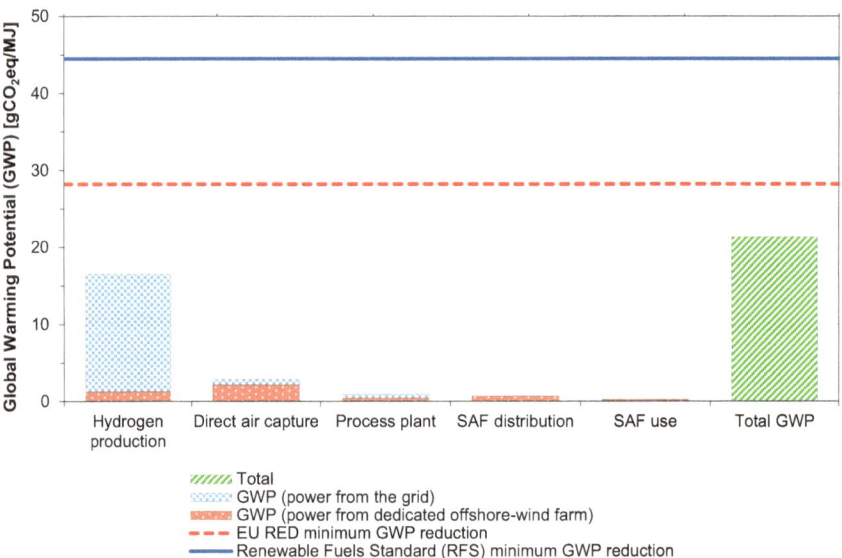

Fig. 11 Exemplary lifecycle emissions of an aviation powerfuel (FT-SPK using electricity from wind power plants; **DAC**—Direct-Air-Capture, **GWP**—Global Warming Potential, **SAF**—Sustainable Aviation Fuel,), adopted from [73]

The emissions factor of the electricity used to produce aviation powerfuels largely determines their lifecycle CO_{2e} emissions. If the power would be supplied from a coal-fired power plant, the powerfuels' life-cycle CO_{2e} emissions would be higher than those of fossil-based kerosene. If power from renewable sources (e.g., wind mills, photovoltaic systems) is used instead, far lower life-cycle CO_{2e} emission values can be achieved. However, it seems questionable to achieve net-zero emissions without additional CO_2 sequestration (e.g., to mitigate residual emissions from the construction of wind or photovoltaic power plants) [70–73] on the short term; on the long term this seems to be possible if defossilation of our overall economy progresses.

3.3 Water Vapor and Aerosol Climate Impacts

The increased hydrogen content of aviation powerfuels slightly shifts the stoichiometric ratio between the final oxidation products CO_2 and H_2O toward H_2O. While CO_2 emissions marginally decrease, H_2O emissions increase slightly. For flights within the stratosphere, this would also slightly increase the climate impact of water vapor emission.

The reduction in soot emissions does not only affect contrail formation, but also reduces aerosol-related climate impacts from soot. Aviation powerfuels are virtually free of sulfur, hence also effects of sulfate emissions are reduced or even avoided. In which direction (more warming/more cooling) this changes the net climate impact of direct and indirect aerosol effects still needs to be investigated [55].

4 Conclusion

In 2015, a global consensus was reached to limit global warming below 2 °C in order to prevent detrimental effects of anthropogenic climate change. International aviation climate mitigation goals currently focus on the abatement of CO_2 emissions. Aviation powerfuels are considered a key measure to achieve these mitigation goals. However, the climate impact of aviation does not only result from CO_{2e} emissions, but—beside other effects of minor importance—also contrails and indirect NO_x effects contribute substantially to the total climate impact from aviation. Against this background, this chapter quantifies and describes aviation climate terms first. Then, the current knowledge on the effect of aviation powerfuels on the overall climate impact from aviation is summarized.

By decreasing order of magnitude, the individual contributions to the total effective radiative forcing (ERF) from aviation are contrails, emissions of CO_{2e}, indirect effects of NO_x and—to a substantially smaller extent—water vapor in the stratosphere as well as direct and indirect aerosol effects.

- The climate impact of contrails is determined by atmospheric and aircraft-related parameters. While the first group includes ice saturation of the ambient air, potential wind gradients and the solar zenith angle, the latter primarily depends on the aircraft's propulsive efficiency, fuel and engine/turbine parameters (soot and ultrafine aqueous particle emissions).
- CO_2 emissions from aviation are primarily caused by fuel combustion in the engines/turbines and emissions from crude extraction and refining.
- Indirect NO_x climate effects are to a large extend determined by the engine's/ turbines thrust setting and combustor design, as well as the flight route; fuel properties do not substantially influence overall NO_x emissions.
- Emissions of water vapor become climate-relevant for flights in the stratosphere due to the increased atmospheric residence time in this atmospheric layer.
- Direct and indirect aerosol climate effects are caused by soot and sulfate aerosol emissions. They cause warming and cooling climate effects; however, especially in the case of indirect aerosol effects the magnitude of the climate impact is highly uncertain.

The use of aviation powerfuels has significant effects on contrail formation and lifecycle CO_{2e} emissions. For the contrail-related climate impact, aviation powerfuels reduce the emissions of soot particles (at least in the soot emission regimes of most present-day aircraft engines/turbines) and in the end shorten the contrail lifetime and reduce the associated net climate impact of contrails. A quantification of the contrail climate mitigation potential of renewably sourced kerosene is still ongoing, existing studies range between 30% and 60%. However, those studies would require market shares of alternative aviation fuels of around 50% or more, and current market shares are below 1%. As just a small fraction of all flights causes contrail formation, targeting those flights specifically with renewably sourced kerosene might allow for a faster reduction in aviation's climate impact while their market matures.

In principle, aviation powerfuels can reduce the CO_{2e} emissions of aviation by more than 80% and in theory on a longer perspective by 100%. However, this is only realistic if the electricity used for fuel production originates from renewable sources of energy, such as wind power or solar radiation used by wind mills and photovoltaic power plants. Electricity provision with higher CO_{2e} emissions can even result in lifecycle CO_{2e} emissions clearly above the values for fossil-based kerosene.

Other aviation climate terms remain largely unaffected by the use of aviation powerfuels (indirect NO_x effects) or are substantially smaller than the aforementioned climate terms. As contrails and CO_{2e} emissions account for a large share of aviation's climate impact, aviation powerfuels allow for a substantial reduction of aviation's overall climate impact and are not limited to CO_{2e} emissions.

In the future, targeting flight routes with a particularly high contrail-climate impact might enable faster reductions in aviation's climate impact. Further studies and flight experiments are required to improve the understanding of the necessary requirements of a targeted fuel use.

References

1. IPCC (2018) Summary for policymakers. Global warming of 1.5°C. An IPCC special report on the impacts of global warming of 1.5°C above pre-industrial levels and related global greenhouse gas emission pathways, in the context of strengthening the global response to the threat of climate change, sustainable development, and efforts to eradicate poverty 2018
2. Intergovernmental Panel on Climate Change (ed) (2023) Climate change 2021—the physical science basis. Cambridge University Press
3. United Nations (2015) Paris agreement: COP21
4. IATA (2021) Resolution on the industry's commitment to reach net zero carbon emissions by 2050
5. ICAO, ICAO welcomes new net-zero 2050 air industry commitment, 18 Oct 2022. Available from: https://www.icao.int/Newsroom/Pages/ICAO-welcomes-new-netzero-2050-air-ind ustry-commitment.aspx
6. Lee DS, Fahey DW, Skowron A, Allen MR, Burkhardt U, Chen Q et al (1994) The contribution of global aviation to anthropogenic climate forcing for 2000 to 2018. Atmos Environ 2021(244):117834. https://doi.org/10.1016/j.atmosenv.2020.117834
7. Transport (2023) In: Intergovernmental Panel on Climate Change (ed) Climate change 2022—mitigation of climate change. Cambridge University Press, pp 1049–1160
8. Annex III (2023) Tables of historical and projected well-mixed greenhouse gas mixing ratios and effective radiative forcing of all climate forcers. In: Intergovernmental Panel on Climate Change (ed) Climate change 2021—the physical science basis. Cambridge University Press; 2023, pp 2139–2152
9. Kärcher B (2018) Formation and radiative forcing of contrail cirrus. Nat Commun 9(1):1824. https://doi.org/10.1038/s41467-018-04068-0
10. Teoh R, Engberg Z, Schumann U, Voigt C, Shapiro M, Rohs S et al (2023) Global aviation contrail climate effects from 2019 to 2021
11. Teoh R, Schumann U, Voigt C, Schripp T, Shapiro M, Engberg Z et al (2022) Targeted use of sustainable aviation fuel to maximize climate benefits. Environ Sci Technol. https://doi.org/10.1021/acs.est.2c05781
12. Appleman H (1953) The formation of exhaust condensation trails by jet aircraft. Bull Am Meteor Soc 34(1):14–20. https://doi.org/10.1175/1520-0477-34.1.14
13. Schumann U (1996) Über Bedingungen zur Bildung von Kondensstreifen aus Flugzeugabgasen. metz 5(1):4–23. https://doi.org/10.1127/metz/5/1996/4
14. Bock L, Burkhardt UNN (2016) Reassessing properties and radiative forcing of contrail cirrus using a climate model. J Geophys Res Atmos 121(16):9717–9736. https://doi.org/10.1002/201 6JD025112
15. Schumann U, Heymsfield AJ (2017) On the life cycle of individual contrails and contrail cirrus. Meteorol Monogr 58:3.1–3.24. https://doi.org/10.1175/AMSMONOGRAPHS-D-16-0005.1
16. Schripp T, Anderson B, Crosbie EC, Moore RH, Herrmann F, Oßwald P et al (2018) Impact of alternative jet fuels on engine exhaust composition during the 2015 ECLIF ground-based measurements campaign. Environ Sci Technol 52(8):4969–4978. https://doi.org/10.1021/acs. est.7b06244
17. Bräuer T, Voigt C, Sauer D, Kaufmann S, Hahn V, Scheibe M et al (2021) Airborne measurements of contrail ice properties—dependence on temperature and humidity. Geophys Res Lett 48(8). https://doi.org/10.1029/2020GL092166
18. Voigt C, Kleine J, Sauer D, Moore RH, Bräuer T, Le Clercq P et al (2021) Cleaner burning aviation fuels can reduce contrail cloudiness. Commun Earth Environ 2(1). https://doi.org/10.1038/s43247-021-00174-y
19. Schripp T, Anderson BE, Bauder U, Rauch B, Corbin JC, Smallwood GJ et al (2022) Aircraft engine particulate matter emissions from sustainable aviation fuels: results from ground-based measurements during the NASA/DLR campaign ECLIF2/ND-MAX. Fuel 325:124764. https://doi.org/10.1016/j.fuel.2022.124764

20. Brem BT, Durdina L, Siegerist F, Beyerle P, Bruderer K, Rindlisbacher T et al (2015) Effects of fuel aromatic content on nonvolatile particulate emissions of an in-production aircraft gas turbine. Environ Sci Technol 49(22):13149–13157. https://doi.org/10.1021/acs.est.5b04167
21. Bräuer T, Voigt C, Sauer D, Kaufmann S, Hahn V, Scheibe M et al (2021) Reduced ice number concentrations in contrails from low-aromatic biofuel blends. Atmos Chem Phys 21(22):16817–16826. https://doi.org/10.5194/acp-21-16817-2021
22. Kärcher B, Burkhardt U, Bier A, Bock L, Ford IJ (2015) The microphysical pathway to contrail formation. J Geophys Res Atmos 120(15):7893–7927. https://doi.org/10.1002/2015JD023491
23. Kleine J, Voigt C, Sauer D, Schlager H, Scheibe M, Jurkat-Witschas T et al (2018) In situ observations of ice particle losses in a young persistent contrail. Geophys Res Lett 45(24). https://doi.org/10.1029/2018GL079390
24. Bier A, Burkhardt U (2022) Impact of parametrizing microphysical processes in the jet and vortex phase on contrail cirrus properties and radiative forcing. J Geophys Res Atmos 127(23). https://doi.org/10.1029/2022JD036677
25. Vázquez-Navarro M, Mannstein H, Kox S (2015) Contrail life cycle and properties from 1 year of MSG/SEVIRI rapid-scan images. Atmos Chem Phys 15(15):8739–8749. https://doi.org/10.5194/acp-15-8739-2015
26. Schumann U, Baumann R, Baumgardner D, Bedka ST, Duda DP, Freudenthaler V et al (2017) Properties of individual contrails: a compilation of observations and some comparisons. Atmos Chem Phys 17(1):403–438. https://doi.org/10.5194/acp-17-403-2017
27. Wang Z, Bugliaro L, Jurkat-Witschas T, Heller R, Burkhardt U, Ziereis H et al (2023) Observations of microphysical properties and radiative effects of a contrail cirrus outbreak over the North Atlantic. Atmos Chem Phys 23(3):1941–1961. https://doi.org/10.5194/acp-23-1941-2023
28. Bock L, Burkhardt UNN (2016) The temporal evolution of a long-lived contrail cirrus cluster: Simulations with a global climate model. J Geophys Res Atmos 121(7):3548–3565. https://doi.org/10.1002/2015JD024475
29. Burkhardt U, Bock L, Bier A (2018) Mitigating the contrail cirrus climate impact by reducing aircraft soot number emissions. npj Clim Atmos Sci 1(1). https://doi.org/10.1038/s41612-018-0046-4
30. Niklaß M, Gollnick V, Lührs B, Dahlmann K, Froemming C, Grewe V et al (2017) Cost-benefit assessment of climate-restricted airspaces as an interim climate mitigation option. J Air Transp 25(2):27–38. https://doi.org/10.2514/1.D0045
31. Teoh R, Schumann U, Stettler MEJ (2020) Beyond contrail avoidance: efficacy of flight altitude changes to minimise contrail climate forcing. Aerospace 7(9):121. https://doi.org/10.3390/aerospace7090121
32. Bräunling G (2015) Flugzeugtriebwerke: Grundlagen, Aero-Thermodynamik, ideale und reale Kreisprozesse, thermische Turbomaschinen, Komponenten, Emissionen und Systeme. Springer Vieweg, Berlin
33. Quante G, Bullerdiek N, Bube S, Neuling U, Kaltschmitt M (2023) Renewable fuel options for aviation—A system-wide comparison of drop-in and non drop-in fuel options. Fuel 333:126269. https://doi.org/10.1016/j.fuel.2022.126269
34. ICAO (2022) CORSIA default life cycle emissions values for CORSIA eligible fuels
35. Bauen A, Harris A, Sim C, Gudde N, Prussi M, Scarlat N (2022) CORSIA lower carbon aviation fuels: an assessment of the greenhouse gas emission reduction potential. Appl Sci 12(22):11818. https://doi.org/10.3390/app122211818
36. concawe (2017) Estimating the marginal CO_2 intensities of EU refinery products: prepared for the CONCAWE Refinery Management Group, Brussels
37. Abella JP, Bergerson JA (2012) Model to investigate energy and greenhouse gas emissions implications of refining petroleum: impacts of crude quality and refinery configuration. Environ Sci Technol 46(24):13037–13047. https://doi.org/10.1021/es3018682
38. IEA (2019) The future of hydrogen: seizing today's opportunities
39. Cetinkaya E, Dincer I, Naterer GF (2012) Life cycle assessment of various hydrogen production methods. Int J Hydrogen Energy 37(3):2071–2080. https://doi.org/10.1016/j.ijhydene.2011.10.064

40. Ruether J, Ramezan M, Grol E (2005) Life-cycle analysis of greenhouse gas emissions for hydrogen fuel production in the United States from LNG and coal

41. Kothari R, Buddhi D, Sawhney RL (2008) Comparison of environmental and economic aspects of various hydrogen production methods. Renew Sustain Energy Rev 12(2):553–563. https://doi.org/10.1016/j.rser.2006.07.012

42. Hsu CS, Robinson PR, Others (eds) (2017) Springer handbook of petroleum technology: with 754 figures and 282 tables. Springer International Publishing, Cham

43. Masnadi MS, El-Houjeiri HM, Schunack D, Li Y, Englander JG, Badahdah A et al (2018) Global carbon intensity of crude oil production. Science 361(6405):851–853. https://doi.org/10.1126/science.aar6859

44. Joos F, Roth R, Fuglestvedt JS, Peters GP, Enting IG, von Bloh W et al (2013) Carbon dioxide and climate impulse response functions for the computation of greenhouse gas metrics: a multi-model analysis. Atmos Chem Phys 13(5):2793–2825. https://doi.org/10.5194/acp-13-2793-2013

45. Ciais P, Sabine C, Bala G, Bopp L, Brovkin V, Canadell J et al (2013) Carbon and other biogeochemical cycles. In: Climate change 2013: the physical science basis. Contribution of working group I to the fifth assessment report of the intergovernmental panel on climate change

46. Niklaß M, Dahlmann K, Grewe V, Maertens S, Plohr M, Scheelhaase J et al (2019) Integration of non-CO_2 effects of aviation in the EU ETS and under CORSIA. Final Report. Dessau-Roßlau

47. Niklaß M (2019) Ein systemanalytischer Ansatz zur Internalisierung der Klimawirkung der Luftfahrt. Dissertationsschrift, Hamburg

48. Szopa S, Naik V, Adhikary B, Artaxo P, Berntsen T, Collins WD, Fuzzi S, Gallardo L, Kiendler-Scharr A, Klimont Z (2023) Short-lived climate forcers. In: Intergovernmental panel on climate change (ed) Climate change 2021—the physical science basis. Cambridge University Press, pp 817–922

49. Voigt C, Jurkat T, Schlager H, Schäuble D, Petzold A, Schumann U (2012) Aircraft emissions at cruise and plume processes. In: Schumann U (ed) Atmospheric physics. Springer Berlin Heidelberg, Berlin, Heidelberg, pp 675–692

50. Lefebvre AH, Ballal DR (2010) Gas turbine combustion: Alternative fuels and emissions. In: Lefebvre AH, Ballal DR (eds), 3rd edn. Taylor & Francis, Boca Raton

51. Nicol D, Malte PC, Lai J, Marinov NN, Pratt DT, Corr RA (1992) NO_x sensitivities for gas turbine engines operated on lean-premixed combustion and conventional diffusion flames. In: Volume 3: coal, biomass and alternative fuels; combustion and fuels; oil and gas applications; cycle innovations. American Society of Mechanical Engineers

52. Fuglestvedt J (1999) Climatic forcing of nitrogen oxides through changes in tropospheric ozone and methane; global 3D model studies. Atmos Environ (1994) 33(6):961–977. https://doi.org/10.1016/S1352-2310(98)00217-9

53. Holmes CD, Prather MJ, Søvde OA, Myhre G (2013) Future methane, hydroxyl, and their uncertainties: key climate and emission parameters for future predictions. Atmos Chem Phys 13(1):285–302. https://doi.org/10.5194/acp-13-285-2013

54. Dörnbrack A (2012) The atmosphere: vast, shallow, and full of subtleties. In: Schumann U (ed) Atmospheric physics. Springer Berlin Heidelberg, Berlin, Heidelberg, pp 3–16

55. Eastham SD, Fritz T, Sanz-Morère I, Prashanth P, Allroggen F, Prinn RG et al (2022) Impacts of a near-future supersonic aircraft fleet on atmospheric composition and climate. Environ Sci Atmos 2(3):388–403. https://doi.org/10.1039/d1ea00081k

56. Brasseur GP, Gupta M, Anderson BE, Balasubramanian S, Barrett S, Duda D et al (2016) Impact of aviation on climate: FAA's Aviation Climate Change Research Initiative (ACCRI) phase II. Bull Am Meteor Soc 97(4):561–583. https://doi.org/10.1175/BAMS-D-13-00089.1

57. Penner JE, Zhou C, Garnier A, Mitchell DL (2018) Anthropogenic aerosol indirect effects in cirrus clouds. J Geophys Res Atmos 123(20):11652–11677. https://doi.org/10.1029/2018JD029204

58. Hahn V, Meerkötter R, Voigt C, Gisinger S, Sauer D, Catoire V et al (2023) Pollution slightly enhances atmospheric cooling by low-level clouds in tropical West Africa. Atmos Chem Phys 23(15):8515–8530. https://doi.org/10.5194/acp-23-8515-2023

59. Penner JE, Chen Y, Wang M, Liu X (2009) Possible influence of anthropogenic aerosols on cirrus clouds and anthropogenic forcing. Atmos Chem Phys 9(3):879–896. https://doi.org/10.5194/acp-9-879-2009

60. Zhou C, Penner JE (2014) Aircraft soot indirect effect on large-scale cirrus clouds: Is the indirect forcing by aircraft soot positive or negative? J Geophys Res Atmos 119(19). https://doi.org/10.1002/2014JD021914

61. Pitari G, Iachetti D, Di Genova G, de Luca N, Søvde O, Hodnebrog Ø et al (2015) Impact of coupled NO_x/aerosol aircraft emissions on ozone photochemistry and radiative forcing. Atmosphere 6(6):751–782. https://doi.org/10.3390/atmos6060751

62. Gettelman A, Chen C (2013) The climate impact of aviation aerosols. Geophys Res Lett 40(11):2785–2789. https://doi.org/10.1002/grl.50520

63. Kapadia ZZ, Spracklen DV, Arnold SR, Borman DJ, Mann GW, Pringle KJ et al (2016) Impacts of aviation fuel sulfur content on climate and human health. Atmos Chem Phys 16(16):10521–10541. https://doi.org/10.5194/acp-16-10521-2016

64. Righi M, Hendricks J, Sausen R (2013) The global impact of the transport sectors on atmospheric aerosol: simulations for year 2000 emissions. Atmos Chem Phys 13(19):9939–9970. https://doi.org/10.5194/acp-13-9939-2013

65. Fuglestvedt JS, Shine KP, Berntsen T, Cook J, Lee DS, Stenke A et al (2010) Transport impacts on atmosphere and climate: Metrics Atmos Environ (1994) 44(37):4648–4677. https://doi.org/10.1016/j.atmosenv.2009.04.044

66. Myrhe G, Shindell D, Bréon F-M, Collins W, Fuglestvedt J, Huang J et al (2013) Anthropogenic and natural radiative forcing. climate change 2013: the physical science basis. In: Contribution of working group I to the fifth assessment report of the intergovernmental panel on climate change 2013

67. ASTM, Standardization news: fueling the future of aviation, 10 Nov 2023. Available from: https://sn.astm.org/features/fueling-future-aviation-ja23.html

68. Florian Wolters (2020) Einfluss alternativer drop-in Flugturbinenkraftstoffe auf das Emissionsverhalten des globalen Luftverkehrs. Ph.D. Thesis. Ruhr-Universität Bochum, Bochum

69. Moore RH, Thornhill KL, Weinzierl B, Sauer D, D'Ascoli E, Kim J et al (2017) Biofuel blending reduces particle emissions from aircraft engines at cruise conditions. Nature 543(7645):411–415. https://doi.org/10.1038/nature21420

70. Schmidt P, Weindorf W, Roth A, Batteiger V, Riegel F (2016) Power-to-liquids: potentials and perspectives for the future supply of renewable aviation fuel. Dessau-Roßlau

71. Falter C, Batteiger V, Sizmann A (2016) Climate impact and economic feasibility of solar thermochemical jet fuel production. Environ Sci Technol 50(1):470–477. https://doi.org/10.1021/acs.est.5b03515

72. Micheli M, Moore D, Bach V, Finkbeiner M (2022) Life-cycle assessment of power-to-liquid kerosene produced from renewable electricity and CO_2 from direct air capture in Germany. Sustainability 14(17):10658. https://doi.org/10.3390/su141710658

73. Rojas-Michaga MF, Michailos S, Cardozo E, Akram M, Hughes KJ, Ingham D et al (2023) Sustainable aviation fuel (SAF) production through power-to-liquid (PtL): a combined techno-economic and life cycle assessment. Energy Convers Manage 292:117427. https://doi.org/10.1016/j.enconman.2023.117427

74. König DH (2016) Techno-ökonomische Prozessbewertung der Herstellung synthetischen Flugturbinentreibstoffes aus CO_2 und H_2. Dissertation

Powerfuels and Alternative Fuels in the Maritime Sector

Larissa Fink, Shaghayegh Kazemi Esfeh, Jorgen Depken, Sören Ehlers, and Martin Kaltschmitt

Abstract The maritime sector accounts for about 3% of global GHG emissions and largely depends on fossil fuels like heavy fuel oil (HFO), marine diesel oil (MDO), and marine gas oil (MGO). These fuels contribute to significant emissions of greenhouse gases (CO_2, CH_4) and pollutants such as particulate matter (PM), sulfur oxides (SO_2), and nitrogen oxides (NOx). To address these emissions, various regulations have been implemented by nations, regional bodies like the EU, and internationally by the International Maritime Organization (IMO), which sets global standards for safety, security, and environmental performance. While MARPOL Annex VI has regulated NOx, SO_2, and energy efficiency since 2005, reducing GHG emissions has gained more focus recently. Adopting alternative fuels is considered the most viable strategy for reducing vessel GHG emissions. Various alternative fuels, each with unique benefits, challenges, and infrastructure requirements, are explored in this article, including liquefied natural gas (LNG), liquefied petroleum gas (LPG), methanol, ammonia, hydrogen, and bio-based or electricity-based diesel fuels. Their properties and technical applications, such as fuel storage and propulsion system compatibility, are discussed, along with regulatory and suitability considerations for reducing GHG emissions in the maritime industry.

Keywords LNG · LPG · Methanol · Ammonia · Hydrogen · Fuel Storage · Propulsion · Bunkering

L. Fink (✉) · M. Kaltschmitt
Hamburg University of Technology (TUHH), Institute of Environmental Technology and Energy Economics (IUE), Hamburg, Germany
e-mail: larissa.fink@tuhh.de

M. Kaltschmitt
e-mail: kaltschmitt@tuhh.de

S. K. Esfeh · J. Depken · S. Ehlers
Institute for Maritime Energy Systems, German Aerospace Center e.V. (DLR), Geesthacht, Germany

905

1   Introduction

Approximately 80% of globally traded goods are transported by ship, with seaborne trade reaching approximately 11×10^9 t in the year 2019 [1]. Related to the year 2021, this transportation was carried out by a fleet of about 99,800 ships of 100 gross tons and above. The average age of these vessels stood at 21.6 years. The maritime industry predominantly relies on fuels of fossil origin, such as heavy fuel oil (HFO), marine diesel oil (MDO), or marine gas oil (MGO) [2]. The combustion of these fuels within the respective propulsion system results in the release of greenhouse gas (GHG) emissions, including carbon dioxide (CO_2) and methane (CH_4), as well as particulate matter (PM), sulfur oxide (SO_2), and nitrogen oxide (NO_x) emissions [3].

To mitigate GHG as well as other emissions from shipping, various rules and regulations are adopted by both individual states as well as associations of states, including the European Union (EU). For ships operating internationally, the International Maritime Organization (IMO) serves as the global authority to set standards regarding safety, security, and environmental performance. While the regulation of nitrogen oxides (NO_x) and sulfur oxide (SO_2) emissions as well as energy efficiency requirements for new ships is covered by MARPOL Annex VI, which entered into force back in the year 2005 and has been updated and expanded since then, the regulation of GHG emissions to achieve significant emission reductions has seen a growing emphasis in recent years [4].

Concerning GHG emissions, the maritime sector is responsible for roughly 3% of global GHG emissions. The first IMO strategy to achieve a reduction in GHG emissions came into force in the year 2018 and was revised in 2023. This strategy aims to expedite the reduction of GHG emissions from shipping with the goal to achieve a net-zero GHG emissions level by or around the year 2050. Additionally, interim GHG reduction targets have been set, aiming for a reduction of at least 20% by the year 2030 and at least 70% by 2040. Furthermore, the target for the year 2030 involves the adoption of technologies and fuels that enable zero or near-zero GHG emissions by a share of at least 5% of the total fuel uptake in the maritime sector [5, 6].

To comply with a ship's GHG emission limits, different improvement areas including specific measures and their GHG emission reduction potential have been assessed [7]. Table 1 presents an overview of these respective improvement areas.

As given in Table 1, the adoption of alternative fuels and energy sources is the only improvement area with the potential to achieve a complete (100%) reduction in vessel GHG emissions [7, 8]. However, a relative wide range of alternative fuels can be considered to be used within the maritime industry, each characterized by different benefits and challenges, depending on their specific properties. Furthermore, requirements related to the production, distribution, and bunkering infrastructures as well as the respective propulsion technologies may differ for a use of different alternative fuels options, and potentially also requiring extensive on-board modifications.

Table 1 Improvement areas for the reduction of GHG emissions from shipping

Improvement Area	GHG Mitigation Measure	Assumed GHG Reduction Potential (Vessel Level)
Logistics	Speed reduction, optimization of routes and vessel utilization	>20%
Hydrodynamics	Hull form optimization, hull coating, hull cleaning	5–15%
Machinery	Efficiency improvements, waste-heat recovery, alternative propulsion systems (e.g., battery hybridization, fuel cells)	5–20%
Fuels and energy sources	Use of alternative fuels (e.g., liquefied natural gas (LNG), liquefied petroleum gas (LPG), methanol, hydrogen, ammonia), electrification, use of wind energy or nuclear power	0–100%
Exhaust gas treatment	Carbon capture from the exhaust gas and permanent storage of the carbon	0–90%

Against this background, various alternative fuel options intended to be used for the reduction of GHG emissions within the maritime sector are discussed and examined in more detail. Specifically, the alternative fuels liquefied natural gas (LNG), liquefied petroleum gas (LPG), methanol, ammonia, and hydrogen as well as diesel fuels produced electricity based or from biogenic sources are analyzed. These alternatives are first introduced, and then, their essential properties are discussed. Subsequently, their application within the maritime sector is examined more closely from a technical perspective, focusing on aspects of fuel storage, the compatibility with the respective propulsion system, and the fuel bunkering. Following that, their use is considered in the context of overarching aspects, including regulatory considerations and further aspects of their fundamental suitability and availability.

2 Alternative Fuel Options

In shipping, different alternative fuels are already utilized (to a very small extend), namely liquefied natural gas (LNG), liquefied petroleum gas (LPG), methanol, hydrogen, and diesel fuels produced from biogenic sources. Therefore, vessels using these fuels so far make up about 1.8% of the current global maritime fleet. The predominant alternative fuels used in shipping, LNG, LPG, methanol, and diesel fuels from biogenic sources, are carbon-based and thus result necessarily in a release of carbon dioxide (CO_2) emissions when combusted; when using diesel fuels from biogenic sources, these CO_2 emissions are treated as neutral to the climate because during plant growth the same amount of CO_2 has been removed from the atmosphere. To counter these emissions, carbon-free fuels like ammonia and hydrogen are also under consideration for their use within the maritime sector.

The properties of a fuel can significantly influence its handling aspects and the respective technical, safety, and environmental requirements throughout the entire supply chain, from its production to its use as a fuel aboard a specific ship. For instance, the boiling point or energy density have an impact on the fuel tank. Fuel-related properties such as the minimum ignition energy or the auto-ignition temperature influence the combustion process (i.e., the use of the respective fuel within an internal combustion engine). Risks and hazards associated with the handling of the fuels, which are increased, for example, by wide flammability limits in air and a low flashpoint, must be determined individually and mitigated by the respective safety measures.

Thus, such alternative fuels can exhibit fuel properties that deviate significantly from the currently used/conventional maritime fuels like heavy fuel oil (HFO). Table 2 provides a comparison of these alternative fuels' properties to HFO, given that HFO is the predominant fuel type used in international shipping so far [7].

As Table 2 presents, with the exception of methanol and diesel fuels, all alternative fuels listed have a negative boiling point. Therefore, they are gaseous under standard conditions for temperature and pressure (STP) (i.e., a temperature of 20 °C and a pressure of 1 bar) resulting in a comparably low density. Given the constraint available volume of a ship, fuels with high volumetric energy densities are generally preferred for a use in ships. Liquefying these gases increases their volumetric energy density; however, liquefaction significantly changes the respective requirements, e.g., for fuel storage. Further aspects of the various alternative fuel options are discussed in more detail below. Besides such aspects, further characteristics of the alternative fuels considered here are subsequently described below.

2.1 Liquefied Natural Gas

Liquefied natural gas (LNG) as a marine fuel mainly consist of methane. This hydrocarbon is described by the molecular formula CH_4 and thus consists of one carbon atom (C) and four hydrogen atoms (H). Methane can be produced by exploiting fossil resources on- and/or offshore (natural gas fields), biogenic resources (e.g., by producing biogas and a subsequent upgrading to biomethane) or by using electricity to produce hydrogen through water electrolysis to be combined downstream with a respective carbon feedstock like CO_2 for the synthesis of so-called synthetic natural gas (SNG).

In the following, a description of how LNG is handled especially for marine applications, its associated GHG emissions, as well as an overview of its current utilization and infrastructure status is given.

Handling. At standard temperature and pressure (STP), natural gas/methane is in a gaseous state and thus has a very low volumetric energy density. To facilitate the transport and storage of natural gas, liquefaction enables a volume reduction by a factor of 600. To keep natural gas/methane in its liquid form at atmospheric

Table 2 Properties of HFO and alternative marine fuels (based on [9–19]; standard temperature and pressure (STP): 20 °C and 1 bar; HFO—heavy fuel oil; LNG—liquefied natural gas; LPG—liquefied petroleum gas)

Properties	HFO	Methane LNG	Propane (P), Butane (B) LPG	Methanol	Ammonia	Hydrogen	Diesel Fuels
Lower heating value (MJ/kg)	41	50	46 (P, B)	20	19	120	43
Volumetric energy density (MJ/l) (atmospheric pressure, liquefied gases)	40	21	27	16	13	9	34
Boiling point (1 bar) (°C)	350–500	−162	−42 (P) −0.5 (B)	65	−33	−253	150–380
Density at STP (kg/m³)	960–1,010	0.70	1.87 (P) 2.51 (B)	790	0.72	0.08	800–900
Density liquid phase (kg/m³)	960–1,010	450	493	790	696	71	800–900
Minimum ignition energy (mJ)	10	0.29	0.25 (P, B)	0.14	8–680	0.017	0.23
Auto-ignition temperature (°C)	>300	537	470 (P) 287 (B)	470	650	560	250
Flammability limits in air (Vol.%)	1–5	5–15	2–11 (P) 1–9 (B)	6–37	15–28	4–75	1–5
Flashpoint (°C)	>60	−188	−104 (P) −60 (B)	12	132	/	>60

pressure, a temperature of at least −162 °C is required [20]. Thus, among the alternative fuels mentioned above, LNG shows after liquid hydrogen the second-lowest storage temperature. To realize under stable and reliable conditions such deep storage temperatures below ambient conditions, a suitable insulation is needed to avoid or at least to minimize heat input from the environment. For this purpose, polyurethane foam is a commonly used insulation material for the respective tank. This reduces the amount of boil-off gas (BOG); boil-off gas is defined as liquefied gas vaporized unintentionally by heat, e.g., from outside the storage system.

Compared to conventional fuels, LNG poses an increased ignition hazard due to its lower minimum ignition energy while having further flammability limits in air. Additionally, the flashpoint of LNG is very low, necessitating increased safety measures, such as, e.g., double-walled pipes [9].

GHG emissions. Methane is a greenhouse gas that has a significant impact on the earth's radiation balance, with a warming effect. Over a 100-year period, its global warming potential is estimated to be about 28 times higher than that of CO_2 (GWP100). When considering a 20-year period, impact on global warming is approximately 84 times higher than that of CO_2 (GWP20) [21]. Against this background, so-called methane slip, the escape of gaseous methane into the atmosphere, needs to be fully avoided and/or minimized along the overall fuel supply as well as the use including the downstream flue gas processing.

Within the maritime context, methane is referred to as a low-carbon fuel, as its combustion leads to lower airborne CO_2 emissions compared to conventional liquid hydrocarbon marine fuels, such as HFO, MDO, or MGO. Nevertheless, if the methane comes from fossil sources (i.e., natural gas), the CO_2 emissions released lead to an increase in GHG emissions within the atmosphere and thus contribute to the anthropogenic greenhouse effect. If, on the other hand, methane is produced via biological fermentation processes or electricity-based synthesis processes, the CO_2 released during methane combustion was previously removed from the atmosphere (in a timely manner). Therefore, a closed CO_2 cycle is possible and typically given; thus, these CO_2 emissions are not counted as GHG emissions.

However, during the combustion of a hydrocarbon also, other emissions in addition to CO_2 are produced. On the one hand, further GHG emissions arise from nitrogen-based emissions (e.g., N_2O) and unburned methane (e.g., slip stream from a use within an internal combustion engine); and on the other hand, further locally effective emissions such as NO_x and particle matter (PM) emissions arise. N_2O, for example, is a greenhouse gas, which supports a depletion of ozone within the atmosphere. Over a 100-year period, its global warming potential is estimated to be about 265 times higher than that of CO_2. When considering a 20-year period, impact on global warming is approximately 264 times higher than that of CO_2 [21].

Usage and infrastructure. Besides being a maritime fuel, LNG is a globally shipped commodity, so both LNG transport vessels as well as LNG compatible port infrastructure are available in many ports worldwide. With regard to its use as a maritime fuel, LNG has firstly been used on LNG carriers by burning the boil-off gas (BOG) generated during the voyage. In the year 2023, LNG was the most widely used alternative fuel in international shipping with more than 1,000 ships in operation and more than 800 ships on order (status 2023) [7]. Nevertheless, ships operated on "green" LNG based either on biological or technical processes are not yet realized. Therefore, compared to the use of HFO potentially a GHG reduction is possible but still the ship operation is characterized by the release of substantial GHG emissions.

2.2 Liquefied Petroleum Gas

Liquefied petroleum gas (LPG) is gas mixture that consists mainly of propane (C_3H_8) and butane (C_4H_{10}). Today, LPG is primarily produced as a by-product during the refining of crude oil and the processing of natural gas. LPG production based on biogenic feedstocks or on electricity together with a carbon source within Power-to-X (PtX) pathways is not envisaged for the time being [9]. In the following, a description of LPG handling for marine applications, its associated GHG emissions, as well as an overview of its current utilization and infrastructure status is given.

Handling. At standard conditions, propane and butane are in a gaseous state. Their liquefaction to LPG can be achieved by both a reduction in temperature to -42 °C (fully-refrigerated) and by an increasing in pressure to 18 bar (fully-pressurized). In addition, a simultaneous reduction in temperature and increase in pressure for liquefaction is also possible (semi-refrigerated/semi-pressurized). Typically, fully-pressurized storage tanks are smaller in size than fully-refrigerated storage tanks.

The main safety concerns with regard to LPG utilization include the density of LPG vapors, heavier than air, requiring leak detectors and ventilation. Furthermore, LPG has a lower minimum ignition energy and broader flammability limits compared to conventional fuels like HFO, making leaks and spills an increased risk [9].

GHG emissions. Both propane and butane are greenhouse gases by themselves. However, their global warming potential is comparatively low. Propane has a GWP20 value of 0.072 and a GWP100 value of 0.02. The GWP20 of butane is 0.022, and the GWP100 is 0.006 [22]. As LPG is a carbon-containing fuel, its combustion in internal combustion engines also results in CO_2 emissions, which increase the amount of CO_2 within the atmosphere, as LPG available on the global energy markets is exclusively produced from fossil sources of energy.

Furthermore, the combustion of LPG also produces other GHG emissions such as methane (CH_4) and nitrous oxide (N_2O) having both a higher GWP than CO_2. Like LNG, LPG is referred to as a low-carbon marine fuel because its use results in lower CO_2 emissions compared to conventional fuels like HFO. This dependency can be described by the average carbon to hydrogen ratio being for methane at 1–4, for propane/butane at 1–2.7/2.5 and in average for HFO 1–2 or even lower; i.e., the longer the carbon chain length of the fuel the higher the relative share of CO_2 emissions.

Usage and infrastructure. LPG is a globally traded and shipped commodity; i.e., both LPG transport vessels and LPG port infrastructure are available worldwide. In the year 2023, only 91 internationally operating LPG vessels used LPG as a marine fuel by burning/utilizing the boil-off gas (BOG) generated during the respective voyage. However, another 96 ships are on order that can use LPG as an alternative fuel (status 2023) [7].

2.3 Methanol

Methanol is a basic alcohol with the molecular formula CH_3OH. This molecule is highly volatile and can be produced based on fossil fuel resources, biogenic resources or based on electricity and a carbon source in respective power-to-methanol (PtM) conversion pathways. In the following, selected aspects of methanol handling, its GHG emissions, and its current use and infrastructure status are explained.

Handling. Methanol remains in a liquid state under standard temperature and pressure conditions, making its handling processes familiar to those accustomed to conventional marine fuels. However, one distinct characteristic of methanol is its significantly lower flash point. This attribute places methanol within the category of low flash point fuels, necessitating the implementation of adapted fuel safety measures for its storage and usage. Another challenge associated with methanol combustion is a nearly invisible flame. This poses a potential hazard during incidents of unintended ignition, making it imperative to use special fire detectors [9].

GHG emissions. Since methanol is a carbon-based fuel, CO_2 emissions are necessarily produced during combustion. Whether these CO_2 emissions are contributing to an increase in GHG emissions within the atmosphere depends on the feedstock used during methanol production. If methanol is produced from fossil fuel sources, the resulting CO_2 emissions are considered to increase the CO_2 inventory within the atmosphere. If biogenic resources are used or methanol is produced using electricity and green CO_2 (e.g., from biogas plants, from bioethanol plants, from ambient air), a closed CO_2 cycle can be assumed, and thus, the CO_2 emissions are considered neutral to global climate.

Burning methanol produces additional GHG emissions, namely methane and nitrous oxide. Since the contribution of these gases to global warming is higher than that of CO_2, they cannot be neglected even when assuming a closed CO_2 cycle (biogenic or electricity-based methanol). By using a fuel cell instead of an internal combustion engine, these additional GHG emissions can be avoided.

Usage and infrastructure. Methanol is a major bulk chemical globally traded as a commodity. Thus, both methanol transport vessels as well as a proper infrastructure to handle methanol established in various ports worldwide are available. In addition, conventional fuel infrastructure can most likely be adapted to methanol with only little additional effort [9].

In the beginning, mainly international methanol carriers used methanol as fuel. However, the commercial fleet of methanol-fueled ships is increasing. In the year 2022, the first container ships fueled by methanol have been ordered. A total of 27 international ships could use methanol as fuel in 2023. Another 151 ships are on order (status 2023) [7].

2.4 Ammonia

Ammonia is a carbon-free molecule with the molecular formula NH_3. It can mainly be produced either from fossil sources of energy or electricity based. In the following, a description of ammonia handling, its GHG emissions, and its current use, and infrastructure status within the maritime industry is given.

Handling. Ammonia is a toxic gas which, like LPG, can be liquefied by both a change in temperature and/or a change in pressure. At standard temperature conditions, ammonia is converted into a liquid state by pressurization to 10 bar. At atmospheric pressure, the liquefaction requires a temperature of -33 °C [23]. Since ammonia is corrosive, stress corrosion cracking is a safety hazard.

The use of ammonia in internal combustion engines is possible but technically challenging. Ammonia is characterized by a comparatively low flame temperature and flame velocity. Also, its higher resistance to auto-ignition can lead to instabilities within the combustion process [24].

GHG emissions. Since ammonia is a carbon-free fuel, no CO_2 emissions are produced during combustion. In addition, ammonia combustion does not produce methane emissions being a further greenhouse gas. However, because ammonia contains nitrogen, the combustion can lead to an increased level of nitrogen oxides (NO_x) and nitrous oxide (N_2O) emissions [21]. Since N_2O contributes significantly more toward climate change compared to CO_2, there is the strong need to minimize these emissions; this is, e.g., possible by optimizing the design of the respective engine. By using a fuel cell as an alternative propulsion system, N_2O emissions could be completely avoided.

Usage and infrastructure. Ammonia is a well-known compound mainly used in the chemical industry to produce nitrogen-based fertilizer for agriculture [23]. It is transported as a global commodity, and existing LPG infrastructure can usually be used to handle ammonia. So far, no ships are in operation and in order that use ammonia as a marine fuel. However, in 2022, a Japanese shipping company has received an approval-in-principle for an ammonia bulk carrier being operated with ammonia. Delivery of such a ship could probably take place in 2026 [25].

2.5 Hydrogen

Hydrogen is a carbon-free molecule with the molecular formula H_2. It is the lightest known element on earth and can be produced based on fossil fuel energy, from biological feedstocks or from electrical energy. So far, most of the hydrogen used globally is produced based on fossil fuel energy (e.g., natural gas, coal). All other options play only a minor role for the time being. In the following, selected aspects of hydrogen handling, its GHG emissions, and its current use and infrastructure status are explained.

Handling. Hydrogen prevails in a gaseous state under ambient temperature and pressure conditions. Therefore, the gas is characterized by a low volumetric energy density. Thus, when storing and/or transporting hydrogen, the primary objective must be to increase the volumetric energy density. Such an increase can be realized based on two main categories of measures: physical treatment (also called direct hydrogen storage) and chemical treatment (also called material-based hydrogen storage).

- Physical treatment is based on the principle of changing the temperature and/or pressure; i.e., the chemical structure of the hydrogen is not modified at all.
- Hydrogen treatment on a chemical-basis is achieved by combining hydrogen molecules with other molecules, forming for example ammonia, synthetic natural gas or methanol, or by connecting it to the surface of other (solid) materials, such as metal hydrides [26]. Another option of chemical hydrogen treatment are liquid organic hydrogen carriers (LOHCs), where hydrogen is bound to an organic carrier molecule through a process typically facilitated by elevated temperatures and specific catalysts.

Hydrogen shows a high flammability and explosivity, which is why it is considered to pose greater operational risks and hazards than conventional fuels.

Three options for hydrogen storage (based on physical treatment and chemical treatment) for use as a marine fuel are outlined below.

- Pressurized hydrogen. Hydrogen can be stored in a gaseous state under pressure to increase the volumetric energy density. Pressures of 200–500 bar are conceivable for maritime applications; despite this high-pressure level, a significant volume is needed, which may lead to a reduction in the available volume of a ship for the transportation of payloads. Due to hydrogen embrittlement, suitable materials have to be selected.
- Liquefied hydrogen. Under atmospheric pressure, hydrogen can be liquefied at (very) low temperatures of -253 °C (20 K). Compared to conventional fuels, suitable materials must be selected for both the tank and the insulation due to these extremely low temperatures. In addition, the operational handling of hydrogen is more complex, for example due to the produced boil-off gas (BOG) resulting in a pressure increase within the tank that has to be handled by appropriate measures.
- Liquid organic hydrogen carriers (LOHCs). From an operational handling perspective, LOHCs behave quite similar to diesel fuels. Therefore, LOHCs can be handled like regular diesel fuel and the respective existing infrastructure can be utilized. As hydrogen is chemically bound to the carrier fluid in case of LOHCs, most of the operation hazards associated with pure hydrogen are mitigated by LOHCs and only occur during the relatively short period when pure hydrogen is present (i.e., between dehydrogenation of the LOHCs and the actual hydrogen use) [27]. On-board a vessel, hydrogen is released from the LOHC+ (organic oil enriched with hydrogen) using heat, and then used in a fuel cell or internal combustion engine, while the dehydrogenated LOHC (LOHC−) is stored in another tank. The next time LOHC+ is bunkered, and the LOHC− is loaded from the ship.

This LOHC− then needs to be re-enriched with hydrogen again in a respective production facility, releasing heat, to be used again [28].

GHG emissions. Since hydrogen is a carbon-free fuel, no CO_2 emissions are produced during combustion. Analogous to ammonia, the combustion of hydrogen does not produce methane emissions, but does produce nitrous oxide (N_2O) emissions also contributing to global climate change. By using a fuel cell instead of an internal combustion engine, N_2O emissions as well as other emissions like nitrogen oxides (NO_x) can be avoided, and only water is produced.

Usage and infrastructure. Hydrogen is currently used, for example, in refineries and for the production of ammonia or methanol. So far, hydrogen is usually produced close to where it is used; i.e., there is only limited experience with the transport of hydrogen in larger amounts over longer distances. Given its nascent adoption in the maritime sector, infrastructure for hydrogen is currently in an early stage only and available only on a limited extent and on a small scale. Nonetheless, in the year 2023, five international ships have been in operation using hydrogen as fuel. Another five ships are on order (status 2023) [7].

2.6 Diesel Fuels

Diesel fuels produced from fossil fuel sources, including marine gasoil (MGO) and marine diesel oil (MDO), are used as a conventional fuel in both deep-sea shipping as well as inland shipping [29]. However, diesel fuels meeting the existing fuel standards can also be produced from energy crops and/or biomass residues as well as based on electricity and a "green" carbon source. Subsequently, such "green" diesel fuels are described, including their production and possible feedstocks [9, 30].

- **Hydrotreated vegetable oil (HVO)**. HVO is produced from organic raw materials such as used cooking oil, animal fats, free fatty acids, and/or virgin vegetable oil through hydrotreating and refining. The production of a fuel meeting the given standards is typically realized in several process steps. First, the feedstock is hydrogenated to remove oxygen from the molecule by forming water and to saturate the existing double bounds. This results in alkenes defined by the respective feedstock in terms of their carbon length. Thus, to meet the given fuel standards, a catalyst-driven isomerization step accompanied by a hydrogenation is necessary. Finally, the resulting mixture needs to be split in the various products.
- **Fischer–Tropsch (FT) diesel**. For the production of FT diesel, synthesis gas (mainly hydrogen and carbon monoxide) is provided in a first step. Solid biofuels can be used as feedstock through thermo-chemical gasification. Alternatively and/ or additionally, hydrogen can be produced by an electricity-based electrolysis of water. Together with a "green" carbon source [e.g., CO_2 from the air through direct air capture (DAC)], a synthesis gas can be provided based on the so-called reverse water–gas shift reaction. Within the subsequent Fischer–Tropsch synthesis, the

synthesis gas reacts in presence of a catalyst to form hydrocarbon chains of various lengths as well as water. These chains are then separated into fractions, such as gasoline, diesel, or kerosene; additionally, an isomerization step in between might be necessary to improve the fuel characteristics.

In addition to the aforementioned "green" diesel fuels HVO and FT diesel, fatty acid methyl ester (FAME) can also be used as a diesel blend or diesel fuel substitute. FAME can be provided from vegetable oil by adding methanol in a catalyst-controlled transesterification. Due the fact that the chain length of the resulting hydrocarbons is defined by the fatty acids—and thus by the used vegetable oil—the given fuel standards for fossil diesel fuel cannot completely met. Nevertheless, FAME was already used in shipping as a blend in the year 2022.

In the following, a description of the handling of such "green" diesel fuels meeting the fuel standards (HVO and FT diesel), their GHG emissions, and their current use and infrastructure status with a special focus on the maritime industry is given.

Handling. Diesel fuels from fossil sources are conventional fuels in shipping. Therefore, handling from bunkering up to the subsequent use within the propulsion system is standard. The risks and hazards that arise are also well known and minimized by existing and highly developed safety requirements. Since HVO and FT diesel produced from biomass or electricity-based meet the diesel fuel standards fully, the handling procedures used for fossil diesel fuels can be taken over. HVO/FT diesel can be used either as a blend with fossil diesel fuels or exclusively.

GHG emissions. Diesel fuels are carbon-based fuels producing necessarily carbon dioxide (CO_2) emissions when they are burned. Furthermore, burning diesel fuels produces additional GHG emissions like methane (CH_4) and nitrous oxide (N_2O). If biogenic sources are used or the diesel fuels are produced using electricity, a closed CO_2 cycle is assumed and the CO_2 emissions are not counted as GHG emissions; however, this does not apply to non-CO_2 greenhouse gases.

Usage and infrastructure. Since fossil diesel fuels are conventionally used in shipping, global infrastructure for transport, storage and bunkering is already fully available and in operation. Due to a full compliance with diesel fuel specifications, HVO and FT diesel can be used without adjustments to the existing infrastructure. In the year 2022, HVO was tested in shipping both as a blend with conventional fuels and as a standalone fuel.

3   Use of Alternative Fuels On-Board Ships

For the alternative marine fuel options previously described, the subsequent sections address their utilization aboard ships. Due to their specific properties, the technical requirements can differ from those of conventional fuels. These requirements are analyzed below, categorized into aspects with regard to fuel storage, propulsion systems, and bunkering.

3.1 Fuel Storage

Adopting alternative fuels in the maritime sector necessitates not only different technical storage systems but also introduces storage requirements that can influence overall ship design and operation. A key parameter that determines aspects of fuel storage, like the storage volume, is the energy density of a fuel. Therefore, below, the influence of the energy density is described first, followed by the individual storage conditions of the alternative fuels.

3.1.1 Energy Density

With regard to the energy density of a fuel, it can be distinguished between gravimetric and volumetric energy density. Related to this, the volumetric energy density is typically the decisive factor for fuels to be stored on ships. As a general rule the lower the volumetric energy density of a fuel, the more space/volume is required for its storage on-board reducing the space for payload more or less accordingly. Since, under standard conditions, gases like methane, propane, butane, ammonia, and hydrogen have significantly lower volumetric energy densities compared to liquid fuels, they need to be liquefied to improve the volumetric energy density. Liquefaction of gases can be done either by a temperature reduction at ambient pressure or a pressure increase at ambient temperature or by a combination of both influencing parameters.

- Fully-refrigerated. Liquefaction of a gas only realized by temperature reduction at ambient pressure is called "fully-refrigerated".
- Fully-pressurized. Liquefaction of a gas performed only by pressure increase at ambient temperature is called "fully-pressurized".
- Semi-refrigerated or semi-pressurized. An adjustment of temperature and pressure to liquefy a gas is called "semi-refrigerated" or "semi-pressurized".

If liquefaction is done mainly or only by a reduction of temperature, so-called boil-off gas (BOG) occur due to heat input from the surrounding/from the ambient environment. This boil-off gas results in an increase of the tank pressure and thus needs to be handled. Furthermore, suitable materials should be selected for storage temperatures below ambient temperatures, adequate materials should be selected resisting embrittlement at the corresponding temperatures. In addition, insulation may be required, for example from materials with low thermal conductivity such as polyurethane foam (PU foam) and/or insulation concepts such as vacuum insulation.

Despite the liquefaction of gaseous fuels at ambient conditions, their volumetric energy density is lower than the volumetric energy density of conventional fuels like heavy fuel oil (HFO) or marine gasoil (MGO). In addition, the tank system reduces both volumetric (tank geometry, insulation thickness) and gravimetric (tank material and tank wall thickness) energy density. Figure 1 provides an overview

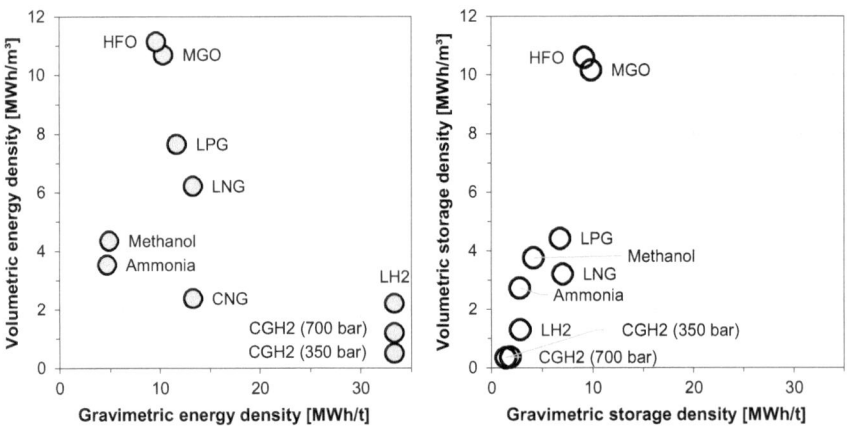

Fig. 1 Overview of volumetric and gravimetric energy density of marine fuels with (right) and without the tank system (left) [31]

of the volumetric and gravimetric energy density of various marine fuels with and without their respective tank systems.

Figure 1 shows that all alternative fuel options considered have lower volumetric energy densities compared to conventional HFO and MGO fuels. Taking into account the tank system, these differences are amplified depending on the fuel option considered. This means that when switching to an alternative fuel option, either more space must be made available for the tank system compared to HFO and MGO fuels, leading to a reduction in cargo capacity/payload, or adjustments might be required in terms of the vessel's route, speed, or bunker frequency to either reduce its specific fuel consumption or bunker fuel more frequently.

3.1.2   Storage Conditions

Individual storage conditions of the different fuel options considered here are explained in detail below. The storage condition is mainly determined by pressure, temperature, and aggregate state of the fuel to be stored. Table 3 presents the storage conditions for the alternative fuels considered.

Liquefied natural gas (LNG). Current LNG fuel tanks store LNG mainly semi-refrigerated/semi-pressurized because a variable internal tank pressure requires simplified boil-off gas (BOG) management equipment for pressure accumulation. Typical boil-off losses of onshore LNG storage tanks range from 0.023 to 0.03%/d, while LNG tanks on LNG carriers range from 0.1 to 0.15%/d [35]. This range depends among other things on the tank size and the insulation thickness. Usually LNG tanks are made of nickel alloys or stainless steel insulated by PU foam [36].

Table 3 Overview of the storage conditions of HFO fuel and alternative fuels (based on [15, 32–34]; HFO—heavy fuel oil; LNG—liquefied natural gas; LPG—liquefied petroleum gas; LOHC—liquid organic hydrogen carriers)

Fuel		State of Aggregation	Pressure (bar)	Temperature (°C)	H_2-Content (wt.%)
HFO		Liquid	Atmospheric	40	~13.5
LNG	(Fully-refrigerated)	Liquid	Atmospheric	−162	25.1
	(Semi-refrigerated)		10	Up to −125	
LPG	(Fully-refrigerated)	Liquid	Atmospheric	−30 to −48	18.3
	(Fully-pressurized)		18	Ambient	(propane) 17.3 (butane)
Methanol		Liquid	Atmospheric	Ambient	12.5
Ammonia	(Fully-refrigerated)	Liquid	Atmospheric	−34	17.7
	(Fully-pressurized)		10	Ambient	
Hydrogen	(Pressurized)	Gaseous	200–500	Ambient	100
	(Fully-refrigerated)	Liquid	Atmospheric	−253	
	(LOHC)			Ambient	6.2
Diesel fuels		Liquid	Atmospheric	Ambient	~12.7

Liquefied petroleum gas (LPG). LPG can be liquefied and stored either by reducing the temperature or by increasing the pressure. Furthermore, a combination of temperature reduction and pressure increase is possible. Since tank wall thickness increases with both growing tank volume and pressure, larger tanks are mainly fully-refrigerated tanks. On the other side, smaller tanks are mainly fully-pressurized, since on the one hand no insulation is required and on the other hand no boil-off gas is produced during storage [37].

Methanol. Since methanol is hygroscopic (i.e., it absorbs moisture from the ambient air) and polar (i.e., electrically conductive), there is a risk of corrosion. Because of these material-related properties, methanol fuel tanks are made of stainless or carbon steel [38]. Additionally, well-developed guidelines exist for the arrangement, installation, control, and monitoring of machinery, equipment, and systems on ships using methanol as a fuel to minimize the risk to the ship, crew, and the environment; main requirements for the fuel containment system and the fuel supply system concern safety measures in case of leakage from the tank [39].

Ammonia. Ammonia is usually transported and stored in a liquid form to reduce its volume. As a liquid, it is relatively safe in terms of flammability. But ammonia is highly toxic; i.e., storage must comply with specific safety regulations. Storage conditions for ammonia are very similar to those for liquefied petroleum gas (LPG), and existing LPG tankers can be specially equipped to be used for transporting ammonia as cargo. Since ammonia has corrosive properties, storage tanks are made of stainless steel [36].

Hydrogen. Hydrogen being the smallest molecule within the periodic system can easily diffuse into and through a broad spectrum of materials. This can lead to hydrogen embrittlement of the materials or a loss of hydrogen quantity (i.e., diffusion losses). Therefore, special requirements are placed on materials with contact to hydrogen. Three forms of storage can be distinguished.

- Compressed hydrogen. Pressure vessels for the storage of compressed hydrogen are typically divided into four types.

    Type I is a full pressure metal container designed for pressures up to 200 bar, which has a high dead weight.
    Type II pressure vessels have smaller wall thicknesses than type I tanks and are additionally reinforced by a fiber winding in the cylinder area for pressure resistance. This makes the tank lighter on the one hand and more resilient on the other, so that pressures of up to 300 bar are possible. However, the manufacturing costs are higher.
    Type III pressure tanks are completely encased in a carbon fiber composite material, allowing pressures of up to 700 bar. Due to the carbon fiber, the manufacturing costs are higher compared to Type II pressure vessels.
    Type IV pressure vessels are no longer made of metal but of a thermoplastic completely wrapped with a fiber composite material. As a result, Type IV pressure vessels have the lowest mass but the highest production costs. Storage pressures up to 700 bar are possible [40].
    Type V pressure tanks are made of carbon fiber laminate without the internal polymer gas barrier used in Type IV pressure vessels. Since ensuring the structural properties at high pressures while preventing gas leaks is a technical challenge, Type V pressure tanks are not currently used in commercial applications (2023) [41].

- Liquefied hydrogen. Due to the very low temperature of liquid hydrogen (ca. − 253 °C), storage tanks are usually installed with a vacuum insulation. This vacuum insulation is combined with specific insulating materials (e.g., multi-layer insulation, loose filling material like perlite). Therefore, a liquid hydrogen storage system usually consists of an inner tank, an insulation, a vacuum chamber, and an outer tank. Unavoidable boil-off losses from liquid hydrogen storage tanks are basically their main disadvantage. A boil-off gas (BOG) handling system is required to allow for a use or a re-liquefaction of the BOG [28]. Typical hydrogen boil-off losses range from 0.1 to 1%/d, depending on the tank size and the insulation installed [35].
- Liquid organic hydrogen carriers (LOHC). Since LOHC is liquid at ambient conditions and has similar properties to petroleum-based fuels, it can be easily handled [42]. LOHC can be stored at ambient pressure and temperature for a long period without any losses. A drawback is that LOHC requires a return loop of the carrier material. If LOHC is supposed to be transported, only one tank is

needed, containing either LOHC− (dehydrogenated LOHC) or LOHC+ (hydrogenated LOHC). However, if LOHC is supposed to be used on-board, LOHC+ and LOHC− must be stored separated from each other. Thus, they must be stored in separate tanks or in one tank with a suitable flexible membrane in between. This leads either to an additional space requirement or a more complex tank system [26]. Furthermore, a dehydrogenation system and, if hydrogen is used in a fuel cell, a purification of the dehydrogenated hydrogen is required. For the latter, heat is required at a relatively high-temperature level (up to approx. 300 °C) [43].

Diesel fuels. Since hydrotreated vegetable oil (HVO) and Fischer–Tropsch (FT) diesel comply with the valid diesel fuel specifications, their storage properties predominantly correspond to the properties of diesel fuels from fossil sources; i.e., existing storages for fossil diesel fuels can be used without any adjustments. However, with an average of 39 MJ/kg, FT diesel currently often has a lower energy density than fossil diesel fuels and HVO (43 MJ/kg) [30].

3.2 Propulsion Systems

The fuel and the propulsion system of a ship must be compatible. Therefore, the selection of an alternative fuel has an influence on the propulsion system and the selection of an alternative propulsion system influences the selection of the alternative fuel. Other factors affecting the choice of propulsion system and fuel are the ship type, the power demand, and the operating profile.

International shipping typically uses 2-stroke-based, 4-stroke-based, and diesel-electric propulsion systems. Here, the chemical energy stored in the fuel is converted into both mechanical (propulsion) energy and electrical energy (required for on-board systems) by means of internal combustion engines. Low-speed, 2-stroke diesel engines are characterized by higher efficiencies and are generally used for propulsion. Medium-speed, 4-stroke diesel engines, on the other hand, have higher power densities and are typically used to provide electrical energy (diesel generators). Diesel-electric propulsion systems use a diesel generator to provide electrical energy, which is used in combination with an electric engine for both propulsion and on-board electrical power supply [31].

With the advancement of alternative fuels, alternative propulsion systems are also being investigated. In the following, first these propulsion systems will be introduced in general, and subsequently detailed in conjunction with a specific alternative fuel option.

3.2.1 Alternative Propulsion Systems

In addition to the 2-stroke-based, 4-stroke-based, and diesel-electric propulsion systems mentioned above, alternative propulsion systems are investigated, respectively, already in use in the context of alternative fuels. Thus, dual-fuel engines, gas engines, and fuel cells are described below being such alternatives.

Dual-fuel engines. Dual-fuel engines can be both 2-stroke and 4-stroke engines that run on an alternative fuel. Ignition is provided on demand by a small amount of pilot fuel (e.g., diesel fuel). The main feature of dual-fuel engines, however, is that they can be operated with two different types of fuel simultaneously or separately. In the case of separate operation, switching between operation with alternative or conventional fuel is possible without load interruption. This results in a high degree of flexibility for the operation of the ship. In addition, dual-fuel engines can also be used when the availability of alternative fuels is limited. Besides the flexibility in terms of multiple usable fuel options, one disadvantage of dual-fuel engines lies in the possible compromises in engine performance when the requirements for using the alternative and conventional fuel vary [31, 44].

Gas engines. Gas engines, for example for LNG, are 4-stroke engines whose ignition is usually provided by external ignition mechanisms, such as spark plugs. Because gas engines are specifically designed for gas operation, they have higher efficiencies with lower exhaust emissions compared to dual-fuel engines. Currently, however, dual-fuel engines dominate as ship propulsion technologies, with gas engines being used as an alternative [44].

Fuel cells. Fuel cells convert the chemical energy stored in a fuel directly into electrical energy. This is the reason why they usually allow for higher efficiencies compared to internal combustion engines. Further advantages include their low levels of emissions, noise, and vibration. For the use of fuel cells on-board a ship, other factors must also be considered. For example, the start-up time of the system should be reasonable and at least comparable to conventional power generation. Since fuel cell technology requires electrification of on-board power distribution, hybridization with additional power storage components can serve to meet these requirements when fuel cells alone cannot fulfill these tasks [45]. Since fuel cells represent a promising solution for maritime applications, various technologies, in particular polymer electrolyte fuel cells (PEMFC) and solid oxide fuel cells (SOFC), but also molten carbonate fuel cells (MCFC) and direct methanol fuel cells (DMFC), have been investigated as alternatives to internal combustion engines. An overview of selected properties of the fuel cells mentioned is given in Table 4 [8, 46–48].

PEMFC can be classified in two categories based on their operating temperature: low-temperature (LT) PEMFC, which operates around 60 to 80 °C, and high-temperature (HT) PEMFC working above 120 °C and up to 200 °C. LT-PEMFCs require high-purity hydrogen and have a low tolerance for impurities. Supplying such high-purity hydrogen poses a challenge and causes potentially high operating costs. In contrast, the HT-PEMFC is more tolerable to impurities, resulting

Table 4 Overview of characteristics of different fuel cell types (based on [48–51]; PEMFC—polymer electrolyte fuel cells; LT—low temperature; HT—high temperature; SOFC—solid oxide fuel cells; MCFC—molten carbonate fuel cells; DMFC—direct methanol fuel cells)

Item	PEMFC	SOFC	MCFC	DMFC
Temperature (°C)	60–80 (LT) 110–180 (HT)	500–1,000	650–200	80
Electrolyte	Water-based polymer membrane	Porous ceramic material	Molten carbonate salt	Based on polymeric membranes (derived from PEMFC)
Typical fuel	Hydrogen	Hydrogen, methanol, hydrocarbons, ammonia	Hydrogen, methanol, hydrocarbons, ammonia	Methanol
Electrical efficiency (%)	45–55	50–60	43–55	20–25
Lifetime (h)	60,000–80,000	20,000–80,000	15,000–30,000	>5,000 (Polyfuel membrane)

in most likely reduced operating costs. Therefore, hydrogen carriers with reformer system (i.e., ammonia or methanol combined with crackers or LOHC combined with dehydrogenation systems) are better suitable options for HT-PEMFC.

Fuel cells like SOFC and MCFC operate at higher temperatures than PEMFC-HT and are therefore not dependent on pure hydrogen; they can also be operated with other fuels such as methanol, LNG, or ammonia. However, they are also characterized by certain disadvantages, such as longer start-up times due to the high operating temperatures, making them rather unsuitable for rapid load changes. Also, they tend to have a shorter lifetime than other fuel cell systems.

DMFCs offer the advantage of using methanol as a fuel, which is liquid at ambient pressure and temperature; this simplifies fuel storage significantly. However, the electrical efficiency of DMFCs is lower compared to the other fuel cell technologies mentioned.

However, although fuel cells allow for higher efficiencies compared to internal combustion engines (ICE), the latter are expected to remain the dominant marine propulsion technology in the foreseeable future. A reason for this is the lower technology readiness level of fuel cells compared to internal combustion engines. For ships that already use alternative fuels, dual-fuel engines are mainly used due to their high flexibility regarding fuel use.

3.2.2 Propulsion Systems for Alternative Fuels

The subsequent sections provide a detailed description of the propulsion systems explained above in the context of the alternative fuel options considered here.

Liquefied natural gas (LNG). LNG is currently the most widely used alternative fuel in shipping, and corresponding propulsion technologies are thus already available on the market. While the first ships to run on LNG used steam turbines for propulsion, dual-fuel engines currently dominate the market. Dual-fuel engines are available as both 2-stroke and 4-stroke engines and can inject LNG under high or low pressures. One challenge with burning methane is the so-called methane slip, as methane has a higher greenhouse effect than CO_2; the problem is that a small share of the gaseous fuel is emitted without oxidation. Commonly, self-ignition engines (diesel engines) show a lower methane slip compared to spark-ignition engines (gasoline engines) typically characterized by 0.2–0.3 g/kWh. Measures to reduce the methane slip include further optimization of the internal engine design and/or using exhaust gas after treatment systems (e.g., oxidation catalysts).

As an alternative to dual-fuel engines, dedicated gas engines for LNG are also available on the market characterized by typically higher investment costs compared to dual-fuel engines. With regard to fuel cells, solid oxide fuel cells (SOFC) can also be operated with LNG, among other fuels. Yet, due to their higher investment costs and relative short lifespans, they are not commercially available. One significant advantage of using LNG with SOFC is the full elimination of the methane slip, and, due to the high concentration of the CO_2 produced, it can be separated relatively easily using carbon capture. This means, if such technical approaches are pursued and the captured CO_2 is stored permanently, a notable decrease in GHG emissions could be possible, even when using fossil-based LNG [7, 31, 44].

Liquefied petroleum gas (LPG). Propulsion technologies for the use of LPG are already available on the market and are usually based on the established LNG propulsion concepts. The most common are 2-stroke diesel engines, but 4-stroke gasoline engines or gas turbines are also considered for commercial use [31].

Methanol. The shipping industry is increasingly relying on ships that run on methanol. Although the supply of methanol does not currently allow for a large-scale conversion of the existing vessel fleets, the methanol sector has taken important steps with new ships. Methanol-powered 2-stroke engines are commercially available, and some 2-stroke engines can be converted to use methanol as a fuel. Methanol-powered 4-stroke engines are also available, with the range of 2- and 4-stroke engines expected to increase in the coming years. Fuel cells can either be operated directly with methanol (SOFC, MCFC, and DMFC) or the hydrogen contained within the methanol can be cracked and used in a pure form in fuel cells. Methanol reformers, which are currently under development, are able to produce hydrogen on demand at the point of use to avoid complexity and high costs associated with the logistics of a pure hydrogen fuel [52]. However, the technological maturity of fuel cells that operate using methanol as fuel is currently low, and DMFCs have a low electrical efficiency. As of today, commercial availability of methanol-powered fuel cells for ships is therefore expected from around 2030.

As methanol is a carbon-based fuel, its use results necessarily in CO_2 emissions. However, if methanol is used in combination with a fuel cell propulsion system, capturing and storing the resultant CO_2 becomes easier and more straightforward compared to a use of methanol in internal combustion engines. If such approaches are pursued, it offers the potential to mitigate GHG emission, even when using methanol from fossil energy resources [2, 31, 53].

Ammonia. Ammonia is currently not used as an alternative fuel in shipping, and as a result, no respective propulsion technologies are commercially available so far. The use of ammonia in internal combustion engines is conceivable; but most likely dual-fuel engines will be used within a first development step due to the poor combustion properties of ammonia. However, if ammonia is used in a dual-fuel engine, the fuel gas–air mixture is ignited by using a pilot fuel (such as diesel fuel) requiring an additional tank system in addition to the actual ammonia tank system. Also, the corrosive properties of ammonia must be taken into account when selecting materials, requiring adaption to existing combustion engines.

The first commercially available ammonia combustion engines are expected to be 2-stroke engines disposable in the market by 2024. One of the primary challenges with ammonia internal combustion engines ensuring a complete ammonia combustion to minimize emissions such as nitrogen oxides, nitrous oxide, and ammonia slip. These emissions can be significantly reduced through exhaust after treatment or entirely eliminated using fuel cells. In such systems, ammonia can either be directly utilized in high-temperature fuel cells or indirectly by cracking it to provide only the hydrogen of the ammonia molecules to the fuel cell system. However, a commercial use of ammonia fuel cells in ships is only expected after the year 2030 [2, 31, 53].

Hydrogen. To utilize hydrogen in ship propulsion systems, it may need to be converted into a gaseous state through evaporation, dehydrogenation, or cracking. The first commercial dual-fuel engines for hydrogen and diesel as a pilot fuel are already available on the market [54]. Further engine suppliers have announced the development of 4-stroke and 2-stroke engines for the use of hydrogen as an alternative marine fuel.

In comparison, the development of fuel cells using hydrogen as a fuel is already more advanced. Several demonstration projects have been successfully executed, and a company intends to produce hydrogen fuel cells for maritime applications on a large scale [31, 53].

Diesel fuels. Since hydrotreated vegetable oil (HVO) and Fischer–Tropsch (FT) diesel comply with the diesel fuel specifications, ship engines for fossil diesel fuels can usually be operated with HVO and FT diesel up to 100% without modifications. This is also because large marine engines are often specifically designed to handle distillate fuels with a wide range of properties; i.e., slightly different fuel properties usually do not require modifications of the propulsion system [30].

3.3   Bunkering

The term "bunkering" is commonly used in the shipping and maritime industry to describe the refueling of ships with various types of fuel. It involves the transfer of fuel from a storage facility or tanker to a vessel's fuel tanks. Ship operators usually prefer to bunker as seldom as possible to minimize interruptions of their ordinary operation. They also favor high fuel transfer rates to shorten the bunkering time and thus the (expensive) time spent within a port. Four different bunker methods are distinguished by the facility supplying the fuel.

- Port-to-ship. If a fixed land-based infrastructure is used to supply the respective fuel, the process is called port-to-ship. This process can deliver the highest volume and is typically characterized by the shortest bunker durations. However, since the ship has to dock at the bunkering facility, where loading or off-loading operations are usually not possible, additional time in the harbor and possibly maneuvering is required [55].
- Truck-to-ship. If the fuel is supplied by a truck, the bunker method is called truck-to-ship. This process allows for a high degree of flexibility, but only small bunker volumes and transfer rates are available. The truck usually can interfere with loading/unloading operations of the payload [56].
- Ship-to-ship. If the fuel is supplied by a ship, this is referred to as ship-to-ship fueling. High volumes of up to 20,000 m^3 are possible for ship-to-ship bunkering. Since the bunkering infrastructure is mobile, bunkering operations can be conducted widely independent from a specific location. This allows for simultaneous loading/off-loading operations for the transported goods to take place and thus reducing time within a certain port [10].
- Container-to-ship. During the container-to-ship process, interchangeable containers with the fuel are loaded on-board the ship and then serve as the fuel tank. The existing port infrastructure can be used during the container-to-ship process to load the fuel container on-board. But these containers will then take up cargo space on-board (i.e., payload is reduced), since they will remain there until the fuel is consumed [56].

All bunkering methods outlined above can be applied for the alternative fuel options. However, since the alternative fuels vary in their properties, the challenges associated with bunkering them vary. Moreover, some bunker infrastructure is already partially in place.

Liquefied natural gas (LNG). Bunkering of LNG is already available in many ports and a growing fleet of LNG bunker ships provide access to large bunker quantities and fast fuel transfer rates. Components for the bunkering facility, like hoses or emergency release coupling (ERC), are available off-the-shelf, reducing CAPEX for new bunkering facilities. Although, except for some test projects, only fossil LNG is used, the LNG bunkering infrastructure can also be used for LNG produced from biomass/biogas or synthetic LNG from renewable electricity and "green" CO_2 as these two options are basically chemically identical to fossil LNG.

Liquefied petroleum gas (LPG). Since LPG can be liquefied by both pressure and temperature changes, bunkering may require applications to bridge the pressure and temperature differences. LPG terminals already exist, and a conversion into a bunkering station is considered to be technically feasible. Truck-to-ship bunkering is considered proven technology, as on-road LPG refueling stations are already supplied by truck. LPG bunkering vessels are not established so far, but could be used in the future due to their advantages if more ships run on LPG [31].

Methanol. For methanol, global logistics already exists, and methanol is available in more than a hundred ports worldwide. Currently, there are several bunkering suppliers around the world interested in supplying ships with methanol as a fuel [57]. Since methanol is liquid at ambient conditions, it is expected that bunkering methanol is similar to bunkering conventional marine fuels and that the existing infrastructure can be used for methanol with only minor modifications [10]. Methanol is characterized by a higher flammability than existing fuels and is toxic for humans upon contact; these fuel characteristics might enforce specific requirements within the bunkering infrastructure. Additionally, due to its lower energy density compared to conventional marine fuels, larger storage tanks and higher transfer rates are necessary, if the same amount of fuel energy is to be stored or transferred within a the time period [58].

Since methanol contains a carbon atom, the energetic use/the oxidation produces necessarily CO_2. To achieve a closed carbon cycle, a carbon capture system with an integrated interim storage can be installed on-board. The captured and stored CO_2 needs to be offloaded during bunkering. This aspect of the bunkering process introduces the need for new technologies and adds complexity to the operation.

Ammonia. Similar to LPG, ammonia can be liquefied through both pressure and temperature adjustment; i.e., the bunkering process might require mechanisms to bridge the difference in pressure and temperature. Beside this challenge which can be handled technically, the high toxicity of ammonia poses a far greater challenge for bunkering; this is especially true for the event of a spill. Although ammonia is already integrated in international shipping as a bulk cargo and as a refrigerant, handling ammonia as a fuel places additional demands on the overall bunker system on-board as well as within the respective harbor and requires appropriate preventive measures. Ports are often located near metropolitan areas, whereas the industry producing and using ammonia is usually more isolated. To safeguard people living close to harbors from any harm due to ammonias toxicity or even its unpleasant odor, it is essential to avoid leakages completely. Even in the event of an accident, the release of substantial amounts of ammonia into the environment must be prevented safely. These requirements are technologically solvable but increase the respective investment and operational expenditure.

For example, Hansen and Martini [59] performed a CFD analysis to assess safety distances for the ammonia bunkering process. Their results show that for a comparably low wind speed of 2 m/s (2 bft) harmful concentrations can still be detected in more than 250 m distance. Such analysis will need to be conducted for each bunkering facility individually and the minimum safety distance might vary from port to port.

Hydrogen. Hydrogen can be used and bunkered in different conditions. Thus, the requirements for the bunkering process differ and existing experience from other areas or with other fuels can be used.

- Compressed hydrogen. Transfer of compressed hydrogen is known from hydrogen refueling stations (HRS) for cars and trucks, where the transfer rate is limited to about 1 kg/min [60]. To cover at such a rate the energy (hydrogen) demand from larger seagoing ships would result in refueling times of several weeks. Alternatively, many nozzles had to be connected to the receiving ship [58]. Another possibility is to load hydrogen tanks, e.g., in standard containers on-board. Due to these challenges and the huge storage volume, most likely such solutions are only feasible for small ships, if at all.
- Liquefied hydrogen. The most challenging bunkering process needs to be realized for liquid hydrogen at cryogenic temperatures. These low temperatures embrittle most materials [61]; adequate materials need to be identified. Additionally, a ship within the water will always move slightly. The connection between the bunkering facility and the ship must compensate this site-specific motion being very challenging at cryogenic temperatures for the used materials [58].
- Liquid organic hydrogen carriers (LOHC). LOHC have very similar properties to the fuels used today in shipping. Thus, it is expected that a large part of the existing bunker infrastructure can also be used for LOHC. In contrast to the commonly used fuels, LOHC are not burned on-board; i.e., only hydrogen is released. The remaining LOHC- has to be unloaded from the ship again, so it can be used to attach hydrogen to it again [58]. During the bunkering process, the ship must therefore be loaded with LOHC+ and unloaded with LOHC− at the same time. This makes the bunkering process more complex and, if the process cannot be performed simultaneously, could lead to longer bunkering times.

Diesel fuels. Since hydrogenated vegetable oil (HVO) and Fischer–Tropsch (FT) diesel from biomass or electricity-based production meet the specifications for diesel fuel, the existing infrastructure can be used, including for bunkering. Because of the structural similarity to fossil diesel, the risks and hazards to be considered in the bunkering process are also similar and thus well known.

4   Comparison and Development

Alternative and carbon-free fuels are promising options for defossilization and decarbonization of shipping. However, the properties of the alternative fuels considered differ from the properties of conventional fuels resulting in individual technical requirements for storage, propulsion, and bunkering. However, for the actual use of alternative fuels, it is not only the technical feasibility that needs to be taken into account. Therefore, within this section, alternative fuels for shipping are considered systemically based on selected overarching factors. A respective overview of individual advantages and disadvantages for each alternative fuel mentioned is given in Table 5.

Table 5 illustrates that each alternative fuel option has its own set of advantages and disadvantages. Therefore, no single alternative fuel option can be regarded as the "optimal fuel" or "silver bullet" for a defossilization of the maritime sector. Furthermore, the suitability of a fuel option for a specific application is influenced by factors such as the regulatory framework, requirements of a fuel option concerning the actual shipping operation, as well as general aspects of the potential availability of alternative fuels over time for the years to come. These factors are discussed in more detail below.

4.1 Regulatory Framework

A uniform regulatory framework is required for the safe integration of marine fuel systems on-board ships. Such frameworks are, of course, already in place for conventional marine fuels such as heavy fuel oil (HFO) and marine diesel oil (MDO). If alternative fuels are used, it is also essential to ensure a safety level comparable to that of conventional fuels. However, due to the differences in physio-chemical properties between alternative and conventional fuels, various risks and hazards arise depending on the specific fuel option. Consequently, the development and implementation of individual and tailored safety measures, regulations, and standards are necessary.

However, since the use of alternative fuels within the maritime sector is currently still limited, comprehensive regulations and standards are not yet fully available for all of the alternative fuels considered. For ships operating internationally, the International Maritime Organization (IMO) is responsible to set the respective legal framework to be valid on a global scale. In contrast, for inland vessels operating nationally, the legal framework is set by the responsible national bodies. Additionally, at the European level, the European Committee for the Development of Standards in the Field of Inland Navigation (CESNI) is also the responsible institution to define the respective legal framework.

Overall, uniform regulations and standards are crucial to facilitate the use of alternative fuels in shipping. The currently valid regulatory framework for the alternative fuel options being considered here are further described below.

Liquefied natural gas (LNG). For the alternative fuel options considered here, the most extensive legal framework is so far set available for LNG. The use of liquid methane as a maritime fuel is governed by international regulations such as the International Code of Safety for Ship Using Gases or Other Low-Flashpoint Liquids as Fuel (IGF Code). However, bunkering regulations for LNG in ports may differ due to different national and local legislations and frame conditions. In certain cases, even special permissions are required for its use. There is ongoing effort to harmonize the currently existing landscape of LNG regulations. In addition to maritime applications,

Table 5 Advantages and disadvantages of alternative fuels [9]

Fuel Type	Advantages	Disadvantages
Liquefied natural gas (LNG)	• Technology availability • Infrastructure availability • Rules and regulations in place • Experience with handling as cargo	• Carbon-containing composition • Risk of methane slip • Low storage temperature required (high CAPEX, complex handling)
Liquefied petroleum gas (LPG)	• Technology availability • Experience with handling as cargo	• Carbon-containing composition • Propane being heavier than air poses safety risks • Rules and regulations not yet available (currently an IMO working item)
Methanol	• Liquid state at ambient temperature • Ease of operational handling • Technology availability • Rules and regulations in place • Experience with handling as cargo	• Toxicity • Carbon-containing composition
Ammonia	• Carbon-free composition • Experience with handling as cargo • Higher volumetric energy density than hydrogen	• High toxicity • Restrictions from IMO IGC Code on use as fuel for gas carriers • Engine technology still under development
Hydrogen (compressed hydrogen)	• Carbon-free composition • High combustion efficiency • Highly purified form of hydrogen available • Solely water-emissions if used in a fuel cell	• Safety risks due to flammability and explosiveness • High volume and storage weight • Rules and regulations not yet available
Hydrogen (liquefied hydrogen)	• Carbon-free composition • Higher energy density than compressed hydrogen • Highly purified form of hydrogen available • Solely water-emissions if used in a fuel cell	• Safety risks due to flammability and explosiveness • Complex storage required due to its cryogenic nature (high CAPEX, complex handling) • Rules and regulations not yet available

(continued)

Table 5 (continued)

Fuel Type	Advantages	Disadvantages
Hydrogen (liquid organic hydrogen carriers, LOHC)	• Liquid state at ambient temperature • Ease of operational handling • Potential to utilize existing conventional fuel infrastructure	• Potential threat to marine ecosystems • Low energy density • Enhanced logistical complexity (parallel infrastructure for LOHC+ and LOHC−)
Diesel fuels (biomass feedstock or electricity-based production)	• Technology availability • Infrastructure availability • Rules and regulations adoptable from fossil diesel fuels and road transport	• Carbon-containing composition • Availability of sustainable biomass and sustainable carbon

the European Standard of Technical Regulations for Inland Waterways (ES-TRIN) also provides construction standards for LNG-powered ships in European inland shipping.

Liquefied petroleum gas (LPG). The International Code for the Construction and Equipment of Ships Carrying Liquefied Gases in Bulk (IGC Code) is the legal framework for the transport of LPG by gas tankers. The IGC Code does not prohibit the use of LPG as a fuel, but only regulates the use of methane (boil-off gas of Cargo LNG) as a fuel. If LPG is to be used as fuel on other types of ships, this falls under the aforementioned IGF Code. While the IGF Code contains rules for the use of LNG as a fuel, it does currently not contain any regulations for using LPG as a fuel. However, in the year 2023 the International Maritime Organization (IMO) published interim guidelines for the safety of ships using LPG [31, 62].

Methanol. The regulatory framework for the use of methanol on-board of ships is currently being established. While methanol has been used as a fuel for road-based transportation for some time, its use in shipping is much more recent. The International Maritime Organization (IMO) has drafted interim guidelines, which are expected to be approved and incorporated into the IGF Code in due course [57].

Ammonia. There are currently no established rules for the use of ammonia as a marine fuel. While the transportation of ammonia is governed by the IGC Code, which prohibits its use as a marine fuel, this Code can be seen as a regulatory basis for to future regulations on the use of ammonia as a marine fuel. However, due to the toxicity of ammonia, safety measures must be developed that achieve an equivalent level of safety as conventional fuels [63]. The IMO's Sub-Committee on Carriage of Cargoes and Containers (CCC 9) is already working on the draft Interim Guidelines for the safety of ships using ammonia as fuel. The draft is expected to be finalized in 2024 [64].

Hydrogen. As of now, there are no regulations in place for the on-board storage of hydrogen as a fuel. Analogous to ammonia, the Sub-Committee on Carriage of

Cargoes and Containers (CCC 9) of the IMO is working on the draft Interim Guidelines for the safety of ships using hydrogen as fuel. In 2024, the draft is expected to be finalized [64].

For inland waterway vessels powered by hydrogen, Lloyd's Register has prepared a proposal for technical provisions to complement ES-TRIN. This proposal has already been presented to the European Committee for the Development of Standards in the Field of Inland Navigation (CESNI) [31].

- Compressed hydrogen. Currently, no specific standards exist for the use of compressed hydrogen as a marine fuel. However, existing regulations for compressed natural gas (CNG) could serve as a foundation for a more specific assessment of hydrogen.
- Liquefied hydrogen. The IGC Code and IGF Code relate to the storage of liquefied gases on-board ships. The tank regulations for the storage of liquefied gases also apply in principle to liquefied hydrogen. However, due to the specific characteristics of hydrogen, including (very) low storage temperatures, additional considerations are required. The Interim Recommendations for Carriage of Liquid Hydrogen in Bulk is an interim recommendation of the IGC Code that currently permits the carriage of hydrogen as cargo under a pilot project (Australia–Japan), but it is not yet part of the actual IGC Code [65].
- Liquid organic hydrogen carriers (LOHC). LOHCs share properties similar to conventional marine fuels, allowing them to be handled like regular diesel oil and utilized in existing infrastructure. Therefore, the existing rules applicable to diesel fuels can potentially be applied to LOHCs on ships [27]. However, such provision have not yet been formalized in international regulations [63].

Diesel fuels. Regulatory frameworks for diesel fuels from fossil origin are already in place for the maritime sector. However, there are currently no marine standards that cover specific diesel fuels from biomass or electricity-based production. However, standards for land-based diesel fuels can be taken into account. This means HVO typically meets the diesel fuel standards EN 590 (Standard from the European Committee: description of physical properties for automotive diesel fuel) and ASTM D975 (American standard: standard specification for diesel fuel oils). For Fisher–Tropsch (FT) diesel, the Standard EN 15940 (Test method for paraffinic diesel fuel) can be taken into account [30, 66].

Table 6 summarizes the current status of the regulatory frameworks for the various alternative fuel types based on the aforementioned explanations.

4.2   Suitability for Ship Operations

The shipping sector and its energy demands can be divided into different categories; i.e., one alternative fuel cannot be generally viewed as the optimal fuel for all ships and their specific driving profiles. Firstly, ships can be divided into seagoing and inland waterway vessels according to their registration. In addition, sea-going vessels can

Table 6 Current status of regulatory frameworks for alternative marine fuels (CNG—compressed natural gas; LOHC—liquid organic hydrogen carriers)

Fuel Type	Status of Regulatory Framework
Liquefied natural gas (LNG)	• Most comprehensive regulatory framework in place among considered alternative fuels, mainly under IGF Code • Yet, port bunkering rules differ by local and national laws; harmonization efforts ongoing
Liquefied petroleum gas (LPG)	• IGC Code oversees LPG transport on gas tankers, allowing its use as fuel • IGF Code addresses LNG but not LPG; interim LPG guidelines issued by IMO in 2023
Methanol	• IMO's interim guidelines for methanol expected to join IGF Code
Ammonia	• IGC Code, which handles ammonia transport, currently bars it as fuel but could underpin future regulations
Hydrogen	• No existing regulations for on-board hydrogen storage; a technical proposal aligned with ES-TRIN for hydrogen-powered inland navigation vessels is under CESNI review
Compressed	• While no specific standards for compressed hydrogen exist, CNG regulations might serve as a foundation
Liquefied	• IGC and IGF Codes address liquefied gases, but liquefied hydrogen's unique characteristics need further assessment • A specific pilot project under interim IGC guidelines allows hydrogen as cargo, but it is not yet an official part of the IGC Code
LOHC	• Similarity of LOHCs to conventional marine fuels suggests existing fuel rules could apply, but this is not yet officially recognized in regulations
Diesel fuels (biomass feedstock or electricity-based production)	• No specific marine fuel standards. For HVO, EN 590 and ASTM D975 can be taken into account. For FT diesel ,EN 15940 can be taken into account

be divided into short-sea vessels and deep-sea vessels according to their (average) voyage distance [67]. As the route and size of ships increase, the requirements for the amount of energy contained in the fuel also increase. This is to ensure that there is no simultaneous need for excessively large fuel storage systems aboard these ships. The characteristics of different ship categories are explained below, along with potential alternative fuel options suitable for each category.

Inland waterway vessels. Inland waterway vessels travel on inland waters such as rivers, canals, and lakes. As a result, inland waterway vessels are often located close to cities and therefore people, as well as in areas that are typically environmentally controlled. There are many different types of ships operating on such inland waterways characterized by very different energy requirements and driving profiles.

In the future, different types of alternative fuels could prove to be advantageous for different types of ships. Gaseous hydrogen is promising for inland vessels with short journeys and lower performance, such as ferries. The first demonstration projects have already been implemented. Inland vessels with longer journeys and higher performance, for example motor freight vessels, require typically higher energy densities. Renewable diesel fuels, methanol, and liquid hydrogen are considered to be particularly promising for these types of ships.

The use of fuel cells instead of internal combustion engines in inland shipping is particularly interesting due to their zero emissions operation. CO_2 emissions can be avoided by using carbon capture and storage and non-CO_2 GHG emissions from internal combustion engines [i.e., methane (CH_4) and nitrous oxide (N_2O)] as well as other locally active emissions [e.g., nitrogen oxides (NO_x) and particulate emissions] can be reduced to zero [68].

Short-sea vessels. Short-sea vessels operate on short voyages with very frequent port calls and usually spend most of their time in areas which are under (strong) environmental control. The current new technologies used in short-sea, offshore, and passenger segments are renewable diesel fuels and methanol produced from both fossil and biogenic sources. Liquid hydrogen could also become a promising fuel for these ships in the future; on the one hand, liquid hydrogen has higher energy densities than gaseous hydrogen, and on the other hand, fuel cells have lower emissions than internal combustion engines predominantly used with biofuels and methanol. Since short-sea vessels travel longer distances, usually across nations, than inland waterway vessels, several bunkering stations are required to switch to an alternative fuel. This requires either increased fuel availability or dual-fuel propulsion systems. On the regulatory level, international standards for the ship and, if possible, uniform national standards for the bunkering station are also required for cross-national routes.

Deep-sea vessels. Deep-sea, large vessels navigate extended routes and, as a result, demand fuels with global availability. Considering the lengthy journeys, it is essential that the on-board fuel possesses high energy density, so excessive cargo space (payload) is not consumed by storage. When fuels like LNG, diesel, methanol, and LPG—produced from renewable sources—become globally accessible, they could serve as potential options. The prospect of using ammonia as a fuel is also under discussion, with its energy density being similar to methanol. Furthermore, ammonia's absence of carbon could make it more cost-effective in the future, compared to other electrically-produced fuels that contain carbon. High-temperature fuel cell systems, which can operate on various fuels, offer potential for reducing emissions relative to conventional internal combustion engines. However, the feasibility of these systems for deep-sea vessels remains to be demonstrated [2, 69].

4.3 Availability

Alternative fuels can be produced from a variety of sources, fossil, biogenic, or electricity-based. If fossil sources of energy are used, hardly any GHG emissions can be saved, or GHG emissions may even be higher compared to conventional fossil fuels. Therefore, alternative fuels produced from biomass and/or based on electricity from renewable sources of energy should be used to mitigate GHG emissions. The GHG reduction potential of biogenic fuels varies, depending among other things on the feedstock origin and the respective production pathway. Biogenic fuels can achieve GHG reductions of up to 80% compared to conventional fuels. However, this value is only a general indication, as e.g., significantly lower GHG reductions are also possible. Electricity-based fuels, when produced from renewable energy sources and combined with sustainably ("green") sourced carbon, are typically regarded as having the highest potential for specific GHG emission reductions, making them the preferable choice.

However, regardless of their specific potential to mitigate GHG emissions, for biogenic and electricity-based alternative fuels to achieve such reductions on a system level in shipping, they must be both available and actually utilized within the maritime sector. Depending on the driving profile of the ships, these fuels have to be available globally. The availability of biogenic and electricity-based fuels for shipping is therefore discussed below.

Availability aspects of fuels from biogenic feedstock. The availability of fuels from biogenic raw materials for shipping is influenced by a wide range of factors over time. However, to mention two primary factors in a general sense, it depends, on the one hand, on the total amount of sustainable biofuels that can be produced globally and, on the other hand, on the proportion that the maritime sector can demand and utilize.

Several estimations of the potential of biofuels in shipping have been conducted in the past. One study [70] projected that between 500 and $1,300 \times 10^6$ t (oil equivalent) of sustainable biofuels, biomethane, biomethanol, and biofuel oils, could be produced annually by the year 2050; sustainable biofuels is defined to be a fuel where the biogenic raw material comes from sources other than food and feed. To achieve this, the roughly 5,000 facilities available worldwide in the year 2023 for the production of advanced biofuels with an average capacity of 11×10^6 t/a (oil equivalent) must be significantly expanded.

The energy demand of global shipping amounted to around 300×10^6 t of oil equivalent in the year 2022. This corresponds to a share of shipping of around 3% of total global energy demand. Assuming that international shipping will be defossilized mainly through the use of fuels from biogenic raw materials by 2050, around 250×10^6 t/a (oil equivalent) of biofuels would be required in the year 2050; resulting in a share of 20–50% of the biofuels available globally for shipping in the year 2050 (Fig. 2) [70].

The study concludes that biofuels in shipping can make an important contribution to the reduction of GHG emissions. In the long term, however, it will probably not be possible to meet the energy demand of shipping exclusively with biofuels. Reasons

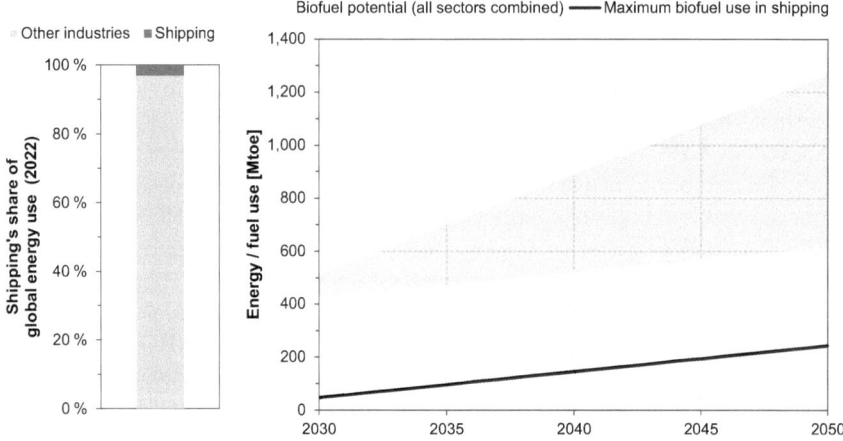

Fig. 2 Estimation of the global biofuel potential including the maximum amount used in international shipping [70]

for this include limitations in the expansion of production capacities for sustainable biofuels and competition with other sectors, such as aviation [70].

Availability aspects of electricity-based fuels. Electricity-based fuels are generally seen as a rather long-term solution for reducing GHG emissions in shipping, as their availability is currently very limited. In order for electricity-based fuels to become available in large quantities at low cost, the infrastructure to provide renewable electricity (e.g., by wind turbines and photovoltaic systems), the necessary electrolysis capacities for the production of hydrogen as well as the CO_2 provision capacities and the downstream processing units must be installed, respectively, significantly expanded. However, various limitations, such as the availability of important materials for electrolyzers (such as nickel, copper, and platinum) or labor shortages, could hamper and delay such a rapid expansion. In addition, the demand for electricity-based (hydrocarbon) fuels in other sectors, combined with the limited availability of electricity-based fuels also in the years to come due to limitations during the ramp-up phase, may lead to rising prices. Therefore, the production of electricity-based fuels will most likely be limited in the 2030s and probably also in the 2040s [71].

Until electricity-based fuels become available in shipping in large quantities at low cost, other alternative fuel options will be needed to reduce GHG emissions swiftly and effectively. In addition to speed reductions, machinery efficiency improvements, and hydrodynamic optimizations, ideas for the transition period also include increased use of the aforementioned low-emission fuels (LNG and LPG from fossil sources), so-called blue fuels (the combination of fossil fuels with carbon capture and storage technologies) and different fuel options based on sustainable biomass [71].

5 Conclusion

To limit global warming, a swiftly and comprehensive reduction in greenhouse gas (GHG) emissions is necessary. Today, through the combustion of fossil fuels, shipping is responsible for a share of about 3% of global GHG emissions. Therefore, the objective is to achieve climate-neutrality in the maritime sector by 2050. On the one hand, this applies to international shipping, which is addressed by the International Maritime Organization (IMO) within the respective GHG reduction strategy. On the other hand, this is true for inland shipping, whose emission reduction targets are set by national and/or transnational targets.

In shipping, there are various strategies such as adopting lower speeds and enhancing efficiencies to reduce GHG emissions. However, to truly achieve climate-neutrality, the use of alternative fuels is essential. Alternative fuels that are either in use or demonstrate potential for shipping include liquefied natural gas (LNG), liquefied petroleum gas (LPG), methanol, ammonia, hydrogen, and diesel fuels produced from electricity or biomass.

Compared to conventional marine fuels, such as heavy fuel oil (HFO) or marine gasoil (MGO), the handling of alternative fuels can differ significantly. This is related to the properties of the fuels, which have an impact on storage, propulsion, bunkering operations, as well as the safety on-board. When storing the alternative fuels on the respective ship, one challenge is their lower volumetric energy density compared to conventional fuels. This leads to larger tanks required and thus lower transport capacities or shorter bunkering frequencies and route adjustments. In the field of propulsion technology for ships, research is currently being conducted into fuel cells, among other things able to reduce additional GHG emissions like methane (CH_4) and nitrous oxide (N_2O) as well as locally active emissions like nitrogen oxides (NO_x) and particulate matter compared to internal combustion engines. However, combustion engines are expected to remain the dominant propulsion technology in the foreseeable future. For the bunkering process, the corresponding infrastructure is required, which is currently not available at all or not available nationwide. Furthermore, when alternative fuels are used, their specific risks must be taken into account and reduced by, for example, safety-related measures.

As the alternative fuels have specific advantages and disadvantages, it seems likely that different alternative fuels will be promising locally and depending on the type of ship. International, deep-sea vessels need especially high energy densities; therefore, they will probably use liquefied natural gas (LNG), liquefied petroleum gas (LPG), methanol, diesel fuels (produced from electricity or biomass), or ammonia in the future. Inland navigation vessels have shorter routes, lower power demands, and frequent start and stop times. Furthermore, they are closer to cities and people. Therefore, especially methanol and hydrogen are promising as alternative fuels under these circumstances.

In addition, other factors have an influence on the future use of alternative fuels in shipping. Uniform regulations and standards, for example, can facilitate their widespread use. At present, however, there are only regulations for LNG and provisional regulations for methanol and LPG. Furthermore, for hydrotreated vegetable oil (HVO) and Fischer–Tropsch (FT) diesel, standards can be derived from existing regulations for fossil diesel fuels or regulations for biomass-based fuels in road transport. Another factor is the global and widespread availability of alternative fuels. Currently, only alternative fuels from fossil sources, such as LNG, LPG, and methanol, as well as selected diesel fuels produced from biomass are available in smaller quantities. Since electricity-based fuels achieve the highest GHG emission reductions, but are not currently available in large quantities, they represent a long-term option. Fuels produced from biomass are thus a short- to medium-term option to reduce GHG emissions from shipping.

References

1. UNCTAD (2019) UNCTAD United Nations conference on trade and development, review of maritime transport 2019. United States of America, New York
2. DNV (2022) Maritime forecast to 2050: energy transition outlook 2022
3. Umweltbundesamt (2022) Seeverkehr: Luftschadstoffe, Energieeffizienz und Klimaschutz. https://www.umweltbundesamt.de/themen/verkehr/emissionsstandards/seeverkehr-luftschad stoffe-energieeffizienz#luftverunreinigung-durch-seeschiffe. Accessed 4 Aug 2023
4. International Maritime Organization (IMO) (2023) International convention for the prevention of pollution from ships (MARPOL). https://www.imo.org/en/about/Conventions/Pages/Intern ational-Convention-for-the-Prevention-of-Pollution-from-Ships-(MARPOL).aspx. Accessed 28 Sept 2023
5. Annex 1, Resolution MEPC.377(80): 2023 IMO strategy on reduction of GHG emissions from ships
6. IMO International Maritime Organization (2023) IMO strategy on reduction of GHG emissions from ships. https://www.imo.org/en/OurWork/Environment/Pages/2023-IMO-Strategy-on-Reduction-of-GHG-Emissions-from-Ships.aspx. Accessed 14 Sept 2023
7. DNV (2023) Maritime forecast to 2050: energy transition outlook 2023
8. Rivarolo M, Rattazzi D, Magistri L, Massardo AF (2021) Energy conversion and management, p 224
9. Bureau Veritas (2022) White paper: alternative fuels outlook for shipping—an overview of alternative fuels from a well-to-wake perspective
10. Ramboll Deutschland GmbH (2021) Abschlussbericht: bunker guidance für alternative Kraftstoffe in deutschen Seehäfen
11. Vermeire MB (2021) Everything you need to know about marine fuels
12. Beuth Verlag GmbH (2018) Deutsche norm: DIN ISO 8217 Mineralölerzeugnisse—Kraft-und Brennstoffe (Klasse F)—Anforderungen an Schifffahrtsbrennstoffe
13. McAllister S, Chen JY, Fernandez-Pello AC (2011) Fundamentals of combustion processes. Springer, New York
14. Altex Energy Ltd (2015) Safety data sheet heavy fuel oil
15. Yeo S-J, Kim J, Lee WJ (2022) J Clean Prod, p 330
16. Younglove BA, Ely JF (1987) J Phys Chem Refer Data 16:4
17. Lagowski JJ (2007) Synthesis and reactivity in inorganic, metal-organic, and nano-metal chemistry 37:2
18. Lin CY (2013) Energies, p 6
19. Gautam R, Kumar S (2020) Energy Rep, p 6

20. Balcombe P, Brierley J, Lewis C, Skatvedt L, Speirs J, Hawkes A, Staffell S (2019) Energy Conv Manag, p 182
21. Myhre G, Shindell D, Bréon FM, Collins W, Fuglestvedt J, Huang J, Koch D, Lamarque JF, Lee D, Mendoza B, Nakajima T, Robock S, Alan B, Stephens S, Graeme B, Takemura T, Zhang H (2013) Climate change 2013: the physical science basis—anthropogenic and natural radiative forcing. Cambridge University Press, Cambridge
22. Stausholm T (2013) IPCC includes GWPs for hydrocarbons in new report. https://hydrocarb ons21.com/ipcc-includes-gwps-for-hydrocarbons-in-new-report/. Accessed 23 Oct 2023
23. Kobayashi H, Hayakawa A, Somarathne K, Okafor E (2019) Proceed Combust Instit, p 37
24. Mohammadpour A, Mazaheri K, Alipoor A (2022) Int J Hydrogen Energy, p 47
25. Wygand J (2022) Ammoniak-Bulker für K-line. https://www.thb.info/rubriken/international/ detail/news/ammoniak-bulker-fuer-k-line.html. Accessed 20 Nov 2023
26. Rheine Hydrogen Integration Network of Excellence (2021) RH2INE kickstart study scenario building: sub-study hydrogen containment systems
27. Teichmann D, Arlt W, Wasserscheid P, Freymann R (2011) Energy Environ Sci, p 4
28. Green Shipping Programme (2022) Infrastructure for liquid organic hydrogen carrier (LOHC). https://greenshippingprogramme.com/pilot/infrastructure-for-liquid-organic-hydrogen-car rier-lohc/. Accessed 1 March 2022
29. IRENA (2021) A pathway to decarbonise the shipping sector by 2050 (AbuDhabi, 2021)
30. Laursen R, Barcarolo D, Patel H, Dowling M, Penfold M, Faber J, Király J, van der Veen R, Pang E, van Grinsven A (2023) Update on potential of biofuels in shipping. Lisbon
31. Ramboll Deutschland GmbH (2022) Abschlussbericht: Kraftstoffanalyse in der Schifffahrt nach segmenten. Rostock
32. Atilhan S, Park S, El-Halwagi MM, Atilhan M, Moore M, Nielsen RB (2021) Curr Opin Chem Eng, p 31
33. Total marine fuels global solutions (2019) Safety data sheet according to regulation (EC) No. 1907/2006: marine distillate fuel (DMA/DFA)
34. LNG Bunkering Course (2015) Section 3.2: properties of LNG. https://www.onthemosway.eu/ wp-content/uploads/2015/06/PRESENTATION-3-%E2%80%93-PROPERTIES-OF-LNG. pdf. Accessed 4 Sept 2023
35. Meyer-Larsen N, Knischka RM, Dreyer M, Kramer H, Arendt F, Baumann M, Würsig G (2021) Die Rolle der maritimen Wirtschaft bei der Etablierung einer deutschen Wasserstoffwirtschaft. Bremerhaven/Bremen
36. Wärtsilä (2020) What does an ammonia-ready vessel look like? https://www.wartsila.com/ media/news/01-12-2020-what-does-an-ammonia-ready-vessel-look-like--2825961. Accessed 4 Sept 2023
37. Clarksons (2023) How is LPG transported? https://www.clarksons.com/glossary/lpg-transp ort/. Accessed 4 Sept 2023
38. Sustainable Ships (2023) The state of methanol as marine fuel 2023: a techno-economic assess-ment for the use of methanol as marine fuel. https://www.sustainable-ships.org/stories/2023/ methanol-marine-fuel. Accessed 4 Sept 2023
39. RCLASS Indian Register of Shipping (2018) Guidelines on methanol fueled vessels
40. Xu Z, Zhao N, Hillmansen S, Roberts C, Yan Y (2022) Energies 15:17
41. Air A, Shamsuddoha M, Gangadhara Prusty B (2023) Comp B Eng, p 253
42. Niermann M, Beckendorff A, Kaltschmitt M, Bonhoff K (2019) Hydr Energy, p 44
43. Rheine Hydrogen Integration Network of Excellence (2021) RH2INE kickstart study scenario building: sub-study hydrogen bunkering scenarios
44. Meyer F (2019) Schiff and hafen: gasmotoren für die schifffahrt
45. van Biert L, Godjevac M, Visser K, Aravind PV (2016) J Power Sour, p 327
46. Shakeri N, Zadeh M, Nielsen JB (2020) IEEE Electr Mag, p 8
47. Rosil RE, Sulong AB, Daud W, Zulkifley MA, Husaini T, Rosli MI, Majlan EH, Haque MA (2017) Int J Hydr Energy, p 42
48. Ahmed AA, Al Labadidi M, Hamada AT, Orhan MF (2022) Membranes 12:12
49. Elkafas AG, Rivarolo M, Gadducci E, Magistri L, Massardo AF (2023) Processes, p 11

50. Ganley JC (2020) Intermediate temperature direct ammonia fuel cells.
51. Campanari S, Guandalini G (2020) Stud Surf Sci Catal 179:179
52. Methanol Institute (2023) Methanol as hydrogen carrier. https://www.methanol.org/fuel-cells/. Accessed 14 Sept 2023
53. DVV Media Group GmbH (2022) Schiff and Hafen: LNG and future fuels: Report 2022/2023 (Hamburg, 2022)
54. Anglo Belgian Corporation (ABC) (2023) Hauptantriebsmotoren. https://www.abc-eng ines.com/de/markets/schiffsantrieb-und-hilfsaggregate/product-solutions/hauptantriebsmotor en--52. Accessed 20 Nov 2023
55. EMSA European Maritime Safety Agency (2018) Guidance on LNG bunkering to port authorities and administrations. https://www.emsa.europa.eu/publications/inventories/dow nload/5104/3207/23.html. Accessed 10 Aug 2022
56. CPL Competence in Ports and Logistics and ISL Institut für Seeverkehrswirtschaft und Logistik (2015) LNG-Marktentwicklungs-und Nachfragepotenzialanalyse für die Schifffahrt und weitere LNG-affine Verkehrsträger in Bremerhaven und Bremen (Bremerhaven)
57. DNV (2020) Methanol as a potential alternative fuel for shipping: a brief talk with Chris Chatterton of the Methanol Institute. https://www.dnv.com/maritime/advisory/afi-update/Met hanol-as-a-potential-alternative-fuel-for-shipping-A-brief-talk-with-Chris-Chatterton.html. Accessed 18 Jan 2022
58. van Hoecke L, Laffineur L, Campe R, Perreault P, Verbruggen SW, Lenaerts S (2021) Energy Environ Sci, p 14
59. Hansen OR, Martini R (2020) Hydrogen and ammonia infrastructure: safety and risk informa- tion and guidance. https://static1.squarespace.com/static/5d1c6c223c9d400001e2f407/t/5eb 553d755f94d75be877403/158894183
60. Reddi K, Elgowainy A, Rustagi N, Gupta E (2017) Int J Hydr Energy, p 42
61. Depken J, Dyck A, Roß L, Ehlers S (2022) Energies, p 15
62. Marine Regulations News (2023) IMO published interim guidelines for the safety of ships using LPG fuels. https://www.marineregulations.news/imo-published-interim-guidelines-for- the-safety-of-ships-using-lpg-fuels/. Accessed 11 Sept 2023
63. Ohle L, Allolio F, Schäfer J (2022) Legal study: regulatory framework for a German–Australian hydrogen bridge (Berlin)
64. DNV (2023) IMO CCC 9: work on interim guidlines for ammonia and hydrogen as fuel. https://www.dnv.com/news/imo-ccc-9-work-on-interim-guidelines-for-ammonia-and- hydrogen-as-fuel-247849
65. DNV (2021) Handbook for hydrogen-fuelled vessels (2021)
66. DNV (2023) Use of biofuels in international shipping. https://www.dnv.com/news/use-of-bio fuels-in-international-shipping-240298. Accessed 30 Oct 2023
67. Zerta M, Diehl L, Landinger H, Moll J, Klemm P, Sattler G (2023) Maritime Wasserstoffan- wender und ihr anteil am H2-bedarf deutschlands (Lauenburg/Elbe, Ottobrunn)
68. Zentralkommission für die Rheinschifffahrt (ZKR) (2022) Roadmap der ZKR zur Verringerung der Emissionen in der Binnenschifffahrt (Strasbourg Cedex)
69. Pape M (2020) Briefing: decarbonising maritime transport—The EU perspective
70. DNV (2023) Exploring the potential of biofuels in shipping. https://www.dnv.com/expert-story/ maritime-impact/Exploring-the-potential-of-biofuels-in-shipping.html. Accessed 4 Sept 2023
71. Mærsk Mc-Kinney Møller Center for Zero Carbon Shipping (2023) Will renewable electricity availiability limit e-fuels in the maritime industry?

Power-to-Chemicals: Defossilization of the Chemical Industry via PtX

Stefan Bube, Steffen Voß, Nils Bullerdiek, and Martin Kaltschmitt

Abstract The chemical industry predominantly relies on fossil fuel-based feedstocks for both material and energetic uses, leading to significant CO_2 emissions. Power-to-Chemicals (PtC) offers a promising solution to defossilize material uses by producing primary chemicals through electricity-derived synthesis gas and various downstream conversion technologies. This chapter provides an overview of the current status of fossil fuel use in the chemical sector and explains various defossilization options for the different requirements. Key PtC pathways, including the production of ammonia, methanol, olefins, and BTX aromatics, are outlined and discussed. The chapter also discusses the synergies and challenges in deploying PtC technologies alongside PtX-based fuel production, emphasizing the need for a collaborative cross-sectoral approach to scale-up and market adoption.

Keywords Power-to-Chemicals (PtC) · Chemical sector · Defossilization · Primary chemicals · Ammonia · Methanol · Olefins · BTX aromatics

1 Introduction

The chemical industry produces a wide variety of products that are applied in almost all areas of the overall economy and serve numerous demands in everyday lives. They are used directly, for instance, to produce plastics or textiles, or indirectly to enhance and support the production of other bulk materials such as metal or timber. Furthermore, several chemical products are almost irreplaceable in their end applications (e.g., in pharmaceuticals or electronic devices) [1, 2].

Today, the production of chemical products relies predominantly on fossil fuel-based, carbon-containing feedstocks (natural gas, crude oil, hard coal) and is therefore associated with greenhouse gas (GHG) emissions. Especially in organic chemistry, these feedstocks serve not only for energy purposes but also as a material/

S. Bube (✉) · S. Voß · N. Bullerdiek · M. Kaltschmitt
Hamburg University of Technology (TUHH), Institute of Environmental Technology and Energy Economics (IUE), Hamburg, Germany
e-mail: stefan.bube@tuhh.de

molecular basis, with (fossil) carbon being bound in the chemical product composition. Consequently, with those carbon-containing products remaining in use, significant segments of the chemical industry cannot be fully decarbonized, unlike other sectors (for instance, where direct electrification is a viable alternative to carbon-based fuels). Since the carbon bound in the molecular composition is typically emitted into the atmosphere as CO_2 at the end of the final product's lifecycle (e.g., by burning plastic waste in waste incineration plants), this carbon must be sourced from nonfossil origins to mitigate climate-impacting GHG emissions.

The overall production structure in the chemical industry can be compared to a "river delta" where the main stream comprises of a few main chemicals, so-called primary chemicals. These serve as the basis to produce a variety of subsequent products and product groups, which in turn also serve for multiple further downstream production pathways and secondary products. As the processing pathways branch out downstream, a vast and interconnected production network is formed, resulting in a wide range of final chemical products. While the energetic use of fossil fuel-based feedstock is required throughout the entire chemical process chain, the material conversion of these feedstocks into chemicals mainly takes places in the production of primary chemicals. As these material carbon requirements are inherently "non-decarbonizable" and yet represent a substantial portion of the overall demand for fossil fuel-based feedstocks in the chemical industry, defossilizing this industry by producing primary chemicals based on nonfossil "green" carbon is pivotal in the broader effort to defossilize the chemical industry.

Defossilization in the chemical industry can be based on biomass as a "green" carbon and energy carrier as well as electricity from renewable sources of energy[1] as a "green" energy source. Although biomass is already utilized in the chemical industry, e.g., for producing bioethanol, its increased use is constrained by a limited availability of sustainable biomass and further restricted by competition with other uses (e.g., food and fodder, building material, biodiversity/nature conservation demands). In contrast, renewable electricity can be used alongside or combined with biomass-based production processes through Power-to-X (PtX) technologies. While renewable electricity is currently also limited (due to insufficient power generation capacities), there is a greater global potential compared to biomass cultivation, which typically involves fewer conflicts in the use of arable land [3, 4]. Presently, PtX products are mainly seen to be used as "green" fuels and thus for energetic purposes (particularly in transportation). However, their material use—specifically as primary chemicals derived from Power-to-Chemicals (PtC) processes—is also crucial to defossilize "hard-to-abate" sectors that depend material wise on hydrogen and carbon.

Against this background, in this article, applications and potentials of Power-to-X (PtX) pathways to defossilize the chemical industry, focusing on primary chemicals, are analyzed. First an overview of the chemical industry is provided, including current energy requirements, a classification of potential defossilization options, and related

---

[1] Electricity from renewable sources of energy is in the following referred to as "renewable electricity."

PtX applications. Following this, the current feedstocks and production routes of primary chemicals are presented in more detail, and various PtX-based defossilization options are discussed. Subsequently, the scale of PtX capacities required in the chemical industry is estimated and compared to demands in aviation and shipping. In this context, additional aspects are discussed on how a utilization of PtX in the chemical industry may affect the use of PtX to produce fuels in other sectors and vice versa.

2 Background

The following section presents the current use of fossil fuel-based hydrocarbons in the chemical sector and describes defossilization options for different sector demands. Since this article focuses on the material use of PtX products, hydrogen-containing, and organic chemicals, which are conventionally produced based on natural gas, crude oil (petrochemistry), or coal, are primarily discussed.

2.1 Demands of Fossil Resources and Resulting GHG Emissions

In addition to metals and inorganic building materials (e.g., cement, gypsum), mainly plastics or fertilizers are among the world's most produced and constantly growing bulk materials. In particular, the demand for plastics has risen above average in recent years [2]. In addition to applications within the packaging and textile sectors, their importance in special applications such as electronics and renewable energy production has also increased. Plastics always consist molecularly of hydrocarbons, forming polymers in different constellations with various possible heteroatoms. As of today, the material origin of the carbon and hydrogen is almost exclusively fossil (i.e., from crude oil, natural gas, or hard coal). Concerning the use of such fossil hydrocarbons, about 90% are accounted for by the petrochemical industry, a subsector of the chemical industry being historically defined as petroleum (oil) and natural gas processing chemistry [2]. The use of hard coal for the production of chemicals is mainly realized in China due to the political will to use predominantly domestic fossil resources. As the demand for chemical products grows, so does the need for feedstock as well as the GHG emissions associated with their production and disposal if fossil resources are continued to be used.

Within organic chemistry, fossil fuel-based feedstocks serve two main demands. On the one hand, the feedstock provides the molecular/atomic building blocks of produced chemicals (material use). On the other hand, most of the process energy required for conversion and processing is obtained via feedstock oxidation/

combustion (energetic use). Thus, demand reductions through improved efficiencies of the conversion processes and energy provision are possible through future technological developments and optimization measures; but according to current knowledge, they are also clearly limited. The efficiency of material feedstock use is at least restricted by the stoichiometric feedstock demand, while the energy efficiency is limited, among others, by the heat integration potential, subject to process-related operational and economic restrictions. Many of the processes used today have been tested and developed for decades already and are thus largely optimized; typically, there is only little additional optimization potential exploitable [1, 2].

Regarding fossil feedstock demand, the chemical sector is the largest industrial energy consumer worldwide [1]. In 2017, the global chemical industry accounted for an energy demand of around 12 PWh, of which almost 60% was required for feedstock energy [2]. As Fig. 1 depicts, the chemical industry accounts for approx. 14% of the global demand in crude oil and approx. 8% in natural gas [2, 5]. Compared to that, the share of coal used for the chemical sector is rather low in most parts of the world. However, in China, the non-power usage of coal for a conversion into fuels and chemicals has increased strongly during the last decades. Nevertheless, less than 5% of the coal demand in China is used within chemical production processes to provide products for a material use [6].

Although slight efficiency improvements are expected in the future, projections expect an increase in the worldwide demand for the 18 most energy-intensive chemicals by more than 280% until 2050 [7]. In the EU, the chemical industry needed 589 TWh of energy in 2020; i.e., the energy demand of this industry sector has remained almost constant since 2015 (590 TWh) [8]. That means, that the substantial demand reduction of 19% between 1990 and 2005 (1.2%/a) could not be sustained within the last decade.

While the chemical industry has the highest energy demand of all industrial sectors, it is only the third largest CO_2 emitter on a global scale. This is because only half of the energy carriers demanded by this industry sector are directly used energetically for the generation of heat or electricity. Beside this, CO_2 emissions occur from adjusting the carbon content of the feedstock to the carbon content of the product. And, what is quite important to note, the materially bonded carbon in

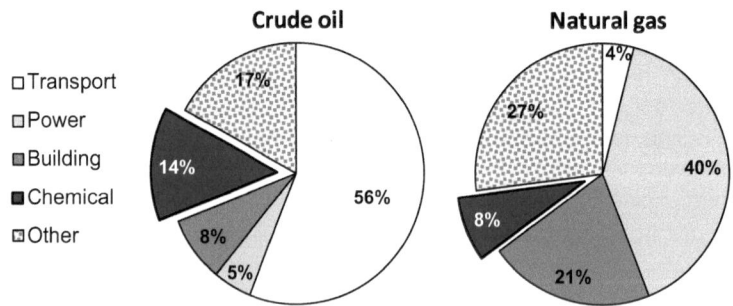

Fig. 1 Demands of crude oil and natural gas for different sectors in 2017 (according to [2])

the products is transferred to other sectors where these products from the chemical industry are used/further processed/recycled. The related fossil CO_2 emissions only occur after the products' life cycles end through combustion or decomposition outside of the producing chemical sectors (indirect emissions from the chemical industry) [2]. Nevertheless, the chemical industry is responsible for approximately 7% of global anthropogenic global GHG emissions and 5.5% when only counting CO_2 emissions [1, 2]. Only the direct CO_2 emissions of chemical production summed up in 2021 to around 925 Mt [9].

Summarizing the above-described demands and GHG emissions, Fig. 2 shows the fossil feedstock for chemical production and the GHG emissions from the major primary chemicals that account for around two-thirds of the total energy demand of the chemical sector [2]. However, emission data show considerable variations in the literature. Uncertainties arise, especially from different allocation approaches regarding producing high-value chemicals (HVCs) from methanol. Further, indirect CO_2 emissions from the conversion or combustion in subsequent sectors can also be relocated to the fossil carbon feedstock bound in the organic chemicals' physical structure.

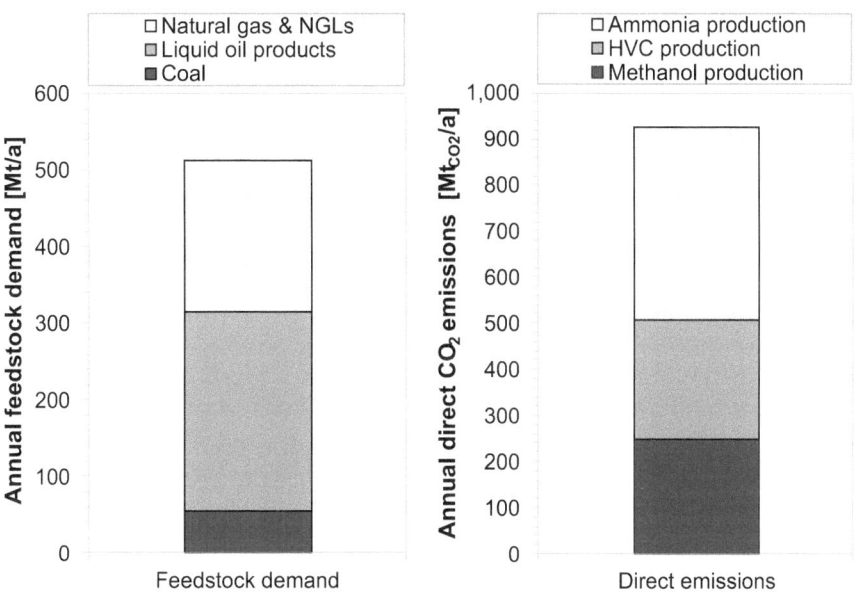

Fig. 2 Major demands and emissions of the chemical industry (feedstock data from [1] and emission data from [9]; HVC—high-value chemicals, NGLs—natural gas liquids)

2.2 Options for Defossilization

As fossil fuel-based feedstocks are used in the chemical industry to fulfill various requirements, different defossilization options in terms of resources and technologies arise depending on the application. Therefore, different terms are used in the context of replacing fossil demands with alternative renewable approaches. In this context, the terms defossilization (also defossilation) and decarbonization are commonly used. Assuming that the original material and energy sources stem from carbon-based fossil feedstock, the terms can be defined as follows:

- Defossilization describes the substitution of fossil carbon-based feedstock and energies through renewable alternatives.
- Decarbonization is a type of defossilization that describes the substitution of fossil carbon-based feedstock and energies through renewable and carbon-free alternatives.

Primarily defossilization can be based on biomass or renewable electricity. Biomass, comparable to fossil energy resources, includes both material and energy supply while using renewable electricity provides only energy. However, renewable electricity can be utilized to provide material streams by driving processes, which, through methods such as electrochemical or thermochemical processes, separate or produce the required substances from non-fossil sources. Examples of these include water splitting through electrolysis for H_2 production, the provision of CO_2 from the air (direct air capture [DAC]) through adsorption-based cycle processes, or the provision of N_2 through cryogenic air separation.

Concerning the primary energy used, defossilization can also be referred to as "electrification." However, in most cases, this term relates to the end application's electrification (e.g., replacing a diesel engine with an electric engine). In the context of this article, only defossilization using renewable electricity – i.e., excluding biogenic energy sources – is considered. Thereby, Fig. 3 illustrates different defossilization pathways. The various demands of the sector are discussed separately in the following. First, the demands and the supply of fossil resources are described. Hereafter, potential non-fossil substitution options are explained and categorized according to the abovementioned terminology. The fossil supply pathway (Fig. 3, left) represents the prevailing supply state today, while renewable supply (Fig. 3, right) is not yet widely used or deployed on a comparable scale.

Defossilization of electricity demands. The demand for and the use of electricity in chemical production plants mainly depends on the feedstock and the desired product as well as on the production process. Electrochemical production processes in which electricity is used directly for material conversion are among the largest point consumers of electricity. An example of this is the energy-intensive chlor-alkali electrolysis currently used to produce chlorine from brine. In addition, almost all chemical production processes require electricity to provide mechanical energy. The latter mainly refers to crushing by grinding or pressing as well as transporting materials (gas, liquid or solid) through conveying equipment such as compressors,

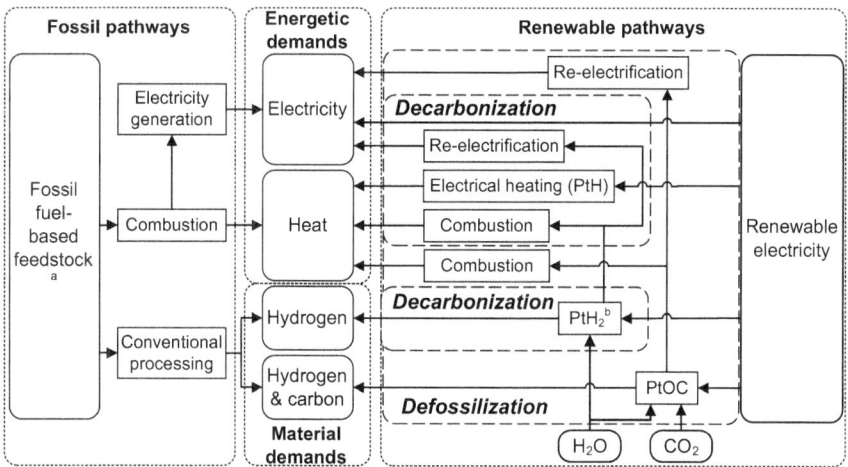

Fig. 3 Defossilization options for different demands of the chemical industry (biogenic resources not included; [a]additional carbon-free feedstocks for a.o., oxidation required, [b]further carbon-free hydrogen carriers such as ammonia (PtNH$_3$) possible; PtH—power-to-heat, PtOC—power-to-organic chemicals)

pumps, or conveyor belts. Compressors usually have significant energy requirements in thermochemical processes since they are used for transporting of feedstock, intermediates, and products as well as increasing gas flow pressure up to operating conditions. With the help of compressors used in cooling circuits, electricity can also be used for cooling, especially at temperatures far below ambient temperature. Furthermore, mostly significantly lower electricity demands arise from process control and operation—i.e., from the supply of measuring and control instruments and the process control system. While the formerly described electricity demands correlate directly with the production output, here, a constant—i.e., continuous and load-independent—supply is necessary.

As Fig. 3 shows, the current electricity supply from fossil fuel-based feedstocks occurs via combustion of the respective fossil fuels within thermal power plants. Since an economically competitive production of chemical products usually requires a cost-effective supply of fossil fuels and, in most cases, a corresponding supply infrastructure already exists to cover further energy and material demands, electrical energy is commonly produced on-site, optionally coupling-out steam for heat supply. In addition, it is common to purchase electricity from the power grid, which, depending on the electricity mix—i.e., the share of renewable energy within the overall delivered electricity—can already lead to partial defossilization; e.g., in 2023 roughly half of the electricity from the German public grid stems from renewable sources of energy. Explicit defossilization is possible through the exclusive purchase of renewable electricity, biomass-derived electricity, or through the use of electricity-based energy carriers for re-electrification. If carbon-free energy carriers are used (like H$_2$ or NH$_3$), it can also be referred to as decarbonization. From a technical point of view, in the context of direct electricity use from the strongly fluctuating renewable sources of energy (e.g., wind power or solar radiation), the volatile availability compared to constant or controllable electricity supply from storable energy carriers must be considered. However, for reasons of efficiency, the direct utilization of renewable electricity is advantageous in most of the known application cases so far.

Defossilization of heat demands. Heat transfer is a process engineering unit operation used in almost all chemical processes. Whereas the environment can usually be used directly or indirectly as a heat sink for cooling, for heating, a higher temperature level must usually be generated by the input of primary or secondary energy. This energy can be provided to a certain amount from internally available "waste" heat—for example, from exothermic chemical reactions or material process flows that are to be cooled (heat integration). In addition, external energy demands for heating (external heating) are commonly required.

Heat requirements in chemical production plants can arise from a huge variety of different processes. The main heat consumers here are often thermal separation processes, such as rectification, desorption, drying, or crystallization [10]. Further requirements arise, among other things, from heating endothermic reactions and process streams up to operating temperatures. Especially when handling highly viscous fluids or liquids with melting points above ambient temperatures, trace heating is usually necessary. The required temperature level is of particular importance for the provision of heat. Apart from such specific applications, such as the trace heating of pipelines, heat is conventionally provided via the combustion of carbon-based fossil fuels. In most cases, the released thermal energy is transferred to a heat transfer medium such as steam or hot liquid utilities (e.g., thermo oil, hot water), which takes up the thermal energy at a specific temperature level and transfers it to the respective heat requirement. In the case of process steam, different pressures are used for this purpose. Heat demands at a temperature level above 400 °C are mainly realized via direct firing/combustion.

The defossilization of heat demands can be achieved via various technical alternatives. Direct heating and heat transfer media utilizing electrical heaters (ohmic resistors) can be a promising or at least a possible solution, especially when a sufficiently large heat transfer surface is available. High-temperature heat pumps can also be used for the highly efficient, renewable provision of heat. However, usability strongly depends on the temperature level, the available quantity of waste heat, and the required temperature level. Temperature increases of 70–90 K can be implemented based on heat pumps; so far, the technical limit is at a maximum temperature clearly below 200 °C [11, 12]. Both cases, ohmic resistors and heat pumps, contribute to decarbonization.[2]

Such an electricity-based heat production is also known as Power-to-Heat (PtH). But so far, only minor extents of electricity are used for heating (e.g., for electrical trace heating) due to economic constraints (electrical energy is typically clearly

---

[2] In terms of conversion technology, this can also be referred to as "electrification."

more expensive compared to competing energy carrier like, e.g., natural gas or wood pellets).

Besides the direct use of electricity, heat defossilization can be realized using combustion processes operated on electricity-based fuels or biomass. Depending on the molecular composition of the fuel, it can also be classified as decarbonization—for example, if heat is supplied by a combustion of hydrogen or ammonia.

Defossilization of material demands. The manufacturing of different primary chemicals or certain intermediate products in the chemical industry requires the supply of the respective molecular/elemental building blocks. For the production of hydrogen or hydrogen- and carbon-containing products, the necessary atoms are today mainly provided by carbon-based fossil feedstock.

Carbon-free hydrogen demands in the chemical sector—i.e., a demand of pure hydrogen—arise, e.g., in ammonia production and the hydrotreatment of oil within a "classical" crude-oil refinery (or a biorefinery producing HEFA-fuel from used cooking oil). The conversion pathway for producing pure hydrogen from fossil resources strongly depends on the feedstock. Today, primarily natural gas is converted to hydrogen via steam reforming, partial oxidation, or autothermal reforming. Besides natural gas, other carbon-based fluids can also be reformed using these technologies. A subsequent water–gas shift reaction is commonly used to maximize the hydrogen yield. In addition, hydrogen occurs as a by-product in selected industrial processes, such as chlor-alkali electrolysis or refinery processes. Decarbonization of the hydrogen supply can be realized via water-electrolysis based on renewable electricity (Power-to-Hydrogen; PtH_2). Reforming electricity-based organic energy carriers is also possible but only reasonable if hydrogen production does not occur at the point of use, which may make the supply of hydrogen derivate through inexpensive infrastructure economically viable.

For the production of organic products, the building components hydrogen and carbon are required. Today, the carbon-based feedstock is therefore processed in the "conventional processing" (Sect. 4). Since carbon is a molecular component of these products and thus physically irreplaceable, this production area cannot be decarbonized under any circumstances. Therefore, the carbon here must be provided from renewable ("green") carbon sources, in terms of electricity-based defossilization mainly as carbon dioxide (CO_2). Various sources and related technologies are available for this purpose (Carbon Capture and Utilization, CCU).

The provision of carbon dioxide and hydrogen as well as the conversion to the respective (organic) products based on renewable electricity is referred to as Power-to-Chemicals (PtC) or, more specifically, Power-to-Organic Chemicals (PtOC). The possible production pathways for this are discussed and classified in detail in Sect. 4.

From a technological point of view, there are various possibilities to defossilize the existing applications of fossil fuel-based feedstocks. Either the current application technologies can be adapted (e.g., by replacing burners with electric heating units) or the energy carriers can be replaced upstream (e.g., by substituting conventional (fossil) natural gas by synthetic natural gas (SNG) from renewable energies). In general, direct electrification, if possible, is the most efficient approach from an energy system point of view, as it involves only few conversion steps and electrical energy is pure exergy. In comparison, electricity-based fuels can use existing infrastructure on the application side but involve more conversion steps and thus more conversions losses on the upstream side. While the defossilization of energetic demands can be realized via several renewable pathways, the material demands can only be defossilized via PtX pathways, when renewable electricity is used (Fig. 3). The primary chemicals, that form the basis for almost all organic downstream chemical products, account for the main material feedstock demands. These primary chemicals are described in Sect. 3.

3 Primary Chemicals

The main energy and feedstock requirements within the chemical industry occur upstream from producing large-volume chemicals. These so-called primary chemicals are intermediates for most downstream processes and a huge variety of different products. Depending on the respective chemical end product, the production chain can involve various conversion steps and several intermediate products. Figure 4 shows a schematic diagram of the terms used along the production chain in this paper.

The primary feedstock is processed upstream within the chemical sector, mainly for the production of primary chemicals (also referred to as basic, bulk, or key chemicals). These primary chemicals are typically produced on a large-scale and traded internationally to satisfy global market demands [13]. Solid, liquid, and/or gaseous substances arising during processing are referred to as intermediates (e.g., synthesis gas). These primary chemicals form the basis for most energy-containing products in the chemical and the petrochemical industry [2]. The further processing of these molecules, down to the target chemical products, is called downstream production or downstream processing. The products processed during these sometimes very branched and multistage pathways are also referred to as intermediate or semifinished products. The subsequent chemical products (including advanced bulk chemicals, fine chemicals, and specialties) mainly leave the chemical sector for further

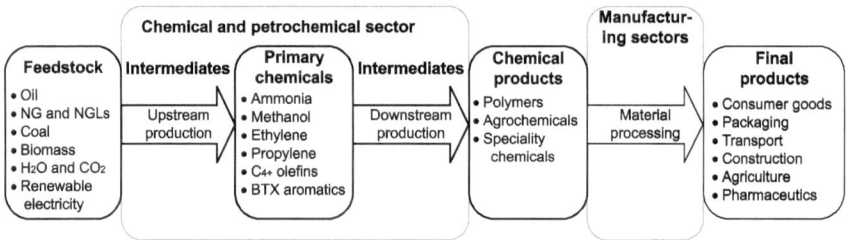

Fig. 4 Terminology along the production chain of the chemical and manufacturing sector (BTX—benzene, toluene, and xylenes, NG—natural gas, NGL—natural gas liquid)

processing within subsequent downstream sectors, whereby molecular composition and structure remain mostly unchanged. Material processing in the so-called manufacturing sector enables the use of the final products outside or, in some cases, within the chemical sector. Examples are processing plastic granulates to plastic packaging or mixing and applying fertilizers on carrier materials for the subsequent use in agricultural and horticulture.

Fossil carbon-based feedstock is used along the production chain in almost all process steps to provide process energy. In contrast, the material use only occurs in the upstream conversion to provide primary chemicals. Along the rest of the process chain, these form the material basis for the provision of the various downstream products. Thus, the "renewable" production of primary chemicals can play a decisive role in defossilizing the chemical industry's material demands. Thereby, the provision of "green" primary chemicals enables additionally the defossilization of significant parts of the respective downstream sector. The primary chemicals under consideration are explained briefly below.

3.1 Synthesis-Based Chemicals

Ammonia (NH_3). Ammonia is a toxic colorless gas that can be liquefied under moderate conditions (boiling point: 20 °C at 7.5 bar/−33 °C at 1 bar [14]). The molecule comprises 82% nitrogen and 18% hydrogen by mass. Since no carbon is in its molecular composition, ammonia is the only inorganic primary chemical used today within our global industrializes society. Today, ammonia is produced within technical syntheses plants, accounting for more than a quarter of the global hydrogen demand. In 2020, the global annual demand accounted for around 185 Mt, making ammonia the most widely mass-produced synthesis-based chemical [15]. Today's production relies mainly on the use of natural gas and hard coal (the latter especially in China with around 85%), as a fossil fuel-based feedstock (see Fig. 5) . The production accounts for around 1.5% of global energy consumption (8.6 EJ in 2020) and is responsible for 1.2–1.8% of the global CO_2 emissions (450 to 500 Mt CO_2 in 2020) [15, 16]. The average CO_2 emissions related to the production of ammonia amount to approx. 2.5 t CO_2/t NH_3. On the application side, most ammonia is used to produce fertilizer to support food production to feed the still growing world population. In addition, ammonia is processed into building blocks for various industrial applications or used directly as a coolant or in exhaust gas and wastewater treatment, among other things [17, 18]. Despite challenging fuel and reforming properties, ammonia is also a potential carbon-free hydrogen carrier and fuel [17] .

Methanol (CH_3OH). Methanol is the smallest and first representative of the homologous series of alcohols. Methanol consists of one carbon atom, three hydrogen atoms, and a hydroxyl group (OH-group). The molecule has a mass of 32 g/mol and thus consists of 50% oxygen. The hydroxyl group is the functional group of

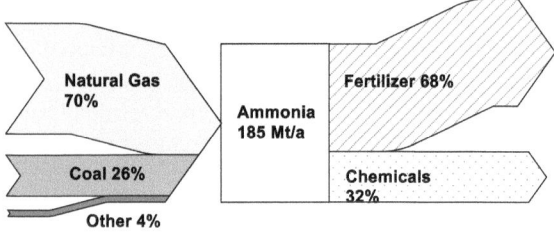

Fig. 5 Up- and downstream flows of the global ammonia value chain (data based on [15, 19], time reference 2019–2021)

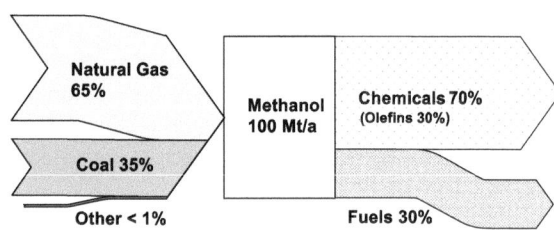

Fig. 6 Up- and downstream flows of the methanol value chain (data based on [21], time reference 2020)

the methanol molecule and is mainly responsible for its polarity and versatile reactive properties. Methanol is liquid under ambient conditions (boiling point: 65 °C at 1.0 bar) [20]. Due to the significant global demand growth, methanol is one of the most widely produced organic primary bulk chemicals worldwide characterized by a production of around 100 Mt/a [21] (see Fig. 6). Like ammonia, so far it is produced predominantly from natural gas, whereas only in China mainly hard coal is used as a feedstock. Technically, methanol is almost exclusively produced via synthesis processes. Downstream, the alcohol is used, e.g., directly as a chemical industry solvent or adsorbent (e.g., Rectisol® process). However, the majority is used in the chemical industry as a feedstock for various downstream products; for example, substantial quantities are used for the production of olefins and formaldehyde. Since about 30% of the produced methanol is converted, mainly on-site, into ethylene and propylene, methanol production is sometimes not fully balanced in this respect.

3.2   High-Value Chemicals

The collective term HVC includes light olefins and aromatics since their market prices are typically significantly higher than those of other primary chemicals. The main reason is that these chemicals are mainly produced from crude oil, which tends to be more expensive than natural gas and hard coal used for ammonia and methanol production [2]. Additionally, olefins can be a downstream product of methanol. In this case, costs must be higher compared to methanol as additional processing is required.

Olefins (C_nH_{2n}). Olefins are unsaturated hydrocarbons, also known as alkenes. Olefins form a group of chemicals to which various substances and isomers are grouped. The most important olefins as primary chemicals are short-chain olefins gaseous under standard conditions (e.g., ethylene [C_2-olefin], propylene [C_3-olefin], butene, butadiene [C_4-olefins]). Due to the double bond and the associated chemical properties, these olefins are particularly used to carry out addition reactions (electrophilic, nucleophilic, or radical) in subsequent processing steps. Polymers (e.g., polyethylene, polypropylene) or monomers (e.g., 1,2-dichloroethane used as a starting material for polyvinyl chloride) are typically produced downstream. In addition, olefins are used to produce other important organic compounds, such as acetaldehyde, acetone, and ethylene oxide, which are available as chemicals for manifold subsequent reactions. Most olefins today are produced via steam cracking of natural gas liquids, liquefied petroleum gas, and naphtha. The feedstock is decisive for the composition of the olefin mixture. Olefins are also produced worldwide as by-products, partly in crude oil processing directly after thermal cracking plants, or they are also produced in smaller quantities by dehydration of methanol or ethanol and dehydrogenation of butanes [22–24]. The combined global ethylene and propylene production is around 255 Mt/a, while the demand for ethylene is generally higher (around 156 Mt/a) than for propylene [2, 25]. The production capacity for butadiene, as one of the C_4 olefins, is about 18 Mt/a and thus significantly lower [26].

BTX aromatics. Benzene, toluene, and xylenes form the substance group of BTX aromatics. The basic molecular structure is a benzene ring (C_6H_6). Toluene is built by an additional methyl group (C_7H_8) and xylenes (C_8H_{10}) by two methyl or one ethyl group. These aromatic compounds are thermally stable but still chemically reactive. All BTX aromatics are liquid under normal conditions but evaporate under moderate temperatures of 80–145 °C. BTX aromatics are present in crude oil or light oils from coal pyrolysis, from which they are obtained via separation processes [27–29]. BTX aromatics are converted into various intermediates and end products, such as solvents, plastics, dyes, perfumery, pharmaceuticals, surfactants, and synthetic rubbers [30]. Large quantities are also fuel components in gasoline. Global production of BTX aromatics is approximately 110 Mt/a [2].

4 Production Pathways

The production of primary chemicals from fossil fuel-based feedstock can be carried out via different process pathways. If the desired product is already inherently present within the feedstock, then the processing primarily consists of separating it from the feedstock mixture (e.g., removal from crude oil). However, usually the target molecules are not or not in the required quantity present in the feedstock. In this case, the desired product need to be generated from rather complex molecular building blocks (conversion) or based on a synthesis process from a so-called synthesis gas (syngas) consisting of very simple molecules (e.g., carbon monoxide [CO], carbon

Fig. 7 PtX-based
substitution options for
primary chemical production

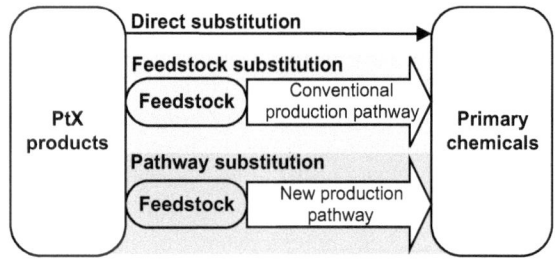

dioxide [CO_2], nitrogen [N_2], hydrogen [H_2]). Defossilization of such a production via electricity-based chemicals can be enabled via different substitution options generally categorized in Fig. 7.

- The simplest case is the "direct substitution" of conventional produced chemicals by Power-to-X (PtX) products[3]; i.e., the product of the PtX process is a primary chemical. This is the case, for example, for electricity-based ammonia production ($PtNH_3$).
- A second option is the substitution of the fossil fuel-based feedstock ("feedstock substitution") by PtX products. In this case, the currently applied conversion pathway remains unchanged but the electricity-based substitute (renewable feedstock) is processed instead of the fossil fuel-based feedstock. For example, electricity-based SNG can substitute conventional (fossil) natural gas used as a feedstock, e.g., for methanol production.
- The third option is the conversion of a PtX product in a new or currently not primarily used production pathway. This so-called pathway substitution requires both, the buildup of a PtX production on the one hand, and the construction of the new downstream conversion pathway on the other. An example of pathway substitution is aromatics production via a Methanol-to-Aromatic (MtA) process from electricity-derived methanol instead of a production based on crude oil.

A more detailed overview of electricity-based pathways for primary chemical defossilization is depicted in Fig. 8. The pathways shown are further described in the following subsections. However, the figure cannot present a complete overview due to the large number of branches and options according to the current state of research.

---

[3] Power-to-X (PtX) in this context includes the conversion of electricity into hydrogen as well as the first synthesis step to a more complex molecule. The further processing of that molecule is then—in the context of this paper—not allocated to the PtX process.

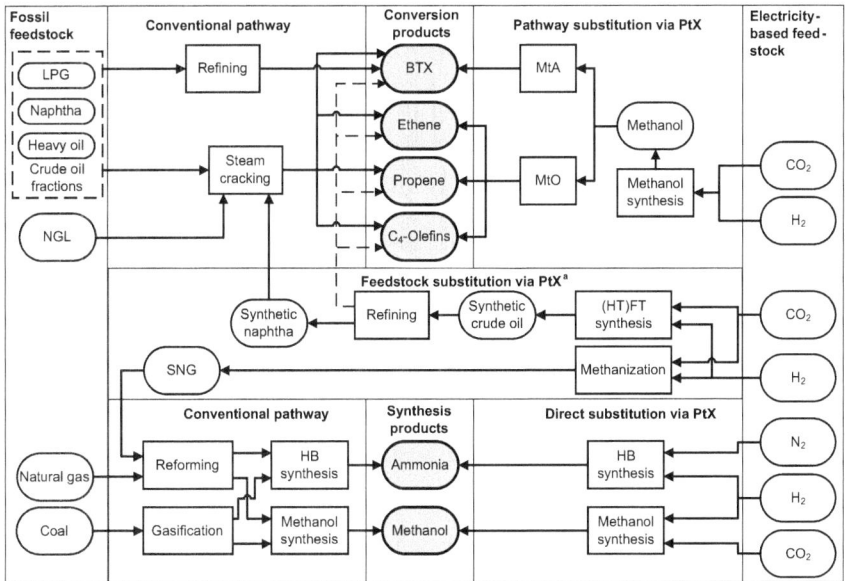

Fig. 8 Electricity-based defossilization pathways for primary chemical production ([a]Dotted line is to be classified differently as a direct substitution; BTX—benzene toluene xylene, FT—Fischer–Tropsch, HB—Haber–Bosch, HT—high temperature, LPG—liquefied petroleum gas, MtA—methanol-to-aromatics, MtO—methanol-to-olefins, NGL—natural gas liquids)

4.1 Ammonia Production

Today, about 90% of the globally realized ammonia production is derived from synthesis processes. The synthesis gas, a mixture of N_2 and H_2, is almost exclusively provided from natural gas and to a smaller extend by hard coal. However, the synthesis gas can also be provided from electricity-based processes.

The Haber–Bosch synthesis was first realized commercially in 1913 and has been continuously developed since then, although the basic synthesis process is still applied in all plants under operation today [31]. The basic chemical reaction is shown in (1). Since the reaction is exothermic and volume-reducing, low temperatures and high pressures shift the equilibrium to the product site [32]. However, the required activation energy is high, requiring high reaction temperatures well above 500 °C.

$$0.5\,N_2 + 1.5\,H_2 \rightleftharpoons NH_3 \mid \Delta H_R = -46\,\text{kJ/mol} \tag{1}$$

The application of mostly iron-based catalysts (Fe_2O_3) allows the necessary reaction temperatures to be reduced to 350–500 °C. This enables conversion rates close to the chemical equilibrium of 15–35%. However, operation pressures from 150 to

350 bar are required. The applied synthesis gas ratio lies close to the stoichiometrically required composition ($H_2/N_2 = 3$). Besides the ammonia-building reaction, almost no side reactions occur, producing pure ammonia.

Although this synthesis reaction is the core of the ammonia production, the synthesis loop is embedded in the overall concept of the plant. Thus, it has to be discussed in the context of synthesis gas supply differing substantially between fossil hydrocarbon-based and electricity-based processes.

Fossil fuel-based ammonia synthesis. In ammonia production from fossil hydrocarbons, the feedstock is, on the one hand, used to supply the physically required hydrogen (40% of the feedstock energy used [15]), and on the other hand, to provide the process energy that is mainly needed for heat generation and for operating auxiliary system compounds (e.g., compressors). Additionally, air is used for hard coal gasification or natural gas reforming (oxidation). Multistage reforming and water–gas shift (WGS) reactors are commonly used to enhance the hydrogen yield. Thereby, the production of hydrogen causes approximately 90% of the CO_2 emissions when derived from fossil fuel-based feedstock [16]. The resulting synthesis gas is an oxygen-free (below 10 ppm [31]) mixture of mainly N_2, CO_2, H_2, and traces of CO. CO_2 is then separated from the gas stream to increase the partial pressure of the reactants and to avoid side reactions and catalyst poisoning within the synthesis reactor. The remaining CO in the mixture has to be reduced below a concentration of 5 ppm since it poisons the iron catalyst. Therefore, it is converted into methane being inert for the applied catalyst. The pure synthesis gas is then fed to the synthesis loop. In general, non-reactive components lower the partial pressure of the reactants. Thus, inert gases like CH_4 and Ar are kept below 10–15% in the synthesis loop by applying purge streams.

Depending on the feedstock and the technology, different plant concepts are realized differing in synthesis gas production and synthesis loop design. Today's most efficient plants are operated on natural gas and reach energetic efficiencies of around $66\%_{LHV}$ [33]. The largest plant show production capacities of more than 3,000 t/d [17].

Electricity-based ammonia synthesis. The defossilization of ammonia production is enabled by "direct substitution" and "feedstock substitution," according to the abovementioned substitution options.

- "Feedstock substitution" is primarily considerable via electricity-based SNG, which can be utilized at conventional (existing) natural gas-converting ammonia plants. However, no decarbonization is achieved in this case, and renewable carbon is still required for intermediate SNG production. Furthermore, the enlarged process chain increases the energetic effort compared to "direct substitution".
- In direct electricity-based ammonia production (Power-to-Ammonia: $PtNH_3$), synthesis gas generation is carried out via air separation, electrochemical hydrogen production, and subsequent mixing. Therefore, process complexity is clearly reduced compared to conventional ammonia plants that commonly require various synthesis gas conditioning and cleaning steps. However, especially hydrogen

production is highly energy intensive. For nitrogen supply from air, the processes of cryogenic air separation (air separation unit: ASU), pressure swing adsorption (PSA), and membrane processes are mainly discussed. For large-scale production in particular, cryogenic air separation units offer advantages such as high nitrogen purity and lower specific energy consumptions [33–35]. Compared to the total energy demand, where H_2 production accounts for more than 90%, the share of N_2 supply is less than 2%. Therefore, the efficiency of the overall process is mainly determined by the electrolysis efficiency.

4.2 Methanol Production

Today methanol is almost exclusively produced via low-temperature methanol synthesis, which has been state-of-technology since the 1960s. As with ammonia synthesis, the synthesis gas is today mainly provided from natural gas or from hard coal. Unlike ammonia, the methanol molecule contains carbon and oxygen instead of nitrogen. As for all organic products, the inherent carbon demand means decarbonization is impossible.

The synthesis gas is a mixture of CO, CO_2, and H_2 in a composition slightly above the stoichiometric ratio. Methanol formation occurs primarily via the reactions represented by (2) and (3), which are coupled through the water–gas-shift reaction (4).

$$CO + 2H_2 \rightleftharpoons CH_3OH \mid \Delta H_R = -90.8 \text{ kJ/mol} \tag{2}$$

$$CO_2 + 3H_2 \rightleftharpoons CH_3OH + H_2O \mid \Delta H_R = -49.2 \text{ kJ/mol} \tag{3}$$

$$CO + H_2O \rightleftharpoons CO_2 + H_2 \mid \Delta H_R = -40.9 \text{ kJ/mol} \tag{4}$$

The equilibrium-limited methanol formation is thermodynamically favored by low temperatures and high pressures [32]. The operation pressure and temperature are 40–100 bar and 200–300 °C. The reaction is heterogeneously catalyzed with copper oxide (CuO)- and zinc oxide (ZnO)-based active sites. The synthesis enables selectivities above $99.9\%_C$. However, per-pass conversion is limited, lying close to the reaction equilibrium [20, 36, 37].

Fossil fuel-based methanol synthesis. In general, synthesis gas production for methanol synthesis involves process steps similar to fossil fuel-based ammonia production. However, since N_2 is not involved in the methanol formation reactions, the presence of N_2 within the synthesis gas should be avoided. Therefore, gasification of hard coal and reforming of natural gas are made with pure oxygen or water instead of air. Synthesis gas conditioning is commonly carried out via water–gas-shift reaction and CO_2 separation. The required synthesis gas cleaning depends mainly on the feedstock composition. In the end, a H_2 and CO-rich synthesis gas is fed into the methanol reactor. Per-pass conversion from 40 to 80% can be achieved, depending on the reactor concept and the applied operating conditions. Since carbon is incorporated in the product, less CO_2 is emitted by the overall production process. However, due to the unfavorable C-to-H ratio concerning the target product, hard coal-based production still produces high specific emissions especially during the synthesis gas provision step.

Electricity-based methanol synthesis. The defossilization of methanol production with electricity-based products can occur via feedstock or direct substitution.

- "Feedstock substitution" can be carried out in the same way as described above for ammonia using electricity-based SNG where large parts of the carbon are bound within the feedstock SNG can be found within the product methanol.
- "Direct substitution" with electricity-based synthesis gas requires, in addition to H_2 production, a supply of "green" CO_2. Since methanol synthesis enables the direct conversion of CO_2, an appropriate mixing of both gases is sufficient for synthesis gas production. However, a prior reverse water–gas shift reaction can be applied if a CO-rich synthesis gas for conventional synthesis is to be provided. In general, the CO_2-rich synthesis gas results in a lower per-pass conversion than the CO-rich synthesis gas [36]. In addition, water formation increases with CO_2 conversion, reducing the methanol concentration in the raw methanol and leading to higher catalyst deactivation [20]. Therefore, slightly adapted catalysts are used for CO_2-converting synthesis, but the operating conditions are within the range of conventional methanol synthesis.

Comparing both defossilization approaches, "direct substitution" without the intermediate formation of methane ("feedstock substitution") is less complex and more efficient due to the considerably shortened process chain. The utilization of "feedstock substitution" could, however, become applicable when existing natural gas infrastructures are converted to "green" SNG among others used by existing methanol production plants.

4.3 Olefin Production

Olefin production, especially primary chemicals like C_2- to C_4-olefins, is today dominated by steam cracking processes of various, but mainly crude oil-based, feedstocks. Steam cracking is a well-known high-temperature process to break down hydrocarbons by thermal decomposition into smaller, more valuable ones.

Typically, within an overall steam cracking process a hydrocarbon-steam mixture is preheated to a temperature of 500–650 °C before the mixture is fed into a fired reactor. Here, the mixture is heated to 750–875 °C within the reactor in a very short time period of 0.1–0.5 s; this results in the desired breaking of carbon–carbon bonds [38]. The initiation reaction exemplarily for ethane is described by (5). The resulting radicals can force various subsequent propagation and termination reaction steps (6)–(9), resulting in a partly very diverse product mixture, depending on the respective feedstock and the actual reaction conditions (e.g., residence time, partial pressure, temperature, catalytic effects). The amount of steam controls the partial pressure of the hydrocarbons in the mixture. A high steam content reduces the probability that larger chains are formed and thus increases the share of short-chain primary products [22]. Since steam cracking is strongly endothermic, the overall process requires significant amounts of heat; for example, ethane cracking is characterized by a thermal energy demand of around 4,900 kJ/kg. In today's steam crackers, this heat is commonly provided via combustion in (natural) gas-fired reactors. The product spectrum resulting from this cracking process is then fed to a cold section downstream of the reactor section, where the respective mixture is separated into various components or groups of components (i.e., single olefins and alkane by-products). The necessary rectification system is operated at low temperatures of around—100 °C and high pressures of 10–25 bar, which also significantly impacts the energy requirements of the entire plant [22, 39].

$$\text{Initiation } C_2H_6 \rightarrow CH_3^{\cdot} + CH_3^{\cdot} \tag{5}$$

$$\text{Propagation } C_2H_6 + CH_3^{\cdot} \rightarrow C_2H_5^{\cdot} + CH_4 \tag{6}$$

$$C_2H_5^{\cdot} \rightarrow C_2H_4 + H^{\cdot} \tag{7}$$

$$C_2H_6 + H^{\cdot} \rightarrow C_2H_5^{\cdot} + H_2. \tag{8}$$

$$\text{Termination } C_2H_5^{\cdot} + CH_3^{\cdot} \rightarrow C_3H_8 \tag{9}$$

Today's steam cracker plants utilize various types of feedstock requiring different upstream treatments described below. Subsequently, electricity-based olefin production via direct, feedstock substitution, and pathway substitution are further described.

Fossil fuel-based olefin production. Various feedstocks are available for fossil fuel-based olefin production. In general, all hydrocarbon feedstocks including a carbon–carbon bond are suitable. The feedstock utilized mainly depends on the local availability and the given product demand. For example, an important feedstock is naphtha being a hydrocarbon mixture in a boiling range from 30 to 200 °C [39]. Compared to the second most used feedstock, ethane, naphtha steam cracking leads to a higher yield of higher olefins and aromatics due to its longer average chain length. Naphtha steam crackers are particularly used in Europe, as the longer-chain olefins and aromatics are also very important for the local industry to produce specialty or fine chemicals. However, steam cracking of lighter gases, such as ethane and propane, from wet natural gas or refinery off-gases and liquefied petroleum gas (LPG) from crude oil refineries are also used. Since worldwide a rising proportion of ethane steam crackers is in operation based on wet natural gas processing, a high share of ethylene is produced as the end product. Thus, the deficiency of propylene has to be overcome by more propylene-selective production routes. The ethylene yield varies between 25 and 53 wt.% and is mainly dependent on the feedstock. Common feedstock conversions start at 65 wt.% and can reach almost complete conversion [22].

Propane dehydrogenation, metathesis reaction of ethylene and 2-butenes, and the Methanol-to-Olefin (MtO) production can achieve selective propylene production. The first two alternatives can utilize a crude oil refining- or steam cracker-based feedstock. MtO is today only applied to a significant amount in China in coal-to-olefin (CtO) plants (approx. 30 Mt_{MeOH}/a [21]). Since methanol can also be derived based on renewable electricity (Sect. 4.2), it is discussed as an alternative pathway in electricity-based olefin production.

Electricity-based olefin production. The electricity-based production can predominantly be achieved through Fischer–Tropsch (FT) synthesis and methanol synthesis. While methanol-based olefin production is only possible via MtO processes, FT-based olefin production can be carried out on different pathways. Unlike methanol synthesis, FT synthesis does not yield a single product but a product spectrum containing various hydrocarbons and oxygenates with different chain lengths representing a kind of synthetic crude oil.

FT technologies are typically distinguished between high-temperature Fischer–Tropsch synthesis (HTFT) and low-temperature Fischer–Tropsch synthesis (LTFT).

- HTFT operates at reaction temperatures of 300–400 °C using iron catalysts to primarily produce short-chain hydrocarbons with relatively high levels of unsaturated components.
- LTFT predominantly generates long-chain n-paraffins and is operated at temperatures of 200–300 °C using cobalt or iron catalysts [36, 40].

Another difference to methanol synthesis is that especially for LTFT, direct CO_2 conversion is state-of-research and not commercially available. Therefore, a prior CO_2 reduction via a reverse water–gas shift (RWGS) reaction is required within a respective reactor (10) [41, 42]. The primary FT reaction (11) is strongly exothermic and volume-decreasing.

$$CO_2 + H_2 \rightleftharpoons CO + H_2O \mid \Delta H_R = 42 \text{ kJ/mol} \tag{10}$$

$$n\,CO + (2n+1)H_2 \rightarrow C_nH_{2n+2} + nH_2O \mid \Delta H_R \approx n(-150 \text{ kJ/mol}) \tag{11}$$

FT synthesis has been commercially used for several decades to produce liquid fuels from natural gas and hard coal, making it a state-of-the-art technology. Additionally, the diverse product spectra offer various options for the production or co-production of chemicals [40]. However, defossilization via "direct substitution" with FT-based olefins (Fig. 8, dotted line) is only possible to a minor extent since, even in HTFT, the C_2 to C_4 olefin fraction is typically below 25 wt.% [40]. Furthermore, the subsequent separation is challenging mainly due the fact that many components with very similar boiling ranges are produced. Besides rectification, the required separation steps can be among others etherification with methanol or removal of oxygenates by polar extraction [40]. Today, the technologies for producing linear olefins from FT syncrude are niche applications only applied for C_3 and higher olefins. Extraction processes have been developed for applications based on HTFT syncrude rather than LTFT.

A more yielding defossilization pathway via electricity-based FT syncrude can be carried out via "feedstock substitution" since FT syncrude can be refined similarly to fossil crude to produce different fuel fractions. The resulting naphtha fraction can also be fed to a steam cracker and thus be defossilized via "feedstock substitution." However, like in fossil naphtha cracking, FT naphtha cracking is constrained in the manageability of the product spectrum and the associated by-product formation, such as methane, aromatics, and hydrogen.

Another potentially more selective defossilization pathway is via MtO processing; i.e., the conversion of electricity-derived methanol via dehydration into short-chain olefins. Olefin generation from methanol is today only carried out in a significant amount in China based on hard coal-derived methanol and is here therefore considered as "pathway substitution" (instead of "feedstock substitution"). The process allows a combined ethylene and propylene production with a total olefin selectivity of up to 90% [43, 44]. The ratio of propylene to ethylene can be varied depending on the operating conditions and the catalyst used; furthermore, almost pure propylene production is achieved in so-called Methanol-to-Propylene (MtP) processes. Catalysts, like protonated ZSM-5 and SAPO-34, are commonly used, with SAPO-34 showing higher selectivity for ethylene and propylene but requiring more extensive regeneration due to coke formation [37, 45]. The generic olefin formation reaction (12) allows a maximum yield of 44 wt.%. The reaction is exothermic and volume-increasing. Therefore, low operating pressures (<5 bar) are commonly applied. The reaction temperature typically lies between 350 and 500 °C to overcome the required activation energy and to reduce the production of longer olefins.

$$n\text{CH}_3\text{OH} \rightarrow \text{C}_n\text{H}_{2n} + n\text{H}_2\text{O} \quad \text{with } 2 \leq n \leq 6 \tag{12}$$

Besides the pathways described above, the direct synthesis of olefins from CO or CO_2-based synthesis gas represents another promising production route. However, the synthesis still suffers from low selectivity due to high CO_2 formation and is still at the research stage [46, 47]. The same applies to the oxidative coupling of methane (OCM) [48], and the direct electrochemical conversion of CO_2 and H_2O especially to ethylene [49].

4.4 Aromatics Production

As in olefin production, aromatics can be produced, among other pathways, by steam cracking, mainly based on naphtha (Sect. 4.3). Aromatics yields of 10–15 wt.% are achieved in naphtha steam cracking [22]. Again, this naphtha can be supplied either from fossil hydrocarbons or electricity-based.

Irrespective of the feedstock basis, the naphtha can also be reformed in a catalytic process. Catalytic reforming is already established in typical crude oil refineries and is part of the refining, along with fractionation and further treatment steps. This reforming step leads to a so-called reformate. Reformate is a mixture of branched alkanes and aromatics. This chemical mixture is today mainly used as a blending component to increase the octane number of gasolines. However, the BTX aromatics are also separated from the reformate, on the one hand, to comply with fuel specifications and, on the other hand, to make them available to the chemical sector as high-quality primary chemicals.

In catalytic reforming, hydrocarbons are catalytically aromatized and isomerized at temperatures of around 500 °C and elevated pressures of 7–25 bar. A lower pressure—respectively, a lower hydrogen partial pressure, since the reaction occurs within a hydrogen atmosphere—is applied if the regeneration occurs continuously in the realized reactor concept. This is because increased coke formation on the catalyst occurs during the rearrangement of hydrocarbon components or ring closure, which can be minimized by applying a higher hydrogen partial pressure. These deposits can be removed from the catalyst based on a continuous regeneration concept by burning off the carbon from the solid catalyst particles. Hydrogen is one of the by-products of catalytic reforming and can also be recovered.

The resulting product mixture can then be further separated by distillation or extraction to supply pure aromatics. Depending on the naphtha used, its composition and the reforming intensity, an aromatics content of about 65–82 wt.% is achieved. In addition, there are fractions of saturated or unsaturated linear or branched hydrocarbons [27].

Fossil fuel-based aromatic production. Since the concentration of BTX aromatics in most crude oils is relatively low and boiling points are close to other crude oil components, direct separation of BTX aromatics is technically challenging. However, the separation can also occur in higher proportions after the catalytic reforming, where the concentration has already increased. If fossil naphtha is used, the naphtha can also be further separated into a light and a heavy fraction prior to reforming to provide a more specific feed mixture for reforming and simplify the subsequent product separation of the various aromatics. Benzene can then be separated from the light naphtha reforming, whereas the C_8 aromatics such as xylenes in particular are separated from the heavy naphtha reformate.

In addition to production from petroleum-based derivatives, an aromatics stream can also be obtained from the light oil of low-temperature or high-temperature carbonization of coal. In this process, up to 94 vol.% aromatics might be contained within the light oil to be collected within a recovery section. These crude aromatics must then be further purified, as the product still might contain too light components, thiophenes and other sulfur compounds. For this purpose, a hydrotreatment step is carried out, which can also incorporate a hydrodealkylation.

Benzene can also be selectively produced from pure toluene via hydrodealkylation or disproportion. In hydrodealkylation, the methyl group is separated from the benzene ring by utilizing hydrogen. This catalytic, exothermic reaction takes place at temperatures of 575–650 °C and pressures between 25 and 60 bar. In the disproportionation reaction, a rearrangement of the methyl group occurs within two toluene molecules. In addition to benzenes, another main product, xylene, is formed. This reaction is also carried out in a hydrogen atmosphere without active consumption. The reaction conditions of the exothermic reaction are milder (350–525 °C and 10–50 bar) [27].

$$\text{Hydrodealkylation} \quad C_6H_5CH_3 + H_2 \rightarrow C_6H_6 + CH_4 \tag{13}$$

$$\text{Disproportionation} \quad 2C_6H_5CH_3 \rightarrow C_6H_6 + C_6H_5(CH_3)_2 \tag{14}$$

Electricity-based aromatic production. The defossilization options for the production of BTX aromatics are similar to that of olefin defossilization. The main process routes start from electricity-based FT or methanol synthesis. However, compared to the "direct substitution" of olefins, the straight-run concentrations of aromatics achievable through FT synthesis are significantly lower and, even on the basis of a HTFT process, typically below 2 wt.% [40]. Due to this low proportion of aromatics within the direct process route, "feedstock substitution" of naphtha can be carried out. In this process, the FT naphtha can also be processed equivalent to the crude oil-based naphtha via a catalytic reforming step explicitly to aromatics or via steam cracking to a mixture of olefins and aromatics. Due to the high share of paraffinic components within the FT naphtha in contrast to crude oil-based naphtha, the feedstock is not ideally usable for catalytic reforming. This means that the reformer must establish severe process conditions to achieve the isomerization and cyclization reactions, which in turn leads to efficiency losses [40].

Instead of more or less conventional production pathways via FT products, the conversion of electricity-derived methanol into aromatics (Methanol-to-Aromatics: MtA) represents an alternative production pathway so far not fully state-of-technology; i.e., most MtA projects are still in the laboratory and an early demonstration scale [30]. The technology is based on the Methanol-to-Gasoline (MtG) concept able to produce an aromatic yield of 30–40 wt.% [30, 37]. However, the selectivity has to be increased for on-purpose production. In MtA, bifunctional catalysts, typically Zn- and Ga-modified ZSM-5 zeolites, are applied. The acidic zeolite forces methanol conversion to olefins, while incorporated metal promotes dehydrogenation functionality [30]. The reaction occurs at temperatures above 400 °C and is favored by high pressures. Major challenges for an industrial application are the low single-pass selectivity (<80 wt.%) and the short and unstable operation due to fast coke formation. To further extend aromatic yields also, two reactor plant concepts are developed. Based on this approach, the first commercial plant was launched in 2013 in China with a production capacity of 30,000 t/a [30].

In addition to the pathways described above, the direct synthesis of aromatics from CO or CO_2-based synthesis gas might be a more direct and efficient production pathway. However, the processes are still under development, and the commercial utilization of the technology is therefore not expected in the near future [50, 51].

5 Systemic Placement

The following section discusses the integration of PtX pathways in the chemical industry under broader system aspects. It focuses on potential competitions and synergy effects with requirements of GHG-neutral PtX fuels for aviation and shipping.

To effectively defossilize global economies, a substantial extension of renewable energy facilities is essential, primarily photovoltaic systems and wind turbines. The potential for generating renewable electricity from solar radiation and wind energy is significantly higher than from hydropower, biomass, and geothermal energy. Additionally, at optimal locations, electricity generation from solar and wind is more cost-effective than other options like biomass. Also, unlike biomass, there is no competition with food, fodder, or raw material markets, which underlines the advantages of electricity-based defossilization strategies by PtX applications.

Despite their considerable potential, the temporal and economic efforts required to harness these renewable energy sources for electricity provision, such as building photovoltaic systems, wind power plants, and their related infrastructure, might mean that in the short to medium term, the overall demand for renewable electricity might not be met completely. Therefore, renewable electricity should be used in the most efficient way and associated with the highest possible GHG savings. Given the disadvantages in energy efficiency of PtX compared to other defossilization options, priority should be placed on demand-side transformation wherever technically feasible (e.g., decarbonization in road transportation through direct electrification). In contrast, in sectors where materials or energy carriers based on carbon are indispensable in the foreseeable future—so-called hard-to-abate sectors—a "green" transformation must focus on the production-side. Primary chemical production and fuel supply in aviation and shipping are currently the largest hard-to-abate sectors. For this reason, PtX is one of the few applicable defossilization options alongside the already limited biomass utilization and, in case of the chemical industry, direct material recycling.

The defossilization of primary chemical and fuel production in these three sectors through PtX largely includes same technologies and intermediate products, like hydrogen, methanol, and Fischer–Tropsch crude. Consequently, scaling up the necessary infrastructure and production processes would simultaneously benefit all three sectors. Therefore, PtX developments in these different sectors should not be considered separately. Synergies could be leveraged, and challenges reduced if the PtX

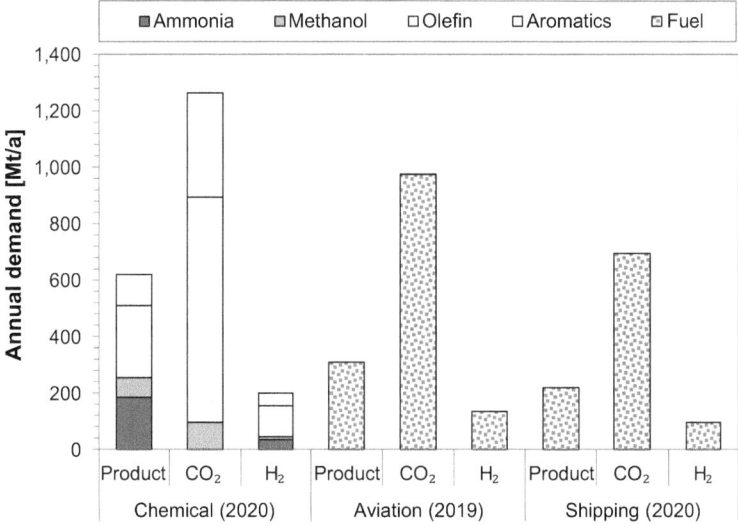

Fig. 9 Demand of PtX products and feedstock by sector assuming complete defossilization via PtX technologies (product data according to [2, 52–56];[4] methanol demand reduced by methanol demand for olefin production)

requirements of these sectors are viewed and aligned from an overarching, holistic system perspective.

Current demands. Alongside the chemical industry, PtX technologies are seen as pivotal options for a defossilization in aviation and shipping. Figure 9 shows the current chemical and fuel demands as well as a basic estimation of the resulting CO_2 and H_2 amounts, for a hypothetical complete defossilization of primary chemical or fuel production through PtX.[5] The current global demand for primary chemicals (600 Mt/a in2020 [5]) exceeds the demand for fuels in aviation (310 Mt/a in 2019 [52]) and shipping (220 Mt/a in 2020 [53, 54]). The demands of the individual sectors are so immense that, even if they are only partially covered by PtX, would require a large-scale expansion of PtX technology in every sector. Synergies may arise if the financial burdens of technology rollout (mainly resulting from uncertainties and increased expenditures for first-of-its-kind plants) could be shared across these

---

[4] The year 2019 has been designated as the reference year for aviation, as in 2020, unlike in the chemistry and shipping sectors, significant impacts of the COVID-19 pandemic were observed.

[5] The values solely relate to material requirements to produce primary chemicals based on stoichiometric calculations. For the H_2 and CO_2 demands in aviation and shipping, it is assumed that only liquid hydrocarbon fuels are used (–CH_2– units); it is also based on stoichiometric calculations. Thus, all values represent minimum estimates based on current demands. However, the actual demands are expected to be in the same order of magnitude since most processes show high selectivities or the potential to recycle by-products within the respective production concept (Sect. 4).

sectors. However, splitting the costs between the sectors requires a parallel ramp-up of PtX in the respective sectors.

Demand developments. All three sectors are expected to see considerable growth in the decades to come. The chemical industry anticipates an ongoing significant growth (compound annual growth rate [CAGR] of 7%/a from 2023 to 2030 [57]) [2]. Regarding material demands, there is no decoupling from this expected growth, as the needed materials will increase with production increases. Also in the aviation sector a CAGR of roughly 3–5%/a [58] is expected until 2050. The growth in international shipping demand is anticipated to increase maritime fuel consumption by 5% until 2030 and 43% until 2050 without proper countermeasures [59]. In aviation and maritime, specific fuel demands can be further reduced through technological and operational improvements. Nevertheless, such advancements are most likely insufficient to fully counterbalance the growth, which would result in rising fuel demands in the sectors.

The growth of these sectors is coupled with a rise in the already high demand for defossilization, and possibly the demand in PtX products. Since producing PtX products is expected to be significantly more expensive than producing fossil fuels in the medium term, increasing shares of PtX products may also result in increasing average costs for products and services in these sectors. This could also have an effect on the forecasted growth. Considering the significant needs for defossilization and the extremely limited time for transformation, it is uncertain whether the projected growth in these sectors can align with achieving climate goals.

Availability of renewable electricity. The production of sustainable PtX products requires large amounts of renewable electricity. From a technical perspective, the global potential to produce renewable electricity (primarily from solar radiation and wind energy) is virtually unlimited. Technical potentials are estimated in the range of up to 10,000 EJ_{el}/a (2,800 PWh_{el}/a) and more [60]. The electrical energy needed to produce the total H_2 demands depicted in Fig. 9, which forms the majority of the PtX energy demands (typically more than 90%, unless CO_2 needs to be provided through DAC units), is approximately 23 PWh. Thus, in principle, there is no technical limitation in meeting current and future PtX demands due to the absolute energy potential available despite the additional energy requirements from other sectors. The main limitations so far arise from the challenge to ensure that the required energy can be made available sufficiently fast under existing framework conditions (see ramp-up aspects below). As of 2021, renewable energy installations worldwide have reached a total capacity of 3.3 TW_{el} [61]. In 2020, approximately 7.5 PWh of renewable electricity were generated worldwide from the installed facilities (3.0 TW_{el}). This total amount of energy currently used in all sectors corresponds to the amount required solely to defossilize aviation using PtX.

Material feedstock availability. The material feedstocks for chemical and fuel production via PtX are H_2O and either CO_2 or N_2. From a global perspective, H_2O is not considered a limiting resource, as seawater desalination can supply deionized water with energy requirements well below 1% of the specific energy needed for water electrolysis. Similarly, N_2, constituting 78vol.% of ambient air, can be provided, among others, by cryogenic air separation or pressure swing adsorption, both of which are state-of-the-art technologies. Similar to seawater desalination, the energy demand for N_2 provision is significantly lower than that required for H_2 production (<5% based on the H_2 demand for ammonia production). The main constraints regarding material feedstock availability for PtX largely pertain to carbon. To defossilize carbon-based energy carriers and chemicals through PtX, only (sustainable) CO_2 can be used, which, when released back into the atmosphere, does not contribute to anthropogenic CO_2 level increase. This primarily includes biogenic CO_2 and CO_2 captured from the atmosphere. The latter is possible through DAC technologies. However, due to the low concentration of CO_2 in ambient air (ca. 420 ppm), DAC entails (very) high specific energy, construction, and area demands, which consequently leads to high economic efforts.

Biogenic CO_2 point sources mainly exist in conversion/treatment processes where biomass is utilized either materially (e.g., biofuel production) or energetically (e.g., combustion of solid biofuels). In a particularly pure and more or less readily usable form, CO_2 is generated within bioethanol factories and biomethane plants. Today, those sustainable or "green" CO_2 sources account for about 87 Mt_{CO2}/a worldwide [56]. However, further substantial amounts of CO_2 can be made available from existing biomass use, with comparatively little effort. In the EU alone, an additional potential to supply CO_2 from biomass combustion and (currently) unprocessed biogas is estimated to about 440 Mt_{CO2}/a [62]. In contrast to these numbers, the combined CO_2 demand of the considered hard-to-abate sectors, if completely defossilized via PtX, adds up to about 3,000 Mt_{CO2}/a. Even if significant amounts of non-fossil CO_2 would be provided from such biogenic sources that are not yet tapped, the challenge of supplying sufficient CO_2 even increases if CO_2 must be supplied from diverse locations (increasing decentralization) and if CO_2 sources with rather high CO_2 concentrations ("low hanging fruits") decrease. Thus, ensuring a sufficient and cost-effective supply of sustainable CO_2 could be (severely) constrained in the short and medium term, especially as suitable centralized CO_2 sources gradually deplete and the supply of CO_2 increasingly shifts to DAC, which is still in its early stages of technological development.

Ramp-up of renewable electricity generation. Most of the world's committed GHG reduction targets require complete defossilization of all sectors by 2045 or by 2060 at the latest. To achieve this, the production and utilization of renewable, electricity-based fuels and chemicals must continuously increase. This necessitates expanding infrastructure and facilities, especially for providing renewable electricity and non-fossil CO_2. Furthermore, there must also be a scale-up or establishment of PtX plants on a global level. In this context, the development of a significant electrolyzer capacity is particularly important, as large-scale manufacturing in this area is still evolving. The annual increase in renewable electricity net capacity—i.e., referring to the actual amount of electricity generated—is projected to rise from about 290 GW_{el} in 2021 to between 450 and 600 GW_{el} by 2027 (90% of this growth is attributed to additional capacities in wind and solar power) [61]. Achieving full decarbonization in primary chemical and fuel production for aviation and shipping sectors would necessitate an addition of more than 100 GW_{el} per year in net capacity. However, this requirement is only part of a broader need for substantial amounts of renewable electricity for defossilization across all economic sectors.

Ramp up of electrolyzer capacity. As of today, the global H_2 production amounts to around 120 Mt/a, with only about 1% derived from renewable sources [63, 64]. The cross-sectoral H_2 demand shown in Fig. 9 exceeds the current total H_2 production by a factor of more than 3 alone (430 Mt_{H2}/a in relation to 120 Mt_{H2}/a). The electrolyzer capacity needed for this cross-sectoral H_2 production amounts to around 5 TW_{el}.[6] Considering the currently installed electrolyzer capacity of about 0.7 GW_{el} in 2021 [64], such an expansion into the terawatt-scale appears quite ambitious. However, electrolyzer manufacturing capacities are currently rising. While in 2022, global electrolyzer manufacturing capacities only reached about 10.3 GW_{el}/a, the announced capacities exceed 35 GW_{el}/a for 2024 and 134 GW_{el}/a for 2030. Again, such production capacities might not be fully dedicated to PtX chemicals and aviation and shipping fuels since a demand for H_2 will likely increase in other sectors as well (e.g., electricity sector, heat sector). Therefore, particularly in the short and medium term, the necessary expansion, especially of renewable power sources and electrolysis capacity, could lag behind the demand for defossilization. Such a gap may also result in more competition among PtX applications, which could cause a use of PtX products primarily in sectors with specific incentives for adopting these higher-priced products.

PtX incentives. In the short and medium term, PtX products are likely to be limited and more expensive than fossil fuel- or biomass-based products. The sector in which PtX products are adopted will largely depend on the willingness of users to pay. This, in turn, is also influenced by incentives and regulations. In the EU, existing regulations predominantly concentrate on the direct, combustion-related release of CO_2 (e.g., EU RED II [65] or EU ETS [66], ReFuelEU Aviation), which currently places greater regulatory pressure on a defossilization of carbon-containing energy usage than on carbon-containing material production. Thus, from a regulatory perspective, a near term use of PtX products as fuels in aviation or shipping is more likely than their integration in the chemical industry. However, given that future regulations might include the material use of fossil fuel-based products or at least affect the respecting downstream sectors (e.g., waste incineration), it is also likely that the production of chemicals will be supported by regulation to be defossilized in the years to come.

Learning curve effects. As PtX technologies are currently employed on a small and semi-commercial scale only, cost reductions are expected to occur as technology development and deployment progress. However, certain key technologies of the PtX process chains have already undergone technology learning curves in the past, such as renewable electricity generation from solar radiation or wind energy. Hence,

---

[6] Assuming an average electrolyzer utilization of 4000 full load hours per year.

further cost reductions in these areas are expected only to a lesser extent. Moreover, the downstream synthesis and conversions in PtX pathways are comprised of thermochemical processes. Such processes have already been extensively employed at commercial scale in the chemical industry for decades, which also reduces the likelihood of substantial further cost reductions for these technologies. The most substantial cost reductions can most likely be expected for electrolyzers, driven by both further technology learning and mass-market electrolyzer production. However, a significant portion of PtX production costs is determined by electricity costs. Consequently, while electrolyzer costs remain a notable investment share of a PtX facility, further cost reduction may have limited additional impact on total PtX production costs as technology advances.

Additional options. Besides the utilization of sustainable biomass and PtX technologies, material recycling of chemical and final products can lower the need to supply additional carbon in the chemical industry (and therefor indirectly contribute to its defossilization). The direct (mechanical or chemical) recycling of predominantly plastics can significantly reduce the demand for primary chemicals, especially methanol, olefins, and aromatics, which are currently used in large quantities for plastic production. Since the chemical structure of materials remains largely unchanged in mechanical and chemical recycling,[7] these processes are significantly more energy-efficient, especially compared to electricity-based production of "new" molecules via PtX. However, both mechanical and chemical recycling preferably require a complete and pure return of end products from their end-use. This is particularly limited by the often complex material combinations in end products and by current limitations in waste separation and improper disposal practices.

Where direct recycling is not possible, the chemical industry also has the option to use CO_2 emitted by incineration of the final products, mainly generated centrally by energetic waste utilization, to expand the available CO_2 sources for PtX. This also results in a closed carbon cycle. However, it should be noted that this "indirect" material recycling is considerably more complex and energy-intensive than direct recycling. As PtX products would be used as fuels in aviation and shipping (i.e., the carbon is released as CO_2 into the atmosphere during combustion), recycling or carbon capture[8] is not applicable to close carbon cycles directly. Therefore, additional defossilization measures in aviation and shipping, beyond PtX (or biomass-based fuels), necessitate demand-side transformations, such electrification of propulsion systems or the use of carbon-free fuels like H_2. However, from a technical and infrastructure perspective, implementing such technologies at scale is only expected in the longer term.

Multiple product plants. Since the production of electricity-based chemicals and fuels often entails similar processes and intermediates, individual facilities could potentially serve both the chemical sector and the demands of aviation and shipping.

---

[7] In chemical recycling, typically chemical or thermal depolymerization occurs, with the respective monomers remaining unchanged.

[8] The use of carbon capture on large ships is not entirely ruled out.

Furthermore, the parallel production of different chemicals and fuels in more complex facility networks is possible, similar to today's crude oil refineries. PtX refineries, using processes like Fischer–Tropsch or methanol synthesis, could produce both fuel components and chemicals through various downstream processes, with the flexibility to adjust product distributions. Thus, beyond technical synergies, such concepts could offer advantages by enabling demand-optimized production depending on specific and variable market requirements.

Overall, the chemical industry, aviation, or shipping alone probably already exhibit a potential demand for PtX products that is high enough to roll out a global large-scale PtX infrastructure in the coming decades. Nonetheless, an early cross-sectoral adoption of PtX could help to reduce risks and costs associated with PtX technology deployment. This means that infrastructure and production facilities could be efficiently shared, serving various sectors by large-scale PtX refineries. However, adverse consequences may still arise as high cross-sectoral demands could occur and result in PtX product shortages. This could particularly be the case during ramp-up phases when renewable energy, "green" CO_2, and electrolyzer capacities are likely to be limited. In such cases, the use and early integration of PtX products would probably shift to the sectors with the greatest willingness of end consumers to pay for PtX products or sectors with the most effective (regulatory) incentives.

6 Summary

The chemical industry currently relies predominantly on fossil fuel-based feedstocks. These feedstocks serve dual purposes: They are used as atomic/molecular building blocks to produce chemical products (material use) or to generate process energy (energetic use). The energetic use results in the direct release of carbon dioxide (CO_2) emissions, whereas the material use leads to an indirect release of CO_2 emissions after the product's lifecycle (typically by incineration).

Applying PtX for chemical production (PtC) offers a promising opportunity especially to defossilize the material use of fossil fuel-based feedstocks. Thereby, all primary chemicals can be produced based on electricity-derived synthesis gas through different synthesis and subsequent conversion technologies being largely state-of-the-art. Ammonia can be directly produced via PtC by the electricity-based provision of N_2 and H_2, employing the Haber–Bosch synthesis also used in conventional (fossil fuel-based) production. The substitution of fossil fuel-based methanol can also be accomplished directly by synthesizing methanol from electricity-based synthesis gas. Based on CO_2 and H_2, in this respect the direct conversion of CO_2-rich synthesis gas appears advantageous over CO-rich synthesis gas. Beside "direct substitution" via PtC, ammonia and methanol can additionally be defossilized through the use of electricity-based methane ("feedstock substitution"), although "direct substitution" is technologically more efficient. However, considering the long operational lifespan of chemical production plants, which can range up to 50 years (with an average of 25 years worldwide and 40 years in Europe) [15], "feedstock substitution" could also be relevant in the medium term.

Currently, olefins are primarily produced through the cracking of naphtha separated from fossil crude oil. Defossilization via PtX can be achieved through "feedstock substitution" with synthetic naphtha produced through high-temperature Fischer–Tropsch (HTFT) synthesis. Small olefin amounts are also directly generated in the FT product, which can be utilized for "direct substitution." However, due to the wide product distribution, the olefin content for a targeted olefin production is generally insufficient. A more selective and demand-specific production of ethylene, propylene, and higher olefins can be achieved through Methanol-to-Olefin (MtO) processes using electricity-based methanol ("pathway substitution"). BTX aromatics, which are currently obtained through naphtha reforming or cracking, can also be defossilized through similar substitution approaches based on synthetic naphtha. A more selective production of BTX aromatics is approached via Methanol-to-Aromatics (MtA) processes; however, these processes are currently between demonstration and large-scale implementation.

The main technologies and pathways to produce primary chemicals via PtX, along with intermediates or end products, are basically identical to a fuel production based on PtX routes. Hence, regarding technology rollout and feedstock supply, PtC in the chemical industry encounters similar challenges as PtX-based fuel production in aviation and shipping. Consequently, synergies could emerge if a holistic cross-sectoral approach to scale-up PtX is pursued across these sectors, such as jointly developing infrastructure and production plants to facilitate a rapid market ramp-up. Nevertheless, adverse effects could also arise, as all three sectors may compete for the limited PtX products, particularly during ramp-up phases. This underscores the importance of advancing a collaborative cross-sectoral PtX deployment approach.

References

1. Levi PG, Cullen JM (2018) Mapping global flows of chemicals: from fossil fuel feedstocks to chemical products
2. IEA (2018) International energy agency. The future of petrochemicals. OECD
3. Quante G, Bullerdiek N, Bube S, Neuling U, Kaltschmitt M (2023) Renewable fuel options for aviation: a system-wide comparison of drop-in and non drop-in fuel options. Fuel 333:126269. https://doi.org/10.1016/j.fuel.2022.126269
4. Kleinschmitt C, Fragoso García J, Franke K, Teza D, Seidel L, Ebner A et al (2022) Global potential of renewable energy sources. Fraunhofer ISI
5. Pales AF, Levi P (2018) The future of petrochemicals: towards more sustainable plastics and fertilisers
6. IEA—International Energy Agency (2022) Coal 2022: analysis and forecast to 2025
7. IEA, ICCA, Dechema (2013) Technology roadmap: energy and GHG reductions in the chemical industry via catalytic processes. Paris
8. The European Chemical Industry Council (2023) Energy consumption in the EU27 chemical industry. Data from Eurostat. https://cefic.org/a-pillar-of-the-european-economy/facts-and-fig ures-of-the-european-chemical-industry/energy-consumption/

9. IEA—International Energy Agency (2023) Direct CO_2 emissions increase, technological innovation is needed to get on track to Net Zero. https://www.iea.org/fuels-and-technologies/chemicals

10. Sattler K (2012) Thermische trennverfahren: grundlagen, auslegung, apparate, 3rd edn. Wiley, Weinheim

11. Dr. Cordin Arpagaus (2018) Hochtemperatur Wärmepumpen: literaturstudie zum Stand der Technik, der Forschung, des Anwendungspotenzials und der Kältemittel

12. Danish Energy Agency and Energinet (2020) Technology data—industrial process heat

13. Chemical Market Forecast (2023) Bulk chemicals. https://chemicalmarketforecast.com/bulk-chemicals/

14. Ammonia AM (2000) 1. Introduction. In: Ullmann's encyclopedia of industrial chemistry. Wiley, p 184

15. IEA—International Energy Agency (2021) Ammonia technology roadmap, 4th edn. Paris

16. Royal Society (2020) Ammonia: zero-carbon fertiliser, fuel and energy store policy briefing. Royal Society, London

17. International Renewable Energy Agency and Ammonia Energy Association (2022) Innovation outlook: renewable ammonia. Abu Dhabi Brooklyn

18. Dolan RH, Anderson JE, Wallington TJ (2021) Outlook for ammonia as a sustainable transportation fuel. Sustain Energy Fuels 5(19):4830–4841

19. TechSciResearch (2023) Global ammonia market 2012–2026. https://refindustry.com/news/market-research/global-ammonia-market-2012-2026/

20. KGaA (2012) Ullmann's encyclopedia of industrial chemistry: methanol. Wiley, Weinheim

21. International Renewable Energy Agency (IRENA) and the Methanol Institute (2021) Innovation outlook: renewable methanol. Abu Dhabi

22. Zimmermann H, Walzl R (2012) Ethylene. In: Ullmann's encyclopedia of industrial chemistry: methanol. Wiley, Weinheim

23. Zimmermann H (2000) Propene. In: Ullmann's encyclopedia of industrial chemistry. Wiley

24. Dahlmann M, Grub J, Löser E (2012) Butadiene. Ullmann's encyclopedia of industrial chemistry: methanol. Wiley, Weinheim, pp 1–24

25. CHEMANALYST (2023) Decode the future of ethylene. https://www.chemanalyst.com/industry-report/ethylene-market-638

26. GlobalData (2023) Butadiene industry installed capacity. https://www.globaldata.com/store/report/butadiene-market-analysis/

27. Folkins HO (2000) Benzene. In: Ullmann's encyclopedia of industrial chemistry. Wiley, p 291

28. Ziegler-Sylakakis K, Fabri J, Graeser U, Simo TA (2012) Toluene. Ullmann's encyclopedia of industrial chemistry: methanol. Wiley, Weinheim, pp 1–12

29. Ziegler-Skylakakis K, Fabri J, Graeser U, Simo TA (2012) Xylenes. Ullmann's encyclopedia of industrial chemistry: methanol. Wiley, Weinheim, pp 1–20

30. Li T, Shoinkhorova T, Gascon J, Ruiz-Martínez J (2021) Aromatics production via methanol-mediated transformation routes. ACS Catal 11(13):7780–7819. https://doi.org/10.1021/acscatal.1c01422

31. Ammonia AM (2012) 2. Production processes. In: Ullmann's encyclopedia of industrial chemistry: methanol. Wiley, Weinheim

32. De Heer J (1957) The principle of Le Châtelier and Braun: University of Colorado. J Chem Educ 1957(34):375–380

33. Morgan ER (2020) Techno-economic feasibility study of ammonia plants powered by offshore wind [Dissertation]. University of Massachusetts Amherst

34. Böcker N, Grahl M, Tota A, Häussinger P, Leitgeb P, Schmücker B (2000) Nitrogen. In: Ullmann's encyclopedia of industrial chemistry. Wiley, pp 1–27

35. Rouwenhorst KHR, Krzywda PM, Benes NE, Mul G, Lefferts L (2000) Ammonia, 4. Green ammonia production. In: Ullmann's encyclopedia of industrial chemistry. Wiley, pp 1–20

36. Dieterich V, Buttler A, Hanel A, Spliethoff H, Fendt S (2020) Power-to-liquid via synthesis of methanol, DME or Fischer–Tropsch-fuels: a review. Energy Environ Sci 13(10):3207–3252. https://doi.org/10.1039/D0EE01187H

37. Bertau M, Offermanns H, Plass L, Schmidt F, Wernicke H-J (2014) Methanol: the basic chemical and energy feedstock of the future. Springer, Berlin
38. Gholami Z, Gholami F, Tišler Z, Vakili M (2021) A review on the production of light olefins using steam cracking of hydrocarbons. Energies 14(23):8190. https://doi.org/10.3390/en1423 8190
39. Ren T, Patel M, Blok K (2006) Olefins from conventional and heavy feedstocks: energy use in steam cracking and alternative processes. Energy 31(4):425–451. https://doi.org/10.1016/j. energy.2005.04.001
40. de Klerk (2011) Fischer-tropsch refining. [Place of publication not identified]
41. Rezaei E, Dzuryk S (2019) Techno-economic comparison of reverse water gas shift reaction to steam and dry methane reforming reactions for syngas production. Chem Eng Res Des 144:354–369. https://doi.org/10.1016/j.cherd.2019.02.005
42. Wolf A, Jess A, Kern C (2016) Syngas production via reverse water-gas shift reaction over a Ni-Al$_2$O$_3$ catalyst: catalyst stability, reaction kinetics, and modeling. Chem Eng Technol 39(6):1040–1048. https://doi.org/10.1002/ceat.201500548
43. Gogate MR (2019) Methanol-to-olefins process technology: current status and future prospects. Petrol Sci Technol 37(5):559–565. https://doi.org/10.1080/10916466.2018.1555589
44. Hongxing L, Zaiku X, Guoliang Z (2013) The progress of SINOPEC methanol-to-olefins (S-MTO) technology: Shanghai research institute of petrochemical technology SINOPEC, China, Dresden
45. Kianfar E (2019) Comparison and assessment of zeolite catalysts performance dimethyl ether and light olefins production through methanol: a review. Rev Inorg Chem 39(3):157–177. https://doi.org/10.1515/revic-2019-0001
46. Wang S, Wang P, Shi D, He S, Zhang L, Yan W et al (2020) Direct conversion of syngas into light olefins with low CO$_2$ emission. ACS Catal 10(3):2046–2059. https://doi.org/10.1021/acs catal.9b04629
47. Zhao S, Li H, Wang B, Yang X, Peng Y, Du H et al (2022) Recent advances on syngas conversion targeting light olefins. Fuel 321:124124. https://doi.org/10.1016/j.fuel.2022.124124
48. Gambo Y, Jalil AA, Triwahyono S, Abdulrasheed AA (2018) Recent advances and future prospect in catalysts for oxidative coupling of methane to ethylene: a review. J Ind Eng Chem 59:218–229. https://doi.org/10.1016/j.jiec.2017.10.027
49. Sisler J, Khan S, Ip AH, Schreiber MW, Jaffer SA, Bobicki ER et al (2021) Ethylene electrosynthesis: a comparative techno-economic analysis of alkaline vs membrane electrode assembly vs CO$_2$–CO–C$_2$H$_4$ tandems. ACS Energy Lett 6(3):997–1002. https://doi.org/10.1021/acsene rgylett.0c02633
50. Kasipandi S, Bae JW (2019) Recent advances in direct synthesis of value-added aromatic chemicals from syngas by cascade reactions over bifunctional catalysts. Adv Mater 31(34):e1803390. https://doi.org/10.1002/adma.201803390
51. Yang X, Su X, de Chen ZT, Huang Y (2021) Direct conversion of syngas to aromatics: a review of recent studies. Chin J Catal 41(4):561–573. https://doi.org/10.1016/S1872-2067(19)63346-2
52. WEF (2020) Clean skies for tomorrow: sustainable aviation fuels as a pathway to net-zero aviation
53. IEA (2023) International shipping: energy consumption in international shipping by fuel in the Net Zero Scenario, 2010–2030. https://www.iea.org/reports/international-shipping
54. Martin Placek (2023) Global energy consumption by shipping 2019–2070, by fuel type. https://www.statista.com/statistics/1105953/shipping-break-down-by-fuel-forecast/
55. IEA (2022) Renewables 2022: analysis and forecast to 2027
56. Schröder J, Naumann K (2022) Monitoring erneuerbarer Energien im Verkehr. DBFZ Deutsches Biomasseforschungszentrum gemeinnützige GmbH
57. Grand View Research (2023) Market analysis report: petrochemicals market size, share and trends analysis report by product (ethylene, propylene, butadiene), by region (North America, Europe, Asia Pacific, Latin America, Middle East, Africa), and segment forecasts, 2023–2030. Report ID: 978-1-68038-560-1https://www.grandviewresearch.com/industry-analysis/petrochemical-market

58. IATA (2021) Resolution on the industry's commitment to reach net zero carbon emissions by 2050. Press Release No: 66
59. International Renewable Energy Agency (2021) A pathway to decarbonise the shipping sector by 2050. Abu Dhabi
60. BP (2021) Statistical review of world energy: 2021, 70th edn
61. IEA—International Energy Agency (2022) Renewables. https://www.iea.org/reports/renewables-2022
62. Rodin V, Lindorfer J, Böhm H, Vieira L (2020) Assessing the potential of carbon dioxide valorisation in Europe with focus on biogenic CO_2. J CO2 Utiliz 41:101219. https://doi.org/10.1016/j.jcou.2020.101219
63. International Energy Agency (2019) The future of hydrogen. Paris
64. IRENA (2023) Hydrogen. https://www.irena.org/Energy-Transition/Technology/Hydrogen
65. European Union (2018) Directive (EU) 2018/2001 of the European parliament and of the council: RED II
66. The European Parliament and of the Council (2003) Directive 2003/87/EC: EU ETS

Systems-Related Aspects

Sustainability Aspects of Powerfuels and Their Certification

Thomas Bock

Abstract Given the crucial role powerfuels are poised to play in a future defossilized energy system and in achieving ambitious climate targets, it is vital to ensure that they live up to their "promised" or envisioned sustainability characteristics. Against this background, the purpose of this chapter is to both describe important sustainability characteristics of powerfuels and show how they are considered in the current landscape of sustainability certification schemes. After an overview of the current regulatory situation for powerfuels and the basic functioning of sustainability certification schemes, different sustainability aspects for powerfuels are analysed in more detail with a focus on those aspects and criteria that are considered most relevant based on ongoing sustainability discussions and their inclusion in major regulatory and voluntary frameworks. Importantly, the chapter will discuss how each of the sustainability aspects is currently reflected in major framework certification frameworks that are expected to be particularly relevant in the years to come.

Keywords Sustainability Characteristics · Certification Schemes · Additionality · LULUC · Water Use · Social Impact · Chain-of-Custody

1 Introduction

The large-scale utilization of fuel options and energy carriers produced from renewable energy sources will play a crucial role in a future defossilized energy system and in achieving ambitious climate targets. In this context, powerfuels are rapidly gaining importance. A growing number of governments consider "green" hydrogen and derivatives as key pillars in their strategies to reduce greenhouse gas (GHG) emissions. The contribution of liquid powerfuels (e.g. kerosene, diesel) will be particularly crucial in those sectors considered notoriously "hard-to-abate," such as aviation, shipping, or heavy road transport.

T. Bock (✉)
ISCC System GmbH, Cologne, Germany
e-mail: thomas.bock@easa.europa.eu

977

With so much importance placed on powerfuels, it is vital to ensure that they live up to their "promised" or envisioned sustainability characteristics. This includes, notably, the delivery of sufficient GHG emissions savings compared to fossil fuel options that are commonly used today. At the same time, it is essential to prevent negative environmental and social effects related to the production of powerfuels as well as lock-in effects of technologies that should be phased out from a wider system perspective.

In this context, the question arises as to how such sustainability aspects, whether implemented through regulations or voluntary commitments, can be effectively implemented and verified along powerfuel supply chains on a global scale. This is where sustainability certification comes into play. Sustainability certification has been used for well over a decade to help ensure the sustainable production of biomass and biofuels and has proved to be an essential piece of the puzzle in the success of the EU's Renewable Energy Directive (RED) and uptake of sustainable, renewable fuels more generally. Sustainability certification is also expected to play a crucial role in ensuring the sustainability characteristics of powerfuels. As in the case of biofuels, the EU Renewable Energy Directive foresees a fundamental role for sustainability certification for the scale-up of powerfuels to ensure that they adhere to specific sustainability criteria. Furthermore, with a growing commitment of larger companies to use powerfuels in the future, they increasingly begin to turn to sustainability certification to be able to properly differentiate between powerfuels that can be considered sustainable and meet defined sustainability criteria and those that cannot.

Against this background, the purpose of this chapter is to both describe important sustainability characteristics of powerfuels and show how they are considered in the current landscape of sustainability certification schemes. Therefore, first a brief overview over the current regulatory situation with regard to sustainable powerfuels is provided. Subsequently, the basic functioning of sustainability certification is explained, with a focus on the different actors involved in the respective systems as well as the certification process itself. Following this, different sustainability aspects for powerfuels are analysed in more detail with a focus on those aspects/criteria that are considered most relevant based on ongoing sustainability discussions and their inclusion in major regulatory and voluntary frameworks. Importantly, it will be discussed how each of the sustainability aspects is currently reflected in major certification frameworks that are expected to be particularly relevant in the years to come.

2 Policies for Sustainable Powerfuels—Regulatory Overview

Regulatory frameworks play an essential role in the promotion and uptake of sustainable products. They provide regulatory certainty and thus certainty for producers and investors, as well as, in many cases, incentives and support schemes to aid in developing nascent markets for sustainable products that would not be viable or competitive otherwise.

This also holds true for renewable fuels. Over the past decade, regulatory frameworks and support have played a crucial role in the development and increased uptake of sustainable biofuels for the transport sector, such as in the EU or the United States. Mandates around growing renewable fuel volumes and/or lowering GHG emissions from fuel use were put in place, alongside different forms of governmental support schemes to help scale-up sustainable biofuels. These regulatory frameworks are generally based on clear definitions for the sustainability characteristics of biofuels (e.g. in terms of feedstock used, greenhouse gas emissions savings achieved and further sustainability requirements met). Many of these frameworks also feature comprehensive certification frameworks to ensure that the respective renewable fuels meet all the sustainability provisions required by law.

2.1 Powerfuels in Global Regulatory Frameworks

Given their currently still nascent market state, comprehensive regulatory frameworks will be key in steering the scale-up of sustainable powerfuels. Unlike for sustainable biofuels, the question as to what constitutes sustainable powerfuels is not yet as conclusively considered and defined in existing regulations. However, this is set to change as governments policies focus increasingly includes "green" hydrogen and its derivatives as key pillars of effective GHG emissions reduction strategies. As of today, sustainable powerfuels are already considered in a variety of governmental schemes and programmes, including legislative frameworks, hydrogen strategies, and different kinds of "roadmaps."

- Inflation Reduction Act (United States). A major example includes the Inflation Reduction Act (IRA) in the United States, which passed the U.S. Congress in the summer of 2022. Among many other provisions, the IRA created a set of tax credits/incentives intended to accelerate the development of clean energy solutions, including "green" hydrogen. The hydrogen production tax credit of the IRA is considered to be particularly important for "green" hydrogen development insofar as it will be highest when the hydrogen can demonstrate very low life cycle GHG emissions [1].
- Hydrogen Industry Development Plan (China). Constituting another major development, the Hydrogen Industry Development Plan drafted by the Chinese government lays out a medium- and long-term development plan for "green" hydrogen. Concretely, the strategy aims for the production of 100–200 kilotons of renewable hydrogen by 2025 and envisages its use in a variety of different sectors, including transport and industry. The Chinese government's strategy is particularly relevant given that China is currently the largest producer and consumer of hydrogen [2].
- Energy Security Strategy (United Kingdom). In the wake of Russia's invasion of Ukraine, the United Kingdom launched its Energy Security Strategy, increasing its

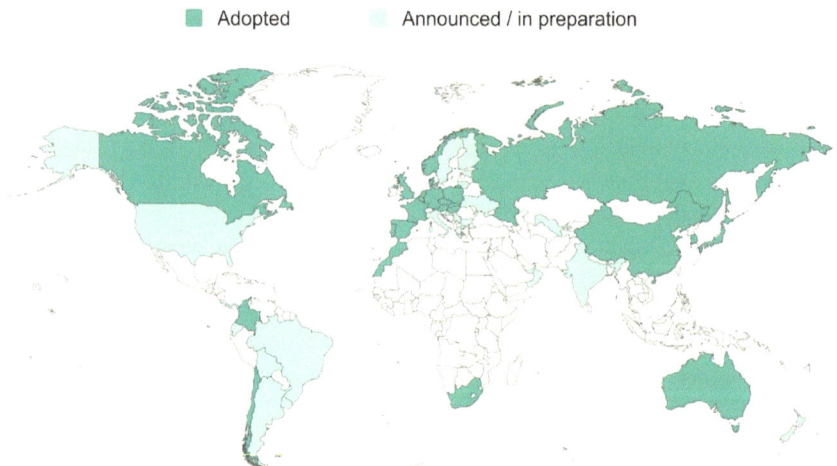

Fig. 1 Overview of adopted and announced hydrogen strategies. Adapted from IEA global hydrogen review 2022

ambition for low-emission hydrogen production (among which "green" hydrogen) to 10 GW by 2030. In 2019 already, Australia, as another major economy, had put in place its National Hydrogen Strategy, focussing on the promotion of "clean" hydrogen.

Besides the programmes and roadmaps mentioned above, at the time of publication of the International Energy Agency's 2022 Global Hydrogen Review, 25 countries plus the European Commission had already adopted national hydrogen strategies, with more than 20 governments in the process of developing their own national hydrogen strategies, as depicted in Fig. 1.[1]

2.2   *Powerfuels in the EU Context*

The growing importance of "green" hydrogen and its derivatives is also recognized in the European Union (EU) and considered in the European Commission's vision for the EU's energy and industrial future, formulated predominantly in the EU strategy

---

[1] An overview over different countries' hydrogen policies is given in the IEA's Global Hydrogen Review 2022, accessible via https://iea.blob.core.windows.net/assets/c5bc75b1-9e4d-460d-9056-6e8e626a11c4/GlobalHydrogenReview2022.pdf.

on hydrogen[2] and the EU strategy for energy system integration.[3] Importantly, the focus is on how "green" hydrogen can help decarbonize the EU economy in a cost-effective way, in line with the European Green Deal. The European Commission has set the ambitious goal of aiming for the production of 10 million tonnes of "green" hydrogen by 2030 in the EU as well as the import of an additional 10 million tonnes by the same year [3, 4].

EU Renewable Energy Directive (EU RED). In the EU, the EU Renewable Energy Directive (EU RED) is the central regulatory framework for the development of renewable energy across all sectors of the EU economy. It was first introduced in 2009 and, among other aspects, set the overarching European target for renewable energy as well as included further provisions to ensure the uptake of renewable fuels in the transport sector as well as heating and cooling sectors. Notably, it also first defined sustainability requirements for biomass cultivation and biomass-derived products to ensure that their production and use truly leads to reduced greenhouse gas emissions, but also avoids other negative environmental impacts, including deforestation, the damaging of high carbon stock areas, or areas with high biodiversity. In essence, only biofuels and bioliquids that met the defined sustainability and GHG emissions savings requirements were eligible to be counted towards the EU's and its Member States' renewable energy targets. This principle remains in place even as the RED undergoes regular revisions. It is important to note that the sustainability requirements defined by the Renewable Energy Directive (RED) apply globally. This means that if a renewable fuel ends up in the EU market and is intended to contribute to the EU's renewable energy targets, companies involved in the fuel's value chain, regardless of their location or in which country they produce in, must comply with the sustainability requirements specified in the RED.

The RED was revised in 2018 (now commonly referred to as RED II) and is legally binding since June 2021. Notably, it sets an increased overall EU target of 32% for the consumption of energy from renewable sources, with a dedicated 14% sub-target for renewable energy consumed in road and rail transport [5]. It is in the RED II that the so-called "renewable fuels of non-biological origin," or "RFNBOs" for short, were first introduced.

RNFBOs are defined as "liquid or gaseous fuels which are used in the transport sector […], the energy content of which is derived from renewable sources other than biomass,"[4] including therefore notably "green" hydrogen and its derivatives. The provisions on RFNBOs in RED II are critical as they define the criteria that "green" hydrogen and its derivatives must meet in order to be classified as RFNBOs and thus be eligible to be counted towards the relevant renewable energy quotas in

---

[2] The European Commissions' communication regarding its hydrogen strategy can be accessed via https://eur-lex.europa.eu/legal-content/EN/TXT/?uri=CELEX:52020DC0301.

[3] More information on the EU strategy on energy system integration is available on the European Commission's website and can be accessed via https://energy.ec.europa.eu/topics/energy-systems-integration/eu-strategy-energy-system-integration_en.

[4] Definition as per RED II. The RED II legislative text can be accessed via https://eur-lex.europa.eu/legal-content/EN/TXT/?uri=uriserv:OJ.L_.2018.328.01.0082.01.ENG&toc=OJ:L:2018:328:TOC.

the EU. Notably, these requirements relate to renewable electricity sourcing, eligible CO_2 sources for RFNBO production as well as the necessity for RFNBOs to achieve an overall GHG emissions reduction of at least 70% compared to a baseline value for fossil fuels (fossil fuel baseline) as defined in the RED II.

EU Renewable Energy Directive (EU RED): Delegated Acts. Given the inherent complexities of and potential pitfalls associated with the accounting of renewable electricity sourcing and the overall methodology to calculate greenhouse gas emissions of RFNBOs, the RED II empowered the European Commission to adopt two Delegated Acts. These complement the basic requirements for RFNBOs already laid down in the RED II itself. Both Delegated Acts were originally required to be adopted by December 2021, yet were significantly delayed and only formally came into force in July 2023. The delays can at least in part be explained by the fact that the provisions in both Delegated Acts needed to be a balancing act between putting in place effective sustainability safeguards and incentivising the market-ramp-up for "green" hydrogen. The European Commission therefore repeatedly delayed the adoption of the Delegated Acts to balance the different interests:

- Conditions for "renewable" hydrogen. Briefly summarized, the first Delegated Act further specifies the conditions under which hydrogen produced from electricity can be considered renewable and thus qualify as RFNBO—and consequently, other fuels derived from this hydrogen, such as kerosene and diesel, can also be classified as RFNBOs. To ensure that RFNBO use ultimately leads to specified minimum reductions in GHG emissions compared to the use of fossil fuels, the Delegated Act relies on three key requirements, namely (1) additionality, (2) temporal correlation, and (3) geographical correlation. Importantly, hydrogen production is required to use additional and newly installed renewable electricity capacity, rather than cannibalize existing renewable electricity supply (this is what is referred to with the term "additionality"). Moreover, referred to the other two requirements, hydrogen production must be linked to actual renewable electricity production by featuring a temporal and geographical component—i.e. hydrogen production needs to happen somewhat near the renewable electricity source (e.g. a wind farm) and at a time when that source is actually generating renewable electricity.[5]
- Calculation of RFNBO GHG emissions. The second Delegated Act specifies how GHG emissions of RFNBOs are to be calculated and verified, as only those RFNBOs which meet the GHG emissions saving threshold of 70% (compared to the fossil fuel comparator) defined in the RED II are eligible for counting towards the EU's renewable energy target. Importantly, the calculation of GHG emissions from RFNBO production and use comprises their entire life cycle, including upstream emissions, emissions in connection with electricity supply, process emissions, and emissions from transport and combustion. The Delegated Act provides further provisions on the use of carbon dioxide (CO_2) for RFNBO

---

[5] Chapter 4.1 goes into more detail on renewable electricity consumption, its implications for sustainability and its coverage under certification.

production. In that regard, it specifies which carbon sources can be associated with avoided emissions—i.e. captured and recycled CO_2—so that the resulting powerfuel has a chance of meeting the 70% GHG emissions saving threshold. It also puts certain restrictions on the continued use of fossil-based carbon and biogenic carbon from practices considered "unsustainable."[6]

Fit for 55/RED III. In July 2021, the European Commission released the so-called "Fit for 55" package, a set of regulatory proposals and revisions designed to achieve the European Green Deal target of a net reduction in GHG emissions of 55% by 2030 compared to 1990 levels. As part of the package, the European Commission proposed another revision of the RED (oftentimes referred to as RED III), on which a provisional agreement was reached by EU institutions in April 2023. The EU co-legislators agreed to raise the EU's binding renewable energy target for 2030 from 32% to a minimum of 42.5% by 2030. For the transport sector, the RED III provides EU Member States the possibility to choose between two different options: they may either set a binding target of a 14.5% reduction of GHG emissions intensity in transport from the use of renewables by 2030 or set a binding target of at least 29% share of renewable within the final consumption of energy in the transport sector by 2030. Furthermore, the provisional agreement also stipulates that the industry sector would need to increase their use of renewable energy by 1.6% annually.

Underpinning the overall increased ambition around renewable energy use, RED III also brings about major implications for "green" hydrogen and powerfuels as a whole. For transport, there is a minimum requirement of a 1% share of RFNBOs in the share of renewable energies supplied to the sector in 2030. For industrial sectors, the RED III will mandate that 42% of hydrogen used in the sector must be RFNBOs, with that share rising to 60% by 2035 [6]. This is a major change insofar as the RED II currently in force focusses on RFNBO use in the transport sector only. In contrast, the RED III not only broadens the scope of RFNBO use to include industry sectors (e.g. metal industry, cement and chemicals production), but also extends the application of RFNBO sustainability provisions. This means the stringent criteria for classifying hydrogen as "green" will also be applicable to its use in these sectors.

Central role of RED sustainability framework. Independently of its regular revisions, the RED and its related Delegated Acts will remain central in terms of the promotion of and setting the rules for renewable energy in the EU in general and renewable fuels and powerfuels in particular. This is to a large degree also because many EU regulations and instruments refer to the sustainability requirements and framework laid down in the RED. For instance, the EU Emissions Trading Scheme (EU ETS), a cornerstone of EU climate policy, allows for renewable fuels and liquids to be accounted with zero GHG emissions, providing these fuels meet the sustainability provisions laid down in the RED. As another example, two regulatory proposals from the Fit for 55 package, the ReFuelEU Aviation and the FuelEU Maritime, largely refer to the sustainability requirements in the RED with regard to

---

[6] Chapter 4.2 goes into more detail on carbon sources, their implications for sustainability and the coverage under certification.

the renewable fuels being considered eligible under these regulations. This is particularly important as both regulatory proposals consider powerfuels as key pillars in reaching the respective proposals' objectives of reducing GHG emissions in two sectors generally considered "hard-to-abate." The ReFuelEU Aviation proposal for instance, aimed at increasing levels of sustainable aviation fuels (SAF) in the aviation fuel supply, envisions a continuously increasing sub-target of synthetic kerosene as SAF volumes increase, with up to 35% of total aviation fuel supply by 2050.

With the provisions for sustainable production and greenhouse gas emissions savings of renewable fuels and powerfuels established in the RED and associated Delegated Acts, the question turns to how these requirements can be verified along oftentimes complex and globally spanning supply chains, while keeping the process practical, flexible and administrative burden for market participants at a manageable level. For the purpose of proving that all RED sustainability requirements are met by renewable fuels that are to be counted towards renewable energy quotas in the EU, the RED therefore requires economic operators to use certification schemes. The following chapter will explore sustainability certification in more detail.

3   Sustainability Certification of Renewable Fuels

In the following, an overview of sustainability certification of renewable fuels is provided. First, the basics of how sustainability certification is set up is introduced. The section then explores the "ecosystem" of organizations involved, followed by more information on how the actual certification process is conducted.

3.1   Basics of Sustainability Certification

In this subsection, the use of sustainability certification in regulatory and voluntary contexts is introduced. In addition, the general set-up and main objectives of certification are briefly discussed.

Sustainability Certification under Regulatory Frameworks. Sustainability certification has become an integral pillar for a growing number of regulatory frameworks aimed at promoting the use of renewable energy and renewable fuels in particular. Increasingly, regulators have been making use of sustainability certification schemes to ensure that only those fuels that comply with defined sustainability criteria set out in key legislative pieces are allowed to count towards renewable fuel targets and benefit from governmental support schemes or incentives.

Prominent examples include the EU's Renewable Energy Directive (EU RED), which sets sustainability requirements and greenhouse gas saving targets for renewable fuels brought to the EU market; the International Civil Aviation Organization's

(ICAO) CORSIA[7] scheme, which sets similar requirements for sustainable aviation fuels (SAF) deemed eligible under the scheme; and the Japanese government's feed-in-tariff (FIT) system, applicable to economic operators processing and trading biomass intended for power generation in Japan. In 2022, the US Inflation Reduction Act (IRA) joined the ranks, as it includes, among many other provisions, a SAF blender's tax credit, provided the fuel achieves GHG emissions reductions beyond 50% on a life cycle basis. One way to prove the fuel's compliance with the IRA provisions is to certify it under a certification scheme recognized under CORSIA.

Certification in Voluntary Markets. In addition to its use under regulatory frameworks, sustainability certification is also increasingly used in non-regulated, voluntary markets. This is predominantly the case for sectors in which a corresponding regulatory framework is not (or not yet) in place, such as in the chemical industry, where the use of renewable and circular feedstock, underpinned by robust certification, has experienced a surge in recent years. Increasingly, it is also widely recognized industry standards for GHG accounting or corporate emissions target setting, such as the Greenhouse Gas Protocol (GHGP) or the Science-Based Targets Initiative (SBTi), that refer to established sustainability certification schemes as an effective way for companies to obtain credible GHG emissions information associated with the materials and products they handle. As more and more companies commit to ambitious climate targets, they increasingly turn to sustainability certification schemes to ensure that the products and fuels they source not only achieve substantial GHG emissions reductions, but also avoid further negative environmental or social effects of their production.

Set-Up of Certification Schemes. Sustainability certification schemes oftentimes vary substantially with regard to their nature, scope, and set-up. In terms of their governance structure, some schemes are set up and managed by one company only (e.g. a fuel producer) or by a group of producers with a very specific focus and common interest (e.g. a group of feedstock producers). Others are based on a broad multi-stakeholder structure that includes companies from across the entire value chain, but also non-governmental organizations (NGOs), research institutes, regulators, and other organizations with a focus on sustainability issues dealt with under the respective certification scheme.[8]

Furthermore, certification schemes may be applicable for a particular region only (e.g. Latin America) or have a global relevance. They may cover only one feedstock (e.g. sugarcane) or certain feedstock types (e.g. agricultural crops), or may be applicable to a wide variety of multiple feedstock types (e.g. crops, wastes, residues, and renewable electricity). They may require economic operators to comply with sustainability requirements defined by a particular regulation only or may go beyond those

---

[7] The acronym "CORSIA" stands for "Carbon Offsetting and Reduction Scheme for International Aviation." CORSIA is a carbon offset scheme intended to ensure carbon–neutral growth of international aviation from 2020 onwards.

[8] For more information on the set-up of certification schemes in the renewable fuel/biofuel space, please also see Feige, A., Pforte, L.; "Sustainability Certification in the Aviation Industry" in Kaltschmitt et al. (2018) "Biokerosene—Status and prospects."

to set additional requirements. They may cover only specific parts of the supply chain (e.g. the feedstock production step) or cover the whole value chain from feedstock production all the way until the final fuel is supplied to the market. Finally, certification schemes may be recognized by competent authorities under one or multiple regulations, in that economic operators can become certified with them to show compliance with the regulations' sustainability stipulations.

Main Objectives of Certification. Though their concrete set-up and application may differ from case to case, certification schemes generally aim to ensure three vital sustainability aspects of renewable fuels, including powerfuels, as laid out in the following:

- Sustainable Feedstock Production. Renewable fuels can be produced from a wide range of different feedstocks, including crops, wastes, and residues, or in the case of powerfuels, renewable electricity and, if needed, a carbon source. To ensure that feedstock production does not lead to unintended negative environmental consequences, many (regulatory) frameworks put in place certain provisions, i.e. in terms of which feedstocks are eligible and the manner in which they should be cultivated or produced. For primary biomass, this often refers to the areas on which feedstock may or may not be cultivated (e.g. exclusion of biomass cultivation on lands classified as high carbon stock or highly biodiverse). For electricity, this may include ensuring its renewable origin and additionality (Chap. 3.1); for a carbon source, this may also include ensuring its renewable origin [e.g. biogenic or atmospheric (Chap. 3.2)].
- Traceability through Supply Chains. Supply chains for renewable fuel include feedstock production, different processing and refining steps, as well as transportation and distribution of raw materials, intermediates and final fuels. Traceability describes the ability to identify and trace the origin, processing history, distribution, and location of these materials and products as they move through supply chains. This is particularly important in view of the fact that renewable fuel supply chains can oftentimes be complex, globally spanning and involve comingling of sustainable with non-sustainable fuels at different supply chain stages (Chap. 4.7).
- Verified GHG Emissions Reductions. Providing assurance that a renewable fuel truly achieves GHG emissions reductions compared to its conventional, fossil-based counterpart is crucial. In that regard, certification schemes provide a robust, standardized framework for how GHG emissions are to be consistently calculated and verified along the fuel supply chain.

Through allowing the verification of the sustainability and GHG emissions reductions of renewable fuels and tracking these characteristics along supply chains, certification schemes cover the key sustainability metrics that many governments aim to ensure in their regulations.

Relevance in the EU. In the EU, certification schemes have been a key piece in the sustainability architecture for renewable fuels under the RED since 2009. The EU RED uses a co-regulation approach, which allows the control of compliance with the

Directive's sustainability provisions by private certification schemes. These schemes must be recognized by the European Commission. Concretely, the RED requires that economic operators prove their compliance with all applicable sustainability, traceability, and GHG emissions requirements via the use of these certification schemes. After more than a decade, there is a well-established ecosystem surrounding the certification of compliance with RED requirements in place. Therefore, given the extraordinary importance and sophistication of the sustainability architecture for sustainable fuels established under the EU RED, for the EU and beyond, the certification ecosystem, its different entities, and their functions are described in more detail in the next chapter.[9] The European Commission will make use of and build on this existing infrastructure for ensuring the sustainability and eligibility of RFNBOs as per the EU RED as well.

3.2   The Sustainability Certification "Ecosystem"

Sustainability certification comprises more than just the certification scheme itself and can rather be understood as a complete "ecosystem" that encompasses different types of organizations, each with their specific roles and responsibilities. In its legislative text, the EU RED defines the requirements that a renewable fuel must comply with in order to be considered "EU RED eligible" and be able to be counted towards an EU Member State's renewable energy target (and thus benefit from any incentives or support schemes that Member State has put in place). Economic operators are required to use EU-recognized certification schemes for demonstrating fuel or product compliance with the respective sustainability requirements. Figure 2 illustrates the sustainability certification architecture, or "ecosystem," in place under the EU RED using the example of a powerfuel supply chain for synthetic kerosene.

In the following, the different elements and organizations of the certification "ecosystem" as depicted above, as well as their roles and responsibilities, are described in more detail.

Regulatory Authority. In the sustainability certification context, the European Commission acts as the overarching regulatory authority. It manages the assessment, approval, and continuous monitoring of all certification schemes recognized under the EU RED. The European Commission also keeps regular contact with all certification schemes and the EU Member State authorities responsible for RED implementation, as well as steps in for clarification whenever there is uncertainty about the applicability of RED provisions in a given situation.

---

[9] Other major regulatory frameworks, such as ICAO's CORSIA for sustainable aviation fuels, function very similarly in their set-up and approach to sustainability certification compared to the EU RED. Different approaches to certification are, however, taken in major North American frameworks for renewable fuels, such as California's Low Carbon Fuel Standard (LCFS) and the US EPA's Renewable Fuel Standard (RFS).

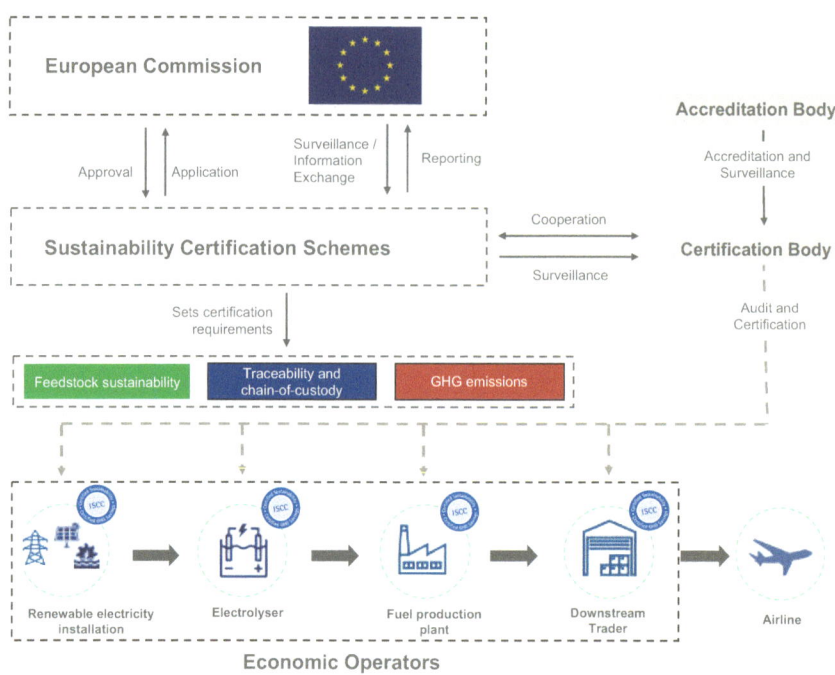

Fig. 2 The EU RED sustainability certification "ecosystem"

Economic Operators. Economic operators operate in supply chains governed by a particular regulatory framework, such as under the EU RED. They include feedstock producers (e.g. renewable electricity installations), processing units (e.g. electrolysers or production plants for synthetic fuels), traders, and storage units. Importantly, economic operators can be located in both EU and non-EU countries—the EU RED provisions for sustainability and certification are applicable on a global basis. Each economic operator that takes part in the supply chain and handles (i.e. receives, stores, processes, and forwards) sustainable product will need to be certified—as indicated in Fig. 1 with a certification logo next to each economic operator. Economic operators are audited by certification bodies against the requirements set by certification schemes, with regard to feedstock sustainability criteria, their traceability and chain-of-custody system, and greenhouse gas emissions.

Certification Schemes. Certification schemes define sets of rules and requirements, on the basis of which certificates of conformity or compliance are issued to economic operators. Under regulatory frameworks such as the EU RED, owners of certification schemes draft and develop certification schemes based on and in line with the applicable regulatory framework. In essence, the sustainability provisions of the legislative framework are taken and implemented as part of a comprehensive auditing and certification framework, with the aim of it being both robust and practical in its widespread application and verification.

In regulated markets, certification schemes generally need to be recognized by the respective competent authorities before they can be used by economic operators. Under the EU RED, the European Commission as the competent authority on EU level has currently formally recognized 14 certification schemes[10] [7]. A re-recognition of certification schemes is typically required after a defined period of time (i.e. every five years under the EU RED) or following a major update of legislation (such as was required when the RED II entered into force in 2021).

Importantly, recognition of a certification scheme by the European Commission implies that EU Member State authorities shall generally accept evidence from suppliers of renewable fuel that a certified batch of fuel meets the EU RED criteria. This is provided that fuel suppliers provide proof of sustainability documents to the authorities that the batches of fuel meet that recognized certification scheme's criteria. As a result, Member State authorities generally do not require additional evidence on RED compliance from fuel suppliers for any given batch of fuel, effectively harmonizing documentation requirements and reducing administrative burden across EU Member States.

Certification Bodies. While certification schemes, based on the regulatory framework, define the criteria that economic operators must comply with in order to become certified, it is not the certification schemes themselves that conduct the audits to check and verify compliance. Instead, owners of certification schemes generally approve and work with so-called certification bodies for that purpose (third-party certification principle).

Certification bodies are independent, third-party organizations that handle the certification process, including the assessment of the individual economic operator's compliance with the requirements of the certification schemes. Certification bodies are authorized by the certification scheme's owners to issue certificates of compliance to economic operators, provided of course the operators comply with the requirements defined under the certification scheme. Commonly, certification bodies are required to work in conformance with relevant ISO standards. In the context of the EU RED, for instance, certification bodies must conduct audits following the ISO/IEC 17065 standard for product certification and ISO/IEC 17021 for management system certification in particular.

Accreditation Bodies. Certification bodies themselves may be subject to audits by their competent national authority (i.e. the competent authority of the country the certification body is registered in). However, their main "license to operate" is generally derived from the certification body's accreditation by so-called accreditation bodies.

Accreditation refers to the independent, third-party evaluation of a conformity assessment body (i.e. a certification body) against recognized standards, conveying formal demonstration of its impartiality and competence to carry out specific conformity assessment tasks (such as certification). Accreditation bodies are established

---

[10] In the context of the EU RED, certification schemes recognized by the European Commission are generally referred to as "voluntary schemes."

in many economies with the primary purpose of ensuring that conformity assessment bodies are subject to oversight by an authoritative body [8]. In the context of the sustainability certification "ecosystem," including the EU RED, accreditation bodies have the essential role of continuously verifying that certification bodies work in conformance with relevant ISO standards and the requirements set by the certification schemes the certification bodies are cooperating with.

Difference in Voluntary Markets. Sustainability certification in voluntary, i.e. non-regulated markets, works in a very similar way compared to regulated markets. The only significant difference is that, here, there is no overarching authority (such as the European Commission for EU RED purposes) that would need to approve certification schemes first. All other aspects remain the same, i.e. the certification scheme sets requirements, which are then verified by ISO-accredited certification bodies. The ISCC PLUS and RSB Advanced Fuels standards, which are part of the analysis in Chap. 3, work in this manner.

3.3   The Sustainability Certification Process for Powerfuels

In this chapter, the sustainability certification process for powerfuels will be explained in more detail. Figure 3 depicts the process as to how a regulatory sustainability framework is translated into a certification system and its audit procedures, to be used by certification bodies in their audits of economic operators.

Regulatory Framework Translation. The three components of translating a regulatory framework into system documents and audit procedures are further explained in the following.

- Regulatory Framework. In regulated markets, the regulatory framework provides the basic sustainability provisions to be fulfilled by-products and/or economic operators subject to that framework. Under the EU RED, for instance, this pertains to requirements for feedstock sustainability, traceability, and chain-of-custody as well as greenhouse gas emissions of EU RED compliant products. Further, the EU RED sets requirements for certification schemes and certification bodies operating thereunder, including with regard to their governance structure, auditing frameworks, independence, and reporting obligations.

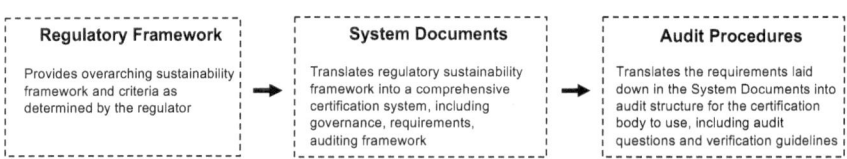

Fig. 3 Process of translating a regulatory framework into a certification system and audit procedures

- System Documents. An essential element of a certification scheme is the set of so-called system documents. These documents translate a regulatory framework, e.g. those EU RED provisions related to sustainability and functioning of EU RED certification systems, into a comprehensive certification system "on the ground." System documents lay down the general provisions of the certification system, including for instance its governance structure, the basic registration, and certification process as well as the processes for dealing with complaints and grievances. Importantly, the system documents also define the different types of economic operators that can be certified (generally referred to as "system users"). In a powerfuels context, these system user types include renewable electricity installations, CO_2 provisioning units, further processing units (which cover electrolysers as well as subsequent synthesis plants), and trading units. In essence, every economic operator participating in powerfuel supply chains can be classified according to one or several of these system user types and thus be "operationalized" for certification. The system documents, crucially, then define the certification requirements for each system user type, with regard to sustainability of (feedstock) production, traceability of sustainable materials, and greenhouse gas emissions calculation and verification.
- Audit Procedures. Based on the system documents, most certification schemes provide so-called audit procedures or checklists. These essentially translate the requirements laid down in the system documents into practical guidelines for the certification body to go through during the actual audit—both in the form of straightforward "Yes or No" questions as well as more elaborate questions (e.g. on the nature of the economic operator's management system). This ensures full transparency and comparability across audits and that all relevant requirements are being checked. The audit procedures also contain additional verification guidance, for instance on which documentation the auditor could check to substantiate the answer to each particular audit question.

Certification Process. Figure 4 illustrates an economic operator's exemplary certification process.[11] Please note that while the basic order of steps is set and must be respected in all cases (e.g. the audit has to be conducted successfully before a certificate can be issued), certification processes can still be different from case to case. This is partly due to the differences in certification bodies' audit approaches as well as due to the sheer variety in set-ups that may be certified (i.e. different types of economic operators, products, production processes, regional specificities, etc.). In most cases (e.g. under EU RED certification), sustainability certification must be conducted at site or facility level (e.g. at the level of the electrolysis plant or trading unit).

Certification Body (CB) Selection and Registration. Certification starts with the registration process. In a first step, a certification contract between the economic operator to be certified and a certification body approved by the certification scheme is

---

[11] Based on the certification process applied under the International Sustainability and Carbon Certification (ISCC) scheme (www.iscc-system.org).

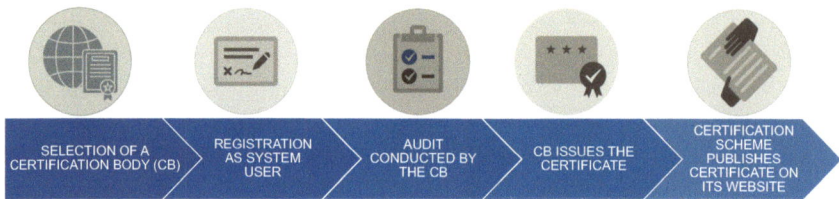

Fig. 4 Typical sustainability certification process (CB—certification body)

concluded. Importantly, each specific site (e.g. production facility or trading unit) will generally need to register separately and undergo its own certification. Certification schemes often approve and work with multiple certification bodies to allow for sufficient availability of qualified auditors and adequate coverage of markets and regions. Major sustainability certification systems, such as ISCC or RSB, usually work with a number of different certification bodies that operate in different countries. ISCC, for instance, collaborates with more than 50 certification bodies that operate in over 120 countries. After a certification contract is concluded, the economic operator registers with the certification scheme, thereby becoming an official system user of that scheme.

Risk Assessment. The certification body will then conduct a risk assessment of the specific supply chain set-up. This assessment has the aim of determining the risk that the system user may not be compliant with some of the requirements set by the certification scheme. Given the diversity of scenarios that can potentially be subject to certification, it is crucial that the certification body takes the basic requirements from the certification scheme's system documents and applies them to the individual scenario at-hand following an in-depth analysis. Naturally, an electrolysis plant has vastly different risks of "non-compliance" compared to a downstream trading unit that only stores and trades final powerfuels. For instance, an electrolysis plant may be situated in a region with high water stress, and with a thus elevated risk of non-compliance with sustainability requirements related to water use (Chap. 4.5). In contrast, a trading unit may have no issue in this regard but may trade different kinds of sustainable and non-sustainable fuels, leading to higher risk of double counting (Chap. 4.7).

The risk assessment is essential insofar as its result determines the audit "intensity." Certification must constantly strike a proper balance. It must provide a reasonable degree of assurance that the economic operator complies with all requirements, yet at the same time keep the efforts, costs, and practicability of an audit to a manageable level. This is true for both certification bodies and economic operators. Therefore, it is necessary and appropriate that the audit intensity level corresponds to the risk level determined by the certification body—i.e. the higher the risk level as determined by the certification body, the more in-depth the audit needs to be (e.g. in terms of documents reviewed, personnel interviewed, suppliers checked). The risk assessment ultimately results in a verification plan that the certification body will use to guide its auditing for that economic operator.

Remote and On-Site Audit. Guided by the verification plan resulting from the risk assessment, the certification body will then, more concretely, evaluate the economic operator's compliance with the certification schemes' requirements. For most certification schemes active in renewable fuels certification, this primarily concerns the economic operator's (e.g. electrolyser's) management system, traceability system, and GHG emissions calculation. These elements are listed below with examples of key questions:

- Management system—Are responsibilities of personnel clear and transparently documented with regard to compliance with all applicable requirements?
- Traceability system—Are adequate tracking systems, such as mass balancing, in place to track all flows of sustainable material in line with all applicable requirements?
- GHG emissions calculation—Does the GHG emissions calculation follow all applicable requirements; is it transparent and an accurate representation of the economic operator's processes?

Importantly, the audit process generally consists of a mix of desk-based assessment and on-site assessment. During the up-front desk assessment, the certification body will first do an in-depth document check and evaluate aspects such as the system user's employee handbook, traceability and mass balance system, and GHG emissions calculation. Where applicable and part of the certification scope, the certification body will also verify compliance with certain feedstock sustainability requirements, such as confirming that feedstock did not come from high carbon stock lands (e.g. via remote sensing tools such as satellite imagery). The desk assessment generally needs to be followed up by an on-site assessment in most cases. On-site assessments are essential insofar as they allow for "ground-truthing" and substantiating the results from the desk-based assessments, as well as interviews with the economic operator's employees. Crucially, as mentioned before, the certification body needs to conduct the audit based on the audit procedures, ensuring all relevant requirements are checked and verified.

If no non-conformities were found or minor non-conformities were corrected by the economic operator in time, the audit is considered successful, and the certification body issues the certificate to the economic operator. It is only from this point onwards that the economic operator is in the position to handle (i.e. receive, store, process and/or forward) sustainable material as per the rules of the certification scheme.

Certificate Issuance. Each certificate is issued for a defined period of time (most commonly for a period of 12 months), after which the economic operator has to undergo another audit and show continued compliance with the relevant requirements in order to become recertified. It is only during initial audits that system users will go through the complete audit process as outlined above, as they will stay registered with the certification scheme and assigned to their certification body up to and including for recertification. During the initial audit, the audit focus will generally centre around the question whether responsibilities, knowledge and processes are clear so that the economic operator is in the position to comply with all relevant requirements *in* *the upcoming certification period* (e.g. 12 months). This is because, at this point, no sustainable material has yet been handled by the economic operator which the certification body could check and verify retrospectively. In contrast, the audit focus during recertification audits will rather be on verifying whether the economic operator complied with all relevant requirements during *the previous certification period* (e.g. 12 months). For this, the certification body will audit the economic operator's sustainability claims and underlying movements of sustainable materials during that period.

In most cases, after the certification process is completed, the certification body is required to send the certificate, the filled-out audit procedures as well as an additional summary audit report to the certification scheme. The scheme then uploads the certificate and the summary audit report to its certificate database. This database is publicly available and can therefore be accessed by suppliers, buyers, competent authorities, and other third parties, enhancing overall transparency and credibility of the certification process.

Before handling (i.e. receiving, storing, processing, and/or forwarding) any sustainable material, economic operators must be in possession of valid certification. Additionally, they can only source from or sell to operators that are certified themselves. While this ensures that only successfully certified operators are involved in sustainable supply chains, it also allows those operators a certain flexibility: they can switch between suppliers and buyers of sustainable material (provided those are certified) and are not bound to just one single supply chain.

At the same time, it is important to keep in mind that a certificate only puts the economic operator in the *position* to handle sustainable material as per the certification standard. It does not automatically mean that each and every batch of product (such as a powerfuel) coming out of a facility can be considered sustainable. Sustainability of products on a batch level is shown via sustainability documents such as sustainability declarations (Chap. 4.7).

4 Sustainability Aspects of Powerfuels and Their Certification

Generally speaking, sustainability aspects of powerfuels encompass all environmental, social, and economic aspects pertaining to the production and use of powerfuels. However, aspects of economic sustainability of powerfuels, such as their contribution to local or regional economies, are not discussed in this chapter. While these aspects are also important, the focus of sustainability certification is primarily on specific supply chains, where a broader economic dimension is generally not evaluated nor compliance with respective criteria certified. Thus, this chapter is focussed on the environmental and social sustainability aspects of powerfuel production through the lens of sustainability certification, as these are commonly the focus of interest in relevant frameworks (including, for instance, the EU RED). Therefore, a set of major sustainability aspects of powerfuels are discussed in more detail, which are chosen because of their relevance and prominence in both important regulatory frameworks and certification standards (see list below). Concretely, the following seven aspects will be considered:

- electricity consumption,
- carbon sources,
- greenhouse gas (GHG) emissions,
- land use and land use change,
- water use,
- social impact,
- traceability and chain-of-custody.

For each of the aforementioned aspects, key sustainability considerations are discussed. The analysis examines how each of these sustainability factors is incorporated into three essential frameworks relevant to powerfuels.

- **EU Renewable Energy Directive (EU RED).** The first framework considered as part of this analysis is the EU Renewable Energy Directive (EU RED). The EU RED is currently applicable in its revised form (RED II, Chap. 1). Importantly, the content of the two Delegated Acts are considered as well, in that they specify in much greater detail the requirements for powerfuels (or "RFNBOs" in EU RED wording) production. Crucially, all certification schemes that will be recognized under the EU RED for the certification of powerfuels must implement each and every requirement set out in the EU RED and the Delegated Acts. This includes the EU RED certification standards of major certification schemes, including ISCC (for its ISCC EU standard) and RSB (for its RSB EU RED standard).
- **ISCC PLUS.** In addition, the ISCC PLUS standard is considered. ISCC PLUS is one of the three major certification standards of ISCC (International Sustainability and Carbon Certification).[12] ISCC is currently among the largest sustainability certification schemes for renewable fuels globally. The ISCC PLUS standard was specifically designed for use in voluntary, i.e. non-regulated markets, and has been certifying the production of "green" hydrogen and derivatives since 2018.[13]
- **RSB Advanced Fuels standard.** Thirdly, the RSB Advanced Fuels standard is considered. This standard has been developed and is maintained by the RSB (Roundtable on Sustainable Biomaterials).[14] The RSB, too, is a sustainability certification scheme and has been playing a significant role in the certification of renewable fuels over the past decade. Similar to ISCC PLUS, the RSB Advanced Fuels standard is designed to be applicable to voluntary markets.

The EU RED as well as the ISCC PLUS and RSB Advanced Fuels standards have been chosen for two main reasons. Firstly, the EU RED (as regulatory framework) as

---

[12] Please find the official website of ISCC under https://www.iscc-system.org.

[13] Please find more information about the ISCC PLUS standard under https://www.iscc-system.org/certification/iscc-certification-schemes/iscc-plus/.

[14] Please find the official website of the RSB under https://rsb.org.

well as ISCC PLUS and the RSB Advanced Fuels standard (as voluntary certification systems) cover powerfuels in the first place. Secondly, all three frameworks are set to play a major role for sustainable powerfuel production in the foreseeable future. The EU RED and its associated Delegated Acts are of particular importance not only because they apply to one of the largest economic blocs in the world, but also because they represent the most comprehensive set of sustainability requirements for powerfuels production to-date. ISCC (with ISCC PLUS) and the RSB (with the RSB Advanced Fuels standard) are not only major certification schemes for renewable fuels in general, but are in essence among the very few certification schemes heavily involved in the certification of fuels for those "hard-to-abate" transport sectors that will require significant volumes of powerfuels in particular, i.e. aviation, shipping, and heavy-duty transport.

4.1 Electricity Consumption

As the chemical energy stored in the different powerfuel options stems predominantly from electricity, one of the most essential and widely discussed sustainability aspects of powerfuel production pertains to electricity generation. In particular, the essential concepts to consider are renewability, additionality, temporal, and geographical correlation of electricity consumption, as discussed in the following.

Renewability. If powerfuels are to live up to their potential in terms of reducing GHG emissions in transport and a variety of other end-uses, the energy for their production must come from renewable and sustainable sources. This also and especially applies to the electricity used to produce electrolysis-based hydrogen, which must likewise come from sustainable sources. Among the limited number of regulatory frameworks and standards that so far consider powerfuels and set criteria for electricity consumption, there is broad consensus that this requires the electricity to come from exclusively renewable sources.[15] Table 1 gives a summary of the requirements for the renewability of the electricity consumption in the three frameworks analysed.

ISCC PLUS and the RSB Advanced Fuels standard base their electricity consumption requirements on the definition of renewable energy included in the EU RED. Consequently, they consider electricity generated from virtually all renewable non-fossil sources, except notably nuclear sources (which is not included in the EU RED definition) and biogenic sources. Renewable electricity generated from biomass is explicitly excluded for the production of powerfuels as per the EU RED and consequently the ISCC PLUS and RSB Advanced Fuels standard. This may be in part because the generation of electricity from biomass and subsequent power-to-liquid (PtL) synthesis would entail significant efficiency losses compared to conversion of that biomass via biomass-to-liquid (BtL) pathways [9], as well as due to general

---

[15] Please also refer to DENA's report on "Global Harmonisation of Hydrogen Certification," Chap. 4.1, for an overview over how different regulations and standard consider criteria for renewable electricity.

Table 1 Comparison of requirements around the renewability of electricity consumed (EAC—energy attribute certificate; PPA—power purchase agreement; RE—renewable electricity)

Framework	Requirements for electricity consumption—renewability
EU RED	• Electricity consumption for powerfuels production must come from 100% renewable sources, excluding nuclear and biogenic sources • Renewability of electricity to be proven via Power Purchase Agreements (PPAs) or direct connection of processing unit with renewable electricity installation
ISCC PLUS	• Electricity consumption for powerfuels production must come from 100% renewable sources, excluding nuclear and biogenic sources • Renewability of electricity to be proven via Energy Attribute Certificates (EACs), PPAs in combination with EACs, or direct connection of processing unit with RE installation
RSB advanced fuels	• Electricity consumption for powerfuels production must come from 100% renewable sources, excluding nuclear and biogenic sources • Renewability of electricity to be proven via PPAs in combination with EACs, or direct connection of processing unit with RE installation

concerns around both the sustainability and limited availability of biomass sources [10].

Renewable electricity from wind, solar, hydro, geothermal, and ocean energy (ocean currents, tides) is considered eligible under many frameworks, including those analysed more closely in this chapter. Nevertheless, there are sustainability concerns for these forms of renewable electricity generation as well. For instance, wind farms may pose a risk for bird species, and bigger photovoltaic systems need non-negligible amounts of land. In particular, renewable electricity from hydropower may be viewed somewhat critically, due to a range of potentially detrimental effects on water courses and related ecosystems.[16]

The EU RED as well as ISCC PLUS and the RSB Advanced Fuels standard require the producer (e.g. the electrolyser) to prove the renewability of the electricity they consume to produce powerfuels. If electricity for powerfuels production is sourced from the grid, as will oftentimes be the case, the EU RED and the RSB explicitly require Power Purchase Agreements (PPAs), whereas ISCC PLUS leaves the option of using either PPAs or Energy Attribute Certificates (EACs). Briefly put, a PPA is a long-term contract under which a company agrees to purchase electricity directly from a renewable energy generator. This allows for close "linking" of the renewable electricity (RE) producer and consumer. In contrast, EACs are typically generated by an RE producer for a given amount of RE produced and added to the grid, and may be sold relatively flexibly to different consumers, without a fixed long-term contract between the producer and those consumers in place.

All three frameworks also consider the option of a direct connection via a direct line between a RE installation (e.g. a wind park) and the powerfuels production facility, in which case no PPA or similar certificates are required.

---

[16] For more information, please refer to the German Federal Agency for Conversation's webpage, accessible via https://www.bfn.de/wasserkraft (in German only).

While some regulatory frameworks and certification standards currently only require a proof of renewability when it comes to electricity consumption, there are some frameworks and standards, including those analysed more closely in this chapter, that go further and set even stricter requirements. Notably, this concerns the aspects of additionality, temporal correlation, and geographical correlation, as explained in the following.

Additionality. A second and closely connected concept is that of additionality. Additionality requires that the production of powerfuels incentivizes new renewable electricity generation capacity to come online. Additionality is important because if renewable electricity generation capacity were not to keep up with growing powerfuels production capacity, existing renewable electricity capacity would simply be diverted ("cannibalized") rather than contribute to decarbonizing the electricity grid. While a focus on additionality is crucial at a time when much of the electricity grid still relies on electricity produced from fossil (and nuclear) sources, it is important to note that this will become less and less important as the share of renewables in the grid mix increases [11].

Table 2 gives a summary of the requirements for the additionality of the renewable electricity consumption under the EU RED as well as lSCC PLUS and the RSB Advanced Fuels standard.

To ensure the additionality of renewable electricity generation, the Delegated Act under the EU RED requires that the RE installations have come into operation no

Table 2 Comparison of requirements around the additionality of electricity consumed (RE—renewable electricity)

Framework	Requirements for electricity consumption—additionality
EU RED	• RE installations must be new, i.e. start of operation no earlier than 36 months before the start of operation of powerfuel production facility • RE installations must be unsupported, i.e. they have not received net operating or investment support (not applicable if direct connection of powerfuel production facility to RE installation) • If powerfuels production facility came online before01 January 2028, it is exempt from the additionality requirement until 01 January 2038 • Exemption from additionality requirements if the proportion of renewable energy in the bidding zone exceeded 90% during last year • The powerfuel production facility improves grid stability (during an imbalance settlement, RE generation installations using RE sources were redispatched downwards, with the electricity consumed for powerfuel production reducing the need for redispatching) • The power grid is sufficiently decarbonized (<18 gCO$_2$eq/MJ)
ISCC PLUS	• No direct requirement for additionality, but requirement for transparency around additionality (EACs must contain information about any support received and start date of operation of RE installation)
RSB Advanced Fuels	• RE installations must be new (start of operation no earlier than 36 months before the start of operation of powerfuel production facility) • If powerfuel production facility came online before 01 January 2027, it is exempt from additionality requirements

earlier than 36 months before the electrolyser. This holds true irrespective of whether the electrolyser is connected via a direct line to the RE installation or whether it sources electricity from the grid (with the respective PPAs with RE installations in place). The RSB Advanced Fuels standard implements the same requirement. ISCC PLUS does not currently feature a direct requirement on additionality, but does require transparency around this aspect, as it requires EACs to contain information about the RE installation's start date of operation as well as any support (financial or otherwise) it has received.

According to the EU RED Delegated Act, RE installations must not have received any financial support, both in terms of capital expenditure (CAPEX) and operational expenditure (OPEX). However, there are exceptions for support received prior to repowering, support for grid connection, fully repaid support, and support received for research, testing, and demonstration of powerfuels (RFNBO) production. Importantly, this does not apply to RE installations that have a direct connection to the powerfuel production facility. Neither ISCC PLUS nor the RSB standard have implemented such a requirement, though ISCC PLUS requires transparency around this aspect (see above).

Regarding the additionality requirement, the Delegated Act includes a grandfathering clause for first movers. Electrolysers that start operation before 01 January 2028 are exempted from the additionality requirement until 01 January 2038. While heavily debated among EU institutions, market operators, and environmental groups, this temporary exemption may support a quicker market-ramp-up of a still nascent electrolyser market. The RSB Advanced Fuels standard includes a similar exemption, though refers to 01 January 2027 and thus an earlier reference date instead.

Importantly, the Delegated Act describes three scenarios in which renewable electricity sourcing does not need to take into account additionality considerations. The first scenario, sometimes referred to as "sunset clause," considers bidding zones where electricity from renewable sources already represents the dominant share (more than 90%)—electrolysers located in that bidding zone can therefore generally count all electricity sourced as fully renewable, provided a certain number of full load hours of electrolyser production is not exceeded. The second scenario describes a situation in which powerfuels production improves grid stability and leads to renewable electricity sources being taken full advantage of. In essence, electricity can be considered fully renewable if, during an imbalance settlement period, RE installations were redispatched downwards and the electricity consumed for powerfuels production reduced the need for redispatching by a corresponding amount. A third scenario considers a situation in which the power grid is sufficiently decarbonized, defined as the bidding zone having an emissions intensity below 18 gCO_2eq/MJ (the rationale being that an emissions intensity of below 18 gCO_2eq/MJ would allow hydrogen produced from such electricity to still meet the 70% GHG emissions savings required under the EU RED (Chap. 3.3). The first two scenarios described above are also considered in the RSB Advanced Fuels standard, while none of the scenarios are currently explicitly covered by ISCC PLUS.

Temporal Correlation. A third important aspect of renewable electricity consumption pertains to temporal correlation. The aim of temporal correlation is that the production of powerfuels, particularly "green" hydrogen, takes place at times when RE installations actually generate renewable electricity. While temporal correlation is not a relevant requirement if the powerfuels production facility is directly connected to an RE installation, it is relevant in almost all other situations where renewable electricity is sourced from the power grid.

Table 3 gives a summary of the requirements surrounding the temporal correlation of the renewable electricity consumption and powerfuels production under the EU RED as well as ISCC PLUS and the RSB Advanced Fuels standard.

Until 31 December 2029, the Delegated Act under the EU RED requires a monthly correlation, i.e. powerfuels production must take place during the same calendar month as the renewable electricity produced under the respective PPAs. This correlation window is narrowed from January 2030 onwards, from which point an hourly correlation is required. In the Delegated Act, the European Commission acknowledges the current nascent state of the industry, with the need for more flexible rules at first. At the same time, the commission argues that the synchronization of electricity generation and hydrogen production should become stricter as markets, infrastructures and needed technologies develop.

Similar to the additionality requirement, the requirement for temporal correlation has led to intensive discussions among stakeholders in the EU. This is because while a strict coupling of renewable electricity production to powerfuels production helps

Table 3 Comparison of requirements around the temporal correlation of electricity consumption and powerfuel production (EAC—Energy Attribute Certificate; PPA—Power Purchase Agreement; RE—renewable electricity)

Framework	Requirements for electricity consumption—temporal correlation
EU RED	• Powerfuel production takes place in the same calendar **month** as the contracted RE sources generate renewable electricity (applicable until December 2029) • Powerfuel production takes place in the same calendar **hour** as the contracted RE sources generate renewable electricity (applicable from January 2030) • Option of using electricity from storage facilities that were charged during the same calendar month (until December 2029) or calendar hour (from January 2030) as the electricity generated by the contracted RE installations • Extra compliance route: Powerfuel production takes place during a one-hour period where the day-ahead price of the concerned bidding zone is less than 20 €/MWh or lower than 0.36 times the price of an allowance for one tonne of CO_2eq
ISCC PLUS	• Currently no requirement for temporal correlation
RSB Advanced Fuels	• Powerfuel production takes place in the same calendar month as the contracted RE sources generate renewable electricity • Option of using renewable electricity from storage facilities that were charged during the same calendar month as the electricity generated by the contracted RE installations

ensure the fuel's "green" characteristics, it also likely leads to higher production costs. On the contrary, procuring electricity freely at power markets allows for much needed flexibility for powerfuel producers (particularly in operating electrolysers), allowing them to benefit from price signals and possibly resulting in lower production costs—whereas the carbon intensity in both the electricity grid and the resulting fuel may be higher [12]. As a result, the requirements needed to be set in a way that ensures high environmental integrity of electricity sourcing and powerfuel production while also being practical and flexible enough to not risk economic viability of powerfuel projects and suffocate a nascent market.

The RSB Advanced Fuels standard adopted the temporal correlation requirement from the Delegated Act and requires a monthly correlation, though without foreseeing a switch to an hourly correlation after December 2029. ISCC PLUS does not currently feature a temporal correlation requirement.

As an additional option, the EU Delegated Act also considers the use of storage facilities. Concretely, sourcing of electricity for powerfuels production can be considered compliant with the temporal correlation requirement under the EU RED if the storage is located behind the same network point as the powerfuels production facility and is charged at the time of generation of the RE installations contracted via PPAs (i.e. in line with the temporal correlation requirement in effect at that time). This option is also considered by the RSB in its Advanced Fuels standard.

Alternatively, the Delegated Act allows for an extra compliance route, in which powerfuels production facilities may source renewable electricity during a one-hour period where the day-ahead price of the concerned bidding zone is less than 20 €/MWh or lower than 0.36 times the price of an allowance for one tonne of CO_2eq during the relevant period under the EU ETS. The rationale is that at times of low electricity prices, fossil-based electricity generation is not economically viable and additional demand from powerfuels production facilities would therefore trigger additional demand for renewable electricity production. Neither ISCC PLUS nor the RSB Advanced Fuels standard consider this extra compliance route in their framework.

Geographical Correlation. Another important aspect of electricity consumption is referred to as geographical correlation. Introducing a geographical correlation requirement aims to ensure the production of powerfuels occurs in places where renewable electricity is available. Similar to temporal correlation, the geographical correlation requirement is particularly relevant if the powerfuels production facility sources renewable electricity from the grid. Table 4 gives a summary of the requirements surrounding the geographical correlation of the renewable electricity consumption and powerfuels production under the EU RED as well as ISCC PLUS and the RSB Advanced Fuels standard.

In the framework of the EU RED, the Delegated Act requires that the RE installation and the electrolyser connected via PPAs are located in the same bidding zone. Given that bidding zones are generally designed to avoid grid congestion within this zone, the assumption is that with this requirement any grid congestion between the electrolyser and the RE installation will be avoided. However, the Delegated Act

Table 4 Comparison of requirements around the geographical correlation of electricity consumption and powerfuel production (EAC—Energy Attribute Certificate; PPA—Power Purchase Agreement; RE—renewable electricity)

Framework	Requirements for electricity consumption—geographical correlation
EU RED	• RE installation and powerfuel production facility must be located in the same bidding zone • RE installation and powerfuel production facility are located in an interconnected bidding zone. This is provided electricity prices on the day-ahead market in the RE installation's bidding zone are equal to or higher than in the powerfuels production facility's bidding zone • RE installation is located in an offshore bidding zone interconnected with the powerfuels production facility's bidding zone
ISCC PLUS	• Currently no requirement for geographical correlation
RSB Advanced Fuels	• Currently no requirement for geographical correlation. However, requirement that the risk of grid congestion is avoided, which amounts to an implicit requirement in this regard

also allows electrolysers to use electricity from RE installations located in neighbouring bidding zones, provided the electricity price there is equal to or higher than in the bidding of the electrolyser on the day-ahead market. Here, the rationale is that additional electricity demand from the electrolyser would help reduce grid congestion. Finally, the electrolyser may also source electricity from renewable electricity installations located in an offshore bidding zone, provided the two bidding zones are interconnected.

Neither ISCC PLUS nor the RSB Advanced Fuels standard have so far integrated direct requirements related to geographical correlation into their respective frameworks. The RSB, however, requires that grid congestion between the RE installation and the powerfuels production facility is avoided, which amounts to an implicit requirement in this regard.

Audit and Verification. The verification of compliance with the requirements with regard to renewability, additionality, temporal, and geographical correlation is done in the framework of audits conducted by certification bodies, as discussed in Chap. 2. In the framework of these audits, powerfuel production facilities will have to provide adequate documentation to substantiate their compliance with the respective requirements. This includes, but is not limited to, PPAs, EACs, documentation by the energy office of the country the RE installations are located in, and certain databases (e.g. Eurostat's SHARES).

Looking ahead, a significant challenge will be the fact that frameworks such as the EU RED, but also important standards such as the ISCC PLUS or RSB Advanced Fuels standard, were developed with an EU-centric view of renewable electricity systems. As a prominent example, all three frameworks heavily rely on concepts such as bidding zones. This is a concept clearly defined and commonly applied in the EU, but not necessarily in non-EU countries, in which electricity markets may be set up quite differently. The Delegated Act under the EU RED acknowledges that certain important concepts and definitions that apply within the EU may not exist outside the

EU territory. In such cases, it specifies that "equivalent concepts" may be used for the purpose of verification. According to the European Commission, these could be similar market regulations, the physical characteristics of the electricity grid (notably the level of interconnection) or, named as a last resort, the country. In practice, however, much of the decision as to how to interpret EU-centric requirements in non-EU countries will be left to the EU-recognized certification schemes. Accordingly, it must be taken care that there will be harmonized interpretation and application of the EU RED provisions by certification schemes. This is particularly important given that the EU expects a significant share of powerfuels used in the EU to be produced outside its territory.

4.2 Carbon Sources

The production of powerfuels, with the exception of fuels that do not contain any carbon (e.g. hydrogen and ammonia), requires a carbon source for the synthesis process. The type of carbon source and the way in which it is provided are major aspects in considering the sustainability performance of a powerfuel. Three different types of carbon sources for powerfuels production can be distinguished:

- Fossil-Based Carbon Sources. Fossil-based carbon sources mainly include industrial CO_2 point sources, such as cement or coal plants. While industrial point sources may be abundantly available in the short- to medium-term, the use of carbon from existing fossil sources entails the risk that non-sustainable economic practices will be continued for a longer time period than is technically necessary. This is also referred to as "lock-in effect," in which non-sustainable practices are maintained even though technical alternatives are on the horizon or already available.
- Biogenic Carbon Sources. Biogenic carbon can be obtained from biomass via different routes, including notably through fermentation to biogas, fermentation to bioethanol, or the combustion of biomass. From a sustainability perspective, the use of biogenic CO_2 is insofar considered preferable because biomass is a renewable raw material that, at least in principle, implies a closed carbon cycle. It is, however, necessary to consider both the type of biomass and the way it was cultivated or generated. Cultivation of crop biomass, for instance, may incur sustainability risks if it is generated using unsustainable farming practices (e.g. through extensive use of pesticides or improper handling of chemicals and waste) or takes place on lands with high carbon stocks or high levels of biodiversity. These risks may be exacerbated particularly for large-scale agricultural production, which may oftentimes be necessary to supply sufficiently large biogenic point sources, such as bioethanol production plants. Use of biogenic wastes, residues, and by-products (e.g. from agriculture or forestry) may be the preferred option, though only if the respective materials truly sit at the end of the waste hierarchy (i.e. are not diverted from a potentially higher form of use). Consequently, there is

also a risk of "lock-in" effects for biogenic carbon sources if non-sustainable agricultural practices or use of non-sustainable biogenic waste streams are extended longer than necessary. Furthermore, the use of CO_2 from biogenic sources should consider and avoid diverting CO_2 from existing and potentially more sensible use cases. As an example, CO_2 from bioethanol production is oftentimes used as additive for mineral water and soft drinks. Diverting this CO_2 as input for powerfuels production would mean that alternative supply routes would have to be established for the original use cases [9].

- Atmospheric Carbon. The extraction of CO_2 from ambient air via direct air capture (DAC) technologies is widely considered as the optimal long-term option for supplying sufficiently large quantities of CO_2 in a sustainable manner. In this case, a closed carbon cycle can be established where CO_2 emitted into the atmosphere from combustion processes is captured from the air and re-integrated into products, such as synthesized powerfuels. While the low concentration of CO_2 in the atmosphere compared to other point sources as well as the comparatively low technology readiness level of DAC in general still pose significant challenges for the scale-up and use of atmospheric carbon, overall sustainability risks are much lower than for either fossil-based or biogenic carbon. Of particular relevance for sustainably operating DAC plants will be the type of process energy used, i.e. this energy (in the form of electricity and/or heat) will also need to come from sustainable, renewable sources.

Table 5 gives a summary of the requirements on carbon sources for powerfuels production under the EU RED as well as ISCC PLUS and the RSB Advanced Fuels standard.

In principle, the Delegated Act established under the EU RED as well as ISCC PLUS and the RSB Advanced Fuels standard all consider fossil-based CO_2 captured from industrial point sources, CO_2 from biogenic sources as well as atmospheric CO_2 captured from the air. The Delegated Act and the RSB standard additionally consider CO_2 from geological sources, which is not currently considered under ISCC PLUS. Importantly, all three frameworks set a requirement that CO_2 is not eligible if it has been deliberately produced just to serve as feedstock for powerfuel production.

Under the EU RED, fossil-based CO_2 can be taken from industrial point sources, provided it has been taken into account upstream in an effective carbon pricing system as well as is incorporated in the chemical composition of the fuel before 2041 (this cut-off date is set to 2036 instead if the CO_2 is produced from the combustion of fuels for electricity generation). Neither ISCC PLUS nor the RSB Advanced Fuels standard currently feature cut-off dates for eligibility of fossil-based carbon or consider coverage under carbon pricing systems.

While CO_2 derived from biogenic sources is generally eligible under the EU RED, the Delegated Act specifies two important provisions in this regard. Firstly, only that biogenic CO_2 is eligible which is derived from biofuels that comply with the sustainability and GHG saving criteria set under the EU RED—thereby aiming to exclude potential biomass sources cultivated in unsustainable ways. Neither ISCC PLUS nor the RSB Advanced Fuels standard currently require a proof of sustainability of the

Table 5 Comparison of requirements for carbon sources (RFNBOs—renewable fuels of non-biological origin; RCFs—recycled carbon fuels)

Framework	Requirements for carbon sources
EU RED	*Valid carbon sources* • Fossil-based CO_2 captured from industrial activities listed in the 2003 EU ETS Directive, provided that • CO_2 is taken into account upstream in an effective carbon pricing system • CO_2 is incorporated in the chemical composition of the fuel before 2041 • If CO_2 is captured from industrial activities for electricity production, it must be incorporated in the chemical composition of the fuel before 2036 • Biogenic CO_2 from the combustion of biofuels, bioliquids, or biomass fuels compliant with the EU RED sustainability and GHG saving criteria • CO_2 capture must not have received credits for emissions savings from the CO_2 capture and replacement • CO_2 captured from the combustion of RFNBOs or recycled carbon fuels (RCFs) • Atmospheric CO_2 captured from the air • Naturally released CO_2 from geological sources *Further requirements* • CO_2 stemming from a fuel that was *deliberately* combusted to produce the CO_2 is not eligible • CO_2, the capture of which has received an emissions credit under other provisions of law, is not eligible
ISCC PLUS	*Valid carbon sources* • Fossil-based CO_2 captured from industrial processes which use fossil sources deliberately to produce electricity, heat, or materials (e.g. cement, iron, steel, petrochemical industry) • Biogenic CO_2 originating from biomass • Atmospheric CO_2 captured from the air *Further requirements* • CO_2 deliberately produced just to serve as carbon source for powerfuel production is not eligible
RSB Advanced Fuels	*Valid carbon sources* • Fossil-based CO_2 captured from industrial processes which use fossil sources deliberately to produce electricity, heat, or materials (e.g. cement, iron, steel, petrochemical industry) • Biogenic CO_2 originating from biomass • Atmospheric CO_2 captured from the air • CO_2 captured from geological sources *Further requirements* • CO_2 stemming from a fuel that was deliberately combusted to produce the CO_2 is not eligible • CO_2, the capture of which has received an emissions credit under other provisions of law, obligatory calculation or voluntary disclosure, is not eligible

source providing the biogenic CO_2. Secondly, biogenic CO_2 can only be considered in the EU RED if it has not already received credits for emissions savings from upstream CO_2 capture and replacement (CCR), e.g. as part of CCR at the biogas or bioethanol production facility. The RSB Advanced Fuels standard features a similar requirement but goes even further by including non-biogenic CO_2 as well. It also extends the requirement insofar that the CO_2 shall not have been credited under any other provision of law, obligatory calculation, or voluntary disclosure.

For atmospheric carbon, neither the EU Delegated Act nor ISCC PLUS or the RSB Advanced Fuels standard set further requirements so far.

Audit and verification. In order to prove type and eligibility of the carbon source they are using, producers of powerfuels will need to provide adequate documentation to their certification bodies in the framework of an audit. This could for instance primarily include documentation on the combusted inputs in a processing facility to substantiate the biogenic fraction of the CO_2, as well as proof of certification of the biomass entering a biogas facility.

4.3   Greenhouse Gas (GHG) Emissions

One of the primary reasons for the promotion of powerfuels is their potential to substantially reduce GHG emissions, particularly in transport sectors generally considered hard-to-abate, namely aviation, shipping, and heavy road transport.

It is therefore of extraordinary importance that GHG emissions from the production of powerfuels are calculated in a comprehensive, credible, and consistent manner. Methodologies as to how GHG emissions from powerfuels production and use are to be determined differ between (regulatory) frameworks. In many major frameworks (e.g. under the EU RED), a life cycle assessment (LCA) approach is applied. LCA is an approach used to assess the potential environmental impacts of a product, process, or system over its entire life cycle, from the extraction of raw material through to the end of its life [13]. In the realm of renewable fuels, LCA is often understood as focussing on the GHG emissions of a fuel across its entire life cycle of production and use.

LCAs can be extensive and complex to conduct. Under regulatory frameworks and standards, the aim is generally to put in place a methodology that is credible and produces robust and comparable results across different economic operators obliged under those frameworks. At the same time, the methodology must be simple enough so that it can be applied by economic operators at a manageable level of complexity. In addition, the methodology should be practicable enough to allow for auditing and verification by certification bodies. Many LCA methodologies applied under different frameworks therefore work with a variety of simplifications [14]. This holds true for the three frameworks analysed more closely in this chapter—the EU RED, the ISCC PLUS, and the RSB Advanced Fuels standard. In fact, both

ISCC PLUS and the RSB standard strongly model their approach after the EU RED methodology.

The EU RED features a well-to-wheel (WTW) approach, which includes emissions from the electricity input all the way up to and including the combustion of powerfuels. Both direct emissions and indirect emissions are considered. Direct emissions are those emissions generated during combustion of the powerfuel, whereas indirect emissions are those that occur upstream—namely emissions from the supply of inputs to the production process (including *their* production and transport to the powerfuel production facility), emissions from the powerfuel production process as well as from the transport and distribution of the final fuel. Importantly, the EU RED methodology accounts for the three major greenhouse gases, namely carbon dioxide (CO_2), methane (CH_4), and nitrous oxide (N_2O).

For simplification purposes, the EU RED methodology focusses on the material and energy flows of an individual powerfuel supply chain. In a first step, GHG emissions for all relevant inputs and outputs are calculated, along all supply chain elements that make up the powerfuel production chain. The results are then converted to a uniform reference value—also referred to as "functional unit"—generally indicated in gCO_2eq/MJ. Through this, the GHG emissions intensity associated with a given unit of energy, e.g. for a powerfuel, can be determined.

Finally, this GHG emissions intensity is compared to a fossil reference value in order to determine the GHG emissions *saving* that the particular powerfuel achieves. Many frameworks set a minimum emission saving threshold that a powerfuel must achieve in order to be considered eligible and/or "sustainable." Both the EU RED and the RSB standard, for instance, require a powerfuel to achieve at least 70% GHG emissions saving to be considered compliant (with the fossil reference value being 94 gCO_2eq/MJ in both cases). In practice, this means that a powerfuel must not exceed specific CO_2 emissions of 28.2 gCO_2eq/MJ.

Figure 5 illustrates an exemplary, simplified powerfuel supply chain and likely inputs and outputs that must be considered within a life cycle GHG emissions calculation. The life cycle GHG emissions for hydrogen generally only include the emissions from electricity production and those generated through operation of the electrolysis plant itself. In contrast, calculating the life cycle GHG emissions of powerfuels such as synthetic methanol or synthetic kerosene requires the consideration of additional processing and transportation steps as well as the emissions resulting from the required carbon source.

Under the EU RED, the GHG emissions methodology for powerfuels follows a similar logic as the methodology that has been applied for biofuels for well over a decade. Still, the European Commission introduced a separate formula with which GHG emissions from powerfuels are to be determined for EU RED purposes, as shown in the following equations:

In a first step, emissions from the supply of inputs (e_i) have to be determined. Here, the EU RED methodology distinguishes between three different factors—elastic inputs, rigid inputs, and the inputs' existing use or fate.

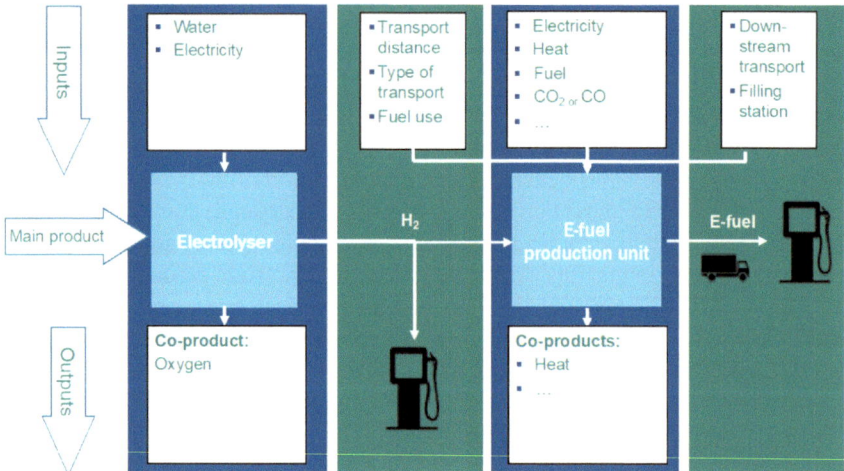

Fig. 5 Simplified depiction of typical inputs and outputs relevant for GHG calculation for powerfuels

Elastic inputs ($e_{ielastic}$) are generally those inputs whose supply can be increased to meet extra demand. Prominent examples for elastic inputs include hydrogen, electricity, and petroleum products. Importantly, electricity that is sourced in compliance with the requirements for renewable electricity set out in the Delegated Act (Chap. 3. 1) is accounted for with zero emissions. Electricity that does not meet all requirements of the Delegated Act must be considered not fully renewable. In this case, the Delegated Act provides certain grid emission factors per country.

Rigid inputs (e_{irigid}) are generally those inputs whose supply *cannot* be increased to meet extra demand. These inputs are often essential to a production process, and potential substitutes not readily available. A prominent example for a rigid input is flue gas.

Emissions from *inputs' existing use or fate* (e_{ex-use}) are generally those emissions that are avoided when carbon is used as input for powerfuel production. They can be considered as avoided (and therefore be subtracted) in the emissions calculation only if the CO_2 that was captured and used for powerfuel production comes from one of the carbon sources deemed eligible under the EU RED (Chap. 3.2). A simple example would be CO_2 from direct air capture to produce powerfuels. In contrast, the respective Delegated Act clearly defines two types of scenarios in which CO_2 cannot be considered as avoided. This includes scenarios in which the CO_2 was deliberately produced as well as scenarios in which the CO_2 already received emissions credits under other provisions of law.

Emissions from *processing* (e_p) include all those GHG emissions directly related to the processing steps along value chain, e.g. at the electrolyser and downstream conversion elements. In essence, these are atmospheric emissions from the processing itself, as well as any emissions stemming from waste treatment and leakages.

Emissions from *transport and distribution* (e_{td}) include all those GHG emissions that occur due to the storage, transportation, and distribution of a powerfuel. For instance, this may include emissions from truck transport of powerfuels to a refuelling station.

Emissions from *combusting the fuel* (e_u) refers to direct GHG emissions, i.e. those that occur as a result of the powerfuel's combustion (e.g. in a conventional internal combustion engine).

Emissions *savings from carbon capture and geological storage* (e_{ccs}) can be accounted for, provided they are stored in accordance with the EU's Directive on geological storage of carbon dioxide. This may, for instance, include tailpipe carbon capture and storage on a ship combusting synthetic methanol.

Importantly, GHG emissions that occur from the construction of infrastructure and equipment (i.e. electricity generation capacity and hydrogen production) can be significant, but are generally not accounted for under the three analysed frameworks. Table 6 gives an overview over the GHG calculation formula for RFNBOs/powerfuels under the EU RED.

Table 6 GHG calculation formula for RFNBOs/powerfuels under the EU RED

$E = e_i + e_p + e_{td} + e_u - e_{ccs}$	(Eq. 1)
$e_i = e_{i\,elastic} + e_{i\,rigid} - e_{ex\text{-}use}$	(Eq. 2)
with	
E	Total emissions from the use of the fuel in gCO_2/MJ
e_i	Emissions from supply of inputs
$e_{i\,elastic}$	Emissions from elastic inputs
$e_{i\,rigid}$	Emissions from rigid inputs
$e_{ex\text{-}use}$	Emissions from inputs' existing use or fate
e_p	Emissions from processing
e_{td}	Emissions from transport and distribution
e_u	Emissions from combusting the fuel
e_{ccs}	Emissions savings from carbon capture and geological storage

Audit and verification. The audit and verification of GHG calculations with respect to their availability, completeness, and correctness is a crucial part of sustainability certification. Economic operators will first need to calculate their GHG emissions following the applicable methodology (e.g. the one outlined above). During the audit, the certification body will then check the GHG calculation with respect to their completeness and accuracy, verifying in particular all relevant material and energy inputs as well as outputs. Only if the calculation is to be found a fair representation of the economic operator's processes can a certificate be issued to the operation. Once verified, the economic operator is allowed to use that GHG value for the upcoming certification period and can forward this value as part of the overall sustainability documentation sent to its customer/recipient (Chap. 4.7). This approach aims to ensure that only audited and verified GHG values are used and forwarded between economic operators in a powerfuel supply chain.

Table 7 gives a summary of the requirements for calculating GHG emissions for powerfuels production under the EU RED as well as ISCC PLUS and the RSB Advanced Fuels standard.

Table 7 Requirements for GHG emissions calculations of powerfuels (RE—renewable electricity)

Framework	Requirements for GHG emissions calculation
EU RED	*Minimum GHG saving required from powerfuels* • 70% (reference value: 94 gCO$_2$eq/MJ) *System boundary of GHG calculation* • Well-to-wheel (from electricity production to powerfuel combustion) • Emissions from construction of infrastructure and equipment not considered
ISCC PLUS	Methodology largely based on and aligned with EU RED methodology *Minimum GHG saving required from powerfuels* • Not currently defined *System boundary of GHG calculation* • Well-to-wheel (from electricity production to powerfuel combustion) • Emissions from construction of infrastructure and equipment not considered
RSB Advanced Fuels	Methodology largely based on and aligned with EU RED methodology *Minimum GHG saving required from powerfuels* • 70% (reference value: 94 gCO$_2$eq/MJ) *System boundary of GHG calculation* • Well-to-wheel (from electricity production to powerfuel combustion) • Emissions from construction of infrastructure and equipment not considered

4.4   Land Use and Land Use Change

Renewable electricity consumption, CO_2 sources, and the life cycle GHG impact of powerfuels are the sustainability aspects most discussed and considered as part of prevalent regulatory frameworks and certification standards. Other sustainability aspects are decidedly less discussed and integrated yet are still important to consider as part of a holistic sustainability examination of powerfuels. The first of these aspects is land use and land use change, as discussed in the following.

For biomass production, land use has long been among the key sustainability aspects addressed as part of regulatory frameworks and certification standards. For instance, since its inception, the EU RED has sought to avoid certain types of land use change that are associated with biofuel production—it notably excluded biofuel feedstocks that were cultivated on areas considered to have high carbon stocks or high biodiversity value after January 2008.

For powerfuels, no such criteria are laid down, neither in the EU RED nor in the more detailed Delegated Acts. The EU RED does, however, allow certification schemes recognized thereunder to go beyond the legally mandated requirements and implement such criteria on a voluntary basis. The lack of mandatory criteria around land use in the EU RED may at least in part be explained by the fact that renewable electricity generation for powerfuel production requires significantly lower land area per energy unit than biofuels made from primary biomass (i.e. crops). Still, electricity generation can be associated with significant land use implications, particularly from solar (e.g. ground-mounted photovoltaic systems) or wind energy (e.g. larger wind parks). This may become particularly relevant in view of the substantial expansion of renewable electricity production that will be needed to supply future powerfuel production capacities. At the same time, it is important to take into account that electricity production from solar or wind energy does not necessarily always pose a risk to highly valuable areas or lands suitable for crop production, as particularly suitable locations for electricity generation based on solar energy may be found in desert-like regions [16]. Compared to the land requirements of installations generating renewable electricity, the land requirements for the provision of CO_2, the electrolysis plant, and subsequent conversion units are generally negligible.

Other than the EU RED, the RSB Advanced Fuels standard features an explicit land-related sustainability requirement. Concretely, it forbids renewable electricity generation from areas identified as "no-go areas" (i.e. those that areas have high carbon stock or high conservation value). A similar requirement is not yet integrated under ISCC PLUS.

Audit and verification. The audit and verification of requirements around land use and land use change are most commonly conducted by making use of remote sensing, including satellite imagery. For instance, for ISCC audits certification, bodies often make use of a tool called GRAS (Global Risk Assessment Services), which allows for the mapping of areas and facilities to be certified in relation to high carbon stock or highly biodiverse areas.[17] An overlap or even just proximity of certification areas with highly valuable lands will inform the certification body's risk assessment in advance of certification audits, as explained in Chap. 2.3. Both certification schemes and certification bodies have built up extensive experience around the audit and verification of land use and land use change in the framework of biomass and biofuels certification. This experience can be utilized for the certification of powerfuels in the future.

Table 8 summarizes the requirements around land use for powerfuels production under the EU RED as well as ISC PLUS and the RSB Advanced Fuels standard.

Table 8 Requirements around land use (RE—renewable electricity)

Framework	Requirements for land use
EU RED	• Currently no requirements around land use and land use change for powerfuel production
ISCC PLUS	• Currently no requirements around land use and land use change for powerfuel production
RSB Advanced Fuels	• No production of renewable electricity from areas identified as "no-go areas" or those with high conservation value after 01 January 2008

4.5   Water Use

Another important sustainability aspect pertains to water use for the production of powerfuels. Water is the basis for electrolysis-based hydrogen production and thus indispensable for the provision of powerfuels. Water use is considered in most life cycle GHG methodologies; however, the GHG emissions from water use are generally negligible. Water use should not primarily be viewed in terms of its contribution to a powerfuel's life cycle GHG emissions value, but rather in terms of its sustainable use as part of the fuel's production. Overall, and especially when compared to the production processes of other renewable fuels derived from cultivated crops, the water consumption for powerfuel production can be considered insignificant [15], as depicted in Fig. 6.

However, it is important to consider the production location in which the water is sourced. Regions that are under high water stress are at particular risk when it comes to unsustainable water use. As a general rule, electrolysers and powerfuel production facilities should therefore be located in regions where water is sufficiently available and does not lead to water scarcity for other uses, particularly for human consumption.

Neither the EU RED nor the Delegated Acts include any provisions for water use when it comes to the production of powerfuels. In contrast, the RSB Advanced Fuels

---

[17] The GRAS system can be accessed via https://www.gras-system.org.

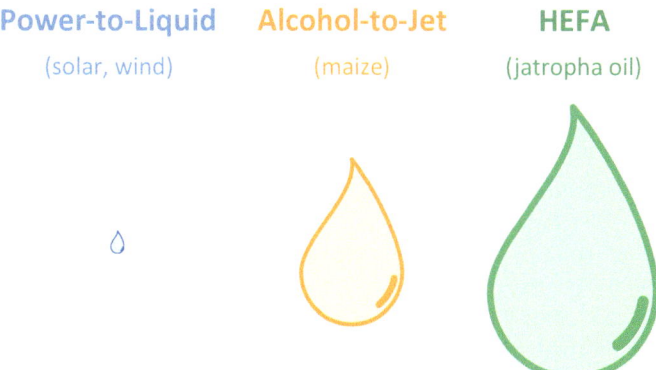

Fig. 6 Water consumption of powerfuels compared to selected biofuels. PtL water demand is about 4 L of water per kg jet fuel. Based on [16]

standard requires a documented water management plan, that puts in context and justifies the water use of the economic operator (e.g. the operator of an electrolyser) as part of the region it is located in, both with regard to an efficient use of water as well as to maintaining water quality. It also requires the respect of existing water rights by local and indigenous communities, in order to foreclose any social conflict around water availability. ISCC PLUS does not have requirements around water use in place so far.

One way of operationalizing the metric of water use in the framework of sustainability certification could be the use of the so-called Water Stress Index (WSI) according to *Falkenmark* [15]. This index refers to the availability of fresh water per capita per year. A value between 1,000 and 1,700 m^3 per capita would imply water scarcity, a value between 500 and 1,000 m^3 per capita water stress, and a value below 500 m^3 per capita would indicate absolute water stress. In practice, certification schemes could set a certain max value of water stress which should not be exceeded by the economic operator subject to certification.

Table 9 summarizes the requirements around water use for powerfuels production under the EU RED as well as ISCC PLUS and the RSB Advanced Fuels standard.

Table 9 Requirements around water use

Framework	Requirements for water use
EU RED	• Currently no requirements around water use for powerfuels production
ISCC PLUS	• Currently no requirements around water use for powerfuels production
RSB Advanced Fuels	• Respect of existing water rights • Documented water management plan for using water efficiently and maintaining water quality

4.6   Social Impact

Environmental sustainability aspects are the clear focus of virtually all regulatory frameworks and voluntary certification standards. At the same time, it is important to ensure that the production of renewable fuels, including powerfuels, does not lead to negative social effects in the countries where it is produced. Negative social impact from fuels production could include, for instance, unsafe working conditions at production facilities, lack of compliance with human or labour rights (e.g. child labour, denial of right to unionize) or putting additional pressure on food security in regions that can already be considered food-insecure.

The EU RED does not include any direct requirements pertaining to social impact of powerfuels production. This may at least in part be explained by the view that this could be seen as too much interference by third countries and may be legally challenged, for instance at level of the World Trade Organization (*DENA*). Importantly, however, the EU RED does allow recognized certification schemes to implement respective requirements on a voluntary basis if they wish to do so.

The RSB Advanced Fuels standard includes a set of requirements around social impact for economic operators. Most notably, this includes criteria for safe working conditions and compliance with human and labour rights (including the promotion of rural development and ensuring fair employment conditions). ISCC PLUS does not currently include a similar requirement. The RSB standard also requires the monitoring and mitigation of food security in particularly food-insecure regions. Overall, however, food security is a considerably less relevant sustainability concern compared with other types of renewable fuels, particularly biofuels produced from crops cultivated on dedicated land areas.

Table 10 summarizes the requirements around social impact for powerfuels production under the EU RED as well as ISCC PLUS and the RSB Advanced Fuels standard.

Table 10   Requirements around social impact of powerfuels production

Framework	Requirements for social impact
EU RED	• Currently no requirements around social impact for powerfuels production
ISCC PLUS	• Currently no requirements around social impact for powerfuels production
RSB Advanced Fuels	• Safe working conditions • Compliance with human and labour rights (e.g. promotion of rural development and fair employment conditions of workers) • Operations should improve food security in food-insecure regions

4.7 Traceability and Chain-of-Custody

Traceability and chain-of-custody is, strictly speaking, not a sustainability aspect in and of itself. Rather, it can be understood as a crucial supporting feature that aims to help ensure compliance with all the sustainability aspects previously discussed.

- The term "traceability" describes the ability to identify and trace the origin, processing history, distribution, and location of products (e.g. sustainably certified renewable fuels) as they move through supply chains [17].
- The term "chain-of-custody" describes the process of transferring, monitoring, and controlling inputs and outputs and related information as they move through the supply chain. In essence, this provides assurance that a given batch of product (e.g. a powerfuel) is associated with a set of specific characteristics (e.g. related to its sustainable production or savings in greenhouse gas emissions) and that the information on these characteristics is also transferred, monitored, and controlled throughout the supply chain [17]. Simply put, traceability is the ability to demonstrate the chain-of-custody [18].

A chain-of-custody (CoC) system forms the basis for any claim that can be made about a certified product, and is essentially comprised of two elements. The CoC standard establishes a set of requirements that provide the necessary controls on the movement of sustainable products for each stage of the product's supply chain. The supporting assurance system, via the use of approved certification bodies, is then used to verify that an economic operator has met the requirements of the CoC standard [18]. In general, all CoC systems aim to fulfil at least the following functions:

- Ensure a custodial sequence along a product's supply chain,
- Ensure that volumes of certified product sold do not exceed volumes of certified bought (at least over a certain defined time frame),
- Serve to carry sustainability claims (e.g. on feedstock sustainability or GHG emissions data) along the supply chain to allow for product sustainability claims at the end of that chain,
- Increase overall transparency in the supply chain.

Broadly speaking, there are three different forms of CoC systems or CoC models as they are commonly referred to, as depicted in Fig. 7.

- Physical Segregation. Physical segregation is a CoC model that requires the certified product to be kept physically separated from non-certified sources along the complete supply chain. Mixing of certified products with non-certified products is not allowed. As a consequence, it can be ensured that a particular product comes from entirely certified sources. The segregation model is commonly used in, though not restricted to, food and agricultural industries, where strict physical traceability of ingredients is required (e.g. for organic cotton or fair trade coffee). A physical segregation model can be further divided into the "identity preserved" and the "segregation" model, the distinction of which is, however, not relevant for powerfuels and shall not be further discussed in this chapter.

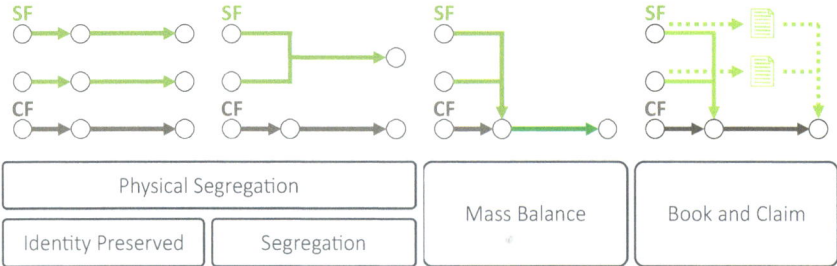

Fig. 7 The three basic forms of CoC systems (SF—sustainable fuel; CF–conventional fuel)

- Mass Balance. The mass balance CoC model, in contrast to physical segregation, allows for the physical mixing of sustainable product with non-sustainable product. At the same time, it ensures that the amount of sustainable product can be tracked as it moves through a production system, while ensuring an appropriate allocation of this amount to outgoing product each supply chain stage. While different forms of mass balance CoC models exist, the ones most commonly applied do not generally provide the assurance that a given percentage of sustainable molecules is actually in the final product. However, incoming and outgoing amounts of sustainable products are generally tracked and reconciled at each stage in the supply chain, so that never more sustainable product is sold than equivalent feedstock has come in. Mass balance is most often used where the segregation of certified and non-certified products is difficult to achieve. This includes industries such as renewable fuels or the chemical industry.
- Book and Claim. The Book and Claim CoC model essentially completely decouples the physical flow of the sustainable product and the associated sustainability value (which may come from the associated GHG emissions saving of the product for instance). Certified and non-certified products are allowed to flow freely through the supply chain, with no further physical traceability for the downstream supply chain. Instead, sustainability certificates, or credits, are generated for a given unit of sustainable product and can then generally be sold and traded on an online platform between economic operators, more or less completely independent of the physical movement of the underlying sustainable product. While it is ensured that a specified volume of sustainable product enters the supply chain at some point, there is no way of knowing where the physical volumes will end up—product claims are instead based on the sustainability certificates bought and retired from the online platform. The application of book and claim is potentially most interesting in industries where it is reasonable to supply sustainable products (commingled with conventional products) in already existing distribution infrastructures or where their supply is still scarce and (geographically) not yet always connected with emerging demand through physical supply chains, such as for sustainable aviation fuels (SAF), marine fuels, or synthetic fuels in general.

Suitability of CoC models. Comparatively, a physical segregation model requires the most logistical and administrative effort as it demands strict physical separation of certified products from non-certified products. This can lead to prohibitive costs given the logistical realities of co-mingled storage, transport, and processing in certain industries, such as in renewable fuels. The mass balance and book and claim models are more suited for renewable fuels and powerfuels instead, as they allow for leveraging existing supply chains and logistics and thus support a gradual increase in the incorporation of renewable fuels in the fuel infrastructure.

Table 11 summarizes the requirements for traceability and chain-of-custody for powerfuels production under the EU RED as well as ISCC PLUS and the RSB Advanced Fuels standard.

Traceability and CoC requirements. For powerfuels and renewable fuels in general, the EU RED, ISCC PLUS, and the RSB Advanced Fuels standard require that, with the exception of feedstock production (as explained below), each economic operator along the supply chain is individually certified in order to ensure full traceability. In practice, this implies that every economic operator, at every supply chain stage, is regularly audited. Further, the three frameworks require that economic operators implement a mass balance CoC model to track sustainability claims as they move through the supply chain alongside the fuels they are associated to. Every audit

Table 11 Requirements for traceability and chain-of-custody (CoC—chain-of-custody)

Framework	Requirements for traceability and chain-of-custody
EU RED	• Every economic operator along the supply chain must be individually certified • Exceptions, i.e. group certification, possible for feedstock producers and storage facilities • Use of mass balance CoC model required • Avoidance of double counting of fuel volumes and GHG emissions savings required
ISCC PLUS	• Every economic operator along the supply chain must be individually certified • Exceptions, i.e. group certification, possible for feedstock producers and storage facilities • Use of mass balance CoC model required • Avoidance of double counting of fuel volumes and GHG emissions savings required
RSB Advanced Fuels	• Every economic operator along the supply chain must be individually certified • Exceptions, i.e. group certification, possible for feedstock producers and storage facilities • Use of mass balance CoC model required • Avoidance of double counting of fuel volumes and GHG emissions savings required

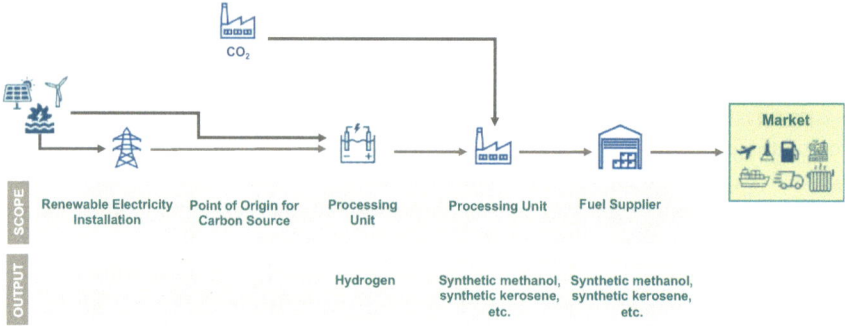

Fig. 8 Exemplary, simplified supply chain for powerfuels

will therefore include the verification of traceability and chain-of-custody requirements set under the respective framework. Figure 8 provides an exemplary, simplified
supply chain for powerfuels.

Renewable electricity installations supply renewable electricity to the first
processing unit of the value chain, i.e. electrolysers. These produce hydrogen, which
can either be used directly or serve as intermediate for the production of hydrogen-
derived powerfuels. For the latter, for many powerfuel options, a carbon source is
needed, which can be supplied from different kinds of sources (Chap. 3.2). The
resulting powerfuels may then be further traded, stored, and ultimately supplied to
the market, e.g. for application in heavy road transport, aviation, or shipping. The
exemplary supply chain is then covered by certification as depicted in Fig. 9.

Each economic operator involved in the powerfuel supply chain is individually
certified and therefore regularly audited. An exception is made for the feedstock
production side—renewable electricity installations do not have to be individually
certified, similar to those units that produce or generate the required carbon. Though
not individually certified, they still need to provide sustainability documentation

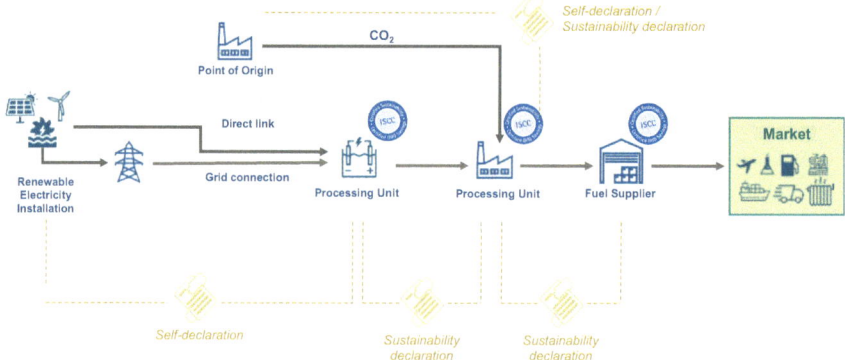

Fig. 9 Certification coverage of exemplary, simplified supply chain for powerfuels

to the subsequent supply chain element, which is checked as part of that element's audit. This sustainability document, referred to as "self-declaration," includes specific information about the type and amount of feedstock supplied. Crucially, this ensures that traceability is given for the complete upstream supply chain without having to necessarily audit every single renewable electricity installation or carbon source individually.

Traceability along the downstream supply chain is ensured via the use of so-called "sustainability declarations." Sustainability declarations are documents that include all relevant sustainability data regarding a batch of product (e.g. on the sustainability of its production or its GHG emissions saving). As part of a mass balance system, they are forwarded by certified economic operators alongside the respective batches of powerfuels they are associated to. As a sustainability declaration moves through the different supply chain stages, the information on it is continuously updated (e.g. additional GHG emissions from further processing steps are included). As the last entity in the supply chain, the fuel supplier will then not only be able to supply a powerfuel to its target market, but also be able to provide a sustainability declaration that includes all needed information to prove compliance of that fuel with the relevant requirements as per the applicable framework (e.g. the EU RED).

While the EU RED, ISCC PLUS, and the RSB Advanced Fuels standard all require a mass balance CoC model currently, a book and claim model could still play an important role in the scale-up of sustainable powerfuels and be integrated in sustainability certification schemes. Book and claim would essentially allow for more flexibility in the still nascent powerfuels market, allowing an easier matching of scarce supply and growing demand in the absence of established physical supply chains. In a certification context, this would require a digital database managed by a trusted party (e.g. by a certification scheme itself) and clear and robust rules around the issuing of sustainability certificates in that database.

Audit and verification. As mentioned before, certification bodies will audit every element that is part of a supply chain element. Feedstock producers' compliance will be checked as part of the audit of the economic operator they are supplying to. For instance, during an audit of an electrolyser, the respective PPAs and/or EACs as concluded with renewable electricity installations need to be made available to the auditor. Specifically with regard to traceability and chain-of-custody requirements, certification bodies will first check that an economic operator has a bookkeeping system in place which allows for the tracking of sustainable product and the associated sustainability data as they move through the operator's custodianship. Under the three frameworks analysed, the sustainability data for renewable fuels that has to be tracked particularly concerns information about GHG emissions as well as sustainability and country of origin of the fuel's feedstock. Furthermore, certification bodies will check that economic operators have correctly forwarded sustainability information to downstream recipients, i.e. via sustainability declarations as explained above.

Of critical importance in the framework of auditing and verification is furthermore the avoidance of double counting. Broadly speaking, the risk of double counting refers to the potential for emissions reductions to be counted more than once towards a climate change mitigation effort. Avoiding double counting is essential for environmental integrity, as double counting would lead to the actual GHG emissions being higher in sum than what individual organizations report them to be and could ultimately undermine the effectiveness of policies. In practice, double counting could, for instance, occur if sustainability declarations were issued twice for one batch of fuel, leading to the powerfuel volume and/or the associated GHG emissions savings to be double counted. It is one of the primary tasks of robust traceability and CoC systems, including the supporting audit and verification under certification schemes, to ensure that no double counting takes place.

Databases, in which sustainability information is entered and tracked digitally, could prove to be an important piece of the puzzle. Under the EU RED, this will be managed and organized via the so-called "Union Database." The database aims at enabling the tracing of sustainable products and will cover the movements of all renewable fuels and their associated sustainability information for EU RED purposes. It is expected that this database will make a vital contribution to the comprehensive monitoring of the production and consumption of renewable fuels and powerfuels, mitigating risks of double counting or irregularities along the respective supply chains.

5   Conclusion

Energy carriers produced from renewable energy will be key in reducing greenhouse gas (GHG) emissions and achieving climate targets in the transport sector. In particular, "hard-to-abate" sectors such as aviation or shipping will rely on liquid fuels from renewable energy for the foreseeable future. Powerfuels, given their basic advantages in terms of scalability of production and potential for high GHG emissions reductions, are poised to play a crucial role in this context.

In their early stage of market development, robust regulatory frameworks and governmental support schemes are crucial for scaling up the production and use of sustainable powerfuels. Presently, many frameworks, programmes, and roadmaps from governments around the world already consider sustainable powerfuels to some degree; however, clear definitions around how to exactly define sustainability in powerfuel production are still emerging and subject to ongoing discussion.

Underpinned by a well-established "ecosystem" of different organizations and implemented via a third-party certification principle, sustainability certification allows to properly differentiate between products that meet defined sustainability criteria and those that do not. Certification is widely used in many major regulated and non-regulated markets today, including for renewable fuels under the EU RED. Sustainability certification can be an equally useful tool to ensure that the production and use of powerfuels achieves sufficient GHG emissions reductions and avoids other negative environmental and social effects.

In this context, seven key sustainability aspects of powerfuels included in important regulatory frameworks and certification standards for powerfuels—namely the framework of the EU Renewable Energy Directive (EU RED), ISCC PLUS, and RSB Advanced Fuels—are analysed. The key characteristics derived from the analysis of the seven sustainability aspects are summarized in Table 12.

Table 12 Overall comparison of key sustainability aspects of powerfuels (CoC—chain-of-custody; GHG—greenhouse gas)

	EU RED	ISCC PLUS	RSB Advanced Fuels
Electricity consumption	Requires electricity consumption to be fully renewable and following principles of additionality, temporal and geographical correlation	Requires electricity consumption to be fully renewable and transparency around additionality	Requires electricity consumption to be fully renewable and following principles of additionality and temporal correlation
Carbon source	Considers all carbon sources (incl. fossil, biogenic, atmospheric); no deliberate production of CO_2	Considers all carbon sources (incl. fossil, biogenic, atmospheric); no deliberate production of CO_2	Considers all carbon sources (incl. fossil, biogenic, atmospheric); no deliberate production of CO_2
GHG emissions	Requires 70% GHG emissions saving of powerfuel; well-to-wheel, i.e. life cycle, approach	Methodology based on EU RED; minimum GHG saving not currently defined, well-to-wheel, i.e. life cycle, approach	Methodology based on EU RED; requires 70% GHG emissions saving of powerfuel; well-to-wheel, i.e. life cycle, approach
Land use (change)	Currently no requirements	Currently no requirements	Protection of areas with high conservation value or those identified as "no-go areas"
Water use	Currently no requirements	Currently no requirements	Respect of water rights, efficient use of water, maintaining water quality
Social impact	Currently no requirements	Currently no requirements	Compliance with human and labour rights, improvement of food security
Traceability/ CoC	Individual certification of every economic operator along supply chain required; use of mass balance CoC model; avoidance of double counting	Individual certification of every economic operator along supply chain required; use of mass balance CoC model; avoidance of double counting	Individual certification of every economic operator along supply chain required; use of mass balance CoC model; avoidance of double counting

In the three frameworks, requirements around electricity consumption are clearly defined already, centring around key principles such as renewability, additionality, and temporal as well as geographical correlation. Similarly, requirements around the eligibility of carbon sources are also reasonably well-established, prominently considering fossil-based, biogenic, and atmospheric carbon sources, with a range of safeguards in place. In addition, the EU Delegated Acts have established a comprehensive methodology for determining the GHG impact of powerfuels, which are key not only under the EU RED but also serve as basis for other certification frameworks—including for ISCC PLUS and the RSB Advanced Fuels standard. Other environmental and socio-economic criteria, including with regard to land use, water use, and social impact, are not yet as widely considered in existing frameworks. Their careful inclusion in regulatory frameworks and certification systems will ultimately be needed to allow for comprehensive sustainability assessments of powerfuels, in line with ambitious sustainable development goals.

At the same time, it is important to keep in mind that the sustainability certification of powerfuels is still a relatively new and rapidly developing field. With most powerfuel production facilities and supply chains still under development, in planning or demonstration phase and not yet fully established, the same applies to sustainability certification of powerfuels. That being said, the extensive experience gained from the certification of renewable fuels will be useful in setting up transparent and robust certification systems for powerfuels going forward. Over the coming years, it will be key to integrate first experiences and learnings to continuously progress towards effective and credible sustainability certification frameworks for powerfuels.

References

1. US Department of Energy (2022) Inflation reduction act summary. https://www.energy.gov/sites/default/files/2022-10/IRA-Energy-Summary_web.pdf. Accessed 28 April 2023
2. International Energy Agency (2022) Hydrogen review 2022. https://iea.blob.core.windows.net/assets/c5bc75b1-9e4d-460d-9056-6e8e626a11c4/GlobalHydrogenReview2022.pdf. Accessed 02 April 2023
3. European Commission: Hydrogen. https://energy.ec.europa.eu/topics/energy-systems-integration/hydrogen_en. Accessed 02 May 2023
4. Deutsche Energie-Agentur (2022) Global harmonisation of hydrogen certification. https://www.dena.de/newsroom/publikationsdetailansicht/pub/report-global-harmonisation-of-hydrogen-certification/. Accessed 12 March 2023
5. Joint Research Centre: Renewable Energy—Recast to 2030 (RED II). https://joint-research-centre.ec.europa.eu/welcome-jec-website/reference-regulatory-framework/renewable-energy-recast-2030-red-ii_en. Accessed 28 March 2023
6. European Commission: RED III proposal. https://eur-lex.europa.eu/resource.html?uri=cellar:dbb7eb9c-e575-11eb-a1a5-01aa75ed71a1.0001.02/DOC_1&format=PDF. Accessed 15 April 2023
7. European Commission: List of recognized voluntary schemes. https://energy.ec.europa.eu/topics/renewable-energy/bioenergy/voluntary-schemes_en. Accessed 15 March 2023
8. International Laboratory Accreditation Cooperation: About ILAC. https://ilac.org/about-ilac/. Accessed 15 March 2023
9. Altmann M, Abdalla N, Astono Y, Fehrenbach H, Krenn P, Schmidt P (2022) Entwicklung von PtX-Nachhaltigkeitsstandards und -Indikatoren. Eine Expertise im Auftrag des PtX Lab Lausitz. Abschlussbericht

10. Gericke N, Thomas S (2022) Certification of green hydrogen: recent efforts and developments in the European Union. https://www.energypartnership.cn/fileadmin/user_upload/china/media_elements/publications/2022/Certification_of_green_hydrogen_EN.pdf. Accessed 02 April 2023

11. Fehrenbach H, Busch M, Bürck S, Bischoff M, Theis S, Reinhardt J, Grahl B (2021) Flächenrucksäcke von Gütern und Dienstleistungen—Ermittlung und Verifizierung von Datenquellen und Datengrundlagen für die Berechnung der Flächenrucksäcke von Gütern und Dienstleistungen für Ökobilanzen—Teilbericht II: FALLBEISPIELE. Dessau-Roßlau: Umweltbundesamt

12. Brauer J, Trüby J, Villavicencio M (2022) Green hydrogen: how grey can it be? Working Paper. Florence School of Regulation. https://fsr.eui.eu/publications/?handle=1814/74850. Accessed 02 April 2023

13. Carvalho F, O'Malley J, Osipova L, Pavlenko N (2023) Key issues in LCA methodology for marine fuels. https://theicct.org/wp-content/uploads/2023/04/Marine-fuels-LCA_final.pdf. Accessed 15 April 2023

14. Majer S, Oehmichen K, Moosmann D et al (2021) Assessment of integrated concepts and identification of key factors and drivers. https://www.regatrace.eu/wp-content/uploads/2021/04/REGATRACE-D5.1.pdf. Accessed 20 April 2023

15. Falkenmark M (1986) Fresh water: time for a modified approach. Ambio 15:192–200

16. Umweltbundesamt (2022) Power-to-liquids: a scalable and sustainable fuel supply perspective for aviation. https://www.umweltbundesamt.de/publikationen/power-to-liquids. Accessed 12 March 2023

17. International Sustainability and Carbon Certification (ISCC) System: ISCC EU 203. Traceability and chain of custody. https://www.iscc-system.org/wp-content/uploads/2022/05/ISCC_EU_203_Traceability_and_Chain-of-Custody-v4.0.pdf. Accessed 20 May 2023

18. ISEAL Alliance: Chain of custody models and definitions. https://www.isealalliance.org/sites/default/files/resource/2017-11/ISEAL_Chain_of_Custody_Models_Guidance_September_2016.pdf. Accessed 02 May 2023

Chain-of-Custody Models for Renewable Fuels: A Comparison of Basic Characteristics

Nils Bullerdiek, Jan Pechstein, Gunnar Quante, and Martin Kaltschmitt

Abstract Powerfuels are expected to play a crucial role in meeting global climate targets, particularly in hard-to-abate sectors like aviation, shipping, or heavy-duty road transportation. Ensuring compliance with established sustainability criteria across the entire value chain and implementing processes to allocate and forward the associated environmental benefits, such as greenhouse gas (GHG) emissions reductions, to end-users is crucial for the large-scale use of powerfuels. This is where so-called chain-of-custody models play a key role. Thus, this chapter provides an overview of the general significance of chain-of-custody models for powerfuels and presents the different available models as well as their specific characteristics. While these chain-of-custody models are not limited to specific sectors or fuel options, they are discussed in the context of the aviation sector and for the use of renewable aviation fuels, including kerosene-type powerfuels.

Keywords Chain-of-Custody · SAF · Book & claim · SAF reporting & accounting

1 Introduction

Powerfuels are expected to play a crucial role in meeting global climate targets, particularly in hard-to-abate sectors like aviation, shipping, or heavy-duty road transportation. Ensuring compliance with established sustainability criteria across the

N. Bullerdiek (✉) · G. Quante · M. Kaltschmitt
Hamburg University of Technology (TUHH), Institute of Environmental Technology and Energy Economics (IUE), Hamburg, Germany
e-mail: nils.bullerdiek@tuhh.de

M. Kaltschmitt
e-mail: kaltschmitt@tuhh.de

J. Pechstein
Lufthansa Aviation Center, Frankfurt Am Main, Germany

© The Author(s), under exclusive license to Springer Nature Switzerland AG 2025
N. Bullerdiek et al. (eds.), *Powerfuels*, Green Energy and Technology,
https://doi.org/10.1007/978-3-031-62411-7_34

entire value chain and implementing processes to allocate and forward the associated environmental benefits, such as greenhouse gas (GHG) emissions reductions, to end-users is crucial for the large-scale use of powerfuels. This is where so-called chain-of-custody models play a key role. Thus, this chapter provides an overview of the general significance of chain-of-custody models for powerfuels and presents the different available models as well as their specific characteristics. While these chain-of-custody models are not limited to specific sectors or fuel options, they are discussed in the context of the aviation sector and for the use of renewable aviation fuels, including kerosene-type powerfuels.

A large-scale use of renewable aviation fuels is vital to reduce air transportation related climate impacts swiftly and extensive [1]. Yet, the dominance of (fossil) kerosene causes path dependencies and lock-in effects that impede a usage of alternative fuels—especially for non-kerosene options [2, 3]. As a result, primarily kerosene-based options have been pursued and approved as renewable fuels for aviation so far [4]. As of today, these are commonly referred to as "sustainable aviation fuel" (SAF) [5]. Although various SAF options have been extensively tested for more than 15 years, a large-scale SAF deployment happens to be complex and time-consuming. Current SAF shares are below 0.1% of total aviation fuel supplies, both globally and within the EU [6–8]. A key issue preventing the upscaling of SAF is the much higher price compared to fossil aviation fuel. It leads to a "chicken-and-egg" situation, where SAF producers and consumers are unwilling and/or unable to carry the cost burden for the upscaling of SAF on their own risk. Especially for nascent fuel options (e.g. power-to-liquid), this can delay technological learning and "economy of scale" effects to reduce production costs [1, 6]. Proper policy options to gradually overcome this chicken-and-egg deadlock have only been pursued in the recent past (e.g. blending mandates, tax incentives, and cross-industry transfers) [9, 10].

To accelerate the deployment of SAF, several actions are required, mainly related to the supply and demand side (e.g. R&D promotion, investment de-risking, feedstock supply, and eligibility criteria) [11–13]. However, also on the use-side of SAF, certain actions are imperative to integrate SAF seamlessly in day-to-day aircraft operations swiftly. A crucial aspect is the development and integration of chain-of-custody models on a larger system level to enable a robust monitoring, reporting, verification and accounting (claiming) of SAF, and its environmental attributes [11, 14, 15]. In principle, a proper chain-of-custody model for SAF serves as a linking element to prove SAF compliance with specific sustainability requirements (e.g. from regulation) and to track, monitor, report, and verify environmental attributes along the fuel supply chain [16]. Particularly in aviation, it should also enable and facilitate overall access to SAF for the stakeholders involved:

- Aircraft operators (airlines) need to be able to access SAF and report its use towards end customers and official authorities in a reliable manner. Also, they need to be able to claim the use of SAF under mandatory greenhouse gas (GHG) reduction schemes (e.g. EU ETS, CORSIA) and voluntary reporting standards (e.g. GHG protocol) in a transparent manner.

- SAF producer/supplier needs to provide information on the sustainability attributes of SAF and proof that predefined sustainability standards are guaranteed throughout the SAF overall supply chain. Also, for the SAF volumes they supply to the market, these need to be recordable for compliance purposes, e.g. to track them as indicators on the success of policy objectives (e.g. within a blending mandate).
- Possible fraud needs to be detectable and avoided, e.g. if sustainability criteria are neglected or SAF volumes are double counted towards policy objectives. Also, so-called additionality of SAF must be trackable. Additionality means that an action leads to an environmental benefit that would not have otherwise occurred. In the context of SAF, loss of such additionality usually occurs when counting SAF against legal obligations in other industries.

Existing GHG schemes involving SAF usually already provide some reporting and accounting processes. In the EU ETS and in CORSIA, for example, airlines may claim SAF usage in their annual emission reports. However, this is often based on basic and convoluted processes, where SAF purchase/blending records need to serve as evidence documents. In some cases, such records then also need to be directly linked with the physically supplied SAF [15, 17–20]. Such processes might serve as short-term solutions, i.e. while SAF volumes and demands are low. But, as SAF market shares increase (as it is necessary to fulfil the valid and legally binding GHG reduction goals),[1] they will most likely reach practical limits by creating too high administrative burden and thus unnecessary costs. Already today (i.e. while the overall SAF availability is still low) the proper reporting and accounting of SAF is often extremely difficult for airlines (e.g. within national blending mandates or in voluntary environmental reports) [21].

Specific chain-of-custody options for SAF are currently being pursued at different economic levels and with various players involved globally. Among them, especially accounting approaches based on the so-called book and claim principle are becoming increasingly popular [7, 22–28]. There are, nevertheless, non-consensual perspectives on the suitability of different chain-of-custody options for SAF among different stakeholders [15, 21]. At the same time, discussions on aviation's full climate impact and non-CO_2-effects gain increasing attention. In this context, it may be pertinent to determine which specific flights use SAF physically, possibly limiting the choice of applicable chain-of-custody models [29–32]. Overall, harmonised chain-of-custody models are still a missing element in current SAF policies, GHG schemes that involve SAF, or voluntary SAF reporting frameworks. Besides, airline customers demonstrate a growing interest in claiming SAF that is used in their upstream value chain. This further emphasises the need for a robust reporting function. So far, airlines and fuel suppliers often need to bridge this gap with individual "best practice" approaches that makes SAF reporting and accounting complex and time-consuming.

---

[1] This means larger SAF volumes will come from various origins (e.g. in terms of feedstock, conversion technologies, production sites, producers, and geographic regions). They may be commingled with fossil aviation fuel multiple times along the overall supply chain and supplied to various airports globally while being used within different policy frameworks by multiple airlines.

Against this background, the aim of this paper is to provide a basic understanding of different chain-of-custody models for SAF, by comparing and analysing their fundamental characteristics, benefits, and limitations. Therefore, firstly different chain-of-custody models are described and compared with one another in general (Sect. 2). Subsequently, the models are transferred to the use case of SAF (Sect. 3). This is followed by an in-depth comparison about specific characteristics, requirements, advantages, and disadvantages that are discussed related to nine specific comparison criteria (Sect. 4).

2 Overview

A chain-of-custody (CoC) can be considered as a sequence of responsibilities for the custodianship of certified materials/products as they move through a supply chain. Its purpose is to ensure that specified characteristics that are claimed for a particular material/product are indeed supplied into the supply chain and delivered to the/an actual output [33]. Different CoC models can be divided into three basic approaches—"physical separation", "mass balance", and "certificate transfer" (Fig. 1) [16].

While all CoC models ensure that the claimed volumes of a certified product match delivered or used volumes of such products (or at least do not exceed those), they have different characteristics, design requirements, advantages, and disadvantages. The overall suitability of a CoC model depends on the certified product itself, but also the characteristics of the system in which it is implemented [16]. The models differ especially in the extent to which certified and non-certified products can be physically mixed/commingled along the product supply chain (Fig. 1). Below, different CoC models are described in detail, considering basic design aspects, monitoring/ supervision aspects, and applicable end user claims.

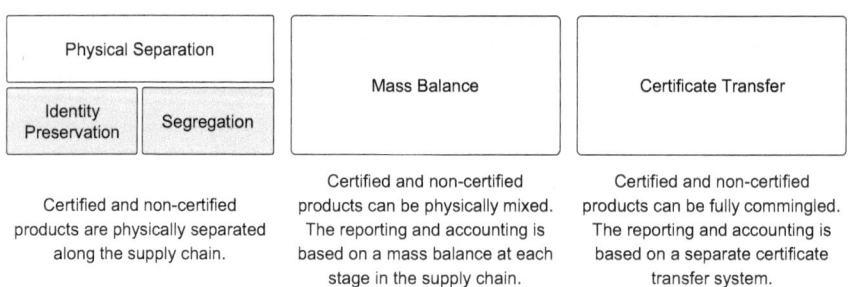

Fig. 1 Different chain-of-custody models

2.1 Physical Separation Approach

In a physical separation model, certified products—i.e. certified according to a certain standard—are physically separated from non-certified products within the respective system limits/system boundaries [33, 34]. Physical separation can be further differentiated into "identity preservation" and "segregation" models, depending on the degree to which certified products may be physically mixed (Fig. 2).

1. **Identity Preservation.** Identity preservation is the most stringent CoC model. At each stage/facility along the supply chain, certified products are not only physically separated from non-certified products, but also from certified products of other origin (e.g. from other production sites). This strict separation allows to trace certified products back to their actual origins (e.g. production site) or at least to the last point of processing, labelling, or certification [16, 33, 35].
2. **Segregation.** Within the system limits of a segregation model, certified and non-certified products are physically separated along the product supply chain. In contrast to an identity preservation model, certified products can be physically mixed amongst each other. Hence, a complete product tracing (tracing of molecules) back to an individual production site is not possible [16, 33, 35].

Monitoring/Supervision. In physical separation CoC models, the administrative flow (i.e. handling, monitoring, verification, and transfer of the certified product attributes (e.g. GHG emission reductions) along the overall supply chain) goes hand in hand with the physical product flow. The corresponding documentation is forwarded between the stages/facilities along the respective provision chain. To ensure a seamless and reliable documentation (i.e. administrative flow), the supply chain stakeholders or economic operators involved are usually registered and certified for the corresponding handling of certified products and audited regularly by (independent) third-party auditors [33, 35, 36].

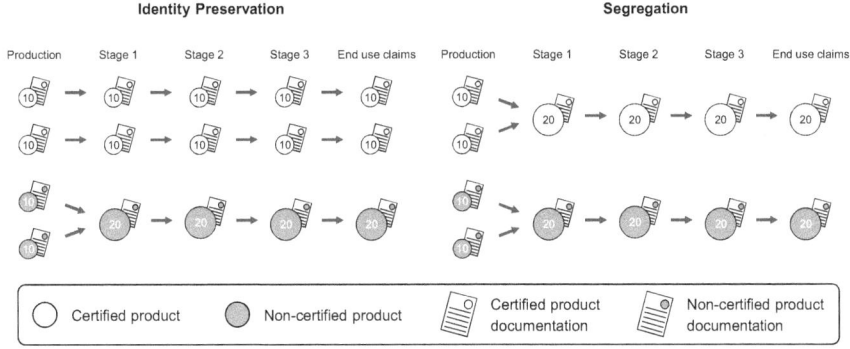

Fig. 2 Identity preservation (left) and segregation (right) chain-of-custody model

Applicable Claims. Within physical separation CoC models, an end user can claim that its physical product is a certified product of a specific certified origin within the respective and agreed system limits (identity preservation) or that its physical product is a certified product from any certified origin within the CoC system limit (segregation).

2.2 Mass Balance Approach

In a mass balance CoC model certified products can be physically mixed with each other as well as with non-certified products. The separation of certified and non-certified products is based on a virtual mass balance by proper input–output accounting at each stage along the supply chain. Within the mathematical/virtual mass balance, the quantity of certified products leaving a supply chain stage (output) may not exceed the quantity of certified products entering that stage (input) within a defined period of time (reconciliation period). Such an input–output balance can be related for example to the mass, volume, or also the energy content of products [33, 37–39]. Until the first point where certified and non-certified products are mixed/blended (i.e. up to the point where certified products first enter the mass balance system limit) the product tracking is usually based on another CoC model (e.g. physical separation) [16]. Mass balance CoC models can be further differentiated into specific mass balance approaches, for example with respect to the mass balance level (tank, site, and group level) or the actual balancing/accounting methodology [16, 33, 35] (Fig. 3).

- **Tank-Level Mass Balance.** Within a mass balance at tank level,[2] the administrative (virtual) separation of certified and non-certified product flows is conducted at tank level; i.e. a single tank represents the balance/accounting limit at a specific stage/facility along the product supply chain. In this case, the mass balance documentation of different tanks at one specific site cannot be mixed with each other [35]. However, at this accounting level, it can be ensured that an end product actually contains a physically share of a certified product [16]. The balancing/accounting methodology of the administrative flow can be strictly linked to the actual physical share of the certified product in an output flow (coupled approach). Alternatively, the balancing/accounting of the administrative flow can be decoupled (Fig. 3). In this case the virtual share of certified products could be allocated to a product output flow in a flexible/free manner, as long as the overall mass balance is consistent within the reconciliation period [16, 33–35].
- **Site-Level Mass Balance.** Within a mass balance at site level, the administrative (virtual) separation of certified and non-certified product flows is conducted at site level; i.e. a single site represents the accounting/balance limit along the product

---

[2] The term "tank level" is not to be confused with the actual fuel level of the tank. In this context, it explicitly refers to the balance limit of the mass balance model.

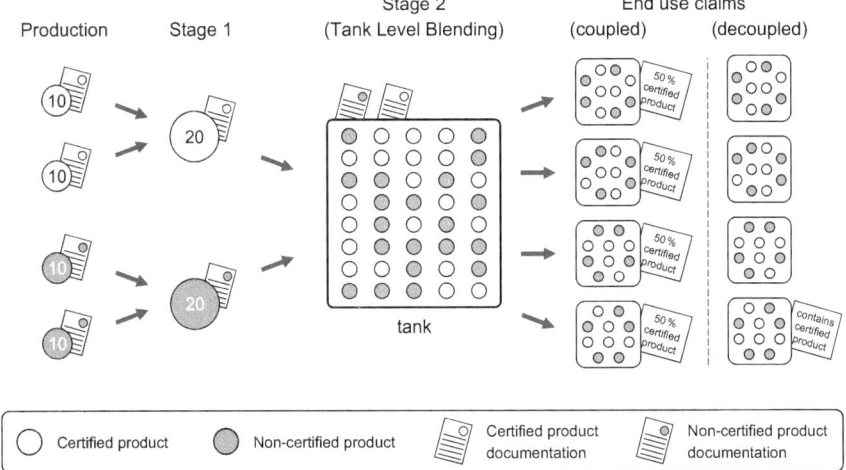

Fig. 3 Coupled and decoupled mass balance chain-of-custody model at tank level (in a coupled mass balance, the administrative flow is coupled to the certified product flow, and in a decoupled mass balance, the administrative flow is decoupled from the certified product flow; a segregation chain-of-custody model is assumed until the point of blending)

chain (Fig. 4) [16]. For example, a specific site (stage/facility) along the product supply chain could be a group of tanks containing specific fuels (e.g. Jet A-1). While no notice would be paid for the individual mass balance of each tank at this site, a proper mass balance accounting is conducted for the entire site as a whole [35].

- **Group Level Mass Balance.** Within a mass balance at group level, the administrative (virtual) separation of certified and non-certified product flows is conducted at a defined group or system level along the product supply. Such groups/systems could, for example, consist of several sites/facilities, a company, a country, a region, or of any other combination of more than one site where product volumes are tracked and balanced [16].

Monitoring/Supervision. Within a mass balance model, the administrative flow is conducted along the product supply chain at every stage/interface on a virtual mass balance by proper accounting processes. Therefore, usually all stakeholders involved in the CoC model are registered, certified accordingly, and audited regularly [35].

Applicable Claims. Within a mass balance CoC model, an end user can claim that a certain amount of certified product from a certified origin[3] was or will[4] be physically

---

[3] The product origin does not necessarily have to be located within the mass balance system limit. This is, for example, the case, if another CoC model is applied up to the point of blending or up to the first supply chain stage within the mass balance system (e.g. a segregation model).

[4] The point of time when a certified product is physical fed into the product supply chain could be delayed due to the reconciliation period.

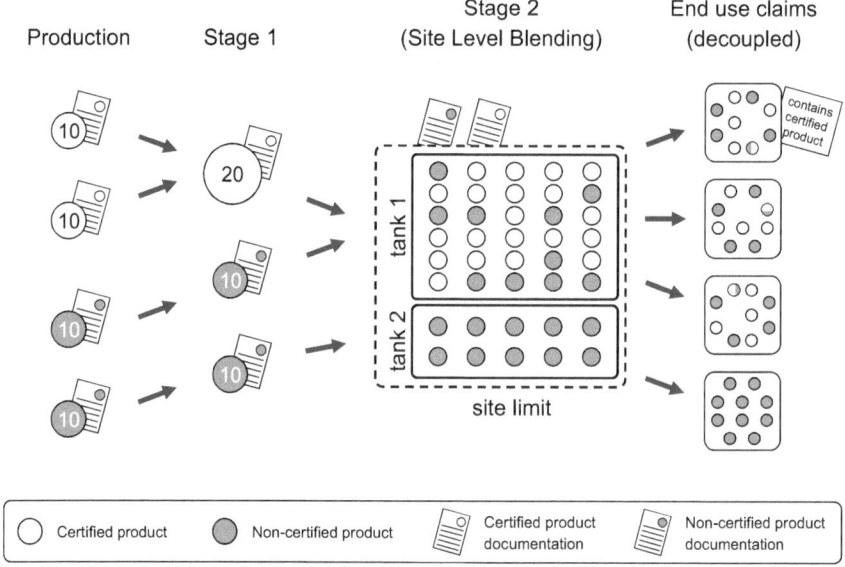

Fig. 4 Mass balance chain-of-custody model at site level (a segregation chain-of-custody model is assumed until the point of blending)

fed into the product supply chain and was, is, or will be physically available at some end customer within the physical product flow limit of the mass balance CoC model. Whether and to what share a certified product is physically present at a specific end customer depends on the further design of the mass balance, such as the mass balance level or the balancing/accounting methodology.

2.3 Certificate Transfer Approach (Book and Claim)

In a certificate transfer model, certified products can be physically mixed with each other as well as with non-certified products (i.e. processed, stored, or transported in a commingled manner in a common logistics infrastructure within the respective system limits). The physical flow of certified products is completely decoupled from the administrative product flow, where the handling, monitoring, verification, and transfer of the certified product attributes are conducted. Within such a CoC model, the information on attributes of certified products is hold in transferable certificates, in some EU frameworks also called "Guarantees of Origin" (GoO). The issuance of certificates needs to be based on a system that ensures equivalence in volumes of physical certified products and corresponding certificates. The issuance, transfer, trade, and redemption of such certificates is based on two main principles: "booking" and "claiming". Thus this CoC model is often also referred to as "book and claim" (Fig. 5) [16, 33, 34, 37, 40, 41].

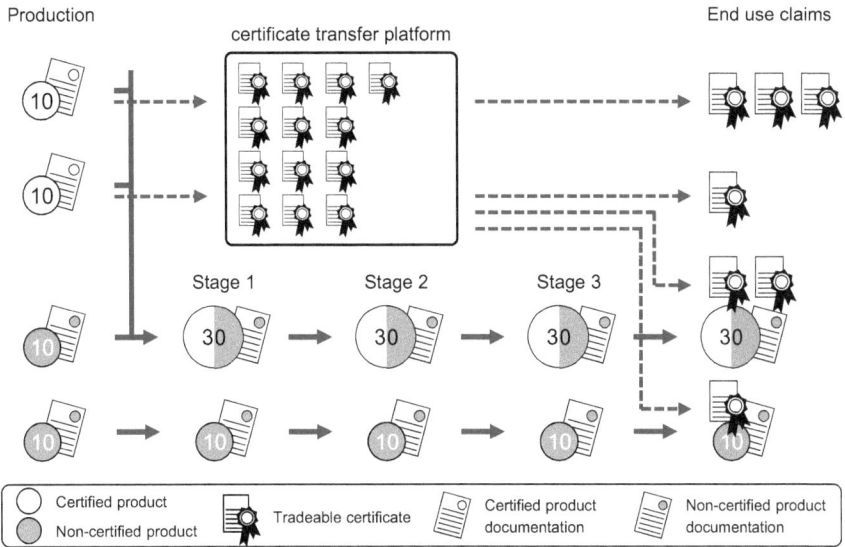

Fig. 5 Certificate transfer chain-of-custody model (book and claim)

1. **Booking.** At a certain point in the product supply chain—most likely the first point practicable within the CoC model system limits—the attributes of certified products are transferred to transferable certificates (usually together with further product information like, e.g. date of issuance and producer information). From that moment on, the certified product attributes are fully transferred to the certificates, while the physical products are completely decoupled from these attributes; i.e. they are fully considered and handled as non-certified products within the existing supply chains and can be handled in a commingled manner.

2. **Claiming.** By redeeming a certificate, the certified attributes are claimed by the respective certificate holder.[5]

Hence, the redemption of a tradeable certificate by a participant proves that an equivalent amount of certified product was physically fed into the further product supply chain and is or will be physically available at some end customer within the physical product provision chain. Depending on the respective product and the specific product supply chain, additional monitoring and control measures might be required. These measures have to ensure that no certified products can leave the respective CoC system limit (unnoticed), once the corresponding certificate have

---

[5] In the context of certificate transfer/book-and-claim systems, it is often stated that the certificate redemption means that a/any physical product is reunited with the certificate attributes, and thus, the physical product is declared as a certified product, irrespective of its actual origin and thus irrespective of whether it is physically a certified or non-certified product. However, in this understanding, this is not the case for a certificate transfer model, but inevitably the case for any mass balance model, regardless of the mass balance design. This is a fundamental and ultimately difference between a mass balance and a certificate transfer model.

been issued and that the certified products are eventually physically used by an end user, so that they indeed substitute the use of a non-certified product.

Monitoring/Supervision. Within a certificate transfer model, the certificate exchange is usually conducted in connection with an IT-based trading infrastructure (certificate registry). Usually not all stakeholders involved within the product supply chain need to participate in the virtual certificate issuance and transfer system. Thus, there is no need for each stakeholder to be registered, certified, and audited accordingly; this is not true for those who are actively involved in the issuance and transfer of certificates. Additionally, a registry administrator might be required for the supervision of the issuing, transfer, and redemption of certificates within the certificate registry. This could be a legitimate body, institution, or authority, who is responsible for the registry operation and corresponding monitoring tasks. Also, this task could be conducted by an organisation monitoring the system already due to a different motivation (e.g. customs authorities). Due to the complete decoupling of certified product attributes and the physical flow of certified products, certificate transfer models are often considered to imply higher risks of fraud compared to other CoC models (e.g. by multiple issuing, selling, counting, or claiming of certificates by end customers) [35, 40].

Applicable Claims. By redeeming a tradeable certificate, an end user can claim that a certain amount of certified product from a certified origin[6] was physically fed into the product supply chain at a certain point in time in the past.[7] This certified product was, is, or will be physically available at some end customer within the physical product supply chain.

2.4   Overall Comparison

The different CoC models are now compared with regards to the physical and administrative product flows and attribute allocation characteristics. Fig. 6 shows the overall models characteristics for the coupling of physical and administrative product flows, the coupling of product attributes to product flows, and the corresponding flexibility in product attribute allocation.

As shown in Fig. 6, the identity preservation model implies the highest level of coupling physical and administrative product flows. For the following CoC models, this coupling characteristic decreases continuously, while the flexibility of product

---

[6] The actual origin of the product does not necessarily have to be within the system limit of the certificate transfer chain-of-custody model. This can, for example, be the case for imports, and if up to the point of blending or up to the first stage in the certificate transfer system limit, another chain-of-custody model is applied, e.g. a segregation chain-of-custody model.

[7] Unlike in the case of mass balance, the injection of certified product into the product supply will most likely always be in the past, once a certificate is available. Otherwise, certificates may have to be issued in advance, which would probably reduce the credibility of a certificate transfer chain-of-custody model and is rather to be considered as unusual.

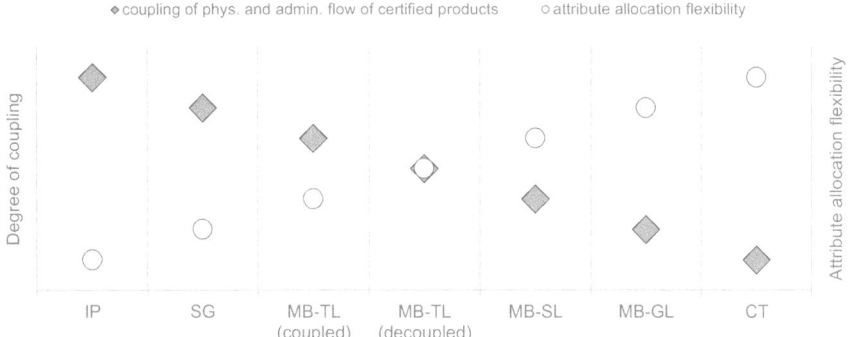

Fig. 6 Chain-of-custody models characteristics/trade-off in terms of product flow coupling and attribute allocation flexibility (CT—certificate transfer; GL—group level; IP—identity preservation; MB—mass balance; SG—segregation; SL—site level; TL—tank level)

attribute allocation increases. Yet, this still requires that a robust monitoring of product flows and compliance with product requirements (e.g. sustainability criteria) is guaranteed. Below, the different CoC models are compared in terms of general logistics/infrastructure, traceability,[8] claiming, and monitoring aspects (Table 1).

Logistics/Infrastructure. Only the mass balance CoC models and the certificate transfer CoC model allow for a mixing/commingling of certified and non-certified products. Thus, a physical separation model requires separate logistics and infrastructure to supply certified products, i.e. a reallocation of existing logistics infrastructure to supply non-certified products or the construction of additional infrastructure. While it depends on the actual product itself and the characteristics of the product supply chain, such separate logistics of certified products are usually associated with (high) additional efforts and costs compared to a mass balance or certificate transfer CoC model. This can also imply environmental downsides, e.g. due to additional logistics related emissions. For all CoC models except for the certificate transfer model, the administrative product flow is always linked to some physical product flow. In other words, delivering certified attributes to an end customer to realise a claim some physical product flow is required in any case. However, in the case of a site or group/system level mass balance CoC model, this physical product flow does not necessarily have to be a physical flow of certified products.

---

[8] Traceability can be considered as the ability to trace the history, application, or location of a product. It enables to follow the movement of a product and its components through specified stages of production, processing, and distribution. Although often considered as interchangeable, the concepts of traceability and chain-of-custody are thus not identical. A chain-of-custody is a chain of responsibility for the custodianship of materials or products as they move through an overall supply chain. Its purpose is to ensure that the specified characteristics that are claimed for a particular material or product (or for the market as a whole) are indeed the ones that are actually delivered in the output [33], p. 25.

Table 1 Comparison of different chain-of-custody models (CP—certified product; NCP—non-certified product; [16, 33–35, 40])

	#	Criterion	Physical separation		Mass balance				Certificate transfer
					Tank level		Site level	Group level	
			Identity preservation	Segregation	Coupled	Decoupled			
Logistics	1	Mixing of CP allowed	No	Yes	Yes	Yes	Yes	Yes	Yes
	2	Mixing of CP and NCP allowed	No	No	Yes	Yes	Yes	Yes	Yes
	3	Separate logistics required for CP	Yes	Yes	No	No	No	No	No
	4	Physical and administrative product flows are linked	Yes	Yes	Yes	Yes	Yes	Yes	No
Traceability	5	Admin. flow allows physical traceability of CP	Yes	Yes	Yes	No	No	No	No
	6	Admin. flow allows physical traceability of CP back to origin	Yes	No	No	No	No	No	No
	7	A physical share of CP products in final products is assured	Yes	Yes	Yes	No	No	No	No

(continued)

Table 1 (continued)

| | # | Criterion | Physical separation | | Mass balance | | | | Certificate transfer |
| | | | Identity preservation | Segregation | Tank level | | Site level | Group level | |
					Coupled	Decoupled			
Claims/monitoring	8	Attributes are assignable to any physical product leaving the system	No	No	No	No	No	No	Yes
	9	Claim of a CP is tied to a physical flow of products	Yes	Yes	Yes	Yes	Yes	Yes	No
	10	Claim of a CP is tied to a physical flow of certified products	Yes	Yes	Yes	No	No	No	No
	11	Monitoring/supervision processes at each supply chain stage	Yes	Yes	Yes	Yes	Yes	Yes	No

Traceability. A true physical traceability of certified products is only given for physical separation models (Table 1). In the case of an identity preservation model, certified products can be traced back to the respective product origin (at least if the product origin lies within the system boundaries of the CoC model). Yet, this is not the case for the segregation model. To some extent, a physical traceability is still possible within a mass balance CoC model at tank level, as the share of certified and non-certified products can be recorded at each stage along the supply chain, and the end product contains at least a share of certified product. For the other mass balance models (site or group/system level), this traceability may only exist to a limited extent, as the physical flow of certified products does not necessarily have to be linked to the administrative flow. For a certificate transfer model, a physical traceability of certified products is not given.

Claims/Monitoring. If it is required that a final product consists only of certified products, this can only be guaranteed based on a physical separation CoC model. If at least a physical share of certified products shall be guaranteed within a final physical product, this could still be realised based on a mass balance CoC model at tank level, even if it is handled in common logistics with non-certified products. In the case of the other mass balance CoC models, it is not guaranteed that a final product used by an end customer contains physical shares of certified products, which excludes a corresponding claim. For these models, it can only be claimed that certified products have been injected into the respective CoC system or the product supply chain and that they are physically present at some end customer of the product supply chain eventually. For all CoC models except for group/system level mass balance models with wide system limits and a certificate transfer model, the monitoring and supervision tasks (registration, certification, auditing, etc.) are basically carried out at all stages along the product supply chain by the stakeholders involved.

3   Application in the Context of SAF

In the following the chain-of-custody (CoC) models described above are applied to the use case of sustainable aviation fuel (SAF). Therefore, an exemplary aviation fuel supply constellation is considered, consisting of two SAF production sites, two conventional (fossil fuel-based) refineries, three fuel supply companies, two airports, and five airlines with one aircraft each (Fig. 7).

In all supply cases SAF production site A (company A) produces neat SAF from lipid feedstock (e.g. used cooking oil), while at SAF production site B (company B) electricity-based SAF (PtL-SAF) is produced. So far, neat (synthetic) SAF must be blended with specification compliant (fossil) aviation fuel to completely meet the standardised aviation fuel requirements. After proper blending, a SAF blend can be seamlessly used in existing aviation fuel and aircraft infrastructure as a "drop-in" fuel [4, 42]. Here, the blending takes place at the respective SAF production site. For reasons of simplicity, a blending ratio of 50% of neat SAF (related to

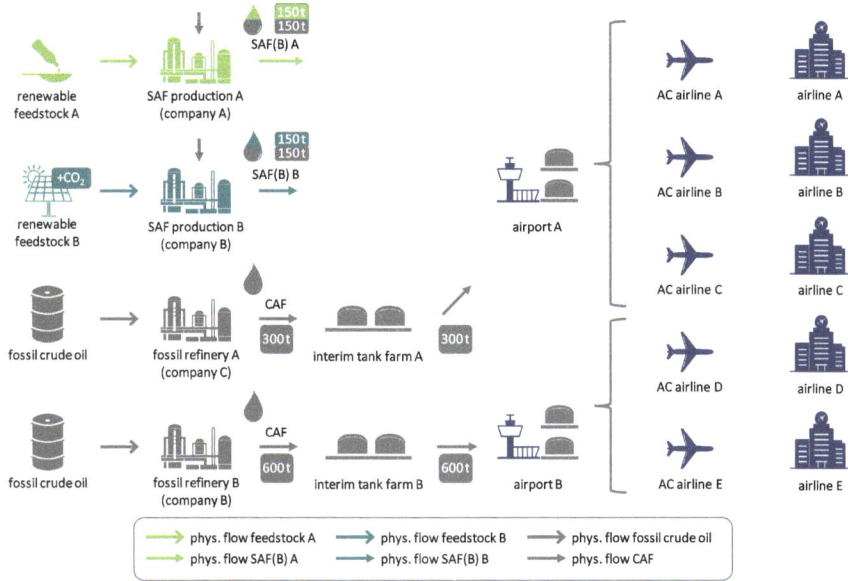

Fig. 7 Example of a SAF supply constellation (AC—aircraft; CAF—conventional aviation fuel; SAF—sustainable aviation fuel; SAF(B)—sustainable aviation fuel blend)

mass) is assumed. Two conventional (fossil fuel-based) refineries A and B provide both airports with conventional aviation fuel (CAF) via an intermediate tank farm. Refinery A (company C) supplies airport A, and refinery B (company B) supplies airport B only. At each airport, two fuel tanks are available. Aircraft A, B, and C uplift fuel at airport A, while aircraft D and E uplift fuel at airport B. All aircraft uplift an exemplary aviation fuel amount of 300 t. The physical product flows of both SAF blends are directed to airport A. Depending on the CoC model, this supply chain constellation is partly modified.

3.1 Identity Preservation and Segregation Model

In case of an identity preservation CoC model, both SAF blends are physically completely separated from other fuels, although they are drop-in capable (Fig. 8, top). Thus, for both SAF blends separate storage and fuelling infrastructure is required along the entire supply chain (i.e. a reallocating of existing fuel infrastructure and/ or the construction of additional infrastructure). As shown in Fig. 8 (top), each SAF supply chain must be certified and audited within this CoC model. Assuming aircraft A uses SAF blend A only and aircraft B uses SAF blend B only, only airlines A and B can account for the use of SAF and at all. As the claiming of SAF attributes is strictly linked to a physical SAF usage within the identity preservation model, airline

A can only account for the attributes of SAF blend A, and airline B can only claim the attributes of SAF blend B.

In case of a segregation CoC model, SAF blends A and B can be commingled (Fig. 8, bottom). Still, they are physically separated from CAF, which requires separate storage/refuelling infrastructure and separate logistics up to the aircraft. It is assumed that aircraft A and B both use a mixture of SAF blend A and B. Hence, in terms of SAF claiming, both airlines can account for the attributes of the fuel

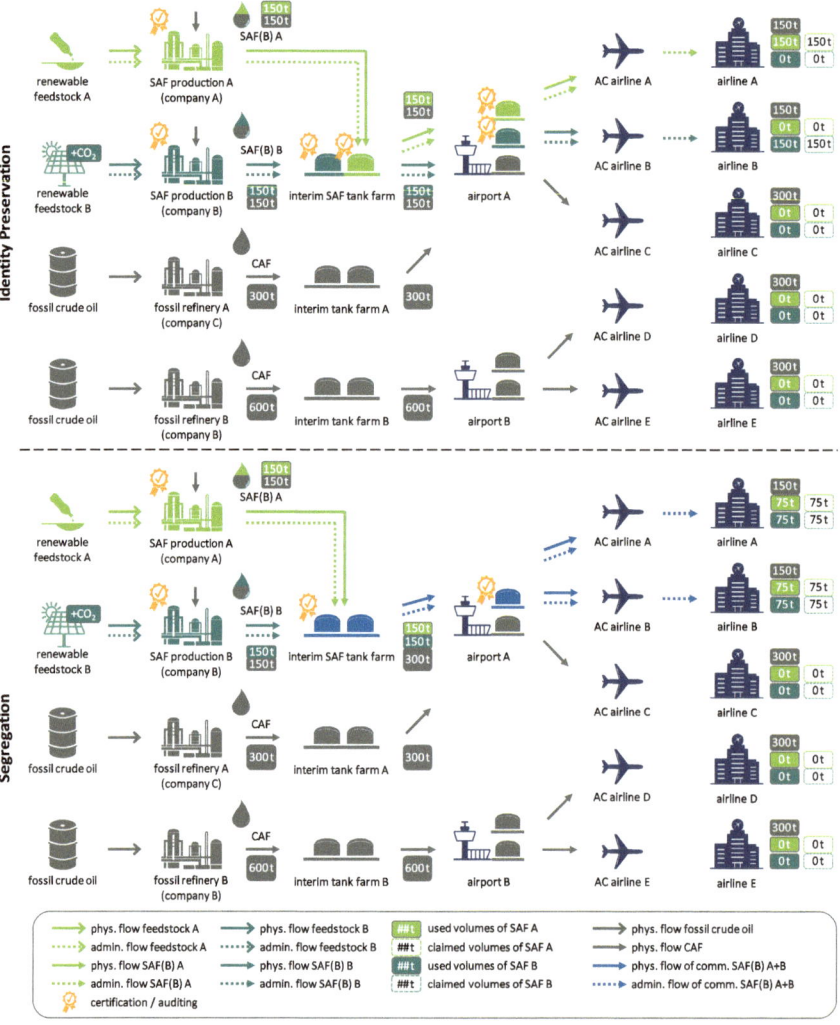

Fig. 8 Exemplary identity preservation and segregation CoC model for SAF (AC—aircraft; CAF—conventional aviation fuel; CoC—chain-of-custody; comm.—commingled; SAF—sustainable aviation fuel; SAF(B)—sustainable aviation fuel blend)

mixture. However, a flexible attribute allocation within the administrative flows is not possible. Thus, airlines that do not operate at airport A or aircraft not being refuelled from the respective "SAF tank" at airport A cannot claim any SAF attributes at all.

Based on the above CoC models, an airline can claim that it physically used a certified SAF product from a specific certified origin (identity preservation) or any certified origin (segregation) in its own aircraft.

3.2 Tank- and Site-Level Mass Balance Model

In case of a tank-level mass balance, the balance limit includes single/specific tanks along the supply chain stages (Fig. 9, top). Both SAF blends can be commingled with CAF along the supply chain but need to be handled in the designated fuel tanks. Thus, a separate transport from interim tank farm A to airport A is still required. All aircraft being supplied with aviation fuel from the corresponding "SAF tank" at airport A obtain physical shares of both SAF options. Considering a decoupled balancing/accounting methodology, the claim of SAF attributes does not need to be strictly coupled/linked to the physical SAF shares. In this case, in contrast to a segregation model, airline A could claim all SAF attributes although aircraft B obtains SAF physically as well. As it is assumed that the accountable volume of SAF cannot exceed the total amount of aviation fuel uplifted (i.e. both SAF and CAF). Depending on the mass balance design, such accounting could also be limited to the volumes of neat SAF that are uplifted by an airline. Without being refuelled from the specific "SAF tank", airline C cannot claim any SAF. Not having physical access to the SAF supply chain, airline D and E cannot access any SAF attributes either.

In case of a mass balance at site level, the balance limit includes the entire site in the supply chain (Fig. 9, bottom). Hence, both SAF options can be completely commingled with CAF along the fuel supply chain at each supply chain stage. In contrast to the tank-level mass balance model, separate fuel logistics from interim tank farm A to airport A are not required. The SAF attributes, i.e. the administrative flow, are not coupled to physical SAF shares. Hence, for example, only airline A could claim SAF attributes, while also aircraft B and C are physically supplied with SAF to equal shares. Even in an extreme supply case, if no SAF is physically delivered to airport A, airline A, B, and C could still claim SAF attributes within the site-level mass balance model. This could, for example, occur, if SAF blend A and B are physically stored in one specific tank at the interim tank farm, while airport A is supplied with pure CAF of another tank from the interim tank farm A. Still, without being connected to the mass balance system and the administrative SAF stream, airline D and E cannot claim any SAF attributes.

Based on the above mass balance models, an airline can claim that a certain amount of certified SAF product was/will be physically fed into the aviation fuel supply chain and was, is, or will be physically available for use in some aircraft

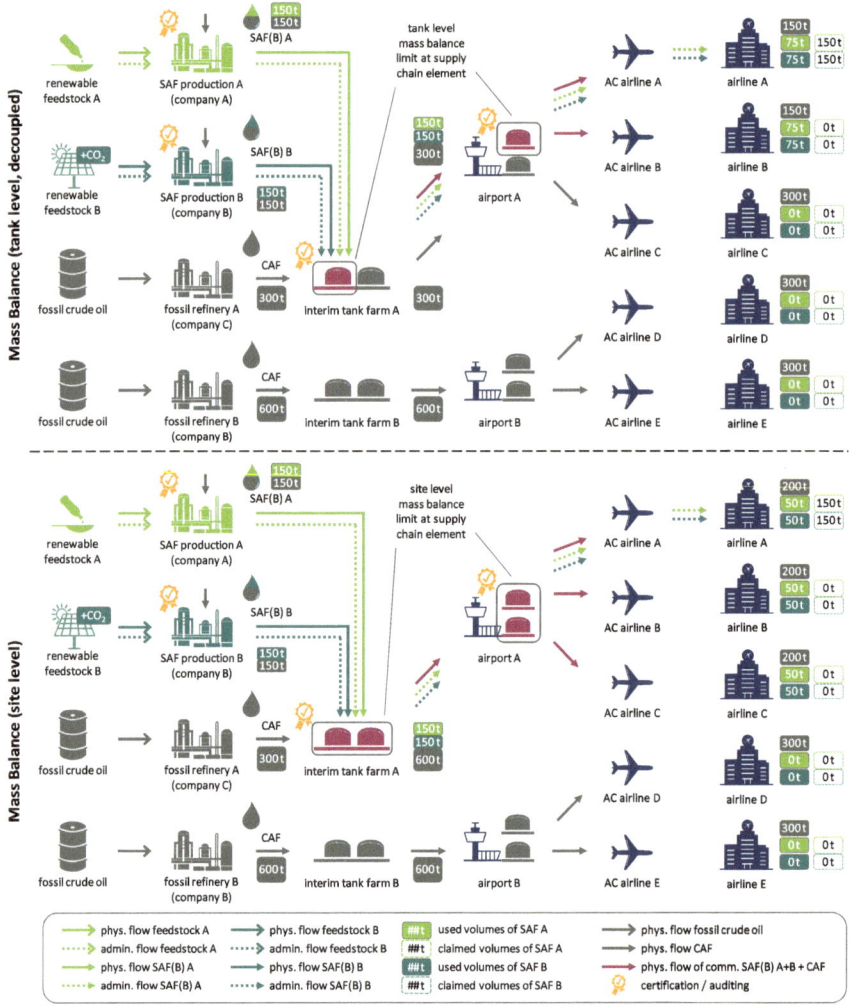

Fig. 9 Exemplary tank-level (decoupled) and site-level mass balance CoC model for SAF (AC—aircraft; CAF—conventional aviation fuel; CoC—chain-of-custody; comm.—commingled; SAF—sustainable aviation fuel; SAF(B)—sustainable aviation fuel blend)

within the physical SAF product flow—these are not necessarily aircraft operated by the claiming airline.

3.3 Supply Chain Level Mass Balance Models

In addition to the mass balance models discussed above, two group/system-level mass balance models are considered offering a wider mass balance limit. Here, the assumed balance limit includes the entire aviation fuel supply chain (Fig. 10). Thus, this mass balance model is considered as a "supply chain level mass balance". Two approaches are differentiated: a company supply chain level as well as a system supply chain level mass balance.

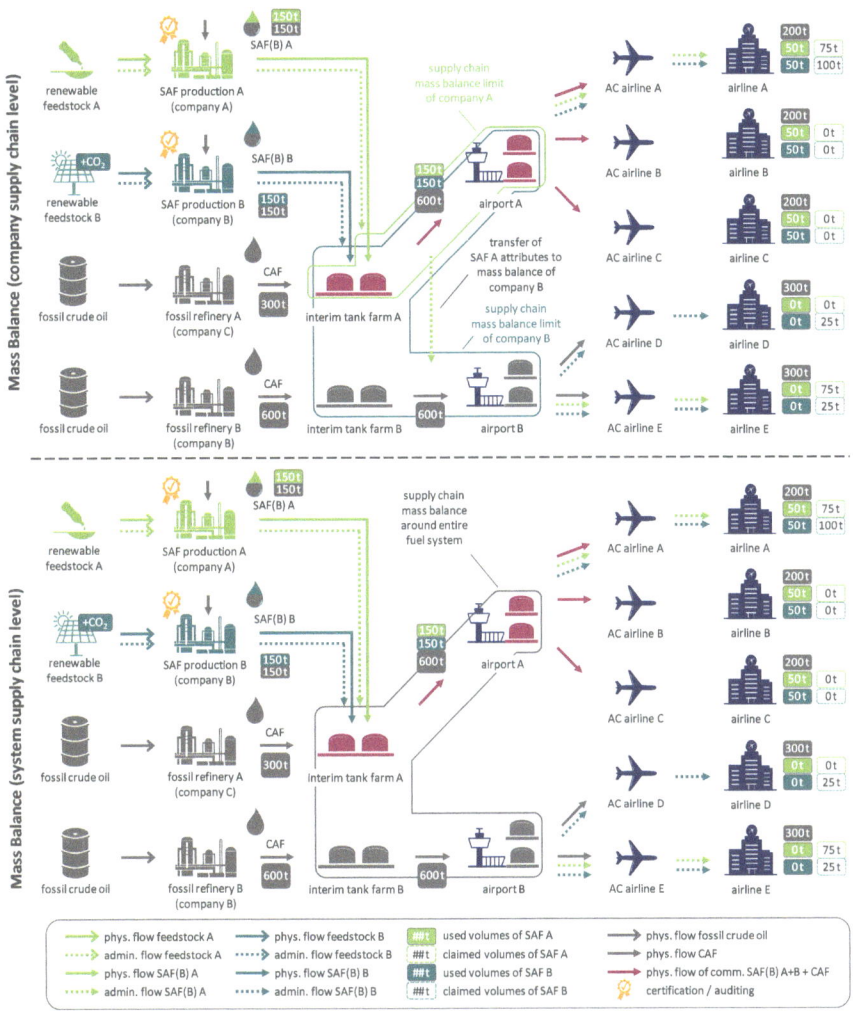

Fig. 10 Exemplary company and system supply chain level mass balance CoC model for SAF (AC—aircraft; CAF—conventional aviation fuel; CoC—chain-of-custody; comm.—commingled; SAF—sustainable aviation fuel; SAF(B)—sustainable aviation fuel blend)

For the company supply chain level mass balance, it is assumed that each SAF supplying company manages a SAF mass balance (i.e. proper virtual input–output accounting of SAF and SAF attributes), around its entire aviation fuel supply chain (SAF and CAF supply; Fig. 10, top). Following this approach, the balance limit of SAF producer A (company A) includes the aviation fuel supply chain from interim tank farm A up to airport A. In contrast, the balance limit of SAF producer B (company B) additionally includes the aviation fuel supply chain from interim tank farm B up to airport B. Within this model, airline D and E could access SAF attributes without a physically supply of SAF—if they are supplied with aviation fuel from company B. Assuming fuel supplying companies are authorised to transfer/trade their mass balance SAF stocks among one another, airline D and E could also access attributes of SAF blends A, if the SAF attributes are traded/transferred between company A and B in advance (Fig. 10, top). Overall, SAF can be seamlessly commingled and supplied with CAF throughout the entire fuel supply chain. Separate fuel logistics are not required within this mass balance model. As shown in Fig. 10, by increasing the mass balance limit, the necessity for supply chain certification and auditing decreases compared to the previously described mass balance models. From a theoretical point of view, the certification and auditing of supply chain elements could be limited to SAF producers/supplier.

For the system supply chain level mass balance, it is considered that the balance limit includes the entire "system", i.e. aviation fuel supply chain, but independently from individual fuel supplying companies (Fig. 10, bottom). This could, for example, be the European aviation fuel supply system. Analogously to a certificate transfer CoC model (Sect. 3.4), the balance limit should begin at the first point practicable, and the stock of SAF attributes needs to be managed in an overall accounting system.[9] In this case, basically all airlines whose aircraft are supplied with aviation fuel within the respective mass balance limit can claim SAF attributes. SAF can be seamlessly commingled and supplied with CAF throughout the entire fuel supply chain. Separate fuel logistics are not required within this mass balance model.

In case of the above supply chain mass balance CoC models, an airline can claim that a certain amount of certified SAF product was/will be physically fed into the aviation fuel supply chain and was, is, or will be physically available for use in some aircraft within the physical SAF product flow—these are not necessarily aircraft operated by the claiming airline.

---

[9] Such a mass balance model would require clarification of several design details (e.g. how SAF attributes are transferred to a virtual mass balance administration system, by whom and in which manner the attributes are forwarded to airlines). Yet, clarifying such details is not subject of this paper. Here the key focus is to assess the overall and basic characteristics of different chain-of-custody models for SAF.

3.4 Certificate Transfer Model (Book and Claim)

In case of a certificate transfer CoC model (book and claim; Fig. 11), both SAF blends can be fed into the existing fuel system and handled in a commingled manner at the first point practicable. A mixture of SAF blend A, B, and CAF from refinery A is supplied to airport A. There, each aircraft obtains shares of both SAF options physically. Yet, the administrative flow for the tracking, reporting, and accounting of SAF is completely decoupled from the physical flows and the SAF attributes transferred onto certificates. The certificates may be purchased, transferred, and redeemed by each airline (or theoretically all stakeholders) with registry access. Hence, also airline D and E can claim SAF in a flexible manner, independently from the fuel supply conditions at airport B.

In theory, an airline could access and acquire all certificates being available in the central registry. In practice, however, certain limits are likely to be specified for the certificate handling and acquiring within such a CoC model. Depending on the model design, i.e. in terms of stakeholder access to certificates, a certificate purchase and direct claiming of SAF attributes does not need to be limited to stakeholders like fuel producers, fuel suppliers, or airlines necessarily. Further stakeholders, such as airline corporate customers, could theoretically participate in certificate trading directly. While some systems of this kind exist in the market already [26, 27], it is essential to carefully assess which stakeholders should participate and under what conditions, taking into account the requirements of existing regulation and established reporting

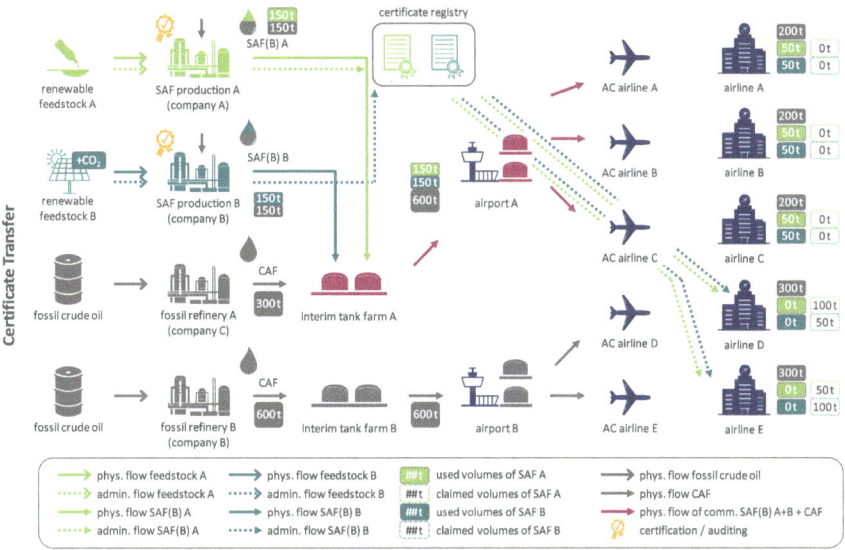

Fig. 11 Exemplary certificate transfer CoC model (book and claim) for SAF (AC—aircraft; CAF—conventional aviation fuel; CoC—chain-of-custody; comm.—commingled; SAF—sustainable aviation fuel; SAF(B)—sustainable aviation fuel blend)

frameworks. However, while indeed especially a certificate transfer model allows for such flexible/free allocation options, in this case the nature of the CoC would shift from a genuine SAF reporting and accounting model for airlines to an offsetting instrument for third parties.

Based on the certificate transfer model discussed above, an airline can claim that a certain amount of certified SAF product was physically fed into the aviation fuel supply chain at a certain point in time and was, is, or will be physically available for use at some aircraft within the physical product supply chain—this is not necessarily an aircraft of the claiming airline itself.

4   Comparison and Discussion

Each chain-of-custody (CoC) model comes with its own specifics, particularly in the context of sustainable aviation fuel (SAF). Therefore, in the following their main characteristics are analysed and discussed in the context of SAF in more detail. All CoC models as described in Sect. 3 are considered. First, the comparison criteria are defined as well as the overall comparison approach is explained (Sect. 4.1). This is followed by the actual criteria-based comparison of the considered CoC models (Sect. 4.2) and a discussion on suitability aspects considering current SAF market conditions (Sect. 4.3).

4.1   Approach and Comparison Criteria

For the comparison of the different CoC models, a completely objective and measurable assessment is practically impossible. In practice, certain comparison/assessment criteria are usually considered more important than others—depending on individual viewpoints or considered requirements. Also, certain criteria are challenging to quantify in a reliable manner. Therefore, a qualitative and basic comparison of main CoC model characteristics is carried out based on a plus-minus classification as a basic and straightforward, but therefore transparent, approach. Hereby, a CoC model is categorised by "++" for the most beneficial characteristic, by "o" for a rather average/neutral characteristic, and by "− −" for the most adverse characteristic with respect to a comparison criterion and in comparison with other CoC models. For this qualitative comparison, nine criteria are considered (Table 2).

Within this ordinal scale comparison, a ranking order is derived for the different CoC models by comparing their individual degree of compliance with regard to each of the comparison criterion shown above.

Table 2 CoC model comparison criteria (CoC—chain-of-custody; MRVA—monitoring, reporting, verification and accounting; SAF—sustainable aviation fuel)

	Sub-level criteria	Explanation
1	Integrity efforts	A CoC model needs a sufficient level of integrity, e.g. to minimise SAF sustainability concerns as far as possible. This requires, amongst others, adequate fraud protection mechanisms. The efforts to achieve a sufficient level of integrity for a CoC model should be as low as reasonably practicable
2	Complexity of operational handling	A CoC model should lead to a simplification of operational and administrative complexity, i.e. it should offer a straightforward application for all operators involved (e.g. fuel producers/suppliers, airlines, end customers, official verifiers, authorities, etc.)
3	Capability of scope expansion	A CoC model should be as flexible as reasonably practicable to expand its scope (system boundary) in a seamless manner after its first implementation, e.g. to involve further SAF producers and supply regions
4	Implementation efforts	The efforts and thus costs for the implementation of a CoC model should be as low as reasonably practicable. This includes direct efforts for the setup of the CoC model itself, but also indirect efforts resulting from the fact that a CoC model may require further system requirements and conditions (e.g. a separate fuel infrastructure)
5	Operating efforts	The efforts and thus costs for the operation of a CoC model should be as low as reasonably practicable. This includes direct efforts (e.g. administrative usage costs and costs for audits), but also indirect efforts, e.g. resulting from the fact that a CoC model may require certain system requirements and conditions to be maintained (e.g. a separate fuel infrastructure)
6	Intra-sectoral access to SAF	In addition to a high overall CoC scope/system limit, access to the CoC model and thus access to SAF attributes should be as seamless as reasonably practicable. In this case this refers to intra-sectoral access, i.e. airlines that physically use aviation fuel directly for their services and would like to access certified SAF attributes, within a given scope of a CoC model

(continued)

Table 2 (continued)

Sub-level criteria		Explanation
7	Extra-sectoral access to SAF	In addition to a broad CoC system boundary, it could hypothetically be intended to provide access to SAF via a CoC model to extra-sectoral stakeholders. In this case, this refers to SAF access for non-aviation stakeholders/companies that typically do not handle or use aviation fuel for their services or products but use air transport services as customers (in terms of both passengers and cargo)
8	CO_2-emissions by SAF logistics	The application of a CoC model should minimise the negative impact on the overall CO_2eq life cycle emissions of a SAF fuel batch. In this respect the main influence of a CoC model most likely results from its specific requirements or flexibility in terms of physical fuel logistics and thus logistics emissions, i.e. emissions for transport of SAF from a production site to an airport/aircraft
9	MRVA of physical fuel properties	Certain conditions could require that information about physical shares of SAF and/or its chemical properties are available/traceable up to a certain stage in the physical fuel supply chain (e.g. up to a dedicated fuel supply region with multiple airports, up to a specific airport (airport tank farm level) or—in a most advanced case—up to a specific aircraft refuelling). Reasons for this could be, for example, influencing non-CO_2-effects[10] or local air quality[11] by a targeted and non-commingled physical supply of SAF. In such a case, at best, a CoC model offers the ability to record the associated information on such effects and seamlessly reflect it within its monitoring, reporting, verification, and accounting (MRVA) process

---

[10] Several SAF options allow for a reduction of aviation related "non-CO_2-effects" [43]. Main non-CO_2-effect is currently seen to be the formation of persistent cirrus clouds from contrails and nitrogen oxide (NO_X) emissions. Other than CO_2-effects, the estimated climate impact of non-CO_2-effects depends on a combination of factors, such as the type of fuel (chemical properties) and propulsion technology, emission location (e.g. cruising altitude and lateral/longitudinal position of aircraft) as well as the time of the emissions and atmospheric/weather conditions [29, 30, 44]. If non-CO_2-effects would to be reduced by SAF usage (e.g. in addition or as an alternative to operational measures, i.e. redirection of flights), a specific non-commingled physical SAF supply would probably be required. However, the full extent of these non-CO_2 effects, along with respective mitigation measures, have yet to be conclusively assessed on a scientific basis.

[11] Several SAF options can lead to lower particle and thus pollutant emissions when combusted [45, 46]. If pollutant emissions are to be reduced at specific regions or locations (airports) by the use of SAF, a specific non-commingled physical SAF supply is probably required.

4.2 Analysis and Results

Based on the plus-minus classification approach and comparison criteria described above, the different CoC models are compared and analysed below. All CoC models presented in Sect. 3 are considered. The overall qualitative assessment or rating CoC models are shown in Table 3.

In the following, the results shown in Table 3 are discussed. According to Table 3, the certificate transfer (CT) model shows overall high benefits across all comparison criteria applied. It is ranked most beneficial or equally beneficial with different mass balance (MB) models with respect to multiple criteria. It shows the highest advantageousness as a CoC model for SAF especially with regard to the abilities for scope expansion, minimisation of CO_2-emissions due to SAF logistics and SAF accessibility. Amongst all CoC models, a theoretical extra-sectoral access to SAF attributes is only possible based on a certificate transfer (CT) model. However, regarding the

Table 3 Assessment of CoC model characteristics in the context of SAF (CSCL—company supply chain level; CT—certificate transfer; de—decoupled; SAF—sustainable aviation fuel; IP—identity preservation; MB—mass balance; MRVA—monitoring, reporting, verification, and accounting; PS—physical separation; SL—site level; SG—segregation; SSCL—system supply chain level; TL—tank level)

Category/sub-level criteria		Physical separation		Mass balance				Certificate transfer
		IP	SG	TL-de	SL	CSCL	SSCL	
1	Integrity efforts	++	+	o	−	− −	− −	− −
2	Operational handling complexity	− −	−	o	o	+	++	++
3	Capability of scope expansion	− −	−	o	o	+	+	++
4	Implementation efforts	− −	−	o	+	+	+	+
5	Operating efforts	− −	−	o	+	+	+	+
6	Intra-sectoral access to SAF	− −	−	o	o	+	+	++
7	Extra-sectoral access to SAF	− −	− −	− −	− −	−	−	+
8	CO_2-emissions by SAF logistics	− −	−	o	o	+	+	++
9	MRVA of physical fuel properties	++	+	o	−	− −	− −	− −

(++) most beneficial characteristic; (+) rather beneficial characteristic; (o) average/neutral characteristic; (−) rather adverse characteristic; (− −) most adverse characteristic of a specific chain-of-custody model with respect to the specific comparison criterion and in comparison with the other chain-of-custody models criteria "integrity efforts", "MRVA capability of non-CO_2-effects", and "MRVA capability of local emissions", it represents the least degree of compliance. This is due to the complete/absolute decoupling of SAF attributes and physical aviation fuel product flows within a certificate transfer (CT) model.

In contrast to a certificate transfer (CT) model, the characteristics of both physical separation models (identity preservation (IP) and segregation (SG)) is basically reverse. They show the highest benefits for those criteria where the characteristics of a certificate transfer (CT) model are least beneficial. Most likely, they lead to the highest costs/efforts and lowest flexibilities in SAF accounting and SAF logistics. Their advantages lie in high integrity as well as their (model-immanent) capability to trace information on non-commingled SAF supply and integrate such information in the monitoring, reporting, verification, and accounting (MRVA) processes.

The mass balance (MB) models are basically ranked between the physical separation models and the certificate transfer (CT) model. Overall, for the mass balance (MB) models, the benefits with regard to the comparison criteria increase with wider balance limits and an increased decoupling of the physical product flows and the administrative SAF flows. Especially the tank and even more the site-level mass balance (MB) models represent a compromise of a physical separation model and a certificate transfer (CT) model.

Overall, the characteristics of a supply chain level mass balance (MB) model are already more comparable to a certificate transfer (CT) model. Yet, several (fundamental) differences still remain between both models. To claim SAF within a mass balance (MB) model, usually the corresponding SAF attributes need to be merged with a corresponding amount of aviation fuel (unless a mass balance (MB) model would explicitly be designed otherwise). Thus, the amount of SAF that can be claimed by an airline cannot (usually) exceed the overall amount of aviation fuel uplifted by the airline within the mass balance (MB) models balance limit. In contrast, this restriction is (usually) not applicable for a certificate transfer (CT) model, as certificates can be handled and traded independently from any physical product flow. However, depending on the actual design of both CoC models, those differences may largely be eliminated by means of a specific rule setting.

Still, a further difference remains with respect to scope expansion and SAF accessibility. For example, if a system supply chain level mass balance (MB) limit was designed around the European aviation fuel supply system, an airline would need to uplift aviation fuel within these system boundaries to gain access to the respective SAF attributes. This is not the case within a certificate transfer (CT) model, unless its system boundaries are explicitly tied to specified geographies (the specifics of this depend on the prevailing regulatory requirements). Furthermore, even in the case of a mass balance (MB) model at system supply chain level, extending the scope/balance limit will always require some sort of physical aviation fuel supply extension (in the form of CAF or SAF). This difference to a certificate transfer (CT) model would only be eliminated if the balance of the mass balance (MB) model would include the global aviation fuel supply system completely from the beginning. The scope of a certificate transfer (CT) model is practically not limited to any balance limit and only defined by the extent to which stakeholders have access to the registry or certificates (independent from their route network or fuel/SAF supply situation).

Following the above considerations and assessment, it can be argued that, in the particular case of SAF, a CoC model with a higher degree of decoupling of physical and administrative product flows offers a greater benefit to the considered criteria over a chain-of-custody model with a tighter coupling of such flows (presupposed that sufficient integrity and fraud protection can also be obtained with reasonable effort).

4.3 Discussion on Suitability Aspects

According to the above analysis, for the reporting and accounting of sustainable aviation fuel (SAF), each chain-of-custody (CoC) model implies individual benefits and limitations. Thus, drawing conclusive aspects about a model's suitability needs a consideration of general framework conditions and requirements for a CoC model in the context of SAF.

As of today, SAF is at a nascent commercial stage. Limited amounts of SAF are available at few locations only. Bringing together the limited initial SAF supply with initial SAF demands from airlines can require considerable efforts compared to logistics of global commodities such as conventional aviation fuel. Given the current SAF market situation, requirements such as a minimum time and effort of implementation, a high accounting flexibility, a wide access to limited SAF supplies and low access barriers, avoidance of designated SAF logistics, or even the ability to further stimulate an overall demand for SAF (consumer demand) seem to be crucial criteria for a proper CoC model. In this case, most likely, a "book and claim" approach based on a certificate transfer (CT) model meets the above criteria best of all CoC models. Its scope is merely defined by the ability for stakeholders to access certificates (independent from their route network or fuel/SAF supply situation) and not limited to any other specific balance limit. In contrast, under the current market circumstances of SAF, physical separation models are most likely significantly less suitable. They would likely imply higher costs, a more limited access to SAF attributes, and less flexibility in both SAF accounting and SAF logistics. Mass balance (MB) models represent a compromise of the preceding models, while this rather applies to a tank- and site-level mass balance (MB) and less to a supply chain level mass balance (MB) model.

However, based on scientific progress in the future, additional requirements might arise if the fuel composition proves suitable to influence non-CO_2 effects. Should the resulting effects need to be captured within the monitoring, reporting, verification, and accounting (MRVA) framework of a CoC model, there could be a requirement to trace information about the chemical composition of the fuel up to a specific point in the supply chain (both for sustainable and fossil fuel). Moreover, the CoC model would have to be capable of managing information about the effects resulting from its physical use throughout its MRVA process. In such a case, CoC models that already imply a coupling of the physical and the administrative flow of SAF within the model's MRVA processes would be beneficial in this respect. In this case (theoretically), only physical separation CoC models seem applicable to allow for such a coupled supply of SAF and tracking of chemical compositions up to a specific point of delivery. Regarding the mass balances (MB) models, a coupling of physical and administrative SAF flow tracking of chemical fuel properties could at most be accomplished by a mass balance (MB) model at tank level. However, certain CoC models, especially those based on physical separation, may impair efficient SAF logistics and impede the trading of SAF as a global commodity. Therefore, even in cases of targeted physical SAF usage, considering other CoC models, such as the certificate transfer (CT) model, which offers enhanced logistical and MRVA flexibility for SAF, still seems appropriate. In this case, should the need arise, these CoC models might require further conceptual refinement to effectively monitor, report, verify, and account for the effects of physical SAF usage within the supply chain.

5 Conclusion

Powerfuels are essential for achieving global climate targets, especially in hard-to-abate sectors like aviation, shipping, and heavy-duty road transportation. Ensuring compliance with sustainability criteria throughout the value chain and implementing chain-of-custody (CoC) models to allocate environmental benefits is pivotal for their widespread use. This is particularly crucial in regions like the EU, where significant shares of powerfuels are expected to be imported in the medium to long term. Given that parts of the powerfuels supply chain will then be located outside the EU, ensuring compliance with established eligibility criteria becomes more challenging, underscoring the need for robust CoC models. In this context, this chapter provides an overview of different chain-of-custody (CoC) options for powerfuels and their specific characteristics, requirements, and both advantageous and disadvantageous implications using the example of aviation and sustainable aviation fuels (SAF). Seven CoC models are compared against nine comparison criteria based on a plus-minus classification approach.

- The certificate transfer (CT) model is found to have high overall benefits and is ranked favourably in multiple criteria. It is particularly advantageous for sustainable aviation fuel (SAF) regarding scope expansion, CO_2 emission reduction in SAF logistics, and SAF accessibility. However, without further adjustments, it has lower compliance with integrity efforts and monitoring, reporting, verification, and accounting (MRVA) capability of physical fuel properties due to the complete decoupling of SAF attributes and physical fuel flows.
- On the other hand, the physical separation models (identity preservation and segregation) have strengths in integrity and tracing non-commingled SAF supply, but they are associated with higher costs, logistical efforts, and less flexibility in SAF accounting and logistics. Already today, and especially in the medium

to long term, these CoC models may only meet the requirements of global SAF logistics to a (quite) limited extend.

- The mass balance (MB) models fall between the physical separation models and the certificate transfer model. They offer increasing benefits with wider balance limits and greater decoupling of physical and administrative flows. The tank- and site-level mass balance models serve as a compromise between physical separation and certificate transfer approaches.

The analysis shows that in the particular case of SAF a CoC model with a high degree of decoupling of physical and administrative product flows offers greater benefits to the considered criteria over a CoC model with a tighter coupling of both flows. This leads to the fact that a certificate transfer (CT) model shows an overall high compliance across all comparison criteria applied—especially with regard to the abilities for scope expansion, minimisation of SAF logistics emissions and SAF accessibility.

However, in terms of suitability aspects, this comparison alone is not sufficient. The actual suitability of a specific CoC model is determined by various individual criteria, whose prioritisation and importance (weighting) will most likely be strongly determined by the existing system and market circumstances over the course of time. Given the current (nascent) SAF market situation, a CoC model with a high degree of decoupling the physical and administrative product flows most likely offers a greater suitability over a CoC model with a tighter coupling of such flows. Thus, especially a certificate transfer (CT) model can be considered as the means of choice. While higher efforts to establish proper fraud protection may arise for this CoC model option, the advantages in terms of accounting flexibility, SAF logistics, and SAF access are higher compared to all other CoC models. Even in case of a direct comparison between a certificate transfer (CT) model and a mass balance (MB) model at system supply chain level, the certificate transfer (CT) model still offers more flexibility. However, in order for those benefits to be operationalised and exploitable in practice, a uniform, preferably global SAF standard is urgently required for the certification and emission reporting of SAF, which should be strongly supported on a policy level. Such a standard should be implemented at the broadest geographical level possible, ideally on global level. This could be initiated in regions like the EU, followed by a progressive expansion.

Yet, additional requirements could arise in future. One of such requirements could be that information about physical shares of SAF and/or its chemical properties need to be available/traceable up to a certain stage in the physical fuel supply chain and a CoC model need to offer the ability to handle information on the resulting effects (e.g. due to non-CO_2-effetcts) seamlessly within its MRVA process. In this context, due to logistical and flexibility considerations, it would be advantageous to also base the MRVA of SAF on flexible CoC concepts, like certificate transfer (CT) models or supply chain level mass balance (MB) models. Nonetheless, given the potential limitations of these CoC models in capturing information related to physical SAF usage—at least in their current conceptual designs—conceptual adjustments to such

CoC models and the integration of an additional mechanism tailored for the MRVA of such effects would be required.

Considering the increasing global prevalence of SAF-related regulation, it is essential that SAF MRVA concepts or SAF CoC models, respectively, are designed in a future-oriented manner and future-proof. Both must anticipate and align with such regulatory developments to ensure compatibility with both current and upcoming regulatory frameworks and compliance requirements. This foresight is required to ensure that SAF CoC models seamlessly integrate with future regulations and remain relevant in an evolving regulatory landscape. Moreover, it is crucial to design CoC models from a holistic system perspective, addressing overarching stakeholder needs that serve the transformation of the aviation industry rather than (short-term) niche MRVA requirements.

Acknowledgements This article has received funding from the European Union's Horizon 2020 research and innovation programme under grant agreement No. 957824.

References

1. WEF (2020) Joint policy proposal to accelerate the deployment of sustainable aviation fuels in Europe. A clean skies for tomorrow publication. White Paper. World Economic Forum (WEF), Köln
2. Bullerdiek N, Quante G, Bube S, Neuling U, Kaltschmitt M (2022a) Non drop-in Kraftstoffe im Luftverkehr—Ein gesamtsystemischer Vergleich von Nutzungs- und Einsatzmöglichkeiten. Institut für Umwelttechnik und Energiewirtschaft (IUE); Technische Universität Hamburg (TUHH). Hamburg, Berlin
3. Bullerdiek N, Voß S, Neuling U, Kaltschmitt M (2022) Direct alcohol versus alcohol-to-jet SPK utilisation in commercial aviation: an energetic-operational analysis. Int J Sustain Aviat 8(3):1. https://doi.org/10.1504/IJSA.2022.10046511
4. Kramer S, Andac G, Heyne J, Ellsworth J, Herzig P, Lewis KC (2022) Perspectives on fully synthesized sustainable aviation fuels: direction and opportunities. Front Energy Res 9:782823. https://doi.org/10.3389/fenrg.2021.782823
5. MPP (2022) Making net-zero aviation possible. An industry-backed, 1.5°C-aligned transition strategy. Mission Possible Partnership (MPP)
6. CST, WEF (2020) Sustainable aviation fuels as a pathway to net-zero aviation. Insight report. Clean skies for tomorrow (CST); McKinsey & Company; World Economic Forum (WEF). Genf
7. CST, WEF (2021a) Powering sustainable aviation through consumer demand: the clean skies for tomorrow sustainable aviation fuel certificate (SAFc) framework. Insight Report. Clean Skies for Tomorrow (CST); World Economic Forum (WEF)
8. EC (2020) Sustainable aviation fuels virtual roundtable. ReFuelEU Aviation Background Paper. European Commission (EC)
9. Bullerdiek N, Neuling U, Kaltschmitt M (2021) A GHG reduction obligation for sustainable aviation fuels (SAF) in the EU and in Germany. J Air Transp Manag 92:102020. https://doi.org/10.1016/j.jairtraman.2021.102020
10. CST, WEF (2021b) Guidelines for a sustainable aviation fuel blending mandate in Europe. Insight Report. Clean Skies for Tomorrow (CST); World Economic Forum (WEF). Genf
11. CST, WEF (2021c) Sustainable aviation fuel policy toolkit. Insight Report. Clean Skies for Tomorrow (CST); World Economic Forum (WEF). Genf

12. Ghatala F (2020) Sustainable aviation fuel policy in the United States. A pragmatic way forward. Hg. v. Atlantic Council, Washington D.C.
13. van Muijden J, Stepchuk I, Boer AI, Kogenhop O, Rademaker ER, van der Sman ES, Kos J, Posada Duque JA, Palmeros Parada M (2021) Final results alternative energy and propulsion technology literature study. Deliverable D1.1 of the TRANSCEND project. NLR-CR-2020-026
14. EASA (2019) Sustainable aviation fuel 'monitoring system'. Grant Agreement EASA.2015.FC21. European Union Aviation Safety Agency (EASA). Köln
15. EC (2021) Proposal for a regulation of the European parliament and of the council on ensuring a level playing field for sustainable air transport. Document 52021PC0561. European Commission (EC), Brussels
16. ISEAL (2016) Chain of custody models and definitions: a reference document for sustainability standards systems, and to complement ISEAL's sustainability claims good practice guide. Version 1.0. ISEAL Alliance, London
17. EC (2022) The monitoring and reporting regulation: general guidance for aircraft operators. MRR Guidance document No. 2. Updated Version, 31 January 2022. European Commission (EC), Brussels
18. IATA (2019) An airline handbook on CORSIA. International Air Transport Association (IATA), Montréal
19. ICAO (2018) First edition to the international standards and recommended practices. In: Environmental protection. Annex 16. Volume IV. Carbon offsetting and reduction scheme for international aviation (CORSIA). International Civil Aviation Organization (ICAO), Montréal
20. Pechstein J, Bullerdiek N, Kaltschmitt M (2020) A "book and claim"-approach to account for sustainable aviation fuels in the EU-ETS: development of a basic concept. Energy Policy 136:111014. https://doi.org/10.1016/j.enpol.2019.111014
21. DfT (2022) Sustainable aviation fuels mandate. Summary of consultation responses. Department for Transport (DfT), London
22. AirGO (2021) Carbon neutral air travel: AirGO provides first SAF book and claim scheme for business aviation. AirGO. https://airgo.de/co2-neutral-private-jet-flights-with-book-and-claim-and-saf/. Accessed 13 May 2022
23. Avfuel NO (2021) Avfuel SAF, powered by neste MY sustainable aviation fuel. Frequently asked questions. Avfuel; Neste Oyj. https://www.avfuel.com/portals/0/Documents/FAQs_Avfuel_Neste_FINAL.pdf. Accessed 13 May 2022
24. FFI (2022) Consensus statement on fuelling flight: response to the ReFuelEU proposal. Fuelling Flight initiative (FFI), Brussels. https://www.transportenvironment.org/wp-content/uploads/2022/02/Aviation_Feb2022-1.pdf. Accessed 13 May 2022
25. Jet Aviation (2021) Jet aviation partners with SkyNRG to implement global 'book and claim' service for SAF. Jet Aviation. https://www.jetaviation.com/company/perspectives/jet-aviation-partners-with-skynrg-to-implement-global-Book-and-claim-service-for-sustainable-aviation-fuel. Accessed 13 May 2022
26. RSB (2021) Book and claim for SAF—FAQs. Roundtable on Sustainable Biomaterials (RSB). Genf. https://rsb.org/wp-content/uploads/2021/11/21-11-08-Book-Claim-FAQ-.pdf. Accessed 13 May 2022
27. Shell (2022) Shell, Accenture and Amex GBT launch one of the world's first blockchain powered digital book-and-claim solutions for scaling sustainable aviation fuel (SAF). Shell International Petroleum Co. Ltd.
28. Smith D, Greene S, Lewis A, Betts K, Bateman A (2021) Sustainable aviation fuel greenhouse gas emission accounting and insetting guidelines. Smart Freight Centre (SFC); MIT Center for Transportation & Logistics, Amsterdam
29. EU Parliament (2022) Revision of the EU emissions trading system for aviation ***I. P9TA(2022)0230. European Parliament (EU Parliament), Strasbourg
30. Lee DS, Fahey DW, Skowron A, Allen MR, Burkhardt U, Chen Q, Doherty SJ, Freeman S, Forster PM, Fuglestvedt J, Gettelman A, León RR, Lim LL, Lund MT, Millar RJ, Owen B, Penner JE, Pitari G, Prather MJ, Sausen R, Wilcox LJ (2021) The contribution of global aviation to anthropogenic climate forcing for 2000 to 2018. Atmosp Environ 244:117834. https://doi.org/10.1016/j.atmosenv.2020.117834

31. Matthes S, Lim L, Burkhardt U, Dahlmann K, Dietmüller S, Grewe V, Haslerud AS, Hendricks J, Owen B, Pitari G, Righi M, Skowron A (2021) Mitigation of non-CO_2 aviation's climate impact by changing cruise altitudes. Aerospace 8(2):36. https://doi.org/10.3390/aerospace8020036

32. Niklaß M (2019) Ein systemanalytischer Ansatz zur Internalisierung der Klimawirkung der Luftfahrt. Dissertation. Technische Universität Hamburg (TUHH), Hamburg

33. ISO (2020) Chain of custody: general terminology and models. ISO 22095:2020. First Edition 2020–2010. International Organization for Standardization (ISO), Genf

34. Dehue B, Meyer S, Hamelinck C (2007) Towards a harmonised sustainable biomass certification scheme. Ecofys, Utrecht

35. IPIECA (2010) Chain of custody options for sustainable biofuels. In: International petroleum industry environmental conservation association (IPIECA). London

36. Goovaerts L, Pelkmans L, Sheng Goh C, Junginger M, Joudrey J, Chum H, Smith CT, Stupak I, Cowie A, Dahlman L, Englund O, Goss A (2013) Strategic inter-task study: monitoring sustainability certification of bioenergy. A cooperation between IEA Bioenergy Task 40, Task 43 and Task 38. IEA Bioenergy

37. Bowe S, Girbig P (2021) Nachweissysteme für erneuerbare Energien. Bericht im Rahmen des Projekts GO4Industryry, gefördert durch das Bundesministerium für Umwelt, Naturschutz und nukleare Sicherheit (FKZ: UM20DC003). Green Gas Advisors, Berlin

38. FONAP (2022) Handelsoptionen von Palmöl (Handelsmodelle). Forum Nachhaltiges Palmöl (FONAP), Bonn. https://www.forumpalmoel.org/zertifizierung/handelsmodelle. Accessed 28 March 2022

39. ISCC (2020) ISCC CORSIA 203 traceability and chain of custody. Version 1.0. ISCC System GmbH, Köln

40. Bullerdiek N, Buse J (2019) Konzeptionierung einer Anrechnungsmethodik im Emissionshandel. Interne Dokumentation der Ergebnisse aus Aufgabe 8 (not published). Forschungs- und Demonstrationsvorhaben zum Einsatz von erneuerbarem Kerosin am Flughafen Leipzig/Halle (DEMO-SPK). Technische Universität Hamburg (TUHH); Adeptus Green Management GmbH, Hamburg

41. Drillisch J (2001) Quotenmodell für regenerative Stromerzeugung. Ein umweltpolitisches Instrument auf liberalisierten Elektrizitätsmärkten. München: Oldenbourg-Industrieverl. (Schriften des Energiewirtschaftlichen Instituts, Bd. 57)

42. IEA (2021) Progress in commercialization of biojet/sustainable aviation fuels (SAF): technologies, potential and challenges. IEA Bioenergy Task 39. International Energy Agency (IEA), Paris

43. Voigt C, Kleine J, Sauer D, Moore RH, Bräuer T et al (2021) Cleaner burning aviation fuels can reduce contrail cloudiness. Commun Earth Environ 2(1):174. https://doi.org/10.1038/s43247-021-00174-y

44. UBA (2019) Umweltschonender Luftverkehr. lokal—national—international. Umweltbundesamt (UBA), Dessau-Roßlau

45. Schripp T, Grein T, Zinsmeister J, Oßwald P, Köhler M, Müller-Langer F, Hauschild S, Marquardt C, Scheuermann S, Zschocke A, Posselt D (2021) Technical application of a ternary alternative jet fuel blend: chemical characterization and impact on jet engine particle emission. Fuel 288:119606. https://doi.org/10.1016/j.fuel.2020.119606

46. Schripp T, Herrmann F, Oßwald P, Köhler M, Zschocke A, Weigelt D, Mroch M, Werner-Spatz C (2019) Particle emissions of two unblended alternative jet fuels in a full scale jet engine. Fuel 256:115903. https://doi.org/10.1016/j.fuel.2019.115903

A Holistic Approach to Sustainability of Powerfuels

Anita Demuth, Nils Fuchs, Harry Lehmann, and Jessica Nagamichi

Abstract Powerfuels produced with renewable electricity are considered a promising option for greenhouse gas mitigation in various sectors. Global scenarios indicate an accelerated and substantially increasing demand for powerfuels to achieve international climate targets. Consequently, a new industrial sector is expected to emerge in the coming decades with links into several transport and manufacturing industries. Past experiences, such as with biofuels, have demonstrated the importance of adopting a holistic perspective when dealing with the implications of a new sector to avoid adverse effects on both people and the planet. Therefore, the objective of this chapter is to define a framework that encompasses the opportunities and risks associated with powerfuels for people, the planet, and prosperity—the three main areas of action next to "peace" and "partnerships" that the United Nations Sustainable Development Goals are categorized against. This framework not only identifies minimum criteria for powerfuels but also considers potential socioeconomic co-benefits at global, regional, and local levels. The approach derives the impacts based on the concept of planetary boundaries and globally shared values, as manifested in international agreements on climate, prosperity, and human rights. The findings of this work reveal that existing sustainability definitions for powerfuels, as exemplified at the European level, seem insufficient to avoid negative impacts. Consequently, there is a need to further advance the sustainability discourse, particularly at the political level.

Keywords Planetary Boundaries · GHG Mitigation · SDG · Socio-Economic Aspects · Holistic Sustainability Standards

A. Demuth (✉) · N. Fuchs · H. Lehmann · J. Nagamichi
PtX Lab Lausitz, Cottbus, Germany
e-mail: anita.demuth@z-u-g.org

© The Author(s), under exclusive license to Springer Nature Switzerland AG 2025
N. Bullerdiek et al. (eds.), *Powerfuels*, Green Energy and Technology,
https://doi.org/10.1007/978-3-031-62411-7_35

1 Introduction

The effects of anthropogenic climate change are already evident globally today. By the end of the twenty-first century, the risks from global warming due to greenhouse gas (GHG) emissions will continue to increase [1]. Therefore, GHG emissions must be drastically reduced. In its sixth Assessment Report (AR6), the Intergovernmental Panel on Climate Change (IPCC) presents different scenarios of how much global warming will increase as a result of GHG emissions by the year 2100. Hydrogen and its derivatives are considered to play an important role in the emissions pathways that are consistent with the Paris Climate Agreement [2]. In particular, the transport sector and parts of industry will have a strong demand for green hydrogen and its derivatives in 2050. This is the outcome of a report by the International Energy Agency (IEA) in scenarios based on net-zero emissions by 2050 [3]. However, due to high production costs and alternatives, such as direct electrification, application should primarily be limited to specific industries, such as steel and chemistry, along with segments within the transport sector, notably aviation, heavy-duty land transport, and shipping. In these sectors, synthetic fuels are often the only or most sustainable option to reduce GHG emissions.

Electrolysis builds the basis for the production of powerfuels. The European Union (EU) has set an ambitious goal within the framework of the REPowerEU plan to produce 10 million tons of renewable hydrogen annually by 2030, aiming to meet its climate targets. This objective corresponds to an electrolysis capacity of 100 gigawatts (GW) by 2030, depending on factors, such as full load hours and efficiency [4]. To achieve this, capacity must increase by a factor of 550 [5]. Additionally, the EU plans to import another 10 million tons of renewable hydrogen from outside Europe. These statistics clearly indicate that both the supply side and the overall demand, as well as the necessary infrastructure, will undergo substantial expansion. Consequently, a new industrial sector is emerging, transcending individual countries and giving rise to new global trade chains. One contributing factor to this development is the varying production potentials and demands across countries.

The number of bilateral trade relationships in the powerfuels sector is steadily increasing [6]. It is evident that developing countries in particular aspire to become exporters. These nations perceive powerfuel production as an opportunity for economic development and value creation. However, the rapid market upscaling and strong export orientation pose risks of neglecting impacts on people, the environment, and resources. Furthermore, the focus on exports may come at the expense of national decarbonization goals [7]. Therefore, it is crucial to identify and mitigate existing risks in the early phase of market upscaling. By establishing a framework and defining standards, a sustainable development pathway can be paved for the powerfuels sector, aligning with the concept of a just transition [8]. This pathway ensures responsible and sustainable resource utilization while mitigating potential socioeconomic and environmental risks.

As of now, a regulatory framework defining sustainable powerfuels has only been established at the EU level. Within the Renewable Energy Directive II (RED II), the European Commission adopted two Delegated Acts (DAs) in February 2023 [9, 10]. These acts establish the requirements for domestic production and import of renewable fuels of non-biological origin (RFNBOs).

- The first Delegated Act (DA) pertains to the requirements for electricity used in the production of RFNBOs. There are fundamentally two options for sourcing electricity: direct connection via transmission lines between the renewable energy (RE) facility and the powerfuels production facility, or sourcing electricity from the grid.
- The second DA pertains to the requirements for emissions reduction. Under this act, RFNBOs are required to exceed a minimum threshold of 70% GHG savings compared to the fossil comparator. The act also establishes the methodology for calculating and attributing these savings, especially on how to attribute the carbon source.

The two criteria for electricity sourcing and GHG reduction represent two important aspects for the sustainable market ramp-up of powerfuels. However, the market ramp-up of powerfuels encompasses additional environmental and socioeconomic risks that are not addressed in the standard. Past examples demonstrate that the positive effects of an emerging industry must be carefully balanced against the associated risks in order to generate a positive impact on the planet.

In the past two decades, the international focus has been on biofuels to achieve emission reductions in sectors, such as transportation. A key criticism of first-generation biofuels was the "Food, Energy and Environmental Trilemma" [11], a situation in which too many different vested interests competed for the same limited resources. This called into question the benefits of biofuels and earned them the ambiguous reputation as a "redeemer and work of the devil" [12] at the same time. In addition to competition for resources, biofuels have also been criticized for certain production methods [13, 14]. One of the main rationales for biofuels, namely their environmental benefits, is challenged by the indirect emissions caused by indirect land-use change [15–17]. In tandem with these impacts, biofuels also have negative consequences in the social sphere due to their high land resource requirements, leading to issues of access to land and water [18, 19] or land rights [20, 21].

The debate on biofuels over the last two decades shows that neglecting the potentially negative impacts of fuels declared "green" by industry and policymakers can induce strong and fundamental rejection for the utilization of such fuels in society—slowing down the overall development dynamics of this industry with corresponding negative economic impacts. The wind power sector has experienced similar challenges in Europe. Wüstenhagen et al. [22] show how social acceptance had been neglected in the first two decades but has led in some countries to many barriers for achieving successful projects at the implementation level. The rapid growth of the powerfuels industry implied in the scenarios on the one hand and the lack of standards for a sustainable path on the other hand therefore justify the need to formulate a holistic sustainability approach.

Previous approaches to sustainability assessment of powerfuels have been limited to the three pillars of sustainability—environment, economy, and society. However, the three sustainability dimensions are interdependent and should not be examined separately [23]. The following work breaks down the rigid view and provides a framework for assessing opportunities and risks to people, planet, and prosperity. The approach is based on the scientific concept of planetary boundaries (PBs), which defines thresholds of human activity without causing permanent harm to the planet [24]. In addition, globally shared values in the form of international treaties on climate, prosperity, and human rights are also utilized. This provides a framework that identifies not only the minimum criteria of powerfuels, but also potential co-benefits at the global, regional, and local levels.

The first section of this work defines the specific risks powerfuels pose for overflowing PBs, highlighting the urgency of formulating effective safeguards. Following this, the focus shifts to examining the socioeconomic and political risks associated with powerfuels, along with forecasting and harnessing potential co-benefits in these domains. The subsequent section addresses the challenge of putting ambitious sustainability criteria into practice, emphasizing the importance of acceptance, transparency, and participation. Finally, the work concludes with final considerations drawn from the comprehensive analysis conducted throughout the study.

2 Impact of Powerfuels on the Planet

2.1 Powerfuels Specific Risks for Overflowing Planetary Boundaries

The main reason for the intended use of powerfuels is their potential for GHG emissions reduction compared to energy options of fossil origin. However, powerfuels also have further impacts on the environment. Depending on the design of the technology and the associated production factors, there may be risks involved. Particularly because the widespread utilization of powerfuels results in a new industrial sector, there may be far-reaching consequences on the planet. Therefore, environmental risks must be reviewed. In order to contextualize environmental risks associated with powerfuels, it is necessary to adopt a framework that takes into account the global impact of human activities. One such framework is that of planetary boundaries, which was originally proposed by [24] and has since been further developed by [25, 26]. The resulting environmental risks associated with the use of powerfuels are therefore derived using this concept.

The concept of planetary boundaries is based on the assumption that human activities negatively impact the Earth system. It defines the "safe operating space" within which humanity can operate without destabilizing the Earth system. Nine planetary boundaries form the framework for classification:

1. climate change,
2. stratospheric ozone depletion,
3. ocean acidification,

4. biosphere integrity (previously "biodiversity loss"),
5. biogeochemical flows,
6. land-system change,
7. freshwater use,
8. atmospheric aerosol loading,
9. novel entities.

All nine planetary boundaries are interconnected, with climate change and biosphere integrity being the core due to their fundamental importance to the Earth system through environmental impacts. While resource use itself is not one of the planetary boundaries, this is only legitimate up to a certain limit. Passing this limit, the environmental impacts resulting from resource use would also be catastrophic and contribute to the destabilization of the Earth system [27]. Therefore, resource use plays an essential role in sustainability considerations and is regarded as the 10th Planetary Boundary in the following. Furthermore, stratospheric ozone depletion and biogeochemical flows (formerly known as "nitrogen and phosphorus cycles") will not be evaluated in this paper.

The assessment of the risk of powerfuels within planetary boundaries is based on a holistic input–output analysis along the value chain. The basis for the quantification of the input are the production factors for powerfuels. Emissions and waste products also represent the output. Table 1 summarizes the inputs and outputs.

Climate Change and Ocean Acidification. Climate change and its overwhelming impact on the Earth system is posing by far the greatest challenge of the twenty-first century [28]. Therefore, it is the core planetary boundary and the center of the global environmental problem. Also, it is closely linked to the goal of the Paris Climate Agreement to limit global warming to well below 2 °C. This target was set based on the findings of the IPCC, which also, as with the planetary boundary, base their findings on the amount of carbon dioxide (CO_2) in the atmosphere. Ocean acidification forms another planetary boundary. However, ocean acidification is directly caused by GHG emissions and therefore climate change [29]. Thus, the influence of powerfuels on both of these planetary boundaries will be examined in the following.

Although the main argument for the utilization of powerfuels is the reduction of CO_2 emissions, there is a risk of negative impacts on climate change if the production is not based on renewable energy (RE). In addition, depending on the fuel produced, usually also CO_2 is required for the production of powerfuels, which might also cause negative climate change impacts.

Table 1 Inputs and outputs along powerfuels value chain (RE—renewable energy)

Input	Output
Plant operation input (e.g., RE, water, CO_2)	Emissions
Human capital	Waste and dissipative losses (materials)
Land (infrastructure and plants)	
Materials and resources	

By shifting to powerfuels, sectors, such as industry and transportation, which are currently still highly dependent on fossil, emission-intensive energy resources, can reduce CO_2 emissions significantly. Therefore, based on the IEA net-zero scenario, [30] it is assumed that green hydrogen and its derivatives can save 7 Gt of CO_2 emissions annually worldwide by 2050, which is a significant reduction compared to the approximately 36.8 Gt of CO_2 emissions worldwide in the year 2022 [31]. However, this is based on the assumption that electricity from RE is used to produce the fuels, which has almost no and significantly less CO_2 emissions than fossil fuels when considering the life cycle assessment (LCA) [32]. However, a look at the global share of solar and wind shows that in 2021 they together accounted for only 12% of electricity generation [33]. Thus, the production of powerfuels would be associated with undesirable CO_2 emissions as electricity generation continues to rely heavily on fossil resources for the most part. To minimize the risk of climate change in the long term, only RE should be used for generation.

Besides the input electricity, the CO_2 intensity of powerfuels strongly depends on the CO_2 source used for fuel production [34]. If the CO_2 is of fossil origin, as it would be for example in the case of reusing CO_2 from industrial sources, such as cement production, this leads to a flow of originally stored CO_2 into the atmosphere and consequently contributes to an increase in atmospheric carbon stock. Hence, despite the use of RE, there would be a negative climate impact in the end. On the other hand, there is the alternative of a closed CO_2 cycle: If CO_2 from sustainable biomass or direct air capture (DAC) is used, no additional CO_2 is emitted into the atmosphere during the subsequent combustion of the powerfuels. Instead, a form of CO_2 recycling occurs, which had already been present in the atmosphere prior to being utilized in the production of powerfuels.

Biosphere Integrity. Alongside climate change, biosphere integrity is the second core planetary boundary due to its interdependency to other planetary boundaries and its irreversible consequences if exceeded persistently. The planetary boundary of changes in biosphere integrity centers on preserving the Earth's ecosystems and biodiversity. It highlights the importance of avoiding substantial biodiversity loss and degradation of ecosystems resulting from human activities. This boundary pertains to the well-being and functionality of ecosystems, the conservation of species diversity, and the safeguarding of ecological processes that sustain life on the planet. At the international level, the International Union for Conservation of Nature (IUCN) plays a pivotal role in international conventions on nature conservation and biodiversity [35]. Their Nature 2030 programme is based on the concept developed by [36], which argues that biosphere builds the foundation of economies and societies and therefore of all Sustainable Development Goals (SDGs).

Although the utilization of powerfuels is less likely to have a negative impact on the biosphere than fossil fuels (indirectly, e.g., through GHG reduction), their impact must be closely examined, as this boundary has already been exceeded. To avoid risks as far as possible, the geographical location is essential—not only in the site decision of a plant, but along the entire value chain. Therefore, there is close correlation with the planetary boundary of land-system change. Even the extraction of the resources needed for machinery, plants and infrastructure can have negative effects on the biosphere. Since there is an extra planetary boundary resource input, this section is limited to further risks associated with powerfuels. In particular, there are risks associated with the construction and operation of plants (including renewable electricity plants) and the construction and use of infrastructure, as well as with a lack of treatment of waste and end products.

To mitigate risks, IUCN has designed a framework that provides project developers, especially in the RE sector, with a framework for monitoring [37]. IUCN distinguishes between three categories of impacts on biodiversity and ecosystems—direct, indirect, and cumulative impacts:

- Construction and operation of powerfuels facilities and infrastructure can have a direct impact on habitat loss, habitat fragmentation, species disturbance, and mortality.
- In addition, there can be indirect influences from increased human activities (agriculture, tree cutting, etc.) in the area as people migrate to previously undisturbed areas for their jobs.
- Cumulative impacts can result from interactions with other sectors. An example would be the mining threat to biodiversity due to increased demand for raw materials for the production of the powerfuel plants as well as the RE generation facilities [38].

The impact of pollution represents a significant facet of direct impacts. Pollution, in this particular context, can manifest in various forms including dust, light, noise, vibration, as well as solid/liquid waste. Especially in the case of solid/liquid waste pollution due to lack of recycling and inadequate treatment of waste or end products, there is a lack of guidelines. Consequently, this situation leads to unregulated release of materials and substances into the environment. For instance, the disposal of cooling water, necessary for facilities like concentrated solar power (CSP), can give rise to algal blooms and fish mortality upon discharge into freshwater bodies. Moreover, the utilization of desalination plants to obtain process water for electrolysis results in the production of brine,[1] which when disposed in sea or fresh water can have detrimental effects on marine organisms. The increased salinity and temperature levels due to disposing untreated brine may lead to reduced dissolved oxygen content, which can negatively impact a wide range of organisms.

Land-System Change. The relationship between planetary boundary land-system change and planetary boundary biosphere integrity is deeply interconnected. Many of the risks posed by powerfuels to biodiversity and ecosystems arise from changes in land systems. This section focuses on other aspects of land-system change, which aligns with the updated concept of planetary boundaries from 2015 [25]. This update aimed to establish a clearer distinction between the two planetary boundaries. As a

---

[1] Brine from seawater desalination is comprised of concentrated salts, minerals, and dissolved organic compounds. The precise composition can vary, depending on the desalination technique employed and the specific characteristics of the seawater.

result, the planetary boundary land-system change was specifically defined to encompass biogeophysical processes, particularly carbon sequestration and the subsequent climate change mitigation effect facilitated by forests. The process of removing CO_2 from the atmosphere, often referred to as "sink" by the United Nations Framework Convention on Climate Change (UNFCCC), is integral to this definition [39]. Human activities thus have a significant impact on natural sinks, such as land use, land-use change, and forestry (LULUCF).

In addition to the risks associated with land use that exhibit strong connections to biodiversity and ecosystems, as described in the preceding section, land-system change resulting from powerfuels can bring forth two main impacts. Firstly, there is the risk that areas serving as crucial carbon sinks can be damaged due to the production of powerfuels. Secondly, the converted areas may not only hold high ecological importance but also possess cultural significance.

The construction of powerfuel plants and infrastructure in regions that serve as significant carbon sinks can result in the unintended release of CO_2 [40]. This not only undermines the intended positive climate impact of powerfuels but also disrupts the carbon cycle within the climate system. Peatlands and forests, in particular, play crucial roles as carbon sinks. Forests, which encompass approximately 31% of the global land surface [41], should receive special protection measures.

The land system is not only important from an environmental perspective but also has cultural significance (Sect. 3.1). This is particularly the case in areas inhabited by indigenous peoples. Due to marginalization and project developments in their areas, their culture and lives face a high level of vulnerability [42]. Taking into account that indigenous territories contain about 80% of the world's biodiversity [43], it is crucial to assess the potential consequences associated with the construction and operation of plants and infrastructure.

Freshwater Use. Freshwater is a finite and essential resource for life on Earth. It sustains ecosystems, supports agriculture, and meets the basic needs of people worldwide. However, the availability and quality of freshwater resources are increasingly being threatened by human activities [44]. Human water consumption has been steadily increasing due to population growth, urbanization, industrialization, and changing consumption patterns. Therefore, the planetary boundary for freshwater use refers to the maximum amount of water that can be consumed from freshwater sources without causing significant ecological disruption or depleting the resource beyond its capacity for renewal. This boundary is essential for maintaining the health of aquatic ecosystems, ensuring water availability for future generations, and preventing water scarcity-related conflicts.

Similar to other industrial sectors, the production of powerfuels necessitates water. However, when compared to fuel alternatives based on biomass, the water intensity of powerfuel production is relatively low [45]. Nevertheless, in regions facing existing or future water scarcity, the utilization of water entails significant risks. The Water Risk Atlas by the World Resources Institute (WRI) identifies three distinct risk categories: physical quantity, physical quality, and regulatory and reputational risks [46].

Many countries that are expected to become major producers of powerfuels in the future due to favorable conditions are already grappling with water scarcity issues. Just the electrolysis production step itself requires 8.9 L of water to produce 1 kg of green hydrogen (stoichiometric) [45]. The EU alone has plans to produce 10 million tons of green hydrogen annually by 2030 and import another 10 million tons [4]. Consequently, this translates to a water demand of approximately 180 million m^3. Particularly in regions with limited surface and groundwater resources, powerfuel production can intensify water withdrawals, thereby increasing the risk of water stress. The elevated risk of water stress is also accompanied by the potential for prolonged droughts, which can impact humans, ecosystems, and biodiversity at various levels.

Quality risks primarily arise from the discharge of untreated wastewater into the environment. In the case of powerfuels, this can occur when untreated process water is discharged into the surroundings. High temperature differentials can have a detrimental impact on ecosystems. Additionally, risks can emerge from the consequences of water treatment. Water treatment, particularly in the form of seawater desalination, plays a crucial role in powerfuel production in regions experiencing water scarcity. This process generates the waste product brine. Disposing brine in sea or fresh water can have detrimental effects on aquatic ecosystems, as described in the section Biosphere Integrity.

Regulatory and reputational risks can particularly exist in countries with poor access to drinking water and sanitation. In these regions, the risk of conflicts over water usage is already high, and the utilization of water for powerfuel production can further exacerbate these tensions. Moreover, these conflicts can result in a lack of social acceptance for the affected powerfuel facilities.

Atmospheric Aerosol Loading. Although the production of powerfuels itself is not directly linked to aerosol emissions, these usually occur during powerfuel combustion. On top of CO_2, aircraft engines emit very small particulates, other gases and water. At high altitudes, these emissions affect atmospheric physical and chemical properties, resulting in an increase in greenhouse gases and the potential formation of contrails. This depends very much on the chemical composition and quality of the fuels. Quantitative estimations are still subject to non-negligible uncertainties, but this climate effect is overall most likely bigger than the change of climate by the GHG emissions itself [47]. In this context, the use of synthetic kerosene options is expected to have a reducing effect on these so-called non-CO_2 effects.

Novel Entities (formerly known as "Chemical Pollution"). The transformation to a climate-neutral and resource-efficient future must not be accompanied by new risks and environmental problems. This means taking precautionary measures to keep chemical pollution as low as possible. Known problems with persistent chemicals must not be exacerbated. But also new mass-produced chemicals, such as hydrogen, must be examined with their effects, e.g., on the climate.

"This is of concern because hydrogen is a small molecule known to easily leak into the atmosphere, and the total amount of emissions (e.g., leakage, venting, and purging) from existing hydrogen systems is unknown. Therefore, the effectiveness of hydrogen as a decarbonization strategy, especially over timescales of several decades, remains unclear" [48].

Resource Use. Natural resources, such as raw materials (biomass, metals, and nonmetallic minerals), water, and land, are the basis of our living. Global materials extraction has increased and accelerated in the last decades to more than 90 Gt in 2017 [49]. Scenarios made by the United Nations Environment Programme (UNEP) and the Organization for Economic Co-operation and Development (OECD) estimate that this could further double to around 160–180 Gt by mid-century [49–51]. In addition, more and more raw materials are getting scarce. More and more materials are being added to the EU's list of critical raw materials. The decisive factors for the classification are, on the one hand, the economic importance of the respective raw material within the EU and, on the other hand, the supply risk.

The sustainable use of raw materials is therefore fundamental to achieving the environmental and socioeconomic targets set out in the United Nations' (UN) Agenda 2030. This is in particular also a question of equitable access to resources. Some countries and regions have set targets for reducing resource use. The EU aims to achieve a circular economy until 2050 as proposed by the European Commission in March 2022 [52]. One finding of all recent scenarios is that without a resource turnaround, the climate targets cannot be achieved either. Another finding is that several materials (e.g., copper, steel, sand, chromium.) are scarce if we calculate the total amount needed for a global transition [53].

In order to measure resource use, a life cycle assessment (LCA) of the materials and land area used for the production of a unit is needed. There are different calculation methods for this, such as "Total Material Requirement" (TMR) and "Material Input per Service Unit" (MIPS). With the help of such an LCA, one also gets indications for the reduction of the required resources. In principle, the resource-saving technique is to be preferred.

The criticality of the use of individual raw materials in certain technologies can be estimated by calculating the cumulative demand in the trend formation until 2045 or 2050 on the basis of scenarios and comparing this with the known global reserves [53].

Due to the fast increasing demand for raw materials, production methods that are increasingly harmful to the environment and human health are being used. This leads, for example, to large-scale water use in copper production and creates problems in other areas of the countries.

2.2 Formulating Safeguards to Minimize the Planetary Risks

Despite the desired positive impact of powerfuels, the portrayal of risks within the planetary boundaries highlights the multifaceted impact on the planet. The previous section identified general planetary risks. Based on the risks, both safeguards and possible benefits are derived in the following with regard to sustainable Power-to-Liquid (PtL) kerosene. The safeguards are based on the principle of "do no harm" to avoid negative externalities, and the benefits follow the principle of "doing good" to generate additional positive impacts. The safeguards can therefore be seen as minimum criteria, which, in contrast to the benefits, must be strictly adhered to in the sense of the planetary boundaries. The scientific basis for the classification is the report of Altmann et al. [54]. Their report defines both an ambitious (long-term) standard and a standard with a trade-off between rapid market ramp-up and sustainability (short-term). Table 2 illustrates the requirements of both standards.

The main reason for using powerfuels is their contribution to the goal of climate neutrality through GHG emission reduction. Therefore, in order to significantly reduce CO_2 emissions, only renewable electricity should be used as electricity input (for all production steps). Besides fossil energy sources, nuclear energy is not an option due to negative impacts on humans and the environment [55]. Since RE currently account for a small share of electricity generation in most countries, additional plants for the generation of powerfuels should be built. In this way, the risk of compromising the achievement of climate targets within the power sector can be eliminated. Particularly in countries where parts of the population are not yet connected to the power supply, the focus should initially be on connecting the affected households to the power grid. Powerfuels projects in the affected regions promise to contribute to a secure power supply for households by overscaling RE plants. However, the degree of feasibility or widespread adoption of this possibility in the affected regions is yet to be determined.

For the production of powerfuels, CO_2 is required. To avoid additional CO_2 emissions, it is important that the CO_2 does not originate from fossil resources. This includes the utilization of CO_2 from industrial point sources, where fossil resources are utilized. Instead, it is preferable to utilize CO_2 from sustainable biogenic sources or employ DAC technology. Capturing CO_2 from the ambient air using DAC allows for a closed carbon cycle when the produced fuel is combusted. However, this technology is not yet commercially mature, making it primarily relevant in the medium and long term. Nonetheless, it represents the most promising solution as it guarantees a closed CO_2 cycle and helps to avoid lock-in effects. Additionally, water is produced as a byproduct in this process. Its utilization in the preceding electrolysis stage reduces the overall water demand during powerfuel production [45].

As all economic activities, the production of powerfuels requires land and therefore inevitably has an impact on biosphere and land use. Therefore, the principle is to minimize the impact on the biosphere and land use. For this reason, the amount of land used should be as small as possible. Furthermore, only a thorough impact assessment can measure and avoid direct as well as indirect and cumulative impacts. Overall, it is difficult to define universally applicable standards because each environment has its own specific characteristics and must therefore be considered in a differentiated manner. However, the better performance of powerfuels compared to biofuels in terms of land use should not be neglected [56]. In addition, the combination of land use can generate positive effects. The joint use of land for agricultural purposes on the one hand and electricity generation by photovoltaics (PV)—also

Table 2 Short-term and long-term criteria for sustainable PtL kerosene from [54] (DA—delegated act; DAC—direct air capture; IUCN—international union for conservation of nature; SDG—sustainable development goal)

Requirements	Short-term criteria	Long-term criteria
CO_2 source and GHG reduction	DAC	DAC
	Biogenic sources that are not associated with significant negative effects*	Biogenic sources that are not associated with significant negative effects
	Industrial point sources in the EU emission trading system until 2035**	
	Geothermal springs***	
Electricity	Renewability: compliance with the RED II DA, and	Renewability: fully renewable and the energy source is associated with limited land and resource consumption, and
	Additionality: compliance with the RED II DA	Additionality: compliance with the RED II DA and the renewable electricity plants are initiated by Power-to-Liquid plant operators themselves and are located in close proximity (max. 150 km distance) to relieve the grid
Land use or land-use change	No conversion of high nature value areas (HNV) according to IUCN with reference to criterion in ISO 13065	No conversion of high nature value areas (HNV) according to IUCN with reference to criterion in ISO 13065
Energy efficiency	Cumulative primary energy consumption: <3.5 MJ/MJ_{fuel}	Cumulative primary energy consumption: <3 MJ/MJ_{fuel}
Water availability	Water scarcity, as defined by UN Water in SDG indicator 6.4.2, does not exceed 60% at the site	Water scarcity, as defined by UN Water in SDG indicator 6.4.2, does not exceed 40% at the site
	Completely from seawater desalination	Completely from seawater desalination
		If water scarcity, as defined by UN Water in SDG indicator 6.4.2, at the site exceeds 60%, then additional supply of desalinated water to the public
Social standards	Do no harm: where applicable, the environmental, social, and governance criteria of the European Investment Bank (EIB) shall be met, and	Do no harm: where applicable, the environmental, social, and governance criteria of the European Investment Bank (EIB) shall be met, and

(continued)

Table 2 (continued)

Requirements	Short-term criteria	Long-term criteria
	Positive effects: permanent creation of a qualified workforce	Positive effects: permanent creation of a qualified workforce, and
		Where the regional supply of electricity/energy and/or water is underdeveloped, the project makes a relevant contribution to the respective supply of the regional population

*No credits for emission reductions through Carbon Capture and Utilization (CCU); biomass fulfills the sustainability requirements of RED II.
**CO_2 was previously taken into account in the process chain in the form of GHG emissions pricing and is chemically incorporated into the fuel before 2036.
***CO_2 was previously emitted naturally.

called agrivoltaics—on the other hand not only reduces the land-use footprint, but can also bring environmental benefits [57].

Water is not only fundamental for sustaining life on Earth but also an indispensable resource for powerfuels production. Currently, numerous regions are already grappling with water stress. It is imperative that powerfuels do not exacerbate water stress levels any further. Therefore, their production should be limited to regions where there is a guarantee of no significant negative impact on the water balance. An exception to this is the utilization of water from seawater desalination plants. Using seawater has no adverse effect on water stress levels, enabling powerfuels production in these regions. However, the environmental impact of seawater desalination plants themselves should be carefully monitored. Furthermore, if more water is desalinated than is required for powerfuels production, the water surplus can contribute to ensuring a secure supply, such as for households or agriculture. It should also be noted that the water footprint of powerfuels is considerably lower compared to biofuels [45].

Lowering the footprint of resources and materials, will help in all sectors of sustainability. Subsequently, it is necessary to replace the current linear, open-loop model with a model of the economy that is as circular as possible. Today's idea of circular economy—as a model of production and consumption [58]—includes newly designed products and services which allow sharing, leasing, reusing, repairing, refurbishing, remanufacturing, and recycling of existing materials and products as long as possible.

3 Impact of Powerfuels on People and Prosperity

3.1 *Risks, Safeguards, and Co-benefits*

Powerfuels do not only have an impact on the planet, but also come with opportunities and risks for people and prosperity, especially in producing countries. In fact, environmental integrity and human well-being—understood in a broader sense as comprising (mental and physical) health, happiness, and material well-being—are inseparably linked. Thus, building upon the planetary boundary framework, Rockström et al. [26] recently proposed an assessment accounting for Earth system resilience and human well-being in an integrated framework. By defining and quantifying safe and just Earth system boundaries (ESBs), the scientists consider (interspecies, intergenerational, and intragenerational) justice implications of crossing safe planetary boundaries and argue, for example, that exceeding the boundaries puts human livelihoods for current and future generations at risk. A recent reinforcement of this interdependency between environmental integrity and human well-being is the recognition of the human right to a clean, healthy, and sustainable environment, which was adopted by the UN General Assembly (GA) in 2022 with resolution A/76/L.75. In the resolution, the General Assembly notes the relation between the right to a clean, healthy, and sustainable environment with other rights and existing international law (operative paragraph 2). Notably, related human rights include the rights to life, health, food, water and sanitation, and development [59]. Reference is also made to the 2030 Agenda for Sustainable Development (preambular paragraph 4).

Consequently, any assessment of the impacts generated by powerfuels must also review risks for people and prosperity (socioeconomic risks). Following up on a definition by the IPCC, such risks are understood as "the potential for adverse consequences for human […] systems" [60]. Depending on the focus, they can have impacts on different levels, i.e., on the level of the state, of the companies involved or of the affected local communities and individuals. While risks related to powerfuel projects are initially relevant at the production sites and thus at local or regional level, the broader context and implications of powerfuel policies and partnerships are mainly of relevance at state level. It should be noted, however, that powerfuel policies ultimately also affect the respective populations.

Nonetheless, there is also a compelling argument to be made that powerfuels could support inclusive human development, including by contributing to economic prosperity and social equity (next to environmental integrity) [61]. In the following chapters, socioeconomic as well as political risks related to powerfuels will thus be analyzed, and—derived from these risks—minimum criteria for their mitigation proposed ("do no harm"). To assess potential socioeconomic and political risks associated with powerfuels, the Sustainable Development Goals (SDGs), complemented by relevant norms and principles of international law as well as justice considerations, serve as benchmarks. Risks, in this regard, are understood as potential adverse impacts on SDGs, human rights, and justice attainment. Furthermore, as the powerfuels market is only an emerging market, insights from other RE development processes will be drawn on. Subsequently, the potential of powerfuels in contributing to human well-being will be explored as well as criteria aimed at harnessing these (co-)benefits ("doing good").

3.2 Socioeconomic Risks at Project Level

At project level, powerfuels come with risks for affected individuals and communities, particularly in terms of human rights violations and livelihood infringements. On the one hand, there are powerfuel-specific risks to human well-being, some of which center around the resources required for their production, such as water and land. If not managed appropriately, potential conflicts over people's access to these resources could have negative impacts on the quality of life and livelihoods of affected communities [23]. On the other hand, risks can stem more broadly from business-related conduct. In general, any powerfuel project should be preceded by a thorough impact assessment geared to the specific (local) context.

3.2.1 Land

As powerfuel production plants require land, particularly in case of production of electricity from RE sources, conflicts over land can arise if parts of the powerfuel value chain are developed in abuse of land rights or in areas in which (informal) land rights are contested. According to [62], land rights abuses are among the most frequently reported abuses in the RE sector. Unfair land acquisitions as well as forced resettlements are thus a real risk which could be exacerbated by powerfuel production and which can have significant impacts on the social, economic, and cultural well-being of affected people, e.g., if sources of income are lost [63]. It should be noted that vulnerable and marginalized communities are especially exposed to this risk [23]. Indigenous peoples in particular are affected, as they many times lack full legal ownership of their land, and as their livelihood and well-being often depend disproportionately on it [64]. For example, as is recognized in both the International Labour Organization (ILO) Convention No. 169 on Indigenous and Tribal Peoples as well as in the United Nations Declaration on the Rights of Indigenous Peoples (UNDRIP), the lands and territories of indigenous peoples are of special importance for their cultures and spiritual values (Art. 13 ILO Convention No. 169, Art. 25 UNDRIP).

Land-use change is another potential risk resulting from powerfuel production, as it cannot only have an impact on the environment, but also clash with the needs of the local population [64]. This is particularly true as land does not only have environmental, but also social, economic, cultural, and spiritual value [64]. Disputes over land-use, for example, may erupt when land that could be used for agricultural purposes is converted [65], or when interferences with local landscapes are unwanted

[23]. As consequence, this might also lead to limited social acceptance of powerfuel projects [23].

As safeguards against the abovementioned risks, powerfuel projects should firstly not violate formal or informal land rights [64]. Forced resettlements in particular should be avoided from the outset. If this is not possible, the European Investment Bank (EIB), for example, stipulates in its Environmental and Social Standards that they should be minimized and adverse impacts on rights holders mitigated [63]. Secondly, local stakeholders should be involved in any land-related questions. For indigenous peoples in particular, this is stipulated in Article 10 UNDRIP, which holds that they shall not be forcibly removed from their lands or territories and that no relocation shall take place without their free, prior, and informed consent (FPIC) and after agreement on just and fair compensation. It further states that indigenous peoples concerned shall be consulted prior to approving any project affecting their lands and other territories, including in connection with the utilization or exploitation of water or other resources (Art. 32 UNDRIP). The principle of FPIC should thus serve as benchmark. In general, procedural rights, such as access to information, public participation, and access to justice, should be guaranteed to all project-affected communities and stakeholders in all matters related to the project (not only in respect to land issues), not least as a means to increase the potential for social acceptance of powerfuel projects.

3.2.2 Access to Water, Energy, and Materials

Other essential resources required for powerfuel production are water, electricity, and (raw) materials. As a consequence of the need for water, powerfuels can aggravate regional water stress, especially in arid regions. This can result in competition for water, especially if water needs for powerfuel production are met at the expense of drinking or sanitary water supplies, the agriculture, or ecosystems [61, 66]. Furthermore, as was noted above, the discharge of untreated wastewater into the environment as well as the disposal of brine in sea or fresh water may have detrimental effects on water quality.

Water supply and access to water are issues of fundamental socioeconomic importance: According to interpretation by the Committee on Economic, Social and Cultural Rights (CESCR), the (human) right to water thus "clearly falls within the category of guarantees essential for securing an adequate standard of living, particularly since it is one of the most fundamental conditions for survival" [67]. Although not recognized explicitly in the first generation of human rights instruments, it is said to be derived from the right to life, the right to an adequate standard of living, including adequate food [68]. Furthermore, it is essential for the fulfillment of other human rights, such as the right to the highest attainable standard of health. The right to water includes the right to access (affordable) water in sufficient quality and amount to meet vital human needs, including drinking, food production, and sanitation [68]. Ensuring availability and sustainable management of water and sanitation for all is also enshrined as one of the goals of the 2030 Agenda (SDG 6).

As safeguard, the production of powerfuels must thus ensure to not result in additional pressure on water availability nor to infringe on water quality. Furthermore, water prices should not increase [64]. In their study on Power-to-X (PtX) sustainability criteria, Altmann et al. [54] thus propose the following guardrail: If water scarcity according to SDG Indicator 6.4.2 at the production site is above 40% (above 60% for the short-term criteria), water must not be taken from freshwater reserves but must be procured through seawater desalination. If water scarcity is above 60%, in addition to using water from seawater desalination, a certain amount of desalinated water must be made available to the public at socially acceptable prices. In the latter case, this could even generate benefits for local communities.

In terms of electricity, the high demand for green electricity related to powerfuel production will require rapidly increasing renewable energy electricity. However, if additional renewable energy electricity is used for powerfuel production, this could exacerbate energy inequality in countries lacking universal access to electricity supply: should the green electricity capacity only fuel green hydrogen or powerfuel production while ignoring local needs, this could hamper access to energy services for local economic development and productive uses [23]. In this regard, SDG 7—the goal to ensure access to affordable, reliable, sustainable, and modern energy for all—will be of particular importance. This holds true especially when considering that according to the 2022 Energy Progress Report, the world is still not on track to achieve SDG 7, with 733 million people globally having no access to electricity [69]. While access to energy is not a human right, it is important for the fulfillment of a number of human rights, such as the right to an adequate standard of living, in particular with regard to housing. Furthermore, it could be argued that an entitlement to accessing (clean) energy may be derived from the right to benefit from scientific progress.

It must thus be ensured that powerfuel production does not weaken the efforts to eliminate energy poverty, e.g., by bypassing local populations in terms of access to clean energy. To prevent such a scenario, new installations of renewable power capacity could be overscaled so that the local population and economy benefit from the additional renewable electricity capacity [65].

Finally, the transformation to a climate-neutral and resource-efficient economy will first lead to an increase in the consumption of materials and raw materials [53]. Even today, some of them are in short supply or their supply is critical. Care should be taken to ensure that these shortages do not lead to distribution struggles at regional or international level. If commodity prices rise due to their increasing scarcity, it is also important to avoid a situation where only wealthy regions would be capable to afford this transformation.

3.2.3 Business-Related Conduct

Regarding business-related behavior, it is essential to acknowledge that globalized economic structures often result in deficiencies in the protection of human rights. Complex supply and value chains in various industries involve production processes that repeatedly result in fatal accidents and serious human rights violations [70]. In fact, a large number of registered business-related human rights complaints has been associated with activities in the extractive and energy sector. According to an empirical analysis by Kamminga [71], nearly a third of registered human rights complaints related to business activities between 2005 and 2014 are related to activities in this sector. Especially in the case of large-scale projects, the consequences often entail the destruction of livelihoods, forced resettlements, and the repression of protests. More recently, human rights violations have also been reported in connection to RE projects. The Business & Human Rights Resource Centre [62] reports nearly 200 allegations of human rights violations related to RE projects between 2010 and 2020. It follows that powerfuel projects cannot be considered as necessarily exempt from business-related human rights risks.

As respect for fundamental human rights is non-negotiable, compliance with human rights requirements along the entire value chain must be a core safeguard for sustainable powerfuel production. This refers, in particular, to the rights enshrined in the Bill of Rights (Universal Declaration of Human Rights, International Covenant on Civil and Political Rights, International Covenant on Economic, Social and Cultural Rights) and to the core labor standards of the International Labour Organization (ILO). While the primary addressee for the protection of human rights is the state, the international community has recognized that companies also have a social responsibility to respect human rights, especially where state protection is lacking. This responsibility is reflected in the Guiding Principles on Business and Human Rights, which were endorsed by the UN Human Rights Council in 2011. In order to close gaps in human rights protection, they address, in addition to the state's duty to protect, a corporate responsibility to respect human rights as well as remedial measures against negative impacts of corporate activities. Minimum guardrails for powerfuels should thus build on these principles and include, for example, the establishment of grievance mechanisms for workers and affected communities, as well as access to remedy in case of human rights violations.

It should be noted in this regard, that for companies involved in powerfuels, violating human rights cannot only lead to reputational risks; with stricter national legislation on supply chain due diligence, corporations are increasingly facing legal sanctions for ignoring basic human rights [61]. In Germany, for example, the act on corporate due diligence obligations in supply chains, which aims at making inhumane working conditions and exploitation harder, has recently come into force. A supply chain law is currently also being discussed at EU level. Furthermore, many banks are now incorporating environmental, social, and governance (ESG) criteria into their lending practices. The lending strategy and objectives of the European Investment Bank (EIB), for instance, are underpinned by the promotion of sustainable development. Its environmental and social standards include standards on stakeholder engagement, involuntary resettlement, labor rights, and cultural heritage [63]. As safeguard (do no harm requirements) against socioeconomic risks related to powerfuels, Altmann et al. [54] even propose in their PtX sustainability standard that the ESG criteria of the European Investment Bank (EIB) shall be met where applicable (Table 2).

3.3 Socioeconomic and Political Risks at Policy and Partnership Level

Risks associated with powerfuels can also have impacts on the level of the state. These risks relate to the broader context of powerfuel policies and partnerships. Assessing them thus requires taking into account justice implications, particularly in terms of justice between states (intragenerational justice). Such justice implications are also reflected in Sustainable Development Goal (SDG) 17: "Partnerships for the Goals".

With numerous hydrogen strategies setting out plans to achieve climate neutrality by using hydrogen and its derivatives adopted at the national level, the powerfuel market is expected to grow massively. As mentioned above, many states in the Global South possess a considerable powerfuel production potential and could develop into exporters [6]. Meanwhile, according to the International Renewable Energy Agency (IRENA) [6], Japan, parts of Europe, as well as some other countries will need to import green hydrogen and its derivatives to satisfy domestic demand. International trade in green hydrogen and powerfuels will thus play an important role. While this is perceived as an opportunity for economic and social development in producing countries, it can also result in injustice and inequalities, particularly if most power-fuels are exported [23]. Scita et al. [72] indicate that there is a risk that industrialized nations decarbonize their domestic economies at the expense of resources and labor of developing countries. Eberhardt [73] for example argues, that German-backed green hydrogen projects abroad already follow colonial patterns, in that resources are appropriated while negative impacts like land conflicts are outsourced. These effects, referred to as "green colonialism" or "energy colonialism", should be carefully evaluated considering the fact that developing countries have only contributed a small share of global warming compared to industrialized nations, yet bear a disproportionate burden of environmental and climate change impacts [23, 74]. It is thus argued that industrialized nations should bear a greater responsibility for climate change mitigation measures. This is also reflected in the international environmental law principle of common but differentiated responsibilities, which holds that while all states need to take responsibility for global environmental problems, there is a need to recognize differences in levels of economic development between states as well in levels of contribution to the problem. If importing countries are privileged over exporting countries, this could risk exacerbating imbalances. Furthermore, using RE installations for producing hydrogen or its derivatives only for exports risks delaying the energy transition in the exporting countries, and locking-in fossil fuel energy generation [61]: If green electricity is used for export-oriented powerfuel production, more fossil fuels are burned to cover domestic needs [73].

Another risk pertains to the high investment requirements of powerfuel production and trade. On the one hand, these might lead to an exclusion of countries of the Global South from the global energy structures in the first place, as many developing countries cannot afford the expensive energy transition technologies and transport infrastructure that are needed [6]. This, in turn, might widen disparities between wealthier and poorer countries, impeding the achievement of a just and equitable global energy transition [6]. On the other hand, the high investment requirements will have major public finance implications, especially for states in the Global South. As massive investments by the private and public sector alike are required for the ramp-up of powerfuel projects, the PtX Hub [61] assumes that due to market uncertainties and current cost structures, major private investments will only be made if there is adequate (long-term) public support, e.g., in the form of investment grants and loans. It should be noted in this context, that due to higher investment risks, financing costs are considerably higher in the Global South than in industrial countries. This discrepancy is especially evident in projects with high initial investments [66]. Should large-scale projects be promoted as part of the development of export capacities, this risks leading to a significant increase in public debt. Large-scale projects could further become stranded assets [75]. While powerfuel production and trade can also generate budget revenues from taxes or export duties [61], it is likely that public expenditure will exceed revenues initially. It would also have to be ensured that private operators of powerfuel production facilities or infrastructure pay taxes locally [75]. However, as Eberhardt [73] notes, projects for hydrogen exports are often located in "special economic zones" with tax break mechanisms, resulting in a reduction in the share of public revenue.

To promote future equality of different nations and to limit possible risks of further power imbalances while unlocking potential benefits, negotiations between importing and exporting countries must be conducted on an equal basis. Future trade and investment agreements must be fair and equitable and ensure benefits for both sides of the value chain, including opportunities for the development of local populations and domestic industries [76]. This is all the more important as the hydrogen trade and, by extension, the powerfuel trade, is set to grow considerably with a likelihood of new energy trading relations emerging [6]. Appropriate policy frameworks are thus needed [76], which take into account conditions under which powerfuels can be produced and traded without compromising the well-being of present and future generations. These must be embedded in broader, just, and sustainable energy transition strategies. Furthermore, public support mechanisms could be used as a way to ensure compliance with such (basic) sustainability requirements [61] as outlined above. This would contribute to mitigating powerfuel project-specific risks. As an extra incentive, going beyond minimum requirements could be rewarded with special support.

3.4 Forecasting and Harnessing Socioeconomic and Political Co-benefits

With numerous hydrogen strategies formulated or planned at the national level and corporate strategies to achieve climate neutrality using hydrogen, many policymakers, researchers, and companies expect the supply of hydrogen and its derivatives to grow massively. These emerging markets and supply chains offer enormous opportunities for economic and social development, in addition to the social and environmental risks already mentioned. There are a range of policy goals that are expected to be achieved by the means of hydrogen and its derivatives that go far beyond climate action. In this section, the German discourse on these products and markets is being examined and the most prevailing forecasted additional benefits discussed in terms of their probability of being harnessed. However, this represents a singular case study and its interpretation may differ significantly in alternative national contexts.

The German Government points out co-benefits in its national hydrogen strategy [77] related to international cooperation, mainly the creation of "opportunities for sustainable growth and development". German companies can hope for sales markets for hydrogen technologies, if a domestic market is built and experiences from flagship projects being displayed internationally. For the partnering exporting countries, the hydrogen and powerfuels value chains are expected to contribute to the decarbonization and economic development and as a stimulus "to rapidly expand their capacities for generating RE", adding that the assumption of "these will, after all, also benefit local markets" [77].

There was a broad consensus among the participants of a German stakeholder survey that the national Government should maintain and—if possible—strengthen the competitiveness of Germany as an industrial location country in the case of prioritizing hydrogen applications. There was disagreement, for example, on the question of whether prioritization should also be aimed at the German labor market in order to maintain qualified jobs here or even create new ones [78].

Furthermore, some stakeholders anticipate that this emerging industry will have significant positive effects on regional (sub-national) development. For example, with the Structural Strengthening Act and the agreed funding for it, new industries are being developed in the areas that are heavily shaped by lignite mining and power generation based on it. One of the three regions is Lusatia and projects like the PtX Lab Lausitz are being funded with the goal to establish a hydrogen-based industry. The assumption is that this will help to achieve a "just transition".

3.4.1 Industrial and Infrastructure Policy

The Sustainable Development Goal (SDG) number 9 seeks to build resilient infrastructure, promote sustainable industrialization, and foster innovation. Even though the goal is aimed more at safeguarding by emphasizing the resilience of economies through infrastructure, diversification of the industry, and innovation, it puts emphasis on inclusiveness. SDG target 9.1. formulates that infrastructure should be developed "with a focus on affordable and equitable access for all" [79]. "Structural policy" can be defined as the totality of all economic policy measures that serve to steer economic structural change in socially desired directions. The term "green industrial policy" has become established in English language and international cooperation and emphasizes steering toward ecologically sustainable economic activity, while at the same time taking economic and social goals into account [80].

Industrial policy is often linked to infrastructure policy, because well-developed infrastructure is generally seen as a sign of a good business location, associated with a high standard of living and better job opportunities. In Germany, different regions have entered into competition which will become one of the main hubs for green hydrogen and powerfuels production and transport. Even though it is intended by funding schemes for hydrogen-related projects, that competition arises—but not in the sense of regional competition, rather in technological and scientific matters. Nevertheless, this competitive approach in Germany has implications for structural policies: latent distribution conflicts in the political multi-level system and between the involved industrial sectors [81]. The establishment of powerfuel plants and the distribution and supply chains imply the construction of infrastructure to provide the inputs electricity, hydrogen, and CO_2. Pipelines and, in some cases, new power lines and power generation plants need to be built.

Whether this will work for the powerfuels value chain and potential regional hubs relies on various factors that are related to the dynamics of the national energy transition, both in comparison with those in other countries and geographical regions. One prerequisite is the speed of diffusion of electrolysis technology, the catalytic conversion, and carbon capture technologies. The scientists of the "Ariadne" project developed a probabilistic technology diffusion model assuming that electrolysis grows as fast as photovoltaics and wind energy. The German expansion target is at the upper end of the range of possibilities, whereas the EU target, on the other hand, lies outside the historical possibility range and thus requires a significantly higher diffusion rate than has ever been achieved for energy technologies [5].

From a global perspective, not all potential hydrogen countries, despite high technical potentials, have equal starting conditions for a realistic, medium-term successful development of hydrogen economies and/or are only limitedly oriented toward [82]. The specific strengths, weaknesses, threats, and opportunities need to be examined from a national, regional, or sub-national perspective. One of the strategic questions that countries with large solar and wind power potential should answer for themselves is whether to envisage green hydrogen as a new export commodity or as a stepping stone toward a diversified and knowledge-based economy [83]. The first would imply that export infrastructure needs to be built. IRENA sees a big chance for Africa to set of infrastructure for the shipping sector, in particular in Nigeria, Morocco, Namibia, and South Africa. "Bunkering facilities for new fuels in the shipping sector go together with the transformation of ports and the construction of export facilities. Actions in this direction have already begun" [6]. The second strategy would include creating value chains with higher value added, such as the full powerfuels value chains locally. This would involve establishing catalytic conversion and refinery facilities to produce powerfuels. However, it may also require developing logistical infrastructure for exporting the powerfuels. Nevertheless, this strategy would result in a larger portion of the overall profits going to the country involved in production. Additionally, in the aviation and the shipping sector it is currently much debated how powerfuels could be certified and how their environmental attributes could be allocated and traded (on a virtual basis) without the need to physically transport the product to the respective customers. Particularly during the nascent stages of powerfuels development and initial commercial ramp-up, this would reduce the need for early investments into physical transport infrastructure as prerequisite to a participation in the powerfuels value chain.

Economic theory, in particular theory of foreign trade and of industrial economics, is largely based on questions of how to spur economic growth and recommended different industrial policies to governments over time. Andreoni and Chang [84] present a brief history of industrial policy theories and practices. The debated issues were the logic of infant industry versus the theory of comparative advantage; export promotion versus import substitution; and state failure versus market failure. They state that industrial policy was a "taboo" for a few decades but is now back on the agenda. They criticize conventional economic growth models and the so-called new Structural Economics for their simplicity and for neglecting structural interdependencies that unfold in a disproportionate way across sectors and production activities. Therefore, they stress the importance of "rethinking industrial policy as packages of interactive measures operating across different policy domains and levels of interventions, and implemented and enforced through a variety of institutions—policy governance" [84]. This is of particular importance with regard to public environmental goods, such as climate change mitigation and the technologies to achieve this. Green structural policy must therefore necessarily be embedded in international negotiation processes (e.g., on sustainability standards for powerfuels), financial transfer mechanisms as well as international energy and climate partnerships [80].

The theoretical models can help industrial, infrastructure, and regional policy planning, but the results need to be verified over time and specifically for the regional and industrial context. There are first insights into the specific case of powerfuels around the globe available through case studies in the 4th roadmap of the Kopernikus project "P2X". The authors analyze the extent to which PtX technologies can be used and thus contribute to the energy transition in these countries. The four case studies clearly show that each country faces its own challenges for entering the green hydrogen/PtX economy. This is reflected both in the national strategies, for example in terms of local value creation, national defossilization targets, and/or national use or export. However, challenges are also evident in strategic collaborations with countries, such as in international cooperation and at the regional level.

3.4.2 Job Creation

One much discussed potential socioeconomic benefit of powerfuel production is the enhancement of human development through job creation. The promotion of regional hydrogen and powerfuel hubs is expected to have positive employment effects in the concerned regions and communities [6, 61, 85]. In general, it is estimated that the transition to renewables will create more employment, with the number of job losses in fossil fuels compensated by the number of new jobs in renewables [85]. Economic models and tools help to predict to which extent employment effects can be achieved, for example, the JOBS Models from Argonne National Laboratory [86]. In the latter, data of the installation of on-site equipment is modeled into an impact on the supply chain and then into induced economic activity. Powerfuel projects will probably create jobs during the installation phase as well as during operation [61]. The Air Transport Action Group (ATAG) estimates a job creation potential of Sustainable Aviation Fuel (SAF) production worldwide of up to 14 million jobs (in an aggressive SAF deployment scenario), of which 2.5 million jobs would be created in Europe [87].

But, what is the probability that this will occur? And, how would those job creation effects unfold? Existing structural realities affect the employment opportunities during the energy transition and leave them unevenly distributed across different regions. IRENA found two sets of situations as particularly shaping the job creation opportunities: first, varying depths, strengths, and diversity of national supply chains; and secondly, the different degrees to which economies depend on the production and sale of fossil fuels [88]. One example for this in the German context is that it is likely that specialist firms have to travel to the new production area for powerfuels which does not have a direct employment effect on the local population, if there are no specialists on site.

Cremonese et al. [23] point out that local job creation could be bypassed unless appropriate capacity-building and skills-creation programs exist. Regarding the often skill-specific requirements of RE projects, value creation does not necessarily accrue to local populations due to lack of capacity and skill development at the local level to match these [23]. As a consequence, capacity-building, re-skilling programs, and other educational opportunities must be deployed, as well as appropriate policies addressing these structural realities must be introduced early in the transition. "Maximizing the employment benefits of the energy transition while minimizing its costs requires a comprehensive policy framework to support a just transition" summarizes IRENA [88]. The authors recommend policymakers to anticipate and address the temporal, spatial, educational, and sectoral misalignments. The latter refers to the fact that new industries need different raw materials and resources than more traditional industries. This is true for the shift from fossil fuels to powerfuels. Moreover, their proposal for a jobs indicator that measures the socioeconomic footprint of the energy transition as relative performance could be adopted in and adapted to powerfuels policies of national governments.

On a product and project level, contributions to employment creation can be incorporated in sustainability standards, too. In the mentioned one for Power-to-Liquid (PtL) kerosene [54], the "permanent creation of qualified workforce" is included as minimum criterion, while acknowledging that the International Labour Organization (ILO) standards as safeguards generally apply in the countries that ratified them. Additional possible positive effects are included as variable bonus points in a score-card principle. This may include skilled job positions filled by women or trainings on PtL production offered to the local population in the producing country/region.

3.4.3 Values-Based Foreign Trade Policy

Countries like Germany will find it difficult to meet their future PtX demand domestically. Therefore, partnerships with countries that have a high generation potential are indispensable. Countries with high generation potential are often developing countries. Especially here, cooperation under a high and uniform sustainability standard is of great importance so that mutual added value is generated for both parties involved. This is to ensure that no neo-colonial structures are created and that added value is generated in the producing countries, too.

Quitzow et al. [89] outline six policy dimensions that European policymakers should consider when engaging in the development of international partnerships within the emerging hydrogen economy. The related "policy objectives may be mutually reinforcing, but, depending on the context, they may also conflict". By explicitly matching the policy dimension "green industrial development" to Europe, and, respectively, "just transitions" to partner countries, the authors highlight questions of global economic and environmental justice and potentially exploitative relationships. This raises the question whether by the means of a green industrial policy a more just international economy and trade relationships can be achieved. For example, if it can be expected from the implementation of green industrial policies to actively contribute to mitigating socioeconomic injustices and violent conflicts between countries and within countries. Or, whether democratic movements should be supported against autocratic, often corrupt elites in partner countries through the means of powerfuels markets and strategies?

A geopolitical strategy coming from the United States called "friend-shoring" intends to cover those states follow the US understanding of open markets but are also committed to respecting labor and environmental standards [90]. This concept is not novel to foreign affairs, international trade, and development cooperation departments, although it has gained renewed momentum, particularly in Europe, due to emerging economic prospects in climate and energy technologies. Additionally, the Russian conflict with Ukraine since March 2022 has further propelled this idea. In the course of a German stakeholder dialogue on hydrogen, it became apparent on several occasions that the participants perceived the Russian war of aggression as a caesura in the European post-war order. They came to the unanimous conclusion that European foreign energy policy should be re-assessed and re-oriented. "At the heart of the discussion was the fundamental question of the extent to which Germany should pursue a values-based foreign energy policy or to the requirements of an interest-based energy foreign energy policy" [78].

The new term "hydrogen diplomacy" unites the idea of opportunities for strategic relationships with like-minded countries to promote European and US-American values and interests abroad [89]. However, the feasibility of such a strategy remains afflicted with uncertainty considering many factors including the geographical distribution of raw materials worldwide. Furthermore, such a strategy could be criticized itself as neo-colonial.

4 Getting Ambitious Sustainability Criteria into Practice

4.1 Acceptance, Transparency, and Participation

A holistic sustainability approach for powerfuels will most likely ensure greater acceptance by society and thus serve as a necessary, but not sufficient prerequisite for the market ramp-up of new technologies. The German economy, in particular, has witnessed the consequences of market instability and policy challenges arising from a lack of sufficient focus on sustainability at the outset, specifically in the development of RE, such as bioenergy and wind energy. This experience has highlighted the importance of addressing sustainability from the beginning to maintain social acceptance and ensure long-term success in these sectors. "Past energy and development endeavors have shown that success and the achievement of sustainability standards do not only depend on policy frameworks and technological advancements but are also influenced by social acceptance and support" [23].

In a collaborative handbook of German industry leaders, academia, and governmental stakeholders on acceptance strategies in energy-intensive industries [91], the authors stress the importance of aiming for "constructive acceptance" in opposition of stopping at "passive acceptance". For constructive acceptance, it is essential to initiate active co-designing of transformation strategies and strengthening the critical-constructive perspective, including making it more audible. The handbook offers a range of examples for civil dialogue formats that go beyond the one-off public consultation binding in German licensing procedures for large industrial installations.

Similarly, in a survey among German stakeholders of hydrogen technologies, the majority expects public discussions, for example, about the safety of hydrogen technologies and sees constructive critique in general as necessary for a successful market ramp-up [82]. In an earlier version of this "roadmap", the researchers found that since synthetic kerosene has so far mainly been part of an expert discussion, measures are needed for a transparent transfer to society. After all, a special feature of aviation is its symbolic power and high potential for emotional reactions, which is demonstrated by the current moral and value discussions. For example, during that time the German Government's support for Lufthansa during the COVID-19 pandemic as well as the rather new "flight shame" due to individual responsibility for climate change was discussed [92]. In another German stakeholder dialogue on hydrogen with representatives from business, science, politics, and organized civil society, many stakeholders affirmed, too, that the participation of citizens in the planning of generation plants and infrastructure measures is necessary and expedient [78]. Infrastructure measures necessitate specific attention to the social dimension, as analyzed in Sect. 3.4.

Bridge technologies and transition strategies, such as the use of biofuels in aviation or of "low-carbon fuels" made from blue hydrogen, will only be able to find acceptance and trust along the way if they adhere to strict sustainability criteria and set out binding roadmaps that outline the transition to major use of electricity-based fuels and new climate-friendly propulsion technologies. In a transitional period, the indicators for the sustainability of powerfuels may also be less ambitious compared to what is imperative in the long run to achieve full climate neutrality and true sustainability. This would allow the market ramp-up of the new hydrogen-based fuels. But, even in a transitional period, demonstration plants as well as public funding schemes should apply a rather comprehensive and ambitious set of sustainability criteria/ indicators in order to set an example for the private sector and to test the defined criteria. Constructive acceptance for powerfuels needs to be safeguarded against also among other European societies and political stakeholders as well as in the producing countries and communities in the Global South.

For individual early movers, a lapse with regard to sustainability aspects could severely undermine the credibility of the industry as a whole. Also, NGOs in Europe and the Global South have already expressed fundamental doubts whether a hydrogen industry truly leads to net positive societal benefits, in particular in the exporting countries [73]. With this respect, it should be in the industry's best interest to ensure that all corporates within the value chain respect the in the article proposed safeguards and to avoid "green washing". A report of the German Institute of Development and Sustainability (IDOS) highlights the interlinkages: Environmental, economic and also social goals should be linked systematically with each other, entirely in the spirit of the SDGs and the vision of a just transition. This could increase the social acceptance of environmental policies that are initially associated with costs for companies and households, especially if they are implemented through multi-actor partnerships and also in Germany or in the EU itself. At the same time, this could contribute to partner countries increasingly anchoring environmental and climate policy elements in their own national development strategies and not perceiving them as an external agenda of donor countries [80].

Governance mechanisms on powerfuels projects have to be established early in order to guarantee participation in and transparency of political decisions. This could include grievance mechanisms, stakeholder mappings based on analyses of land and water rights, social impact assessments, hearings, interactive co-operative workshops, protected whistle-blowing channels, or the nomination of ombudsmen [61]. Particular attention should be paid to including disadvantaged groups in the respective country and enabling them to participate. This is especially, but not exclusively, important in fragile states where conflicts over financial and environmental resources are smoldering.

The recently proclaimed feminist development and foreign affairs policies of the German Government provides a basis for policy coherence for such participatory principles for powerfuels projects that are usually assigned to the fields of energy, industrial, and/or transport policies. While women and girls are the largest disadvantaged group globally, other groups are addressed, too, that are being discriminated against and oppressed due to their gender, their age, their religion, their race, or for some other reason [93]. While equality is recognized as a fundamental human right and is included in various sets of sustainability criteria, the reality is that true equality has not yet been achieved in any country worldwide. Therefore, specific safeguards should be incorporated to make sure that equal access to abovementioned participatory formats is provided to all groups in the countries and sub-national regions where PtX projects and policies are implemented. And, co-benefits for a more just distribution of resources, political and economic power within nations, both in the Global North and Global South, as well as between nations are achievable.

4.2 Implementation of Holistic Sustainability Standards

The preceding analysis has shown that a range of academic and political stakeholders in Germany and beyond promote a systemic and holistic definition of sustainability standards for powerfuels and its fast implementation. National governments in the EU and outside the EU could set up own ambitious sustainability standards to make a case for a systemic view on sustainability and hydrogen-based fuels and seek harmonization with other regulatory environments over time—without compromising too much. Countries that have the economic power to provide subsidies and/or that procure powerfuels for its own consumption (in the case of Germany through H2Global and H2Uppp) could do this under the conditions of high standards of sustainability to provide strong signals for sustainability to the market.

The EU has the potential to establish itself as the leading market for sustainable and green powerfuels by adopting a common standard and exclusively procuring powerfuel products that meet this standard. It is of high importance that the holistic approach on sustainability will be integrated into EU regulation as well as voluntary certification schemes. So far, the proposed criteria for defining powerfuels in the EU, so-called RFNBOs, relate only to the electricity supply and a minimum GHG reduction with a GHG accounting methodology. In the medium-term, the legal minimum requirements in the EU and the criteria for eligible fuels within the market-based instrument CORSIA for international aviation need to be extended and adjusted dynamically, in order to push conditions to establish itself as a green lead market. Voluntary standards can support this regulatory development by increasing the ambitions faster within their certification systems which would be beneficiary for the front runners of the market as well as society and the environment as a whole.

Adapting standards could also happen in form of detailing and interpreting general ESG standards like the one related to in the proposed PtL sustainability standard by Altmann et al. [54]—the ESG-guidelines from the European Investment Bank (EIB). Established voluntary sustainability standard systems for alternative fuels could consider offering specific premium powerfuels certification that go beyond the expected minimum legal requirements laid out in the delegated act for RFNBOs in RED II. Another area where EU regulation could incorporate a systemic sustainability criteria for powerfuels is the EU-taxonomy for sustainable activities as well as in existing and future funding programs for powerfuel market penetration and research and development like EU-Life and Horizon Europe.

On an international level, there are further possibilities to include the set of sustainability criteria to ensure policy coherence. It is important for powerfuels policies and production to align not only with the UN Agenda 2030 (SDGs), but also with the

National Determined Contributions (NDCs) and the long-term strategies of countries as agreed on in the Paris Agreement. Further alignment is needed with energy, infrastructure, and trade policies. Öko-Institut proposes to seek the incorporation of its set of criteria into international trade regulations from the World Trade Organization (WTO) and in the Energy Charter Treaty [64]. In addition to integrating the sustainability criteria in the aviation sector—in particular in the international GHG offsetting mechanisms for international aviation, CORSIA—the criteria need to be integrated also in the shipping sector on policy, project, and product levels. For example, sustainability criteria could be incorporated as minimum requirements for eligible fuels to align with the International Maritime Organization's (IMO) climate goals. These criteria may also be considered for inclusion in the carbon mechanism currently being discussed at the IMO level. In the European Union, the definition of minimum sustainability criteria for renewable fuels of non-biological origin (RFNBO) should be expanded by the criteria mentioned in this chapter. The sustainability criteria are applied already to the aviation and the maritime transport sector—for both, RFNBO mandates where introduced.

The choice of partner countries and multilateral trade agreements for powerfuels should also be done in a smart way by state actors and by business to ensure a holistic approach on a sustainable powerfuels value chain. The EU could set up a kind of plurilateral "sustainable powerfuels club" with ambitious standards that can slowly expand memberships (expanding the idea of a "hydrogen club" from [76]). Currently, an expansion with a sustainability assessment component for the global map of potentials for green hydrogen production from Fraunhofer IEE[2] is under way. It can lead entrepreneurial and political decision-making on future bilateral or multilateral cooperation agreements.

Furthermore, in the aviation and shipping sector a so-called book & claim system is being discussed that relates to a specific chain-of-custody model for the monitoring, reporting, verification, and accounting of the certified powerfuels. A book & claim mechanism could be established in regulated or voluntary markets, on a national, regional, or international level. The biggest benefits would probably arise from implementing it on a global level because it could allow GHG-reduced and rather simple supply chain of powerfuels without geographical limits. Some stakeholders doubt that under the bottom line, such a system would have a positive effect for the climate mitigation and the biosphere, mainly due to risks of double-counting, fraud, and adverse incentives for fuel producers and consumers. Here, too, it is essential that a holistic set of sustainability criteria is established as mandatory minimum criteria for the traded certificates of sustainable aviation, respectively, maritime fuel.

---

[2] https://maps.iee.fraunhofer.de/ptx-atlas; PtX Lab Lausitz is supporting the development of this additional feature of the PtX Atlas.

5   Final Considerations

Climate-neutral powerfuels have to be used as efficiently as possible. For their production, green hydrogen must first be produced from renewable electricity. This is then used to produce powerfuels in further synthesis steps—again with a significant energy input. Each conversion step is associated with energy losses. The issue must be how to produce and supply synthetic fuels as efficiently and sustainably as possible for those sectors that cannot resort to alternatives. In the transport sector, therefore, powerfuels should be dedicated to aviation and shipping, possibly also to heavy-duty road transportation. High efficiency in the direct use of renewable energy electricity is often the more cost-effective climate protection option compared to the production of synthetic fuel options. "Initially, hydrogen should be used primarily for 'no regret applications', as long as it is not foreseeable when and at what prices electricity-based energy sources will be available. In particular, these are applications in which direct electrification is, in the next decades not possible: the use of hydrogen in industry (ammonia, steel) and of powerfuels in petrochemistry and in long-distance aviation and shipping" [94].

Fraunhofer ISI describes the prioritization of usage sectors in a report as a "hierarchical principle" that should be applied by producing and consuming countries of hydrogen/powerfuels: "This comprises four stages: (1) The 'energy efficiency first' principle to minimize demand, (2) Priority to decarbonizing the electricity sector, (3) Priority use of alternatives based on RE sources with similar services but lower environmental impacts (e.g., direct use of electricity, sustainable biomass/biofuels/biogas taking their limited availability into account), and (iv) Use of hydrogen and its synthesis products if the first three stages have been exploited as far as this is reasonable" [66]. A recent study on the hydrogen governance in Germany points out additional economic benefits: "the prioritization of individual consumption sectors, as planned by the federal government, can reduce the financing costs of the market ramp-up" [81].

In this article, a range of research and data gaps were identified, in particular in the area of socioeconomic risks and co-benefits. The transmission belts between climate policy goals and other policy goals of promoting powerfuels are based on assumptions and wishful strategies. They could become reality if the preconditions and specific local contexts are taken into account. Development cooperation has a long track record of partly painful learnings on the challenges of applying technologies and social innovation concepts in different countries. Since the powerfuel value chains that are expected to evolve are of international character and involve ties between developing, emerging, and industrial countries, these learnings should be taken seriously by policymakers, scholars, entrepreneurs, and other stakeholders. Further case studies on both the expectations and the environmental, socioeconomic, and political realities in the different countries need to be examined.

A global pipeline of research projects is needed for accompanying the market ramp-up of powerfuels. In majority, they should be composed in an equal or almost equal quantity of researchers from the Global South and the Global North to safeguard against the bias in the assessment of positive and negative effects of powerfuels development. Setting up implementation and study projects in this manner has been a standard practice in development cooperation for a while. This approach not only enhances ownership among local communities where the project is implemented but also ensures that local knowledge and perspectives are given equal importance alongside the perspectives from industrial countries (which often come from a predominantly white perspective) during project assessments.

Also, the interlinkages between the environmental, economic, social, and political dimensions in the context of powerfuels need to be evaluated in more depth—ideally from scholars around the world as well as from stakeholders that can provide valuable nontraditional knowledge, too. Several authors from reports on green hydrogen and the derivatives see the need for further research in that regard, and some believe that the challenges of importing these are partially underestimated [66]. IRENA summarized it as follows: "A greater understanding of the multidimensional nature of global threats and vulnerabilities will make it possible to foresee and defuse certain risks that may come with the deployment of hydrogen on a major scale" [6].

Another significant research gap is the question of the future scarcity of raw materials. It is increasingly being discussed more openly and prominently in the context of existing and new technological solutions and innovations. Raw materials and resources are in limited supply, and the environmental impacts of their extraction are increasingly visible. At present, there is a lack of technology-specific studies focusing on the critical raw materials associated with powerfuels.

Finally, further technological advancements are needed to achieve a fully sustainable carbon resource for the production of powerfuels. In terms of sustainability, the preferred technical solution is the capture of CO_2 ambient air (direct air capture, DAC). Currently, DAC technology has only been realized in pilot plants, and the necessary production capacity for DAC plants still needs to be developed and scaled-up significantly. Although DAC technology is currently at a relatively low level of technological maturity and is only available on a small scale in the market, it is considered to have significant potential from a sustainability perspective. This technology ensures that no additional CO_2 is released into the atmosphere during the production and consumption of powerfuels, as long as the processes are conducted solely using renewable energies, making it a favorable option.

Acknowledgements We would like to thank our colleagues Christoph Menzel (former), Felix Schmermer, Anja Paumen, and Balthasar Kirchgäßner for their contributions to the scientific project work within the scope of the topic area "Sustainability of PtX" of the PtX Lab Lausitz. We thank the participants of the expert workshop on the "Introduction of an ambitious sustainability standard for PtL kerosene" (15th of March 2023) for their contributions to the working group on sustainability criteria and certification.

References

1. Magnan AK, Pörtner H-O, Duvat VK, Garschagen M, Guinder VA, Zommers Z, Hoegh-Guldberg O, Gattuso J-P (2021) Estimating the global risk of anthropogenic climate change. Nat Clim Change
2. IPCC (2023) Climate change 2023: synthesis report
3. IEA (2022) Global energy and climate model. Int Energy Agency
4. European Commission (2023) Hydrogen. https://energy.ec.europa.eu/topics/energy-systems-integration/hydrogen_en. Accessed 26 May 2023
5. Odenweller A, George J, Müller VP, Verpoort P, Gast L, Pflüger B, Ueckerdt F (2022) Wasserstoff und die Energiekrise: fünf Knackpunkte. Kopernikus-Projekt Ariadne, Potsdam
6. IRENA (2022) Geopolitics of the energy transformation: the hydrogen factor. Int Renew Energy Agency
7. Corporate Europe Observatory and Transnational Institute, Assessing EU plans to import hydrogen from North Africa, Morocco, 2022
8. OECD (2022) CAF and European commission. In: Latin American economic outlook 2022: towards a green and just transition. OECD Publishing, Paris
9. European Commission (2023) Delegated regulation on Union methodology for RFNBOs. https://energy.ec.europa.eu/publications/delegated-regulation-union-methodology-rfnbos_en. Accessed 26 May 2023
10. European Commission (2023) Delegated regulation for a minimum threshold for GHG savings of recycled carbon fuels and annex. https://energy.ec.europa.eu/publications/delegated-regulation-minimum-threshold-ghg-savings-recycled-carbon-fuels-and-annex_en. Accessed 26 May 2023
11. Tilman D, Socolow R, Foley JA, Hill J, Larson E, Lynd L, Pacala S, Reilly J, Searchinger T, Somerville C, Williams R (2009) Beneficial biofuels: the food, energy, and environment trilemma. Science
12. Selbmann K, Ide T (2015) Between redeemer and work of the devil: the transnational Brazilian biofuel discourse. Energy Sustain Develop
13. Gerbens-Leenes PW, Hoekstra AY, Van der Meer T (2009) The water footprint of energy from biomass: a quantitative assessment and consequences of an increasing share of bio-energy in energy supply. Ecol Econ
14. Uusitalo V, Väisänen S, Havukainen J, Havukainen M, Soukka R, Louranen M (2014) Carbon footprint of renewable diesel from palm oil, jatropha oil and rapeseed oil. Renew Energy
15. Fargione J, Hill J, Tilman D, Polasky S, Hawthorne P (2008) Land clearing and the biofuel carbon debt. Science
16. Gawel E, Ludwig G (2011) The iLUC dilemma: how to deal with indirect land use changes when governing energy crops? Land Use Policy
17. Searchinger T, Heimlich R, Houghton RA, Dong F, Elobeid A, Fabiosa J, Tokgoz S, Hayes D, Yu T-H (2008) Use of U.S. croplands for biofuels increases greenhouse gases through emissions from land-use change. Science
18. Cotula L, Dyer N, Vermeulen S (2008) Fuelling exclusion? The biofuels boom and poor people's access to land. IIED, London
19. German L, Schoneveld GC (2011) Social sustainability of EU-approved voluntary schemes for biofuels: implications for rural livelihoods. CIFOR, Bogor
20. Obidzinski K, Rubeta A, Komarudin H, Andrianto A (2012) Environmental and social impacts of oil palm plantations and their implications for biofuel production in Indonesia. Ecol Soc
21. Vermeulen S, Lorenzo C (2010) Over the heads of local people: consultation, consent, and recompense in large-scale land deals for biofuels projects in Africa. J Peasant Stud
22. Wüstenhagen R, Wolsink M, Bürer MJ (2007) Social acceptance of renewable energy innovation: an introduction to the concept. Energy Policy
23. Cremonese L, Mbungu G, Quitzow R (2023) The sustainability of green hydrogen: an uncertain proposition. Int J Hydr Energy

24. Rockström J, Steffen W, Noone K, Persson Å, Chapin FS, Lambin EF, Lenton TM, Scheffer M, Folke C, Schellnhuber HJ, Nykvist B, de Wit CA, Hughes T, van der Leeuw S, Rodhe H, Sörlin S et al (2009) A safe operating space for humanity. Nature
25. Steffen W, Richardson K, Rockström J, Cornell SE, Fetzer I, Bennett EM, Biggs R, Carpenter SR, de Vries W, de Wit CA, Folke C, Gerten D, Heinke J, Mace GM, Persson LM, Ramanathan V, Reyers B, Sörlin S (2015) Planetary boundaries: guiding human development on a changing planet. Science
26. Rockström J, Gupta J, Qin D, Lad SJ, Andersen JF, McKay DIA, Bai X et al (2023) Safe and just Earth system boundaries. Nature
27. Dittrich M, Limberger S, Vogt R, Keppner B, Leuser L, Schoer K (2013) Vorstudie zu Ansätzen und Konzepten zur Verknüpfung des "Planetaren Grenzen" Konzepts mit der Inanspruchnahme von abiotischen Rohstoffen/Materialien. Umweltbundesamt, Dessau-Roßlau
28. McNutt M (2013) Climate change impacts. Science
29. National Oceanic and Atmospheric Administration (2020) Ocean acidification. https://www.noaa.gov/education/resource-collections/ocean-coasts/ocean-acidification. Accessed 12 May 2023
30. Hydrogen Council and McKinsey and Company (2021) Hydrogen for net-zero. Hydrogen Council
31. IEA (2023) CO_2 emissions in 2022. https://www.iea.org/reports/co2-emissions-in-2022. Accessed 24 Aug 2023
32. Bruckner T, Fulton L, Hertwich E, McKinnon A, Perczyk D, Roy J, Schaeffer R, Schlömer S, Sims R, Smith P, Wiser R (2014) Annex III: technology-specific cost and performance parameters. In: Climate change 2014: mitigation of climate change. Contribution of working group III to the fifth assessment report of the intergovernmental panel on climate change. Cambridge University Press, Cambridge
33. EMBER (2023) Global electricity review 2023. https://ember-climate.org/insights/research/global-electricity-review-2023/. Accessed 24 Aug 2023
34. Ueckerdt F, Bauer C, Dirnaichner A, Everall J, Sacchi R, Luderer G (2021) Potential and risks of hydrogen-based e-fuels in climate change mitigation. Nat Clim Change
35. Hurley I, Tittensor DP (2020) The uptake of the biosphere integrity planetary boundary concept into national and international environmental policy. Global Ecol Conser
36. Stockholm Resilience Centre (2016) The SDGs wedding cake. https://www.stockholmresilience.org/research/research-news/2016-06-14-the-sdgs-wedding-cake.html. Accessed 11 May 2023
37. Bennun L, van Bochove J, Ng C, Fletcher C, Wilson D, Phair N, Carbone G (2021) Mitigating biodiversity impacts associated with solar and wind energy development. Guidelines for project developers. IUCN and The Biodiversity Consultancy, Gland
38. Sonter LJ, Dade MC, Watson JEM, Valenta RK (2020) Renewable energy production will exacerbate mining threats to biodiversity. Nat Commun
39. UNFCCC (2023) Land use, land-use change and forestry (LULUCF). https://unfccc.int/topics/land-use/workstreams/land-use--land-use-change-and-forestry-lulucf. Accessed 17 May 2023
40. Umweltbundesamt (2023) Emissionen der Landnutzung, -änderung und Forstwirtschaft. https://www.umweltbundesamt.de/daten/klima/treibhausgas-emissionen-in-deutschland/emissionen-der-landnutzung-aenderung#bedeutung-von-landnutzung-und-forstwirtschaft. Accessed 17 May 2023
41. FAO and UNEP (2020) The state of the world's forests 2020. Forests, biodiversity and people. FAO, Rome
42. Yap MLM, Watene K (2019) The sustainable development goals (SDGs) and indigenous peoples: another missed opportunity? J Hum Develop Capabil
43. WWF (2020) Recognizing indignhous peoples' land interests is critical for people and nature. https://www.worldwildlife.org/stories/recognizing-indigenous-peoples-land-interests-is-critical-for-people-and-nature. Accessed 17 May 2023
44. Vörösmarty CJ, McIntyre PB, Gessner MO, Dudgeon D, Prusevich A, Green P, Glidden S, Bunn S, Sullivan CA, Reidy Liermann C, Davies PM (2010) Global threats to human water security and river biodiversity. Nature

45. Batteiger V, Schmidt P, Ebner K, Habersetzer A, Moser L, Weindorf W, Rakscha T (2022) Power-to-liquids: a scalable and sustainable fuel supply perspective for aviation. Umweltbundesamt, Dessau-Roßlau
46. WRI (2023) Water risk atlas. https://www.wri.org/applications/aqueduct/water-risk-atlas/. Accessed 19 May 2023
47. EASA (2020) Updated analysis of the non-CO_2 climate impacts of aviation and potential policy measures pursuant to the EU. Emissions Trad Syst Direct 30(4)
48. Ocko IB, Hamburg SP (2022) Climate consequences of hydrogen emissions. Atmosp Chem Phys
49. International Resource Panel (2019) Global resources outlook 2019: natural resources for the future we want. UNEP, Nairobi
50. Hatfield-Dodds S, Schandl H, Newth D, Obersteiner M, Cai Y, Baynes T, West J, Havlik P (2017) Assessing global resource use and greenhouse emissions to 2050, with ambitious resource efficiency and climate mitigation policies. J Clean Product
51. OECD (2019) Global material resources outlook to 2060. Economic drivers and environmental consequences. OECD Publishing, Paris
52. European Parliament (2023) How the EU wants to achieve a circular economy by 2050. https://www.europarl.europa.eu/news/en/headlines/society/20210128STO96607/how-the-eu-wants-to-achieve-a-circular-economy-by-2050. Accessed 12 June 2023
53. Günther J, Lehmann H, Nuss P, Purr K (2019) Resource-efficient pathways towards greenhouse-gas-neutrality—RESCUE. Umweltbundesamt, Dessau-Roßlau
54. Altmann M, Schmidt P, Krenn P, Astono YS, Fehrenbach H, Abdalla NJ (2022) Development of PtX sustainability standards and indicators: an expertise commissioned by the PtX Lab Lausitz. LBST and ifeu, Munich
55. Wealer B, Bauer S, Göke L, von Hirschhausen C, Kemfert C (2019) High-priced and dangerous: nuclear power is not an option for the climate-friendly energy mix. In: DIW weekly report
56. Luderer G, Pehl M, Arvesen A, Gibon T, Bodirsky BL, de Boer HS, Fricko O, Humpenöder F, Iyer G, Mima S, Mouratiadou I, Pietzcker RC, Popp A, van den Berg M, van Vuuren D, Hertwich EG (2019) Environmental co-benefits and adverse side-effects of alternative power sector decarbonization strategies. Nat Commun
57. Barron-Gafford GA, Pavao-Zuckerman MA, Minor RL, Sutter LF, Barnett-Moreno I, Blackett DT, Thompson M, Dimond K, Gerlak AK, Nabhan GP, Macknick JE (2019) Agrivoltaics provide mutual benefits across the food–energy–water nexus in drylands. Nat Sustain
58. Bourguignon D (2016) Closing the loop. New circular economy package. European Parliamentary Research Service, Brussels
59. OHCHR, UNEP and UNDP (2023) What is the right to a healthy environment
60. IPCC (2022) Climate change 2022: impacts, adaptation and vulnerability. In: Contribution of working group II to the sixth assessment report of the intergovernmental panel on climate change. Cambridge University Press, Cambridge
61. Ptx Hub (2020) PtX.Sustainability dimensions and concerns. Towards a conceptual framework for standards and certification. PtX Hub, Berlin
62. Business and Human Rights Resource Centre (2020) Renewable energy and human rights benchmark. Key findings from the wind and solar sectors
63. European Investment Bank (2022) Environmental and social standards. EIB, Luxembourg
64. Heinemann C, Mendelevitch R (2021) Sustainability dimensions of imported hydrogen, Freiburg: Öko-Institut e.V
65. Morgen S, Schmidt M, Steppe J, Wörlen C (2022) Fair green hydrogen: chance or chimera in Morocco, Niger and Senegal? Arepo GmbH and Rosa-Luxemburg-Stiftung, Berlin
66. Wietschel M, Bekk A, Breitschopf B, Boie I, Edler J, Eichhammer W, Klobasa M, Marscheider-Weidemann F, Plötz P, Sensfuß F, Thorpe D, Walz R (2020) Opportunities and challenges when importing green hydrogen and synthesis products. Fraunhofer Institute for Systems and Innovation Research, Karlsruhe
67. Committee on Economic, Social and Cultural Rights, General Comment No. 15 (2002). The right to water (arts. 11 and 12 of the International Covenant on Economic, Social and Cultural Rights), 2002.

68. Benvenisti E (2010) Water, right to, international protection. Max Planck Encyclopedia of International Law
69. The World Bank (2022) Tracking SDG 7: the energy progress report 2022. https://www.worldbank.org/en/topic/energy/publication/tracking-sdg-7-the-energy-progress-report-2022. Accessed 09 June 2023
70. Deutsches Institut für Menschenrechte (2023) Wirtschaft und Menschenrechte. https://www.institut-fuer-menschenrechte.de/themen/wirtschaft-und-menschenrechte. Accessed 02 June 2023
71. Kamminga MT (2016) Company responses to human rights reports: an empirical analysis. Bus Hum Rights J
72. Scita R, Raimondi PP, Noussan M (2020) Green hydrogen: the holy grail of decarbonisation? An analysis of the technical and geopolitical implications of the future hydrogen economy. Fondazione Eni Enrico Mattei, Milano
73. Eberhardt P (2023) Germany's great hydrogen race. The corporate perpetuation of fossil fuels, energy colonialism and climate disaster. Corporate Europe Observatory, Brussels
74. Popovich N, Plumer B (2021) Who has the most historical responsibility for climate change? The New York Times. https://www.nytimes.com/interactive/2021/11/12/climate/cop26-emissions-compensation.html. Accessed 05 June 2023
75. Bread for the World and Misereor (2022) Auswirkungen der deutschen Klimapolitik auf den Globalen Süden. Ein Impuls der Bank Entwicklungszusammenarbeit. https://www.brot-fuer-die-welt.de/fileadmin/mediapool/blogs/Fuenfgelt_Joachim/201022_Impulspapier_Entwicklungsbank_Auswirkungen-Dt-Klimapolitik.pdf. Accessed 07 June 2023
76. Villagrasa D (2022) Green hydrogen: key success criteria for sustainable trade and production. In: A synthesis based on consultations in Africa and Latin America. Bread for the World and Heinrich Böll Foundation
77. BMWi (2020) The national hydrogen strategy. BMWi, Berlin
78. acatech; DECHEMA (2023) Wasserstoff-Kompass. Ergebnisse des Stakeholder-Dialogs. acatech and Dechema, Berlin
79. United Nations General Assembly (2017) Work of the statistical commission pertaining to the 2030 agenda for sustainable development, A/RES/71/313
80. Altenburg T, Bauer S, Brandi C, Brüntrup M, Malerba D, Never B, Pegels A, Stamm A, To J, Volz U (2022) Ökologische Strukturpolitik. Ein starker Baustein für die deutsche Entwicklungszusammenarbeit. German Institute of Development and Sustainability (IDOS), Bonn
81. Walker B (2022) Governance der deutschen Wasserstoffwirtschaft. Standort
82. Ausfelder F, Bauer F, Cadavid Isaza A, de la Rua Lope C, Deutz S, Fröhlich T, Gawlick J, Hamacher T et al (2022) Optionen für ein nachhaltiges Energiesystem mit Power-to-X-Technologien. 4. Roadmap des Kopernikus-Projektes P2X Phase II. Florian Ausfelder and Dinh Du Tran, Germany
83. Albaladejo M, Altenburg T, Fokeer S, Wenck N, Schwager P (2022) Green hydrogen: fuelling industrial development for a clean and sustainable future. https://iap.unido.org/articles/green-hydrogen-fuelling-industrial-development-clean-and-sustainable-future. Accessed 11 June 2023
84. Andreoni A, Chang HJ (2019) The political economy of industrial policy: structural interdependencies, policy alignment and conflict management. Struct Change Econ Dyn
85. OHCHR (2022) Renewable energy and the right to development: realizing human rights for sustainable development. https://www.ohchr.org/sites/default/files/2022-05/2022-05-22-Renewable-Energy.pdf. Accessed 09 June 2023
86. Argonne National Laboratory (2023) JOBS models. Energy systems and infrastructure analysis. https://www.anl.gov/esia/jobs-models. Accessed 22 May 2023
87. ATAG (2021) Waypoint 2050. Air Transport Action Group (ATAG), Geneva
88. IRENA (2020) Measuring the socio-economics of transition: focus on jobs. Int Renew Energy
89. Quitzow R, Mewes C, Thielges S, Tsoumpa M, Zabanova Y (2022) Building partnerships for an international hydrogen economy. Entry-points for European policy action. Friedrich-Ebert-Stiftung, Berlin

90. Maihold G (2022) Die neue Geopolitik der Lieferketten. "Friend-shoring" als Zielvorgabe für den Umbau von Lieferketten. Stiftung Wissenschaft und Politik, Berlin
91. Kompetenzzentrum Klimaschutz in energieintensiven Industrien (ed) (2023) Akzeptanzstrategien in den energieintensiven Industrien. Aus der Praxis für die Praxis, KEI, Cottbus
92. Ausfelder F, Bauer F, Cadavid Isaza A, de la Rua C, Dura HFT, Gawlick J et al (2021) Optionen für ein nachhaltiges Energiesystem mit Power-to-X-Technologien. 3. Roadmap des Kopernikus-Projektes P2X Phase II. Florian Ausfelder and Hanna Dura, Frankfurt
93. BMZ (2023) Feminist development policy. For just and strong societies worldwide. Federal Ministry for Economic Cooperation and Development (BMZ), Bonn
94. Ueckerdt F, Günther C, Rehfeldt M, Gils HC, Pfluger B, Knodt M, Bauer C, Luderer G, Odenweller A, Kemmerzell J, Verpoort P (2021) Cornerstones of an adaptable hydrogen strategy: summary. Kopernikus-Projekt Ariadne, Potsdam

Market Introduction and Ramp-Up of Powerfuels under Political and Regulatory Aspects

Peter Kasten

Abstract This chapter explores the challenges and opportunities associated with the market introduction and ramp-up of power-to-liquid (PtL) fuels, essential for achieving climate neutrality in the transport sector. The focus is on the political and regulatory landscape in Europe and Germany influencing PtL fuel deployment. Challenges include technical and economic factors, including the current status of PtL fuel production technologies, high production costs, and the global scope of the future PtL fuel market. The need for certification of renewable and sustainable PtL fuels, crucial for ensuring environmental and climate integrity as well as for defining PtL production standards, is also addressed. The importance of long-term investment security, incentive schemes for PtL production and use, and a stable regulatory framework is emphasized. The regulatory and policy design in Europe and Germany is scrutinized, focusing on policies such as the Renewable Energy Directive, the German GHG Emission Reduction Quota, ReFuelEU Aviation and FuelEU Maritime Regulations, Emission Trading System 2, and the funding mechanisms H2Global and European Hydrogen Bank. The analysis concludes that while power-to-liquid (PtL) fuels are vital for a climate-neutral society, significant scaling in the transport sector faces major market barriers. Ambitious and long-term quota systems, supplemented by funding mechanisms for the first production plants, are essential to incentivize PtL fuel development. The ReFuelEU Aviation and the FuelEU Maritime Regulation are identified as suitable tools for a future market for PtL fuels in aviation and maritime transport, due to their long-term outlook. However, the RED and, consequently, the GHG quota in Germany are lacking in these long-term market incentives. The analysis also emphasizes the need for stable certification systems for PtL fuels that are transferable between global markets, to ensure seamless global PtL fuel markets.

Keywords Market ramp-up · Regulatory Framework · EU RED · EU ETS · ReFuelEU Aviation · Policy Design

P. Kasten (✉)
Oeko-Institut e.V., Berlin, Germany
e-mail: p.kasten@oeko.de

1 Introduction

In the transition towards "green" energies in the energy sector, climate-neutral alternative fuels are pivotal. Achieving carbon neutrality in transportation is particularly challenging but crucial in realizing this overall objective. Alternative, climate-neutral fuels are essential if the transport sector is to become climate-neutral within the next 25–30 years. The development of production capacities and a market for such fuels are relevant elements for the imminent transformation of the transport sector and its energy supply.

Power-to-liquid (PtL) fuels often emerge as a key element in the transformation towards climate-neutral sectors like aviation and shipping. These synthetic fuels are produced through various processes, utilizing electricity generated from renewable sources of energy, water, and typically carbon dioxide (CO_2). PtL fuels are pivotal for a seamless transition, as they can, depending on the specific PtL fuel type, be integrated into existing transportation and fuel infrastructure with minimal changes.

However, achieving climate neutrality in the transport sector is a complex endeavour, accompanied by numerous challenges. There is a common misperception in public discussions that hydrogen and PtL fuels are already technically sufficiently mature and only require suitable policy incentives for immediate utilization. In reality, the market ramp-up of PtL fuels requires specific market incentives to overcome the principal challenges to create a large-scale PtL fuel market.

Against this background, this chapter therefore addresses the challenges for the market introduction of PtL fuels and the existing regulatory framework that can play a key role in stimulating the market ramp-up of renewable PtL fuels.[1] The most important challenges with regard to a market introduction for PtL fuels are presented as a basis for exploring possible policy designs, and also key regulations to support this market ramp-up EU and Germany are described.

2 Challenges and Opportunities for a PtL Fuel Deployment

The perception in the public discussion is often that hydrogen and PtL fuels are technically available and "only" the appropriate policy incentives are needed for an immediate demand for these. In reality, however, the market ramp-up requires market incentives to overcome the main challenges for a PtL fuel market. The most important challenges are presented below as a basis for discussing possible policy designs.

---

[1] In the context of this chapter, unless we explicitly discuss sustainability aspects and the climate impact of PtL fuels, we assume that PtL fuels are renewable fuels, i.e. produced based on renewable energy.

2.1   Technical Status Quo of PtL Fuel Production

Various technologies and process routes exist for the production of hydrogen, the provision of CO_2 from various potential sustainable CO_2 sources, and conversion of hydrogen and CO_2 into PtL fuels. Although there are currently numerous projects aiming to build PtL plants, none of these routes is yet available along the entire process chain on an industrial production scale. Technical challenges include, for example, scaling of the reverse water–gas shift reaction and CO_2 capture from ambient air, advancing the technological maturity of new processes such as co-electrolysis, and reducing the specific demands for scarce and expensive materials such as iridium and platinum [1]. Furthermore, the technologies available for the production of PtL fuels are currently not in high volume production. Automated production processes and large-scale upstream value chains from the mining of the required materials (e.g. iridium) to the production of the precursors of hydrogen manufacture and CO_2 capture must first be developed and established over several years. A long-term planning perspective for hydrogen and PtL fuel use is therefore a major prerequisite for the development of the respective value chains. In the initial market phase in particular, it is equally important that new experiences and advances in production are made with each production facility, and the costs of producing PtL fuels are thereby continuously reduced. In addition to the technical development, it is evident that the timespan from the investment decision to the construction and commissioning of the technical facilities alone will take several years. The beginning of industrial production of PtL fuels in relevant quantities will therefore in all likelihood only be possible from around the year 2030 [2].

2.2   Economics of PtL Fuel Production

Current production costs of PtL fuels are estimated at more than 2 €/L of fuel [3]. The process chain of PtL fuel production consists of several energy-intensive processes and conversion steps with associated "conversion losses" ("energy losses"), resulting in relatively low overall production efficiencies. This relatively high energy input for the production of a specific amount of PtL fuel is a main factor that leads to the rather high production costs compared to both fossil fuels as well as sustainable biofuels used so far. Even if the costs of PtL fuel production can be significantly reduced in the future through technical progress and efficiency gains, lower prices for renewable electricity, the scaling of plants and an ongoing development of the value chain, and higher costs compared to fossil fuels and sustainable biofuels are usually still assumed in the medium and long term. For the year 2030, studies assume CO_2 reduction costs of up to 200 €/tCO_2 and more using PtL fuels [4]. In contrast, the Emission Trading System 2 (ETS2), which will apply in the EU from 2027 onwards, provides for countermeasures to slow down or prevent further price increases from a CO_2 price of 45 €/tCO_2. Consequently, in order to create a market for PtL fuels, appropriate additional incentivizing policies are needed in the short term and may in all likelihood still be essential in the medium and longer term.

The cost structure of PtL fuel production is also worth noting with regard to its market ramp-up. Especially for PtL, production plants are capital-intensive [5]. For this reason, plant utilization and capital procurement costs are relevant components on the overall production costs of PtL fuel. Thus, PtL fuel production is characterized by large investment sums that are amortized over periods of 20 years or more. In addition, Power Purchase Agreements (PPA) must be signed for the provision of electricity from new renewable electricity capacities over the whole lifetime of a PtL plant in order to be classified as a renewable PtL fuel and to be eligible for EU support schemes. Both aspects highlight the importance of providing a high level of investment security to facilitate the development of production capacities.

2.3   Geographical Scope of the Future PtL Fuel Market

Liquid fuels are characterized by the fact that they are easy to store and transport. For the value chain from crude oil extraction to refuelling in transport vehicles, a global transport infrastructure with low transport costs already exists for a global fossil fuel market. This transportation and distribution infrastructure is basically also applicable for PtL logistics, especially if hydrocarbons like kerosene or diesel are produced via PtL processes, which are compatible with the current infrastructure. With the exception of the initial pilot projects, a global market for PtL fuels can therefore be expected. Optimal production sites with low costs for renewable power generation can thus be developed for PtL fuel production at an early stage.

2.4   Definition and Certification of Renewable and Sustainable PtL Fuels

Besides the origin of the CO_2 required for the production of PtL fuels, the greenhouse gas (GHG) mitigation effect of PtL fuels depends largely on the emissions associated with the provision of the electricity that is used for PtL production, due to the high level of electrical energy input required [6]. Given a sufficiently high share of renewable electricity in the PtL production, PtL fuels can have a (significantly) lower climate impact than fossil fuels. However, in the worst case scenario with unfavourable high shares of fossil power generation in the grid, they can also be associated with higher GHG emissions than their fossil fuel equivalents. In addition, negative as well as positive sustainability impacts can occur in terms of the land and water used and with regard to social effects at fuel production sites.

As a result, a definition of the framework conditions under which PtL fuels are considered sustainable and renewable (eligibility criteria) must be established for developing a PtL fuel market. The definition of renewable and sustainable PtL fuels has an impact on the production costs of PtL fuels and on the type of potential production sites. The challenge in defining these criteria is to make fuel certification as simple as possible while maintaining environmental integrity and to keep the criteria the same over a long period of time in order to offer adequate planning and investment security. Ideally, certification systems will be set up that are similar worldwide to enable a global market.

The EU adopted its certification rules for renewable PtL fuels in summer 2023, and the first fuels certified as renewable PtL can come onto the market in the EU. At present, however, the proposed and already adopted certification rules are different in most global markets. Past experience with the market ramp-up of biofuels also shows that the definition of renewable fuels can change over time. For example, market restrictions can be set if—contrary to initial expectations—fuels have no or a negative sustainability effect. With regard to the market ramp-up, it is therefore important to note that the definition of sustainable and renewable PtL fuels can change over time.

2.5 Integration into a Cross-Sector Strategy for Reaching Climate Neutrality

Three pillars of the transition for climate protection that have been (significantly) progressed are: (I) a renewable electricity supply, (II) the electrification of energy demands, and (III) increased efficiency in energy use. The use of renewable hydrogen and PtL fuels is the fourth pillar and so far the least developed of the energy transition for climate protection. These four pillars thus represent the foundation for the climate protection efforts not only in the transport sector but also in other parts of the economy and society. The market ramp-up of PtL fuels should therefore not be looked at independently of the developments and markets in the other sectors.

The design of the political and regulatory framework of the PtL fuel ramp-up should therefore adhere to various basic premises from an overall cross-sectoral perspective.

- Climate protection scenarios with a focus on energy efficiency and the electrification of applications not previously supplied with electricity have lower overall economic costs across sectors [7] and in the transport sector [8] than those with high shares of hydrogen and PtL fuel use. In land transportation, technical alternatives with lower overall future costs are available, such as battery-powered and direct-electric vehicles, and possibly also fuel cell technology. For this reason, incentive instruments for the PtL fuel ramp-up in land transportation, which necessitate their use in current combustion engine technologies, are not effective as this would hinder the transformation to zero-emission vehicles. These instruments—such as crediting the GHG emission reduction of fuels to the CO_2 emission standards of vehicles—are therefore not considered appropriate.

- There are economic sectors in which the previously mentioned strategies such as increasing electrification and energy efficiency are not technically and economically feasible. For such applications, there is no alternative to the use of hydrogen and PtL fuels. These include, for example, shipping and aviation as well as steel production. The available quantity of renewable hydrogen products will be low until around 2030. As a commodity, these will probably only be available in the later part of the 2030s. The early development of the supply chains and transport infrastructures for the above-mentioned applications is therefore crucial for the long-term development towards climate neutrality. The support measures for PtL fuels (and hydrogen) should therefore consider which requirements will arise in the long term and should stimulate or not hinder the necessary developments [9].

2.6 Overview of Challenges and Opportunities for PtL Fuel Deployment

In summary, PtL fuels will not be available in larger quantities until around 2030 and may not become a commodity until later in the 2030s. Long-term investment security is of high importance to kick-start the high upfront investments of the production plants. In all likelihood, in a similar way to fossil fuels, a global market for PtL fuels will develop, for which a clear definition is required as to the production criteria for which a PtL fuel is certified as sustainable and renewable. The prerequisite for a market ramp-up is also that incentive instruments are created that compensate for the higher production costs compared to other fuel options. These instruments should be integrated into an overall cross-sector strategy for climate neutrality. Transformation processes towards zero-emission vehicles and the development of long-lasting transport and utilization infrastructures for hydrogen in all sectors of the economy should not be hindered by the support of PtL fuels in transport. In an early phase of hydrogen use in particular, it seems essential to allocate the scarce quantity of hydrogen and its derivatives to specific applications with the help of support instruments.

3 Current Regulatory and Policy Design in Europe and Germany

In the EU and Germany, there is an incentive system for climate protection in transport. The most important regulations to support the market ramp-up of PtL fuels are as follows.

3.1 Renewable Energy Directive

With the revision of the Renewable Energy Directive (RED)[2] in 2023, this directive is the main policy instrument for the market for renewable fuels at EU level. It has two functions relating to the market ramp-up of PtL fuels. First, it defines the minimum level of demand for renewable fuels in Europe up to 2030, and second, it specifies the requirements for certification as renewable fuels in the EU.

- The RED sets a target for 2030 that at least 29% of final energy used in transport should come from renewable sources or that the GHG intensity in transport should be reduced by at least 14.5%. Various types of biofuels and electricity from different renewable sources are eligible options for meeting the target. PtL fuels are part of the renewable fuels of non-biological origin (RFNBO), which can also be used to meet the target. Some types of biofuels are subject to crediting limits, and advanced biofuels and RFNBO are specifically incentivized via a joint minimum energy share of 1% in 2025 and 5.5% in 2030. Both fuel types are also eligible for double crediting towards the target. Additionally, a separate 2030 energy share target of 1% is set for RFNBO, and an indicative RFNBO energy share target of 1.2% exists for fuels used in maritime shipping. The use of renewable electricity can also be accounted for multiple times. The real minimum share of renewable fuels to be achieved is therefore less than 29%.
- With regard to the compliance options, the RED defines the conditions under which RFNBOs are considered renewable fuels. While it is specified that RFNBO must have at least 70% less GHG emissions than a reference value for fossil fuels. The methodology for determining the GHG emissions of RFNBO and the regulation when the electricity used for the production of RFNBO is considered renewable and therefore emission free were published in summer 2023.[3]

3.2 GHG Emission Reduction Quota in Germany

As a directive, the RED provides the framework for the support of renewable fuels in the EU Member States. However, the implementation into law is carried out at national level, so that the legislation resulting from the RED differs between EU Member States, in some cases significantly. In Germany, the GHG emission reduction quota (in short: GHG quota)[4] is the implementation of the RED. The GHG quota obligates fuel suppliers in land-based transport to reduce the GHG emissions of their fuels by 25% by 2030 compared to the fossil reference value. They can achieve this by using the previously mentioned RED compliance options (eligible fuels). As stipulated by the RED, there are certain crediting limits or minimum energy shares to be achieved for biofuels. Emission reductions from the use of electricity

---

[2] (EU) 2023/2413.

[3] (EU) 2023/1184 and (EU) 2023/1185.

[4] §§37a-H of https://www.gesetze-im-internet.de/bimschg/.

in electric vehicles can also be included. For the purpose of crediting, the suppliers can apply a multiplier of 2 for advanced biofuels and RFNBOs, i.e. power-to-X fuels (e.g. hydrogen, synthetic methane, synthetic kerosene / diesel, etc.). In addition, as electricity generated from renewable sources of energy will count with a multiplier of 3 towards the quota, a mechanism exists to increase the ambition level of the GHG quota as soon as fuel suppliers meet the GHG quota mainly via the crediting of emission reductions from electricity use in transport. This ensures that also renewable fuels are needed to meet the overall target.

The target of the GHG quota will increase annually, starting from 2022. These annual target values are very ambitious and go far beyond the minimum ambition level defined by the RED. For non-compliance with the GHG quota targets, there is a penalty of 600 €/tCO$_2$eq of non-achieved reduction. This defines the maximum achievable price incentive for PtL fuels under the GHG quota. In 2022, the prices for emission reductions in the GHG quota are above 400 €/tCO$_2$eq.

In addition to the GHG quota, in 2021 Germany also introduced a minimum energy quota for the use of PtL fuel in aviation. This is to be met for the first time in 2026 (0.5%) and rises to 2% in 2030.

The GHG quota will be revised on the basis of the amended RED (see Chap. 4), and the targets set will be adjusted if necessary.

3.3 ReFuelEU Aviation Regulation and FuelEU Maritime Regulation

The ReFuelEU Aviation Regulation and FuelEU Maritime Regulation are both new regulatory instruments in the EU. Both regulations have been approved by the EU Parliament and the EU Council but have not yet been published as legal texts at the time of writing.

The ReFuelEU Aviation Regulation provides for a minimum quota for renewable fuels in aviation to be met by fuel suppliers. This is to increase from 2% (2025) through 6% (2030) to 70% in 2050. PtL fuels can be used for compliance with the targets of the quota. In addition, there is a continuously increasing specific sub-quota for synthetic fuels (including PtL fuels) from 1.2% (combined average target for 2030 and 2031) to 35% in 2050. In a transitioning period, these quotas can be fulfilled as a weighted average across the EU until 2034. Afterwards, the quotas must be fulfilled as minimum shares at each EU airport (Union Airports).

The FuelEU Maritime Regulation is a new instrument for emissions reduction in maritime transport. Similar to the RED and the German GHG quota, a reduction pathway is provided for maritime GHG emission intensity against a fossil reference value. The targeted emission reduction for ship operators increases continuously from 2% (2025) through 6% (2030) to 80% (2050). Fossil fuels can be used to reduce emissions. However, if an indicative share for RFNBO fuels of 1% is not reached until 2031 (1%), a binding quota of 2% shall apply from 2034.

3.4 Emission Trading System 2

In addition to the existing European emissions trading system, a second emissions trading system for the heating sector and road transport (ETS2[5]) will start from 2026. The quantity of emissions allowances in this emissions trading system will decrease over time (emission reduction of 42% by 2030 compared to 2005 levels) in order to provide an incentive to reduce emissions. It can be assumed that the prices for CO_2 emissions in the two sectors will increase over time. However, mechanisms are foreseen for the early years of the system in order to reduce a possible rapid increase in the cost of emission allowances. Specific price stabilization mechanisms are also integrated into the ETS2 for the case of the CO_2 price exceeding 45 €/tCO_2. In Germany, a pricing instrument for CO_2 emissions in the transport sector already exists in the form of the "Brennstoffemissionshandelsgesetz" (BEHG). It remains to be seen how the BEHG will be integrated into the ETS2 and what price level will be reached in Germany compared to the EU.

3.5 H2Global/European Hydrogen Bank

H2Global[6] is a German funding instrument for the rollout of hydrogen and PtL fuel production. The mechanism offers long-term contracts (minimum 10 years) for the supply of hydrogen and PtL fuels through a bidding mechanism (lowest price) for tendered production volumes. On the demand side, short-term purchase contracts are established, which are also determined by a bidding process (highest price). The difference in cost between the price of supply and the purchase price at the demand side is covered by the state. This long-term purchase mechanism is intended to make the first production plants for hydrogen and PtL financially feasible. With the European Hydrogen Bank,[7] the EU intends to use the same instrument in future to promote the first hydrogen and PtL production plants.

4 Main Policy Design to Ramp-Up the Market for PtL Fuels in Europe and Germany

The above-mentioned existing and proposed instruments are expected to constitute the main framework for the ramp-up of PtL fuel use in Europe and Germany.

Quota systems like the effective RED and the GHG quota define a minimum market for the products with which the targets of the quotas can be achieved. The

---

[5] (EU) 2023/959.

[6] https://www.h2global-stiftung.com/.

[7] https://climate.ec.europa.eu/news-your-voice/news/innovation-fund-announces-two-upcoming-calls-proposals-2023-10-23_en.

obligated fuel suppliers give priority to those options for meeting the quota that have low GHG emission reduction costs. However, among the possible compliance options, PtL fuel is currently a rather expensive choice for meeting the targets of the RED and GHG quotas and is likely to remain so in the medium to long term. The incentive for a long-term market for PtL fuels therefore depends heavily on the extent to which other options are available and eligible for crediting towards the quota, the general ambition level, and the specific targets for PtL fuels.

The amended RED and the currently valid GHG quotas implemented in Germany therefore have a similar impact on the PtL market ramp-up in the transport sector. The GHG quota in Germany has a similar, but probably slightly less ambitious target level than the amended RED. It guarantees—by means of the target adjustment mechanism for heavy use of emission reduction certificates from electricity usage in transport— that advanced biofuels or PtL fuels must be used to meet the target, and PtL fuel therefore competes with advanced biofuels for being used for compliance. A similar effect occurs in the RED as a result of the joint sub-quota for advanced biofuels and RFNBO. The GHG quota targets are only set until 2030, and fuel suppliers can assume that the GHG quota will continue with increasing ambition levels beyond 2030. However, due to the rather short time frame of the targets, this system only offers a moderate degree of investment security. The prerequisite for this is that PtL fuels can be made available in larger quantities and at a price comparable to modern biofuels.

Generally, quota systems also result in the polluters of GHG emissions paying the costs of renewable fuels, as it can be assumed that the obligated fuel suppliers will pass on the costs incurred by the GHG quota to the users. High ambition levels lead to rather high prices for meeting the targets—as visible in 2022 in the GHG quota in Germany. Accordingly, such quota systems can increase fuel costs for users and can also have an effect on the availability of hydrogen and sustainable biomass or the cost of their use in other sectors. These sectoral quota systems therefore interact significantly with policies and can potentially make the use of these energy carriers in other sectors very expensive, if the ambition level is set too high. However, a positive effect is that the quotas do not hinder the transformation to new powertrain systems in road transport and that investments in new technologies will be made.

The specific PtL quota for aviation in Germany, which came into force in 2021, can be seen as positive for a market ramp-up of PtL fuels, as it provides an incentive for PtL fuels to be developed for aviation that will rely on this fuel in the long term. The specific quota for aviation thus results in production facilities being targeted at an early stage in order to produce certified fuels for aviation. Similar to the German GHG quota, however, there is only moderate investment security due to the short time frame for targets.

The new specific sustainable fuel quota system for aviation of the EU including a sub-quota for synthetic fuels makes sense as the quota system extends to the whole EU and it ensures that PtL fuels for aviation are targeted as part of the market ramp-up. The FuelEU Maritime Regulation, on the other hand, is a suitable approach to phase in domestic GHG emission mitigation measures for maritime shipping, but the option to use fossil fuels for compliance is not suitable for bringing renewable fuels into use in the short and medium term. However, the indicative target for PtL fuels also supports the potential demand for PtL fuels in this regulation. Despite the rather low target levels, the targets of the two regulations ensure—different to the German fuel quotas—that the long-term prospects for the energy supply of shipping and aviation are clear and thus provide for fairly high investment security.

The ETS2 will support renewable fuels in their market ramp-up. However, it is unlikely that carbon price levels will be reached to incentivize additional volumes of renewable fuels beyond those required for compliance with the quota systems. The price in the ETS2 that is created on the market and politically feasible is unlikely to be sufficient for additional volumes of renewable fuels being marketed. The long-term investment security is increased though as it can be assumed that a carbon price will still exist in the medium and long term.

The implementation of the first PtL production plants is crucial in the short term to initiate the scaling and industrialization of PtL production. Given that these plants may face cost disadvantages within the quota systems in the EU and Germany in the medium term, providing initial financial support through instruments like H2Global and the European Hydrogen Bank is a rational approach for PtL production financing.

5 Conclusion

PtL fuels are crucial for a climate-neutral society. However, they have some market barriers that make very short-term deployment in the transport sector at a relevant scale unlikely. High production costs and upfront investments for PtL fuel production indicate that high investment security is a key element for the market ramp-up of PtL fuels. From a technical point of view, PtL fuels could become available in larger quantities around 2030, and a global market for PtL fuels will emerge in the long term only. Another key element for the design of incentive instruments for a PtL fuel market is the definition of the production conditions under which PtL fuels are sustainable and renewable and the development of certification systems to prove these properties. The EU published their certification regulation in summer 2023. However, the different certification systems in the various hydrogen and PtL fuel markets make it difficult to establish a global system for these energy sources.

The quota systems for renewable fuels used in the EU and Germany, if designed ambitiously and with a long-term perspective, can provide an adequate incentive for a market for PtL fuels to develop. In all likelihood, in these quota systems PtL fuels will compete with advanced biofuels for market shares to comply with the targets. The recently developed specific quotas for aviation and maritime shipping are also appropriate additions to European legislation to facilitate the future use of PtL fuels in these applications. The CO_2 pricing element of ETS2 is a complementary factor for the uptake of PtL fuel production. Funding mechanisms for initial production facilities such as H2Global and the European Hydrogen Bank also support the industrialization and scaling of PtL production.

References

1. Minke C, Suermann M, Bensmann B, Hanke-Rauschenbach R (2021) Is iridium demand a potential bottleneck in the realization of large-scale PEM water electrolysis? Int J Hydr Energy 46:23581–23590. https://doi.org/10.1016/j.ijhydene.2021.04.174
2. Nationale Plattform Zukunft der Mobilität (2020) Arbeitsgruppe 1: Werkstattbericht Alternative Kraftstoffe. Klimawirkungen und Wege zum Einsatz Alternativer Kraftstoffe. Bundesministerium für Verkehr und digitale Infrastruktur. https://www.plattform-zukunft-mobilitaet.de/wp-content/uploads/2020/12/NPM_AG1_Werkstattbericht_AK.pdf
3. Grahn M, Malmgren E, Korberg A, Taljegard M, Anderson J, Brynolf S, Hansson J, Ridjan Skov I, Wallington T (2022) Review of electrofuel feasibility—cost and environmental impact. Progr Energy 4:7937. https://doi.org/10.1088/2516-1083/ac7937
4. Matthes F, Heinemann C, Hesse T, Kasten P, Mendelevitch R, Seebach D (2020) Wasserstoff sowie wasserstoffbasierte Energieträger und Rohstoffe. Eine Überblicksuntersuchung. Öko-Institut e.V. https://www.oeko.de/fileadmin/oekodoc/Wasserstoff-und-wasserstoffbasierte-Brennstoffe.pdf
5. Agora Verkehrswende (2023) E-Fuels zwischen Wunsch und Wirklichkeit. Was strombasierte synthetische Kraftstoffe für die Energiewende im Verkehr leisten können—und was nicht. https://www.agora-verkehrswende.de/fileadmin/user_upload/103-E-Fuels_v2.pdf
6. Kasten P (2020) E-Fuels im Verkehrssektor. Kurzstudie über den Stand des Wissens und die mögliche Bedeutung von E-Fuels für den Klimaschutz im Verkehrssektor: Öko-Institut e.V. https://www.oeko.de/fileadmin/oekodoc/E-Fuels-im-Verkehrssektor-Hintergrundbericht.pdf
7. Sensfuß F et al (2021) Langfristszenarien für die Transformation des Energiesystems in Deutschland 3. Kurzbericht: 3 Hauptszenarien. Bundesministerium für Wirtschaft und Energie. https://langfristszenarien.de/enertile-explorer-wAssets/docs/LFS_Kurzbericht_final_v5.pdf
8. Mottschall M, Kasten P, Kühnel S, Minnich L (2019) Sensitivitäten zur Bewertung der Kosten verschiedener Energieversorgungsoptionen des Verkehrs bis zum Jahr 2050. Abschlussbericht. Umweltbundesamt (Texte, 114/2019). https://www.umweltbundesamt.de/publikationen/sensitivitaeten-zur-bewertung-der-kosten
9. Matthes F, Braungardt S, Bürger V, Göckeler K, Heinemann C, Hermann H et al (2021) Die Wasserstoffstrategie 2.0 für Deutschland. Untersuchung für die Stiftung Klimaneutralität. Öko-Institut e.V. https://www.oeko.de/fileadmin/oekodoc/Die-Wasserstoffstrategie-2-0-fuer-DE.pdf

Importing Powerfuels to Europe: Options and Challenges

Martin Wietschel, Natalia Pieton, Jana Thomann, and Barbara Breitschopf

Abstract Hydrogen and other power-based fuels will be needed in the future to achieve ambitious climate targets. Due to different production costs and production potentials in different countries, international trade will emerge, with Europe importing powerfuels. The challenges and opportunities for Europe are discussed in this chapter.

Keywords Powerfuel Import · Country Potentials · Energy Partnerships · Investment Risks · Hydrogen Economy

1 Trading Hydrogen and Other Power-Based Fuels Internationally Will Be Necessary to Achieve Climate Protection Goals

Today's hydrogen industry is a sector without significant trading activities. Only 5% of the globally produced hydrogen is transported and traded at the moment [1]. This trade is mainly conducted by industrial gas suppliers directly supplying large-scale industrial consumers. Most of the hydrogen currently produced is either from these industrial gas suppliers or produced on-site by the respective industrial enterprises. For instance, in refineries, hydrogen is produced based, e.g., on natural gas via steam methane reforming and subsequently used for crude oil refining (e.g., hydrofiner). This is why there is currently little need for an expanded national or international hydrogen transportation infrastructure. However, this situation is very likely to change in the future.

M. Wietschel (✉) · J. Thomann · B. Breitschopf
Fraunhofer Institute for Systems and Innovation Research (Fraunhofer ISI), Karlsruhe, Germany
e-mail: martin.wietschel@isi.fraunhofer.de

N. Pieton
Fraunhofer Research Institution for Energy Infrastructures and Geothermal Systems (Fraunhofer IEG), Cottbus, Germany

© The Author(s), under exclusive license to Springer Nature Switzerland AG 2025
N. Bullerdiek et al. (eds.), *Powerfuels*, Green Energy and Technology,
https://doi.org/10.1007/978-3-031-62411-7_37

Decarbonizing the energy system will ultimately take place to a large extent by generating power based on renewable energies and substituting fossil fuel energy sources, such as natural gas or hard coal. This electrical power can be used within the energy system either directly or indirectly by using it to produce hydrogen or synthetic hydrocarbons. Riemer et al. [2] have conducted a meta-analysis to explore how the global demand for hydrogen could develop in the future by re-evaluating more than 40 current energy system and hydrogen scenarios. Compared to today's figures, a large proportion of these scenarios indicates a significant increase in hydrogen demand, resulting in a global hydrogen share of 4–11% in the final energy consumption in 2050 (in absolute terms, the demand ranges from 14 to 55 EJ or 4 to 15 PWh). Some scenarios predict significantly higher values of up to 23% (79 EJ or 22 PWh). However, there are regional differences. In the European Union (EU), there is a higher demand range; hydrogen demand here is expected to be between 4 and 14% of final energy demand (1–4 EJ or 0.3–1 PWh) [2]. In this context, the ambition of the European Commission (EC) as well as the governments of the various member states to reduce greenhouse gas (GHG) emissions is identified as a key influencing factor. Hydrogen demand increases for all regions in line with the ambition to reduce GHG emissions, especially if the target exceeds 80% in 2050 compared to 1990.[1]

Hydrogen and other power-based fuels are characterized by a high flexibility of use, but their production also requires a high specific energy input compared to their energy content (e.g., heating value), because of the conversion steps required within the overall process chain and the associated conversion losses. Therefore, producing power-based fuels requires substantial expansion of renewables-based electricity generation. Not every country has sufficient potential to expand their domestic renewables-based electricity generation to cover the energy demand needed to produce power-based fuels. This is limited by the country's renewable energy potentials, costs and also factors like the social acceptance of such an expansion. Germany and the EU are worth mentioning here as examples. Importing hydrogen and hydrogen-based powerfuels, such as methanol and ammonia, forms an important pillar of the hydrogen strategies of both Germany and the EU. The majority of studies assume future import shares of more than 50% for Germany and the EU [3]. It is therefore very likely that hydrogen and other power-based fuels will be traded internationally to a greater extent in the future and that this will be of special interest to Germany and the EU.

The EU's hydrogen strategy expresses the clear need to import hydrogen and hydrogen-based powerfuels [4]. Its strategy paper includes the following three guiding principles:

- "Clean" hydrogen offers new opportunities for re-designing the EU's energy partnerships with both neighboring countries and regions and its international, regional and bilateral partners, promoting supply diversification and helping to establish stable and secure supply chains.

---

[1] See Riemer et al. [2] for worldwide figures and Wietschel et al. [3] for EU and German figures.

- Given the rise of promising resource opportunities in the Middle East and in North Africa (MENA), the EU's hydrogen strategy focuses on the MENA region. Taking into account natural resources, physical interconnections and technological development, the EU's eastern neighbors, particularly the Ukraine, and southern neighbors should also be regarded as priority partners.
- Cooperation should range from research and innovation to regulatory policy, direct investments and an undistorted and fair trade in hydrogen, hydrogen-based power-based fuels and the associated technologies and services.

With the publication of the REPowerEU plan in May 2022, the EC has set a target of 10 Mt (334 TWh) of renewable hydrogen from domestic production and 10 Mt of renewable hydrogen from imports by 2030 [5].

2 Importing Green Hydrogen and Other Power-Based Fuels Could Be Economically Attractive on a Cost Basis

Several studies have analyzed the economics of importing hydrogen. They look at the production and supply costs of imported synthesized hydrocarbon-based fuels based on renewable electricity generation [6–11]. Electricity costs, electrolyzer efficiency, investments in the electrolyzers and their annual full-load hours (AFLHs) have the biggest impact on the costs of producing hydrogen. In countries with favorable conditions for renewable energy, such as North Africa, the electricity costs of installations using renewable sources of energy (e.g., wind, solar) are much lower than in central Europe, even less than half in some cases. With more than 4,000 annual full-load hours (electrolysis) of hydrogen production, these countries clearly offer higher annual full-load hours than any EU country. Accordingly, the production costs of hydrogen and other power-based fuels in countries with such favorable climatic conditions for renewables are significantly lower than those in central Europe. Furthermore, costs for transporting hydrogen and synthetic fuels from these countries into the EU and also to Germany are low in relation to the production costs (Fig. 1).

These favorable conditions suggest that it could be economically more efficient to import hydrogen and synthetic fuels to central Europe rather than produce them domestically. However, as shown in Fig. 1, the assumed cost of capital has a major influence on the costs. The cost of capital reflects different investment risks and policy action could be required to offset the risks during the early market development phase of hydrogen and other powerfuels (see the discussion in Chap. 6).

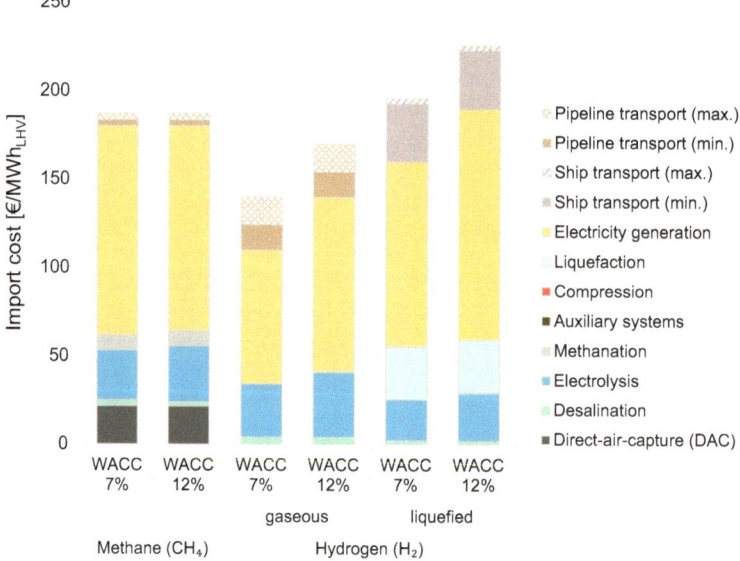

Fig. 1 Costs of importing hydrogen [gaseous (H2) and liquid (LH2)] and methane (gaseous) from the MENA region to the EU in 2030 including transport costs and calculated with 7 and 12% WACC (weighted average cost of capital). *Source* Lux et al. [8]

3 Several Countries Are Promising Candidates for Producing Green Powerfuels

A global Power-to-X (PtX) atlas has been compiled [12] that illustrates the estimated power-based fuel production potential of different areas. Generally applicable criteria for the potential area analysis for systems using renewable sources of energy (RE) form a rough estimate of future area potentials. The PtX atlas only considers regions in which sustainable energy production is possible. In concrete terms, this means that sites are excluded where there could be conflicts, e.g., with nature conservation demands. The same applies to inland regions where large-scale use of electrolysis would lead to water stress. Ensuring a constant supply of water presents a huge challenge in some regions and may require seawater desalination plants. Areas of fertile and managed agricultural land where food can be grown are also excluded, especially with regard to the use of photovoltaic (PV) ground-mounted systems.

Taking these criteria into account, the PtX atlas shows that it would be possible in the long term to produce a total of around 109,000 TWh/year. of liquid "green" hydrogen, or alternatively 87,000 TWh/year. of "green" hydrogen-based synthesized fuels. The PtX atlas identifies ten countries with the largest potential for powerfuel production outside of Europe: USA, Australia, Argentina, Russia, Egypt, Canada, Mexico, Libya, Chile and Saudi Arabia. A large share of this potential is found in regions able to generate electricity using both wind and solar energy, which enables a high utilization of the electrolyzers. Australia and the USA also offer significant solar-only potential. Wind-only sites in the USA, Canada and Russia constitute another significant source of potential [12].

The potentials identified in the PtX atlas are significantly higher than the demand for hydrogen in the different scenarios, which ranges between 4,000 and 15,000 TWh in 2050 (Chap. 1).

However, it will only be possible to exploit a certain proportion of this total potential due to constraining factors, which include, for example:

- the pace (rate) at which renewable and power-based fuel generation facilities can be expanded,
- the development of a proper transportation infrastructure (Chap. 5),
- political stability and its impact on investment dynamics (Chap. 6),
- access to low-cost capital (Chaps. 2, 6),
- the need to also meet domestic energy demands,
- the application of sustainability criteria.

Some of these factors are taken into account in the PtX atlas and the potential production volumes that could be achieved still amount to 69,000 TWh/year. of hydrogen or 57,000 TWh/year. of PtL [12]. Thus, even including limiting factors, there is a potentially high supply of power-based fuels.

However, attention should be drawn to the debate about the sustainability of hydrogen and its synthesis products [4, 13–17]. In some studies, very strict requirements are set that would significantly lower the hydrogen production potential and could significantly increase the costs. It is the task of policymakers to find the right balance between sustainability requirements and economic efficiency, especially during the phase of market development.

4 The War in Ukraine Has Forced a Reassessment of Energy Partnerships and Highlights the Importance of Diversification

In addition to Russia, countries like Kazakhstan, Morocco, Saudi Arabia and Ukraine are potentially relevant partner countries in the German and European hydrogen strategies based, among other things, on good economic conditions for producing and transporting hydrogen. Russia's war in Ukraine has raised doubts about the formerly assumed certainty that close trade relations mean a stable energy supply. This assumption was largely based on the belief that trade always induces mutual dependence, and that any doubt or uncertainty would be smaller for the economically strong West than for its partners, for whom the inflow of foreign currency from trading commodities was considered essential. Even at the height of the Cold War, for instance, the USSR continued to provide a reliable supply of raw materials to the West. The longer such trade relations ran, the more the partners involved were perceived as reliable [18, 19].

In order to strengthen the resilience of the energy system, key strategic assumptions must be critically reviewed. The future topic of hydrogen is particularly affected by this. The strategic decisions pending here will have long-term economic and political impacts, particularly since the agreements and infrastructure investments made today create path dependencies. Clear criteria should guide the evaluation of potential supplier countries and greater weighting should be assigned to political risks. In the past, technical availability and price as well as political stability played the dominant role when selecting energy suppliers. Russia's war against Ukraine highlights the importance of supply sovereignty and of the careful evaluation of possible partners—in addition to their reliability, their systemic and political resilience should be considered to a greater extent. Geopolitical considerations should also play a more important role in hydrogen strategies, for example, for the EU as a counterweight to China within the MENA region.

Finally, value-based criteria should be given greater attention. Assessments should include the implications of a commodity partnership for fostering the rule of law and democracy, sustainable development, as well as promoting peace and the protection of human rights. In the past, commodity partnerships with autocratic countries often mainly benefited small elite groups close to the country's ruler, who were able to cement their status using the inflow of foreign currency. The wider population, on the other hand, were often beneficiaries of trade revenues to a very limited extent, and, in some cases, even had to fear negative impacts, such as environmental destruction and distributional conflicts. Numerous recognized criteria already exist in the field of development cooperation that are used to assess whether a state is a reliable partner worthy of support. These aspects should also be integrated into the search for suitable cooperation partners in the energy sector. In this respect, the future topic of hydrogen offers a major opportunity to put trading relations on a new strategic footing that considers national interests and is guided by values. Governments need to be aware that this will inevitably lead to conflicting and contradictory goals but a systematic and transparent assessment can provide a valuable foundation for political decision-making [18, 19].

The recent energy crisis due to the war between Russia and Ukraine has underscored the significance of diversifying the suppliers of energy imports and that this will play an important role in the future as well. Diversification leads to higher import costs and takes more time to develop the respective supply systems, but reduces the economic risk by lowering dependency. Hydrogen liquefaction and its transport by ship raise the overall costs (Chap. 2) and the construction of production and transport capacities requires time—but these factors contribute to greater diversification and protection against too few suppliers exercising too much market power Wietschel et al. [18].

5 International Infrastructure Must Be Established to Transport Hydrogen

For larger quantities of hydrogen, transportation options via pipeline and ship are currently being examined. Hydrogen is gaseous at ambient pressure and temperature. To transport it by pipeline, it is first compressed to over 30 bar [20], depending on pipeline distance, recompression necessity and quantities. To transport hydrogen by ship, it is either liquefied (LH_2) or chemically converted into hydrogen-based powerfuels like ammonia (NH_3), liquid organic hydrogen carriers (LOHC), synthetic natural gas (SNG), synthetic methanol (MeOH) or into synthetic Fischer–Tropsch (FT) fuels [21]. The Fischer–Tropsch fuels are also often referred to as Power-to-Liquids (PtL) fuels, synfuels or e-fuels, which include kerosene, diesel and naphtha [22].

The transport costs for hydrogen (products) depend on different factors, such as:

- the transport pathway [gaseous H_2, liquid H_2, liquid NH_3, methanol (CH_3OH, MeOH), FT-SynFuels, LOHC],
- the product quantity to be transported and
- the transport distance Wietschel et al. [21].

Recently published studies have examined the costs of transporting hydrogen and other powerfuels (e.g., [20, 23–25]). These studies show that the distance for transporting larger hydrogen quantities (7–85 TWh_{H2}/year.) is decisive for the transport pathway and cost range:

I. For distances <3,500 km: pipeline transport (new and repurposed) is the most cost-effective option (max. 10–20 €/MWh_{H2}).
II. For distances from 3,500 to 7,700 km: pipeline transport (repurposed) is the most cost-effective option (10–20 €/MWh_{H2}).
III. For distances >10,000 km: shipping is the most cost-effective option (20–50 €/MWh_{H2}).

Pipelines are the most cost-effective option for an average transport distance of up to 5,000 km including both new and repurposed pipelines [21]. The further the transportation distance, the more compressor stations are needed to compensate for the pressure loss within the pipeline [26]. As a result, pipeline transportation costs depend strongly on the distance involved. The costs of transportation using repurposed pipelines have been estimated as half those of newly built pipelines [27]. Up to a transport distance of 7,000–8,000 km, repurposed pipelines have the lowest costs of all transportation pathways, calculated as 20 €/MWh_{H2} in the studies mentioned above.

Effective and secure long-term transportation of hydrogen in repurposed natural gas pipelines has not yet been technically proven [28]. As a result, the operational costs for maintenance and monitoring are still unclear. In Germany, for example, injecting up to 10 vol.% hydrogen into the natural gas networks is permitted under German regulations [22]. Higher injection rates are currently being tested in pilot projects to study the material behavior of the gas pipelines and feasible purification technologies [28].

In order to ensure natural gas pipelines are fully hydrogen-ready, the existing compressors and valves need to be replaced. Furthermore, each natural gas pipeline needs to be examined individually, considering the material used and its operational history in order to assess the hydrogen embrittlement risk and to avoid hydrogen leakage. More knowledge is required about the technical aspects of natural gas pipeline repurposing [18, 19].

Newly built pipelines have the second lowest hydrogen transportation costs (15 €/ MWh$_{H2}$) for distances up to 3,500 km and capacities over 50 TWh/year. [25, 27]. Even though the investments required for repurposing existing natural gas pipelines are half those for building new hydrogen pipelines [27], it may still be necessary to construct new pipelines if infrastructure does not exist for either natural gas or electricity. Germany has been operating pipelines to transport fossil hydrogen for many years and therefore has the know-how to construct and operate hydrogen pipelines [29, 30]. In addition, connecting countries via new pipelines can help to establish strong energy partnerships. However, the construction of new pipelines requires significant investments, typically ranging from 2.5 to 4.0 M€/km [20, 31, 32]. The uncertain market environment is a barrier to potential investors, and there may be challenges related to social acceptance, such as Not In My Backyard (NIMBY) concerns. Additionally, lengthy planning and approval processes can inhibit the development of pipeline infrastructure.

Given their large renewable energy potentials at lower cost, it is obvious that especially the MENA region as well as Ukraine, Russia, Kazakhstan, Turkey and Norway are within the economically viable pipeline distance and therefore potential candidates for importing hydrogen via pipeline into the EU [33]. Nevertheless, the financing costs for hydrogen imports are uncertain for countries that are politically unstable and geopolitically significant.

For hydrogen transportation over long distances (>7,700 km), shipping is the most cost-effective pathway with costs between 20 and 50 €/MWh$_{LH2}$ [20, 23–25]. Since the conversion and reconversion of hydrogen into liquid H$_2$, NH$_3$, LOHC and methanol (MeOH) are the main cost driver of hydrogen transportation, distance becomes less relevant. Thus, the costs for transporting liquid H$_2$ and hydrogen carriers by ship are less dependent on the distance covered. For the EU, many potential export countries like Chile, Australia, South Africa, or Namibia, are beyond the range of an economically viable pipeline transport, so that shipping becomes necessary and cost-effective [33]. Shipping has higher costs compared to pipeline transport over distances smaller than 7,500 km, but is an important addition, as it contributes to supplier diversification and thus to greater security of supply. In general, shipping allows greater flexibility in transport routes to avoid politically insecure regions and flexibly adapt destinations to market price dynamics [20]. Hence, shipping is of greater interest in an established hydrogen market.

Different technical solutions already exist to transport energy carriers and chemicals by ship, such as gas tankers, crude oil tankers, bulk chemicals tankers and container ships [34].

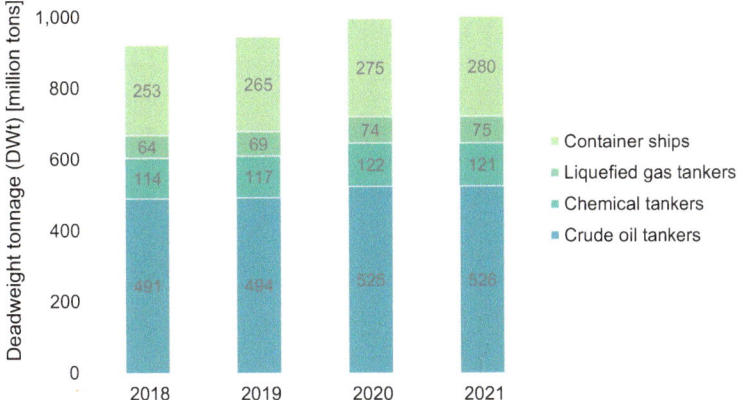

Fig. 2 Development of the world trade fleet by ship type between 2018 and 2021 (Data: [36])

- Gas tankers have the technical advantage to transport compressed gases as liquids, such as LNG, LPG or NH_3, under constant pressure and temperature.
- Oil tankers are characterized by their ability to transport large capacities between 60,000 DWt^2 (Panamax) and more than 320,000 DWt (ultra large crude carrier, ULCC) [34].
- The configurations of tankers for bulk chemicals allow the transportation of high standard, corrosive or hazardous liquids in bulk, such as nutritional oil, lubricants or methanol [35].
- Container ships transport all kinds of manufactured goods in standardized containers.

Figure 2 shows the development of the world trade fleet by ship between 2018 and 2021.

Roughly 1,000 Mt of goods were shipped in 2021 including cargo, fuel, water, food and personnel. Among the above named ship types, crude oil tankers shipped more than 50% and gas tankers 7% of the traded goods [36]. In order to reach climate neutrality by 2045, the goal is to drastically reduce the use of fossil gases and crude oil [37, 38]. Therefore, the transportation of crude oil and fossil gases is expected to decrease gradually until 2045. Based on the average ship lifetime of 30 years [39], it is assumed there will be sufficient capacities to transport Fischer–Tropsch products and liquefied gases for the decades to come. To understand which ship type is suited to which energy carrier, it is necessary to examine the technical characteristics of liquid H_2, liquid NH_3, methanol (MeOH), FT-SynFuels and LOHC.

- Hydrogen has a significantly higher energy density when liquefied and requires less space for transportation in a liquid state. To liquefy hydrogen, it needs to be

---

[2] The maximum loading capacity is expressed as deadweight tonnage DWt.

cooled down to -253 °C [40]. Therefore, to transport hydrogen by ship, liquefaction plants have to be built and operated that require high energy input, leading to significant investments and high operational costs dominated by energy costs. Although liquid H_2 is more difficult to transport and store than (liquid) ammonia and LOHC, it does not require complex reconversion steps. So far, only a small 90 t prototype ship has been built and put into operation [41]. However, no relevant infrastructure exists for shipping liquid H_2, neither liquefaction plants, ships at scale, nor harbor infrastructure including regasification plants, storage and hydrogen pipelines. Ships for larger transport volumes must still be developed and built. Therefore, international shipping of liquid H_2 is not expected before 2030 [41].

- NH_3 production is a well-known process, where hydrogen and nitrogen react in the Haber-Bosch process to form ammonia [24]. The reconversion, however, is currently still being developed [40] and is not expected to be feasible on a large scale before 2030. Transport infrastructures are globally established for fossil fuel-based liquid NH_3, where NH_3 is liquefied at -33 °C and shipped via gas tankers or, especially in the USA, pipelines [42]. Due to a growing global population and the rising demand for fertilizers, NH_3 demand is likely to increase in the future [38]. Therefore, it is expected that additional NH_3 transport capacities will be required to provide sufficient NH_3 quantities for energy and nutrition purposes. In parallel, to reach climate neutrality in 2045, the demand for LNG and LPG will decrease continuously [37, 43] and thus, LNG and LPG tankers might become available to transport NH_3. It is therefore necessary to whether LNG and LPG ships can be made NH_3-ready.

- Liquid organic hydrogen carriers (LOHCs) are hydrocarbons (e.g., benzyl toluene) that are hydrogenated to store hydrogen and later dehydrogenated to release the hydrogen [44]. This process is proven at industrial scale and is economically interesting if the heat generated during the conversion can be integrated and used to compensate the heat required for the reconversion. As, e.g., the LOHC benzyl toluene is an oil, it can be transported in chemical tankers or potentially even in container or crude oil ships.

- Natural gas is currently used for heating purposes, power generation and production of fossil fuel-based hydrogen. Transporting hydrogen via SNG is a controversial subject as this can be both an enabler and an inhibitor of a hydrogen ramp-up [21]. On the one hand, abundant infrastructures are already established for natural gas which could be used immediately to transport SNG (e.g., pipelines, LNG tanker, LNG terminals), and assets that currently run on natural gas could continue to be operated in the future. On the other hand, continuing the use of natural gas pipelines to transport SNG hinders their repurposing into hydrogen pipelines and running or investing into SNG infrastructures might lead to be locked in investments. In addition, a climate-neutral carbon source is needed to produce SNG, competing with carbon sources for FT-SynFuels.

- Methanol is synthesized from hydrogen and a carbon source and is already a widely used bulk chemical in the chemical industry. At present, it is globally traded and transported in chemical tankers. Methanol demand will increase significantly in the future (Big 5, IRENA shipping, [24]), as it can be used as a feedstock for the production of synthetic high value chemicals (HVC) and synthetic fuels (diesel, kerosene and naphtha) as well as being used directly as a transportation fuel. Therefore, the existing transport capacities for methanol are likely to be extended in the future. In more recent studies, methanol is no longer considered as a means for transporting hydrogen because it has sufficient potential applications of its own and its reconversion is generally cost-intensive.

- Hydrogen and a carbon source can be synthesized within the FT-process to a FT-crude and then further processed to FT-SynFuels, particularly synthetic diesel, kerosene and naphtha [22]. These synthetic oils can replace their fossil equivalents, which implies that the existing infrastructure for crude oil and oil products, such as oil pipelines and crude oil tankers, can continue to be used for synthetic diesel, kerosene and naphtha. FT-crude and methanol are competing feedstocks for the production of synthetic fuels [22, 45]. From an infrastructure perspective, FT-crude is preferred, as it has a very similar chemical composition to crude oil. Moreover, FT-crude has a higher energy density compared to methanol, requiring lower transport capacities.

To achieve climate targets, the ships used to transport hydrogen and hydrogen-based powerfuels must be greenhouse gas-neutral. It is expected that fossil fuels will be predominantly used for ship propulsion by 2030. The International Maritime organization targets to reduce the total annual GHG emissions from international shipping by at least 50% by 2050 compared to 2008 [46]. It is currently challenging to use liquid H_2, NH_3 and LOHC as fuel for cargo ships. Using liquid H_2 and NH_3 as propulsion fuel requires sufficient capacities for a high-weight fuel cell on board. NH_3 could be used with a combustion engine, which is currently still under development, but the combustion of NH_3 generates NO_x emissions. There are plans to launch the first commercial propulsion system for LOHC in 2025 [44].

Furthermore, terminals have to be established to load and unload ships carrying liquid H_2, NH_3 and LOHC. When repurposing existing LNG terminals, the insulation plays a key role for the use of liquid H_2 and liquid NH_3 [40]. Hydrogen liquefies at $-253\,°C$, natural gas at $-161\,°C$ and NH_3 at $-33\,°C$. As NH_3 liquefies at milder temperatures compared to natural gas, LNG terminals could potentially be repurposed into liquid NH_3 terminals. Repurposing LNG terminals into liquid hydrogen terminals would require more electrical power and higher insulation. Loading and unloading methanol, FT-crude and LOHC are not expected to present a technical challenge.

Even though pipelines offer the most cost-effective transport of hydrogen over distances up to 7,500 km, developing hydrogen pipeline infrastructure is capital intensive and therefore represents a barrier to potential investors. Liquefied ammonia (NH_3), natural gas (NG), crude oil and crude oil products as well as fossil-based methanol are already transported over long distances by ship or pipeline so that infrastructures already exist for their synthetic counterparts. Therefore, it is expected that importing hydrogen will take place mainly via hydrogen-based powerfuels in the next few years and no significant hydrogen transport by pipeline is expected before

2030. Potential investors need planning certainty to realize the development of a hydrogen pipeline infrastructure and the associated large-scale import of hydrogen at transportation costs of less than 20 €/MWh$_{H2}$. Hydrogen transportation infrastructure is a crucial puzzle to solve before a successful hydrogen ramp-up can take place [47].

6 Investment Risks Can Be a Significant Obstacle

Particularly in developing and emerging countries, there may be certain risks that can hinder possible investments in the construction of plants for hydrogen production. The DESERTEC project is a prominent example where such risks have hindered the realization of energy import projects. DESERTEC was an initiative aiming to generate green electricity in energy-rich locations, such as the MENA region, and, in addition to meeting domestic demand, to export electricity via high-voltage direct current cables by constructing transnational power grids. However, the goal of meeting a significant portion of Europe's electricity supply in this way has since been abandoned. Some studies explored the failure of the project [48–50]. The challenges identified in this context can be differentiated into:

- planning uncertainty due to political events (civil wars, etc.),
- fear of terrorist attacks (power lines, plants),
- breach of contract due to political changes (change of power within MENA countries) and
- more difficult communication with MENA countries due to political uncertainties.

The production and utilization of hydrogen requires substantial investments in different types of infrastructure, particularly in plants to generate electricity from renewable sources of energy, water electrolysis plants, as well as conversion and application technologies. For private equity investors, the expected return is decisive, while for the provision of debt capital, the default risk is relevant [51]. High risks raise the costs of financing [52] which can significantly increase the specific production costs.

Investments in hydrogen generation are exposed to a variety of different risks and uncertainties in the same way as investments in renewable energy generation [53].

- Risks at the macro-level include financial sector development, political stability, sovereignty and economic stability.
- Risks at the meso-level include the reliability of energy policy, risks resulting from policy instruments, experiences of the financial and energy sector with the respective technology, societal acceptance.
- At the micro-level, there are risks that pertain to specific aspects of the project, such as technology and yield risks, financial structuring and conditions of the project.

In general, investors in energy projects in emerging and developing countries have to raise more expensive equity and debt capital, and thus have to pay more than for similar projects in industrialized countries. The difference in the cost of capital of renewable energy investments is mainly driven by macro-level risks that could be compensated by adapting the structure of the projects [53].

An overview from the International Energy Agency (IEA) on the cost of capital is intended to create greater transparency. It compares and monitors the average cost of capital after taxes for five countries: Brazil, India, Indonesia, Mexico and South Africa, taking into account the local currency [54]. The overview shows that, for a solar energy project with 100 MW of installed capacity, Brazil had the highest cost of capital in the year 2021 at 13%. This was lower in South Africa (10.25%), followed by Indonesia (10%), India and Mexico (both 9.75%). Compared to 2019, the cost of capital had even risen slightly in some countries, while investment in renewable energies should be increasing rapidly. In comparison, the cost of capital for larger plants in industrialized countries is often in the low single-digit percentage range—in Germany, the cost of capital was just 1.7% for solar energy projects in the year 2017. These huge differences are due to macro-economic effects, such as economic and political risks and sovereignty issues. In such cases, guarantees or interim financing of the project by international companies might have a mitigating effect on the cost of capital. Such project structuring might even help to reduce the levelized cost of hydrogen (LCOH) and thus increase the potential import volume of hydrogen.

A benchmark that is often used for this purpose is the weighted average cost of capital (WACC), which was originally a company-specific measure, but can be applied to project appraisals as well.

As the risk exposure of projects also depends on the individual countries, an evaluation based on Damodaran [55] can be used. This approach allows the estimation of a risk premium, the Equity Risk Premium (ERP), which represents the risks in the equity market and can be determined for all countries included in Moody's Rating as a standard for credit risk ratings [56].

Secure demand (guaranteed purchase of electricity) created by various bundles of policy instruments, which influenced the risks involved and thus the financing costs, was decisive for the expansion of renewable energies and thus also for investments and financing [57]. If policy measures such as premiums, quotas or certificates in the countries importing hydrogen generated a similarly reliable demand as has been partly accomplished in the expansion of renewable energies, this would reduce the market risk and make the market more attractive to investors. If, in addition, corresponding guarantees from the importing countries reduced the existing country or counterparty risk in the exporting countries, and if development banks served the capital market as enablers, investment risks could be reduced and the supply of hydrogen increased, so that the market as a whole could grow significantly [58].

7 There is a High Degree of Uncertainty Regarding Future Price Developments

As shown in Chap. 2, a number of studies have analyzed the economic viability and import potentials for hydrogen and hydrogen-derived synthesized fuels but do not include statements about market and price development. Typically, they are limited to the analysis of production and transportation costs. However, market prices are decisive for any realistic estimation of future development. These are based on the marginal costs of production including transport plus markups for profits, risks, sales, warranties and R&D costs. A government can also influence prices through taxes and/or levies. Especially at the beginning of market take-off as is the case for an international market for hydrogen and other powerfuels, countries will take different approaches to stimulating supply and demand. The strategic behavior of market players, price agreements and lack of competition also have a decisive influence on the price-finding and adjustment procedure, as is the case with today's oil and gas prices. Scarcity pricing can occur when demand is high but supply is low (as has been the case at times for crude oil), or pricing can be based on the prices of other energy sources (indexation, such as for natural gas). It is also relevant whether different market and price regions have emerged, as for natural gas, or whether a quasi-global market exists, such as for crude oil, although, even here, there are also product differentiations and different contracts. Given such possible developments, it is conceivable that analyses employing approaches that only refer to production costs are potentially underestimating the actual development and volatility of prices [21]. Figure 2 illustrates price formation in perfect markets and shows the possible factors influencing price.

Current uncertainty about price developments is a major barrier to investment in powerfuels. To mitigate these uncertainties, their reasons and origin need to be addressed. Based on the above outline, they are grouped into two main categories—uncertainty about production costs and uncertainty about market structures.

Uncertainty about production costs. The uncertainty regarding production costs has its roots in technical risks, yield risks and uncertain input costs or prices (e.g., for energy). Uncertainty about input prices depends on their general availability and demand. These factors increase investment and operational expenditure as well as the costs of capital.

Unclear market structures. In a fully liberalized and functioning market, demand and supply determine the price, see Fig. 3; the market acts as a price-setting mechanism and the resulting price is called the market price. In contrast, on a regulated market, prices are set "exogenously" based on normative aspects and plans by an authority. This procedure is often referred to as administrative price setting. Following characteristics of markets exist:

- Market with a price-setting mechanism under full competition. If information about quantities and prices is transparent and many actors participate in both the

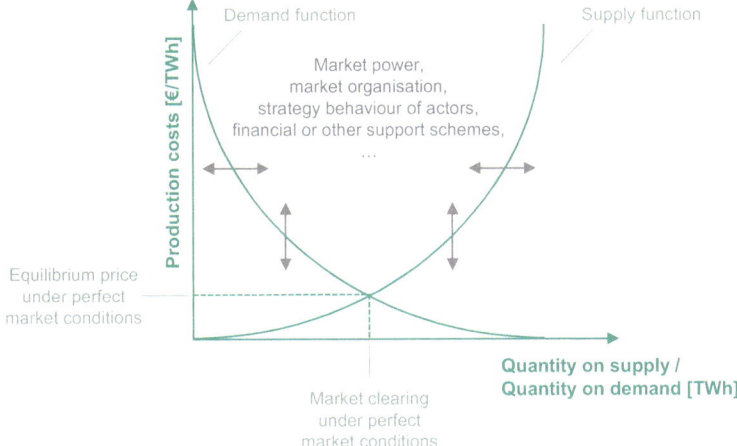

Fig. 3 Formation of prices in markets

supply and the demand side of the market, there is full competition. In such a case, the resulting market price is close to the minimum of the average costs per unit [e.g., per unit of hydrogen and here the levelized cost of hydrogen (LCOH)]. Taxes on production or products or grants might shift this price up or down, but even with taxes or grants, the price mechanism is unchanged. The minimum of average costs account for grants or taxes. In such a market setting, uncertainty arises, if the amount of hydrogen supplied and demanded is unknown or at least uncertain and decisions regarding hydrogen production depend on hydrogen demand and vice versa (coordination problem). Investors can easily enter and exit the market if production takes place on a small scale. However, in the case of investments in the production and industrial usage of hydrogen, substantial upfront capital is required, which entails a considerable level of risk from an investor's perspective. In combination with an uncertain environment entailing high cost of capital, large investments are at a comparative disadvantage to low upfront capital investments, as the present value of their revenues is low [59]. This acts like a market barrier, and both suppliers and industrial demanders are reluctant to invest. To overcome this coordination failure and avoid a lack of investments, the state has to interfere in the market and mitigate this barrier as in the case of renewables.

- Market price mechanism under imperfect competition. In situations where the number of market participants is limited, there is a lack of transparent information, or a natural monopoly exists (due to high infrastructure costs), competition is imperfect and the market price-setting mechanism results in a price that is significantly higher than the minimum average costs of production (here: hydrogen production), because the price is set at a level, at which marginal revenue equals marginal prices. The resulting price is above that of a competitive market. Such a situation opens the door to bilateral agreements, for example, with price indexing, or even gaming strategies that might lead to huge deviations of the agreed price from the minimum of average costs, while at the same time the quantity offered is smaller than under full competition. To avoid this market failure and reduce extreme price distortions from the competitive optimum, governments have to interfere and set a regulatory framework as in the case of district heating, for example.

- Regulated market. In regulated markets, governments control and regulate the prices, supply and demand of the respective good (here: hydrogen and other power-based fuels) to varying degrees. In a very strictly regulated market, prices are set administratively by the government. In a less regulated market, the government establishes a price-setting mechanism, for example, tenders for production plants or networks and ensures transparent prices and supply characteristics. In addition to price-setting mechanisms and controls, market access of third parties to production sites or networks might be elements of regulation as in the case of district heating [60]. Furthermore, clear product and production standards that are certified and controlled by authorized and trustworthy organizations can be used to increase transparency.

8 The Hydrogen Economy Must Be Integrated into the Overarching Governance Structures for Transforming the Energy System

Generating electricity to produce hydrogen in Europe and in developing countries is linked with environmental impacts that are enhanced by the low efficiencies across the entire chain from producing to using the hydrogen and/or other powerfuels. Although these impacts are not easy to quantify and have to be taken into account in renewable power generation, they provide arguments for a hierarchy principle that minimizes the renewable electricity capacities needed—even if costs continue to decrease. This implies that a hydrogen economy must be integrated into the general governance structures for transforming the energy system via a hierarchical principle. This hierarchical principle should—analogous to the energy efficiency first principle—minimize the cost of hydrogen production, even if the costs for electricity from renewable sources of energy, electrolysis and the production of synthetic energy carriers continue to decrease. This principle applies to the entire system of supplying and consuming energy, i.e., it covers both producing and consuming countries, and comprises the following four stages (seen [58]):

- The "energy efficiency first" principle recently introduced into European policy must be a strong guiding principle when expanding energy supply and therefore the RE capacities in a country. This refers to both PtX import and export regions. This also applies to the direct use of renewable energies to decarbonize the power sector and the demand sectors as well as to RE capacities used to produce the hydrogen itself.

- The second stage in the hierarchical principle implies priority given to renewable energies in the continued expansion of the electricity sector, i.e., the principle of avoidance of emissions through switching to renewable energies. Fossil fuels should be phased out as quickly as possible to make room for clean power generation. This second stage is especially relevant for potential hydrogen-producing countries that have not yet defossilized their own power sector.
- The third stage of the hierarchical principle gives priority to alternative energy carriers based on renewable energy sources. These alternative energy carriers entail no or small changes in habits and processes when used, and with lower environmental impacts than fossil fuels and higher transformation efficiency than hydrogen. They include the direct use of electricity, in particular, and the use of sustainable biomass, biofuels and biogas as well. However, when using the latter, one should account for their limited availability.
- Applications where none of the above three stages apply rely on hydrogen and synthesis products. This still opens up a global market of 100,000–700,000 M€ for the hydrogen economy.

This four-stage principle, which can be regarded as an extension to the "energy efficiency first" principle, should be implemented in the governance structures of the energy system transformation in those countries that consume and/or produce hydrogen. It is an important prerequisite for sustainable development.

9 Local Economic Competencies Are Required to Successfully Produce Hydrogen and Its Synthesis Products

A global hydrogen economy requires the development of new and the enhancement of existing local technical and economic competencies. This is the only way to ensure that a globally networked green hydrogen economy not only contributes to the global energy transition, but also to sustainable and innovative economic development in the countries producing the hydrogen. This is a major driver for establishing new energy technologies, especially in emerging economies and developing countries. In addition, creating jobs and expanding local value creation potentials in sustainable sectors are key elements of national development strategies [58]. Local hydrogen value creation concerns the following elements:

- electricity generation from renewable sources of energy (e.g., based on wind or solar energy),
- hydrogen production (e.g., electrolyzers),
- hydrogen conversion (compression, liquefaction),
- hydrogen storage,
- hydrogen transport (e.g., by pipeline, ship) and
- hydrogen application (e.g., ammonia production, iron and steel production).

CO_2 extraction and conversion into powerfuels are additional elements for the production of hydrogen-based powerfuels.

Assuming an ambitious scenario, developing a hydrogen and fuel cell economy within the EU could have economic effects of up to 35,000 M€ and create 38,500 direct jobs and more than 70,000 indirect ones (i.e., full-time equivalents) by 2030 [61]. By 2050, the hydrogen economy could reach a volume ranging from 100,000 M€ worldwide, which is equivalent to today's steel market, up to almost 700,000 M€, about one third of today's oil market. The size of the market depends on where alternatives are available in the form of direct electricity use, such as freight transport [62]. However, as is the case with renewable energy expansion, other technologies or energy sources might become redundant and be abandoned, and workers in these sectors might lose their jobs [63]. The shrinking demand for fossil fuels might be replaced by synthetic hydrogen-based fuels and imply lower employment in one region of the world and higher employment in another. However, since the value chain of synthetic fuels requires more (sophisticated) processes, a net increase in (qualified) jobs can be expected at the global level. The global value chains for hydrogen are still in their early stages, with only a few active players so far. In this context, there is significant potential for local value creation in those countries that could produce hydrogen. Currently, numerous emerging economies and developing countries are not only working on strategies to use different hydrogen application technologies in the transport sector or industry, for example, but are also actively assessing the possibilities for local value creation by establishing a local hydrogen economy [27, 58, 64].

References

1. Monopolkommission (2021) Energie 2021: Wettbewerbschancen bei Strombörsen, E-Ladesäulen und Wasserstoff nutzen 8. Sektorgutachten der Monopolkommission gemäß § 62 EnWG 2021
2. Riemer M, Zheng L, Pieton N, Eckstein J, Kunze R, Wietschel M (2022) Future hydrogen demand: a cross-sectoral, multiregional meta-analysis. In: HYPAT Working Paper 04/2022. Fraunhofer ISI, Karlsruhe
3. Wietschel M, Zheng L, Arens M, Hebling C, Ranzmeyer O, Schaadt A, Hank C et al (2021) Metastudie Wasserstoff: Auswertung von Energiesystemstudien. Studie im Auftrag des Nationalen Wasserstoffrats. Fraunhofer ISI, Karlsruhe
4. European Commission (2020) A hydrogen strategy for a climate-neutral Europe, Communication from the Commission to the European Parliament, the Council, the European Economic and Social Committee and the Committee of the Regions. European Commission, Brussels
5. European Commission (2022) REPowerEUPlant. Communication from the Commission to the European Parliament, the Council, the European Economic and Social Committee and the Committee of the Regions. European Commission, Brussels
6. Deutsch M, Maier U, Perner J, Unteutsch M, Lövenich A (2018) Die zukünftigen Kosten strombasierter synthetischer Brennstoffe. Agora Energiewende/Frontier Economics, Berlin
7. Hampp J, Düren M, Brown T (2021) Import options for chemical energy carriers from renewable sources to Germany

8. Lux B, Gegenheimer J, Franke K, Sensfuß F, Pfluger B (2021) Supply curves of electricity-based gaseous fuels in the MENA region. Comput Ind Eng 12:107647. https://doi.org/10.1016/j.cie.2021.107647

9. Pfennig M, Gerhardt N, Pape C, Böttger D (2017) Mittel- und langfristige Potenziale von PtL- und H2-Importen aus internationalen Vorzugsregionen. Teilbericht im Rahmen des Projektes: Klimawirksamkeit Elektromobilität—Entwicklungsoptionen des Straßenverkehrs unter Berücksichtigung der Rückkopplung des Energieversorgungssystems in Hinblick auf mittel- und langfristige Klimaziele. Teilbericht. Fraunhofer-Institut für Windenergie und Energiesystemtechnik IWES, Kassel

10. Prognos (2020) Kosten und Transformationspfade für strombasierte Energieträger. https://www.bmwi.de/Redaktion/DE/Downloads/Studien/transformationspfade-fuerstrombasierte-energietraeger.pdf?blob=publicationFile. Accessed 11 Oct 2020

11. Timmerberg S, Kaltschmitt M (2019) Hydrogen from renewables: supply from North Africa to Central Europe as blend in existing pipelines—potentials and costs. Appl Energy 237:798–809

12. Pfennig M, von Bonin M, Gerhardt N (2021) PTX-Atlas: Weltweite Potenziale für die Erzeugung von grünem Wassrstoff und kliamneutralen synthetsichen Kraft- und Brennstoffen. Teilbericht im Rahmen des Projektes: DeV-KopSys. Fraunhofer-Institut für Energiewirtschaft und Energiesystemtechnik (Fraunhofer IEE)

13. GIZ (ed) (2022) Potenziale für die Produktion von grünem Wasserstoff -Übersicht über Beurteilungskriterien aus entwicklungspolitischer Sicht und Abgleich mit aktuellen Potenzialatlanten. Deutsche Gesellschaft fürInternationale Zusammenarbeit (GIZ) GmbH

14. Heinemann C, Mendelevitch R (2021) Sustainability dimensions of imported hydrogen

15. Nationaler Wasserstoffrat (2021) Nachhaltigkeitskriterien für Importprojekte von erneuerbarem Wasserstoff und PtX-Produkten

16. Thomann J, Edenhofer L, Hank C, Lorych L, Marscheider-Weidemann F, Stamm A, Thiel Z, Weise F (2022) Background paper on sustainable green hydrogen and synthesis products. In: HYPAT working paper 01/2022. Fraunhofer ISI, Karlsruhe

17. Sachverständigenrat für Umweltfragen (Hrsg.) (2021) Wasserstoff im Klimaschutz: Klasse statt Masse. Stellungnahme

18. Wietschel M et al (2022) War in Ukraine: Implications for the European and German strategies for importing hydrogen and synthesis product. In: HYPAT policy paper. Fraunhofer ISI, Karlsruhe

19. Wietschel M, Ragwitz M, Stamm A, Marscheider-Weidemann Frank, Löschel A (2022) Import von Wasserstoff und Syntheseprodukten in einer veränderten Welt. Hg. v. HYPAT. Berliner Dialogforum. https://hypat.de/hypat-wAssets/docs/new/publikationen/Berliner-Dialogforum_Import-von-Wasserstoff-und-Syntheseprodukten-in-einer-veraenderten-Welt_20-Juni-2022.pdf

20. Wang A, Jens J, Mavins D, Moultak M, Schimmel M, van der Leun K (2021) European hydrogen backbone. Analysing future demand, supply, and transport of hydrogen. Creos, DESFA, Elering, Enagás, Energinet, Eustream, FGSZ, Fluxys Belgium, Gas Connect. https://gasforclimate2050.eu/wp-content/uploads/2021/06/EHB_Analysing-the-future-demand-supplyand-transport-of-hydrogen_June-2021_v3.pdf. Accessed 02 Oct 2021

21. Wietschel M et al (2021) Importing hydrogen and hydrogen derivatives: from costs to prices. In: HYPAT working paper 01/2021. Fraunhofer ISI, Karlsruhe

22. Arndt C, Neuling U, Vorsatz M, Prause J (2021) Konzeptionelle und technische Ausgestaltung einer Entwicklungsplattform für Power-to-Liquid-Kraftstoffe. Abschlussbericht. Hg. v. DLR, Griesemann Gruppe, Technische Universität Hamburg und UE. https://www.now-gmbh.de/wp-content/uploads/2021/08/EPP_Abschlussbericht.pdf. Accessed 01 March 2022

23. European Commission (2021) Assessment of hydrogen delivery options. In: Science for policy briefs. https://ec.europa.eu/jrc/sites/default/files/jrc124206_assessment_of_hydrogen_delivery_options.pdf. Accessed 03 Oct 2021

24. IEA (2021) Global hydrogen review 2021. International Energy Agency. https://iea.blob.core.windows.net/assets/e57fd1ee-aac7-494d-a351-f2a4024909b4/GlobalHydrogenReview2021.pdf. Accessed 05 Oct 2021

25. IRENA (2022) Global hydrogen trend to meet the 1.5°C climate goal. Part II. Technology review of hydrogen carriers
26. Wang A, van der Leun K, Peters D, Buseman M (2020) European hydrogen backbone. How a dedicated hydrogen infrastructure can be created. https://gasforclimate2050.eu/wp-content/uploads/2020/07/2020_European-Hydrogen-Backbone_Report.pdf
27. AHP (2020) African hydrogen partnership. https://www.afr-h2-p.com/. Accessed 12 Nov 2020
28. IEA (2019) The future of hydrogen. seizing today's opportunities. https://iea.blob.core.windows.net/assets/8ab96d80-f2a5-4714-8eb5-7d3c157599a4/English-Future-Hydrogen-ES.pdf. Accessed 07 Aug 2023
29. Air Liquide (2016) *Wasserstoffanlagen*. Technische Gase von Air Liquide in Deutschland. https://industrie.airliquide.de/wasserstoffanlagen. Accessed 06 May 2021
30. ECSPP (2023) Chemical pipeline networks and infrastructure in Europe. https://chemicalparks.eu/europe/pipeline-networks. Accessed 07 Aug 2023
31. André J, Auray S, Brac J, Wolf D, Maisonnier G, Ould-Sidi MM, Simonnet A (2013) Design and dimensioning of hydrogen transmission pipeline networks. Eur J Oper Res 229(1):239–251. https://doi.org/10.1016/j.ejor.2013.02.036
32. van Gerwen R, Eijgleaar M, Bosma T (2019) Hydrogen in the electricity value chain. In: Group technology and research position paper 2019. Hg. v. DNVGL. https://www.dnv.com/Publications/hydrogen-in-the-electricity-value-chain-225850. Accessed 07 Aug 2023
33. Breitschopf B, Thomann J, Garcia JF, Kleinschmitt C, Hettesheimer T, Neuner F et al (2022) Import von Wasserstoff und Wasserstoffderivaten. Exportländer. Hg. v. HYPAT. ISI; IEG; RUB; DIE; GIZ (HYPAT Working Paper, 02). https://www.hypat.de/hypat-wAssets/docs/new/publikationen/HyPAT_Working_Paper_02-2022_Import_Wasserstoff_und_Derivate_Exportlaender.pdf. Accessed 22 Aug 2022
34. Marinekommando (Hg.) (2021) Maritime Abhängigkeit Deutschlands. Jahresbericht 2021. https://www.bundeswehr-journal.de/2020/maritime-abhaengigkeit-deutschlands-jahresbericht-2020/. Accessed 29 Oct 2022
35. Applications (1997) Duplex stainless steels. In: Gunn RN (ed) Microstructure, properties and applications; based … on the keynote papers from the Glasgow and Beaune conferences. Abington, Cambridge, pp 175–186. https://www.sciencedirect.com/science/article/pii/B9781855733183500177
36. Marinekommando (Hg.) (2020) Maritime Abhängigkeit Deutschlands. Jahresbericht 2020. https://www.bundeswehr-journal.de/2020/maritime-abhaengigkeit-deutschlands-jahresbericht-2020/. Accessed 29 Oct 2022
37. ARIADNE (2022) Szenarien zur Klimaneutralität: Vergleich der Big 5-Studien. Datenanhang, Version 1.0. https://ariadneprojekt.de/news/big5-szenarienvergleich/. Accessed 15 Sept 2022
38. IEA (2021) Ammonia technology roadmap. https://www.iea.org/reports/ammonia-technology-roadmap. Accessed 31 Oct 2022
39. Marinekommando (Hg.) (2019) Maritime Abhängigkeit Deutschlands. Jahresbericht 2019. https://www.bundeswehr.de/resource/blob/156014/fa1039c05301b9c63ad642c683880778/jahresbericht-marinekommando-2019-data.pdf. Accessed 29 Oct 2022
40. FfE (Hg.) (2022) LNG-terminals im kontext rückläufiger Gasverbräuche. München. https://www.ffe.de/veroeffentlichungen/lng-terminals-im-kontext-ruecklaeufiger-gasverbraeuche-2/. Accessed 31 Oct 2022
41. SCI4climate.NRW (2021) Wasserstoffimporte: Bewertung der Realisierbarkeit von Wasserstoffimporten gemäß den Zielvorgaben der Nationalen Wasserstoffstrategie bis zum Jahr 2030
42. Gezerman AO (2021) Industrial scale ammonia pipeline transfer system and exergy analysis. In: Kemija u industriji, pp 11–12. https://doi.org/10.15255/KUI.2020.080
43. IEA (2021) World energy outlook 2021. Hg. v. IEA. https://iea.blob.core.windows.net/assets/888004cf-1a38-4716-9e0c-3b0e3fdbf609/WorldEnergyOutlook2021.pdf. Accessed 19 Nov 2021
44. Hydrogenious LOHC Technologies (2021) Auf neuem Wege zur sicheren Null-Emissions-Schifffahrt. https://hydrogenious.net/. Accessed 29 June 2022

45. Höhlein B, Grube T, Biedermann P, Bielawa H, Erdmann G, Schlecht L et al (2003) Methanol als Energieträger. Schriften des Forschungszentrums Jülich. Hg. v. Forschungszentrum Jülich GmbH. https://juser.fz-juelich.de/record/32762/files/Energietechnik_28.pdf. Accessed 17 Sept 2022
46. International Maritime Organization (IMO) (Hg.) (2018) Initial IMO strategy on reduction of GHG emissions from ships. Note by the international maritime organization. London. https://unfccc.int/sites/default/files/resource/250_IMO%20submission_Tal anoa%20Dialogue_April%202018.pdf. Accessed 29 Oct 2022
47. Roland Berger (2021) Hydrogen transportation: the key to unlocking the clean hydrogen economy. Roland Berger
48. Looney RL (2018) Handbook of transitions to energy and climate security. Verlag Routledge
49. Schmitt TM (2018) Why did Desertec fail? An interim analysis of a large-scale renewable energy infrastructure project from a social studies of technology perspective. Local Environ 23(7):747–776. https://doi.org/10.1080/13549839.2018.1469119l
50. Stegen KS, Gilmartin P, Carlucci J (2012) Terrorists versus the sun: Desertec in North Africa as a case study for assessing risks to energy infrastructure. in Risk management: a journal of risk, crisis and disaster. Palgrave Macmillan, Basingstoke
51. Gerhard M, Rüschen T, Sandhövel A (eds) (2015) Finanzierung Erneuerbarer Energien, 2nd ed. Frankfurt-School-Verlag, Frankfurt am Main
52. Agora Energiewende (2018) Die energiewende im stromsektor: stand der dinge, p 74
53. Breitschopf B, Alexander-Haw A (2022) Auctions, risk and the WACC. How auctions and other risk factors impact renewable electricity financing costs, in Energy Strategy Reviews
54. IEA (2022) Cost of capital observatory: tracking the cost of capital for clean energy projects in emerging and developing economies. https://www.iea.org/data-and-statistics/data-tools/cost-of-capital-observatory-data-explorer. Accessed 03 Oct 2022
55. Damodaran A (2019) Country risk: determinants, measures and implications—the 2019 edition
56. Breitschopf B, Lux B, Neuner F, Wietschel M (2022) Länderrisiken adäquat berücksichtigen—Wirtschaftlichkeit von Wasserstoffimporten. In BWK, Jahrgang 74, Heftnummer 5–6
57. Polzin F, Migendt M, Täube FA, von Flotow P (2015) Public policy influence on renewable energy investments: a panel data study across OECD countries. Energy Policy 80:98–111
58. Wietschel M et al (2020) Opportunities and challenges when importing green hydrogen and synthesis products. In: Policy Paper 03-2020, Fraunhofer ISI
59. Breitschopf B, Winkler J (2019) The EU 2030 renewable energy vision—can it be more ambitious? Adv Environ Stud 3(1):164–178
60. Billerbeck et al (2023) Policy frameworks for district heating: a comprehensive overview and analysis of regulations and support measures across Europe, in Energy Policy
61. E4tech (2019) Value added of the hydrogen and fuel cell sector in Europe: supporting European growth and competitiveness—study on value chain and manufacturing competitiveness analysis for hydrogen and fuel cells technologies. Study by E4tech on behalf of the Fuel Cells and Hydrogen
62. Eichhammer W, Oberle S, Händel M, Gnann T, Wietschel M, Lux B (2019) Etude sur les Opportunites et Priorites du Power-to-X au Maroc. Studie gefördert von Bundesministerium für Wirtschaft und Energie. Durchgeführt vom Fraunhofer ISI im Rahmen der Deutsch-Marokkanischen Energiepartnerschaft. https://www.giz.de/de/weltweit/57157.html. Accessed 12 Nov 2020
63. Breitschopf B, Held A, Resch G (2016) A concept to assess the costs and benefits of renewable energy use and distributional effects among actors: the example of Germany. Energy Environ 27(1):55–81. https://doi.org/10.1177/0958305X16638572
64. ESMAP (2020) Green hydrogen in developing countries. World Bank, Washington, DC. http://documents1.worldbank.org/curated/en/953571597951239276/pdf/Green-Hydrogen-in-Developing-Countries.pdf. Accessed 12 Nov 2020
65. IRENA (2021) A pathway to decarbonise the shipping sector by 2050. Hg. v. International Renewable Energy Agency, Abu Dhabi. https://www.irena.org/publications/2021/Oct/A-Pat hway-to-Decarbonise-the-Shipping-Sector-by-2050. Accessed 07 Aug 2023

PtL Fuels and Biofuels: A Dream Team?

Agneev Mukherjee, Maria Grahn, Julia Hansson, Thijmen Boter, Martin Junginger, Henrik Rådberg, Timothy J. Wallington, Sierk de Jong, and Robert De Kleine

Abstract This chapter reviews the pros and cons of Power-to-Liquid (PtL) fuels and fuels produced from biomass (biofuels). Possible integrated production and deployment pathways for these fuels are described as this will improve the potential for these fuels. An updated comparison of the economics of biofuels, PtL fuels and combinations of them as well as an overview of important environmental issues is presented. Finally, some industry actors provide their views on the possibilities for PtL fuels versus biofuels in various sectors. Overall, it can be concluded that the combination of biofuels and PtL fuels could—at least temporarily—be a 'dream team', providing a transitional trade-off between costs, land use, and electricity use.

Keywords Power-to-liquid · Biofuels · PtL vs. biofuels · Biomass · PtL potential · PtL possibilities

A. Mukherjee · T. Boter · M. Junginger (✉)
Utrecht University, Copernicus Institute of Sustainable Development, Utrecht, Netherlands
e-mail: h.m.junginger@uu.nl

A. Mukherjee
VITO, Materials & Chemistry, Mol, Belgium

M. Grahn · J. Hansson
Chalmers University of Technology, Department of Mechanics and Maritime Sciences, Maritime Environmental Sciences, Gothenburg, Sweden

J. Hansson
IVL Swedish Environmental Research Institute, Sustainable Society, Gothenburg, Sweden

H. Rådberg
Preem AB, Stockholm, Sweden

T. J. Wallington · R. De Kleine
Ford Motor Company, Research and Advanced Engineering, Dearborn, MI, USA

S. de Jong
SkyNRG BV, Amsterdam, Netherlands

© The Author(s), under exclusive license to Springer Nature Switzerland AG 2025   1127
N. Bullerdiek et al. (eds.), *Powerfuels*, Green Energy and Technology,
https://doi.org/10.1007/978-3-031-62411-7_38

1   Introduction

An effective transition to climate neutrality and broad decarbonisation of the global economy requires an extensive generation and utilisation of renewable electricity to substitute fossil fuels. There are many sectors, however, which are hard-to-electrify, and therefore are likely to require alternative methods for defossilisation. These include industries such as chemicals and cement, as well as parts of the transport sector, including heavy-duty road, maritime and aviation. In the coming years and decades, these sectors will continue to heavily rely on the use of carbon-based liquid energy sources with high energy densities. To achieve climate neutrality in these sectors as well, the required fuels will need to be supplied based on renewable energy.

While there are numerous technology routes for producing a range of such renewable fuels, these can be broadly segmented into two categories—biofuels and Power-to-Liquid (PtL) fuels. Technologies in these two categories can have the same fuel output but differ in the feedstock source—biofuels are made from biomass, while PtL fuels are based on electricity to produce hydrocarbon fuels out of carbon dioxide (CO_2) and water via electrolytic hydrogen (H_2). Within this basic classification, both of these categories encompass an umbrella of diverse technology routes, often offering multiple options for producing the same type of fuel. As an example, bio-methanol can be produced thermochemically (via gasification), biochemically (via anaerobic digestion), or via chemical digestion of wood pulp. PtL methanol can also be produced by several conversion pathways—the CO_2 can be captured from the atmosphere or from a point source and combined with hydrogen produced via electrolysis or photoelectrochemistry, for instance. A combination of biomass-based and PtL technologies can also be applied—e.g. by combining hydrogen from electrolysis with captured biogenic CO_2 generated from biofuel production.

On the face of it, this range of choices presents a dilemma, as selecting the 'best' route involves a range of techno-economic, environmental and other considerations. In reality, though, such optimisation may be irrelevant or unnecessary. None of the individual methods are likely to meet future renewable fuel demands by themselves, and since they basically lead to an identical end product (e.g. methanol in the above case), serious consideration should instead be given to examining how different biomass-based and PtL-based production pathways for a particular fuel can be combined in an efficient way.

Against this background, in this article, the pros and cons of biofuels and PtL fuels are reviewed, both in general and for a selection of specific conversion routes. This is followed by an explanation of possible synergies between the various technology options and an overview of some of the integrated production and deployment pathways. Finally, potential roadblocks in ramping up such pathways to a large scale are discussed, along with possible solutions.

2 Comparison of Biofuels and PtL Fuels

In this section, an analysis and discussion of the distinct characteristics, advantages and disadvantages of biofuels and PtL fuels is given.

Biofuels—pros and cons. With regard to the final fuel as an end product, biofuels are essentially the same type of fuels as those produced by PtL routes, except that they are derived from biomass instead of carbon dioxide (CO_2) and hydrogen (H_2). A major advantage they have over PtL fuels is a high technological maturity for certain routes. First generation biofuel technologies, such as ethanol production from sugarcane or corn, or the production of fatty acid methyl esters (FAME) from soy or palm oil, also called 'biodiesel', have been commercially available for decades. First generation biomass feedstocks are often questioned from a sustainability perspective (Sect. 5), prompting the quest for second generation (from lignocellulosic/waste biomass) and third generation (from algal biomass) alternatives. While not as advanced as their first generation counterparts, some second generation biofuel routes—such as the anaerobic digestion of wet biomass to obtain methane or (hydrotreated) pyrolysis oil suitable for co-processing in oil refineries—are almost technologically mature. Others, like making bio-oil using hydrothermal liquefaction, are relatively immature, but still better developed than many of the PtL routes. Biofuels, therefore, are a good option for meeting part of the need for renewable fuels in the immediate future, rather than decades from now.

The higher technological maturity of biofuels is reflected in their smaller price differential with conventional fuels, with biofuels often estimated to be available at half the price or less than that of PtL fuels [1–3]. However, these prices still need further reduction through technology advancements and the use of carbon credits to approach the price levels of conventional fuels, but closing this gap is conceivably easier than for PtL fuels.

One major drawback of biofuels is the associated feedstock limitations. Most commercially produced biofuels are 'first generation' biofuels, which are made from feedstocks like corn, sugarcane and vegetable oils [4–6]. These feedstocks are generally not held to be sustainable in terms of environmental or social impact. It is also unlikely that the vast quantities of renewable fuels needed can be made from these first generation feedstocks without severely affecting food supplies or leading to deforestation and biodiversity losses. Many second and third generation feedstock-based biofuels are still in the nascent stages of commercial deployment (e.g. lignin-based gasoline or algae-based biofuels), with either the feedstock availability or the conversion technologies posing challenges to their production at scale and at competitive prices.

Biofuels have also gained an increasingly negative reputation as not actually being as low carbon or environmentally friendly as claimed. As a result, their deployment is increasingly being subjected to political and regulatory scrutiny, causing hesitation on the part of many stakeholders to proceed with biofuel deployment at scale.

Power-to-Liquid (PtL) fuels—pros and cons. The main attraction of PtL fuels is that the feedstock (CO_2) availability is in principle unlimited, given the opportunity to capture CO_2 from ambient air, which mitigates one of the main limitations of biofuels. To contextualise this, the annual global liquid biofuel production of around 170 billion litres is less than 3% of the world's liquid fuels consumption [7, 8] and feedstock limitations make bridging this gap sustainably a huge challenge. In contrast, atmospheric CO_2 and hydrogen from water are both far larger sources, and PtL fuels can therefore in theory solve the liquid fuel demands by themselves.

In practice, the deployment of PtL fuels is constrained by the availability of land, equipment, and most critically, electricity from renewable energy sources. The last of these constraints can be better illustrated by considering some numbers. The global annual liquid fuel consumption corresponds to approximately 215 EJ. Meeting this demand entirely with, for instance, PtL methanol (LHV of 20.1 MJ/kg) will require the production of 9.5 billion tons of methanol per year. Presently, PtL methanol production requires around 9–10 MWh of electricity per ton of output [1], which translates into around 324 EJ of electricity from renewable energies for meeting the global liquid fuel demand annually. Additional energy is required for capturing the CO_2 to be used as feedstock; assuming a CO_2 requirement of 1.46 t CO_2 per ton of methanol [9] and an optimistic total energy requirement of 8.8 GJ per ton CO_2 captured from ambient air [10] means that a further 122 EJ (34 PWh) will be required. This increases the total renewable electricity requirement to 446 EJ (124 PWh) annually. If this is compared to the 2.3 EJ (8.3 PWh) of renewable electricity actually generated worldwide in 2021 [11], it becomes evident that current infrastructural capacities for providing renewable electricity are far from the scales necessary for a comprehensive defossilisation of annual liquid fuel demands by PtL fuels.

Another challenge for PtL fuels is their reliance on a combination of technologies with low technology readiness levels (TRL). For example, combining direct air capture (DAC) with hydrogen generated from PEM electrolysis and using Fischer–Tropsch synthesis to produce jet fuel means a commercially unproven combination of three technologies, with the first two in this combination being relatively immature technologies [12–14]. This means that many PtL fuel pathways are still technologically immature today, and the rapid upscaling and deployment of renewable fuels that is needed is unlikely to happen using these routes by themselves.

This technological immaturity is also reflected in the high production costs of PtL fuels (Sect. 4). Sustainable aviation PtL fuels, e.g. were estimated to be at least six times costlier than conventional jet fuel in 2022 [15, 16]. Other fuels, such as PtL gasoline and PtL methanol, are comparatively less expensive, but still on average at least three times as expensive as fossil diesel [17]. This price or cost difference is too big to be compensated by carbon credits alone, nor can it be expected to be borne by airlines or by passengers completely.

Comparison. The main characteristics and aspects with regard to biofuels and PtL fuels considered above are summarised in the following table.

To summarise, PtL fuels show plenty of promise, but it will most likely still take several decades before they can make a meaningful contribution to global fuel supply for hard-to-abate transportation sectors. In the interim, it is advisable to conduct a more thorough examination of biofuels or the potential of combining biofuel and PtL production pathways.

3   Combining Biofuels and PtL Fuels

As shown in the previous sections, biofuels and PtL fuels have somewhat contrasting advantages and disadvantages. This, and the fact that the same product can be made by either route, makes combining production pathways for biofuels and PtL fuels an interesting strategy to counterbalance drawbacks of one route with the advantages of the other. This combination can be achieved by two main ways:

- **Separate production, combined deployment.** The first option to combine biofuel and PtL fuel production pathways involves producing an identical fuel (e.g. methanol) by the two routes separately and then combining their deployment. Understandably, the production of many biofuels is quite location-dependent—feedstock availability not only directly affects production costs but also limits the scalability of the biofuel plant itself. At first glance, PtL fuels would appear to be less location-dependent, given that direct air capture (DAC) and water electrolysis only require water and renewable electricity, which would allow them to be produced at many locations. However, in reality economic considerations dictate that, at least in the near-term, the required CO_2 source will more likely be from a specific point source, which makes PtL fuel production in such cases also location-dependent. The availability of low-cost, abundant renewable electricity, land, labour and other factors will impose additional constraints on PtL locations. It follows, therefore, that some locations are likely to be better suited for biofuels production and others to produce PtL fuels. As these fuels are fungible, it therefore makes sense to consider producing them in a location-specific way and deploying them together to meet the market demands. Since the fuels produced can be chemically identical regardless of the origin of the carbon and hydrogen present in them, the same distribution infrastructure can, if needed, be used to supply them to the point of use. This approach is explored further in Sect. 3.1.
- **Integrated production pathways.** The second option to combine biofuel and PtL fuel production pathways involves the integration of their production. One of the biggest factors affecting the production of both biofuels and PtL fuels is the feedstock price (Sect. 4); small fluctuations in this can severely impact the output fuel price. This is the motivation behind considering production pathways integrating the routes. The logic here is that in integrated production, a part of the feedstock comes from biomass while the rest comes from captured biogenic or atmospheric CO_2. This means that a rise or fall in the price of either biomass or captured CO_2 has a much smaller effect on the overall fuel price than if either is

used as the sole feedstock. The ways in which this integrated production can be achieved are detailed in Sect. 3.2.

The above approaches to combine biofuels and PtL fuels are presented in further detail below.

3.1 Separate Production, Combined Deployment

The role that biofuels can play in the renewable liquid fuels market globally has been debated and analysed for years now. In 2010, Fischer et al. [18] estimated that if drastic land use conversion was permitted and advanced second generation biofuel technology chains were developed to the fullest, the biofuel potential in the EU in 2030 would amount to around 5.2 EJ, or about one-third of the EU transport sector's energy demand. A similar study from 2011, considering a global scale, estimated that biofuels could meet only a limited share, ranging from 10 to 52%, of the 2006 global liquid fuel consumption [19]. This shows that even under aggressive assumptions, biofuels are only likely to suffice for a limited share of the global liquid fuel demand, which, in addition to transport, also includes use in sectors like industrial heating and electricity generation. A further aggravating factor is that a lot of the biofuel volumes that could be produced may not be available at practically economical prices [20]. It may therefore make most sense to produce biofuels where they can be produced under economically more viable conditions, due to factors such as proximity to low-cost biomass (e.g. agro-residues) or the possibility of using low capital cost brownfield projects.

Meeting the remainder of the liquid fuel demand will therefore require an alternative source, and this can be PtL fuels. Like biofuels, the cost of PtL fuel production is highly location-dependent. Firstly, if the CO_2 used is captured from a point source, such as the flue gas from a cement or steel plant, then the higher CO_2 concentration lowers the cost of capture. As an illustration, the CO_2 capture cost from certain industrial facilities can be as low as 22.5 €/t CO_2, while the cost from direct air capture is likely to be above 135 €/t CO_2 [21] (assuming an exchange rate of 0.90 €/ US\$). Secondly, the cost of producing the required hydrogen for PtL fuels through renewable electricity is highly location-specific. For instance, in the case of solar PV, the global electricity costs ranged from a 5th percentile cost of 0.026 €/kWh to a 95th percentile cost of 0.11 €/kWh in 2021, suing the same exchange rate [22]. Again, besides the costs, the availability of renewable electricity itself is often a challenge, with the share of electricity supplied from renewable sources varying by country from 0 to 100% (Fig. 1) [23].

As Figs. 2 and 3 show, the areas of high biomass availability are often those lacking in other renewable electricity resources (linked to low H_2/PtL production cost). For example, northern Africa has a high solar energy potential but low biomass potential stock, while the situation is reverse in Europe. This makes clear that rather than attempting a one-size-fits-all solution, it would make more sense to produce biofuels

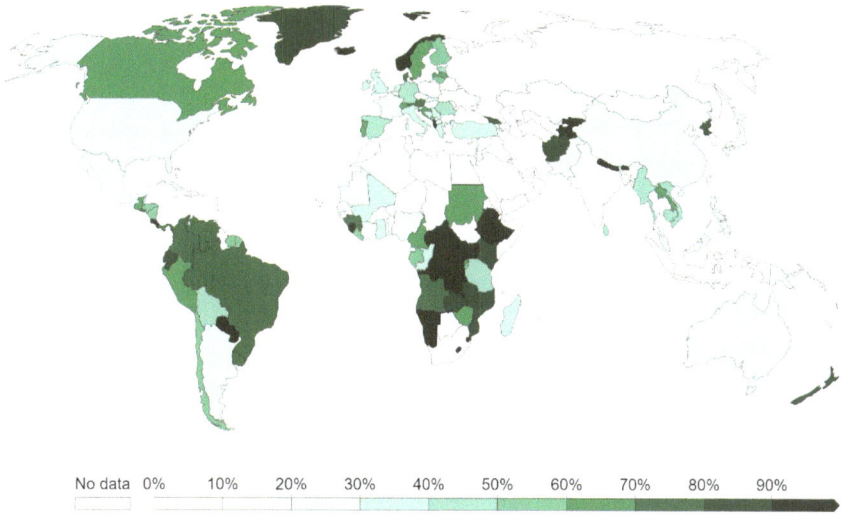

Fig. 1 Share of renewables in global electricity production by country (2022) [23]

in areas of plentiful low-cost biomass and PtL fuels in areas of low-cost renewable electricity availability, preferably from sustainable CO_2 point sources.

3.2 Integrated Production Pathways

The integrated production of biofuels and PtL fuels, in this chapter called bio-H_2-fuels, can be implemented in various ways. One option is using electrolysis hydrogen to upgrade biofuels. During biofuel production processes, both biochemical (e.g. anaerobic digestion) and thermochemical (e.g. biomass gasification with water–gas shift process), a portion of the feed carbon content is converted into excess CO_2 rather than into the desired fuel product. Inserting additional hydrogen into the production process can prevent the loss of carbon in the form of CO_2 and allows for the production of higher fuel volumes—and thus for a more efficient use of the carbon feedstocks (Fig. 4).

A second approach is capturing the biogenic CO_2 from biofuels production and combining this with electrolysis-generated hydrogen. The important difference with the aforementioned option lies in the inclusion of a carbon capture step and a separate reactor. These additions raise the capital costs of the process but avoid the problem of side reactions and resulting lowered efficiency that would occur if hydrogen were added directly to the main reactor.

A third way that is subtly different from the above two approaches is using electrolytic hydrogen to improve the quality rather than the yield of biofuels. One of the major examples of this is the upgrading of pyrolysis bio-oil. Pyrolysis bio-oil has a

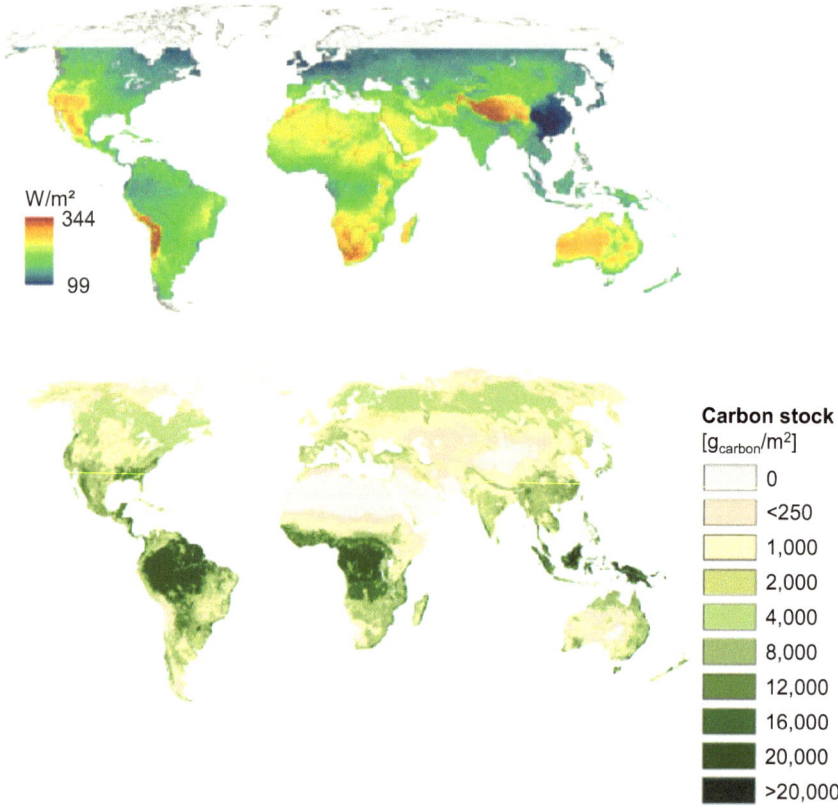

Fig. 2 Solar energy potential by region (top; based on [24]) and biomass potential stock by region (bottom; based on [25])

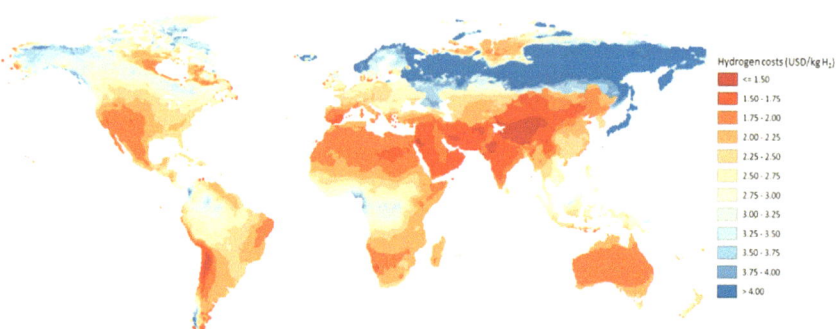

Fig. 3 Cost of hydrogen production using hybrid solar PV and wind systems in 2030 by region [26]

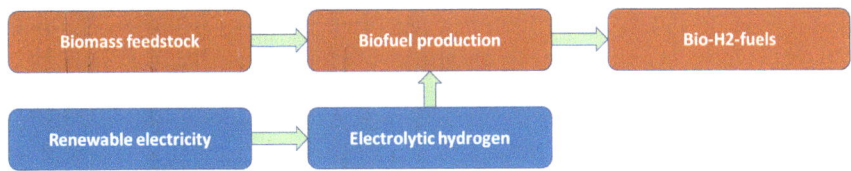

Fig. 4 Using hydrogen to increase biofuel yields, where bio-H_2-fuels are a mixture of actual biofuels and the additional yield (based on [17])

high proportion of oxygenates, whose reduction is essential for its use as a commercial fuel [27]. This reduction can be accomplished by the addition of hydrogen to the bio-oil. If electrolytic hydrogen is used instead of hydrogen produced from fossil resources, a product that is of higher quality from an environmental perspective is produced.

4   Overview of Economics of Biofuels, PtL Fuels and Combinations

The production costs of biofuels, bio-H_2-fuels and PtL fuels result from different energy conversion steps based on different key components. A literature review of future cost and efficiency data on these components for PtL fuels show that estimates vary considerably, where dominant factors impacting production costs are the costs for electrolysers and electricity [17]. A brief description of the routes studied in this article is given in Table 1.

Estimated production costs for biofuels, bio-H_2-fuels and PtL fuels in the near-term (approximately 5–10 years) and long-term (approximately 20–30 years), based on Grahn et al. [17] and Mukherjee et al. [28] and authors' structured calculations are presented in Fig. 5. These estimates assume a biomass cost of 7 €/GJ for solid biomass and 1.2 €/GJ for biogas substrate, and an electricity cost of 50 €/MWh, in

Table 1 Comparison of pros and cons of biofuels and PtL fuels

	Biofuels	PtL fuels
Pros	• High technological maturity (for some routes)	• No feedstock limitation
	• Lower production cost than corresponding PtL fuel	
Cons	• Feedstock limitations, mainly to limited availability of land	• Low technological maturity for some parts of the production chain
	• Sustainability concerns	• Higher production cost than corresponding biofuel
		• Will require considerable amounts of renewable electricity

line with [17]. A plant size of 100 PJ/a of fuel produced annually has been assumed for all the fuels, in order to ensure an equitable comparison [17] (Table 2).

From Fig. 5, it is evident that, as a group, biofuels are less expensive than either bio-H_2-fuels or PtL fuels—especially in the near-term. However, it can also be seen that due to their already relatively high technological maturity, their potential cost reductions in the long-term are quite limited, and therefore bio-H_2-fuels can possibly reach a similar cost range compared to biofuels over time. As a group, PtL fuels are in general the costliest fuel options, but their costs show the largest variation of any fuel group depicted, with the most expensive PtL fuel being jet fuel produced through CO_2 hydrogenation via methanol (CO_2–H_2-MT-Jet). Among the PtL fuels, liquefied methane, or ammonia, are expected to have the lowest production cost. The bio-H_2-fuels are, as might be expected, intermediate in cost terms. For instance, the near-term cost of methanol produced by this hybrid route is around 33.6 €/GJ, roughly midway between the 27.6 €/GJ of bio-methanol and the 43.2 €/GJ of PtL methanol.

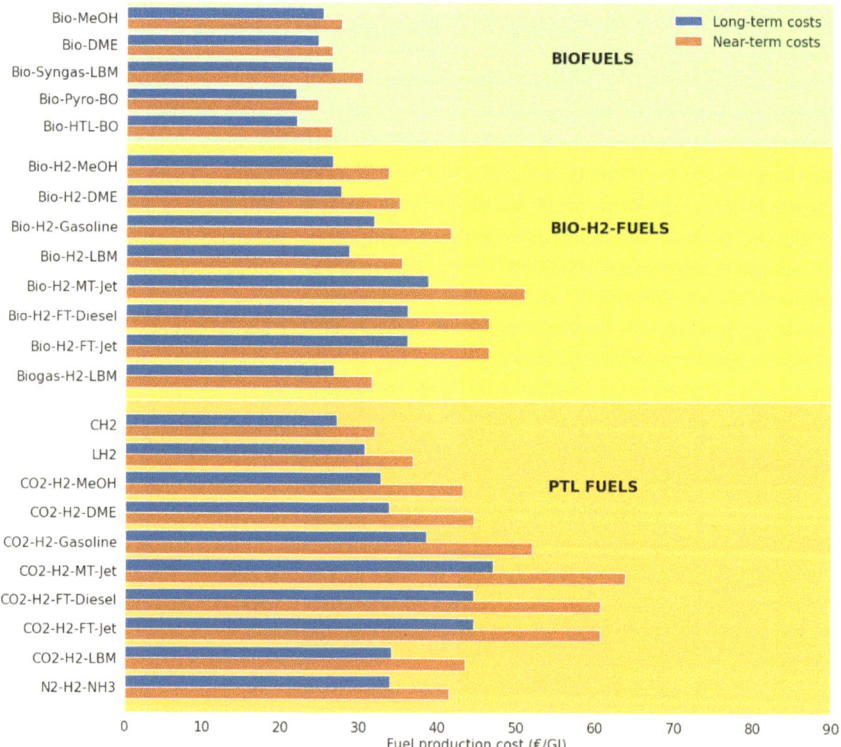

Fig. 5 Estimated production costs of selected fuels based on a literature review ([17, 28]) and authors' calculations (top section: biofuels; middle section: bio-H_2-fuels; bottom section: PtL fuels; these estimates assume a biomass cost of 7 €/GJ for solid biomass and 1.2 €/GJ for biogas substrate, and an electricity cost of 50 €/MWh, in line with [17])

Table 2 Brief description of the biofuels, bio-H_2-fuels and PtL fuels (CO_2–H_2-fuels) routes studied

Abbreviation	Description
Bio-MeOH	Methanol produced from syngas made by thermochemical gasification of biomass
Bio-DME	Dimethyl ether produced via catalytic dehydration of methanol synthesised from syngas made by thermochemical gasification of biomass
Bio-Syngas-LBM	Liquefied biomethane produced via methanation of syngas made by thermochemical gasification of biomass
Bio-Pyro-BO	Bio-oil produced via fast pyrolysis of biomass
Bio-HTL-BO	Bio-oil produced via hydrothermal liquefaction of biomass
Bio-H_2-MeOH	Methanol produced from syngas made by thermochemical gasification of biomass + hydrogen from water electrolysis
Bio-H_2-DME	Dimethyl ether produced via catalytic dehydration of methanol synthesised from syngas made by thermochemical gasification of biomass + hydrogen from water electrolysis
Bio-H_2-Gasoline	Gasoline produced from methanol produced from syngas made by thermochemical gasification of biomass + hydrogen from water electrolysis
Bio-H_2-LBM	Liquefied methane gas produced via methanation of syngas made by thermochemical gasification of biomass + hydrogen from water electrolysis
Bio-H_2-MT-Jet	Jet fuel produced from methanol produced from syngas made by thermochemical gasification of biomass + hydrogen from water electrolysis
Biogas-H_2-LBM	Liquefied biogas produced by anaerobic digestion of biomass + hydrogen from water electrolysis
Bio-H_2-FT-Jet	Jet fuel produced via Fischer–Tropsch synthesis of syngas made by thermochemical gasification of biomass + hydrogen from water electrolysis
Bio-H_2-FT-Diesel	Diesel produced via Fischer–Tropsch synthesis of syngas made by thermochemical gasification of biomass + hydrogen from water electrolysis
CH_2	Compressed hydrogen from water electrolysis
LH_2	Liquefied hydrogen from water electrolysis
CO_2–H_2-MeOH	CO_2 hydrogenated to methanol using hydrogen from water electrolysis
CO_2–H_2-DME	CO_2 hydrogenated to DME via methanol intermediate using hydrogen from water electrolysis
CO_2–H_2-Gasoline	CO_2 hydrogenated to gasoline via methanol intermediate using hydrogen from water electrolysis
CO_2–H_2-MT-Jet	CO_2 hydrogenated to jet fuel via methanol intermediate using hydrogen from water electrolysis
CO_2–H_2-FT-Diesel	CO_2 hydrogenated to diesel using Fischer–Tropsch synthesis with hydrogen from water electrolysis
CO_2–H_2-FT-Jet	CO_2 hydrogenated to jet fuel using Fischer–Tropsch synthesis with hydrogen from water electrolysis

(continued)

Table 2 (continued)

Abbreviation	Description
CO_2–H_2-LBM	CO_2 hydrogenated to liquefied methane gas using hydrogen from water electrolysis
N_2–H_2–NH_3	Nitrogen hydrogenation to ammonia using hydrogen from water electrolysis

The plant size has a major impact on the fuel production costs, with larger plants being able to achieve lower costs due to economies of scale. The plant size of 100 PJ annually assumed for plants here may be impractically large, especially for the biofuel plants, whose biomass collection radius and corresponding transport costs rise with increasing biomass inputs. To put this into context, a 100 PJ bio-methanol plant using wood pellets (LHV around 17.5 GJ/t [29]) may require around 12 million tons of feedstock annually, which is more than half of the entire EU wood pellet consumption in 2022 [30]. While the UK-based Drax power plant uses over 7 million tonnes of wood pellets each year at a single location, this requires intercontinental trade of biomass and advanced logistics [31]. Using lower calorific value biomass like straw (LHV of 14.5 GJ/t [32]) will need even larger quantities of feedstock. Smaller plants are therefore likely to be required, but these will probably not achieve the production cost ranges outlined above. For instance, if the biofuel plant size is reduced to 10 PJ/a, all other conditions remaining unchanged, then the near-term cost of bio-methanol rises from the 27.6 €/GJ mentioned above to 43.7 €/GJ, a significant increase.

It should also be noted that the estimated production costs depend on the assumptions made with the total capital investment (TCI) being a major factor for all of the fuels, biomass costs being important for the biomass-based fuels, and electricity prices being important for the bio-H_2-fuels and PtL fuels. Therefore, a sensitivity analysis was conducted (Figs. 6, 7 and 8) to assess the effect of:

- a 50% increase or reduction in the TCI for each fuel option;
- a reduction in biomass cost to 2 €/GJ or an increase to 10 €/GJ (0–2.4 €/GJ for anaerobic digestion feedstock) for the bio- and bio-H_2-fuels;
- a reduction in electricity cost to 10 €/MWh or an increase to 85 €/MWh for the bio-H_2-fuels and PtL fuels.

These ranges correspond to the range of values mentioned in [30] and represent realistic ranges for the expected fluctuation of these parameters in the near future.

From Fig. 6, it can be seen that changes in the feedstock cost have a greater effect on fuel prices for bio-methanol, bio-DME and liquefied biomethane made from syngas, while for pyrolysis and HTL, the effect of changes in the capital investment is roughly similar in magnitude if both the near and long-term scenarios are considered. It can also be seen that the effect of changes in the capital investment is larger for the near-term case for all fuels. This is understandable, since one of the main differences between the near-term and long-term scenarios is that the capital costs are expected to reduce over time, making them a smaller component of the fuel costs and therefore affecting these costs less if they fluctuate.

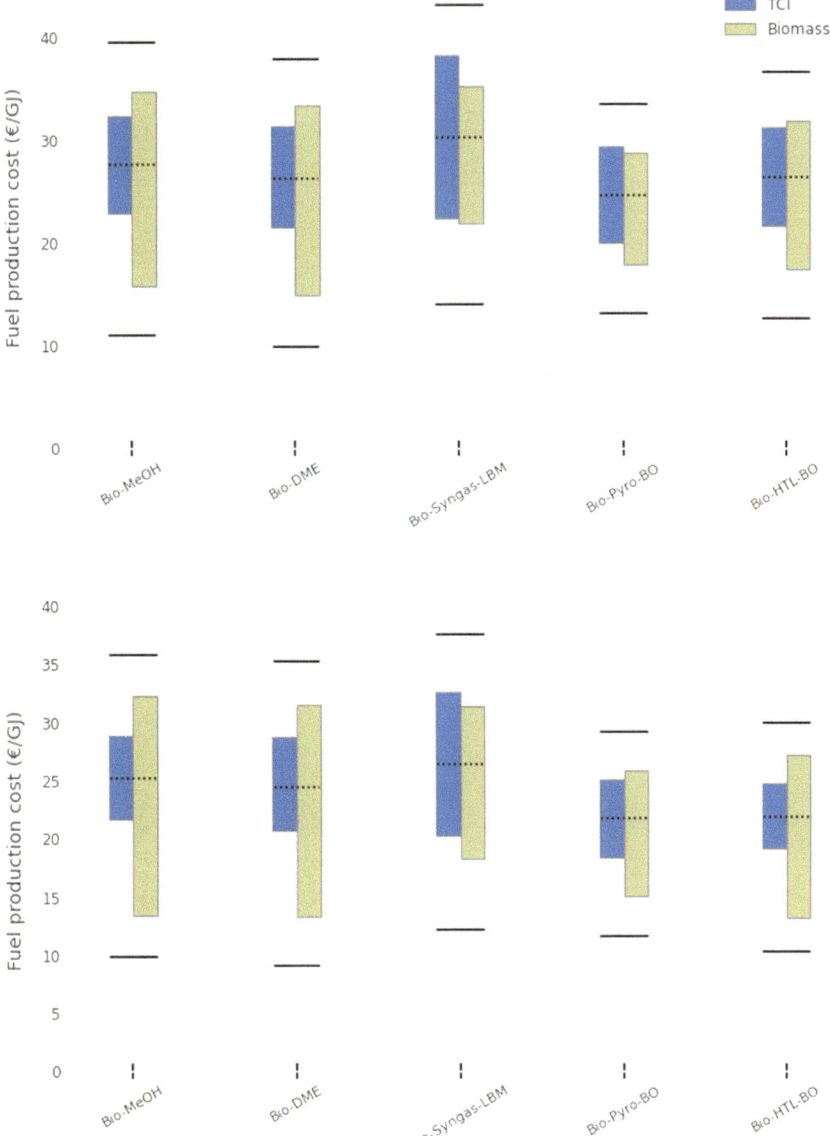

Fig. 6 Sensitivity of estimated fuel production costs of biofuels to changes in total capital investment cost (TCI) and biomass cost (top: near-term plants; bottom: long-term plants; the dashed horizontal lines represent the base cost of the fuels, while the solid horizontal lines represent the cost range if both TCI and biomass costs are simultaneously minimised or maximised)

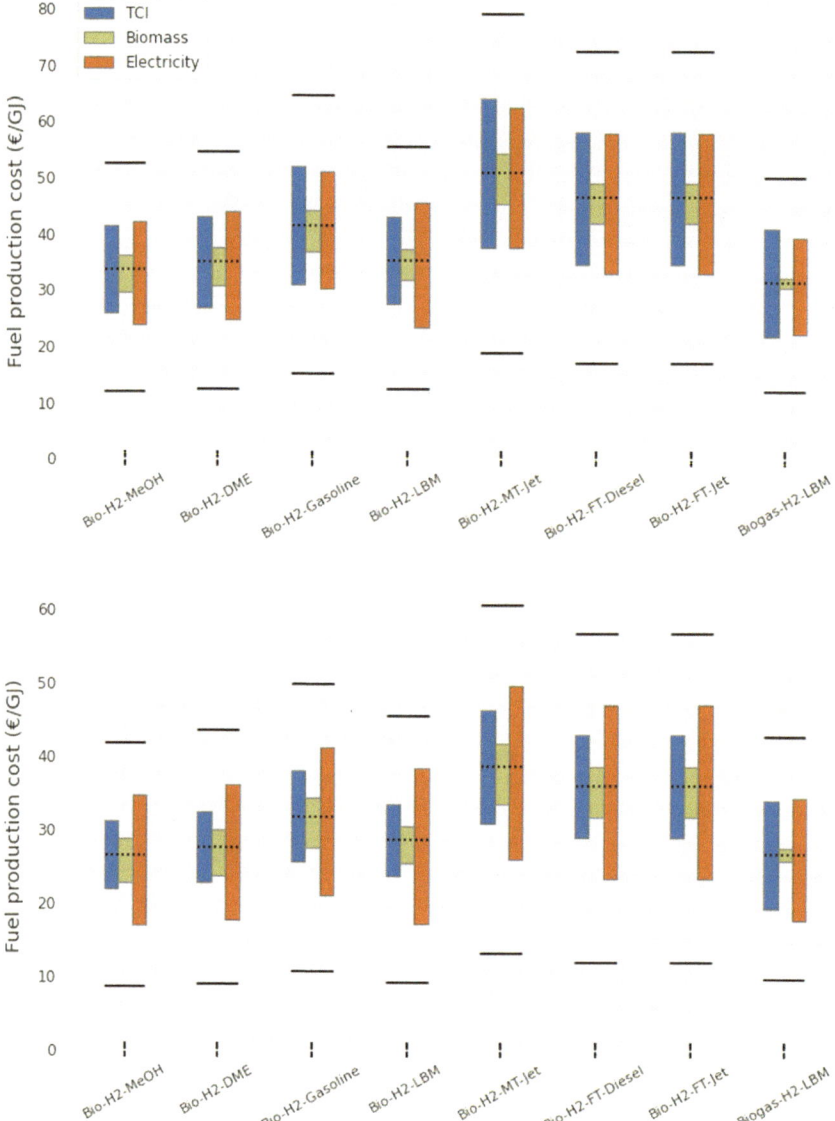

Fig. 7 Sensitivity of estimated fuel production costs of bio-H₂-fuels to changes in total capital investment cost (TCI), biomass cost and electricity prices (top: near-term plants; bottom: long-term plants; the dashed horizontal lines represent the base cost of the fuels, while the solid horizontal lines represent the cost range if TCI, electricity prices, and biomass costs are simultaneously minimised or maximised)

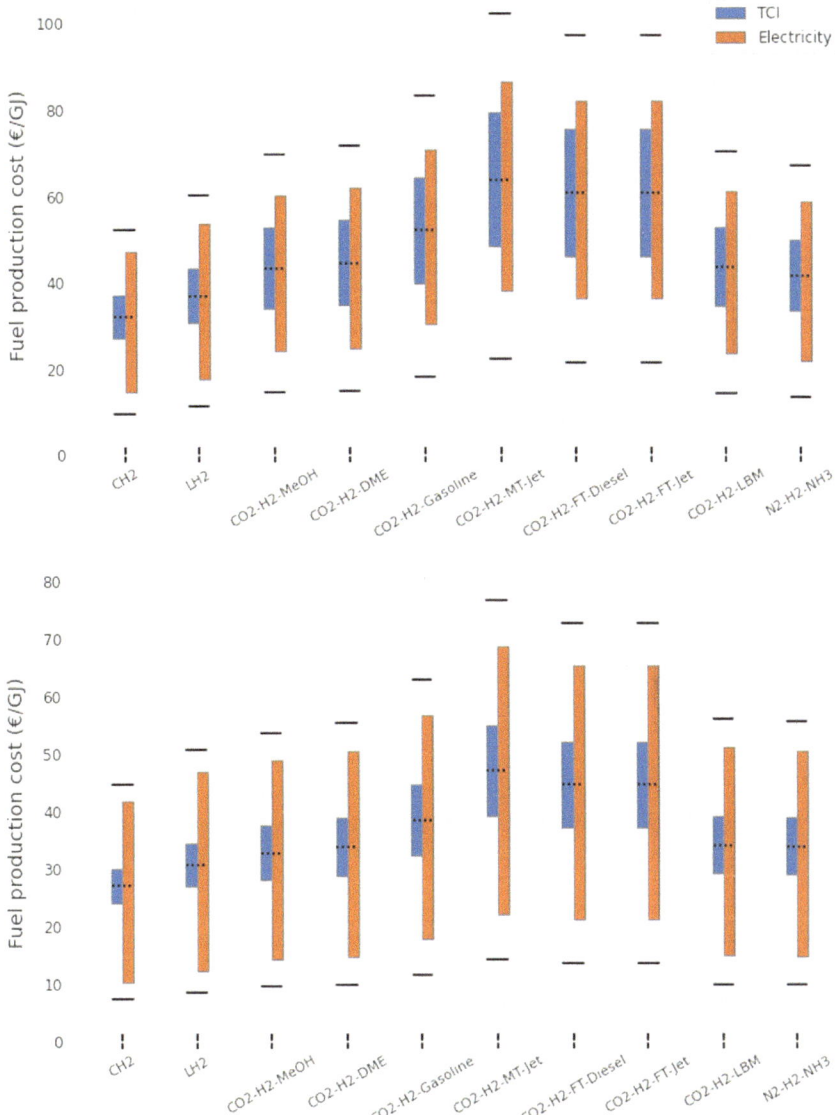

Fig. 8 Sensitivity of estimated fuel production costs of PtL fuels to changes in total capital investment cost (TCI) and electricity prices (top: near-term plants; bottom: long-term plants; the dashed horizontal lines represent the base cost of the fuels, while the solid horizontal lines represent the cost range if both TCI and electricity prices are simultaneously minimised or maximised)

Finally, if both the TCI and biomass costs are assumed to be minimised, then the fuels become highly cost-competitive, with prices approaching 10 €/GJ. Conversely, a pessimistic scenario with both terms maximised pushes prices beyond 35 €/GJ for certain fuels. A noticeable fact here is that since bio-oil production using pyrolysis is both more technically mature and has a higher biomass-to-fuel efficiency than the overall production pathways via methanol or DME synthesis, it is less susceptible to changes in TCI and biomass costs. Therefore, methanol and DME can potentially become cheaper than bio-oil if sufficient cost reductions in TCI and biomass are achieved.

For Fig. 7, one of the takeaways for bio-H_2-fuels is similar to that obtained for the biofuels, which is that changes in capital investment lead to a greater impact for near-term plants than their long-term counterparts. That apart, biomass costs lead to only small changes in the fuel costs, the exiguity of the change being especially noticeable for the fuels using anaerobic digestion, i.e. the two fuels to the right-most of Fig. 7. In contrast, electricity is a massive contributor to the fuel costs and therefore to variability in the costs as well.

Finally, Fig. 8 shows the sensitivity analysis for the PtL fuels considered. Electricity costs are massive cost contributors for every PtL fuel, making the availability of cheap renewable electricity crucial to achieve cost competitiveness of these fuels. Once again, the impact of capital investment is largely proportional to the level of technological maturity of the fuels, with relatively mature technologies like hydrogen and ammonia being less susceptible to these factors compared to less mature options like the CO_2 hydrogenation to jet fuel routes.

5 Overview of Environmental Impacts of Biofuels, PtL Fuels and Combinations

The environmental impacts of biofuels have been assessed since their advent in the 1990s and have been described extensively in the literature [33]. They also are typically the first type of bioenergy of which the environmental impacts were regulated by laws and regulations, such as the first EU Renewable Energy Directive (RED) which came into force in 2009 [34].

Such environmental impacts can be separated into those that are related to the provision of the feedstock (e.g. sugarcane or rapeseed) and fuel production (well-to-tank), and impacts that are related to fuel combustion (tank-to-wheel/wake). The former is related to general impacts of agriculture and forestry, such as impacts on biodiversity (e.g. due to deforestation and loss of habitat, and use of pesticides and herbicides), soil due to soil erosion and water (e.g. leaching of fertilisers). Often a difference is made between first generation feedstocks, which typically are food or fodder crops (sugar, starch or oil crops) grown on agricultural lands, and second generation feedstocks, which are usually derived from residue and waste streams, such as lignocellulosic material, waste streams from food processing and manure.

The production-related environmental impacts of second generation biofuels are generally far lower, as land use related impacts are either absent (secondary and tertiary waste streams), or less pronounced (e.g. lignocellulosic crops often have lower impacts than food crops).

Potential environmental impacts associated with fuel production are linked to factors such as energy consumption, which in turn depend on the electricity (and heat) source used. On the one hand, the impacts of tailpipe emissions are largely similar to the combustion of conventional fossil fuels, e.g. with regard to nitrogen oxide (NO_x) emissions (and related eutrophication), carbon monoxide (CO), particulate matter and polycyclic aromatic compounds (PACs) (and related toxic properties for the environment and humans). On the other hand, biofuels have a significant advantage over fossil fuels in that they do not contain sulphur. This characteristic is especially important, for instance, for marine applications, as fossil fuels (here especially bottom-distillates) typically contain substantial amounts of sulphur, which is increasingly restricted near coastal areas due to environmental regulations.

Greenhouse gas (GHG) emissions. The most commonly assessed environmental impact related to renewable fuels, climate change, is linked to the emissions of greenhouse gases (GHGs) [35]. The GHG emissions are typically assessed over the entire production chain up until the point of combustion. In Fig. 9, estimated GHG emissions are shown for various biofuels over the entire production chain up until the point of combustion (the biogenic tailpipe CO_2 emissions are not shown as they are considered carbon neutral). Median emissions from most sugar, starch and oil crops typically range between 30 and 80 gCO_2/MJ_{fuel}, with the notable exception of sugarcane ethanol which has a lower range. Palm oil-based biodiesel can also in principle achieve emissions below 30 gCO_2/MJ_{fuel}, but due to high risks of indirect land use change (not shown in Fig. 9) and competition with food and fodder purposes, its use is highly controversial from an environmental perspective. In general, for first generation biofuels, land use change (direct and indirect) can add significantly to the overall GHG emissions, e.g. by carbon emission due to deforestation or carbon emissions from peat soils after drainage (not included in Fig. 9). In contrast, most lignocellulosic crops and residue-based fuels are able to reach (median) values between 10 and 25 gCO_2/MJ_{fuel}.

Figure 10 shows the estimated GHG emissions of different PtL fuels under different assumptions for the GHG intensity of the electricity mix used. For each PtL fuel, the electricity mix, i.e. the share of different energy sources, used is the major determinant of the emissions. If wind energy-based electricity (10 $gCO_{2,eq}$/kWh) is assumed to be used, the GHG emissions for PtL fuels can be as low as 5 $gCO_{2,eq}/MJ$. On the other hand, if the expected global electricity carbon intensity for 2030 (277 $gCO_{2,eq}$/kWh) is considered, then the GHG emissions footprint can exceed 180 $gCO_{2,eq}/MJ$ and hence be double as high as for fossil fuels. Thus, the range of GHG emissions for the different PtL fuels depends on the carbon intensity of the electricity source, with all the fuels achieving low GHG emissions if a green source is used. However, if a carbon-intense source is used, the disparities in the GHG emissions become significant. For instance, in the near-term scenario using the

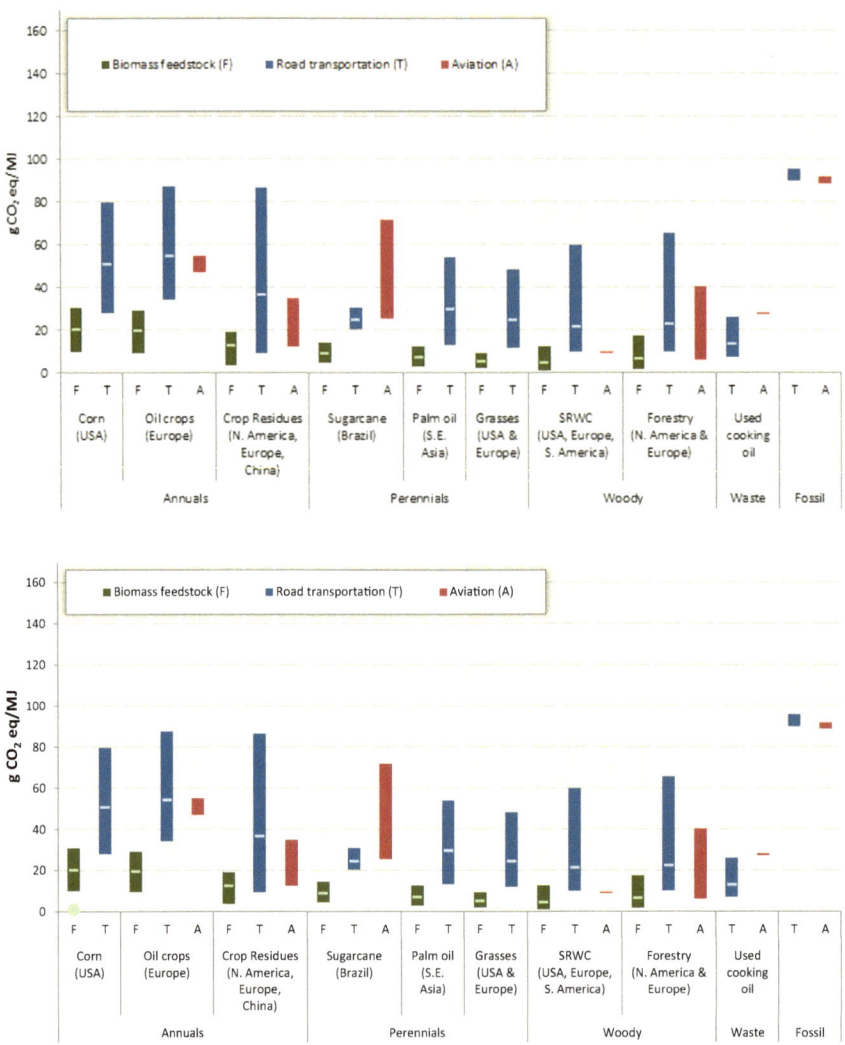

Fig. 9 Greenhouse gas (GHG) emissions for different biofuels. *Source* [36]; SRWC—short rotation woody crops

global electricity mix, ammonia has a footprint of 137 $gCO_{2,eq}$/MJ, while PtL jet has a footprint of 186 $gCO_{2,eq}$/MJ.

A general conclusion to be drawn from Fig. 10 is that PtL fuels only make sense as a renewable fuel option from an environmental perspective (with regard to GHG emissions) if low-carbon electricity (preferably from renewable sources) is used in their production. Indeed, if the current global electricity grid mix is used, then the use of PtL fuels is actually associated with higher overall GHG emissions compared to fossil fuels (Fig. 9 for the emissions factor of fossil fuels).

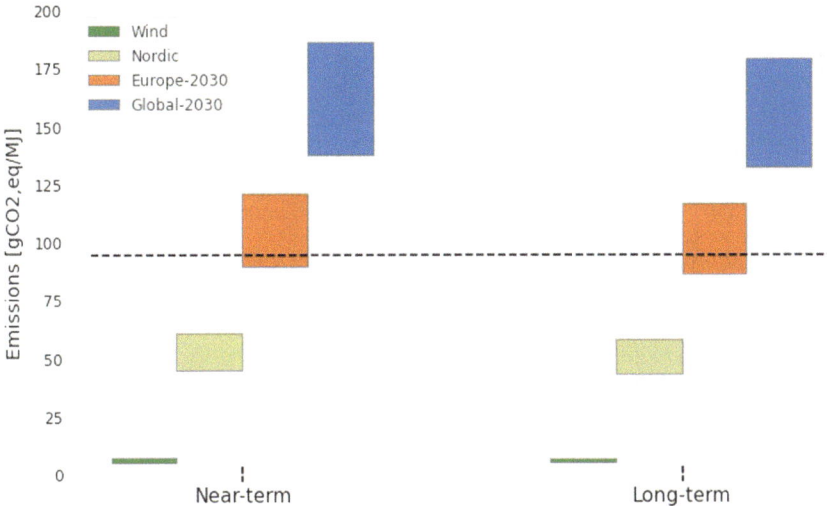

Fig. 10 Estimated variation in GHG emissions of various PtL fuels depending on the electricity mix assumed (the range for each box is because several PtL fuels are included in the assessment; the dashed line is the EU RED-II fossil fuel comparator of 94 $gCO_{2,eq}$/MJ)

Intuitively, the GHG emissions of bio-H_2-fuels can be expected to be between biofuels and PtL fuels, given that they use both biomass and electricity as feed. An assessment of bio-H_2-methanol shows that this is indeed the case. On the one hand, if the use of low-carbon footprint biomass is considered (5 $gCO_{2,eq}$/MJ) coupled with wind electricity (10 $gCO_{2,eq}$/kWh), it results in a long-term bio-H_2-methanol carbon footprint of around 6 $gCO_{2,eq}$/MJ—comparable with the best case biofuels and PtL fuels. On the other hand, high carbon footprint biomass (30 $gCO_{2,eq}$/MJ) coupled with global grid electricity (277 $gCO_{2,eq}$/kWh) gives a long-term bio-H_2-methanol footprint of 88 $gCO_{2,eq}$/MJ, a value higher than bio-methanol (63 $gCO_{2,eq}$/MJ), lower than PtL methanol (132 $gCO_{2,eq}$/MJ), and roughly comparable with fossil fuels.

There is yet a lack of studies estimating the GHG emissions from bio-H_2-fuels that will depend on the GHG emissions for the corresponding biofuels and electricity used. However, [37] provides an estimate for some bio-H_2-fuels based on lignocellulosic forest residues. These estimates are for Swedish conditions of about 5–9 $gCO_{2,eq}$/MJ fuel, depending on the process pathway, which include methane, diesel and jet fuel. The estimates assume a Swedish electricity mix corresponding to GHG emissions of 7 $gCO_{2,eq}$/MJ electricity. This emissions factor for Swedish electricity is equal to 26 $gCO_{2,eq}$/kWh electricity, which is much lower than the 180 $gCO_{2,eq}$/kWh electricity emissions factor estimated for the EU electricity mix in 2030. In a different study, Brynolf et al. [38] provide a very rough estimate for a range of bio-H_2-fuels based on forest residues and assuming zero GHG emissions from the electricity used resulting in 5.3–16 $gCO_{2,eq}$/MJ fuel. The GHG emissions for corresponding PtL fuels, also assuming zero emissions from electricity, are 0.9–6.7 $gCO_{2,eq}$/MJ fuel [38].

In summary, due to the low specific GHG emissions achievable both a use of biofuels and PtL fuels can achieve substantial GHG emission reductions compared to fossil references—which is in best cases in the order of magnitude of more than 90%. However, (indirect) land use impacts and high carbon grid electricity may currently still cause more mediocre emission reductions. Bio-H_2-fuels scores roughly comparable to the separate production of biofuels and PtL fuels in terms of GHG emissions but may result in more efficient use of land and resources.

Additional emission categories. Apart from climate change, there are also other environmental impacts to consider. In terms of other potential environmental impacts, it has been indicated that PtL fuels used for shipping may lead to lower impacts on acidification, particulate matter and photochemical ozone formation compared to conventional fossil fuels [39]. However, there is a risk that PtL fuels may have a higher impact on human health (through higher toxicity impacts). The reason is the electricity production (wind power assumed) which may give rise to several potentially harmful substances [39]. However, the toxicity impacts are uncertain and need to be further assessed.

Water demand. Fresh water is a scarce resource and increased water use can be critical in some regions. Water is required in the electrolyser as well as in some carbon capture processes and fuel synthesis related purposes. Water is also key when growing biomass and in the biofuel production processes. There is a lack of scientific papers which directly assess and compare water consumption for PtL fuel production with water consumption to produce other fuels. However, water consumption in the electrolyser depends on the type of electrolyser; Bhandari et al. [40] conclude that alkaline electrolysers consume around 10–12 L and PEM electrolysers around 18 L of water per kg H_2, while the stoichiometric water demand amounts to about 9 L of water per kg H_2. In [14], the water demand for the fuel synthesis of PtL jet fuel using the Fischer–Tropsch process is estimated to 1.3–1.4 L per litre of PtL jet fuel which according to the authors is lower than the Fischer–Tropsch bio-jet pathway. In [41], the water demand for PtL jet fuels via Fischer–Tropsch and methanol is estimated to be higher (around 3.8–4.1 L per litre of jet fuel) but still found considerable lower than for similar bio-jet fuel pathways.

Land use. Since the amount of land on Earth is limited and the competition for land, in a long-term perspective, imposes a risk of, e.g. increased food prices and biodiversity losses, the area-efficiency of biofuels and PtL fuels is an important aspect to consider. Winther-Rennuit-Mortensen et al. [42] assess hypothetical cases on how much land area would be required if all fossil fuels were substituted by PtL fuels, biomass or by electricity. They find that PtL fuels (1540 EJ) from solar and wind demand 1–12% and 11–31% of Earth's land surface, respectively. Substituting fossil fuels globally with biomass would require less energy expenditure (843 EJ). However, providing this quantity of biomass would still demand 5–134% of Earth's land surface. The former assumes only the use of relatively area efficient biomass in the form of sugarcane (110 MJ/m^2), while the latter assumes rapeseed (4 MJ/m^2).

Schmidt et al. [14] conclude that the specific land area needed to produce PtL fuels appears to be lower than the specific land requirements for the production of biofuels. A similar result is also seen in [43] where a factor of 3, 4 and 10 more land is needed for the production of imported palm oil-based HVO diesel, grass-based biomethane and rapeseed oil-based biodiesel, respectively, compared to PtL fuels production. Lai et al. [44], however, show higher specific land use for two PtL fuel production pathways (using either alkaline or PEM electrolysis) compared to two biofuels pathways, that are based on forest residues or black liquor, given that both these biomass-based resources are judged as waste and therefore not assumed to require any additional land.

Finally, Boter [45] compares several food crop- and lignocellulosic crop-based routes with three PtL fuels, both in terms of land use and renewable electricity use. In terms of land use, the PtL routes typically require 30–75 ha/kt of SAF, representing a difference of approximately a factor of 10–20 compared to the 700–900 ha/kt of SAF for the biofuels routes (see Fig. 11).

However, when assessing the amount of renewable electricity required, an inverse pattern emerges; the biofuels require (net) 0–7 GWh/kt of sustainable aviation fuels (SAF), whereas all PtL routes demand about 30 GWh/kt. This assessment shows that PtL SAF is significantly more land-efficient compared to dedicated crops. However, as biofuels like bio-ethanol are currently produced in substantial amounts in regions like the EU, the US and Brazil, where the electricity grid is not yet fully renewable, intermediate Bio-H_2 hybrid strategies may provide an attractive transition pathway.

For example, in the bio-H_2 alcohol to jet (AtJ+) case (see Fig. 11), the concentrated CO_2 released during fermentation is upgraded with hydrogen to bio-H_2 SAF. This case shows significantly less land use than the default AtJ route (400 instead of 700 ha/kt of SAF), with an electricity requirement that is still a factor of 4–5 lower than 100% PtL fuels. As such, the combination of biofuels and PtL fuels could— at least temporarily—be a 'dream team', providing a transitional trade-off between costs, land use and electricity use.

6 Potential Roadblocks and Industry Reflections

In the following section, potential roadblocks in ramping up biofuels and PtL pathways are discussed considering three industry viewpoints, namely automotive, aviation, and fuel producer perspective.

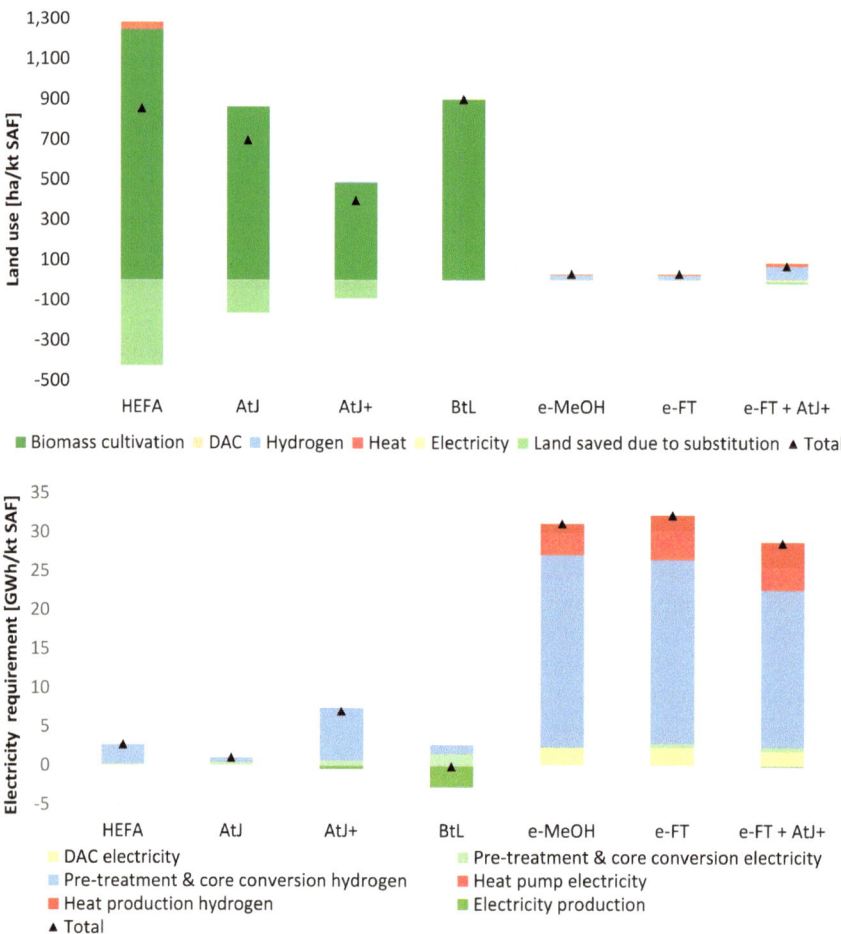

Fig. 11 Comparison of bio-based (HEFA, AtJ, BTL), PtL (e-MeOH, e-FT), and hybrid bio-H_2 (AtJ+, e-FT+AtJ+) sustainable aviation fuels (SAF) production in terms of land use and electricity use. *Source* [45]. Acronyms used: HEFA= hydro-processed esters and fatty acids, AtJ= alcohol to jet, BTL= biomass to liquids, MeOH= methanol, FT= Fischer-Tropsch, DAC= direct air capture of CO_2

6.1 PtL Fuels and Biofuels in Road Transport—An Automotive Perspective

The automotive industry is undergoing a profound transformation to a lower emission future. Governments, regulatory agencies and manufacturers have set aggressive targets for the introduction of zero emission vehicles [46]. The Zero-Emissions Vehicle Declaration at the COP 26 meeting had more than 200 signatories including 14 automotive manufacturers and 30 countries and included a call for working 'towards all sales of new cars and vans being zero emission globally by 2040 and by no later than 2035 in leading markets'. In the U.S., an executive order was issued by the president which set a U.S. target of 50% zero-emissions vehicle (ZEV) sales including battery electric, plug-in hybrid electric or fuel cell electric vehicles in the U.S. by 2030 [47]. California has adopted a rule requiring 100% of new car sales to be zero-emission vehicles (ZEVs) by 2035 [48]. The EU has introduced CO_2 emission reduction targets for new cars and vans equivalent to a ban on the sale of new petrol and diesel cars by 2035 [49]. Global automotive firms have announced major shifts in their future car and van product portfolios towards battery electric vehicles. The major push for electric vehicles raises the question of what role is left for PtL fuels and biofuels to play?

PtL and biofuels could complement electrification and hence play an important role in future transportation because of two considerations. First, while electrification is well suited to light-duty vehicles such as cars and vans, the electrification of medium- and heavy-duty vehicles for many applications is somewhat more challenging because of the large amounts of energy and hence heavy batteries needed by these vehicles. As an example, it has been estimated that a Class X heavy-duty electric vehicle with 300-mile range would require a 6.4 tonne battery and would have approximately 20% lower freight capacity than a conventional diesel vehicle [50]. The battery mass and payload penalty would be larger for heavy-duty electric vehicles with ranges greater than 300-miles. While electrification may become a more realistic option also for medium- and heavy-duty vehicles via,e.g. improved battery capacities, PtL and biofuels could be used for medium- and heavy-duty vehicles where electrification yet is more impractical.

Second, PtL and biofuels that are fungible with conventional fuels offer perhaps the only mechanism for substantially decreasing the emissions of the existing vehicle fleet which largely utilise internal combustion engines. Fungible PtL fuels and biofuels, either blended with conventional fossil fuel blendstocks or as complete fuels, by definition meet the specifications of conventional fuels and can be used in internal combustion engine vehicles in lieu of their fossil fuel equivalent. It will take decades for electric vehicles to replace most of the existing internal combustion engine vehicles in the light-duty vehicle fleet, even with the anticipated rapid ramp-up in their development and production. During at least the next 10–20 years, internal combustion engine vehicles will comprise the majority of the on-road light-duty vehicle fleet. Given increased supply PtL and biofuels offer a means to reduce the CO_2 emissions from conventional vehicles, while electric vehicles penetrate the on-road fleet.

Fuelling/charging infrastructure, as well as compliance with vehicle emissions standards, is critical considerations for road transport. In addition to maintaining the existing liquid fuel infrastructure, a large effort is underway to build an electric vehicle charging infrastructure. Widespread use of PtL and biofuels that are not fungible with conventional fuels would require: (i) a third fuelling infrastructure additional to the conventional liquid and the new electric charging infrastructure and (ii) additional vehicle engine and after treatment system development and testing to ensure robust operation and compliant criteria emissions at the same time as vehicle manufacturers are rapidly redeploying personnel and financial resources to electric vehicles. The cost and complexity of an additional fuelling infrastructure and additional vehicle development are significant barriers against the widespread use in road transportation of PtL fuels and biofuels that are not fungible with conventional fuels.

6.2 PtL Fuels and Biofuels in Aviation—An Aviation Perspective

Before the COVID-19 pandemic, aviation accounted for roughly 1 Gt CO_2 emissions per annum, equivalent to approximately 3% of global greenhouse gas emissions [51]. The International Air Transport Association (IATA) expects passenger numbers to be at pre-COVID-19 levels in 2024 [52]. Further industry growth is projected; if the industry remains unabated, the industry could account for almost a quarter of global anthropogenic emissions by 2050 [51]. The industry has set the target to cap net aviation CO_2 emissions from 2020 onwards and pledged to achieve net-zero carbon emissions from their operations by 2050, bringing air transport in line with the objectives of the Paris agreement to limit global warming to 1.5 °C [53].

Sustainable Aviation Fuels (SAF), i.e. kerosene-type renewable hydrocarbon fuels, are an important measure to reduce GHG emissions in aviation [51, 53–55]. In IATA's roadmap, 65% of emission reductions towards net zero is supposed to come from SAF [53]. The remainder is to come from infrastructure and efficiency improvements (3%), novel propulsion methods (13%) and offsets and carbon capture (19%). The role of novel propulsion methods such as hybrid-, electric- or hydrogen-powered aircraft is deemed relatively limited as large-scale deployment could take 10–20 years and will initially be limited to shorter-range and smaller aircraft [54, 55]. Moreover, aircraft refuelling based on electricity or hydrogen will require significant adaptation to global fuelling infrastructure, airport logistics, aircraft design and airline operating models (in the case of electricity, recharging affects turnaround times).

There are a variety of pathways to produce SAF. As of today, the most mature and predominant technology for SAF production is via the Hydroprocessed Esters and Fatty Acids (HEFA) route, in which SAF is produced from (waste) fat and oil feedstocks. Given the (sustainable) scalability of such lipid-based feedstocks is limited, the contribution of HEFA SAF to aviation GHG reduction targets has a ceiling that might possibly be reached in a few years already. Alternatives for SAF production include SAF from advanced biofuel pathways based on cellulosic biomass (through catalytic and biochemical pathways) and PtL that are more scalable, and therefore key to achieving the aviation industry's emission goals.

Two key barriers to the commercialisation of advanced biofuels and PtL pathways at this moment include the high production costs and lower technological readiness compared to HEFA-based SAF. Policy measures to stimulate the production and uptake of SAF are therefore key. Multiple jurisdictions are currently designing dedicated policy mechanisms for SAF, most notably in the EU and the U.S.

The EU is anticipating a SAF blending mandate (as part of ReFuelEU Aviation), which starts in 2025 with a dedicated sub-mandate for synthetic aviation fuels including PtL fuels from 2030 onwards. These blending targets are progressively moving up to 70% SAF by 2050 of which 35% synthetic SAF including PtL [56]. With the inclusion of the sub-mandate for PtL, the EU is essentially pushing for a combination of bio-based pathways and synthetic pathways, recognising it takes multiple technology platforms to meet the set volume targets.

The U.S. government is targeting 20% lower aviation emissions by 2030. It has also launched the U.S. SAF Grand Challenge—a government-wide commitment by U.S. to scale up the production of SAF to 35 billion gallons per year by 2050 (roughly 100 million tonnes), implying 100% SAF use in 2050 [57]. The incentive package of the U.S. consists of different instruments, including renewable fuel credits (under the Renewable Fuel Standard), tax credits and state incentives (e.g. the Californian Low Carbon Fuel Standard).

6.3 PtL Fuels and Biofuels in Transport—A Fuel Producer's Perspective

Decarbonisation of transport is driven both by regulation and economic benefits. Besides the recently updated CO_2 emission reduction targets for new cars and light commercial vehicles in the EU [49], the corresponding regulation for heavy-duty vehicle is also being revised [58]. Electrification to a rather large degree is needed to fulfil these regulations, under which biofuels cannot be used to account for the CO_2 emission reductions. Biofuels come under the RED (EU) 2018/2001 [59]. The EU focus is set to direct the use of advanced biofuels and electrofuels (PtL fuels) for transport to the hard-to-decarbonise aviation and maritime sectors. Biofuels in road transport will remain important and the ambition in RED III is increased, but at the same time, electrification will cause the total volumes of liquid fuels for road transport in Europe to decline, a decline that mostly will hit the fossil fuels while biofuel volumes for road transport will remain rather constant until 2035 [60]. In parallel, biofuel and PtL fuel demand for aviation and shipping are predicted to increase, driven by Europe's Green Deal and ReFuelEU Aviation.

Refiners recognise the advantages of PtL fuels, as they can be integrated into existing infrastructure and combustion engines without requiring major modifications. This compatibility allows refiners to leverage their expertise and existing assets while gradually transitioning towards a low-carbon future. However, the high associated production cost, mainly arising from investment capital in electrolyser and operating costs for the purchased electricity, as well as the scalability of PtL fuels (mainly due to power duty limitations) will be main obstacles for many years, also in the 2030s. If it wasn't for the aviation fuel providers obligation to supply increasing percentages of PtL fuels needed to fulfil EU regulations, starting in 2025 and significantly increasing in 2030, PtL fuels would not be on the priority list for a refiner today.

Biofuels have been a familiar presence in the European fuel market for several years. In the 2030s, fuel refiners will continue to view biofuels as a viable option for defossilisation. Advances in technology and research have led to the development of lignocellulosic-based biofuels, which boast improved sustainability and further reduces environmental impacts, while fats and oils, including crop-based biofuels, will still be commonly used, and accompanied by an increased portion of climate-adapted crops, such as cover-crops, that also benefit, e.g. soil organic carbon and erosion.

Biofuels and PtL fuels will co-exist and as electrification of transport evolves, their use will shift more and more towards hard-to-electrify sectors, including aviation, marine, chemical industry and as use in materials.

7 Conclusions

Transitioning to a climate-neutral global economy necessitates substituting fossil fuels with renewable electricity. However, sectors like chemicals, cement, heavy-duty road transport, maritime and aviation are challenging to electrify and will continue to depend on carbon-based energy sources for the years to come. To defossilise these sectors, pathways to produce biofuels and Power-to-Liquid (PtL) fuels must come into play. While both overall pathways can produce chemically identical fuels, the required feedstocks differ; biofuels are derived from biomass, whereas PtL fuels are based on the use of electricity to convert CO_2 and water into higher hydrocarbon fuels. Given the multiple technological pathways available to produce a similar fuel type, such as bio-methanol and PtL methanol, the challenge is not to merely select an optimal route based on techno-economic and environmental factors. Instead, the focus should be on integrating various biomass-based and PtL-based production pathways efficiently, as no single method can meet future renewable fuel demands alone.

In this context, this article examines advantages and disadvantages of biofuels and PtL fuels, both in general and for a selection of specific conversion routes. This is followed by assessing potential synergies among the various technological approaches and an overview of integrated production and deployment pathways. Finally, the potential challenges of scaling up these pathways are addressed, accompanied by potential solutions. The following key aspects can be concluded:

- PtL fuels generally have higher production costs than corresponding biofuel production pathway. The main factor contributing to both is the total capital investment cost, and both depend heavily on economies of scale. The costs of PtL fuels are also strongly dependent on the cost of electricity, while biofuels costs are also sensitive to biomass feedstock costs. Production cost ranges are substantial, and are highly dependent on local conditions, such as the cost and availability of electricity. Given the currently still low technological maturity of PtL fuels, their longer-term production cost estimates are also uncertain. Combined biofuels and PtL fuels (Bio-H_2-fuels) are intermediate in cost terms but can possibly reach a similar cost range compared to biofuels over time.

- Due to low GHG emissions both biofuels and PtL fuels can achieve substantial GHG emission reductions compared to fossil references—in best cases more than 90%. However, (indirect) land use impacts and high carbon grid electricity may cause more mediocre emission reductions. Combined biofuels and PtL fuels (Bio-H_2-fuels) scores roughly comparable to the separate production of biofuels and PtL fuels in terms of GHG emissions but may result in more efficient use of land and resources. Other sustainability issues also need to be considered such as water use and tailpipe emissions and related environmental impacts.
- PtL fuels show plenty of promise, but it will take decades, or at least years, before they can make a meaningful contribution to the global supply of fuels for hard-to-abate transportation sectors, mainly due to high costs and limited availability of baseload renewable electricity. Meanwhile, the potential of combining biofuel and PtL production pathways as well as continued use of sustainable biofuels need to be further explored.

Overall, the combination of biofuels and PtL fuels could—at least temporarily—be a 'dream team', providing a transitional trade-off between costs, land use, and electricity use. More specifically, while PtL fuels hold significant potential for the future, their current challenges, especially in terms of cost and technological development, cannot be ignored. Biofuels and PtL fuels both present opportunities for substantial GHG reduction, but it is also crucial to address associated sustainability and environmental concerns. As the transition towards sustainable transport advances, the synergy of biofuel and PtL production routes, paired with an ongoing commitment to sustainable biofuels, will be central to achieving global climate targets.

References

1. IRENA (2021) Innovation outlook: renewable methanol. International Renewable Energy Agency (IRENA); Methanol Institute. https://www.irena.org/-/media/Files/IRENA/Agency/Publication/2021/Jan/IRENA_Innovation_Renewable_Methanol_2021.pdf
2. Dietrich RU et al (2018) Cost calculations for three different approaches of biofuel production using biomass, electricity and CO_2. Biomass Bioenergy 111:165–173. https://doi.org/10.1016/j.biombioe.2017.07.006
3. IRENA (2021) Reaching zero with renewables: biojet fuels. International Renewable Energy Agency. https://www.irena.org/-/media/Files/IRENA/Agency/Publication/2021/Jul/IRENA_Reaching_Zero_Biojet_Fuels_2021.pdf
4. Seabra JEA (2021) 1. Biofuels in the Global Energy Mix. Sustainable Development Solutions Network, Fondazione Eni Enrico Mattei. https://roadmap2050.report/biofuels/biofuels-in-the-global-energy-mix/

5. OECD-FAO (2021) Biofuels. In: OECD-FAO agricultural outlook 2021–2030: organisation for economic co-operation development. Food and Agricultural Organization. https://www.oecd-ilibrary.org/sites/89d2ac54-en/index.html?itemId=/content/component/89d2ac54-en

6. IEA (2020) Transport biofuels. International Energy Agency

7. IEA (2019) Global biofuel production in 2019 and forecast to 2025—charts—data and statistics. https://www.iea.org/data-and-statistics/charts/global-biofuel-production-in-2019-and-forecast-to-2025. Accessed 24 March 2023

8. EIA (2020) Short-term energy outlook. U.S. Energy Information Administration (EIA). https://www.eia.gov/outlooks/steo/report/global_oil.php. Accessed 24 March 2023

9. Pérez-Fortes M, Schöneberger JC, Boulamanti A, Tzimas E (2016) Methanol synthesis using captured CO_2 as raw material: techno-economic and environmental assessment. Appl Energy 161:718–732. https://doi.org/10.1016/j.apenergy.2015.07.067

10. Lebling K, Leslie-Bole H, Byrum Z, Bridgwater L (2023) 6 things to know about direct air capture. World Resources Institute. https://www.wri.org/insights/direct-air-capture-resource-considerations-and-costs-carbon-removal. Accessed 31 May 2023

11. IEA (2021) Global energy review 2021: analysis. International Energy Agency. https://www.iea.org/reports/global-energy-review-2021

12. Erans M, Sanz-Pérez ES, Hanak DP, Clulow Z, Reiner DM, Mutch GA (2022) Direct air capture: process technology, techno-economic and socio-political challenges. Energy Environ Sci 15(4):1360–1405. https://doi.org/10.1039/D1EE03523A

13. Pinsky R, Sabharwall P, Hartvigsen J, O'Brien J (2020) Comparative review of hydrogen production technologies for nuclear hybrid energy systems. Prog Nucl Energy 123:103317. https://doi.org/10.1016/j.pnucene.2020.103317

14. Schmidt P, Weindorf W, Roth A, Batteiger V, Riegel F (2016) Power-to-liquids: potentials and perspectives for the future supply of renewable aviation fuel. German Environment Agency. https://www.umweltbundesamt.de/sites/default/files/medien/377/publikationen/161005_uba_hintergrund_ptl_barrierrefrei.pdf

15. Zhou Y, Searle S, Pavlenko N (2022) Current and future cost of e-kerosene in the United States and Europe. International Council on Clean Transportation. https://theicct.org/wp-content/uploads/2022/02/fuels-us-europe-current-future-cost-ekerosene-us-europe-mar22.pdf

16. Freire Ordóñez D, Halfdanarson T, Ganzer C, Shah N, Dowell NM, Guillén-Gosálbez G (2022) Evaluation of the potential use of e-fuels in the European aviation sector: a comprehensive economic and environmental assessment including externalities. Sustain Energy Fuels 6(20):4749–4764. https://doi.org/10.1039/D2SE00757F

17. Grahn M et al (2022) Review of electrofuel feasibility—cost and environmental impact. Progr Energy 4(3):032010. https://doi.org/10.1088/2516-1083/ac7937

18. Fischer G et al (2010) Biofuel production potentials in Europe: sustainable use of cultivated land and pastures, Part II: land use scenarios. Biomass Bioenergy 34(2):173–187. https://doi.org/10.1016/j.biombioe.2009.07.009

19. Cai X, Zhang X, Wang D (2011) Land availability for biofuel production. Environ Sci Technol 45(1):334–339. https://doi.org/10.1021/es103338e

20. Bioenergy I (2020) Annual report 2020. IEA Bioenergy. https://www.ieabioenergy.com/wp-content/uploads/2021/04/IEAB-Annual-Report-2020.pdf

21. Baylin-Stern A, Berghout N (2020) Is carbon capture too expensive? International Energy Agency. https://www.iea.org/commentaries/is-carbon-capture-too-expensive. Accessed 31 May 2023

22. IRENA (2022) Renewable power generation costs in 2021. International Renewable Energy Agency. https://www.irena.org/-/media/Files/IRENA/Agency/Publication/2022/Jul/IRENA_Power_Generation_Costs_2021.pdf?rev=34c22a4b244d434da0accde7de7c73d8

23. Share of electricity production from renewables. Our World in Data. https://ourworldindata.org/grapher/share-electricity-renewables. Accessed 31 May 2023

24. Oakleaf JR et al (2019) Mapping global development potential for renewable energy, fossil fuels, mining and agriculture sectors. Sci Data 6(1):101

25. Erb KH et al (2018) Unexpectedly large impact of forest management and grazing on global vegetation biomass. Nature 553(7686):73–76. https://doi.org/10.1038/nature25138
26. IEA (2021) Global hydrogen review 2021. International Energy Agency. https://iea.blob.core. windows.net/assets/5bd46d7b-906a-4429-abda-e9c507a62341/GlobalHydrogenReview2021. pdf
27. Dimitriadis A et al (2021) Bio-based refinery intermediate production via hydrodeoxygenation of fast pyrolysis bio-oil. Renew Energy 168:593–605. https://doi.org/10.1016/j.renene.2020. 12.047
28. Mukherjee A, Bruijnincx P, Junginger M (2023) Techno-economic competitiveness of renewable fuel alternatives in the marine sector. Renew Sustain Energy Rev 174:113127. https://doi. org/10.1016/j.rser.2022.113127
29. Visser L, Hoefnagels R, Junginger M (2020) Wood pellet supply chain costs: a review and cost optimization analysis. Renew Sustain Energy Rev 118:109506. https://doi.org/10.1016/j.rser. 2019.109506
30. Watak T (2023) Czapp explains: EU forestry biomass. Czapp. https://www.czapp.com/analyst-insights/czapp-explains-eu-forestry-biomass/. accessed 31 May 2023
31. Drax (2023) Drax receives 100th biomass cargo from dedicated US export facility. https:// www.drax.com/press_release/drax-receives-100th-biomass-cargo-from-dedicated-us-export-facility/. Accessed 31 May 2023
32. Biomass Statistics: Straw (2018) Ea energy analyses. https://ens.dk/sites/ens.dk/files/Statistik/ metode_halm.pdf
33. Martin M, Røyne F, Ekvall T, Moberg Å (2018) Life cycle sustainability evaluations of bio-based value chains: reviewing the indicators from a Swedish perspective. Sustainability 10:2. https://doi.org/10.3390/su10020547
34. European Commission (2009) Directive 2009/28/EC of the European Parliament and of the Council of 23 April 2009 on the promotion of the use of energy from renewable sources. http:// data.europa.eu/eli/dir/2009/28/oj
35. Bioenergy and sustainability: bridging the gaps. Scientific committee on problems of the environment, vol 72. https://bioenfapesp.org/scopebioenergy/index.php/chapters/
36. van der Hilst F, Hoefnagels R, Junginger M, Londo M, Shen L, Wicke B (2018) Biomass provision and use, sustainability aspects. In: Meyers RA (ed) Encyclopedia of sustainability science and technology. Springer, New York, pp 1–30
37. Furusjö E, Mesfun S, Samavati M, Larsson A, Gustafsson G (2022) Bio-electro fuels: hybrid technology for improved resource efficiency. https://f3centre.se/en/research/bio-electro-fuels-technology-that-can-offer-improved-resource-efficiency/
38. Brynolf S et al (2022) Review of electrofuel feasibility—prospects for road, ocean, and air transport. Progr Energy 4(4):042007. https://doi.org/10.1088/2516-1083/ac8097
39. Kanchiralla FM, Brynolf S, Malmgren E, Hansson J, Grahn M (2022) Life-cycle assessment and costing of fuels and propulsion systems in future fossil-free shipping. Environ Sci Technol 56(17):12517–12531. https://doi.org/10.1021/acs.est.2c03016
40. Bhandari R, Trudewind CA, Zapp P (2024) Life cycle assessment of hydrogen production via electrolysis: a review. J Clean Product 85:151–163. https://doi.org/10.1016/j.jclepro.2013. 07.048
41. Batteiger V et al (2022) Power-to-liquids: a scalable and sustainable fuel supply perspective for aviation. Umwelt Bundesamt
42. Winther-Rennuit-Mortensen A, Dalgas-Rasmussen K, Grahn M (2023) How replacing fossil fuels with electrofuels could influence the demand for renewable energy and land area. Smart Energy
43. Gray N, McDonagh S, O'Shea R, Smyth B, Murphy JD (2021) Decarbonising ships, planes and trucks: an analysis of suitable low-carbon fuels for the maritime, aviation and haulage sectors. Adv Appl Energy 1:100008. https://doi.org/10.1016/j.adapen.2021.100008
44. Lai YY, Karakaya E, Björklund A (2022) Employing a socio-technical system approach in prospective life cycle assessment: a case of large-scale Swedish sustainable aviation fuels. Front Sustain 3:676. https://doi.org/10.3389/frsus.2022.912676

45. Boter T (2023) Bio-SAF versus e-SAF: land-use efficiency of conversion routes for sustainable aviation fuel production in the EU. A land-use comparison of SAF production routes using biomass, renewable hydrogen and direct air capture. Master, Energy Science, Utrecht University. https://studenttheses.uu.nl/handle/20.500.12932/43990

46. The Declaration. Accelerating to Zero Coalition. https://acceleratingtozero.org/the-declaration/. Accessed 02 June 2023

47. The White House (2021) FACT SHEET: president Biden announces steps to drive American leadership forward on clean cars and trucks. The White House

48. California enacts world-leading plan to achieve 100 percent zero-emission vehicles by 2035, cut pollution. State of California

49. European Union (2023) Regulation (EU) 2023/851 of the European Parliament and of the Council of 19 April 2023 amending regulation (EU) 2019/631 as regards strengthening the CO_2 emission performance standards for new passenger cars and new light commercial vehicles in line with the Union's increased climate ambition

50. He X et al (2021) Life-cycle greenhouse gas emission benefits of natural gas vehicles. ACS Sustain Chem Eng 9(23):7813–7823. https://doi.org/10.1021/acssuschemeng.1c01324

51. Making Net-Zero Aviation Possible (2022) Mission possible partnership. https://www.mckinsey.com/~/media/mckinsey/industries/aerospace%20and%20defense/our%20insights/decarbonizing%20the%20aviation%20sector%20making%20net%20zero%20aviation%20possible/making-net-zero-aviation-possible-full-report.pdf

52. IATA (2022) Air passenger numbers to recover in 2024. IATA

53. IATA (2022) Our commitment to fly net zero by 2050. IATA. https://www.iata.org/en/programs/environment/flynetzero/. Accessed 02 June 2023

54. World Economic Forum (2020) Clean skies for tomorrow: sustainable aviation fuels as a pathway to net-zero aviation. World Economic Forum. https://www3.weforum.org/docs/WEF_Clean_Skies_Tomorrow_SAF_Analytics_2020.pdf

55. Harris J, Danicourt J, Papania A, Kim A (2023) Will plans to decarbonize the aviation industry fly? Bain and Company. https://www.bain.com/insights/will-plans-to-decarbonize-the-aviation-industry-fly/. Accessed 02 June 2023

56. Fit for 55 (2023) Parliament and council reach deal on greener aviation fuels. European Parliament

57. U.S. Department of Energy (2023) Sustainable aviation fuel grand challenge. U.S. Department of Energy. https://www.energy.gov/eere/bioenergy/sustainable-aviation-fuel-grand-challenge. Accessed 02 June 2023

58. European Union (2023) Proposal for a regulation of the European Parliament and of the Council amending Regulation (EU) 2019/1242 as regards strengthening the CO_2 emission performance standards for new heavy-duty vehicles and integrating reporting obligations, and repealing Regulation (EU) 2018/956

59. European Commission (2018) Directive (EU) 2018/2001 of the European Parliament and of the Council of 11 December 2018 on the promotion of the use of energy from renewable sources (recast). https://eur-lex.europa.eu/legal-content/EN/TXT/PDF/?uri=CELEX:32018L2001

60. Afonso M (2021) The impact of EU road transport decarbonization on biofuels, G&O, and sugar: an outlook toward 2050. Rabobank. https://research.rabobank.com/far/en/documents/103727_Rabobank_The-Impact-of-EU-Road-Transport-Decarbonization-on-Biofuels-GO-and-Sugar_Afonso_Oct2021.pdf